Mammalian Genomics

Mammalian Genomics

Edited by

A. Ruvinsky

The Institute for Genetics and Bioinformatics
University of New England
Armidale
Australia

and

J.A. Marshall Graves

Comparative Genomics Group
Research School of Biological Sciences
Australian National University
Canberra
Australia

CABI Publishing

CABI Publishing is a division of CAB International

CABI Publishing
CAB International
Wallingford
Oxfordshire OX10 8DE
UK

Tel: +44 (0)1491 832111
Fax: +44 (0)1491 833508
E-mail: cabi@cabi.org
Website: www.cabi-publishing.org

CABI Publishing
875 Massachusetts Avenue
7th Floor
Cambridge, MA 02139
USA

Tel: +1 617 395 4056
Fax: +1 617 354 6875
E-mail: cabi-nao@cabi.org

A catalogue record for this book is available from the British Library,
London, UK.

Library of Congress Cataloging-in-Publication Data
Mammalian genomics / edited by R. Ruvinsky and J. Marshall Graves.
     p. ; cm.
     Includes bibliographical references and index.
     ISBN 0-85199-910-7 (alk. paper)
     1. Mammals--Genetics. 2. Genomics.
     [DNLM: 1. Chromosomes, Mammalian--genetics. 2. Chromosome
Mapping. 3. Computational Biology--methods. 4. Evolution, Molecular.
5. Gene Expression Regulation. 6. Genomics--methods.] I. Ruvinsky,
Anatoly. II. Graves, Jennifer A. Marshall. III. Title.

     QL738.5.M3593 2004
     572.8'619--dc22

                                        2004009115

ISBN 0 85199 910 7

Typeset in 9/11pt Souvenir by Columns Design Ltd, Reading.
Printed and bound in the UK by Biddles Ltd, King's Lynn.

Contents

Contributors vii

Preface ix

Part 1. Organization of the Mammalian Genome

1 Linkage mapping 1
C. Moran and J.W. James

2 Mapping genomes at the chromosome level 23
B.P. Chowdhary and T. Raudsepp

3 Mapping genomes at the molecular level 67
F. Galibert and N.E. Cockett

4 DNA sequence of the human and other mammalian genomes 87
D. Vaiman

Part 2. Expression of the Mammalian Genome

5 The transcriptome 117
A. Verger and M. Crossley

6 The proteome 153
M.B. Datto and T.A. Haystead

7 The epigenome: epigenetic regulation of gene expression in mammalian 179
species
E. Whitelaw and D. Garrick

8 Regulation of genome activity and genetic networks in mammals 201
V. VanBuren and M.S.H. Ko

9 Inducing alterations in the mammalian genome for investigating the functions 221
of genes
J.-L. Guénet

Part 3. Evolution of the Mammalian Genome

10   A comparative analysis of mammalian genomics: prokaryote and eukaryote   263
        perspectives
        M.I. Bellgard and T. Gojobori

11   Elements and mechanisms of genome change   279
        R.J. O'Neill, G.C. Ferreri and M.J. O'Neill

12   DNA sequence evolution and phylogenetic footprinting   301
        E.T. Dermitzakis and A. Reymond

13   Evolution of the mammalian karyotype   317
        F. Pardo-Manuel de Villena

14   Comparative gene mapping, chromosome painting and the reconstruction of   349
        the ancestral mammalian karyotype
        O. Serov, B.P. Chowdhary, J.E. Womack and J.A. Marshall Graves

Part 4. Genome Analysis and Bioinformatics

15   Bioinformatics: from computational analysis through to integrated systems   393
        M.I. Bellgard

16   Genetic databases   411
        V. Brusic and J.L.Y. Koh

17   Gene predictions and annotations   429
        R. Guigó and M.Q. Zhang

Part 5. The Fruits of Mammalian Genomics

18   Genomic research and progress in understanding inherited disorders in   449
        humans and other mammals
        D.R. Sargan and A.I. Agoulnik

19   Pharmacogenomics   477
        W.W. Weber and J.M. Rae

20   Genome scanning for quantitative trait loci   507
        B.J. Hayes, B.P. Kinghorn and A. Ruvinsky

21   Mammalian population genetics and genomics   539
        L. Chikhi and M. Bruford

Index   **585**

The colour plate section can be found preceding p. 263

Contributors

Agoulnik, A.I., Department of Obstetrics and Gynecology, Baylor College of Medicine, Houston, TX 77030, USA.

Bellgard, M.I., Centre for Bioinformatics and Biological Computing, Murdoch University, Perth, WA 6150, Australia.

Bruford, M., Biodiversity and Ecological Processes Group, School of Biosciences, Cardiff University, Main Building, Cardiff CF10 3TL, UK.

Brusic, V., Institute for Infocomm Research, 21 Heng Mui Keng Terrace, Singapore 119613, Singapore.

Chikhi, L., UMR 5174 Evolution et Diversité Biologique, Bat. IV R3, Université Paul Sabatier, 118 Route de Narbonne, 310062 Toulouse Cédex 4, France.

Chowdhary, B.P., Department of Veterinary Anatomy & Public Health, College of Veterinary Medicine, Texas A&M University, Mail Stop 4458, College Station, TX 77843-4458, USA.

Cockett, N.E. Department of Animal Health, Utah State University, Logan, UT 84322-4815, USA.

Crossley, M., Department of Molecular and Microbial Biosciences, University of Sydney, NSW 2006, Australia.

Datto, M.B. Department of Pathology, Duke University, Medical Center, Durham, NC 27719, USA.

Dermitzakis, E.T., Population and Comparative Genomics, The Wellcome Trust Sanger Institute, Wellcome Trust Genome Campus, Hinxton, Cambridge CB10 1SA, UK.

Ferreri, G.C., Department of Molecular and Cell Biology, U-2131, University of Connecticut, Storrs, CT 06269, USA.

Galibert, F., Faculte de Medecine 'Genetique et developpement', Université de Rennes 1, UMR-CNRS, Rennes 35043, France.

Garrick, D., MRC Molecular Haematology Unit, Institute of Molecular Medicine, John Radcliffe Hospital, Oxford OX3 9DS, UK.

Guénet, J.L., Unite de Genetique des Mammiferes, Institut Pasteur, 25 rue de Doctor Roux, Paris, Cedex 15, 75724, France.

Gojobori, T., Center for Information Biology, National Institute of Genetics, Mishima, Shizuoka-ken 411-8540, Japan.

Guigó, R., Institut Municipal d'Investigacio Medica, Centre de Regulacio Genomica, Universitat Pompeu Fabra, Barcelona 08003, Spain.

Hayes, B.J., AKVAFORSK, Institute of Aquaculture Research, Agricultural University of Norway, Ås 1432, Norway.

Haystead T.A., Department of Pharmacology and Cancer Biology, Duke University and Serenex Inc., Durham, NC 27710, USA.

James, J.W., Centre for Advanced Technologies in Animal Genetics and Reproduction, Faculty of Veterinary Science, University of Sydney, NSW 2006, Australia.

Kinghorn, B.P., Institute for Genetics and Bioinformatics, University of New England, Armidale, NSW 2351, Australia.

Ko, M.S.H., National Institutes of Health, National Institute of Aging, Laboratory of Genetics, Developmental Genomics and Aging Section, Baltimore, MD 21224, USA.

Koh, J.L.Y., Institute for Infocomm Research, 21 Heng Mui Keng Terrace, Singapore 119613, Singapore.

Marshall Graves, J.A., Comparative Genomics Group, Research School of Bilogical Sciences, The Australian National University, Canberra, ACT 2601, Australia.

Moran, C., Centre for Advanced Technologies in Animal Genetics and Reproduction, Faculty of Veterinary Science, University of Sydney, NSW 2006, Australia.

O'Neill, M.J., Department of Molecular and Cell Biology, U-2131, University of Connecticut, Storrs, CT 06269, USA.

O'Neill, R.J., Department of Molecular and Cell Biology, U-2131, University of Connecticut, Storrs, CT 06269, USA.

Pardo-Manuel de Villena, F., Department of Genetics & Lineberger Comprehensive Cancer Center, Campus Box 7264, University of North Carolina at Chapel Hill, Chapel Hill, NC 27599-7264, USA

Rae, J.M., University of Michigan School of Medicine, 1301b, MSRB III, 1150 West Medical Center Drive, Ann Arbor, MI 48109-0632, USA.

Raudsepp, T., Department of Animal Sciences, College of Agriculture and Life Sciences, Texas A&M University, Mail Stop 4458, College Station, TX 77843-4458, USA.

Reymond, A., Division of Medical Genetics, University of Geneva Medical School, 1 Rue Michel-Servet, Geneva 1211, Switzerland.

Ruvinsky, A., Institute for Genetics and Bioinformatics, University of New England, Armidale, NSW 2351, Australia.

Sargan, D.R. Department of Clinical Veterinary Medicine, University of Cambridge, Madingley Road, Cambridge CB3 0ES, UK.

Serov, O.L., Institute of Cytology and Genetics, Russian Academy of Sciences, Novosibirsk 90, Russia.

Vaiman, D., INRA Department of Animal Genetics and U361, INSERM, Pavillon Baudelocque, Hôpital Cochim, 123 Boulevard de Port-Royal, 75014 Paris, France.

VanBuren, V., National Institutes of Health, National Institute of Aging, Laboratory of Genetics, Developmental Genomics and Aging Section, Baltimore, MD 21224, USA.

Verger, A., Department of Molecular and Microbial Biosciences, University of Sydney, NSW 2006, Australia.

Weber, W.W., University of Michigan School of Medicine, 1301b, MSRB III, 1150 West Medical Center Drive, Ann Arbor, MI 48109-0632, USA.

Whitelaw, E., School of Molecular and Microbial Biosciences, Biochemistry Building G08, University of Sydney, Sydney, NSW 2008, Australia.

Womack, J.E., Department of Veterinary Pathobiology, College of Veterinary Medicine, Texas A&M University, College Station, TX 778430-4458, USA.

Zhang, M., Watson School of Biological Sciences, Cold Spring Harbor Laboratory, 1 Bungtown Road, Cold Spring Harbor, NY 11724, USA.

Preface

Hardly any other area of science has experienced such dramatic development as genomics during the last 15–20 years. While genomics of many pro- and eukaryotic species progressed rapidly, mammalian genomics, despite the obvious complexity, was in the centre of development. Over the last several years, the Human Genome Project abandoned its initial academic status and became a human endeavour comparable with the flight to the moon and other extraordinary human achievements.

Data from other mammalian genomes such as mouse and rat have been published recently; others for dog, cattle, chimpanzee and several more will soon follow. Incredible and unprecedented opportunities for the comparative analysis of genomes brought by these huge international efforts have already started to yield new results, and will certainly do much more in the near future.

This explosion of genomic research has led to a quite widely expressed view that the genomic era is behind us, and we are entering a post-genomic era. Although terminology and attempts to categorize scientific development are usually subjective, it seems obvious to the editors of this book that the genomic era is far from being over, and there are extraordinary opportunities to expand this very productive way of generating deep understanding of the essential core of life.

This vision is reflected in the structure of the book, which is composed of five linked sections. The first section is devoted to the structure of the mammalian genome and covers topics central to genomics projects. It contains four chapters, covering linkage maps, cytogenetic and physical maps, and the molecular anatomy of the human genome. The second section describes functional aspects of genomics, topics often called functional genomics. There are five chapters in this section covering different 'omes' including transcriptome, proteome, epigenome, regulation of genome activity and genetic networks, as well as the analysis of gene functions by genome manipulation. These fields of research provide essential information on the physiological output of the genome.

Evolution of the mammalian genome forms the topic of the third section of the book. Chapter 10 considers the mammalian genome as a part of broad genomic landscape, and the following chapter is focused on mechanisms of genomic changes at the molecular level. Chapter 12 is about the evolution of DNA sequences and their usefulness in phylogenetic footprinting. The next chapter considers major features of mammalian genome evolution, including a broad set of classical cytogenetic phenomena considered from the modern point of view. Finally Chapter 14 concentrates on comparative chromosome mapping and painting, evolution of sex chromosomes and reconstruction of the karyotype of a common mammalian ancestor.

A shorter section on mammalian genomics and bioinformatics is included to highlight the critical importance of an informatic approach to building genomic knowledge. The first of its three chapters considers bioinformatics tools commonly used in genomic research, and the second chapter mainly describes genome databases. Chapter 17 concentrates on important issues of gene prediction and annotation. Several years ago specialized bioinformatics chapters were rare components of biologically oriented books. Not any more! This is a clear sign of further integration of bioinformatic methods, approaches and ideology into mainstream biology.

The final section of the book, called 'The Fruits of Mammalian Genomics', seeks to demonstrate the diverse contributions of mammalian genomics. Chapter 18 describes the progress achieved in using genomic information to understand inherited diseases in mammalian species. Dramatic developments of the last decade in pharmacogenomics and personalized medicine are discussed in the next chapter. Then in Chapter 20 the focus of attention is shifted to genome scanning and QTL mapping, which has become a high priority of mapping efforts in humans as well as domestic and laboratory mammals. The final chapter showcases recent developments in population genetics, which is steadily transforming to population genomics, thus increasing the power of classical population genetics analysis.

All 21 chapters of this book are the result of international collaborations between many researchers working in different areas of mammalian genetics and genomics. This includes not only the authors but also many people, who made a great contribution reviewing chapters and providing very useful advice. The editors are very much in debt to all of them and are very happy to acknowledge the friendly contributions of the following people: S. Agulnik, C. Babinet, R. Baker, N. Belyaev, G. Blobel, F. Calafell, A. Clark, L. Croft, M. Eldridge, V. Filippov, W. Flood, J. Gibson-Brown, T. Itoh, S. Lehnert, K. Mitchelhill, G. Montgomery, J.J. Panthier, T. Petes, N. Petrovsky, M. Switonski, I. Tammen and J. van der Werf. In a project like this, it is hard to avoid mistakes, omissions and unnecessary repeats. The editors have tried to avoid such errors but we accept full responsibility for any that escaped our attention.

This book is addressed to a wide audience of researchers and advanced students interested in mammalian genomics. We hope that the publication provides not only the latest information in this field of modern biology but also creates a novel insight into mammalian genomics.

Anatoly Ruvinsky
Jenny Graves
20 September 2004

1 Linkage Mapping

C. Moran and J.W. James

Centre for Advanced Technologies in Animal Genetics and Reproduction, Faculty of Veterinary Science, University of Sydney, NSW 2006, Australia

---

Introduction	2
Molecular Biology of Mammalian Recombination	2
Double-stranded break formation and processing	3
Peculiarities of Meiosis in Mammals with Implication for Linkage Mapping	4
Male meiosis in mammals	4
Female meiosis in mammals	4
Distribution of Chiasmata	4
Models of crossover distribution	5
DNA composition and crossing over	5
Sex Differences in Chiasma Distribution in Mammals	6
Marsupials	6
Eutherians	7
Comparison of map lengths in eutherian mammals	8
A Brief History of Linkage Mapping	8
Interference	9
Detection of Linkage	10
Specific crosses	10
LOD approach	10
Estimation of recombination rate	11
Mapping Functions	13
The Haldane function	13
The Kosambi function	14
Comparison of mapping functions	14
Current State of Linkage Maps in Mammals	15
Why Make Linkage Maps?	15
QTL and complex trait mapping	16
High resolution linkage disequilibrium mapping of QTLs	17
The ultimate QTL mapping resource	17
Concluding Remarks	18
Acknowledgements	18
References	19

Introduction

In eukaryotes, genes are located on chromosomes and thus genes on the same chromosome are physically linked. This physical linkage affects the co-segregation of alleles at pairs of loci on the same chromosome and is manifested as genetic linkage in families and crosses. During meiosis, there is exchange of material between homologous chromosomes during gamete formation. If the gene loci are far apart, the variants at two loci are likely to be split by this process of recombination, whereas if the loci are close together, the parental combination of alleles at the different loci is less likely to be separated. Genetic linkage maps are plots reflecting the frequency of occurrence of genetically detectable exchanges or crossovers among progeny produced during meiotic recombination. While linkage maps provide a true picture of the order of gene loci along a chromosome, the separation of loci in linkage maps reflects the distribution of crossing over along chromosomes and the factors which control and regulate this process rather than the physical separation of the loci along the DNA molecule.

A full molecular understanding of recombination in mammals, and their resultant linkage maps, will require identification of all gene loci whose products are involved, structurally and enzymically, in recombination or which regulate the actions and expression of these loci. Further, we would need to recognize all variants across all these loci impinging on their function. We are still a long way from identifying and characterizing all these structural and functional loci in mammals or any eukaryote, let alone their polymorphic variants causing variation in recombination. There is also growing evidence that the gross molecular composition (GC content) of a chromosome or chromosomal region can influence recombination (Fullerton *et al.*, 2001), so it is not only the recombinational mechanism we must understand better but the recombinational substrate, namely the chromosomes. Currently, there is no molecular explanation for the substantially different patterns of recombination between chromosomes of similar size and structure, between the sexes nor for the fact that the pattern of

sex differences differs dramatically between species of eutherian mammals and marsupials. However, even if we knew all the loci involved and their variants, it is clear that the pattern of meiotic crossing over is not genetically hardwired. The substantial intra-individual variation in the distribution and number of meiotic crossovers illustrates that there is an inherently stochastic element to the occurrence and positioning of crossovers along chromosomes and thus some inherent unpredictability about genetic maps.

Molecular Biology of Mammalian Recombination

Although many of the molecular details of meiosis and crossing over have been discovered in yeast, in general the process is similar at the ultrastructural and molecular level in mammals, where homologues of most of the yeast genes functioning in meiosis have been discovered. Hassold *et al.* (2000) and Cohen and Pollard (2001) have reviewed meiotic recombination and early meiosis in mammals, and Roeder (1997) has provided an overview of meiosis. The Gene Ontology section of the Yeast Genome Database (http://db.yeastgenome.org) provides an excellent overview of the complex molecular processes and genes involved in meiotic recombination.

During prophase 1 of meiosis, sister chromatids replicate and enter the leptotene stage. A proteinaceous backbone called an axial element (encoded by the synaptonemal complex protein 3 (*SCP3*) locus) joins the sister chromatids (Yuan *et al.*, 2000). During the next stage, zygotene, numerous proteinaceous foci called early stage recombination nodules become associated with the axial elements and become the first points of interaction between homologous chromosomes. Only a small number of these early stage recombination nodules will go on to become sites of homologous recombination. As the axial elements pair to form lateral elements, they are zippered together by the central element (encoded by the synaptonemal complex protein 1, *SCP1*) (Liu *et al.*, 1996) to form the synaptonemal complex. The recombination nodules then 'mature' as various gene products are expressed within them, ultimately generating recombinant chromosomes. It is unknown why most early stage recombination nodules regress and only some mature. If we knew, we would be much closer to understanding some important biological phenomena, like interference.

Double stranded break formation and processing

Following the discovery in the yeast, *Saccharomyces cerevisiae*, that meiotic recombination involved double stranded breaks (Sun *et al.*, 1989), the double stranded break repair model has provided the molecular basis for understanding meiotic recombination. Many of the gene products involved have been identified, usually first in yeast with the mammalian orthologues identified later. Dudas and Chovanec (2003) comparatively review genes and gene products involved in double-strand break repair in homologous recombination in yeast and humans.

The enzyme making the double stranded break, a topoisomerase, is encoded by the *SPO1* (Sporulation 1) locus, so named since it was discovered to be essential for sporulation in yeast. This topoisomerase is present within the recombination nodule but may have additional functions than simply making the double stranded breaks (Romanienko and Camerini-Otero, 2000). During repair of the double stranded break, the paired cut and uncut homologous duplex DNA must be converted into a four-stranded branched intermediate or joint molecule, containing Holliday junctions on either side of heteroduplex DNA. First exonuclease digestion exposes single-stranded 3′ termini in the cut molecule. These invade the uncut homologous double stranded molecule and initiate repair synthesis followed by branch migration to produce two Holliday junctions (see Figure 6 in Roeder, 1997). Resolution of these Holliday junctions results in a reciprocal crossover in half the cases (Roeder, 1997). Numerous gene products, at least 11, have been implicated in the production of double stranded breaks, generation of joint molecules and resolution of the Holliday junctions, and undoubtedly many more

remain to be discovered. This overview will attempt to cover the roles of the major players, many of which also have important roles in DNA repair in somatic cells.

The Mre11–Rad50–Nbs1 protein complex of mammals, also vitally important in repairing cellular damage (Petrini, 1999), is responsible for the exonuclease digestion producing the 3′ single stranded tails that invade the uncut homologous duplex and produce the joint molecule. However, there is evidence from mutations at these loci that the complex is also required for proper formation of the double stranded breaks.

The *RAD51* and *DMC1* (Disruption of Meiotic Control 1) gene products co-localize in zygotene recombination nodules where they catalyse invasion of the single-stranded tails into the uncut DNA duplex. Mutations at both loci result in disruption of the normal pattern of meiotic recombination where non-sister strand exchanges are favoured over sister chromatid exchanges. Yoshida *et al.* (1998) have also shown that *DMC1* is necessary for homologous pairing of chromosomes in mouse spermatogenesis. In mammals it is noteworthy that the number of *RAD51* foci on leptotene chromosomes far exceeds the number of recombination sites in mouse germ cells (Cohen and Pollard, 2001). It is not known what determines which of these zygotene recombination nodules matures into a crossover nor how they interact with each other. However, as will be explained later, there is good biological evidence from both cytological and genetic studies that such interactions must occur, as recombination events on a chromosome do not occur independently of each other. There clearly is some form of interaction but its mode of action is still a molecular mystery.

Several proteins have been implicated in the maintenance of pairing of homologous chromosomes in bivalents necessary for their regular segregation at anaphase 1 of meiosis and at least one reflects the positions of crossovers. Foci of the *MLH1* (MutL homologue 1) gene product, detected by immunochemistry, are found only on pachytene bivalents and are believed to accurately reflect the position of chiasmata (Anderson *et al.*, 1999; Frönicke *et al.*, 2002). Mice deficient for *MLH1* have regular synapsis of chromosomes in male meiosis up until pachytene, but subsequently desynapse. Thus the *MLH1* gene product is required for maintenance of meiotic pairing. Interestingly the *MLH1* gene product, like many other gene products involved in meiosis, is multifunctional, known also to be a DNA mismatch repair enzyme and has even been implicated in hereditary non-polyposis cancer (Bronner *et al.*, 1994).

Peculiarities of Meiosis in Mammals with Implication for Linkage Mapping

Linkage maps differ substantially between male and female mammals, reflecting substantial differences in the regulation and control of the meiotic process between males and females. The timing and duration of meiosis differ dramatically between male and female mammals and so it is probably not surprising that the patterns of recombination and the resultant linkage maps differ also. Furthermore Hunt and Hassold (2002) have reviewed data on mammalian gene knockouts at 12 loci involved in synapsis and recombination and found that in seven cases these mutations caused male sterility due to prophase arrest when females retained some or even close to normal fertility. Male meiosis seems less robust to mutational disturbance than female meiosis.

Male meiosis in mammals

The products of male meiosis are countless millions of sperm cells. Since so many gametes are produced, male meiosis is dependent on a stock of gonadal stem cells called spermatogonia, which divide to regenerate the stem cell population throughout the life of the animal as well as providing terminally differentiating cells via meiosis to form spermatids. Although the early stages of meiosis can be detected prior to puberty, male meiosis is initiated and occurs continuously throughout sexual maturity and thus in many mammalian species spermatogenesis may continue for many years. Studies in mice have established that male meiosis takes about 13 days for completion (Nebel *et al.*, 1961) and a similar duration is likely in most mammals.

Female meiosis in mammals

Female meiosis is dramatically different from that in the male. Not only is there only one gamete produced per meiotic division but meiosis occurs in two relatively brief and separate phases in females, with the phases in some species separated by many years. The first phase is during early fetal development when the primordial germ cells divide mitotically to produce 30,000–75,000 oogonia. All oogonia then proceed into meiosis, reaching the diplotene stage of the first meiotic division shortly after the time of birth. In female mice, this occurs at day 5 after birth (Silver, 1995). The oogonia then become arrested at this dictyate stage, acquiring a coating of follicle cells to comprise a follicle. With the onset of puberty, hormones stimulate individual follicles or groups of follicles to mature and complete meiosis during each oestrus cycle. The first meiotic division and first polar body formation is completed prior to ovulation. The second meiotic division then stalls at metaphase II, until fertilization triggers its completion and the generation of the gamete or maternal pronucleus and second polar body.

These differences in the staging of meiosis in male and female mammals mean that the developmental control of the process is quite different. Although many of the same gene products implicated in meiosis and recombination have been demonstrated at the appropriate developmental stages in male and female mammals, the control of expression of these genes and the effect of mutations at these loci show some difference between the sexes. It is therefore not surprising given these dramatic developmental differences between male and female meiosis that fine control of distribution of crossing over differs between the sexes.

Sharp (1984) has shown that female meiosis in marsupials is similar to that in placental mammals except that entry into the dictyate stage takes place in female pouch young rather than *in utero*.

Distribution of Chiasmata

For many years, the only direct way to observe the distribution of crossovers was to use cytologically amenable material from species like newts with lampbrush chromosomes or grasshoppers, where pachytene and diplotene stages could be easily observed by light microscopy. Especially in male meiosis, chiasmata could be easily seen and recorded. In a series of elegant experiments, Shaw (1971, 1972, 1974) demonstrated genetic control of variation in distribution of chiasmata in grasshopper species by selecting for increased and decreased total number of chiasmata in males, in the process estimating realized heritability of total chiasma number. Thus chiasma number was shown to be genetically variable and potentially subject to natural selection. Shaw and Knowles (1976) subsequently mapped the distribution of diplotene chiasmata in grasshoppers using a computerized digitizer in an early attempt to electronically quantify chiasma distribution, although accurate recognition of some chromosomes remained a challenge in these studies.

Mammalian meiosis in general is not amenable to such analysis by light microscopy and it took the advent of electron microscopy and the discovery of the synaptonemal complex for progress on the details of early stages of mammalian meiosis to begin. The recognition by Carpenter (1975) of recombination nodules was a crucial step in the mapping of chiasma distribution in species which did not have meiosis favourable for light microscopic analysis.

Models of crossover distribution

Many observations relate crossover frequency to length of synaptonemal complex rather than to DNA content of chromosomes. The observations by Bojko (1983) and Wallace and Hulten (1985) that the lengths of human pachytene chromosomes in oocytes are approximately double those in spermatocytes can be correlated with the higher observed recombination frequency in human females than males. Lynn et al. (2002) have taken this comparison of crossover frequency with synaptonemal complex length a step further by comparing synaptonal complex length with DNA content and map length in human chromosomes in spermatocytes. For example, chromo-

somes 21 and 22, have similar DNA contents (39 Mb and 43 Mb, respectively), but quite different male map lengths (54 cM and 70 cM, respectively). Conversely, human chromosome 16 (98 Mb and 106 cM) and chromosome 19 (67 Mb and 104 cM) have very similar male map lengths but quite different DNA contents. In each case, the map length is best predicted by the synaptonemal complex axis length.

Kleckner et al. (2003) have attempted to develop models to explain this correlation between chromosome axis length and crossover frequency. The development of such models is challenging since it is not completely clear whether the length of the axes determines the frequency of crossing over or whether the recombination process determines the synaptonemal complex length or whether there is some other factor which coordinately determines both features. Kleckner et al. (2003) favour the first hypothesis since loss-of-function mutations in *Spo11* prevent development of double stranded breaks and thus of recombination, but do not affect homologue axis lengths or pachytene meiotic structures in pachytene.

DNA composition and crossing over

There is evidence that nucleotide composition of chromosomal regions called isochores, which are relatively large regions of uniform GC content (Zhang and Zhang, 2003), affects crossing over. Holmquist (1992) found more chiasmata in GC-rich chromosomal bands and Eyre-Walker (1993) also showed a significant positive correlation between GC content and chiasma density. Recently Fullerton et al. (2001) have analysed the correlation between GC content and recombination for all individual human chromosomes and found highly significant positive correlations for nine of the 23 chromosomes – many of the non-significant correlations were probably due to paucity of data for those chromosomes. Eisenbarth et al. (2000) have found evidence for a sudden switch in recombination pattern on human chromosome 17q in the neurofibromatosis type 1 gene associated with a transition from low GC content to high GC content. In the low (37.2%) GC content L1 isochore, very strong linkage disequilibrium was observed, whereas in an equivalent sized region of the high (51%) H2 isochore, no linkage disequilibrium was detectable. From this, it was inferred that recombination was more frequent in the high GC content region, consistent with the earlier finding of Holmquist (1992) and Eyre-Walker (1993). Eisenbarth *et al.* (2000) speculated on the possible necessity of taking GC content into account in designing mapping studies. More markers would be required in GC-rich isochores and less in GC-poor regions.

Sex Differences in Chiasma Distribution in Mammals

Haldane (1922) long ago predicted that crossing over would be less frequent in the heterogametic sex, based primarily on data from *Drosophila* species with heterogametic males which lacked recombination and silkworm species where the heterogametic females lacked recombination. Dunn and Bennet (1967) challenged the validity of this generalization, especially for mammals, although they did find some evidence even in the literature up to 1967 for higher levels of recombination in female mammals, although there were also some cases of higher male recombination. How well has Haldane's generalization survived the subsequent test of time in mammals? Although it holds reasonably true for eutherian mammals, marsupials appear to be a general exception and, even within eutherians, there are numerous exceptions for specific chromosomes and chromosomal regions to the prediction that males will have lower recombination frequency. Perversely, there is even evidence that the female marsupials can have an overall higher chiasma frequency but have lower overall rates of recombination and much shorter maps since the crossovers are located telomerically and thus contribute little to recombination. Clearly reality is much more complex than the original simple prediction by Haldane.

Marsupials

Few studies have been made of meiosis in marsupials, with Sharp (1984) providing a comprehensive early analysis. He observed male and female meiosis in a number of species of Australian marsupials by light microscopy, a more feasible procedure in marsupials which generally have smaller chromosome counts and more easily recognizable chromosomes than eutherians. Chiasmata counts were obtained for males from 33 species of marsupials, including kangaroos, possums and wombats. For three species, *Perameles gunnii*, the eastern barred bandicoot, *Isodon obesulus*, the southern brown bandicoot, and *Sarcophilus harrissii*, the Tasmanian devil, chiasma distributions were determined for all chromosomes in males. Chiasmata were found to be frequent near telomeres and rare near centromeres and were influenced by intra-arm interference. Female meiosis was observed in a more limited set of samples from *Macropus giganteus*, the eastern grey kangaroo, *Petauris breviceps*, the sugar glider, *Trichosurus vulpecula*, the common brushtail possum, and *Sminthopsis crassicaudata*, the fat-tailed dunnart. Male *T. vulpecula* and *P. breviceps* were found to have a lower chiasma frequency per cell (18.14 and 14.9, respectively) than females (28 and 31.7, respectively), although Sharp expressed some concern about the accuracy of the estimates of female chiasma frequencies.

Subsequently Bennett *et al.* (1986) described extreme differences in the linkage maps and the meiotic behaviour of chromosomes between males and females of *S. crassicaudata*. Pairs of loci displaying no or very little recombination in females displayed very frequent recombination in males. Females were found to have strictly terminally localized chiasmata, whereas males had frequent interstitial chiasmata. Thus, it could be concluded that while the number of chiasmata in female marsupials could be as high as or even higher than in male marsupials, their strictly terminal localization in females meant that they were having no or little impact on recombination, hence explaining the shrinkage of the female maps in marsupials. van Oorschot *et al.* (1992) also observed similar severely reduced recombination in females of the South American marsupial, *Monodelphis domestica*.

Zenger *et al.* (2002) have recently reported the first comprehensive genetic linkage map for a marsupial, the tammar wallaby, *Macropus eugenii*, and compared sex-specific recombination rates. They found significantly higher male recombination, with the female map about 78% as long as the male linkage map. The deficiency in female recombination was restricted to eight intervals on the map and when these were removed from the analysis, male and female map lengths were otherwise not different. These intervals of low female recombination tended to be interstitial but the positions were not sufficiently well characterized to make reliable generalizations about their location.

Eutherians

Meiotic chromosomes of eutherian mammals are less amenable to the type of cytological analysis applied to marsupials since they are more numerous, smaller and more difficult to recognize. Nevertheless Polani (1972) showed that chiasma frequency was higher in female than in male mice. Subsequently Davisson *et al.* (1989) reported that the overall rate of recombination in male mice is about 50–85% of that in females, consistent with the chiasma frequency difference. However, as in other species, there are exceptional chromosomes and chromosomal regions where there are no differences or male recombination even exceeds that in females. Gorlov *et al.* (1994) analysed the distribution of chiasmata along two easily recognizable chromosomes, 1 and 14, in male and female meiosis in the mouse and compared them with male and female genetic map data. For both chromosomes, they found that there was no significant difference in chiasma frequency between males and females, but that there were highly significant differences in chiasma distribution between the sexes. Males were found to have more terminally localized chiasmata than females which tended to have more chiasmata in subterminal regions than males and a more uniform distribution across the chromosome.

Modern techniques in cytology and electron microscopy are now being applied to the analysis of crossover distribution in mammals. Chromosome-specific probes permit unambiguous recognition of individual chromosomes, and antibodies against the *MLH1* foci

on synaptonemal complexes, whose number and positions are strongly correlated with chiasmata at metaphase 1, identify the sites of exchanges. For example, Barlow and Hulten (1998) and Lynn *et al.* (2002) have mapped the sites of exchanges in meiotic chromosomes of human males, obtaining detailed data on the number and distribution of crossovers for some chromosomes. They found, respectively, means of 50.6 and 46.2–52.8 for the numbers of *MLH1* foci per cell in males. Tease *et al.* (2002) also used similar techniques to analyse female meiosis in humans, observing a mean of 70.3 foci. The number of crossovers per female germ cell is 1.4 times as great as the number in male germ cells, agreeing well with the ratio of 1.6:1 for the ratio of human female map length (44 Morgans) to male map length (27 Morgans) (Broman *et al.*, 1998). Lynn *et al.* (2002) reported the distribution of crossovers in males for chromosomes 1 and 16, which have interstitial centromeres, and for chromosomes 21 and 22 with terminal centromeres. The crossovers were distributed preferentially at both telomeres of the submetacentric chromosomes, with a minimum at the position of the centromere, and at the distal telomere of the acrocentric chromosomes. Tease *et al.* (2002) reported no such preferential telomeric distribution of chiasmata in females where there was a tendency for a more uniform distribution across the chromosomes and relatively fewer crossovers at the telomeres. For chromosome 21, for which data were available from both Lynn *et al.* (2002) and Tease *et al.* (2002), a more precise comparison can be made between male and female crossover distribution. For males, the most common site of crossovers is at the telomere of the long arm. For females, cells with single chiasmata show two peaks in the long arm, with no crossovers in the short arm or near the telomeres of the long arm. For cells with two *MLH1* foci on chromosome 21, almost all crossovers are on the long arm, but are displaced towards the centromere and the telomere. Broman *et al.* (1998) have examined the outcome of these differences in chiasma distribution on genetic maps. For almost all chromosomes examined, males showed an expansion of their maps telomerically with respect to females. Perhaps more surprisingly, almost all chromosomes showed a substantial excess of female recombination in the vicinity of the centromeres, often by a factor of 8 with respect to the sex-averaged distance.

One particularly intriguing observation by Tease *et al.* (2002) relates to the occurrence of multiple crossovers on a bivalent. For some chromosomes, including 13, 18, 21 and X, the number of *MLH1* foci ranged from one to five per bivalent. For each of these chromosomes, the mean length of the synaptonemal complexes in the diplotene bivalents increased more or less linearly with each additional crossover, again emphasizing the important relationship between length of synaptonemal complex and frequency of crossover mentioned earlier. However, again it is not clear which comes first – more crossovers cause a longer synaptonemal complex or vice versa.

Anderson *et al.* (1999) pioneered the use of *MLH1* foci for mapping the positions of crossovers on the synaptonemal complex using mice. In a subsequent study, Frönicke *et al.* (2002) have mapped *MLH1* foci on all chromosomes in male mice, individually identifying every chromosome by multicolour fluorescence *in situ* hybridization and producing detailed maps of single and multiple crossover positions for all chromosomes. The number of *MLH1* foci per cell reported by Frönicke *et al.* (2002) is 22.9, less than half the number reported for human males, and consistent with the much shorter total map length reported for mice compared with humans. For almost every chromosome, they found a strong telomeric bias in the distribution of chiasmata. Further, they showed that in mice as in humans, the length of the synaptonemal complex, rather than DNA content per chromosome or mitotic length, is by far the best predictor of the number of crossovers.

Comparison of map lengths in eutherian mammals

Mapping of *MLH1* foci has not yet occurred in other mammalian species. However, for many species, such as the human (Li *et al.*, 1998), pig (Archibald *et al.*, 1995; http://www.thearkdb.org/), dog (Mellersh *et al.*, 1997; Ostrander *et al.*, 2001), cattle (Barendse *et al.*, 1997; Kappes *et al.*, 1997; Barendse and Fries, 1999) and sheep (Maddox *et al.*, 2001), comprehensive genetic maps permit comparisons of male and female recombination rates. In the pig for example, the female map is about 25% longer than the male map. Most female maps for individual chromosomes display the expected greater length, in most cases due to telomeric expansion of the female maps, presumably due to a greater occurrence of sub-telomeric chiasmata in females. Similarly in humans and dogs, the female maps are longer than the male maps.

While longer female maps tend to be the rule among studied species, there are important exceptions that illustrate the dangers of such generalizations about higher rates of recombination in female eutherian mammals. For cattle, there is no significant difference in the size of the male and female maps. Sheep provide an even more dramatic exception. Maddox *et al.* (2001) have shown that the overall male autosomal map length (3807 cM) of sheep is considerably and significantly greater than the female map (3145 cM), with 23 of 26 autosomes showing higher male recombination. The autosomal female map length in sheep is only 82% of that in males, contrasting dramatically with humans where the ratio is 160%.

Individual chromosomes also provide exceptions to these general trends, even within species. For example the male map of chromosome 1 in the pig is considerably larger than the female map (Archibald *et al.*, 1995), completely running against the trend of elevated female recombination. This exception is attributable primarily to a short interstitial interval in females of about 2–5 CM which is so grossly expanded as to imply an obligate chiasma within it in males. Such recombinational hotspots have been described in many species. In sheep, female recombination for chromosome 24 is substantially higher than male recombination, contrary to the trend for other chromosomes.

A Brief History of Linkage Mapping

Genetic linkage was first recognized early in the 20th century shortly after the rediscovery of Mendel's laws of inheritance. Exceptions to the Mendelian predictions for independent assortment of alleles in dihybrid crosses began to crop up which could not be easily explained away by epistasis. It was apparent that there was a tendency in some cases for alleles at different loci to stay together in the original parental configurations. Analysis of data initially on eye colour and wing shape loci in *Drosophila* led Morgan in 1910 and 1911 (Morgan 1910, 1911a,b) to make the critical suggestion that genes are sometimes physically located on the same chromosome and therefore would not segregate independently of each other, but that they could recombine. Recombination between chromosomes was required to break up the parental combinations of alleles at the different loci and to generate new chromosomal combinations. Sturtevant, Morgan's undergraduate student, then conceived of using recombination frequency data to describe the relative physical positions of genes on a chromosome, producing the first ever linkage map of six loci on the X chromosome of *Drosophila melanogaster* by examining the frequency of recombination between pairwise combinations of the loci (Sturtevant, 1913). It is fascinating to read Sturtevant's account of how he created this first ever linkage map when he should have been doing his undergraduate homework. The first, albeit rather incomprehensible, report of mammalian linkage, in mice, was published by Haldane *et al.* (1915). This paper was written while Haldane was on active military duty in France and after his co-author, Sprunt, had been killed.

Definitive proof of the connection between chromosomal exchange and genetic recombination was quite slow in coming. Creighton and McClintock (1931) finally provided the definitive proof of chromosomal exchange in genetic recombination using an elegant combination of genetic and cytological analysis in maize. By using two genetic marker loci, *colorless* seed and *waxy* endosperm, in conjunction with chromosomal markers on maize chromosome 9 (knob at one end and a translocation at the other), they were able to demonstrate that recombination of the genetic markers was always associated with chromosomal exchange of the cytological markers.

Interference

After Sturtevant's initial stroke of genius, it was realized subsequently that more than two loci could be tested simultaneously for linkage. For the simple case of three loci on a chromosome, it was apparent that double crossovers could occur (one recombination event between the first pair of loci and another between the second pair of loci). However, another important observation was then made, with important repercussions for constructing linkage maps. Recombination events on the same chromosome were not necessarily independent of each other. If they were, the frequency of double crossovers would be predicted simply as the product of the single crossover frequencies. However, in most cases, fewer double recombinations were found than would be predicted on this basis. Crossovers appeared to interfere with each other. Muller (1916), another student of Morgan's, introduced the concept of interference ($I = 1 - $ (observed frequency of double recombinants)/(predicted frequency of double recombinants)) to quantify this phenomenon. The ratio of observed to expected double recombinants is also called the coefficient of coincidence, C. Generally interference is close to zero for very widely separated loci but approaches 1 for close loci. Why don't crossovers occur close together? We still await a full molecular explanation for the cross-talk between recombination sites, although there are plenty of observational data on chiasma distribution showing that double chiasmata are differently distributed along chromosomes from single chiasma and thus that the crossovers affect each other. Sym and Roeder (1994) showed that mutation of the *zip1* gene, part of the central region of the synaptonemal complex and involved in synapsis, abolishes interference in the yeast *Saccharomyces cerevisiae*. The *SCP1* locus of mammals is functionally similar to *zip1*, although it may not be homologous, and it is interesting to speculate whether it could be similarly implicated in interference in mammals. It is clear that mutation of structural components of the synaptonemal complex could interfere with synaptonemal complex formation and this could be reflected in interference levels. However, there is interest in identifying molecules which might communicate relative positions of crossover sites. Ross-Macdonald and Roeder (1994) have identified a potential communication molecule encoded by the *msh4* locus where mutation almost completely abolishes interference but does not affect synaptonemal complex formation.

Detection of Linkage

The detection of linkage between a pair of loci requires that at least one of the parents of a family be double heterozygotes, *AaBb*. If data are being combined between families, the phase of heterozygosity must also be known or a probability assigned to the alternative phases. In other words, did the parent arise from one gamete containing *AB* and the other *ab*, or was it from *Ab* and *aB*, as this will clearly affect the gametes transmitted from the parent to their offspring?

Specific crosses

In many species, it is possible to set up specific crosses to test for the presence of linkage. These have been given various names like testcross or intercross and are described in elementary genetics textbooks. If inbred lines are being crossed to produce heterozygous F_1 parents in whose progeny crossing over can be detected, only one phase is possible. The accumulation and interpretation of data is very straightforward in these cases.

LOD approach

In many species, humans most obviously, it is difficult or impossible to arrange matings to be completely informative for linkage studies. Some families will be informative and others not. In some families, both parents will be informative and, of course, the linkage phase can vary from individual to individual, further complicating the interpretation. In these cases, the best possible use must be made of data from these less than ideal family sources. With the wide application of multiple DNA markers in genome scans, the problem of aggregating data from these various heterogeneous sources

is apparent, even in outbred domestic animal species where crosses can be made at will. Although wide crosses can be made in domestic animals, like pigs and cattle, to substantially increase the heterozygosity in the F_1, the high allelic diversity of the DNA markers means that each individual family must be treated separately. Morton (1955) devised a statistically efficient method for aggregating data from many families to test for the presence of linkage which is still widely used today. The LOD score (logarithm of odds) essentially compares the probability of a data set given linkage at some particular level with the probability of the data given no linkage, that is a recombination frequency equal to 0.5. The *a priori* probability of autosomal linkage for humans assumed by Morton in his original paper is 0.05, based on the fact that humans have about 20 pairs of autosomes.

Numerous software packages have implemented Morton's method or refinements of it for estimating linkage. These packages generally estimate the LOD scores for a range of possible recombination frequencies and choose the recombination frequency giving the peak LOD score as the best estimate. CRI-MAP (Green *et al.*, 1990) is a widely used example.

When data from many loci must be combined into multi-marker maps, guidelines for the systematic construction of human maps have been developed by Keats *et al.* (1991) for a sequential procedure involving the identification of framework markers whose position and order is very strongly supported statistically followed by the positioning of less informative markers around these framework loci. These guidelines have been subsequently adopted for other species, for example the pig (Archibald *et al.*, 1995).

Morton (1955) in fact devised a form of the then recent (Wald, 1947) sequential probability ratio test in which tests are made sequentially as families accumulate, with the testing procedure designed to provide a balance between Type I errors (finding apparent linkage when there is none) and Type II errors (failing to detect genuine linkage). The odds value for any family is $P(\theta)/P(\frac{1}{2})$ where $P(\theta)$ is the probability of observing the family data on the assumption of a recombination rate of θ, while $P(\frac{1}{2})$ is the probability on the assumption of no linkage. Then the LOD score for the family is $\log_{10}(P(\theta)/P(\frac{1}{2}))$. LOD scores are then added as each new family becomes available. If α is the Type I and β the Type II error rate, end points $A = \log_{10}((1 - \beta)/\alpha)$ and $B = \log_{10}(\beta/(1 - \alpha))$ are defined and families are added until the total LOD score falls outside the limits A and B, linkage being accepted if LOD > A and independent segregation if LOD < B. Cases of apparent linkage will then consist of true linkage situations and Type I errors. If π is the power of the test (the probability of declaring linkage when loci are linked with recombination rate θ) and φ is the prior probability of linkage, the posterior probability that apparent linkage is not real is $[\alpha(1 - \varphi)/\{\alpha(1 - \varphi) + \pi\}]$. On the assumption that $\varphi = 0.05$ and for what he considered reasonable values of θ and number of families, Morton suggested that to keep the posterior probability of a Type I error satisfactorily low the value of A should be about 3. Current use of the LOD score limit of 3 does not depend on Morton's argument.

An alternative criterion based on LOD scores for non-sequential tests would use the non-sequential likelihood-ratio test, namely that if θ is the maximum likelihood estimate of recombination rate then $2\ln(P(\theta)/P(\frac{1}{2}))$ is distributed approximately as a χ^2 with one degree of freedom. A LOD of 3 thus corresponds to a χ^2 of 13.82 which has a probability of 0.0002.

Estimation of recombination rate

As noted before, it is necessary for one parent to be doubly heterozygous in order to detect recombination. When the two dominant alleles and the two recessive alleles are each on the same chromosome, the loci are said to be in coupling, while if there is one dominant and one recessive on each chromosome the loci are said to be in repulsion. With co-dominant loci this distinction is not possible, but the alleles at each locus may be arbitrarily assigned. It is simply a matter of distinguishing the haplotypes in the doubly heterozygous parent.

In the classical case of a testcross in which an F_1 (AB/ab or Ab/aB) was mated to double recessive homozygotes (ab/ab) the phenotypes of the progeny directly reflect the gametes from the F_1 parent, assuming no viability differ-

ences. Assuming coupling, the number of recombinant gametes equals the number of single dominant progeny, and the recombination rate is simply the proportion of recombinant progeny. If the progeny numbers are n_{AB}, n_{Ab}, n_{aB} and n_{ab} totalling N, the estimated recombination rate is

$$\hat{r} = (n_{Ab} + n_{aB})/N$$

and the sampling variance is

$$\sigma_{\hat{r}}^2 = r(1 - r)/N$$

which is estimated by substituting $\hat{r}$ for r in giving a standard error. The test of significance for linkage, when $r < \frac{1}{2}$ is, however, obtained by using $r = \frac{1}{2}$ in the variance formula, giving

$$\chi^2 = 4N(\hat{r} - 0.5)^2$$

which is also the value found by comparing observed and expected numbers of recombinants and non-recombinants in the usual way.

The amount of information for estimation of r present in the data is the reciprocal of the sampling variance, or $N/\{r(1 - r)\}$, being much greater for close than for loose linkage.

There is in this simple case no need for maximum likelihood (ML) estimation as simple gamete counting is available, but it may be applied as follows. The probabilities of the four gametes are $P_{AB} = P_{ab} = (1 - r)/2$, and $P_{Ab} = P_{aB} = r/2$, so that the likelihood is

$$L = (1 - r)^{n_{AB} + n_{ab}} r^{n_{Ab} + n_{aB}}/2^N$$

and thus the log likelihood is

$$LL = (n_{AB} + n_{ab})\ln(1 - r) + (n_{Ab} + n_{aB}) \ln r - N\ln 2$$

from which the ML estimate is $(n_{Ab} + n_{aB})/N$ as above, and from the second derivative of LL the variance of the ML estimate is $r(1 - r)/N$.

Biased segregation does not cause bias in the estimate of recombination rate because the ratio of parental to recombinant gametes is the same for each allele. However, if the genotypes differ in viability the simple estimate described above will be biased. Severe cases will be noticed because of differences in numbers between the two types of non-recombinant or recombinant genotypes, but less severe cases may well pass unnoticed. A simple χ^2 test of equality of numbers of the same type can be made if desired. If the cross is made in only one way, there are only three degrees of freedom among the four classes, but four parameters to be estimated (three relative viabilities and r). If both coupling and repulsion backcrosses are made, differential viability will produce different estimates of r in the crosses from the above formula. Suppose that relative to AB/ab the viabilities of Ab/ab, aB/ab and ab/ab genotypes are u, v and w. Then in the coupling cross the ratio of recombinant to parental types is

$$[(u + v)r/2]/[(1 + w)(1 - r)/2]$$

while in the repulsion cross the corresponding ratio is

$$[(1 + w)r/2]/[(u + v)(1 - r)/2]$$

because the genotypes which are recombinant and parental are reversed. However, the product of these two ratios is $r^2/(1 - r)^2$ which is unaffected by differential viability, thus allowing a clean estimate of r.

It is common practice to ignore all but very obvious viability differences, which probably has much greater justification for the DNA markers now used than for the traditional recessive markers.

There is greater complexity in estimating r in other types of mating. For example, instead of back-crossing an F_1 (AB/ab) to the double recessive (ab/ab) an F_2 could be produced. In this case the progeny numbers do not directly reflect gamete production. Assuming dominance at both loci the probabilities for the four observable phenotypes are

A-B-	A-bb	aaB-	aabb
$(2 + Q)/4$	$(1 - Q)/4$	$(1 - Q)/4$	$Q/4$

where $Q = (1 - r_M)(1 - r_F)$, r_M and r_F being the male and female recombination rates, respectively. It should be noted that Q is the estimable parameter in this case. Only on the assumption that $r_M = r_F$ can a unique recombination rate be estimated.

If the four progeny types are present in numbers n_1, n_2, n_3 and n_4, the log likelihood is

$$LL = n_1 \ln(0.5 + Q/4) + (n_2 + n_3)\ln([1 - Q]/4) + n_4 \ln(Q/4)$$

Maximizing LL with respect to Q leads to the quadratic equation

$$2n_4 + (n_1 - 2n_2 - 2n_3 - n_4)Q - (n_1 + n_2 + n_3 + n_4)Q^2 = 0.$$

The estimate $\hat{Q}$ is the appropriate root of this equation. The sampling variance of the estimate obtained from the second derivative of LL is

$$\sigma_Q^2 = 2Q(1 - Q)(2 + Q)/(N(1 + 2Q))$$

where N is the total number of progeny.

If we assume the recombination rate is the same in both sexes we can estimate it as

$$\hat{r} = 1 - \sqrt{Q}$$

and its sampling variance as

$$\sigma_r^2 = \sigma_Q^2/(dQ/dr)^2 = \frac{r(2 - r)(3 - 2r + r^2)}{2N(3 - 4r + 2r^2)}.$$

Sampling variances are very useful for comparing the relative precision of different experimental designs. For example, we can use the variance formulae for backcross and F_2 experiments of equal size to assess their efficiencies with recessive markers. The relative efficiency of F_2 compared with backcross experiments depends on whether the cross is in coupling or repulsion since this changes the distribution of F_2 phenotypes and thus the sampling variance. The relative efficiency of F_2 experiments is:

$$\text{Coupling} \quad \frac{2(1 - r)(3 - 4r + 2r^2)}{(2 - r)(3 - 2r + r^2)}.$$

$$\text{Repulsion} \quad \frac{2r(1 + 2r^2)}{(2 + r^2)(1 + r)}.$$

If the genes are co-dominant, the precision is the same for coupling and repulsion since all the genotypes can be distinguished, and the relative efficiency of an F_2 experiment turns out to be

$$\text{Co-dominant} \quad \frac{2(1 - 3r + 3r^2)}{1 - 2r + 2r^2}$$

It is instructive to compare these efficiencies for a few values of r, as in Table 1.1.

For recessives the F_2 is half as efficient as a backcross for unlinked loci, and for coupling approaches the backcross in efficiency as linkage becomes tight, whereas the repulsion design gives essentially no information when linkage is tight. On the other hand, with co-dominant genes the F_2 is always at least as efficient as the backcross and approaches an efficiency of 2 with tight linkage. The greater efficiency of the F_2 arises because meioses in both parents contribute information, whereas in the backcross only meioses in the AB/ab parent are informative. In the F_2 there is some loss of

Table 1.1. Relative efficiencies of different designs as affected by recombination rate.

Design	Recombination rate				
	0.5	0.4	0.2	0.1	0.05
Coupling	0.50	0.59	0.77	0.88	0.94
Repulsion	0.50	0.35	0.18	0.09	0.05
Co-dominant	1.00	1.08	1.53	1.78	1.89

efficiency because the progeny do not all give information on the gametes transmitted by both parents. Of course, in the backcross design sex-specific recombination rates are estimable.

Van Arendonk and Van der Beek (1991) studied the efficiency of linkage estimation in full sib families by simulating samples from matings of different parental genotypes and different recombination rates. Not surprisingly, the most informative family type was a mating of $A_1A_2B_1B_2$ by $A_3A_4B_3B_4$ in which all gametes can be identified. Knowledge of the parental linkage phase is equivalent to two extra progeny when the recombination rate is 0.05, but with $r=0.3$ was equivalent to an extra 10–15 progeny. Thus for tight linkage it may be better to increase progeny numbers rather than to genotype grandparents in order to establish linkage phase in parents.

In some situations, genotype data will be available only for sires and progeny in half-sib families, the dams being either unavailable or considered too expensive to genotype. Gomez-Raya (2001) considered estimation of recombination rates in this case, paying special attention to the fact that many progeny will be uninformative. By developing ML procedures to estimate recombination rate including uninformative progeny, assuming known gene frequencies in the dam population, as the basis for assessing the effect of discarding them from the analysis, Gomez-Raya showed that ignoring uninformative offspring can lead to biased estimation of map distance and reduction of LOD score in analysing half-sib family data.

Mapping Functions

If two or indeed any even number of crossovers occur between a pair of marker loci, then in the absence of other information, no recombina-

tion event will be observable. It will appear that no crossover has occurred. If three or more odd number of crossovers occur, then only one recombination event will be apparent and the additional crossovers will not be added to the true tally. Mapping functions attempt to better estimate true crossover frequency between a pair of loci by estimating the number of these undetectable multiple crossovers and adding them to the actual tally of observed recombination events between the loci. This adjusted frequency is called a map distance. In practice, we really only need to worry about the effect of double crossovers, since higher order multiple crossovers are normally going to be very rare. Indeed even double crossovers are so rare in practice that their occurrence in real data sets is usually taken as indicative of genotyping errors. Standard linkage analysis programs such as CRI-MAP (Green et al., 1990) can be set to flag all double recombinants as potential genotyping errors.

The Haldane function

J.B.S. Haldane devised the first mapping function in 1919 (Haldane, 1919). He assumed that the distribution of crossovers was random and approximated a Poisson distribution. This is equivalent to assuming that there is no crossover interference, a situation in which the observed recombination frequency most severely underestimates the actual crossover frequency. Thus for a triplet of loci in the order ABC, he assumed $r_{AC} = r_{AB} + r_{BC} - 2 r_{AB} r_{BC}$.

The map distance or estimated true frequency of crossover events is estimated as $d = -\frac{1}{2}\ln(1 - 2r)$ where r is the observed frequency of recombination events. Haldane introduced the unit of map length, the Morgan, in this 1919 publication.

However, the assumption of no interference is clearly not valid for mammals, where interference has been shown to extend for very long distances. Kwiatkowski et al. (1993) showed for a very large sample of human chromosome 9 that interference was still very strong even for markers separated by 20 cM – double crossovers were 100-fold less common than predicted, given no interference, for 10 cM intervals and 80-fold less common for 20 cM intervals.

The Kosambi function

Kosambi (1944) developed a new mapping function which partly accounts for interference, as data from many organisms had shown the importance of interference, especially for close loci. For three loci in the order ABC, the relationship, $r_{AC} = r_{AB} + r_{BC} - 2C$ r_{AB} r_{BC}, takes account of interference. If C is 0, there is complete interference and recombination frequencies are additive as map lengths. If C is 1, there is no interference, as assumed for the Haldane function. The problem of course is how do we choose the value of C, as this will vary for loci close together versus loci far apart and may well vary for different chromosomal regions in the genome. The Kosambi mapping function, $d = \frac{1}{4}$ $\ln[(1+2r)/(1-2r)]$, sets the coefficient of coincidence, C, equal to 2r. Thus for closely linked loci, interference is strong but for distant loci it decreases towards zero.

Comparison of mapping functions

Figure 1.1 shows the relationship between recombination frequency and map distance for the Kosambi and Haldane mapping functions. It can be seen that for recombination frequencies of less than about 10%, it makes little difference which mapping function is used or indeed whether any mapping function is used to transform recombination frequency. Numerous other more complicated mapping functions have been developed. Crow (1990) has provided an amusing and informative overview of the history, development and application of these functions. Liu (1998) provides a thorough coverage of eight functions, although only the Haldane and Kosambi functions tend to be used in practice. Indeed, as genetic maps become more complete and recombination frequencies between adjacent markers average only a few per cent or less, application of mapping function transforma-

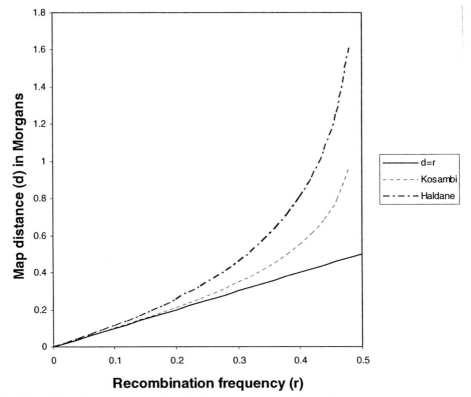

Fig. 1.1. Relationship between recombination frequency and map distance for several mapping functions.

tions to recombination frequencies becomes less necessary as interference is usually complete and recombination frequencies are in fact additive for such closely linked loci.

However, for less complete maps, Lathrop *et al.* (1984) pointed out that mapping functions can be very helpful in multi-point mapping. Estimation of the amount of interference may be difficult, so that the assumption of a known mapping function (e.g. Haldane or Kosambi) can increase precision, perhaps at the expense of some bias because of use of an incorrect function. It may be especially useful in the case of adding new loci to well-established maps. Suppose we have three loci, A, B and C which are well located, and a new locus D to be placed on the map. Using an assumed mapping function we can let the map location of D vary from well to the left of A to well to the right of C, computing the likelihood of the data for each location, and thus finding the ML estimate of the map location of D, conditional on the assumed function.

When several linkage maps are available, based on overlapping sets of markers, the construction of a combined or consensus map is a useful way to integrate the various maps. There is no generally accepted way of doing this, with subjective decisions being commonly made. If all the raw data are available, the several maps can be integrated using a software package, but all data would not normally be available, and in some situations, such as maps based on recombination and physical maps, it would be impossible. Yap *et al.* (2003) proposed a graph-theoretic method of combining maps which would give gene orders for those loci which were in the same order on all maps, and would indicate those loci which needed to have their order resolved. This could be useful in guiding further research, but would not avoid the need for interim decisions to be made.

Current State of Linkage Maps in Mammals

Commencing in the 1980s in humans, there has been explosive growth of linkage and other genetic maps in a range of mammalian species, including humans, mouse, rat, cattle, sheep, goats, pigs, rabbits, dogs, cats and horses. Much of this growth in genetic mapping arose from the seminal paper of Botstein *et al.* (1980) when the first genetic map based on DNA markers, specifically restriction fragment length polymorphisms, was developed for humans. It soon became apparent that molecular biology could provide an effectively unlimited number of genetic markers in any species and the race was on to construct comprehensive maps. Marker dense linkage maps, mainly based on microsatellites, are now available for these species (Table 1.2).

Why Make Linkage Maps?

Genetic linkage maps, like road maps, are a means to an end. Their primary purpose is to help locate, identify and exploit genes that are of specific utility. Human geneticists are interested in mapping genes and mutations predisposing to disease to assist in assessing health risks, to provide genetic counselling and for developing therapies. Plant and animal breeders are interested in identifying genes and mutations affecting productivity, product quality and disease resistance, usually with the objective of increasing the frequency of favourable alleles in breeding programmes.

The distribution of effects of mutations shows an L-shape (gamma distribution), with a very large number of mutations of small effect and a very small number of mutations of large effect. Whilst the mapping of genes of largest effect is quite straightforward and the mapping of mutations of smallest (infinitesimal) effect is quite impossible, a genetic marker map provides a valuable resource, in conjunction with appropriate family and pedigree resources, for mapping many of the genes lying at intermediate positions between these extremes. Hayes and Goddard (2001) have estimated the distribution of gene effects in two important species, namely cattle and pigs, and have concluded that genes responsible for about half of the observed variation are mappable in resources of realistic size.

If mutations have a sufficiently large effect, then their segregation in families and crosses can be directly observed. In these cases, the co-segregation of alleles at the gene of interest and the

Table 1.2. Web sites providing genetic maps and/or genomic information for a range of mammalian species.

Human, mouse, rat, cow, pig, dog and other species	National Center for Biotechnology Information Mapviewer	http://www.ncbi.nlm.nih.gov/ mapview/
Mouse	Mouse Genome Informatics	http://www.informatics.jax.org/
Human	The Genome Database	http://gdb.mirror.edu.cn/
Rat	RATMAP: the rat genome database	http://ratmap.gen.gu.se/
Rat	Rat Genome database	http://rgd.mcw.edu/
Rat	The Wellcome Trust Centre for Human Genetics Rat Mapping Resources	http://www.well.ox.ac.uk/rat_mapping_resources/
Cattle	BOVMAP	http://locus.jouy.inra.fr/cgi-bin/bovmap/intro2.pl
Cattle, pig, sheep, deer, horse and others	ARKdb	http://www.thearkdb.org/
Cattle, pig, horse, sheep	U.S. Livestock Genome Mapping Projects	http://www.genome.iastate.edu/
Horse	Horse Genome Project	http://www.uky.edu/Ag/Horsemap/
Dog	The FHCRC Dog Genome Project	http://www.fhcrc.org/science/dog_genome/
Goat	GOATMAP	http://locus.jouy.inra.fr/cgi-bin/lgbc/mapping/goatmap/Goatmap/main.pl
Kangaroo	The Kangaroo Genome project	http://kangaroo.genome.org.au/
Vertebrates including sheep, cattle, goat, pig, horse and dog	The Genome Web	http://www.hgmp.mrc.ac.uk/GenomeWeb/
Rabbit	Rabbitmap	http://locus.jouy.inra.fr/cgi-bin/lgbc/mapping/common/intro2.pl?BASE=rabbit

marker loci can be treated little differently from the co-segregation of alleles at pairs of marker loci, assuming the gene of interest does not have a substantial impact on the fitness of homozygotes or heterozygotes for the mutation. In these straightforward cases, linkage analysis packages, like CRIMAP (Green et al., 1990), can be used to map the gene of interest. In human genetic studies, where small families are the rule, homozygosity mapping (Lander and Botstein, 1987) is an effective way of quickly locating a gene to a chromosome interval where positional candidate loci can then be investigated. This approach has also been successfully applied to the mapping of domestic animal mutations.

QTL and complex trait mapping

For mutations lying between the extremes of large effect and infinitesimal effect, there is a subset of mutations whose effects are too small for their segregation to be observed directly amongst the background noise of segregation of other loci and random environmental factors but large enough to be detected by co-segregation with genetic markers. Geldermann (1975) introduced the term quantitative trait locus for such genes and described how they could be detected. Human geneticists tend to use the term complex trait mapping, since many of the traits which interest them are all-or-none binary traits, such as presence or absence of a disease, but are nevertheless multifactorial. There are far too many QTL or complex trait research projects on many species throughout the world in domestic animals and in human medical genetics to attempt to summarize here, but specific examples can be found at the web sites listed in Table 1.3. QTL mapping is dealt with more fully in Chapter 20.

Table 1.3. Sample of web sites for quantitative trait loci in mammalian species.

		URL	Traits
Proteome Centre Rostock	Rat and mouse models for human disease	http://qtl.pzr.uni-rostock.de/	Multiple sclerosis, rheumatoid arthritis
Rat QTL	Rat models for human disease	http://ratmap.gen.gu.se/qtler/	Diabetes, metabolic disorders, hypertension and cardiovascular diseases, arthritis, encephalomyelitis, behaviour, cancer
Jackson Laboratory Heart, Blood, Lung and Sleep Disorders Center	Mouse models for human disease	http://pga.jax.org/qtl/	Atherosclerosis, cholesterol, gallstones HDL cholesterol, hypertension
Online Combined QTL Map of Dairy Cattle Traits	Cattle	http://www.vetsci.usyd.edu.au/reprogen/QTL_Map/	Milk, protein, fat, somatic cell count

High resolution linkage disequilibrium mapping of QTLs

High resolution mapping in future will rely increasingly not on the detection of linkage, but on the occurrence of linkage disequilibrium, that is non-random associations between pairs of loci detectable at the population level rather than the family or pedigree level. There is insufficient opportunity for enough recombination to accumulate over the course of a two or three generation pedigree to map QTLs with anything but very coarse precision, generally no better than about 20 cM resolution. However, appropriately designed resources can exploit the combined historical accumulation of recombination by utilizing linkage disequilibrium to map genes of interest to much higher resolution and therefore assist in their identification and use. Cardon and Abecasis (2003) have described preliminary progress in the human HapMap project which eventually aims to use many thousands of closely spaced single nucleotide polymorphism (SNP) markers to precisely map and identify loci affecting complex trait phenotypes in humans such as susceptibility to common diseases like asthma and diabetes. This same approach is also being used in domestic animals.

For inbred model organisms such as mice, various schemes are used to accumulate the additional recombination required to map QTLs at higher resolution than is possible in an F_2 cross. The development of congenic lines via many generations of backcrossing is one strategy for accumulating the additional recombination, but this requires very extensive phenotyping and genotyping and the creation of a large number of congenics – usually one or two per QTL detected in an F_2 screen. Darvasi and Soller (1995) proposed an efficient strategy for accumulation of additional recombination by the development of what they termed advanced intercross lines or AILs, where a cross is taken to F_{10} or beyond, but with sufficient numbers of breeding animals that genetic drift will be relatively inconsequential. No genotyping or phenotyping is required in the intermediate generations. Map resolution of QTL positions can be improved by a factor of five or more. Iraqi (2000) has reported successful application of an AIL including positional cloning of the relevant genes.

The ultimate QTL mapping resource

The Complex Trait Consortium was established in 2002 with the objective of generating a generally available mouse QTL mapping resource based on a very large series of recombinant inbred lines from crosses between eight inbred lines of mice (Williams *et* al., 2002; Threadgill et al., 2002). The strains would be crossed A×B up to G×H, then the progeny AB×CD, EF×GH, then ABCD×EFGH and finally ABCDEFGH would be sib-mated for 20 generations to produce the RI lines containing genetic contributions from all eight parents. About 1000 recombinant inbred lines are planned. A core set of 50–100 lines would be kept for routine coarse level map-ping and also for generating large numbers of recombinant inbred crosses (RIX) which would be generated when required and would permit more reliable phenotyping due to their higher levels of heterozygosity. The consortium aims to achieve a mapping resolution for QTLs of 0.1 cM and to be able to detect QTLs with effects greater than about 0.25 SD. By exploit-ing historical haplotype breakpoints already existing in the inbred lines, the consortium believes that QTL mapping resolution down to 100 kbp will be possible, greatly narrowing the search for candidate genes underlying QTLs to about five loci. An enormous advantage of recombinant inbred strains is that they only need genotyping once, but they can be phe-notyped to any necessary degree of precision for any conceivable and measurable trait later. The problem of course is cost, estimated to be about US$50 million, and determining if there are more economically competitive methods of achieving the same results. Mappers of other species of mammals can only look on with envy at these possibilities available to mouse QTL mappers and assume that much functional information extracted from mice can be extrapolated to other species from such mouse resources.

Concluding Remarks

The cellular process of meiotic recombination, underlying genetic linkage, is becoming better understood at the molecular level in mammals, building on fundamental discoveries made in simpler eukaryotes. However, some fundamen-tally important aspects of meiotic crossing over and linkage, such as interference, are still not

at all understood at the molecular level, even though interference was recognized almost 90 years ago. One must expect that recognition of all the genes and gene products involved in meiosis and crossing over, which will eventu-ally flow from genome sequences, will eventu-ally throw light even on this question.

Genetic linkage maps have a more impor-tant role than ever to play in the era of func-tional genomics. Much, indeed probably most, of the biological variation of interest to medical, agricultural and biological scientists is due to segregation of genes whose effects are individually too small to be detected directly. Providing their effects are not too small, co-segregation with genetic markers provides the only means for recognizing these genes and their variants. New linkage map-ping strategies, involving exploitation of historical recombination, namely linkage disequilibrium mapping, or breeding strate-gies to accumulate higher levels of recombi-nation, promise to map QTLs at far better precision than currently possible and will lead more directly to the genes and their variants underlying important QTLs.

Acknowledgements

Chris Moran wishes to acknowledge the long-term financial support of Australian Pork Limited (formerly the Pig Research and Development Corporation) which has sup-ported his genetic mapping studies. He wishes to thank Professor Peter Sharp for providing access to his unpublished PhD thesis on mar-supial meiosis. Finally he wishes to dedicate this paper to his PhD supervisor, Dr David Shaw, for introducing him many years ago to the beauty of meiosis and the challenges of the study of recombination. Some of the material in this chapter was used by John James for a course given at Wageningen Agricultural University, and he thanks Professor E.W. Brascamp for giving him this opportunity. He also wishes to thank Reprogen for giving him a scientific home.

References

Anderson, L.K., Reeves, A., Webb, L.M. and Ashley, T. (1999) Distribution of crossovers on mouse chromosomes using immunofluorescent localization of MLH1 protein. *Genetics* 151, 1569–1579.

Archibald, A.L., Haley, C.S., Brown, J.F., Couperwhite, S., McQueen, H., Nicholson, D., Coppieters, W., Van de Weghe, A., Stratil, A., Wintero, A.-K., Fredholm, M., Larsen, N.J., Nielsen, V.H., Milan, D., Woloszyn, N., Robic, A., Dalens, M., Riquet, J., Gellin, J., Caritez, J.-C. *et al.* (1995) The PiGMaP consortium linkage map of the pig (*Sus scrofa*). *Mammalian Genome* 6, 157–175.

Barendse, W. and Fries, R. (1999) Genetic linkage mapping, the gene maps of cattle and the lists of loci. In: Fries, R. and Ruvinsky, A. (eds) *The Genetics of Cattle.* CAB International, Wallingford, UK.

Barendse, W., Vaiman, D., Kemp, S.J., Sugimoto, Y., Armitage, S.M., Williams, J.L., Sun, H.S., Eggen, A., Agaba, M. and Aleyasin, S.A. (1997) A medium-density genetic linkage map of the bovine genome. *Mammalian Genome* 8, 21–28.

Barlow, A.L. and Hulten, M.A. (1998) Crossing over analysis at pachytene in man. *European Journal of Human Genetics* 5, 350–358.

Bennett, J.H., Hayman, D.L. and Hope, R.M. (1986) Novel sex differences in linkage values and meiotic chromosome behaviour in a marsupial. *Nature* 323, 59–60.

Bojko, M. (1983) Human meiosis VIII. Chromosome pairing and formation of the synaptonemal complex in oocytes. *Carlsberg Research Communications* 48, 457–483.

Botstein, D., White, R.L., Skolnick, M. and Davis, R.W. (1980) Construction of a genetic linkage map in man using restriction fragment length polymorphisms. *American Journal of Human Genetics* 32, 314–331.

Broman, K.W., Murray, J.C., Sheffield, V.C., White, R.L. and Weber, J.L. (1998) Comprehensive human genetic maps: individual and sex-specific variation in recombination. *American Journal of Human Genetics* 63, 861–869.

Bronner, C.E., Baker, S.M., Morrison, P.T., Warren, G., Smith, L.G., Lescoe, M.K., Kane, M., Earabino, C., Lipford, J., Lindblom, A., Tannergard, P., Bollag, R.J., Godwin, A.R., Ward, D.C., Nordenskjold, M., Fishel, R., Kolodner, R. and Liskay, R.M. (1994) Mutation in the DNA mismatch repair gene homologue hMLH1 is associated with hereditary non-polyposis colon cancer. *Nature* 368, 258–261.

Cardon, L.R. and Abecasis, G.R. (2003) Using haplotype blocks to map human complex trait loci. *Trends in Genetics* 19, 135–140.

Carpenter, A.T. (1975) Electron microscopy of meiosis in *Drosophila melanogaster* females: II. The recombination nodule – a recombination-associated structure at pachytene? *Proceedings of the National Academy of Sciences USA* 72, 3186–3189.

Cohen, P.E. and Pollard, J.W. (2001) Regulation of meiotic recombination and prophase I progression in mammals. *BioEssays* 23, 996–1009.

Creighton, H.S. and McClintock, B. (1931) A correlation of cytological and genetical crossing-over in *Zea mays. Proceedings of the National Academy of Sciences USA* 17, 492–497.

Crow, J.F. (1990) Mapping functions. *Genetics* 125, 669–671.

Darvasi, A. and Soller, M. (1995) Advanced intercross lines, an experimental population for fine genetic mapping. *Genetics* 141, 1199–1207.

Davisson, M.T., Roderick, T.H. and Doolittle, D.P. (1989) Recombination percentages and chromosomal assignments. In: Lyon, M.F. and Searle, A.G. (eds) *Genetic Variants and Strains of the Laboratory Mouse.* Oxford University Press, pp. 432–505.

Dudas, A. and Chovanec, M. (2003) DNA double-strand break repair by homologous recombination. *Mutation Research* 566, 131–167.

Dunn, L.C. and Bennett, D. (1967) Sex differences in recombination of linked genes in animals. *Genetical Research Cambridge* 9, 211–220.

Eisenbarth, I., Vogel, G., Krone, W., Vogel, W. and Assum, G. (2000) An isochore transition in the *NF1* gene region coincides with a switch in the extent of linkage disequilibrium. *American Journal of Human Genetics* 67, 873–880.

Eyre-Walker, A. (1993) Recombination and mammalian genome evolution. *Proceedings of the Royal Society London B. Biological Sciences* 252, 237–243.

Frönicke, L., Anderson, L.K., Wienberg, J. and Ashley, T. (2002) Male mouse recombination maps for each autosome identified by chromosome painting. *American Journal of Human Genetics* 71, 1353–1368.

Fullerton, S.M., Carvalho, A.B. and Clark, A.G. (2001) Local rates of recombination are positively correlated with GC content in the human genome. *Molecular Biology and Evolution* 18, 1139–1142.

Geldermann, H. (1975) Investigations on inheritance of quantitative characters in animals by gene markers. I. Methods. *Theoretical and Applied Genetics* 46, 319–330.

Gomez-Raya, L. (2001) Biased estimation of the recombination fraction using half-sib families and informative offspring. *Genetics* 157,1357–1367.

Gorlov, I.P., Zhelezova, A.I. and Gorlova, O.Y. (1994) Sex differences in chiasma distribution along two marked mouse chromosomes: differences in chiasma distribution as a reason for sex differences in recombination frequency. *Genetical Research* 64, 161–166.

Green, P., Falls, K. and Crooks, S. (1990) Documentation for CRI-MAP, Version 2.4 (St Louis: Washington University School of Medicine).

Hassold, T., Sherman, S. and Hunt, P. (2000) Counting cross-overs: characterising meiotic recombination in mammals. *Human Molecular Genetics* 9, 2409–2419.

Haldane, J.B.S. (1919) The mapping function. *Journal of Genetics* 8, 299–309.

Haldane, J.B.S. (1922) Sex ratio and unisexual sterility in hybrid animals. *Journal of Genetics* 12, 101–109.

Haldane, J.B.S., Sprunt, A.D. and Haldane, N.M. (1915) Reduplication in mice. *Journal of Genetics* 5, 133–136.

Hayes, B. and Goddard, M.E. (2001) The distribution of the effects of genes affecting quantitative traits in livestock. *Genetics Selection Evolution* 33, 209–229.

Holmquist, G.P. (1992) Chromosome bands, their chromatin flavours, and their functional features. *American Journal of Human Genetics* 51, 17–37.

Hunt, P.A. and Hassold, T.J. (2002) Sex matters in meiosis. *Science* 296, 2181–2183.

Iraqi, F. (2000) Fine mapping of quantitative trait loci using advanced intercross lines of mice and positional cloning of the corresponding genes. *Experimental Lung Research* 26, 641–649.

Kappes, S.M., Keele, J.W., Stone, R.T., McGraw, R.A., Sonstegard, T.S., Smith, T.P.L., Lopez-Corrales, N.L. and Beattie, C.W. (1997) A second-generation linkage map of the bovine genome. *Genome Research* 7, 235–249.

Keats, B.J.B., Sherman, S.L., Morton, N.E., Robson, E.B., Buetow, K.H., Cartwright, P.E., Chakravarti, A., Francke, U., Green, P.P. and Ott, J. (1991) Guidelines for human linkage maps: an international system for human linkage maps. *Genomics* 9, 557–560.

Kleckner, N., Storlazzi, A. and Zickler, D. (2003) Co-ordinate variation in meiotic pachytene SC length and total crossover/chiasma frequency under conditions of constant DNA length. *Trends in Genetics* 19, 623–628.

Kosambi, D.D. (1944) The estimation of map distances from recombination values. *Annals of Eugenics* 12, 172–175.

Kwiatkowski, D.J., Dib, C., Slaugenhaupt, S.A., Povey, S., Gusella, J.F. and Haine, J.L. (1993) An index marker map of chromosome 9 provides strong evidence for positive interference. *American Journal of Human Genetics* 53, 1279–1288.

Lander, E.S. and Botstein, D. (1987) Homozygosity mapping: a way to map human recessive traits with the DNA of inbred children. *Science* 236, 1567–1570.

Lathrop, G.M., Lalouel, J.M., Julier, C. and Ott, J. (1984) Strategies for multilocus linkage analysis in humans. *Proceedings of the National Academy of Sciences USA* 81, 3443–3446.

Li, W.T., Fann, C.S.J. and Ott, J. (1998) Low-order polynomial trends of female-to-male map distance ratios along human chromosomes. *Human Heredity* 48, 266–270.

Liu, B. (1998) *Statistical Genomics: Linkage, Mapping and QTL Analysis*. CRC Press, Boca Raton, Florida.

Liu, J.G., Yuan, L., Brundell, E., Bjorkroth, B., Daneholt, B. and Hoog, C. (1996) Localization of the N-terminus of SCP1 to the central element of the synaptonemal complex and the evidence for direct interactions between the N-termini of SCP1 molecules organized head-to-head. *Experimental Cell Research* 226, 11–19.

Lynn, A., Koehler, K.E., Judis, L., Chan, E.R., Cherry, J.P., Schwartz, S., Seftel, A., Hunt, P.A. and Hassold, T.J. (2002) Covariation of synaptonemal complex length and mammalian meiotic exchange rates. *Science* 296, 2222–2225.

Maddox, J.F., Davies, K.P., Crawford, A.M., Hulme, D.J., Vaiman, D., Cribiu, E.P., Freking, B.A., Beh, K.J., Cockett, N.E., Kang, N., Riffkin, C.D., Drinkwater, R., Moore, S.S., Dodds, K.G., Lumsden, J.M., van Stijn, T.C., Phua, S.H., Adelson, D.L., Burkin, H.R., Broom, J.E., Buitkamp, J., Cambridge, L., Cushwa,

W.T., Gerard, E., Galloway, S.M., Harrison, B., Hawken, R.J., Hiendleder, S., Henry, H.M., Medrano, J.F., Paterson, K.A., Schibler, L., Stone, R.T. and van Hest, B. (2001) An enhanced linkage map of the sheep genome comprising more than 1000 loci. *Genome Research* 11, 1275–1289.

Mellersh, C.S., Langston, A.A., Acland, G.M., Fleming, M.A., Ray, K., Wiegand, N.A., Francisco, L.V., Gibbs, M., Aguirre, G.D. and Ostrander, E.A. (1997) A linkage map of the canine genome. *Genomics* 46, 326–336.

Morgan, T.H. (1910) The method of inheritance of two sex-limited characters in the same animal. *Proceedings of the Society for Experimental Biology and Medicine* 8, 17.

Morgan, T.H. (1911a) Random segregation versus coupling in Mendelian inheritance. *Science* 34, 384.

Morgan, T.H. (1911b) An attempt to analyze the constitution of the chromosomes on the basis of sex-limited inheritance in *Drosophila. Journal of Experimental Zoology* 11, 365.

Morton, N.E. (1955) Sequential tests for the detection of linkage. *American Journal of Human Genetics* 7, 277–318.

Muller, H.J. (1916) The mechanism of crossing-over. *American Naturalist* 50, 193–221.

Nebel, B.R., Amarose, A.P. and Hackett, E.M. (1961) Calendar of gametogenic development in the prepubertal male mouse. *Science* 134, 832–833.

Ostrander, E.A., Galibert, F. and Mellersh, C.S. (2001) Linkage and radiation hybrid mapping in the canine genome. In: Ruvinsky, A. and Sampson, J.J. (eds) *Genetics of the Dog.* CAB International, Wallingford, UK.

Petrini, J.H.G. (1999) The mammalian Mre11–Rad50–Nbs1 protein complex: integration of functions in the cellular DNA-damage response. *American Journal of Human Genetics* 64, 1264–1269.

Polani, P.E. (1972) Centromere localization at meiosis and the position of chiasmata in the male and female mouse. *Chromosoma* 64, 241–254.

Roeder, G.S. (1997) Meiotic chromosomes: it takes two to tango. *Genes and Development* 11, 2600–2621.

Romanienko, P.J. and Camerini-Otero, R.D. (2000) The mouse Spo11 gene is required for meiotic chromosome synapsis. *Molecular Cell* 6, 975–987.

Ross-Macdonald, P. and Roeder, G.S. (1994) Mutation of a meiosis-specific *MutS* homolog decreases crossing over but not mismatch correction. *Cell* 79,1069–1080.

Sharp, P.J. (1984) Cytological studies in Australian marsupials. PhD thesis, University of Adelaide, Adelaide, Australia.

Shaw, D.D. (1971) Genetic and environmental components of chiasma control. I. Spatial and temporal variation in *Schistocerca* and *Stethophyma. Chromosoma* 34, 281–301.

Shaw, D.D. (1972) Genetic and environmental components of chiasma control. II. The response to selection in *Schistocerca. Chromosoma* 37, 297–308.

Shaw, D.D. (1974) Genetic and environmental components of chiasma control. 3. Genetic analysis of chiasma frequency variation in two selected lines of *Schistocerca gregaria* Forsk. *Chromosoma* 46, 365–374.

Shaw, D.D. and Knowles, G.R. (1976) Comparative chiasma analysis using a computerised optical digitiser. *Chromosoma* 59, 103–127.

Silver, L.M. (1995) *Mouse Genetics. Concepts and Applications.* Oxford University Press, New York and Oxford.

Sturtevant, A.H. (1913) The linear arrangement of six sex-linked factors in *Drosophila*, as shown by their mode of association. *Journal of Experimental Biology* 14, 43–59.

Sun, H., Treco, D., Schultes, N.P. and Szostak, J.W. (1989) Double-strand breaks at an initiation site for meiotic gene conversion. *Nature* 338, 87–90.

Sym, M. and Roeder, G.S. (1994) Crossover interference is abolished in the absence of a synaptonemal complex protein. *Cell* 79, 283–292.

Tease, C., Hartshorne, G.M. and Hulten, M.A. (2002) Patterns of recombination in human fetal oocytes. *American Journal of Human Genetics* 70, 1469–1479.

Threadgill, D.W., Hunter, K.W. and Williams, R.W. (2002) Genetic dissection of complex and quantitative traits: from fantasy to reality via a community effort. *Mammalian Genome* 13, 175–178.

Van Arendonk, J.A.M. and Van der Beek, S. (1991) Estimation of recombination rates between markers from segregating populations. *Proceedings of the Australian Association of Animal Breeding and Genetics* 9,144–150.

van Oorschot, R.A., Porter, P.A., Kammerer, C.M. and VandeBerg, J.L. (1992) Severely reduced recombination in females of the South American marsupial *Monodelphis domestica. Cytogenetics and Cell Genetics* 60, 64–67.

Wald, A. (1947) *Sequential Analysis*. John Wiley and Sons, New York.

Wallace, B.M.N. and Hulten, M.A. (1985) Meiotic chromosome pairing in the normal human female. *Annals of Human Genetics* 49, 215–226.

Williams, R.W., Broman, K.W., Cheverud, J.M., Churchill, G.A., Hitzemann, R.W., Hunter, K.W., Mountz, J.D., Pomp, D., Reeves, R.H., Schalkwyk, L.C. and Threadgill, D.W. (2002) A collaborative cross for high-precision complex trait analysis. *1st Workshop Report of the Complex Trait Consortium*: Sept 2002. www.complextrait.org/Workshop1.pdf

Yap, I.V., Schneider, D., Kleinberg, J., Matthews, D., Cartinhour, S. and McCouch, S.R. (2003) A graph-theoretic approach to comparing and integrating genetic, physical and sequence-based maps. *Genetics* 165, 2235–2247.

Yoshida, K., Kondoh, G., Matsuda, Y., Habu, T., Nishimune, Y. and Morita, T. (1998) The mouse *RecA*-like gene *Dmc1* is required for homologous chromosome synapsis during meiosis. *Molecular Cell* 1, 705–718.

Yuan, L., Liu, J.G., Zhao, J., Brundell, E., Daneholt, B. and Hoog, C. (2000) The murine *SCP3* gene is required for synaptonemal complex assembly, chromosome synapsis and male fertility. *Molecular Cell* 5, 73–83.

Zenger, K.R., McKenzie, L.M. and Cooper, D.W. (2002) The first comprehensive genetic linkage map of a marsupial: the tammar wallaby (*Macropus eugenii*). *Genetics* 162, 321–330.

Zhang, C.T. and Zhang, R. (2003) An isochore map of the human genome based on the Z curve method. *Gene* 317, 127–135.

2 Mapping Genomes at the Chromosome Level

B.P. Chowdhary[1,2*] and T. Raudsepp[1]

[1]Department of Veterinary Anatomy & Public Health, College of Veterinary Medicine and [2]Department of Animal Sciences, College of Agriculture & Life Sciences, MS 4458, Texas A&M University, College Station, TX 77843, USA

Introduction	23
Mammalian Cytogenetics: Launchpad for Chromosome Mapping	24
Comparative Cytogenetics and Gene Mapping: a Brief Overview	25
Standard Karyotypes and Nomenclature Systems as Prerequisites for Chromosome Mapping	26
Physical Mapping: an Overview of Chromosome Mapping Approaches	27
Cytogenetic Mapping: Fluorescence *in situ* Hybridization (FISH)	27
Somatic Cell Hybrid (SCH) Panels and Synteny Mapping	32
Construction of an SCH panel	32
Methods for SCH panel analysis	33
Synteny maps in different mammalian species	34
Radiation Hybrid (RH) Mapping	39
Genotyping an RH panel	40
Data analysis	40
RH mapping in different mammals	41
Future Prospects	45
Acknowledgements	46
References	46

Introduction

All gene maps relate in one way or another to the arrangement of genetic markers on chromosomes. They provide us with the fundamental information about how genomes are organized. Gene maps are divided into three major categories: first the *genetic linkage maps* – discussed at length in Chapter 1 – and secondly, the *physi-cal maps* – that originate from three major approaches, i.e. somatic cell hybrid (SCH) analysis, radiation hybrid (RH) analysis and *in situ* hybridization (ISH). The past decade has witnessed the emergence of a third group of maps that are more refined and sophisticated than genetic linkage and physical maps. These maps, conveniently designated as *molecular maps*, refer to the study of the organization of genomes

* Correspondence: bchowdhary@cvm.tamu.edu

© CAB International 2005. *Mammalian Genomics* (eds A. Ruvinsky and J. Marshall Graves) at the DNA level. The maps provide information on the assembly of bacterial artificial chromosome (BAC) contigs through end-sequencing and fingerprinting and also elaborate map information at the sequence level (see Chapter 3). In this chapter, we will only focus on physical maps.

Physical maps are also commonly referred to as *chromosome maps* because they encompass techniques where mapping can be directly visualized on the chromosomes, e.g. by ISH (Gall and Pardue, 1969; Pardue and Gall, 1969). Even using the SCH (Kucherlapati and Ruddle, 1975; Ruddle, 1981; Kao, 1983) and RH techniques (Goss and Harris, 1975; Cox *et al.*, 1990), where chromosomes are mostly fragmented and analysis is nowadays done by polymerase chain reaction (PCR), the chromosomal component of individual hybrid cell lines can be readily visualized and the mapping data can be associated with individual chromosomal fragments. This advantage is not present in linkage maps (Morton, 1955) where proportion of meiotic recombination in a family material helps to: (i) detect co-segregation of markers on the same chromosome and (ii) deduce their relative order. The markers can be allocated neither to specific chromosomes, nor to chromosomal regions.

Since the inception of genome analysis programmes in humans and rodents almost three decades ago, *chromosome maps* have played a critical role in the development of gene maps in over 40 additional mammalian species. Among these:

- *Synteny maps* have been critical in detecting groups of markers located on the same chromosome (Minna *et al.*, 1976; Ruddle, 1981; Kao, 1983).
- *In situ* hybridization (Gall and Pardue, 1969; Pardue and Gall, 1969), or its fluorescence version FISH (Pinkel *et al.*, 1986a,b), has facilitated the allocation of these groups to specific chromosomes, chromosomal regions or bands.
- *RH mapping* (Goss and Harris, 1975; Cox *et al.*, 1990), a revolutionary technique that overcomes the drawbacks of various mapping approaches, has provided a common platform for integration of all types of maps into a consensus map. Most importantly, the technique facilitates physical ordering of mapped markers.

In the ensuing discussion we will provide a brief description of each of the mapping techniques, followed by information on the current status of gene maps generated using these approaches in different mammalian species. However before expounding on this, we would like to: (i) provide some background information on how chromosome studies or cytogenetics influenced physical gene mapping, and (ii) how during the early gene mapping era, cytogenetic comparisons triggered discussions on cross-species genome comparisons.

Mammalian Cytogenetics: Launchpad for Chromosome Mapping

The origins of mammalian cytogenetics may be accredited to a range of attempts by various researchers during the early 1900s to ~1950 to identify correct chromosome number in different species. Understandably, the detection of correct chromosome number in humans by Tjio and Levan (1956) was more publicized than the identification of correct chromosome number in some other species. However, the real burst of activity in chromosome studies is ascribed to two accidental discoveries: the hypotonic treatment (Hsu, 1952; Ford and Hamerton, 1956) and the colchicine/colcemid-mediated arrest of cell cultures (Sumner *et al.*, 1973). These discoveries revolutionized the way by which crisp and clear chromosome preparations could be obtained (van der Ploeg, 2000). Consequently, during the 1960s, researchers started studying chromosomes of any species they could lay their hands on (Hsu and Benirschke, 1967). Among them, laboratory and domesticated animals were obviously the primary choices (Gustavsson, 1980).

When cytogenetic analysis in different species was evolving, a number of laboratories were also finding ways to map genes. Intense efforts were being made to ascertain whether or not two (or a group of) enzyme-producing genetic markers were located on the same chromosome. The next apparent pursuit was to find the chromosome where they were located. Researchers were therefore confronted with two major tasks:

1. Develop a way by which syntenic loci could be identified, and
2. Somehow tag the syntenic loci/groups to specific chromosomes.

The ability to make clean metaphase chromosome preparations and to be able to visualize them separately under the microscope provided a major impetus to achieve both goals.

Strategies were devised whereby random assortment of chromosomes or chromosomal fragments could be obtained to infer togetherness or synteny between markers. The result was the development of the SCH technique (Ruddle, 1981; Kao, 1983). Though the technique relied considerably on important advancements related to the generation of a transformed cell line, gene transfer approaches and approaches concerning cell fusion and growth of hybrid cells, the concept basically originated from chromosomal analysis of transformed cell lines. Each line had a different chromosome number even though they were from the same species. Researchers wished to get a large assortment of cell lines with a different combination of chromosomes from one species. Somatic cell hybrids were the answer.

Next, the ability to make chromosome preparations on slides also made researchers surmise approaches whereby they could 'see' the location of the gene of interest directly on the chromosome. The result was the discovery of the ISH technique (Gall and Pardue, 1969; Pardue and Gall, 1969). Lastly, the discovery of the RH mapping approach (Goss and Harris, 1975) is often considered as an offshoot of SCH mapping that, despite budding concurrently, blossomed almost two decades later when desired reagents to facilitate analysis were available. In brief, the ability to see the chromosomes under the microscope was a major instigation to develop gene maps in different species.

Over the years, mammalian cytogenetics has progressed enormously. Starting with chromosome analysis primarily in humans (Ford and Hamerton, 1956; Tjio and Levan, 1956), and then in laboratory species (Ford and Hamerton, 1956; Levan *et al.*, 1962), farm and domestic animals (Evans *et al.*, 1973) and wild animal species (Hsu and Bernischke, 1967), chromosome analysis now covers over 200 mammalian species spanning ~15 orders/suborders (Hsu and Bernischke, 1967; Richard *et*

al., 2003). In a variety of these species, detailed chromosome analysis has been carried out using different banding techniques. However, in a number of others, chromosome studies are still in a 'primitive' stage. It is very easy to distinguish this disparity and correlate it with gene mapping efforts in a species. In species where gene mapping has progressed (or is progressing) rapidly, the chromosomes are typically well characterized, karyotype analysis has been carried out in the population, an international standard depicting arrangement of chromosomes from a cell is available and, in many cases, schematic drawings of individual chromosomes (ideograms) and a description of individual bands (nomenclature system) is also in place. Thus, in general, the more advanced the gene map of a species, the better characterized are their chromosomes, thereby stressing the significance of detailed chromosome information in facilitating the rapid evolution of gene mapping programmes.

Comparative Cytogenetics and Gene Mapping: a Brief Overview

The advent of banding techniques during the early 1970s marked a new era in chromosome analysis (Caspersson *et al.*, 1968; Gustavsson, 1980). The techniques not only facilitated chromosome identification, but also provided the possibility to compare chromosomes across species (Dutrillaux, 1979; Gibson, 1986; Baker *et al.*, 1987; Sumner, 1990). It enabled researchers to investigate whether or not similar banding patterns between chromosomes of different species reflect similar genetic content. This represented the initial transition of comparative cytogenetics to comparative chromosome/gene mapping.

One of the first amalgamations of *comparative cytogenetics* and *gene mapping* was seen in studies focused on the mammalian X chromosome. Almost identical banding patterns and gene content of the X chromosome among a large cross-section of mammals suggested that at least some long-term evolutionary conservation of chromosome structure and gene content had occurred among mammals (Ohno *et al.*, 1964; Ohno, 1967; Pathak and Stock, 1974). Likewise, cross-species chromosome banding comparisons between autosomes also revealed conservation of chromosome organization among species within individual mammalian order/family, e.g. carnivores (Wayne et al., 1987a,b), bovids (Evans et al., 1973; Gallagher and Womack, 1992) and cetaceans (Arnason, 1977). Surprisingly, such similarities were also observed between distantly related species, e.g. rodents and primates (Sawyer, 1991), lagomorphs and primates (Dutrillaux et al., 1980; Grafodatskii and Biltueva, 1987), rodents and carnivores (Kiel et al., 1985; Grafodatskii and Biltueva, 1987), human and cat (Nash and O'Brien, 1982), human and mouse (Sawyer and Hozier, 1986), etc. In some cases, the data were strongly supported by the presence of the same genes, e.g. in humans and mouse (Lalley et al., 1978; Sawyer and Hozier, 1986). Thus cross-species similarities in banding patterns in the euchromatic regions (where most of the genes are located) of certain chromosomal segments across species, in general, reflected homology in their genetic content (Baker et al., 1987; Womack, 1987; Sumner, 1990; Sawyer et al., 1992).

The early to mid-1990s is heaving with reports indicating that similarities in banding patterns, in most cases, reflected similarities in gene content and gene order. This principally seemed to hold good for some groups of closely (as well as distantly) related species. However, in some of the other related/unrelated groups of species, partial or whole chromosome banding similarities did not yield similar results.

During the late 1990s, the importance of comparative cytogenetics started declining, particularly in species with rapidly evolving gene maps. Moreover, emergence of newer approaches such as comparative chromosome painting or Zoo-FISH (Scherthan et al., 1994; Chowdhary et al., 1998) seemed to provide a quicker and more reliable mode of discerning cross-species chromosome/genome similarities than was available only by banding.

Standard Karyotypes and Nomenclature Systems as Prerequisites for Chromosome Mapping

Karyotypes represent arrangement of all the chromosomes from a cell on the basis of a universally agreed layout scheme that is specific for each species. In most cases chromosomes are classified into groups (e.g. meta-/submetacentric and acrocentric) and then arranged based on their decreasing size within each group. Using banding techniques, homologous chromosomes are paired and presented together. Once a chromosome arrangement for a species is agreed upon by researchers around the world, it is referred to as a standard karyotype. Such karyotypes are extremely important as global platforms for chromosome identification and numbering within a species. Standard karyotypes are generally supplemented with a universally accepted nomenclature system wherein individual regions and chromosome bands are numerically recognized. This markedly facilitates pinpointing specific regions on the chromosomes.

Chromosome mapping and standard karyotype/nomenclature are closely connected. All gene mapping efforts will be in vain unless identified linkage or syntenic groups are assigned to specific chromosomes. Also, once a gene is mapped to a chromosome, it must be allocated to a specific region or band to indicate its position on the chromosome. Hence in species where genome analysis is carried out, each chromosome must: (i) bear a number; (ii) have a fixed position on the karyotype; and (iii) have an elaborate nomenclature system for individual regions and bands. This will enable every researcher around the world to associate a numbered chromosome of a species with a pre-determined size, structure, banding pattern and gene content.

Presently, standard karyotypes with nomenclature descriptions are available for ~15 species. Incidentally, all these species have strong to evolving gene mapping programmes. The most prominent among these is humans where several resolutions of banded standard karyotypes and ideograms are available. The most commonly cited human standard is the ISCN 1985 (An International System for Human Cytogenetic Nomenclature, 1985). In view of rapid progress in cytogenetic analysis and gene mapping, this system was revised in 1991 and 1995 for higher resolution and for specific applications to gene mapping and cancer cytogenetics (ISCN, 1991, 1995). To accommodate the growing need of computerized ideograms in genomic applications, Francke (1994) provided digitized and differentially shaded ideograms for individual human chromosomes.

Nomenclature systems are also available in mouse (Nesbitt and Francke, 1973), rat (Levan, 1974; Satoh *et al.*, 1989), Chinese hamster (Ray and Mohandas, 1976), a range of livestock/farm species, i.e. cattle, sheep, goat (ISCNDB, 2000), river buffalo (Iannuzzi *et al.*, 1990), pig (Gustavsson, 1988; Yerle *et al.*, 1991), horse (ISCNH, 1997), dog (Reimann *et al.*, 1996), Chinese raccoon dog (Pienkowska *et al.*, 2002a), cat (Cho *et al.*, 1997), donkey (Raudsepp *et al.*, 2000) and rabbit (Hayes *et al.*, 2002). Incidentally, other than rabbit, raccoon dog and donkey, genome projects are expanding rapidly among other mammals.

Genome programmes have also been initiated in some other mammalian species, e.g. American mink (Kuznetsov *et al.*, 2003). However the scope of these programmes is limited. Either very few genes are mapped in these species or the genomes/chromosomes of these species are only compared by Zoo-FISH. For these species G-banded standard karyotypes are available. Though these karyotypes are useful for preliminary comparisons, they do not provide detailed information/nomenclature on individual chromosomes. Hence, before the launch of any expanded genome programme in these species, development of a comprehensive nomenclature will be essential.

Physical Mapping: an Overview of Chromosome Mapping Approaches

Physical alignment of gene maps to individual chromosomes or chromosome regions is an integral part of any gene mapping initiative. This is because gene maps developed using various analytical approaches must be validated by testing the proposed order of loci against the one visualized on the chromosomes. Historically, *genetic linkage maps* reflect meiotic recombination between markers along the chromosomes. These maps do not tag the markers to a specific chromosomal location, and the map distances (Kosambi centiMorgan; cM; Kosambi, 1944) simply reveal recombination rates between markers – not real physical distances (Ott, 1985). Hence traditionally, these maps are not referred to as physical maps.

Compared with this, synteny and cytogenetic maps, and lately RH maps too, are conventionally referred to as *physical chromosome maps*, and are therefore briefly introduced below.

- *Synteny map:* In addition to delineating coexistence of loci on the same chromosome, synteny maps have the potential to discern their relative physical order if the hybrid cells are adequately characterized.
- In *RH mapping*, similar characterization of individual cell lines can be done only to a much lesser extent than done for somatic cell hybrids. For RH mapping, chromosomes are fractionated by irradiation and the physical closeness of markers on a chromosome is deduced through their co-segregation in the hybrid cell panel. Hence these maps are also physical maps.
- *Cytogenetic maps* are perceptibly the most classical form of physical maps wherein ISH shows the location of a marker directly on the chromosome.

In brief, the three types of maps – synteny, RH and cytogenetic – are the *pillars of physical mapping*. The approaches used to develop them have been extensively used in different mammalian species. In the following text we will provide a brief overview of each of the techniques, summarize their methodological considerations and discuss their individual uses and drawbacks. A synopsis of gene maps developed using these approaches in different mammalian species will be presented at the end of the section. Finally, the prospective use of each of the mapping approaches for future chromosome mapping efforts in mammals will be summarized.

Cytogenetic Mapping: Fluorescence *in Situ* Hybridization (FISH)

In situ hybridization is a technique that is widely used in several branches of biology. With specific reference to chromosome mapping, the technique permits direct visualization of the location of DNA markers on the chromosomes (*in situ* – Latin 'in its original place'). The method relies on the Watson–Crick base pairing complementarity principle and the location thus provides information on the molecular constitution of the chromosome at the hybridization site. There are two major components of chromosomal ISH, i.e. chromosomes, that are the targets, and probes, that are DNA segments of various lengths. Usually, the *target* is either metaphase or prometaphase chromosomes, but in cases where high resolution physical mapping is conducted, the target could be chromatin fibre either from interphase cells or from mechanically stretched cellular DNA. The *probes*, however, vary considerably in size as well as origin, for example from a few base pairs-long telomeric or centromeric repeat sequences to 1–2 Mb-long DNA segments cloned in YACs (yeast artificial chromosomes). When the probe DNA and the target chromosomes are from the same species the hybridization is referred to as *homologous*. However, when DNA probes and target chromosomes originate from different species, sometimes evolutionarily distant mammalian species, the hybridization is referred to as *heterologous*.

Depending on the type of ISH, the DNA probe can be labelled radioactively or non-radioactively. Radioactive ISH (RISH), first introduced by Gall and Pardue (1969) and John *et al.* (1969), was extensively used during the 1970s and early 1980s to map a range of nucleic acid probes to specific chromosomal regions. The approach, however, became obsolete following the rapid evolution of non-radioactive methods to label and hybridize DNA to fixed chromosomal targets (Manning *et al.*, 1975; Langer *et al.*, 1981; Pinkel *et al.*, 1986a,b; Lawrence *et al.*, 1988; Lichter *et al.*, 1991; Trask, 1991a,b). Because of significantly higher resolution, rapid turnaround time and safety, the new approach – referred to as fluorescence *in situ* hybridization or FISH – became the method of choice in ensuing years. Briefly, the approach entails labelling probe DNA with nucleotides tagged with biotin, digoxigenin (DIG), di-/tri-nitrophenol, or other labelling molecules. The hybridization is then detected with a variety of reporter molecules that have affinity for the labels. The detection can either be carried out *enzymatically* or with the help of *fluorochrome conjugates*. Lately, labelling can be done directly with fluorescent labelled nucleotides, thus culminating in the need for reporter molecules and related post-hybridization steps to visualize the signals. Increase in the number of spectrally distinct flu-orochromes and development of sophisticated computer imaging systems nowadays allows simultaneous detection of hybridization of two, three or even multiple probes (reviewed by Lichter, 1997; Levsky and Singer, 2003).

Over the years, FISH is said to have become 10,000 times more sensitive (Trask, 2002). However, general experience shows that FISH localization of sequences <2 kb in size is indeed cumbersome. There are very few groups that have effectively (with only partial success) carried out band-specific localization of 1–2 kb cDNA probes (see Trask *et al.*, 1993; Chaudhary *et al.*, 1997; Thomsen *et al.*, 1998). Conversely, targets larger than 10 kb, such as those cloned in λ phages, cosmids, BACs and YACs, can be detected with >90% efficiency (Trask *et al.*, 1993).

The FISH process itself, as well as the post-hybridization washing, can vary considerably depending upon the type and size of the DNA probes used. Details on individual aspects/components of FISH are described elsewhere (Lichter *et al.*, 1991; Trask, 1991a,b, 2002; Lichter and Cremer, 1992; Levsky and Singer, 2003). Below we briefly discuss the major uses of ISH/FISH in developing physical gene maps in different mammalian species.

1. *Chromosomal assignment of linkage and syntenic groups.* One of the early principal uses of ISH in humans and mouse was to align individual syntenic and linkage groups to specific chromosomes (Kozak and Ruddle, 1976; Green *et al.*, 1984; Naylor *et al.*, 1984; Mamula *et al.*, 1985; McKusick, 1991; Korenberg *et al.*, 1992). This work was accomplished rather quickly, i.e. by the early 1980s, because the maps in the two species were developing at a rapid pace. In other species such as rat and domestic animals, however, it took almost another 15 years before individual linkage and syntenic groups could be allocated to specific chromosomes (Szpirer *et al.*, 1998). To a large extent this is attributed to the late start of genome programmes in these species. For example, in cattle and pigs, all linkage/syntenic groups were allocated to specific chromosomes by 1995 (Ellegren *et al.*, 1994; Mezzelani *et al.*, 1994, 1995; Eggen and Fries, 1995; Masabanda *et al.*, 1996; Robic *et al.*, 1996; Rohrer *et al.*, 1996; Yerle *et al.*, 1996). Compared with this, a similar alignment in the horse was not completed until recently (Milenkovic *et al.*, 2002; Chowdhary *et al.*, 2003) because organized efforts to map the equine genome started in 1995–1996 after the 1st International Equine Gene Mapping Workshop (October 1995, Lexington, Kentucky).

In rabbit and cat, where syntenic groups were available much earlier than other domestic species, the chromosomal alignment proceeded much faster in the beginning, but was then delayed probably due to a shift in mapping priorities. In rabbit, the synteny mapping programme virtually came to a standstill after initial success during the early 1980s (Echard *et al.*, 1981, 1982). In fact presently there are more FISH than synteny mapped loci in rabbit (Hayes *et al.*, 2002; Zijlstra *et al.*, 2002). In dog, linkage, syntenic and RH mapped groups were almost concurrently assigned to specific chromosomes. In part this is attributed to: (i) lack of a universally accepted standard karyotype in the dog and (ii) difficulty in identifying/distinguishing individual dog chromosomes due to high diploid number. FISH mapping of chromosome-specific clones (Breen *et al.*, 2001) and comparative chromosome painting information (Breen *et al.*, 1999a; Yang *et al.*, 1999) was significant in resolving this problem and assigning linkage/syntenic/RH groups to specific chromosomes. In sheep, goat and buffalo, FISH mapping was instrumental in transferring cattle gene mapping data (synteny and linkage groups) to individual chromosomes of these species (Hayes *et al.*, 2000; Iannuzzi *et al.*, 2001a; Di Meo *et al.*, 2003). This has by default transferred the syntenic groups developed recently in river buffaloes (El Nahas *et al.*, 1996b, 2001; Iannuzzi *et al.*, 2001a,b). Similarly, ISH has helped to assign synteny groups to specific chromosomes in marsupials (Watson *et al.*, 1992; Spencer *et al.*, 1991a,b,c) and monotremes (Watson *et al.*, 1991).

2. *Single locus mapping.* Radioactive ISH was the main source of chromosome mapping in humans for almost 10 years (1980–1990). During the late 1980s, FISH was introduced, and it soon became the main source of chromosomal mapping. The first single-copy gene mapped by FISH was human thyroglobulin, *TG* (Landegent *et al.*, 1985) and the results

confirmed the isotopic ISH data. Since then thousands of genes, expressed sequence tags (ESTs; Korenberg *et al.*, 1995), sequence-tagged sites (STSs), microsatellites, clones from BAC/YAC contigs (McPherson *et al.*, 2001), etc. have been assigned to human chromosomes by FISH, resulting in exponential growth in the number of publications and reviews during these years (reviewed by Trask, 2002; Levsky and Singer, 2003).

In mouse and rat this number is significantly lower. The Mouse Genome Informatics website (MGI: http://www.informatics.jax.org/) shows that up until now ~1300 markers have been FISH mapped to all mouse chromosomes. Compared with this, ~600 loci (550 autosomal and 50 sex chromosome) have been FISH mapped in rat (RatMap: http://ratmap.gen. gu.se/ShowSingleCitation.htm?citno=904). In other species such as cattle (BovMap database: http://locus.jouy.inra.fr/cgi-bin/bovmap/ intro2.pl), pig (PigMap database: http://www. thearkdb.org/browser?species=pig&objtype=s tats, Robic *et al.*, 1996; Pinton *et al.*, 2000), dog (Breen *et al.*, 2001; Thomas *et al.*, 2001) and horse (Milenkovic *et al.*, 2002; Chowdhary *et al.*, 2003), the number ranges between 300 and 500 loci. In sheep (http://www.thearkdb. org/browser?species=sheep&objtype=stats, Di Meo *et al.*, 2002, 2003), goats (Schibler *et al.*, 1998) and buffaloes (El Nahas *et al.*, 2001; Iannuzzi *et al.*, 2001a,b), this number ranges between 150 and 250 loci.

In rabbit (Hayes *et al.*, 2002; Zijlstra *et al.*, 2002) and donkey (Raudsepp *et al.*, 2001, 2002b), the number of FISH-mapped loci ranges between 40 and 65, while very few markers have been localized in cat chromosomes (Beck *et al.*, 2001; Fujino *et al.*, 2001a,b). Besides these species, a handful of loci (up to 20) are also FISH assigned in other mammals, e.g. sika deer (Bonnet *et al.*, 2001), reindeer (Prakash *et al.*, 1996), American mink (Biltueva *et al.*, 1996; Christensen *et al.*, 1998) and marsupials/monotremes (Sinclair and Graves, 1991; Spencer *et al.*, 1991a,b; Watson *et al.*, 1991, 1993; Sturm *et al.*, 1994; Toder *et al.*, 1996). In the latter group, ISH or FISH has predominantly contributed to the understanding of differences in sex chromosome evolution (Wilcox *et al.*, 1996; Waters *et al.*, 2001).

Overall, the mammalian FISH assignments have been critical in providing information on the spatial/physical distribution of loci on the chromosomes. This is not only important for validating gene order deduced using different mapping approaches, but is equally valuable for studying the comparative organization of different genomes, and hence their evolution. FISH localizations have also been crucial in verifying the order of loci proposed by, for example, synteny, linkage and RH mapping. These approaches propose an algorithm-based order that needs to be corroborated by other mapping approaches, of which FISH is the most accurate. Finally, as mentioned above, in species with high chromosome number and uniform morphology, e.g. dogs ($2n=78$) or cattle ($2n=60$), FISH has been key in facilitating chromosome identification (Hayes *et al.*, 2000; Breen *et al.*, 2001). The significance of FISH in human and mouse gene mapping can be gauged from the fact that despite availability of sequence data, researchers even now tend to carry out FISH localizations to verify the origin of their genomic clone before proceeding with further analysis (McPherson *et al.*, 2001).

FISH probes are primarily used in species of origin, i.e. human probes on human metaphase spreads. However, during recent years, these probes have also been used *across closely related* species primarily for comparative mapping or, in rare cases, to initiate development of their gene maps. For example, numerous human genomic clones were used to map orthologues in a range of primates (Stankiewicz *et al.*, 2001; Carbone *et al.*, 2002; de Pontbriand *et al.*, 2002, etc.). Similarly, cattle λ, cosmid and BAC clones have been extensively used to map the same genes/markers in sheep, goat and buffaloes (Prakash *et al.*, 1997; Iannuzzi *et al.*, 2001a,b; Di Meo *et al.*, 2002). During recent years, dog genomic clones have also been utilized to carry out comparative mapping with other canids by FISH mapping dog genes and microsatellites in raccoon dog, red fox, silver fox and arctic fox (Park, 1996; Pienkowska *et al.*, 2002b; Szamalek *et al.*, 2002; Rogalska-Niznik *et al.*, 2003; Szczerbal *et al.*, 2003). Next, following initiation of organized efforts to map the horse genome, equine genomic clones have been

extensively used to understand the comparative organization of the genomes of other equids, e.g. donkey, Hartmann's mountain zebra, common zebra, kiang and onager (Raudsepp *et al.*, 2001, 2002b; Myka *et al.*, 2003a,b). It is interesting that despite divergence in chromosome number (32 in Hartmann's zebra to 66 in Przewalskii horse), and evidence of evolutionary re-organization in some genomic regions (Raudsepp and Chowdhary, 1999, 2001), all extant equids interbreed giving viable offspring.

3. *Complementing contig mapping and genome sequencing projects.* Lately FISH has been used to verify the relative order of overlapping BACs assembled into contigs while developing BAC fingerprinting maps. The latter are needed to obtain assortment of BACs for building a minimum tiling path over a specific chromosomal region (Korenberg *et al.* 1999; McPherson *et al.*, 2001) (The BAC Resource Consortium, 2001: http://bacpac. chori.org/mapped-clones.htm). More than 10,000 BAC clones have been mapped at 300 kb resolution to human chromosomes to provide integration of RH and BAC contig maps with the finished sequence and also to help with the characterization of genes altered by disease-related chromosomal aberrations. A similar contribution by FISH could also be expected for the animal genomes (cattle and pig) where sequencing projects are soon expected to be initiated.

4. *Multicolour FISH.* With the rapid expansion of cytogenetic maps in some of the mammalian species (primarily human, mouse and domestic animals), researchers confronted the problem to precisely orient/order loci that showed hybridization signals on the same chromosomal band. In some instances the order could be resolved by co-hybridization of the probes, provided the signals were physically apart from each other by ~3–5 Mbp (Plate 1A). However in cases where the hybridization signals virtually overlapped, it became difficult to assess which of the two loci is proximal and which one is distal. This was overcome by the development of the double colour FISH approach. Hopman *et al.* (1986) demonstrated the first successful use of the approach by simultaneously hybridizing mercurated total human DNA and a biotinylated

mouse satellite DNA probe to metaphase chromosomes and interphase nuclei of human–mouse hybrid cells. The results were detected with FITC (fluorescein isothiocyanate) and TRITC (tetramethylrhodamine isothiocyanate) fluorescence. Since then, the approach has evolved enormously and today two or more labels and flurophores (Nederlof et al., 1989, 1990) can be used to decipher physical arrangement of closely located loci on metaphase/pro-metaphase chromosomes (reviewed by Fauth and Speicher, 2001; Liehr and Claussen, 2002; Levsky and Singer, 2003) and interphase chromatin (see below). The multicolour FISH approach has been most frequently used in humans because of the greater number of FISH localizations carried out in this species. Compared with this, the technique has been less frequently used in rodents. Among other mammalian species, the technique has been used in pigs (Chowdhary et al., 1995; Quilter et al., 2002), cattle (Schlapfer et al., 1998; Gallagher et al., 1999; McShane et al., 2001), horse (Raudsepp et al., 1999; Chowdhary et al., 2002, 2003; Raudsepp et al., 2002a,b), dog (Breen et al., 1999b) and Chinese raccoon dog (Szczerbal et al., 2003). Very recently, the technique found critical use in inferring the relative order of Y specific loci in the horse (Raudsepp et al., 2004).

5. *Interphase and fibre-FISH.* The resolution of the metaphase/prometaphase FISH permits physical orientation of loci separated by 1–3 Mb (Lichter et al., 1990; Trask, 1991a). However, to order loci separated by distances in the range of few hundred kb or less, FISH to extended chromatin in interphase cells has emerged as the technique of choice. Using this approach, probes merely 25–50 kb apart could be readily resolved from each other (Lawrence et al., 1988, 1990, 1992; Trask et al., 1989, 1993) (Plate 1B). However, if a set of probes is separated by more than 750 kb, the reliability of the deduced order reduces because the chromatin fibre is fairly decondensed in the interphase stage and therefore tends to twist on average every 1 Mb. As expected, the interphase mapping technique has been most frequently used in humans, followed by mouse and rats (e.g. see Nath and Johnson, 2000; Kraan et al., 2002; Liehr and Claussen, 2002). In other mammals, interphase FISH has been

used so far for fine mapping only in pigs (Chowdhary et al., 1995; Quilter et al., 2002). However, recently it was crucial to develop a physically ordered map of the euchromatic region of the horse Y chromosome (Raudsepp et al., 2004a). Using this approach, a number of overlapping markers in the RH map were linearly oriented on the chromosome. Thus, the approach provides one of the quickest and easiest ways to physically order closely located loci/DNA clones.

A further improvement in the resolution of chromosome mapping by FISH has been reported through the use of fibre-FISH. The technique enables the researcher to distinguish probes separated by even 1–2 kb (Plate 1C). Probes more than 500 kb apart are relatively less suitable to be studied using this approach because mechanically stretched DNA fibres tend to break beyond the ~500–700 kb level (Parra and Windle, 1993; Heiskanen et al., 1994, 1995, 1996; Senger et al., 1994; Palotie et al., 1996). None the less, in some instances, when the fibres are not stretched too much, probes located within a span of ~1000 kb can be analysed. Thus, fibre-FISH has proven to be well suited as an adjunct physical mapping tool (reviewed by Kraan et al., 2002). In humans, it has been used to find transcriptional orientation of clones, and in detection of minor chromosomal rearrangements (Palotie et al., 1996). In some instances, it has been used for positional cloning of disease genes (*CLN5;* Laan et al., 1996). The use of the technique in mouse has been occasional. Recently it was used for high resolution mapping of mouse collagen genes on MMU10 (Sallinen et al., 2001). The results were compared with homologous human data to detect both similarities and differences concerning the chromosomal distribution, order, transcriptional orientations, and intergenic distances of the collagen genes in the two species.

In domesticated animals, the technique has been used in pigs to study the organization of major histocompatibility complex (Sjoberg et al., 1997) and to order subclones from the porcine erythropoietin gene (Liu et al., 1998). The studies delineated the order of individual clones and also estimated physical distances (in kb) between them. In cattle, the approach has been used to study the organization of the bovine aromatase gene (Brunner et al., 1998) and a family of bovine signal transducers and activators of transcription, STATs (Seyfert et al., 2000). The latter study revealed extended intragenic sequence homogeneity in the STAT locus between cattle and mouse. Very recently, fibre-FISH was used extensively to generate an ordered physical map of the euchromatic region of the human (Rottger et al., 2002) and horse (Raudsepp et al., 2004a) Y chromosomes.

Somatic Cell Hybrid (SCH) Panels and Synteny Mapping

The term synteny originates from two Greek words: 'syn' meaning the same or together, and 'tene' meaning the thread. Thus synteny describes togetherness of two or more genes on the same chromosome 'thread', irrespective of demonstrable linkage between them. A synteny map represents one of the most basic forms of a gene map wherein all loci located on the same chromosome of a particular species are catalogued. Synteny maps have been developed in a variety of mammalian species and are inarguably the foundations of physical gene maps. The maps have also played an important role as a bridge between physical and genetic linkage maps. We will first discuss how SCH panels are developed, elaborate various approaches used to analyse the panels and then provide an overview of synteny mapping in a range of mammalian species, with special focus on primates, rodents and domesticated animals.

Construction of an SCH panel

Somatic cell genetics has been the key to the development of hybrid cell panels commonly used for generating synteny maps (Ruddle, 1981). Over the years, very little has changed in terms of the approaches used to develop these panels in different species. In essence, the procedure involves: (i) fusion of cell lines from two species (one donor and the other recipient); (ii) isolation of an assortment of hybrid cell clones with different combinations of donor chromosomes; and (iii) analysis of individual clones for the presence or absence of markers.

Markers co-segregating in the majority of the hybrid cell lines are considered to be syntenic. A number of methodological breakthroughs in somatic cell genetics during the 1950s and 1960s, e.g. the possibility of growing mammalian cells in vitro, the generation of biochemically well-defined mutant cell lines and the discovery of systems that could induce fusion of cell lines from two different species (Kucherlapati and Ruddle, 1975; Kao, 1983), were central to the development of SCH panels.

Briefly, a biochemically deficient cell line of one species (recipient) is fused with normal somatic cells of the species for which the map has to be constructed (donor). Recipient cell lines are usually transformed, traditionally deficient for a certain gene, e.g. hypoxanthine–guanine phosphoribosyl transferase (HPRT$^-$) or thymidine kinase (TK$^-$), and mainly have a rodent origin (mouse, Chinese hamster, etc.). These lines are well established, preserve long-term karyotypic stability and are immortalized in special growth media to compensate their deficiency. Contrary to this, the donor cell lines are normal. The fusion between the donor and the recipient cell lines is in part facilitated by 'transformation' of the latter. However, it is also mediated by a viral (e.g. Sendai virus) and/or chemical agent (e.g. polyethylene glycol). Following fusion, the newly formed hybrid cells contain both parental genomes. During subsequent divisions, chromosomes from the donor cell are preferentially but randomly lost. Thus, each hybrid cell line retains the entire rodent (recipient) genome and segregates the donor chromosomes in different combinations. The donor chromosome carrying the gene deficient in recipient genome is preferentially retained in all hybrid cells (Fig. 2.1).

The hybrid cell clones are characterized for content and representation of donor chromosomes/genome because the precision of synteny mapping depends largely on how well the hybrid clones are characterized cytogenetically. For this, chromosome preparations from individual hybrid cell lines are obtained. During the early years, the chromosomes were analysed using differential staining and banding techniques (Kao, 1983). However, during recent years, ISH with labelled donor genomic DNA (Boyle et al., 1990), chromosome-specific short interspersed DNA elements (SINEs) (Zijlstra et al., 1996) and chromosome-specific painting probes

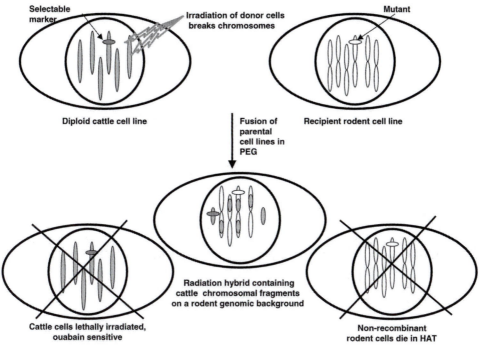

Fig. 2.1. Illustration showing an outline of the development of a radiation hybrid panel. In principle, the approach is similar to somatic hybrid cell panel generation and analysis.

(Archidiacono *et al.*, 1994a,b) have been used to precisely identify donor chromosomes in hybrid cells. Very recently, a cocktail of differentially labelled chromosome-specific paints were used for a single step hybridization and detection of all donor chromosomes using spectral karyotyping (Matsui *et al.*, 2003).

A collection of hybrid cell lines (commonly referred to as a 'panel'), each containing different combinations of donor chromosomes, is used to ascertain synteny between different loci. A range of approaches has been used during the past 30 years to analyse somatic cell hybrid panels for developing synteny maps. The approaches have been summarized below in the order they evolved during these years.

Methods for SCH panel analysis

Enzyme electrophoresis

This is one of the most extensively used initial approaches to detect synteny between markers in somatic cell hybrid panels. The method is based on the fact that variants of the same enzyme – isozymes – in different species have different physical properties (e.g. electrophoretic mobility) due to which they can be readily distinguished (Westerveld *et al.*, 1971; Kao, 1983). Electrophoresis analysis of cell lysates from cultures of individual SCH cell lines shows the presence or absence of donor isozyme(s), thus indicating the retention or absence of the chromosome(s) carrying the corresponding gene(s). Co-retention of a pair or group of markers suggests synteny, which is statistically verified against results from clones showing dis-concordance between the markers. Normally, a concordance of ~70% or above is accepted as an indicator of synteny between the tested markers. Because the method requires analysis of gene products in hybrid cell lysates, this approach could detect synteny only between functional genes. Early synteny maps in a number of species, e.g. human (Minna *et al.*, 1976; Kao, 1983), cattle (Womack and Cummins, 1984; Womack and Moll, 1986) and pig (Forster *et al.*, 1980; Leong *et al.*, 1983), were developed using this approach.

Southern blotting

Following advances in recombinant DNA techniques (Berg *et al.*, 1974; Berg, 1977; Cohen, 1977) and filter hybridization (Southern, 1975), analysis of SCH panels witnessed a new era wherein all types of markers – genes with or without products in the lysates and even non-coding sequences – could be rapidly placed on synteny maps. DNA from hybrid clones was isolated, cleaved with a rare cutting restriction enzyme (e.g. *Eco*RI) and transferred to nitrocellulose membranes. The membranes were hybridized with radiolabelled gene-specific probes and retention of genes of interest in the panel was detected by autoradiography. Compared with the enzymatic method, this approach was more sensitive and allowed mapping of all types of markers.

PCR

The advent of PCR (Saiki *et al.*, 1985; Mullis *et al.*, 1986; Scharf *et al.*, 1986) added a new phase in SCH analysis. It enabled quick and high-throughput synteny mapping of any kind of marker for which primers could be developed. DNA from individual cell hybrid clones in the panel is used as a template to obtain amplification product from the primers. A positive amplification product indicated the presence of the marker in the clone. Concordance of markers in a range of hybrid clones indicates synteny. In some laboratories, this approach continues to be the approach of choice even today due to the ease with which a new marker can be added to the synteny map. During recent years, heterologous (non-species-specific) primers, following due optimization to curtail recipient DNA amplification, have been successfully used to detect synteny between evolutionarily conserved genes across species (Venta *et al.*, 1996; Lyons *et al.*, 1997, 1999; Jiang *et al.*, 1998).

Irrespective of the approach used to analyse the panel, synteny between markers is determined on the basis of their retention concordance or discordance. Correlation values >70% are usually accepted as evidence for synteny between two markers (Chevalet and Corpet, 1986). The results simply illustrate whether or not a given set of loci are present on the same chromosome. They do not indicate the chromosome unless: (i) one of the syntenic markers is assigned to a specific chromosome using other approaches (e.g. ISH) or (ii) donor chromosomes of all hybrid cell lines are adequately characterized. In fact, well characterized SCH panels considerably facilitate assignment of markers to specific chromosomal segments or bands thus permitting them to be physically ordered.

Synteny maps in different mammalian species

Humans

Development and analysis of SCH panels represents a milestone in facilitating analysis of the human genome in an era that was equipped with far fewer techniques, instruments and methods. Initiated during the early 1970s, the human synteny maps in fact galvanized development of physical maps as well as comparative maps (Westerveld *et al.*, 1971; Kucherlapati and Ruddle, 1975; Minna *et al.*, 1976; McKusick and Ruddle, 1977). By the mid-1980s, over 600 genes were mapped using this approach and all syntenic groups were assigned to individual chromosomes (Washington *et al.*, 1994; Roberts *et al.*, 1996). The development of synteny maps in parallel with the linkage maps was very useful because the two approaches complemented each other and provided key verification to mapping data. Today, with the completion of the human genome sequencing project (http://www.ornl. gov/TechResources/Human_Genome/project/50yr.html), SCH panels are no longer needed for actual mapping of genes/markers. However, recently generated and cytogenetically characterized monochromosome and regional SCH mapping panels (Drwinga *et al.*, 1993; Leonard *et al.*, 1997, 1998; Marzella *et al.*, 2000; Tanabe *et al.*, 2000) are proving invaluable for detailed characterization of disease-, cancer- and development-related structural rearrangements and also for fine mapping deletion and translocation breakpoints on the chromosomes (Matsui *et al.*, 2003). Some of the human × rodent SCH panels are so specialized that they even enable mapping of differentially expressed genes. For example, panels with clones containing either active or inactive X chromosomes have been extremely useful in determining human X chromosome inactivation regional profiles (Brown *et al.*, 1997; Carrel *et al.*, 1999).

Other primates

Concurrent with the development of the human synteny map, SCH panels were also constructed for a variety of non-human primates: chimpanzee (Finaz *et al.*, 1975; Rebourcet *et al.*, 1975; Ishizaki *et al.*, 1977; Archidiacono *et al.*, 1994a), gorilla (Khan *et al.*, 1976; Khan, 1987), orangutan (Khan, 1987), gibbon (Turleau *et al.*, 1983), baboon (Thiessen and Lalley, 1986, 1987), rhesus and African green monkeys (Khan, 1987) and grey mouse lemur (Cochet *et al.*, 1982). During recent years, SCH panels were generated for black spider monkey (Moreira *et al.*, 1997; Canavez *et al.*, 1998, 1999a; Seuanez *et al.*, 2001), where to date almost 80 loci have been mapped using this approach (Seuanez *et al.*, 2001).

Interestingly, initial mapping of human orthologues on SCH panels provided one of the first comparative maps between humans and primates. These mapping experiments were valuable in the identification of evolutionarily conserved synteny groups in these species and were also critical to detect rearrangements between their karyotypes. Even today, when comparative genome analysis in primates is dominated by comparative chromosome painting (also referred to as Zoo-FISH), FISH mapping and comparative sequence analysis, synteny mapping has not lost momentum in non-human primates. For example, recent studies showing rearrangements between black spider monkey and human or other primate karyotypes are based primarily on the results of SCH analysis (Canavez *et al.*, 1999b; Seuanez *et al.*, 2001).

Rodents

Mouse and rat genomes are the second best studied mammalian genomes after human (Carver and Stubbs, 1997). However, unlike humans, synteny mapping has never been the central force facilitating the development of gene maps in the two species. This is evident from the fact that in both species SCH panels were constructed much later than in humans and some of the other mammals: in mouse (Bucan *et al.*, 1986; Williamson *et al.*, 1995; Sabile *et al.*, 1997) and rat (Szpirer *et al.*, 1984; Yeung *et al.*, 1993; Walter *et al.*, 1996). Despite a relatively low number of SCH-mapped loci in these species, synteny mapping data have played an important role in generating some of the earliest comparative maps between rodents and humans (Minna *et al.*, 1976; Bucan *et al.*, 1986; Yeung *et al.*, 1993), and mouse and rat (Szpirer *et al.*, 1984).

Domesticated animals

After human and primates, SCH panel and synteny mapping information is most abundant in the main livestock species (cattle, pig, sheep, horse, river buffalo) and some domestic animals (cat, dog).

Cattle. Cattle is probably the second species after humans where SCH analysis played a critical role in the development of initial synteny and comparative maps. The first reports about synteny mapping of enzyme-producing loci appeared in the mid-1980s (Womack and Cummins, 1984; Womack and Moll, 1986; Fries *et al.*, 1989), when the generation of the first cattle × hamster SCH panel was reported (Dain *et al.*, 1984; Womack and Cummins, 1984). This was later followed by the identification of 26 of the 29 autosomal synteny groups, each marked by at least one gene (Fries *et al.*, 1989). Following developments in human genomics, the enzymatic approach of analysis in cattle was soon replaced by Southern blotting (Adkison *et al.*, 1988; Gallagher *et al.*, 1991; Rogers *et al.*, 1991) and PCR (Guérin *et al.*, 1994a,b). The latter particularly increased the output of synteny mapping and allowed assignment of all kinds of markers: microsatellites (Ferretti *et al.*, 1994; Guérin *et al.*, 1994b; Vaiman *et al.*, 1994), genes and ESTs (Ma *et al.*, 1998; Goldammer *et al.*, 2002b). The panel thus became a key resource for developing the cattle gene map and carrying comparisons with the rapidly expanding human and mouse maps (Womack, 1993, 1994, 1996; Womack and Kata, 1995). Detailed cytogenetic characterization of some hybrid cell panels

(Konfortov *et al.*, 1998) partly improved mapping efficiency and accuracy.

At present, when more accurate and increased throughput physical mapping methods (RH, FISH, BAC contigs) are available, and sequencing the whole cattle genome by the end of 2005 has become a reality (http://www.genome.gov/Pages/Research/Sequencing/SeqProposals/BovineSEQ.pdf), one may anticipate that SCH mapping may not be needed any more in the future. On the contrary, the Texas A&M SCH panel is still in use for quick initial chromosomal assignment of new markers (Goldammer *et al.*, 2002a,b; Reed *et al.*, 2002). The cattle RH mapping website (http://bovid.cvm.tamu.edu/cgi-bin/rhmapper.cgi) displays a note stating that before RH analysis 'all markers must be genotyped on the 31 clone bovine–hamster SCH panel'.

Pig. As in most mammalian species, SCH genetics in pigs (reviewed by Chowdhary, 1998) began with the mapping of enzyme genes, first to the X chromosome (Forster *et al.*, 1980; Leong *et al.*, 1983) and then to the autosomes (Ryttman *et al.*, 1986, 1988). Soon thereafter, Southern blot analysis replaced the enzymatic method (e.g. Harbitz *et al.*, 1990; Poulsen *et al.*, 1991; Thomsen *et al.*, 1991). After a brief hiatus of ~4–5 years, SCH analysis in pigs once again gained momentum (Rettenberger *et al.*, 1994; Zijlstra *et al.*, 1994; Yerle *et al.*, 1996), primarily due to switchover to analysis by PCR. The latter almost revolutionized physical assignment of genes, ESTs, microsatellites and anonymous DNA segments to the porcine genome. In 1998, synteny mapping of more than 228 loci was reported (Wintero *et al.*, 1996; Yerle *et al.*, 1996; Fridolfsson *et al.*, 1997; Chowdhary, 1998). Today, with around 700 loci mapped by SCH analysis, this figure has increased by more than threefold (Lahbib-Mansais *et al.*, 1999, 2000; Davoli *et al.*, 2000).

Large-scale mapping and detailed characterization of two of the SCH panels (Rettenberger *et al.*, 1994; Robic *et al.*, 1996) enabled accurate assignment of markers either to a chromosomal arm or, in some cases, even to specific chromosomal bands. Though highly efficient and rapid physical mapping

approaches (RH, FISH) are currently available, SCH analysis continues to be an active contributor to porcine genome analysis. This is evident from the fact that all the recent large-scale comparative mapping projects have exploited SCH panels for initial assignment of new markers and for anchoring RH linkage groups to chromosomes (Cirera *et al.*, 2003; Lahbib-Mansais *et al.*, 2003).

Sheep. Several SCH panels have been generated in sheep. These include three sheep–Chinese hamster (Cianfriglia, 1979; Saidi-Mehtar and Hors-Cayla, 1981; Saidi-Mehtar *et al.*, 1981; Burkin *et al.*, 1993), one sheep–mouse (Tucker *et al.*, 1981) and one sheep–American mink (Kuznetsov *et al.*, 1998) panels. These panels have been continuously used for quick chromosomal assignment of ovine genes and polymorphic markers (Burkin *et al.*, 1993, 1998; Tabet-Aoul *et al.*, 2000; Hotzel and Cheevers, 2002). At present, over 200 loci have been mapped using this approach, leading to 130 syntenic groups on all autosomes and the X chromosome (Tabet-Aoul *et al.*, 2000).

Following characterization of individual hybrid cell lines, the sheep–hamster SCH panel was used for flow-sorting sheep chromosomes and generating chromosome-specific paints (Burkin *et al.*, 1997). Attempts have also been made to develop single sheep chromosome panels by selectively capturing individual sheep chromosomes and fusing them with Chinese hamster auxothrophs (Burkin *et al.*, 1993). Characterization of the panels has also been done by PCR with markers of known chromosomal location (Tabet-Aoul *et al.*, 2000). Despite all advances, synteny mapping in sheep has not progressed in the way it did in cattle and pig. Part of this is attributed to the overall slow pace of the sheep genome analysis programme and availability of the cattle genome as a template for the development of the gene map in sheep.

Horse. The first synteny group in the horse can be traced back to indirect analysis carried out in a mule–mouse hybrid cell panel (Deys, 1972). Three genes were then assigned to the X chromosome. After that, four more SCH panels were constructed (Lear *et al.*, 1992;

Williams *et al.*, 1993; Bailey *et al.*, 1995; Raney *et al.*, 1998; Shiue *et al.*, 1999). It was interesting to note that in all the panels, large submetacentric chromosomes were preferentially lost. A similarly poor retention of large submetacentric chromosomes (e.g. SSC2) has been reported for a pig–mouse SCH panel (Zijlstra *et al.*, 1996). In most of the other species, as well as in other porcine panels, loss of donor species chromosomes is usually found to be random (Robic *et al.*, 1996; Yerle *et al.*, 1996). Of the three equine RH panels, only one has as yet been characterized using FISH with equine genomic DNA (Lear *et al.*, 1992).

By 1995, only nine syntenic groups were established in the horse using enzymatic analysis of lysates (Williams *et al.*, 1993) or by PCR (Bailey *et al.*, 1995). Unfortunately, none of the groups could be assigned to specific chromosomes. During later years, the panel created at UC Davis (Shiue *et al.*, 1999) became the main contributor to the equine synteny map. With around 450 markers analysed on this panel, the resource has been central in assigning synteny groups (comprising genes and microsatellites) to individual equine autosomes (Caetano *et al.*, 1999a,b; Shiue *et al.*, 1999; Lindgren *et al.*, 2001a,b) and the sex chromosomes (Shiue *et al.*, 2000). The resource has also significantly contributed to the development of the horse–human comparative map (Caetano *et al.*, 1999b; Chowdhary *et al.*, 2003). As in some of the other species, the potential use of the resource in the future seems limited. Part of this is attributed to availability of other faster mapping approaches; but the other part is attributed to depleting DNA reserves from the original panel.

River buffalo. To date ~75 loci have been assigned to the river buffalo gene map using buffalo–hamster hybrid cell lines (de Hondt *et al.*, 1991; El Nahas *et al.*, 1996a,b, 2001). The panel consists of 37 hybrid clones and has been used for assigning loci by PCR (de Hondt *et al.*, 1997; Oraby *et al.*, 1998; El Nahas *et al.*, 1999; Othman and El Nahas, 1999), by isozyme electrophoresis (de Hondt *et al.*, 1991; El Nahas *et al.*, 1999), by restriction enzyme digestion (El Nahas *et al.*, 1996a) and by immunofluorescence (El Nahas *et al.*, 1996b).

The map thus generated is primarily useful for comparative purpose – mainly to align the river buffalo genome to cattle and human genomes.

Cat. Development of the cat gene map during the late 1980s marked an important step in comparative genomics. Being one of the first and better developed gene maps in any of the domesticated animals, it provided interesting insights into cross-species genome conservation and rearrangements. The map was almost exclusively based on SCH analysis (O'Brien and Nash, 1982; Berman *et al.*, 1986; O'Brien *et al.*, 1997a,b). Today the cat synteny map comprises ~105 genes and shows a high degree of evolutionary synteny conservation with human. Almost half of the 258 described feline genetic diseases have human homologues. Extensive synteny conservation between the two species thus makes cat an important model animal to study human genetic disorders (O'Brien *et al.*, 2002) and at the same time treatment regimes for the same conditions in the cat.

Dog. The complexity of the canine karyotype (76 acrocentric autosomes) has been an impediment to progress in dog genome analysis. This basic problem extends also to somatic cell genetics. Thus far the only dog–Chinese hamster SCH panel generated for mapping the dog genome was built using microcell-mediated transfer of canine chromosomes into rodent cell lines (Langston *et al.*, 1997). Following selection, the final mapping panel includes 43 microcell hybrid cell lines and has been characterized by PCR using primers for 181 microsatellites and 27 genes, and also by FISH with dog genomic DNA on chromosome spreads from individual cell lines.

Synteny mapping of candidate genes for canine diseases has been helpful in assigning candidate genes to specific chromosomes. The information is useful in quickly sifting through the existing gene mapping data to identify informative markers that could be used for more focused/detailed analysis, or to devise strategies for generating a new set of markers (see Fredholm, 1997). With the availability of the RH panel in dog (Vignaux *et al.*, 1999a,b; Research Genetics), the use of SCH panel has reduced considerably. Nevertheless, it is still available as a useful resource for rapid chromosome assignment of markers.

Other mammals

SCH panels, and occasional synteny maps, have been constructed for a number of other species from different mammalian orders. The primary focus of these studies is to generate comparative information across diverse mammalian orders. This information will be vital to develop an improved understanding about the evolution of various mammalian genomes from a common ancestor.

Common shrew. At least two shrew–rodent panels are currently available and have been used for mapping of more than 30 genes to eight out of the ten shrew chromosomes (Pack et al., 1995; Malchenko et al., 1996; Matyakhina et al., 1996, 1997). These assignments show a high degree of evolutionary synteny conservation between shrew and human. The findings are in overall (with a few exceptions) agreement with the human–shrew Zoo-FISH data (Dixkens et al., 1998).

American mink. Initially, an American mink–Chinese hamster hybrid cell panel with 39 clones was constructed (Rubtsov et al., 1981a,b) and checked for the possibility of mapping enzyme genes (*LDHA*, *LDHB*, *MDH1*, *G6PD*, *PGD*). In the following years 28 hybrid clones from this panel were successfully used to map over ten enzyme genes to specific mink chromosomes (Rubtsov et al., 1981a,b, 1982; Matveeva et al., 1987; Khlebodarova et al., 1995). Following this, a mink–mouse SCH panel with 23 hybrid clones was generated (Pack et al., 1992) and used for assigning mink synteny groups by either enzyme electrophoresis or Southern blotting (Pack et al., 1992; Khlebodarova et al., 1995). To date about 110 genes have been mapped to all American mink autosomes and the X chromosome using this approach (Kuznetsov et al., 2003). Overall, the information has been useful to compare gene associations between the mink genome and genomes of other species. Together with subsequently generated Zoo-FISH data, these findings reveal a high degree of evolutionary

conservation between mink and human syntenic groups (Khlebodarova et al., 1995; Hameister et al., 1997; Kuznetsov et al., 2003).

Rabbit. During the 1970s and early 1980s, rabbit emerged as an important laboratory animal to study various genetic/biological phenomena. This led to the generation of a rabbit–hamster SCH panel in France that was used to identify synteny between loci. Consequently, four enzyme loci were mapped to the X chromosome and four to two of the autosomes (Echard et al., 1981, 1982). Since then the panel has not made any major contributions towards the expansion of the rabbit gene map. Presently the rabbit gene map comprises ~115 loci, most of which are assigned by *in situ* hybridization and genetic linkage mapping approaches (RabbitMap Database: http://locus.jouy.inra.fr/cgi-bin/lgbc/mapping/common/intro2.pl?BASE=rabbit).

Indian muntjac. As far as is known, only two SCH panels were generated for Indian muntjacs. This muntjac–mouse panel was used to map three enzyme loci (*HPRT*, *G6PD*, *PGK*) to the X chromosome (Shows et al., 1976). Later, a muntjac–rat SCH panel was developed and used to study induced chromosome aberrations. This study demonstrated the presence of several rearrangement 'hot spots' on chromosomes 1 and X (Polianskaia and Fridlianskaia, 1991).

Marsupials. SCH hybrid analysis has proven an invaluable tool for comparing synteny conservation of genes between the three mammalian subclasses: eutherians, marsupials and monotremes. Hybrid cell panels for grey kangaroo–mouse (Dawson and Graves, 1984), red-necked wallaby–mouse (Sykes and Hope, 1985), platypus–Chinese hamster (Watson and Graves, 1988) and South American opossum–Chinese hamster, mouse, mink or vole (Nesterova et al., 1994, 1997) have been constructed for developing synteny maps. Common to all these panels is extreme segregation and fragmentation of marsupial/monotreme chromosomes. Therefore, before practical use of individual hybrid cell lines for synteny mapping, careful characterization of individual cell lines was carried out by enzymatic expression and cytogenetic and/or molecular methods (Watson and Graves, 1988; Nesterova *et al.*, 1994, 1997). The panels proved extremely useful for synteny mapping of genes from the rather conserved mammalian X chromosome, and for revealing similarities and differences between eutherian and marsupial/monotreme X chromosomes (Spencer *et al.*, 1991a,b,c; Watson *et al.*, 1991, 1992, 1993; Watson and Graves, 1998). Synteny assignments of autosomal (*GPI, GOT1, 6PGD, LDHA, LDHB*) and X chromosome genes (*PGK, G6PD*) on the opossum–eutherian panels represent the first gene mapping data in this South American marsupial species (Nesterova *et al.*, 1997).

Overall, it seems that the era of developing gene maps using SCH hybrid panels is gradually coming to an end. This is primarily due to the availability of faster and efficient approaches such as RH mapping and FISH that provide more accurate and detailed information than simply syntenic association with another locus obtained through SCH panel analysis. This is clearly evident in the analysis of animal genomes where, due to a pressing need to rapidly expand the gene maps, the current method of choice is RH analysis. However, in species where detailed genome analysis is presently not essential, existing SCH panels will continue to be extremely valuable tools to detect physical association between loci. It is our view point that during coming years, construction of new SCH panels will be rare among mammals.

Radiation Hybrid (RH) Mapping

Radiation hybrid (RH) mapping is basically an SCH technique with the difference that before fusion of cell lines, the whole or partial genome of the species of interest is exposed to high (lethal) doses of X-ray irradiation that cause fragmentation of donor chromosomes (Fig. 2.1). The dosage may range from as low as 3000 rad (Gyapay *et al.*, 1996; McCarthy *et al.*, 1997) to as high as 50,000 rad (Lunetta *et al.*, 1996). The irradiated cells are fused with rodent recipient cells and the fragments are recovered in various combinations in the hybrid cells. Thus, the resultant panel of hybrid cell lines each contains a different combination of chromosomal fragments from the donor species genome (Goss and Harris, 1975; Cox *et al.*, 1990; Walter and Goodfellow, 1993). The method was first suggested by G. Pontecorvo (Pontecorvo, 1971), and was initially applied for constructing monochromosomal panels to map single human chromosomes, e.g. HSAX, HSA1 (Goss and Harris, 1975, 1977a,b). However, the discovery of the approach and the entire concept of mapping seemed way ahead of technological developments. It therefore comes as no surprise that the advent of PCR and recombinant techniques more than a decade later resulted in the 'rediscovery' of RH mapping, first with the development of single human chromosomes RH panels, e.g. HSA21 and X (Benham *et al.*, 1989; Cox *et al.*, 1990), and then immediately thereafter also with the generation of whole genome RH (WGRH) panels (Walter and Goodfellow, 1993; Walter *et al.*, 1994).

Presently, RH mapping is the method of choice in a range of genome analysis programmes in animals as well as plants. One of the key factors in favour of this approach is the possibility of mapping and physically ordering all types of markers (polymorphic as well as non-polymorphic) that could be genotyped by PCR. This emerged as a major advantage over SCH mapping where markers could be chromosomally assigned but not ordered, and over genetic linkage mapping by analysis of family material where only polymorphic markers could be mapped and ordered. Let us examine the approaches used to characterize the hybrid cell lines for their donor genome content and then look into the ways adopted for the statistical analysis of the results.

Characterization of RH cell lines

In order to assess the overall representation of the donor genome or the genome of interest in the panel, and to assess irradiation-caused fragmentation of chromosomes, RH panels are usually analysed by typing a panel of evenly distributed markers of known chromosomal location. It is estimated that, on average, marker retention frequencies of around 20% should be sufficient to permit determination of accurate marker order (Barrett, 1992; Jones,

1996). Lately, FISH with labelled genomic DNA (Walter and Goodfellow, 1993; McCarthy *et al.*, 1997; Chowdhary *et al.*, 2002) or SINE sequences (Yerle *et al.*, 1998) from the donor species for which the RH panel is constructed have been used on metaphase spreads from hybrid cell lines, for direct visualization of retained donor chromosome fragments and to assess the size of these fragments. Groups working with RH mapping in primates have used also *Alu*-PCR-amplified DNA from individual cell lines as FISH probes on normal metaphase spreads from donor species to identify retained donor chromosome fragments (Marzella *et al.*, 2000).

Genotyping an RH panel

A typical WGRH panel comprises ~90 hybrid cell lines, e.g. human 3000 rad panel – 93 clones (Gyapay *et al.*, 1996), cattle 5000 rad panel – 91 clones (Womack *et al.*, 1997), horse 5000 rad panel – 92 clones (Chowdhary *et al.*, 2002), etc. In some cases, this number may go as high as 130 to even 160 cell lines, e.g. dog 5000 rad panel – 126 clones (Priat *et al.*, 1998). Obviously, the more the number of cell lines, the greater is the accuracy of mapping loci and the higher is the power of resolving physical order of markers. The dose/degree of irradiation also determines how many cell lines must be obtained to generate an informative panel. In general, the greater the irradiation dosage, the more is the fragmentation of the chromosomes. Due to this, fewer fragments are retained in the resultant hybrid cells. Hence more hybrid cell lines are needed for complete representation of the donor genome in the panel (Goss and Harris, 1975; Cox *et al.*, 1990; Walter and Goodfellow, 1993; Walter *et al.*, 1994; Leach and O'Connell, 1995; Jones, 1996). With regards to single chromosome RH panels, the number of hybrid cell lines may range from ~20 to 40, depending on the size of the chromosome, e.g. HSA1 or HSAX requires more cell lines than HSA20–22 (Benham *et al.*, 1989; Leach and O'Connell, 1995). However, these estimates are further compounded with the dosage of irradition.

RH cell lines were initially analysed using enzyme assays (Goss and Harris, 1975,

1977a,b). Later, just for a short time, Southern blot hybridization was used to detect the presence or absence of markers in individual cell lines (Benham *et al.*, 1989). However, with the discovery of PCR in 1989/90 (Saiki *et al.*, 1985; Mullis *et al.*, 1986), almost everyone switched to this new, more accurate and efficient approach. The latter is nowadays almost exclusively the method of choice to map all kinds of DNA sequences – genes, ESTs, polymorphic markers and even anonymous DNA fragments provided the primers of the marker specifically amplify only donor DNA in the recipient DNA background. As in SCH analysis, retention of markers in different RH cell lines is scored based on duplicate typing results ('1' for positive amplification; '0' for negative amplification; '2' for questionable amplification). In some human and mouse laboratories and also in groups working on domestic animals, the entire process is significantly accelerated by duplexing the primers based on the size of their amplification product (Raudsepp *et al.*, 2002a), followed by the use of fluorescent primers for automated capture of results (Ranade *et al.*, 2001).

Data analysis

Analysis of RH typing results is a statistical procedure that reflects the probability of association between two loci. The basic hypothesis here is that the closer two loci are on a chromosome, the greater are the chances that they will not be separated due to irradiation-induced breakages, hence lesser will be the distance between them. Due to this statistical nature, the RH map defined represents relative likelihood of one order versus another at a threshold of 1000:1 odds (Cox *et al.*, 1990). Distances between loci, usually expressed in centiRays (cR), depend mainly on the irradiation dosage used, though other factors such as the number and density of markers on the map also have an influence (Cox *et al.*, 1990; Walter and Goodfellow, 1993; Matise *et al.*, 2002). The relationship of distance between markers and radiation dosage is expressed by a simple formula: a distance of 1 cR between two markers corresponds to ~1% frequency of breakage between them after exposure to an irradiation dosage of N rad of irradiation. Distances estimated by RH are suggested by some to be proportional to physical distances (kilobase pair, kbp), provided irradiation dosage is taken into account. For example, 1 cR = 30 kb (at 6500 rad) and 1 cR = 55 kb (at 9000 rad) (Walter and Goodfellow, 1993). However, the accuracy/validity of these estimates is still a matter of debate.

A number of statistical programs/packages, e.g. RHMAP (Boehnke, 1992; Lunetta *et al.*, 1995), RHMAPPER (Slonim *et al.*, 1997), MULTIMAP (Matise *et al.*, 1994), have been used during the past decade for analysing the genotyping data. At the RH mapping webpage (http://compgen.rutgers.edu/rhmap/#programs) one can find nine different statistical analysis packages with comments by Tara C. Matise for the choice of a particular program over another. The crucial factor in the decision making is the number of markers to be mapped and the size of the mapped region (i.e. the marker density).

RH mapping in different mammals

Precision and efficiency of RH mapping depend not only on the resolution of the panel but also on how many markers are already mapped and ordered in the species investigated. Thus, radiation hybrid mapping is most advanced for species with well-developed or rapidly expanding maps. This is evident from the fact that RH panels and maps have been constructed first in human, mouse and rat, followed by the main domestic and livestock species. A list and general description of available commercial and public WGRH panels, together with the current genome-wide RH mapping projects, can be found on http://compgen.rutgers.edu/rhmap/. Below we provide a brief overview of RH mapping efforts in different mammalian species.

Human

The RH mapping resources in humans are indisputably the most developed among mammals. In addition to the three main commercially available WGRH panels, i.e. Genebridge4, StanfordG3 and StanfordTNG4 (http://comp-gen.rutgers.edu/rhmap/), monochromosomal panels for almost all human chromosomes are also available (e.g. HSA3, Siden *et al.*, 1992; HSA5, McPherson *et al.*, 1997; HSA6, Pappas *et al.*, 1995; HSA11, James *et al.*, 1994; HSA12, Raeymaekers *et al.*, 1995, etc.). It is noteworthy that there are panels even for small chromosomal regions, e.g. HSAXp22 (Heuertz *et al.*, 1995). Collectively, the panels have contributed immensely to the rapid generation of high resolution RH maps for individual human chromosomes. These maps (Hudson *et al.*, 1995; Gyapay *et al.*, 1996; Stewart *et al.*, 1997; Deloukas *et al.*, 1998; Agarwala *et al.*, 2000) have acted as important tools for: (i) the study of complex genetic traits; (ii) positional cloning of disease genes; and (iii) cross-referencing or comparison of the human genome with other mammalian genomes. Additionally, the maps have played an important role in the assembly of the draft sequence of the human genome by validating the order of the sequenced contigs (Schuler *et al.*, 1996; Lander *et al.*, 2001; Olivier *et al.*, 2001; Venter *et al.*, 2001; Matise *et al.*, 2002). With the availability of the assembled complete human (and mouse) genome sequences, the role of RH panels in gene identification and localization seems to be limited in these species.

Primates

Non-human primates provide powerful models for the study of various human diseases. Consequently, large-scale genome analysis projects have been initiated in these species. Since RH analysis provides one of the most efficient ways to rapidly develop ordered physical maps, WGRH panels have been constructed in some species such as baboon (Research Genetics, 3000 rad) and rhesus macaque (Murphy *et al.*, 2001, 5000 rad). Until now, no RH mapping information has been published for baboon. The rhesus macaque (*Macaca mulatta*, MMA) panel comprises 93 clones, and has been analysed to develop RH maps for MMA7 and MMA9. The results show a high degree of concordance with available human and baboon maps (Murphy *et al.*, 2001). It is anticipated that during the coming years RH panels will be available for some more non-human primates.

As far as is known, no RH panels have been constructed for the genomes of great apes, however, an acellular adaptation of the RH method – the HAPPY mapping – was used for fine mapping of chimpanzee and gorilla homologues of HSA2 to identify evolutionary breakpoints (de Pontbriand *et al.*, 2002). In this mapping approach, DNA from any source is irradiated or sheared; the resulting fragments are diluted to produce aliquots containing approximately one haploid genome equivalent. Minute amounts of DNA are then distributed into microtitre wells so that the markers to be PCR tested on these aliquots are either absent or present in one to three copies only. Linked markers tend to be found together in an aliquot and final analysis of the typing results is the same as for RHs (Dear and Cook, 1993). The main advantage of the HAPPY method over RH is that in the former there is no need for time-consuming construction of RH panels.

Rodents

The success of WGRH mapping in humans paved the way for the development of similar resources in mouse and rat. Construction of the T31 mouse–hamster 3000 rad panel (McCarthy *et al.*, 1997; Flaherty and Herron, 1998) provided a 271 marker first-generation WGRH map in the mouse (McCarthy *et al.*, 1997). Soon thereafter, a more comprehensive whole genome map comprising 2486 loci (van Etten *et al.*, 1999), a refined map of MMU17 including 75 microsatellite markers (Schalkwyk *et al.*, 1998) and a high-resolution map for MMU5 comprising 138 markers (Tarantino *et al.*, 2000) were published using the panel. As in humans, RH mapping in mouse is playing a key role in the annotation and assembly of the whole genome sequence data by providing physically ordered markers along the genome.

A 3000 rad rat–hamster RH panel (T55) was among the first created for the laboratory rat (Watanabe *et al.*, 1999). This panel has been crucial in providing high resolution framework maps altogether containing over 5000 microsatellites and 11,000 ESTs/genes (Watanabe *et al.*, 1999; Scheetz *et al.*, 2001; Wallace *et al.*, 2002). These maps significantly facilitated integration of genetic linkage and

RH maps of the rat genome (Steen *et al.*, 1999), thus providing one of the essential tools for rat disease and functional genomics. As in mouse, the physically ordered map is providing a scaffold for large-scale genome sequencing efforts currently underway, and for developing comparative maps in relation to human, mouse and other mammalian genomes. This is in part evident from recent effort where human, mouse and rat sequence data, gene assembly and RH map information were coupled and used for the computer-simulated generation of a comparative framework map for the three species with 2155 homologous UniGenes (Kwitek *et al.*, 2001).

Domestic animals

Cattle. Four WGRH panels generated using different dosages of irradiation are currently available for genome mapping in cattle (5000 rad cattle–hamster, Womack *et al.*, 1997; 3000 rad cattle–hamster, Williams *et al.*, 2002; 12,000 rad cattle–hamster, Rexroad *et al.*, 2000; 7000 rad cattle–hamster, unpublished, see Liu *et al.*, 2002). Published literature shows that the 5000 rad panel (Womack *et al.*, 1997) has been most extensively used, and has served as the key contributor to genome mapping efforts in cattle. A large assortment of medium to high density RH maps for several individual bovine chromosomes are all based on this panel among which the first parallel RH map between BTA19 and HSA17 is worth mentioning (Yang and Womack, 1998; Yang *et al.*, 1998). Over the years, the panel has been widely used to integrate physical and genetic linkage maps of various cattle chromosomes (e.g. Band *et al.*, 1998; Rexroad and Womack, 1999; Amarante *et al.*, 2000; Antoniou *et al.*, 2002; Ashwell *et al.*, 2002). The first-generation WHRH and comparative map in cattle was also generated using the same panel (Band *et al.*, 2000). The map comprised over 1000 loci (768 genes and 319 microsatellite markers) and provided one of the first detailed whole genome comparisons between cattle and human genomes. Recently, the panel was used to develop a high density comprehensive RH and comparative map for some of the cattle chromosomes (BTA18, 103 markers, Goldammer *et al.*, 2002a; BTA15 and BTA29, Larkin *et al.*, 2003). The latter study generated RH and comparative maps for BTA15 and BTA29 that identified conserved segmental boundaries and revealed a complex patchwork shuffling of conserved syntenic segments across cattle, human and mouse genomes.

Compared with the 5000 rad panel, other cattle WGRH panels have not yet been extensively used. The 3000 rad panel (Williams *et al.*, 2002) developed in Edinburgh recently provided a 1200 marker whole genome outline map for all bovine autosomes and the sex chromosomes. Additionally, the panel was used to obtain a detailed RH map for the proximal part of BTA1 (25 markers; Drogemuller *et al.*, 2002) and for BTA15 (94 markers; Gautier *et al.*, 2002). The 7000 rad panel (P. Mariani, Y. Sugimoto and C.W. Beattie, unpublished), however, has been used mainly to develop a physically ordered map of 59 microsatellite and STS markers on the cattle Y chromosome (Liu *et al.*, 2002).

It is suggested that the 3000–7000 rad RH panels in cattle have been extremely useful in developing the first-generation physical maps. However, their mapping resolution is insufficient for fine-mapping and positional cloning in defined chromosomal sub-regions of interest (Weikard *et al.*, 2002). Therefore, in order to improve the mapping resolution, a 12,000 rad WGRH panel was constructed (Rexroad *et al.*, 2000). The power of this panel was tested by comparative mapping of 18 markers from the BTA1 5000 rad map (Rexroad *et al.*, 2000) and constructing high-resolution RH maps for BTA6 (Weikard *et al.*, 2002) and BTA13 (Schlapfer *et al.*, 2002). The results show a clear improvement in the resolution of the maps because markers located on the same location in the 5000 rad map could readily be physically ordered in the 12,000 rad panel. The latter is therefore being used to obtain the second-generation RH and comparative map of the cattle genome that is expected to be published soon (Everts-van der Wind *et al.*, 2004). This map, along with others, will play a pivotal role in the assembly of cattle whole genome sequence data anticipated to be obtained by the end of 2005 (http://www.genome.gov/Pages/Research/Sequencing/SeqProposals/BovineSEQ.pdf).

Pig. As far as is known, four different WGRH panels have been constructed in the pig: a 7000 rad pig–hamster panel (Yerle *et al.*, 1998), 3000 rad pig–hamster panel (Research Genetics), 5000 rad pig–mouse panel (Hamasima *et al.*, 2000) and 12,000 rad pig–hamster panel (Yerle *et al.*, 2002). During the past 5–6 years these panels, in particular the 7000 rad panel (referred to as IMpRH or INRA-Minnesota Porcine Radiation Hybrid panel), has been extensively used for mapping individual chromosomes and chromosomal regions, and for developing a WGRH map. The IMpRH panel contributed to the isolation of the RN mutation by facilitating the development of a detailed map over the candidate region of SSC15 (Robic *et al.*, 1999). The panel has also been instrumental in integrating linkage and physical maps of the pig genome (e.g. whole genome Hawken *et al.*, 1999; SSC13, van Poucke *et al.*, 2001; SSC4q, Stratil *et al.*, 2001; SSC15, Robic *et al.*, 2001; SSC1q, Sarker *et al.*, 2001).

Very recently, a large number of porcine ESTs, i.e. 64 ESTs (Karnuah *et al.*, 2001), 48 ESTs (Wang *et al.*, 2001), 1058 ESTs (Rink *et al.*, 2002), 243 ESTs (Cirera *et al.*, 2003) and orthologues of human genes (Lahbib-Mansais *et al.*, 2003) were mapped using RH analysis. This includes publication of the first-generation RH comparative map of porcine and human genomes involving 489 markers and covering 98% of the assembled human genome. The study identified 60 breakpoints and 90 microrearrangements between pig and human chromosomes (Rink *et al.*, 2002). Overall, the findings of all these research groups represent important contributions to the pig–human comparative map.

Compared with the IMpRH panel, other pig RH panels have not been exploited equally. The 5000 rad panel developed in Japan (Hamasima *et al.*, 2000) was used to generate a medium-resolution RH and comparative map between SSC8 and HSA4 by typing 46 genes and 11 microsatellites (Jiang *et al.*, 2002). The commercially available 3000 rad panel (Research Genetics) has led to the development of a high-resolution RH and comparative map of SSC2 (Rattink *et al.*, 2001) and a map of the porcine X and Y chromosomes (Quilter *et al.*, 2002). Recently a

12,000 rad panel (IMNpRH2) was constructed (Yerle *et al.*, 2002) with the aim of complementing the previous 7000 rad panel constructed by the same groups. Resolutions of the two panels were compared by constructing framework maps in the 2.4 Mb region of SSC15 containing the RN gene. The results show that the resolution of IMNpRH2 is two to three times better than that of the 7000 rad panel (Yerle *et al.*, 2002). The panel was also used to orient two sequence-ready BAC/PAC contigs over SSC6q1.2 (Martins-Wess *et al.*, 2003).

Horse. Two horse–hamster radiation hybrid panels have been made in the horse: a 3000 rad (Kiguwa *et al.*, 2000) and a 5000 rad panel (Chowdhary *et al.*, 2002). The 5000 rad panel comprising 92 hybrid cell lines has been extensively used first to obtain comprehensive RH and comparative maps for ECA11 (Chowdhary *et al.*, 2002) and ECAX (Raudsepp *et al.*, 2002a) and very recently to develop the first-generation RH and comparative map of the equine genome (Chowdhary *et al.*, 2003). This map, comprising a total of 730 markers (258 type I and 472 type II), is the first map in horse that incorporates type I as well as type II markers, integrates synteny, cytogenetic and meiotic maps into a consensus map and provides the most detailed genome-wide information to date on the comparative organization of the equine genome in relation to humans. Very recently, a 1.4 Mb resolution map of ECA17 comprising a total of 75 loci was reported (Lee *et al.*, 2003). This map provides one of the best marker coverages for any of the equine autosomes. Of greater interest is the most recent generation of gene maps for the sex chromosomes in the horse. The X chromosome RH map comprises 169 markers, of which 136 are genes and the remaining markers are microsatellites (Raudsepp *et al.*, 2004b). With an average spacing of markers at ~800 kb intervals, the map is one of the most comprehensive among domesticated animals. Lastly, a 24 marker RH map of the Y chromosome focusing primarily on the euchromatic region has also been generated (Raudsepp *et al.*, 2004a). Thus RH mapping is playing a key role in rapid expansion of the gene map of horse.

Cat. So far, only one WGRH panel has been constructed for the domestic cat, i.e. a cat–hamster 5000 rad panel comprising 93 hybrid clones (Murphy *et al.*, 1999a). Initial testing of the quality of this panel was done by typing 54 genome-wide markers, including those from the sex chromosomes. The panel was useful for comparative studies of cat and human sex chromosomes (Murphy *et al.*, 1999b) revealing a high degree of gene order conservation between feline and human X and Y chromosomes. The first-generation WGRH map reported later (Murphy *et al.*, 2000) provided an average density of one marker every 4.9 Mb interval. The map includes >400 gene-specific loci that helped to identify the presence of 100 conserved segments between cat and human. Owing to RH mapping, the physical and genetic linkage maps have been integrated in the cat (Sun *et al.*, 2001). This has been instrumental in highlighting chromosomal regions that have discrepancies between the two maps. More recently, a second-generation integrated genetic linkage and RH map of the cat was reported (Menotti-Raymond *et al.*, 2003). This map contains a total of 834 markers, including 579 genes and 255 polymorphic loci and reveals approximately 110 conserved segments ordered between the human and cat genomes.

Dog. Dog is a widely recognized model animal for various human genetic diseases/disorders. Discovery of canine genes responsible for these traits can be considerably facilitated through a high resolution genome map comprising polymorphic markers and functional genes. To date, two WGRH panels are reported in dog: a 5000 rad (Vignaux *et al.*, 1999a,b) and a 3000 rad panel (T72, Research Genetics). The 5000 rad or RHDF5000-2 panel has been used by the wider dog gene mapping community. Using this panel, the first-generation WGRH map consisting of 400 markers – 218 genes and 182 microsatellites – was reported 7 years ago (Priat *et al.*, 1998). Subsequently, several attempts were made to expand the RH map and to integrate it with genetic linkage and cytogenetic maps. By adding 200 new loci to the first map, Mellersh *et al.* (2000) reported an integrated map of the dog genome. A year later, a composite RH, meiotic and FISH map with 1800 markers and with >90% genome coverage was released (Breen *et al.*, 2001). This was very recently topped by a 3270 marker map including microsatellites, genes, ESTs, BAC ends and STSs (Guyon *et al.*, 2003). This map has an average intermarker distance of ~1 Mb, and provides 325 highly informative well-spaced markers for canine whole genome scan. The map was also instrumental in the identification of 85 conserved fragments with the human genome, of which few were new. Over the years, the 5000 rad panel has also been used to generate a refined RH map over a small region on CFA12 known to harbour the gene for autosomal recessive narcolepsy (Li *et al.*, 2001). As yet, only two studies have reported the use of the 3000 rad commercial panel to develop integrated RH, cytogenetic and comparative maps for CFA5 (Thomas *et al.*, 2001) and CFAX (Spriggs *et al.*, 2003).

Other mammals

As far as is known, the only non-domestic mammal for which a RH panel has been constructed is the American mink (Knud Christensen, http://www.ihh.kvl.dk/htm/kc/mink/radiat.htm). The initial panel consists of 123 mink–hamster RH cell lines. The quality of the panel has been tested by PCR analysis with canine, ermine, otter, polecat and porcine primers. Primers for two mink-specific SINEs were used to test the presence or absence of mink DNA in individual RH clones. The panel has not yet been actively used for mapping the mink genome.

Non-mammalian vertebrates

The generation of RH panels for two non-mammalian species – chicken and zebrafish – is a relatively recent development. The two species represent important models for studying genetics of growth, development and reproduction, and their gene maps contribute to our understanding of vertebrate genome/chromosome evolution.

The feasibility of creating WGRH panels for non-mammalian species was first shown by Kwok and colleagues (Kwok *et al.*, 1998, 1999) by generating zebrafish–hamster and chicken–hamster RH panels using different radiation doses (1000, 3000, 5000 and 10,000 rad). However, PCR characterization of the hybrid clones did not show significant variation in retention frequencies at different irradiation dosages. Interesting observations emerged after characterizing hybrid clones using FISH analysis – hybridizing chicken or zebrafish genomic DNA to metaphase chromosomes from hybrid cell lines, as well as, hybridizing RH clone DNA to normal chicken or zebrafish metaphase spreads. It appeared that zebrafish chromosome fragments were integrated into the hamster genome, while chicken fragments, several of which were from macrochromosomes, were retained primarily as independent entities and not integrated into the hamster genome. Additionally, the chicken microchromosomes were preferentially retained. Recently a 6000 rad panel comprising 90 chicken × hamster RH clones was produced and characterized with 46 genome-wide markers (Morisson *et al.*, 2002). The panel is considered as a valuable resource for integrating chicken physical and genetic maps.

The 94 hybrid cell 3000 rad zebrafish–hamster panel (Kwok *et al.*, 1998) was used for constructing the first RH map of the zebrafish genome by typing a set of 1275 STSs, genes and ESTs (Geisler *et al.*, 1999). Simultaneously, panels – a zebrafish–mouse 5000 rad and 4000 rad panel consisting of 93 cell lines each – were established and used for mapping 1053 sequence length polymorphism markers, STSs and cloned genes to the zebrafish genome (Hukriede *et al.*, 1999).

Future Prospects

Chromosome maps are integral to the development of gene maps in any species. The three approaches described in this chapter have been extensively used in a range of mammalian species to expand their gene maps. SCH mapping, the first of the three techniques introduced for chromosome mapping, is soon to become a technique of the past. In most of the species with developed (human, mouse, rat, etc.) or rapidly developing (cattle, pig, horse, cat, dog, etc.) gene maps, this technique has already served its purpose. In human, mouse, dog, chicken and rat, whole genome sequence information is available, while in cattle, pig, cat and horse, either sequencing is soon expected to begin or high resolution RH maps are under construction. Hence, SCH mapping will be of little value in all these species. Perhaps, in some of the species where gene mapping efforts are still rudimentary, this approach may be of significance in establishment of syntenic groups and their assignment to specific chromosomes. Alternatively, depleting resources of the stock DNA from individual somatic hybrid cell lines, and availability of better, faster and more efficient mapping methods such as RH, will gradually eclipse the use of SCH panels during the coming years. FISH is a technique which promises to stay longer in genome analysis of mammals. This application should not be confused with applications related to cytogenetic analysis and diagnostics. There will constantly be a need to visualize and confirm the presence/location of a gene or DNA marker on the chromosome, unless alternative accurate methods are available. Lastly, RH analysis is a technique here to stay. The techniques will be extensively used in species that are rapidly moving towards generating high resolution maps and have a sufficiently large stock of markers. We anticipate development of RH panels in some more mammals (primates, domestic animals as well as a few wild animal species). One of the advances to observe during coming years is the automated typing of RH panels using fluorescent markers. Such a set-up will definitely encourage the use of this approach as the main mapping method.

Acknowledgements

This work was funded by grants from the Texas Higher Education Board (ARP 010366-0162-2001; ATP 000517-0306-2003 B.P.C., J.E.W.), NRICGP/USDA Grant 2003-03687 (B.P.C.), NRICGP/USDA Grant 2000-03510 (L.C.S.), Texas Equine Research Foundation (B.P.C., L.C.S.), Link Endowment (B.P.C., L.C.S.), American Quarter Horse Association and the Dorothy Russell Havemeyer Foundation.

References

Adkison, L.R., Leung, D.W. and Womack, J.E. (1988) Somatic cell mapping and restriction fragment analysis of bovine alpha and beta interferon gene families. *Cytogenetics and Cell Genetics* 47, 62–65.

Agarwala, R., Applegate, D.L., Maglott, D., Schuler, G.D. and Schaffer, A.A. (2000) A fast and scalable radiation hybrid map construction and integration strategy. *Genome Research* 10, 350–364.

Amarante, M.R., Yang, Y.P., Kata, S.R., Lopes, C.R. and Womack, J.E. (2000) RH maps of bovine chromosomes 15 and 29: conservation of human chromosomes 11 and 5. *Mammalian Genome* 11, 364–368.

Antoniou, E., Gallagher, D., Jr, Taylor, J., Davis, S., Womack, J. and Grosz, M. (2002) A comparative map of bovine chromosome 25 with human chromosomes 7 and 16. *Cytogenetics and Genome Research* 97, 128–132.

Archidiacono, N., Marzella, R., Finelli, P., Antonacci, R., Jones, C. and Rocchi, M. (1994a) Characterization of chimpanzee–hamster hybrids by chromosome painting. *Somatic Cell and Molecular Genetics* 20, 439–442.

Archidiacono, N., Antonacci, R., Forabosco, A. and Rocchi, M. (1994b) Preparation of human chromosomal painting probes from somatic cell hybrids. *Methods in Molecular Biology* 33, 1–13.

Arnason, U. (1977) The relationship between the four principal pinniped karyotypes. *Hereditas* 87, 227–242.

Ashwell, M.S., Sonstegard, T.S., Kata, S. and Womack, J.E. (2002) A radiation hybrid map of bovine chromosome 27. *Animal Genetics* 33, 75–76.

Bailey, E., Graves, K.T., Cothran, E.G., Reid, R., Lear, T.L. and Ennis, R.B. (1995) Synteny-mapping horse microsatellite markers using a heterohybridoma panel. *Animal Genetics* 26, 177–180.

Baker, R.J., Qumsiyeh, M.B. and Hood, C.S. (1987) Role of chromosomal banding patterns in understanding mammalian evolution. In: Genoways, H.H. (ed.) *Current Mammology*. Plenum Press, New York, pp. 67–96.

Band, M., Larson, J.H., Womack, J.E. and Lewin, H.A. (1998) A radiation hybrid map of BTA23: identification of a chromosomal rearrangement leading to separation of the cattle MHC class II subregions. *Genomics* 53, 269–275.

Band, M.R., Larson, J.H., Rebeiz, M., Green, C.A., Heyen, D.W., Donovan, J., Windish, R., Steining, C., Mahyuddin, P., Womack, J.E. and Lewin, H.A. (2000) An ordered comparative map of the cattle and human genomes. *Genome Research* 10, 1359–1368.

Barrett, J.H. (1992) Genetic mapping based on radiation hybrid data. *Genomics* 13, 95–103.

Beck, T.W., Menninger, J., Voigt, G., Newmann, K., Nishigaki, Y., Nash, W.G., Stephens, R.M., Wang, Y., de Jong, P.J., O'Brien, S.J. and Yuhki, N. (2001) Comparative feline genomics: a BAC/PAC contig map of the major histocompatibility complex class II region. *Genomics* 71, 282–295.

Benham, F., Hart, K., Crolla, J., Bobrow, M., Francavilla, M. and Goodfellow, P.N. (1989) A method for generating hybrids containing nonselected fragments of human chromosomes. *Genomics* 4, 509–517.

Berg, P. (1977) Genetic engineering: challenge and responsibility. *Ambio* 6, 253–260.

Berg, P., Baltimore, D., Boyer, H.W., Cohen, S.N., Davis, R.W., Hogness, D.S., Nathans, D., Roblin, R., Watson, J.D., Weissman, S. and Zinder, N.D. (1974) Letter: potential biohazards of recombinant DNA molecules. *Science* 185, 303.

Berman, E.J., Nash, W.G., Seuanez, H.N. and O'Brien, S. J. (1986) Chromosomal mapping of enzyme loci in the domestic cat: GSR to C2, ADA and ITPA to A3, and LDHA-ACP2 to D1. *Cytogenetics and Cell Genetics* 41, 114–120.

Biltueva, L.S., Sablina, O.V., Beklemisheva, V.R., Shvets, Y., Tkachenko, A., Dukhanina, O., Lushnikova, T.P., Vorobieva, N.V., Graphodatsky, A.S. and Kisselev, L.L. (1996) Localization of rat K51 keratin-like locus (Krt10l) to human and animal chromosomes by *in situ* hybridization. *Cytogenetics and Cell Genetics* 73, 209–213.

Boehnke, M. (1992) Multipoint analysis for radiation hybrid mapping. *Annals of Medicine (Helsinki)* 24, 383–86.

Bonnet, A., Thevenon, S., Claro, F., Gautier, M. and Hayes, H. (2001) Cytogenetic comparison between Vietnamese sika deer and cattle: R-banded karyotypes and FISH mapping. *Chromosome Research* 9, 673–687.

Boyle, A.L., Lichter, P. and Ward, D.C. (1990) Rapid analysis of mouse–hamster hybrid cell lines by *in situ* hybridization. *Genomics* 7, 127–130.

Breen, M., Thomas, R., Binns, M.M., Carter, N.P. and Langford, C.F. (1999a) Reciprocal chromosome painting reveals detailed regions of conserved synteny between the karyotypes of the domestic dog (*Canis familiaris*) and human. *Genomics* 61, 145–155.

Breen, M., Langford, C.F., Carter, N.P., Holmes, N.G., Dickens, H.F., Thomas, R., Suter, N., Ryder, E.J., Pope, M. and Binns, M.M. (1999b) FISH mapping and identification of canine chromosomes. *Journal of Heredity* 90, 27–30.

Breen, M., Jouquand, S., Renier, C., Mellersh, C.S., Hitte, C., Holmes, N.G., Cheron, A., Suter, N., Vignaux, F., Bristow, A.E., Priat, C., McCann, E., Andre, C., Boundy, S., Gitsham, P., Thomas, R., Bridge, W.L., Spriggs, H.F., Ryder, E.J., Curson, A., Sampson, J., Ostrander, E.A., Binns, M.M. and Galibert, F. (2001) Chromosome-specific single-locus FISH probes allow anchorage of an 1800-marker integrated radiation-hybrid/linkage map of the domestic dog genome to all chromosomes. *Genome Research* 11, 1784–1795.

Brown, C.J., Carrel, L. and Willard, H.F. (1997) Expression of genes from the human active and inactive X chromosomes. *American Journal of Human Genetics* 60, 1333–1343.

Brunner, R.M., Goldammer, T., Furbass, R., Vanselow, J. and Schwerin, M. (1998) Genomic organization of the bovine aromatase encoding gene and a homologous pseudogene as revealed by DNA fibre FISH. *Cytogenetics and Cell Genetics* 82, 37–40.

Bucan, M., Yang-Feng, T., Colberg-Poley, A.M., Wolgemuth, D.J., Guenet, J.L., Francke, U. and Lehrach, H. (1986) Genetic and cytogenetic localisation of the homeo box containing genes on mouse chromosome 6 and human chromosome 7. *EMBO Journal* 5, 2899–2905.

Burkin, D.J., Morse, H.G., Broad, T.E., Pearce, P.D., Ansari, H.A., Lewis, P.E. and Jones, C. (1993) Mapping the sheep genome: production of characterized sheep × hamster cell hybrids. *Genomics* 16, 466–472.

Burkin, D.J., O'Brien, P.C., Broad, T.E., Hill, D.F., Jones, C.A., Wienberg, J. and Ferguson-Smith, M.A. (1997) Isolation of chromosome-specific paints from high-resolution flow karyotypes of the sheep (*Ovis aries*) *Chromosome Research* 5, 102–108.

Burkin, D.J., Broad, T.E., Lambeth, M.R., Burkin, H.R. and Jones, C. (1998) New gene assignments using a complete, characterized sheep–hamster somatic cell hybrid panel. *Animal Genetics* 29, 48–54.

Caetano, A.R., Lyons, L.A., Laughlin, T.F., O'Brien, S.J., Murray, J.D. and Bowling, A.T. (1999a) Equine synteny mapping of comparative anchor tagged sequences (CATS) from human chromosome 5. *Mammalian Genome* 10, 1082–1084.

Caetano, A.R., Shiue, Y.L., Lyons, L.A., O'Brien, S.J., Laughlin, T.F., Bowling, A.T. and Murray, J.D. (1999b) A comparative gene map of the horse (*Equus caballus*). *Genome Research* 9, 1239–1249.

Canavez, F., Moreira, M.A., Bonvicino, C.R., Parham, P. and Seuanez, H.N. (1998) Comparative gene assignment in *Ateles paniscus chamek* (*Platyrrhini, Primates*) and man: association of three separate human syntenic groups and evolutionary considerations. *Chromosoma* 107, 73–79.

Canavez, F., Moreira, M.A., Bonvicino, C.R., Olicio, R. and Seuanez, H.N. (1999a) Gene assignment in the spider monkey (*Ateles paniscus chamek*—APC): APE-MYH7 to 2q; AR-GLA-F8C to the X chromosome. *Journal of Heredity* 90, 460–463.

Canavez, F.C., Moreira, M.A., Bonvicino, C.R., Parham, P. and Seuanez, H.N. (1999b) Evolutionary disruptions of human syntenic groups 3, 12, 14, and 15 in *Ateles paniscus chamek* (Platyrrhini, primates). *Cytogenetics and Cell Genetics* 87, 182–188.

Carbone, L., Ventura, M., Tempesta, S., Rocchi, M. and Archidiacono, N. (2002) Evolutionary history of chromosome 10 in primates. *Chromosoma* 111, 267–272.

Carrel, L., Cottle, A.A., Goglin, K.C. and Willard, H.F. (1999) A first-generation X-inactivation profile of the human X chromosome. *Proceedings of the National Academy of Sciences USA* 96, 14440–11444.

Carver, E.A. and Stubbs, L. (1997) Zooming in on the human–mouse comparative map: genome conservation re-examined on a high-resolution scale. *Genome Research* 7, 1123–1137.

Caspersson, T., Farber, S., Foley, G.E., Kudynowski, J., Modest, E.J., Simonsson, E., Wagh, U. and Zech, L. (1968) Chemical differentiation along metaphase chromosomes. *Experimental Cell Research* 49, 219–222.

Chaudhary, R., Wintero, A.K., Fredholm, M. and Chowdhary, B.P. (1997) FISH mapping of seven cDNA sequences in the pig. *Chromosome Research* 5, 545–549.

Chevalet, C. and Corpet, F. (1986) Statistical decision rules concerning synteny or independence between markers. *Cytogenetics and Cell Genetics* 43, 132–139.

Cho, K.W., Youn, H.Y., Watari, T., Tsujimoto, H., Hasegawa, A. and Satoh, H. (1997) A proposed nomenclature of the domestic cat karyotype. *Cytogenetics and Cell Genetics* 79, 71–78.

Chowdhary, B.P. (1998) Cytogenetics and physical chromosome maps. In: Rothschild, M.F. and Ruvinsky, A. (eds) *The Genetics of the Pig*. CAB International, Wallingford, UK, pp. 199–264.

Chowdhary, B.P., de la Sena, C., Harbitz, I., Eriksson, L. and Gustavsson, I. (1995) FISH on metaphase and interphase chromosomes demonstrates the physical order of the genes for GPI, CRC, and LIPE in pigs. *Cytogenetics and Cell Genetics* 71, 175–178.

Chowdhary, B.P., Raudsepp, T., Fronicke, L. and Scherthan, H. (1998) Emerging patterns of comparative genome organization in some mammalian species as revealed by Zoo-FISH. *Genome Research* 8, 577–589.

Chowdhary, B.P., Raudsepp, T., Honeycutt, D., Owens, E.K., Piumi, F., Guérin, G., Matise, T.C., Kata, S.R., Womack, J.E. and Skow, L.C. (2002) Construction of a 5000(rad) whole-genome radiation hybrid panel in the horse and generation of a comprehensive and comparative map for ECA11. *Mammalian Genome* 13, 89–94.

Chowdhary, B.P., Raudsepp, T., Kata, S.R., Goh, G., Millon, L.V., Allan, V., Piumi, F., Guérin, G., Swinburne, J., Binns, M., Lear, T.L., Mickelson, J., Murray, J., Antczak, D.F., Womack, J.E. and Skow, L.C. (2003) The first-generation whole-genome radiation hybrid map in the horse identifies conserved segments in human and mouse genomes. *Genome Research* 13, 742–751.

Christensen, K., Lomholt, B., Hallenberg, C. and Nielsen, K.V. (1998) Mink 5S rRNA genes map to 2q in three loci suggesting conservation of synteny with human 1q. *Hereditas* 128, 17–20.

Cianfriglia, M. (1979) Cytogenetic characterization of Chinese hamster–sheep somatic cell hybrids. *Bollettino della Società Italiana di Biologia Sperimentale (Napoli)* 55, 2201–2207.

Cirera, S., Jorgensen, C.B., Sawera, M., Raudsepp, T., Chowdhary, B.P. and Fredholm, M. (2003) Comparative mapping in the pig: localization of 214 expressed sequence tags. *Mammalian Genome* 14, 405–426.

Cochet, C., Creau-Goldberg, N., Turleau, C. and De Grouchy, J. (1982) Gene mapping of *Microcebus murinus* (Lemuridae): a comparison with man and *Cebus capucinus* (Cebidae). *Cytogenetics and Cell Genetics* 33, 213–221.

Cohen, S.N. (1977) Recombinant DNA: fact and fiction. *Science* 195, 654–657.

Cox, D.R., Burmeister, M., Price, E.R., Kim, S. and Myers, R.M. (1990) Radiation hybrid mapping: a somatic cell genetic method for constructing high-resolution maps of mammalian chromosomes. *Science* 250, 245–250.

Dain, A.R., Tucker, E.M., Donker, R.A. and Clarke, S.W. (1984) Chromosome mapping in cattle using mouse myeloma/calf lymph node cell hybridomas. *Biochemical Genetics* 22, 429–439.

Davoli, R., Bigi, D., Fontanesi, L., Zambonelli, P., Yerle, M., Zijlstra, C., Bosma, A.A., Robic, A. and Russo, V. (2000) Mapping of 14 expressed sequence tags (ESTs) from porcine skeletal muscle by somatic cell hybrid analysis. *Animal Genetics* 31, 400–403.

Dawson, G.W. and Graves, J.A. (1984) Gene mapping in marsupials and monotremes. I. The chromosomes of rodent–marsupial (*Macropus*) cell hybrids, and gene assignments to the X chromosome of the grey kangaroo. *Chromosoma* 91, 20–27.

de Hondt, H.A., Bosma, A.A., den Bieman, M., de Haan, N.A. and van Zutphen, L.F.M. (1991) Gene mapping in river buffalo (*Bubalus bubalis* L.). *Genetics, Selection, Evolution, Supplement 1*, 104s–108s.

de Hondt, H.A., Gallagher, D.S., Oraby, H.A., Othman, O.E., Bosma, A.A., Womack, J.E. and El-Nahas, S.M. (1997) Gene mapping in the river buffalo (*Bubalus bubalis*): five syntenic groups. *Journal of Animal Breeding and Genetics* 114, 79–85.

de Pontbriand, A., Wang, X.P., Cavaloc, Y., Mattei, M.G. and Galibert, F. (2002) Synteny comparison between apes and human using fine-mapping of the genome. *Genomics* 80, 395–401.

Dear, P.H. and Cook, P.R. (1993) Happy mapping: linkage mapping using a physical analogue of meiosis. *Nucleic Acids Research* 21, 13–20.

Deloukas, P., Schuler, G.D., Gyapay, G., Beasley, E.M., Soderlund, C., Rodriguez-Tome, P., Hui, L., Matise, T.C., McKusick, K.B., Beckmann, J.S., Bentolila, S., Bihoreau, M., Birren, B.B., Browne, J., Butler, A., Castle, A.B., Chiannilkulchai, N., Clee, C., Day, P.J., Dehejia, A., Dibling, T., Drouot, N., Duprat, S., Fizames, C., Bentley, D.R. *et al.* (1998) A physical map of 30,000 human genes. *Science* 282, 744–746.

Deys, B.F. (1972) Demonstration of X-linkage of G6PD, HGPRT, and PGK in the horse by means of mule–mouse cell hybridization. PhD thesis, University of Leiden, Leiden.

Di Meo, G.P., Perucatti, A., Incarnato, D., Ferretti, L., Di Berardino, D., Caputi Jambrenghi, A., Vonghia, G. and Iannuzzi, L. (2002) Comparative mapping of twenty-eight bovine loci in sheep (*Ovis aries*, 2n = 54) and river buffalo (*Bubalus bubalis*, 2n = 50) by FISH. *Cytogenetics and Genome Research* 98, 262–264.

Di Meo, G.P., Perucatti, A., Gautier, M., Hayes, H., Incarnato, D., Eggen, A. and Iannuzzi, L. (2003) Chromosome localization of the 31 type I Texas bovine markers in sheep and goat chromosomes by comparative FISH-mapping and R-banding. *Animal Genetics* 34, 294–296.

Dixkens, C., Klett, C., Bruch, J., Kollak, A., Serov, O.L., Zhdanova, N., Vogel, W. and Hameister, H. (1998) ZOO-FISH analysis in insectivores: 'evolution extols the virtue of the status quo'. *Cytogenetics and Cell Genetics* 80, 61–67.

Drogemuller, C., Bader, A., Wohlke, A., Kuiper, H., Leeb, T. and Distl, O. (2002) A high-resolution comparative RH map of the proximal part of bovine chromosome 1. *Animal Genetics* 33, 271–279.

Drwinga, H.L., Toji, L.H., Kim, C.H., Greene, A.E. and Mulivor, R.A. (1993) NIGMS human/rodent somatic cell hybrid mapping panels 1 and 2. *Genomics* 16, 311–314.

Dutrillaux, B. (1979) Chromosomal evolution in primates: tentative phylogeny from *Microcebus murinus* (Prosimian) to man. *Human Genetics* 48, 251–314.

Dutrillaux, B., Viegas-Pequignot, E. and Couturier, J. (1980) Great homology of chromosome banding of the rabbit (*Oryctolagus cuniculus*) and primates, including man. *Annales de Genetique* 23, 22–25.

Echard, G., Gellin, J., Benne, F. and Gillois, M. (1981) The gene map of the rabbit (*Oryctolagus cuniculus* L.) I. Synteny between the rabbit gene loci coding for HPRT, PGK, G6PD, and GLA: their localization on the X chromosome. *Cytogenetics and Cell Genetics* 29, 176–183.

Echard, G., Gellin, J., Benne, F. and Gillois, M. (1982) The gene map of the rabbit, *Oryctolagus cuniculus* l. II. Analysis of the segregation of 11 enzymes in rabbit × hamster somatic cell hybrids: two syntenic groups, LDHB-TPI and LDHA-ACP2. *Cytogenetics and Cell Genetics* 34, 289–295.

Eggen, A. and Fries, R. (1995) An integrated cytogenetic and meiotic map of the bovine genome. *Animal Genetics* 26, 215–236.

El Nahas, S.M., Oraby, H.A., de Hondt, H.A., Medhat, A.M., Zahran, M.M., Mahfouz, E.R. and Karim, A.M. (1996a) Synteny mapping in river buffalo. *Mammalian Genome* 7, 831–834.

El Nahas, S.M., Ramadan, H.A., Abou-Mossallem, A.A., Kurucz, E., Vilmos, P. and Ando, I. (1996b) Assignment of genes coding for leukocyte surface molecules to river buffalo chromosomes. *Veterinary Immunology and Immunopathology* 52, 435–443.

El Nahas, S.M., de Hondt, H.A., Soussa, S.F. and Hassan, A.M. (1999) Assignment of new loci to river buffalo chromosomes confirm the nature of chromosomes 4 and 5. *Journal of Animal Breeding and Genetics* 116, 21–28.

El Nahas, S.M., de Hondt, H.A. and Womack, J.E. (2001) Current status of the river buffalo (*Bubalus bubalis* L.) gene map. *Journal of Heredity* 92, 221–225.

Ellegren, H., Chowdhary, B., Johansson, M. and Andersson, L. (1994) Integrating the porcine physical and linkage map using cosmid-derived markers. *Animal Genetics* 25, 155–164.

Evans, H.J., Buckland, R.A. and Sumner, A.T. (1973) Chromosome homology and heterochromatin in goat, sheep and ox studied by banding techniques. *Chromosoma* 42, 383–402.

Everts-van der Wind, A., Kata, S.R., Band, M.R., Rebeiz, M., Larkin, D.M., Everts, R.E., Green, C.A., Liu, L., Natarajan, S., Goldammer, T., Lee, J.H., McKay, S., Womack, J.E. and Lewin, H.A. (2004) A 1463 gene cattle-human comparative map with anchor points defined by human genome sequence coordinates. *Genome Research* 14, 1424–1437.

Fauth, C. and Speicher, M.R. (2001) Classifying by colors: FISH-based genome analysis. *Cytogenetics and Cell Genetics* 93, 1–10.

Ferretti, L., Leone, P., Pilla, F., Zhang, Y., Nocart, M. and Guérin, G. (1994) Direct characterization of bovine microsatellites from cosmids: polymorphism and synteny mapping. *Animal Genetics* 25, 209–214.

Finaz, C., Cochet, C., de Grouchy, J., Van Cong, N., Rebourcet, R. and Frezal, J. (1975) Gene localization in the chimpanzee (*Pan troglodytes*). Comparison with the factor mapping of man (*Homo sapiens*). *Annals of Genetics* 18, 169–177.

Flaherty, L. and Herron, B. (1998) The new kid on the block – a whole genome mouse radiation hybrid panel. *Mammalian Genome* 9, 417–418.

Ford, C.E. and Hamerton, J.L. (1956) The chromosomes of man. *Nature* 178, 1020–1023.

Forster, M., Stranzinger, G. and Hellkuhl, B. (1980) X-chromosome gene assignment of swine and cattle. *Naturwissenschaften* 67, 48–49.

Francke, U. (1994) Digitized and differentially shaded human chromosome ideograms for genomic applications. *Cytogenetics and Cell Genetics* 65, 206–218.

Fredholm, M. (1997) State of the art and prospects for gene mapping in the dog. *AgBiotechnology News and Information* 9, 267N–272N.

Fridolfsson, A.K., Hori, T., Wintero, A.K., Fredholm, M., Yerle, M., Robic, A., Andersson, L. and Ellegren, H. (1997) Expansion of the pig comparative map by expressed sequence tags (EST) mapping. *Mammalian Genome* 8, 907–912.

Fries, R., Beckmann, J.S., Georges, M., Soller, M. and Womack, J. (1989) The bovine gene map. *Animal Genetics* 20, 3–29.

Fujino, Y., Mizuno, T., Masuda, K., Ohno, K., Satoh, H. and Tsujimoto, H. (2001a) Assignment of the feline Fas (TNFRSF6) gene to chromosome D2p13→p12.2 by fluorescence *in situ* hybridization. *Cytogenetics and Cell Genetics* 95, 122–124.

Fujino, Y., Mizuno, T., Masuda, K., Ohno, K., Satoh, H. and Tsujimoto, H. (2001b) Assignment of the feline Fas ligand gene (TNFSF6) to chromosome F1q12→q13 by fluorescence *in situ* hybridization. *Cytogenetics and Cell Genetics* 94, 92–93.

Gall, J.G. and Pardue, M.L. (1969) Formation and detection of RNA–DNA hybrid molecules in cytological preparations. *Proceedings of the National Academy of Sciences USA* 63, 378–883.

Gallagher, D.S., Jr and Womack, J.E. (1992) Chromosome conservation in the Bovidae. *Journal of Heredity* 83, 287–298.

Gallagher, D.S., Threadgill, D.W., Jackson, A.P., Parham, P. and Womack, J.E. (1991) Somatic cell mapping of bovine clathrin light chain genes: identification of a new bovine syntenic group. *Cytogenetics and Cell Genetics* 56, 154–156.

Gallagher, D. S., Jr, Schlapfer, J., Burzlaff, J.D., Womack, J.E., Stelly, D.M., Davis, S.K. and Taylor, J.F. (1999) Cytogenetic alignment of the bovine chromosome 13 genome map by fluorescence *in-situ* hybridization of human chromosome 10 and 20 comparative markers. *Chromosome Research* 7, 115–119.

Gautier, M., Hayes, H. and Eggen, A. (2002) An extensive and comprehensive radiation hybrid map of bovine chromosome 15: comparison with human chromosome 11. *Mammalian Genome* 13, 316–319.

Geisler, R., Rauch, G.J., Baier, H., van Bebber, F., Brobeta, L., Dekens, M.P., Finger, K., Fricke, C., Gates, M.A., Geiger, H., Geiger-Rudolph, S., Gilmour, D., Glaser, S., Gnugge, L., Habeck, H., Hingst, K., Holley, S., Keenan, J., Kirn, A., Knaut, H., Lashkari, D., Maderspacher, F., Martyn, U., Neuhauss, S., Haffter, P. et al. (1999) A radiation hybrid map of the zebrafish genome. *Nature Genetics* 23, 86–89.

Gibson, L.J. (1986) A creationist view of chromosome banding and evolution. *Origins* 13, 9–35.

Goldammer, T., Kata, S.R., Brunner, R.M., Dorroch, U., Sanftleben, H., Schwerin, M. and Womack, J.E. (2002a) A comparative radiation hybrid map of bovine chromosome 18 and homologous chromosomes in human and mice. *Proceeding of the National Academy of Sciences USA* 99, 2106–2111.

Goldammer, T., Dorroch, U., Brunner, R.M., Kata, S.R., Womack, J.E. and Schwerin, M. (2002b) Identification and chromosome assignment of 23 genes expressed in meat and dairy cattle. *Chromosome Research* 10, 411–418.

Goss, S.J. and Harris, H. (1975) New method for mapping genes in human chromosomes. *Nature* 255, 680–684.

Goss, S.J. and Harris, H. (1977a) Gene transfer by means of cell fusion I. Statistical mapping of the human X-chromosome by analysis of radiation-induced gene segregation. *Journal of Cell Science* 25, 17–37.

Goss, S. J. and Harris, H. (1977b) Gene transfer by means of cell fusion. II. The mapping of 8 loci on human chromosome 1 by statistical analysis of gene assortment in somatic cell hybrids. *Journal of Cell Science* 25, 39–57.

Grafodatskii, A.S. and Biltueva, L.S. (1987) Homology of G-banded mammalian chromosomes. *Genetika* 23, 93–103.

Green, L., Van Antwerpen, R., Stein, J., Stein, G., Tripputi, P., Emanuel, B., Selden, J. and Croce, C. (1984) A major human histone gene cluster on the long arm of chromosome 1. *Science* 226, 838–840.

Guérin, G., Eggen, A., Vaiman, D., Nocart, M., Laurent, P., Bechet, D. and Ferrara, M. (1994a) Further characterization of a somatic cell hybrid panel: ten new assignments to the bovine genome. *Animal Genetics* 25, 31–35.

Guérin, G., Nocart, M. and Kemp, S.J. (1994b) Fifteen new synteny assignments of microsatellites to the bovine genome. *Animal Genetics* 25, 179–181.

Gustavsson, I. (1980) Banding techniques in chromosome analysis of domestic animals. *Advances in Veterinary Science and Comparative Medicine* 24, 245–289.

Gustavsson, I. (1988) Standard karyotype of the domestic pig. *Hereditas* 109, 151–157.

Guyon, R., Lorentzen, T.D., Hitte, C., Kim, L., Cadieu, E., Parker, H.G., Quignon, P., Lowe, J.K., Renier, C., Gelfenbeyn, B., Vignaux, F., DeFrance, H.B., Gloux, S., Mahairas, G.G., Andre, C., Galibert, F. and Ostrander, E.A. (2003) A 1-Mb resolution radiation hybrid map of the canine genome. *Proceedings of the National Academy of Sciences USA* 100, 5296–5301.

Gyapay, G., Schmitt, K., Fizames, C., Jones, H., Vega-Czarny, N., Spillett, D., Muselet, D., Prud'homme, J.F., Dib, C., Auffray, C., Morissette, J., Weissenbach, J. and Goodfellow, P.N. (1996) A radiation hybrid map of the human genome. *Human Molecular Genetics* 5, 339–346.

Hamasima, N., Suzuki, H. and Mitsuhashi, T. (2000) The preliminary report on establishment of 113 porcine radiation hybrid panel cells. *Plant and Animal Genome Conference VIII, 9–12 January, 2000, San Diego, CA, USA*, p. 150.

Hameister, H., Klett, C., Bruch, J., Dixkens, C., Vogel, W. and Christensen, K. (1997) Zoo-FISH analysis: the American mink (*Mustela vison*) closely resembles the cat karyotype. *Chromosome Research* 5, 5–11.

Harbitz, I., Chowdhary, B., Thomsen, P.D., Davies, W., Kaufmann, U., Kran, S., Gustavsson, I., Christensen, K. and Hauge, J.G. (1990) Assignment of the porcine calcium release channel gene, a candidate for the malignant hyperthermia locus, to the 6p11–q21 segment of chromosome 6. *Genomics* 8, 243–248.

Hawken, R.J., Murtaugh, J., Flickinger, G.H., Yerle, M., Robic, A., Milan, D., Gellin, J., Beattie, C.W., Schook, L.B. and Alexander, L.J. (1999) A first-generation porcine whole-genome radiation hybrid map. *Mammalian Genome* 10, 824–830.

Hayes, H., Di Meo, G.P., Gautier, M., Laurent, P., Eggen, A. and Iannuzzi, L. (2000) Localization by FISH of the 31 Texas nomenclature type I markers to both Q- and R-banded bovine chromosomes. *Cytogenetics and Cell Genetics* 90, 315–320.

Hayes, H., Rogel-Gaillard, C., Zijlstra, C., De Haan, N.A., Urien, C., Bourgeaux, N., Bertaud, M. and Bosma, A.A. (2002) Establishment of an R-banded rabbit karyotype nomenclature by FISH localization of 23 chromosome-specific genes on both G- and R-banded chromosomes. *Cytogenetics and Genome Research* 98, 199–205.

Heiskanen, M., Karhu, R., Hellsten, E., Peltonen, L., Kallioniemi, O.P. and Palotie, A. (1994) High resolution mapping using fluorescence *in situ* hybridization to extended DNA fibres prepared from agarose-embedded cells. *Biotechniques* 17, 928–929, 932–933.

Heiskanen, M., Hellsten, E., Kallioniemi, O.P., Makela, T.P., Alitalo, K., Peltonen, L. and Palotie, A. (1995) Visual mapping by fibre-FISH. *Genomics* 30, 31–36.

Heiskanen, M., Kallioniemi, O. and Palotie, A. (1996) Fibre-FISH: experiences and a refined protocol. *Genetic Analysis* 12, 179–184.

Heuertz, S., Smahi, A., Sanak, M., Holvoet-Vermaut, L. and Hors-Cayla, M.C. (1995) Fine deletion mapping of the p22 region of the human X chromosome using a radiation-induced hybrid panel. *Cytogenetics and Cell Genetics* 69, 7–10.

Hopman, A.H., Wiegant, J., Raap, A.K., Landegent, J.E., van der Ploeg, M. and van Duijn, P. (1986) Bi-color detection of two target DNAs by non-radioactive *in situ* hybridization. *Histochemistry* 85, 1–4.

Hotzel, I. and Cheevers, W.P. (2002) A maedi-visna virus strain K1514 receptor gene is located in sheep chromosome 3p and the syntenic region of human chromosome 2. *Journal of General Virology* 83, 1759–1764.

Hsu, T.C. (1952) Mammalian chromosomes *in vitro*. The karyotype of man. *Journal of Heredity* 43, 172.

Hsu, T.C. and Benirschke, K. (1967) *Mammalian Chromosomes*. Springer, Berlin.

Hudson, T.J., Stein, L.D., Gerety, S.S., Ma, J., Castle, A.B., Silva, J., Slonim, D.K., Baptista, R., Kruglyak, L., Xu, S.H. *et al.* (1995) An STS-based map of the human genome. *Science* 270, 1945–1954.

Hukriede, N.A., Joly, L., Tsang, M., Miles, J., Tellis, P., Epstein, J.A., Barbazuk, W.B., Li, F. N., Paw, B., Postlethwait, J.H., Hudson, T.J., Zon, L.I., McPherson, J.D., Chevrette, M., Dawid, I.B., Johnson, S.L. and Ekker, M. (1999) Radiation hybrid mapping of the zebrafish genome. *Proceedings of the National Academy of Sciences USA* 96, 9745–9750.

Iannuzzi, L., Di Meo, G.P., Perucatti, A. and Ferrara, L. (1990) The high resolution G- and R-banding pattern in chromosomes of river buffalo (*Bubalus bubalis* L.). *Hereditas* 112, 209–215.

Iannuzzi, L., Gallagher, D.S., Di Meo, G.P., Schlapfer, J., Perucatti, A., Amarante, M.R., Incarnato, D., Davis, S.K., Taylor, J.F. and Womack, J.E. (2001a) Twelve loci from HSA10, HSA11 and HSA20 were comparatively FISH-mapped on river buffalo and sheep chromosomes. *Cytogenetics and Cell Genetics* 93, 124–126.

Iannuzzi, L., Di Meo, G.P., Hayes, H., Perucatti, A., Incarnato, D., Gautier, M. and Eggen, A. (2001b) FISH-mapping of 31 type I loci (Texas markers) to river buffalo chromosomes. *Chromosome Research* 9, 339–342.

ISCN (1985) An International System for Human Cytogenetic Nomenclature ISCN 1985. Report of the Standing Committee on Human Cytogenetic Nomenclature. *Birth Defects Original Article Series* 21, 1–117.

ISCN (1991) *Guidelines for Cancer Cytogenetics*. S. Karger, New York.

ISCN (1995) *An International System for Human Cytogenetic Nomenclature*. S. Karger, Basel.

ISCNDB (2001) International System for Chromosome Nomenclature of Domestic Bovids (ISCNDB 2000) Cribiu, E.P., Di Berardino, D., Di Meo, G.P., Eggen, A., Gallagher, D.S., Gustavsson, I., Hayes, H., Iannuzzi, L., Popescu, C.P., Rubes, J., Schmutz, S., Stranzinger, G., Vaiman, A. and Womack, J. (committee). *Cytogenetics and Cell Genetics* 92, 283–299.

ISCNH (1997) International system for cytogenetic nomenclature of the domestic horse. Bowling, A.T., Breen, M., Chowdhary, B.P., Hirota, K., Lear, T., Millon, L.V., Ponce de Leon, F.A. Raudsepp, T. and Stranzinger, G. (commitee). *Chromosome Research* 5, 433–443.

Ishizaki, K., Omoto, K. and Sekiguchi, T. (1977) Confirmation of the assignment of the chimpanzee thymidine kinase and galactokinase genes to chromosome 19. *Human Genetics* 37, 231–234.

James, M.R., Richard, C.W., 3rd, Schott, J.J., Yousry, C., Clark, K., Bell, J., Terwilliger, J.D., Hazan, J., Dubay, C., Vignal, A. *et al.* (1994) A radiation hybrid map of 506 STS markers spanning human chromosome 11. *Nature Genetics* 8, 70–76.

Jiang, Z., Priat, C. and Galibert, F. (1998) Traced orthologous amplified sequence tags (TOASTs) and mammalian comparative maps. *Mammalian Genome* 9, 577–587.

Jiang, Z., He, H., Hamasima, N., Suzuki, H. and Verrinder, G. (2002) Comparative mapping of *Homo sapiens* chromosome 4 (HSA4) and *Sus scrofa* chromosome 8 (SSC8) using orthologous genes representing different cytogenetic bands as landmarks. *Genome* 45, 147–156.

John, H.A., Birnstiel, M.L. and Jones, K.W. (1969) RNA–DNA hybrids at the cytological level. *Nature* 223, 582–587.

Jones, H.B. (1996) Hybrid selection as a method of increasing mapping power for radiation hybrids. *Genome Research* 6, 761–769.

Kao, F.T. (1983) Somatic cell genetics and gene mapping. *International Review of Cytology* 85, 109–146.

Karnuah, A.B., Uenishi, H., Kiuchi, S., Kojima, M., Onishi, A., Yasue, H. and Mitsuhashi, T. (2001) Assignment of 64 genes expressed in 28-day-old pig embryo to radiation hybrid map. *Mammalian Genome* 12, 518–523.

Khan, M.P., Pearson, P.L., Wijnen, L.L., Doppert, B.A., Westerveld, A. and Bootsma, D. (1976) Assignment of inosine triphosphatase gene to gorilla chromosome 13 and to human chromosome 20 in primate–rodent somatic cell hybrids. *Cytogenetics and Cell Genetics* 16, 420–421.

Khan, P.M. (1987) Isozymes as bioprobes for genetic analysis of nonhuman primates. *Genetica* 73, 25–36.

Khlebodarova, T.M., Malchenko, S.N., Matveeva, N.M., Pack, S.D., Sokolova, O.V., Alabiev, B.Y., Belousov, E.S., Peremislov, V.V., Nayakshin, A.M., Brusgaard, K. *et al.* (1995) Chromosomal and regional localization of the loci for IGKC, IGGC, ALDB, HOXB, GPT, and PRNP in the American mink (*Mustela vison*): comparisons with human and mouse. *Mammalian Genome* 6, 705–709.

Kiel von, K., Hameister, H., Somssich, I.E. and Adolph, S. (1985) Early replication banding reveals a strongly conserved functional pattern in mammalian chromosomes. *Chromosoma* 93, 69–76.

Kiguwa, S.L., Hextall, P., Smith, A.L., Critcher, R., Swinburne, J., Millon, L., Binns, M.M., Goodfellow, P.N., McCarthy, L.C., Farr, C.J. and Oakenfull, E.A. (2000) A horse whole-genome-radiation hybrid panel: chromosome 1 and 10 preliminary maps. *Mammalian Genome* 11, 803–805.

Konfortov, B.A., Jorgensen, C.B., Miller, J.R. and Tucker, E.M. (1998) Characterisation of a bovine/murine hybrid cell panel informative for all bovine autosomes. *Animal Genetics* 29, 302–306.

Korenberg, J.R., Yang-Feng, T., Schreck, R. and Chen, X.N. (1992) Using fluorescence *in situ* hybridization (FISH) in genome mapping. *Trends in Biotechnology* 10, 27–32.

Korenberg, J.R., Chen, X.N., Adams, M.D. and Venter, J.C. (1995) Toward a cDNA map of the human genome. *Genomics* 29, 364–370.

Korenberg, J.R., Chen, X.N., Sun, Z., Shi, Z.Y., Ma, S., Vataru, E., Yimlamai, D., Weissenbach, J.S., Shizuya, H., Simon, M.I., Gerety, S.S., Nguyen, H., Zemsteva, I.S., Hui, L., Silva, J., Wu, X., Birren, B.W. and Hudson, T.J. (1999) Human genome anatomy: BACs integrating the genetic and cytogenetic maps for bridging genome and biomedicine. *Genome Research* 9, 994–1001.

Kosambi, D.D. (1944) The estimation of map distances from recombination values. *Annals of Eugenics* 12, 172–175.

Kozak, C.A. and Ruddle, F.H. (1976) Sexual and parasexual approaches to the genetic analysis of the laboratory mouse, *Mus musculus*. *In Vitro* 12, 720–725.

Kraan, J., von Bergh, A.R., Kleiverda, K., Vaandrager, J.W., Jordanova, E.S., Raap, A.K., Kluin, P.M. and Schuuring, E. (2002) Multicolor Fibre FISH. *Methods in Molecular Biology* 204, 143–153.

Kucherlapati, R.S. and Ruddle, F.H. (1975) Mammalian somatic hybrids and human gene mapping. *Annals of Internal Medicine* 83, 553–560.

Kuznetsov, S.B., Larkin, D.M., Kaftanovskaia, E.M., Ivanova, E.V., Astakhova, N.M., Cheriakene, O.V. and Zhdanova, N.S. (1998) Chromosome localization and analysis of synteny analysis of some genes in swine, cattle, and sheep (Artiodactyla). *Genetika* 34, 1200–1204.

Kuznetsov, S.B., Matveeva, N.M., Murphy, W.J., O'Brien, S.J. and Serov, O.L. (2003) Mapping of 53 loci in American mink (*Mustela vison*). *Journal of Heredity* 94, 386–391.

Kwitek, A.E., Tonellato, P.J., Chen, D., Gullings-Handley, J., Cheng, Y.S., Twigger, S., Scheetz, T.E., Casavant, T.L., Stoll, M., Nobrega, M.A., Shiozawa, M., Soares, M.B., Sheffield, V.C. and Jacob, H.J. (2001) Automated construction of high-density comparative maps between rat, human, and mouse. *Genome Research* 11, 1935–1943.

Kwok, C., Korn, R.M., Davis, M.E., Burt, D.W., Critcher, R., McCarthy, L., Paw, B.H., Zon, L.I., Goodfellow, P.N. and Schmitt, K. (1998) Characterization of whole genome radiation hybrid mapping resources for non-mammalian vertebrates. *Nucleic Acids Research* 26, 3562–3566.

Kwok, C., Critcher, R. and Schmitt, K. (1999) Construction and characterization of zebrafish whole genome radiation hybrids. *Methods in Cell Biology* 60, 287–302.

Laan, M., Isosomppi, J., Klockars, T., Peltonen, L. and Palotie, A. (1996) Utilization of FISH in positional cloning: an example on 13q22. *Genome Research* 6, 1002–1012.

Lahbib-Mansais, Y., Dalias, G., Milan, D., Yerle, M., Robic, A., Gyapay, G. and Gellin, J. (1999) A successful strategy for comparative mapping with human ESTs: 65 new regional assignments in the pig. *Mammalian Genome* 10, 145–153.

Lahbib-Mansais, Y., Leroux, S., Milan, D., Yerle, M., Robic, A., Jiang, Z., Andre, C. and Gellin, J. (2000) Comparative mapping between humans and pigs: localization of 58 anchorage markers (TOASTs) by use of porcine somatic cell and radiation hybrid panels. *Mammalian Genome* 11, 1098–1106.

Lahbib-Mansais, Y., Tosser-Klopp, G., Leroux, S., Cabau, C., Karsenty, E., Milan, D., Barillot, E., Yerle, M., Hatey, F. and Gellin, J. (2003) Contribution to high-resolution mapping in pigs with 101 type I markers and progress in comparative map between humans and pigs. *Mammalian Genome* 14, 275–288.

Lalley, P.A., Minna, J.D. and Francke, U. (1978) Conservation of autosomal gene synteny groups in mouse and man. *Nature* 274, 160–163.

Landegent, J.E., Jansen in de Wal, N., van Ommen, G.J., Baas, F., de Vijlder, J.J., van Duijn, P. and Van der Ploeg, M. (1985) Chromosomal localization of a unique gene by non-autoradiographic *in situ* hybridization. *Nature* 317, 175–177.

Lander, E.S., Linton, L.M., Birren, B., Nusbaum, C., Zody, M.C., Baldwin, J., Devon, K., Dewar, K., Doyle, M., FitzHugh, W., Funke, R., Gage, D., Harris, K., Heaford, A., Howland, J., Kann, L., Lehoczky, J., LeVine, R., McEwan, P., McKernan, K. *et al.* (2001) Initial sequencing and analysis of the human genome. *Nature* 409, 860–921.

Langer, P.R., Waldrop, A.A. and Ward, D.C. (1981) Enzymatic synthesis of biotin-labeled polynucleotides: novel nucleic acid affinity probes. *Proceedings of the National Academy of Sciences USA* 78, 6633–6637.

Langston, A.A., Mellersh, C.S., Neal, C.L., Ray, K., Acland, G.M., Gibbs, M., Aguirre, G.D., Fournier, R.E. and Ostrander, E.A. (1997) Construction of a panel of canine–rodent hybrid cell lines for use in partitioning of the canine genome. *Genomics* 46, 317–325.

Larkin, D.M., Everts-van der Wind, A., Rebeiz, M., Schweitzer, P.A., Bachman, S., Green, C., Wright, C.L., Campos, E.J., Benson, L.D., Edwards, J., Liu, L., Osoegawa, K., Womack, J.E., de Jong, P.J. and Lewin, H.A. (2003) A cattle–human comparative map built with cattle BAC-ends and human genome sequence. *Genome Research* 13, 1966–1972.

Lawrence, J.B., Villnave, C.A. and Singer, R.H. (1988) Sensitive, high-resolution chromatin and chromosome mapping *in situ*: presence and orientation of two closely integrated copies of EBV in a lymphoma line. *Cell* 52, 51–61.

Lawrence, J.B., Singer, R.H. and McNeil, J.A. (1990) Interphase and metaphase resolution of different distances within the human dystrophin gene. *Science* 249, 928–932.

Lawrence, J.B., Carter, K.C. and Gerdes, M.J. (1992) Extending the capabilities of interphase chromatin mapping. *Nature Genetics* 2, 171–172.

Leach, R.J. and O'Connell, P. (1995) Mapping of mammalian genomes with radiation (Goss and Harris) hybrids. *Advances in Genetics* 33, 63–99.

Lear, T.L., Trembicki, K.A. and Ennis, R.B. (1992) Identification of equine chromosomes in horse × mouse somatic cell hybrids. *Cytogenetics and Cell Genetics* 61, 58–60.

Lee, E.-J., Raudsepp, T., Kata, S.R., Adelson, D., Womack, J.E., Skow, L.C. and Chowdhary, B.P. (2004) A 1.4-Mb interval RH map of horse chromosome 17 provides detailed comparison with human and mouse homologues. *Genomics* 83, 203–215.

Leonard, J.C., Drwinga, H.L., Kim, C.H., Toji, L.H., Bender, P.K., Mulivor, R.A. and Beck, J.C. (1997) Regional mapping panels for chromosomes 3, 4, 5, 11, 15, 17, 18, and X. *Genomics* 46, 530–534.

Leonard, J.C., Toji, L.H., Bender, P.K., Beiswanger, C.M. and Beck, J.C. (1998) Regional mapping panels for chromosomes 8, 13, 21, and 22. *Genomics* 51, 17–20.

Leong, M.M., Lin, C.C. and Ruth, R.F. (1983) The localization of genes for HPRT, G6PD, and alpha-GAL onto the X-chromosome of domestic pig (*Sus scrofa domesticus*). *Canadian Journal of Genetics and Cytology* 25, 239–245.

Levan, A., Hsu, T.C. and Stich, H.F. (1962) The idiogram of the mouse. *Hereditas (Lund)* 48, 677–687.

Levan, G. (1974) Nomenclature for G-bands in rat chromosomes. *Hereditas* 77, 37–52.

Levsky, J.M. and Singer, R.H. (2003) Fluorescence in situ hybridization: past, present and future. *Journal of Cell Science* 116, 2833–2838.

Li, R., Faraco, J.H., Lin, L., Lin, X., Hinton, L., Rogers, W., Lowe, J.K., Ostrander, E.A. and Mignot, E. (2001) Physical and radiation hybrid mapping of canine chromosome 12, in a region corresponding to human chromosome 6p12–q12. *Genomics* 73, 299–315.

Lichter, P. (1997) Multicolor FISHing: what's the catch? *Trends in Genetics* 13, 475–479.

Lichter, P. and Cremer, T. (1992) Chromosome analysis by non-isotopic in situ hybridization. In: *Human Cytogenetics: A Practical Approach*. IRL Press, Oxford, pp. 157–192.

Lichter, P., Tang, C.J., Call, K., Hermanson, G., Evans, G.A., Housman, D. and Ward, D.C. (1990) High-resolution mapping of human chromosome 11 by in situ hybridization with cosmid clones. *Science* 247, 64–69.

Lichter, P., Boyle, A.L., Cremer, T. and Ward, D.C. (1991) Analysis of genes and chromosomes by nonisotopic in situ hybridization. *Genetic Analysis, Techniques and Applications* 8, 24–35.

Liehr, T. and Claussen, U. (2002) Current developments in human molecular cytogenetic techniques. *Current Molecular Medicine* 2, 283–297.

Lindgren, G., Breen, M., Godard, S., Bowling, A., Murray, J., Scavone, M., Skow, L., Sandberg, K., Guérin, G., Binns, M. and Ellegren, H. (2001a) Mapping of 13 horse genes by fluorescence in-situ hybridization (FISH) and somatic cell hybrid analysis. *Chromosome Research* 9, 53–59.

Lindgren, G., Swinburne, J.E., Breen, M., Mariat, D., Sandberg, K., Guérin, G., Ellegren, H., and Binns, M.M. (2001b) Physical anchorage and orientation of equine linkage groups by FISH mapping BAC clones containing microsatellite markers. *Animal Genetics* 32, 37–39.

Liu, W.S., Harbitz, I., Gustavsson, I. and Chowdhary, B.P. (1998) Mapping of the porcine erythropoietin gene to chromosome 3p16–p15 and ordering of four related subclones by fibre-FISH and DNA-combing. *Hereditas* 128, 77–81.

Liu, W.S., Mariani, P., Beattie, C.W., Alexander, L.J. and Ponce De Leon, F.A. (2002) A radiation hybrid map for the bovine Y chromosome. *Mammalian Genome* 13, 320–326.

Lunetta, K.L., Boehnke, M., Lange, K. and Cox, D.R. (1995) Experimental design and error detection for polyploid radiation hybrid mapping. *Genome Research* 5, 151–163.

Lunetta, K.L., Boehnke, M., Lange, K. and Cox, D.R. (1996) Selected locus and multiple panel models for radiation hybrid mapping. *American Journal of Human Genetics* 59, 717–725.

Lyons, L.A., Laughlin, T.F., Copeland, N.G., Jenkins, N.A., Womack, J.E. and O'Brien, S.J. (1997) Comparative anchor tagged sequences (CATS) for integrative mapping of mammalian genomes. *Nature Genetics* 15, 47–56.

Lyons, L.A., Kehler, J.S. and O'Brien, S.J. (1999) Development of comparative anchor tagged sequences (CATS) for canine genome mapping. *Journal of Heredity* 90, 15–26.

Ma, R.Z., van Eijk, M.J., Beever, J.E., Guérin, G., Mummery, C.L. and Lewin, H.A. (1998) Comparative analysis of 82 expressed sequence tags from a cattle ovary cDNA library. *Mammalian Genome* 9, 545–549.

Malchenko, S.N., Koroleva, I.V., Brusgaard, K., Matyakhina, L.D., Colonin, M.G., Pack, S.D., Searle, J.B., Borodin, P.M., Serov, O.L. and Bendixen, C. (1996) Chromosome localization of the gene for growth hormone in the common shrew (*Sorex araneus*). *Hereditas* 125, 243–245.

Mamula, P.W., Heerema, N.A., Palmer, C.G., Lyons, K.M. and Karn, R.C. (1985) Localization of the human salivary protein complex (SPC) to chromosome band 12p13.2. *Cytogenetics and Cell Genetics* 39, 279–284.

Manning, J.E., Hershey, N.D., Broker, T.R., Pellegrini, M., Mitchell, H.K. and Davidson, N. (1975) A new method of *in situ* hybridization. *Chromosoma* 53, 107–117.

Martins-Wess, F., Milan, D., Drogemuller, C., Vobeta-Nemitz, R., Brenig, B., Robic, A., Yerle, M. and Leeb, T. (2003) A high resolution physical and RH map of pig chromosome 6q1.2 and comparative analysis with human chromosome 19q13.1. *BMC Genomics* 4, 20.

Marzella, R., Viggiano, L., Miolla, V., Storlazzi, C.T., Ricco, A., Gentile, E., Roberto, R., Surace, C., Fratello, A., Mancini, M., Archidiacono, N. and Rocchi, M. (2000) Molecular cytogenetic resources for chromosome 4 and comparative analysis of phylogenetic chromosome IV in great apes. *Genomics* 63, 307–313.

Masabanda, J., Kappes, S.M., Smith, T.P., Beattie, C.W. and Fries, R. (1996) Mapping of a linkage group to the last bovine chromosome (BTA27) without an assignment. *Mammalian Genome* 7, 229–230.

Matise, T.C., Perlin, M. and Chakravarti, A. (1994) Automated construction of genetic linkage maps using an expert system (MultiMap): a human genome linkage map. *Nature Genetics* 6, 384–390.

Matise, T.C., Porter, C.J., Buyske, S., Cuttichia, A.J., Sulman, E.P. and White, P.S. (2002) Systematic evaluation of map quality: human chromosome 22. *American Journal of Human Genetics* 70, 1398–1410.

Matsui, S., Faitar, S.L., Rossi, M.R. and Cowell, J.K. (2003) Application of spectral karyotyping to the analysis of the human chromosome complement of interspecies somatic cell hybrids. *Cancer Genetics and Cytogenetics* 142, 30–35.

Matveeva, N.M., Khlebodarova, T.M., Karasik, G.I., Rubtsov, N.B., Serov, O.L., Sverdlov, E.D., Broude, N.E., Modyanov, N.N., Monastyrskaya, G.S. and Ovchinnikov, Yu, A. (1987) Chromosomal localization of the gene coding for alpha-subunit of Na$^+$,K$^+$-ATPase in the American mink (*Mustela vison*). *FEBS Lett* 217, 42–44.

Matyakhina, L.D., Colonin, M.G., Pack, S.D., Borodin, P.M., Searle, J.B. and Serov, O.L. (1996) Chromosome localization of the loci for PEPA, PEPB, PEPS, IDH1, GSR, MPI, PGM1, NP, SOD1, and ME1 in the common shrew (*Sorex araneus*). *Mammalian Genome* 7, 265–267.

Matyakhina, L.D., Koroleva, I.V., Malchenko, S.N., Bendixen, C., Cheryaukene, O.V., Pack, S.D., Borodin, P.M., Serov, O.L. and Searle, J.B. (1997) Chromosome location of sixteen genes in the common shrew, *Sorex araneus* L. (Mammalia, Insectivora). *Cytogenetics and Cell Genetics* 77, 201–204.

McCarthy, L.C., Terrett, J., Davis, M.E., Knights, C.J., Smith, A.L., Critcher, R., Schmitt, K., Hudson, J., Spurr, N.K. and Goodfellow, P.N. (1997) A first-generation whole genome-radiation hybrid map spanning the mouse genome. *Genome Research* 7, 1153–1161.

McKusick, V.A. (1991) Current trends in mapping human genes. *FASEB Journal* 5, 12–20.

McKusick, V.A. and Ruddle, F.H. (1977) The status of the gene map of the human chromosomes. *Science* 196, 390–405.

McPherson, J.D., Apostol, B., Wagner-McPherson, C.B., Hakim, S., Del Mastro, R.G., Aziz, N., Baer, E., Gonzales, G., Krane, M.C., Markovich, R., Masny, P., Ortega, M., Vu, J., Vujicic, M., Church, D.M., Segal, A., Grady, D.L., Moyzis, R.K., Spence, M.A., Lovett, M. and Wasmuth, J.J. (1997) A radiation hybrid map of human chromosome 5 with integration of cytogenetic, genetic, and transcript maps. *Genome Research* 7, 897–909.

McPherson, J.D., Marra, M., Hillier, L., Waterston, R.H., Chinwalla, A., Wallis, J., Sekhon, M., Wylie, K., Mardis, E.R., Wilson, R.K., Fulton, R., Kucaba, T.A., Wagner-McPherson, C., Barbazuk, W.B., Gregory, S.G., Humphray, S.J., French, L., Evans, R.S., Bethel, G., Whittaker, A. *et al.* (2001) A physical map of the human genome. *Nature* 409, 934–941.

McShane, R.D., Gallagher, D.S., Jr, Newkirk, H., Taylor, J.F., Burzlaff, J.D., Davis, S.K. and Skow, L.C. (2001) Physical localization and order of genes in the class I region of the bovine MHC. *Animal Genetics* 32, 235–239.

Mellersh, C.S., Hitte, C., Richman, M., Vignaux, F., Priat, C., Jouquand, S., Werner, P., Andre, C., DeRose, S., Patterson, D.F., Ostrander, E.A. and Galibert, F. (2000) An integrated linkage-radiation hybrid map of the canine genome. *Mammalian Genome* 11, 120–130.

Menotti-Raymond, M., David, V.A., Chen, Z.Q., Menotti, K.A., Sun, S., Schaffer, A.A., Agarwala, R., Tomlin, J.F., O'Brien, S.J. and Murphy, W.J. (2003) Second-generation integrated genetic linkage/radiation hybrid maps of the domestic cat (*Felis catus*). *Journal of Heredity* 94, 95–106.

Mezzelani, A., Solinas Toldo, S., Nocart, M., Guérin, G., Ferretti, L. and Fries, R. (1994) Mapping of syntenic groups U7 and U27 to bovine chromosomes 25 and 12, respectively. *Mammalian Genome* 5, 574–576.

Mezzelani, A., Zhang, Y., Redaelli, L., Castiglioni, B., Leone, P., Williams, J.L., Toldo, S.S., Wigger, G., Fries, R. and Ferretti, L. (1995) Chromosomal localization and molecular characterization of 53 cosmid-derived bovine microsatellites. *Mammalian Genome* 6, 629–635.

Milenkovic, D., Oustry-Vaiman, A., Lear, T.L., Billault, A., Mariat, D., Piumi, F., Schibler, L., Cribiu, E. and Guérin, G. (2002) Cytogenetic localization of 136 genes in the horse: comparative mapping with the human genome. *Mammalian Genome* 13, 524–534.

Minna, J.D., Lalley, P.A. and Francke, U. (1976) Comparative mapping using somatic cell hybrids. *In Vitro* 12, 726–733.

Moreira, M.A., Canavez, F., Parham, P. and Seuanez, H.N. (1997) Assignment of TCF1, TGM1, CALM1, CKB, THBS1, B2M, and FES in *Ateles paniscus chamek* (Platyrrhini, Primates). *Cytogenetics and Cell Genetics* 79, 92–96.

Morisson, M., Lemiere, A., Bosc, S., Galan, M., Plisson-Petit, F., Pinton, P., Delcros, C., Feve, K., Pitel, F., Fillon, V., Yerle, M. and Vignal, A. (2002) ChickRH6: a chicken whole-genome radiation hybrid panel. *Genetics, Selection and Evolution* 34, 521–533.

Morton, N.E. (1955) Sequential tests for the detection of linkage. *American Journal of Human Genetics* 7, 277–318.

Mullis, K., Faloona, F., Scharf, S., Saiki, R., Horn, G. and Erlich, H. (1986) Specific enzymatic amplification of DNA *in vitro*: the polymerase chain reaction. *Cold Spring Harbor Symposia on Quantitative Biology* 51, 263–273.

Murphy, W.J., Menotti-Raymond, M., Lyons, L.A., Thompson, M.A. and O'Brien, S.J. (1999a) Development of a feline whole genome radiation hybrid panel and comparative mapping of human chromosome 12 and 22 loci. *Genomics* 57, 1–8.

Murphy, W.J., Sun, S., Chen, Z.Q., Pecon-Slattery, J. and O'Brien, S.J. (1999b) Extensive conservation of sex chromosome organization between cat and human revealed by parallel radiation hybrid mapping. *Genome Research* 9, 1223–1230.

Murphy, W.J., Sun, S., Chen, Z., Yuhki, N., Hirschmann, D., Menotti-Raymond, M. and O'Brien, S.J. (2000) A radiation hybrid map of the cat genome: implications for comparative mapping. *Genome Research* 10, 691–702.

Murphy, W.J., Page, J.E., Smith, C., Jr, Desrosiers, R.C. and O'Brien, S.J. (2001) A radiation hybrid mapping panel for the rhesus macaque. *Journal of Heredity* 92, 516–519.

Myka, J.L., Lear, T.L., Houck, M.L., Ryder, O.A. and Bailey, E. (2003a) Conservation of a Robertsonian chromosome polymorphism in the Equidae. *13th North American Colloquium on Animal Cytogenetics and Gene Mapping July 13–17, Louisville, Kentucky, USA*, p. 14.

Myka, J.L., Lear, T.L., Houck, M.L., Ryder, O.A. and Bailey, E. (2003b) A comparative gene map for the onager, *Equus hemionus onager*. *13th North American Colloquium on Animal Cytogenetics and Gene Mapping July 13–17, Louisville, Kentucky, USA*, p. 21.

Nash, W.G. and O'Brien, S.J. (1982) Conserved regions of homologous G-banded chromosomes between orders in mammalian evolution: carnivores and primates. *Proceedings of the National Academy of Sciences USA* 79, 6631–6635.

Nath, J. and Johnson, K.L. (2000) A review of fluorescence *in situ* hybridization (FISH): current status and future prospects. *Biotechnic and Histochemistry* 75, 54–78.

Naylor, S.L., Zabel, B.U., Manser, T., Gesteland, R. and Sakaguchi, A.Y. (1984) Localization of human U1 small nuclear RNA genes to band p36.3 of chromosome 1 by *in situ* hybridization. *Somatic Cell and Molecular Genetics* 10, 307–313.

Nederlof, P.M., Robinson, D., Abuknesha, R., Wiegant, J., Hopman, A.H., Tanke, H.J. and Raap, A.K. (1989) Three-color fluorescence *in situ* hybridization for the simultaneous detection of multiple nucleic acid sequences. *Cytometry* 10, 20–27.

Nederlof, P.M., van der Flier, S., Wiegant, J., Raap, A.K., Tanke, H.J., Ploem, J.S. and van der Ploeg, M. (1990) Multiple fluorescence *in situ* hybridization. *Cytometry* 11, 126–131.

Nesbitt, M.N. and Francke, U. (1973) A system of nomenclature for band patterns of mouse chromosomes. *Chromosoma* 41, 145–158.

Nesterova, T.B., Isaenko, A.A., Matveeva, N.M., Shilov, A.G., Rubtsov, N.B., Vorobeva, N.V., Rubtsov, N.V., VandeBerg, J.L. and Zakiian, S.M. (1994) Prospects for obtaining a mapping panel for somatic cell marsupial–rodent hybrids for the short-tailed opossum (*Monodelphis domestica*). *Genetika* 30, 1516–1524.

Nesterova, T.B., Isaenko, A.A., Matveeva, N.M., Shilov, A.G., Rubtsov, N.B., Vorobieva, N.V., Rubtsova, N.V., VandeBerg, J.L. and Zakian, S.M. (1997) Novel strategies for eutherian × marsupial somatic cell hybrids: mapping the genome of *Monodelphis domestica*. *Cytogenetics and Cell Genetics* 76, 115–122.

O'Brien, S.J. and Nash, W.G. (1982) Genetic mapping in mammals: chromosome map of domestic cat. *Science* 216, 257–265.

O'Brien, S.J., Cevario, S.J., Martenson, J.S., Thompson, M.A., Nash, W.G., Chang, E., Graves, J.A., Spencer, J.A., Cho, K.W., Tsujimoto, H. and Lyons, L.A. (1997a) Comparative gene mapping in the domestic cat (*Felis catus*). *Journal of Heredity* 88, 408–414.

O'Brien, S.J., Wienberg, J. and Lyons, L.A. (1997b) Comparative genomics: lessons from cats. *Trends in Genetics* 13, 393–399.

O'Brien, S.J., Menotti-Raymond, M., Murphy, W.J. and Yuhki, N. (2002) The Feline Genome Project. *Annual Review of Genetics* 36, 657–686.

Ohno, S. (1967) *Sex Chromosomes and Sex-linked Genes*. Springer, Berlin.

Ohno, S., Becak, W. and Becak, M.L. (1964) X–autosome ratio and the behavior pattern of individual X-chromosomes in placental mammals. *Chromosoma* 15, 14–30.

Olivier, M., Aggarwal, A., Allen, J., Almendras, A.A., Bajorek, E.S., Beasley, E.M., Brady, S.D., Bushard, J.M., Bustos, V.I., Chu, A., Chung, T.R., De Witte, A., Denys, M.E., Dominguez, R., Fang, N.Y., Foster, B.D., Freudenberg, R.W., Hadley, D., Hamilton, L.R., Jeffrey, T.J. *et al.* (2001) A high-resolution radiation hybrid map of the human genome draft sequence. *Science* 291, 1298–1302.

Oraby, H.A., El Nahas, S.M., de Hondt, H.A. and Abdel Samad, M.F. (1998) Assignment of PCR markers to river buffalo chromosomes. *Genetics, Selection and Evolution* 30, 71–78.

Othman, O.E. and El Nahas, S.M. (1999) Synteny assignment of four genes and two microsatellite markers in river buffalo (*Bubalus bubalis* L.). *Journal of Animal Breeding and Genetics* 116, 161–168.

Ott, J. (1985) *Analysis of the Human Genetic Linkage*. The Johns Hopkins University Press, Baltimore and London.

Pack, S.D., Bedanov, V.M., Sokolova, O.V., Zhdanova, N.S., Matveeva, N.M. and Serov, O.L. (1992) Characterization of a new hybrid mink–mouse clone panel: chromosomal and regional assignments of the GLO, ACY, NP, CKBB, ADH2, and ME1 loci in mink (*Mustela vison*). *Mammalian Genome* 3, 112–118.

Pack, S.D., Kolonin, M.G., Borodin, P.M., Searle, J.B. and Serov, O.L. (1995) Gene mapping in the common shrew (*Sorex araneus*; Insectivora) by shrew-rodent cell hybrids: chromosome localization of the loci for HPRT, TK, LDHA, MDH1, G6PD, PGD, and ADA. *Mammalian Genome* 6, 784–787.

Palotie, A., Heiskanen, M., Laan, M. and Horelli-Kuitunen, N. (1996) High-resolution fluorescence *in situ* hybridization: a new approach in genome mapping. *Annals of Medicine* 28, 101–106.

Pappas, G.J., Thompson, E., Burgess, A., Greenwood, A. and Trent, J.M. (1995) Generation and molecular cytogenetic characterization of a radiation-reduction hybrid panel for human chromosome 6. *Cytogenetics and Cell Genetics* 69, 201–206.

Pardue, M.L. and Gall, J.G. (1969) Molecular hybridization of radioactive DNA to the DNA of cytological preparations. *Proceedings of the National Academy of Sciences USA* 64, 600–604.

Park, J.P. (1996) Shared synteny of human chromosome 17 loci in Canids. *Cytogenetics and Cell Genetics* 74, 133–137.

Parra, I. and Windle, B. (1993) High resolution visual mapping of stretched DNA by fluorescent hybridization. *Nature Genetics* 5, 17–21.

Pathak, S. and Stock, A.D. (1974) The X chromosomes of mammals: karylogical homology as revealed by banding techniques. *Genetics* 78, 703–714.

Pienkowska, A., Szczerbal, I., Makinen, A. and Switonski, M. (2002a) G/Q-banded chromosome nomenclature of the Chinese raccoon dog, *Nyctereutes procyonoides procyonoides* Gray. *Hereditas* 137, 75–78.

Pienkowska, A., Schelling, C., Opiola, T., Rozek, M. and Barciszewski, J. (2002b) Canine 5S rRNA: nucleotide sequence and chromosomal assignment of its gene cluster in four canid species. *Cytogenetics and Genome Research* 97, 187–190.

Pinkel, D., Gray, J.W., Trask, B., van den Engh, G., Fuscoe, J. and van Dekken, H. (1986a) Cytogenetic analysis by *in situ* hybridization with fluorescently labelled nucleic acid probes. *Cold Spring Harbor Symposia on Quantitative Biology* 51, 151–157.

Pinkel, D., Straume, T. and Gray, J.W. (1986b) Cytogenetic analysis using quantitative, high-sensitivity, fluorescence hybridization. *Proceedings of the National Academy of Sciences USA* 83, 2934–2938.

Pinton, P., Schibler, L., Cribiu, E., Gellin, J. and Yerle, M. (2000) Localization of 113 anchor loci in pigs: improvement of the comparative map for humans, pigs, and goats. *Mammalian Genome* 11, 306–315.

Polianskaia, G.G. and Fridlianskaia, I.I. (1991) [The chromosome variability of the Indian muntjac in somatic cell hybrids.] *Tsitologiia* 33, 86–94.

Pontecorvo, G. (1971) Induction of directional chromosome elimination in somatic cell hybrids. *Nature* 230, 367–369.

Poulsen, P.H., Thomsen, P.D. and Olsen, J. (1991) Assignment of the porcine aminopeptidase N (PEPN) gene to chromosome 7cen–q21. *Cytogenetics and Cell Genetics* 57, 44–46.

Prakash, B., Kuosku, V., Olsaker, I., Gustavsson, I. and Chowdhary, B.P. (1996) Comparative FISH mapping of bovine cosmids to reindeer chromosomes demonstrates conservation of the X-chromosome. *Chromosome Research* 4, 214–217.

Prakash, B., Olsaker, I., Gustavsson, I. and Chowdhary, B.P. (1997) FISH mapping of three bovine cosmids to cattle, goat, sheep and buffalo X chromosomes. *Hereditas* 126, 115–119.

Priat, C., Hitte, C., Vignaux, F., Renier, C., Jiang, Z., Jouquand, S., Cheron, A., Andre, C. and Galibert, F. (1998) A whole-genome radiation hybrid map of the dog genome. *Genomics* 54, 361–378.

Quilter, C.R., Blott, S.C., Mileham, A.J., Affara, N.A., Sargent, C.A. and Griffin, D.K. (2002) A mapping and evolutionary study of porcine sex chromosome genes. *Mammalian Genome* 13, 588–594.

Raeymaekers, P., Van Zand, K., Jun, L., Hoglund, M., Cassiman, J.J., Van den Berghe, H. and Marynen, P. (1995) A radiation hybrid map with 60 loci covering the entire short arm of chromosome 12. *Genomics* 29, 170–178.

Ranade, K., Chang, M.S., Ting, C.T., Pei, D., Hsiao, C.F., Olivier, M., Pesich, R., Hebert, J., Chen, Y.D., Dzau, V.J., Curb, D., Olshen, R., Risch, N., Cox, D.R. and Botstein, D. (2001) High-throughput genotyping with single nucleotide polymorphisms. *Genome Research* 11, 1262–1268.

Raney, N.E., Graves, K.T., Cothran, E.G., Bailey, E. and Coogle, L. (1998) Synteny mapping of the horse using a heterohybridoma panel. Paper presented at: Plant & Animal Genome VI (San Diego, CA, USA).

Rattink, A.P., Faivre, M., Jungerius, B.J., Groenen, M.A. and Harlizius, B. (2001) A high-resolution comparative RH map of porcine chromosome (SSC) 2. *Mammalian Genome* 12, 366–370.

Raudsepp, T. and Chowdhary, B.P. (1999) Construction of chromosome-specific paints for meta- and submetacentric autosomes and the sex chromosomes in the horse and their use to detect homologous chromosomal segments in the donkey. *Chromosome Research* 7, 103–114.

Raudsepp, T., and Chowdhary, B.P. (2001) Correspondence of human chromosomes 9, 12, 15, 16, 19 and 20 with donkey chromosomes refines homology between horse and donkey karyotypes. *Chromosome Research* 9, 623–629.

Raudsepp, T., Kijas, J., Godard, S., Guérin, G., Andersson, L. and Chowdhary, B.P. (1999) Comparison of horse chromosome 3 with donkey and human chromosomes by cross-species painting and heterologous FISH mapping. *Mammalian Genome* 10, 277–282.

Raudsepp, T., Christensen, K. and Chowdhary, B.P. (2000) Cytogenetics of donkey chromosomes: nomenclature proposal based on GTG-banded chromosomes and depiction of NORs and telomeric sites. *Chromosome Research* 8, 659–670.

Raudsepp, T., Mariat, D., Guérin, G. and Chowdhary, B.P. (2001) Comparative FISH mapping of 32 loci reveals new homologous regions between donkey and horse karyotypes. *Cytogenetics and Cell Genetics* 94, 180–185. Comparative mapping in equids: the asine X chromosome is rearranged compared to horse and Hartmann's mountain zebra. *Cytogenetics and Genome Research* 96, 206–209.

Raudsepp, T., Kata, S.R., Piumi, F., Swinburne, J., Womack, J.E., Skow, L.C. and Chowdhary, B.P. (2002a) Conservation of gene order between horse and human X chromosomes as evidenced through radiation hybrid mapping. *Genomics* 79, 451–457.

Raudsepp, T., Lear, T.L. and Chowdhary, B.P. (2002b) Comparative mapping in equids: the asine X chromosome is rearranged compared to horse and Hartmann's mountain zebra. *Cytogenetics and Genome Research* 96, 206–209.

Raudsepp, T., Santani, A., Wallner, B., Kata, S.R., Ren, C., Zhang, H.B., Womack, J.E., Skow, L.C. and Chowdhary, B.P. (2004a) A detailed physical map of the horse Y chromosome. *Proceedings of the National Academy of Sciences USA* 101, 9321–9326.

Raudsepp, T., Lee, E.J., Kata, S.R., Brinkmeyer, C., Mickelson, J.R., Skow, L.C., Womack, J.E. and Chowdhary, B.P. (2004b) Exceptional conservation of horse-human gene order on X chromosome revealed by high-resolution radiation hybrid mapping. *Proceedings of the National Academy of Sciences USA* 101, 2386–2391.

Ray, M. and Mohandas, T. (1976) Proposed banding nomenclature for the Chinese hamster chromosomes (*Cricetulus griseus*). *Cytogenetics and Cell Genetics* 16, 83–91.

Rebourcet, R., van Cong, N., Frezal, J., Finaz, C., Cochet, C. and de Grouchy, J. (1975) Chromosome no. 1 of man and chimpanzee: identity of gene mapping for three loci: PPH, PGM1, and Pep-C. *Humangenetik* 29, 337–340.

Reed, K.M., Ihara, N., Mariani, P., Mendoza, K.M., Jensen, L.E., Bellavia, R., Ponce De Leon, F.A., Bennett, G.L., Sugimoto, Y. and Beattie, C.W. (2002) High-resolution genetic map of bovine chromosome 29 through focused marker development. *Cytogenetics and Genome Research* 96, 210–216.

Reimann, N., Bartnitzke, S., Bullerdiek, J., Schmitz, U., Rogalla, P., Nolte, I. and Ronne, M. (1996) An extended nomenclature of the canine karyotype. *Cytogenetics and Cell Genetics* 73, 140–144.

Rettenberger, G., Fredholm, M. and Fries, R. (1994) Chromosomal assignment of porcine microsatellites by use of a somatic cell hybrid mapping panel. *Animal Genetics* 25, 343–345.

Rexroad, C.E., 3rd and Womack, J.E. (1999) Parallel RH mapping of BTA1 with HSA3 and HSA21. *Mammalian Genome* 10, 1095–1097.

Rexroad, C.E., 3rd, Owens, E.K., Johnson, J.S. and Womack, J.E. (2000) A 12,000 rad whole genome radiation hybrid panel for high resolution mapping in cattle: characterization of the centromeric end of chromosome 1. *Animal Genetics* 31, 262–265.

Richard, F., Lombard, M. and Dutrillaux, B. (2003) Reconstruction of the ancestral karyotype of eutherian mammals. *Chromosome Research* 11, 605–618.

Rink, A., Santschi, E.M., Eyer, K.M., Roelofs, B., Hess, M., Godfrey, M., Karajusuf, E.K., Yerle, M., Milan, D. and Beattie, C.W. (2002) A first-generation EST RH comparative map of the porcine and human genome. *Mammalian Genome* 13, 578–587.

Roberts, T., Mead, R.S. and Cowell, J.K. (1996) Characterisation of a human chromosome 1 somatic cell hybrid mapping panel and regional assignment of 6 novel STS. *Annals of Human Genetics* 60, 213–220.

Robic, A., Riquet, J., Yerle, M., Milan, D., Lahbib-Mansais, Y., Dubut-Fontana, C. and Gellin, J. (1996) Porcine linkage and cytogenetic maps integrated by regional mapping of 100 microsatellites on somatic cell hybrid panel. *Mammalian Genome* 7, 438–445.

Robic, A., Seroude, V., Jeon, J.T., Yerle, M., Wasungu, L., Andersson, L., Gellin, J. and Milan, D. (1999) A radiation hybrid map of the RN region in pigs demonstrates conserved gene order compared with the human and mouse genomes. *Mammalian Genome* 10, 565–568.

Robic, A., Jeon, J.T., Rey, V., Amarger, V., Chardon, P., Looft, C., Andersson, L., Gellin, J. and Milan, D. (2001) Construction of a high-resolution RH map of the human 2q35 region on TNG panel and comparison with a physical map of the porcine homologous region 15q25. *Mammalian Genome* 12, 380–386.

Rogalska-Niznik, N., Szczerbal, I., Dolf, G., Schlapfer, J., Schelling, C. and Switonski, M. (2003) Canine-derived cosmid probes containing microsatellites can be used in physical mapping of Arctic fox (*Alopex lagopus*) and Chinese raccoon dog (*Nyctereutes procyonoides procyonoides*) genomes. *Journal of Heredity* 94, 89–93.

Rogers, D.S., Gallagher, D.S. and Womack, J.E. (1991) Somatic cell mapping of the genes for anti-mullerian hormone and osteonectin in cattle: identification of a new bovine syntenic group. *Genomics* 9, 298–300.

Rohrer, G.A., Alexander, L.J., Hu, Z., Smith, T.P., Keele, J.W. and Beattie, C.W. (1996) A comprehensive map of the porcine genome. *Genome Research* 6, 371–391.

Rottger, S., Yen, P.H. and Schempp, W. (2002) A fibre-FISH contig spanning the non-recombining region of the human Y chromosome. *Chromosome Research* 10, 621–635.

Rubtsov, N.B., Radzhabili, S.I., Gradov, A.A. and Serov, O.L. (1981a) Isolation and characterization of somatic cell hybrids of the Chinese hamster and American mink. *Tsitologija i Genetika* 15, 54–58.

Rubtsov, N.B., Radjabli, S.I., Gradov, A.A. and Serov, O.L. (1981b) Chromosome localization of three syntenic gene pairs in the American mink (*Mustela vison*). *Cytogenetics and Cell Genetics* 31, 184–187.

Rubtsov, N.B., Radjabli, S.I., Gradov, A.A. and Serov, O.L. (1982) Chromosome localization of the genes for isocitrate dehydrogenase-1, isocitrate dehydrogenase-2, glutathione reductase, and phosphoglycerate kinase-1 in the American mink (*Mustela vison*). *Cytogenetics and Cell Genetics* 33, 256–260.

Ruddle, F.H. (1981) A new era in mammalian gene mapping: somatic cell genetics and recombinant DNA methodologies. *Nature* 294, 115–120.

Ryttman, H., Thebo, P., Gustavsson, I., Gahne, B. and Juneja, R.K. (1986) Further data on chromosomal assignments of pig enzyme loci LDHA, LDHB, MPI, PEPB and PGM1, using somatic cell hybrids. *Animal Genetics* 17, 323–333.

Ryttman, H., Thebo, P. and Gustavsson, I. (1988) Regional assignments of NP and MPI on chromosome 7 in pig, *Sus scrofa*. *Animal Genetics* 19, 197–200.

Sabile, A., Poras, I., Cherif, D., Goodfellow, P. and Avner, P. (1997) Isolation of monochromosomal hybrids for mouse chromosomes 3, 6, 10, 12, 14, and 18. *Mammalian Genome* 8, 81–85.

Saidi-Mehtar, N. and Hors-Cayla, M.C. (1981) Sheep gene mapping by somatic cell hybridization. III. Synteny between pyruvate kinase M2 (PKM2) and nucleoside phosphorylase (NP) in domestic sheep. *Annales de Genetique (Paris)* 24, 148–151.

Saidi-Mehtar, N., Hors-Cayla, M.C. and Cog, N.V. (1981) Sheep gene mapping by somatic cell hybridization: four syntenic groups: ENO1-PGD, ME1-PGM3, LDHB-PEPB-TPI, and G6PD-PGK-GALA. *Cytogenetics and Cell Genetics* 30, 193–204.

Saiki, R.K., Scharf, S., Faloona, F., Mullis, K.B., Horn, G.T., Erlich, H.A. and Arnheim, N. (1985) Enzymatic amplification of beta-globin genomic sequences and restriction site analysis for diagnosis of sickle cell anemia. *Science* 230, 1350–1354.

Sallinen, R., Latvanlehto, A., Kvist, A.P., Rehn, M., Eerola, I., Chu, M.L., Bonaldo, P., Saitta, B., Bressan, G.M., Pihlajaniemi, T., Vuorio, E., Palotie, A., Wessman, M. and Horelli-Kuitunen, N. (2001) Physical mapping of mouse collagen genes on chromosome 10 by high-resolution FISH. *Mammalian Genome* 12, 340–346.

Sarker, N., Hawken, R.J., Takahashi, S., Alexander, L.J., Awata, T., Schook, L.B. and Yasue, H. (2001) Directed isolation and mapping of microsatellites from swine chromosome 1q telomeric region through microdissection and RH mapping. *Mammalian Genome* 12, 524–527.

Satoh, H., Yoshida, M.C. and Sasaki, M. (1989) High resolution chromosome banding in the Norway rat, *Rattus norvegicus*. *Cytogenetics and Cell Genetics* 50, 151–154.

Sawyer, J.R. (1991) Highly conserved segments in mammalian chromosomes. *Journal of Heredity* 82, 128–133.

Sawyer, J.R. and Hozier, J.C. (1986) High resolution of mouse chromosomes: banding conservation between man and mouse. *Science* 232, 1632–1635.

Sawyer, J.R., Johnson, M.P. and Miller, O.J. (1992) Traditional and molecular cytogenetics. *Journal of Reproductive Medicine* 37, 485–498.

Schalkwyk, L.C., Weiher, M., Kirby, M., Cusack, B., Himmelbauer, H. and Lehrach, H. (1998) Refined radiation hybrid map of mouse chromosome 17. *Mammalian Genome* 9, 807–811.

Scharf, S.J., Horn, G.T. and Erlich, H.A. (1986) Direct cloning and sequence analysis of enzymatically amplified genomic sequences. *Science* 233, 1076–1078.

Scheetz, T.E., Raymond, M.R., Nishimura, D.Y., McClain, A., Roberts, C., Birkett, C., Gardiner, J., Zhang, J., Butters, N., Sun, C., Kwitek-Black, A., Jacob, H., Casavant, T.L., Soares, M.B. and Sheffield, V.C. (2001) Generation of a high-density rat EST map. *Genome Research* 11, 497–502.

Scherthan, H., Cremer, T., Arnason, U., Weier, H.U., Lima-de-Faria, A. and Fronicke, L. (1994) Comparative chromosome painting discloses homologous segments in distantly related mammals. *Nature Genetics* 6, 342–347.

Schibler, L., Vaiman, D., Oustry, A., Giraud-Delville, C. and Cribiu, E.P. (1998) Comparative gene mapping: a fine-scale survey of chromosome rearrangements between ruminants and humans. *Genome Research* 8, 901–915.

Schlapfer, J., Gallagher, D.S., Jr, Burzlaff, J.D., Womack, J.E., Stelly, D.M., Taylor, J.F. and Davis, S.K. (1998) Comparative mapping of bovine chromosome 13 by fluorescence *in situ* hybridization. *Animal Genetics* 29, 265–272.

Schlapfer, J., Stahlberger-Saitbekova, N., Comincini, S., Gaillard, C., Hills, D., Meyer, R.K., Williams, J.L., Womack, J.E., Zurbriggen, A. and Dolf, G. (2002) A higher resolution radiation hybrid map of bovine chromosome 13. *Genetics Selection and Evolution* 34, 255–267.

Schuler, G.D., Boguski, M.S., Stewart, E.A., Stein, L.D., Gyapay, G., Rice, K., White, R.E., Rodriguez-Tome, P., Aggarwal, A., Bajorek, E., Bentolila, S., Birren, B.B., Butler, A., Castle, A.B., Chiannilkulchai, N., Chu, A., Clee, C., Cowles, S., Day, P.J., Dibling, T., Drouot, N., Dunham, I., Duprat, S., East, C., Hudson, T.J. *et al.* (1996) A gene map of the human genome. *Science* 274, 540–546.

Senger, G., Jones, T.A., Fidlerova, H., Sanseau, P., Trowsdale, J., Duff, M. and Sheer, D. (1994) Released chromatin: linearized DNA for high resolution fluorescence *in situ* hybridization. *Human Molecular Genetics* 3, 1275–1280.

Seuanez, H.N., Lima, C.R., Lemos, B., Bonvicino, C.R., Moreira, M.A. and Canavez, F.C. (2001) Gene assignment in *Ateles paniscus chamek* (Platyrrhini, Primates). Allocation of 18 markers of human syntenic groups 1, 2, 7, 14, 15, 17 and 22. *Chromosome Research* 9, 631–639.

Seyfert, H.M., Pitra, C., Meyer, L., Brunner, R.M., Wheeler, T.T., Molenaar, A., McCracken, J.Y., Herrmann, J., Thiesen, H.J. and Schwerin, M. (2000) Molecular characterization of STAT5A- and STAT5B-encoding genes reveals extended intragenic sequence homogeneity in cattle and mouse and different degrees of divergent evolution of various domains. *Journal of Molecular Evolution* 50, 550–561.

Shiue, Y.L., Bickel, L.A., Caetano, A.R., Millon, L.V., Clark, R.S., Eggleston, M.L., Michelmore, R., Bailey, E., Guérin, G., Godard, S., Mickelson, J.R., Valberg, S.J., Murray, J.D. and Bowling, A.T. (1999) A synteny map of the horse genome comprised of 240 microsatellite and RAPD markers. *Animal Genetics* 30, 1–9.

Shiue, Y.L., Millon, L.V., Skow, L.C., Honeycutt, D., Murray, J.D. and Bowling, A.T. (2000) Synteny and regional marker order assignment of 26 type I and microsatellite markers to the horse X- and Y-chromosomes. *Chromosome Research* 8, 45–55.

Shows, T.B., Brown, J.A. and Chapman, V.M. (1976) Comparative gene mapping of HPRT, G6PD, and PGK in man, mouse, and muntjac deer. *Cytogenetics and Cell Genetics* 16, 436–439.

Siden, T.S., Kumlien, J., Schwartz, C.E. and Rohme, D. (1992) Radiation fusion hybrids for human chromosomes 3 and X generated at various irradiation doses. *Somatic Cell and Molecular Genetics* 18, 33–44.

Sinclair, A.H. and Graves, J.A. (1991) Gene mapping in marsupials: detection of an ancient autosomal gene cluster. *Genomics* 9, 581–586.

Sjoberg, A., Peelman, L.J. and Chowdhary, B.P. (1997) Application of three different methods to analyse fibre-FISH results obtained using four lambda clones from the porcine MHC III region. *Chromosome Research* 5, 247–253.

Slonim, D., Kruglyak, L., Stein, L. and Lander, E. (1997) Building human genome maps with radiation hybrids. *Journal of Computational Biology* 4, 487–504.

Southern, E.M. (1975) Long range periodicities in mouse satellite DNA. *Journal of Molecular Biology* 94, 51–69.

Spencer, J.A., Sinclair, A.H., Watson, J.M. and Graves, J.A. (1991a) Genes on the short arm of the human X chromosome are not shared with the marsupial X. *Genomics* 11, 339–345.

Spencer, J.A., Watson, J.M. and Graves, J.A. (1991b) The X chromosome of marsupials shares a highly conserved region with eutherians. *Genomics* 9, 598–604.

Spencer, J.A., Watson, J.M., Lubahn, D.B., Joseph, D.R., French, F.S., Wilson, E.M. and Graves, J.A. (1991c) The androgen receptor gene is located on a highly conserved region of the X chromosomes of marsupial and monotreme as well as eutherian mammals. *The Journal of Heredity* 82, 134–139.

Spriggs, H.F., Holmes, N.G., Breen, M.G., Deloukas, P.G., Langford, C.F., Ross, M.T., Carter, N.P., Davis, M.E., Knights, C.E., Smith, A.E., Farr, C.J., McCarthy, L.C. and Binns, M.M. (2003) Construction and integration of radiation-hybrid and cytogenetic maps of dog chromosome X. *Mammalian Genome* 14, 214–221.

Stankiewicz, P., Park, S.S., Inoue, K. and Lupski, J.R. (2001) The evolutionary chromosome translocation 4;19 in *Gorilla gorilla* is associated with microduplication of the chromosome fragment syntenic to sequences surrounding the human proximal CMT1A-REP. *Genome Research* 11, 1205–1210.

Steen, R.G., Kwitek-Black, A.E., Glenn, C., Gullings-Handley, J., Van Etten, W., Atkinson, O.S., Appel, D., Twigger, S., Muir, M., Mull, T., Granados, M., Kissebah, M., Russo, K., Crane, R., Popp, M., Peden, M., Matise, T., Brown, D.M., Lu, J., Kingsmore, S., Tonellato, P.J., Rozen, S., Slonim, D., Young, P., Jacob, H.J. *et al.* (1999) A high-density integrated genetic linkage and radiation hybrid map of the laboratory rat. *Genome Research* 9, AP1–8, insert.

Stewart, E.A., McKusick, K.B., Aggarwal, A., Bajorek, E., Brady, S., Chu, A., Fang, N., Hadley, D., Harris, M., Hussain, S., Lee, R., Maratukulam, A., O'Connor, K., Perkins, S., Piercy, M., Qin, F., Reif, T., Sanders, C., She, X., Sun, W.L., Tabar, P., Voyticky, S., Cowles, S., Fan, J.B., Cox, D.R. *et al.* (1997) An STS-based radiation hybrid map of the human genome. *Genome Research* 7, 422–433.

Stratil, A., Reiner, G., Peelman, L.J., Van Poucke, M. and Geldermann, H. (2001) Linkage and radiation hybrid mapping of the porcine gene for subunit C of succinate dehydrogenase complex (SDHC) to chromosome 4. *Animal Genetics* 32, 110–112.

Sturm, R.A., Francis, D.I., Cassady, J.L. and Graves, J.A. (1994) Identification of a marsupial OTF1 gene: cross-species STS analysis and *in situ* cross-hybridization to *Macropus eugenii* chromosomes 3/4 and 5. *Cytogenetics and Cell Genetics* 65, 272–275.

Sumner, A.T. (1990) *Chromosome Banding*. Unwin Hyman Ltd, London.

Sumner, A.T., Evans, H.J. and Buckland, R.A. (1973) Mechanisms involved in the banding of chromosomes with quinacrine and Giemsa. I. The effects of fixation in methanol–acetic acid. *Experimental Cell Research* 81, 214–222.

Sun, S., Murphy, W.J., Menotti-Raymond, M. and O'Brien, S.J. (2001) Integration of the feline radiation hybrid and linkage maps. *Mammalian Genome* 12, 436–441.

Sykes, P.J. and Hope, R.M. (1985) Provisional mapping of the gene for a cell surface marker, GA-1, in the red-necked wallaby *Macropus rufogriseus*. *Australian Journal of Biological Sciences* 38, 365–376.

Szamalek, J.M., Szczerbal, I., Rogalska-Niznik, N., Switonski, M., Ladon, D. and Schelling, C. (2002) Chromosomal localization of two keratin gene families in the karyotype of three species of the family Canidae. *Animal Genetics* 33, 404–405.

Szczerbal, I., Kubickova, S., Schelling, C., Dolf, G., Schlapfer, J. and Rubes, J. (2003) Application of dual colour FISH for localisation of canine microsatellite markers in the Chinese raccoon dog (*Nyctereutes procyonoides procyonoides*) genome. *Journal of Applied Genetics* 44, 71–76.

Szpirer, C., Szpirer, J., Van Vooren, P., Tissir, F., Simon, J.S., Koike, G., Jacob, H.J., Lander, E.S., Helou, K., Klinga-Levan, K. and Levan, G. (1998) Gene-based anchoring of the rat genetic linkage and cytogenetic maps: new regional localizations, orientation of the linkage groups, and insights into mammalian chromosome evolution. *Mammalian Genome* 9, 721–734.

Szpirer, J., Levan, G., Thorn, M. and Szpirer, C. (1984) Gene mapping in the rat by mouse–rat somatic cell hybridization: synteny of the albumin and alpha-fetoprotein genes and assignment to chromosome 14. *Cytogenetics and Cell Genetics* 38, 142–149.

Tabet-Aoul, K., Schibler, L., Vaiman, D., Oustry-Vaiman, A., Lantier, I., Saidi-Mehtar, N., Cribiu, E.P. and Lantier, F. (2000) Regional characterization of a hamster–sheep somatic cell hybrid panel. *Mammalian Genome* 11, 37–40.

Tanabe, H., Nakagawa, Y., Minegishi, D., Hashimoto, K., Tanaka, N., Oshimura, M., Sofuni, T. and Mizusawa, H. (2000) Human monochromosome hybrid cell panel characterized by FISH in the JCRB/HSRRB. *Chromosome Research* 8, 319–334.

Tarantino, L.M., Feiner, L., Alavizadeh, A., Wiltshire, T., Hurle, B., Ornitz, D.M., Webber, A.L., Raper, J., Lengeling, A., Rowe, L.B. and Bucan, M. (2000) A high-resolution radiation hybrid map of the proximal portion of mouse chromosome 5. *Genomics* 66, 55–64.

Thiessen, K.M. and Lalley, P.A. (1986) New gene assignments and syntenic groups in the baboon (*Papio papio*). *Cytogenetics and Cell Genetics* 42, 19–23.

Thiessen, K.M. and Lalley, P.A. (1987) Gene assignments and syntenic groups in the sacred baboon (*Papio hamadryas*). *Cytogenetics and Cell Genetics* 44, 82–88.

Thomas, R., Breen, M., Deloukas, P., Holmes, N.G. and Binns, M.M. (2001) An integrated cytogenetic, radiation-hybrid, and comparative map of dog chromosome 5. *Mammalian Genome* 12, 371–375.

Thomsen, P.D., Bosma, A.A., Kaufmann, U. and Harbitz, I. (1991) The porcine PGD gene is preferentially lost from chromosome 6 in pig × rodent somatic cell hybrids. *Hereditas* 115, 63–67.

Thomsen, P.D., Wintero, A.K. and Fredholm, M. (1998) Chromosomal assignments of 19 porcine cDNA sequences by FISH. *Mammalian Genome* 9, 394–396.

Tjio, J.H. and Levan, A. (1956) The chromosome number in man. *Hereditas* 42, 1–6.

Toder, R., Wilcox, S.A., Smithwick, M. and Graves, J.A. (1996) The human/mouse imprinted genes IGF2, H19, SNRPN and ZNF127 map to two conserved autosomal clusters in a marsupial. *Chromosome Research* 4, 295–300.

Trask, B.J. (1991a) Gene mapping by *in situ* hybridization. *Current Opinion in Genetics and Development* 1, 82–87.

Trask, B.J. (1991b) Fluorescence *in situ* hybridization: applications in cytogenetics and gene mapping. *Trends in Genetics* 7, 149–154.

Trask, B.J. (2002) Human cytogenetics: 46 chromosomes, 46 years and counting. *Nature Reviews in Genetics* 3, 769–778.

Trask, B., Pinkel, D. and van den Engh, G. (1989) The proximity of DNA sequences in interphase cell nuclei is correlated to genomic distance and permits ordering of cosmids spanning 250 kilobase pairs. *Genomics* 5, 710–717.

Trask, B.J., Allen, S., Massa, H., Fertitta, A., Sachs, R., van den Engh, G. and Wu, M. (1993) Studies of metaphase and interphase chromosomes using fluorescence *in situ* hybridization. *Cold Spring Harbor Symposia on Quantitative Biology* 58, 767–775.

Tucker, E.M., Dain, A.R., Wright, L.J. and Clarke, S.W. (1981) Culture of sheep × mouse hybridoma cells *in vitro*. *Hybridoma* 1, 77–86.

Turleau, C., Creau-Goldberg, N., Cochet, C. and de Grouchy, J. (1983) Gene mapping of the gibbon. Its position in primate evolution. *Human Genetics* 64, 65–72.

Vaiman, D., Imam-Ghali, M., Moazami-Goudarzi, K., Guérin, G., Grohs, C., Leveziel, H. and Saidi-Mehtar, N. (1994) Conservation of a syntenic group of microsatellite loci between cattle and sheep. *Mammalian Genome* 5, 310–314.

van der Ploeg, M. (2000) Cytochemical nucleic acid research during the twentieth century. *European Journal of Histochemistry* 44, 7–42.

van Etten, W.J., Steen, R.G., Nguyen, H., Castle, A.B., Slonim, D.K., Ge, B., Nusbaum, C., Schuler, G.D., Lander, E.S. and Hudson, T.J. (1999) Radiation hybrid map of the mouse genome. *Nature Genetics* 22, 384–387.

van Poucke, M., Yerle, M., Tuggle, C., Piumi, F., Genet, C., Van Zeveren, A. and Peelman, L.J. (2001) Integration of porcine chromosome 13 maps. *Cytogenetics and Cell Genetics* 93, 297–303.

Venta, P.J., Brouillette, J.A., Yuzbasiyan-Gurkan, V. and Brewer, G.J. (1996) Gene-specific universal mammalian sequence-tagged sites: application to the canine genome. *Biochemical Genetics* 34, 321–341.

Venter, J.C., Adams, M.D., Myers, E.W., Li, P.W., Mural, R.J., Sutton, G.G., Smith, H.O., Yandell, M., Evans, C.A., Holt, R.A., Gocayne, J.D., Amanatides, P., Ballew, R.M., Huson, D.H., Wortman, J.R., Zhang, Q., Kodira, C.D., Zheng, X.H., Chen, L., Skupski, M. *et al.* (2001) The sequence of the human genome. *Science* 291, 1304–1351.

Vignaux, F., Hitte, C., Priat, C., Chuat, J.C., Andre, C. and Galibert, F. (1999a) Construction and optimization of a dog whole-genome radiation hybrid panel. *Mammalian Genome* 10, 888–894.

Vignaux, F., Priat, C., Jouquand, S., Hitte, C., Jiang, Z., Cheron, A., Renier, C., Andre, C. and Galibert, F. (1999b) Toward a dog radiation hybrid map. *Journal of Heredity* 90, 62–67.

Wallace, C.A., Ali, S., Glazier, A.M., Norsworthy, P.J., Carlos, D.C., Scott, J., Freeman, T.C., Stanton, L.W., Kwitek, A.E. and Aitman, T.J. (2002) Radiation hybrid mapping of 70 rat genes from a data set of differentially expressed genes. *Mammalian Genome* 13, 194–197.

Walter, L., Klinga-Levan, K., Helou, K., Albig, W., Drabent, B., Doenecke, D., Gunther, E. and Levan, G. (1996) Chromosome mapping of rat histone genes H1fv, H1d, H1t, Th2a and Th2b. *Cytogenetics and Cell Genetics* 75, 136–139.

Walter, M.A. and Goodfellow, P.N. (1993) Radiation hybrids: irradiation and fusion gene transfer. *Trends in Genetics* 9, 352–356.

Walter, M.A., Spillett, D.J., Thomas, P., Weissenbach, J. and Goodfellow, P.N. (1994) A method for constructing radiation hybrid maps of whole genomes. *Nature Genetics* 7, 22–28.

Wang, C., Hawken, R.J., Larson, E., Zhang, X., Alexander, L. and Rutherford, M.S. (2001) Generation and mapping of expressed sequence tags from virus-infected swine macrophages. *Animal Biotechnology* 12, 51–67.

Washington, S.S., Bowcock, A.M., Gerken, S., Matsunami, N., Lesh, D., Osbourne-Lawrence, S.L., Cowell, J.K., Ledbetter, D.H., White, R.L. and Chakravati, A. (1994) A somatic cell hybrid map of human chromosome 13. *Genomics* 18, 486–495.

Watanabe, T.K., Bihoreau, M.T., McCarthy, L.C., Kiguwa, S.L., Hishigaki, H., Tsuji, A., Browne, J., Yamasaki, Y., Mizoguchi-Miyakita, A., Oga, K., Ono, T., Okuno, S., Kanemoto, N., Takahashi, E., Tomita, K., Hayashi, H., Adachi, M., Webber, C., Davis, M., Kiel, S., Knights, C., Smith, A., Critcher, R., Miller, J., James, M.R. *et al.* (1999) A radiation hybrid map of the rat genome containing 5,255 markers. *Nature Genetics* 22, 27–36.

Waters, P.D., Duffy, B., Frost, C.J., Delbridge, M.L. and Graves, J.A. (2001) The human Y chromosome derives largely from a single autosomal region added to the sex chromosomes 80–130 million years ago. *Cytogenetics and Cell Genetics* 92, 74–79.

Watson, J.M. and Graves, J.A. (1988) Gene mapping in marsupials and monotremes, V. Synteny between hypoxanthine phosphoribosyltransferase and phosphoglycerate kinase in the platypus. *Australian Journal of Biological Sciences* 41, 231–237.

Watson, J.M., Spencer, J.A., Riggs, A.D. and Graves, J.A. (1991) Sex chromosome evolution: platypus gene mapping suggests that part of the human X chromosome was originally autosomal. *Proceedings of the National Academy of Science, USA* 88, 11256–11260.

Watson, J.M., Spencer, J.A., Graves, J.A., Snead, M.L. and Lau, E.C. (1992) Autosomal localization of the amelogenin gene in monotremes and marsupials: implications for mammalian sex chromosome evolution. *Genomics* 14, 785–789.

Watson, J.M., Frost, C., Spencer, J.A. and Graves, J.A. (1993) Sequences homologous to the human X- and Y-borne zinc finger protein genes (ZFX/Y) are autosomal in monotreme mammals. *Genomics* 15, 317–322.

Wayne, R.K., Nash, W.G. and O'Brien, S.J. (1987a) Chromosomal evolution of the Canidae. I. Species with high diploid numbers. *Cytogenetics and Cell Genetics* 44, 123–133.

Wayne, R.K., Nash, W.G. and O'Brien, S.J. (1987b) Chromosomal evolution of the Canidae. II. Divergence from the primitive carnivore karyotype. *Cytogenetics and Cell Genetics* 44, 134–141.

Weikard, R., Kuhn, C., Goldammer, T., Laurent, P., Womack, J.E. and Schwerin, M. (2002) Targeted construction of a high-resolution, integrated, comprehensive, and comparative map for a region specific to bovine chromosome 6 based on radiation hybrid mapping. *Genomics* 79, 768–776.

Westerveld, A., Visser, R.P., Meera Khan, P. and Bootsma, D. (1971) Loss of human genetic markers in man–Chinese hamster somatic cell hybrids. *Nature: New Biology* 234, 20–24.

Wilcox, S.A., Watson, J.M., Spencer, J.A. and Graves, J.A. (1996) Comparative mapping identifies the fusion point of an ancient mammalian X–autosomal rearrangement. *Genomics* 35, 66–70.

Williams, H., Richards, C.M., Konfortov, B.A., Miller, J.R. and Tucker, E.M. (1993) Synteny mapping in the horse using horse–mouse heterohybridomas. *Animal Genetics* 24, 257–260.

Williams, J.L., Eggen, A., Ferretti, L., Farr, C.J., Gautier, M., Amati, G., Ball, G., Caramorr, T., Critcher, R., Costa, S., Hextall, P., Hills, D., Jeulin, A., Kiguwa, S.L., Ross, O., Smith, A.L., Saunier, K., Urquhart, B. and Waddington, D. (2002) A bovine whole-genome radiation hybrid panel and outline map. *Mammalian Genome* 13, 469–474.

Williamson, P., Holt, S., Townsend, S. and Boyd, Y. (1995) A somatic cell hybrid panel for mouse gene mapping characterized by PCR and FISH. *Mammalian Genome* 6, 429–432.

Wintero, A.K., Fredholm, M. and Davies, W. (1996) Evaluation and characterization of a porcine small intestine cDNA library: analysis of 839 clones. *Mammalian Genome* 7, 509–517.

Womack, J.E. (1987) Comparative gene mapping: a valuable new tool for mammalian developmental studies. *Developmental Genetics* 8, 281–293.

Womack, J.E. (1993) The goals and status of the bovine gene map. *Journal of Dairy Science* 76, 1199–1203.

Womack, J.E. (1994) Chromosomal evolution from the perspective of the bovine gene map. *Animal Biotechnology* 5, 123–128.

Womack, J.E. (1996) The bovine gene map: a tool for comparative candidate positional cloning. *Archivos de Zootecnia* 45, 151–164.

Womack, J.E. and Cummins, J.M. (1984) Comparative gene mapping: interferon sensitivity (IFREC) and cytoplasmic superoxide dismutase (SOD1) are linked in cattle. *Cytogenetics and Cell Genetics* 37, 612.

Womack, J.E. and Kata, S.R. (1995) Bovine genome mapping: evolutionary inference and the power of comparative genomics. *Current Opinion in Genetics and Development* 5, 725–733.

Womack, J.E. and Moll, Y.D. (1986) Gene map of the cow: conservation of linkage with mouse and man. *Journal of Heredity* 77, 2–7.

Womack, J.E., Johnson, J.S., Owens, E.K., Rexroad, C.E., 3rd, Schlapfer, J. and Yang, Y.P. (1997) A whole-genome radiation hybrid panel for bovine gene mapping. *Mammalian Genome* 8, 854–856.

Yang, F., O'Brien, P.C., Milne, B.S., Graphodatsky, A.S., Solanky, N., Trifonov, V., Rens, W., Sargan, D. and Ferguson-Smith, M.A. (1999) A complete comparative chromosome map for the dog, red fox, and human and its integration with canine genetic maps. *Genomics* 62, 189–202.

Yang, Y.P. and Womack, J.E. (1998) Parallel radiation hybrid mapping: a powerful tool for high-resolution genomic comparison. *Genome Research* 8, 731–736.

Yang, Y.P., Rexroad, C.E., 3rd, Schlapfer, J. and Womack, J.E. (1998) An integrated radiation hybrid map of bovine chromosome 19 and ordered comparative mapping with human chromosome 17. *Genomics* 48, 93–99.

Yerle, M., Galman, O. and Echard, G. (1991) The high-resolution GTG-banding pattern of pig chromosomes. *Cytogenetics and Cell Genetics* 56, 45–47.

Yerle, M., Echard, G., Robic, A., Mairal, A., Dubut-Fontana, C., Riquet, J., Pinton, P., Milan, D., Lahbib-Mansais, Y. and Gellin, J. (1996) A somatic cell hybrid panel for pig regional gene mapping characterized by molecular cytogenetics. *Cytogenetics and Cell Genetics* 73, 194–202.

Yerle, M., Pinton, P., Robic, A., Alfonso, A., Palvadeau, Y., Delcros, C., Hawken, R., Alexander, L., Beattie, C., Schook, L., Milan, D. and Gellin, J. (1998) Construction of a whole-genome radiation hybrid panel for high-resolution gene mapping in pigs. *Cytogenetics and Cell Genetics* 82, 182–188.

Yerle, M., Pinton, P., Delcros, C., Arnal, N., Milan, D. and Robic, A. (2002) Generation and characterization of a 12,000-rad radiation hybrid panel for fine mapping in pig. *Cytogenetics and Genome Research* 97, 219–228.

Yeung, R.S., Hino, O., Vilensky, M., Buetow, K., Szpirer, C., Szpirer, J., Klinga-Levan, K., Levan, G. and Knudson, A.G. (1993) Assignment of 22 loci in the rat by somatic hybrid and linkage analysis. *Mammalian Genome* 4, 585–588.

Zijlstra, C., Bosma, A.A. and de Haan, N.A. (1994) Comparative study of pig–rodent somatic cell hybrids. *Animal Genetics* 25, 319–327.

Zijlstra, C., Bosma, A.A., de Haan, N.A. and Mellink, C. (1996) Construction of a cytogenetically characterized porcine somatic cell hybrid panel and its use as a mapping tool. *Mammalian Genome* 7, 280–284.

Zijlstra, C., de Haan, N.A., Korstanje, R., Rogel-Gaillard, C., Piumi, F., van Lith, H.A., van Zutphen, L.F. and Bosma, A.A. (2002) Fourteen chromosomal localizations and an update of the cytogenetic map of the rabbit. *Cytogenetics and Genome Research* 97, 191–199.

Web addresses

Mouse Genome Informatics: http://www.informatics.jax.org/

RatMap database: http://ratmap.gen.gu.se/ShowSingleCitation.htm?citno=904

BovMap database: http://locus.jouy.inra.fr/cgi-bin/bovmap/intro2.pl

PigMap database: http://www.thearkdb.org/browser?species=pig&objtype=stats

SheepMap database: http://www.thearkdb.org/browser?species=sheep&objtype=stats

RabbitMap: http://locus.jouy.inra.fr/cgibin/lgbc/mapping/common/intro2.pl?BASE=rabbit

The BAC Resource Consortium, 2001: http://bacpac.chori.org/mapped-clones.htm

Human Genome Sequencing Project: (http://www.ornl.gov/TechResources/Human_Genome/project/50yr.html

Cattle Genome Sequencing Proposal: (http://www.genome.gov/Pages/Research/Sequencing/SeqProposals/BovineSEQ.pdf

Cattle Radiation Hybrid Mapping Website: (http://bovid.cvm.tamu.edu/cgi-bin/rhmapper.cgi

Computer Programs for RH Mapping: http://compgen.rutgers.edu/rhmap/#programs

Radiation Mapping Information Page: http://compgen.rutgers.edu/rhmap/

Mink chromosome mapping by Knud Christensen: http://www.ihh.kvl.dk/htm/kc/mink/radiat.htm

3 Mapping Genomes at the Molecular Level

F. Galibert[1] and N.E. Cockett[2]

[1]CNRS/*Université de Rennes 1, France;* [2]*Utah State University, Logan, USA*

Introduction	67
Integrated Maps	68
Cloning Systems	68
Cosmids	69
Fosmids	70
Bacteriophage P1 cloning system	70
Bacterial artificial chromosome (BAC) system	71
P1-derived artifical chromosome (PAC) system	72
Yeast artificial chromosome (YAC) system	72
Screening Large Insert Libraries	73
Contig Assembly	74
Chromosome-specific Libraries	75
Sequencing of the Entire Genome: Current Approaches and Technologies	76
The Shotgun Approach	77
Novel Sequencing Methods and Strategies	77
Sequencing the Human Genome and Other Large Genomes	79
Where Do We Go From Here?	79
Conclusions	81
References	81

Introduction

Genome maps are schematic representations in which markers that correspond to specific sequences of known or unknown function are linearly dispersed at a distance representative of their respective position in the genome. There are two types of genome maps, designated according to the assembly process used for their construction: (i) linkage or genetic maps (Chapter 1) and radiation hybrid (RH) maps (Chapter 2), in which the order of markers is determined on the basis of statistical analysis (Cox *et al.*, 1990), and (ii) physical maps (Chapter 2), which are not based on statistical analysis. Although this difference in terms of data processing is clear, the term 'physical map' is often used to describe an RH map (Deloukas *et al.*, 1998). The main consequence of these differences in data processing is that the addition of a new set of markers to a linkage or RH map can and frequently does change the order and distances between previously mapped markers whereas no such modification should follow the addition of a new set of markers to a physical map unless mistakes were made in its original construction. The reason for this is simple: in linkage and RH maps, the order and distances between markers on the map depend to some extent on the algorithm used to construct the map and on the parameters used to evaluate the data set. Thus, for a given set of data (vectors) there can be several statistically sound graphical representations (Hitte et al., 2003). Often, the various orders of markers that are statistically sound correspond to simple local inversions of the positions of certain groups of markers. However, in some cases an entire set of markers may be mapped to the correct position, but the order of these markers completely reversed with respect to the rest of the map. Alternatively, a set of markers may be mapped such that it appears to extend beyond the end of the chromosome. Unfortunately, there is generally no basis on which to choose the best order of markers during map construction. Agarwala et al. (2000) recently developed an approach in which the same set of RH data is processed automatically five times, with a non-parametric method used for three of these calculations and a parametric method used for the other two calculations, and the parameters set changed each time. Such an approach represents a major step forward as any discrepancy between the five proposed maps provides an early warning of potential problems or weaknesses in the construction of the map. This warning encourages the researcher to search for a better solution or to indicate potential weaknesses in construction to users of the map. Thus, this difference between linkage or RH maps and physical maps is not simply semantics. Users should keep in mind the fact that two maps may have markers in different positions, yet both maps are statistically sound.

Integrated Maps

Integrated maps are produced by merging maps of the same genome that have been constructed independently. Generally, maps of different natures are integrated. For example, a physical map may be merged with a linkage map or an RH map, or a linkage map with an RH map (Mellersh et al., 2000). Integrated maps may also be produced by merging different maps generated by similar data processing methods.

Whatever the nature of the maps, integration is made possible when a set of common markers or anchor loci is mapped on to each map that is to be integrated. The objective is to allow one to move from map to map rather than placing all markers on a single map. Integrated maps make it easier to move from a genetic locus at which a particular trait has been located to the cloning of the corresponding gene. As both linkage maps and RH maps are based on a statistical treatment of the data, integrated maps combining these two types of map should if possible be anchored on markers that have also been physically located on metaphase chromosome spreads by fluorescence in situ hybridization (FISH), as shown in Fig. 3.1 (Breen et al., 2001).

Cloning Systems

One of the keys to developing a genome map is the ability to capture, propagate and manipulate large fragments of the genome via cloning and then systematically organizing these fragments into overlapping segments or contigs. This term, derived from the word 'contiguity', was invented in the late 1980s to designate a series of recombinant clones – principally cosmids at the time – that overlap each other. It is also used to describe a series of overlapping short nucleotide sequences obtained by shotgun analysis.

Development of cloning systems that allow the insertion of large exogenous DNA fragments (30 kb or more) has been critical to the advancement of genome maps in complex species such as mammals. Systems that facilitate whole-genome mapping include cosmids (Collins and Hohn, 1978), fosmids (Kim et al., 1992), the bacteriophage P1 system (Sternberg, 1990), bacterial artificial chromosomes or BACs (Shizuya et al., 1992), P1-derived artificial chromosomes or PACs (Ioannou et al., 1994), and yeast artificial chromosomes or YACs (Burke et al., 1987). These systems are

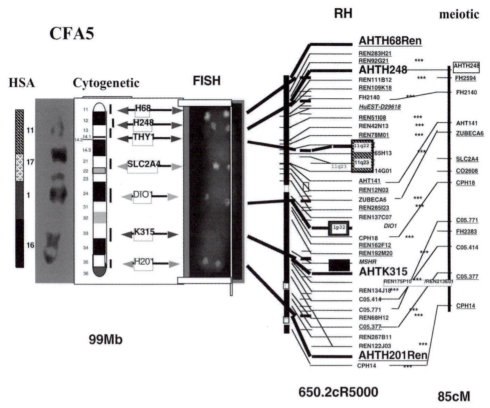

Fig. 3.1. An integrated map of canine chromosome 5 (CFA5) with RH, meiotic, FISH and cytogenetic maps shown from right to left (Breen *et al.*, 2001). Markers localized on two or more maps, indicated by horizontal bars, served as anchors and allow movement from one map to another.

described below in order of increasing insert size and several characteristics of the cloning systems are summarized in Table 3.1.

Cosmids

The original cosmid vector, pJC703 (Collins and Hohn, 1978), contained the lambda phage *cos* (cohesive end) site and the ColE replicon, as well as lambda sequences necessary for packaging (Hohn, 1983). After ligation of exogenous DNA into the *Bam*HI cloning site within the rifampicin resistance (*rpoB*) gene, the resulting recombinant clones were packaged into lambda bacteriophage particles using the packaging system described by Hohn and Murray (1977). Early studies of the ColE1 cosmid indicated that there was an inverse relationship between size of the inserted DNA and

copy number (Collins and Hohn, 1978; Little and Cross, 1985) and that the clones were often unstable. Modifications of the pJC703 vector into the pJB8 series prevented formation of clones with small or multiple inserts (Ish-Horowicz and Burke, 1981). In order to improve the stability of the recombinant cosmid and increase DNA yield, the ColE replicon was replaced with the lambda phage replicon (Little and Cross, 1985), creating the loric (or 'lambda-origin cosmid') vector. It was suggested that the increased yield was due to more efficient 'rolling circle' replication of the lambda replicon. Efficiency of packaging hybrid cosmids is about 10^5 hybrid clones per µg of foreign DNA (Collins and Hohn, 1978).

There are drawbacks of the cosmid system that have limited its application to mammalian genome libraries. First and foremost, cosmid clones contain comparatively small inserts

Table 3.1. Comparisons of large insert cloning systems.

Cloning system	Common vector	Vector size	Maximum insert size	Selection for recombinants	Transformation efficiency	Stability[a]	Chimerism[b]
Cosmid	pJB8	5.4 kb	45 kb	Ampicillin	10^5 clones/μg insert	Unstable	5–40%
Fosmid	pFOS1	9.7 kb	45 kb	Chloramphenicol	Not available	Stable	5%
P1	pAd10sacBII	16.0 kb	100 kb	Kanamycin	10^5 clones/μg vector	Stable	5%
BAC	pBeloBAC11	7.4 kb	350 kb	Chloramphenicol	10^6 clones/μg vector	Very stable	1%
PAC	pCYPAC2	18.8 kb	300 kb	Kanamycin	10^5 clones/μg vector	Very stable	1%
YAC	pYAC4	11.4 kb	1 Mb	Tryptophan/uracil	10^3 clones/μg vector	Very unstable	10–30%

[a]Stability is defined as the ability of inserts to remain intact after repeated cell generations.
[b]Chimerism is defined as the cloning of non-contiguous genomic DNA into the same clone.

(around 35–45 kb) by virtue of its dependency on a phage packaging system, because there are limits on the length of DNA that can be maintained in the lambda phage head. There are also minimum size restrictions because of the packaging requirement for a full or nearly full head. Frequent instability of the cosmid clones has been observed, detected as rearrangements and deletions in the inserts after 30 or more cell generations. The use of *rec*⁻ host strains such as DH5alphaMCR reduced the original estimates of rearrangements of 40% to <5% (Yokobata *et al.*, 1991), but *in vivo* site-specific recombination can still occur (Kim *et al.*, 1992). Also, the system appears to have a non-random cloning bias (Harrison-Lavoie *et al.*, 1989) with certain regions, such as highly repetitive sequences, difficult or impossible to maintain in *Escherichia coli* at high copy number possibly due to the DNA repair functions of the phage. In addition, there may be an absence of restriction enzyme sites in a region, creating a fragment that is too large to be cloned (Harrison-Lavoie *et al.*, 1989).

Fosmids

The fosmid plasmid pFOS1 was engineered by Kim *et al.* (1992) to include the *E. coli* F-factor replicon, which is used for plasmid DNA replication and segregation after *in vitro* packaging. This replicon limits the number of clones maintained in bacteria to a single or very few copies. Therefore, the stability of inserted DNA in fosmids (95%) is improved over that of cosmids (Kim *et al.*, 1992). However, the fosmid system allows cloning of inserts in the same size range as

cosmids (35–45 kb) because it also relies on phage packaging for introducing the clone into a bacterial cell. The small insert size has limited the application of fosmids in mammalian genome mapping, although a gorilla fosmid library with an average insert size of 30 kb was recently constructed using liver tissue (Kim *et al.*, 2003).

Bacteriophage P1 cloning system

The bacteriophage P1 cloning system was originally developed by Sternberg (1990). In this system, DNA is inserted into the cloning site of a bacteriophage P1-based vector (originally pAd10) and the ligation product is then packaged *in vitro* into the P1 phage in a two stage process (Pierce and Sternberg, 1992). In stage 1, the packaging site (*pac*) in the vector DNA is cleaved by the pacase extract and then one of the *pac* ends is brought into the empty phage prohead and the head is filled with DNA. In stage 2, the resulting DNA is packaged into phage particles by converting it to an infectious particle with the addition of phage tails. The resulting phage lysate is used to infect an *E. coli* host.

Two replicons in the P1 vector allow alternative replication modes for the plasmid within the cell. When the P1 plasmid replicon is active, the plasmid is stably maintained as a single copy. Inactivation of the *lacI*q repressor of the host cell by adding isopropyl β-D-thiogalactopyranoside (IPTG) to the growth media allows induction of the P1 lytic replicon and subsequent increase in the number of plasmids (about 100–200 copies) within 30–40 min. Each clone can yield several μg of plasmid DNA after alkaline lysis extraction. Thus, the clone can be stabilized at just a single copy per cell under the P1 plasmid replicon until large amounts are needed and IPTG is added to the growth media.

Because there was no selection against non-recombinants in the pAd10 system, non-recombinant vectors after ligation were estimated at 10–20%. Pools of P1 clones were rapidly overgrown by the non-recombinants because non-recombinant clones grow significantly faster, with up to 80% of the clones in a pool having no insert. To minimize the recovery of clones without insert, Pierce *et al.* (1992) added the *Bacillus amyloliquefaciens sacB* gene and a synthetic *E. coli* promoter into the pAd10 vector. When expressed, the *sacB* gene produces the enzyme levansucrase that catalyses the hydrolysis of sucrose to levan, which is highly toxic to *E. coli* cells. By placing the *Bam*HI cloning site between the promoter and *sacB*, insertion into the cloning site prevents *sacB* expression and permits growth of plasmid-containing cells on media containing sucrose. A small proportion of clones that lack insert still survive, but these are aberrant vectors that have lost the *sacB* gene and, therefore, can survive in the presence of sucrose.

Other changes to the pAd10 vector (Pierce and Sternberg, 1992; Pierce *et al.*, 1992) provided mechanisms for characterization of the insert. The addition of rare-cutter restriction enzyme sites (*Not*I, *Sal*I and *Sfi*I) flanking the *Bam*HI cloning site allows excision of the insert with minimum fragmentation. The addition of Sp6 and T7 promoters allows the analysis of the cloned insert ends via sequencing and the generation of RNA probes. The resulting pAd10sacBII vector (Pierce *et al.*, 1992) and its derivatives are now used almost exclusively for P1 libraries.

Cloning efficiency of the P1 system is about 10^5 clones per µg of vector. The maximum insert size is about 95–100 kb (depending on the size of the vector) because of limits on the DNA that can be contained within the P1 phage head, but these limits are higher than those of the lambda phage head used in the cosmid and fosmid systems. Average insert size is about 80 kb. The P1 system has been further adapted by Rao *et al.* (1992) who replaced the *in vitro* P1 packaging system with that of the bacteriophage T4, while retaining the P1 vector. The larger DNA capacity within the T4 head allows an increased insert size of up to 122 kb.

Bacterial artificial chromosome (BAC) system

The BAC vector is a derivative of the *E. coli* fertility (F-factor) plasmid, which maintains its copy number at one to two per cell through the expression of the *parA* and *parB* genes. The first BAC vector that was constructed, called pBAC108L, was developed from pMBO1331 (Shizuya *et al.*, 1992). It contained a chloramphenicol resistance marker (Cm^R), *Hin*dIII and *Bam*HI cloning sites, rare-cutter restriction enzyme sites for excision of the cloned inserts, T7 and Sp6 promoter sites, and the bacteriophage lambda *cosN* site and P1 *loxP* sites. Later modification by Kim *et al.* (1996) eliminated the *Eco*RI site from the Cm^R gene and added an *Eco*RI site to the multiple cloning site, thereby allowing the cloning of DNA fragments generated with *Eco*RI, as well as *Hin*dIII and *Bam*HI.

Recombinant BAC molecules are introduced into bacterial host cells by electroporation (Leonardo and Sedivy, 1990), thus avoiding the phage packaging requirements and limitations of the cosmid and P1 cloning systems. The cloned BAC insert can be up to 350 kb in size, with an average insert size around 150 kb. Improved transformation efficiency due to electroporation results in about 10^6 transformants per µg of DNA. When combined with cell strains that lack recombination functions, thereby eliminating *in vivo* site-specific recombination, the BAC system provides highly stable transformants, with essentially no rearrangements of the inserts even after 100 generations (Ioannou *et al.*, 1994). Because the recombinant BACs exist as supercoiled circular plasmids, the DNA can be isolated by alkaline lysis extraction with minimal shearing.

The original pBAC108L vector did not allow selection for recombinants/non-recombinants and, therefore, libraries made with pBAC108L contained up to 90% vectors without inserts (Shizuya *et al.*, 1992). Rather, non-recombinants were detected by lack of hybridization to probes containing species-specific repetitive elements. However, vector modifications by Kim *et al.* (1996) produced pBeloBAC11 and allowed blue/white colour screening for non-recombinants/recombinants based on the alpha-complementation of β-galactosidase. Frengen *et al.* (1999) developed the pBACe3.6

vector that contains multiple cloning sites within the *sacB* gene and, therefore, positive selection for recombinant clones on sucrose-containing medium, similar to the P1 system.

As mentioned above, a significant advantage of the BAC system is the stability of large inserts originally resulting from their very low copy number. However, the limit on copy number also limits DNA recovery. Modifications by Wild *et al.* (2002) to the pBeloBAC11 vector and the DH10B host allow the replication of about 100 copies of the vector per host cell. The resulting pBAC/oriV vector includes the *oriV*/TrfA replicon system and can be conditionally induced with L-arabinose to increase copy number by 30- to 100-fold when large amounts of the clones are needed.

Chimeric BAC clones occur at a very low level (about 1%) because of a strong size bias during the bacterial transformation. Those clones that are chimeric usually consist of one large and one small fragment (Osoegawa *et al.*, 2000), and are detectable by mapping the clone ends to different genomic regions.

As noted above, BAC clones are stable, have high transformation efficiency, and produce large amounts of DNA. These attributes have produced a system that is commonly used for end sequencing, shotgun sequencing, fingerprinting and positional cloning. Based on these characteristics, the BAC cloning system has become the method of choice for most mammalian genome projects.

P1-derived artificial chromosome (PAC) system

Ioannou *et al.* (1994) developed a large insert cloning system, subsequently referred to as PAC, that combines features of the bacteriophage P1 with the *E. coli* F-factor plasmid used in the BAC system. Manipulations of the Pierce *et al.* (1992) BAC pAd10sacBII plasmid produced the pCYPAC2 vector, in which the 13.4 kb adenovirus stuffer fragment, the packaging signal (pac) and the second *loxP* site were deleted. A modified pUC19 plasmid was inserted into the *Bam*HI cloning site of the pAd10SacBII vector, between the *E. coli* promoter and the *sacB* gene. The insertion of the pUC19 disrupts *sacB* gene expression, thereby

eliminating its selection against non-recombinant vectors when propagating large amounts of the native vector for further manipulation. However, the pUC plasmid can be subsequently eliminated by double digestion of the vector with *Sca*I and *Bam*HI, re-establishing the *sacB* selection against non-recombinant clones. Exogenous DNA is then ligated into the *Bam*HI cohesive ends.

Presence of the F-factor in the vector allows the clones to be introduced into bacterial cells (usually DH10B) by electroporation, thereby eliminating packaging problems and phage head restrictions on insert size. Insert sizes in PAC clones are typically 150 kb and can theoretically be >300 kb. However, this system is not as widely used as the BAC.

Yeast artificial chromosome (YAC) system

The YAC system, first developed by Burke *et al.* (1987), involves the cloning of exogenous DNA fragments up to 1 Mb or more into linear artificial chromosomes that are maintained in *Saccharomyces cerevisiae*.

The pYAC vector, derived from pBR322, is a circular plasmid that can replicate in *E. coli* and contains several key sequence modules (Burke and Olson, 1991). These modules include the *CEN4* centromere sequence that is necessary for mitotic and meiotic function in the yeast cell, the *ARS1* or autonomous replication sequence that prevents integration of the YAC into a natural yeast chromosome, a cloning site that contains *SUP4*, an interruptible marker for insertion, two *TEL* sequences that are derived from *Tetrahymena* and required for telomere formation, and two yeast selectable markers, *TRP1* and *URA3*, which are on opposite sides of the cloning site.

Several pYAC vectors have been developed that differ primarily in the cloning site but the most commonly used vector is pYAC4. The *Eco*RI cloning site in pYAC4 is placed within *SUP4*, an ochre-suppressor allele of a *tRNA*[Try] gene. Insertion of exogenous DNA into the cloning site results in disruption of *SUP4* expression and, therefore, ochre suppression, resulting in red colonies, as opposed to white colonies in which *SUP4* is active.

Digestion of the pYAC4 vector with *Bam*HI and *Eco*RI cuts the plasmid into three parts, including the two arms and a throwaway region that separates the two *TEL* sequences when circularized. Upon digestion, the *TEL* sequences seed the formation of functional telomeres. The left arm contains *CEN4* and *TRP1*, whereas the right arm contains *URA3*. *TRP1* and *URA3* are the wild-type alleles for the genes that are mutated in yeast strains bearing the auxotrophic markers *ura3* and *trp1*; selection for transformation of the vector involves complementation of these auxotrophic markers in suitable hosts.

Incomplete digestion of genomic DNA is performed using a combination of *Eco*RI and *Eco*RI methylase (Larin *et al.*, 1997; Sanchez *et al.*, 2002). Size fractionation of high molecular weight DNA is performed either in solution through sucrose gradient or in agarose by pulsed-field gel electrophoresis (PFGE). Insert size is larger on average when prepared in agarose because shear forces are minimized. However, partial degradation of DNA occurs when melting agarose and subsequent β-agarase treatment, probably due to metal ion cofactors found in commercial agarose or DNA denaturation. The presence of polyamines during the melting step protects DNA in agarose from degradation (Larin *et al.*, 1991) but transformation efficiency is reduced if polyamines are present during the transformation procedure. The DNA can also be removed from the agarose by electroelution into dialysis bags (Strong *et al.*, 1997; Osoegawa *et al.*, 1998).

The size-fractionated DNA is then ligated to de-phosphorylated vector arms. The ligation products are transformed into yeast spheroplasts, which are made from yeast cells, usually AB1380, after the addition of lyticase (Burgers and Percival, 1987) that removes the yeast cell walls. The transformed spheroplasts are then embedded in agar on selective medium that lacks sorbitol, uracil and tryptophan. Because the yeast cells must grow within a supportive agar matrix, it is difficult to use colony lifts directly from the surface of transformation plates. Instead, the YAC clones are picked into microtitre plates and can then be transferred to filters for screening by hybridization.

The insert size maintained in YACs is the largest of all cloning systems, thereby providing genome coverage with the least number of clones. However, the advantage of large insert size is offset by several well known drawbacks of the YAC system. Transformation efficiency of yeast spheroplasts by electroporation is about 10–100 times lower than transformation of bacteria, with about 10^3 yeast transformants per μg of vector. Of particular concern for genome mapping, YAC libraries contain high levels of chimeric clones (40–60% in the original system, decreasing to 10–30% with modifications). These chimeric inserts are due to ligation of non-contiguous sequences into a single YAC vector or from *in vivo* recombination following co-transformation of multiple YACs into the same yeast cell. Recombination of the cloned insert is typically detected by fluorescence *in situ* hybridization (FISH) or restriction enzyme digestion. After cloning, there is up to 2% insert instability, resulting in small deletions and/or rearrangements of the cloned insert after several generations of growth. Also, certain regions of the genome are unclonable in YACs, estimated to occur every 1–2 Mb in the human genome (Schlessinger, 1990). Separation of exogenous DNA from yeast chromosomes requires extensive technical manipulation and the yields of insert DNA can be quite low, given that the YACs are maintained within the cells as single copies. Smith *et al.* (1990) added a conditional centromere to the YAC vector that can be turned on and off by changing the carbon source, thereby permitting copy number to increase by 10- to 20-fold without loss of stability. However, this modified vector (pCGGS966) has not been used that extensively for YAC library construction.

Screening Large Insert Libraries

Colony hybridization and polymerase chain reaction (PCR) amplification of isolated DNA are the two most common methods for screening large insert libraries (Campbell and Choy, 2002). For colony hybridization, a set of clones is replicated on to filters in an ordered array. The cells on the filters are lysed and then the filters are hybridized to a labelled probe, such as overlapping oligonucleotides (overgos; Han *et al.*, 2000) or probes specific to a chromosomal region. While colony hybridization allows the screening of large numbers of clones in parallel, there are often multiple false positive/negatives. Also, large amounts of the insert DNA must be present in each spot on the filter in order to have positive hybridization, which makes screening of low copy number clones difficult (i.e. cosmids, fosmids, YACs). An advantage of screening by colony hybridization is that sequence information on the probe is not necessary, but this is much less of a hindrance now than when the cloning systems were first devised. Also, preparation of DNA is only needed for positive clones, as opposed to PCR-based screening (described below).

PCR-based screening protocols are based on the positive amplification of a clone using unique sequence primers. However, rather than testing the primers on each individual clone which would require hundreds of thousands of PCRs, pooling strategies have been developed. A common strategy involves combining equal aliquots of prepared DNA across plates, rows and columns (such as Libert *et al.*, 1993; Shepherd *et al.*, 1994). Based on 96- or 386-well formats, up to 6144 clones are included in any one pool. The pools are then screened for the presence of a particular sequence and in just a few rounds of PCR, the complete address (plate/column/row) of each positive clone is revealed. Pooling is done after plating out the library, with all clones having a unique address. This approach requires significant investment in DNA preparation, but procedures for using crude cell lysates or cultures have been published (Bloem and Yu, 1990; Kwiatkowski *et al.*, 1990; Campbell and Choy, 2001).

Emerging automated systems allow improved efficiency for both colony hybridization and PCR-based screening (Meldrum, 2000), with robotics now available for plaque and colony picking, filter preparation, DNA extraction and DNA pooling.

Contig Assembly

A contig of cosmids, BACs or YACs corresponds to a physical map covering part or all of a genome or a chromosome (Fig. 3.2). It is characterized by its depth or coverage: the mean number of clones found at any position. BAC contigs

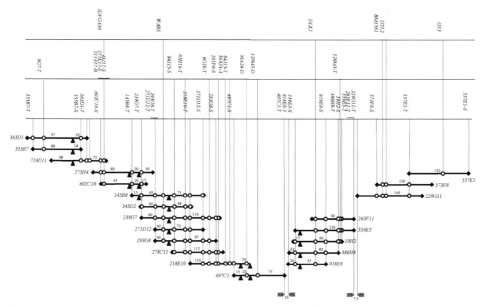

Fig. 3.2. An ovine BAC contig spanning the *callipyge* locus on ovine chromosome 18 (Segers *et al.*, 2000). STSs (sequence-tagged sites) used for contig assembly are shown at the top of the figure. Microsatellite markers (black) and genes (grey) are on the top horizontal line. STS from the orthologous region in the bovine (Shay *et al.*, 2001) are on the second horizontal line. STS developed from ovine BAC ends are on the third horizontal line. The arrows flank the gaps bridged by long-range PCR. Triangles mark the positions of *Not*I restriction sites. Estimates of fragment sizes obtained by *Not*I restriction are indicated above the corresponding BAC clones.

are now preferred to cosmid or YAC contigs as they are thought to provide the best compromise between the length of inserted sequences and stability. The methods used to construct contigs from cosmids, BACs or YACs were developed several years ago for the physical cloning of disabled genes and as a first step in the hierarchical genome sequencing approach. Two basic approaches are used for developing contigs. In the first approach, short nucleotide sequences, called markers, are amplified by PCR from each individual clone of the library or from pools of clones, according to the number of clones to assemble and the size of the map under construction, reflecting the fact that clones sharing a marker overlap at the sequence marker location (Capela *et al.*, 1999). The second approach, popularized by the physical mapping of the 100 Mb *Caenorhabditis elegans* genome (Coulson *et al.*, 1995), is based on the clone restriction fingerprinting method (Coulson *et al.*, 1986). In this method, clones are digested by one or more restriction enzymes, successively or in combination, and the products separated on the basis of size by acrylamide or agarose gel electrophoresis. The restriction fragments making up a clone are identified and it is assumed that clones sharing a sufficient number of fragments overlap in the region encompassed by those fragments. This approach is particularly useful for the construction of large maps as it is easily automated. It also offers the advantage of providing an accurate assessment of the length of the map. The main disadvantage of this approach is that a large number of clones are required for coverage compared with the PCR-based method.

The recent development of capillary sequencers that have high-throughput analytical capability for DNA fragment analysis, and appropriate software (see ftp.sanger.ac.uk), has overcome the technical problems that were previously associated with the construction of contigs. However, contig construction of mammalian genomes and of other genomes of similar size and larger still represents a challenge, in terms of cost-effectiveness. There is also need for construction of contigs corresponding to a specific region of interest, which requires the selection of appropriate BACs or clones corresponding to the region of interest from an entire genomic library. Elegant solutions to this problem can be found in the use of flow-sorted chromosomes or the microdissection of metaphase chromosomes (Claussen *et al.*, 1997, see below).

Another strategy that has been successfully applied to the construction of regional maps is chromosome walking, as reviewed by Ragoussis and Olavesen (1997). To begin chromosome walking, a large insert library is first screened with markers previously assigned to the region of interest. Positive BAC clones are end-sequenced using universal sequencing primers. Primers specific to the end sequences are designed and then used in subsequent rounds of library screening. Overlapping clones are identified in each screening and arrayed into the expanding contig. While the chromosome walking strategy is effective, it is labour- and time-consuming and should therefore be limited to the construction of small regional maps.

Chromosome-specific Libraries

While whole genome libraries are now available for a wide range of mammalian species, chromosome-specific libraries provide an alternative for direct examination of a particular chromosomal segment or chromosome.

DNA used for construction of chromosome-specific libraries is usually obtained by flow-sorting (Vooijs *et al.*, 1993) or microdissection of banded chromosomes (Scalenghe *et al.*, 1981; Ludecke *et al.*, 1989). Libraries can also be generated by using flow-sorted chromosomes derived from monochromosomal hybrid cell lines (Krumlauf *et al.*, 1982; McCormick *et al.*, 1993) or interspecies cell hybrids (Rohme *et al.*, 1984). Unfortunately, the amount of starting material is often limited with these methods of chromosome isolation, particularly with microdissection, yet several hundred thousand chromosomes are needed to produce the 100 ng quantities needed for cloning. The microscale cloning methods developed by Scalenghe *et al.* (1981) for *Drosophila melanogaster* polytene chromosomes and later adapted by Rohme *et al.* (1984) for mammalian chromosomes are now widely used for handling chromosome preparations. These methods use nanolitre reaction volumes, thereby reducing the amount of initial starting chromosomal material.

In addition, techniques to amplify the DNA within the isolated chromosomes have been developed, most commonly using ligated oligonucleotide adaptors (Jinno et al., 1992; Vooijs et al., 1993) or degenerate oligonucleotide primed (DOP) PCR (Guan et al., 1993; Christian et al., 1999). In a comparison of both systems using microdissected chromosomal DNA from rice and wheat, Zhou et al. (2000) found yields to be the same for the two systems but the oligonucleotide adaptors produced larger fragments than the DOP-PCR (0.3–2.0 kb compared with 0.3–1.4 kb). Also, the low annealing temperature and degenerate primers used in DOP-PCR makes it difficult to eliminate exogenous DNA contamination. It has been demonstrated by Guan et al. (1993) that topoisomerase I treatment of the chromosome prior to DOP-PCR improves efficiency of amplification, allowing the creation of sufficient product from a single microdissected chromosome. This enzyme catalyses the relaxation of supercoiled, condensed DNA (Liu and Wang, 1987). Amplification of Alu or L1 primers has also been used to increase the amount of starting material, but this procedure results in fragments that are flanked by repetitive sequences and, therefore, can have very limited complexity.

All of these amplification methods result in very small sized amplicons (300–2000 bp). Therefore, libraries prepared from amplified chromosomal material are not suitable for development of contigs. However, these amplicons can be cloned into plasmid vectors, creating 'microclones', which can then be used as complex chromosome-specific probes for subsequent screening of whole-genome libraries (Zimmer et al., 1997). In this way, a chromosome-specific contig can be produced. These amplified products are also widely used as whole chromosome painting probes (such as Goldammer et al., 1996; Ponce de Leon et al., 1996; Zimmer et al., 1997; Raudsepp and Chowdhary, 1999) and for isolation of microsatellites used in linkage analysis (such as Ambady et al., 1997, 2001, 2002; Korstanje et al., 2003).

Sequencing of the Entire Genome: Current Approaches and Technologies

During the course of 2003, many scientific meetings around the world, particularly those devoted to molecular biology, genetics or genomics, celebrated the 50th anniversary of Watson and Crick's 'DNA double helix' model proposed in a short paper published in Nature (Watson and Crick, 1953). The annual CSH (Cold Spring Harbor) meeting on 'DNA Sequencing and Genome Analysis' was formally elevated to a Quantitative Biology Meeting (number LXVIII), to celebrate both the 50th anniversary of the double helix model and the release of an almost complete human genome sequence (International Human Genome Sequencing Consortium, 2001; Venter et al., 2001; NIH release 'builds 33'). Fair enough, but what about the publication almost 25 years ago in the Proceedings of the National Academy of Sciences, USA of the 'chain termination method' by Sanger et al. (1977). Astonishing as it may seem, this method was not fully appreciated at the time of its publication. The chemical degradation method proposed a few months previously by Maxam and Gilbert (1977) was generally considered more robust and potentially more useful for determining the sequence of double-stranded DNA fragments. Sanger's method has since become the power horse of the genomic era and is capable of determining the sequences of hundreds of genomes of all sizes from thousands to billions of nucleotides, from all kinds of living organisms, ranging from viruses to mammals (see http://www.ncbi.nlm.nih.gov/). Even more surprising, the Sanger method has not changed since its inception despite improvements in the quality of the reagents, the development of various cloning vectors, including the M13 series of Joshua Messing (Sanger et al., 1980; Messing et al., 1981), the advent of PCR (Mullis et al., 1986), almost complete automation and the development of many algorithms for storing, manipulating and comparing billions of strings of nucleotides.

The Shotgun Approach

In 1998, Craig Venter proposed a radical new approach to the sequencing of the human genome, which was received with scepticism, bitterness and criticism. Conceptually, this strategy was not new as it had been used for many years worldwide for the sequencing of thousands of BAC clones. What was new was its scale. The shotgun approach was developed at the time of the conception and construction of the M13 vector system, and the genome of the phage lambda was the first large DNA molecule to be sequenced by this method (Sanger *et al.*, 1982). It consists of a double-stranded DNA molecule, almost 50,000 base pairs in length. The method used to determine the phage lambda genome sequence is now known as the whole genome shotgun (WGS) approach. Following a logical trend, the shotgun approach was applied to ever longer molecules and, in 1995, Fleischmann *et al.*, from The Institute for Genomic Research (TIGR), published the sequence of the *Haemophilus influenzae Rd* genome. This sequence covered some 1,830,137 base pairs and was assembled from over 11 million bases, determined from nearly 24,000 random reads plus gap closures and finishing. This was the first of many bacterial genomes to be sequenced by the chain termination method, using the shotgun approach (see http://www.tigr.org/). Needless to say, this achievement necessitated the development of new algorithms and ever more powerful bioinformatic platforms. At this stage, there was no reason to believe that the limit had been reached in terms of the size of molecules that could be sequenced by the whole shotgun approach. With increasing automation and sequencing capabilities due to the development of high-throughput capillary machines and new algorithms, it is not surprising that the idea of a WGS approach for sequencing the human genome emerged. A milestone in this trend was the WGS analysis of the *Drosophila* genome (Adams *et al.*, 2000).

In 1997, Weber and Myers had already suggested sequencing the human genome by a WGS strategy.

This trend towards increasingly large projects was not new in biology. In the early 1980s scientists involved in genome mapping moved from projects restricted to a portion of the chromosome encompassing a few megabases to the whole genome approach. After many ups and downs, the WGS approach is now widely accepted, not as the unique, self-sufficient strategy first proposed by its fiercest proponents, but as the most efficient way to obtain a draft sequence. There is no doubt that the WGS analysis of the human genome has revolutionized our way of thinking about large-scale sequencing projects. The WGS approach made it possible to analyse the mouse genome in record time (Mouse Genome Sequencing Consortium, 2002). Today, not even the most vocal detractors of the WGS strategy would propose a hierarchical clone-by-clone strategy for a new mammalian genome sequence project. The human genome sequencing programme made use of both strategies and, despite an atmosphere of conflict, each approach benefited from the other. It is also clear that without the WGS initiative, we would still be working on the human genome sequencing programme, which was originally scheduled to end in 2005. Furthermore, any new WGS assembly benefits from the development of ever more powerful algorithms and from the alignment of sequences with those of genomes already sequenced and analysed. Finally, the goals set for new mammalian genome sequencing projects differ from those of previous projects, and are limited to a draft rather than a finished sequence.

Novel Sequencing Methods and Strategies

Sanger's chain termination method remains the only sequencing method for use in small and large sequencing projects. Does this mean that no other option is available, either now or in the near future? This may well prove to be the case, despite several advances in this area.

In 1989, Mirzabekov proposed a new sequencing strategy based on the hybridization of a DNA template to a series of overlapping oligonucleotides of known sequences (Khrapko

Box 3.1. The whole shotgun strategy

Sequencing read lengths vary depending upon several parameters but 600–800 nucleotides corresponds to a good estimate. To sequence much larger fragments or even a whole genome, essentially two strategies have been designed.

1. The hierarchical approach in which collections of subfragments obtained by enzymatic restriction are mapped to get a unique contigs from which a minimal set of subfragments can be selected and sequenced, thus limiting sequence redundancy.

2. The shotgun approach in which the DNA is randomly fragmented by sonication or shearing. Following fragmentation and enzymatic end repair the DNA fragments are ligated to a plasmid vector and a bacterium host transformed to produce a library. Clones taken at random from the library are then sequenced from both ends using two universal primers. At this stage a shotgun is characterized by its depth, i.e. the cumulative length of sequence determined divided by the length of the fragment or genome to be sequenced. For example, with an estimated size of 4 Mb a 10x shotgun would correspond to the assembly of about 60,000 reads with a mean size of 650 nucleotides. The resulting sequences are assembled in a unique contig representing the whole fragment by sequence comparison using appropriate bioinformatic programs. The final stage or 'polishing stage' corresponds to the elimination of gaps, single strandedness and of all possible errors.

Except for the very early sequencing projects (see for example Galibert *et al.*, 1978) in which the hierarchical strategy was used, sequencing projects have rapidly moved to using the two approaches sequentially. For example, the construction of a BAC map covering an entire genome or chromosome is followed by a shotgun strategy to sequence a minimal set of BACs. The change that occurred along the years was the size of the DNA fragment or genome that was directly shotgunned. It should be noted that C. Venter did not invent or introduce the whole shotgun approach, rather he totally eliminated the hierarchical step. The possibility of increasing the size of the shotgun projects was dependent upon the development of robots adapted to high-throughput projects and of bioinformatic programs that solve two major problems. One is a quantitative problem regarding the capacity to store, compare and retrieve millions of reads corresponding to billions of nucleotides. The second problem is related to the presence of numerous repeat sequences that are often longer than the mean read length, complicating correct assembly (Myers *et al.*, 2000, 2002; Istrail *et al.*, 2004).

et al., 1991). This strategy only really became practicable with the advent of the oligonucleotide array technology. However, this approach suffered from two main problems. For technical reasons, arrays of short oligonucleotides (8-mers) were first selected. Unfortunately the amount of information that such octanucleotides contain is too small in comparison with the size of the DNA molecules to be sequenced. Secondly, oligonucleotides with deviated base composition (high versus low GC content) have large differences in melting temperature (T_m) making hybridization in parallel on a single array nearly impossible. Thus, with the exception of re-sequencing, polymorphism analysis and comparison of the sequences of orthologous genes in closely related species, this method has been little used (Hacia *et al.*, 1998; Pe'er *et al.*, 2003). Does this mean that sequencing by hybridization will for ever be limited to re-sequencing or polymorphism analysis? This is hard to say but it is worth mentioning here that the tremendous development in micro-array construction would nowadays allow researchers, through *in situ* oligonucleotide synthesis, to get arrays with several thousands of features and thus arrays with a full set of 10mer oligonucleotides.

Other methods were proposed during the 1990s, but none has been successful. The main reason for this is probably the high efficiency, reliability and cost-effectiveness of Sanger's method, making it hard, if not impossible for any new method based on a different principle to supercede it.

Sequencing the Human Genome and Other Large Genomes

A complete or almost complete human genome sequence was released in April 2003, and a high-quality WGS draft sequence of the *Mus musculus* genome (C57BL/6J strain) has been assembled by the Mouse Genome Sequencing Consortium (2002). A complete final sequence will be generated by this public consortium in 2004/2005, by means of a more traditional map-based approach for filling in the gaps. A draft sequence covering 90% of the estimated 2.8 Gb of the rat genome *(Rattus norvegicus)* was released in June 2003 and is available via the Internet (http://hgsc.bcm.tmc. edu/projects/rat/). As for the mouse, the strategy for sequencing the rat genome and obtaining a complete final sequence combines the whole shotgun and hierarchical approaches. In the future, this dual approach is likely to become the general rule for projects aiming to determine the complete sequence of any large genome.

It is beyond the scope of this chapter to summarize everything that has been learned from these studies and from the sequencing of the genomes of several model organisms (see http://www://ncbi.nih.gov/genomes/). Instead, we will recap some of the findings that were least anticipated when the Human Genome Project was launched.

At the start of the Human Genome Project in 1988, estimates of the number of genes encoded by the human genome and other mammalian genomes were in the range of 70,000 and estimates exceeding 100,000 were not uncommon. These values have decreased over time and, in 2002, based on comparison with the sequence of the model fish *Rubripes fluvialis*, it was suggested that the human genome would contain only 30,000 genes (Roest Crollius *et al.*, 2000). Even today, with the completed human genome sequence and draft genome sequences for the mouse and rat, the exact number of genes encoded by the human genome and other mammalian genomes is not precisely known. However, it is clear that this number is astonishingly small – probably around 28,000 – not much larger than the number of genes in the *C. elegans* genome, in which 20,000 genes have been

identified, or in the *Drosophila* genome, which contains some 15,000 identified genes, and smaller than the 29,454 predicted genes encoded by the genome of the model plant *Arabidopsis thaliana* (*Arabidopsis* Genome Initiative, 2000).

The importance of alternative splicing has now also been recognized. The most recent estimates suggest that this process probably affects up to 60% of the transcripts of a mammalian genome (Modrek and Lee, 2002). Thus, although the number of genes has decreased over time, the number of resulting proteins and their variants has remained high. We also now know that there is an editing process that reprogrammes primary transcripts, expanding the number of protein products, but this process is currently poorly understood and its importance probably underestimated (Hoopengardner *et al.*, 2003). Interestingly, mammalian genomes are very similar, with almost all genes having orthologues, often in similar environments, as shown by sequence alignments and genome map comparisons indicating the conservation of large blocks of genes. Surprisingly, there also seems to be a high level of conservation in gene organization and sequences between distantly related organisms (Chervitz *et al.*, 1998; Carrascosa *et al.*, 2001; Lawrence, 2002). Although most of the gene repertoire making a mammal is now known, it is currently impossible, and is likely to remain so for some time, to identify the key features of a mammalian genome and to understand how such similar genomes can direct the development of different species. In other words, what makes a mouse a mouse and an elephant an elephant? The determination of the human genome sequence and the anticipated completion of the mouse and rat genome sequences is clearly no more than the end of the first phase of the genomic revolution that began 50 years ago.

Where Do We Go From Here?

Even as early as the late 1980s, as James Watson and others were pushing for a human genome programme, with the sequencing of the entire human genome considered the 'holy grail', it was recognized that the sequencing of other genomes would be necessary if we were to understand the human sequence. It was therefore decided that the genomes of the mouse and the rat and of divergent species from major phyla should be sequenced in parallel. Much remains to be learned from the sequencing of other genomes. It is therefore encouraging that other genomes, such as the chimpanzee, dog, chicken and bovine genomes, are currently being sequenced (and more will come) and that the sequencing of a number of bacterial and lower eukaryote genomes is planned. For economic reasons, it has been decided that new mammalian genome projects will not aim for a perfect, clean complete sequence. Sequencing efforts will instead be limited to a 6× shotgun, as for the canine genome project (http://www.genome.wi.mit.edu/media/2003/pr_03_tasha.thml). It will not be possible to produce a finished continuous sequence with this level of redundancy. Instead, a draft sequence of about 150,000 contigs will be generated that could be combined to give a few thousand supercontigs with the aid of BAC end-paired sequences. The production, in parallel, of a dense RH map will not only facilitate the assembly of the contigs into supercontigs, but will also make it possible to order the supercontigs – a necessary step if we are to understand genome rearrangements and to compare synteny. It should, however, be stressed that this intermediate level of sequencing is not appropriate for the smaller genomes of lower eukaryotes, for which obtaining a perfect finished sequences is both highly desirable on scientific grounds and economically viable.

Although comparison of the genomes of mammals and other organisms is an essential step in understanding the functioning of individual genes and of the genome as a whole, it is not sufficient in itself. We need to go one step further, into the new disciplines such as proteomics, metabolomics and other 'omics' approaches (see Chapters 5, 6 and 7). However, to tackle the problem of phenotypic polymorphism and to identify the genes and alleles governing complex traits that define an individual and make it unique, we need to analyse the genetic diversity within species. This involves the evaluation of sequence variation, mostly in the form of single nucleotide polymorphisms (SNPs) and short insertions

and deletions, that characterize the various alleles and regulatory sequences. In the genomes of humans and other mammals, this genetic polymorphism affects up to one nucleotide per thousand. We also need to analyse the combinations of gene alleles present in a given individual. This problem has been approached by collecting several hundred thousand human SNPs and carrying out linkage disequilibrium studies to identify regions of the genome that may contain susceptibility genes. Such studies have led, for example, to the recent identification of a functional haplotype of PAD14 associated with rheumatoid arthritis (Susuki et al., 2003). However, with the exception of a few human populations, such as Icelanders and Askenazi Jews who live in a geographical or cultural environment resulting in limited gene flow, linkage disequilibrium studies in humans require the analysis of very large numbers of SNPs in a large study population. As a consequence of an important gene flow, the genetic structure of the human populations is such that the mean proportion of genetic differences between individuals from different human populations only slightly exceeds that between unrelated individuals from a single population (Rosenberg et al., 2002). SNPs located in close proximity tend to be inherited in blocks and constitute a haplotype. Studies have revealed considerable variability in local haplotype structure, with some haplotypes extending for only a few kilobases and others for more than 100 kb (Daly et al., 2001). A recent elegant, detailed analysis of chromosome 21, based on chromosomes isolated from 24 ethnically diverse individuals, showed that a large number of SNPs was required to identify all the common haplotypes. With this information in hand, blocks of limited haplotype diversity can be defined, making it possible to identify a much smaller subset of SNPs for linkage studies (Patil et al., 2001).

The dog genome, which the National Institutes of Health (NIH) has selected as a new mammalian genome project, is particularly suited for analysis of the relationship between genotype and phenotype. A 6× WGS project is already underway and should be completed by October 2004. Dogs originated from domestication of the wolf, more than 15,000 years ago (Leonard *et al.*, 2002; Savolainen *et al.*, 2002). They have been subjected to intensive breeding, particularly during the last three centuries, resulting in the creation of many well-characterized breeds. The 350 or so breeds registered by kennel and breed clubs around the world provide a reservoir of phenotypic and behavioural traits unequalled in any other mammal. Furthermore, most dog breeds are susceptible to a large number of genetic diseases that are also observed in the human population (Ostrander *et al.*, 2000). Analysis of the genetic and allelic basis of this extraordinary diversity and the differences observed between breeds represents a major challenge to scientists in the next few years. Resolving these issues in the dog will undoubtedly provide answers to similar questions in humans.

Conclusions

The progress made in the analysis of genomes of all sizes and from all branches of the tree of life could not have been foreseen a mere decade ago. This remarkable achievement has been made possible by the synergy of new approaches and methods including bioinformatics and automation, which have proved to be instrumental. It also owes much to the human wisdom underlying decisions to seek the financial support without which nothing is possible in genomics, and the willingness of the scientific community worldwide to work openly and in close collaboration.

It is often said that we have now entered the post-genomic era. This statement is erroneous in two ways. First, with the large number of genomes already sequenced, including the human genome, it would seem more correct to say that we are passing from the sequencing era to the post-sequencing era. Secondly, it would clearly be a mistake to end genome sequencing activities now, without sequencing other genomes. This is because our understanding of the meaning and significance of the sequences already determined is dependent on comparison with a large number of other sequences. One case in point is the identification of all the genes within a mammalian genome, and within the human genome in particular. Furthermore, life is so diverse that we cannot claim to have an exhaustive view if we rely purely on analyses carried out to date.

Acknowledgement

The authors would like to thank the CNRS and the University of Rennes (FG) and Utah State University (NC) for their support and their colleagues for critical reading and discussion.

References

Adams, M.D., Celniker, S.E., Holt, R.A., Evans, C.A., Gocayne, J.D., Amanatides, P.G., Scherer, S.E., Li, P.W., Hoskins, R.A., Galle, R.F. *et al.* (2000) The genome sequence of *Drosophila melanogaster*. *Science* 287, 2185–2195.

Agarwala, R., Applegate, D.L., Maglott, D., Schuler, G.D. and Schaffer, A.A. (2000) A fast and scalable radiation hybrid map construction and integration strategy. *Genome Research* 3, 350–364.

Ambady, S., Mendiola, J.R., Louis, C.F., Janzen, M.A., Schook, L.B., Buoen, L., Lunney, J.K., Grimm, D.R. and Ponce de Leon, F.A. (1997) Development and use of a microdissected swine chromosome 6 DNA library. *Cytogenetics and Cell Genetics* 76, 27–33.

Ambady, S., Kappes, S.M., Park, C., Ma, R.Z., Beever, J.E., Lewin, H.A., Smith, T.P.L., Beattie, C.W., Basrur, P.K. and Ponce de Leon, F.A. (2001) Development and mapping of microsatellites from a microdissected BTA11-specific DNA library. *Animal Genetics* 32, 152–155.

Ambady, S., Cheng, H.H. and Ponce de Leon, F.A. (2002) Development and mapping of microsatellite markers derived from chicken chromosome-specific libraries. *Poultry Science* 81, 1644–1646.

Bloem, L.J. and Yu, L. (1990) A time-saving method for screening cDNA or genomic libraries. *Nucleic Acids Research* 18, 2830.

Breen, M., Jouquand, S., Renier, C., Mellersh, C. S., Hitte, C., Holmes, N.G., Chéron, A., Suter, N., Vignaux, F., Bristow, A.E., Priat, C., McCann, E., Andre, C., Boundy, S., Gitsham, P., Thomas, R., Bridge, W., Spriggs, H.F., Ryder, E.J., Curson, A., Sampson, J., Ostrander, E.A. Binns, M. and Galibert, F. (2001)

Chromosome specific-locus FISH probes allow anchorage of an 1800-marker integrated radiation-hybrid/linkage map of the domestic dog genome to all chomosomes. *Genome Research* 11, 1784–1795.

Burgers, P.M. and Percival, K.J. (1987) Transformation of yeast spheroplasts without cell fusion. *Analytical Biochemistry* 163, 391–397.

Burke, D.T. and Olson, M.V. (1991) Preparation of clone libraries in yeast artificial chromosome vectors. *Methods in Enzymology* 194, 251–270.

Burke, D.T., Carle, G.F. and Olson, M.V. (1987) Cloning of large segments of exogenous DNA into yeast by means of artificial chromosome vectors. *Science* 236, 806–812.

Campbell, T.N. and Choy, F.Y.M. (2001) Large-scale colony screening and insert orientation determination using PCR. *BioTechniques* 30, 32–34.

Campbell, T.N. and Choy, F.Y.M. (2002) Approaches to library screening. *Journal of Molecular Microbiology and Biotechnology* 4, 551–554.

Capela, D., Barloy-Hubler, F., Gatius, M.T., Gouzy, J. and Galibert, F. (1999) A high-density physical map of *Sinorhizobium meliloti* 1021 chromosome derived from bacterial artificial chromosome library. *Proceedings of the National Academy of Sciences USA* 96, 9357–9362.

Carrascosa, J.L., Llorca, O. and Valpuesta J.M. (2001) Structural comparison of prokaryotic and eukaryotic chaperonins. *Micron* 32, 43–50.

Chervitz, S.A., Aravind, L., Sherlock, G., Ball, C.A., Koonin, E.V., Dwight, S.S., Harris, M.A., Dolinski, K., Mohr, S., Smith, T., Weng, S., Cherry, J.M. and Botstein, D. (1998) Comparison of the complete protein sets of worm and yeast: orthology and divergence. *Science* 282, 2022–2028.

Christian, A.T., Garcia, H.E. and Tucker, J.D. (1999) PCR *in situ* followed by microdissection allows whole chromosome painting probes to be made from single microdissected chromosomes. *Mammalian Genome* 10, 628–631.

Claussen, U., Senger, G. and Chudoba, I. (1997) In: Dear, P. (ed.) *Genome Mapping, a Practical Approach*. IRL Press, Oxford, UK.

Collins, J. and Hohn, B. (1978) Cosmids: a type of plasmid gene-cloning vector that is packageable *in vitro* in bacteriophage lambda heads. *Proceedings of the National Academy of Sciences USA* 75, 4242–4246.

Coulson, A., Sulston, J., Brenner, S. and Karn, J. (1986) Toward a physical map of the nematode *Caenorhabditis elegans*. *Proceedings of the National Academy of Sciences USA* 83, 7821–7825.

Coulson, A., Huynp, C., Kozono, Y. and Shownkeen, R. (1995) The physical map of the *Caenorhabditis elegans* genome. *Methods in Cell Biology* 48, 533–550.

Cox, D.R., Burnmeister, M., Price, E.R., Kim, S. and Myers, R.M. (1990) Radiation hybrid mapping: a somatic cell generic method for constructing high resolution maps of mammalian chromosome. *Science* 250, 245–250.

Daly, M.J., Rioux, J.D., Schaffner, S.F., Hudson, T.J. and Lander, E.S. (2001) High-resolution haplotype structure in the human genome. *Nature Genetics* 29, 229–232.

Deloukas, P., Schuler, G.D., Gyapay, G., Beasley, E.M., Soderlund, C., Rodriguez-Tome, P., Hui, L., Matisse, T.C., McKusick, K.B., Beckmann, J.S., Bentolila, S., Bihoreau N., Birren, B.B., Browne, J., Butler, A., Castle, A.B., Chiannilkulchia, N., Clee, C., Day, P.J., Dehejita, A. *et al.* (1998) A physical map of 30,000 human genes. *Science* 282, 744–746.

Fleischmann, R.D., Adams, M.D., White, O., Clayton, R.A., Kirkness, E.F., Kerlavage, A.R., Bult, C.J., Tomb, J.F., Dougherty, B.A., Merrick, J.M. *et al.* (1995) Whole-genome random sequencing and assembly of *Haemophilus influenzae* Rd. *Science* 269, 496–512.

Frengen, E., Weichenhan, D., Zhao, B., Osoegawa, K., van Geel, M. and de Jong, P.J. (1999) A modular, positive selection bacterial artificial chromosome vector with multiple cloning sites. *Genomics* 58, 250–253.

Galibert, F., Mandart, E., Fitoussi, F., Tiollais, P. and Charnay, P. (1979) Nucleotide sequence of the hepatitis B virus genome (subtype ayw) cloned in *E. coli. Nature* 281, 646–650.

Goldammer, T., Weikard, R., Brunner, R.M. and Schwerin, M. (1996) Generation of chromosome fragment specific bovine cDNA sequences by microdissection and DOP-PCR. *Mammalian Genome* 7, 291–296.

Guan, X.-Y., Trent, J.M. and Meltzer, P.S. (1993) Generation of band-specific painting probes from a single microdissected chromosome. *Human Molecular Genetics* 2, 1117–1121.

Hacia, J.G., Makalowski, W., Edgemon, K., Erdos, M.R., Robbins, C.M., Fodor, S.P., Brody, L.C. and Collins, F.S. (1998) Evolutionary sequence comparisons using high-density oligonucleotide arrays. *Nature Genetics* 18, 155–158.

Han, C.S., Sutherland, R.D., Jewett, P.B., Campbell, J.L., Meincke, L.J., Tesmer, J.G., Mundt, M.O., Fawcett, J.J., Kim, U.-J., Deaven, L.L. and Doggett, N.A. (2000) Construction of a BAC contig map of chromosome 16q by two-dimensional overgo hybridization. *Genome Research* 10, 714–721.

Harrison-Lavoie, K.J., John, R.M., Porteous, D.J. and Little, P.F.R. (1989) A cosmid clone map derived from a small region of human chromosome 11. *Genomics* 5, 501–509.

Hitte, C., Lorentzen, T.D., Guyon, R., Kim, L., Cadieu, E., Parker, H.G., Quignon, P., Lowe, J.K., Gelfenbeyn, B., Andre, C., Ostrander, E.A. and Galibert, F. (2003) Comparison of MultiMap and TST/CONCORDE for constructing radiation hybrid maps. *Journal of Heredity* 94, 9–13.

Hohn, B. (1983) DNA sequences necessary for packaging of bacteriophage lambda DNA. *Proceedings of the National Academy of Sciences USA* 80, 7456–7460.

Hohn, B. and Murray, K. (1977) Packaging recombinant DNA molecules into bacteriophage particles *in vitro*. *Proceedings of the National Academy of Sciences USA* 74, 3259–3263.

Hoopengardner, B., Bhalla, T., Staber, C. and Reenan, R. (2003) Nervous system targets of RNA editing identified by comparative genomics. *Science* 301, 832–836.

International Human Genome Sequencing Consortium (2001) Initial sequencing and analysis of the human genome. *Nature* 409, 860–921.

Ioannou, P.A., Amemiya, C.T., Garnes, J., Kroisel, P.M., Shizuya, H., Chen, C., Batzer, M.A. and de Jong, P.J. (1994) A new bacteriophage P1-derived vector for the propagation of large human DNA fragments. *Nature Genetics* 6, 84–89.

Ish-Horowicz, D. and Burke, J.F. (1981) Rapid and efficient cosmid cloning. *Nucleic Acids Research* 9, 2989–2998.

Istrail, S., Sutton, G.G., Florea, L., Halpern, A.L., Mobarry, C.M., Lippert, R., Walenz, B., Shatkay, H., Dew, I., Miller, J.R., Flanigan, M.J., Edwards, N.J., Bolanos, R., Fasulo, D., Halldorsson, B.V., Hannenhalli, S., Turner, R., Yooseph, S., Lu, F., Nusskern, D.R. *et al.* (2004) Whole-genome shotgun assembly and comparison of human genome assemblies. *Proceedings of the National Academy of Sciences USA* 101, 1916–1921.

Jinno, Y., Harada, N., Yoshiura, K., Ohta, T., Tohma, T., Hirota, T., Tsukamot, K., Deng, H.X., Oshimura, M. and Niikawa, N. (1992) A simple and efficient amplification method of DNA with unknown sequences and its application to microdissection/microcoding. *Journal of Biochemistry (Tokyo)* 12, 75–80.

Khrapko, K.R., Lysov, Yu.P., Khorlin, A.A., Ivanov, I.B., Yershov, G.M., Vasilenko, S.K., Florentiev, V.L. and Mirzabekov, A.D. (1991) A method for DNA sequencing by hybridization with oligonucleotide matrix. *DNA Sequencing* 1, 375–388.

Kim, C.-G., Fujiyama, A. and Saitou, N. (2003) Construction of a gorilla fosmid library and its PCR screening system. *Genomics* 82, 571–573.

Kim, U.J., Shizuya, H., de Jong, P.J., Birren, B. and Simon, M.I. (1992) Stable propagation of cosmid sized human DNA inserts in an F factor based vector. *Nucleic Acids Research* 20, 1083–1085.

Kim, U.J., Birren, B.W., Slepak, T., Mancino, V., Boysen, C., Kang, H.L., Simon, M.I. and Shizuya, H. (1996) Construction and characterization of a human bacterial artificial chromosome library. *Genomics* 34, 213–218.

Korstanje, R., Gillissen, G.F., Versteeg, S.A., van Oost, B.A., Bosma, A.A., Rogel-Gaillard, C., van Zuthphen, L.F.M. and van Lith, H.A. (2003) Mapping of rabbit microsatellite markers using chromosome-specific libraries. *Journal of Heredity* 94, 161–169.

Krumlauf, R., Jeanpierre, M. and Young, B.D. (1982) Construction and characterization of genomic libraries from specific chromosomes. *Proceedings of the National Academy of Sciences USA* 79, 2971–2975.

Kwiatkowski, T.J., Zoghbi, H.Y., Ledbetter, S.A., Ellison, K.A. and Chinault, A.C. (1990) Rapid identification of yeast artificial chromosome clones by matrix pooling and crude lysate PCR. *Nucleic Acids Research* 18, 7191.

Larin, Z., Monaco, A.P. and Lehrach, H. (1991) Yeast artificial chromosome libraries containing large inserts from mouse and human DNA. *Proceedings of the National Academy of Sciences USA* 88, 4123–4127.

Larin, Z., Monaco, A.P. and Lehrach, H. (1997) Construction of large insert yeast artificial chromosome libraries. *Molecular Biotechnology* 8, 147–153.

Lawrence, J.G. (2002) Shared strategies in gene organization among prokaryotes and eukaryotes. *Cell* 110, 407–413.

Leonard, J.A., Wayne, R.K., Wheeler, J., Valadez, R., Guillen, S. and Vila, C. (2002) Ancient DNA evidence for old world origin of new world dogs. *Science* 298, 1613–1616.

Leonardo, E.D. and Sedivy, J.M. (1990) A new vector for cloning large eukaryotic DNA segments in *Escherichia coli*. *Bio/Technology* 8, 841–844.

Libert, F., Lefort, A., Okimoto, R., Womack, J. and Georges, M. (1993) Construction of a bovine genomic library of large yeast artificial chromosomes clones. *Genomics* 18, 270–276.

Little, P.F.R. and Cross, S.H. (1985) A cosmid vector that facilitates restriction enzyme mapping. *Proceedings of the National Academy of Sciences USA* 82, 3159–3163.

Liu, L.F. and Wang, J.C. (1987) Supercoiling of the DNA template during transcription. *Proceedings of the National Academy of Sciences USA* 84, 7024–7027.

Ludecke, H.-J., Senger, G., Claussen, U. and Horsthemke, B. (1989) Cloning defined regions of the human genome by microdissection of banded chromosomes and enzymatic amplification. *Nature* 338, 348–350.

Maxam, A.M. and Gilbert, W. (1977) A new method for sequencing DNA. *Proceedings of the National Academy of Sciences USA* 74, 560–564.

McCormick, M.K. Buckler, A., Bruno, W., Campbell, E., Shera, K., Torney, D., Deaven, L. and Moyzis, R. (1993) Construction and characterization of a YAC library with a low frequency of chimeric clones from flow-sorted human chromosome 9. *Genomics* 18, 553–558.

Meldrum, D. (2000) Automation for genomics, part one: preparation for sequencing. *Genome Research* 10, 1081–1092.

Mellersh, C.S., Hitte, C., Richmann, M., Vignaux, F., Priat, C., Jouquand, S., Werner, P., Andre, C., DeRose, S., Patterson, D.F., Ostrander, E.A. and Galibert, F. (2000) An integrated linkage-radiation hybrid map of the canine genome. *Mammalian Genome* 11, 120–130.

Messing, J., Crea, R. and Seeburg, P.H. (1981) A system for shotgun DNA sequencing. *Nucleic Acids Research* 9, 309–321.

Modrek, B. and Lee, C.J. (2003) Alternative splicing in the human, mouse and rat genomes is associated with an increased frequency of exon creation and/or loss. *Nature Genetics* 34, 177–180.

Mouse Genome Sequencing Consortium (2002) Initial sequencing and comparative analysis of the mouse genome. *Nature* 420, 520–562.

Mullis, K., Faloona, F., Scharf, S., Sarki, R., Hom, G. and Erlich, H. (1986) Specific enzymatic amplification of DNA *in vitro*: the polymerase chain reaction. *Cold Spring Harbor Symposia on Quantitative Biology* 51, 263–273.

Myers, E.W., Sutton, G.G., Delcher, A.L., Dew, I.M., Fasulo, D.P., Flanigan, M.J., Kravitz, S.A., Mobarry, C.M., Reinert, K.H., Remington, K.A., Anson, E.L., Bolanos, R.A., Chou, H.H., Jordan, C.M., Halpern, A.L., Lonardi, S., Beasley, E.M., Brandon, R.C., Chen, L., Dunn, P.J., Lai, Z., Liang, Y., Nusskern, D.R., Zhan, M., Zhang, Q., Zheng, X., Rubin, G.M., Adams, M.D. and Venter, J.C. (2000) A whole-genome assembly of *Drosophila*. *Science* 287, 2196–2204.

Myers, E.W., Sutton, G.G., Smith, H.O., Adams, M.D. and Venter, J.C. (2002) On the sequencing and assembly of the human genome. *Proceedings of the National Academy of Sciences USA* 99, 4145–4146.

Osoegawa, K., Woon, P.Y., Zhao, B., Frengen, E., Tateno, M., Catanese, J.J. and de Jong, P.J. (1998) An improved approach for construction of bacterial artificial chromosome libraries. *Genomics* 52, 1–8.

Osoegawa, K., Tateno, M., Woon, P.Y., Frengen, E., Mammoser, A.G., Catanese, J.J., Hayashizaki, Y. and de Jong, P.J. (2000) Bacterial artificial chromosome libraries for mouse sequencing and functional analysis. *Genome Research* 10, 116–128.

Ostrander, E.A., Galibert, F. and Patterson, D.F. (2000) Canine genetics comes of age. *Trends in Genetics* 16, 117–124.

Patil, N., Berno, A.J., Hinds, D.A., Barrett, W.A., Doshi, J.M., Hacker, C.R., Kautzer, C.R., Lee, D.H., Marjoribanks, C., McDonnough, D.P. *et al.* (2001) Blocks of limited haplotype diversity revealed by high-resolution scanning of human chromosome 21. *Science* 294, 1719–1723.

Pe'er, I., Arbili, N., Liu, Y., Enck, C., Gelfand, C. and Shamir, R. (2003) Advanced computational techniques for re-sequencing DNA with polymerase signaling assay arrays. *Nucleic Acids Research* 31, 5667–5675.

Pierce, J.C. and Sternberg, N.L. (1992) Using bacteriophage P1 system to clone high molecular weight genomic DNA. *Methods in Enzymology* 216, 549–574.

Pierce, J.C., Sauer, B. and Sternberg, N. (1992) A positive selection vector for cloning high molecular weight DNA by the bacteriophage P1 system: improved cloning efficacy. *Proceedings of the National Academy of Sciences USA* 889, 2056–2060.

Ponce de Leon, F.A., Ambady, S., Hawkins, G.A., Kappes, S.M., Bishop, M.D., Robl, J.M. and Beattie, C.W. (1996) Development of a bovine X chromosome linkage group and painting probes to assess cattle, sheep and goat X chromosome segment homologies. *Proceedings of the National Academy of Sciences USA* 92, 3450–3454.

Ragoussis, J. and Olavesen, M.G. (1997) In: Dear, P. (ed.) *Genome Mapping, a Practical Approach*. IRL Press, Oxford, UK.

Rao, V.B., Thaker, V. and Black, L.W. (1992) A phage T4 *in vitro* packaging system for cloning long DNA molecules. *Gene* 113, 25–33.

Raudsepp, T. and Chowdhary, B.P. (1999) Construction of chromosome-specific paints for meta- and sub-metacentric autosomes and the sex chromosomes in the horse and their use to detect homologous chromosomal segments in the donkey. *Chromosome Research* 6, 103–114.

Roest Crollius, H., Jaillon, O., Bernot, A., Dasilva, C., Bouneau, L., Fischer, C., Fizames, C., Wincker, P., Brottier, P., Quetier, F., Saurin, W. and Weissenbach, J. (2000) Estimate of human gene number provided by genome-wide analysis using *Tetraodon nigroviridis* DNA sequence. *Nature Genetics* 25, 235–238.

Rohme, D., Fox, H., Herrmann, B., Frischauf, A.-M., Edstrom, J.-E., Mains, P., Silver, L.M. and Lehrach, H. (1984) Molecular clones of the mouse *t* complex derived from microdissected metaphase chromosomes. *Cell* 36, 783–788.

Rosenberg, N.A., Pritchard, J.K., Weber, J.L., Cann, H.M., Kidds, K.K., Zhvotovsky, L.A. and Feldman, M.W. (2002) Genetic structure of human populations. *Science* 298, 2381–2385.

Sanchez, C.P., Preuss, M. and Lanzer, M. (2002) Construction and screening of YAC libraries. *Methods in Molecular Medicine* 72, 291–304.

Sanger, F., Nicklen, S. and Coulson, A.R. (1977) DNA sequencing with chain-terminating inhibitions. *Proceedings of the National Academy of Sciences USA* 74, 5463–5467.

Sanger, F., Coulson, A.R., Barrel, B.G., Smith, A.J. and Roe, B.A. (1980) Cloning in single stranded bacteriophage as an aid to rapid DNA sequencing. *Journal of Molecular Biology* 143, 161–178.

Sanger, F., Coulson, A.R., Hong, G.F., Hill, D.F. and Petersen, G.B. (1982) Nucleotide sequence of bacteriophage lambda DNA. *Journal of Molecular Biology* 162, 729–773.

Savolainen, P., Zhang, Y.P., Luo, J., Lundeberg, J. and Leitner, T. (2002) Genetic evidence for an East Asian origin of domestic dogs. *Science* 298, 1610–1613.

Scalenghe, F., Turco, E., Edstrom, J.E., Pirrotta, V. and Melli, M. (1981) Microdissection and cloning of DNA from a specific region of *Drosophila melanogaster* polytene chromosomes. *Chromosoma* 82, 205–216.

Schlessinger, D. (1990) Yeast artificial chromosomes: tools for mapping and analysis of complex genomes. *Trends in Genetics* 6, 254–258.

Segers, K., Vaiman, D., Berghmans, S., Shay, T., Meyers, S., Beever, J., Cockett, N., Georges, M. and Charlier, C. (2000) Construction and characterization of an ovine BAC contig spanning the callipyge locus. *Animal Genetics* 21, 352–359.

Shay, T.L., Berghmans, S., Segers, K., Meyers, S., Beever, J.E., Womack, J.E., Georges, M., Charlier, C. and Cockett, N.E. (2001) Fine mapping and construction of a bovine contig spanning the ovine callipyge locus. *Mammalian Genome* 12, 141–149.

Shepherd, N.S., Pfrogner, B.D., Coulby, J.N., Ackerman, S.L., Vaidyanathan, G., Sauer, R.H., Balkenhol, T.C. and Sternberg, N. (1994) Preparation and screening of an arrayed human genomic library with the P1 cloning system. *Proceedings of the National Academy of Sciences USA* 19, 2629–2633.

Shizuya, H., Birren, B., Kim, U.-J., Mancino, V., Slepak, T., Tachiiri, Y. and Simon, M. (1992) Cloning and stable maintenance of 300-kilobase-pair fragments of human DNA in *Escherichia coli* using an F-factor-based vector. *Proceedings of the National Academy of Sciences USA* 89, 8794–8797.

Smith, D.R., Smyth, A.P. and Moir, D.T. (1990) Amplification of large artificial chromosomes. *Proceedings of the National Academy of Sciences USA* 87, 8242–8246.

Sternberg, N. (1990) Bacteriophage P1 cloning system for the isolation, amplification and recovery of DNA fragments as large as 100 kilobase pairs. *Proceedings of the National Academy of Sciences USA* 87, 103–107.

Strong, S.J., Ohta, Y., Litman, G.W. and Amemiya, C.T. (1997) Marked improvement of PAC and BAC cloning is achieved using electroelution of pulsed-field gel-separated partial digests of genomic DNA. *Nucleic Acids Research* 25, 3959–3961.

Susuki, A., Yamada, R., Chang, X., Tokuhiro, S., Sawada, T., Susuki, M., Nagasaki, M., Nakayama-Hamada, M., Kawaida, R., Ono, M. *et al.* (2003) Functional haplotypes of *PADI4*, encoding citrullinating enzyme peptidylarginine deiminase 4, are associated with rheumatoid arthritis. *Nature Genetics* 34, 395–402.

Venter, J.C. *et al.* (2001) The sequence of the human genome. *Science* 291, 1304–1351.

Vooijs, M., Yu, L.-C., Tkachuk, D., Pinkel, D., Johnson, D. and Gray, J.W. (1993) Libraries for each human chromosome, constructed from sorter-enriched chromosomes by using linker-adapted PCR. *American Journal of Human Genetics* 52, 586–597.

Watson, J.D. and Crick, F.H.C. (1953) Molecular structure of nucleic acids. *Nature* 171, 737–738.

Weber, J.L. and Myers, E.W. (1997) Human whole-genome shotgun sequencing. *Genome Research* 7, 401–409.

Wild, J., Hradecna, Z. and Szybalski, W. (2002) Conditionally amplifiable BACs: switching from single-copy to high-copy vectors and genomic clones. *Genome Research* 12, 1434–1444.

Yokobata, K., Trenchak, B. and de Jong, P.J. (1991) Rescue of unstable cosmids by *in vitro* packaging. *Nucleic Acids Research* 19, 403.

Zhou, Y., Wang, H., Wei, J., Cui, L., Deng, X., Wang, X. and Chen, Z. (2000) Comparison of two PCR techniques used in amplification of microdissected plant chromosomes from rice and wheat. *BioTechniques* 28, 766–774.

Zimmer, R., King, W.A. and Verrinder Gibbons, A.M. (1997) Generation of chicken X-chromosome painting probes by microdissection for screening large-insert genomic libraries. *Cytogenetics and Cell Genetics* 78, 124–130.

4 DNA Sequence of the Human and Other Mammalian Genomes

D. Vaiman*

INRA Department of Animal Genetics and U361, INSERM, Pavillon Baudelocque, Hôpital Cochin, 123 Boulevard de Port-Royal, 75014 Paris, France

Introduction	87
The past and present of genomic sequencing	88
The future	88
The Targets of Comprehensive Sequencing Projects	89
Sequencing strategies	90
Some general features of finished and ongoing projects	90
Humans	90
Mice	91
Chimpanzee	92
Other mammals	93
Mammalian Genome Components	94
Mononucleotides	94
Base pairs and their usage: periodic and aperiodic di- and trinucleotides	94
Meta-elements of the functional genome	96
Gene Expression Regulation and Genome Sequencing	107
The promoter model. What is a promoter?	107
Interspecific promoter comparisons	107
Bioinformatics and experimental validation	108
Long-range regulation of gene expression	109
Conclusion	110
References	110

Introduction

In the last few years, the 'new era' of genetics and genomics has become a commonplace topic of conversation following the completion of the human genomic sequence (Lander *et al.*, 2001; Venter *et al.*, 2001). The turmoil accompanying the achievement of organizing almost 3 billion base pairs led to this frequently heard 'new era' leitmotiv. However, this wears

*Correspondence: vaiman@cochin.inserm.fr

the mask of insufficient analysis. Indeed, what appears now as the most significant way of putting in order the bunch of collected data is the use of the results of evolutionary time and speciation as a sieve for distinguishing the true jewellery from the rubbish called either junk DNA, or, quoting the British geneticist Richard Dawkins, selfish DNA (Dawkins, 1976).

Actually, in parallel with two independent human genome sequences, the mouse genomic sequence completion has been a very important breakthrough (Waterston *et al.*, 2002), not of course because mice are more relevant study targets than human beings, but because two fairly complete (although modestly called 'drafts' by their authors) mammalian genomes are today available for extensive studies.

The cornucopia of data from both human and mouse genomes will undoubtedly, in the near future, yield a very important impetus to a new discipline: comparative eukaryotic sequence analysis. This emerging area of knowledge will rest on bioinformatics and provide a wide collection of data, which will differ according to the pairs of species under scrutiny. Phylogenetically distant species (such as mice and humans, for the class Mammalia, as well as fugu and humans for comparisons inside vertebrates) will indicate the position and edges of coding sequences, as well as the most general regulatory elements. In contrast, comparing genomes of close species, such as humans and chimpanzees, will pinpoint the most relevant tiny differences that make man a man. In 2002, A. Sidow proposed a provocative title for a short review paper in *Cell* (Sidow, 2002), entitled 'Sequence first. Ask questions later'. In this paper, the author addresses the evolutionary constraints affecting biological molecules, either by maintaining protein domains and DNA sequences unchanged, or by allowing subtle changes, sometimes able to lead to decisive speciation events. In this respect, the study of Enard and co-workers (2002a,b) on the two amino acid substitutions of the *FOXP2* gene, involved in language acquisition, and separating the human from the chimpanzee lineage is striking. This discovery of specific, evolutionarily relevant 'language mutations' is somehow reminiscent of Richard Goldsmith's theory of macromutations and hopeful monsters.

The past and present of genomic sequencing

In the past, research on mammalian genomes consisted of the building of genetic and physical maps (Botstein *et al.*, 1980; Donis-Keller *et al.*, 1987; Weissenbach *et al.*, 1992; Lander and Schork, 1994), which started with the tedious collection of rare restriction fragment length polymorphisms (RFLPs) and microsatellite markers in the late 1980s, and continued with the construction of dense maps made of thousands of polymorphic sequences. These tools became so efficient that the discovery of almost any locus, either Mendelian or underlying quantitative variations (quantitative trait loci, QTLs; Knott and Haley, 1992), is now possible in humans, mice and more recently in various domestic species (see Chapter 20). The sequencing capacity of many 'Genome Centres' around the world, exploited at 'full-gear' for the human and mouse projects, is now under-employed. Despite the extensive cost of whole genome sequencing projects, many mammalian geneticists would like their favourite species to constitute the future target for a comprehensive genomic sequencing project. Such consideration leads directly to the question of the benefit of such extensive endeavour (estimated in 1997 to be comprised in a range of US$220–900 million, Weber and Myers, 1997; even though these projects are certainly less costly today, they still represent a huge investment), while the possibility of comparing human and mice sequences has already yielded a huge resource for data mining. The major point in this question relies on whether the human and mouse genomes are adequate models for the other ~4000 existing mammalian species.

The future

Clearly, the major challenge of the coming years on the 'genome sequencing' front will be accurate analysis of the collected data and, as mentioned above, the capability to ask the right questions, because answers may lie in information already obtained, but hidden in huge databases. One of the major questions will undoubtedly be seeking an understanding of how mammalian genomes are regulated to allow tissue differentiation, body development and cell response to external stimuli.

Comparative information exists not only for coding sequences which are the most easily picked up from the genome, both by aligning mass sequencing expressed sequence tag (EST) information with genomic sequences and by interspecific comparisons of the most conserved chromosome fragments, but also for the more elusive regulatory elements which could represent over 15% of the sequence submitted to a strong selection pressure (Waterston *et al.*, 2002a). For this, new bioinformatic approaches begin to be developed and will for sure expand enormously in the years to come, for instance, by identifying important conserved promoter elements in the 5′ region of genes; such approaches have been called phylogenetic footprinting (Gumucio *et al.*, 1992; Lenhard *et al.*, 2003). In this book, Chapter 12 specifically relates to these techniques.

Now, with the availability of the mouse genome sequence, the time may have come to start asking the right questions. Most of these questions could be directed towards the intricate pathways of gene regulation, for which concerted long-range sequencing will certainly change the way of observing the genomic landscape, thus changing simultaneously our understanding of genome evolution, and medical practice towards an individual-directed medicine approach.

One can predict that unsuspected ways by which genomic expression is modulated will be discovered, such as shown recently by the transcriptional modulation achieved in complex genomes over wide chromosome domains (Cremer and Cremer, 2001; Caron *et al.*, 2001; Cosseddu *et al.*, 2004). In a recent study, it was shown that HSA21 contains many non-coding human/mouse conserved sequences (Dermitzakis *et al.*, 2002, see also Chapter 12). The biological significance (gene regulation, DNA compaction) of these sequences remains an open question but the identification and understanding of such genome regions will be a priority concern in the years to come. Another important issue of gene regulation might be brought about by the systematic identification of CpG islands in mammalian genomes, strengthened by interspecific conservation. Indeed CpG islands are the targets for methylation in CmG dinucleotides, which generally suppress transcription very efficiently. Therefore, the systematic identification of interspecifically conserved CpG islands will make it possible to evaluate experimentally distant methylation regulatory (DMR) elements, able to act upon side chromosomal regions.

In summary, the publication of 'draft' genome sequences has considerably modified the way we think about the subject of mammalian genomes, which cannot anymore be viewed as a descriptive topic. On the contrary, it is the inter- and intra-specific dynamics of mammalian genome evolution that have been opened for study. To use a metaphor, our time may correspond to the relatively abrupt passage from stills photography to moving pictures. Trying to predict the consequences of complete genome sequences upon the science of genomics, only a couple of years after their publication and in the limited space of a book chapter, is a difficult challenge for any writer, especially if all aspects are to be dealt with. An author can either be overwhelmed by the huge volume of data, if he attempts to be comprehensive, or appear strikingly 'old-fashioned' if he only sticks to the 'classical' way of presenting genomes found in textbooks. I have therefore chosen to focus on some relevant features of recent research pathways in each subchapter, without any claim for exhaustive coverage. This chapter is divided into three parts. In the first part, I describe the technical strategies of the whole genome projects that are finished or underway, and the interpretation of these data on a large chromosomal scale. In the second part, I will detail the basic components of mammalian genomes, starting from the nucleotide level, to the genes and gene families. In the last part, I outline the potential of comparative sequencing of genomes or ESTs for the understanding of gene expression regulation.

The Targets of Comprehensive Sequencing Projects

The initial impetus for sequencing whole mammalian genomes was provided in the 1980s (see for instance Hood *et al.*, 1987). Two draft human sequences were produced as early as

2001, far ahead of the most optimistic predictions (Lander et al., 2001; Venter et al., 2001), and were followed 1 year later by a mouse genome sequence (Waterston et al., 2002). While two preliminary genome sequences are now available for mammals, some technical problems remain, such as obtaining the most reliable alignment of the sequences, and are still the subject of bioinformatic research projects (Couronne et al., 2003; Jaffe et al., 2003; Mullikin and Ning, 2003), while the systematic sifting of the data collected has already started.

Sequencing strategies

The sequencing of the human genome relied on two different strategies: whole genome sequencing (WGS) and sequencing following the establishment of a contig map (see Chapter 3). The first strategy is derived from the methods used for sequencing bacterial genomes, which are about 1000 times smaller than mammalian genomes, which roughly encompass 3 billion bp. All the genome is cut physically or enzymatically into fragments of ~2000– ~10,000 bp, in a random fashion. All the fragments are cloned and sequenced from both ends ('shotgun' sequencing) and assembled making an extensive use of computer time. To achieve statistically sufficient redundancy by this strategy, almost 15 billion bp (over five genome equivalents) had to be sequenced (Venter et al., 2001).

In contrast, the alternative strategy rests first on a subdivision of the genome into large-insert genomic fragments (100–200 kb in bacterial artificial chromosome (BAC) vectors; Shizuya et al., 1992), followed by the establishment of contigs, a series of overlapping DNA inserts, which, extended to the whole genome, constitute a physical scaffold of several tens of thousands of clones. This contig organization is mainly carried out by establishing a fluorescent 'fingerprint' of each BAC. This fingerprint makes it possible to align BACs according to the similarities present in their banding patterns (see for instance Depatie et al., 2000; Ding et al., 2001). Each clone is then the target for shotgun sequencing, involving at least fivefold coverage of the BAC total sequence (hierarchical shotgun, Green, 2001). While the first approach appears tremendously straightfor-

ward, it is actually severely complicated by the fact that mammalian genomes are replete with repetitive sequences, sometimes several kilobases long, which can rapidly poison the nascent sequence alignments. Therefore, this approach has been used in combination with sequencing results present in public databases for humans, which has been of considerable help to complete alignments of WGS data programs.

There is still considerable controversy about the ability of the WGS-only approach to construct a draft sequence of a mammalian genome (Waterston et al., 2002b) as the private project of Celera Genomics (Venter et al., 2001) used sequencing data from the public genome initiative (Lander et al., 2001). The major question resides in the type of data used, either pre-assembled into an organized draft genome, or rough sequences (Green, 2002; Myers et al., 2002). If pre-assembled data were used, the publication of a human sequence by the Celera team did not prove that the assembly of a complex genome sequence is possible by the WGS approach alone. However, in any event, the sequences available from the private effort provide a complementary data set to the International Human Genome Consortium sequence, leading the way to the discovery of new polymorphisms (SNPs, single nucleotide polymorphisms). By their number and possibilities of automated genotyping, these additional data may later constitute a major tool for identifying disease genes at the genome level from population analysis by linkage disequilibrium.

Some general features of finished and ongoing projects

Humans

While many key features of the human genome had already been discovered or estimated by various approaches, the first two drafts of the human genome brought several useful confirmations (Lander et al., 2001; Venter et al., 2001). One direct issue addressed has been the GC content which is indicative of the presence of CpG islands suggested earlier as reliable landmarks for housekeeping genes (Larsen et al., 1992). Variation in CpG content has been associated with the existence of defi-nite genomic fractions called isochores (Bernardi *et al.*, 1988), accessible by density centrifugation, and which are classified in L (light, >43% G+C), H1/H2 (heavy, 43–48% G+C) or H3 (heavy, >48% G+C) isochores. Interestingly, CpG islands, regions rich in these nucleotides in the vicinity of housekeeping genes, which are correlated with gene density, are in variable numbers on the chromosomes with the highest value on HSA19, known as a particularly gene-rich chromosome. In the Lander paper, the authors state that 'strict' iso-chores (i.e. particularly homogenous chromo-some regions) are not apparent compared with random sequences. In fact, a comprehensive analysis of base pair compositions along the genome reveals clearly the existence of chro-mosome heterogeneities in GC contents by sliding a 100 kb window over the draft genome sequences (Pavlicek *et al.*, 2002). The iso-chores were also examined by 50 kb windows in the Celera paper (Venter *et al.*, 2001), show-ing that H3 isochores represent 273.9 kb, H1 and H2 isochores represent 202.8 kb, while L isochores (G+C <43%) represent 1078.6 kb. The existence of isochores appears therefore confirmed by the human genome sequence (Clay and Bernardi, 2002).

In both human genome papers, the issue of regional chromosomal duplication is exten-sively addressed. In duplication events, blocks of 1–200 kb are transferred from one to several (or one) other position(s) in the genome. These events have been shown to amount to 3.3% of the whole genome (1.5% interchro-mosomal and about 2% intrachromosomal). In the Celera paper, 1077 of these duplicated blocks are identified, contain 10,310 pairs of genes, and are represented by chromosome segments in a very complex figure matching each chromosome with all the others sharing at least one duplicated segment (see Fig. 4.1). This intricate figure may not be much simpli-fied, as it directly represents the historical dynamics of mammalian genomes.

Mice

Rodent genome evolution seems to present several very peculiar features, leading to a rapid accumulation of mutations, either by

'population' effects at work for small animals which can rapidly constitute geographical iso-lates, or by rodent-specific molecular mecha-nisms (Guénet and Bonhomme, 2003). This means that the simultaneous availability of the human and mouse genomes is a route towards the most conserved features common to pla-cental mammals. Synteny disruptions have been used to estimate the number of shared chromosome segments between the two species. From small sets of data, and an exten-sive use of statistics, it has been estimated that roughly 170 syntenic segments are conserved between primates and rodents (Nadeau and Taylor, 1984). This estimation has not changed much with the increasing density of genetic and physical maps of the mouse and human

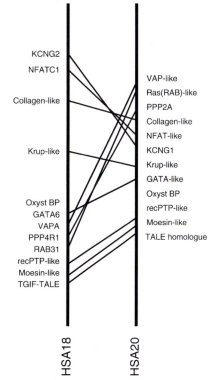

Fig. 4.1. A local example of chromosome fragment duplication. This figure illustrates the similarity between two chromosome segments located on human chromosomes 18 and 20 (from Venter *et al.*, 2001). Inside a DNA segment where a series of genes is present, many synteny breaks occur. Many segment duplications, followed by translocations and point mutation, constituted the basic material from which mammalian genomes are made.

genomes. For instance, the recent use of more sophisticated tools, such as radiation hybrids, identified about 200 regions of conserved synteny by mapping 3600 coding sequences (Hudson *et al.*, 2001). Sequencing of human/ mouse genomes increased the estimation to 217 segments of conserved synteny, using 558,000 orthologous landmarks. Most of the difference between 170 and 217 originates from rearrangements occurring inside synteny blocks that were not detectable in the early studies.

The mouse genome possesses a slightly higher G+C content than the human genome (42% vs. 41%), although the X chromosome is richer in CpG in humans compared with mice. This higher C+G frequency in the mouse autosomes veils the actual existence of human genomic regions presenting an extremely high G+C content, the distribution being wider in humans than in mice. CpG islands, which have been defined as the characteristic genomic tag for housekeeping genes, are far less represented in mice (15,500 vs. 27,000 in humans), maybe as a direct consequence of the lack of mouse genome regions containing an extremely high G+C percentage.

Anyway, the 10,000 or so interspecifically conserved CpG may be reasonably suspected of being genuine functional CpG islands (Waterston *et al.*, 2002a).

The comprehensive sequence of mouse and human genomes is also an invaluable tool for understanding the differences in nature and content of repetitive DNA elements in both species. Repetitive elements (about 50% of the genome sequences) present apparent specificity at the Order level inside the Class Mammalia. For instance, as described below (section on 'Dispersed elements'), ruminant short interspersed DNA elements (SINEs) have been shown to be shared by other Artiodactyla such as pigs, but also to display similarities with more distantly related species, such as Cetacea (Shimamura *et al.*, 1997; Nikaido *et al.*, 2001). Therefore, while the comparison of repetitive elements *per se* will probably not raise useful data for uncovering the evolutionary pathways linking primates and rodents, it gives nevertheless instant pictures of two functionally complete mammalian genomes, and thus describes two 'equilibrium states' for repetitive DNA elements.

At the gene level, comparisons of human and mice genomes have made it possible to detect unambiguously the most relevant conserved elements in coding sequences and the variation in nucleotide conservation within these sequences (Waterston *et al.*, 2002a). To achieve this comparison, a sample of 3165 human mRNAs was collected and aligned against their mouse counterparts. Each gene was divided into seven regions; 200 bp upstream (5' untranslated region, UTR), first exon, first intron, middle exons assembled in one, last intron, last exon and 200 bp of 3' UTR (Fig. 4.2). On average, nucleotide identity is 75% in the 5' UTR, 67% in the introns, 85% in the exons, and 75% in the 3' UTR. There are striking peaks of similarity at the first and last exon bases, which are the conserved acceptor and donor (GT/AG) splice sites. Gene structure was examined by comparison of a sample of 1506 genes. In 86% of the cases, the number of exons was identical between the two species, representing 10,061 pairs of orthologous exons. The differences are mainly due to the insertion/deletion of single exons.

Chimpanzee

While whole genome sequencing efforts are difficult to achieve to date, owing to the extreme expense of such programmes, a preliminary alternative is now envisaged for several domestic species, for which BAC libraries organized in contigs are available, such as cattle or pigs. The project consists of the systematic sequencing of BAC ends. Such limited programmes (60,000 sequences of 500 bp each, i.e. 30 Mbp, or about 1% of the genome) can provide useful data about the general level of sequence conservation, and, by anchoring sequences to one of the existing mammalian completed maps, may give an extensive panorama of gene order and organization in many mammalian genomes. Such an approach has been recently implemented for analysing the chimpanzee genome (Fujiyama *et al.*, 2002). In this study, over 75,000 BACs were sequenced from both ends, and aligned against the human genome. In this particular case, due to our close parentage to chimpanzees (a divergence time of only about 6 million years), all the sequences could be

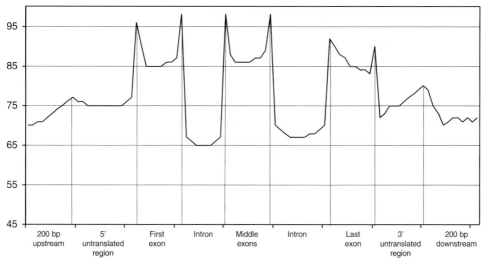

Fig. 4.2. Alignment of two 'synthetic' paradigmatic genes from the mouse and the human genome. The middle exons are 'summarized' in a unique sequence. Clearly, the highest similarity scores are observed at the exon/intron borders, where very conserved donor/acceptor splice site are located. Similarly, exons are much more conserved than introns. Regulatory elements in the proximal promoter and 5' region are globally more conserved than introns, which are much closer to intergenic sequences (redrawn from Waterston *et al.*, 2002a).

aligned to the human genome. Several conclusions have been drawn from this extensive analysis. First, the authors confirmed the level of sequence divergence generally agreed upon for human/chimpanzee (i.e. 1.5%). Moreover, one of the most interesting outputs of this study was the systematic search for rearrangements by comparison of the distance separating the two BAC ends in the human sequence versus the BAC size, some of these differences possibly accounting for bases of species specificities between two very close mammalian species. The authors rediscovered the previously known inversion separating human and chimpanzee chromosomes between 12p12 and 12q15.

Other mammals

A similar approach has been proposed for multi-species comparisons between vertebrates (Chen *et al.*, 2001). In this study, roughly 451,000 rat sequences provided by a whole genome shotgun sequencing approach have been aligned against human genome sequences. This number is about 60 times less than the number of sequences requested for a shotgun sequencing of a complete mammalian

genome, which consisted of almost 30 million sequence reads for the human genome sequence generated by Celera Genomics (Venter *et al.*, 2001). The authors stress that one of the major problems in identifying genes by this method is the difficulty in distinguishing between *bona fide* genes and low-abundance repetitive elements. Among the 451,000 sequence reads, 3% are homologous to transcribed regions of the human genome, consistent with the estimated overall conservation of 90% of the genes between humans and rodents, and with the number of coding sequences in mammalian genomes (roughly 3%). The possibility to adapt BLAST parameters for aligning sequences between rodents and humans may implicate that between species where mutations are accumulating more slowly (such as artiodactyla and primates), the same approach could be successfully used, with a cost far less than that of complete genome sequencing. It would be an accessible alternative for many species of economic interest, as suggested in a recent review, where vertebrate genome megabase sequencing is described as a 'backbone' for comparative genomics, allowing parallel comparisons with many species, given the condition that BACs will consistently consti-tute the major targets for sequencing pro-grammes (Thomas and Touchman, 2002).

Mammalian Genome Components

Mononucleotides

Edwin Chargaff demonstrated long ago the striking equality of all genomes between A and T nucleotides, on the one hand, and of G and C nucleotides on the other hand. This result is now trifling for many children after elementary school. It is due to the strict base-pairing of DNA. Aside from this strict parity, the propor-tion of A and T versus C and G is fixed for a given species, but very variable between species. Vertebrates have a genome clearly richer in A or T than in G or C (in humans, around 48–49% of GC for the gene-rich chro-mosomes HSA19 or HSA22, and 39–40% for the gene-poor chromosomes such as HSA13 or HSA18). In contrast, some eukaryotic para-sites, such as *Leishmania major*, possess an excess of GC/AT bases (over 60% of GC). In 2001, Shiori and Takahata demonstrated the existence of local deviations in mononucleotide frequencies, in both prokaryotic and human genomes (Shioiri and Takahata, 2001). In 1997, Hardison and co-workers performed a systematic study of base pair composition on available contigs from HSA6, HSA21 and HSA22. They attributed the skews in mononu-cleotidic frequencies to errors accumulating during replication, as these deviations were specifically positioned at precise chromosome locations that might therefore correspond to multiple replication origins. Mechanisms of DNA replication may thus provide clues about the fast accumulation of nucleotide substitu-tions for some specific DNA sites.

The extensive sequence data now available allow a more comprehensive analysis of nucleotide evolution by a systematic compari-son of human and mouse genomes. In a recent study, Hardison and co-workers demonstrated that modifications in genomic sequences do not appear randomly in mammalian genome evolution (Hardison *et al.*, 2003). Six criteria were chosen for this purpose, three resulting from the double set of data:

- The study of nucleotide substitution in cod-ing regions.
- The study of nucleotide substitution in transposable elements pre-dating human/mouse speciation.
- Nucleotide substitution in non-repetitive or ancestral repeats.

and three from the human data set alone:

- SNP density.
- Insertion of transposable elements.
- Recombination rate.

The major outcome of this study, which rests on a very large genomic basis (coding, non-coding, and repetitive sequences) and on a comprehensive mechanistic base (nucleotide substitutions, deletions and insertions), was the surprising observation that broad chromoso-mal regions (in the megabase range) vary either rapidly or slowly, but often in a con-certed fashion, where all possible mechanisms of DNA alteration are involved. This observa-tion can be correlated with recent papers indi-cating a concerted level of expression along wide chromosome regions (see below). Therefore, contrary to what was supposed pre-viously, it appears that both for gene expres-sion regulation and for genome evolution, mechanisms at work may involve very large chromosomal regions, by ways that are not yet satisfactorily elucidated.

Base pairs and their usage: periodic and aperiodic di- and trinucleotides

Biases in dinucleotide frequencies

Recent years have seen the development of novel statistical approaches in search for biases in nucleotide composition, i.e. di- or tri-nucleotide associations that could not be pre-dicted only by the average base composition of mammalian genomes. Major insights have been provided as to the specific occurrences of nucleotides between various eukaryotes at the genome scale, including human sequences from chromosomes 21 and 22, which were published before the rest of the genome sequence (Gentles and Karlin, 2001). In this study, the authors have used a simple formula aimed at evaluating the distortion between expected and observed dinucleotide frequencies, defined as:

$$\rho^*xy = f^*xy/f^*xf^*y$$

where f^*x and f^*y are the frequencies of the two nucleotides under scrutiny, while f^*xy is the frequency of the dinucleotide. When ρ^*xy is under 0.78, or above 1.23, the bias is considered significant. This measure, carried out on windows of 50 kb, makes it possible to define a genome signature, characteristic of the species under scrutiny. The authors carried out a systematic comparison of genomic sequences from five eukaryotes for the ten possible dinucleotide pairs: *Drosophila melanogaster*, *Caenorhabditis elegans*, *Homo sapiens*, *Arabidopsis thaliana* and *Saccharomyces cerevisiae*, and *Mus musculus* for a limited part. The most striking feature of mammalian genomes is the systematic under-representation of the CG dinucleotide with ρ^*xy values ranging from 0.18 to 0.31 only. In contrast, GC dinucleotides are normally represented. To a lesser extent, TA dinucleotides are also under-represented (ρ^*xy comprised between 0.66 and 0.75), but this feature is shared with all other eukaryotes (with the exception of the parasite *Plasmodium falciparum*). The low level of CG dinucleotides may be interpreted as a result of the importance of these sequences for gene regulation in vertebrate genomes, where DNA methylases exist and modulate tissue-specific expression, together with other mechanisms of histone acetylation/methylation, both mechanisms being probably related (Hashimshony et al., 2003). Therefore, in the normal biology of DNA, CpG can be methylated to CmG, which can then be mutated in TG, suggesting a regular mechanism for their disparity (Ohno and Yomo, 1991). This interpretation would also be consistent with the normal representation of CpG in other eukaryotic genomes where DNA methylation does not play any role in modulating gene expression, such as *Caenorhabditis* and *Drosophila*. The slight under-representation of TA could be due to the very general function of TATA boxes in promoters, or AATAAA polyadenylation signals in the 3′ end sequence of genes which could suggest that a selective pressure acts to eliminate such sequences when they arise randomly.

Specific dinucleotides and microsatellite markers

Mammalian genome analysis in the 1990s rested for a large part upon apparently boring sequences, made of stretches of short polynucleotide elements (1–5 bases), microsatellites. Microsatellites occur once every 2–30 kb in mammalian genomes, according to the criteria used for their definition (Schlötterer, 2000). Some specific repeated sequences have been related to a biological role. This is clearly the case for CpG dinucleotides which, when repeated, can constitute HpaII tiny fragment (HTF) islands. In this case, they are the major target for DNA methylation mechanisms, themselves generally correlated with tissue-specific gene silencing. However, these associations mask the fact that most combinations of mono-, di- or trinucleotide repeats could not be clearly associated with a defined function. Indeed, the most frequent microsatellites in mammalian genomes (i.e. $(TG)_n$ repeats and $(A)_n$ repeats) could not be clearly assigned to a specific biological function although several studies suggest that these repeats may contribute to a putative biological role, maybe as transcription enhancers (Hamada et al., 1984). As noted above, TG microsatellites may be derived from CpG islands having lost their regulatory function, followed by their chemical modification, from CpG to CmG and then to TG (Ohno and Yomo, 1991). This hypothesis is considered controversial today however.

The real reason why microsatellite sequences have been the subject of numerous studies (over 16,000 references in the Medline database) is in fact their utility as markers, spread relatively evenly along mammalian chromosomes, polymorphic, numerous and easy to analyse in segregation, a set of properties which renders them the most advantageous tools for localizing any gene on the genetic map by pedigree analysis. Genomic properties of microsatellite markers will be developed in the next chapter.

Trinucleotides

A special mention should be made of trinucleotides and more specifically of trinucleotide repeats, which have been associated with many dominant neurological diseases. These disorders comprise a series of spinocerebellar ataxias (SCAs), Huntington disease, dentalorubral-palidoluysian atrophy, Machado–Joseph disease, spinal and bulbar muscular atrophy and fragile X mental retardation. In each case, the disease is caused by the expansion of a trinucleotide repeat, from below 100 repeats to over 1000, depending on the disease locus considered (Margolis and Ross, 2001; Ranum and Day, 2002). An intermediate level of repetition is associated with a special state of 'pre-mutation', which increases dramatically the risk of expansion and mutation in the following generation. The repeat most generally corresponds to a polyglutamine CAG coding sequence, but can also sometimes be located in introns or untranslated regions.

Some other diseases have been associated with other repeat expansions. For instance, a polyalanine expansion in the *FOXL2* gene is one of the various causal mutations of the blepharophimosis ptosis epicanthus syndrome (De Baere, 2003). In mice, the *Sry* gene product, the major determinant of the male sex, includes a large tail with a series of 15 $(CAG)_n$-encoded polyglutamines (from 2 to 13 repeats) separated by series of FHDHH (Phe-His-Asp-Phe-Phe) amino acids. This remarkable 'tail' is specific to several but not all mouse species and has been shown to play an essential role in mouse male sex determination in these species (Bowles *et al.*, 1999). This polyglutamine region is absent from most mammalian *SRY* genes, including that of humans, suggesting that in these cases its function may be carried out by another protein, physically interacting with *SRY*.

Meta-elements of the functional genome

In textbooks, mammalian genomes are defined by their size and composition in terms of non-repetitive and repetitive elements. Classically, the size of mammalian genomes was estimated at approximately 3 billion bp, although this number can go up to 6 billion bp in some species such as the aardvark (*Orycteropus afer*). This size was previously estimated by a simple division between the amount of DNA

obtained and the number of original cells used to prepare it. The sequencing of human and mice genomes reduces this estimate somewhat to about 2.91 billion bp for the euchromatic portion of the human genome (Venter *et al.*, 2001) and slightly less for the mouse genome, 2.5 billion bp (Waterston *et al.*, 2002a). Clearly the division in chromosomes, whose number varies in mammals from $2n = 6$ (muntjac, Wurster and Benirschke, 1970) to $2n = 92$ (Rhinocerotidae, Wurster and Benirschke, 1968) and higher affects neither the total amount of DNA nor the way its expression is regulated (see below). Statistically speaking, the fairly complete sequencing of two phylogenetically distant mammalian species did not much affect the way scientists see how the genome is compartmentalized. Basically, following the historical use of different responses to dyes, chromosomes have been divided into regions of euchromatin (easily stained) and heterochromatin (more difficult to stain). Heterochromatin (around 20% of the total DNA), composed of satellite DNA elements (tandemly repeated units of 200–6000 bp), is the major component of centromeric regions, where these repetitive elements are closely associated with specific proteins. Several of these factors are constitutive (centromere proteins A, B, C, G and H, *CENPA*, *CENPB*, *CENPC*, *CENPG*, *CENPH*), while some others are transiently present (over ten different proteins). In a recent paper, it has been proposed that satellite DNA adopts a regular helical conformation (Gilbert and Allan, 2001). In this study, sucrose gradient centrifugation was used to show that centromeric satellite DNA is separated from the bulk of the cell's DNA. The authors suggest that centromeric DNA adopts a helical conformation consistent with the established model of a 30 nm fibre in chromosomal DNA. Euchromatin represents 80% of the DNA, and is considered as the most biologically active part of the DNA. When genomic sequence is spoken about, it is the euchromatin that is generally referred to, since the repetitive status of centromeric heterochromatin renders long-range sequence alignment impractical.

On the other hand, textbooks propose a division of mammalian genomes into compartments composed of dispersed or tandemly repeated sequences, intergenic non-coding sequences and genes (Fig. 4.3). The advent of partial and then complete sequence elements has made it possible to address the composition of mammalian genomes in a more sophisticated way and has allowed for a more descriptive comparison, based upon the actual data and not only on a sampling of these data. In particular, the extensive comparison of the mouse and human genome sequences carried out in the 'mouse genome' paper (Waterston *et al.*, 2002a), constitutes a good starting point to address comprehensively the fundamental structure of mammalian genomes. In this paper, genomic comparisons are carried out at various scales, from the 'chemical' GC composition of the chromosome, and identification of CpG islands, through a comprehensive analysis of the repeats and of their origins in rodents and primates, to end by an enlightening vision of genes and cross-species gene conservation.

Repetitive sequences

Mammalian genomes are replete with repetitive sequences, which may be classified into two main categories, dispersed or interspersed elements, and tandem repeats, also called satellites. Dispersed repetitive elements are distributed more or less regularly over the chromosomes (Fig. 4.4). They comprised LINEs (long interspersed DNA elements), SINEs (short interspersed DNA elements), retrovirus-like elements with LTRs (long tandem repeats), and DNA transposons.

Dispersed elements. LINEs are also described as 'autonomous' as they contain an RNA polymerase II-binding site and two open reading frames (ORFs) encoding genes able to replicate them and to insert them in the genome. Although their total size is about 6000 bp, they are most often truncated, only a small fraction

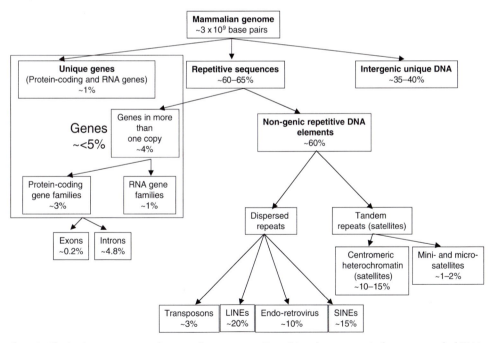

Fig. 4.3. The basic components of mammalian genomes. Repetitive elements are in fact composed of DNA segments, which are only duplicated, but also of many sequences present thousands of times in the genome. Experiments performed at the end of the 1960s, based upon the reassociation time of DNA in solution, indicated that mammalian genomes are indeed made of three groups of sequences: highly repetitive, which re-associate very rapidly, owing to the high probability of two complementary strands meeting each other, moderately repetitive, and unique. Further analysis based upon comprehensive DNA sequencing strengthened this view of mammalian genomic DNA.

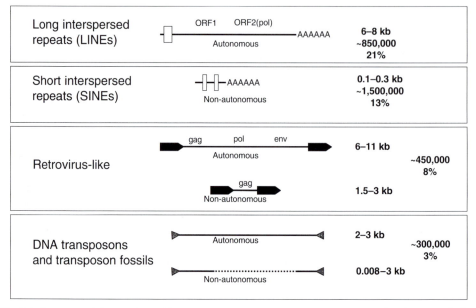

Fig. 4.4. Summary of dispersed repetitive sequences. Some are autonomous, meaning that they encode the genes needed to duplicate and copy them from one chromosome location to another, some are non-autonomous, meaning that they depend upon the machinery of homologous replicating repetitive DNA elements to move in the genome.

having kept their total length and being thus fully functional. They represent several hundreds of thousands of copies in mammals (about 21% of the genome in humans). Although their structure indicates a loss of function for most LINEs, their role in genome evolution is no longer downplayed. In a recent study, it was shown that human L1 LINE retrotransposase is expressed in male germ cells in transgenic mice (Ostertag *et al.*, 2002), while the natural promoter restricts its expression to testis and ovary. In mouse lines constructed with the human L1 and expressing the transgene at a high level, *de novo* insertions are observed in over 1% of the offspring. This high insertion frequency substantiates the idea of a high mutagenic potential of random LINE insertions (Ovchinnikov *et al.*, 2001). In addition to its use in self-promotion, the LINE machinery is readily used by other components of the nuclear environment, SINEs being the most documented in this respect. However, they extend their action to processed retropseudogenes (Esnault *et al.*, 2000). Such pseudogenes are derived from reverse transcription of transcripts, followed by random genomic insertion. Consequently, they do not contain promoters and introns, but are terminated by a poly(A) tail, unlike pseudogenes derived from gene duplication followed by mutation accumulation. In this study, the authors demonstrate that both LINE ORFs are required for the formation of processed pseudogenes, while retroviral genome elements are not able to perform the same task. In conclusion, coincident clues suggest a very important role for LINEs in shaping and modifying mammalian genomes.

SINEs are much shorter and depend upon the LINEs machinery for duplicating and moving in the genome. They represent up to 15% of the mammalian DNA and often originate from the reverse transcription and insertion of specific tRNA molecules. In this case, they consist of three regions, a tRNA-related region, a tRNA-unrelated region, and an AT-rich region, often associated with a $(TG)_n$ microsatellite (see below) in Artiodactyla (Varvio and Kaukinen, 1993; Vaiman *et al.*, 1994). Mammalian genomes also contain retrovirus-like elements, either autonomous or non-autonomous according to their gene content. In humans they amount to 8% of the genome.

Finally, transposons, both complete (autonomous) and deleted (non-autonomous), represent about 3% of mammalian genomes.

In mammals, SINEs have been particularly useful to study phylogeny, particularly in ruminants (Lenstra *et al.*, 1993; Okada and Hamada, 1997). The existence of cetacean DNA sequence, although limited, revealed the existence of SINE elements that had first been detected in ruminants. In 1999, Shimamura and co-workers discovered a huge family of SINEs derived from tRNA[Glu], common to a new 'super-order' of mammals: Cetartiodactyla (Shimamura *et al.*, 1999). Analysis of this SINE enabled the authors to reach interesting conclusions about the evolution of this large group of mammals. It showed that camelids (Tylopoda) diverged first from stem Cetartiodactyla. Following this divergence, LINE families (CHRS) derived from tRNA appeared in the lineage of the other Cetartiodactyla (suiforms, ruminants, cetaceans and hippopotamuses). Subsequent divergence created CHR-1, which is absent from suiforms. Finally in ruminants only, one of the most common SINEs, Bov-tA, was built from Bov-A and CHR-2. Its duplication conducted to Bov-A2, parts which are used in the typical ruminant LINE Bov-B. In the pig lineage, PRE-1 is derived from a subfamily of CHRS, CHRS-S.

This study emphasizes the interest of studying repetitive elements, which are less conserved than coding sequences, to understand mammalian evolution.

Other dispersed elements, endovirus (retrovirus) or transposons are much less numerous in mammalian genomes and their remodelling function seems more limited than that of LINEs.

Satellite elements.

Satellites and minisatellites. Satellite DNA, previously identified as the major DNA constituent of centromeres, gained its name from the fact that it can be physically separated from the bulk of the DNA when physical or biological methods are used (e.g. density centrifugation, cytogenetics, which in the early days of metaphase chromosome analysis made it possible to see chromosome fragments appearing separated from the end of chromosome arms, or restriction enzyme digestion, followed by gel electrophoresis). Whatever the method used, it

appears that one characteristic feature of satellite DNA is the head-to-tail repetition of a common sequence of nucleotides, unchanged or only slightly modified. Satellite DNA is composed of thousands of repetitive units ranging from 100 to 10,000 bp, where in coordination with several protein factors, it participates in the centromere structure, as previously described (see section on meta-elements of the functional genome). When the repeated units range from 10 to 100 bp, these are referred to as minisatellites, which were once used as highly polymorphic genomic markers of the VNTR class (variable number of tandem repeats), for which polymorphism can be studied either by a polymerase chain reaction (PCR) with primers surrounding the repeat, or by hybridization with a specific probe in Southern blots (Gill *et al.*, 1985), and are still in use for some forensic studies.

Microsatellites. Other satellite markers, microsatellites, have now taken over from minisatellites for genomic fingerprinting, as well as for individual identification. In microsatellites, the repeated motif ranges generally from 1 to 6 bp. Microsatellites have been, and still remain, indispensable tools for building genetic maps and to map relevant genes by positional cloning approaches. In mammals, the most studied is composed of TG repeats, which are generally polymorphic when the number of repeated units exceeds 10 or 12, which is the case for about 50,000 of such sequences in mammalian genomes although striking differences have been reported between humans and rodents, long before whole genome sequencing programmes achieved their goals (Beckman and Weber, 1992). Sequencing programmes have shown that microsatellites are about three times more frequent in mice than in humans. For instance, TG repeats are present at a frequency of 62 per Mb in mice, while there are about 20 per Mb in humans. Explanations for this increased number are not currently proposed. From sequencing data, it appears that the density of microsatellites in cattle and pigs is more similar to that of humans than of mice. Another unexplained observation is the long average size of rodent microsatellites. Again, from this point of view, it appears that mice constitute an exception rather than a rule, compared with other mammalian species, such as ruminants (Vaiman et al., 1994), pigs (Rohrer et al., 1996) or horses (Godard et al., 1997). A comprehensive genome-wide analysis of 1–6 bp microsatellites has been recently performed from the human genomic sequences (Subramanian et al., 2003). The authors took into account all the repeats of more than 12 bp. The density ranges from about 12,000 bp per Mb (various autosomes and the X) to 20,351 bp per Mb (HSA19). Variations were observed according to the nature of the repeats. Poly(A) repeats are the vastly predominant type of mononucleotide repeat in all chromosomes (3500–7500 bp per Mb). Amongst dinucleotide repeats, AC and AT are prominent and display a fairly homogenous density. Except for dinucleotide repeats, the skew towards an increased microsatellite density is striking for HSA19, and to a lesser extent for HSA17 and HSA22.

Microsatellite composition: a comprehensive study of microsatellite composition through various phylogenetic units has been carried out recently by Toth and co-workers (Toth et al., 2000). This study took into account simple sequence repeats (microsatellites) in eukaryotes from yeast and fungi to mammals. It therefore represents a compendium of the different features of microsatellites across a wide sampling of the living world. In exons, microsatellites are logically almost exclusively tri- or hexanucleotides, which do not break ORFs. Another striking feature is the specific occurrence of CCG and ACG repeats in vertebrate intergenic regions, which are absent in introns, but relatively abundant in exons. The authors correlate these facts to the highly mutable CpG dinucleotide, present in both these trinucleotides which, after methylation to CmG, may easily be changed into TG. It is therefore concluded that either intergenic regions are sufficiently unmethylated to maintain the repeats, or that an active mechanism is maintaining these sequences. The latter hypothesis is substantiated by the rarity of CCG and ACG in intergenic regions from species which do not methylate their DNA such as Drosophila and Caenorhabditis. Except for these examples, it is remarkable that the study fails to uncover explanations truly applicable to

take into account microsatellite distribution and choice in various genomes. For instance, vertebrate genomes appear to favour $(TG)_n$ microsatellites, while Caenorhabditis, as well as some insects, appear to favour $(TC)_n$ microsatellites (Estoup et al., 1995). Similarly, within mammals, differences in the preferred motif are often found between primates and rodents for tri- and tetranucleotides.

Microsatellite evolution: although microsatellites are apparently not subject to a high selection pressure, their flanking sequences can still be sufficiently conserved to be amplified interspecifically by PCR, as long as the species are not separated by too large a phylogenetic distance. This property has been used to build the sheep and goat genetic maps from bovine microsatellite markers (Vaiman et al., 1996; Maddox et al., 2001). Similarly, human and chimpanzee microsatellite flanking sequences are very well conserved. This has constituted the grounds for a recent study comparing the length of orthologous microsatellite loci covering a distance of 5.1 Mb of genomic DNA (Webster et al., 2002). The analysis revealed that inside a specific genome, mutation rates are highly variable, an observation that may be related to the above-mentioned study of Hardison and co-workers, which suggests a high variability in specific genomic regions even between close species. Besides the evolution of microsatellite flanking sequences, another important issue is the origin of microsatellite alleles, which have mainly been studied on TG repeats (the major bricks of vertebrate genetic map construction). Polymorphism of microsatellites is represented by length polymorphism, detectable by electrophoresis after PCR. Indeed, between locus-specific framing primers, the number of repeats may vary (Weber, 1990), and this variation can form the basis for allele determination. The mutation rate of microsatellites has been estimated at $\sim 10^{-4}$ in mammals (see for instance Dib, 1996). Microsatellite alleles have been proposed to arise from two possible mechanisms, polymerase slippage and unequal crossing over, the first being seemingly predominant (Schlötterer and Tautz, 1992). More precisely, two mechanisms of either stepwise mutation or all-allele mutation have been considered, experimental and theoretical data involving essentially the first model (Valdes et al., 1993; Kimmel and

Chakraborty, 1996). Equilibrium must be found not to allow overlong alleles. The regulation of microsatellite allele length may also be obtained by a stepwise mechanism, involving counter-selection of the largest alleles, although mathematical modelling indicates that such a selective mechanism may not be necessary (Kruglyak *et al.*, 1998).

Gene families. For a long time, it has been known that some genes are organized in families over the genome, these families sometimes being far apart (Hood, 1977). Basically, such families originate from duplications of chromosome fragments, first in syntenic blocks; these blocks may subsequently be dispersed to different chromosomes (Kasahara, 1997). Examples of such gene families abound, one of the most well known being the haemoglobin gene cluster (Forget, 1980). Haemoglobin is a tetrapolypeptide composed of four polypeptide chains (two alpha chains, and two beta chains in the adult). Alpha chains are clustered on human chromosome 16 (alpha-1, alpha-2, theta, zeta as well as three pseudogenes, respectively, similar to alpha-1, alpha-2 and zeta), while beta chains (five different 'beta-type' polypeptides, beta, delta,

epsilon1, gamma A, gamma G, one pseudogene and a regulator of gamma haemoglobin) are clustered on human chromosome 11. At the protein level, alpha and beta chains share 61 identities amongst ~145 amino acids, indicating unambiguously their common origin (Fig. 4.5). Haemoglobins appear themselves to be derived from the unique globin gene present in the most primitive vertebrates, the Agnatha (lampreys).

Other examples of gene families are given by regulatory factors. For instance, the *HMG*-box gene family (Fig. 4.6) encompasses *SRY* (the primary determinant of the male sex in mammals) and several genes resembling *SRY*, the *SOX* family (for *SRY-HMG* box, Pevny and Lovell-Badge, 1997). Some of these genes, such as *SOX9* or *SOX3*, appear to be involved in sex determination, while others have completely different functions, such as *SOX10*, which is involved in the development of neural crest-derived melanocytes and glia (Mollaaghababa and Pavan, 2003). There are at least 20 *SOX* genes, spread over the whole genome, and sharing similarity essentially in the *HMG* box. From their dispersion, it can be deduced that local gene duplication has not played a major role for the development of the family.

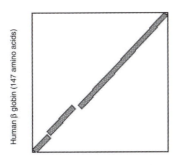

```
Query:   4   LTPEEKSAVTALWGKV--NVDEVGGEALGRLLVVYPWTQRFFESFGDLSTPDAVMGNPKV   61
             L+P +K+ V A WGKV +   E G EAL R+ + +P T+ +F  F      D   G+ +V
Sbjct:   3   LSPADKTNVKAAWGKVGAHAGEYGAEALERMFLSFPTTKTYFPHF------DLSHGSAQV   56

Query:  62   KAHGKKVLGAFSDGLAHLDNLKGTFATLSELHCDKLHVDPENFRLLGNVLVCVLAHHFGK  121
             K HGKKV A ++ +AH+D++    + LS+LH  KL VDP NF+LL + L+  LA H
Sbjct:  57   KGHGKKVADALTNAVAHVDDMPNALSALSDLHAHKLRVDPVNFKLLSHCLLVTLAAHLPA  116

Query: 122   EFTPPVQAAYQKVVAGVANALAHKY  146
             EFTP V A+  K +A V+  L   KY
Sbjct: 117   EFTPAVHASLDKFLASVSTVLTSKY  141
```

Fig. 4.5. An illustration of functional gene duplication provided by the haemoglobin family. The alpha and beta chains of the protein are located on different chromosomes (HAS16 and HAS11, respectively). The alignment was made using the BLAST program from the NCBI site http://www.ncbi.nlm.nih.gov/BLAST/.

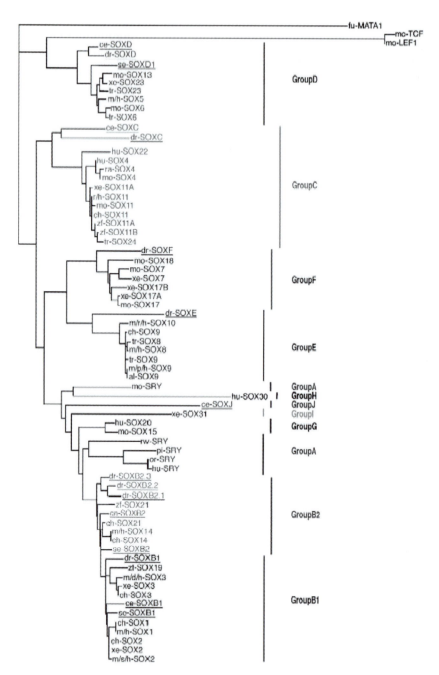

Fig. 4.6. An example of a moderately repetitive gene family, the *SOX* genes, which share a common element called the HMG box. This family presents many genes that are present in many vertebrates. The species are indicated by the letters preceding the gene denomination: al = *Alligator mississipiensis*, ce = *Caenorhabditis elegans*, ch = *Gallus gallus*, dr = *Drosophila melanogaster*, d = *Sminthopsis macroura* (marsupial), fu = *Saccharomyces cerevisiae*, h = *Homo sapiens*, mo or m = *Mus musculus*, or = *Pongo pygmaeus*, pi or p = *Sus scrofa*, r = *Rattus norvegicus*, tw = *Macropus eugenii* (tammar wallaby), sh or s = *Ovis aries*, tr = *Oncorhynchus mykiss* (rainbow trout), se = *Strongylocentrotus pupuratus* (sea urchin), xe = *Xenopus laevis*, zf = *Danio rerio*.

Gene families can contain an enormous number of related genes. The study of gene superfamilies opens vistas towards an understanding of some important specific pathways of genome evolution, both within and between species. Gene superfamilies are composed of one or several clusters of related gene sequences, some of which may have lost their function by the accumulation of mutations (pseudogenes). The evolutionary pathways of such families have been the subject of debate between two models: concerted evolution and birth-and-death evolution. According to the first model, genes inside clusters are homogenized by interlocus recombination or gene conversion. In the second model, clusters of genes and pseudogenes are inherited from an ancestral species, while, later, genes inside the cluster may evolve towards pseudogenes and lose their function. Two examples of such huge gene families are provided by: (i) the 'immunoglobulin' gene family, which *sensu lato* comprises various molecules involved in immunity: the major histocompatibility complex (MHC), T-cell receptors (TCRs) and immunoglobulin (Hunkapiller and Hood, 1989) and (ii) the olfactory receptor (OR) gene family (Mombaerts, 1999; Buck, 2000). A third multigene family, very different in terms of evolutionary constraints, ribosomal genes, will also be discussed.

Among the three major branches comprising the 'immunoglobulin' superfamily, the MHC has been the topic of many studies. It is characterized by extensive polymorphism, which can now be studied directly at the sequence level in several mammalian species, including humans, mice, pigs and cattle. This gene complex spans about 3 Mb in mammals and harbours the highest gene density in the genome. MHC genes are divided into three classes, class II, class III and class I, in this order along human chromosomes. MHC molecules distinguish between self- and non-self-antigens, and present foreign peptides to the TCR of T lymphocytes. The biological bases for MHC polymorphism maintenance are a subject of debate (Parham and Ohta, 1996). One hypothesis resides in the advantage conferred to heterozygotes, which can respond to a higher number of antigens than homozygotes (Hughes and Nei, 1988; Hughes *et al.*, 1994).

By studying six primate species, it was possible to date the birth of MHC alleles in both Old World (Platyrrhini) and New World (Catharrhini) monkeys (Pionkivska and Nei, 2003) This study demonstrated also that the F locus, the most ancestral, diverged from previous MHC class I between 66 and 46 million years ago; a major diversification of class I loci followed 49–35 million years ago. The C locus appeared in apes by gene duplication 28–21 million years ago. In humans the three MHC class I loci, A, B and C could be dated between 19 and 10 million years ago. The authors state that the phylogenetic trees observed for this genetic system are consistent with the birth-and-death model.

Another interesting example of genome evolution at the level of gene families is given by the OR superfamily. ORs are located in olfactory cell membranes, and contain seven transmembrane domains (Buck and Axel, 1991). Intriguingly, OR genes display a monoallelic pattern of expression, thus conferring a functional identity to each olfactory neurone (Chess *et al.*, 1994). OR genes are divided into two major groups: class I share similarity with 'fish-like' OR, while class II appear to be mammal-specific (Freitag *et al.*, 1998; Glusman *et al.*, 2000). ORs constitute the largest existing gene superfamily in mammalian genomes. In mice, where olfaction plays a vital biological role (as for most mammals except primates), it is comprised of 1296 genes, of which 20% are pseudogenes (Glusman *et al.*, 2001; Zozulya *et al.*, 2001; Zhang and Firestein, 2002). The family is distributed in 27 clusters on all chromosomes except MMU12 and MMUY. Human chromosomes devote as much genomic space as those of mice to OR genes (all chromosomes except HSA20 and HSAY), however, almost two-thirds of the OR genes have been lost by mutations or frame disruptions, consistent with the birth-and-and death model. Roughly 900 genes have been discovered in the human genome sequence release (Glusman *et al.*, 2001; Zozulya *et al.*, 2001). As in mice, they are organized in clusters of six or more, for 80% of them. Expansions in the number of OR genes have been shown, by analysis of mice and human genomic sequences, to be related to a very general duplication mechanism, sometimes involving large chromosome regions, encountered in mammals but apparently not in fish. The monoallelic expression of ORs requires highly specific regulation mechanisms, some of which are still unknown. As for the immunoglobulin superfamily, there is a clear association between function and gene locus organization in order to reach an accurate gene regulation, and to ascertain the expression of a single allele per neurone. This regulation mechanism is at work even when multiple transgenes are integrated into the cell nucleus (Serizawa et al., 2000), suggesting the titration of a specific factor (Reed, 2000; Lane et al., 2001). Three models have been suggested to account for the monoallelic OR expression (Fig. 4.7, Kratz et al., 2002). In the first hypothesis, each gene possesses a specific promoter to which specific regulatory factors bind, resulting in the activation of this specific gene only.

Alternatively, a spatial modification of the DNA would position a locus control region (LCR) in contact with the proximal promoter of one gene. As the LCR would be able to make contact with only a subset of genes from the cluster, the problem would be simplified to the choice of only one of these genes as being the neurone-specific expressed OR. Finally, genomic somatic recombination has been put forward as a way to reduce the number of ORs to only one in a given gene cluster. However, there are no experimental data to substantiate this hypothesis. Comparative sequence analysis between active mouse and human OR genes may be a promising way to start elucidating the challenging question of the exquisite regulation of OR genes. While binding sites for several transcription factors have been discovered directly in the human 3' regions, genomic comparisons have also been

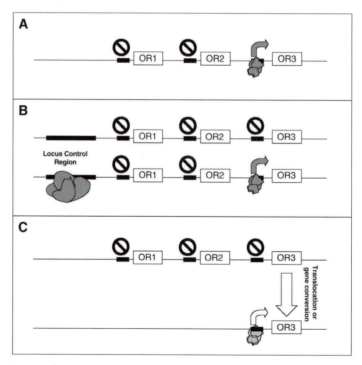

Fig. 4.7. Three hypotheses to explain the selective expression of genes inside the largest mammalian gene family, the olfactory receptor (OR) family. In the figure, three OR genes are represented; only OR3 is transcribed in the tissue. (A) Each gene possesses a different and highly specific promoter. (B) Upper part, transcription factors are absent; lower part, a distant control element (LCR) binds specific regulatory factors, that are susceptible to interact with other factors bound to the promoter of the transcribed gene. (C) Chromosome rearrangements are positioning the specific gene in the vicinity of a specific regulator. The other genes may be lost in this somatic recombination event.

performed with orthologous mouse genome sequences to obtain a better resolution of these important sites. In the particular case of ORs, however, the intricacies and number of active mice genes could be deceptive to perform useful comparisons. Another interesting source of comparative genomics information about the OR gene family originates from the dog species, which like mice belongs to the group of macrosmic species. A description of dog OR genes can be found in Sargan *et al.* (2001).

Ribosomal genes are a family of very similar genes spread in clusters along mammalian genomes. In this case, in contrast to the immunoglobulin superfamily and the 'OR' complexes, there is clearly an evolutionary advantage to limiting the divergence of these gene complexes, since they enable the organism to adapt rapidly by increasing drastically its translational response following a given stimulus. It is therefore reasonable to assume that ribosomal genes are subject to concerted evolution (Arnheim *et al.*, 1980). It is suggested that genetic exchanges of DNA occur in the nucleolar organizer regions between non-syntenic loci. In this case, concerted evolution favours the maintenance of a group of genes in a homogeneous state. In conclusion, the comparative sequence evolution of ribosomal genes, on the one hand, and of the immunoglobulin or ORs gene families, on the other hand, may present two different faces of multigene evolution, according to the biology and the role of the family under scrutiny.

When a complete genome sequence is not available (i.e. for most species) it may be interesting to be able to construct comprehensive inventories of large gene families, for which an approach has been proposed (Fuchs *et al.*, 2002). The methodology is based upon the design of a set of degenerated oligonucleotides, and upon their use in PCR reactions starting from the species DNA. The authors used the OR superfamily as an example, designed 20 primer pairs from the exonic regions of 127 OR genes. After PCR from human genomic DNA, the fragments were cloned, and arrayed on nylon membranes. Hybridization fingerprinting of each clone made it possible to cluster similar or identical clones and to choose a subset of 924 clones that were sequenced, yielding

358 new ORs, comprising genes and pseudogenes, in the expected proportion. This is a good example of the possible transfer of information from genomes sequenced with high throughput technologies, towards other genomes; the proposed method could make it possible to readily characterize gene families.

Unique sequences

Designing PCR primers is a good illustration that mammalian genomes in fact contain many repetitions; indeed, if all the genome was made of random sequence, then 16 nucleotides would find a unique match in a complete mammalian genome as $4^{16} = 4.294.967.296$, that is to say that two 8-mer oligonucleotide primers (with a calculated annealing temperature of about 20–25°C) would recognize a specific sequence, and would therefore be more than sufficient for achieving specific DNA amplification in PCRs. In view of the experimental facts, a specific PCR amplification needs in fact two 20-mers, theoretically much more than necessary for a specific hybridization. This discrepancy relates to the redundancy which saturates mammalian genomes. In fact, gene families (whatever their size) are much more frequent than unique genes.

Genes. The question 'how many genes are unique?' in mammalian genomes has been very difficult to answer, with estimates varying over a fourfold range. This was illustrated recently in three papers from the same issue of *Nature Genetics*, where the number of genes was estimated between 28,000 and 120,000 according to the method employed (Ewing and Green, 2000; Liang *et al.*, 2000; Roest Crollius *et al.*, 2000). The sequences of humans and mice have now provided consistent estimates converging on 30,000–35,000 protein-coding genes in mammalian genomes, considerably lower than most estimates of the 1990s. This result was a surprise when these values were compared with those of *Drosophila* (16,000 genes) or *Caenorhabditis* (19,000 genes), placing mammalian or vertebrate 'complexity' just twofold more than that of insects or nematodes (*C. elegans* Sequencing Consortium, 1998; Adams *et al.*, 2000). Two arguments may mitigate this surprise. First, there is an extreme tendency of vertebrate genes to use combinatorial processes to generate variation, which could be much more efficient than in invertebrates. For instance, the *WT1* gene, associated with the Wilm's tumour, encodes at least 24 protein isoforms resulting from alternative splicing, alternative promoting and RNA editing (see Wagner *et al.*, 2003, for a recent review). Secondly and maybe more important, estimates fluctuate about the number of non-protein-coding genes but appear to range from 10 to 50% of the active cell machinery. The actual number of this ever-growing class of genes is difficult to predict *ab initio*, therefore specific computer programs have been designed to identify them. For instance, starting from the RIKEN collection of mouse cDNA, a very recent systematic sequencing of the mouse transcriptome was carried out. The sequencing was carried out from 246 cDNA libraries. From the 3′ end of cDNA, 1,442,236 reads were obtained and 60,770 full-length cDNA clones were completely sequenced. After assembly, the cDNA reads were grouped in 70,000 transcription units (Carninci *et al.*, 2003). The recent development of a stringent informatic method of screening made it possible to identify 4280 putative non-coding transcripts (Numata *et al.*, 2003). With less stringent procedures, the amount of non-coding RNA can be estimated at up to 50% of the coding genome. These RNA-coding genes may therefore increase the number of mammalian genes to 60,000, a number close to 'ancient' (1990) 'guesstimates' of the number of human genes. The differences between this measure and the protein-coding gene number estimated in genome papers, around 35,000, may at least partially originate from non-coding RNAs which are nevertheless *bona fide* active genes. Another recently improved annotation of the human chromosome 22 sequence extrapolates that the human genome may contain 29,000–36,000 coding genes, as many as 21,300 pseudogenes, and 1500 RNA genes acting as antisense in regulating gene expression (Collins *et al.*, 2003). In summary, despite all the existing sequencing data, gene number in mammalian or vertebrate genomes remains an unanswered question, depending on various extrapolations.

A more relevant question may be the assignment of genes to cellular function. One possible way to address this issue may be given by the comparison of the human and yeast genome *Saccharomyces cerevisiae*, the first eukaryote for which the genome sequence was made available. As mentioned above, estimates of gene number in humans range between 26,000 and 35,000, estimates based either upon direct gene predictions or EST assemblies along the human genome sequence, or by more indirect means. The yeast genome comprises about 6000 genes, 50% more than an *Escherichia coli* bacterium (roughly 4000 genes). Qualitative differences between mammals and yeast can be found in specific gene functions, such as immunity and self-recognition, neural system function and development, intercellular contacts and recognition, developmental organization and homeostasis. These new functions are undoubtedly novel to metazoans, and may have needed the development of complex gene families, such as the above-mentioned immunoglobulin superfamily. In the Celera human genome sequence paper (Venter *et al.*, 2001), a comprehensive study of gene expansions was carried out between humans and other sequenced eukaryotes, *Caenorhabditis elegans* and *Drosophila melanogaster*, considered as representative of nematodes and insects, respectively. The metabolic enzymes cytochromes P450 are less numerous in humans than in other sequenced metazoans. In contrast, GAPDH, involved mainly in anaerobic glycolysis, is present 46 times in human genomes (three in *Drosophila*, four in *Caenorhabditis*). Similarly, ribosomal proteins are 8–10 times more frequent in human genomes. While these differences are genuine, they may be relevant to the evolutionary history of each species and may not have a true adaptative value.

Intergenic unique sequences. This category of genome sequences probably belongs to the least studied of DNA compartments. While genes or repetitive sequences have characteristic features that are able to make them popular targets for study, intergenic DNA sequences, although representing a significant fraction of the genome, are apparently not considered as a relevant subject for genetic analyses. The study of sequence conservation of these regions is now possible between humans and mice. Soon, it should be possible to categorize them between simple 'barrier' regions or regions containing important regulatory elements. Further genome knowledge will undoubtedly transform many of these intergenic sequences into important regulatory elements. To illustrate this, an example can be drawn from the *IGF2* gene. *IGF2* encodes a growth factor presenting a quite general trophic function in mammalian cells. While four alternative promoters have been described in the past, recently, Constancia and co-workers identified a new promoter 5' to those previously defined and called P0, for this reason. P0 is responsible for the placenta-specific expression of one *IGF2* isoform. The knockout of this isoform results only in a decreased growth of the placenta structure called labyrinth, causing a decreased placental growth followed some days later by fetal growth retardation (Constancia *et al.*, 2002). These experiments demonstrate that alternative, still unknown promoters may reside in intergenic regions.

Another example of an important intergenic region is provided by the callipyge phenotype in sheep that causes a hypermuscularity of the hind quarters in ewes and rams. The phenotype is the result of the regulation of at least four genes according to a complex imprinting phenomenon (reviewed in Georges *et al.*, 2003). In contrast to the wide-ranging chromosomal effect of the mutation (over 400 kb), the responsible mutation has been identified as a single point mutation located outside any coding region on sheep chromosome 18 (Freking *et al.*, 2002; Smit *et al.*, 2003). Large-scale sequencing was the only way to compare systematically a carrier chromosome against a non-carrier chromosome, and to identify the mutation. The result was obtained independently by two research teams. This example illustrates the importance of some ill-defined intergenic regions in achieving the precise regulation of mammalian gene expression, and is probably the mark of many such regulations yet to be discovered in mammalian genomes.

Gene Expression, Regulation and Genome Sequencing

The promoter model. What is a promoter?

For 25 years, gene expression regulation has been explored against the background of the operon theory of Lwoff, Monod and Jacob (Shafrir, 1996). This theory rests on the initial grounds that regulatory factors (transcription factors, TFs), interact physically with specific DNA regions, generally located upstream of the first codon of the gene (in the so-called 5' region, or promoter region). Although intended for describing bacterial gene regulation, this model has been successfully applied to understanding the profile of gene expression of hundreds of eukaryotic genes. However, besides the simple idea of a non-coding promoter playing the role of a bandmaster for the 'following' gene, it appears that the definition of such a promoter is not trivial. Most generally, 2–5 kb of DNA 5' from the ATG (first codon of the ORF of a given gene) are experimentally studied by a standard panel of techniques, including: (i) serial deletions of the promoter placed in front of a reporter gene, whose expression is checked in transfected cells; and (ii) detection of TF–promoter protein–DNA interactions by 'DNase footprinting' (see for instance Cappabianca *et al.*, 1999; Saluz and Jost, 1993; Rippe *et al.*, 2001) or gel-retardation (EMSA, electrophoretic mobility shift assay; Fried and Crothers, 1981). Experimental ways of exploring promoter function have also been achieved using more 'functional' approaches, such as transgenesis. However, the delimitation of a promoter remains a problem, since several genes, such as *WT1* or *SOX9* (Pfeifer *et al.*, 1999), may possess huge regulatory regions. In the case of *SOX9*, translocations located up to 1 Mbp 5' from the ATG codon result in the same phenotype as a mutation in the gene coding sequence itself, i.e. a rare genetic disease affecting bone and cartilage, campomelic dysplasia.

Interspecific promoter comparisons

Apart from these difficulties in unambiguously defining promoter regions, the effort of sequencing in parallel both humans and mice provides new opportunities to identify putative targets for transcription factors. While the sequencing programmes were still in progress, several preliminary studies indicated the hopes and problems of such new approaches

(Hardison *et al.*, 1997; Loots *et al.*, 2000; Wassermann *et al.*, 2000). The extensiveness of available sequences drove the construction of bioinformatic tools aimed at the systematic cross-species screening of promoter regions. In a recent study, Lenhard and co-workers, proposed a Web-based interface to make direct use of the possibilities of interspecific promoter data, ConSite, (http://www.phylofoot.org/). The site provides a graphical representation of promoter sequence similarities, and is connected to a TF-binding site database (Wingender *et al.*, 2001; Lenhard *et al.*, 2003). In addition to these bioinformatic tools, systematic approaches are carried out to discover conserved regions around genes (not only in promoters) excluding coding regions. In 2001, Levy and co-workers analysed a set of 502 transcripts present in the OMIM database, in order to identify conserved regions in the different gene elements (exons, introns, 5 kb upstream region, Levy *et al.*, 2001). While exon conservation reaches 79% between humans and mice, the score is around 20% for upstream regions, and 10–12% for introns, with the highly significant presence of TF-binding sites in both upstream elements and introns. This study also statistically demonstrated the physical clustering of binding sites in promoters, confirming the general view of a complex set of regulatory factors binding jointly to the promoter. Furthermore, it made it possible to classify genes according to the set of TFs interacting cooperatively with their promoters, opening doorways to complex patterns of concerted gene regulation networks (Wagner, 1997). Indeed, while footprinting techniques involve incubating promoter sequences with nuclear extracts, and testing the protection from DNase actions (for instance) of the complexed DNA (Cappabianca *et al.*, 1999), phylogenetic footprinting will achieve the same result by aligning promoter regions between the human and mouse (or other) genomes.

It is now clear that the differences in expression levels between two different mammalian genomes (e.g. humans and mice) are not achieved by single-point mutations inside the ORF of genes, but are rather the result of a precise and specific tuning of gene action. The orchestration of developmental gene expression is finely adjusted by modifications to regulatory sequences of the genome. This adjustment may be achieved by subtle alterations to TF-binding sites. Such fluctuations have been shown in the promoters of genes belonging to gene families, leading to the concept of statistically defined binding sites (Tronche *et al.*, 1997, for the example of hepatocyte nuclear factor-binding sites; see Stormo, 2000, for a review).

Bioinformatics and experimental validation

As described above, one way of improving the detection of relevant regulatory sequences is to complement bioinformatic data with data from comparative genomics, a very important issue that is the topic of Part 4 in this book. In several cases, comparative genomics has been used to detect regulatory elements. For instance, Loots and co-workers focused on HSA5q31, which contains a cluster of three biomedically important cytokines (interleukin-4 (IL-4), IL-13 and IL-5, Loots *et al.*, 2000). These cytokines are accompanied by 23 other genes in an ~1 Mb chromosomal region, and all but one are conserved in the same relative order on mouse chromosome 11. Comparative sequence analysis of this region allowed the authors to screen for DNA elements of more than 100 bp and 70% base identity between mice and humans. Two hundred and fifty-five DNA elements could be identified by this approach, 155 of which were shown to belong to coding sequences, while the other 90 were defined as non-coding. Of these, 46% were in introns, 9% lay within 1 kb of identified 5′ or 3′ transcripts, and 45% lay in intergenic regions (i.e. at least 1 kb away from any known gene). One of these sequences, *CNS-7* (for conserved non-coding sequence-7), is located between *GMCSF* (granulocyte–macrophage colony-stimulating factor) and *IL-3*, and had been previously identified as a common enhancer for the two cytokines (Osborne *et al.*, 1995). By PCR and sequencing with degenerated primers, 70% of the sequences were shown to be conserved in mammals, suggesting a role as a regulatory element (Li *et al.*, 1999). The authors focused their study on *CNS-1*, located in the 13 kb separating *IL-4* and *IL-13*. This sequence is the largest conserved element (401 bp), and was either conserved or deleted experimentally in a 450 kb yeast artificial chromosome (YAC) construct, before transgenesis in FVB mice. The authors could observe a marked increase in CD4+ cells expressing *IL-4* and *IL-13* as compared with mice where the unmodified transgene alone had been introduced. This increase was much less pronounced in transgenic mice where the putative regulatory element has been deleted. The authors were able to demonstrate that *CNS-1* regulates some of the genes in a 120 kb interval on HSA5q31, although other genes of the same region were not regulated, illustrating the still poorly understood complexity of long-range gene regulation in vertebrate genomes (Loots *et al.*, 2000).

These studies provide landmark results indicating that interspecific genomic comparisons may compensate for the fact that TF-binding sites, generally only 4–10 bp long, occur much too frequently by random chance alone (one every 1000 bp on average) to be all relevant. If a single species is scrutinized, many non-biologically relevant sites appear, as clearly shown from the monospecific analysis of the human beta-globin promoter, this tendency clearly disappearing when the human promoter is compared with that of other mammals. Comparison with chicken, on the other hand, fails to provide evidence for the existence of conserved regulatory elements, despite the fact that a significant (although weaker than intermammal) overall genomic conservation of expression levels between mammals and birds has been found (Cossedu *et al.*, 2004).

Long-range regulation of gene expression

From Loots and co-workers, we have seen that small DNA elements, most of them still unidentified, are able to modulate gene expression. How general is this phenomenon in vertebrate genomes? Recent studies in *Drosophila*, humans and domestic species indicate that long-range regulation of gene expression does exist in metazoan genomes. While the underlying mechanisms are still not well understood, they probably involve chromatin remodelling

over long genomic distances. In a paper by Caron and co-workers (2001), it was found that regions of high gene expression can exist across wide regions over 2 Mb in length, which have been called RIDGEs (region of increased gene expression) by the authors. In this paper, over 2.5 million SAGE sequences (serial analysis of gene expression), from 12 different tissues, were assembled in clusters and positioned on human chromosomes.

The biological reasons underlying these long-range variations of gene expression are not yet well understood, nevertheless, two trails may help to address the important question of long-distance gene regulation: the existence of chromosome territories localized at specific positions inside the cell nucleus, and the existence of genes modulating the overall level of gene expression. The existence of chromosome territories was hypothesized very early (Schardin *et al.*, 1985; for a historical review, Cremer *et al.*, 1993). This concept is consistent with the hypothesis that nuclear regions, although not visibly distinguishable, are not transcriptionally equivalent. The view proposed by Cremer and co-workers proposes that giant chromosomal loops protrude into the interchromatin compartment (IC), where transcription factors are concentrated, more generally near the centre of the nucleus. This view is somewhat supported at the level of entire chromosomes, since in humans, HSA19 (gene-rich, and with a high transcriptional activity, Caron *et al.*, 2001) localizes to the centre of the interphase nucleus, as revealed by fluorescent chromosome painting, while HSA18 (of equivalent size but gene-poor, and with a low transcriptional activity) localizes in the periphery of the nucleus.

The other trail lies in the recent discovery of specific genes (QTLs) affecting the overall level of gene transcription, for organisms as diverse as humans, maize and mice. In this study, the authors considered genomic transcription levels as quantitatively segregating characters accessible to genome scanning experiments. Several regions could be identified in this study (Schadt *et al.*, 2003), and in particular a QTL associated with obesity was identified in a murine cross. Such an approach suggests that the modulation of gene expression at the whole genome level is now accessible to genetic dissection.

Conclusion

Summarizing recent advances in mammalian genomics and sequencing is an almost impossible challenge, particularly in the limited space of a book chapter. However, after this overview it is clear that the next emerging field of genome biology can be summarized in a single word, 'regulation'. This word emphasizes what the future will be about: identifying as yet unknown genomic regions that influence the tuning of gene expression. By gene expression, I do not intend to mean solely the level of mRNA expression, but also the way the pre-mRNA will be spliced and/or edited to determine which protein will be the final end-product.

It clearly appears, therefore, that the true value of mammalian sequencing programmes will lie in the possibility of performing inter-specific comparative sequence analyses. Today, two genomes from very divergent mammalian species are available, which only give access to elements highly conserved by evolutionary pressure. In the not too distant future, other mammalian genomes will be completely sequenced and will provide insights into much more subtle differences. Besides the help to biomedical research, these achievements will certainly constitute efficient tools to crack the code that made us similar to other mammals, and yet different from them all, along continuous threads of 220 million years of mammalian evolutionary history.

References

Adams, M.D., Celniker, S.E., Holt, R.A., Evans, C.A., Gocayne, J.D., Amanatides, P.G., Scherer, S.E., Li, P.W., Hoskins, R.A., Galle, R.F., George, R.A., Lewis, S.E., Richards, S., Ashburner, M., Henderson, S.N., Sutton, G.G., Yandell, M.D., Zhang, Q., Chen, L.X. *et al.* (2000) The genome sequence of *Drosophila melanogaster*. *Science* 287, 2185–2195.

Arnheim, N., Krystal, M., Schmickel, R., Wilson, G., Ryder, O. and Zimmer, E. (1980) Molecular evidence for genetic exchanges among ribosomal genes on nonhomologous chromosomes in man and apes. *Proceedings of the National Academy of Sciences USA* 77, 7323–7327.

Beckman, J.S. and Weber, J.L. (1992) Survey of human and rat microsatellites. *Genomics* 12, 627–631.

Bernardi, G., Mouchiroud, D., Gautier, C. and Bernardi, G. (1998) Compositional patterns in vertebrate genomes: conservation and change in evolution. *Journal of Molecular Evolution* 28, 7–18.

Botstein, D., White, R.L., Skolnick, M. and Davis, R.W. (1980) Construction of a genetic linkage map in man using restriction fragment length polymorphisms. *American Journal of Human Genetics* 32, 314–331.

Bowles, J., Cooper, L., Berkman, J. and Koopman, P. (1999) Sry requires a CAG repeat domain for male sex determination in *Mus musculus*. *Nature Genetics* 22, 405–408.

Buck, L. and Axel, R. (1991) A novel multigene family may encode odorant receptors: a molecular basis for odor recognition. *Cell* 65, 175–187.

Buck, L.B. (2000) The molecular architecture of odor and pheromone sensing in mammals. *Cell* 100, 611–618.

Cappabianca, L., Thomassin, H., Pictet, R. and Grange, T. (1999) Genomic footprinting using nucleases. *Methods in Molecular Biology* 119, 427–442.

Carninci, P., Waki, K., Shiraki, T., Konno, H., Shibata, K., Itoh, M., Aizawa, K., Arakawa, T., Ishii, Y., Sasaki, D., Bono, H., Kondo, S., Sugahara, Y., Saito, R., Osato, N., Fukuda, S., Sato, K., Watahiki, A., Hirozane-Kishikawa, T., Nakamura, M. *et al.* (2003) Targeting a complex transcriptome: the construction of the mouse full-length cDNA encyclopedia. *Genome Research* 13, 1273–1279.

Caron, H., van Schaik, B., van der Mee, M., Baas, F., Riggins, G., van Sluis, P., Hermus, M.C., van Asperen, R., Boon, K., Voute, P.A., Heisterkamp, S., van Kampen, A. and Versteeg, R. (2001) The human transcriptome map: clustering of highly expressed genes in chromosomal domains. *Science* 291, 1289–1292.

C. elegans Sequencing Consortium (1998) Genome sequence of the nematode *C. elegans*: a platform for investigating biology. *Science* 282, 2012–2018.

Chen, R., Bouck, J.B., Weinstock, G.M. and Gibbs, R.A. (2001) Comparing vertebrate whole-genome shotgun reads to the human genome. *Genome Research* 11, 1807–1816.

Chess, A., Simon, I., Cedar, H. and Axel, R. (1994) Allelic inactivation regulates olfactory receptor gene expression. *Cell* 78, 823–834.

Clay, O. and Bernardi, G. (2002) Isochores: dream or reality? *Trends in Biotechnology* 20, 237.

Collins, J.E., Goward, M.E., Cole, C.G., Smink, L.J., Huckle, E.J., Knowles, S., Bye, J.M., Beare, D.M. and Dunham, I. (2003) Reevaluating human gene annotation: a second-generation analysis of chromosome 22. *Genome Research* 13, 27–36.

Constancia, M., Hemberger, M., Hughes, J., Dean, W., Ferguson-Smith, A., Fundele, R., Stewart, F., Kelsey, G., Fowden, A., Sibley, C. and Reik, W. (2002) Placental-specific IGF-II is a major modulator of placental and fetal growth. *Nature* 417, 945–948.

Cosseddu, G.M., Perez-Enciso, M., Fellous, M. and Vaiman, D. (2004) Interspecific chromosome-wide transcription profiles reveal the existence of mammalian-specific and species-specific chromosome domains. *Journal of Molecular Evolution* (in press).

Couronne, O., Poliakov, A., Bray, N., Ishkhanov, T., Ryaboy, D., Rubin, E., Pachter, L. and Dubchak, I. (2003) Strategies and tools for whole-genome alignments. *Genome Research* 13, 73–80.

Cremer, T. and Cremer, C. (2001) Chromosome territories, nuclear architecture and gene regulation in mammalian cells. *Nature Review Genetics* 2, 292–301.

Cremer, T., Kurz, A., Zirbel, R., Dietzel, S., Rinke, B., Schrock, E., Speicher, M.R., Mathieu, U., Jauch, A., Emmerich, P. *et al.* (1993) Role of chromosome territories in the functional compartmentalization of the cell nucleus. *Cold Spring Harbor Symposia on Quantitative Biology* 58, 777–792.

Dawkins, R. (1976) *The Selfish Gene.* Oxford University Press, Oxford.

De Baere, E., Beysen, D., Oley, C., Lorenz, B., Cocquet, J., De Sutter, P., Devriendt, K., Dixon, M., Fellous, M., Fryns, J.P., Garza, A., Jonsrud, C., Koivisto, P.A., Krause, A., Leroy, B.P., Meire, F., Plomp, A., Van Maldergem, L., De Paepe, A., Veitia, R. and Messiaen, L. (2003) FOXL2 and BPES: mutational hotspots, phenotypic variability, and revision of the genotype-phenotype correlation. *American Journal of Human Genetics* 72, 478–487.

Depatie, C., Lee, S.H., Stafford, A., Avner, P., Belouchi, A., Gros, P. and Vidal, S.M. (2000) Sequence-ready BAC contig, physical, and transcriptional map of a 2-Mb region overlapping the mouse chromosome 6 host-resistance locus Cmv1. *Genomics* 66, 161–174.

Dermitzakis, E.T., Reymond, A., Lyle, R., Scamuffa, N., Ucla, C., Deutsch, S., Stevenson, B.J., Flegel, V., Bucher, P., Jongeneel, C.V. and Antonarakis, S.E. (2002) Numerous potentially functional but non-genic conserved sequences on human chromosome 21. *Nature* 420, 578–582.

Dib, C., Faure, S., Fizames, C., Samson, D., Drouot, N., Vignal, A., Millasseau, P., Marc, S., Hazan, J., Seboun, E., Lathrop, M., Gyapay, G., Morissette, J. and Weissenbach, J. (1996) A comprehensive genetic map of the human genome based on 5,264 microsatellites. *Nature* 380, 152–154.

Ding, Y., Johnson, M.D., Chen, W.Q., Wong, D., Chen, Y.J., Benson, S.C., Lam, J.Y., Kim, Y.M. and Shizuya, H. (2001) Five-color-based high-information-content fingerprinting of bacterial artificial chromosome clones using type IIS restriction endonucleases. *Genomics* 74, 142–154.

Donis-Keller, H., Green, P., Helms, C., Cartinhour, S., Weiffenbach, B., Stephens, K., Keith, T.P., Bowden, D.W., Smith, D.R., Lander, E.S. *et al.* (1987) A genetic linkage map of the human genome. *Cell* 51, 319–337.

Enard, W., Przeworski, M., Fisher, S.E., Lai, C.S., Wiebe, V., Kitano, T., Monaco, A.P. and Paabo, S. (2002a) Molecular evolution of *FOXP2*, a gene involved in speech and language. *Nature* 418, 869–872.

Enard, W., Khaitovich, P., Klose, J., Zollner, S., Heissig, F., Giavalisco, P., Nieselt-Struwe, K., Muchmore, E., Varki, A., Ravid, R., Doxiadis, G.M., Bontrop, R.E. and Paabo, S. (2002b) Intra- and interspecific variation in primate gene expression patterns. *Science* 12, 340–343.

Esnault, C., Maestre, J. and Heidmann, T. (2000) Human LINE retrotransposons generate processed pseudogenes. *Nature Genetics* 24, 363–367.

Estoup, A., Garnery, L., Solignac, M. and Cornuet, J.M. (1995) Microsatellite variation in honey bee (*Apis mellifera* L.) populations: hierarchical genetic structure and test of the infinite allele and stepwise mutation models. *Genetics* 140, 679–695.

Ewing, B. and Green, P. (2000) Analysis of expressed sequence tags indicates 35,000 human genes. *Nature Genetics* 25, 232–234.

Forget, B.G. (1980/81) Structure and organization of the human globin genes. *Texas Report on Biology and Medicine* 40, 77–86.

Freking, B.A., Murphy, S.K., Wylie, A.A., Rhodes, S.J., Keele, J.W., Leymaster, K.A., Jirtle, R.L. and Smith, T.P. (2002) Identification of the single base change causing the callipyge muscle hypertrophy phenotype, the only known example of polar overdominance in mammals. *Genome Research* 12, 1496–1506.

Freitag, J., Ludwig, G., Andreini, I., Rossler, P. and Breer, H. (1998) Olfactory receptors in aquatic and terrestrial vertebrates. *Journal of Comparative Physiology [A]* 183, 635–650.

Fried, M. and Crothers, D.M. (1981) Equilibria and kinetics of lac repressor–operator interactions by polyacrylamide gel electrophoresis. *Nucleic Acids Research* 9, 6505–6525.

Fuchs, T., Malecova, B., Linhart, C., Sharan, R., Khen, M., Herwig, R., Shmulevich, D., Elkon, R., Steinfath, M., O'Brien, J.K., Radelof, U., Lehrach, H., Lancet, D. and Shamir, R. (2002) DEFOG: a practical scheme for deciphering families of genes. *Genomics* 80, 295–302.

Fujiyama, A., Watanabe, H., Toyoda, A., Taylor, T.D., Itoh, T., Tsai, S.F., Park, H.S., Yaspo, M.L., Lehrach, H., Chen, Z., Fu, G., Saitou, N., Osoegawa, K., de Jong, P.J., Suto, Y., Hattori, M. and Sakaki, Y. (2002) Construction and analysis of a human–chimpanzee comparative clone map. *Science* 295, 131–134.

Gentles, A.J. and Karlin, S. (2001) Genome-scale compositional comparisons in eukaryotes. *Genome Research* 11, 540–546.

Georges, M., Charlier, C. and Cockett, N. (2003) The callipyge locus: evidence for the trans interaction of reciprocally imprinted genes. *Trends in Genetics* 19, 248–252.

Gilbert, N. and Allan, J. (2001) Distinctive higher-order chromatin structure at mammalian centromeres. *Proceedings of the National Academy of Sciences USA* 98, 11949–11954.

Gill, P., Jeffreys, A.J. and Werrett, D.J. (1985) Forensic application of DNA 'fingerprints'. *Nature* 318, 577–579.

Glusman, G., Yanai, I., Rubin, I. and Lancet, D. (2001) The complete human olfactory subgenome. *Genome Research* 11, 685–702.

Glusman, G., Bahar, A., Sharon, D., Pilpel, Y., White, J. and Lancet, D. (2002) The olfactory receptor gene superfamily: data mining, classification, and nomenclature. *Mammalian Genome* 11, 1016–1023.

Godard, S., Vaiman, D., Oustry, A., Nocart, M., Bertaud, M., Guzylack, S., Meriaux, J.C., Cribiu, E.P. and Guerin, G. (1997) Characterization, genetic and physical mapping analysis of 36 horse plasmid and cosmid-derived microsatellites. *Mammalian Genome* 8, 745–750.

Green, E.D. (2001) Strategies for the systematic sequencing of complex genomes. *Nature Review Genetics* 2, 573–583.

Green, P. (2002) Whole-genome disassembly. *Proceedings of the National Academy of Sciences USA* 99, 4143–4144.

Guénet, J.L. and Bonhomme, F. (2003) Wild mice: an ever-increasing contribution to a popular mammalian model. *Trends in Genetics* 19, 24–31.

Gumucio, D.L., Heilstedt-Williamson, H., Gray, T.A., Tarle, S.A., Shelton, D.A., Tagle, D.A., Slightom, J.L., Goodman, M. and Collins, F.S. (1992) Phylogenetic footprinting reveals a nuclear protein which binds to silencer sequences in the human gamma and epsilon globin genes. *Molecular and Cellular Biology* 12, 4919–4929.

Hamada, H., Seidman, M., Howard, B.H. and Gorman, C.M. (1984) Enhanced gene expression by the poly(dT–dG)·poly(dC–dA) sequence. *Molecular and Cellular Biology* 4, 2622–2630.

Hardison, R.C., Oeltjen, J. and Miller, W. (1997) Long human–mouse sequence alignments reveal novel regulatory elements: a reason to sequence the mouse genome. *Genome Research* 7, 959–966.

Hardison, R.C., Roskin, K.M., Yang, S., Diekhans, M., Kent, W.J., Weber, R., Elnitski, L., Li, J., O'Connor, M., Kolbe, D., Schwartz, S., Furey, T.S., Whelan, S., Goldman, N., Smit, A., Miller, W., Chiaromonte, F. and Haussler, D. (2003) Covariation in frequencies of substitution, deletion, transposition, and recombination during eutherian evolution. *Genome Research* 13, 13–26.

Hashimshony, T., Zhang, J., Keshet, I., Bustin, M. and Cedar, H. (2003) The role of DNA methylation in setting up chromatin structure during development. *Nature Genetics* 34, 187–192.

Hood, L. (1977) The evolution of multigene families. *Advances in Pathobiology* 6, 51–67.

Hood, L.E., Hunkapiller, M.W. and Smith, L.M. (1987) Automated DNA sequencing and analysis of the human genome. *Genomics* 1, 201–212.

Hudson, T.J., Church, D.M., Greenaway, S., Nguyen, H., Cook, A., Steen, R.G., Van Etten, W.J., Castle, A.B., Strivens, M.A., Trickett, P., Heuston, C., Davison, C., Southwell, A., Hardisty, R., Varela-Carver, A., Haynes, A.R., Rodriguez-Tome, P., Doi, H., Ko, M.S., Pontius, J. *et al.* (2001) A radiation hybrid map of mouse genes. *Nature Genetics* 29, 201–205.

Hughes, A.L. and Nei, M. (1988) Pattern of nucleotide substitution at major histocompatibility complex class I loci reveals overdominant selection. *Nature* 335, 167–170.

Hughes, A.L., Hughes, M.K., Howell, C.Y. and Nei, M. (1994) Natural selection at the class II major histocompatibility complex loci of mammals. *Philosophical Transactions of the Royal Society of London B: Biological Sciences* 346, 359–366.

Hunkapiller, T. and Hood, L. (1989) Diversity of the immunoglobulin gene superfamily. *Advances in Immunology* 44, 1–63.

Jaffe, D.B., Butler, J., Gnerre, S., Mauceli, E., Lindblad-Toh, K., Mesirov, J.P., Zody. M.C. and Lander, E.S. (2003) Whole-genome sequence assembly for mammalian genomes: Arachne 2. *Genome Research* 13, 91–96.

Kasahara, M. (1997) New insights into the genomic organization and origin of the major histocompatibility complex: role of chromosomal (genome) duplication in the emergence of the adaptive immune system. *Hereditas* 127, 59–65.

Kimmel, M. and Chakraborty, R. (1996) Measures of variation at DNA repeat loci under a general stepwise mutation model. *Theoretical Population Biology* 50, 345–367.

Knott, S.A. and Haley, C.S. (1992) Maximum likelihood mapping of quantitative trait loci using full-sib families. *Genetics* 132, 1211–1222.

Kratz, E., Dugas, J.C. and Ngai, J. (2002) Odorant receptor gene regulation: implications from genomic organization. *Trends in Genetics* 18, 29–34.

Kruglyak, S., Durrett, R.T., Schug, M.D. and Aquadro, C.F. (1998) Equilibrium distributions of microsatellite repeat length resulting from a balance between slippage events and point mutations. *Proceedings of the National Academy of Sciences USA* 95, 10774–10778.

Lander, E.S. and Schork, N.J. (1994) Genetic dissection of complex traits. *Science* 265, 2037–2048. Review. Erratum in: *Science* 266, 353.

Lander, E.S., Linton, L.M., Birren, B., Nusbaum, C., Zody, M.C., Baldwin, J., Devon, K., Dewar, K., Doyle, M., FitzHugh, W., Funke, R., Gage, D., Harris, K., Heaford, A., Howland, J., Kann, L., Lehoczky, J., LeVine, R., McEwan, P., McKernan, K. *et al.*: International Human Genome Sequencing Consortium (2001) Initial sequencing and analysis of the human genome. *Nature* 409, 860–921.

Lane, R.P., Cutforth, T., Young, J., Athanasiou, M., Friedman, C., Rowen, L., Evans, G., Axel, R., Hood, L. and Trask, B.J. (2001) Genomic analysis of orthologous mouse and human olfactory receptor loci. *Proceedings of the National Academy of Sciences USA* 98, 7390–7395.

Larsen, F., Gundersen, G., Lopez, R. and Prydz, H. (1992) CpG islands as gene markers in the human genome. *Genomics* 13, 1095–1107.

Lenhard, B., Sandelin, A., Mendoza, L., Engstrom, P., Jareborg, N. and Wasserman, W.W. (2003) Identification of conserved regulatory elements by comparative genome analysis. *Journal of Biology* 2, 13.

Lenstra, J.A., van Boxtel, J.A., Zwaagstra, K.A. and Schwerin, M. (1993) Short interspersed nuclear element (SINE) sequences of the Bovidae. *Animal Genetics* 24, 33–39.

Levy, S., Hannenhalli, S. and Workman, C. (2001) Enrichment of regulatory signals in conserved non-coding genomic sequence. *Bioinformatics* 17, 871–877.

Li, Q., Harju, S. and Peterson, K.R. (1999) Locus control regions: coming of age at a decade plus. *Trends in Genetics* 15, 403–408.

Liang, F., Holt, I., Pertea, G., Karamycheva, S., Salzberg, S.L. and Quackenbush, J. (2000) Gene index analysis of the human genome estimates approximately 120,000 genes. *Nature Genetics* 25, 239–240.

Loots, G.G., Locksley, R.M., Blankespoor, C.M., Wang, Z.E., Miller, W., Rubin, E.M. and Frazer, K.A. (2000) Identification of a coordinate regulator of interleukins 4, 13, and 5 by cross-species sequence comparisons. *Science* 288, 136–140.

Maddox, J.F., Davies, K.P., Crawford, A.M., Hulme, D.J., Vaiman, D., Cribiu, E.P., Freking, B.A., Beh, K.J., Cockett, N.E., Kang, N., Riffkin, C.D., Drinkwater, R., Moore, S.S., Dodds, K.G., Lumsden, J.M., van Stijn, T.C., Phua, S.H., Adelson, D.L., Burkin, H.R., Broom, J.E. *et al.* (2001) An enhanced linkage map of the sheep genome comprising more than 1000 loci. *Genome Research* 11, 1275–1289.

Margolis, R.L. and Ross, C.A. (2001) Expansion explosion: new clues to the pathogenesis of repeat expansion neurodegenerative diseases. *Trends in Molecular Medicine* 7, 479–482.

Mollaaghababa, R. and Pavan, W.J. (2003) The importance of having your SOX on: role of SOX10 (dagger) in the development of neural crest-derived melanocytes and glia. *Oncogene* 22, 3024–3034.

Mombaerts, P. (1999) Seven-transmembrane proteins as odorant and chemosensory receptors. *Science* 286, 707–711.

Mullikin, J.C. and Ning, Z. (2003) The phusion assembler. *Genome Research* 13, 81–90.

Myers, E.W., Sutton, G.G., Smith, H.O., Adams, M.D. and Venter, J.C. (2002) On the sequencing and assembly of the human genome. *Proceedings of the National Academy of Sciences USA* 99, 4145–4146.

Nadeau, J.H. and Taylor, B.A. (1984) Lengths of chromosomal segments conserved since divergence of man and mouse. *Proceedings of the National Academy of Sciences USA* 81, 814–818.

Nikaido, M., Matsuno, F., Abe, H., Shimamura, M., Hamilton, H., Matsubayashi, H. and Okada, N. (2001) Evolution of CHR-2 SINEs in cetartiodactyl genomes: possible evidence for the monophyletic origin of toothed whales. *Mammalian Genome* 12, 909–915.

Numata, K., Kanai, A., Saito, R., Kondo, S., Adachi, J., Wilming, L.G., Hume, D.A., Hayashizaki, Y. and Tomita, M. (2003) Identification of putative noncoding RNAs among the RIKEN mouse full-length cDNA collection. *Genome Research* 13, 1301–1306.

Ohno, S. and Yomo, T. (1991) The grammatical rule for all DNA: junk and coding sequences. *Electrophoresis* 12, 103–108.

Okada, N. and Hamada, M. (1997) The 3′ ends of tRNA-derived SINEs originated from the 3′ ends of LINEs: a new example from the bovine genome. *Journal of Molecular Evolution* 44, Suppl. 1, S52–S56.

Osborne, C.S., Vadas, M.A. and Cockerill, P.N. (1995) Transcriptional regulation of mouse granulocyte–macrophage colony-stimulating factor/IL-3 locus. *Journal of Immunology* 155, 226–235.

Ostertag, E.M., DeBerardinis, R.J., Goodier, J.L., Zhang, Y., Yang, N., Gerton, G.L. and Kazazian, H.H., Jr (2002) A mouse model of human L1 retrotransposition. *Nature Genetics* 32, 655–660.

Ovchinnikov, I., Troxel, A.B. and Swergold, G.D. (2001) Genomic characterization of recent human LINE-1 insertions: evidence supporting random insertion. *Genome Research* 11, 2050–2058.

Parham, P. and Ohta, T. (1996) Population biology of antigen presentation by MHC class I molecules. *Science* 272, 67–74.

Pavlicek, A., Paces, J., Clay, O. and Bernardi, G. (2002) A compact view of isochores in the draft human genome sequence. *FEBS Letters* 511, 165–169.

Pevny, L.H. and Lovell-Badge, R. (1997) Sox genes find their feet. *Current Opinion in Genetics and Development* 7, 338–344.

Pfeifer, D., Kist, R., Dewar, K., Devon, K., Lander, E.S., Birren, B., Korniszewski, L., Back, E. and Scherer, G. (1999) Campomelic dysplasia translocation breakpoints are scattered over 1 Mb proximal to SOX9: evidence for an extended control region. *American Journal of Human Genetics* 65, 111–124.

Piontkivska, H. and Nei, M. (2003) Birth-and-death evolution in primate MHC class I genes: divergence time estimates. *Molecular Biology and Evolution* 20, 601–609.

Ranum, L.P.W. and Day, J.W. (2002) Dominantly inherited, non-coding microsatellite expansion disorders. *Current Opinion in Genetics and Development* 12, 266–271.

Reed, R.R. (2000) Regulating olfactory receptor expression: controlling globally, acting locally. *Nature Neuroscience* 3, 638–639.

Rippe, R.A., Brenner, D.A. and Tugores, A. (2001) Techniques to measure nucleic acid–protein binding and specificity. Nuclear extract preparations, DNase I footprinting, and mobility shift assays. *Methods in Molecular Biology* 160, 459–479.

Rohrer, G.A., Alexander, L.J., Hu, Z., Smith, T.P., Keele, J.W. and Beattie, C.W. (1996) A comprehensive map of the porcine genome. *Genome Research* 6, 371–391.

Roest Crollius, H., Jaillon, O., Bernot, A., Dasilva, C., Bouneau, L., Fischer, C., Fizames, C., Wincker, P., Brottier, P., Quetier, F., Saurin, W. and Weissenbach, J. (2000) Estimate of human gene number provided by genome-wide analysis using *Tetraodon nigroviridis* DNA sequence. *Nature Genetics* 25, 235–238.

Saluz, H.P. and Jost, J.P. (1993) Approaches to characterize protein–DNA interactions *in vivo*. *Critical Reviews in Eukaryotic Gene Expression* 3, 1–29.

Sargan, D.R., Sampson, J. and Binns, M.M. (2001) Molecular genetics of the dog. In: Ruvinsky, A. and Sampson, J. (eds) *The Genetics of the Dog*. CAB International, Wallingford, UK, pp. 139–157.

Schadt, E.E., Monks, S.A., Drake, T.A., Lusis, A.J., Che, N., Colinayo, V., Ruff, T.G., Milligan, S.B., Lamb, J.R., Cavet, G., Linsley, P.S., Mao, M., Stoughton, R.B. and Friend, S.H. (2003) Genetics of gene expression surveyed in maize, mouse and man. *Nature* 422, 297–302.

Schardin, M., Cremer, T., Hager, H.D. and Lang, M. (1985) Specific staining of human chromosomes in Chinese hamster × man hybrid cell lines demonstrates interphase chromosome territories. *Human Genetics* 71, 281–287.

Schlötterer, C. (2000) Evolutionary dynamics of microsatellite DNA. *Chromosoma* 109, 365–371.

Schlötterer, C. and Tautz, D. (1992) Slippage synthesis of simple sequence DNA. *Nucleic Acids Research* 20, 211–215.

Serizawa, S., Ishii, T., Nakatani, H., Tsuboi, A., Nagawa, F., Asano, M., Sudo, K., Sakagami, J., Sakano, H., Ijiri, T., Matsuda, Y., Suzuki, M., Yamamori, T., Iwakura, Y. and Sakano, H. (2000) Mutually exclusive expression of odorant receptor transgenes. *Nature Neuroscience* 3, 687–693.

Shafrir, E. (1996) Monod, Jacob and Lwoff, introducers of new dimensions in cellular genetics and molecular biology. *Israeli Journal of Medical Science* 32, 162.

Shimamura, M., Yasue, H., Ohshima, K., Abe, H., Kato, H., Kishiro, T., Goto, M., Munechika, I. and Okada, N. (1997) Molecular evidence from retroposons that whales form a clade within even-toed ungulates. *Nature* 388, 666–670.

Shimamura, M., Abe, H., Nikaido, M., Ohshima, K. and Okada, N. (1999) Genealogy of families of SINEs in cetaceans and artiodactyls: the presence of a huge superfamily of tRNA(Glu)-derived families of SINEs. *Molecular Biology and Evolution* 16, 1046–1060.

Shioiri, C. and Takahata, N. (2001) Skew of mononucleotide frequencies, relative abundance of dinucleotides, and DNA strand asymmetry. *Journal of Molecular Evolution* 53, 364–376.

Shizuya, H., Birren, B., Kim, U.J., Mancino, V., Slepak, T., Tachiiri, Y. and Simon, M. (1992) Cloning and stable maintenance of 300-kilobase-pair fragments of human DNA in *Escherichia coli* using an F-factor-based vector. *Proceedings of the National Academy of Sciences USA* 89, 8794–8797.

Sidow, A. (2002) Sequence first. Ask questions later. *Cell* 111, 13–16.

Smit, M., Segers, K., Carrascosa, L.G., Shay, T., Baraldi, F., Gyapay, G., Snowder, G., Georges, M., Cockett, N. and Charlier, C. (2003) Mosaicism of solid gold supports the causality of a noncoding A-to-G transition in the determinism of the callipyge phenotype. *Genetics* 163, 453–456.

Stormo, G.D. (2000) DNA binding sites: representation and discovery. *Bioinformatics* 16, 16–23.

Subramanian, S., Mishra, R.K. and Singh, L. (2003) Genome-wide analysis of microsatellite repeats in humans: their abundance and density in specific genomic regions. *Genome Biology* 4, R13.

Thomas, J.W. and Touchman, J.W. (2002) Vertebrate genome sequencing: building a backbone for comparative genomics. *Trends in Genetics* 18, 104–108.

Toth, G., Gaspari, Z. and Jurka, J. (2000) Microsatellites in different eukaryotic genomes: survey and analysis. *Genome Research* 10, 967–981.

Tronche, F., Ringeisen, F., Blumenfeld, M., Yaniv, M. and Pontoglio, M. (1997) Analysis of the distribution of binding sites for a tissue-specific transcription factor in the vertebrate genome. *Journal of Molecular Biology* 266, 231–245.

Vaiman, D., Mercier, D., Moazami-Goudarzi, K., Eggen, A., Ciampolini, R., Lepingle, A., Velmala, R., Kaukinen, J., Varvio, S.L., Martin, P. *et al.* (1994) A set of 99 cattle microsatellites: characterization, synteny mapping, and polymorphism. *Mammalian Genome* 5, 288–297.

Vaiman, D., Schibler, L., Bourgeois, F., Oustry, A., Amigues, Y. and Cribiu, E.P. (1996) A genetic linkage map of the male goat genome. *Genetics* 144, 279–305.

Valdes, A.M., Slatkin, M. and Freimer, N.B. (1993) Allele frequencies at microsatellite loci: the stepwise mutation model revisited. *Genetics* 133, 737–749.

Varvio, S.L. and Kaukinen, J. (1993) Bovine microsatellites: racial differences and association with SINE-elements. *Experientia. Supplementum* 67, 437–443.

Venter, J.C., Adams, M.D., Myers, E.W., Li, P.W., Mural, R.J., Sutton, G.G., Smith, H.O., Yandell, M., Evans, C.A., Holt, R.A., Gocayne, J.D., Amanatides, P., Ballew, R.M., Huson, D.H., Wortman, J.R., Zhang, Q., Kodira, C.D., Zheng, X.H., Chen, L., Skupski, M. *et al.* (1997) The sequence of the human genome. *Science* 291, 1304–1351.

Wagner, A. (1997) A computational genomics approach to the identification of gene networks. *Nucleic Acids Research* 25, 3594–3604.

Wagner, K.D., Wagner, N. and Schedl, A. (2003) The complex life of WT1. *Journal of Cell Science* 116, 1653–1658.

Wasserman, W.W., Palumbo, M., Thompson, W., Fickett, J.W. and Lawrence, C.E. (2000) Human–mouse genome comparisons to locate regulatory sites. *Nature Genetics* 26, 225–228.

Waterston, R.H., Lindblad-Toh, K., Birney, E., Rogers, J., Abril, J.F., Agarwal, P., Agarwala, R., Ainscough, R., Alexandersson, M., An, P., Antonarakis, S.E., Attwood, J., Baertsch, R., Bailey, J., Barlow, K., Beck, S., Berry, E., Birren, B., Bloom, T., Bork, P. *et al.* Mouse Genome Sequencing Consortium (2002a) Initial sequencing and comparative analysis of the mouse genome. *Nature* 420, 520–562.

Waterston, R.H., Lander, E.S. and Sulston, J.E. (2002b) On the sequencing of the human genome. *Proceedings of the National Academy of Sciences USA* 99, 3712–3716.

Weber, J.L. (1990) Informativeness of human (dC–dA)n·(dG–dT)n polymorphisms. *Genomics* 7, 524–530.

Weber, J.L. and Myers, E.W. (1997) Human whole-genome shotgun sequencing. *Genome Research* 7, 401–409.

Webster, M.T., Smith, N.G. and Ellegren, H. (2002) Microsatellite evolution inferred from human–chimpanzee genomic sequence alignments. *Proceedings of the National Academy of Sciences USA* 99, 8748–8753.

Weissenbach, J., Gyapay, G., Dib, C., Vignal, A., Morissette, J., Millasseau, P., Vaysseix, G. and Lathrop, M. (1992) A second-generation linkage map of the human genome. *Nature* 359, 794–801.

Wingender, E., Chen, X., Fricke, E., Geffers, R., Hehl, R., Liebich, I., Krull, M., Matys, V., Michael, H., Ohnhauser, R., Pruss, M., Schacherer, F., Thiele, S. and Urbach, S. (2001) The TRANSFAC system on gene expression regulation. *Nucleic Acids Research* 29, 281–283.

Wurster, D.H. and Benirschke, K. (1968) The chromosomes of the great Indian rhinoceros (*Rhinoceros unicornis* L.). *Experientia* 24, 511.

Wurster, D.H. and Benirschke, K. (1970) Indian muntjac, *Muntiacus muntjac*: a deer with a low diploid chromosome number. *Science* 168, 1364–1366.

Zhang, X. and Firestein, S. (2002) The olfactory receptor gene superfamily of the mouse. *Nature Neuroscience* 5, 124–133.

Zozulya, S., Echeverri, F. and Nguyen, T. (2001) The human olfactory receptor repertoire. *Genome Biology* 2, 0018.1–0018.12.

5   The Transcriptome

A. Verger* and M. Crossley*

Molecular and Microbial Biosciences, University of Sydney, NSW 2006, Australia

Introduction	118
Gene Regulatory Elements: *Cis*-acting Promoter and Enhancer Sequences	119
Promoters and enhancers	119
The structure of the core promoter	120
Locus control regions	122
The RNA Polymerase II Complex and General Transcription Factors	122
Core RNA polymerase II	122
Basal/general transcription factors	123
Mediator complexes	124
Chromatin in Transcriptional Regulation	124
Chromatin remodelling complexes	124
Chromatin modifying enzymes	126
Cooperation between chromatin remodelling factors and modifying complexes	129
DNA methylation	130
Sequence-specific DNA-binding Transcription Factors	131
Sequence-specific DNA-binding domains	131
Activation and repression domains	133
Regulating Gene Expression	134
Developmental and tissue-specific gene expression	134
Regulating the regulators	135
Nuclear compartmentalization and gene activity	137
The Non-coding RNA World	137
Dosage compensation in mammals	137
Gene silencing by the non-coding RNA *Air*	138
RNA interference	138
Mechanism of RNA interference	139
Double-stranded RNA and heterochromatin	139
Post-transcriptional Regulation and Control of Splicing	140
'Gene expression factories'	140
Splicing generates diversity	142

*Correspondence: A.Verger@mmb.usyd.edu.au, M.Crossley@mmb.usyd.edu.au

© CAB International 2005. *Mammalian Genomics*
(eds A. Ruvinsky and J. Marshall Graves)

Conclusion   142
   Current methodologies, results, limitations   143
   Past limitations and future prospects   143
References   143

Introduction

The ability of multicellular organisms to control the output of their genomes and determine which genes are expressed at a given time, in a particular organ and in response to different physiological stimuli is central to many biological processes, including tissue differentiation, organogenesis, metabolic control and disease. Gene output also needs to respond to external stimuli, as it must in single-celled organisms, but additionally patterns of gene expression must be carefully coordinated during complex developmental histories. The expression of some genes will remain fixed for long periods once different cell types have terminally differentiated in the adult organism but other genes will continue to respond to physiological and environmental stimuli. Some genes involved in ongoing processes such as haematopoiesis will continue to be dynamically regulated throughout life. Although mammals have a very extensive range of cell types, the same genetic information is contained within every cell (except specialized cells of the immune system), so sophisticated strategies (Fig. 5.1) are required to ensure the proper differential and controlled regulation of gene expression during the mammalian life cycle.

The total number of genes present in the mammalian genome, estimated at around 30,000 protein-coding genes, and perhaps several thousand non-coding RNA transcripts, further highlights the complexity of regulation required (Ewing and Green, 2000; Roest Crollius *et al.*, 2000; Kapranov *et al.*, 2002; Collins *et al.*, 2003). The absolute size of the genome and the necessity for it to be compacted into the nuclear space further complicates the process of regulating gene expression. In simple terms, it appears that the primary point of gene control usually occurs at the level of transcriptional initiation. Major questions that arise, therefore, are how does the molecular machinery required to transcribe a gene, the RNA polymerase and its associated proteins identify the beginning of each gene against the vast landscape of highly packaged DNA sequence, what triggers the initiation of transcription, and how is this regulated?

The coordinated assembly of active transcription complexes (Fig. 5.1, step 1) requires

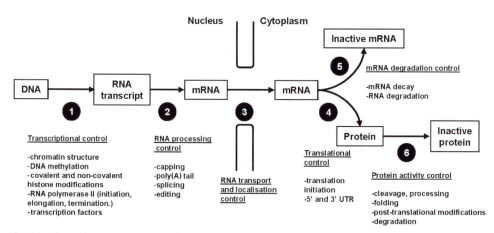

Fig. 5.1. The major steps in gene regulation.

the interplay of multiple *cis*-acting promoter and enhancer sequence elements, a large number of sequence-specific DNA-binding proteins, and recruited chromatin remodelling and modifying factors (reviewed in Lemon and Tjian, 2000; Orphanides and Reinberg, 2002). Subsequent steps in gene expression, such as the mRNA processing reactions of capping, splicing and polyadenylation (step 2 in Fig. 5.1) occur co-transcriptionally. Interestingly, the processes of RNA transcription and processing appear to be linked (reviewed in Maniatis and Reed, 2002; Proudfoot *et al.*, 2002; Reed, 2003). Thus, whereas Fig. 5.1 shows a simple linear assembly line, recent studies suggest that a complex and extensively coupled network has evolved to coordinate the steps involved in gene expression. The extensive coupling is consistent with a model in which the different molecular machines involved in transcription and RNA processing are tethered to each other to form 'gene expression factories' (see below).

In this chapter, selected examples of gene regulatory mechanisms will be discussed with a particular focus on recent progress in elucidating the mechanisms by which the control of specific protein-coding genes is achieved (essentially steps 1 and 2, Fig. 5.1). For a more detailed analysis of additional aspects of gene regulation, the reader is directed to some excellent reviews (Lee and Young, 2000; Lemon and Tjian, 2000; Naar *et al.*, 2001; Roth *et al.*, 2001; Zhang and Reinberg, 2001; Cosma, 2002; Lachner and Jenuwein, 2002; Narlikar *et al.*, 2002; Orphanides and Reinberg, 2002; Fischle *et al.*, 2003; Gaston and Jayaraman, 2003) and to other chapters of this book.

Gene Regulatory Elements: *Cis*-acting Promoter and Enhancer Sequences

Regulation of gene expression at the level of transcriptional initiation is achieved primarily through the binding of regulatory proteins, called transcription factors (see below), to the promoter, enhancer or additional control regions that direct the expression of the (usually) adjacent gene. The regulatory transcription factors directly or indirectly facilitate the binding of RNA polymerase immediately upstream of the gene to be transcribed. There are three nuclear RNA poly-

merases in mammals, each responsible for the transcription of a different set of genes. RNA polymerase I transcribes only the rRNA genes (reviewed in Grummt, 2003), RNA polymerase II transcribes predominantly protein-coding genes, resulting in the production of mRNA (reviewed in Woychik and Hampsey, 2002), and RNA polymerase III transcribes predominantly tRNA genes and genes encoding small RNAs (reviewed in Schramm and Hernandez, 2002). This chapter will focus on regulation of RNA polymerase II transcription. In this section, the DNA elements involved will be summarized, the basic process of RNA polymerase II transcription will be described, and some of the essential proteins will be discussed.

Promoters and enhancers

Metazoan genes contain highly structured regulatory DNA sequences that direct complex patterns of expression in many different cell types during development (reviewed in Levine and Tjian, 2003). There are at least two features common to most promoters of protein-coding genes: the core promoter element occupied by RNA polymerase and associated proteins and additional sequences bound by regulatory proteins, termed transcription factors. These latter sequences include enhancers, silencers, insulators and proximal promoter elements (Fig. 5.2A). A typical animal gene may contain several enhancers and these can be located either 5' or 3' of the gene or within its introns. Each enhancer may be responsible for a subset of the total gene expression pattern, possibly for expression in a single cell type or tissue. A typical enhancer might be around 500 bp in length and may contain in the order of ten binding sites for at least three different sequence-specific DNA-binding transcription factors (reviewed in Levine and Tjian, 2003).

The core promoter is localized to the ~100 nucleotides surrounding the transcription start site, which is defined as the first transcribed nucleotide (Fig. 5.2A). The core promoter defines the site over which a large multiprotein transcription complex assembles. This complex, called the pre-initiation complex (PIC), contains the basal or general transcription factors (GTFs),

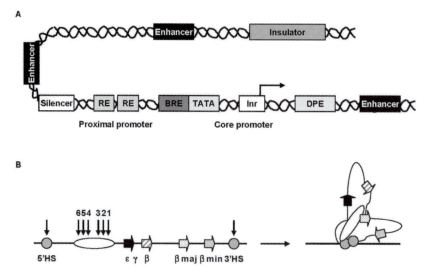

Fig. 5.2. Gene regulatory elements. (A) A typical mammalian regulatory module. A complex arrangement of multiple clustered enhancer modules interspersed with silencer and insulator elements which can be located 10–50 kb either upstream or downstream of a composite core promoter containing the TATA box (TATA), initiator sequences (Inr), downstream promoter elements (DPE) and TFIIB recognition element (BRE). RE, responsive element. Adapted from Levine and Tjian (2003). (B) Looping interactions in the murine *β-globin* locus. A schematic representation of the murine *β-globin* locus is shown. Vertical arrows indicate DNase I-hypersensitive sites (HS) present in the locus control region (LCR) and at the 5′ and 3′ ends of the locus. Horizontal arrows indicate the four *globin* genes. Chromosome conformation capture (3C) analysis demonstrated interactions between 5′HS, LCR, active *globin* genes and 3′HS, leading Tolhuis *et al.* (2002) to propose that all those sites cluster to form an 'active chromatin hub'.

in addition to RNA polymerase II itself, and comprises the core transcription machinery (Fig. 5.3). After formation of the PIC, the next step in transcriptional initiation is localized melting of the duplex promoter DNA to form an open complex. Formation of the first phosphodiester bond can then occur. Promoter clearance follows and, when approximately 10–40 nucleotides have been added to the transcript, a transition to processive elongation occurs. Transcriptional elongation continues until termination signals are reached (reviewed in Lee and Young, 2000).

The structure of the core promoter

Usually, at least one of several important sequence motifs, including the TATA box, initiator (Inr), and downstream core promoter element (DPE), is found within the core promoter (Fig. 5.2A). These motifs each play roles in assembling and aligning the initiation complex in anticipation of transcription (reviewed in

Butler and Kadonaga, 2002; Smale and Kadonaga, 2003).

The TATA element is a DNA sequence located 25–30 bp upstream from the transcription start site. This element is the binding site for a DNA-binding protein termed the TATA-binding protein (TBP). The consensus DNA sequence for the TATA element is TATAAA, but some derivatives of this sequence can also function as a TATA box (Singer *et al.*, 1990). TBP is associated with up to 14 accessory proteins termed TAFs (TBP-associated factors) and the complex of TBP and its TAFs is termed TFIID (reviewed in Albright and Tjian, 2000). Binding of TFIID to the TATA element is an important step in the formation of the PIC. It appears that TBP is the predominant TATA box-binding protein, but it is also important to note that there are a small number of TBP-related factors (TRFs) that are thought to be involved in the regulation of genes in specific tissues (reviewed in Hochheimer and Tjian, 2003).

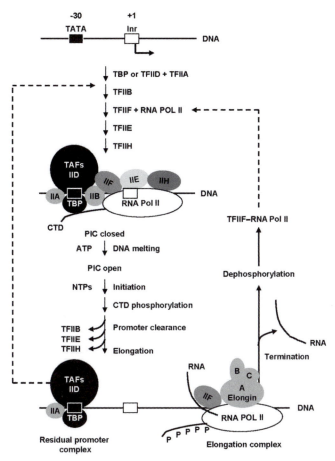

Fig. 5.3. Order of assembly of the basal polymerase II factors. Promoter recognition is achieved by binding of TBP to the TATA box. DNA-bound TBP recruits TFIIA and TFIIB. TFIIB, in turn, provides the docking site for a pre-formed complex of TFIIF bound to the RNA polymerase II. TFIIE and TFIIH are recruited last through interactions with the RNA polymerase II enzyme. After the pre-initiation complex (PIC) assembly, the C-terminal domain (CTD) of the RNA polymerase II is phosphorylated by TFIIH. As the RNA polymerase II moves away (elongation) from the promoter, TFIIB, TFIIF and TFIIH are released into solution and the elongator complex is recruited.

A second element found in the core promoter is the initiator element or Inr. The Inr surrounds the transcription start site itself and is pyrimidine-rich, but has no clearly defined consensus sequence. The Inr element is a DNA element distinct from the TATA element but it seems to have a related function in that it also contributes to the nucleation and proper placement of the PIC. Promoters may contain either TATA or Inr elements, both, or neither (reviewed in Butler and Kadonaga, 2002; Smale and Kadonaga, 2003). A variety of factors has been found to interact with the Inr element. In some instances, particular subunits of the TFIID complex appear to make contact with the Inr element (Oelgeschlager *et al.*, 1996) but purified RNA polymerase II and the sequence-specific DNA-binding factors TFII-I and YY1 have also been reported to recognize the Inr and contribute to transcriptional initiation (Cheriyath *et al.*, 1998; Weis and Reinberg, 1997).

A third core promoter element is the downstream promoter element (DPE). This is a 7-nucleotide sequence positioned about 30 nucleotides downstream from the transcription start site (Fig. 5.2A). The DPE has been shown to be required for the binding of TFIID to a subset of promoters that do not contain TATA elements (reviewed in Kadonaga, 2002). Other promoters may contain additional elements, such as the TFIIB recognition element (BRE) that is located immediately upstream of some TATA boxes (Lagrange et al., 1998), and binds a protein termed TFIIB (discussed further below). TFIIB interacts with TFIID and thus also may help facilitate formation and positioning of the PIC. As can be seen from the discussion above, there are several different ways in which TFIID can be positioned at the core promoter and accordingly, promoters may differ in which subset of these elements are found.

Certain promoters, often referred to as TATA-less GC-rich promoters, seem to be quite different in that these promoters contain a number of GC-rich elements instead of or in addition to some of the elements described above. GC-rich promoters are a subset of so-called CpG islands – sequences ranging in size from 0.5 to 2 kbp which have a higher proportion of CpG dinucleotides than other regions of the genome. So-called 'housekeeping genes', such as those encoding basic metabolic enzymes that are required in all cell types, often have GC-rich promoters, though such promoters are also sometimes found associated with tissue-restricted genes (for example, the human α-globin gene has a GC-rich promoter (Fischel-Ghodsian et al., 1987)). Interestingly, GC-rich promoters often generate a cluster of different start points of transcription, perhaps suggesting that in the absence of a TATA box, the precise positioning of RNA polymerase does not occur. Nevertheless, although the start point may be imprecise, there is no evidence that these promoters are less efficient than the typical TATA-containing promoters.

Locus control regions

Locus control regions (LCRs) are another interesting type of genetic regulatory element similar to enhancers in that they consist of multiple activator binding sites. However, the two elements differ in that classical enhancers are orientation- and distance-independent, though in some assays their effect can depend on the site of integration into native chromatin, apparently because local chromatin structure can dominate the function of the enhancer (this variation has been termed position effect variegation (PEV), see Chapter 7 for more details). In contrast, LCRs stimulate transcription independent of their site of integration in native chromatin, although their effects are limited by orientation and distance (reviewed in Li et al., 1999).

The particular potency of LCRs to activate transcription is commonly ascribed to a dominant chromatin opening activity that functions in a position-independent fashion. Two new methods have recently been developed to analyse the β-globin locus and its LCR and suggest that the formation of specific chromosome structural features is involved in gene activation (reviewed in Dekker, 2003). Briefly, the murine β-globin locus contains four globin-like genes that are differentially expressed during development, an LCR which is characterized by multiple transcription factor-binding sites and DNase I-hypersensitive sites (HS), as well as potential boundary elements which are located at both ends of the locus (5'HS and 3'HS) (Fig. 5.2B). Using the chromosome conformation capture technique (3C), Tolhuis and co-workers (Tolhuis et al., 2002) compared the conformation of the active and repressed states of the locus. This analysis revealed interactions between 5'HS and the LCR, the active globin genes and 3'HS and clearly demonstrates that distant elements can physically interact to control gene activity, as predicted by the original looping models (Fig. 5.2B) (reviewed in Bulger and Groudine, 1999). Similar analyses of other genomic regions will probably reveal that sophisticated chromosome structures are involved in controlling gene expression.

The RNA Polymerase II Complex and General Transcription Factors

Core RNA polymerase II

Mammalian RNA polymerase II is made up of 12 subunits (Table 5.1 and Fig. 5.3). There is evidence that different subunits play specific roles in determining start site selection, tran-

Table 5.1. RNA polymerase II and the general transcription factors.

Factor	Subunit	Features
RNA polymerase II	12 subunits (hRPB1–12)	
TFIID	TBP and TAFs	Promoter selection, binds DPE
TFIIA	TFIIα, TFIIβ, TFIIγ	Stabilizes TBP–DNA interaction. Interacts with numerous activators
TFIIB	TFIIB	Selection of transcription start sites, binds BRE
TFIIF	RAP74 and RAP30	Stabilizes the pre-initiation complex, DNA wrapping ability
TFIIE	TFIIEa, TFIIEb	Plays a role in melting of promoter DNA, stimulates the CTD kinase and ATPase activities of TFIIH
TFIIH	P62, p52, MAT1, p34, XPD/ERCC2, XPB/ERCC3, Cdk7, cyclin H	DNA-dependent ATPase, ATP-dependent helicase and CTD kinase. Phosphorylates the CTD of RNA polymerase II

Adapted from Lee and Young (2000).

scriptional elongation rates and interactions with transcriptional activators (reviewed in Lee and Young, 2000). The largest subunit of the RNA polymerase II contains a unique C-terminal repeat domain (CTD) that consists of tandem repeats of a consensus heptapeptide sequence Tyr-Ser-Pro-Thr-Ser-Pro-Ser (reviewed in Majello and Napolitano, 2001). The functions of the CTD are closely associated with the phosphorylation state of the domain. The CTD exists in at least two phosphorylation states. RNA polymerase II molecules lacking phosphate on the CTD are found in initiation complexes, while elongating polymerase molecules contain heavily phosphorylated CTDs (Dahmus, 1996). The switch in CTD phosphorylation states that occurs between initiation and elongation appears to allow the RNA polymerase II molecule to switch co-factors. For example, proteins involved in initiation (such as the mediator, discussed below) are tightly associated with RNA polymerase II molecules that lack phosphate on their CTDs (Sun *et al.*, 1998). In contrast, an elongator complex and various RNA processing factors become associated with RNA polymerase II molecules that possess hyperphosphorylated CTDs (Otero *et al.*, 1999).

Basal/general transcription factors

In addition to TFIID, a number of other basal or general transcription factors (GTFs) are required for specific promoter binding and initiation by RNA polymerase II, including TFIIA, TFIIB, TFIIE, TFIIF and TFIIH (Table 5.1 and Fig. 5.3) (reviewed in Gill, 2001). A pre-initiation complex containing the GTFs and RNA polymerase II can be assembled in a stepwise fashion on promoters *in vitro* (Fig. 5.3). Of the six GTFs mentioned above, TFIID, which contains the TBP protein, is usually thought to be the first to bind to the core promoter and to facilitate the assembly of the remaining components. In the current model of PIC assembly (Fig. 5.3), TFIIA and TFIIB contact TBP and DNA, thereby increasing the stability and sequence specificity of TBP binding. Next, a pre-formed complex of TFIIF and RNA polymerase II is recruited followed by TFIIE and TFIIH. TFIIH is required for DNA melting at the transcriptional start site and to enable RNA polymerase II to initiate transcription and clear the promoter (Fig. 5.3).

The current view is that the regulation of transcriptional initiation largely reflects the frequency with which the GTFs and RNA polymerase are recruited to the core promoter (Ptashne and Gann, 1997). Additionally, certain regulatory proteins could have a direct influence on the conformation of particular GTFs and/or DNA structure (Chi and Carey, 1996). Other studies support the idea that some regulatory proteins stimulate promoter clearance, thereby accelerating the entrance of new RNA polymerases so that multiple mRNAs can be produced in quick succession (Sandaltzopoulos and Becker, 1998).

Mediator complexes

Another large protein complex termed the mediator is also found at the core promoter in addition to the subunits of RNA polymerase II itself and the GTFs (reviewed in Naar et al., 2001; Boube et al., 2002). The Mediator proteins appear to link RNA polymerase II to sequence-specific gene regulatory proteins that we discuss below. Mediator complexes were initially purified from yeast as proteins that helped stimulate specific transcriptional activity in reconstituted transcription assays. Importantly, the yeast mediator complex was also found to interact with the CTD of RNA polymerase II and to stimulate TFIIH-dependent phosphorylation of the CTD (Myers et al., 1998). Yeast mediator complexes may be recruited by several different sequence-specific regulatory proteins (Myers et al., 1999), and transmit activator signals to the basal transcriptional apparatus.

A number of metazoan complexes distantly related to the yeast mediator have now been isolated. These include TRAP/SMCC (TR-associated protein/SRB/MED-containing factor), ARC (activator-recruited cofactor), DRIP (vitamin D receptor-interacting protein), NAT (negative regulator of activated transcription), CRSP (cofactor required for SP1 activation) and a human Sur2-containing complex (reviewed in Lee and Young, 2000). SMCC, for example, was isolated by immunopurification on the basis of its ability to mediate activation by a Gal4-activator fusion protein (Gu et al., 1999). ARC was identified as a cofactor required for synergistic activation by the sequence-specific DNA-binding protein and transcriptional activator Sp1 and SREBP-1a activators (Naar et al., 1999), whereas DRIP was originally identified as a complex of proteins that interacts specifically with ligand-bound vitamin D receptor, which is itself a sequence-specific DNA-binding activator protein (Rachez et al., 1998).

The functional similarity of these different complexes and the identity of specific subunits indicate that they are related but the complexes are not identical. It is possible that each is derived from a single in vivo complex, and that differences in biochemical fractionation account for the differences in composition observed. More likely perhaps is the possibility that different mediator complexes exist in different cells, or complexes of different subunit composition are recruited by different sequence-specific DNA-binding proteins.

Chromatin in Transcriptional Regulation

Mammalian DNA must be tightly packaged in order to fit within the nucleus. The first level of packaging involves the winding of about 150 bp of DNA around an eight subunit core of proteins termed histones (Fig. 5.4A). The histone core and its DNA is referred to as a nucleosome. Strings of nucleosomes can exist as an extended 10 nm fibre, reminiscent of beads on a string, or can be further condensed to become a coiled 30 nm fibre. This can then be further wound into a large solenoid structure which is thought to be looped together to form the chromosome proper (reviewed in Horn and Peterson, 2002). Chromatin is the term used to describe the complex of DNA, histones and additional proteins that exists naturally in the nucleus.

For a long period histones were regarded as having merely a passive role in packaging. It was known that the packing of genes and promoters into chromatin is an impediment to PIC formation, transcriptional initiation and also RNA polymerase II progression, but the dynamic nature of chromatin and the central importance of chromatin remodelling and histone modifications has only recently been recognized (reviewed in Zhang and Reinberg, 2001; Felsenfeld and Groudine, 2003; Fischle et al., 2003). The two main classes of regulatory proteins implicated in the regulation of chromatin packaging are the ATP-dependent chromatin remodelling proteins (Table 5.2), and the histone modifying enzymes, such as histone acetyltransferases (HATs) (Table 5.3), deacetylases (HDACs) (Table 5.4) and methyltransferases (HMTs).

Chromatin remodelling complexes

Remodelling involves the breaking and reforming of histone–DNA contacts and results in the mobilization of nucleosomes on the chromatin template. Several different remodelling complexes have been identified (reviewed in

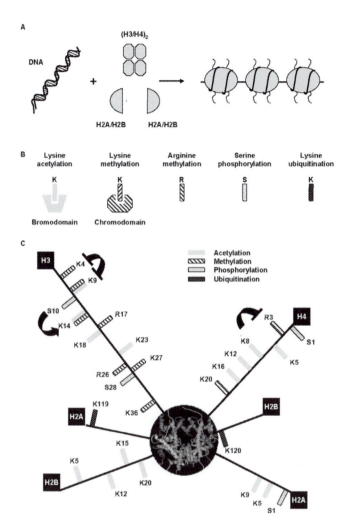

Fig. 5.4. The histone code. (A) Diagrammatic representation of the nucleosome. (B) Types and patterns of histone covalent modifications and interacting domains. These modifications include acetylation at lysine (K), phosphorylation at serine (S), methylation at lysine (K) and arginine (R) and ubiquitination at lysine (K). The two classes of domains that interact with specific modified residues are bromodomains, which interact with acetylated lysines, and chromodomains, which interact with methylated lysines. (C) Patterns of histone tail modifications. Adapted from Felsenfeld and Groudine (2003). Each modification is coded as indicated and the position of the modified amino acid labelled. Note that methylation of K79 H3 (not indicated in the figure) is unusual as it takes place in the core particle. The structure of the nucleosome core particle is indicated (Luger *et al.*, 1997). Some examples of modification that influence the modification of other sites are given. Thus, methylation of H3 K4 impairs the methylation of H3 K9 and vice versa. Acetylation of H4 K8 inhibits the methylation of H4 R3 whereas phosphorylation of H3 S10 induces the acetylation of H3 K14. For more detailed examples of cross-talk, see Fischle *et al.* (2003).

Becker and Horz, 2002; Narlikar *et al.*, 2002). These complexes can be divided into three main classes based on the identity of their catalytic ATPase subunits (Table 5.2). They also contain additional subunits that may affect regulation, efficiency and specificity.

Chromatin remodelling factors apparently act by catalysing fluidity in the position and

Table 5.2. ATP-dependent chromatin remodelling factors.

ATPase family	Complex	Species	Subunits
ISWI	ISWI1	Yeast	ISWI1, p110, p105, p74
	ISWI2	Yeast	ISWI2, p140
	ACF	*Drosophila, Xenopus*, human	Acf-1, ISWI (hSNF2H)
	WICH	*Xenopus*, mouse	WSTF, hSNF2H
	CHRAC	*Drosophila*, human	Acf-1, ISWI (hSNF2H), CHRAC16, CHRAC14
	NURF	*Drosophila*	NURF301, ISWI, NURF55, NURF38
	RSF	Human	Rsf-1 (p325), hSNF2H
	NCoR-C	Human	Tip5, hSNF2H
	ISWI-D	*Xenopus*	p195, ISWI
SWI2/SNF2	SWI/SNF	Yeast, *Drosophila*, mouse, human	Brg1/Brm, ~10 subunits depending on the species
	RSC	Yeast	Sth1, ~15 polypeptides complex
INO	INO80	Yeast	12 polypeptides complex
CHD	CHD1	Yeast, human	CHD1
	NuRD	*Drosophila, Xenopus*, human	CHD4 (Mi-2), MTA2, MBD3, HDAC1/2, RbAp48/46

Adapted from Vaquero *et al.* (2003). ISWI, imitation switch; ACF, ATP-utilizing chromatin assembly and remodelling factor; WICH, WSTF (Williams syndrome transcription factor)-ISWI chromatin remodelling complex; CHRAC, chromatin accessibility complex; NURF, nucleosome remodelling factor; RSF, remodelling and spacing factor; NcoR-C, nuclear receptor co-repressor complex; SWI/SNF, switch/sucrose-non-fermentation; RSC, remodels the structure of chromatin; INO80, inositol80; CHD1, chromodomain-helicase-DNA binding protein 1; NuRD, nucleosome remodelling and deacetylase repressor complex.

conformation of nuclesomes in an ATP-dependent manner. They are thought to do this by catalysing the interconversion between various chromatin states via an activated intermediate consisting of the remodelling factor and a nucleosome with weakened histone–DNA contacts (reviewed in Narlikar *et al.*, 2002). As this model posits only that remodelling complexes increase the rate of interconversion between chromatin states, an important implication is that the action of remodelling complexes does not in itself specify whether the resulting chromatin state is positive or negative for transcription. Indeed, genome-wide analysis of the effects of loss of the Swi2 remodelling factor in yeast indicates that this factor has positive roles in transcription at some genes and negative roles at others (Holstege *et al.*, 1998).

Chromatin modifying enzymes

When the structure of the nucleosome was solved by X-ray crystallography (Richmond *et al.*, 1984; Luger *et al.*, 1997), it became apparent that the termini of the histone subunits, the so-called histone tails, protruded outside the nucleosome and it was hypothesized that these might either influence the general level of packaging by mediating nucleosome–nucleosome contacts or be instrumental in forming the binding sites for additional proteins (Table 5.3 and Fig. 5.4). There is now considerable evidence that this is essentially correct and that the histone tails are of vital importance to the regulation of gene expression. It has become apparent that a 'histone code' exists (Strahl and Allis, 2000; Jenuwein and Allis, 2001), whereby different modifications to the tails of the different subunits do indeed alter packaging and form the docking sites for enzymes that ultimately lead to either activation or repression of particular genes (Figs 5.4 and 5.5). For example, the co-activator CREB-binding protein (CBP) is an enzyme that activates gene expression in part by acetylating lysines within the histone tails (Bannister and Kouzarides, 1996; Ogryzko *et al.*, 1996).

Table 5.3. Histone acetyl transferase factors.

HAT family	HAT enzyme	Species	Complex	Specificity	Function
GNAT	Gcn5	Yeast to human	SAGA, ADA, ADA2	H3, H2B	Co-activator
	PCAF	Human, mouse	PCAF	H3, H4	Co-activator
	Hat1	Yeast	HatB	H4, H2B	Histone deposition, silencing
	Elp3	Yeast to human	Elongator	H3, H4	Transcriptional elongation
	Hpa2	Yeast		H3, H4	Unknown
	ATF-2	Human, mouse		H2B, H4	Transcription factor
MYST	Sas2	Yeast		H4	Silencing
	Sas3	Yeast	NuA3	H3, H4, H2A	Silencing
	MORF	Human		H4, H3	Unknown
	TIP60	Human	TIP60	H4, H3, H2A	Co-activator. DNA repair, apoptosis
	Esa1	Human	NuA4	H4, H2A	Cell cycle progression
	MOF	*Drosophila*	MSL	H4	X chromosome dosage compensation
	HB01	Human	HB01	H3, H4	DNA replication
	MOZ	Human	AML1	H3, H4, H2A	Transcriptional activation
p300/CBP	p300	Human, mouse		H2A, H2B, H3, H4	Co-activator
	CBP	Human, mouse		H2A, H2B, H3, H4	Co-activator
NR	ACTR	Human, mouse		H3, H4	NR co-activator
	SRC-1	Human, mouse			NR co-activator
	TIF2	Human, mouse			NR co-activator
TAFII250	TAFII250	Yeast to human	TFIID	H3, H4, H2A	RNA polymerase II complex
TFIIIC	TFIII220				
	TFIII110	Human	TFIIIC	H3, H4, H2A	RNA polymerase III complex
	TFIII90				
Nut1	Nut1	Human	Mediator	H3, H4	Co-activator of RNA polymerase II

Adapted from Vaquero *et al.* (2003). GNAT, GCN5-related *N*-acetyltransferase; MYST, MOZ, Ybf2/Sas3, Sas2, Tip60; NR, nuclear receptor.

In general, the acetylation of histones is linked to transcriptional activation because histone acetylation decreases internucleosome interaction and the interaction of nucleosome tails with linker DNA, thereby allowing greater accessibility (Fig. 5.5), but acetylation also has additional effects. Acetylated lysine can be recognized and bound by activating proteins that contain bromodomains (Fig. 5.4B). Such proteins include SWI/SNF components and one of the TBP-associated factors, TAF1 (formerly TAF$_{II}$-250 (Tora, 2002)), and CBP itself (reviewed in Zeng and Zhou, 2002). Thus, acetylation of lysine can theoretically lead to sustained gene activation. For example, the recruitment of CBP might lead to the acetylation of lysines within the promoter; this may facilitate the recruitment or retention of SWI/SNF through binding of the acetyl lysine to its bromodomain-containing subunits. SWI/SNF may then loosen the chromatin packaging, and the acetyl lysines may contribute to the recruitment of TAF1 and further CBP to ensure that the gene remains permanently on as the cells terminally differentiate. It is a hallmark of mammalian

Table 5.4. Histone de-acetylase factors.

HDAC group	HDACs members	Complex	TSA sensitivity	NAD+ dependence	Localization	Function
Class I (Rpd3)	HDAC1 HDAC2 HDAC3 HDAC8	Sin3, NuRD, CoREST (HDAC1, 2) N-CoR, SMRT (HDAC3)	Yes	No	Nuclear and ubiquitous	Involved in a variety of functions such as transcriptional repression and cell differentiation
Class II (Hda1)	HDAC4 HDAC5 HDAC6 HDAC7 HDAC9 HDAC10 HDAC11	P97/PLAP (HDAC6)	Yes	No	Nuclear and cytoplasmic, tissue-specific	Transcriptional repression, microtubule
Class III (Sir2)	SIRT1 SIRT2 SIRT3 SIRT4 SIRT5 SIRT6 SIRT7	?	No	Yes	Nuclear, cytoplasmic and mitochondrial	Silencing, gene repression, DNA repair, microtubule

Adapted from Vaquero *et al.* (2003). TSA, trichostatin; NAD, nicotinamide dinucleotide.

gene expression that genes turned on during differentiation must often be kept on thereafter for the life of the cell and epigenetic marks such as the acetylation of histones could facilitate the establishment of an active chromatin state so that the gene remains on.

Histone modifications have also been associated with repression of gene expression. The first histone-modifying enzymes associated with repression were the histone deacetylase or HDAC enzymes (Table 5.4) (Taunton *et al.*, 1996). The deacetylation of lysine will restore the positive charge and possibly thereby allow strong binding of the histone tails to DNA, including nucleosome–nucleosome contacts and ultimately a tightening of chromatin structure (Fig. 5.5). Recently another effect of deacetylation has become apparent. Deacetylated lysine can be targeted by methyltransferase enzymes and it appears that HDACs and certain methyltransferases can cooperate in a pathway to repression (reviewed in Rice and Allis, 2001). Moreover, just as acetylated lysine can recruit activating proteins that contain bromodomains, such as certain SWI/SNF subunits, methylated lysine residues can be bound by repressive proteins that contain so-called chromodomains, such as polycomb and heterochromatin protein

1 (HP1) (Fig. 5.4B) (reviewed in Rice and Allis, 2001).

HP1 is known to recognize methylated lysine 9 on the tail of the histone subunit, histone H3 (Jacobs *et al.*, 2001). Methylated lysine 9 also recruits additional methyltransferase enzyme and HP1 binds the methyltransferase too, so the repressive state can be maintained and may even spread to neighbouring regions of chromatin. Additionally, HP1 has been shown to display RNA-binding activity (Muchardt *et al.*, 2002) and there is some evidence that non-coding RNA molecules may further contribute to setting up a silenced chromatin state, at least in plants and yeast (see below).

Interestingly, it has also become clear that as well as histones being modified by acetylases and methylases, other gene regulatory proteins such as sequence-specific DNA-binding proteins can also be modified and their activity regulated by such modifications (reviewed in Gamble and Freedman, 2002). Acetylation and methylation occur on lysine and arginine residues which are positively charged and found abundantly in histone tails and in DNA-binding proteins, so it is perhaps not surprising that this modification

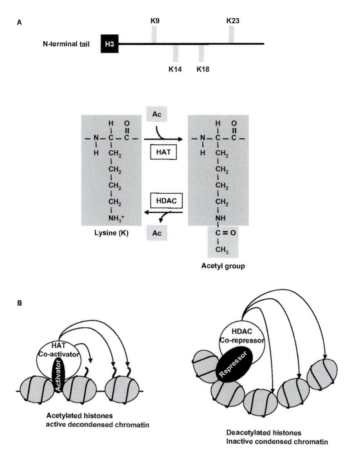

Fig. 5.5. Histone acetylation and deacetylation. (A) The acetylated lysines of the H3 tail are indicated as in Fig. 5.4C. The ε-amino group of lysine residue is modified by acetyl-CoA as indicated by histone acetyl-transferase complexes (HATs) and de-modified by histone deacetylase complexes (HDACs). (B) Acetylated histones correspond to active decondensed chromatin whereas deacetylated histones correspond to inactive condensed chromatin.

plays a central role in the regulation of gene expression.

While acetylation is readily reversible by deacetylase enzymes, there is as yet no evidence for demethylases that remove the methyl groups from lysine and arginine residues. Thus methylation of histones can presumably only be reversed by rounds of DNA replication and repackaging with new, unmodified histones (Ahmad and Henikoff, 2002a,b). Thus methylation appears to be a relatively permanent mark and is likely to be ideal for maintaining chromatin states in terminally differentiated non-dividing cells that remain relatively static over long periods of time during the life cycle of mammals.

Cooperation between chromatin remodelling factors and modifying complexes

There is strong evidence to support the common-sense notion that ATP-dependent remodellers and covalent modifiers work together to regulate gene expression. A functional link was first suggested by genetic studies in yeast (Roberts and Winston, 1997). In mammalian cells, chromatin immunoprecipitation (ChIP) experiments have demonstrated that both BRG1 (the ATPase subunit of the human SWI/SNF complex) and CBP are present on promoters bound by the oestrogen receptor DNA-binding activator (DiRenzo *et al.*, 2000; Shang *et al.*, 2000). Direct physical interactions between the ATP-dependent remodellers and chromatin modifiers have been observed and may act to increase binding to the chromatin template. Alteration of the chromatin by one complex could make it a better substrate for the other complex. For example, remodelling of the nucleosomes by ATP-dependent remodellers may increase the accessibility of the histone N termini for acetylation or deacetylation. Alternatively, as described above, ATP-dependent remodellers, particularly those containing bromodomains, would be expected to bind more strongly to, or dissociate more slowly from, nucleosomes having N termini acetylated at specific positions.

It appears that there is no obligate order for function of ATP-dependent remodellers and covalent modifiers that is general for all promoters. Indeed, on the yeast HO promoter, the recruitment of ATP-dependent chromatin complexes precedes that of histone acetylase complexes (Cosma et al., 1999; Krebs et al., 1999). The reverse order of complex recruitment has been observed on the interferon-β promoter and for retinoic acid-induced transcription (Agalioti et al., 2000; Dilworth et al., 2000). The precise order seems to depend upon the nature of the promoter, the complement of transcription factors present and the chromatin structure in which the promoter resides. Moreover, the ordered recruitment of factors is probably influenced by the speed with which a gene must be activated (reviewed in Cosma, 2002).

DNA methylation

Modification of proteins by methylation is one way of influencing gene expression but modification of DNA by methylation is another. In mammals, cytosine residues followed directly by guanine residues are substrates for methylation and are converted to 5 methyl cytosine. 5 Methyl cytosine is prone to deaminating to form thymine and over evolutionary time many ^{5Me}CpG dinucleotides have become TpG, leaving the mammalian genome poor in CpG sites (reviewed in Jones and Baylin, 2002). Nevertheless there are some regions that are not depleted in CpG dinucleotides, and these regions are termed CpG islands.

They are often found over the promoter and first exon of broadly expressed genes. One hypothesis for the existence of CpG islands is that the regions are regulatory regions that have been protected from methylation by the presence of transcription factors or an associated phenomenon such as the process of active transcription (Macleod et al., 1994). The inheritance of CpG islands after reproduction depends on their under-methylation in germ cells, accordingly 'housekeeping' genes that are expressed in all cell types including germ cells are often observed to contain CpG regions (reviewed in Jones and Baylin, 2002).

Overall there seems to be a connection between methylation and gene expression in that unmethylated DNA is associated with actively transcribed genes, whereas heavily methylated DNA is associated with silent chromatin (reviewed in Jones and Baylin, 2002). Our understanding of the link between methylation and gene repression is still incomplete but there is some evidence that a ^{5Me}CpG-binding protein, termed MeCP2, contributes to gene silencing. Most interesting is the observation that MeCP2 associates with HDACs and in some circumstances it may be that methylation of DNA leads to the recruitment of MeCP2 and HDAC enzymes that function to silence gene expression (Nan et al., 1997, 1998). On the other hand, there are also likely to be instances where methylation, rather than triggering gene repression, is the final result of a complex process of repression. As mentioned above, one path to repression involves the removal of acetyl groups from lysines which then allows lysine methylation. It is now known that in some organisms lysine methylation of histones is linked to cytosine methylation of DNA (Tamaru et al., 2003) and it would not be surprising if histone and DNA methylation were also found to be associated in mammals (reviewed in Li, 2002).

At present relatively little is known about changes in DNA methylation at specific genes during mammalian development, and the mechanisms by which methylation is targeted and controlled remain uncertain. It is notable that in model experimental organisms such as yeast, Drosophila and nematodes, methylation of DNA is not thought to occur at high levels and this fact has probably hindered efforts to understand how methylation contributes to the regulation of gene expression. In mammals, the observation that the methylase enzymes are required for proper development (reviewed in Li, 2002) and that aberrant patterns of methylation are found in various cancer cells, together with the realization that methylation is important for many epigenetic processes, such as imprinting, X-inactivation, retroviral and transgene silencing, argues strongly that methylation has a very important role to play in mammalian development and physiology (reviewed in Li, 2002).

DNA methylation, like histone methylation, seems to be a fairly stable epigenetic mark. Methylation can be reversed by DNA replication, as new DNA strands are built from unmethylated components, and there is some evidence that active demethylation of DNA can also occur during development (reviewed in Dean *et al.*, 2003) but in general methylation patterns seem to be relatively stable and may contribute to the maintenance of gene expression patterns in long-lived organisms such as mammals.

As we have discussed, the regulated recruitment of chromatin modifying enzymes is the major new theme underlying our current understanding of gene control. Below we discuss the features of the regulatory proteins that control the recruitment of these enzymes and/or RNA polymerase and its associated cofactors to specific genes. These regulatory proteins must themselves be able to recognize specific target genes and they do this by binding specific promoter and enhancer elements.

Sequence-specific DNA-binding Transcription Factors

Sequence-specific DNA-binding domains

So far we have introduced the proteins involved in opening chromatin and those involved in the subsequent assembly of the PIC ready for the initiation of transcription. Controlling these two processes at particular promoters usually determines whether or not a gene will be expressed. The gene-specific regulatory proteins that thus control gene expression are the sequence-specific DNA-binding transcription factors.

Many insights into the mechanisms by which transcription factors regulate gene expression have been gleaned from very detailed examinations of the structure of individual transcription factors (reviewed in Harrison, 1991; Pabo and Sauer, 1992; Luscombe *et al.*, 2000). Transcription factors must perform two main functions: first they must specifically recognize and bind to their particular target genes, and secondly they must activate or repress transcription. Most transcription factors seem to have evolved to have a modular structure, with distinct and separable domains dedicated to DNA binding and transcriptional regulation. The first activity sequence-specific recognition seems to have been solved by evolution in only a small number of ways. While there may be as many as 2000 different transcription factors encoded in mammals (Tupler *et al.*, 2001), they can be classified into perhaps as few as 20 different groups based on the structures of their DNA-binding domains (Table 5.5) (reviewed in Harrison, 1991; Pabo and Sauer, 1992; Luscombe *et al.*, 2000). Whilst it may be reasonably easy to generate a protein that binds DNA, devising a structure that is highly specific – able to bind some sites with high affinity but not others – is perhaps a more difficult problem. This may explain why the same DNA recognition domains have been used repeatedly together with different activation and repression domains.

The ways in which different DNA-binding domains make specific contacts with DNA vary between different families, but some general principles emerge. The major groove of DNA presents a more varied surface for molecular recognition than the minor groove, and the majority of transcription factors bind in the major groove (reviewed in Marmorstein and Fitzgerald, 2003). Arginine residues in DNA-binding domains are particularly important in making contacts with guanines in DNA but many other amino acids also make specific contacts with the exposed surfaces of the bases or more general stabilizing contacts with the phosphate sugar backbone of DNA. DNA contact residues are often presented on α-helices as the size of an α-helix is such that it fits into the major groove (Jones *et al.*, 1999).

Table 5.5. Examples of sequence-specific transcription factors based on the structural DNA-binding domain.

Basic domains				
Class	Leucine zipper (bZIP)	Helix–loop–helix (bHLH)	bHLH-ZIP	
Examples	Jun, Fos, CREB	MyoD, E2A	c-Myc, Mad/Max	
Zinc-coordinating domains				
Class	C4 of NR	C4	C2H2	
Examples	GR, ER, RXR	GATA-4	Sp1, BKLF, YY1	
Helix–turn–helix (HTH)				
Class	Homeodomain	Paired box	wHTH	
Examples	MATα2, Oct-1	Pax-1	HNF-3, ETS	
β-Scaffold with minor groove contacts				
Class	Rel homology region	STAT	P53	MADS-box	HMG
Examples	NF-KB, NFAT	STAT1	P53	SRF, MEF2	UBF, TCF-1

The most common domain found in transcription factors is the C_2H_2 or classical zinc finger domain (Wolfe *et al.*, 2000; Tupler *et al.*, 2001; Krishna *et al.*, 2003). This is a small (around 30 residue) domain. In the reducing atmosphere of the nucleus, disulphide bonds cannot be used to stabilize small protein modules, and folding around metal ions such as zinc provides a good alternative. In classical zinc fingers, one zinc atom is bound by two cysteine and two histidine residues (Table 5.5). The domain folds to produce a β-hairpin followed by an α-helix. A small number of residues within the α-helix dictate the DNA-binding specificity. These residues are distinct from the cysteines and histidines required to maintain the structure, thus it is possible to find zinc fingers with a very wide array of different specificities. Nevertheless, zinc fingers are small and typically only contact three or four contiguous bases in the major groove. In order to obtain greater specificity several fingers are linked together. Thus some zinc finger proteins, such as erythroid Krüppel-like factor (EKLF), have three zinc fingers and bind sites made up of about 9 bp, while other proteins have two zinc fingers and bind shorter recognition elements (reviewed in Wolfe *et al.*, 2000; Matthews and Sunde, 2002; Krishna *et al.*, 2003). Some zinc finger proteins have as many as 30 repeated zinc finger domains but it is unlikely these proteins recognize sequences of 90 bp, and the full biological function of these proteins remains to be elucidated.

Other DNA-binding domains commonly found in mammalian transcription factors include homeodomains (reviewed in Pabo and Sauer, 1992). There are several hundred homeodomain proteins encoded in the human genome (Tupler *et al.*, 2001). The homeodomain consists of three α-helices and includes a derivative of the helix–turn–helix motif first recognized in bacterial DNA-binding proteins. Typically the third helix in homeodomains, the so-called recognition helix, docks into the major groove of DNA and mediates sequence-specific binding. Different homeodomain proteins recognize different sites but in general these proteins recognize motifs with the core TAAT so these proteins are not as versatile as zinc finger proteins. Homeodomain proteins were initially identified as proteins controlling segmental positioning during development in organisms such as *Drosophila*, and many mammalian homologues have also been found to have critical developmental roles (reviewed in Kmita and Duboule, 2003).

Treble clef finger domains also merit attention (reviewed in Krishna *et al.*, 2003). These are also zinc-binding domains but have a different topology from classical C_2H_2 zinc fingers (Table 5.5). They typically use four cysteine residues to bind a single zinc atom, and sequence-specific DNA-binding is mediated by an α-helix within the domain. Proteins such as GATA-1 and the nuclear receptor family contain treble clef finger domains. GATA-1 predominantly uses its C-terminal zinc finger to contact DNA and recognizes a relatively short site with the core GATA (Martin and Orkin, 1990), while nuclear receptors typically dimerize and recognize palindromic or repeat sites with the core sequence related to GGTCA (reviewed in Renaud and Moras, 2000).

Another common class of dimerizing DNA-binding domain is the basic-leucine zipper (reviewed in Vinson *et al.*, 2002). Leucine zippers are short (up to about 40 amino acids) α-helices that come together to form a coiled-coil structure. There is a basic region at the end of the pair of helices and the two helices contact the major groove of DNA in a configuration often referred to as the 'scissors grip'. Each α-helix recognizes about 3 bp, so leucine zipper recognition sites generally span around 6 bp. Transcription factors of this family can exist as homodimers, such as the c-Jun/c-Jun homodimer, or members of the same family can heterodimerize, such as c-Jun and c-Fos (reviewed in Vinson *et al.*, 2002). In general, members of the same family recognize similar half-sites. Consequently the dimers, be they homo- or heterodimers, generally recognize palindromic sites. Jun/Fos family proteins recognize TGANTCA sequences (where N is any nucleotide) (Glover and Harrison, 1995) and members of the C/EBP family bind sequences of the general form TGCGCA (Miller *et al.*, 2003).

As can be seen from the above discussion, mammalian transcription factors tend to recognize relatively short motifs, such as GATA, etc. The concatenation of domains, as occurs in zinc finger proteins, or the dimerization of proteins serves to increase the length and therefore specificity of the target site but still the sites bound remain relatively short compared with the size of mammalian genomes. Most transcription factors can recognize sites of about 6 bp, and such sites will occur many times by chance in the human genome so a single transcription factor-binding site will not be sufficiently specific to direct regulation to a single target gene. To solve the specificity problem it seems that transcription factors often work combinatorially (reviewed in Wolberger, 1998; Marmorstein and Fitzgerald, 2003). Thus the combination of GATA sites and CACCC boxes can be used as a more specific composite recognition element. In some cases, the two DNA-binding domains involved physically interact and this interaction can increase the rate or affinity of binding (reviewed in Verger and Duterque-Coquillaud, 2002). Thus many genes expressed in erythroid cells contain closely spaced GATA and CACCC elements and bind GATA-1 and CACCC box proteins such as EKLF (Orkin, 1996). Similarly, genes expressed in megakaryocyte cells contain GATA elements closely juxtaposed to GGAA sites, that are bound by the protein Fli-1, and cooperative binding between GATA-1 and Fli-1 leads to synergistic activation of some megakaryocytic promoters (Eisbacher *et al.*, 2003; Starck *et al.*, 2003).

Activation and repression domains

Whereas DNA-binding domains fall into a relatively small number of clearly defined structural categories, activation and repression domains have proved to be significantly more diverse and it can be argued that they have defied classification. Relatively sophisticated structures are required to distinguish one DNA sequence from another, but it appears that much simpler motifs suffice for protein–protein interactions that underlie gene activation and repression.

Accordingly, it is possible that most activation and repression domains do not have any permanent tertiary structure but consist of a series of very short motifs that dock with partners, and sometimes only adopt structure during the binding process (see for example Uesugi *et al.*, 1997). Consistent with this view, very little progress has been made defining the three-dimensional structures of activation and repression domains. Small motifs within the domains mediate critical specific contacts and thereby facilitate the recruitment of chromatin modifying complexes or RNA polymerase–associated complexes like the mediator. Indeed, a short Pro-X-Asp-Leu-Ser motif that is found in the repression domains of transcription factors, including the EKLF-related protein basic Krüppel-like factor (BKLF), binds to a specific pocket in the co-repressor protein CtBP (Turner and Crossley, 1998, 2001; Nardini *et al.*, 2003), and CtBP then recruits HDACs and methylases to turn genes off (Shi *et al.*, 2003). Likewise, activation can be mediated through short motifs. For example, nuclear receptor proteins use small activation domains to dock to short Leu-X-X-Leu-Leu motifs that are presented as α-helical regions within various histone acetylase enzymes (Heery et al., 1997; Torchia et al., 1997; Glass and Rosenfeld, 2000). Nuclear receptors also recruit SWI/SNF complexes (Deroo and Archer, 2001; Hassan et al., 2001) as do small domains within transcription activators as divergent as C/EBPβ, c-Myc and MyoD (Hassan et al., 2001; Sullivan et al., 2001).

Chromatin remodelling complexes are also recruited by transcriptional repressors, such as the NuRD complex, which contains an ATP-dependent chromatin remodeller, Mi-2, and an HDAC (Khochbin et al., 2001). Indeed, upon T-cell activation, the DNA-binding protein Ikaros recruits NuRD to regions of hetero-chromatin (Kim et al., 1999). The trans-criptional co-repressor Kap-1 can also target NuRD to specific promoters to repress gene ex-pression (Schultz et al., 2001). Taken together, these targeting mechanisms probably serve to initiate a cascade of events at a given promoter that results in local alteration of chromatin structure to facilitate formation of an active or repressed state.

The simple picture that emerges is that activation domains contain short motifs that recruit either chromatin remodelling com-plexes, such as SWI/SNF, or histone acety-lases like CBP, or contact the mediator, one of the general transcription factors or RNA polymerase itself to facilitate transcription. In several instances, a single protein, such as EKLF, may contribute to several of these recruitment steps. That is, EKLF is known to recruit the histone acetylase CBP as well as a SWI/SNF-related chromatin remodelling com-plex (Armstrong et al., 1993; Zhang et al., 2001). Repression domains may also recruit remodellers but generally recruit HDACs and/or histone methyltransferases.

Regulating Gene Expression

The general states and mechanisms associated with active or silent genes are discussed above but how are patterns of gene expression con-trolled over time or in response to particular

stimuli? The complex pattern of gene expres-sion that occurs during development largely reflects the different sets of transcription factors that operate in different parts of the embryo at different stages of development. Likewise the activation and repression of genes in response to environmental or physiological stimuli depend on transcription factors that respond to different external conditions. It is estimated that there are around 2000 different sequence-specific DNA-binding proteins encoded in the human genome (Tupler et al., 2001) and con-sequently it is easy to imagine that by using combinatorial control mechanisms and the distinct properties of different transcription factors, it is possible to generate almost limitless different patterns of gene expression.

Developmental and tissue-specific gene expression

Development can be viewed as a process whereby cells differentiate into distinct cell types, or at a more macroscopic level form dif-ferent organs. Some general principles of how development is orchestrated by transcription factors have been illuminated by studying gene regulation in different organs. Considerable progress has been made in understanding the generation of the eye, of muscle cells and of blood cells, particularly erythroid cells. Experiments in Drosophila have demonstrated that a single gene, termed Pax6, encodes a DNA-binding protein that functions as a 'mas-ter regulator' for eye formation (reviewed in Baker, 2001; Pichaud and Desplan, 2002). During development, cells destined to give rise to eyes express Pax6 and it presumably then turns on the other genes necessary to form an eye. When Pax6 is mis-expressed in Drosophila legs, it causes the production of eyes at this unlikely anatomical position (Halder et al., 1995). Thus the formation of different tissues reflects the presence of different 'master regula-tors' in different parts of the embryo during development.

Homologues of Pax6 are also implicated in eye formation in mammals, as shown by natu-rally occurring mutations in the human Pax6 gene that lead to a condition termed aniridia (Prosser and van Heyningen, 1998). In mammals, however, there is no evidence that expression of *Pax6* alone can generate a fully functioning eye in a novel location, and indeed the idea of true master regulators in mammals is somewhat controversial (Mathers and Jamrich, 2000). It is likely that the chromatin and DNA modifications detailed above that occur during development leave a permanent mark on key genes so that even the subsequent expression of high level regulatory proteins cannot easily initiate full cascades of events needed for ectopic tissue formation.

Nevertheless, there is some evidence to support the 'master regulator' concept in mammals. The muscle cell-specific transcription factor MyoD is capable of transforming non-muscle cells into muscle at least in cell culture systems. Knockout experiments in mice have confirmed that MyoD and related proteins are important for muscle formation *in vivo* (reviewed in Puri and Sartorelli, 2000; Pownall *et al.*, 2002). MyoD and its relatives can be thought of as 'master regulators' that turn on the sets of genes that coordinate muscle development and differentiation. The restriction of these activities to precise locations within the embryo is essentially what defines which areas will become muscle.

Other tissues and cell types also depend on tissue-specific DNA-binding transcription factors. Studies on genes expressed in immature erythroid cells showed that most if not all the promoters and enhancers identified contained binding sites for the transcription factor GATA-1. GATA-1 is expressed in the blood islands during development and is critical for activating the transcription of genes such as the globin genes, and haem biosynthetic genes (reviewed in Orkin, 1998).

A number of interesting points have emerged from more detailed analysis of these systems of cellular differentiation. In some instances, families of structurally related transcription factors work sequentially in a pathway of differentiation. For instance, the GATA-1-related protein GATA-2 is important for early blood cell development and then GATA-1 is required for the terminal differentiation of erythroid cells (Weiss *et al.*, 1997; Orkin, 1998). In the case of muscle cells, the protein MyoD is required early on, and later MyoD-related proteins come into action

(Pownall *et al.*, 2002). In the case of eye development, *Pax6* regulates a gene called *eyeless* that encodes a protein of the same family as *Pax6* (Pichaud and Desplan, 2002).

It is not clear why related regulatory proteins should work sequentially in a single pathway but one explanation is that transcription factors often regulate their own expression and when gene duplications occur during evolution the family members then regulate each others' expression. Consequently a hierarchy or cascade of transcriptional initiation events required to coordinate complex processes can be developed.

Another outcome of work on tissue-specific and developmental gene expression is that, although the concept of 'master regulators' is attractive and useful in some instances, in general, single factors are unable to drive specific differentiation pathways. Rather, sets of transcription factors tend to be associated with the generation of different tissues (reviewed in Levine and Tjian, 2003). As mentioned above, sometimes these sets include members of single protein families, but different families also collaborate to drive differentiation (reviewed in Sieweke and Graf, 1998). For instance, it appears that GATA and CACCC box-binding proteins work together to drive erythroid gene expression. Thus both GATA-1 and EKLF appear to be required for globin gene expression in red blood cells (reviewed in Cao and Moi, 2002). The GATA-1 protein is not structurally related to the EKLF protein but both proteins appear to recruit similar cofactors, such as the chromatin remodelling complex SWI/SNF, and the histone acetylase CBP. It is not certain why two different factors are needed but using combinations of proteins does increase the specificity of control (reviewed in Wolberger, 1998).

Regulating the regulators

As mentioned, the changing patterns of gene expression during development are coordinated by changes in the transcription factor complement in different parts of the embryo. Obviously, each transcription factor must itself be activated by an upstream transcription factor, and cascades of regulation do exist, but if every transcription factor required a separate master to regulate it, one would need an infinite supply of such factors. How does biology overcome this problem? The short answer is that many transcription factors are regulated at levels other than at the transcriptional level. That is, in early development some transcription factors may not be regulated at all, they may be expressed stochastically but direct particular cell fates only when their chance expression coincides with a particular external growth factor. Others may be tightly regulated but at a level subsequent to transcription. Thus they may be expressed widely but lie dormant in the cell awaiting an external growth factor signal for activation. The key issue here is not whether the protein factor is present, but whether or not it is activated by a particular extracellular signal.

In short, many (and perhaps even all) developmentally important and other transcription factors are subject to post-transcriptional regulation. Post-translational modifications to transcription factors that alter their activity include phosphorylation of serine, threonine, and sometimes tyrosine residues, acetylation of lysine, methylation of lysine or arginine, and ubiquitination or sumoylation of lysine (reviewed in Gamble and Freedman, 2002; Muratani and Tansey, 2003; Verger et al., 2003).

One good example of transcription factor regulation concerns the leucine zipper protein cAMP-response element-binding protein or CREB. In response to an increase in the levels of cyclic AMP, protein kinase A is activated and phosphorylates CREB at serine 133. This alters the surface of CREB and a new binding face for CBP is presented. CBP is recruited and then acetylates histones in the region, thereby facilitating transcription (Radhakrishnan et al., 1997).

One of the best characterized pathways of signal transduction involves the signal transducer and activator of transcription or STAT family of proteins (reviewed in Leonard and O'Shea, 1998). Members of the family respond to signalling by interleukins, cytokines and hormones such as prolactin. Growth factor binding triggers the dimerization of particular receptors. Kinases associated with the cytoplasmic tails of the receptors mediate reciprocal phosphorylation of these tails on tyrosine residues. The phosphotyrosines are then recog-

nized by the STAT transcription factors, through their SH2 phosphotyrosine-binding domains. The STAT transcription factors are brought into proximity to the kinases by binding the receptor and are themselves phosphorylated on key tyrosine residues. They then homodimerize via their SH2 domains; this unmasks a nuclear localization signal and activates them so they are competent to bind to their target sites in DNA (reviewed in Leonard and O'Shea, 1998). In this way the signal produced by the presence of an external growth factor is passed through to the nucleus and results in changes in gene expression. Among the target genes activated by STATs are proteins of the suppressor of cytokine signalling (SOCS) family (reviewed in Kile and Alexander, 2001). These proteins are implicated in terminating the signal by binding to the receptors and/or the STAT proteins to either inhibit their phosphorylation or facilitate their ubiquitination and ultimate breakdown.

The STAT proteins are unusual in two respects; first they can be found near the cytoplasmic membrane and are directly phosphorylated by the relevant receptor or associated kinase, rather than being targeted after a lengthy cascade of phosphorylation. Secondly, they are phosphorylated on tyrosine. Tyrosine phosphorylation is relatively uncommonly found in transcriptional activators. There is some evidence that STAT proteins appeared relatively late on the evolutionary scene as they are not present in lower organisms such as yeast, but are found in mammals (reviewed in Leonard and O'Shea, 1998).

Growth factors are not the only small molecules that provide signals that activate dormant receptors present in the cytosol. Perhaps the best characterized hormones are the steroids that are known to activate members of the nuclear receptor family, including the androgen, oestrogen and glucocorticoid receptors (reviewed in Xu et al., 1999). Although the details vary depending on which family member is studied (some nuclear receptors are primarily in the nucleus, others are in the cytoplasm and some occupy both cytoplasmic and nuclear compartments), the general principle is that the proteins rest in a dormant state complexed with proteins such as the chaperone protein Hsp90. Upon ligand binding, the proteins are released and concentrate in the nucleus, where they dimerize on their recognition elements and activate transcription by recruiting histone acetylase enzymes (reviewed in Xu *et al.*, 1999).

The related proteins, the thyroid hormone and retinoic acid receptors, work similarly but in this case a proportion of the protein remains bound to its DNA target site even in the absence of its ligand. In this state it recruits co-repressor complexes that contain HDAC enzymes (reviewed in Glass and Rosenfeld, 2000). Thus these receptors mediate a profound switch in gene expression. In the absence of their hormone they actively repress their target genes by means of recruited HDAC enzymes (together with proteins of the N-CoR and Sin family). After binding ligand, however, they release the repressor complex and bind a co-activator complex that contains histone acetylases as well as other components such as arginine methylases (reviewed in Glass and Rosenfeld, 2000). These latter proteins serve to activate the expression of the target genes (reviewed in McKenna and O'Malley, 2002).

The exact details of how the very large repressor complex is disassembled so that the activator complex can take its place is still under investigation but there is good evidence that components of the proteosome are required for activation in response to steroid and related hormones. The proteosome is a large protein complex that specifically degrades the proteins it targets (reviewed in Muratani and Tansey, 2003). It may be that protein degradation is required to actually dismantle the complex, despite the fact that intuitively this would seem an energy-expensive way of facilitating a genetic switch (see for example Reid *et al.*, 2003).

Nuclear compartmentalization and gene activity

Another important step in the process by which genes are specifically expressed is nuclear organization, the position of genes within the nucleus and compartmentalization of proteins that regulate their expression (reviewed in Isogai and Tjian, 2003). Current evidence supports the idea that the eukaryotic nucleus is functionally divided into heterochromatin compartments that repress transcription, and compartments in which transcription is permitted. For example, the *β-globin* locus is transcribed at very low levels before differentiation of the erythroid lineage and is preferentially associated with repressive compartments near centromeric heterochromatin and the nuclear periphery. After differentiation, *β-globin* gene expression is dramatically enhanced, and this activation is coupled with relocation of the gene to regions of the nucleus distant from the centromere and nuclear periphery (Francastel *et al.*, 2001). Interestingly, p18, one of the two subunits of the erythroid-restricted heterodimer NF-E2, is found in centromeric heterochromatin before differentiation and relocates upon differentiation to the euchromatin compartment to join its partner p45 (Francastel *et al.*, 2001). This study suggests that one mechanism of transcriptional activation entails a relocation of the target gene locus and/or a key transcription factor from one nuclear territory to another.

The Non-coding RNA World

Non-coding RNA (ncRNA) genes produce structural, catalytic or regulatory RNA molecules that function as RNA rather than acting as templates for protein synthesis. Recently, several different systematic screens have identified a surprisingly large number of new ncRNA genes (Table 5.6) (reviewed in Eddy, 2001). They seem to be particularly abundant in roles that require highly specific nucleic acid recognition without complex catalysis, such as in directing post-transcriptional regulation of gene expression or in guiding RNA modifications.

Dosage compensation in mammals

Besides their well-known roles in translation and splicing (rRNA, tRNA and snRNA), ncRNA are now known to play several new roles in gene regulation. Among the more fascinating stories is the discovery that RNAs have roles in chromatin structure and transcriptional regulation (reviewed in Kelley and Kuroda, 2000;

Table 5.6. The different classes of non-coding RNA.

Abbreviations	Definitions
tRNA	Functional RNA – essentially synonymous with non-coding RNA
miRNA	Micro RNA – putative translational regulatory gene family
ncRNA	Non-coding RNA – all RNAs other than mRNA
rRNA	Ribosomal RNA
siRNA	Small intefering RNA – active molecules in RNA interference
snRNA	Small nuclear RNA – includes spliceosomal RNAs
snmRNA	Small non-mRNA – essentially synonymous with small ncRNAs
snoRNA	Small nucleolar RNA – involved in rRNA modification
stRNA	Small temporal RNA–essentially synonymous with miRNA
tRNA	Transfer RNA

Adapted from Eddy (2001).

Tijsterman et al., 2002; Andersen and Panning, 2003; Dykxhoorn et al., 2003). One example is the human *Xist* (X inactive specific transcript) RNA, a 17 kb ncRNA with a key role in dosage compensation and X chromosome inactivation (reviewed in Avner and Heard, 2001). It is believed that *Xist* RNA spreads from its site of transcription to coat the entire chromosome that will become the inactive X and that this spread correlates tightly with the spread of transcriptional silencing (reviewed in Plath et al., 2002). This global silencing is linked to chromatin modifications and it is possible that *Xist* operates by recruiting chromatin modifying enzymes. Briefly, the inactivated X acquires many characteristics of constitutive heterochromatin, including hypoacetylation and methylation of histone H3, replication late in S phase, methylation of CpG dinucleotides, and hypoacetylation of histone H4 (reviewed in Plath et al., 2002). One function of *Xist* RNA could be to recruit a protein complex with histone H3–K9 deacetylase and/or methyltransferase activity, probably the Eed/Enx1 polycomb group complex (Cao et al., 2002; Kuzmichev et al., 2002; Plath et al., 2003; Silva et al., 2003).

Gene silencing by the non-coding RNA *Air*

In some animals, including mammals, a number of genes are expressed differently according to whether they have been inherited from the mother or from the father. This process is termed genomic imprinting (see Chapter 7 for more details). Several large ncRNAs have been found associated with imprinted genes, but their role, if any, has remained enigmatic (reviewed in Andersen and Panning, 2003). A recent study provides the first evidence that, at least in one case, a ncRNA has a direct role in regulating imprinted gene expression *in cis* (Sleutels et al., 2002). In this case, *Igf2r* is an imprinted gene expressed exclusively from the maternal allele (Zwart et al., 2001). The non-coding RNA *Air*, which exhibits imprinted expression exclusively from the paternal allele, overlaps *Igf2r* and is transcribed in the antisense direction through this locus. In this model, the *Air* transcript might recruit a repressor complex to chromatin *in cis*, in a way that is reminiscent of dosage compensation in mammals by *Xist* RNA (Sleutels et al., 2002).

RNA interference

In 1998, Fire and colleagues found that the injection of double-stranded (ds) RNA into *Caenorhabditis elegans* led to an efficient sequence-specific gene silencing (Fire et al., 1998), which is referred to as RNA interference (RNAi). RNAi has been linked to many previously described silencing phenomena such as post-transcriptional gene silencing (PTGS) in plants and quelling in fungi (reviewed in Tijsterman et al., 2002). The RNAi reaction has been recapitulated in *Drosophila melanogaster* embryo extracts, in which it was confirmed that long dsRNA substrates could be cleaved into short interfering dsRNA species (siRNA) of

~22 nucleotides (Zamore *et al.*, 2000) and that the introduction of chemically synthesized 21 nt and 22 nt siRNAs to these extracts led to the degradation of homologous RNA (Elbashir *et al.*, 2001b).

While the natural presence of RNAi had been observed in plants, insects and nematodes, evidence for the existence of RNAi in mammalian cells took longer to establish. Indeed, transfection of long dsRNA molecules (>30 nt) into most mammalian cells causes non-specific suppression of gene expression, as opposed to the gene-specific suppression seen in other organisms. This suppression has been attributed to an antiviral interferon response (Elbashir *et al.*, 2001a). Indeed, in mammals interferon-induced antiviral pathways are activated by dsRNA and extensive cleavage of single-stranded RNA follows (reviewed in Stark *et al.*, 1998). Moreover, while fungi, plants and worms can replicate siRNAs, there is no indication of siRNA replication in mammals (Zamore, 2002). Therefore, siRNA-directed silencing by transfection in mammals is limited by its transient nature. To overcome some of these problems, several groups have developed DNA vector-mediated mechanisms to express substrates that can be converted into siRNA *in vivo* (Brummelkamp *et al.*, 2002; Sui *et al.*, 2002).

The advent of siRNA-directed knockdown has the potential to allow for the determination of the function of many genes. In addition, siRNA-directed gene silencing might allow the silencing of genes that are pathogenic to the host organism. Indeed, many disease-related genes have already been targeted with some success in mammals (reviewed in Dykxhoorn *et al.*, 2003). It is important to note that the siRNA technology is quite different from the antisense technology that uses either DNA or RNA molecules that are complementary to sequences on the target mRNA to inhibit protein production.

Mechanism of RNA interference

Both biochemical and genetic approaches have led to the current model of the RNAi mechanism (Fig. 5.6). In this model, RNAi includes both initiation and effector steps (reviewed in Denli and Hannon, 2003).

Biochemical characterization showed that siRNAs are 21–23 nt dsRNA duplexes with symmetric 2–3 nt 3' overhangs and 5' phosphate and 3' hydroxyl groups (Fig. 5.6) This structure is characteristic of an RNase III-like enzymatic cleavage pattern, which led to the identification of a highly conserved *Dicer* family of RNase III enzymes as the mediators of the dsRNA cleavage (reviewed in Denli and Hannon, 2003; Dykxhoorn *et al.*, 2003). Thus, in the initiation step, input dsRNA is digested into 21–23 nt siRNAs in an ATP-dependent, processive manner. Successive cleavage events degrade the RNA to 19–21 bp each with 2nt 3' overhangs (Fig. 5.6). In the effector step, the siRNA duplexes bind to a nuclease complex to form what is known as the RNA-induced silencing complex or RISC. The duplex siRNA is unwound in an ATP-dependent manner, leaving the antisense strand to guide RISC to its homologous target mRNA for endonucleolytic cleavage. The target mRNA is then cleaved at a single site in the centre of the duplex region between the guide siRNA and the target mRNA, 10 nt from the 5' end of the siRNA (Fig. 5.6).

Double-stranded RNA and heterochromatin

Despite their natural role in PTGS and quelling, or their 'technological' role in mammals, recent work has pointed to an even more central role for dsRNA/RNAi in genome maintenance. Indeed, components of the RNAi machinery, including a member of the RNase III enzyme Dicer (*dcr1*), are required for heterochromatin formation and for targeting of histone H3 lysine 9 methylation in the yeast *S. pombe*, a hallmark histone modification that correlates with large-scale, or chromosome level, repression (Hall *et al.*, 2002; Volpe *et al.*, 2002). The role of RNAi in epigenetic gene silencing appears to be conserved among diverse species, as it is also required for silencing and chromatin modification in *Arabidopsis* (Zilberman *et al.*, 2003). RNA has also been implicated in subnuclear localization of heterochromatic proteins such as HP1 (Muchardt *et al.*, 2002). Whereas these intriguing results await further confirmation in mammals, the mouse HP1 family has been

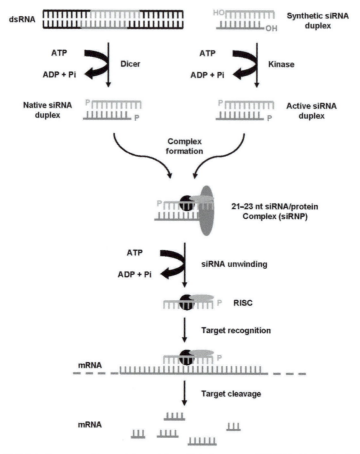

Fig. 5.6. The RNA interference pathway. Adapted from Dykxhoorn *et al.* (2003). Molecular hallmarks of a short interfering (si) RNA include 5′ phosphorylated ends, a 19-nucleotide (nt) duplexed region and 2-nt unpaired and unphosphorylated 3′ ends that are characteristic of RNase III cleavage products. Long double-stranded (ds) RNA is cleaved by the RNase III family member, Dicer, into siRNAs in an ATP-dependent reaction. These siRNAs are then incorporated into the RNA-inducing silencing complex (RISC). Once unwound, the single-stranded antisense guides RISC to mRNA that has complementary sequence, which results in the endonucleolytic cleavage of the target mRNA.

recently implicated in stable epigenetic gene silencing (Ayyanathan *et al.*, 2003) and it seems likely that RNAi-mediated gene silencing will also be found to be important in the stable repression of mammalian genes and/or transposable elements.

Post-transcriptional Regulation and Control of Splicing

Although much regulation involves the opening of chromatin and controlling the rate of transcriptional initiation, there are many exam-ples of regulation that occurs at subsequent stages, such as the level of splicing, mRNA stability, translational initiation and ultimately protein activity (Fig. 5.1, Steps 2–6 and Fig. 5.7).

'Gene expression factories'

As mentioned in the Introduction, recent studies lead to the view that, in contrast to a simple linear assembly line (Fig. 5.1), a complex and extensively coupled network has evolved to coordinate the activities of gene expression

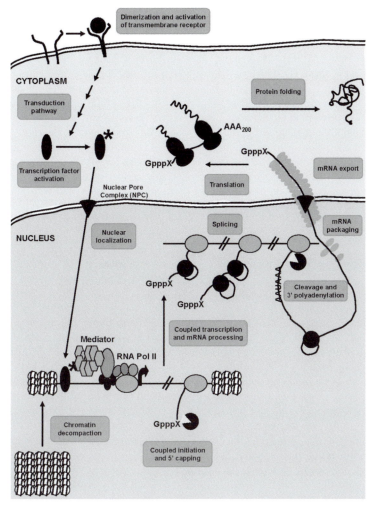

Fig. 5.7. A contemporary view of gene expression. Adapted from Orphanides and Reinberg (2002). Each step regulating gene expression (from transcription to translation) is a subdivision of a continuous process. In this contemporary view, each stage is physically and functionally connected to the next.

machines (Fig. 5.7). Indeed, based on *in vitro* assays showing that each of the major steps, that is, transcription, capping, splicing and polyadenylation, can be carried out in isolation, and because intuitively each of these reactions seemed quite distinct from the others, it had been widely assumed that the machinery responsible for each step was distinct and functioned essentially independently (reviewed in Hirose and Manley, 2000; Maniatis and Reed, 2002; Proudfoot *et al.*, 2002; Reed, 2003). However, numerous studies have provided considerable evidence that this is not the case.

Thus, transcription is clearly coupled with mRNA processing (capping, splicing and polyadenylation). For example, by tight coupling between capping and transcription initiation, rapid capping of the nascent pre-mRNA is ensured, thereby protecting it from degradation. Coupling also plays a critical role in gene expression by tethering machines to each other and to their substrates, a mechanism that dramatically increases the rate and specificity of enzymatic reactions (reviewed in Hirose and Manley, 2000; Maniatis and Reed, 2002; Proudfoot *et al.*, 2002; Reed, 2003). The CTD

of the RNA polymerase II complex plays a critical role in this coupling by recruiting many of
the activities to the transcriptional machinery
(reviewed in Howe, 2002). Throughout the
transcription cycle, the CTD undergoes a variety of covalent and structural modifications
which can, in turn, modulate the interactions
and functions of processing factors during transcription initiation, elongation and termination.
This harmonic integration of transcriptional and
post-transcriptional activities supports a general
model for the coordination of gene expression
by the RNA polymerase II complex within the
nucleus, the so-called transcriptome or tran-
scriptosome.

Splicing generates diversity

Rather than simple on/off regulation, splicing
provides the opportunity to generate a wide
range of different gene products from a single
gene. Although the exact make up of genes
varies, mammalian genes contain on average
around ten exons and it is estimated that perhaps half of all genes undergo some form of
alternative splicing. That is different subsets of
the exons are retained in different transcripts,
although the order of the exons does not
change. Unfortunately, however, the exact
mechanisms that control alternative splicing
are still largely unknown. Moreover, the function of the majority of alternatively spliced
mRNA remains to be determined.

Nevertheless, there are some interesting
examples of differential splicing generating
proteins with distinct functions. For example,
the human Wilm's tumour suppressor gene
(*WT1*) undergoes extensive alternative splicing; however, the only alternative splice conserved among vertebrates is the use of two
alternative 5′ splice sites for exon 9 separated
by 9 nt that encode lysine-threonine-serine
(KTS) (reviewed in Faustino and Cooper,
2003). The +KTS and –KTS isoforms are
expressed at a constant ratio favouring the
+KTS isoform in all tissues and developmental stages that express *WT1* (Haber *et al.*,
1991). The variable KTS region is located in
the DNA-binding domain of *WT1*, between
the third and fourth C_2H_2 zinc fingers. The
+KTS isoform binds DNA only weakly and is

unable to activate targets of the –KTS isoform
(Wilhelm and Englert, 2002). The distinct DNA
binding affinity of these two isoforms appears
to be important for their differential functions
during development. Whereas the –KTS is
believed to play a role in early gonad development, the +KTS appears to be involved in
RNA metabolism and perhaps pre-mRNA
splicing (reviewed in Faustino and Cooper,
2003). Interestingly, inactivation of *WT1* is
responsible for ~15% of Wilm's tumour, a pae-
diatric cancer of the kidney. Three additional
disorders are associated with abnormalities in
WT1 expression: WAGR (Wilms tumour,
aniridia, genitourinary abnormalities, mental
retardation), Denys–Drash syndrome (DDS)
and Frasier syndrome (FS). All three diseases
are characterized by urogenital disorders
involving kidney and gonad development
(Armstrong *et al.*, 1993). Consistent with the
distinct functions performed by the +KTS and
the –KTS isoforms, the majority of individuals
with FS were found to have mutations that
inactivate the downstream 5′ splice site, resulting in a shift to the –KTS form (Melo *et al.*,
2002). The basis for FS is therefore underex-
pression of +KTS, overexpression of –KTS, or
a combination of the two, indicating that the
ratio of the two isoforms is critical, especially
during gonad and kidney development.

Conclusion

Many of the concepts discussed in this chapter
are common to all organisms while others,
such as the long-term regulation achieved by
methylating histones and DNA, may be particularly important in large multicellular organisms such as plants and animals that have
relatively long life times. Furthermore, the
structure of animals, as opposed to plants, their
complex interdependent organ networks and
development suggest a particular requirement
for gene control and it is likely that some
mechanisms are peculiar to larger animals. It is
notable that DNA methylation, although it
occurs in bacteria, where it is important for
restriction modification and defence against
viral pathogens (reviewed in Casadesus and
D'Ari, 2002), seems to play important regulatory roles mostly in higher organisms.

Current methodologies, results, limitations

Although a great deal has been learnt about how genes are differently regulated during mammalian development and in response to different physiological stimuli, our understanding is still too limited to allow us to reliably predict the expression of genes by studying their regulatory sequences or to design regulatory sequences to drive expression in desired tissues. Indeed, even generating systems that allow properly regulated inducible gene expression in experimental applications has proved problematic (reviewed in Gatz, 1995). Although some progress has been made by transplanting tetracycline- or lactose-inducible systems from bacteria, it is still not always routine to generate cell lines where expression of a particular heterologous gene can be reliably controlled artificially and this challenge is even greater when transgenic animals are required.

The seemingly more accessible goal of expressing transgenes constitutively in a given tissue is also far from routinely available (reviewed in Hadjantonakis *et al.*, 2003). This explains the relatively modest progress that has been made towards realizing the goal of gene therapy in humans. Artificial promoters have not been particularly effective and in general the best approach has been to utilize existing regulatory sequences and fuse them to genes to be expressed. The larger the amount of sequence of the regulatory region used, the more likely faithful expression recapitulating the desired pattern will be achieved *in vivo*, but it is likely that some transgenic mouse lines will still show variations from the desired expression pattern. Distant regulatory sequences (termed enhancers) can affect the expression of adjacent genes and it seems that when a foreign gene is introduced into the genome it may be up-regulated by existing enhancers, or down-regulated by silencer elements. For instance, even when relatively large globin regulatory regions are used to drive the expression of transgenes, expression may be observed in unintended regions such as the brain or the transgene may fail to be expressed in red blood cells, or be expressed in only a subset of cells (reviewed in Martin and Whitelaw, 1996; Rakyan *et al.*, 2002, see also Chapter 7 of this book). These are referred to

as positive and negative position effects. There is some evidence for the existence of 'insulator' elements that can protect transgenes from distant regulatory sequences but at this stage these elements are not well characterized and are not in routine use (reviewed in Geyer and Clark, 2002).

An alternative approach to directing gene expression has been to design artificial sequence-specific DNA-binding factors to target endogenous genes of interest to turn them on or off. Much success has been achieved in generating zinc finger proteins with almost any desired specificity, using protein engineering techniques and screening libraries of variants to obtain proteins with optimal binding specificities (Beerli and Barbas, 2002; Bae *et al.*, 2003). When tested *in vitro* these proteins have proved to be very effective at DNA binding and have even been shown to influence the expression of desired target genes in physiologically relevant experimental systems.

Perhaps the most promising approach to regulating gene expression artificially is through the use of short RNAi. The full rules for achieving effective shut-off reliably are not yet certain but it is clear that good silencing can often be achieved for any chosen gene simply by testing a relatively small number of short RNAi molecules targeted against a gene of interest (often as few as three suffices). The full potential of RNAi in mammals is still being explored but if the signal feeds back to the gene to induce methylation and more permanent silencing, then RNAi may prove to be a very effective way of artificially regulating the expression of chosen genes in mammals (reviewed in Dykxhoorn *et al.*, 2003). Indeed, the likelihood that human endogenous retroviruses and repetitive elements are silenced by RNAi-related mechanisms, as are retroviral-like elements in other organisms (reviewed in Zamore, 2002), suggests that RNAi-based strategies will be effective for regulating gene expression in mammalian systems.

Past limitations and future prospects

The regulation of gene expression in bacteria relies on sequence-specific DNA-binding proteins that typically either block or recruit RNA polymerase (reviewed in Lawrence, 2002). The study of mammalian gene regulation revealed similar concepts but it is now clear that multiple sequence-specific DNA-binding proteins are generally necessary for the expression of single genes and these operate through a wide variety of mechanisms, including not only the recruitment of RNA polymerase and its associated factors, but also the recruitment of chromatin remodelling and modifying enzymes. It is also clear that in addition to proteins, RNA molecules are also involved in regulating gene expression. Finally, though we have concentrated on the control of transcriptional initiation, many of the other steps required for making an active protein from a gene template can and often are regulated (Figs 5.1 and 5.7).

Although there has been much progress in understanding the mechanisms that control the output of the transcriptome, there have been some limitations to advancement in mammalian systems. The large size of the mammalian genome causes a number of problems, for instance the fact that regulatory regions may be spread across many kilobases makes it difficult to identify all the *cis*-elements that regulate a single gene, and the mechanisms involved in packaging large amounts of chromatin are clearly complex. Additionally, given that most sequence-specific DNA-binding proteins recognize relatively short motifs *in vitro* and each may be a member of a family of as many as 30 highly related proteins with similar DNA-binding specificity, it is often difficult to determine whether the biochemical activities observed *in vitro* are important *in vivo*. For instance, showing that a particular DNA-binding protein can bind to a regulatory sequence and affect expression in cell culture assays is not always a guarantee that it has this activity *in vivo*.

Conversely, the complexity of mammalian cell protein networks is such that when changes in gene expression are observed in knockout mice with mutations in specific gene regulatory proteins, it is not always possible to determine whether the effects are directly attributable to the original mutation or an indirect consequence of a complex pathway, or whether the full effects are masked by redundancy. Nevertheless, the recent use of chromatin immuno-precipitation experiments that reveal whether or not regulatory proteins are bound at particular control elements in living cells, and the use of micro-array experiments with inducible transcription factors that allow the full complement of gene expression to be analysed over time, so that direct and indirect effects can be identified, are important steps towards a fuller understanding of gene regulatory mechanisms. It is likely that as the pathways regulated by individual regulatory factors and combinations of factors are more fully characterized, useful patterns will emerge and at least some general rules for understanding gene expression and artificially controlling it will become clear.

References

Agalioti, T., Lomvardas, S., Parekh, B., Yie, J., Maniatis, T. and Thanos, D. (2000) Ordered recruitment of chromatin modifying and general transcription factors to the IFN-beta promoter. *Cell* 103, 667–678.

Ahmad, K. and Henikoff, S. (2002a) Epigenetic consequences of nucleosome dynamics. *Cell* 111, 281–284.

Ahmad, K. and Henikoff, S. (2002b) The histone variant H3.3 marks active chromatin by replication-independent nucleosome assembly. *Molecular Cell* 9, 1191–2000.

Albright, S.R. and Tjian, R. (2000) TAFs revisited: more data reveal new twists and confirm old ideas. *Gene* 242, 1–13.

Andersen, A.A. and Panning, B. (2003) Epigenetic gene regulation by noncoding RNAs. *Current Opinion in Cell Biology* 15, 281–289.

Armstrong, J.F., Pritchard-Jones, K., Bickmore, W.A., Hastie, N.D. and Bard, J.B. (1993) The expression of the Wilms' tumour gene, WT1, in the developing mammalian embryo. *Mechanisms of Development* 40, 85–97.

Avner, P. and Heard, E. (2001) X-chromosome inactivation: counting, choice and initiation. *Nature Reviews of Genetics* 2, 59–67.

Ayyanathan, K., Lechner, M.S., Bell, P., Maul, G.G., Schultz, D.C., Yamada, Y., Tanaka, K., Torigoe, K. and Rauscher, F.J., 3rd (2003) Regulated recruitment of HP1 to a euchromatic gene induces mitotically heritable, epigenetic gene silencing: a mammalian cell culture model of gene variegation. *Genes and Development* 17, 1855–1869.

Bae, K.H., Kwon, Y.D., Shin, H.C., Hwang, M.S., Ryu, E.H., Park, K.S., Yang, H.Y., Lee, D.K., Lee, Y., Park, J., Kwon, H.S., Kim, H.W., Yeh, B.I., Lee, H.W., Sohn, S.H., Yoon, J., Seol, W. and Kim, J.S. (2003) Human zinc fingers as building blocks in the construction of artificial transcription factors. *Nature Biotechnology* 21, 275–280.

Baker, N.E. (2001) Master regulatory genes; telling them what to do. *Bioessays* 23, 763–766.

Bannister, A.J. and Kouzarides, T. (1996) The CBP co-activator is a histone acetyltransferase. *Nature* 384, 641–643.

Becker, P.B. and Horz, W. (2002) ATP-dependent nucleosome remodeling. *Annual Review of Biochemistry* 71, 247–273.

Beerli, R.R. and Barbas, C.F., 3rd (2002) Engineering polydactyl zinc-finger transcription factors. *Nature Biotechnology* 20, 135–141.

Boube, M., Joulia, L., Cribbs, D.L. and Bourbon, H.M. (2002) Evidence for a mediator of RNA polymerase II transcriptional regulation conserved from yeast to man. *Cell* 110, 143–151.

Brummelkamp, T.R., Bernards, R. and Agami, R. (2002) A system for stable expression of short interfering RNAs in mammalian cells. *Science* 296, 550–553.

Bulger, M. and Groudine, M. (1999) Looping versus linking: toward a model for long-distance gene activation. *Genes and Development* 13, 2465–2477.

Butler, J.E. and Kadonaga, J.T. (2002) The RNA polymerase II core promoter: a key component in the regulation of gene expression. *Genes and Development* 16, 2583–2592.

Cao, A. and Moi, P. (2002) Regulation of the globin genes. *Pediatric Research* 51, 415–421.

Cao, R., Wang, L., Wang, H., Xia, L., Erdjument-Bromage, H., Tempst, P., Jones, R.S. and Zhang, Y. (2002) Role of histone H3 lysine 27 methylation in Polycomb-group silencing. *Science* 298, 1039–1043.

Casadesus, J. and D'Ari, R. (2002) Memory in bacteria and phage. *Bioessays* 24, 512–518.

Cheriyath, V., Novina, C.D. and Roy, A.L. (1998) TFII-I regulates Vbeta promoter activity through an initiator element. *Molecular and Cellular Biology* 18, 4444–4454.

Chi, T. and Carey, M. (1996) Assembly of the isomerized TFIIA–TFIID–TATA ternary complex is necessary and sufficient for gene activation. *Genes and Development* 10, 2540–2550.

Collins, J.E., Goward, M.E., Cole, C.G., Smink, L.J., Huckle, E.J., Knowles, S., Bye, J.M., Beare, D.M. and Dunham, I. (2003) Reevaluating human gene annotation: a second-generation analysis of chromosome 22. *Genome Research* 13, 27–36.

Cosma, M.P. (2002) Ordered recruitment: gene-specific mechanism of transcription activation. *Molecular Cell* 10, 227–236.

Cosma, M.P., Tanaka, T. and Nasmyth, K. (1999) Ordered recruitment of transcription and chromatin remodeling factors to a cell cycle- and developmentally regulated promoter. *Cell* 97, 299–311.

Dahmus, M.E. (1996) Reversible phosphorylation of the C-terminal domain of RNA polymerase II. *Journal of Biological Chemistry* 271, 19009–19012.

Dean, W., Santos, F. and Reik, W. (2003) Epigenetic reprogramming in early mammalian development and following somatic nuclear transfer. *Seminars in Cellular and Developmental Biology*, 14, 93–100.

Dekker, J. (2003) A closer look at long-range chromosomal interactions. *Trends in Biochemical Sciences* 28, 277–280.

Denli, A.M. and Hannon, G.J. (2003) RNAi: an ever-growing puzzle. *Trends in Biochemical Sciences* 28, 196–201.

Deroo, B.J. and Archer, T.K. (2001) Glucocorticoid receptor-mediated chromatin remodeling *in vivo*. *Oncogene* 20, 3039–3046.

Dilworth, F.J., Fromental-Ramain, C., Yamamoto, K. and Chambon, P. (2000) ATP-driven chromatin remodeling activity and histone acetyltransferases act sequentially during transactivation by RAR/RXR *in vitro*. *Molecular Cell* 6, 1049–1058.

DiRenzo, J., Shang, Y., Phelan, M., Sif, S., Myers, M., Kingston, R. and Brown, M. (2000) BRG-1 is recruited to estrogen-responsive promoters and cooperates with factors involved in histone acetylation. *Molecular and Cellular Biology* 20, 7541–7549.

Dykxhoorn, D.M., Novina, C.D. and Sharp, P.A. (2003) Killing the messenger: short RNAs that silence gene expression. *Nature Reviews of Molecular and Cellular Biology* 4, 457–467.

Eddy, S.R. (2001) Non-coding RNA genes and the modern RNA world. *Nature Reviews Genetics* 2, 919–929.

Eisbacher, M., Holmes, M.L., Newton, A., Hogg, P.J., Khachigian, L.M., Crossley, M. and Chong, B.H. (2003) Protein–protein interaction between Fli-1 and GATA-1 mediates synergistic expression of megakaryocyte-specific genes through cooperative DNA binding. *Molecular and Cellular Biology* 23, 3427–3441.

Elbashir, S.M., Harborth, J., Lendeckel, W., Yalcin, A., Weber, K. and Tuschl, T. (2001a) Duplexes of 21-nucleotide RNAs mediate RNA interference in cultured mammalian cells. *Nature* 411, 494–498.

Elbashir, S.M., Lendeckel, W. and Tuschl, T. (2001b) RNA interference is mediated by 21- and 22-nucleotide RNAs. *Genes and Development* 15, 188–200.

Ewing, B. and Green, P. (2000) Analysis of expressed sequence tags indicates 35,000 human genes. *Nature Genetics* 25, 232–234.

Faustino, N.A. and Cooper, T.A. (2003) Pre-mRNA splicing and human disease. *Genes and Development* 17, 419–437.

Felsenfeld, G. and Groudine, M. (2003) Controlling the double helix. *Nature* 421, 448–453.

Fire, A., Xu, S., Montgomery, M.K., Kostas, S.A., Driver, S.E. and Mello, C.C. (1998) Potent and specific genetic interference by double-stranded RNA in *Caenorhabditis elegans*. *Nature* 391, 806–811.

Fischel-Ghodsian, N., Nicholls, R.D. and Higgs, D.R. (1987) Long range genome structure around the human alpha-globin complex analysed by PFGE. *Nucleic Acids Research* 15, 6197–6207.

Fischle, W., Wang, Y. and Allis, C.D. (2003) Histone and chromatin cross-talk. *Current Opinion in Cell Biology* 15, 172–183.

Francastel, C., Magis, W. and Groudine, M. (2001) Nuclear relocation of a transactivator subunit precedes target gene activation. *Proceedings of the National Academy of Sciences USA* 98, 12120–12125.

Gamble, M.J. and Freedman, L.P. (2002) A coactivator code for transcription. *Trends in Biochemical Sciences* 27, 165–167.

Gaston, K. and Jayaraman, P.S. (2003) Transcriptional repression in eukaryotes: repressors and repression mechanisms. *Cellular and Molecular Life Sciences* 60, 721–741.

Gatz, C. (1995) Novel inducible/repressible gene expression systems. *Methods in Cell Biology* 50, 411–424.

Geyer, P.K. and Clark, I. (2002) Protecting against promiscuity: the regulatory role of insulators. *Cellular and Molecular Life Sciences*, 59, 2112–2127.

Gill, G. (2001) Regulation of the initiation of eukaryotic transcription. *Essays in Biochemistry* 37, 33–43.

Glass, C.K. and Rosenfeld, M.G. (2000) The coregulator exchange in transcriptional functions of nuclear receptors. *Genes and Development* 14, 121–141.

Glover, J.N. and Harrison, S.C. (1995) Crystal structure of the heterodimeric bZIP transcription factor c-Fos–c-Jun bound to DNA. *Nature* 373, 257–261.

Grummt, I. (2003) Life on a planet of its own: regulation of RNA polymerase I transcription in the nucleolus. *Genes and Development* 17, 1691–1702.

Gu, W., Malik, S., Ito, M., Yuan, C.X., Fondell, J.D., Zhang, X., Martinez, E., Qin, J. and Roeder, R.G. (1999) A novel human SRB/MED-containing cofactor complex, SMCC, involved in transcription regulation. *Molecular Cell* 3, 97–108.

Haber, D.A., Sohn, R.L., Buckler, A.J., Pelletier, J., Call, K.M. and Housman, D.E. (1991) Alternative splicing and genomic structure of the Wilms tumor gene WT1. *Proceedings of the National Academy of Sciences USA* 88, 9618–9622.

Hadjantonakis, A.K., Dickinson, M.E., Fraser, S.E. and Papaioannou, V.E. (2003) Technicolour transgenics: imaging tools for functional genomics in the mouse. *Nature Review of Genetics* 4, 613–625.

Halder, G., Callaerts, P. and Gehring, W.J. (1995) Induction of ectopic eyes by targeted expression of the eyeless gene in *Drosophila*. *Science* 267, 1788–1792.

Hall, I.M., Shankaranarayana, G.D., Noma, K., Ayoub, N., Cohen, A. and Grewal, S.I. (2002) Establishment and maintenance of a heterochromatin domain. *Science* 297, 2232–2237.

Harrison, S.C. (1991) A structural taxonomy of DNA-binding domains. *Nature* 353, 715–719.

Hassan, A.H., Neely, K.E., Vignali, M., Reese, J.C. and Workman, J.L. (2001) Promoter targeting of chromatin-modifying complexes. *Frontiers in Bioscience*, 6, D1054–D1064.

Heery, D.M., Kalkhoven, E., Hoare, S. and Parker, M.G. (1997) A signature motif in transcriptional co-activators mediates binding to nuclear receptors. *Nature* 387, 733–736.

Hirose, Y. and Manley, J.L. (2000) RNA polymerase II and the integration of nuclear events. *Genes and Development* 14, 1415–1429.

Hochheimer, A. and Tjian, R. (2003) Diversified transcription initiation complexes expand promoter selectivity and tissue-specific gene expression. *Genes and Development* 17, 1309–1320.

Holstege, F.C., Jennings, E.G., Wyrick, J.J., Lee, T.I., Hengartner, C.J., Green, M.R., Golub, T.R., Lander, E.S. and Young, R.A. (1998) Dissecting the regulatory circuitry of a eukaryotic genome. *Cell* 95, 717–728.

Horn, P.J. and Peterson, C.L. (2002) Molecular biology. Chromatin higher order folding–wrapping up transcription. *Science* 297, 1824–1827.

Howe, K.J. (2002) RNA polymerase II conducts a symphony of pre-mRNA processing activities. *Biochimica Biophysica Acta* 1577, 308–324.

Isogai, Y. and Tjian, R. (2003) Targeting genes and transcription factors to segregated nuclear compartments. *Current Opinion in Cell Biology* 15, 296–303.

Jacobs, S.A., Taverna, S.D., Zhang, Y., Briggs, S.D., Li, J., Eissenberg, J.C., Allis, C.D. and Khorasanizadeh, S. (2001) Specificity of the HP1 chromo domain for the methylated N-terminus of histone H3. *EMBO Journal* 20, 5232–5241.

Jenuwein, T. and Allis, C.D. (2001) Translating the histone code. *Science* 293, 1074–1080.

Jones, P.A. and Baylin, S.B. (2002) The fundamental role of epigenetic events in cancer. *Nature Review Genetics* 3, 415–428.

Jones, S., van Heyningen, P., Berman, H.M. and Thornton, J.M. (1999) Protein-DNA interactions: A structural analysis. *Journal of Molecular Biology* 287, 877–896.

Kadonaga, J.T. (2002) The DPE, a core promoter element for transcription by RNA polymerase II. *Experimental and Molecular Medicine* 34, 259–264.

Kapranov, P., Cawley, S.E., Drenkow, J., Bekiranov, S., Strausberg, R.L., Fodor, S.P. and Gingeras, T.R. (2002) Large-scale transcriptional activity in chromosomes 21 and 22. *Science* 296, 916–919.

Kelley, R.L. and Kuroda, M.I. (2000) Noncoding RNA genes in dosage compensation and imprinting. *Cell* 103, 9–12.

Khochbin, S., Verdel, A., Lemercier, C. and Seigneurin-Berny, D. (2001) Functional significance of histone deacetylase diversity. *Current Opinion in Genetics and Development* 11, 162–166.

Kile, B.T. and Alexander, W.S. (2001) The suppressors of cytokine signalling (SOCS). *Cellular and Molecular Life Sciences* 58, 1627–1635.

Kim, J., Sif, S., Jones, B., Jackson, A., Koipally, J., Heller, E., Winandy, S., Viel, A., Sawyer, A., Ikeda, T., Kingston, R. and Georgopoulos, K. (1999) Ikaros DNA-binding proteins direct formation of chromatin remodeling complexes in lymphocytes. *Immunity* 10, 345–355.

Kmita, M. and Duboule, D. (2003) Organizing axes in time and space; 25 years of colinear tinkering. *Science* 301, 331–333.

Krebs, J.E., Kuo, M.H., Allis, C.D. and Peterson, C.L. (1999) Cell cycle-regulated histone acetylation required for expression of the yeast HO gene. *Genes and Development* 13, 1412–1421.

Krishna, S.S., Majumdar, I. and Grishin, N.V. (2003) Structural classification of zinc fingers: survey and summary. *Nucleic Acids Research* 31, 532–550.

Kuzmichev, A., Nishioka, K., Erdjument-Bromage, H., Tempst, P. and Reinberg, D. (2002) Histone methyltransferase activity associated with a human multiprotein complex containing the Enhancer of Zeste protein. *Genes and Development* 16, 2893–2905.

Lachner, M. and Jenuwein, T. (2002) The many faces of histone lysine methylation. *Current Opinion in Cell Biology* 14, 286–298.

Lagrange, T., Kapanidis, A.N., Tang, H., Reinberg, D. and Ebright, R.H. (1998) New core promoter element in RNA polymerase II-dependent transcription: sequence-specific DNA binding by transcription factor IIB. *Genes and Development* 12, 34–44.

Lawrence, J.G. (2002) Shared strategies in gene organization among prokaryotes and eukaryotes. *Cell* 110, 407–413.

Lee, T.I. and Young, R.A. (2000) Transcription of eukaryotic protein-coding genes. *Annual Review of Genetics* 34, 77–137.

Lemon, B. and Tjian, R. (2000) Orchestrated response: a symphony of transcription factors for gene control. *Genes and Development* 14, 2551–2569.

Leonard, W.J. and O'Shea, J.J. (1998) Jaks and STATs: biological implications. *Annual Review of Immunology* 16, 293–322.

Levine, M. and Tjian, R. (2003) Transcription regulation and animal diversity. *Nature* 424, 147–151.

Li, E. (2002) Chromatin modification and epigenetic reprogramming in mammalian development. *Nature Review Genetics* 3, 662–673.

Li, Q., Harju, S. and Peterson, K.R. (1999) Locus control regions: coming of age at a decade plus. *Trends in Genetics* 15, 403–408.

Luger, K., Mader, A.W., Richmond, R.K., Sargent, D.F. and Richmond, T.J. (1997) Crystal structure of the nucleosome core particle at 2.8 A resolution. *Nature* 389, 251–260.

Luscombe, N.M., Austin, S.E., Berman, H.M. and Thornton, J.M. (2000) An overview of the structures of protein-DNA complexes. *Genome Biology* 1, REVIEWS001.

Macleod, D., Charlton, J., Mullins, J. and Bird, A.P. (1994) Sp1 sites in the mouse aprt gene promoter are required to prevent methylation of the CpG island. *Genes and Development* 8, 2282–2292.

Majello, B. and Napolitano, G. (2001) Control of RNA polymerase II activity by dedicated CTD kinases and phosphatases. *Frontiers in Bioscience* 6, D1358–D1368.

Maniatis, T. and Reed, R. (2002) An extensive network of coupling among gene expression machines. *Nature* 416, 499–506.

Marmorstein, R. and Fitzgerald, M.X. (2003) Modulation of DNA-binding domains for sequence-specific DNA recognition. *Gene* 304, 1–12.

Martin, D.I. and Orkin, S.H. (1990) Transcriptional activation and DNA binding by the erythroid factor GF-1/NF-E1/Eryf 1. *Genes and Development* 4, 1886–1898.

Martin, D.I. and Whitelaw, E. (1996) The vagaries of variegating transgenes. *Bioessays* 18, 919–923.

Mathers, P.H. and Jamrich, M. (2000) Regulation of eye formation by the Rx and pax6 homeobox genes. *Cellular and Molecular Life Sciences* 57, 186–194.

Matthews, J.M. and Sunde, M. (2002) Zinc fingers–folds for many occasions. *International Union of Biochemistry and Molecular Biology Life*, 54, 351–355.

McKenna, N.J. and O'Malley, B.W. (2002) Combinatorial control of gene expression by nuclear receptors and coregulators. *Cell* 108, 465–474.

Melo, K.F., Martin, R.M., Costa, E.M., Carvalho, F.M., Jorge, A.A., Arnhold, I.J. and Mendonca, B.B. (2002) An unusual phenotype of Frasier syndrome due to IVS9 +4C>T mutation in the WT1 gene: predominantly male ambiguous genitalia and absence of gonadal dysgenesis. *Journal of Clinical Endocrinology and Metabolism* 87, 2500–2505.

Miller, M., Shuman, J.D., Sebastian, T., Dauter, Z. and Johnson, P.F. (2003) Structural basis for DNA recognition by the basic region leucine zipper transcription factor CCAAT/enhancer-binding protein alpha. *Journal of Biological Chemistry* 278, 15178–15184.

Muchardt, C., Guilleme, M., Seeler, J.S., Trouche, D., Dejean, A. and Yaniv, M. (2002) Coordinated methyl and RNA binding is required for heterochromatin localization of mammalian HP1alpha. *EMBO Reports* 3, 975–981.

Muratani, M. and Tansey, W.P. (2003) How the ubiquitin-proteasome system controls transcription. *Nature Reviews of Molecular and Cellular Biology* 4, 192–201.

Myers, L.C., Gustafsson, C.M., Bushnell, D.A., Lui, M., Erdjument-Bromage, H., Tempst, P. and Kornberg, R.D. (1998) The Med proteins of yeast and their function through the RNA polymerase II carboxy-terminal domain. *Genes and Development* 12, 45–54.

Myers, L.C., Gustafsson, C.M., Hayashibara, K.C., Brown, P.O. and Kornberg, R.D. (1999) Mediator protein mutations that selectively abolish activated transcription. *Proceedings of the National Academy of Sciences USA* 96, 67–72.

Naar, A.M., Beaurang, P.A., Zhou, S., Abraham, S., Solomon, W. and Tjian, R. (1999) Composite co-activator ARC mediates chromatin-directed transcriptional activation. *Nature* 398, 828–832.

Naar, A.M., Lemon, B.D. and Tjian, R. (2001) Transcriptional coactivator complexes. *Annual Review of Biochemistry* 70, 475–501.

Nan, X., Campoy, F.J. and Bird, A. (1997) MeCP2 is a transcriptional repressor with abundant binding sites in genomic chromatin. *Cell* 88, 471–481.

Nan, X., Ng, H.H., Johnson, C.A., Laherty, C.D., Turner, B.M., Eisenman, R.N. and Bird, A. (1998) Transcriptional repression by the methyl-CpG-binding protein MeCP2 involves a histone deacetylase complex. *Nature* 393, 386–389.

Nardini, M., Spano, S., Cericola, C., Pesce, A., Massaro, A., Millo, E., Luini, A., Corda, D. and Bolognesi, M. (2003) CtBP/BARS: a dual-function protein involved in transcription co-repression and Golgi membrane fission. *EMBO Journal* 22, 3122–3130.

Narlikar, G.J., Fan, H.Y. and Kingston, R.E. (2002) Cooperation between complexes that regulate chromatin structure and transcription. *Cell* 108, 475–487.

Oelgeschlager, T., Chiang, C.M. and Roeder, R.G. (1996) Topology and reorganization of a human TFIID–promoter complex. *Nature* 382, 735–738.

Ogryzko, V.V., Schiltz, R.L., Russanova, V., Howard, B.H. and Nakatani, Y. (1996) The transcriptional coactivators p300 and CBP are histone acetyltransferases. *Cell* 87, 953–959.

Orkin, S.H. (1996) Development of the hematopoietic system. *Current Opinion in Genetics and Development* 6, 597–602.

Orkin, S.H. (1998) Embryonic stem cells and transgenic mice in the study of hematopoiesis. *International Journal of Developmental Biology* 42, 927–934.

Orphanides, G. and Reinberg, D. (2002) A unified theory of gene expression. *Cell* 108, 439–451.

Otero, G., Fellows, J., Li, Y., de Bizemont, T., Dirac, A.M., Gustafsson, C.M., Erdjument-Bromage, H., Tempst, P. and Svejstrup, J.Q. (1999) Elongator, a multisubunit component of a novel RNA polymerase II holoenzyme for transcriptional elongation. *Molecular Cell* 3, 109–118.

Pabo, C.O. and Sauer, R.T. (1992) Transcription factors: structural families and principles of DNA recognition. *Annual Review of Biochemistry* 61, 1053–1095.

Pichaud, F. and Desplan, C. (2002) Pax genes and eye organogenesis. *Current Opinion in Genetics and Development* 12, 430–434.

Plath, K., Mlynarczyk-Evans, S., Nusinow, D.A. and Panning, B. (2002) Xist RNA and the mechanism of X chromosome inactivation. *Annual Review of Genetics* 36, 233–278.

Plath, K., Fang, J., Mlynarczyk-Evans, S.K., Cao, R., Worringer, K.A., Wang, H., de la Cruz, C.C., Otte, A.P., Panning, B. and Zhang, Y. (2003) Role of histone H3 lysine 27 methylation in X inactivation. *Science* 300, 131–135.

Pownall, M.E., Gustafsson, M.K. and Emerson, C.P., Jr (2002) Myogenic regulatory factors and the specification of muscle progenitors in vertebrate embryos. *Annual Review of Cellular and Developmental Biology* 18, 747–783.

Prosser, J. and van Heyningen, V. (1998) PAX6 mutations reviewed. *Human Mutation* 11, 93–108.

Proudfoot, N.J., Furger, A. and Dye, M.J. (2002) Integrating mRNA processing with transcription. *Cell* 108, 501–512.

Ptashne, M. and Gann, A. (1997) Transcriptional activation by recruitment. *Nature* 386, 569–577.

Puri, P.L. and Sartorelli, V. (2000) Regulation of muscle regulatory factors by DNA-binding, interacting proteins, and post-transcriptional modifications. *Journal of Cellular Physiology* 185, 155–173.

Rachez, C., Suldan, Z., Ward, J., Chang, C.P., Burakov, D., Erdjument-Bromage, H., Tempst, P. and Freedman, L.P. (1998) A novel protein complex that interacts with the vitamin D3 receptor in a ligand-dependent manner and enhances VDR transactivation in a cell-free system. *Genes and Development* 12, 1787–1800.

Radhakrishnan, I., Perez-Alvarado, G.C., Parker, D., Dyson, H.J., Montminy, M.R. and Wright, P.E. (1997) Solution structure of the KIX domain of CBP bound to the transactivation domain of CREB: a model for activator:coactivator interactions. *Cell* 91, 741–752.

Rakyan, V.K., Blewitt, M.E., Druker, R., Preis, J.I. and Whitelaw, E. (2002) Metastable epialleles in mammals. *Trends in Genetics* 18, 348–351.

Reed, R. (2003) Coupling transcription, splicing and mRNA export. *Current Opinion in Cell Biology* 15, 326–331.

Reid, G., Hubner, M.R., Metivier, R., Brand, H., Denger, S., Manu, D., Beaudouin, J., Ellenberg, J. and Gannon, F. (2003) Cyclic, proteasome-mediated turnover of unliganded and liganded ERalpha on responsive promoters is an integral feature of estrogen signaling. *Molecular Cell* 11, 695–707.

Renaud, J.P. and Moras, D. (2000) Structural studies on nuclear receptors. *Cellular and Molecular Life Sciences* 57, 1748–1769.

Rice, J.C. and Allis, C.D. (2001) Histone methylation versus histone acetylation: new insights into epigenetic regulation. *Current Opinion in Cell Biology* 13, 263–273.

Richmond, T.J., Finch, J.T., Rushton, B., Rhodes, D. and Klug, A. (1984) Structure of the nucleosome core particle at 7 Å resolution. *Nature* 311, 532–537.

Roberts, S.M. and Winston, F. (1997) Essential functional interactions of SAGA, a *Saccharomyces cerevisiae* complex of Spt, Ada, and Gcn5 proteins, with the Snf/Swi and Srb/mediator complexes. *Genetics* 147, 451–465.

Roest Crollius, H., Jaillon, O., Bernot, A., Dasilva, C., Bouneau, L., Fischer, C., Fizames, C., Wincker, P., Brottier, P., Quetier, F., Saurin, W. and Weissenbach, J. (2000) Estimate of human gene number provided by genome-wide analysis using *Tetraodon nigroviridis* DNA sequence. *Nature Genetics* 25, 235–238.

Roth, S.Y., Denu, J.M. and Allis, C.D. (2001) Histone acetyltransferases. *Annual Review of Biochemistry* 70, 81–120.

Sandaltzopoulos, R. and Becker, P.B. (1998) Heat shock factor increases the reinitiation rate from potentiated chromatin templates. *Molecular and Cellular Biology* 18, 361–367.

Schramm, L. and Hernandez, N. (2002) Recruitment of RNA polymerase III to its target promoters. *Genes and Development* 16, 2593–2620.

Schultz, D.C., Friedman, J.R. and Rauscher, F.J., 3rd (2001) Targeting histone deacetylase complexes via KRAB-zinc finger proteins: the PHD and bromodomains of KAP-1 form a cooperative unit that recruits a novel isoform of the Mi-2alpha subunit of NuRD. *Genes and Development* 15, 428–443.

Shang, Y., Hu, X., DiRenzo, J., Lazar, M.A. and Brown, M. (2000) Cofactor dynamics and sufficiency in estrogen receptor-regulated transcription. *Cell* 103, 843–852.

Shi, Y., Sawada, J., Sui, G., Affar el, B., Whetstine, J.R., Lan, F., Ogawa, H., Luke, M.P. and Nakatani, Y. (2003) Coordinated histone modifications mediated by a CtBP co-repressor complex. *Nature* 422, 735–738.

Sieweke, M.H. and Graf, T. (1998) A transcription factor party during blood cell differentiation. *Current Opinion in Genetics and Development* 8, 545–551.

Silva, J., Mak, W., Zvetkova, I., Appanah, R., Nesterova, T.B., Webster, Z., Peters, A.H., Jenuwein, T., Otte, A.P. and Brockdorff, N. (2003) Establishment of histone h3 methylation on the inactive X chromosome requires transient recruitment of Eed–Enx1 polycomb group complexes. *Development of the Cell* 4, 481–495.

Singer, V.L., Wobbe, C.R. and Struhl, K. (1990) A wide variety of DNA sequences can functionally replace a yeast TATA element for transcriptional activation. *Genes and Development* 4, 636–645.

Sleutels, F., Zwart, R. and Barlow, D.P. (2002) The non-coding Air RNA is required for silencing autosomal imprinted genes. *Nature* 415, 810–813.

Smale, S.T. and Kadonaga, J.T. (2003) The RNA polymerase II core promoter. *Annual Review of Biochemistry* 72, 449–479.

Starck, J., Cohet, N., Gonnet, C., Sarrazin, S., Doubeikovskaia, Z., Doubeikovski, A., Verger, A., Duterque-Coquillaud, M. and Morle, F. (2003) Functional cross-antagonism between transcription factors FLI-1 and EKLF. *Molecular and Cellular Biology* 23, 1390–1402.

Stark, G.R., Kerr, I.M., Williams, B.R., Silverman, R.H. and Schreiber, R.D. (1998) How cells respond to interferons. *Annual Review of Biochemistry* 67, 227–264.

Strahl, B.D. and Allis, C.D. (2000) The language of covalent histone modifications. *Nature* 403, 41–45.

Sui, G., Soohoo, C., Affar el, B., Gay, F., Shi, Y. and Forrester, W.C. (2002) A DNA vector-based RNAi technology to suppress gene expression in mammalian cells. *Proceedings of the National Academy of Sciences USA* 99, 5515–5520.

Sullivan, E.K., Weirich, C.S., Guyon, J.R., Sif, S. and Kingston, R.E. (2001) Transcriptional activation domains of human heat shock factor 1 recruit human SWI/SNF. *Molecular and Cellular Biology* 21, 5826–5837.

Sun, X., Zhang, Y., Cho, H., Rickert, P., Lees, E., Lane, W. and Reinberg, D. (1998) NAT, a human complex containing Srb polypeptides that functions as a negative regulator of activated transcription. *Molecular Cell* 2, 213–222.

Tamaru, H., Zhang, X., McMillen, D., Singh, P.B., Nakayama, J., Grewal, S.I., Allis, C.D., Cheng, X. and Selker, E.U. (2003) Trimethylated lysine 9 of histone H3 is a mark for DNA methylation in *Neurospora crassa*. *Nature Genetics* 34, 75–79.

Taunton, J., Hassig, C.A. and Schreiber, S.L. (1996) A mammalian histone deacetylase related to the yeast transcriptional regulator Rpd3p. *Science* 272, 408–411.

Tijsterman, M., Ketting, R.F. and Plasterk, R.H. (2002) The genetics of RNA silencing. *Annual Review of Genetics* 36, 489–519.

Tolhuis, B., Palstra, R.J., Splinter, E., Grosveld, F. and de Laat, W. (2002) Looping and interaction between hypersensitive sites in the active beta-globin locus. *Molecular Cell* 10, 1453–1465.

Tora, L. (2002) A unified nomenclature for TATA box binding protein (TBP)-associated factors (TAFs) involved in RNA polymerase II transcription. *Genes and Development* 16, 673–675.

Torchia, J., Rose, D.W., Inostroza, J., Kamei, Y., Westin, S., Glass, C.K. and Rosenfeld, M.G. (1997) The transcriptional co-activator p/CIP binds CBP and mediates nuclear-receptor function. *Nature* 387, 677–684.

Tupler, R., Perini, G. and Green, M.R. (2001) Expressing the human genome. *Nature* 409, 832–833.

Turner, J. and Crossley, M. (1998) Cloning and characterization of mCtBP2, a co-repressor that associates with basic Kruppel-like factor and other mammalian transcriptional regulators. *EMBO Journal* 17, 5129–5140.

Turner, J. and Crossley, M. (2001) The CtBP family: enigmatic and enzymatic transcriptional co-repressors. *Bioessays* 23, 683–690.

Uesugi, M., Nyanguile, O., Lu, H., Levine, A.J. and Verdine, G.L. (1997) Induced alpha helix in the VP16 activation domain upon binding to a human TAF. *Science* 277, 1310–1313.

Vaquero, A., Loyola, A. and Reinberg, D. (2003) The constantly changing face of chromatin. *Science of Aging and Knowledge of the Environment* RE4.

Verger, A. and Duterque-Coquillaud, M. (2002) When Ets transcription factors meet their partners. *Bioessays* 24, 362–370.

Verger, A., Perdomo, J. and Crossley, M. (2003) Modification with SUMO: a role in transcriptional regulation. *EMBO Reports* 4, 137–142.

Vinson, C., Myakishev, M., Acharya, A., Mir, A.A., Moll, J.R. and Bonovich, M. (2002) Classification of human B-ZIP proteins based on dimerization properties. *Molecular and Cellular Biology* 22, 6321–6335.

Volpe, T.A., Kidner, C., Hall, I.M., Teng, G., Grewal, S.I. and Martienssen, R.A. (2002) Regulation of heterochromatic silencing and histone H3 lysine-9 methylation by RNAi. *Science* 297, 1833–1837.

Weis, L. and Reinberg, D. (1997) Accurate positioning of RNA polymerase II on a natural TATA-less promoter is independent of TATA-binding-protein-associated factors and initiator-binding proteins. *Molecular and Cellular Biology* 17, 2973–2984.

Weiss, M.J., Yu, C. and Orkin, S.H. (1997) Erythroid-cell-specific properties of transcription factor GATA-1 revealed by phenotypic rescue of a gene-targeted cell line. *Molecular and Cellular Biology* 17, 1642–1651.

Wilhelm, D. and Englert, C. (2002) The Wilms tumor suppressor WT1 regulates early gonad development by activation of Sf1. *Genes and Development* 16, 1839–1851.

Wolberger, C. (1998) Combinatorial transcription factors. *Current Opinion in Genetics and Development* 8, 552–559.

Wolfe, S.A., Nekludova, L. and Pabo, C.O. (2000) DNA recognition by Cys2His2 zinc finger proteins. *Annual Review of Biophysical and Biomolecular Structure* 29, 183–212.

Woychik, N.A. and Hampsey, M. (2002) The RNA polymerase II machinery: structure illuminates function. *Cell* 108, 453–463.

Xu, L., Glass, C.K. and Rosenfeld, M.G. (1999) Coactivator and corepressor complexes in nuclear receptor function. *Current Opinion in Genetics and Development* 9, 140–147.

Zamore, P.D. (2002) Ancient pathways programmed by small RNAs. *Science* 296, 1265–1269.

Zamore, P.D., Tuschl, T., Sharp, P.A. and Bartel, D.P. (2000) RNAi: double-stranded RNA directs the ATP-dependent cleavage of mRNA at 21 to 23 nucleotide intervals. *Cell* 101, 25–33.

Zeng, L. and Zhou, M.M. (2002) Bromodomain: an acetyl-lysine binding domain. *FEBS Letters* 513, 124–128.

Zhang, W., Kadam, S., Emerson, B.M. and Bieker, J.J. (2001) Site-specific acetylation by p300 or CREB binding protein regulates erythroid Kruppel-like factor transcriptional activity via its interaction with the SWI–SNF complex. *Molecular and Cellular Biology* 21, 2413–2422.

Zhang, Y. and Reinberg, D. (2001) Transcription regulation by histone methylation: interplay between different covalent modifications of the core histone tails. *Genes and Development* 15, 2343–2360.

Zilberman, D., Cao, X. and Jacobsen, S.E. (2003) ARGONAUTE4 control of locus-specific siRNA accumulation and DNA and histone methylation. *Science* 299, 716–719.

Zwart, R., Sleutels, F., Wutz, A., Schinkel, A.H. and Barlow, D.P. (2001) Bidirectional action of the Igf2r imprint control element on upstream and downstream imprinted genes. *Genes and Development* 15, 2361–2366.

6 The Proteome

M.B. Datto[1] and T.A.J. Haystead[2]

[1]*Department of Pathology, Duke University and* [2]*Department of Pharmacology and Cancer Biology, Duke University and Serenex Inc., Durham, NC 27710, USA*

The Promise of Proteomics	153
Mass Spectrometry Basics	154
Instrumentation	154
Ion sources	154
Mass analysers	156
Mass spectrometers	157
Protein Identification/Database Searching	157
De novo peptide sequencing	158
Sequence tags	160
Peptide mass fingerprints	160
Uninterpreted MS/MS database searching	161
Proteomic Applications	162
Protein expression profiling	162
The serum proteome	165
Subproteome profiling: the phosphoproteome	167
Proteome mining and the purine-binding proteome	169
Merging Proteomics and Genomics	169
Future Challenges	171

The Promise of Proteomics

Coined only 10 years ago, the term proteomics, in all its various forms and definitions, has captured the imagination of the scientific community. Simply defined, proteomics is the study of the entire complement of proteins expressed by a given genome. This includes not only the quantity of each protein, but also when and where each protein is expressed, including expression in specific organelles, subcellular locations, cell types, tissues, at specific stages of development, or in response to extracellular signals, environmental stimuli or insults. Thus, in contrast to the comparably static genome of an organism, there is no one single proteome, but rather an infinite number of specific spatial and temporal proteomes. Of course, the properties of a protein are not limited to its amino acid sequence. Accordingly, proteomics has grown to include characterizing the array of protein post-translational modification, most notably phosphorylation. In addition to defining the individual properties of each

© CAB International 2005. *Mammalian Genomics* (eds A. Ruvinsky and J. Marshall Graves)

protein, proteomics hopes to characterize how these proteins interact to form higher order functional units involved in essential cellular processes including transcription, DNA repair and signal transduction. Simply put, proteomics has grown to encompass, in a general sense, the understanding of all biology at its most basic protein level. Add to this lofty goal, the rapid technological developments of mass spectrometry, automation and high-throughput analysis and one begins to understand the high hopes of proteomics.

If these goals can be only partially achieved, the impact on our understanding of basic biology and medicine would be profound. The ability to determine the identity and quantity of all phosphorylated proteins (the phosphoproteome) in a given cell in a rapid and reproducible fashion would be an enormous aid in defining kinase-mediated signal transduction pathways. To accurately define the components of large complexes or determine in a systematic way protein–protein interactions (the interactome) will undoubtedly yield benefits in all areas of basic biology. The ability to determine all protein-binding targets of specific drugs (the drug-binding proteome) will yield more effective therapeutics, as well as a better understanding of disease processes. The ability to determine the identity and quantity of all proteins in human serum (the serum proteome) holds the hope of finding disease-associated markers which could aid in diagnosis or prognosis. In a similar fashion, the ability to define the proteome of specific tumour types may lead to a better understanding of the tumorigenic process and subsequently targeted therapeutics. These are just a few of the promises of proteomics. Although none are yet a complete reality, it is these promises that drive the further development of this field.

Mass Spectrometry Basics

Instrumentation

Mass spectrometry (MS) is central to modern proteomics, and recent technological advances in MS have driven the emergence of this field. Although MS has been used in the analysis of small molecules for decades, recent years have seen the development of mass spectrometers that can analyse complex molecules including proteins and peptides with a high degree of precision. These instruments allow for the rapid identification of purified proteins, with sensitivities into the femtomolar range. Thus, the identity of a protein is no longer the limiting factor in a well designed experiment. In addition, mass spectrometers have been developed which can analyse complex peptide mixtures with little front end purification or fractionation, allowing for protein profiling of clinical or experimental samples. Regardless of the specific type of mass spectrometer or its application, all of these instruments have three basic elements; an ion source, a mass analyser and an ion detector (Fig. 6.1).

Ion sources

Mass spectrometers measure the mass to charge ratio (M/Z) of analytes. In the case of peptides, the M/Z ratio is one key feature used to determine the peptide's identity. In order to gain accurate information on the mass and charge of a peptide it must first be ionized in a way that is not only uniform and reproducible, but causes minimal fragmentation or modifications which would lead to changes in mass. In addition, the ionized peptide must be volatized in order to enter the vacuum chamber of the mass analyser. The development of techniques to ionize peptides in this way has been a key technological advance in the development of MS-based proteomics, as recognized by the 2002 Nobel Prize in chemistry. These techniques, termed electrospray ionization (ESI) (Fenn *et al.*, 1989) and matrix-assisted laser desorption/ionization (MALDI) (Karas and Hillenkamp, 1988), allow for the ionization of large, complex molecules without significant chemical change or fragmentation. Each of these ionization techniques has a specific set of applications to which it is best suited, as well as certain limitations.

Electrospray ionization (ESI)

ESI ionizes peptides by spraying a fine mist of peptide solution into a high voltage electric field (Fenn *et al.*, 1989; Wilm and Mann, 1996;

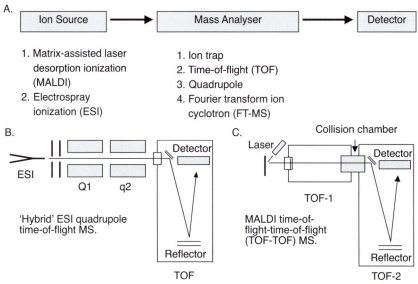

Fig. 6.1. Basic mass spectrometer configurations. (A) Mass spectrometer components. A mass spectrometer consists of three components: an ion source, a mass analyser (or tandem mass analysers) and a detector. Specific ion sources and analysers are listed. (B) ESI quadrupole time-of-flight mass spectrometer, ESI-Q-TOF. A specific mass spectrometer configuration consisting of an ESI ion source, a quadrupole mass analyser (Q1), a collision cell (q2) and a time-of-flight analyser. This instrument can be operated in either MS or MS/MS mode, and is used predominantly for collision-induced dissociation (CID) peptide sequencing. (C) MALDI time-of-flight-time-of-flight mass spectrometer, MALDI-TOF-TOF. A mass spectrometer configuration consisting of a MALDI ion source and two time-of-flight analysers separated by a collision cell. This instrument can be operated both in MS mode for peptide mass fingerprinting and in MS/MS mode for CID peptide sequencing. MS/MS spectra produced from this instrument, however, tend to be more difficult to interpret than those generated by ESI-Q-TOF.

Wilm *et al.*, 1996). This is accomplished by pushing the sample through a very fine (1 mM), metal-lined, glass capillary at flow rates as low as 5 nl/min (Wilm and Mann, 1996; Wilm *et al.*, 1996). As the peptide mist enters the electric field at the orifice of the mass analyser, the solvent evaporates, and the peptides take on an additional positive charge. Thus, ESI of peptides derived from tryptic digestion of larger proteins will contain at least two positive charges; one from the C-terminal lysine or arginine, and one imparted by the ionization technique. As will become clear in the next section, this feature allows for more straightforward '*de novo*' sequencing of peptides.

The exact physical mechanism of ion production remains unknown. However, certain experimental observations have become clear. Not all peptides ionize with the same efficiency. Some peptides ionize well and can be analysed easily by MS. Others are ionized very poorly. In addition, the composition and complexity of the starting sample influences the ionization and ultimate detection of its component peptides. The presence of highly abundant peptides within a mixture will inhibit the ionization and detection of low abundance peptides, through a phenomenon called ion suppression. For these reasons, MS is inherently poorly quantitative.

Since ESI is a liquid-based technology it can be easily coupled to liquid-based fractionation methods. A liquid chromatography system has been designed to elute directly into the MS ESI capillary (Deterding *et al.*, 1991; Davis *et al.*, 1995; McCormack *et al.*, 1997; Davis and Lee, 1998). This allows for the automated partial fractionation of peptides prior to MS analysis. This fractionation increases the ability to detect low abundant peptides from complex mixtures by separating them from higher abundant components. It also eliminates the tedious

work of partially purifying peptides and manually loading peptide solutions into capillaries.
Chromatography can be carried out using any
type of column. The most common for this
application are strong ion exchange and
reverse phase columns. In addition, distinct
chromatography steps can be run in tandem to
further increase peptide resolution prior to MS
analysis. This technology has been termed
MudPIT (multidimensional protein identification technology) (Washburn *et al.*, 2001;
Wolters *et al.*, 2001; Florens *et al.*, 2002). The
degree of resolution capable by MudPIT rivals
the separation and detection methods of conventional protein gel electrophoresis. This resolution dramatically increases the number of
peptides capable of being detected by MS
analysis in a single experimental run and has
become a major advantage of ESI MS instruments.

Matrix-assisted laser desorption/ionization

In contrast to ESI, MALDI ionizes peptides
from a solid crystallized matrix (Karas and
Hillenkamp, 1988). In this technique, a peptide
or peptide mixture is mixed with an energy-
absorbing compound, typically α-cyano-4-
hydrocinnamic acid or 2,5-dihihydroxybenoic
acid. This mixture is then spotted on to a metal
plate and dried to form a matrix – peptide
crystalline solid. To produce peptide ions, the
crystal matrix is irradiated by brief laser pulses.
Peptides derived from tryptic digestion of proteins will usually form singly charged ion
species when ionized by MALDI. MALDI also
produces good peptide 'coverage'. In other
words, in a sample consisting of multiple peptides, MALDI usually produces measurable ion
species from many of the component peptides.
This property of MALDI makes it ideal for peptide mass fingerprinting, a technique for protein identification that relies on good peptide
coverage. Mass fingerprinting is described in
detail in the next section. MALDI is inherently
poorly quantitative, suffering from the same
physical limitations as ESI. In contrast to ESI,
where samples are analysed one at a time,
MALDI lends itself well to high-throughput
analysis. Up to 96 different samples can be
spotted on to a single metal plate. The laser
then ionizes each spot in turn. This process can

be completely automated, from sample
preparation, to spotting, to data analysis, to
database searching and ultimately protein
identification. These features of MALDI make it
the ionization procedure of choice for large-
scale proteomic projects (Ekstrom *et al.*, 2000).

Mass analysers

After ionization, peptides are separated by their
differing mass-to-charge ratio by the mass
analyser. Currently, four types of mass analysers are used in modern mass spectrometers;
time-of-flight (TOF), quadrupole, ion trap and
Fourier transform (FT) ion cyclotron analysers
(Yates, 1998). Each type of analyser has its
own inherent strengths and weaknesses with
varying mass accuracies, sensitivities, and
dynamic ranges. These differences have been
discussed in several excellent reviews
(Burlingame *et al.*, 1998; Yates, 1998;
Aebersold and Goodlett, 2001). The properties
and general function of these analysers will be
discussed only briefly here.
      The time-of-flight mass analyser is the most
conceptually simple of the four types. It functions by determining the time an ion takes to
traverse the length of the analyser
(Chernushevich *et al.*, 2001). This time is a
function of the mass and charge of the ion;
larger ions move more slowly, and higher
charged species move more quickly. At the end
of the ion flight path is an ion detector which
counts the ion current at each M/Z. Quadrupole
mass analysers are the most common analysers
used today. As the name implies, the quadrupole consists of four metal rods which generate
an electromagnetic field. A quadrupole analyser
can allow the transmition of all ions in a sample
or the field can be modulated to allow a stable
flight path for peptide ions of a specific M/Z
ratio (Chernushevich *et al.*, 2001). In this way,
a quadrupole can function as a mass filter,
allowing for the analysis of only a single ion
species. This is a particularly attractive feature
when multiple analysers are run in tandem for
the peptide sequencing application discussed
below. Ion trap analysers function by 'trapping'
ions in a three-dimensional electric field
(Jonscher and Yates, 1997). Ions of specific M/Z
ratio are then selectively ejected from the field

into the detector. Ion trap analysers are quite sensitive and relatively inexpensive, but have a comparably limited mass accuracy. Recent developments in ion trap analysers have led to the generation of two-dimensional analysers which maintain the sensitivity of the 3D ion trap analysers but have an increased mass accuracy (Schwartz et al., 2002). Like the ion trap analysers, Fourier transform (FT) ion cyclotron analysers are also molecular ion trapping devices. FT analysers have the greatest potential of the mass analysers for excellent sensitivity, mass accuracy, resolution, and dynamic range (Marshall et al., 1998; Smith et al., 2001). These analysers, however, are costly and their operation is complex. They are also limited in their ability to fragment peptides to generate sequence information.

Mass spectrometers

In its simplest configuration, a mass spectrometer consists of an ion source, a single mass analyser and a detector (Fig. 6.1). A MALDI-TOF MS is configured in this way. Named for its component parts, a MALDI-TOF consists of a MALDI ion source and a time-of-flight detector (Chernushevich et al., 2001). The most common application of a MALDI-TOF MS is protein identification. To this end, a sample composed of peptide fragments derived from the enzymatic digestion of a purified unknown protein is analysed. What is produced from such an experiment is a TOF spectrum: a plot of ion current intensity at each M/Z ratio. Individual peaks in this spectrum represent individual peptides from the digested protein. Singly charged peptides (the molecular ion species produced by MALDI) produce a characteristic pattern on the TOF spectrum and can, therefore, be easily identified. This pattern consists of a cluster of peaks spaced at 1 Da, arising from the natural occurrence of carbon 13. Thus, a MALDI-TOF MS, when used in this way, produces a peptide fingerprint of the original protein.

Whereas a mass analyser when used alone can produce a peptide fingerprint, mass analysers when used in tandem can give information on the sequence of each individual peptide ion. For the purposes of this discussion, a specific mass spectrometer configuration will be used as an example: the hybrid ESI-quadrupole-TOF MS (Chernushevich et al., 2001; Morris et al., 1996; Morris et al., 1997). This mass spectrometer consists of an ESI ion source linked to a quadrupole mass analyser, a quadrupole collision cell, and finally a time-of-flight mass analyser. This instrument can operate in one of two modes; MS mode and MS/MS mode (Fig. 6.2). MS mode produces a peptide fingerprint as described above. In MS/MS mode the first quadrupole acts as a molecular ion filter to allow the passage of only one specific peptide of a chosen M/Z ratio. This peptide enters a collision cell where it interacts with molecules of the collision gas (typically nitrogen or argon). This collision produces fragmentation preferentially along peptide bonds through a process termed collision-induced dissociation, CID (Hunt et al., 1981, 1986). Thus, in MS/MS mode, a spectrum is obtained that contains all of the fragments produced by a single parent peptide ion. These data can be used to determine the amino acid sequence of the parent ion. Although the above example uses the ESI-QqTOF as an example, many types of tandem mass analyser MS instruments are emerging for the generation and analysis of CID spectrum.

Protein Identification/Database Searching

The most powerful and common use of MS is protein identification. Using MS and MS/MS spectra, there are several approaches to protein identification. With the exception of de novo sequencing, all of these approaches rely heavily on protein database searching. The ability to identify a protein, therefore, directly depends on the completeness of the database. Identification of proteins from databases generated from completely sequenced and well annotated genomes is generally easier than identification from poorly annotated or incomplete genomes. Examples of completely sequenced and well annotated genomes include human (Lander et al., 2001; Venter et al., 2001), mouse (Waterston et al., 2002), Drosophila melanogaster (Adams et al., 2000; Myers et al., 2000), Saccharomyces

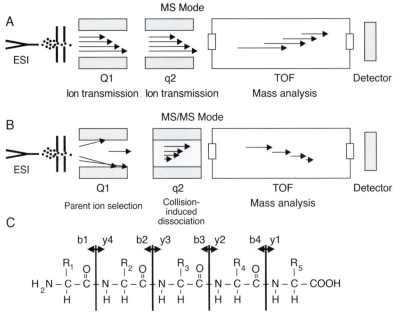

Fig. 6.2. Mass spectrometer operating modes. MS and MS/MS modes of operation using an ESI-Q-TOF mass spectrometer are shown. (A) MS mode. Peptides are ionized using electrospray ionization. They pass through the first quadrupole and collision chamber (Q1 and q2) which are set to allow transmition of all peptide ions. Ions are then separated along the TOF flight tube by their differing mass to charge ratios. An MS spectrum is produced. Individual peaks in this spectrum represent individual peptides. (B) MS/MS mode. From the MS spectrum, an individual peptide ion is selected for collision-induced dissociation (CID). The first quadrupole acts as a mass filter and is set to allow transmission of only the selected ion, termed the parent ion. This parent ion is fragmented in the collision chamber (q2), to produce daughter ions. Daughter ions are analysed in the TOF analyser to produce an MS/MS spectrum. (C) Peptide ion fragmentation patterns. Fragmentation by CID occurs preferentially along peptide bonds. If the additional charge imparted by ESI is maintained on the carboxy terminus after fragmentation, the resulting ion is called a y ion. If the charge remains on the amino terminus, the ion is called a b ion.

cerevisiae (Goffeau *et al.*, 1996) and *Plasmodium falciparum* (Gardner *et al.*, 2002) to name a few. Obviously, if a protein sequence does not occur in the database being searched, identification of this protein would be impossible. On the other hand, if the quality of the spectra is not optimal or the peptides or peptide fragments are not well represented, identification may also be difficult or ultimately ambiguous.

The simplest protein identifications start with purified proteins. Often this is accomplished by one-dimensional or two-dimensional protein gel electrophoresis. The protein band or spot of interest is excised from the gel, digested with trypsin and the peptide fragments eluted. MS and MS/MS analysis is then performed.

De novo peptide sequencing

De novo sequencing is performed by analysing MS/MS spectra (Shevchenko *et al.*, 1997, 2002). As discussed above, MS/MS spectra are produced by the collision of a single peptide with a collision gas. Fragmentation occurs preferentially along peptide bonds, allowing for sequence determination. A nomenclature has been adopted to define the different types of peptide fragments produced by CID (Roepstorff and Fohlman, 1984). If the additional charge imparted by ESI is maintained on the carboxy terminus after fragmentation, the resulting ion is called a y ion. If the charge remains on the amino terminus, it is called a b ion. Sequence information is gained by measuring the difference in mass between adjacent y or b ions. This corresponds to the mass of the lost amino acid. In this way, unambiguous *de novo* peptide sequences are obtained (Fig. 6.3). Using this sequence, protein databases are searched to identify the parent protein. The identification of several peptides from the same protein confirms its identity. A search algorithm called FASTS has been developed that can search both protein and nucleic acid databases using the sequence of several short peptides (Mackey *et al.*, 2002). This program is relatively error tolerant, allowing for cross-species identification as well as the identification of proteins from poorly annotated genomes. *De novo* sequencing is also an excellent means of peptide identification if the starting sample contains a mixture of

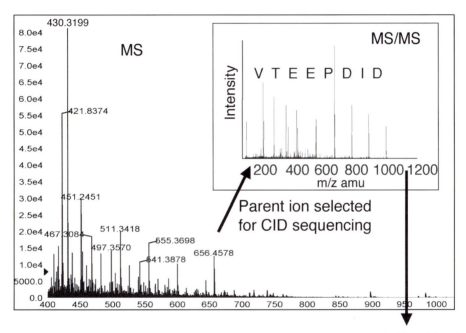

Fig. 6.3. *De novo* protein sequencing by MS/MS. An MS spectrum of the sample is first obtained which contains multiple peptide ions. In this example, several of the most prominent peaks (e.g. 421.83) correspond to proteolytic fragments of trypsin, which was used to cleave the protein sample to produce peptides for analysis. A doubly charged peptide ion (M/Z = 656.46 in this example) is chosen for fragmentation. Fragmentation produces an MS/MS spectrum (inset panel). Peptide sequence is determined by measuring the difference in mass between adjacent y ions. This corresponds to sequentially lost amino acids. Partial or complete sequence is then used to search protein databases from well annotated genomes. In this example a positive identification for a RAD23b homologue was obtained.

proteins. The sequence of individual peptides can be used to determine the identities of the individual protein components.

Although *de novo* sequencing sounds straightforward, interpretation of MS/MS can be very difficult (Shevchenko *et al.*, 1997, 2002). A priori, there is no way to distinguish y ions from b ions in a spectrum. In addition, CID can produce fragmentation at bonds other than peptide bonds, further complicating analysis. Definitive peptide identification comes at the cost of time as well as a need for technical MS and MS/MS interpretation expertise. User intervention is required to select peptide ions for fragmentation, and the MS/MS spectrum for each peptide is collected individually. Often these parent ions can have very low intensity on the MS spectrum, particularly if the starting amount of protein is small. Finding appropriate ions for sequencing in the sea of background on an MS spectrum takes experience. In addition, for reasons that are not understood, some peptides fragment poorly to yield MS/MS spectra that are uninterpretable (Shevchenko *et al.*, 1997, 2002). Although software exists to automatically perform *de novo* sequencing on peptides from CID spectra (Fernandez-de-Cossio *et al.*, 2000; Johnson and Taylor, 2002), analysis of these spectra is often best performed manually. Even with these drawbacks, *de novo* sequencing can be routinely performed on nanogram quantities of proteins purified from silver-stained gels to yield accurate protein identifications (Shevchenko *et al.*, 1997, 2002).

Sequence tags

Often only partial *de novo* sequence can be obtained for a peptide. A database searching technique has been developed which uses this partial sequence, the measured mass of the entire peptide, and the mass of the remaining amino and carboxy ends of the peptide. In addition, if the peptide is a fragment produced from a proteolytic digest, the algorithm will add the condition that the peptide must end at a cleavage site for that enzyme (in the case of trypsin, a lysine or arginine). An example of a peptide sequence tag identification is given in Fig. 6.4. Using these parameters, relatively

high confidence peptide identifications can usually be made (Mann and Wilm, 1994).

Peptide mass fingerprints

Peptide mass fingerprinting is a database-driven protein identification algorithm that uses the masses of peptides produced from the proteolytic digest of a protein to determine the protein's identity (Henzel *et al.*, 1993; James *et al.*, 1993; Mann *et al.*, 1993; Yates *et al.*, 1993). In this technique, the masses of peptides produced by proteolysis are compared with the masses of peptides of theoretical digests of all proteins in a well annotated genome. If there is significant overlap between observed peptide masses and the theoretical peptide masses of any one protein, a positive identification is made. Since this technique relies solely on intact peptide masses, only MS analysis is necessary. This type of analysis can be done using a MALDI-TOF MS and database searching can be performed without user input, allowing for both speed and high-throughput analysis (Henzel *et al.*, 1993; James *et al.*, 1993; Mann *et al.*, 1993; Yates *et al.*, 1993). An example of protein identification by peptide mass fingerprinting is given in Fig. 6.5.

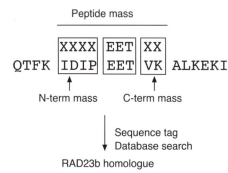

Fig. 6.4. Peptide mass tags. An example of peptide mass tag searching is given using the Rad23b homologue from Fig. 6.3. If only partial sequence was obtained (EET), positive identification can still be made using the measured mass of the entire peptide, and the calculated masses of the remaining amino and carboxy ends. In this example, the peptide was produced from a proteolytic digest using trypsin. This adds the condition that the peptide must end at a trypsin cleavage site: lysine or arginine.

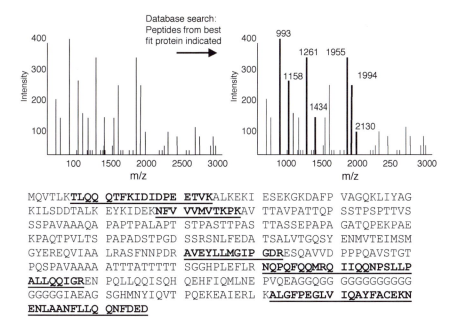

Fig. 6.5. Peptide mass fingerprints. A purified unknown protein is digested with a specific protease. The resulting peptides are analysed by MS to produce a peptide mass fingerprint. This fingerprint is compared with peptide mass fingerprints of theoretical digests of all proteins in a well annotated genome. The Rad23b homologue is used again in this example. Seven tryptic peptides in the experimental MS spectrum match peptides in theoretical peptide mass fingerprint (bold peaks). This represents 25.7% coverage of the entire protein.

Mass fingerprinting has several limitations, the greatest of which is peptide mass redundancy. The mass of a peptide is not unique to that peptide. Two peptides with the same amino acid composition but different primary sequence have the same mass, and therefore cannot be distinguished by conventional MS alone. In addition, any one peptide is not specific to a single protein. Many proteins share domains with very similar or partially identical sequences among protein family members. Larger genomes have a greater chance for redundancy. Fingerprinting is also limited by the fact that observed peptide masses may differ from theoretical masses due to post-translational modifications, gene polymorphisms, or genome sequencing errors. Fingerprinting is also dependent on the purity of the analyte. If a mixture of proteins is present in the original proteolytic digest, the resulting peptide fingerprint would obviously not match to any one protein in the database. Peptide mass finger-printing is also highly dependent on the quality of the MS spectra. Good peptide representation is required for high confidence protein identification (Henzel *et al.*, 1993; James *et al.*, 1993; Mann *et al.*, 1993; Yates *et al.*, 1993).

Uninterpreted MS/MS database searching

A final approach to protein identification is searching of databases with uninterpreted MS/MS data (Griffin *et al.*, 1995; Yates *et al.*, 1995; Perkins *et al.*, 1999; MacCoss *et al.*, 2002). This approach is theoretically similar to peptide mass fingerprinting. An MS/MS spectrum of a peptide is compared against theoretical MS/MS spectra derived from all peptides in a well annotated genome. If a significant number of experimental ion fragments match y ions and b ions derived from the theoretical fragmentation of a specific peptide, an identification can be made. This technique has many of the same limitations as mass fingerprinting, including poor genome annotation, polymorphisms, and genome sequencing errors. In spite of its limitations, uninterpreted MS/MS searching, particularly when coupled with peptide mass fingerprinting, can lead to high confidence protein identification in a completely automated, high-throughput fashion.

Proteomic Applications

Protein expression profiling

One of the primary applications of these powerful new technologies is protein expression profiling. The goal of expression profiling is to determine the identity and quantity of individual proteins in complex mixtures. This has been made possible, in part, by the advances in mass spectrometry that allow for relatively easy identification of proteins, both purified and in mixtures. This, however, has been a mixed blessing. Many studies define specific proteomes by generating large lists of expressed proteins. The increasing size of these lists over the past couple of years is a true testament to the technological advances in this field. However, by the nature of mass spectrometry, no matter how long they are, these lists are always incomplete and often minimally quantitative. In addition, the absence of a detected protein does not mean its true absence from a sample. Even if detected, a protein's expression level is difficult to determine since the ion current for a given peptide is dependent on many complex factors other than peptide abundance. In addition, mass spectrometers are very sensitive, and the detection of a peptide in a proteome may represent contamination rather than its true presence. Thus, it is difficult to answer biological questions from these protein lists. However, if biology is allowed to drive technology and expression profiling starts with a specific biological question, significant findings are often discovered.

Two-dimensional gel electrophoresis

Protein profiling has its origins in two-dimensional protein gel electrophoresis, a technique which separates proteins based on their size and isoelectric point. First described in 1975 by O'Farrell (O'Farrell, 1975), Klose (Klose, 1975), and Scheele (Scheele, 1975), the high resolution two-dimensional gel has formed the foundation of modern proteomics. Using this technique, the protein expression pattern of a sample can be visualized. In addition, differences in expression among samples can be quantitatively compared. In a single 2D gel, hundreds of proteins from complex mixtures can be resolved making it, to the present day, an essential tool in proteome profiling (Rabilloud, 2002). A simple scheme for proteomic profiling by 2D electrophoresis is outlined in Fig. 6.6.

Unfortunately, 2D gels have several limitations. One of the greatest is the technical expertise required to obtain reproducible protein patterns. This limitation is partially solved by the creation of immobilized pH gradients which have greatly improved reproducibility (Bjellqvist et al., 1993; Gorg et al., 2000). Narrow range pH gradients have also increased the resolving power of 2D gels for complex protein mixtures. It remains, however, a labour intensive, technically challenging technique. Intrinsic limitations to 2D electrophoresis also exist. Many proteomes have a very large range of protein expression levels. In the serum proteome, for example, expression of proteins varies by many orders of magnitude from the most abundant protein (serum albumin 50 mg/ml) to the least abundant proteins (in the range of 1 pg/ml or less). This is not a unique problem of serum, but is present to varying degrees in every proteome, including relatively simple proteomes such as yeast (Gygi et al., 2000). Detecting these low abundant, low copy proteins by 2D gel electrophoresis in this context is not possible, even with the most sensitive visualization techniques. Although biochemical fractionation prior to 2D electrophoresis can partially solve this problem, it remains a considerable limitation. Another fundamental assumption when protein profiling by 2D gel electrophoresis is that each spot on the gel represents an individual protein. Co-migration of proteins can occur with highly complex protein mixtures, compromising expression analysis (Gygi et al., 2000).

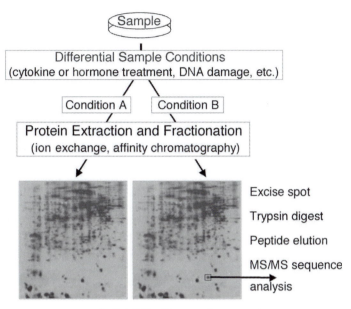

Fig. 6.6. Differential proteome profiling by two-dimensional protein electrophoresis. Differential protein profiling begins with the selection of experimental conditions. Infinite possibilities exist. One example is the comparative profiling of a cell line or tissue in the presence or absence of a specific cytokine or hormone. Protein lysates are prepared for each condition. These lysates are then analysed by 2D gel electrophoresis, and differences in protein expression are visualized by silver staining. Proteins that change in abundance between experimental conditions are excised from the gel, proteolysed with trypsin and identified by MS. Alternatively, whole cell lysates can be fractionated prior to electrophoresis by a variety of chromatographic methods. Each fraction is then analysed by 2D gel electrophoresis. This decreases the complexity of each 2D gel and taken together gives a better representation of the profiled proteome.

Finally, some proteins, including large or hydrophobic proteins or proteins which are poorly soluble, do not enter the first dimension of the 2D gel, leading to their absence in the 2D profile.

Despite the limitations of 2D gel electrophoresis, many proteomes have been profiled in this way including the proteomes of a variety of microorganisms (Shevchenko et al., 1996; Cash, 2000; Nilsson et al., 2000; Bernhardt et al., 2003), human serum (Anderson and Anderson, 1977, 1991; Pieper et al., 2003), cellular organelles (Cordwell et al., 2000; Jung et al., 2000) and a number of different tumour types including adenocarcinoma of the breast (Czerwenka et al., 2001; Gharbi et al., 2002; Dwek and Alaiya, 2003), oesophagus (Soldes et al., 1999), bladder (Gromova et al., 1998; Celis et al., 1999; Gromov et al., 2002), lung (Beer

et al., 2002; Chen et al., 2002a,b), and prostate (Nelson et al., 2000a,b; Ahram et al., 2002; Meehan et al., 2002). The advantage of profiling of these kinds of samples is clear. For example, comparative profiling of neoplastic versus normal tissue could lead to insight into the mechanism of tumorigenesis, or identification of new screening markers or chemotherapeutic targets. In a similar fashion, profiling of serum could eventually lead to the discovery of disease-associated markers. Profiling of microorganisms could lead to a better understanding of their pathogenesis or mechanisms of drug resistance. These ultimate goals, however, have yet to be met. Expression profiling using 2D gel electrophoresis has led mostly to identification of proteins with relatively high expression levels, leaving low expression level proteins largely unevaluated.

Gel-independent profiling

Despite the continuing improvements in 2D gel electrophoresis technology (Herbert, 1999; Rabilloud, 2002), including new and better detection and visualization techniques (Unlu et al., 1997; Tonge et al., 2001), the limitations of 2D gels have provided the impetus for developing gel-independent approaches to large-scale proteomic projects. This work was pioneered in studies of peptide antigens associated with the major histocompatibility complexes (MHCs) of antigen-presenting cells (Henderson et al., 1992; Hunt et al., 1992; Appella et al., 1995). These presented peptides are small, 8–10 amino acids for MHC class I-associated peptides and 10–34 for class II-associated peptides, and represent a snap shot of proteins that are being recognized and presented by the immune system. If this complex mixture of peptides was isolated and subjected directly to MS analysis, the majority of information would be lost due to ion suppression by abundant peptides. This study overcomes this limitation by coupling liquid chromatography directly to MS analysis. This technique is called LC-MS/MS. Using LC-MS/MS, partial fractionation of peptides occurs prior to MS analysis, reducing the number of peptides being ionized and analysed at any one time (Henderson et al., 1992; Hunt et al., 1992; Appella et al., 1995). As peptide ions are detected by MS they are subjected to MS/MS CID sequencing, followed by database searching and ultimately peptide identification.

Additional improvements have since been made in this technique (McDonald and Yates, 2002). LC-MS/MS is now routinely performed with two column chromatography steps with different separation properties (Henderson et al., 1992; Hunt et al., 1992; Appella et al., 1995). Most commonly, these are reverse-phase and strong cation exchange chromatography. The separation of peptides prior to MS analysis provided by tandem chromatography has allowed for the identification of thousands of peptides representing hundreds to thousands of proteins from complex mixtures (Henderson et al., 1992; Hunt et al., 1992; Appella et al., 1995). LC-MS/MS has been widely used for the initial cataloguing of proteins in complex proteomes. Many examples of proteome profiling can be found in the literature, including profiling of human serum (Adkins et al., 2002; Wu et al., 2002), human urine (Davis et al., 2001; Spahr et al., 2001; Vlahou et al., 2001), human membrane proteins (Simpson et al., 2000; Barnidge et al., 2003;), various developmental stages of the malaria parasite (Florens et al., 2002; Lasonder et al., 2002), a variety of yeast subproteomes (Rout et al., 2000; Schafer et al., 2001; Pflieger et al., 2002), proteasomal proteins (Verma et al., 2000), and the nucleolus (Scherl et al., 2002; Dreger, 2003) to name a few.

Proteomic analysis of the malaria parasite demonstrates the type of the valuable information obtainable through gel-independent profiling. The life cycle of malaria is complex, with four different stages present in either the human host or mosquito vector. In two studies Florens et al. and Lasonder et al. set out to define malaria stage-specific protein expression profiles (Florens et al., 2002; Lasonder et al., 2002). In these two studies, 2400 and 1300 different proteins were identified from the estimated 6000 encoded by the malaria genome using LC-MS/MS. As was expected, many proteins were identified that were stage-specific. This information alone provides insight into the potential function of these stage-specific proteins. Stage-specific expression also provides possible targets for anti-malarial drugs, or vaccines. These efforts also identified a number of protein products that were not predicted by analysis of the genome alone, allowing for a more complete annotation of the malaria genome. Finally, the study by Lasonder et al. attempted to quantitate the relative abundance of peptides by integrating ion currents on MS spectra. Quantitative measurements made by mass spectrometry were then checked by reverse transcription-polymerase chain reaction (RT–PCR) of specific gene products. A general correlation was seen between MS quantitation and RNA expression levels for many of the stage-specific proteins studied.

More exact methods for comparative proteomics have been developed. The most widely used of these is isotope-coded affinity tagging (ICAT) (Gygi et al., 1999, 2002; Zhou et al., 2002). ICAT relies on differential incorporation of heavy isotopes into proteins from two different experimental sources. This leaves the proteins from these two sources chemically identical, but with different masses. Ultimately, this allows for a direct comparison of protein expression levels between these two sources by MS. In a typical ICAT experiment, proteins from the two samples for comparison are chemically coupled at cysteine residues with a biotin tag. Proteins from one sample are tagged with biotin containing eight molecules of deuterium (d8-ICAT). Proteins from the other sample are tagged with an entirely ^{2}H-containing biotin (d0-ICAT). The samples are mixed, fractionated or processed depending on the experimental aims, and finally proteolytically digested. The biotinylated peptides are purified by avidin affinity chromatography, and then analysed by LC-MS/MS. Each cysteine-containing peptide will appear as a doublet on the MS spectrum. The space between individual peaks is determined by the size difference of the isotope tag. The smaller peptide is derived from the d0-ICAT-labelled specimen, the larger is derived from the d8-ICAT-labelled specimen. A direct comparison of intensity for these coupled peaks provides the relative ratio of each peptide between the two samples (Fig. 6.7). MS/MS analysis can be automatically performed on these peptides allowing for their identification. Thus, relative quantities and identities of proteins from similar specimens can be determined by a single fully automated technique. Several studies have demonstrated the utility of ICAT in quantitative comparative proteomics (Shiio *et al.*, 2002; Ranish *et al.*, 2003).

The ICAT technique is not without limitations. Not all proteins contain a cysteine residue on a proteolytic fragment that will be amenable to MS analysis. In addition, very low abundance peptides will probably be under-represented due to inefficiency of the chemical reactions at very low concentration of substrate, and non-specific adsorption on to the avidin–Sepharose substrate during peptide purification. However, even with these limitations, ICAT when coupled with multidimensional chromatography can rival gel electrophoresis in the task of comparative expression protein profiling (Gygi *et al.*, 2002).

Isotope tags can also be introduced into proteins through metabolic labelling of viable cells (Ong *et al.*, 2002; Washburn *et al.*, 2002). In this technique, termed SILAC, cells are grown in the presence or absence of an amino acid tagged with heavy isotopes of carbon, for example ^{13}C-labelled arginine, or in the presence of heavy isotope-enriched media (^{15}N media). Isotope-tagged peptides can be distinguished from native peptides by a difference in mass, allowing for the type of quantitation described above.

The serum proteome

Recently, several studies have attempted the proteomic profiling of serum (Adkins *et al.*, 2002; Li *et al.*, 2002; Petricoin *et al.*, 2002a; Wu *et al.*, 2002; Pieper *et al.*, 2003; Poon *et al.*, 2003). Serum is a complex protein mixture that highlights many of the difficulties of protein profiling. The greatest of these difficulties is the high concentration of a single protein product, albumin. The large albumin signal in any profiling approach can mask the presences of less abundant proteins. The second most abundant protein family in serum is the immunoglobins, which contain highly diverse peptides within their variable regions. This presents another unique problem; these diverse sequences would not exist in any database. The high degree of post-translational modification and proteolysis adds another degree of complexity to expression profiling of serum. Finally, the concentrations of the remaining proteins in serum vary over many orders of magnitude. In spite of these hurdles, great progress has been made in defining the serum proteome. In one study (Adkins *et al.*, 2002), 490 different proteins were identified by LC-MS/MS. The identified proteins include some that are present at very low abundance including prostate-specific antigen, a protein currently used for prostate cancer screening.

Other studies have developed profiling techniques which dispense with protein identification and profile merely by comparative MS spectra analysis. These studies rely on a technology termed SELDI, surface-enhanced laser desorption ionization. SELDI uses a surface with specific chemical properties to capture a subset of proteins. These proteins are then ionized in a way similar to MALDI. In one study, serum from women with and without ovarian cancer was analysed (Petricoin *et al.*, 2002b). By simply comparing spectra from these women, a nearly perfect segregation between those with and those without cancer was achieved. Similar

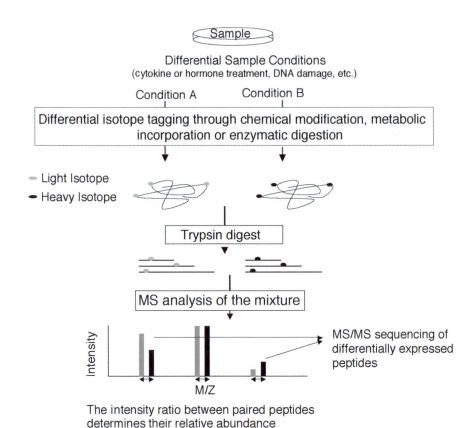

Fig. 6.7. Differential proteome profiling by isotope-coded affinity tagging (ICAT). Protein lysates are prepared from two different experimental conditions. Through chemical modification, enzymatic digestion or metabolic incorporation, proteins from sample A are tagged with a light isotope-containing reagent and proteins from sample B are tagged with a heavy isotope-containing reagent. This leaves the individual proteins between the two samples chemically identical, but of differing masses. The samples are then mixed, fractionated or processed, and proteolytically digested. The resulting peptides are analysed by LC-MS/MS. Each cysteine-containing peptide will appear as a doublet on the MS spectrum. The smaller peptide is derived from the light isotope-labelled specimen, and the larger is derived from the heavy isotope-labelled specimen. A direct comparison of intensities for these coupled peaks provides the relative ratio of each peptide between the two samples. In this example, three peptides are present. The smallest is more abundant in condition A, the middle is equal in quantity between the two conditions and the largest is more abundant in condition B.

results have been reported for prostate cancer (Petricoin *et al.*, 2002a), breast cancer (Li *et al.*, 2002), and hepatocellular carcinoma (Poon *et al.*, 2003). Unfortunately, protein identification is not possible with SELDI. Proteins are not proteolysed when analysed by SELDI, and fragmentation of full-length proteins is inefficient and would yield MS/MS spectra that are impossible to interpret. Therefore, the identity of the proteins that are providing this high degree of accuracy remains unknown. However, due to ion suppression, high abundance proteins in serum are probably being analysed at the expense of the lower abundance proteins. It is likely that the distinguishing factors in serum are a side effect of ovarian cancer, representing the body's reaction to the disease process. High accuracy in identifying a specific disease process in the general population using this approach, therefore, is unlikely. Regardless, these experiments have highlighted the potential of profiling serum in the detection of disease.

Subproteome profiling: the phosphoproteome

Post-translational protein modifications play a central role in regulating protein function. One of the most common protein modifications is phosphorylation. Protein phosphorylation has been described in nearly every cellular process and signalling pathway. Site-specific phosphorylation is a key regulator of many enzymes, either activating or inactivating enzymatic activity. Phosphorylation can also regulate protein–protein interactions and subcellular localization. Phosphorylation can dictate protein stability, regulate transcription, control cell cycle progression, and regulate smooth muscle contraction. This list does not even begin to describe the global nature and importance of phosphorylation in protein function, cellular biology and physiology (Manning *et al.*, 2002). Proteomics has provided the tools to profile the phosphorylation state of all proteins in a given cell or biological sample (the phosphoproteome).

The phosphoproteome is a dynamic entity, constantly changing with changes in cellular environment. An infinite number of phosphoproteomes exist. For example, the phosphoproteome of a rapidly dividing cell under the effect of a growth stimulatory hormone or cytokine is dramatically different from that of a quiescent cell. These phosphorylation differences, in part, are responsible for the different cellular behaviours. The ability to profile these phosphoproteomes can provide insight into the underlying biochemical mechanisms of fundamental cellular processes.

One approach to phosphoproteome profiling is two-dimensional gel electrophoresis of ^{32}P radioactively labelled cell or tissue lysates (Fig. 6.8). In this approach two samples of different experimental conditions are metabolically labelled with ^{32}P. Lysates are prepared from these samples, and fractionated by ion exchange chromatography. Fractions are then resolved on 2D gels and autoradiography is performed. In this way, changes in the phosphorylation status of thousands of proteins can

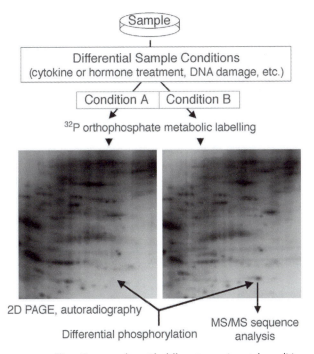

Fig. 6.8. Phosphoproteome profiling. Two samples with different experimental conditions are metabolically labelled with ^{32}P orthophosphate. Protein lysates are analysed by 2D gel electrophoresis and visualized by autoradiography. Differentially phosphorylated proteins are excised from the gel and subjected to sequence analysis by MS/MS.

be simultaneously measured by evaluating changes in ^{32}P incorporation. Differentially phosphorylated proteins are subsequently excised and subjected to sequence analysis by MS/MS. This technique has been successfully used, in our lab, in the identification of several proteins which are differentially phosphorylated in the regulation of smooth muscle contraction (MacDonald et al., 2001a; Wu et al., 1998).

When profiling by ^{32}P metabolic labelling, the specific phosphorylated amino acid within the differentially phosphorylated proteins can usually be determined using Edman degradation sequencing (Aebersold et al., 1991; Boyle et al., 1991). Many examples of phosphorylation site analysis using this technology are present in literature (MacDonald et al., 2000, 2001b; Walker et al., 2001; Borman et al., 2002). Even with the introduction of MS for phosphorylation site identification, Edman degradation remains a widely used approach. Briefly, a ^{32}P-labelled protein is purified and subject to protease digestion. The peptides are separated by reverse phase HPLC, coupled to an inert membrane at their C terminus and subjected to multiple rounds of Edman degradation. Released ^{32}P is measured in each round, and the phosphorylated residue can be inferred if the peptide sequence is known. Even if the identity of the peptide is unknown, the phosphorylated residue can still be determined, given the sequence of the original protein. This is accomplished by performing the above experiment in duplicate with two different proteases. The cycle in which radioactivity is released from the two peptide digests is compared. This usually provides sufficient information to unambiguously determine the phosphorylation site. Software developed for this application dramatically simplifies this approach (Mackey et al., 2003).

The phosphorylation site of differentially phosphorylated proteins can also be determined by MS (Zhang et al., 1998; Mitchelhill and Kemp, 1999). This can be done using a technique called precursor ion scanning, which is based on the formation of phosphate groups on CID fragmentation with a mass of 79 Da (PO_3^-) (Neubauer and Mann, 1999; Hinsby et al., 2003). Phosphoserine and phosphothreonine can also be identified on CID fragmentation by the formation of their β-elimination

products dehydroalanine and dehydroamino-2 butyric acid. Phosphotyrosines are also easily identified on CID fragmentation due to their high stability. Using MALDI-MS, phosphorylated peptides can be identified by treatment of samples with a phosphatase and looking for a 79 Da shift on the MS spectrum. Frequently, however, phosphate groups are released when subjected to MALDI, resulting in coupled 79 Da peaks without the need for phosphatase treatment. This is particularly true for phosphoserine- and phosphothreonine-containing peptides. Careful analysis of MS spectra, therefore, can sometimes yield information on peptide phosphorylation. These topics, as well as other MS approaches to phosphorylation site determination are reviewed by Mitchelhill and Kemp (1999).

Other approaches to profile the phosphoproteome centre on affinity purification of phosphorylated proteins or peptides (Mann et al., 2002). Metal affinity chromatography is an approach that has shown promise (Cao and Stults, 1999; Posewitz and Tempst, 1999; Ficarro et al., 2002). Although initially suffering from a low specificity for phosphorylated peptides, refinements in the technique, including esterification of peptide carboxyl groups, have increased the selectivity of this approach. Proof of its effectiveness has been demonstrated using the yeast phosphoproteome (Ficarro et al., 2002). Other approaches use antibody purification of phosphorylated proteins (Pandey et al., 2000). Sensitive phosphotyrosine antibodies have been developed which can selectively immunoprecipitate a broad range of tyrosine-phosphorylated proteins. These can subsequently be profiled using 2D gel electrophoresis or LC-MS/MS. Recently, some success has also been seen with phosphothreonine and phosphoserine antibody-mediated enrichment (Gronborg et al., 2002). Chemical modification of phosphorylated amino acids within peptides is another promising approach (Meyer et al., 1986; Fadden and Haystead, 1995; Oda et al., 2001; Zhou et al., 2001). In these techniques, phosphorylated peptides are either specifically coupled to biotin or immobilized on beads through β-elimination reactions, or the formation of phosphoramidates. Isolated peptides are then subject to LC-MS/MS for identification. Comparative phosphoproteome profiling using chemical derivitization coupled with differential isotope tagging has been developed (Goshe et al., 2001, 2002). This technique adds a quantitative nature to phosphoproteome profiling in the same way that the ICAT techniques have added a quantitative nature to LC-MS/MS-based proteome profiling.

No method has yet been developed that can completely profile the phosphoproteome. Each technique suffers from the same fundamental problem, incomplete representation of the phosphoproteome with poor sampling of low abundance proteins. This limitation, however, will undoubtedly be overcome as these MS- and non-MS-based techniques continue to develop.

Proteome mining and the purine-binding proteome

Proteome mining is a functional proteomics approach which defines subproteomes by their small molecule-binding properties. Using affinity chromatography with an immobilized natural ligand, a portion of the proteome is captured. This affinity-purified proteome is then analysed for its composition. With the addition of the analytical techniques described above, comparative analysis between various experimental conditions can be made on ligand-specific subproteomes. Proteome mining involves the use of small molecule drug libraries to competitively elute proteins from the captured ligand-specific proteomes. These drug-binding proteomes can provide critical information in understanding the efficacy of existing drugs, or for the discovery and development of new drugs.

In our lab, we have developed an ATP–Sepharose column for this purpose (Graves et al., 2002). ATP when coupled in a specific orientation to a solid support selectively captures purine-binding proteins. This subset of the proteome includes protein kinases, purine-dependent metabolic enzymes, heat shock proteins, and a variety of other proteins whose function is dependent on the binding of ATP. These proteins, taken together, comprise approximately 4% of the entire eukaryotic proteome. This captured proteome is then competitively eluted with

small molecule drug libraries consisting of purine analogues. With the clinical success of gleavac, a BCR-Abl tyrosine kinase inhibitor, (Druker et al., 1996; Mauro et al., 2002) purine analogues have become an attractive drug design, and many purine-based libraries have been developed. Proteins eluted from the ATP–Sepharose column with these drug libraries are then identified by MS/MS. Lead compounds, compounds which elute interesting proteins, are modified to create new related libraries which are again analysed. Through this iterative process, compounds can be developed with very high specificity and affinity for a single target (Fig. 6.9).

Proteome mining can also be used to identify the therapeutic targets of established purine analogue drugs. Using these techniques we have recently determined the binding targets of the anti-malarial drug, chloroquine (Graves et al., 2002). Chloroquine is a purine analogue and, therefore, lends itself well to this analysis. When used to elute targets from ATP–Sepharose bound to proteins extracted from malaria-infected red blood cells, two proteins were identified, quinone reductase 2 and aldehyde dehydrogenase 1. These two proteins can explain both the therapeutic utility and side effects of chloroquine. Binding of chloroquine to QR2 could increase intracellular oxidative stress which is known to be toxic to the malaria parasite. The inhibition of aldehyde dehydrogenase by chloroquine in the retina could lead to a build-up of retinaldehyde in the eye, and subsequent retinopathy. With chloroquine as a lead compound, iterative screening of synthetic drugs has begun to increase the affinity for the therapeutic target and decrease the affinity for the side effect target.

Merging Proteomics and Genomics

The techniques used in mRNA expression profiling are more established than those of proteome profiling. With cDNA microarray techniques, the expression level of all genes in a sample can be determined with relative ease and accuracy. This is largely due to the relative biochemical uniformity of mRNA in comparison with protein and the ability to detect both poorly expressed transcripts and abundant transcripts using the same

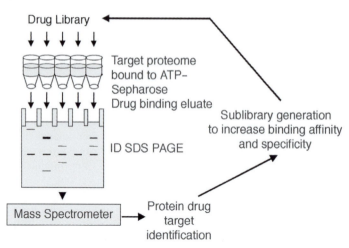

Fig. 6.9. Proteome mining. An ATP–Sepharose column is used to selectively capture purine-binding proteins from a target tissue or cell line. This captured proteome is then competitively eluted with small molecule drug libraries consisting of purine analogues. Proteins eluted from the ATP–Sepharose column for each drug within the library are resolved by protein gel electrophoresis and identified by MS/MS. Lead compounds, compounds which elute interesting proteins, are modified to created new related libraries which are again analysed. Through this iterative process, compounds can be developed with very high specificity and affinity for a single target.

techniques. However, proteins are the functional units of the cell. The abundance, modification, localization and interactions of the proteins within a cell ultimately determine biological behaviour. Thus, with current technologies, mRNA expression profiling gives a more thorough picture of gene expression, but protein profiling takes a more direct look at gene function. The data obtained from these profiling approaches are largely complementary with the strengths of each compensating for their individual weaknesses. Together, these profiling techniques can be used to define the overall state of a cell or biological system. Given enough information, predictions on the biological behaviour of that system can be made. These predictions range from the protein expression patterns of a yeast cell when subjected to specific growth conditions to the malignant potential of histologically identical tumours. These types of global system profiling experiments are just now beginning to be published.

Yeast biologists have been the greatest pioneers of protein expression profiling. Many of the techniques of proteomic profiling, including ICAT, were developed and validated in the yeast system. It is fitting that the most comprehensive combined mRNA and protein expression profil-

ing studies have been done in yeast. In a recent study, Ghaemmaghami *et al.* generated thousands of yeast strains that each have a single tandem affinity purification (TAP) tag inserted into specific open reading frames (ORFs) (Ghaemmaghami *et al.*, 2003). In this yeast library, every ORF is tagged. By quantitative immunoblotting, the expression level of every protein was determined. This was done without MS-based profiling approaches, thus bypassing all of the technical limitations described above. These studies demonstrated that the most abundant proteins are largely the ones identified by LC-MS/MS profiling studies. Proteins present at levels less than 10^{11} copies per cell represent the majority of yeast proteins yet they are poorly represented by traditional protein profiling approaches. This yeast library also allowed for a comprehensive combined mRNA/protein expression profiling of yeast under normal growth conditions. This represents the ultimate combined genomic/proteomic profile in which the expression level of every transcript and protein can be directly compared. These studies revealed that the expression of many genes could not be predicted from mRNA levels alone, although there was a general correlation between mRNA and protein expression levels for genes of all abundances. This finding highlights the complementary nature of genomic and proteomic profiling. In addition to this TAP-tagged resource, a library of yeast strains has been created which have green fluorescent protein (GFP)-tagged ORFs (Huh *et al.*, 2003). This allowed for the localization of all expressed proteins in yeast into 22 distinct subcellular locations providing a detailed and comprehensive view of protein physical interactions within the cell. These two yeast libraries represent an invaluable resource for further understanding basic eukaryotic biology and, when combined with genomic profiling, allow for comprehensive analysis of gene expression and protein localization in yeast

While it is possible to tag all yeast proteins in their genomic locus for the purpose of profiling, protein expression profiling in most experimental conditions relies on the MS and electrophoretic techniques described above. Even though limited, proteomic data obtained from more complex specimens have been shown to complement mRNA expression profiling. These studies, however, are limited in number. In a study of stage I and stage III lung adenocarcinomas, only 20% of analysed proteins had a significant correlation between protein level and mRNA levels. Of these genes, only five showed a significant difference in expression level between stage I and stage III cancer (Chen *et al.*, 2002a,b). In a study of similar design using lipopolysaccharide (LPS)-treated neutrophils in the presence or absence of a p38 map kinase inhibitor, discordance between mRNA and protein expression was also seen. The genes affected by the map kinase inhibitor detected by proteomic analysis were largely distinct from those detected by cDNA microarrays (Fessler *et al.*, 2002). These two studies highlight the complementary nature of the data obtained by protein and mRNA profiling. The benefits of combined profiling will probably become clearer with more experimentation.

Future Challenges

The way researchers study biology continues to evolve. Less than a decade ago, merely identifying an unknown protein was a laborious and uncertain task. After protein identification, the straightforward biochemical approach ruled the day. Individual proteins were identified, and studied *in vitro* and in isolation to determine their functions. Potential protein interactions or hypothesized substrates for kinases or phosphatases were tested one at a time. Expression of proteins was assayed individually, by western blotting, based on best guesses for potential function. The types of questions that could be asked were limited by the available technology.

Recent advances have begun to change the way that science is done. The genomes of many organisms have been completely sequenced and well annotated. Couple this with the relative ease of protein sequencing by MS, and the identity of a protein is no longer a limiting factor in a well designed experiment. An extension of this is the emergence of protein profiling, which allows researchers to look at the expression or phosphorylation of many proteins at once rather than one at a time. Quantitative profiling subsequently emerged, enabling automated directly comparison of the expression of hundreds of proteins between different experimental conditions.

The recent rapid advances in mass spectrometry and genomics suggest that many of the limitations of proteomics today will probably be overcome in the near future. With current technology, low abundant proteins which in part consist of transcription factors, kinases and cell surface receptors, are probably not being detected in large-scale profiling experiments. If recent progress is any indication, mass spectrometers will undoubtedly become more and more sensitive and accurate, allowing for more complete protein profiling. In a similar fashion, experimental designs will become more elegant and imaginative to make the best use of the MS technologies available. As more genomes are completely sequenced and annotated and as computer technology continues to improve, identification of proteins through large-scale database searching will in turn become more reliable.

The true challenge for the future is determining what experimental questions are best answered with these emerging technologies and how to manage the wealth of information obtained in a single large-scale MS-based experiment. The answers to these challenges will probably be found in a return to the basics and a merging of technologies. Interesting candidate proteins will be identified through MS-based protein profiling experiments and validated using the tried and true biochemical and genetic experimental approaches. The merging of genetics and proteomics has great potential. Expression or phosphorylation profiling can identify candidate proteins as important in a specific system. This gene product can then be modified, for example through targeted disruption in the mouse, and the system again studied through large-scale proteomic approaches. These types of experiments could lead to detailed protein information on biochemical pathways or physiological and disease processes that would be impossible using a gene by gene, protein by protein approach.

References

Adams, M.D., Celniker, S.E., Holt, R.A., Evans, C.A., Gocayne, J.D., Amanatides, P.G., Scherer, S.E., Li, P.W., Hoskins, R.A., Galle, R.F. *et al.* (2000) The genome sequence of *Drosophila melanogaster*. *Science* 287, 2185–2195.

Adkins, J.N., Varnum, S.M., Auberry, K.J., Moore, R.J., Angell, N.H., Smith, R.D., Springer, D.L. and Pounds, J.G. (2002) Toward a human blood serum proteome: analysis by multidimensional separation coupled with mass spectrometry. *Molecular and Cellular Proteomics* 1, 947–955.

Aebersold, R. and Goodlett, D.R. (2001) Mass spectrometry in proteomics. *Chemical Reviews* 101, 269–295.

Aebersold, R., Watts, J.D., Morrison, H.D. and Bures, E.J. (1991) Determination of the site of tyrosine phosphorylation at the low picomole level by automated solid-phase sequence analysis. *Analytical Biochemistry* 199, 51–60.

Ahram, M., Best, C.J., Flaig, M.J., Gillespie, J.W., Leiva, I.M., Chuaqui, R.F., Zhou, G., Shu, H., Duray, P.H., Linehan, W.M. *et al.* (2002) Proteomic analysis of human prostate cancer. *Molecular Carcinogenesis* 33, 9–15.

Anderson, L. and Anderson, N.G. (1977) High resolution two-dimensional electrophoresis of human plasma proteins. *Proceedings of the National Academy of Sciences USA* 74, 5421–5425.

Anderson, N.L. and Anderson, N.G. (1991) A two-dimensional gel database of human plasma proteins. *Electrophoresis* 12, 883–906.

Appella, E., Padlan, E.A. and Hunt, D.F. (1995) Analysis of the structure of naturally processed peptides bound by class I and class II major histocompatibility complex molecules. *Experientia. Supplementum* 73, 105–119.

Barnidge, D.R., Dratz, E.A., Martin, T., Bonilla, L.E., Moran, L.B. and Lindall, A. (2003) Absolute quantification of the G protein-coupled receptor rhodopsin by LC/MS/MS using proteolysis product peptides and synthetic peptide standards. *Analytical Chemistry* 75, 445–451.

Beer, D.G., Kardia, S.L., Huang, C.C., Giordano, T.J., Levin, A.M., Misek, D.E., Lin, L., Chen, G., Gharib, T.G., Thomas, D.G. *et al.* (2002) Gene-expression profiles predict survival of patients with lung adenocarcinoma. *Nature Medicine* 8, 816–824.

Bernhardt, J., Weibezahn, J., Scharf, C. and Hecker, M. (2003) *Bacillus subtilis* during feast and famine: visualization of the overall regulation of protein synthesis during glucose starvation by proteome analysis. *Genome Research* 13, 224–237.

Bjellqvist, B., Pasquali, C., Ravier, F., Sanchez, J.C. and Hochstrasser, D. (1993) A nonlinear wide-range immobilized pH gradient for two-dimensional electrophoresis and its definition in a relevant pH scale. *Electrophoresis* 14, 1357–1365.

Borman, M.A., MacDonald, J.A., Muranyi, A., Hartshorne, D.J. and Haystead, T.A. (2002) Smooth muscle myosin phosphatase-associated kinase induces Ca^{2+} sensitization via myosin phosphatase inhibition. *Journal of Biological Chemistry* 277, 23441–23446.

Boyle, W.J., van der Geer, P. and Hunter, T. (1991) Phosphopeptide mapping and phosphoamino acid analysis by two-dimensional separation on thin-layer cellulose plates. *Methods in Enzymology* 201, 110–149.

Burlingame, A.L., Boyd, R.K. and Gaskell, S.J. (1998) Mass spectrometry. *Analytical Chemistry* 70, 647R–716R.

Cao, P. and Stults, J.T. (1999) Phosphopeptide analysis by on-line immobilized metal-ion affinity chromatography-capillary electrophoresis-electrospray ionization mass spectrometry. *Journal of Chromatography A* 853, 225–235.

Cash, P. (2000) Proteomics in medical microbiology. *Electrophoresis* 21, 1187–1201.

Celis, J.E., Ostergaard, M., Rasmussen, H.H., Gromov, P., Gromova, I., Varmark, H., Palsdottir, H., Magnusson, N., Andersen, I., Basse, B. *et al.* (1999) A comprehensive protein resource for the study of bladder cancer: *http://biobase.dk/cgi-bin/celis. Electrophoresis* 20, 300–309.

Chen, G., Gharib, T.G., Huang, C.C., Taylor, J.M., Misek, D.E., Kardia, S.L., Giordano, T.J., Iannettoni, M.D., Orringer, M.B., Hanash, S.M. and Beer, D.G. (2002a) Discordant protein and mRNA expression in lung adenocarcinomas. *Molecular and Cellular Proteomics* 1, 304–313.

Chen, G., Gharib, T.G., Huang, C.C., Thomas, D.G., Shedden, K.A., Taylor, J.M., Kardia, S.L., Misek, D.E., Giordano, T.J., Iannettoni, M.D. *et al.* (2002b) Proteomic analysis of lung adenocarcinoma: identification of a highly expressed set of proteins in tumors. *Clinical Cancer Research* 8, 2298–2305.

Chernushevich, I.V., Loboda, A.V. and Thomson, B.A. (2001) An introduction to quadrupole-time-of-flight mass spectrometry. *Journal of Mass Spectrometry* 36, 849–865.

Cordwell, S.J., Nouwens, A.S., Verrills, N.M., Basseal, D.J. and Walsh, B.J. (2000) Subproteomics based upon protein cellular location and relative solubilities in conjunction with composite two-dimensional electrophoresis gels. *Electrophoresis* 21, 1094–1103.

Czerwenka, K.F., Manavi, M., Hosmann, J., Jelincic, D., Pischinger, K.I., Battistutti, W.B., Behnam, M. and Kubista, E. (2001) Comparative analysis of two-dimensional protein patterns in malignant and normal human breast tissue. *Cancer Detection and Prevention* 25, 268–279.

Davis, M.T. and Lee, T.D. (1998) Rapid protein identification using a microscale electrospray LC/MS system on an ion trap mass spectrometer. *Journal of the American Society of Mass Spectrometry* 9, 194–201.

Davis, M.T., Stahl, D.C., Hefta, S.A. and Lee, T.D. (1995) A microscale electrospray interface for on-line, capillary liquid chromatography/tandem mass spectrometry of complex peptide mixtures. *Analytical Chemistry* 67, 4549–4556.

Davis, M.T., Spahr, C.S., McGinley, M.D., Robinson, J.H., Bures, E.J., Beierle, J., Mort, J., Yu, W., Luethy, R. and Patterson, S.D. (2001) Towards defining the urinary proteome using liquid chromatography-tandem mass spectrometry. II. Limitations of complex mixture analyses. *Proteomics* 1, 108–117.

Deterding, L.J., Moseley, M.A., Tomer, K.B. and Jorgenson, J.W. (1991) Nanoscale separations combined with tandem mass spectrometry. *Journal of Chromatography* 554, 73–82.

Dreger, M. (2003) Subcellular proteomics. *Mass Spectrometry Reviews* 22, 27–56.

Druker, B.J., Tamura, S., Buchdunger, E., Ohno, S., Segal, G.M., Fanning, S., Zimmermann, J. and Lydon, N.B. (1996) Effects of a selective inhibitor of the Abl tyrosine kinase on the growth of Bcr-Abl positive cells. *Nature Medicine* 2, 561–566.

Dwek, M.V. and Alaiya, A.A. (2003) Proteome analysis enables separate clustering of normal breast, benign breast and breast cancer tissues. *British Journal of Cancer* 89, 305–307.

Ekstrom, S., Onnerfjord, P., Nilsson, J., Bengtsson, M., Laurell, T. and Marko-Varga, G. (2000) Integrated microanalytical technology enabling rapid and automated protein identification. *Analytical Chemistry* 72, 286–293.

Fadden, P. and Haystead, T.A. (1995) Quantitative and selective fluorophore labeling of phosphoserine on peptides and proteins: characterization at the attomole level by capillary electrophoresis and laser-induced fluorescence. *Analytical Biochemistry* 225, 81–88.

Fenn, J.B., Mann, M., Meng, C.K., Wong, S.F. and Whitehouse, C.M. (1989) Electrospray ionization for mass spectrometry of large biomolecules. *Science* 246, 64–71.

Fernandez-de-Cossio, J., Gonzalez, J., Satomi, Y., Shima, T., Okumura, N., Besada, V., Betancourt, L., Padron, G., Shimonishi, Y. and Takao, T. (2000) Automated interpretation of low-energy collision-induced dissociation spectra by SeqMS, a software aid for *de novo* sequencing by tandem mass spectrometry. *Electrophoresis* 21, 1694–1699.

Fessler, M.B., Malcolm, K.C., Duncan, M.W. and Worthen, G.S. (2002) A genomic and proteomic analysis of activation of the human neutrophil by lipopolysaccharide and its mediation by p38 mitogen-activated protein kinase. *Journal of Biological Chemistry* 277, 31291–31302.

Ficarro, S.B., McCleland, M.L., Stukenberg, P.T., Burke, D.J., Ross, M.M., Shabanowitz, J., Hunt, D.F. and White, F.M. (2002) Phosphoproteome analysis by mass spectrometry and its application to *Saccharomyces cerevisiae. Nature Biotechnology* 20, 301–305.

Florens, L., Washburn, M.P., Raine, J.D., Anthony, R.M., Grainger, M., Haynes, J.D., Moch, J.K., Muster, N., Sacci, J.B., Tabb, D.L. *et al.* (2002) A proteomic view of the *Plasmodium falciparum* life cycle. *Nature* 419, 520–526.

Gardner, M.J., Hall, N., Fung, E., White, O., Berriman, M., Hyman, R.W., Carlton, J.M., Pain, A., Nelson, K.E., Bowman, S. *et al.* (2002) Genome sequence of the human malaria parasite *Plasmodium falciparum. Nature* 419, 498–511.

Ghaemmaghami, S., Huh, W.K., Bower, K., Howson, R.W., Belle, A., Dephoure, N., O'Shea, E.K. and Weissman, J. S. (2003) Global analysis of protein expression in yeast. *Nature* 425, 737–741.

Gharbi, S., Gaffney, P., Yang, A., Zvelebil, M.J., Cramer, R., Waterfield, M.D. and Timms, J.F. (2002) Evaluation of two-dimensional differential gel electrophoresis for proteomic expression analysis of a model breast cancer cell system. *Molecular and Cellular Proteomics* 1, 91–98.

Goffeau, A., Barrell, B.G., Bussey, H., Davis, R.W., Dujon, B., Feldmann, H., Galibert, F., Hoheisel, J.D., Jacq, C., Johnston, M. *et al.* (1996) Life with 6000 genes. *Science* 274, 546, 563–547.

Gorg, A., Obermaier, C., Boguth, G., Harder, A., Scheibe, B., Wildgruber, R., and Weiss, W. (2000) The current state of two-dimensional electrophoresis with immobilized pH gradients. *Electrophoresis* 21, 1037–1053.

Goshe, M.B., Conrads, T.P., Panisko, E.A., Angell, N.H., Veenstra, T.D. and Smith, R.D. (2001) Phosphoprotein isotope-coded affinity tag approach for isolating and quantitating phosphopeptides in proteome-wide analyses. *Analytical Chemistry* 73, 2578–2586.

Goshe, M.B., Veenstra, T.D., Panisko, E.A., Conrads, T.P., Angell, N.H. and Smith, R.D. (2002) Phosphoprotein isotope-coded affinity tags: application to the enrichment and identification of low-abundance phosphoproteins. *Analytical Chemistry* 74, 607–616.

Graves, P.R., Kwiek, J.J., Fadden, P., Ray, R., Hardeman, K., Coley, A.M., Foley, M., and Haystead, T.A. (2002) Discovery of novel targets of quinoline drugs in the human purine binding proteome. *Molecular Pharmacology* 62, 1364–1372.

Griffin, P.R., MacCoss, M.J., Eng, J.K., Blevins, R.A., Aaronson, J.S. and Yates, J.R., 3rd (1995) Direct database searching with MALDI-PSD spectra of peptides. *Rapid Communications in Mass Spectrometry* 9, 1546–1551.

Gromov, P.S., Ostergaard, M., Gromova, I. and Celis, J.E. (2002) Human proteomic databases: a powerful resource for functional genomics in health and disease. *Progress in Biophysics and Molecular Biology* 80, 3–22.

Gromova, I., Gromov, P., Wolf, H. and Celis, J.E. (1998) Protein abundancy and mRNA levels of the adipocyte-type fatty acid binding protein correlate in non-invasive and invasive bladder transitional cell carcinomas. *International Journal of Oncology* 13, 379–383.

Gronborg, M., Kristiansen, T.Z., Stensballe, A., Andersen, J.S., Ohara, O., Mann, M., Jensen, O.N. and Pandey, A. (2002) A mass spectrometry-based proteomic approach for identification of serine/threonine-phosphorylated proteins by enrichment with phospho-specific antibodies: identification of a novel protein, Frigg, as a protein kinase A substrate. *Molecular and Cellular Proteomics* 1, 517–527.

Gygi, S.P., Rist, B., Gerber, S.A., Turecek, F., Gelb, M.H. and Aebersold, R. (1999) Quantitative analysis of complex protein mixtures using isotope-coded affinity tags. *Nature Biotechnology* 17, 994–999.

Gygi, S.P., Corthals, G.L., Zhang, Y., Rochon, Y. and Aebersold, R. (2000) Evaluation of two-dimensional gel electrophoresis-based proteome analysis technology. *Proceedings of the National Academy of Sciences USA* 97, 9390–9395.

Gygi, S.P., Rist, B., Griffin, T.J., Eng, J. and Aebersold, R. (2002) Proteome analysis of low-abundance proteins using multidimensional chromatography and isotope-coded affinity tags. *Journal of Proteome Research* 1, 47–54.

Henderson, R.A., Michel, H., Sakaguchi, K., Shabanowitz, J., Appella, E., Hunt, D.F., and Engelhard, V.H. (1992) HLA-A2.1-associated peptides from a mutant cell line: a second pathway of antigen presentation. *Science* 255, 1264–1266.

Henzel, W.J., Billeci, T.M., Stults, J.T., Wong, S.C., Grimley, C. and Watanabe, C. (1993) Identifying proteins from two-dimensional gels by molecular mass searching of peptide fragments in protein sequence databases. *Proceedings of the National Academy of Sciences USA* 90, 5011–5015.

Herbert, B. (1999) Advances in protein solubilisation for two-dimensional electrophoresis. *Electrophoresis* 20, 660–663.

Hinsby, A.M., Olsen, J.V., Bennett, K.L. and Mann, M. (2003) Signaling initiated by overexpression of the fibroblast growth factor receptor-1 investigated by mass spectrometry. *Molecular and Cellular Proteomics* 2, 29–36.

Huh, W.K., Falvo, J.V., Gerke, L.C., Carroll, A.S., Howson, R.W., Weissman, J.S. and O'Shea, E.K. (2003) Global analysis of protein localization in budding yeast. *Nature* 425, 686–691.

Hunt, D.F., Buko, A.M., Ballard, J.M., Shabanowitz, J. and Giordani, A.B. (1981) Sequence analysis of polypeptides by collision activated dissociation on a triple quadrupole mass spectrometer. *Biomedical Mass Spectrometry* 8, 397–408.

Hunt, D.F., Yates, J.R., 3rd, Shabanowitz, J., Winston, S. and Hauer, C.R. (1986) Protein sequencing by tandem mass spectrometry. *Proceedings of the National Academy of Sciences USA* 83, 6233–6237.

Hunt, D.F., Henderson, R.A., Shabanowitz, J., Sakaguchi, K., Michel, H., Sevilir, N., Cox, A.L., Appella, E. and Engelhard, V.H. (1992) Characterization of peptides bound to the class I MHC molecule HLA-A2.1 by mass spectrometry. *Science* 255, 1261–1263.

James, P., Quadroni, M., Carafoli, E. and Gonnet, G. (1993) Protein identification by mass profile fingerprinting. *Biochemical and Biophysical Research Communications* 195, 58–64.

Johnson, R.S. and Taylor, J.A. (2002) Searching sequence databases via *de novo* peptide sequencing by tandem mass spectrometry. *Molecular Biotechnology* 22, 301–315.

Jonscher, K.R. and Yates, J.R., 3rd (1997) The quadrupole ion trap mass spectrometer–a small solution to a big challenge. *Analytical Biochemistry* 244, 1–15.

Jung, E., Heller, M., Sanchez, J.C. and Hochstrasser, D.F. (2000) Proteomics meets cell biology: the establishment of subcellular proteomes. *Electrophoresis* 21, 3369–3377.

Karas, M. and Hillenkamp, F. (1988) Laser desorption ionization of proteins with molecular masses exceeding 10,000 daltons. *Analytical Chemistry* 60, 2299–2301.

Klose, J. (1975) Protein mapping by combined isoelectric focusing and electrophoresis of mouse tissues. A novel approach to testing for induced point mutations in mammals. *Humangenetik* 26, 231–243.

Lander, E.S., Linton, L.M., Birren, B., Nusbaum, C., Zody, M.C., Baldwin, J., Devon, K., Dewar, K., Doyle, M., FitzHugh, W. *et al.* (2001) Initial sequencing and analysis of the human genome. *Nature* 409, 860–921.

Lasonder, E., Ishihama, Y., Andersen, J.S., Vermunt, A.M., Pain, A., Sauerwein, R.W., Eling, W.M., Hall, N., Waters, A.P., Stunnenberg, H.G. and Mann, M. (2002) Analysis of the *Plasmodium falciparum* proteome by high-accuracy mass spectrometry. *Nature* 419, 537–542.

Li, J., Zhang, Z., Rosenzweig, J., Wang, Y.Y. and Chan, D. W. (2002) Proteomics and bioinformatics approaches for identification of serum biomarkers to detect breast cancer. *Clinical Chemistry* 48, 1296–1304.

MacCoss, M.J., Wu, C.C. and Yates, J.R., 3rd (2002) Probability-based validation of protein identifications using a modified SEQUEST algorithm. *Analytical Chemistry* 74, 5593–5599.

MacDonald, J.A., Walker, L.A., Nakamoto, R.K., Gorenne, I., Somlyo, A.V., Somlyo, A.P. and Haystead, T.A. (2000) Phosphorylation of telokin by cyclic nucleotide kinases and the identification of *in vivo* phosphorylation sites in smooth muscle. *FEBS Letters* 479, 83–88.

MacDonald, J.A., Borman, M.A., Muranyi, A., Somlyo, A.V., Hartshorne, D.J. and Haystead, T.A. (2001a) Identification of the endogenous smooth muscle myosin phosphatase-associated kinase. *Proceedings of the National Academy of Sciences USA* 98, 2419–2424.

MacDonald, J.A., Eto, M., Borman, M.A., Brautigan, D.L. and Haystead, T.A. (2001b) Dual Ser and Thr phosphorylation of CPI-17, an inhibitor of myosin phosphatase, by MYPT-associated kinase. *FEBS Letters* 493, 91–94.

Mackey, A.J., Haystead, T.A. and Pearson, W.R. (2002) Getting more from less: algorithms for rapid protein identification with multiple short peptide sequences. *Molecular and Cellular Proteomics* 1, 139–147.

Mackey, A.J., Haystead, T.A. and Pearson, W.R. (2003) CRP: cleavage of radiolabeled phosphoproteins. *Nucleic Acids Research* 31, 3859–3861.

Mann, M. and Wilm, M. (1994) Error-tolerant identification of peptides in sequence databases by peptide sequence tags. *Analytical Chemistry* 66, 4390–4399.

Mann, M., Hojrup, P. and Roepstorff, P. (1993) Use of mass spectrometric molecular weight information to identify proteins in sequence databases. *Biological Mass Spectrometry* 22, 338–345.

Mann, M., Ong, S.E., Gronborg, M., Steen, H., Jensen, O.N. and Pandey, A. (2002) Analysis of protein phosphorylation using mass spectrometry: deciphering the phosphoproteome. *Trends in Biotechnology* 20, 261–268.

Manning, G., Whyte, D.B., Martinez, R., Hunter, T. and Sudarsanam, S. (2002) The protein kinase complement of the human genome. *Science* 298, 1912–1934.

Marshall, A.G., Hendrickson, C.L. and Jackson, G.S. (1998) Fourier transform ion cyclotron resonance mass spectrometry: a primer. *Mass Spectrometry Reviews* 17, 1–35.

Mauro, M.J., O'Dwyer, M., Heinrich, M.C. and Druker, B.J. (2002) STI571: a paradigm of new agents for cancer therapeutics. *Journal of Clinical Oncology* 20, 325–334.

McCormack, A.L., Schieltz, D.M., Goode, B., Yang, S., Barnes, G., Drubin, D. and Yates, J.R., 3rd (1997) Direct analysis and identification of proteins in mixtures by LC/MS/MS and database searching at the low-femtomole level. *Analytical Chemistry* 69, 767–776.

McDonald, W.H. and Yates, J.R., 3rd (2002) Shotgun proteomics and biomarker discovery. *Disease Markers* 18, 99–105.

Meehan, K.L., Holland, J.W. and Dawkins, H.J. (2002) Proteomic analysis of normal and malignant prostate tissue to identify novel proteins lost in cancer. *Prostate* 50, 54–63.

Meyer, H.E., Hoffmann-Posorske, E., Korte, H. and Heilmeyer, L.M., Jr (1986) Sequence analysis of phosphoserine-containing peptides. Modification for picomolar sensitivity. *FEBS Letters* 204, 61–66.

Mitchelhill, K. I. and Kemp, B.E. (1999) *Phosphorylation Site Analysis by Mass Spectrometry*, 2nd edn. Oxford University Press, New York.

Morris, H.R., Paxton, T., Dell, A., Langhorne, J., Berg, M., Bordoli, R.S., Hoyes, J. and Bateman, R.H. (1996) High sensitivity collisionally-activated decomposition tandem mass spectrometry on a novel quadrupole/orthogonal-acceleration time-of-flight mass spectrometer. *Rapid Communications in Mass Spectrometry* 10, 889–896.

Morris, H.R., Paxton, T., Panico, M., McDowell, R. and Dell, A. (1997) A novel geometry mass spectrometer, the Q-TOF, for low-femtomole/attomole-range biopolymer sequencing. *Journal of Protein Chemistry* 16, 469–479.

Myers, E.W., Sutton, G.G., Delcher, A.L., Dew, I.M., Fasulo, D.P., Flanigan, M.J., Kravitz, S.A., Mobarry, C.M., Reinert, K.H., Remington, K.A. *et al.* (2000) A whole-genome assembly of *Drosophila. Science* 287, 2196–2204.

Nelson, P.S., Clegg, N., Eroglu, B., Hawkins, V., Bumgarner, R., Smith, T. and Hood, L. (2000a) The prostate expression database (PEDB): status and enhancements in 2000. *Nucleic Acids Research* 28, 212–213.

Nelson, P.S., Han, D., Rochon, Y., Corthals, G.L., Lin, B., Monson, A., Nguyen, V., Franza, B.R., Plymate, S.R., Aebersold, R. and Hood, L. (2000b) Comprehensive analyses of prostate gene expression: convergence of expressed sequence tag databases, transcript profiling and proteomics. *Electrophoresis* 21, 1823–1831.

Neubauer, G. and Mann, M. (1999) Mapping of phosphorylation sites of gel-isolated proteins by nanoelectrospray tandem mass spectrometry: potentials and limitations. *Analytical Chemistry* 71, 235–242.

Nilsson, C.L., Larsson, T., Gustafsson, E., Karlsson, K.A. and Davidsson, P. (2000) Identification of protein vaccine candidates from *Helicobacter pylori* using a preparative two-dimensional electrophoretic procedure and mass spectrometry. *Analytical Chemistry* 72, 2148–2153.

Oda, Y., Nagasu, T. and Chait, B.T. (2001) Enrichment analysis of phosphorylated proteins as a tool for probing the phosphoproteome. *Nature Biotechnology* 19, 379–382.

O'Farrell, P.H. (1975) High resolution two-dimensional electrophoresis of proteins. *Journal of Biological Chemistry* 250, 4007–4021.

Ong, S.E., Blagoev, B., Kratchmarova, I., Kristensen, D.B., Steen, H., Pandey, A. and Mann, M. (2002) Stable isotope labeling by amino acids in cell culture, SILAC, as a simple and accurate approach to expression proteomics. *Molecular and Cellular Proteomics* 1, 376–386.

Pandey, A., Podtelejnikov, A.V., Blagoev, B., Bustelo, X.R., Mann, M. and Lodish, H.F. (2000) Analysis of receptor signaling pathways by mass spectrometry: identification of vav-2 as a substrate of the epidermal and platelet-derived growth factor receptors. *Proceedings of the National Academy of Sciences USA* 97, 179–184.

Perkins, D.N., Pappin, D.J., Creasy, D.M. and Cottrell, J.S. (1999) Probability-based protein identification by searching sequence databases using mass spectrometry data. *Electrophoresis* 20, 3551–3567.

Petricoin, E.F., 3rd, Ornstein, D.K., Paweletz, C.P., Ardekani, A., Hackett, P.S., Hitt, B.A., Velassco, A., Trucco, C., Wiegand, L., Wood, K. *et al.* (2002a) Serum proteomic patterns for detection of prostate cancer. *Journal of the National Cancer Institute* 94, 1576–1578.

Petricoin, E.F., Ardekani, A.M., Hitt, B.A., Levine, P.J., Fusaro, V.A., Steinberg, S.M., Mills, G.B., Simone, C., Fishman, D.A., Kohn, E.C. and Liotta, L.A. (2002b) Use of proteomic patterns in serum to identify ovarian cancer. *Lancet* 359, 572–577.

Pflieger, D., Le Caer, J.P., Lemaire, C., Bernard, B.A., Dujardin, G. and Rossier, J. (2002) Systematic identification of mitochondrial proteins by LC-MS/MS. *Analytical Chemistry* 74, 2400–2406.

Pieper, R., Gatlin, C.L., Makusky, A.J., Russo, P.S., Schatz, C.R., Miller, S.S., Su, Q., McGrath, A.M., Estock, M.A., Parmar, P.P. *et al.* (2003) The human serum proteome: display of nearly 3700 chromatographically separated protein spots on two-dimensional electrophoresis gels and identification of 325 distinct proteins. *Proteomics* 3, 1345–1364.

Poon, T.C., Yip, T.T., Chan, A.T., Yip, C., Yip, V., Mok, T.S., Lee, C.C., Leung, T.W., Ho, S.K. and Johnson, P.J. (2003) Comprehensive proteomic profiling identifies serum proteomic signatures for detection of hepatocellular carcinoma and its subtypes. *Clinical Chemistry* 49, 752–760.

Posewitz, M.C. and Tempst, P. (1999) Immobilized gallium(III) affinity chromatography of phosphopeptides. *Analytical Chemistry* 71, 2883–2892.

Rabilloud, T. (2002) Two-dimensional gel electrophoresis in proteomics: old, old fashioned, but it still climbs up the mountains. *Proteomics* 2, 3–10.

Ranish, J.A., Yi, E.C., Leslie, D.M., Purvine, S.O., Goodlett, D.R., Eng, J. and Aebersold, R. (2003) The study of macromolecular complexes by quantitative proteomics. *Nature Genetics* 33, 349–355.

Roepstorff, P. and Fohlman, J. (1984) Proposal for a common nomenclature for sequence ions in mass spectra of peptides. *Biomedical Mass Spectrometry* 11, 601.

Rout, M.P., Aitchison, J.D., Suprapto, A., Hjertaas, K., Zhao, Y. and Chait, B.T. (2000) The yeast nuclear pore complex: composition, architecture, and transport mechanism. *Journal of Cell Biology* 148, 635–651.

Schafer, H., Nau, K., Sickmann, A., Erdmann, R. and Meyer, H.E. (2001) Identification of peroxisomal membrane proteins of *Saccharomyces cerevisiae* by mass spectrometry. *Electrophoresis* 22, 2955–2968.

Scheele, G.A. (1975) Two-dimensional gel analysis of soluble proteins. Characterization of guinea pig exocrine pancreatic proteins. *Journal of Biological Chemistry* 250, 5375–5385.

Scherl, A., Coute, Y., Deon, C., Calle, A., Kindbeiter, K., Sanchez, J.C., Greco, A., Hochstrasser, D. and Diaz, J.J. (2002) Functional proteomic analysis of human nucleolus. *Molecular Biology of the Cell* 13, 4100–4109.

Schwartz, J.C., Senko, M.W. and Syka, J.E. (2002) A two-dimensional quadrupole ion trap mass spectrometer. *Journal of the American Society of Mass Spectrometry* 13, 659–669.

Shevchenko, A., Jensen, O.N., Podtelejnikov, A.V., Sagliocco, F., Wilm, M., Vorm, O., Mortensen, P., Boucherie, H. and Mann, M. (1996) Linking genome and proteome by mass spectrometry: large-scale identification of yeast proteins from two dimensional gels. *Proceedings of the National Academy of Sciences USA* 93, 14440–14445.

Shevchenko, A., Chernushevich, I., Ens, W., Standing, K.G., Thomson, B., Wilm, M. and Mann, M. (1997) Rapid 'de novo' peptide sequencing by a combination of nanoelectrospray, isotopic labeling and a quadrupole/time-of-flight mass spectrometer. *Rapid Communications in Mass Spectrometry* 11, 1015–1024.

Shevchenko, A., Chernushevic, I., Wilm, M. and Mann, M. (2002) 'De novo' sequencing of peptides recovered from in-gel digested proteins by nanoelectrospray tandem mass spectrometry. *Molecular Biotechnology* 20, 107–118.

Shiio, Y., Donohoe, S., Yi, E.C., Goodlett, D.R., Aebersold, R. and Eisenman, R.N. (2002) Quantitative proteomic analysis of Myc oncoprotein function. *EMBO Journal* 21, 5088–5096.

Simpson, R.J., Connolly, L.M., Eddes, J.S., Pereira, J.J., Moritz, R.L. and Reid, G.E. (2000) Proteomic analysis of the human colon carcinoma cell line (LIM 1215): development of a membrane protein database. *Electrophoresis* 21, 1707–1732.

Smith, R.D., Pasa-Tolic, L., Lipton, M.S., Jensen, P.K., Anderson, G.A., Shen, Y., Conrads, T.P., Udseth, H.R., Harkewicz, R., Belov, M.E. *et al.* (2001) Rapid quantitative measurements of proteomes by Fourier transform ion cyclotron resonance mass spectrometry. *Electrophoresis* 22, 1652–1668.

Soldes, O.S., Kuick, R.D., Thompson, I.A., 2nd, Hughes, S.J., Orringer, M.B., Iannettoni, M.D., Hanash, S.M., and Beer, D.G. (1999) Differential expression of Hsp27 in normal oesophagus, Barrett's metaplasia and oesophageal adenocarcinomas. *British Journal of Cancer* 79, 595–603.

Spahr, C.S., Davis, M.T., McGinley, M.D., Robinson, J.H., Bures, E.J., Beierle, J., Mort, J., Courchesne, P.L., Chen, K., Wahl, R.C. *et al.* (2001) Towards defining the urinary proteome using liquid chromatography-tandem mass spectrometry. I. Profiling an unfractionated tryptic digest. *Proteomics* 1, 93–107.

Tonge, R., Shaw, J., Middleton, B., Rowlinson, R., Rayner, S., Young, J., Pognan, F., Hawkins, E., Currie, I. and Davison, M. (2001) Validation and development of fluorescence two-dimensional differential gel electrophoresis proteomics technology. *Proteomics* 1, 377–396.

Unlu, M., Morgan, M.E. and Minden, J.S. (1997) Difference gel electrophoresis: a single gel method for detecting changes in protein extracts. *Electrophoresis* 18, 2071–2077.

Venter, J.C., Adams, M.D., Myers, E.W., Li, P.W., Mural, R.J., Sutton, G.G., Smith, H.O., Yandell, M., Evans, C.A., Holt, R.A. *et al.* (2001) The sequence of the human genome. *Science* 291, 1304–1351.

Verma, R., Chen, S., Feldman, R., Schieltz, D., Yates, J., Dohmen, J. and Deshaies, R.J. (2000) Proteasomal proteomics: identification of nucleotide-sensitive proteasome-interacting proteins by mass spectrometric analysis of affinity-purified proteasomes. *Molecular Biology of the Cell* 11, 3425–3439.

Vlahou, A., Schellhammer, P.F., Mendrinos, S., Patel, K., Kondylis, F.I., Gong, L., Nasim, S. and Wright, G.L., Jr (2001) Development of a novel proteomic approach for the detection of transitional cell carcinoma of the bladder in urine. *American Journal of Pathology* 158, 1491–1502.

Walker, L.A., MacDonald, J.A., Liu, X., Nakamoto, R.K., Haystead, T.A., Somlyo, A.V., and Somlyo, A.P. (2001) Site-specific phosphorylation and point mutations of telokin modulate its Ca^{2+}-desensitizing effect in smooth muscle. *Journal of Biological Chemistry* 276, 24519–24524.

Washburn, M.P., Wolters, D. and Yates, J.R., 3rd (2001) Large-scale analysis of the yeast proteome by multidimensional protein identification technology. *Nature Biotechnology* 19, 242–247.

Washburn, M.P., Ulaszek, R., Deciu, C., Schieltz, D.M. and Yates, J.R., 3rd (2002) Analysis of quantitative proteomic data generated via multidimensional protein identification technology. *Analytical Chemistry* 74, 1650–1657.

Waterston, R.H., Lindblad-Toh, K., Birney, E., Rogers, J., Abril, J.F., Agarwal, P., Agarwala, R., Ainscough, R., Alexandersson, M., An, P. *et al.* (2002) Initial sequencing and comparative analysis of the mouse genome. *Nature* 420, 520–562.

Wilm, M. and Mann, M. (1996) Analytical properties of the nanoelectrospray ion source. *Analytical Chemistry* 68, 1–8.

Wilm, M., Shevchenko, A., Houthaeve, T., Breit, S., Schweigerer, L., Fotsis, T. and Mann, M. (1996) Femtomole sequencing of proteins from polyacrylamide gels by nano-electrospray mass spectrometry. *Nature* 379, 466–469.

Wolters, D.A., Washburn, M.P. and Yates, J.R., 3rd (2001) An automated multidimensional protein identification technology for shotgun proteomics. *Analytical Chemistry* 73, 5683–5690.

Wu, S.L., Amato, H., Biringer, R., Choudhary, G., Shieh, P. and Hancock, W.S. (2002) Targeted proteomics of low-level proteins in human plasma by LC/MSn: using human growth hormone as a model system. *Journal of Proteome Research* 1, 459–465.

Wu, X., Haystead, T.A., Nakamoto, R.K., Somlyo, A.V. and Somlyo, A.P. (1998) Acceleration of myosin light chain dephosphorylation and relaxation of smooth muscle by telokin. Synergism with cyclic nucleotide-activated kinase. *Journal of Biological Chemistry* 273, 11362–11369.

Yates, J.R., 3rd (1998) Mass spectrometry and the age of the proteome. *Journal of Mass Spectrometry* 33, 1–19.

Yates, J.R., 3rd, Speicher, S., Griffin, P.R. and Hunkapiller, T. (1993) Peptide mass maps: a highly informative approach to protein identification. *Analytical Biochemistry* 214, 397–408.

Yates, J.R., 3rd, Eng, J.K., McCormack, A.L. and Schieltz, D. (1995) Method to correlate tandem mass spectra of modified peptides to amino acid sequences in the protein database. *Analytical Chemistry* 67, 1426–1436.

Zhang, X., Herring, C.J., Romano, P.R., Szczepanowska, J., Brzeska, H., Hinnebusch, A.G. and Qin, J. (1998) Identification of phosphorylation sites in proteins separated by polyacrylamide gel electrophoresis. *Analytical Chemistry* 70, 2050–2059.

Zhou, H., Watts, J.D. and Aebersold, R. (2001) A systematic approach to the analysis of protein phosphorylation. *Nature Biotechnology* 19, 375–378.

Zhou, H., Ranish, J.A., Watts, J.D. and Aebersold, R. (2002) Quantitative proteome analysis by solid-phase isotope tagging and mass spectrometry. *Nature Biotechnology* 20, 512–515.

7 The Epigenome: Epigenetic Regulation of Gene Expression in Mammalian Species

E. Whitelaw[1] and D. Garrick[2]

[1]*School of Molecular and Microbial Biosciences, Biochemistry Building G08, University of Sydney, NSW 2006, Australia;* [2]*MRC Molecular Haematology Unit, Weatherall Institute of Molecular Medicine, John Radcliffe Hospital, Oxford OX3 9DS, UK*

Introduction	179
Epigenetic Modifications of the Genome	180
Topic 1: DNA methylation and its effect on gene expression	180
Topic 2: Chromatin and histone modifications	183
Topic 3: Interactions between different epigenetic signals	186
Epigenetic Phenomena in Mammalian Development	187
Topic 4: Clearing and establishment of epigenetic marks in development	187
Topic 5: Parental imprinting	189
Topic 6: X Chromosome inactivation	192
Topic 7: Metastable epialleles	195
Conclusions and the Future	195
References	196

Introduction

There is a growing realization that the knowledge of the primary DNA sequence alone is insufficient to predict whether or not a gene will be active. In many instances genes can be switched off as a result of changes in the way the DNA is modified. These modifications, termed epigenetic, include both methylation of cytosine residues and modifications to the histone proteins that package the DNA. The patterns of epigenetic modifications vary as one proceeds along a chromosome and they vary from tissue to tissue. Within any particular tissue, these epigenetic marks, once established, are relatively permanent for the life of that

organism, i.e. they are mitotically inherited. This heritability differentiates control of gene expression by epigenetic processes from those that result from transient changes in the concentration of transcription factors. Work over the last decade has greatly increased our knowledge of the chemical nature of epigenetic modifications, and Topics 1–3 of this chapter review the progress made in this area.

While we know a great deal about the nature of the marks themselves, we understand very little about how or why particular areas of chromosomes are ear-marked for particular epigenetic modifications. It is clear, for example, that regions of DNA that are heavily methylated are generally transcriptionally silent, but how this decision is made in the first place remains unclear. One might guess that this a direct consequence of the primary DNA sequence. However, epigenetic changes in the absence of an underlying mutation at the gene of interest do occur, in some cases as a result of DNA mutations in the adjacent gene (Tufarelli et al., 2003) and in other cases as a result of stochastic events in early development (Duhl et al., 1994; Morgan et al., 1999). Topic 4 discusses the timing and nature of the epigenetic changes that take place during mammalian development. Topics 5 and 6 cover two important situations in which epigenetic control of gene expression is known to occur in mammals: parental imprinting and X inactivation. These phenomena, particularly parental imprinting, have been extensively studied and reviewed over the last 20 years and so these sections will be brief. One of the more recent discoveries in mammalian biology is the importance of the stochastic establishment of epigenetic processes during embryonic development, and this is discussed in Topic 7.

Epigenetic Modifications of the Genome

Topic 1. DNA methylation and its effect on gene expression

The methylation of DNA at the C-5 position of cytosine residues is the most extensively studied epigenetic modification of the genome. In vertebrate animals, the majority of methylated cytosine residues are situated within the symmetrical dinucleotide 5'-CpG-3'. In human somatic cells, some 70–80% of all CpG dinucleotides are methylated (Ehrlich et al., 1982). While methylated CpGs are dispersed widely throughout the genome, those that escape methylation exhibit dramatic clustering. These clusters of unmethylated CpG dinucleotides are referred to as CpG islands. Because of the high rate with which methyl-CpG undergoes spontaneous deamination (resulting in TpG), methylated genomes are characterized by overall CpG depletion, and CpG islands, which are devoid of methylation, can be readily identified as short segments of the genome which are G+C rich (>50%) but in which the ratio of observed to expected CpGs is high

(>0.6) (Gardiner-Garden and Frommer, 1987). Computational analysis suggests that the human genome contains approximately 29,000 CpG islands (International Human Sequencing Consortium, 2001) most of which are associated with the 5′ ends of genes, including the promoters of all housekeeping genes as well as those of many tissue-restricted genes. The majority of CpG islands remain unmethylated at all stages of development and in all tissue types, regardless of whether the associated gene is expressed or not (Cross and Bird, 1995). It is not yet understood how these regions remain unmethylated, despite presenting an ideal target for the methylation machinery. While most CpG islands remain unmethylated, the developmentally programmed methylation of specific CpG islands is observed as part of the processes of parental imprinting (Topic 5) and X chromosome inactivation (Topic 6).

The cellular methylation machinery

The genomic methylation pattern that is observed in adult somatic tissues is determined by a combination of two distinct enzymatic processes: maintenance and de novo DNA methylation. Maintenance methylation describes the attachment of a methyl moiety to the unmethylated cytosine within a hemimethylated CpG dinucleotide (one in which only one strand of the CpG duplex is methylated), thereby converting the duplex to the symmetrically methylated form (Fig. 7.1). Hemimethylated sites arise in genomic DNA when a symmetrically methylated CpG dinucleotide undergoes semi-conservative replication. Thus the pattern of genomic DNA methylation is transmitted from parent to daughter nucleus on only a single strand of the DNA duplex. Maintenance methylation reconverts these hemimethylated sites to the symmetrically methylated form ready for the next round of DNA replication. Since the maintenance methylation activity ensures that the methylation pattern will be faithfully transmitted through multiple cellular divisions, DNA methylation is a mitotically stable epigenetic modification. In contrast to maintenance methylation, de novo methylation occurs at CpG dinucleotides that are completely devoid of methyl moieties. The de novo modification of unmethylated sites

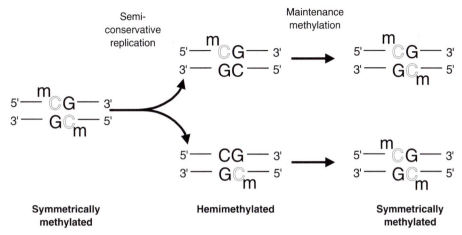

Fig. 7.1. Maintenance methylation of DNA. Semi-conservative replication of a methylated CpG dinucleotide in genomic DNA gives rise to daughter helices that are hemimethylated, as unmethylated cytosines are incorporated into the two nascent strands. The maintenance methyltransferase activity reconverts the hemimethylated sites to the symmetrically methylated form, essentially copying the methylation pattern on to the other strand. The activity of the maintenance methyltransferase thus ensures that, upon a further round of semi-conservative replication, each of the daughter nuclei will again carry the methylation pattern on one strand.

occurs primarily in the early embryo, and the methylation pattern thus established is then maintained in somatic tissues by the action of the maintenance methyltransferase. *De novo* methylation of DNA also occurs in the germline (see Topic 5, Parental imprinting).

Three different DNA methyltransferases have now been characterized in mammals. Dnmt1, the first to be discovered, exhibits a preference for hemimethylated DNA over completely unmethylated DNA (Pradhan *et al.*, 1999) and has been proposed to function as the primary cellular maintenance methyltransferase. Consistent with this proposed role, the targeted inactivation of the *Dnmt1* gene in mice severely impairs methylation at all sequences tested and reduces the level of total genomic methylation to about a third of normal (Li *et al.*, 1992; Lei *et al.*, 1996). The early embryonic lethality observed in mice homozygous for disrupted *Dnmt1* alleles also provided the first direct evidence that DNA methylation was indeed critical for normal mammalian development. Within the nucleus, the Dnmt1 protein is actually localized at DNA replication foci, suggesting that maintenance methylation is coupled to the replication process (Leonhardt *et al.*, 1992).

The residual levels of methylation, and a demonstrable *de novo* methylation activity, in embryonic stem (ES) cells lacking Dnmt1, indicated that other methyltransferase enzymes remained undiscovered. A search for expressed sequence tags (ESTs) bearing homology to bacterial methyltransferase domains led to the discovery of two proteins, Dnmt3a and Dnmt3b, which exhibited many of the properties expected for the putative mammalian *de novo* methyltransferase (Okano *et al.*, 1998). Both genes are most highly expressed in ES cells and early embryos, the stage where most programmed *de novo* methylation is occurring. Moreover, unlike Dnmt1, Dnmt3a and 3b are equally active on hemimethylated and unmethylated DNA *in vitro*. Mouse ES cells lacking both Dnmt3a and 3b lost the ability to methylate incoming unmethylated proviral DNA, confirming that together, these enzymes constitute the cellular *de novo* methyltransferase activity (Okano *et al.*, 1999). Mice which were lacking either or both of these enzymes failed to develop normally.

The cellular pattern of DNA methylation will also be influenced by any DNA demethylating activity. As mentioned above, methylation patterns will be passively removed from DNA over successive cellular generations if there is a failure to methylate the progeny strands following DNA replication. However, the observation in fertilized zygotes that the paternal genome is subject to active demethylation, the loss of methylation in the absence of DNA replication (Mayer et al., 2000), suggests the existence of an enzyme with a catalytic demethylase activity, and several laboratories have endeavoured to isolate the demethylase enzyme. An activity isolated from human cells (which subsequently turned out to be the methyl-binding protein MBD2) was reported to exhibit a robust demethylation activity in vitro (Bhattacharya et al., 1999), but other attempts to demonstrate this property of MBD2 have failed.

The functions of DNA methylation

One of the primary cellular functions for which DNA methylation appears to be essential for normal development is as a regulator of gene expression. For more than 20 years there has been a recognized correlation in vertebrates between the transcriptional repression of a gene and the methylation of associated CpG dinucleotides. Conversely, induced demethylation of local CpG dinucleotides has been shown to accompany transcriptional activation at a number of loci. Repression appears dependent on the location and density of methyl-CpGs relative to the promoter. The methylation-mediated silencing of gene expression is important for a number of highly specialized biological functions during development. These include the processes of parental imprinting, where methylation is required to achieve allele-specific silencing of imprinted genes (discussed in Topic 5) and X-inactivation, where DNA methylation is essential to maintain the repressed state of genes on the inactive X chromosome in females (discussed in Topic 6).

It has also been proposed that an important function of DNA methylation is to silence expression of parasitic DNA elements or retrotransposons and thereby limit their spread throughout the genome. These repetitive elements, which include endogenous retroviruses, LINE (L1) and SINE (Alu) repeat families, account for almost 40% of the human genome. It has been shown that in somatic tissues, where these elements are transcriptionally silent, the promoters are also heavily methylated. The importance of DNA methylation in maintaining the silent state of retrotransposons was confirmed by the observed derepression of the human LINE and SINE families and of the mouse intracisternal A particle (IAP) family when genomic DNA methylation is impaired (Liu et al., 1994; Woodcock et al., 1997; Walsh et al., 1998). The biological significance of methylation-mediated repression of transposons is unclear. The genome defence model proposes that unconstrained transposition could lead to a decrease in genomic stability, both as a result of direct transposition into structural genes and as a result of chromosome rearrangements which may occur due to recombination between repetitive elements at non-allelic sites (Yoder et al., 1997).

As well as its proposed roles in imprinting and X inactivation and in controlling expression of retroviral elements, it has long been suggested that DNA methylation patterns, which vary between different cell types, might contribute to the establishment of tissue-specific patterns of gene expression, and the maintenance of those expression patterns during mitotic cell division (Holliday and Pugh, 1975). The general importance of DNA methylation in regulating the expression of tissue-specific genes is still unclear. For many tissue-restricted genes, correlations between gene repression and DNA methylation have been made in artificial systems involving transgenic constructs or established cell lines (which frequently acquire methylation at sites which are unmethylated in primary tissues) and it is essential to confirm that DNA methylation is involved in gene repression during normal development in vivo. Moreover, the depletion of genomic methylation in Dnmt1-mutant mouse embryos did not result in widescale ectopic expression of tissue-restricted genes (Walsh and Bestor, 1999), suggesting that the methylation-mediated silencing of genes in non-expressing tissues may not be as general as has been proposed.

DNA methylation and transcriptional activity

DNA methylation can interfere with gene transcription by both direct and indirect mechanisms. Direct repression occurs when the presence of the methyl moiety interferes with the binding of transcription factors to their cognate sites within methylated promoters or regulatory elements. Several transcription factors have been identified whose binding is impaired when one or more CpG dinucleotides within the recognition site are methylated. One of the most important examples of methylation-sensitive binding is that of CTCF, a factor required for the activity of several domain boundaries and insulator elements.

DNA methylation can also indirectly influence transcription by a mechanism that involves the recruitment of secondary factors or complexes that exert a repressive effect. These factors are specifically attracted to, rather than repelled by, the methyl-CpG moieties. The first of these factors to be characterized, MeCP2, contains a methyl-binding domain (MBD) which specifically recognizes a symmetrically methylated CpG dinucleotide (Nan *et al.*, 1993). Since the discovery of MeCP2, four other members of the methyl-binding protein family (MBD1–4) have been identified based on the presence of the MBD. Members of the MBD family (MeCP2, MBD1, MBD2 and MBD3) have been directly shown to mediate transcriptional repression at methylated sites. These MBD proteins are themselves associated with large multicomponent corepressor complexes that include histone deacetylases (HDACs) and chromatin remodelling factors, and serve to recruit these complexes to sites of DNA methylation. These complexes cause the formation of a repressive chromatin structure. The effects of histone modifications on gene expression will be further discussed in the next section. More recently it has been shown that the DNA methyltransferases Dnmt1, 3A and 3B are also associated with HDAC activity (Fuks *et al.*, 2000, 2001; Bachman *et al.*, 2001), suggesting that at some methylated loci, the adoption of a repressive chromatin structure may be directly linked to laying down of the methylation pattern itself and therefore not rely on recruitment of HDACs by the MBD proteins.

It is worth noting that while the correlation between DNA methylation and transcriptional silencing is well established, it is often more difficult to determine whether methylation is the primary silencing event or is a secondary modification which occurs at loci that have already been repressed by other means. Certainly there are well characterized examples of methylation occurring at genes which are silenced. For example, during X inactivation in mammalian females, silencing of genes on the inactive X chromosome is achieved well before the appearance of methylation at promoter CpG islands (Lock *et al.*, 1987). Similarly, the *de novo* methylation of infected retroviral elements occurs several days after transcriptional silencing (Niwa *et al.*, 1983). However, both X inactivation and retroviral silencing are weakened when DNA methylation is impaired in somatic cells either genetically or by treating with the demethylating agent 5-azacytidine. Thus, while not the primary silencing event, DNA methylation appears to be essential to lock in and maintain the inactive state, and render the silencing irreversible.

Topic 2. Chromatin and histone modifications

DNA does not exist naked within the eukaryotic nucleus but is packaged into a compacted state referred to as chromatin. The basic unit of chromatin is the nucleosome core particle, in which the DNA double helix wraps itself around an octameric complex containing two molecules each of the histone proteins H2A, H2B, H3 and H4 (Luger *et al.*, 1997). The formation of the nucleosome core particle (which reduces the length of the DNA by approximately sevenfold) represents the first step in the several thousand-fold reduction required to compact the DNA into metaphase chromosomes, although the formation of higher-order structures is less well understood. As well as satisfying the physical constraints of the nucleus, it is now well established that the compaction of DNA into chromatin also presents an opportunity to regulate the expression of genes.

The most obvious effects of chromatin formation on gene expression are the result simply of physical exclusion. Nucleosomes and higher order structures are believed to act as simple physical barriers to prevent DNA-binding proteins from accessing their cognate sites on the DNA helix, thereby blocking the assembly of transcription initiation complexes. Thus, the potential exists to control gene expression by 'remodelling' the histone core particles so as to expose or obscure core regulatory elements. Over the past decade, several evolutionarily conserved protein complexes have been discovered which exhibit these nucleosome remodelling capabilities. All nucleosome remodelling factors characterized so far are large multisubunit complexes, and all contain an ATPase core subunit. For further discussion of chromatin remodelling complexes see Chapter 5 and Table 5.2.

As well as acting as a passive barrier to the transcription apparatus, it is known that the histones that comprise the core particle also carry additional epigenetic information. This epigenetic information exists in the form of different enzyme-catalysed covalent modifications that occur at the exposed N-terminal tails of the nucleosomal histones. Covalent histone modifications detected to date include acetylation, methylation, phosphorylation, ubiquitination and ATP-ribosylation (illustrated in Fig. 7.2). It is clear that there is an enormous num-

ber of possible combinations of these different modifications that can mark the surface of the nucleosome. This observation has led to the proposal of the histone code hypothesis (Turner, 2000; Strahl and Allis, 2000). This hypothesis posits that the total combination of different modifications at a nucleosome constitutes an epigenetic code, which dictates distinct biological consequences for chromatin-associated activities including DNA replication, chromosome segregation and, importantly, gene expression. The pattern of histone modifications is believed to be stably transmitted during mitotic cell division, thereby ensuring that the biological state that it determines (such as the expression state of a gene) is maintained from one cell generation to the next. However, the mechanisms by which histone modifications are transmitted are not fully understood. Although many of the histone modifications have been known to occur for years, it is only relatively recently that headway has been made in understanding the biological con-

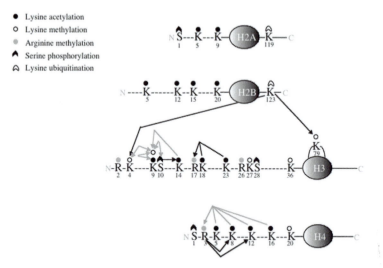

Fig. 7.2. Sites of post-translational modification of histone tails and regulatory interactions between different modifications. The N- and C-terminal tails and globular histone-fold domains are shown for each of the four core histones, with sites of potential modification (lysine acetylation, lysine methylation, arginine methylation, serine phosphorylation and lysine ubiquitination) indicated. Mono-, di- or tri-methylated residues are not distinguished. Two distinct, mutually exclusive modifications (acetylation and methylaton) are possible at H3–K9. Regulatory interactions between different modifications are indicated by arrows, with arrowheads signifying the direction of influence. Black arrows indicate a stimulatory influence and grey arrows indicate a repressive influence. The *trans*-histone regulatory pathway observed in *S. cerevisiae* in which ubiquitination of H2B–K123 upregulates methylation of H3–K4 and H3–K79 (the only modification known to occur within the histone fold domain) is indicated. Note that the figure represents a compilation of observations from different organisms and that not all modifications are seen in all organisms.

sequences of the different modifications, and identifying the mediatory factors that 'read' the code and implement the downstream events.

Histone acetylation

Histone acetylation became the first modification to be associated with a particular functional state when it was observed that hyperacetylated histone isoforms are enriched at transcriptionally active chromatin fragments (Pogo et al., 1966; Hebbes et al., 1988). Further, chromatin regions that are transcriptionally inactive, such as constitutive (pericentromeric) heterochromatin and facultative heterochromatin of the inactive X chromosome (see Topic 6), are characterized by histones that are generally poorly acetylated. Acetylation of lysine residues has been detected at the N terminal tails of all four core histones and acetylation can occur at multiple lysine residues within a single tail (Fig. 7.2). Histone acetylation is believed to affect gene transcription by directly modifying the chromatin structure since lysine acetylation reduces the overall positive charge of the histone tail and so is predicted to weaken the strength of the interaction between the histones and the (negatively charged) DNA (Hong et al., 1993). The resulting decondensation or 'opening' of the chromatin is thought to facilitate the access of transcription factors to their cognate recognition sites. The acetylation of histone lysine residues is generally a very labile modification (with half-lives ranging from minutes to several hours), and the acetylation state of a given histone at any moment is the result of an equilibrium achieved between the constant and opposing activities of histone acetyltransferases (HATs) and HDACs. For more detail see Tables 5.3 and 5.4. Histone acetylation is a modification ideally suited to allow rapid changes in transcription levels for genes that must respond quickly to environmental or hormonal signals.

Histone phosphorylation

Phosphorylation of serine residues has also emerged as an important covalent modification occurring at histone tails. The functional consequences of phosphorylation at Ser10 of histone H3 (H3-S10) have been most closely studied, although phosphorylation also occurs elsewhere (Fig. 7.2). Histone phosphorylation is involved in at least two different nuclear processes. Phosphorylation of H3-S10 occurs on a genome-wide scale during chromosome condensation at mitosis, and is essential for proper chromosome transmission (Wei et al., 1999). However, phosphorylation of H3-S10 also occurs during interphase, when it is targeted to specific regulatory elements and has been associated with the transcriptional activation (Clayton et al., 2000; Nowak and Corces, 2000). Distinct kinases are responsible for histone phosphorylation during these two processes. It is surprising that the same modification is involved in the apparently opposite processes of chromatin condensation (at the onset of mitosis) and chromatin relaxation (during transcriptional activation), suggesting that the ultimate biological outcome in these situations is determined by the interaction between phosphorylation and other modifications or mediatory proteins.

Histone methylation

While histones have long been recognized as substrates for the covalent addition of methyl moieties (Murray, 1964), it is only relatively recently that histone methylation has emerged as a critical modification for determining gene expression status. Methyl groups covalently attached to proteins are thermodynamically more stable than either acetyl or phosphoryl groups. Further, while both of the latter groups can be enzymatically removed from histones (by the actions of HDACs and histone phosphatases), no histone demethylase activity has yet been isolated. Histone methylation is thus well suited to act as a stable and long-term epigenetic mark of gene expression status. Thus far, histone methylation has been detected at arginine and lysine residues (Fig. 7.2), however most is understood about methylation of lysine residues.

There are at least five different lysine residues which can be methylated within the amino tails of histones H3 and H4 (K4, K9, K27 and K36 of histone H3 and K20 of histone H4). A sixth methylatable site (K79) is located not at the amino tails, but within the histone-fold domain of histone H3 (see below). Since each of these lysine ϵ-amino groups can be either mono-, di- or tri-methylated, the total number of possible different methyl-lysine histone isoforms is very large. Unlike methylation of arginine residues, which appears to be associated specifically with transcriptional activation, the methylation of histone lysine residues can dictate diverse functional outcomes ranging from transcriptional activation of euchromatic gene loci to the establishment of silent chromatin domains within pericentromeric heterochromatin. All lysine histone methyltransferases contain the characteristic SET domain which constitutes the catalytic core of the enzyme and most contain other recognizable motifs such as bromo-, chromo- and PHD-domains (reviewed in Kouzarides, 2002). It is likely that these other motifs are critical in determining the substrate specificity of the HMT activity.

The methylation of several histone lysine residues (particularly Lys9, Lys27 and Lys79 of histone H3, as well as Lys20 of histone H4) has been associated with the adoption of a transcriptionally repressed state at the associated chromatin. Trimethylation of histone H3 at Lys27 has been recently recognized as an important modification during the formation of facultative heterochromatin at the inactive X chromosome and is discussed in Topic 6. Methylation of histone H3 at Lys9 is one of the best understood modifications, in that the enzyme responsible for the mark, the mediatory proteins which read it and the functional consequences have all been well characterized. Members of the SUV39 family of SET-containing proteins, which includes the human (SUV39H1) and mouse (Suv39h1 and Suv39h2) proteins are methyltransferases whose activity is highly specific for Lys9 of histone H3 (Rea et al., 2000). Histone H3-K9 methylation creates a high affinity binding site for members of the HP1 family of heterochromatin structural proteins that recognize the epitope through their conserved chromodomains. Mice which are null for the Suv39h HMTs show disrupted centromere function and chromosome instability, confirming the importance of Suv39h, H3-K9 methylation and HP1-protein recruitment for the functional integrity of pericentromeric heterochromatin (Peters et al., 2001).

In contrast, the methylation of other histone lysine residues, specifically Lys4 and Lys36 of histone H3, is associated with transcriptional activation. Recent refinement of antibodies against K4-methylated histone H3 suggest that it is specifically the trimethylated rather than the dimethylated form which is consistently associated with fully activated promoters (Santos-Rosa et al., 2002). The downstream proteins or complexes that are recruited (or perhaps displaced) by this modification to bring about transcriptional activation, are yet to be determined.

Topic 3. Interactions between different epigenetic signals

It is obvious that different epigenetic signals, such as DNA methylation, nucleosome remodelling and covalent histone modifications, do not occur in isolation but rather that they act together to constitute a complete epigenetic code. It is not surprising that there is frequently regulatory cross-talk between the different epigenetic modifications themselves, such that the presence or absence of one signal can influence the acquisition of another signal at the same genetic locus. In this way, it is possible to discern various regulatory cascades in which individual epigenetic changes accumulate in a specified order at a locus to dictate the final state of gene activity.

Interactions between different histone tail modifications

Given the large number of diverse covalent modifications that can take place within the spatially confined histone amino tails, it is not surprising that different modifications can affect each other (Fig. 7.2). One obvious opportunity for cross-talk is where different modifications affect the same histone tail residue. This occurs at Lys9 of histone H3, which can be modified by both acetylation and methylation. Acetylation of H3–K9 prevents methylation of the same residue by the SUV39 HMT (Rea et al., 2000). Regulatory cross-talk also occurs between different residues on the same tail, e.g. phosphorylation at Ser10 prevents methylation of adjacent Lys9. Finally, regulatory interactions have also been observed even between modifications on different histone tails. One such modification that has only recently been identified is ubiquitination. Ubiquitin is a 76-residue peptide which can be covalently attached to a variety of cellular proteins. In yeast it has been shown that ubiquitination of histone H2B at Lys123 (within the C-terminal tail) is essential for methylation to occur at Lys79 and Lys4 of histone H3 (Briggs et al., 2002). Interactions between modifications on different tails have yet to be detected in higher eukaryotes.

Interactions between histone tail modifications and DNA methylation

Over recent years, it has become clear that not only can DNA methylation patterns exert an influence on the accumulation of histone modifications, but conversely that some histone modifications are critical for determining the pattern of DNA methylation. As discussed above, several MBD-containing proteins, which are recruited to methylated DNA, are themselves associated with HDACs. In this way, sites of DNA methylation are also targeted for histone deacetylation, thereby locking in a repressive chromatin state. Further, the DNA methyltransferases Dnmt1, 3A and 3B themselves are also associated with HDACs. DNA methylation can also feed back on histone lysine methylation. It has recently been shown that both the MBD protein MeCP2 as well as the DNA methyltransferases Dnmt1 and Dnmt3A are all associated with a histone H3-K9 methyltransferase activity, suggesting that like histone deacetylation, the repressive modification of H3-K9 methylation may also be coordinately recruited to methylated DNA (Fuks et al., 2003a,b).

Conversely, histone modifications can also affect the pattern of DNA methylation. In both the fungus *Neurospora crassa* and the plant *Arabidopsis thaliana*, mutations in genes encoding histone methyltransferases (*dim-5* and *kryptonite*, respectively) resulted in a dramatic reduction of DNA methylation (Tamaru and Selker, 2001; Jackson et al., 2002) suggesting that at least in these organisms, histone methylation is critical for DNA methylation. The mechanisms by which histone methylation influences DNA methylation are still being elucidated.

Effects of chromatin remodelling on DNA methylation and histone modification

The acquisition of a normal pattern of DNA methylation is also critically dependent on a variety of chromatin remodelling activities. It seems likely that the remodelling activity is required to allow the *de novo* and/or maintenance DNA methyltransferases to gain access to the chromatin. This relationship was first revealed in the plant *A. thaliana* (Jeddeloh et al., 1999) but is also observed in mammals. In mice, targeted disruption of Lsh, a putative chromatin remodeller of the ATPase superfamily, results in reduction of DNA methylation (Dennis et al., 2001) and perinatal lethality. Further, in humans, mutations in an X-linked gene (*ATRX*) encoding another member of the Swi2/Snf2 ATPase superfamily, give rise to changes in the pattern of DNA methylation at several repetitive genomic elements (Gibbons et al., 2000). Functional interactions also exist between ATP-dependent chromatin remodelling and the modification of histone tails. At some promoters, it appears that nucleosome remodelling must take place first in order to allow access of histone modifying enzymes (Cosma et al., 1999). In contrast, at some other promoters the order of regulatory events appears to be reversed (Agalioti et al., 2000).

Epigenetic Phenomena in Mammalian Development

Topic 4. Clearing and establishment of epigenetic marks in development

In general, epigenetic modifications are cleared between generations. Analysis of global methylation levels shows that genomic DNA undergoes dramatic demethylation between fertilization and blastocyst (Monk et al., 1987) (Fig. 7.3). As a result, in the blastocyst most (but not all) epigenetic information inherited from the gametes has been removed, and the cells of the blastocyst are in a pluripotent state, i.e. they have the potential to generate most cell types. Recent experiments using antibodies to methylated cytosine residues have revealed that in a fertilized egg, the paternal genome (i.e. the chromosomes supplied by the sperm)

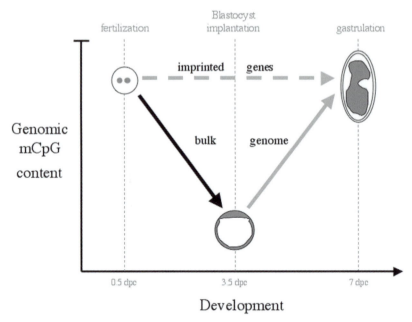

Fig. 7.3. Changes in DNA methylation during early mouse development. The vertical axis indicates the relative genomic content of methylated CpG. The horizontal axis represents a time course of early mouse development (numbers indicate approximate days post-coitum (dpc)). Following fertilization, the bulk of the genomic DNA (mainly at repeat sequences and some unique genes, but excluding imprinted genes) undergoes demethylation (black arrow) and the lowest level of DNA methylation is reached at about the time of implantation of the blastocyst (3.5 dpc). After implantation, significant *de novo* methylation of the bulk genome is observed in the embryo itself (solid grey arrow). In contrast, the methylation patterns at imprinted genetic loci escape this process of demethylation and remethylation (dashed grey arrow), such that the epigenetic distinction between the maternally and paternally inherited alleles is maintained. Figure adapted from Li (2002).

undergoes a particularly rapid demethylation process (Mayer *et al.*, 2000; Oswald *et al.*, 2000). This process, which is independent of DNA replication, obviously involves an active demethylation activity, although the enzyme responsible has not yet been identified (see Topic 1). The maternally inherited set of chromosomes becomes demethylated more slowly. This may involve active and/or passive demethylation. Despite this general clearing of epigenetic information prior to implantation, it should be noted that marks at some specific alleles (called imprinted loci) are not cleared during this stage (see Topic 5). There is also evidence of failure to clear epigenetic information at a group of unusual alleles called metastable epialleles (see Topic 7). Clearing of epigenetic signals also occurs during gametogenesis and this is particularly important for the parentally imprinted genes (see Topic 5)

(Monk, 1987; Howlett and Reik, 1991). However not much is known about the global extent of clearing during gametogenesis.

After clearing, during the period of differentiation of the early post-implantation embryo, the epigenetic marks must be re-established. Following implantation, DNA methylation levels increase rapidly in the tissues of the primitive ectoderm, which gives rise to the embryo itself (Santos *et al.*, 2002) (Fig. 7.3). It is believed that the different epigenetic patterns laid down at this stage are intimately involved in the determination of different cell type. One of the fundamental unresolved issues in this field is how the decisions are made about where (which genetic loci) specific marks are laid down, and what those marks are. While most is known about the dynamics of the clearing and establishment of methylation, there is good reason to believe that it is not the primary event during this process. In *Drosophila*, a complex multicellular organism, only trace amounts of cytosine methylation are detected during early development, and none is found in the adult. Chromatin proteins may turn out to play a more critical role in this process. In the future, an understanding of both the nature and the timing of these mechanisms will be critical if we are ever to be able to culture specific differentiated cell types from pluripotent ES cells *in vitro*, a technology which has broad therapeutic potential.

Topic 5. Parental imprinting

There are approximately 60–100 genes in the mouse genome that are monoallelically expressed in at least some tissue in the adult animal (Morison and Reeve, 1998). This process is called parental imprinting or genomic imprinting. At fertilization, differences exist between the epigenetic states of the paternally and maternally inherited alleles, and it is these epigenetic differences that determine the different expression status of the two alleles. The epigenetic differences between the male and female alleles are actually established in the gametes of the parents (when the alleles are still physically separate) and so are present before the alleles come together at fertilization. The first evidence for genomic imprinting came about during experiments attempting to produce mice from purely maternal (gynogenetic) or paternal (androgenetic) nuclear genomes (McGrath and Solter, 1984). In both cases, the embryos failed to develop beyond implantation. These experiments demonstrated that both the maternal and paternal genomes are required for normal embryonic development. Consistent with this was the finding in humans of clinical disorders associated with uniparental disomy of some chromosomes (Spence *et al.*, 1988).

A selection of mouse genes that are known to be imprinted is shown in Table 7.1. A great deal of research has been carried out on these genes over the last 20 years and many reviews on the subject are available (Hall, 1990; Chaillet, 1994; Reik and Walter, 2001; Sleutels and Barlow, 2002). In summary, some of the genes become epigenetically silenced during male gametogenesis (such as *Igf2*), and some

during female gametogenesis (such as *H19*). Many of the genes found to be imprinted are associated with embryonic and fetal growth. It has been hypothesized that the reason why these genes are controlled such that there is only monoallelic rather than biallelic expression is that the father's genome has evolved to produce large offspring while the mother's genome is trying to suppress embryonic overgrowth so that she herself will survive childbirth (Moore and Haig, 1991). More recently, alternative explanations for the evolution of parental imprinting have been suggested (Pardo-Manuel de Villena *et al.*, 2000; Paldi, 2003). In essence, these models focus on the idea that the parental imprinting is a remnant of gamete-specific marks which are established in order to achieve sex-specific patterns of expression in mature gametes, i.e. depending on if it is a sperm or an egg.

Most of the imprinted genes are clustered at a small number of chromosomal locations (see Table 7.1). It is still not clear why the genes are clustered, but increasing evidence supports the notion that long-range chromatin states are established in these regions (Razin and Cedar, 1994; Khosla *et al.*, 1999). These chromatin states appear to be controlled by small regions of DNA a few kilobases long known as differentially methylated regions (DMRs) (Stöger *et al.*, 1993), so called because they show different levels of DNA methylation between the maternal and paternal alleles. Thus cytosine methylation is emerging as the most likely candidate for the epigenetic mark that distinguishes the imprinted alleles (Barlow, 1994). One of the interesting aspects of parental imprinting is that these marks are not cleared during the period of demethylation that occurs in early development (see Topic 4 and Fig. 7.3), thereby retaining the mark established in the previous generation (Reik and Walter, 2001). What is it about these genes that enables them to avoid erasure at this period in development? Recent progress in the molecular nature of epigenetic marks (see Topics 1 and 2) should help us to answer this question.

One obvious further aspect of imprinting is that the epigenetic marks that have been inherited from the parents must be cleared away during gametogenesis such that the new mark can be re-established according to the sex of

Table 7.1. Imprinted genes in the mouse*.

Imprinted loci	Chr.	Chromosomal region	Repressed allele	Name
Gatm	2	central 2	P	L-arginine: glucine amidino-transferase
Nnat	2	distal 2	M	Neuronatin
Gnas	2	distal 2	P	Guanine nucleotide-binding protein, alpha stimulating
Gnasxl	2	distal 2	M	Guanine nucleotide-binding protein, alpha stimulating, extra large
Nesp	2	distal 2	P	Neuroendocrine secretory protein
Nespas	2	distal 2	M	Neuroendocrine secretory protein antisense
Dlx5	6	centromere to T77H (A3.2)	P	Distal-less homeobox 5
Calcr	6	centromere to T77H (A3.2)	P	Calcitonin receptor
Sgce	6	centromere to T77H (A3.2)	M	Sarcoglycan, epsilon
Peg 10	6	centromere to T77H (A3.2)	M	Paternally expressed gene 10
Neurabin	6	centromere to T77H (A3.2)	P	Neurabin
Pon 3	6	centromere to T77H (A3.2)	P	Paroxonase 3
Pon 2	6	centromere to T77H (A3.2)	P	Paroxonase 2
Asb4	6	centromere to T77H (A3.2)	P	Ankyrin repeat and suppressor of cytokine signalling
Peg1/Mest	6	proximal 6 (distal to A3.2)	M	Mesoderm-specific transcript
Copg2	6	proximal 6 (distal to A3.2)	P	Coatomer protein complex subunit gamma 2
Copg2as	6	proximal 6 (distal to A3.2)	M	Antisense to Copg2
Mit1/lb9	6	proximal 6 (distal to A3.2)	M	Mest-linked imprinted transcript 1
Nap1l5	6	proximal 6 (distal to A3.2)	M	Nucleosome assembly protein 1, like 5.
Zim1	7	proximal 7	P	Imprinted zinc-finger gene 1
Peg3/Pw1	7	proximal 7	M	Paternally expressed gene 3
Usp29	7	proximal 7	M	Ubiquitin-specific processing protease 29
Zim3	7	proximal 7	P	Zinc finger gene 3 from imprinted domain
Zfp264	7	proximal 7	M	Zinc finger gene 264
Snrpn	7	central 7	M	Small nuclear ribonucleoprotein polypeptide N
Snurf	7	central 7	M	Snrpn upstream reading frame
Pwcr1	7	central 7	M	Prader–Willi chromosome region 1
Magel2	7	central 7	M	Magel2
Ndn	7	central 7	M	Necdin
Zfp127/Mkrn3	7	central 7	M	Ring zinc-finger-encoding gene
Zfp127as/Mkrn3as	7	central 7	M	Ring zinc-finger-encoding gene antisense
Frat3	7	central 7	M	Frequently rearranged in advanced T-cell lymphomas.
Ipw	7	central 7	M	Imprinted in Prader–Willi syndrome
Atp10c/Atp10a	7	central 7	P	Aminophospholipid translocase
Ube3a	7	central 7	P	E6-AP ubiquitin protein ligase 3A
Ube3aas	7	central 7	M	Ube3a antisense
Nap1l4/Nap2	7	central 7	P	A cDNA clone isolated from a fetal hepatic library
H19	7	distal 7	P	
Igf2	7	distal 7	M	Insulin-like growth factor type 2

Continued

Table 7.1. Continued.

Imprinted loci	Chr.	Chromosomal region	Repressed allele	Name
Igf2as	7	distal 7	M	Insulin-like growth factor type 2, antisense
Ins2	7	distal 7	M	Insulin 2
Mash2	7	distal 7	P	Mus musculus achaete-scute homologue 2
Kvlqt1	7	distal 7	P	
Kvlqt1as/Lit1	7	distal 7	M	Kvlqt1 antisense
Tapa1/Cd81	7	distal 7	P	Cd 81 antigen
p57KIP2/Cdkn1c	7	distal 7	P	Cyclin-dependent kinase inhibitor 1C
Msuit	7	distal 7	P	Mouse-specific ubiquitously expressed imprinted transcript 1
Slc221l	7	distal 7	P	Solute carrier family 22 (organic cation transporter memter-1 like). Formally known as Impt1, Itm and Orctl2.
Ipl/Tssc3	7	distal 7	P	Imprinted in placenta and liver (Tdag51?)
Tssc4	7	distal 7	P	
Obph1	7	distal 7	P	Oxysterol-binding protein 1
A19	9	9	M	
Rasgrf1	9	9	M	Ras protein-specific guanine nucleotide-releasing factor 1
Zac1	10	10	M	Zinc finger DNA-binding protein
Dcn	10	distal 10	P	Decorin
Meg1/Grb10	11	proximal 11 (A1–A4)	P	Growth factor receptor-bound protein 10
U2af1-rs1	11	proximal 11 (A3.2–4)	M	U2 small nuclear ribonucleoprotein auxiliary factor (U2AF), 35 Da, related sequence 1
Dlk/Pref1	12	distal 12 (E–F)	M	Delta-like 1
Meg3/Gtl2	12	distal 12 (E–F)	P	Gene trap locus 2
Dio3	12	distal 12 (E–F)	M	Deiodinase iodothyronine type 3
Rian	12	distal 12 (E–F)	P	RNA imprinted and accumulated in the nucleus.
Rtl1	12	distal 12 (E–F)	M	Retrotransposone-like gene 1
Mirg	12	distal 12 (E–F)	P	MicroRNA-containing gene
Htr2a	14	distal 14	P	5-Hydroxytryptamine (serotonin) receptor 2 A
Slc38a4/Ata3	15	distal 15	M	Solute carrier family 38, member 4/amino acid transport system A3
Peg13	15	distal 15	M	Paternally expressed gene 13
Slc22a2	17	proximal 17	P	Membrane-spanning transporter protein
Slc22a3	17	proximal 17	P	Membrane-spanning transporter protein
Igf2r	17	proximal 17	P	Insulin-like growth factor type 2 receptor
Igf2ras/Air	17	proximal 17	M	Insulin-like growth factor type 2 receptor antisense RNA
Impact	18	proximal 18 (A2–B2)	M	Homology with yeast and bacterial protein family YCR59c/yigZ
Ins1	19	19	M	insulin 1

M = maternal; P = paternal.
*Data presented in this table were obtained from Beechey et al. (2003).

the individual. The erasure of methylation marks at imprinted loci takes place in primordial germ cells, at around 11.5–12.5 days of mouse development (Lee et al., 2002), although little is known about this process.

Most genes found to be imprinted in the mouse are also imprinted in humans. A number of human diseases are associated with a failure of parental imprinting (e.g. Angelmans, Prader–Willi, and Beckman–Wiederman syndrome). Again, numerous up-to-date reviews on these diseases have been written (Hall, 1990; Buiting et al., 2003).

Topic 6. X Chromosome inactivation

From as early as 1961, it has been recognized that in female mammals, one of the two X chromosomes undergoes a developmentally regulated programme of transcriptional silencing, in order to compensate for the gene dosage difference with the XY male (Lyon, 1961). As in parental imprinting, X inactivation involves establishing a functional distinction between two alleles that coexist within the same nucleus. It is now known that, as with imprinted genes, epigenetic modifications play a crucial role in distinguishing the active (Xa) and inactive (Xi) X chromosome, and the study of X inactivation has contributed to a more general understanding of epigenetic mechanisms of gene regulation. This section will focus particularly on the epigenetic changes associated with X chromosome inactivation. For a more detailed review of the specific aspects of X chromosome counting and choice, the reader is referred to Avner and Heard (2001) and Boumil and Lee (2001).

Xist and the initiation of inactivation

X inactivation occurs early in female embryogenesis, and correlates with the differentiation from totipotent or pluripotent cell lineages. Once established during early differentiation, the silenced state of the inactive X is stably inherited through mitotic cell divisions. In eutherian mammals, X inactivation within embryonic tissues is a random process, in which either the paternally or maternally inherited X can become silenced. In contrast, in some extraembryonic tissues and in marsupial mammals, X inactivation is non-random, with the paternally inherited X being selected for silencing (see below). The first step in the silencing of the future inactive X chromosome involves the upregulated expression of the Xist (X-inactive-specific transcript) gene located within a region of the X chromosome termed the X inactivation centre (XIC). The Xist gene produces a large (17 kb) non-coding transcript that is spliced and polyadenylated, but which is retained within the nucleus (Brown et al., 1992). In undifferentiated cells, Xist is expressed at low levels from both future Xa and Xi chromosomes. However, upon differentiation and the initiation of inactivation, Xist expression is upregulated from thistlST from the future Xa allele is silenced. Xist transcripts accumulate and coat the chromosome arms in cis to the expressed allele, the first distinguishing mark of the future inactive X. The accumulation of the Xist transcript is closely followed by the transcriptional silencing of Xi-linked genes. During the earliest stages of differentiation, silencing of genes on the inactive X is dependent on the continued presence of the Xist RNA and is reversed in its absence. In contrast, later in differentiation, the silent state is stably maintained even in the absence of Xist (Brown et al., 1994), suggesting that X inactivation involves the accumulation of multiple epigenetic changes that render the chromosome stably repressed. These changes culminate in the formation of the condensed heterochromatic state recognized cytologically as the Barr body, which is usually located at the nuclear periphery. The nature of these changes and the relative timing of events in this process have largely been established through observations of X inactivation as XX mouse ES cells are induced to differentiate in vitro (Fig. 7.4).

Early events in X inactivation

Two very recent studies have demonstrated that the first epigenetic mark to follow accumulation of the Xist transcript at the inactive X is the specific enrichment of this chromosome in histone H3 which is trimethylated at Lys27 (Plath et al., 2003; Silva et al., 2003). Trimethylation of H3-K27, which is observed during both imprinted and random X inactiva-

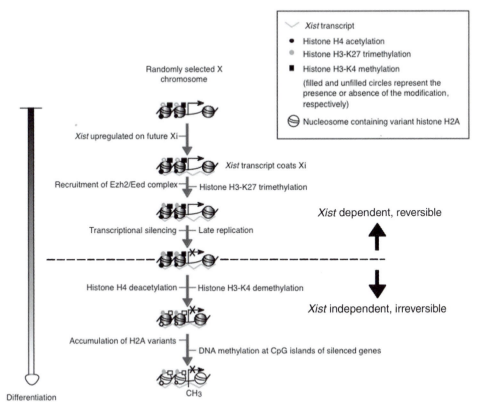

Fig. 7.4. Order of epigenetic changes observed at Xi during random X inactivation. The relative order of events has largely been determined from observations of X inactivation in differentiating XX mouse ES cells. The cartoon represents the chromatin surrounding the promoter of an X-linked gene, which is initially transcriptionally active on Xi (arrow). As differentiation proceeds, the promoter becomes stably silenced (blocked arrow) due to the accumulation of epigenetic changes as described in the text. The dashed grey line indicates the approximate point at which the silenced state of the chromosome becomes irreversible and is no longer dependent on the presence of the Xist transcript.

tion, is catalysed by the SET domain of a polycomb group (PcG) protein called Ezh2 (also called Enx1). A complex containing Ezh2 and another PcG protein (Eed) is transiently recruited to the inactive X in response to Xist RNA accumulation. It should also be noted that two previous studies have also identified methylation of H3 Lys9 (previously shown to mark constitutive heterochromatin) as an early modification of the inactive X chromosome (Boggs *et al.*, 2002; Peters *et al.*, 2002). However, since the H3-K9 and H3-K27 residues are embedded within the same sequence in the histone H3 tail (-ARKS-), recent data suggest that the apparent H3-K9 methylation initially observed on Xi may be

due to cross-reactivity of the anti-methyl H3-K9 serum towards methyl H3-K27. More recent highly specific antisera suggest that di- and trimethylated H3-K9 is located in pericentromeric heterochromatin but not at Xi, while trimethylated H3-K27 is specifically at the inactive X, but not at constitutive heterochromatin (Plath *et al.*, 2003; Silva *et al.*, 2003). Transcriptional silencing of Xi-linked genes is observed early during the inactivation process (Keohane *et al.*, 1996) and so roughly coincides with the accumulation of H3-K27 methylation (Fig. 7.4). Also at around this time, the inactive X is observed to replicate later in S phase relative to its active homologue (Keohane *et al.*, 1996).

Late events in X inactivation

Early during differentiation, transcriptional silencing is dependent on *Xist* expression and is easily reversed if *Xist* is removed. However, as differentiation proceeds, X inactivation becomes irreversible and is no longer dependent on the presence of the *Xist* transcript. Thus, a model is emerging in which the *Xist*-dependent recruitment of the Eed/Ezh2 complex is a temporary signal which establishes an initial level of silencing upon which later epigenetic modifications accrue to convert the chromosome to a heritably silenced state, which is independent of *Xist* and the Eed/Ezh2 complex. These later changes are likely to be important for maintaining the inactive state of the chromosome which has been established.

The accumulation of trimethylation at H3-K27 is followed by a number of other changes at the histone tails on Xi, although the factors responsible for these secondary changes are not fully understood. There is a depletion of methylation of histone H3 Lys4, a marker associated with transcriptional activation (Boggs *et al.*, 2002; Silva *et al.*, 2003) and a global histone deacetylation, as the inactive X becomes generally depleted of acetylated isoforms of histones H2A, H2B, H3 and H4 (Jeppesen and Turner, 1993; Belyaev *et al.*, 1996). This global hypoacetylation is a relatively late event in inactivation, and is observed several days after transcriptional silencing in differentiating XX ES cells (Keohane *et al.*, 1996).

A further epigenetic modification occurring later during the X inactivation process is the accumulation of variant forms of histone H2A (in particular members of the macroH2A1 family) on Xi. In undifferentiated ES cells, macroH2A1 is located away from the X chromosome and is concentrated within a distinct body (the macrochromatin body) usually at the nuclear periphery. Upon differentiation of XX ES cells, the macroH2A1 body becomes relocalized to the inactive X chromosome (Mermoud *et al.*, 1999; Rasmussen *et al.*, 2000). This association of macroH2A1 with Xi is dependent on the *Xist* transcript, and is disrupted when *Xist* is deleted but occurs only well after transcriptional silencing (Fig. 7.4).

DNA methylation also appears to be an important feature of the X inactivation process. The CpG islands associated with X-inactivated genes tend to be heavily methylated on the inactive X, while their homologous counterparts on Xa remain unmethylated (Norris *et al.*, 1991). In differentiating XX ES cells, *de novo* methylation of CpG islands on Xi occurs some days after the upregulation of *Xist* and the observation of late replication (Keohane *et al.*, 1996; Wutz and Jaenisch, 2000), suggesting that DNA methylation is not required to initially establish silencing. Evidence is accumulating that the importance of DNA methylation is to lock in and maintain the repressed state of the inactivated chromosome (Sado *et al.*, 2000).

Imprinted X inactivation

In extraembryonic tissues of eutherian mammals, and all tissues of marsupials, X inactivation is not random, but occurs in an imprinted manner, with the paternally derived X (Xp) being preferentially silenced. This imprinted form of X inactivation is also mediated by *Xist* RNA, but unlike in the embryo proper, in extraembryonic (and marsupial) tissues, *Xist* is expressed exclusively from the paternally inherited X. The mechanistic basis for the imprinted expression of *Xist* in these tissues is yet to be clearly established. While *Xist* is still the catalyst for inactivation, it is becoming clear that there are differences in the nature and order of downstream events involved in random and imprinted X inactivation. Unlike the late association of macroH2A1 with Xi observed during random X inactivation, macroH2A1 associates with the paternal X early during inactivation in extraembryonic tissues (Costanzi *et al.*, 2000). There is also indirect evidence that DNA methylation of CpG islands may be less important during imprinted X inactivation (Sado *et al.*, 2000). The inactive X (Xp) in extraembryonic tissues replicates earlier during S phase than its active (maternally derived) homologue, in contrast to the late replication observed for the inactive X in the embryo itself. Finally, it was originally suggested that the mouse PcG-group protein Eed, now known to catalyse the trimethylation of H3–K27 (see above), was specifically involved in imprinted X inactivation in extraembryonic tissues (Wang *et al.*, 2001). However, as mentioned above, more recent findings have shown that both random and imprinted X inactivation are characterized by the transient recruitment of the Eed/Ezh2 complex and the accumulation of H3-K27 trimethylation (Plath *et al.*, 2003).

Topic 7. Metastable epialleles

Recent research in the mouse has established the existence of a group of alleles where the epigenetic state, at least the level of cytosine methylation, can vary from cell to cell within a single tissue type. Since the epigenetic state at these alleles determines transcriptional activity, this results in mosaic patterns of expression, also known as variegation. One good example of this is the mottled coat colour associated with the Agouti viable yellow (Avy) mouse (Duhl *et al.*, 1994; Morgan *et al.*, 1999). Interestingly, the establishment of the epigenetic state can also vary from mouse to mouse, even in a situation where the mice are genetically identical (Wolff, 1978). This variable expressivity associated with isogenicity suggests that there is some element of stochasticity associated with the establishment of these marks in early development. Alleles of this type are now referred to as metastable epialleles (Rakyan *et al.*, 2002) to emphasize the importance of epigenetic state in determining their activity and the plasticity of that epigenetic state between generations.

As has been found at parentally imprinted genes, the transcriptional activity of these alleles correlates with the level of cytosine methylation (Rakyan *et al.*, 2002). One interesting characteristic of these alleles is that the phenotype of the parent influences, to some extent, the phenotype of the offspring. Careful studies of this effect suggest that this is the result of a failure to completely clear the epigenetic marks between generations (Morgan *et al.*, 1999; Rakyan *et al.*, 2002). This raises the possibility that at least in mice, it is not only DNA that is inherited from the parents. At least at some loci, at some times, the epigenetic modifications are also inherited. The inheritance of epigenetic states between generations has also been observed in *Drosophila* (Cavalli and Paro, 1998; Sollars *et al.*, 2003). These studies raise the interesting possibility that the recent

findings in mice force us to consider the implications of this type of inheritance in mammals. One can imagine situations in which epigenetic inheritance could enable offspring to be better adapted, i.e. fitter, to the predicted environmental conditions. There are good reasons to believe that epigenetic inheritance of this sort would provide a more rapid form of adaptation than mutation (Rutherford and Henikoff, 2003; Sollars *et al.*, 2003).

It remains unclear how many alleles of this type exist in the mouse. To date, all metastable epialleles are associated with the intracisternal A-particle (IAP) family of retrotransposons. While there are 1000–2000 IAP elements scattered throughout the mouse genome, evidence to date suggests that the majority of these are always transcriptionally inactive, and not associated with metastable epialleles.

One of the most important questions that arises from this work is whether or not such alleles exist in humans. Unfortunately, the outbred nature of humans makes it difficult to study in the human population as a whole; however, a number of groups are now investigating this possibility using monozygotic twins. The hypothesis is that some genes will be found where the epigenetic state differs between monozygotic twins (Petronis, 2001).

Conclusions and the Future

The past decade has seen an explosion in our understanding of both the nature and significance of epigenetic information. Modifications of chromosomal DNA that are not accompanied by a change in the primary DNA sequence, including methylation and histone modification, exert a significant impact on both chromosome and gene function. As reviewed in this chapter, these epigenetic changes are fundamental to a number of developmentally important processes in mammals, including parental imprinting and X chromosome inactivation, and the reprogramming of the genome during gametogenesis and embryogenesis. Several important human diseases (including cancer and various mental retardation syndromes) are known to be associated with an abnormal epigenetic profile. Accordingly it is likely that one important future direction for epigenetic research will be the development of new drugs to target epigenetic regulators. As discussed in Topic 7, another important future question will be to understand the extent to which the adult phenotype is influenced by the inheritance between generations of epigenetic as well as genetic information. Finally, a thorough understanding of the epigenetic information which accumulates in somatic tissues and how this information becomes cleared and reprogrammed during development will be fundamental to unlocking the significant medical potential offered by somatic stem cell therapy, and for studies of nuclear cloning.

References

Agalioti, T., Lomvardas, S., Parekh, B., Yie, J., Maniatis, T. and Thanos, D. (2000) Ordered recruitment of chromatin modifying and general transcription factors to the IFN-beta promoter. *Cell* 103, 667–678.

Avner, P. and Heard, E. (2001) X-chromosome inactivation: counting, choice and initiation. *Nature Reviews of Genetics* 2, 59–67.

Bachman, K.E., Rountree, M.R. and Baylin, S.B. (2001) Dnmt3a and Dnmt3b are transcriptional repressors that exhibit unique localization properties to heterochromatin. *Journal of Biological Chemistry* 276, 32282–32287.

Barlow, D.P. (1994) Imprinting: a gamete's point of view. *Trends in Genetics* 10, 194–199.

Beechey, C.V., Cattanach, B.M., Blake, A. and Peters, J. (2003) MRC Mammalian Genetics Unit, Harwell, Oxfordshire. World Wide Web Site – Mouse Imprinting Data and References (http://www.mgu.har.mrc.ac.uk/imprinting/imprinting.html).

Belyaev, N.D., Keohane, A.M. and Turner, B.M. (1996) Differential underacetylation of histones H2A, H3 and H4 on the inactive X chromosome in human female cells. *Human Genetics* 97, 573–578.

Bhattacharya, S.K., Ramchandani, S., Cervoni, N. and Szyf, M. (1999) A mammalian protein with specific demethylase activity for mCpG DNA. *Nature* 397, 579–583.

Boggs, B.A., Cheung, P., Heard, E., Spector, D.L., Chinault, A.C. and Allis, C.D. (2002) Differentially methylated forms of histone H3 show unique association patterns with inactive human X chromosomes. *Nature Genetics* 30, 73–76.

Boumil, R.M. and Lee, J.T. (2001) Forty years of decoding the silence in X-chromosome inactivation. *Human Molecular Genetics* 10, 2225–2232.

Briggs, S.D., Xiao, T., Sun, Z.-W., Caldwell, J.A., Shabanowitz, J., Hunt, D.F., Allis, C.D. and Strahl, B.D. (2002) *Trans*-histone regulatory pathway in chromatin. *Nature* 418, 498.

Brown, C.J. and Willard, H.F. (1994) The human X-inactivation centre is not required for maintenance of X-chromosome inactivation. *Nature* 368, 154–156.

Brown, C.J., Hendrich, B.D., Rupert, J.L., Lafreniere, R.G., Xing, Y., Lawrence, J. and Willard, H.F. (1992) The human XIST gene: analysis of a 17 kb inactive X-specific RNA that contains conserved repeats and is highly localized within the nucleus. *Cell* 71, 527–542.

Buiting, K., Gross, S., Lich, C., Gillessen-Kaesbach, G., el-Maarri, O. and Horsthemke, B. (2003) Epimutations in Prader–Willi and Angelman syndromes: a molecular study of 136 patients with an imprinting defect. *American Journal of Human Genetics* 72, 571–577.

Cavalli, G. and Paro, R. (1998) The *Drosophila* Fab-7 chromosomal element conveys epigenetic inheritance during mitosis and meiosis. *Cell* 93, 505–518.

Chaillet, J.R. (1994). Genomic imprinting: lessons from mouse transgenes. *Mutation Research* 307, 441–449.

Clayton, A.L., Rose, S., Barratt, M.J. and Mahadevan, L.C. (2000) Phosphoacetylation of histone H3 on c-fos- and c-jun-associated nucleosomes upon gene activation. *EMBO Journal* 19, 3714–3726.

Cosma, M.P., Tanaka, T. and Nasmyth, K. (1999) Ordered recruitment of transcription and chromatin remodelling factors to a cell cycle- and developmentally regulated promoter. *Cell* 97, 299–311.

Costanzi, C., Stein, P., Worrad, D.M., Schultz, R.M. and Pehrson, J.R. (2000) Histone macroH2A is concentrated in the inactive X chromosome of female preimplantation mouse embryos. *Development* 127, 2283–2289.

Cross, S.H. and Bird, A.P. (1995) CpG islands and genes. *Current Opinion in Genetics and Development* 5, 309–314.

Dennis, K., Fan, T., Geiman, T., Yan, Q. and Muegee, K. (2001) Lsh, a member of the SNF2 family, is required for genome-wide methylation. *Genes and Development* 15, 2940–2944.

Duhl, D.M., Vrieling, H., Miller, K.A., Wolff, G.L. and Barsh, G. S. (1994) Neomorphic agouti mutations in obese yellow mice. *Nature Genetics* 8, 59–65.

Ehrlich, M., Gama-Sosa, M.A., Huang, L.H., Midgett, R.M., Kuo, K.C., McCune, R.A. and Gehrke, C. (1982) Amount and distribution of 5-methylcytosine in human DNA from different types of tissues of cells. *Nucleic Acids Research* 10, 2709–2721.

Fuks, F., Burgers, W.A., Brehm, A., Hughes-Davies, L. and Kouzarides, T. (2000) DNA methyltransferase Dnmt1 associates with histone deacetylase activity. *Nature Genetics* 24, 88–91.

Fuks, F., Burgers, W.A., Godin, N., Kasai, M. and Kouzarides, T. (2001) Dnmt3a binds deacetylases and is recruited by a sequence-specific repressor to silence transcription. *EMBO Journal* 20, 2536–2544.

Fuks, F., Hurd, P.J., Wolf, D., Nan, X., Bird, A.P. and Kouzarides, T. (2003a) The methyl-CpG-binding protein MeCP2 links DNA methylation to histone methylation. *Journal of Biological Chemistry* 278, 4035–4040.

Fuks, F., Hurd, P.J., Deplus, R. and Kouzarides, T. (2003b) The DNA methyltransferases associate with HP1 and the SUV39H1 histone methyltransferase. *Nucleic Acids Research* 31, 2305–2312.

Gardiner-Garden, M. and Frommer, M. (1987) CpG islands in vertebrate genomes. *Journal of Molecular Biology* 196, 261–282.

Gibbons, R.J., McDowell, T.L., Raman, S., O'Rourke, D.M., Garrick, D., Ayyub, H. and Higgs, D.R. (2000) Mutations in *ATRX*, encoding a SWI/SNF-like protein, cause diverse changes in the pattern of DNA methylation. *Nature Genetics* 24, 368–371.

Hall, J.G. (1990) Genomic imprinting: review and relevance to human diseases. *American Journal of Human Genetics* 46, 857–873.

Hebbes, T., Thorne, A.W. and Crane-Robinson, C. (1988) A direct link between core histone acetylation and transcriptionally active chromatin. *EMBO Journal* 7, 1395–1403.

Holliday, R. and Pugh, J.E. (1975) DNA modification mechanisms and gene activity during development. *Science* 187, 226–232.

Hong, L., Schroth, G.P., Matthews, H.R., Yau, P. and Bradbury, E.M. (1993) Studies of the DNA binding properties of histone H4 amino terminus. Thermal denaturation studies reveal that acetylation markedly reduces the binding constant of the H4 'tail' to DNA. *Journal of Biological Chemistry* 268, 305–314.

Howlett, S.K. and Reik, W. (1991) Methylation levels of maternal and paternal genomes during preimplantation development. *Development* 113, 119–127.

International Human Genome Sequencing Consortium (2001) Initial sequencing and analysis of the human genome. *Nature* 409. 860–921.

Jackson, J.P., Lindroth, A.M., Cao, X. and Jacobsen, S.E. (2002) Control of CpNpG DNA methylation by the KRYPTONITE histone H3 methyltransferase. *Nature* 416, 556–560.

Jeddeloh, J.A., Stokes, T.L. and Richards, E.J. (1999) Maintenance of genomic methylation requires a SWI2/SNF2-like protein. *Nature Genetics* 22, 94–97.

Jeppesen, P. and Turner, B.M. (1993) The inactive X chromosome in female mammals is distinguished by a lack of histone H4 acetylation, a cytogenetic marker for gene expression. *Cell* 74, 281–289.

Keohane, A.M., O'Neil, L.P., Belyaev, N.D., Lavender, J.S. and Turner, B.M. (1996) X inactivation and histone H4 acetylation in embryonic stem cells. *Developmental Biology* 180, 618–630.

Khosla, S., Aitchison, A., Gregory, R., Allen, N.D. and Feil, R. (1999) Parental allele-specific chromatin configuration in a boundary-imprinting-control element upstream of the mouse *H19* gene. *Molecular and Cellular Biology* 19, 2556–2566.

Kouzarides, T. (2002) Histone methylation in transcriptional control. *Current Opinion in Genetics and Development* 12, 198–209.

Lee, J., Inoue, K., Ono, R., Ogonuki, N., Kohda, T., Kaneko-Ishino, T., Ogura, A. and Ishino, F. (2002) Erasing genomic imprinting memory in mouse clone embryos produced from day 11.5 primordial germ cells. *Development* 129, 1807–1817.

Lei, H., Okano, M., Jutterman, R., Goss, K.A. and Jaenisch, R. (1996) *De novo* DNA cytosine methyltransferase activities in mouse embryonic stem cells. *Development* 122, 3195–3205.

Leonhardt, H., Page, A.W., Weier, H.-U. and Bestor, T.H. (1992) A targeting sequence directs DNA methyltransferase to sites of DNA replication in mammalian nuclei. *Cell* 71, 865–873.

Li, E. (2002) Chromatin modification and epigenetic reprogramming in mammalian development. *Nature Reviews Genetics* 3, 662–673.

Li, E., Bestor, T. and Jaenisch, R. (1992) Targeted mutation of the DNA methyltransferase gene results in embryonic lethality. *Cell* 69, 915–926.

Liu, W.M., Maraia, R.J., Rubin, C.M. and Schmid, C.W. (1994) Alu transcripts: cytoplasmic localisation and regulation by DNA methylation. *Nucleic Acids Research* 22, 1087–1095.

Lock, L.F., Takagi, N. and Martin, G.R. (1987) Methylation of the Hprt gene on the inactive X occurs after chromosome inactivation. *Cell* 48, 39–46.

Luger, K., Mader, A.W., Richmond, R.K., Sargent, D.F. and Richmond, T.J. (1997) Crystal structure of the nucleosome core particle at 2.8Å resolution. *Nature* 389, 251–260.

Lyon, M.F. (1961) Gene action in the X-chromosome of the mouse (*Mus musculus* L.). *Nature* 190, 372–373.

Mayer, W., Niveleau, A., Walter, J., Fundele, R. and Haaf, T. (2000) Demethylation of the zygotic paternal genome. *Nature* 403, 501–502.

McGrath, J. and Solter, D. (1984) Completion of mouse embryogenesis requires both the maternal and paternal genomes. *Cell* 37, 179–183.

Mermoud, J.E., Costanzi, C., Pehrson, J.R. and Brockdorff, N. (1999) Histone MacroH2A relocates to the inactive X chromosome after initiation and propogation of X-inactivation. *Journal of Cell Biology* 147, 1399–1408.

Monk, M. (1987) Genomic imprinting. Memories of mother and father. *Nature* 328, 203–204.

Monk, M., Boubelik, M. and Lehnert, S. (1987) Temporal and regional changes in DNA methylation in the embryonic, extraembryonic and germ cell lineages during mouse embryo development. *Development* 99, 371–382.

Moore, T. and Haig, D. (1991) Genomic imprinting in mammalian development: a parental tug-of-war. *Trends in Genetics* 7, 45–49.

Morgan, H.D., Sutherland, H.G., Martin, D.I. and Whitelaw, E. (1999) Epigenetic inheritance at the agouti locus in the mouse. *Nature Genetics* 23, 314–318.

Morison, I.M. and Reeve, A.E. (1998) A catalogue of imprinted genes and parent-of-origin effects in humans and animals. *Human Molecular Genetics* 7, 1599–1609.

Murray, K. (1964) The occurrence of ε-N-methyllysine in histones. *Biochemistry* 3, 10–15.

Nan, X., Meehan, R.R. and Bird, A. (1993) Dissection of the methyl-CpG binding domain from the chromosomal protein MeCP2. *Nucleic Acids Research* 21, 4886–4892.

Niwa, O., Yokota, Y., Ishida, H. and Sugahara, T. (1983) Independent mechanisms involved in suppression of the Moloney leukemia virus genome during differentiation of murine teratocarcinoma cells. *Cell* 32, 1105–1113.

Norris, D.P., Brockdorff, N. and Rastan, S. (1991) Methylation status of CpG-rich bands on active and inactive mouse X chromosomes. *Mammalian Genome* 1, 78–83.

Nowak, S.J. and Corces, V.G. (2000) Phosphorylation of histone H3 correlates with transcriptionally active loci. *Genes and Development* 14, 3003–3013.

Okano, M., Xie, S. and Li. E. (1998) Cloning and characterization of a family of novel mammalian DNA (cytosine-5) methyltransferases. *Nature Genetics* 19, 219–220.

Okano, M., Bell, D.W., Haber, D.A. and Li, E. (1999) DNA methyltransferases Dnmt3a and Dnmt3b are essential for *de novo* methylation and mammalian development. *Cell* 99, 247–257.

Oswald, J., Engemann, S., Lane, N., Mayer, W., Olek, A., Fundele, R., Dean, W., Reik, W. and Walter, J. (2000) Active demethylation of the paternal genome in the mouse zygote. *Current Biology* 10, 475–478.

Paldi, A. (2003) Genomic imprinting: could the chromatin structure be the driving force? *Current Topics in Developmental Biology* 53, 115–138.

Pardo-Manuel de Villena, F., de la Casa-Esperon, E. and Sapienza, C. (2000) Natural selection and the function of genome imprinting: beyond the silenced minority. *Trends in Genetics* 16, 573–579.

Peters, A.H.F.M., O'Carroll, D., Scherthan, H., Mechtler, K., Sauer, S., Schöfer, C., Weipoltshammer, K., Pagani, M., Lachner, M., Kohlmaier, A., Opravil, S., Doyle, M., Sibilia, M. and Jenuwein, T. (2001) Loss of the *Suv39h* histone methyltransferases impairs mammalian heterochromatin and genome stability. *Cell* 107, 323–337.

Peters, A.H.F.M., Mermoud, J.E., O'Carroll, D., Pagani, M., Schweizer, D., Brockdorff, N. and Jenuwein, T. (2002) Histone H3 lysine 9 methylation is an epigenetic imprint of facultative heterochromatin. *Nature Genetics* 30, 77–80.

Petronis, A. (2001) Human morbid genetics revisited: relevance of epigenetics. *Trends in Genetics* 17, 142–146.

Plath, K., Fang, J., Mlynarczyk-Evans, S.K., Cao, R., Worringer, K.A., Wang, H., de la Cruz, C.C., Otte, A.P., Panning, B. and Zhang, Y. (2003) Role of histone H3 lysine 27 methylation in X inactivation. *Science* 300, 131–135.

Pradhan, S., Bacolla, A., Wells, R.D. and Roberts, R.J. (1999) Recombinant human DNA (cytosine-5) methyltransferase. I. Expression, purification, and comparison of *de novo* and maintenance methylation. *Journal of Biological Chemistry* 274, 33002–33010.

Pogo, B.G.T., Allfrey, V.G. and Mirsky, A.E. (1966) RNA synthesis and histone acetylation during the course of gene activation in lymphocytes. *Proceedings of the National Academy of Sciences USA* 55, 6212–6222.

Rakyan, V.K., Blewitt, M.E., Druker, R., Preis, J.I. and Whitelaw, E. (2002) Metastable epialleles in mammals. *Trends in Genetics* 18, 348–351.

Rasmussen, T.P., Mastrangelo, M.-A., Eden, A., Pehrson, J.R. and Jaenisch, R. (2000) Dynamic relocalization of histone macroH2A1 from centrosomes to inactive X chromosome during X inactivation. *Journal of Cell Biology* 150, 1189–1198.

Razin, A. and Cedar, H. (1994) DNA methylation and genomic imprinting. *Cell* 77, 473–476.

Rea, S., Eisenhaber, F., O'Carroll, D., Strahl, B.D., Sun, Z.-W., Schmid, M., Opravil, S., Mechtler, K., Ponting, C.P., Allis, C.D. and Jenuwein, T. (2000) Regulation of chromatin structure by site-specific histone H3 methyltransferases. *Nature* 406, 593–599.

Reik, W. and Walter, J. (2001) Genomic imprinting: parental influence on the genome. *Nature Reviews of Genetics* 2, 21–32.

Rutherford, S.L. and Henikoff, S. (2003) Quantitative epigenetics. *Nature Genetics* 33, 6–8.

Sado, T., Fenner, M.H., Tan, S.S., Tam, P., Shioda, T. and Li, E. (2000) X inactivation in the mouse embryo deficient for Dnmt1: distinct effect of hypomethylation on imprinted and random X inactivation. *Developmental Biology* 225, 294–303.

Santos, F., Hendrich, B., Reik, W. and Dean, W. (2002) Dynamic reprogramming of DNA methylation in the early mouse embryo. *Developmental Biology* 241, 172–182.

Santos-Rosa, H., Schneider, R., Bannister, A.J., Sherriff, J., Bernstein, B.E., Emre, N., Schreiber, S.L., Mellor, J. and Kouzarides, T. (2002) Active genes are tri-methylated at K4 of histone H3. *Nature* 419, 407–411.

Silva, J., Mak, W., Zvetkova, I., Appanah, R., Nesterova, T.B., Webster, Z., Peters, A.H.F.M., Jenuwein, T., Otte, A.P. and Brockdorff, N. (2003) Establishment of histone H3 methylation on the inactive X chromosome requires transient recruitment of Eed–Enx1 polycomb group complexes. *Developmental Cell* 4, 481–495.

Sleutels, F. and Barlow, D.P. (2002) The origins of genomic imprinting in mammals. *Advances in Genetics* 46, 119–163.

Sollars, V., Lu, X., Xiao, L., Wang, X., Garfinkel, M.D. and Ruden D.M. (2003) Evidence for an epigenetic mechanism by which Hsp90 acts as a capacitor for morphological evolution. *Nature Genetics* 33, 70–74.

Spence, J.E., Perciaccante, R.G., Greig, G.M., Willard, H.F., Ledbetter, D.H., Hejtmancik, J.F., Pollack, M.S., O'Brien, W.E. and Beaudet, A.L. (1988) Uniparental disomy as a mechanism for human genetic disease. *American Journal of Human Genetics* 42, 217–226.

Stöger, R., Kubicka, P., Liu, C.G., Kafri, T., Razin, A., Cedar, H. and Barlow, D.P. (1993) Maternal-specific methylation of the imprinted mouse Igf2r locus identifies the expressed locus as carrying the imprinting signal. *Cell* 73, 61–71.

Strahl, B.D. and Allis, C.D. (2000) The language of covalent histone modifications. *Nature* 403, 41–45.

Tamaru, H. and Selker, E.U. (2001) A histone H3 methyltransferase controls DNA methylation in *Neurospora crassa*. *Nature* 414, 277–283.

Tufarelli, C., Stanley, J.A., Garrick, D., Sharpe, J.A., Ayyub, H., Wood, W.G. and Higgs, D.R. (2003) Transcription of antisense RNA leading to gene silencing and methylation as a novel cause of human genetic disease. *Nature Genetics* 34, 157–165.

Turner, B.M. (2000) Histone acetylation and an epigenetic code. *Bioessays* 22, 836–845.

Walsh, C.P. and Bestor, T.H. (1999) Cytosine methylation and mammalian development. *Genes and Development* 13, 26–34.

Walsh, C.P., Chaillet, J.R. and Bestor, T.H. (1998) Transcription of IAP endogenous retroviruses is constrained by cytosine methylation. *Nature Genetics* 20, 116–117.

Wang, J., Mager, J., Chen, Y., Schneider, E., Cross, J.C., Nagy, A. and Magnuson, T. (2001) Imprinted X inactivation maintained by a mouse *Polycomb* group gene. *Nature Genetics* 28, 371–375.

Wei, Y., Yu, L., Bowen, J., Gorovsky, M.A. and Allis, C.D. (1999) Phosphorylation of histone H3 is required for proper chromosome condensation and segregation. *Cell* 97, 99–109.

Wolff, G.L. (1978) Influence of maternal phenotype on metabolic differentiation of agouti locus mutants in the mouse. *Genetics* 88, 529–539.

Woodcock, D.M., Lawler, C.B., Linsenmeyer, M.E., Doherty, J.P. and Warren, W.D. (1997) Asymmetric methylation in the hypermethylated CpG promoter region of the human L1 retrotransposon. *Journal of Biological Chemistry* 272, 7810–7816.

Wutz, A. and Jaenisch, R. (2000) A shift from reversible to irreversible X inactivation is triggered during ES cell differentiation. *Molecular Cell* 5, 695–705.

Yoder, J.A., Walsh, C.P. and Bestor, T.H. (1997) Cytosine methylation and the ecology of intragenomic parasites. *Trends in Genetics* 13, 335–340.

8 Regulation of Genome Activity and Genetic Networks in Mammals

V. VanBuren* and M.S.H. Ko

National Institutes of Health, National Institute on Aging, Laboratory of Genetics, Developmental Genomics and Aging Section, Baltimore, MD 21224, USA

Introduction	201
Contrasting Transcriptional Regulation for Mammalian and 'Lower' Species	202
Basics of Transcription Activation	202
The primary components of transcriptional activation	203
The stochastic component of gene regulation	204
Signalling and Changes in Genome Activity	205
Nuclear receptors	205
Whole Genome Expression Studies	207
ESTs, SAGE and MPSS	207
DNA microarrays	208
Molecular Genetic Networks	209
Reconstruction of network topologies from large-scale data	210
Mammalian gene regulatory networks	212
Analysis of biological networks	212
Knowledge-based Prediction of Protein and DNA Interactions	213
Identification of cis-regulatory elements based on computational motif finding	213
Summary	215
References	216

Introduction

Networks are everywhere. Complex and dynamic functions of cells, and their interactions to form and maintain organisms, are all governed by regulatory networks of genes, proteins and other biological molecules. For example, cell responses to stimuli and maintenance of homeostasis may be governed by a variety of cellular control mechanisms, including control at the phases of transcription, RNA processing, transport and localization of mRNA, translation, mRNA degradation and protein activity control (Fig. 5.1). Transcriptional activity of the genome, however, underlies much of the important regulatory circuitry of the cell. In this chap-

* Correspondence: vincent_vanburen@nih.gov; minoru_ko@nih.gov

© CAB International 2005. *Mammalian Genomics*
(eds A. Ruvinsky and J. Marshall Graves)

ter, we therefore discuss mainly transcriptional regulation by transcription factors and the networks formed by their subsequent regulation of genes encoding transcription factors.

This chapter necessarily provides only an overview of these complex subjects. For a discussion of epigenetic control of transcription, see Chapter 7 in this book. For a comprehensive coverage of transcriptional regulation by transcription factors, see Chapter 5 of this book, Carey and Smale (2000) or Alberts et al. (2002). Discussions of gene network reconstruction may be found in Bower and Bolouri (2001) and Kitano (2001).

Contrasting Transcriptional Regulation for Mammalian and 'Lower' Species

The basic genomic sequences of different species do not tell the story of how different species have such different morphologies, or how mammals have such physiological and morphological complexity when compared with other metazoans such as the nematode Caenorhabditis elegans. For example, genome size does not correlate at all with the complexity of the mRNA population. When animals with >100-fold differences in genome size were compared, most had very similar mRNA population sizes ($35 \pm 10 \times 10^6$ nt), and only Drosophila was markedly different, with an mRNA population size ~twofold lower than the others (Davidson, 2001).

Mammalian species and other higher eukaryotes are special in that their transcriptional regulation is extremely complex (Fig. 8.1). It is likely that information about the regulation of even a single mammalian gene is not complete (Carey and Smale, 2000). This complexity is due to four main factors. First, proteins that regulate transcription can act over thousands of nucleotides in higher eukaryotes, whereas this type of regulation is normally restricted to a region in close proximity to the transcription start site in other species. It is not unusual in higher eukaryotes to find regulatory sequences for a gene spread out over 50,000 nucleotide base pairs, and this greatly complicates efforts to experimentally determine or predict elements important for regulation. Secondly, RNA polymerase II cannot initiate transcription on its own in higher eukaryotes. General transcription factors (GTFs) are required for initiation, thus providing further opportunities for regulation that are unavailable in bacteria. Thirdly, chromatin packing provides still more opportunities for regulation that are not available to bacteria by regulating DNA access to the transcriptional machinery (Alberts et al., 2002). Fourthly, it is thought that there have been at least two complete genome duplications in the early evolution of vertebrates. As an example of the increased complexity of gene regulatory networks that must be considered for mammals, there are four clusters of hox genes on four different chromosomes in amniotes, whereas there is only one hox cluster in the invertebrate chordate Amphioxus (Garcia-Fernandez and Holland, 1994) and one cluster in Strongylocentrotus purpuratus (Martinez et al., 1999).

Bilaterians (bilaterally organized animals) are likely to have members of each known major transcription factor family, and have components of every known signalling pathway (Davidson, 2001). These gene families, however, are differently duplicated and diversified between different clades of bilaterians, which is accompanied by diversifications of their functional roles in development. This leads to the assertion that different bilaterian clades all have a similar transcription factor toolbox, and that the differential morphology between clades is a result of the differing architecture of cis-regulatory elements.

Basics of Transcription Activation

Differences in cellular morphology and function generally depend on differences in gene expression, and are not dependent on changes in the DNA sequence itself. The notable exception to this is the generation of immune system diversity by DNA rearrangement. Here we describe some of the prominent features of mammalian transcription regulation by transcription factors for a single gene, with emphasis on the features of transcriptional regulation that contribute to the complexity of genetic networks. We will also note the role of stochasticity in gene regulatory switches. In a later section ('Molecular Genetic Networks'), we revisit a discussion about factors contributing to the complexity of the network of regulatory control. To avoid much redundancy,

It is not uncommon for eukaryotic enhancers to 50 kb away from the transcriptional start site, whereas regulatory elements in bacteria are normally within a few hundred bases of the start site.

Unlike bacterial promoters, eukaryotic promoters require binding of general transcription factors (GTFs), including TFIID, for RNA polymerase II (RNA pol II) binding and transcription initiation.

Eukaryotic chromatin must be remodeled to allow transcription, permitting regulatory control not found in bacteria.

5-10% of genes in mammalian genomes code for gene regulatory proteins, and at least 2-3 different factors bind multiple motifs within a single regulatory module. There are likely to be ten or more modules per gene.

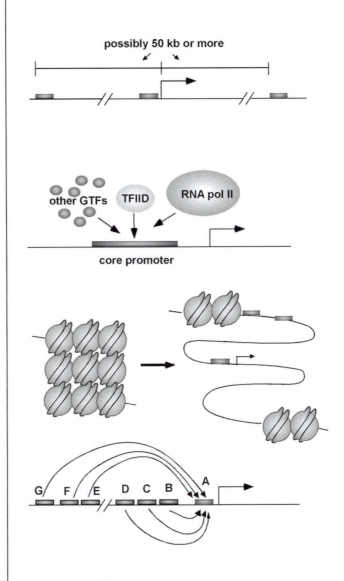

Fig. 8.1. Some ways that complexity arises in mammalian transcriptional regulation networks.

we will point the reader to the earlier Chapter 5, as necessary. As our understanding of mammalian transcription is incomplete, we note that generalizations about this regulation are suspect, and that the components of this regulation are present in various combinations that contribute to the complexity of transcription activity.

The primary components of transcriptional activation

A typical gene model for gene activation and gene inactivation includes a core promoter, a regulatory promoter, and enhancers. The core promoter, which typically lies between −40 and +50 relative to the transcription start site, binds the general transcription machinery and recruits RNA polymerase II to begin transcription at the correct start site. The regulatory promoter, which typically lies somewhere between −500 and −40, together with the enhancer sequences, which lie further upstream or downstream of the transcriptional start site, bind to 'activator' proteins. Animal genes usually include several enhancers, and these are likely to be ~500 bp and have ~10 binding sites for at least three sequence-specific transcription factors, one of which is normally a repressor (Levine and Tjian, 2003). Some of these *trans* activators that bind *cis*-regulatory elements are ubiquitous, while others are expressed in specific cell types. Transcription activation occurs when activator proteins bound to these elements recruit the general transcription machinery to the core promoter (Carey and Smale, 2000).

Genes may be inactivated by establishing an inactive chromatin environment. This can be accomplished by ATP-dependent remodelling and histone deacetylation and by the formation of heterochromatin. Thus, transcription initiation often requires that the chromatin containing a given gene must first decondensate its higher order structures. Remodelling of the chromatin occurs in the presence of ATP-dependent remodelling enzymes and histone acetyltransferases. Locus control regions may help govern accessibility of the promoters over an extended area by controlling chromatin remodelling with the recruitment of remodelling complexes, and thus control access of the transcription factors and transcription machinery to promoters and the transcription start sites. As an alternative to transcriptional inactivation by chromatin inactivation, transcription factors may have a repressive effect on the formation of an initiation complex. It is possible for *cis*-regulatory elements to be overlapping such that binding of a repressor may block binding of an activating factor.

Insulator or boundary elements flank gene regions in the genome and prevent enhancers within the region they flank from communicating with elements outside of that region. Boundary elements are like locus control regions in the sense that they can confer position-independent expression to a gene, but differ in that they do not enhance transcription and can prevent transcription when placed between a promoter and an enhancer. Study of matrix attachment regions (MARs) has been motivated by the idea that their physical interaction along with their proximity to other regulatory elements suggests a possible role similar to boundary elements. These A/T-rich regions are defined by physical attachment to the intranuclear matrix, and so their functions may be heterogeneous. Although they have been found to possess functions similar to boundary elements, they cannot insulate a gene from upstream enhancers (Carey and Smale, 2000).

The regulatory promoter and enhancer regions of a gene process complex information in the regulation of transcription initiation. *cis*-regulatory elements for a gene may be spread over great distances and there may be interdigitation of elements that bind ubiquitously expressed transcription factors with elements that bind factors specifically expressed. Furthermore, some factors may be inhibitory and many may associate only indirectly by binding other *trans* factors rather than binding directly to the *cis*-regulatory element itself. The nucleoprotein structures that comprise these arrays are collectively referred to as the enhanceosome. The enhanceosome recruits the general transcription machinery to the core promoter to form a pre-initiation complex.

The core promoter itself is also subject to regulation. TFIID is the only general transcription factor that can bind to the core promoter specifically and independently. TFIID contains TATA-binding protein (TBP) and ten or more TBP-associated factors ($TAF_{II}s$). TBP binds directly to the TATA DNA motif TATAAA, which is generally found 25–30 bp upstream of the transcription start site. This binding is the nucleation step for the start of transcription. The TATA-binding motif can independently regulate a low level of basal transcription, and may participate in activated transcription when an activator binds to a nearby regulatory element. The overall level of basal transcription is probably related to the affinities of GTFs and the affinity of TFIID (via TBP and other components).

The stochastic component of gene regulation

To build a model of a gene expression network, it is important to consider the behaviour of a single template molecule in a cell. There are generally only two copies of genes, paternally derived and maternally derived, and thus, the network model should be built with this in mind. However, almost all molecular, cell and developmental biology research has been performed using samples consisting of large numbers of cells. These studies nevertheless tend to draw conclusions as if the measurement was made on a single gene copy. Early work to address this issue demonstrated that gene expression regulation is indeed a stochastic event (reviewed in Ko, 1992). Subsequently, similar findings have been reported (e.g. Hume, 2000; de Krom *et al.*, 2002). More recent efforts using modern technology have also supported the stochasticity of gene regulation (Becskei and Serrano, 2000; Blake *et al.*, 2003).

It has been demonstrated both computationally and experimentally that stochastic behaviour of the genetic switch can lead to haploinsufficiency for some genetic loci (Cook *et al.*, 1998; Veitia, 2002) and to noise in biological systems (Cook *et al.*, 1998; Kemkemer *et al.*, 2002; Rao *et al.*, 2002; Magee *et al.*, 2003).

The stochastic nature of genetic switches prompted a search for a genetic switch that can function as a rheostat. A rheostat switch would allow for the graded regulation of transcription, rather than for simple on–off dynamics. An example of this kind of switch has been engineered in the tet-inducible system (Blau and Rossi, 1999). This work not only solved the problem of precisely manipulating gene expression levels experimentally, but also suggests the possibility that such rheostat switches exist in nature.

The problem of stochastic behaviour of genetic switches from the viewpoint of entire genetic networks and circuits is analogous to building electronic circuits with an unreliable switch. Stochastic switches cannot be entirely 'on' when they need to be on, nor can they be entirely 'off' when they need to be off. Only the probability of being on or off is controlled, and the control is not 100% accurate. It is easy to imagine that such a circuit will not function properly. An understanding of how this expression noise is controlled or accommodated in transcriptional switches will be necessary for a complete understanding of gene regulatory network control.

Signalling and Changes in Genome Activity

Transcription may occur at basal or activated levels depending on the combinatorial activity of transcription factors. The activity of many transcription activators and inactivators (or repressors) is controlled by signals transmitted from the cell surface to the nucleus. This signalling may be quite direct and have a lasting duration as with signals from steroid hormones, or may be short-lived and indirect as with transmembrane receptors that initiate signal transduction.

Signals received by cells are important for establishing the developmental plan and for maintenance of homeostasis. Receptor-mediated signal transduction is discussed in Chapter 5 of this volume. Here we will discuss steroid hormone interaction with nuclear receptors as an additional salient example of how signalling mechanisms influence transcription factor activity and thus influence transcription rates. Steroid hormones can trigger expression of both primary and secondary response proteins. In this case, the primary response proteins activate expression of the secondary response proteins, and so our discussion of steroid hormones offers an excellent example of how a network of *cis*-regulatory control elements can receive and process communications from outside the cell.

Nuclear receptors

Steroid hormones are lipid-soluble molecules that act as signals that participate in a variety of cell regulatory pathways and are essential for development and homeostasis in mammals. The steroid hormone receptor is the archetype for a superfamily of receptors called nuclear receptors, a group that also includes retinoid receptors, vitamin D receptors, thyroid hormone receptors, and other orphan nuclear receptors. Steroid hormone receptors are activated by bound steroid, and the activated receptor binds to a hormone responsive element (HRE), thus influencing the transcriptional activity of the gene under its control.

The receiver is composed of a nuclear localization signal, a ligand-binding domain and a DNA-binding domain, which is organized around two zinc ions. Receptors are dimerized with an association between the DNA-binding domains of two receptor monomers. This dimerization is organized such that the DNA-binding domains both have access to the DNA, and the complete ligand-bound receptor complex binds to palindromic HRE sequences (Fig. 8.2). Prior to ligand binding, the unbound receptor complex is bound by a complex of several chaperones, including Hsp90, Hsp56, Hsp70 and p23, that prevent interactions with HRE.

Steroid receptors interact with GTFs and co-activators. Oestrogen receptor (ER) transactivation is enhanced when TBP is overexpressed, and these two proteins are reported to interact *in vitro* (Sadovsky *et al.*, 1995). $TAF_{II}s$ have also been implicated as potential targets for hormone receptors. For example, human $TAF_{II}30$ is required for transactivation by ER (Jacq *et al.*, 1994). There are numerous examples of potential co-activators called transcription intermediary factors (TIFs) that interact with steroid receptors. The ligand-binding domain of ER is known to associate with several proteins in a ligand-dependent manner (e.g. Cavailles *et al.*, 1995). Glucocorticoid

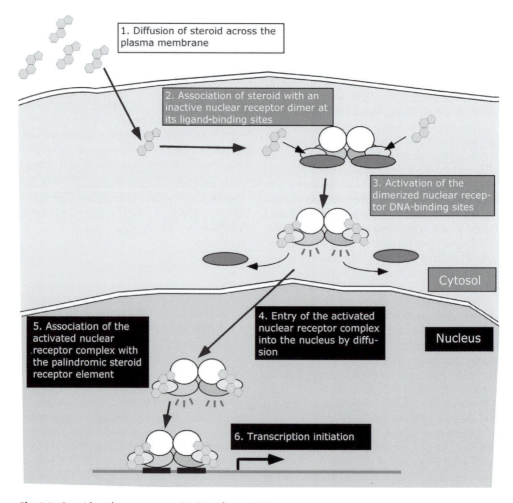

Fig. 8.2. Steroid nuclear receptor activation of transcription.

receptor (GR) binds a ubiquitous protein called GRIP170 in a complex with two GREs, and GRIP170 enhances GR transcription *in vitro* (Eggert *et al.*, 1995). There are many other such examples of potential TIFs, including proteins identified by genetic screens for intermediary factors.

Hormone receptors also interact with specific transcription factors. Many transcription factors have been shown to mediate or enhance the activity of ligand-bound steroid receptors. Some examples illustrate further complexity that is difficult to interpret. One such example is that although the MMTV promoter requires both the CTF1/NF1 and Oct1/OTF1 transcription factors for optimal transactivation by glucocorticoids or progestins, CTF1/NF1 does not bind synergistically, whereas Oct1/OTF1 has been shown to interact directly with receptors on free DNA (Beato *et al.*, 1996).

The DNA-binding domains of steroid hormone receptors typically have multiple transactivation domains which allow the ligand-bound receptor to directly influence the transcription of multiple genes. As some genes under this direct control may code for transcription factors, steroid hormones may have extensive indirect influence over the transcription of entire batteries of genes, underscoring their importance in developmental processes. Cell type profoundly affects the transcriptional control that steroids exert. The importance of cell type for steroid action is attributed to the integrative process of gene activation, where activation is co-dependent on the activated nuclear receptor and other factors (Alberts *et al.*, 2002).

Whole Genome Expression Studies

Profiling the expression of genes on a large scale gives snapshots of cellular status and valuable insight into which genes are important to or have unique expression patterns in a particular cell type, or which genes have enriched or depleted expression in response to a given cue. Despite the large scale of the data collected in this type of expression profiling, most efforts that use microarrays are actually large-scale reductionist studies that aim to single out candidate genes for further functional studies. A relatively smaller constituency of these efforts

tries to assemble complex networks of genetic regulation. None the less, the collection of whole genome expression data is an important step for such a goal. For whole genome expression analysis, four technologies have been used: expressed sequence tag (EST), serial analysis of gene expression (SAGE), massively parallel signature sequencing (MPSS) and DNA microarrays.

ESTs, SAGE and MPSS

ESTs are short sequence reads from either the 5' or 3' end of complementary DNA (cDNA) (Adams *et al.*, 1993; reviewed in Schuler, 1997). Construction of cDNA libraries and large-scale analysis of these libraries forms part of the core of genomics (reviewed in Ko, 2001). Building a gene catalogue from the fundamental gene identification and gene structure information provided by EST projects is a prerequisite for most expression profiling techniques. ESTs give gene signatures, which help define gene structure by alignment with the genome, and they allow for an evaluation of the frequency of randomly selected cDNA clones within a library. The cDNA clones from which ESTs are derived are useful for functional studies of a particular gene, or may be used as part of a collection in large-scale efforts such as spotting on cDNA microarrays or in large-scale *in situ* hybridization.

EST frequency has been used to identify genes expressed in specific cell types and stages. For example, the generation and analysis of ESTs from embryonic tissues lead to the discovery of genes predominantly expressed in mouse embryos and stem cells as well as genes defining developmental potency (Sharov *et al.*, 2003). Publicly available ESTs have been used to identify genes specific to ovary and early embryos (Stanton and Green, 2001). Public databases have also provided tools for such analysis, e.g. Digital Differential Display (http://www.ncbi.nlm.nih.gov/UniGene/info_ddd.shtml).

SAGE is essentially an accelerated version of determining EST frequency from an EST project, though this does not provide the information regarding structures of transcripts and genes (Velculescu *et al.*, 2000). Expressed sequences are cut with restriction enzymes to produce small unique 'tags' representing each expressed sequence. These tags are ligated together with linker sequences into vectors containing approximately 20 tags separated by the linker sequences. Sequencing of these vectors allows the frequency of the expressed sequences to be determined in a high-throughput capacity compared with normal EST sequencing efforts.

SAGE has been used to evaluate expression profiles in mammals in a variety of clinical and basic research investigations and the archived data sets can be used as *in silico* or virtual dot blots (see the database: http://www.ncbi.nlm. nih.gov/SAGE/). For example, SAGE has been performed after adenoviral expression of the oncogenic transcription factor c-MYC in primary human umbilical vein, and it was found that 216 tags are induced and 260 tags are repressed (Menssen and Hermeking, 2002). Of these, 54 induced tags were validated by microarray and reverse transcription–polymerase chain reaction (RT–PCR). The combination of cell cycle genes found to be c-MYC-induced (CDC2-L1, cyclin E-binding protein 1 and cyclin B1) and c-MYC-induced DNA repair genes (BRCA1, MSH2 and APEX) suggests that c-MYC's promotion of cell proliferation is coupled to protection of genome integrity. SAGE has also been used extensively in cardiovascular studies, including studies following the differentiation of monocytes to macrophages (Hashimoto et al., 1999), expression profiles for normal mouse and human heart (Anisimov et al., 2002), profiles of human platelets, human intercranial aneurism, and comparisons of normal and tumour human endothelial cells, to name a few.

MPSS combines non-gel-based signature sequencing with *in vitro* cloning of millions of templates on separate microbeads (Brenner et al., 2000). Signature sequences of 16–20 bases obtained from cDNA libraries provide the expression profiles of cells and tissues. This relatively new technology is beginning to be applied to large-scale gene expression profiling.

DNA microarrays

DNA microarray technology has been leading the way in offering applications and knowledge-seeking strategies in large-scale biology.

Microarrays give a snapshot of gene expression for thousands of transcripts simultaneously. Transcript abundance is normally characterized as a difference in abundance, i.e. fold change, rather than as an absolute measure of transcript counts.

Three types of microarrays are currently used: (i) cDNAs spotted on either glass slides or nylon membranes (Schena et al., 1995); (ii) 25-mer oligonucleotides synthesized by the lithographic technique (Pease et al., 1994, e.g. Affymetrix); and (iii) 60-mer to 80-mer oligonucleotides synthesized *in situ* (Hughes et al., 2001, e.g. Agilent Technologies) or spotted on either a glass slide or nylon membrane. The cDNA-spotted arrays (i) require physical cDNA clones to start with. Given the scarcity of embryonic cDNA clones, an effort to assemble collections of representative mouse transcripts culminated in the construction of the NIA Mouse 15K (Tanaka et al., 2000) and NIA Mouse 7.4K cDNA (VanBuren et al., 2002) collections, which have been freely distributed worldwide. In contrast, oligonucleotide-based arrays, both (ii) and (iii), do not require cDNA clones, but the downstream functional studies are still dependent on the availability of cDNA clones. In this context, fully sequenced full-length cDNA clones become important resources (Strausberg et al., 2000; Okazaki et al., 2002).

Transcripts are labelled, typically with Cy3 and Cy5 fluorophores for two-channel glass arrays, and with the radioisotope phosphorus-33 for one-channel nylon membranes. When transcripts are collected from rare sample tissues such as pre-implantation embryos (Hamatani et al., 2004), they are often linear-amplified by an *in vitro* RNA polymerase reaction, and then hybridized to the microarray (Eberwine et al., 1992). The intensity of the spot is used as a means of quantifying the transcript abundance relative to the abundance of the same transcript from a different sample, which may be measured in the second channel on the same two-channel array, or measured from a different array when using one-channel arrays.

ChIP-chip is used to identify protein–DNA interactions at a large scale

A new technique called ChIP-chip offers a promising experimental platform for comprehensive identification of *cis*-regulatory elements bound by a known transcription factor (Horak and Snyder, 2002). The name is derived from the two principal technologies drawn upon – <u>Ch</u>romatin <u>i</u>mmuno-<u>p</u>recipitation and DNA microarrays (<u>chip</u>). DNA–protein complexes are first fixed with formaldehyde, and the tissue is lysed. DNA in the lysate is then sheared by sonication, and the transcription factor of interest is isolated using an antibody. DNA that was bound to the protein is captured along with the transcription factor. This DNA is extracted, amplified and labelled for hybridization to a microarray. ChIP-chip requires a good antibody to the transcription factor of interest and an *intergenic* DNA microarray, which contains probes to the genomic regions between genes.

For example, ChIP-chip has been used to determine binding sites for the mammalian GATA-1 transcription factor, which is haematopoietic lineage-specific (Horak *et al.*, 2002). GATA-1 is important for regulating expression of erythroid-specific genes, so selected targets could be chosen to reduce the representaive intergenic sequences to a manageable size. In this case, the mammalian β-globin locus was analysed in the chronic myelogenous leukaemia cell line K562. GATA-1 was found to bind a region encompassing the HS2 core element and another region upstream of the γG gene. This approach may be applied to other factors that bind in the β-globin locus or elsewhere, provided an appropriate intergenic microarray is constructed.

Although ChIP-chip potentially offers a comprehensive map of all the binding sites for a known transcription factor, the scaling-up of this technique to a large number of transcription factors may be difficult due to its requirement for high-quality and specific antibodies. Furthermore, the seminal studies using ChIP-chip have been applied to yeast, and much larger chips must be created to cover all potential regulatory regions in a mammalian genome such as the murine genome. Using ChIP-chip for regulatory motif discovery is also faced with several technical challenges, including false positives, motif variability from consensus, non-functional repeats, and that many motifs may be involved in regulatory complex formation (Horak and Snyder, 2002).

Molecular Genetic Networks

The complexity of biochemical and gene regulatory networks *in vivo* and *in vitro* confounds efforts to analyse expression data and reconstruct gene regulatory networks from that analysis. Some of the issues arising from this complexity are: (i) transcript expression and protein activity sometimes do not correlate; (ii) although we readily gather information about how a single gene varies expression between samples, finding absolute expression values so the expression levels of different genes may be compared is not yet a standard practice; (iii) comparison of different cell types may lead to erroneous conclusions due to heterogeneity of some sample collections and volumetric differences between cell types, and (iv) even when the experimental conditions are well described, it is not yet clear how the results from different experimental platforms and methods may be compared. Despite these challenges, many groups have made strides in reconstructing genetic networks from DNA microarray data and other large-scale data sets.

Networks in cells can be visualized in forms depending on the interests of a given investigator and tolerance for complexity. They are logical constructions that are difficult to define precisely because such cellular networks are interconnected with other networks. As the author Jorge Luis Borges is sometimes quoted, 'everything touches everything'. As examples of some logical constructions of networks, we can think about large metabolic networks at the level of a whole cell, or we may think about subnetworks in metabolism, such as the citric acid cycle, or the pathway for synthesis of a particular amino acid, or signalling networks, or we may think about transcriptional regulatory networks. Here we will focus most of our attention on the regulatory networks formed by the dynamic population of transcription factors, with some notice of signalling networks and their connection to genetic networks.

As such, genetic network reconstruction of a complex circuit has been approached by assembling experimental data and previous data from the literature for the lambda bacteriophage lysis–lysogeny decision circuit (McAdams and Shapiro, 1995). In this seminal work, experimentally determined aspects of the decision circuit were mapped in a circuit diagram style borrowed from electrical engineering, and simulation of the circuit made predictions in agreement with known outcomes. As described above, the gene regulatory control by combinatorial *cis*-regulatory elements in mammals makes prospects for simulation of mammalian gene networks significantly more complex. The best-described genetic network in metazoans is that of the endomesoderm specification in the sea urchin *S. purpuratus* (Davidson *et al.*, 2002a,b). The current model for endoderm and mesoderm specification *in S. purpuratus* includes almost 50 genes and presents views both 'from the genome' and 'from the nucleus' (see http://sugp.caltech.edu/endomes/ for the most recent version of the complete network). The view from the genome portrays the entire genetic network, regardless of when components are active. The view from the nucleus gives a picture of only the operational components during a snapshot of the developmental process.

Endo16 has the best-understood *cis*-regulatory control of any member in the endomesoderm specification network (reviewed in Yuh, 2001; Yuh *et al.*, 2001). There are 55 protein-binding sites in the 2300 bp upstream of the Endo16 transcriptional start site. Fifteen different proteins bind to these sites with greater than 10^4 times the affinity that these proteins have for non-specific synthetic double-stranded (poly(deoxyinosine–deoxycytidine) poly(dI–dC)). To determine the activity of different components of this regulatory system, various expression constructs were made with segments of the upstream regulatory system combined with the coding region of the chloramphenicol acetyltransferase (CAT) reporter gene. The smallest unique segments were referred to as 'modules', and a combinatorial design of segments allowed inference of the role of each module in the transcriptional activation decision circuit. For example, these experiments demonstrated that all regulatory activity is mediated through the first module (module A), so that all other modules (modules B–G) modulate transcriptional activity through interactions with module A (as in the cartoon in the bottom panel of Fig. 8.1).

The current understanding of the endomesoderm network was reconstructed from many years of laborious experimentation to elucidate the basic elements of the network (Davidson *et al.*, 2002a). To fully understand genetic network behaviour at this level of complexity, a large amount of information is required: *cis*-regulatory elements must be known for all of the relevant genes, transcription cofactors must be known, and the abundance and effect of transcription factor binding must be understood in the combinatorial milieu of regulatory modules, to name a few of the requisites for a full understanding. This poses a severe challenge to experimentalists, and even with all of this information in hand, simulations should be performed (and may be required) to understand how the network behaves under various conditions (Bower and Bolouri, 2001). For this reason, most preliminary attempts to understand genetic networks are at the level of network topology, or to ask how the network is wired as a first step towards grasping its complex behaviour (e.g. Rzhetsky and Gomez, 2001; Tanay and Shamir, 2001; Wagner, 2001; Clipsham *et al.*, 2002; Featherstone and Broadie, 2002; Sabatti *et al.*, 2002; Albert and Othmer, 2003), whereas studies examining network dynamics have generally been attempted with only a small number of genes (e.g. Mendoza and Alvarez-Buylla, 1998; Watts and Strogatz, 1998; Mendoza *et al.*, 1999; Salazar-Ciudad *et al.*, 2001).

Reconstruction of network topologies from large-scale data

To build accepted models for gene regulatory networks, it has been necessary to gradually construct them using targeted experimentation to elucidate network connection, as described above. It is advantageous, however, to estimate genetic network topologies based on high-throughput experimental approaches. Here we turn the discussion to the question of how networks may be inferred or estimated from large-scale data. *De novo* efforts to predict *cis*-regulatory elements should prove useful in narrowing the experimental search for these elements and their respective transcription factors. Reconstruction of genetic networks from large data sets such as microarray data provides further predictions about the complex interactions of many genes and gene products. Several strategies and algorithms have emerged for this purpose and these efforts typically render a prediction about the topology of a gene network, or the map of the interacting genes, rather than a full rendering of the rates at which the various interactions occur and what the abundances of the participants in these interactions are.

The typical microarray or SAGE experiment produces a list of candidate differentially expressed genes. This information is useful starting material that may be combined with other data for the reconstruction of putative biochemical or genetic networks. Information about regulatory promoters, whether experimentally derived or predicted, can be applied with microarray data towards reconstruction of genetic networks. Protein–protein interaction networks may be combined with microarray expression data for reconstructing signal transduction pathways.

Building genetic networks from the compilation of independent experimental data, using literature searches to build up specific aspects of circuits and larger networks, has been referred to as the bottom-up approach to genetic network reconstruction (Kitano, 2001). Using large-scale technologies such as DNA microarrays for network reconstruction has thus been called the top-down approach. For example, inference of genetic networks from expression data taken from the developing central nervous system (CNS) has been performed with the REVEAL algorithm (Somogyi et al., 2001).

One effort to reconstruct signal cascade networks has combined microarray data with yeast two-hybrid data (Steffen et al., 2002). Yeast two-hybrid experiments aim to detect protein–protein interactions, and a large-scale effort has been applied to yeast to determine such interactions exhaustively. The major caveat of the yeast two-hybrid approach is that some proteins are very 'sticky' and appear to have interactions with very many other proteins. The false-positive rate is high and so many apparent interactions may not be functionally relevant. With this in mind, highly interacting proteins were removed from consid-

eration. This was the first step for signal transduction reconstruction (Steffen et al., 2002). Microarray clustering results were then combined with the remaining protein interaction network to reconstruct signalling pathways.

Providing a detailed reconstruction of which genes have a direct influence over the expression of other genes generally requires microarray analysis of systematic genetic perturbations with the inclusion of data about other aspects of the genetic networks such as the protein phosphorylation states of potential protein intermediates in the network. One method to reconstruct a genetic network with this kind of ordering of influences uses a directed graph model and can reconstruct the network with fewer than n^2 perturbations, where n is the number of genes in the circuit or network of interest (Wagner, 2001). When considering the scale and number of replications required to achieve reliable results, it becomes apparent that such an experimental plan is ambitious. As with all methods that aim to reconstruct networks from microarray data, the accuracy of the reconstruction will be sensitive to false negatives and false positives in the microarray data, so adding additional sources of data to the reconstruction effort will improve reliability of the network model.

Regardless of the method of topology reconstruction, a deeper understanding of genetic networks will require simulations that can help to validate or reject models. Simulations should also provide testable hypotheses and reveal unintuitive aspects of generally correct models of genetic networks. A detailed account of how simulations may be applied to genetic networks is outside the scope of this chapter, and we point the reader to de Jong (2002) for a review of this complex subject. Briefly, methods to simulate genetic networks may be categorized as approaches that use directed graphs, Bayesian networks, Boolean networks, generalized local networks, non-linear differential equations, piece-wise differential equations, partial differential equations, stochastic master equations, or rule-based formalisms. Each method has advantages and disadvantages relating to the coarseness of the simulation resolution and the practical aspects of simulation running time (de Jong, 2002).

Mammalian gene regulatory networks

Gene regulatory networks arise in mammalian regulatory control of transcription where the products of regulated transcription are themselves transcription factors. Transcription factors probably number in the thousands for many mammals and may bind to *cis*-regulatory elements often far from the transcriptional start site. They behave combinatorially, so many transcription factors may be involved in the decision circuit for the transcription of a single gene. Furthermore, cofactors may bind and mediate interaction between regulatory modules and the transcriptional start site, and these cofactors cannot be inferred from the known or predicted *cis*-regulatory elements (Carey and Smale, 2000; Levine and Tjian, 2003).

Networks can be employed for differentiation in the embryo and in adulthood. One example of this kind of redeployment of a genetic network in the adult mammal is the network utilized in myogenesis (reviewed in Arnold and Winter, 1998). Other contributions towards understanding mammalian networks include a mammalian protein–protein database (Suzuki *et al.*, 2003), study of the dynamic regulatory network in rat spinal cord development (Elkon *et al.*, 2003), *in silico* prediction of transcriptional regulators involved in the human cell cycle (Elkon *et al.*, 2003), and network analysis in neural progenitors in mouse (Karsten *et al.*, 2003).

Transcription has multiple outputs

Although protein-coding transcripts have been considered essential for a long time, there is emerging evidence that suggests important functions for non-coding sequences. Recent studies have shown the presence of a large number of transcripts that contain a single exon with no apparent open reading frame (Okazaki *et al.*, 2002; Yelin *et al.*, 2003). Many of these are coded on antisense strands, suggesting possible regulatory functions. Another example comes from introns, which account for up to 95% of the transcriptional output in higher eukaryotes. It has been proposed that higher eukaryotes may utilize intronic output as a form of regulatory control (Mattick and Gagen, 2001). Multitasking in regulatory networks is most effective when nodes in the network have both multiple inputs and multiple outputs. Although introns have traditionally been thought to have no interesting function, there is no direct evidence to support this claim. The issue of potential intron function is confounded by the likelihood that some potential function would be expected to evolve independently for each intron, and whether such functions may exist is further obscured by the lower conservation of sequence in introns relative to exons. Interest in intron function is supported by the good correlation of intron size and sequence complexity with developmental complexity.

Analysis of biological networks

Biological networks are prescribed by evolutionary constraints and are thus non-random. In a random network, the network nodes have a number of connections, or edges, to other nodes that is close to the mean number of connections for all nodes. This situation arises when the edges between nodes are connected randomly, and thus the number of connections per node follows a Poisson distribution (Fig. 8.3). Recent studies have revealed that biological networks have a quite different topology (Oltvai and Barabasi, 2002; Dezso *et al.*, 2003; Ravasz and Barabasi, 2003). Most nodes have very few connections, while a few nodes are very highly connected. This non-random type of network, which follows a power law distribution, is called a scale-free network (Fig. 8.3). Network analysis of this kind offers some general impressions about the networks, and provides some insights about how biological networks might arise from an evolutionary perspective. For example, one property of protein–protein interaction networks that is suggested by their scale-free connectedness is that this kind of network can arise from gene duplications (Dezso *et al.*, 2003). It has been shown that such duplications can promote the formation of hubs of highly interacting proteins, while the majority of proteins have a small number of specific interactions with other proteins.

Scale-free networks are very robust against random 'attacks' or failures of random nodes. If

n=19, e=26   n=19, e=26

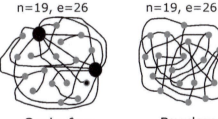

Scale-free   Random

Fig. 8.3. Scale-free and random networks. Although the number of nodes (n) and edges (e) in each of these networks is the same, the configurations of their respective connections are different: scale-free networks have many nodes with very few connections (small circles) and a small number of 'hubs' with a large number of connections (large circles), while random networks follow a Poisson distribution of connections, with most nodes having a number of connections close to the mean number of connections for all nodes.

a random node fails, or loses its connections, the network typically only suffers a small loss of connections. On the other hand, if a hub fails, large segments of the network can become disconnected. An implication for network analysis in medicine is that molecular hubs involved in disease may be promising targets for engineered medicines (Ravasz and Barabasi, 2003).

Although progress in understanding global expression patterns will come in part from network analysis, it is important to recognize that these approaches also require one to examine the subnetwork details to glean a deep understanding of network architecture and function (Bray, 2003).

Knowledge-based Prediction of Protein and DNA Interactions

One approach to finding genes and regulatory sequences essential for execution of a developmental plan is to compare the genomes of closely related species, looking for highly conserved sequences. The converse, looking for genes uniquely expressed in a developmental profile, may yield insights into how that profile generates the unique features of its developmental plan. The two forerunners for this kind of analysis are comparisons between

Drosophila and *Anopheles*, and between *Mus musculus* and *Homo sapiens*. At this time, comparative genomics approaches have mainly been performed at the global level, without a published account detailing a comparison of genes with important developmental roles between genomes. Analysis of the current draft genomes with a focus on embryology has some beginnings in the supplemental materials accompanying the published comparative genomics of the *Drosophila* and *Anopheles* genomes, but we find little else that approaches this problem in the literature. This scarcity of 'comparative embryogenomics' approaches is probably a consequence of our presently poor understanding of large regulatory networks for the specification of development. It is difficult to make comprehensive comparisons of the molecular details of development in different organisms without the corresponding network topologies to use as a conceptual framework of comparison.

Mammalian complexity creates enormous challenges for arriving at an understanding of their genetic networks by experimental investigation. Although ChIP-chip approaches offer a large-scale method to reach this goal, there are limitations in the sensitivity of this approach. It is highly advantageous to approach this problem by exploring computational methods for the prediction of interactions between the *cis* and *trans* elements involved in the regulation. Although such computational approaches are enormously useful, it should be noted that transcriptional cofactors may also be involved in the integration of activation signals and that there is no information about these cofactors present in the regulatory sequence.

Identification of *cis*-regulatory elements based on computational motif finding

Defining *cis*-regulatory elements in higher eukaryotes is a laborious task that requires hunting for these elements over large segments of the genome. It is clearly advantageous to have some way to narrow this search computationally to the most likely candidate sequences before embarking on this large-scale experimental task. Below we discuss selected methods for computational motif finding, and

Web sites with application availability information for these and some other popular methods, where applicable, may be found in Table 8.1. There are two principal strategies that may be applied to the computational prediction of *cis*-regulatory elements: (i) the motifs of these sequences may be predicted *de novo*, or directly from raw sequence information; or (ii) element motifs may be predicted using both the sequence information and some other data source, such as ChIP-chip, as discussed earlier in this chapter ('Whole-genome Expression Studies'). Most efforts that use *de novo* methods may be divided into those that explore regular expressions of the motifs and those that update a position weight matrix to find the best motif candidates (reviewed in Brazma *et al.*, 1998b; Pedersen *et al.*, 1999; Eskin and Pevzner, 2002; Roven and Bussemaker, 2003; Sinha and Tompa, 2003). Another approach is to build a consensus motif by pairwise comparison of regulatory sequences from genes believed to be co-regulated. This approach compares two sequences, finds the most promising motifs, then looks for those motifs in a third sequence, and so on. This method is very fast, but also sensitive to the initial ordering of the comparisons. Below we discuss each of the major *de novo* approaches (regular expressions and position weight matrix updates), and conclude our discussion on computational motif finding with a discussion about computational methods that employ microarray data from ChIP-chip experiments.

Regular expressions

In computer science, a 'regular expression' is a patterned character string that may incorporate defined alternatives for each position in the string. DNA sequence is represented by a string of characters (A, T, G, C) and thus can be analysed computationally by regular expression rules. A regular expression may be as simple as a single sequence without alternatives or wild-card characters, it may be entirely composed of wild-cards, or it may be some combination of fixed characters and wild-cards. Searching for conserved regular expressions in intergenic sequences is a popular approach for finding regulatory sequences (Brazma *et al.*, 1998a; Tompa, 1999, 2001; Bussemaker *et al.*, 2000; van Helden *et al.*, 2000; Keich and Pevzner, 2002). For example, one approach is to look for so-called spaced dyads, which are simply two trinucleotides separated by *n* bases, where *n* can be 0–16. This pattern was chosen because dimerized transcription factors may bind to such sequences in yeast DNA. The prediction technique is to count spaced dyads exhaustively, and determine which dyads are statistically over-represented.

Table 8.1. Computational approaches to cis-regulatory motif finding.

Approach derivative	Application/ approach name	Citation/ laboratory	Web site
Expectation maximization	MEME	TL Baily	http://meme.sdsc.edu/meme/website/intro.html
	Random projection	J Buhler	n/a
	First EM motif finder	C Lawrence	n/a
Gibbs sampling	AlignAce	GM Church	http://atlas.med.harvard.edu/
	Gibbs Motif sampler	JS Liu	http://www.people.fas.harvard.edu/~junliu/index1.html
	Bioprospector	XS Liu	http://bioprospector.stanford.edu/
Consensus	Consensus	G Stormo	http://ural.wustl.edu/~jhc1/consensus/html/Html/main.html
Regular expression enumeration	spaced dyads	J van Helden	http://rsat.ulb.ac.be/rsat/
	WINNOWER	P Pevzner	http://www.cs.ucsd.edu/groups/bioinformatics/software.html
	suffix trie	A Brazma	n/a
	MobyDick	H Bussemaker	http://uqbar.rockefeller.edu/~siggia/projects/
	IUPAC alphabet, Markov background	M Tompa	http://bio.cs.washington.edu/software.html
	2-bit encoding	P Baldi	n/a
Microarray guided motif search	MDSCAN	XS Liu	http://bioprospector.stanford.edu/MDscan/Overview.html
	REDUCE	H Bussemaker	http://bussemaker.bio.columbia.edu/tools/reduce/
	GMEP	M Eisen	http://rana.lbl.gov/EisenSoftware.htm
	MOTIF REGRESSOR	EM Conolon	http://www.techtransfer.harvard.edu/Software/MotifRegressor/

The WINNOWER algorithm is good at finding relatively long motifs with mismatches (e.g. 15 bases with four mismatches) (Eskin and Pevzner, 2002). The difficulty in finding such sequences is that, for the example motif of 15 bases with four allowed mismatches, there may be up to eight differences between any two sequences that fit this motif. WINNOWER solves this problem by identifying non-signal motifs (spurious motifs) and eliminating them from a graph that has edges representing similarity between vertices representing different substrings from the sample sequence. What remains after this removal of non-signal similarities are so-called cliques in the graph that most often can be used to identify the correct motif(s).

Most efforts to search for regular expressions begin with an *a priori* defined pattern or class of patterns and search a region limited to a few hundred bases upstream of the transcription start site (where yeast is the likely model organism for the pilot use of the algorithm). The challenge of these pattern-finding approaches is that the computational time taken to do the search can be prohibitive, especially for mammalian sequences where the potential motif may be many kilobases upstream or downstream of the transcriptional start site. An efficient data structure for searching called a suffix trie has been used to represent all of the intergenic sequences in the yeast genome (Brazma *et al.*, 1998b). Suffix *trees* are efficiently built by pre-processing a sequence T in linear time, so that when a string S is searched in T, the time required is dependent only on the length of S, not the length of T (Gusfeld, 1997). The suffix *trie* is a simpler version of the suffix tree structure, but is more resource demanding.

Position weight matrix

Whereas the methods to reveal motifs as regular expressions are mainly driven by *a priori* patterns, position weight matrices may be used to search for motifs using methods more driven by the sequence data (Stormo, 1990; Lawrence and Reilly, 1990; Bailey and Elkan, 1994, 1995; Neuwald *et al.*, 1995; Hughes *et al.*, 2000; Liu *et al.*, 2001; Buhler and Tompa, 2002). The matrix probabilities are given as

the probability of finding the ith sequence character (row) at the jth position (column) in the putative motif. The strategy applied is to randomly initialize a matrix that represents probabilities of each nucleotide at each position in the putative motif, and then update the matrix according to iterative examinations of the sequence data. This general strategy is applied in two principle variations: expectation maximization and Gibbs sampling. The general problem may be formally defined as finding the probability of a motif probability matrix parameter and the site locations of the motif given the sequence data and the probability of genome background.

Microarrays

Microarray data may be used to guide the search for common motifs important to the co-regulation of specific genes. The basic strategy applied is to identify potential motifs that are best correlated with variation in gene expression. Several applications have been constructed with this general approach, including, for example: REDUCE, GMEP, MDSCAN and MOTIFREGRESSOR (Bussemaker *et al.*, 2001; Chiang *et al.*, 2001; Liu *et al.*, 2002; Roven and Bussemaker, 2003). MDSCAN reports regulatory motif predictions based on sequences acquired from ChIP-chip hybridization studies. One logical difference between using microarrays and *de novo* searching of genomic sequences is that microarray measures narrow the search space and algorithms can be designed to take advantage of this. A second difference is that expression may be correlated with known transcription factor expression, which may be of further aid in the identification of *cis*-regulatory elements.

Summary

Mammalian systems of genetic regulation are one of the most complex and most import areas of study in biology. Several levels of control, including control at the levels of transcription, RNA processing, transport and localization of mRNA, translation, mRNA degradation and protein activity control, contribute to this complexity. Here we focused only on the aspects of transcriptional initiation that involve transcription factors. The combinatorial control of transcription by *cis*-regulatory elements is considered to be the most important contribution to the unique developmental plans of different species (Davidson, 2001). The evolutionary impact of genetic networks is that changes arising in a gene that is downstream or terminal in a genetic network will generally have a smaller impact than changes upstream in the regulatory network. This explains how some evolutionary changes appear to have occurred in a discontinuous fashion. A deeper understanding of genetic networks in mammals will require the combined efforts of detailed genetics and functional analysis, genomics approaches, and computational predictions.

References

Adams, M.D., Soares, M.B., Kerlavage, A.R., Fields, C. and Venter, J.C. (1993) Rapid cDNA sequencing (expressed sequence tags) from a directionally cloned human infant brain cDNA library. *Nature Genetics* 4, 373–380.

Albert, R. and Othmer, H.G. (2003) The topology of the regulatory interactions predicts the expression pattern of the segment polarity genes in *Drosophila melanogaster*. *Journal of Theoretical Biology* 223, 1–18.

Alberts, B., Johnson, A., Lewis, J., Raff, M., Roberts, K. and Walter, P. (2002) *Molecular Biology of the Cell*. Garland Science, New York. 1463 pp.

Anisimov, S.V., Tarasov, K.V., Stern, M.D., Lakatta, E.G. and Boheler, K.R. (2002) A quantitative and validated SAGE transcriptome reference for adult mouse heart. *Genomics* 80, 213–222.

Arnold, H.H. and Winter, B. (1998) Muscle differentiation: more complexity to the network of myogenic regulators. *Current Opinion in Genetics and Development* 8, 539–544.

Bailey, T.L. and Elkan, C. (1994) Fitting a mixture model by expectation maximization to discover motifs in biopolymers. *Proceedings of the International Conference on Intelligent Systems for Molecular Biology* 2, 28–36.

Bailey, T.L. and Elkan, C. (1995) The value of prior knowledge in discovering motifs with MEME. *Proceedings of the International Conference on Intelligent Systems for Molecular Biology* 3, 21–29.

Beato, M., Chavez, S. and Truss, M. (1996) Transcriptional regulation by steroid hormones. *Steroids* 61, 240–251.

Becskei, A. and Serrano, L. (2000) Engineering stability in gene networks by autoregulation. *Nature* 405, 590–593.

Blake, W.J., Kærn, M., Cantor, C.R. and Collins, J.J. (2003) Noise in eukaryotic gene expression. *Nature* 422, 633–637.

Blau, H.M. and Rossi, F.M. (1999) Tet B or not tet B: advances in tetracycline-inducible gene expression. *Proceedings of the National Academy of Sciences USA* 96, 797–799.

Bower, J.M. and Bolouri, H. (2001) *Computational Modeling of Genetic and Biochemical Networks*. The MIT Press, London. 336 pp.

Bray, D. (2003) Molecular networks: the top-down view. *Science* 301, 1864–1865.

Brazma, A., Jonassen, I., Eidhammer, I. and Gilbert, D. (1998a) Approaches to the automatic discovery of patterns in biosequences. *Journal of Computational Biology* 5, 279–305.

Brazma, A., Jonassen, I., Vilo, J. and Ukkonen, E. (1998b) Predicting gene regulatory elements *in silico* on a genomic scale. *Genome Research* 8, 1202–1215.

Brenner, S., Johnson, M., Bridgham, J., Golda, G., Lloyd, D.H., Johnson, D., Luo, S., McCurdy, S., Foy, M., Ewan, M., Roth, R., George, D., Eletr, S., Albrecht, G., Vermaas, E., Williams, S.R., Moon, K., Burcham, T., Pallas, M., DuBridge, R.B., Kirchner, J., Fearon, K., Mao, J. and Corcoran, K. (2000) Gene expression analysis by massively parallel signature sequencing (MPSS) on microbead arrays. *Nature Biotechnology* 18, 630–634.

Buhler, J. and Tompa, M. (2002) Finding motifs using random projections. *Journal of Computational Biology* 9, 225–242.

Bussemaker, H.J., Li, H. and Siggia, E.D. (2000) Regulatory element detection using a probabilistic segmentation model. *Proceedings of the International Conference on Intelligent Systems for Molecular Biology* 8, 67–74.

Bussemaker, H.J., Li, H. and Siggia, E.D. (2001) Regulatory element detection using correlation with expression. *Nature Genetics* 27, 167–171.

Carey, M. and Smale, S.T. (2000) *Transcriptional Regulation in Eukaryotes: Concepts, Strategies, and Techniques*. Cold Spring Harbor Laboratory Press, Cold Spring Harbor, New York. 640 pp.

Cavailles, V., Dauvois, S., L'Horset, F., Lopez, G., Hoare, S., Kushner, P.J. and Parker, M.G. (1995) Nuclear factor RIP140 modulates transcriptional activation by the estrogen receptor. *EMBO Journal* 14, 3741–3751.

Chiang, D.Y., Brown, P.O. and Eisen, M.B. (2001) Visualizing associations between genome sequences and gene expression data using genome-mean expression profiles. *Bioinformatics* 17, Suppl. 1, S49–S55.

Clipsham, R., Zhang, Y.H., Huang, B.L. and McCabe, E.R. (2002) Genetic network identification by high density, multiplexed reversed transcriptional (HD-MRT) analysis in steroidogenic axis model cell lines. *Molecular Genetics and Metabolism* 77, 159–178.

Cook, D.L., Gerber, A.N. and Tapscott, S.J. (1998) Modeling stochastic gene expression: implications for haploinsufficiency. *Proceedings of the National Academy of Sciences USA* 95, 15641–15646.

Davidson, E.H. (2001) *Genomic Regulatory Systems*. Academic Press, San Diego. 261 pp.

Davidson, E.H., Rast, J.P., Oliveri, P., Ransick, A., Calestani, C., Yuh, C.H., Minokawa, T., Amore, G., Hinman, V., Arenas-Mena, C., Otim, O., Brown, C.T., Livi, C.B., Lee, P.Y., Revilla, R., Rust, A.G., Pan, Z., Schilstra, M.J., Clarke, P.J., Arnone, M.I., Rowen, L., Cameron, R.A., McClay, D.R., Hood, L. and Bolouri, H. (2002a) A genomic regulatory network for development. *Science* 295,1669–1678.

Davidson, E.H., Rast, J.P., Oliveri, P., Ransick, A., Calestani, C., Yuh, C.H., Minokawa, T., Amore, G., Hinman, V., Arenas-Mena, C., Otim, O., Brown, C.T., Livi, C.B., Lee, P.Y., Revilla, R., Schilstra, M.J., Clarke, P.J., Rust, A.G., Pan, Z., Arnone, M.I., Rowen, L., Cameron, R.A., McClay, D.R., Hood, L. and Bolouri, H. (2002b) A provisional regulatory gene network for specification of endomesoderm in the sea urchin embryo. *Developmental Biology* 246, 162–190.

de Jong, H. (2002) Modeling and simulation of genetic regulatory systems: a literature review. *Journal of Computational Biology* 9, 67–103.

Dezso, Z., Oltvai, Z.N. and Barabasi, A.L. (2003) Bioinformatics analysis of experimentally determined protein complexes in the yeast *Saccharomyces cerevisiae*. *Genome Research* 13, 2450–2454.

Eberwine, J., Spencer, C., Miyashiro, K., Mackler, S. and Finnell, R. (1992) Complementary DNA synthesis in situ: methods and applications. *Methods in Enzymology* 216, 80–100.

Eggert, M., Mows, C.C. Tripier, D., Arnold, R., Michel, J., Nickel, J., Schmidt, S., Beato, M. and Renkawitz, R. (1995) A fraction enriched in a novel glucocorticoid receptor-interacting protein stimulates receptor-dependent transcription in vitro. *Journal of Biological Chemistry* 270, 30755–30759.

Elkon, R., Linhart, C., Sharan, R., Shamir, R. and Shiloh, Y. (2003) Genome-wide in silico identification of transcriptional regulators controlling the cell cycle in human cells. *Genome Research* 13, 773–780.

Eskin, E. and Pevzner, P.A. (2002) Finding composite regulatory patterns in DNA sequences. *Bioinformatics* 18 Supplement 1, S354–S363.

Featherstone, D.E. and Broadie, K. (2002) Wrestling with pleiotropy: genomic and topological analysis of the yeast gene expression network. *Bioessays* 24, 267–274.

Garcia-Fernandez, J. and Holland, P.W. (1994) Archetypal organization of the amphioxus Hox gene cluster. *Nature* 370, 563–566.

Gusfeld, D. (1997) *Algorithms on Strings, Trees, and Sequences: Computer Science and Computational Biology*. Cambridge University Press, New York. 534 pp.

Hamatani, T., Carter, M.G., Sharov, A.A. and Ko, M.S.H. (2004) Dynamics of global gene expression changes during mouse preimplantation development. *Developmental Cell* 6, 117–131.

Hashimoto, S., Suzuki, T., Dong, H.Y., Yamazaki, N. and Matsushima, K. (1999) Serial analysis of gene expression in human monocytes and macrophages. *Blood* 94, 837–844.

Horak, C.E. and Snyder, M. (2002) ChIP-chip: a genomic approach for identifying transcription factor binding sites. *Methods in Enzymology* 350, 469–483.

Horak, C.E., Mahajan, M.C., Luscombe, N.M., Gerstein, M., Weissman, S.M. and Snyder, M. (2002) GATA-1 binding sites mapped in the beta-globin locus by using mammalian ChIP-chip analysis. *Proceedings of the National Academy of Sciences USA* 99, 2924–2929.

Hughes, J.D., Estep, P.W., Tavazoie, S. and Church, G.M. (2000) Computational identification of cis-regulatory elements associated with groups of functionally related genes in *Saccharomyces cerevisiae*. *Journal of Molecular Biology* 296, 1205–1214.

Hughes, T.R., Mao, M., Jones, A.R., Burchard, J., Marton, M.J., Shannon, K.W., Lefkowitz, S.M., Ziman, M., Schelter, J.M., Meyer, M.R., Kobayashi, S., Davis, C., Dai, H., He, Y.D., Stephaniants, S.B., Cavet, G., Walker, W.L., West, A., Coffey, E., Shoemaker, D.D., Stoughton, R., Blanchard, A.P., Friend, S.H. and

Linsley, P.S. (2001) Expression profiling using microarrays fabricated by an ink-jet oligonucleotide synthesizer. *Nature Biotechnology* 19, 342–347.

Hume, D.A. (2000) Probability in transcriptional regulation and its implications for leukocyte differentiation and inducible gene expression. *Blood* 96, 2323–2328.

Jacq, X., Brou, C., Lutz, Y., Davidson, I., Chambon, P. and Tora, L. (1994) Human TAFII30 is present in a distinct TFIID complex and is required for transcriptional activation by the estrogen receptor. *Cell* 79, 107–117.

Karsten, S.L., Kudo, L.C., Jackson, R., Sabatti, C., Kornblum, H.I. and Geschwind, D.H. (2003) Global analysis of gene expression in neural progenitors reveals specific cell-cycle, signalling, and metabolic networks. *Developmental Biology* 261,165–182.

Keich, U. and Pevzner, P.A. (2002) Subtle motifs: defining the limits of motif finding algorithms. *Bioinformatics* 18, 1382–1390.

Kemkemer, R., Schrank, S., Vogel, W., Gruler, H. and Kaufmann, D. (2002) Increased noise as an effect of haploinsufficiency of the tumor-suppressor gene neurofibromatosis type 1 *in vitro*. *Proceedings of the National Academy of Sciences USA* 99, 13783–13788.

Kitano, H. (2001) Systems biology: toward system-level understanding of biological systems. In: Kitano, H. (ed.) *Foundations of Systems Biology*. The MIT Press, Cambridge, Massachusetts, pp. 1–36.

Ko, M.S. (1992) Induction of a single gene molecule: stochastic or deterministic? *Bioessays* 14, 341–346.

Ko, M.S. (2001) Embryogenomics: developmental biology meets genomics. *Trends in Biotechnology* 19, 511–518.

Lawrence, C.E. and Reilly, A.A. (1990) An expectation maximization (EM) algorithm for the identification and characterization of common sites in unaligned biopolymer sequences. *Proteins* 7, 41–51.

Levine, M. and Tjian, R. (2003) Transcription regulation and animal diversity. *Nature* 424, 147–151.

Liu, X., Brutlag, D.L. and Liu, J.S. (2001) BioProspector: discovering conserved DNA motifs in upstream regulatory regions of co-expressed genes. *Pacific Symposium on Biocomputing* 6, 127–138.

Liu, X.S., Brutlag, D.L. and Liu, J.S. (2002) An algorithm for finding protein–DNA binding sites with applications to chromatin-immunoprecipitation microarray experiments. *Nature Biotechnology* 20, 835–839.

Magee, J.A., Abdulkadir, S.A. and Milbrandt, J. (2003) Haploinsufficiency at the Nkx3.1 locus. A paradigm for stochastic, dosage-sensitive gene regulation during tumor initiation. *Cancer Cell* 3, 273–283.

Martinez, P., Rast, J.P., Arenas-Mena, C. and Davidson, E.H. (1999) Organization of an echinoderm Hox gene cluster. *Proceedings of the National Academy of Sciences USA* 96, 1469–1474.

Mattick, J.S. and Gagen, M.J. (2001) The evolution of controlled multitasked gene networks: the role of introns and other noncoding RNAs in the development of complex organisms. *Molecular Biology and Evolution* 18, 1611–1630.

McAdams, H.H. and Shapiro, L. (1995) Circuit simulation of genetic networks. *Science* 269, 650–656.

Mendoza, L. and Alvarez-Buylla, E.R. (1998) Dynamics of the genetic regulatory network for *Arabidopsis thaliana* flower morphogenesis. *Journal of Theoretical Biology* 193, 307–319.

Mendoza, L., Thieffry, D. and Alvarez-Buylla, E.R. (1999) Genetic control of flower morphogenesis in *Arabidopsis thaliana*: a logical analysis. *Bioinformatics* 15, 593–606.

Menssen, A. and Hermeking, H. (2002) Characterization of the c-MYC-regulated transcriptome by SAGE: identification and analysis of c-MYC target genes. *Proceedings of the National Academy of Sciences USA* 99, 6274–6279.

Neuwald, A.F., Liu, J.S. and Lawrence, C.E. (1995) Gibbs motif sampling: detection of bacterial outer membrane protein repeats. *Protein Science* 4, 1618–1632.

Okazaki, Y., Furuno, M., Kasukawa, T., Adachi, J., Bono, H., Kondo, S., Nikaido, I., Osato, N., Saito, R., Suzuki, H., Yamanaka, I., Kiyosawa, H., Yagi, K., Tomaru, Y., Hasegawa, Y., Nogami, A., Schonbach, C., Gojobori, T., Baldarelli, R., Hill, D.P. *et al.* (2002) Analysis of the mouse transcriptome based on functional annotation of 60,770 full-length cDNAs. *Nature* 420, 563–573.

Oltvai, Z.N. and Barabasi, A.L. (2002) Systems biology. Life's complexity pyramid. *Science* 298, 763–764.

Pease, A.C., Solas, D., Sullivan, E.J., Cronin, M.T., Holmes, C.P. and Fodor, S.P. (1994) Light-generated oligonucleotide arrays for rapid DNA sequence analysis. *Proceedings of the National Academy of Sciences USA* 91, 5022–5026.

Pedersen, A.G., Baldi, P., Chauvin, Y. and Brunak, S. (1999) The biology of eukaryotic promoter prediction – a review. *Computational Chemistry* 23, 191–207.

Rao, C.V., Wolf, D.M. and Arkin, A.P. (2002) Control, exploitation and tolerance of intracellular noise. *Nature* 420, 231–237.

Ravasz, E. and Barabasi, A.L. (2003) Hierarchical organization in complex networks. *Physical Review. E, Statistical Nonlinear, and Soft Matter Physics* 67, 026112.

Roven, C. and Bussemaker, H.J. (2003) REDUCE: an online tool for inferring *cis*-regulatory elements and transcriptional module activities from microarray data. *Nucleic Acids Research* 31, 3487–3490.

Rzhetsky, A. and Gomez, S.M. (2001) Birth of scale-free molecular networks and the number of distinct DNA and protein domains per genome. *Bioinformatics* 17, 988–996.

Sabatti, C., Rohlin, L., Oh, M.K. and Liao, J.C. (2002) Co-expression pattern from DNA microarray experiments as a tool for operon prediction. *Nucleic Acids Research* 30, 2886–2893.

Sadovsky, Y., Webb, P., Lopez, G., Baxter, J.D., Fitzpatrick, P.M., Gizang-Ginsberg, E., Cavailles, V., Parker, M.G. and Kushner, P.J. (1995) Transcriptional activators differ in their responses to overexpression of TATA-box-binding protein. *Molecular and Cellular Biology* 15, 1554–1563.

Salazar-Ciudad, I., Newman, S.A. and Sole, R.V. (2001) Phenotypic and dynamical transitions in model genetic networks. I. Emergence of patterns and genotype–phenotype relationships. *Evolution and Development* 3, 84–94.

Schena, M., Shalon, D., Davies, R.W. and Brown, P.O. (1995) Quantitative monitoring of gene expression patterns with a complementary DNA microarray. *Science* 270, 467–470.

Schuler, G.D. (1997) Pieces of the puzzle: expressed sequence tags and the catalog of human genes. *Journal of Molecular Medicine* 75, 694–698.

Sharov, A.A., Piao, Y., Matoba, R., Dudekula, D.B., Qian, Y., VanBuren, V., Falco, G., Martin, P.R., Stagg, C.A., Bassey, U.C., Wang, Y., Carter, M.G., Hamatani, T., Aiba, K., Akutsu, H., Sharova, L., Tanaka, T.S., Kimber, W.L., Yoshikawa, T., Jaradat, S.A. *et al.* (2003) Transcriptome analysis of mouse stem cells and early embryos. *Public Library of Science Biology* 1, 410–419.

Sinha, S. and Tompa, M. (2003) YMF: a program for discovery of novel transcription factor binding sites by statistical overrepresentation. *Nucleic Acids Research* 31, 3586–3588.

Somogyi, R., Fuhrman, S. and We, X. (2001) Genetic network inference in computational models and applications to large-scale gene expression data. In: Bower, J.M. and Bolouri, H. (eds) *Computational Modeling of Genetic and Biochemical Networks*. The MIT Press, Cambridge, Massachusetts, pp. 336.

Stanton, J.L. and Green, D.P. (2001) Meta-analysis of gene expression in mouse preimplantation embryo development. *Molecular Human Reproduction* 7, 545–552.

Steffen, M., Petti, A., Aach, J., D'Haeseleer, P. and Church, G. (2002) Automated modelling of signal transduction networks. *BioMed Central Bioinformatics* 3, 34.

Stormo, G.D. (1990) Consensus patterns in DNA. *Methods in Enzymology* 183, 211–221.

Strausberg, R.L., Buetow, K.H., Emmert-Buck, M.R. and Klausner, R.D. (2000) The cancer genome anatomy project: building an annotated gene index. *Trends in Genetics* 16, 103–106.

Suzuki, H., Saito, R., Kanamori, M., Kai, C., Schonbach, C., Nagashima, T., Hosaka, J. and Hayashizaki, Y. (2003) The mammalian protein–protein interaction database and its viewing system that is linked to the main FANTOM2 viewer. *Genome Research* 13, 1534–1541.

Tanaka, T.S., Jaradat, S.A., Lim, M.K., Kargul, G.J., Wang, X., Grahovac, M.J., Pantano, S., Sano, Y., Piao, Y., Nagaraja, R., Doi, H., Wood, W.H., 3rd, Becker, K.G. and Ko, M.S. (2000) Genome-wide expression profiling of mid-gestation placenta and embryo using a 15,000 mouse developmental cDNA microarray. *Proceedings of the National Academy of Sciences USA* 97, 9127–9132.

Tanay, A. and Shamir, R. (2001) Computational expansion of genetic networks. *Bioinformatics* 17 Supplement 1, S270–S278.

Tompa, M. (1999) An exact method for finding short motifs in sequences, with application to the ribosome binding site problem. *Proceedings of the International Conference on Intelligent Systems for Molecular Biology* 262–271.

Tompa, M. (2001) Identifying functional elements by comparative DNA sequence analysis. *Genome Research* 11, 1143–1144.

VanBuren, V., Piao, Y., Dudekula, D.B., Qian, Y., Carter, M.G., Martin, P.R., Stagg, C.A., Bassey, U.C., Aiba, K., Hamatani, T., Kargul, G.J., Luo, A.G., Kelso, J., Hide, W. and Ko, M.S. (2002) Assembly, verification, and initial annotation of the NIA mouse 7.4K cDNA clone set. *Genome Research* 12, 1999–2003.

van Helden, J., Rios, A.F. and Collado-Vides, J. (2000) Discovering regulatory elements in non-coding sequences by analysis of spaced dyads. *Nucleic Acids Research* 28, 1808–1818.

Veitia, R.A. (2002) Exploring the etiology of haploinsufficiency. *Bioessays* 24, 175–184.

Velculescu, V.E., Vogelstein, B. and Kinzler, K.W. (2000) Analysing uncharted transcriptomes with SAGE. *Trends in Genetics* 16, 423–425.

Wagner, A. (2001) How to reconstruct a large genetic network from n gene perturbations in fewer than n(2) easy steps. *Bioinformatics* 17,1183–1197.

Watts, D.J. and Strogatz, S.H. (1998) Collective dynamics of 'small-world' networks. *Nature* 393, 440–442.

Yelin, R., Dahary, D., Sorek, R., Levanon, E.Y., Goldstein, O., Shoshan, A., Diber, A., Biton, S., Tamir, Y., Khosravi, R., Nemzer, S., Pinner, E., Walach, S., Bernstein, J., Savitsky, K. and Rotman, G. (2003) Widespread occurrence of antisense transcription in the human genome. *Nature Biotechnology* 21, 379–386.

Yuh, C.H. (2001) A logical model of *cis*-regulatory control in a eukaryotic system. In: Bower, J.M. and Bolouri, H (eds) *Computational Modeling of Genetic and Biochemical Networks*. The MIT Press, Cambridge, Massachusetts, pp. 336.

Yuh, C.H., Bolouri, H. and Davidson, E.H. (2001) *Cis*-regulatory logic in the *endo16* gene: switching from a specification to a differentiation mode of control. *Development* 128, 617–629.

9 Inducing Alterations in the Mammalian Genome for Investigating the Functions of Genes

J.-L. Guénet
Institut Pasteur, Paris, France

Introduction	221
Analysing the Genetic Determinism of Phenotypic Variations: from Phenotype to Genotype	222
Positional cloning of the gene mutated in *obese* mice: an example of forward genetics	222
The experimental production of mutations in the mammalian genome	226
Understanding the genetic determinism of quantitative variations: QTL analysis and its contribution to genome annotation	235
Inducing Engineered Alterations at the Genome Level and Analysing their Consequences at the Organism Level: from Genotype to Phenotype	236
Transgenesis by viral infection of early embryos	236
Transgenesis by *in ovo* injection of a cloned DNA	237
Inducing alterations at the genome level through the *in vitro* manipulation of embryo-derived stem cells	242
Other Strategies of Gene Inactivation	253
Conclusions	253
References	254

Introduction

Analysis of the genome sequence with specialized software utilities allows the identification of genes but does not reveal much concerning their function(s). The observation of sequence homologies with domains that have known functions and/or the analysis of the spatiotemporal expression profile is also informative but this is in general only sufficient to speculate on the function of genes, and it is probably for this reason that, as of today, functional information is available for no more than 30% of the mammalian genes. In fact in most cases, the most reliable way to assess the function of a gene is to specifically address this question through experimentation. Experiments performed with this aim are based on two strategies that are complementary:

1. The first strategy utilizes phenotypic variations in a given population to facilitate the molecular identification of the underlying genetic factor(s). Since the process ensues from phe-

© CAB International 2005. *Mammalian Genomics*
(eds A. Ruvinsky and J. Marshall Graves)

notype to genotype, this strategy is often called 'forward genetics' and is considered a 'bottom-up' approach.

2. The second strategy, termed 'reverse genetics', is a 'top-down' approach. In this situation an organism is genetically modified, introducing specific changes at a targeted locus or misexpressing a gene, and the functional consequences observed through phenotypic analysis.

Both strategies have been widely used in model organisms, in particular the mouse. Here we will review some of their practical and technical aspects.

Analysing the Genetic Determinism of Phenotypic Variations: from Phenotype to Genotype

Some phenotypic variations are the consequence of an allelic difference at a given locus and accordingly are inherited as simple Mendelian traits. Such situations are relatively simple to analyse and in this case, as we shall see, the link between a gene and its function can be established using the now classical strategy of positional cloning. In most instances however, phenotypic variations are controlled by many genes that sometimes interact with each other in a complex network. The identification of these complex genetic factors is more difficult.

Positional cloning of the gene mutated in obese mice: an example of forward genetics

Mouse and rat mutations with deleterious effects are common and the resulting phenotypes often recapitulate more or less faithfully a human syndrome. This is the case for the mouse recessive mutation obese (*ob*), on chromosome 6, which provokes massive obesity sometimes associated with diabetes depending on genetic background (Fig. 9.1). In *ob/ob* mice, obesity is characterized by both an increased number of adipocytes as well as an increase in their size. It was observed over 30 years ago that, when put in parabiosis with wild-type mice of the same strain, *ob/ob* mice ate less and progressively lost weight to the point of achieving almost normal size (Coleman, 1973). This experiment indicated that the abnormal regulation of fat metabolism in *ob/ob* mice could be corrected by a diffusible substance acting like a hormone or cytokine. Obviously, there was much interest in identifying this substance, which had therapeutic potential as a drug to fight obesity, and this was achieved, by positional cloning of the *ob* locus (Zhang *et al.*, 1994). To achieve this aim the *ob* locus was mapped using several crosses and molecular markers of all kinds in order to determine the physical position of the locus on chromosome 6. After scoring the DNA samples collected from a number of offspring representing a total of 1606 meiotic events, the scientists

Fig. 9.1. An obese (*ob/ob*) mutant mouse and its control (*+/ob*). Positional cloning of this mutation led to the discovery of *leptin*.

involved in this project reduced the size of the segment harbouring the *ob* locus to approximately 650 kb of DNA. Then, using the technique of exon trapping (Church *et al.*, 1994), they detected the coding sequences in the DNA segment in question and finally, after a lot of hard bench work, they identified a gene with a 4.52 kb predicted transcript expressed only in white adipose cells. The gene encoded a 16 kDa protein and was called *leptin* (from the Greek *leptos*, which means slim). Since that time the annotation for the obese mutation has been changed for *Lep*[ob]. The *Lep*[ob] mutation is a C→T transition that generates a nonsense (stop) mutation at 105 nt resulting in a non-functional allele. Another allele at the *Lep* locus, *Lep*[ob-2J], with very similar phenotypic effects, was found to be the consequence of a small sized deletion that completely inactivates transcription of the *Lep* gene. This validated leptin *Lep* gene as a candidate for the obese phenotype.

Positional cloning of the *Lep* locus made it clear that this gene, which was not known, had an essential role in the control of satiety, fat metabolism and energy expenditure. After this pioneering experiment a function could be assigned to a genomic sequence. Nowadays, geneticists would say that the genomic sequence at the *Lep* locus is fully *annotated*.

Many genes have been identified by positional cloning, including for example the gene encoding the receptor of leptin (*Lepr*, for *Leptin receptor* – chromosome 4) which, when mutated, produces a phenotype very similar to that seen for *Lep*[ob]. It is anticipated that many other genes will be cloned in the future because the strategy is now standardized and greatly facilitated by the availability of the entire genome sequence for several species. Basically, positional cloning consists of three steps, as follows.

First a high resolution/high density linkage map of the region containing the locus to be cloned (symbolized *Mut*, when dominant and *mut*, when recessive) must be established using as many molecular markers as possible. To achieve such a genetic map, interspecific crosses or intersubspecific crosses are often preferred to intraspecific crosses because inbred strains recently derived from wild mice of the same genus *Mus* but from different species

(*Mus spretus* or *Mus m. castaneus* for example) represent a virtually unlimited reservoir of polymorphisms (Guénet and Bonhomme, 2003). This first step is relatively tedious but, unfortunately, it is unavoidable (Fig. 9.2).

When the segment of DNA harbouring the *Mut* (or *mut*) locus (designated the 'critical interval') is reduced to 0.3 cM or less, then the breeding phase is stopped and the 'bench' or 'molecular' component begins. Since the mouse genome consists of 2.6×10^9 base pairs and spans approximately 1560 cM, a genetic interval of 0.3 cM is theoretically equivalent to 500/600 kb of DNA and contains an average of 5 to 8 genes (Waterston *et al.*, 2002). The sequences of the nearest flanking markers are used as entries into the genome databases to delimit the region of interest and establish a list and the position of candidate genes (Fig. 9.2). The actual molecular size of the segment that corresponds to 0.3 cM is subject to variations because the frequency of recombination within the mouse genome is not evenly distributed, with 'hot' and 'cold' points of recombination scattered throughout. Zhang and co-workers (1994) observed that in one of their crosses, 1 cM of genetic distance corresponded to ~5.8 Mb, which is about three times more than expected. The reasons for such variations in recombination frequency are not known. It is clear however that the strains used in the backcross and the localization of the gene on the chromosome both have an influence.

The purpose of the last step is to identify any changes at the genomic level that might explain the observed phenotypic difference(s). This is sometimes easy, for example when the mutation is a deletion or an insertion, or any change leading to an abnormal transcript. In other instances, identification of the genetic change is more difficult and it is sometimes even impossible. This is the case, for example, when there is no obvious alteration in the coding sequence or when RNA splicing is abnormal because an alien DNA sequence has inserted into an intron, introducing a new splicing site. In the latter two cases, additional experiments are required such as northern blot analysis or reverse transcription–polymerase chain reaction (RT–PCR) to study the transcribed mRNAs in the mutant and control alleles. An alternative approach consists of the

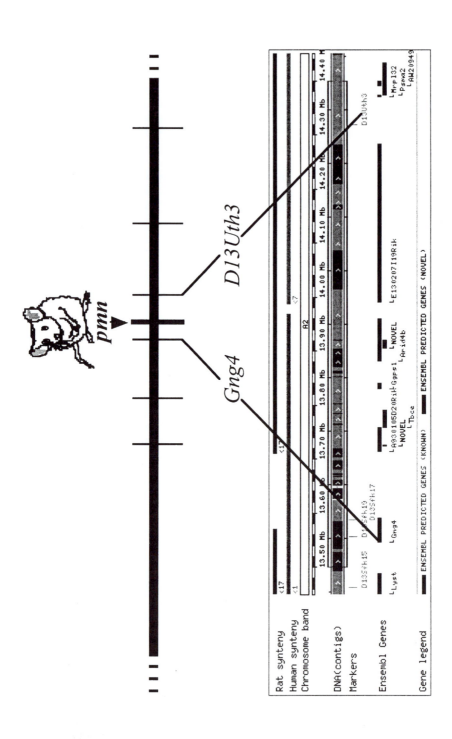

Fig. 9.2. A high resolution genetic map localizes the mutant allele *pmn* between molecular markers *Gng4* and *D13Uth3*. The same molecular markers can then be used as entries to define the molecular interval containing *pmn* in the mouse genome sequence. After sequencing, *Tbce* was found to be the mutant gene in the interval.

rescue of the mutant phenotype by a transgenic copy of the wild-type allele (refer to the section on 'inducing engineered alterations' for details). An example of this so-called transgenic rescue is provided in a case reported by Martin and co-workers (Martin *et al.*, 2002), concerning the progressive motor neuronopathy (*pmn*) mutation. After characterization of the mutant allele through positional cloning, the scientists found that the only consistent difference between the mutant allele and its normal counterpart was a G→T transversion in the last codon of the gene coding for tubulin folding cofactor E, replacing a Trp (W) by a Gly (G) residue. Given that the mutant mRNA was of the expected size and the predicted protein

structure almost normal, the absolute demonstration that the neurological defect observed (axonal degeneration of the motor neurones in the spinal cord) was a direct consequence of the transversion was provided by the rescue of homozygous mutant *Tbce^{pmn}/Tbce^{pmn}* genotypes by a transgenic copy of the wild or *Tbce^+* allele (Fig. 9.3).

The strategy of positional cloning has several advantages. The first and most important is that the phenotype of the mutations, which made their discovery possible, is generally not ambiguous. It is thus relatively easy to associate a given gene with a given phenotype, and accordingly, with a specific function. Another important advantage is that positional cloning

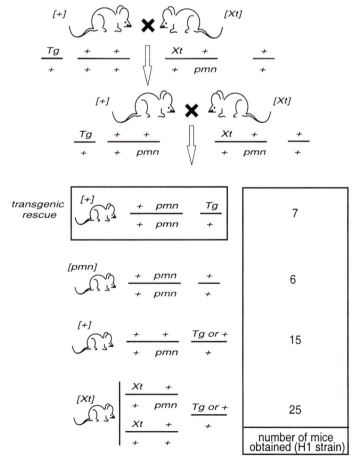

Fig. 9.3. Transgenic rescue of the mutation progressive motor neuronopathy (*Tbce^{pmn}*) by a transgenic copy of the normal allele *Tbce^+*. The seven mice, homozygous for the mutant allele *Tbce^{pmn}*, exhibited a normal phenotype.

sometimes points to genes that could not be suspected *a priori* based only on the pathology of the mutation. The *pmn* mutation is again a good example. In this case, the gene is expressed ubiquitously but the deleterious effects of the mutant allele are observed mainly in the motor neurones with long axons. In fact, no one would have suspected that a missense mutation in *Tbce* could have such specific effects, especially when considering the dramatic effects of null mutations occurring at the same, homologous locus in man which result in the so-called Kenny Caffey or Sanjad–Sakati syndromes (Parvari *et al.*, 2002), a severe disease with growth retardation, craniofacial anomalies, small hands and feet, hypocalcaemia, hypoparathyroidism and early psychomotor retardation. Another advantage of positional cloning is that this strategy exploits an enormous variety of mutations, much wider than those engineered by man (see further, the section dealing with engineered mutations).

Aside from these many advantages, there are however some important limitations to the strategy of positional cloning for gene annotation. The first and most important of these limitations is that, even if the number of mutant alleles increases constantly, the vast majority of mammalian genes do not have mutant alleles or do not have a mutant allele that produces an obvious phenotype to facilitate this strategy. Another important limitation is that not all mutations are easily identifiable at the molecular level. For example, the gene encoding ATP/GTP-binding protein 1 (*Agtpb1*) has several independent mutant alleles in the mouse (the *pcd* alleles), which all result in the same phenotype of Purkinje cell degeneration. Among all these alleles, the *pcd* allele, which was discovered first, displays no change at the level of the coding sequence and if other alleles had not been found subsequently, annotation of this gene would have been very difficult if not impossible. This is also true for a few other mutations.

To bypass these limitations, large-scale mutagenesis projects are now in progress worldwide. In some instances mutagenesis is also performed in conjunction with phenotypic analyses to generate new mutant alleles with effects on specific tissues, organs or functions. Such specific phenotypes may involve hearing, integument morphology, brain and behaviour,

resistance to experimental infections, cardiac physiology, etc. This *phenotype-driven* mutagenesis has generated interesting results and appears to be a very promising strategy to generate animal models for genome annotation.

The experimental production of mutations in the mammalian genome

Spontaneous mutations and their frequency

Mutations are rare events that occur (probably) at random in the genome. At the DNA level, mutations exhibit a great variety of forms, ranging from single base substitution (also known as single nucleotide polymorphism or SNP) to complex alterations such as deletions, insertions, transpositions, inversions, etc. The spontaneous mutation rate is not known with accuracy because only a fraction of these mutations are detected at the phenotypic level. Dominant mutations are, in theory, easier to discover than recessive but they generally occur only once and their detection greatly depends on their phenotype. Some mutations are so severe that they impair reproductive capacity and accordingly cannot even be identified as a heritable trait. Others have a very subtle phenotypic expression and for this reason they may remain undetected. A good example of the latter category is the mouse mutation extra-toes (formerly *Xt*, now *Gli3a*^{Xt}), whose only phenotype is a tiny extra digit at the inner edge of the hind feet! Another major difficulty in the detection of dominant mutations is that they frequently exhibit wide variations in expressivity. A well-known example of a mutation exhibiting variable expressivity is brachyury (*T*) that leads to a reduction in tail size. Sometimes the tail is so dramatically reduced that the (*T/+*) mutant mice do not even have a sacrum. In other instances the tail exhibits only a few kinks at its tip, which may escape detection (Fig. 9.4).

Recessive mutations are different. They are easy to detect, because they are in general observed several times, especially when they occur in an inbred strain. Many recessive mutations, however, are lost by chance, after a few generations of inbreeding. Variations in expressivity also occur with recessive mutations but

Fig. 9.4. Mouse mutations often exhibit large variations in expressivity. All mice on this picture are heterozygous for the same mutation brachyury (*T/+*). They exhibit, however, rather different phenotypes.

this is generally not a serious problem given that many mutant homozygotes are produced by the heterozygous carriers.

Based on the mutant genotypes discovered at a number of loci in the very large mouse breeding colonies at the Jackson Laboratory, the spontaneous mutation rates have been estimated as 10^{-7} to 0.5×10^{-6} per locus per gamete for mutations towards a dominant allele and 0.6 to 0.8×10^{-6} per locus per gamete for mutations towards a recessive allele (Schlager and Dickie, 1967). The same spontaneous mutation rates have also been computed with accuracy at a small number of specific loci on a very large number of gametes in control experiments performed by health physicists testing the mutagenicity of different forms of radiation and by geno-toxicologists testing potentially hazardous chemical compounds in the environment. Whatever the loci concerned, these mutation rates are very low and this explains why mammalian geneticists, like other geneticists, have developed strategies to increase the rates of mutation.

Mutagenesis in the mouse and other mammals

As already indicated, mice have been used during the last four decades as living test tubes for assessing the risk generated by potentially hazardous environmental conditions, and many mutations have been generated and detected as a consequence of these activities. However, mutagenesis *sensu stricto*, where mice are specifically treated with mutagenic substances to purposefully increase the mutation rates, is only recent.

Gametogenesis and experimental mutagenesis. Experimental mutagenesis consists of exposing progenitors of one sex or the other to a mutagenic agent, looking for an increase in the mutation frequency in the successive progeny of the treated animals. In practice, only male progenitors are treated because, in this sex, gametogenesis is a continuous process, starting at puberty and lasting several months or even years. In females, on the contrary, gametogenesis is a cyclic process and the number of germ cells that are potential targets for mutagenesis is dramatically reduced in adult animals.

Spermatogonia are the stem cells of the male germline. When they divide these cells produce one daughter cell, which remains in the pool of germ cells, and one daughter cell that undergoes several cycles of mitosis then enters into the meiotic process and finally differentiates into spermatozoa. Each spermatogonia generates a cluster of around 120/140 spermatozoa and the duration of the spermatogenetic wave, from the stem cell to the spermatozoa, lasts about 13 weeks in the mouse. It is then clear that, depending on the time that elapses between the mutagenic treatment and the mating of the treated animal, the cells that were affected by the mutagen are different. When a male mouse was mated 3 weeks following a mutagenic treatment, the targeted cells were post-meiotic (or haploid and accordingly resting) while when the same male was mated 3 months after the mutagenic treatment the targeted cells were spermatogonia. An important difference is that, when a mutagenic treatment is active on pre-meiotic germ cells, the male becomes a permanent provider of mutations. On the contrary, a mutagenic treatment applied to post-meiotic germ cells only is transient.

The mutagens. Basically there are two kinds of mutagens: radiation and chemicals. Radiation is interesting because the protocol of mutagenesis is, in general, easily standardizable. However, chemicals have become more popular over recent years because they are more efficient and somewhat easier to use.

Radiation as mutagens. There is a great variety of forms of radiation that can be classified in two major groups: (i) corpuscular radiation (protons, α- and β-rays, neutrons, etc.) and (ii) electromagnetic radiation (γ, X and UV rays). All forms of radiation are mutagenic provided their energy, linear energy transfer (or LET) and doses are sufficiently high. In normal laboratory conditions, however, only the X- and γ-rays can be easily and efficiently used for the purpose of mutagenesis. There have been many experiments during the 1950s–1970s that specifically addressed the use of radiation as a mutagen in the laboratory, in particular at the MRC Harwell, UK, and at Oak Ridge National Laboratory in Tennessee, USA. The main conclusion of these numerous experiments is that the most efficient way to induce mutations in the mouse is to irradiate an adult male with a split dose of 2×500 rads (2×5 Gray) of γ-rays (Co^{60} or Cs^{137}) distributed at a 24 h interval. In these conditions, the number of recessive mutations induced in spermatogonia is 30–50 times higher than for the normal control rate (Green and Roderick, 1966).

Irradiation with X- or γ-rays induces all kinds of mutations, from point mutations, resulting in 1 bp substitution or deletion, to more extensive damage such as large deletions

or translocations. However, when these mutations are generated in spermatogonia, those with the most deleterious effects at the cellular level are strongly selected against and in general are not transmitted in the male gametes. This is not the case for the mutations induced in the post-meiotic germ cells, such as spermatids, which are generally capable of fertilizing an oocyte irrespective of the genetic damage they carry.

The chemical mutagens. Ideally, the optimal mutagenic substance has to comply with a number of pre-requisites: the mutagen should be readily available to purchase or easy to prepare, it should be easy to handle (for instance it should be readily soluble in water), it should not be too toxic and, if possible, it should act on pre-meiotic germ cells to produce transmittable mutations. Up to the late 1970s only a handful of chemical substances had these virtues but none appeared to be a mutagen of choice (Vogel and Röhrborn, 1970).

The observation by William Russell and colleagues in 1979 (Russell *et al.*, 1979) that *N*-ethyl-*N*-nitroso-urea (abbreviated ENU) was a potent mutagen in the mouse had a major impact on genetic research and must be considered a milestone in the history of mouse genetics. ENU is an alkylating agent. It is relatively easy to handle although it is light, heat and pH sensitive. It does not dissolve easily in water but adding a few drops of ethanol helps its dissolution. The mutagenic activity of ENU results from its capacity to transfer an ethyl group to oxygen or nitrogen radicals in the DNA molecule provoking mispairing and ultimately leading to base pair substitutions or deletions (Van Zeeland *et al.*, 1989; Vogel and Natarajan, 1995). In fact, the mutagenic activity of ENU results from two components acting in opposite directions: the alkylation of the DNA molecule on one hand and the efficiency of the enzymatic DNA repair mechanisms on the other. In spermatogonia, the ENU-alkylated N atoms are efficiently repaired while ENU-alkylated O atoms are repaired with a much lower efficiency.

Many ENU-induced germline mutations have been analysed at the molecular level and it has been found that, in the great majority of cases, adenine is the target of ENU activity, with the primary genetic alteration being either A-T to T-A transversions or A-T to G-C transitions (Justice *et al.*, 1999, plus personal observation).

The mutagenic activity of ENU has been evaluated using several tests (Russell *et al.*, 1979; Favor, 1986; Lewis *et al.*, 1991, 1992; Favor, 1994; Ashby *et al.*, 1997; Schmezer and Eckert, 1999). In his initial paper, William Russell and colleagues (1979) found 35 confirmed mutations at seven specific loci among 7584 offspring in the treated group (250 mg/kg of body weight) compared with 28 among 531,500 mice in the control group. This indicated a mutation rate 90 times higher than the spontaneous rate and five times higher than for 600 rads (6 Gy) of γ rays.

Plotting the mutation rates computed with the same 'multiple loci' assay to the doses of ENU injected in male mice, Favor and colleagues (1990) observed that the mutation rate for ENU increases linearly with dose, from the threshold dose of 34 mg/kg of body weight up to 400 mg/kg, a dose that seems to be the highest tolerable by the adult mouse. If the dose remains low, say less than 30 mg/kg of body weight, the mutation rates are not significantly different from the rate of spontaneous mutations in the same assay. Favor's computations can be summarized in the following two formulae:

$$MR \times 10^{-5} = 1.2 \qquad \text{for D} < 33.9 \text{ mg/kg}$$

$$MR \times 10^{-5} = 1.2 + 0.4 \times (D - 33.9)$$
$$\text{for D} \geq 33.9 \text{ mg/kg}$$

where MR = mutation rates and D = dose in mg/kg of body weight.

The threshold effect observed by Favor and colleagues may be explained by the fact that when the number of alkylated sites remains low, the repair mechanisms can cope but when it becomes high or very high then they become saturated and mispairing increases in proportion to the dose of mutagen.

This linear dose relationship for induced mutation rates at the seven specific loci demonstrates the extraordinary power of ENU as a mutagen but cannot adequately measure the absolute rate of induced mutation at an 'average' locus in the mouse genome. Lewis *et al.* (1991), for example, computed the number of electrophoretic variants induced at 32 loci, after treatment with increasing doses of ENU

(from 0 to 250 mg/kg of body weight), in DBA/2 or C57BL/6 male mice (Lewis, 1991). The mutation rates again appeared to increase linearly with the dose but were on average 2.6 times lower than for the 'multiple loci' test. This latter observation, which has been reported by many others in the literature with different tests, indicates that the sensitivity of a locus to the mutagenic activity of ENU probably depends on a variety of parameters such as (maybe) its 'molecular' size, the gene structure (density in A-T, number of introns, etc.) and presumably several other unknown parameters (Kiernan *et al.*, 2002).

W. Russell and colleagues (1982a,b) reported that fractionation of the dose had a reducing effect on the mutation rate. This is presumably because a dose of 100 mg/kg of body weight in one dose is more capable of saturating the repair mechanisms than the same dose delivered as 10 weekly injections of 10 mg/kg. In contrast to this, however, the same team (Hitotsumachi *et al.*, 1985) reported 3 years later that three or four injections of 100 mg/kg of body weight, each delivered at weekly intervals, enhanced the mutation rates by a factor 1.8 and 2.2, respectively, compared with a single dose of 250 mg/kg of body weight, without impairing the viability or the fertility of the treated mice. With such a treatment, the maximum mutation rate of 1.5 to 6 $\times$ 10^{-3} per locus can be obtained, which is equivalent to obtaining a mutation in a gene of interest in one out of every 175 to 655 gametes screened and roughly corresponds to 150 times the spontaneous mutation rate. This frequency, established for seven specific loci, was later refined by Bode (1984) in another experimental context. Bode considered that, from a correctly mutagenized male, one can expect to get, on average, one mutation out of 1500 of its gametes. It must however be kept in mind that a given male can produce only a limited number of mutations that are dependent on the number of targets that have been hit by the mutagen. From his experimental data Bode concluded that this number is close to 500.

Mutagenesis in the rat has also been achieved using ENU. In this case the dose must be reduced to 90 mg/kg of body weight and, here again, splitting of the doses has proven more efficient than a single dose. The efficiency of mutagenesis in this species is almost the same as in the mouse although many strains are hypersensitive to the mutagenic treatment.

Protocols for experimental mutagenesis. The induction of mutations in mammals can be achieved either genome wide, i.e. at all loci, or in more or less accurately targeted regions. These protocols do not depend upon the mutagen and can apply to radiation as well as to chemicals. We will review some of the most commonly used protocols.

Genome-wide induction of new mutant alleles. When a male mouse is treated with a mutagenic agent, for example by injecting 250 mg of ENU per gram of body weight, spermatogenesis ceases for a period spanning 10–13 weeks. Then, the surviving spermatogonia repopulate the testis, the sperm concentration rises and the male in question regains fertility and produces sperm from the several different clones of mutagenized spermatogonia. In the sperm population (and later in the embryos), all kinds of mutations are present but, while dominant mutations can be observed directly in the F1 progeny, recessive mutations have to be homozygous to produce a detectable phenotype. This requires two more generations (Fig. 9.5).

Many ENU-induced dominant alleles have been detected in the mouse genome, particularly within the frame of the presently ongoing large-scale mutagenesis programmes (Hrabe de Angelis *et al.*, 2000; Nolan *et al.*, 2000). As already discussed, dominant mutations are sometimes difficult to detect in the progeny of treated males (G1 or F$_1$) because of their subtle phenotypes. They are also difficult to characterize because complementation tests with this category of mutations are not always possible. The efficiency of detection also depends greatly on the phenotyping protocols that are associated with the mutagenesis programme, especially for the mutations with late onset.

Because their deleterious effects are compensated by the presence of a normal allele in heterozygotes, recessive mutations are easier to detect and to preserve. For a genome-wide project, F1 male offspring of mutagenized males are mated to normal wild-type females (G2) then the female offspring of these matings are mated to their presumptive heterozygous father (backcross or G3) and the subsequent progenies are carefully scrutinized for the presence of mutant phenotypes. Each of the recessive mutations carried by the F1 male, son of the mutagenized male, is transmitted to every other female among its offspring and, in the G3 (backcross) progeny, a quarter of the offspring exhibit a mutant phenotype when the

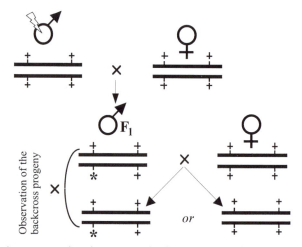

Fig. 9.5. Detection of mutations induced genome-wide after treatment with a mutagen. Dominant mutations can be detected directly in the F$_1$ while recessive mutations require a three generations micropedigree and are observed in the progeny only when both parents are heterozygous.

mating is set between two heterozygous partners. To minimize the probability of not detecting a recessive mutation by chance, each $+/m$?
male must be backcrossed to six $+/m$? females
and at least ten offspring must be scored for
each of these six females. Following such a
protocol, the probability of not detecting a new
ENU-induced recessive mutation is less than
2% at the 95% confidence level (Fig. 9.5).

The genome-wide production of recessive
mutations is a tedious enterprise that requires
both animal care and large sized breeding
facilities. It is however very rewarding since
studies of mutagenesis in many organisms has
indicated that the majority (probably over
90%) of mutations are recessive to wild-type
(Wilkie, 1994).

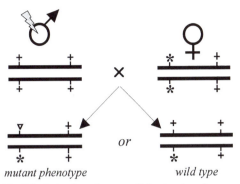

mutant phenotype   *wild type*

Fig. 9.6. Induction of a new allele at a given locus
by mutagenesis. This strategy requires that at least
one viable allele exists at the locus in question.
Identification of the new allele often requires the
use of molecular markers.

*The induction of new mutations at specific
loci.* This strategy is well exemplified in the so-
called 'multiple loci' test where newly ENU-
induced alleles are detected, directly in the F1
progeny of treated wild-type males mated with
females homozygous for a set of recessive
viable alleles. Using this protocol, a large number of new alleles have been induced at the *a*,
Tyrp1 (formerly *b*), *Tyr* (formerly *c*), *p*, *Myo5a*
(formerly *d*), *Bmp5* (formerly *se*) and *Ednrb*
(formerly *s*) loci, which have been investigated
in detail (Rinchik and Carpenter, 1999). Such
a strategy can be applied to any situation
where the production of a series of new alleles
at a given locus might be interesting. It
requires however that at least one viable
recessive allele be available (Fig. 9.6). Bode
(1984) used ENU to produce new alleles at
the *Brachyury* (*T*), *quaking* (*qk*) and *tufted* (*tf*)
loci by mutagenizing $+++/+++$ male mice
and crossing them to females with the genetic
constitution *T qk tf/++ tf*. In the F$_1$ progeny
he found three [*tf*], one [*qk*] and one *t*-inter-
acting (or *t^{int}*) allele out of 5172 offspring. In
other experiments of the same kind Justice
and Bode (1986) and Bode *et al.* (1988) produced several new alleles at the same three
loci, with each new allele exhibiting interesting
properties (Justice and Bode, 1990; Cox *et al.*,
1999). Similarly, Chapman *et al.* (1989) identified four new alleles at the *Dmd^{mdx}* locus
(*mdx^{cv2}*, *mdx^{cv3}*, *mdx^{cv4}* and *mdx^{cv5}*) by
checking for an increase in creatine phosphok-
inase (CPK) plasmatic levels in the female

progeny of ENU-mutagenized $+/Y$ males
crossed to *Dmd^{mdx}/Dmd^{mdx}* homozygous
females. This experiment was undertaken to
control for the possibility that the lack of an
obvious phenotype in *Dmd^{mdx}/Dmd^{mdx}* or
Dmd^{mdx}/Y mice was not simply due to an
allele with an exceptionally weak effect.
Chapman found that all five alleles, the four
ENU-induced and the original one (Bulfield *et
al.*, 1984), had a very a similar pathology.
Interestingly, however, it was later demonstrated that mice homozygous for some of
these alleles exhibited variations in their electro-retinogram (ERG) patterns. In particular,
the original *mdx* and the ENU-induced *mdx^{cv5}*
alleles had a weaker effect than the *mdx^{cv3}*.
This observation indicated that the position of
the mutation in the dystrophin-encoding gene
had a direct consequence on the ERG phenotype (Pillers *et al.*, 1999) and contributed to
the annotation of the different domains of the
Dmd gene.

A variation of the above-mentioned strategy is to analyse the electrophoretic pattern of
enzymatic proteins in an inter-strain F1 hybrid
where one parent (the male in general) has
been mutagenized. Such an 'electrophoretic
multiple loci test' has been successfully used to
identify new mutations at loci coding for enzymatic proteins (Johnson and Lewis, 1981;
Marshall *et al.*, 1983; Lewis *et al.*, 1991,
1992). ENU mutagenesis has also been used
to induce mutant alleles in the gene encoding
the β-chain of haemoglobins (Peters *et al.*,

1986). This strategy has also been used to produce null alleles or functionally different alleles (Charles and Pretsch, 1987; Pretsch et al., 1994).

The production of new alleles at specific loci by ENU mutagenesis might also be very useful for the production of new independent alleles in a positional cloning project. For example, cloning the mouse mutations hotfoot/Lurcher (Grid2[ho]-Grid2[Lc], chromosome 6) (Lalouette et al., 1998), beige (bg, chromosome 13) (Perou et al., 1996), strigosus (Npr3, Chr 15) (Jaubert et al., 1999), satin (sa, chromosome 13) (Hong et al., 2001) or Purkinje cell degeneration (Agtpb1, chromosome 13) (Fernandez-Gonzalez et al., 2002) would not have been an easy task if several alleles had not been available that offered various independent alterations in the same gene. Many genes, which have only one allele with deleterious effects, may lead to an inconclusive situation once cloned and sequenced, simply because the alteration found at the genomic level is a missense mutation, resulting in an amino acid substitution of which the consequence is not easily predictable. In this case, it may be wise to generate new alleles by ENU mutagenesis concurrently with positional cloning.

The condition set above, that at least one recessive and viable mutant allele for the locus of interest exists to facilitate the generation of other alleles with deleterious effects, is not an absolute pre-requisite, it simply makes the experiment easier. If this is not the case, alternative strategies are still possible. Let us suppose that other alleles are desired at the Mut locus, which to date is only characterized by the unviable (or sterile) mutation mut[1]. In this case, several F1 males, heterozygous for many new ENU-induced mutations (among which a potential new mut[2] allele may be found) are produced, then crossed to +/mut[1] females. If, by chance, a mouse with an abnormal [mut] phenotype is detected in the progeny of one of these females, this suggests that the expected new mut[2] allele at the Mut locus has indeed been induced by the treatment. According to the statistical computations made by Bode (1984) in his paper reporting the generation of new alleles at the qk and tf loci, the mutation frequency after a single dose of 250 mg/kg of ENU is one in 1500 gametes and one in five mutagenized males has mutated sperm at any

given locus. This indicates that the production of new alleles at a specific locus by ENU mutagenesis, even if time and breeding space are required, is not completely unrealistic.

The induction of new mutations at specific regions of the genome. Recessive mutations have also been induced in some specific regions of the mouse genome and many strategies have been used for this. Here, we report three of these strategies that may be of interest in the future: the first makes use of *deletions*, the second utilizes *consomic* or *congenic* strains and the last strategy requires a set of overlapping *inversions*.

Using deletions to detect recessive mutations can only be applied to regions where haploidy is compatible with life. Although a complete inventory of these regions is not available we know that, by definition, the imprinted regions must be excluded. The basic principle when using deletions is that, when a mutation is induced in the normal chromosomal segment that faces a deletion, a mutant phenotype (often lethal) is observed when the chromosome carrying the induced mutation and the chromosome carrying the deletion are paired in the same genome. In these conditions the breeding protocol is not straightforward and requires more than one generation since the induced mutation must be kept in the heterozygous state while it is being revealed using the deletion (Fig. 9.7). The deletion strategy has been used many times (Justice et al., 1997; Rinchik and Carpenter, 1999) and is still included in modern mutagenesis programmes (Nolan et al., 2000).

Consomic strains are strains where a complete chromosome has been backcrossed from a donor strain into a recipient or background strain. Such strains are not very common but at least one complete set exists (Nadeau et al., 2000) and this is sufficient for the strategy to be applied (Williams, 1999). Male mice are mutagenized then mated to female mice consomic for a specifically targeted chromosome. F_1 male offspring of this cross are then crossed again to consomic female mice of the same strain. Subsequently a few female offspring from this second cross, with the same 'heterozygous' chromosomal constitution as their father, are selected by microsatellite genotyping

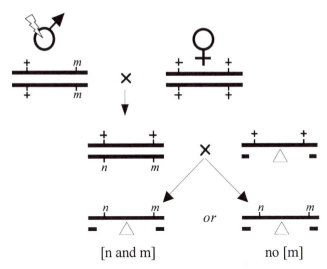

[n and m]   no [m]

Fig. 9.7. Induction of mutations in a region encompassed by a deletion. A male, homozygous for a recessive mutation (*m*), is treated with a mutagen and mated to a wild-type female. Its offspring are then crossed to a female heterozygous for a deletion (△) encompassing the *m* locus and the progeny are carefully scrutinized. If the induced mutation (*n*) is not lethal, two phenotypes [m and n] are observed in the progeny of the mice heterozygous for the deletion. Absence of the expected [m] phenotypes suggests that *n* is lethal.

and backcrossed to the latter. Finally mice homozygous for the 'mutagenesis'-targeted chromosome are produced and carefully checked for the occurrence of mutations. If these mutations have effects compatible with normal life, then one male and one female 'homozygous' for the targeted chromosome are crossed together and a co-isogenic strain is then created for all the mutations induced in the chromosome in question (Fig. 9.8). The strategy that makes use of consomic strains has several advantages. First, by definition, it facilitates the detection of all the mutations that have been induced in the targeted chromosome. This is advantageous for the mutations that have a weak effect or that require sophisticated tests for their detection, for example, behavioural, biochemical or immunological tests. Secondly, the strategy that makes use of consomic strains allows the establishment of a co-isogenic strain where the newly ENU-induced mutations are safely stored before being studied. This is particularly advantageous when populations (not only individuals) are to be compared at the phenotypic level, for example, histocompatibility, susceptibility to infectious diseases, and quantitative trait locus (QTL) analysis. If we consider that there are

about 30,000 genes in the mouse genome (Waterston *et al.*, 2002) and that the mutation frequency is high (an average of one mutation per 1500 gametes), then we expect around 20 new mutant alleles with clear phenotypic effects per spermatozoa and presumably many more phenotypically neutral DNA polymorphisms. The use of consomic strains to store new mutations and neutral polymorphism such as SNPs might then be advantageous. Another advantage of this strategy is that the same co-isogenic strain that is homozygous for the targeted chromosome can be used over and over in similar mutagenesis experiments resulting in the accumulation of new alleles in the targeted chromosome. Congenic strains can be used in this approach, the only difference being that only a piece of chromosome, instead of one entire chromosome, is involved.

The use of a set of inversions, which may be induced by genetic engineering, in a given chromosome is similar in its principles to the use of consomic strains. However, it is more sophisticated in that, when appropriate markers are used, this approach does not require any genotyping for the selection of the progenitors. An example of the strategy was described in detail for an inversion in chromosome 11:

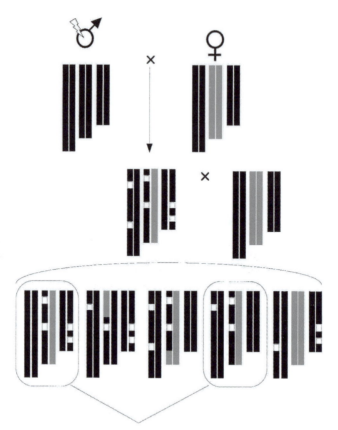

Fig. 9.8. Use of a consomic mouse strain for the collection and accumulation of mutations in an inbred background. A male of an inbred strain is treated with a mutagen then mated with a female of a consomic strain for a given chromosome. The F$_1$ from this cross, which carry several new mutations, are crossed again to the same consomic strain and two mice of the same heterozygous genotype are selected in the progeny and crossed together. If they are viable and fertile, mice homozygous for the 'targeted' chromosome can then be inbred.

Inv(11)8Brd$^{Trp-53-Wnt3}$ (Zheng *et al.*, 1999) and the practical use of this inversion for the isolation of mutation (Kile *et al.*, 2003). In this case all categories of segregants born from the crosses used in the micropedigree are easily recognized by the convenient use of a coat colour marker (K14-Agouti – with yellow ears, tail and ventrum) and a fur texture allele (mutation Rex – *Re*). The inversion induced by the Cre-*lox*P strategy (see further in this chapter) acts as a crossover suppressor and it is lethal when homozygous because mice are *Wnt3*-deficient. New mutations are homozygous in the 'normal' C57BL/6 background and are automatically kept heterozygous in the yellow mice.

Many other strategies have been used to generate and keep mutations in specific areas of the mouse genome that cannot be described in detail here. One could just briefly mention that, using a specially designed strategy, Shedlovsky and her colleagues (1986, 1988) were able to induce and study a dozen new lethal alleles within 2 cM flanking the *T/t* region on mouse chromosome 17.

Phenotype-driven mutagenesis. Bode and colleagues (1988), followed by McDonald and colleagues (1994) were the first to use ENU mutagenesis to induce mutations resulting in phenylketonuria. After treatment with ENU the progeny of treated male mice (G1 and G3) were subjected to the popular Guthrie test, a biochemical test detecting an elevated level of phenylalanine in the blood. In so doing, three loci (*hph1*, *hph2* and *Pah*) were identified by a deleterious allele. It is interesting to note that using such a strategy, the biochemical pathways at work in the catabolism of the amino acid phenylalanine were literally dissected out with at least one mutation at each step, in exactly the same way the bacterial geneticists of the early days investigated metabolism in bacteria (McDonald, 1995).

Nowadays, with the progress in genotyping and phenotyping, several projects have been undertaken where the G1 and G3 progeny of ENU-mutagenized males are systematically phenotyped by a team of specialists against a number of criteria. As 'end-products' of such phenotype-driven mutagenesis, many interesting mutations have been discovered that would probably not have been noticed otherwise. Among these interesting mutations we must mention *Clock*, a mutation that perturbs the circadian rhythm of affected mice (Wilsbacher *et al.*, 2000), and *Lps2*, a mutation resulting in impaired defence mechanisms against viral and bacterial diseases (Beutler *et al.*, 2003). Numerous projects are now in progress in several laboratories where these new mutations are immediately offered to the community of mouse geneticists by way of web sites describing the phenotype. There is no doubt that genome annotation will benefit of all these programmes even if a significant amount of work remains, for example mapping at high resolution and testing for allelism with the already existing mutations.

Genome-driven strategies for the production of targeted mutations. With the development of techniques for the structural analysis of DNA, new strategies have been recently reported that are based on the direct, *in vitro* detection of DNA alterations, at specific loci, in F1 offspring from ENU-mutagenized males. These strategies make use of sensitive techniques that can identify SNPs in control and ENU-treated samples to identify potentially interesting mutations in selected genes. Of course, only a small proportion of the changes that are randomly induced at the DNA level by the chemical mutagen have 'functional' consequences. Beier (2000), for example, calcu-

lated that after mutagenesis under optimal conditions, there would be ten sequence changes per 1000 genetic loci, but only one of these would result in a functional change leading to a phenotypic variant. These new strategies are nevertheless potentially interesting for several reasons. The first and most important reason is that they are applicable to virtually any species, including mammals, where specific gene disruption (knockout) is not yet possible as embryonic stem cells are not yet available. Second is that the identification of specific gene alterations can be achieved using pooled DNA samples and that the bench work can be run concurrently in several different laboratories to increase the efficiency thereby and ultimately lowering the cost of mutagenesis. The final advantage is that in species where sperm cells can easily be deep-frozen, there is no time limit for the identification of mutations. Several laboratories have already published interesting results on this matter on mouse (Coghill *et al.*, 2002) and rat (Zan *et al.*, 2003; Smits *et al.*, 2004).

The strategy developed at the University of Wisconsin (Zan *et al.*, 2003) is particularly clever. Male rats are mutagenized with ENU, then a yeast-based screening assay is applied to DNA samples taken from the F1 that select for various classes of functional mutations. This assay uses gap-repair cloning to integrate either genomic DNA (gDNA) or cDNA for a selected gene between the yeast promoter *ADH1* and the reporter gene *ADE2* to form a chimeric protein. If the DNA from a specific allele contains functional mutations that interfere with translation, then an active *ADE2* chimeric protein is not produced. This is visualized by the growth of small red yeast colonies instead of the large white colonies that are observed when screening wild-type DNA. With this technique, two mutations equivalent to knockout for the breast cancer suppressor genes *BRCA1* and *BRCA2* have been generated in the rat.

Understanding the genetic determinism of quantitative variations: QTL analysis and its contribution to genome annotation

In the previous sections the use of spontaneous or induced mutations for genome annotation was discussed. The strategy was to detect the specific point mutations at the genome level that could explain a phenotypic change, characterized as a Mendelian trait, which would reveal the function of the protein encoded at this locus. Unfortunately most of the phenotypic differences that are observed between any two individuals of the same species are in general not the consequence of a simple change in the coding sequence of one gene. On the contrary, most heritable traits exhibit quantitative variations. For example when a cross is set between two parental inbred strains with dramatically different phenotypes for a given parameter, the resulting F2 offspring exhibit a wide range of phenotypes. Such phenotypic variation can sometimes be even more extreme than the parental lines. For instance the daily water intake for inbred strains of mice varies from 3.8 ± 0.8 ml for strain RIII to 8.3 ± 0.6 ml for strain SEA/Gn. However, in the F2 population from a cross between these two parental strains, all intermediates were observed indicating a rather complex genetic determinism. In the same way it has been found that mice of strain C57BL/6 were extremely resistant to experimental infections with *Trypanosoma cruzi* (the protozoa responsible for Chagas' disease) while mice of strain A were highly susceptible. C57BL/6 mice survive an injection with up to 10^5 infectious particles while A/J mice die after a single injection with ten trypomastigotes. Here again, however, most other inbred strains exhibit an intermediate phenotype of resistance indicating a complex determinism.

The genetic analysis of quantitative or complex traits is much more difficult than the analysis of simple Mendelian traits but this form of analysis undoubtedly represents the genetics of tomorrow. In the frame of this chapter we cannot go into too much detail but a few QTLs have been recently dissected at the molecular level and the availability of the complete genome sequence will certainly contribute to a better understanding of these loci (Korstanje and Paigen, 2002). This will probably improve understanding of sequence variations observed in non-coding sequences and the importance of missense mutations in coding regions of moderately conserved domains. A lot more work will be required to understand the nature and importance of the many complex interactions between genes in the mammalian genome.

Inducing Engineered Alterations at the Genome Level and Analysing their Consequences at the Organism Level: from Genotype to Phenotype

As we have already said, another way to assess the function of genes in the mammalian genome is to produce a variety of alterations at the genome level, followed by careful analysis of their effect(s) in the context of the whole organism. Several techniques designed for the production of such changes have been developed from the mid-1970s and new words, such as *transgenesis* and *transgenics*, have been coined to designate them collectively. Transgenic animals have foreign DNA sequences inserted stably into their genome (Silver, 1995). Transgenic animals can be created by viral infection of early embryos, by micro-injection of DNA fragments into one-cell eggs or through the *in vitro* manipulation of embryo-derived stem cells (ES cells) that are subsequently placed into a blastocyst to form a chimeric mouse.

Transgenesis by viral infection of early embryos

The incorporation of exogenous DNA into the germline through viral infection of mouse embryos was reported for the first time in 1976 (Jaenisch, 1976). Newborns and pre-implantation embryos (4–8 cell stage) were infected with the Moloney leukaemia virus (M-MuLV) and it was observed that infection of pre-implantation embryos, in contrast to infection of newborns, could lead to stable integration of proviral copies into the germline. Infections of embryos with M-MuLV have yielded several strains of mice with stable germline integration of retroviral DNA at distinct chromosomal loci (the *Mov* loci – Jaenisch *et al.*, 1981) and at least one mutation in the gene encoding procollagen, type I, alpha 1 (*Col1a1^{Mov13}*) has been induced using this strategy (Stacey *et al.*, 1988). Infection of embryos with retroviruses or retroviral vectors allows generation of transgenic animals with a reduced number of single intact copies of the provirus integrated into their genome, sometimes into a gene.

Integrations are in general stable, they occur at random and they do not cause any of the chromosomal rearrangements or deletions that often occur with the other classical transgenic techniques (see below).

When viral infection of mouse embryos results in a mutation with a clear-cut phenotype, it is likely that the integration occurred in a gene or in its close vicinity. The DNA of the retrovirus can then be used as a hook to clone DNA sequences flanking the insertion site and this helps in gene annotation.

Viral infection of mouse embryos can also be used to introduce exogenous DNA into embryos or eukaryotic cells. The advantages of this strategy have been reviewed in detail by Nicolas and Rubenstein (1988). An advantage that is particularly noteworthy in the context of this chapter is that all the sequences of the viral genome that are required for its replication, transcription and integration, are grouped in or adjacent to the long terminal repeats (or LTRs) and can be removed from an engineered 'shuttle' virus and provided *in trans* by a 'helper' virus leaving space for foreign DNA inserts up to 8 kb.

Although the use of retroviruses for transgenesis has some advantages, reviewed by Jaenisch (1988), the method has not been widely used for gene annotation because the number of insertion events cannot be controlled and because a high frequency of mosaicism often occurs in the infected embryos. It also has the disadvantage that the expression of foreign genes integrated using a retroviral vector is not guaranteed (Hogan *et al.*, 1994).

Transgenesis by *in ovo* injection of a cloned DNA

The stable insertion of foreign DNAs into the germline through microinjection into the pronuclei of fertilized mouse eggs was reported simultaneously in several laboratories (Brinster *et al.*, 1981; Costantini and Lacy, 1981; Gordon and Ruddle, 1981; Harbers *et al.*, 1981; Wagner *et al.*, 1981a,b). Since these first descriptions, the technique has now become routine in many laboratories. Nowadays, anybody who wants to obtain an animal that is transgenic for a cloned DNA ranging from a few kilobases to a few hundred kilobases in size can order it from a number of companies. Protocols for the generation of transgenic mice and reviews on the subject have been published repeatedly. Among the most popular books dealing with the subject one can cite Hogan and co-workers (1994), Hammes and Schedl (2000) and, more recently, Houdebine (2003).

The production of transgenic mice

The production of transgenic mice is achieved by using a sharpened glass micropipette. A few picolitres of a DNA solution is injected into one of the pronuclei while the egg itself is held to another glass micropipette by suction (Fig. 9.9). In skilled hands, around 10–20% of the microinjected eggs eventually develop into a transgenic animal. Confirmation of the transgenic status is achieved either by Southern blotting or by PCR amplification of the introduced DNA using specific primers and DNA templates prepared from the presumptive transgenic animals. Experience teaches that the integration of the introduced DNA probably occurs at random in the genome of the transgenic animal and that it frequently occurs at the one-cell stage. In this case, the foreign DNA is present in every cell of the transgenic animal and accordingly it is transmitted generation after generation as a new, co-dominant, Mendelian trait (symbolized *Tg*). When the DNA integrates at a later stage of development, the transgenic animal is a mosaic and the detection of the transgene may then be more difficult. The number of copies of the foreign DNA sequence that inte-

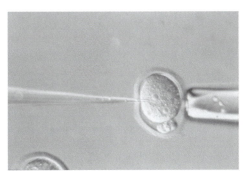

Fig. 9.9. Injection of DNA molecules into a mouse pronucleus (courtesy C. Babinet).

grate into the genome, here again, is not controlled and can range from one to several hundred. Because sticky ends are generated when the foreign DNA is processed for injection, the transgenic copies are generally arranged in head-to-tail arrays with frequent and sometimes extensive rearrangements generated in the flanking regions.

Since the integration occurs in only one chromosome, transgenic animals are *hemizygous* ($Tg/-$) for the DNA segment but (Tg/Tg) homozygotes can be generated through breeding. This however is difficult to achieve and often requires a long and tedious breeding protocol, unless the transgenic insertion(s) can be visualized directly on chromosome preparations by fluorescence *in situ* hybridization (FISH), quantified by quantitative PCR or mapped to a chromosome by setting appropriate crosses (Plate 2). A classical observation is that 7–10% of the transgenic insertions appear to be lethal when in the homozygous state, presumably because a recessive lethal mutation (deletion or gene disruption) was generated at the time of integration.

The uses of transgenic mice for investigating gene functions and regulation

A virtually unlimited number of transgenes can be engineered *in vitro* by the mere apposition of any coding sequence, normal or mutant, with any regulatory elements taken from any gene of any species. When constructing a fusion or chimeric gene for expression in transgenic mice, it is often advantageous to use a cDNA clone to provide the coding sequences rather than the genomic DNA. This is especially true when the coding sequences in question stretch over a very long DNA segment or when it consists of many exons. Unfortunately, the levels of gene expression obtained with cDNA-based constructs are sometimes much lower than those obtained when genomic sequences are used (Brinster *et al.*, 1988). Among the many explanations that can account for this observation, the existence of enhancers in introns is the most likely.

The use of transgenic mice thus appears a very efficient way to assess the function of genes. Here we will consider a few cases that have been selected as didactic examples.

The use of transgenic mice to define the function of genes. Examples of this approach are provided by the *homeogenes* and the *oncogenes*, which both are important for mammalian development. Homeogenes are a family of genes (or isoforms) with a remote ancestral origin and are present in the mouse genome in four *paralogous* clusters (*Hoxa*, *Hoxb*, *Hoxc* and *Hoxd*), located on four different chromosomes. Because the structure of these genes is very similar it was impossible to decide *a priori* whether each of them had a specific function, whether they had an effect because of the copy number (additive effect) or whether some of the isoforms were mere 'back-up' copies set aside (or preserved) by evolution for an unknown purpose. Transgenic mice have been made for some of these homeogenes with an intact coding sequence driven by a regulatory sequence different from the native one (driving ubiquitous expression in general). In most instances, the transgenic embryos in question exhibited severe 'homeotic' transformations indicating that, indeed, most of the homeogenes in the *Hox* clusters had a specific function in the developmental patterning of the mouse embryo, a patterning reminiscent of their function in *Drosophila*, where they were initially discovered (Duboule, 1998).

Transgenic mice have also been created with the coding sequence of intact or mutated oncogenes such as *Abl1*, *Jun*, *Mos*, *Nras*, *Myc*, etc. as well as with the coding sequences of genes whose function was not completely understood, such as those encoding protein p53 or p105-RB, downstream of a variety of regulatory sequences. The subsequent analysis of the transgenic animals has provided an enormous amount of information concerning the role of these oncogenes in the regulation of several basic cellular functions and during the process of malignant transformation. The unique advantage of transgenesis in the case of the homeogenes and oncogenes was to make possible the analysis of the gene function(s) at the level of the whole organism.

The use of transgenic mice to define and characterize the regulatory sequences of genes. While some mammalian genes are constantly and ubiquitously expressed, many others are expressed in a tissue-specific manner during different stages of embryonic development or in the adult organism. Such a selective expression pattern occurs because the genes in question are controlled by regulatory sequences that are frequently, although not always, located upstream of the coding regions. At the beginning of this chapter, when we described the positional cloning of the obese (*Lep^{ob}*) gene that encodes *leptin*, it was mentioned that this cytokine is expressed almost exclusively in the adipocytes. We now know that this highly tissue-specific expression of the *Lep^{ob}* gene is controlled by a 161 bp-long, *cis*-acting, regulatory sequence which is located upstream of exon 1 and which is known as '*minimum promoter*' (He *et al.*, 1995). Unfortunately the regulatory sequences are not yet known for most genes and geneticists have to design experiments to identify them accurately. Genes cloned in their native genomic configuration and introduced into the mouse germline by transgenesis retain, in most instances, their tissue-specific and stage-specific patterns of expression despite integration at random sites.

A popular strategy for characterization of the regulatory sequences is to design a series of transgenes in which the sequence encodes an easy to detect product that is not normally encoded in a mammalian genome (a reporter gene) and to associate it with a variety of DNA sequences that are suggested to be regulatory. These suggested regulatory sequences can be put either upstream of the coding region, at the 5′ end of the coding sequence or, less frequently, downstream of its 3′ end. Among the many genes for which the coding sequence has been used as a reporter, the most popular are the *lacZ* gene from *Escherichia coli* that encodes β-galactosidase (Goring *et al.*, 1987); the *CAT* gene also from *E. coli* that encodes the chloramphenicol acetyltransferase (Overbeek *et al.*, 1985) and the luciferase gene from the firefly (*Photinus pyralis*) (Lira *et al.*, 1990, 1992). Each of these reporter genes has been extensively used and each has specific advantages. The *lacZ* gene is particularly useful for studies of tissue- or position-specific gene expression (Fig. 9.10). The CAT and the luciferase genes are useful because the assays to detect them are easy, sensitive, and quanti-

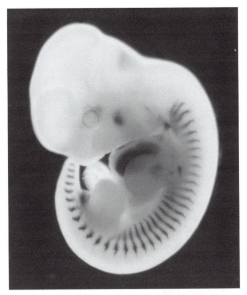

Fig. 9.10. Expression of *lacZ* driven by a muscle-specific promoter (Desmin) (courtesy C. Babinet).

tative. Another reporter gene has been developed more recently that consists of the sequence of the green fluorescent protein (GFP) of the jellyfish *Aequora victoria*. The product of this gene is fluorescent and the analysis of the expression pattern requires no co-factor or specific substrate, only UV light. Transgenic mice with a reporter gene have been extensively used by developmental geneticists and have greatly contributed to the annotation of the non-coding sequences.

The use of transgenic mice to generate tissue- or cell-specific ablations. Transgenes have been designed using tissue-specific regulatory sequences, associated with sequences encoding cytotoxic proteins, to programme the genetic ablation of specific cell types either in the developing embryo or in the adult (Breitman *et al.*, 1989). The most common strategy makes use of sequences encoding toxic proteins such as the A chains of the diptheria toxin (DT-A) or of ricin (R-A), which both block protein synthesis. In this case, the cell death takes place as soon as the transgene is transcribed. These studies indicate that programmed ablation of specific cell types can be stably transmitted through the germline (Breitman *et al.*, 1987).

Another strategy has been developed which relies on the induced intracellular expression of the enzyme thymidine kinase of the herpes simplex virus. This enzyme is not directly toxic to animal cells but, unlike the mammalian thymidine kinase, it can phosphorylate certain nucleoside analogues, such as *acyclovir* or *gancyclovir*, converting them into toxic molecules in dividing cells. In this particular case, the cell death becomes conditional since it depends on both the expression of the gene coding for viral thymidine kinase and the administration of the nucleoside analogues.

These methods of genetic ablation can be used to confirm the tissue specificity of a promoter and, from this point of view, they appear complementary to the methods described above. Unfortunately these methods, particularly the method using the highly toxic DT-A or R-A toxins as cell killing agents, have a major drawback in the sense that eukaryotic cells are extremely sensitive to these toxins. If the regulatory elements used in the transgenic construction are not specific enough or are 'leaky', a background expression of the transgene in cells that are not targeted may result in misleading pathological conditions. Alternative strategies of cell- or tissue-specific ablation based on more specific approaches have been designed (see further, section on engineering conditional knockout mutations in ES cells).

The validation by transgenic complementation of a candidate gene suggested by positional cloning of a mutant allele. As we already mentioned in the previous section, positional cloning of mouse mutations is an efficient strategy to assess the function of genes. However, when there are only two alleles at a given locus, one normal and one mutant, with the mutant being the consequence of a missense mutation (about 60% of the cases), it is sometimes difficult to conclude unambiguously what the gene function is. In this case it is generally necessary to prove that the candidate gene is indeed the 'good' one, either by generating other alleles by mutagenesis or by attempting the rescue of a mutant genotype by transgenic complementation. In this latter case, an appropriate breeding protocol is used to obtain genotypes that are homozygous for the recessive mutation

(*mut/mut*) with an additional normal transgenic copy of the candidate gene ($Tg+^{mut}$). The observation of a normal or nearly normal phenotype for this genotype validates the candidacy (see above, section on positional cloning of the gene mutated in obese mice, where the *pmn* mutation was complemented by a normal allele of *Tbce*).

Engineering transgenic mice for modelling human diseases. Different types of transgenic mice have been designed either to allow scientists to perform experiments that were not possible with normal mice or to model a pathological condition existing in humans and not in mice. A few examples demonstrate the great versatility of this technology.

Transgenic mice susceptible to human infectious diseases. The poliovirus, the causative agent of poliomyelitis, normally only infects primates and cannot infect mice except for type 2 virulent strains. Transgenic animals susceptible to all three poliovirus serotypes have been produced by injecting *in ovo*, the human gene encoding the cellular receptor for the virus (Koike *et al.*, 1991). These transgenic mice, when inoculated, mimic clinical symptoms observed in humans and monkeys and thus represent good models to study the molecular mechanisms of pathogenesis of the poliovirus as well as for testing vaccines against poliovirus infections.

The bacteria *Listeria monocytogenes*, once ingested by humans, can produce severe and sometimes fatal infections. The mechanisms by which the bacteria passes through the intestinal barrier are not totally understood but we know that a surface protein of *Listeria*, called *internalin*, interacts with a host receptor, E-cadherin, to promote entry into intestinal epithelial cells. Murine E-cadherin, in contrast to that in human or guinea pig, does not interact with internalin, excluding the mouse as a model for experimental oral infection with *L. monocytogenes*. In contrast, in transgenic mice expressing human E-cadherin, internalin was found to mediate invasion of enterocytes and crossing of the intestinal barrier (Lecuit *et al.*, 2001). These results illustrate well the value of transgenesis for the understanding of the physiopathology of infections in human.

Transgenic models of human genetic diseases. A mouse model of the human disease osteogenesis imperfecta type II has been produced by injecting *in ovo*, an abnormal mouse pro-α1 (I) collagen gene, homologous to the abnormal human gene (Stacey *et al.*, 1988; Pereira *et al.*, 1993). Soon after birth the animals carrying such a transgene appeared very sick, because of the modification of the extracellular matrix by the abnormal collagen fibres. In this case, with the transgene having a dominant deleterious effect, the mouse strain was difficult to propagate, and accordingly the model was of limited value. Nowadays much better models can be generated using modern techniques of transgenesis as described further.

A transgenic strain has been successfully produced by inserting in the mouse genome both the normal human α-globin gene and the abnormal β^s-globin gene characteristic of sickle cell anaemia (Ryan *et al.*, 1990). These animals were then bred to β-thalassaemic mice to reduce endogenous mouse globin levels. When erythrocytes from these mice were deoxygenated, greater than 90% of the cells displayed the same characteristic sickled shapes as erythrocytes from humans with sickle cell anaemia. Furthermore, when compared with controls, the mice exhibited a series of symptoms that are commonly associated with human sickle cell anaemia. Such models are of great help in the understanding of the pathophysiology of this debilitating disease as well as in the development of new drugs and therapies.

Models resulting from the introduction of large DNA fragments into the germline. Several techniques have been used to make mice transgenic for large DNA fragments. Among these techniques, direct microinjection of purified yeast artificial chromosomes (YACs) or bacterial artificial chromosomes (BACs) *in ovo* appears to be the most popular (Jakobovits *et al.*, 1993; Schedl *et al.*, 1993; Lee and Jaenisch, 1996). Such transgenic mice, when available, are helpful. Several examples documenting the ability of wild-type alleles carried in YACs to complement mutations have been reported. The first one was the rescue of the classical mouse albino mutation after introducing into albino (Tyr^c/Tyr^c) mice a

250 kb YAC covering the mouse tyrosinase (*Tyr*) gene with all its introns and 155 kb of the 5'-flanking region (Schedl *et al.*, 1992).

Original animal models of human genetic diseases have also been established using YAC transgenes. Among these, a model for the Charcot–Marie–Tooth disease type 1A (Huxley *et al.*, 1996), a model for Down syndrome where a cloned fragment of mouse chromosome 16 was added to the mouse genome producing a phenotype somewhat similar to trisomy 21 in humans (Smith *et al.*, 1995), and a model for Huntington disease (Hodgson *et al.*, 1996; Mangiarini *et al.*, 1996).

Transgenesis with BACs or other large chromosomal segments is bound to become a popular technology with the foreseeable development of quantitative genetics in the years to come. This is because, unlike in the case of single Mendelian mutations, the genetic alterations at the genome level that can have a quantitative effect on the phenotype are mostly unknown. Accordingly, BACs containing the DNA segment where a QTL has been identified can be transfected into zygotes and the resulting mice tested for the quantitative trait in question. However, for this system to be applicable, BAC libraries that contain the appropriate alternative alleles at the QTL in question should be available (Heintz, 2001; Biola *et al.*, 2003). Such libraries are now being prepared for different strains of mice and rats.

The use of transgenic technology for the induction of mutations. As mentioned earlier, approximately 10% of transgenic lines carry phenotypically visible insertional mutations. Theoretically, this appears to be an interesting situation considering that the inserted DNA 'tags' the mutated gene and accordingly may allow it to be cloned and characterized (for reviews see Jaenisch, 1988; Gridley *et al.*, 1990; Meisler, 1992). In practice, however, the cloning of DNA flanking the insertion loci has proven rather difficult. The complementary approaches of enhancer trapping and promoter trapping (Gossler *et al.*, 1989; Friedrich and Soriano, 1991; Skarnes *et al.*, 1992), which are alternative strategies used to identify new genes on the basis of the expression patterns of inserted transgenes in ES cells, are discussed further.

The use of transgenic mice encoding antisense RNA to knock-down gene expression. Examples of transgenic mice in which the transgene consists of an antisense cDNA are not very numerous. At least one case has been reported for the gene encoding the myelin basic proteins (MBPs). Several offspring of such a transgenic mouse exhibited a phenotype characteristic of the mutant '*shiverer*' (*Mbp^shi^*) in which the MBPs are defective. In these mice, antisense mRNA was transcribed and was associated with reduced myelination in the central nervous system (Katsuki *et al.*, 1988).

Inducing alterations at the genome level through the *in vitro* manipulation of embryo-derived stem cells

The advent of the technology for introducing extrinsic DNA fragments into the mouse genome has had a profound impact in many areas of biology, in particular for sequence annotation. Unfortunately, the *in ovo* microinjection of an exogenous DNA can add to, but not subtract from or substitute for, genetic material in a directed or targeted manner. This means that, except in a few cases, it is not possible to produce recessive alterations by this procedure. Another drawback is due to the fact that the injected DNA sequence, as a rule, inserts randomly in the genome of the host and, in most cases, in the form of several copies associated in tandem. Such limitations do not apply to other techniques for germline manipulation using ES cells, which have become extremely popular over these last years. Techniques making use of ES cells were developed at the beginning of the 1980s (Evans and Kaufman, 1981; Martin *et al.*, 1981) and have now reached a high level of sophistication.

ES cells and their advantages

ES cells are derived from the inner cell mass of blastocysts. They are cultured *in vitro*, generally on feeder layers of fibroblasts, in tissue culture media supplemented with fetal calf serum, a high concentration of glucose, glutamine and β-mercaptoethanol. To prevent them from differentiating, low concentrations of leukaemia inhibitory factor (LIF) or ciliary neurotrophic factor (CNTF) are added into the culture medium and the cells are replated at a relatively rapid pace. These embryonic cells represent a material of choice for geneticists because they can be manipulated like ordinary somatic cells as long as they are *in vitro* and retain their developmental potential when injected into the cavity of a blastocyst. Provided that they stay karyotypically normal and that they have not been maintained *in vitro* for too long, ES cells are capable of participating in the formation of the germ cell lineage of chimerical embryos (Bradley *et al.*, 1984). It is then possible to use the techniques of somatic cell genetics to isolate ES cell clones with a particular genotype and then 'recycle' them back into the germline. The first experiments of genetic engineering with ES cells were achieved by Gossler and co-workers (1986) and by Robertson and co-workers (1986). These were ground breaking experiments!

In spite of the many attempts performed in several laboratories and for several species, so far ES cells have been isolated only from mouse embryos and from a limited number of inbred strains indicating that the derivation of such cell lines depends, at least in part, upon the genetic constitution of the embryonic cells themselves.

Targeted mutagenesis in ES cells

Since ES cells behave like somatic cells when cultured *in vitro* it is possible, in principle, to induce mutations in their genome then select the mutant cells generated, provided they have acquired a selective advantage over the normal cells when grown in specific culture conditions. It is also possible to transfect these cells with foreign DNA molecules and to select the transfected cells based on the same principle, i.e. when a selective advantage has been conferred to the cells by the transfected heterologous DNA. All these techniques have been used to induce a variety of genomic alterations in ES cells.

A first mouse model of the Lesch–Nyhan syndrome. The Lesch–Nyhan syndrome (OMIM 308000) is a rare but severe X-linked condition in humans. The primary metabolic defect is characterized by the absence or inactivity of the enzyme hypoxanthine phosphoribosyl transferase (HPRT), an essential enzyme for the catabolism of purines. No model for this disease was available until the mid-1980s when Hooper *et al.* (1987) isolated *Hprt⁻* clones of ES cells resulting from spontaneous null mutations occurring in the gene. This was achieved *in vitro* by positive selection of the *Hprt* locus in male hemizygous ES cells, allowing the immediate detection of the rare *Hprt⁻/Y* mutant cells occurring by chance in the colony. A selective advantage was conferred because mutant cells would become insensitive to the poisonous effect of the purine analogue 6-thioguanine added to the culture medium. These investigators found several such cell clones, bred mice from them and succeeded in establishing an *Hprt⁻* strain. To their surprise, however, the mice appeared completely normal and did not exhibit any symptoms reminiscent of the human Lesch–Nyhan syndrome. This was the first mutation generated *in vitro*, in ES cells. Furthermore this was another example that a mouse can in no way be considered *a priori* as a 'human in reduction' and that what is to be learnt from experiments with mice must be considered only as an indication for what may occur in humans.

Retroviruses as mutagenic agents in ES cells: another model for the Lesch–Nyhan syndrome. The isolation of clones of mutant ES cells by *in vitro* selection of a particular genotype has proven successful, although with a very low yield in the case of *Hprt*. Unfortunately this strategy has a relatively narrow range of applications because most mammalian genes have two alleles at each locus and not only one as in the case of X-linked genes in males. In addition, for most genes there is no efficient possibility to select for mutant alleles *in vitro*. Another drawback is the very low rate of spontaneous mutations occurring *in vitro*. Obviously other strategies had to be developed to make the use of ES cells more efficient. The first step in a series of technical refinements came from the use of retroviruses as mutagenic agents and was a consequence of the early observations by Jaenisch and colleagues that we reported earlier. Two major

conclusions of these pioneering experiments were: (i) retroviral vectors can be used as mutagenic agents in mammalian embryonic cells and (ii) they insert into the genome almost at random, without generating extensive chromosomal rearrangements. Based on these observations ES cells have been infected with the M-MuLV and mutant null alleles have been induced at the *Hprt* locus, but this time at higher frequency (Kuehn *et al.*, 1987).

Considering the relatively high efficiency of the technique in terms of proviral integrations, massive infections of ES cells have been performed and embryos homozygous for these potential mutations by insertion have been generated by appropriate crosses. An interesting example is a recessive lethal mutation that was found to be the result of a proviral insertion causing a loss of function of *Nodal*, a transforming growth factor-β (TGF-β)-related gene (Conlon *et al.*, 1994). Another mutation, generated by a similar proviral insertion, was found in our laboratory and affects the process of limb formation (Fig. 9.11).

In most instances retroviral insertions occur in non-coding regions and accordingly have no direct mutagenic effects. In some instances however they disrupt a gene either because they occur within an exon or because they occur in an intron and disorganize the splicing process of the transcript encoded in the neighbouring gene. Here again, the retroviral insertion can be used as a 'hook' or 'tag' to pull out DNA clones containing the flanking gene.

In vitro infection of ES cells with genetically engineered retroviruses has also been used as a way to produce transgenic animals. However the technique has been abandoned for the same reasons as for *in vitro* embryo infection, to the benefit of more efficient strategies.

A third model of the Lesch–Nyhan syndrome generated by targeted homologous recombination in ES cells. A major drawback of the use of retroviral vectors for the production of mutations in ES cells is that one cannot choose the site of integration of the retrovirus and accordingly the mutations generated are of different and unpredictable kinds. From this point of view, *homologous recombination* of extrinsic DNA molecules in ES cells has been another breakthrough.

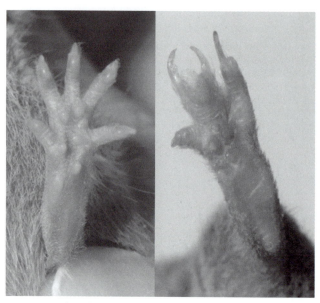

Fig. 9.11. The mouse mutation *dan* (*digitation anormale*) is the consequence of a proviral insertion after experimental infection of an ES cell line.

The principle for the production of mutations by homologous recombination is based on the observation that DNA molecules, once introduced into cells by an appropriate experimental procedure, for example by electroporation, can recombine with the chromosome(s) of these cells either at homologous or at non-homologous (illegitimate) sites. As a consequence, and provided that the transfected DNA molecules are adequately modified by genetic engineering *in vitro*, a homologous recombination event can disrupt and inactivate a gene. The idea that homologous recombination could occur in mammalian cells, and in particular in ES cells, originated from observations made in other eukaryotic organisms, in particular in the yeast *Saccharomyces cerevisiae*, where similar experiments had been successfully conducted. The intimate molecular mechanisms at work in the recombination process are not yet fully understood. It is likely that the mechanisms of homologous recombination overlap with those of illegitimate recombination but a number of experiments indicate that they are not identical (for a review refer to Hooper, 1992). Mammalian cells have the enzymatic 'equipment' necessary for homologous recombination to take place though it

occurs at a frequency much lower than random integration (Smithies *et al.*, 1985; Wong and Capecchi, 1986). At this point it should be noted that the idea to develop such a strategy was audacious if one compares the relatively small size of a cloned DNA that can be handled experimentally, with the gigantic dimensions of a mammalian genome. To increase the yield of homologous recombination, experimental data indicate that the DNA molecule transfected into the ES cell must be linear, as large as technically possible, up to 10 kb and more if possible, and should have the greatest possible length of sequence homology with the target DNA in the ES cell.

The first endogenous mouse gene that was modified by homologous recombination in ES cells was the *Hprt* locus with the aim to produce mutant alleles (Thomas and Capecchi, 1987). For the latter, the experiment consisted of three steps: in the first step a DNA molecule homologous to the *Hprt* targeted region containing a few exons, the intervening introns and some flanking sequence, was cloned. In the second step, one exon in this cloned *Hprt*-DNA molecule was substituted by a piece of DNA of roughly similar size from another origin. The idea underlying this manipulation was that, in the event of successful homologous recombination, the substitution of an exon by another segment of DNA would alter profoundly the coding sequence of the native *Hprt* gene making it unable to transcribe a correct mRNA. Synthesizing this altered DNA molecule *in vitro*, instead of using a mere segment of DNA with no coding capacities as a foreign sequence, the scientists had the clever idea to use a minigene of bacterial origin encoding *neomycin phosphotransferase* (*neo^r*) which can confer neomycin resistance to transfected cells. Finally, in a third step, the ES cells were transfected with these engineered DNA molecules and grown in a culture medium with the antibiotic *neomycin* or, more precisely, with one of its amino glycoside analogues, G418. This drug kills normal mammalian cells but cells synthesizing neomycin phosphotransferase are resistant. In other words, with such culture conditions, only those ES cells that have an engineered DNA molecule integrated in their chromosomes (at a homologous site or elsewhere) could grow. Those rare ES cells clones where a homologous recombination occurred would have reciprocally exchanged a functional copy of the *Hprt* gene for a non-functional one and accordingly would also have acquired the property to resist the toxic effects of thioguanine just like the *Hprt$^-$* mutant cells generated by retroviral infection (Fig. 9.12). The advantages of the technique are twofold. The first is that, after two successive rounds of selection, one with G418, the other one with thioguanine, only clones of ES cells that have undergone homologous recombination at the *Hprt* locus would survive *in vitro*. In other words only the cells where the gene actually targeted has been effectively inactivated, or 'knocked out', would survive. The second advantage is that the mutation frequency by homologous recombination is higher than with any other technique. In the case reported above, for example, one stably transfected ES cell clone out of 150 was found to be a knockout (Capecchi, 1989). This frequency of homologous recombination was considered high enough to adapt the technique to all cases where it was suitable to generate a null allele even if the selection for homologous, as opposed to non-homologous recombination, could not be achieved by the same straight, *in*

vitro selection as in the case we just reported for *Hprt$^-$* cells. Since these early experiments, hundreds of genes have been invalidated using this strategy. Genes invalidated by homologous recombination in ES cells are now collectively designated by the name 'knockout'. The *in vitro* engineered DNA molecule that is used for targeting the homologous native counterpart in the chromosome of the ES cells is designated '*recombination or targeting vector*'. Nowadays, in all knockout experiments, the expected homologous recombination event is confirmed in the manipulated ES cells by PCR amplification before the cells in question are placed into a recipient blastocyst for the production of a chimeric embryo.

Generating a great variety of knockout alleles in ES cells, by homologous recombination.

Most of the knockout mutations that have been generated in mouse ES cells over recent years have resulted from the use of targeting vectors as described above. In this case, after homologous recombination, one of the specific coding sequences of the gene is deleted and replaced by a heterologous DNA that is itself, in most cases, a 'selection cassette'. As a consequence of this substitution the gene is inactivated and, at the same time, the ES cells acquire a selective advantage over a drug and can survive to be positively selected. Several variations on this basic scenario have been used and it is almost impossible to describe them all. However, we can say that selection cassettes making use of bacterial genes encoding resistance to *hygromycin B* or *puromycin* have been used as alternatives to the *neo^r* cassette.

The design of the selection cassettes in the replacement vectors for homologous recombination depends on the nature of the gene that is targeted. If the gene in question is transcriptionally active in the ES cells then the selection cassette is transcribed and positive selection with a drug can operate. If however the gene in question is not expressed in the ES cells or if its expression pattern is unknown, it is then necessary to design a vector that incorporates the necessary regulatory elements.

Replacement vectors allowing positive/negative selection have also been designed by inserting a *neo^r* minigene between two regions of high homology plus, outside of these homol-

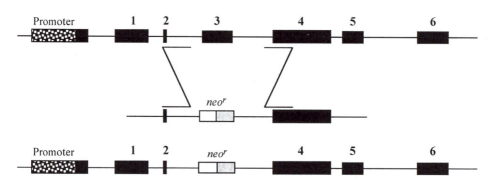

Fig. 9.12. Gene targeting with a *replacement vector*. Recombination events occurring in the regions flanking the *neo^r* cassette result in the deletion of exon 3. The *neo^r* cassette confers a selective advantage to the recombined ES cells.

ogous regions, a gene encoding herpes simplex virus thymidine kinase (HSV*tk*). ES cells that are transfected *in vitro* with such a replacement vector can subsequently be submitted to a double selection: (i) the first one with G418, inducing the destruction of all ES cells that had not integrated at least one copy of the vector; and (ii) the second one with the guanosine analogue gancyclovir that kills all cells where non-homologous events of recombination occurred. The second level of selection is possible because when non-homologous recombination occurs, the HSV*tk* component of the vector is retained while it is deleted after a homologous recombination event (Fig. 9.13).

Aside from replacement vectors, *sequence insertion vectors* have also been used for targeting genes in ES cells (Thomas and Capecchi, 1987). These vectors are engineered

from a cloned gene sequence where a single region of high homology is adjacent to a selectable marker. The plasmid that is used for the cloning process is then linearized by cutting with a restriction enzyme, at a single site, within the homologous region, and recombination occurs with the free homologous extremities of the cloned molecule. The insertion of the vector into the genome of the ES cells has two consequences: (i) as in the case of replacement vectors, it confers to the transfected cells the capacity to resist the toxic effects of a drug added to the culture medium for positive selection; and (ii) it generates a tandem duplication of the same size as the homologous region. Together with the intrusion of some plasmid DNA, insertion of the vector hampers the transcription of a correct mRNA molecule. The advantages of this sort of insertion vector over

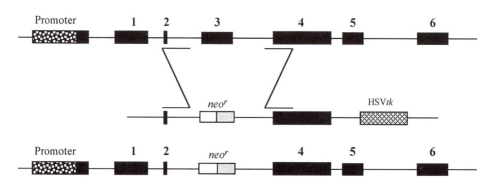

Fig. 9.13. Gene targeting with a *replacement vector* engineered with a positive/negative selection cassette. After homologous recombination, the HSV*tk* cassette is deleted while the *neo^r* cassette replaces exon 3. This confers to the recombined ES cells a selective advantage to G418 and a selective disadvantage to gancyclovir.

the replacement vectors are that homologous recombination events are more frequent, maybe because they require only a single crossover and not two as in the case of replacement vectors (Fig. 9.14).

The techniques for gene invalidation we just reported have been described in detail in several review papers or book chapters (Hooper, 1992; Hasty *et al.*, 2000; Plagge *et al.*, 2000; Babinet and Cohen-Tannoudji, 2001; DeChiara, 2001). The use of replacement vectors with a selectable marker has been and still is very popular for the generation of null alleles because it is relatively straightforward and produces stable, permanent alterations. Unfortunately cases have been reported where the selection cassette alters to some extent the expression of neighbouring genes. Sequence insertion vectors produce null alleles in most cases but the mutation is less stable, with occasional spontaneous reversions to the wild-type resulting from intrachromosomal recombination events (gene conversion). Mutations of that kind sometimes are also leaky because alternative splicing in the duplicated region may skip the extra exon(s) and the cloned sequence.

Generating point mutations in ES cells, by homologous recombination. Strategies making use of replacement or sequence insertion vec-

tors have in common the introduction of extrinsic DNA sequences of various sizes into the ES cells. Although the adverse effects of this contamination of the ES cell genome are mostly unknown and thus totally neglected, they may, none the less, represent a potential drawback. This is why scientists have developed a strategy in two steps that can lead to the introduction of very precise, punctual mutations, such as single base pair changes (missenses or nonsense) in exonic sequences. This strategy is based on a double replacement with positive and negative selection and makes use of *Hprt*⁻ ES cells similar to those cells resulting from the experiments reported above (Hooper *et al.*, 1987). The first replacement vector replaces an exon by a normal, functional (mini)copy of the *Hprt* gene by homologous recombination in the targeted sequence. After this first replacement, ES cells that have undergone homologous recombination are no longer resistant to the poisonous effect of 6-thioguanine but can grow normally in the so-called Littlefield's HAT (hypoxanthine, aminopterin and thymidine) culture medium because the cells have acquired a functional copy of the *Hprt* gene. An *alteration vector* is then used that corresponds perfectly to the sequence of the targeted gene with only one single base pair difference in the targeted exon. The critical part of this vector is synthesized *in vitro*, using a PCR technique of directed muta-

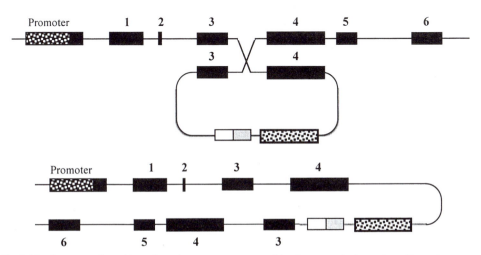

Fig. 9.14. Gene targeting with an *insertion vector*: one recombination event in the region of homology generates a duplication of exon 3 and 4 of the gene, with insertion of two selection cassettes one of them being a *neo*ʳ cassette. These cassettes confer selective advantages to the recombined cells.

genesis that is now routine in most laboratories. After this second replacement, homologously recombined ES cells are killed in HAT medium but survive selection by 6-thioguanine as did the original cells (Stacey *et al.*, 1994) (Fig. 9.15). Finally, and although it has no deleterious effects in the mouse, if the *Hprt⁻* mutation is considered undesirable it can be eliminated by two rounds of sexual reproduction once the offspring of the mutated ES cells are born. An alternative technique using an insertion vector instead of replacement vector, and known as the 'hit-and-run' or 'in-and-out' technique, has also been used for the generation of point mutations in targeted genes (Hasty *et al.*, 1991; Valancius and Smithies, 1991). This technique requires two rare intrachromosomal recombination events to occur and accordingly is less efficient than the technique making use of replacement vectors. Whatever the technique used for the induction of point mutations (replacement or insertion vectors), it must be kept in mind that mice of that kind are not transgenic animals *sensu stricto* because they do not have any extrinsic DNA sequences inserted into their genome.

Knockins are sophisticated knockout mutations. An interesting variation of the technique used for obtaining knockout null mutations has been designed by introducing, via the replacement vector, the coding sequence of reporter genes in-frame with the promoter of the targeted gene. To give an example of the high degree of sophistication for this type of approach, one could refer to an experiment designed to assess the function of the many genes encoding connexins (Filippov *et al.*, 2003). Connexins are expressed in the diverse cell types of the central nervous system and are thought to regulate some of the functional properties exhibited by immature and mature cells. Understanding of the role of specific connexins in these processes required an unambiguous characterization of their spatial and temporal pattern of expression. To reach this aim with connexin 26 (symbol *Gjb2*: for gap junction membrane channel protein beta 2) the scientists generated a reporter allele (*Gjb2^{lacZ}*) by manipulating the gene in such a way that the gene coding for β-galactosidase was expressed from the endogenous *Gjb2* promoter. This resulted in embryonic lethality for the

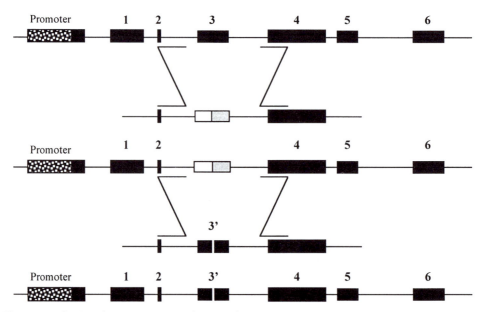

Fig. 9.15. Induction of point mutations with two *replacement vectors* in *Hprt⁻* ES cells. The first replacement vector substitutes a *Hprt* cassette for exon 3 and confers resistance to HAT. The second recombination replaces the *Hprt* cassette by a mutated exon 3 (exon 3′) engineered *in vitro*. The ES cells then become sensitive to HAT but insensitive to 6-thioguanine.

Gjb2^{lacZ}/Gjb2^{lacZ} knockout animals, indicating that an intact copy of the *Gjb2* gene is necessary for embryonic development. However in addition to this information, by using heterozygous mice *Gjb2^{lacZ}/+* the authors could label the tissues known to contain Cx26 (i.e. liver, kidney, skin, cochlea, small intestine, placenta and thyroid gland) and finally demonstrated that both in embryonic and adult brain the expression of *Gjb2* is restricted to the meninges. Such a knockout mutation, where a gene is invalidated by a coding sequence using the same promoter, is designated 'knockin'. The knockin strategy is universal and can be applied to any unknown gene to invalidate it and, at the same time, visualize its expression pattern in the developing embryo or in the adult.

Engineering conditional knockout mutations in ES cells. Depending on their allelic interactions (dominant or recessive) and their homozygous or heterozygous status, mutations induced in ES cells by retroviral insertion or by targeting vectors may affect all the cells of the developing embryo from the earliest stages of development. This precludes the analysis of gene function(s) in later developmental stages or in the adult. It is also a drawback when the targeted gene has a wide expression pattern because, in this case, the analysis of the function(s) of the gene is too complex and sometimes impossible. To bypass these problems, gene targeting strategies have been developed that allow the induction of conditional knockout mutations ('cko' mice). Accordingly, the time of gene expression (the gene is 'switched off' when necessary) or the tissue in which the gene is expressed (the gene is 'switched off' in some cells or tissues but not in others) can be controlled. This strategy, in general, makes use of two mouse strains; one in which a gene of interest is engineered in a specific manner to become a target for inactivation and the other where the time- or tissue-specific expression of the mutation is predetermined. This strategy is known as the Cre-*loxP* system. Discussion of this system will be based on the example of tissue-specific inactivation of *Polb*, the gene encoding (DNA directed) β-polymerase (Gu *et al.*, 1994) in T lymphocytes. Firstly a strain of mice (strain A) in which the normal *Polb* allele was specifically modified by homologous recombination with a replacement vector was generated. The replacement vector was designed in such a way that an essential sequence of the targeted gene, actually its promoter and first exon, became flanked by two specific short sequences known as *loxP* sites (Fig. 9.16), after the recombination event.

These *loxP* sites (short for *locus* of X-ing over P1) each consist of two 13 bp inverted (palindromic) repeats separated by an 8 bp asymmetric spacer region: such a sequence does not normally exist in the mouse genome (Fig. 9.17). The replacement vector was also designed in such a way that it could allow positive/negative selection with a positive selection cassette that is also inside the region flanked by the *loxP* sites. After homologous recombination, the targeted gene *Polb* ended up with two *loxP* sites inserted in the same direct orientation; the first one upstream of its promoter, the other in intron 1. The gene was 'floxed', as specialists would say but, at this point it was still functional and normally transcribed. Nothing changed in the functional genomics of

Fig. 9.16. A *loxP* site (top) consists of two 13 bp palindromic sequences (arrowed) flanking a 8 bp spacer region (boxed). These 8 bp define the directionality of the site. When two *loxP* sites are placed on the same strand and in the same orientation the Cre recombinase deletes the intervening sequence plus one *loxP* site. When the sites are in opposite orientations Cre generates an inversion of the intervening sequence and both *loxP* sites are retained. An *Frt* site (below). The sequence of the site is longer (47 bp) and asymmetrical.

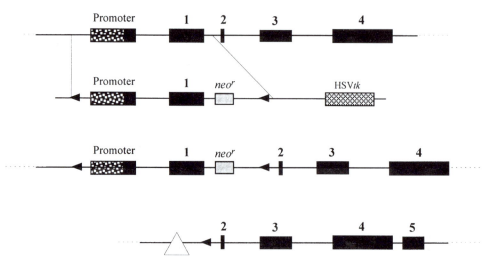

Fig. 9.17. Inducing gene-targeted deletions with the Cre-_loxP_ system. First, a positive/negative selection is used to insert two _loxP_ sites and a positive selection cassette inside the targeted region. At this stage the altered sequence is still functional. When Cre is synthesized, the segment flanked by the two _loxP_ sites is deleted and the cells lose the positive selection cassette.

the region, the mutation was cryptic, or 'premeditated' so to say.

Concurrently another strain of mice (strain B) was produced, which was transgenic for a gene encoding Cre (short for cyclization recombinase), a 38 kDa recombinase protein from bacteriophage P1 that mediates intramolecular or intermolecular site-specific recombination between _loxP_ sites (Sauer, 1993). The Cre encoding transgene in this case was driven by a lymphocyte creatine kinase (_lck_) promoter, specific for T cells. When strain A and strain B were intercrossed, the product of the Cre transgene triggered deletion of the floxed segment, in one or both chromosomes according to the genetic constitution of strain A, but in T cells only. The consequences of the _Polb⁻_ mutation on T cells could then be analysed because mutant mice were viable, while they would have died if the mutation was expressed ubiquitously during development.

Many similar experiments leading to tissue- or cell-specific gene inactivation, have been performed over the last 10 years using either the Cre-_loxP_ system or a similar system known as FLP-_Frt_. The FLP-_Frt_ system makes use of a yeast recombinase with its specific restriction site. With these systems an unlimited number of mutations may be designed keeping in mind that Cre (or FLP) deletes any DNA segment

once it is flanked by two _loxP_ (or _Frt_) sites, provided these sites are oriented the same way. When they are oriented in opposite direction, then Cre generates an inversion of the 'floxed' segment. When the _loxP_ sites are not adjacent, say on two different chromosomes, Cre mediates reciprocal translocations between chromosomes. When these _loxP_ sites are more than two in the same cell, Cre would cut at each site and, under specific conditions, generate a variety of deletions or translocations. Selection can then be applied to retain one cell type and not the others if selection cassettes have been inserted in critical regions.

A similar strategy has been employed using the same strain A (with floxed _Polb_) and another strain (strain C) with the interferon-inducible promoter of the gene _Mx1_ to regulate Cre expression using the _lck_ promoter to drive the expression of Cre in T lymphocytes. After crossing strains A × C, _Polb_ inactivation was induced in adult animals after interferon treatment. In this case inactivation was complete in liver and spleen whilst it was incomplete in other tissues (Kuhn _et al._, 1995).

The Cre-_loxP_ strategy can also be used to regulate the expression of specific genes and proteins in a tissue- or cell-specific way (Fig. 9.18). In this case the _lacZ_ gene is driven by a ubiquitous promoter with a floxed 'stop'

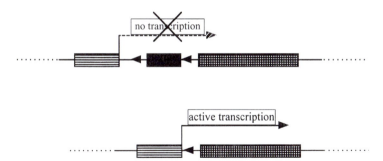

Fig. 9.18. Cre-*lox*P regulation of transcription. A floxed 'stop' sequence (black box) hampers transcription of the gene down stream. When this sequence is deleted transcription resumes.

sequence inserted between the promoter and the *lacZ* coding sequence proper. The 'stop' sequence is a short segment of DNA with several terminator codons that prevents translation of protein. When the floxed 'stop' sequence is deleted by the action of Cre in some specific cells or tissues, then *lacZ* protein is transcribed with the same pattern of cell/tissue specificity (Lakso *et al.*, 1992; Pichel *et al.*, 1993).

Since experiments of conditional targeting all entail the use of mouse strains that synthesize Cre, a specific database of the Cre strains was established (http://www.mshri.on.ca/nagy/cre.htm) to make them freely available.

Induction of mutations with conditional expression: the ultimate refinement of the 'Tet-Off' and 'Tet-On' expression systems. The Cre-*lox*P or FLP-*Frt* strategies allow the conditional induction of gene knockout. With these strategies virtually any gene can be inactivated in a specific tissue or cell type. However once the Cre recombinase has deleted a floxed DNA segment the situation is irreversible as the gene is permanently inactivated (or activated) in all daughter cells. Accordingly this may represent a drawback in experiments where only a transient inactivation (or activation) would be desired. In the same way it may be beneficial in some experiments with transgenic mice to have a transgene expressed only during a certain period but switched off the rest of the time. Unfortunately this is not possible using classical techniques. The 'Tet-Off' and 'Tet-On' inducible expression systems overcame these problems and put the transcription of a given gene under the total control of the experimenter.

In this system the expression of a transgene is dependent on a tetracycline-controlled trans-activator protein and can be regulated, both reversibly and quantitatively, by exposing the transgenic mice to *tetracycline* (Tc) or to one of its derivatives such as *doxycline* (Dox).

In the *Tet-Off* expression system, the tetracycline-controlled trans-activator (tTA) protein is composed of the *Tet* repressor DNA-binding protein (TetR), from the tetracycline resistance operon (*tet*) of *E. coli* transposon *Tn*10 fused to the strong transactivating domain of herpes simplex virus protein VP16. tTA induces transcription of a target gene that is under transcriptional control of a tetracycline-responsive promoter element (TRE). The TRE promoter is composed of a concatemer of seven *tet* operators (tetO) fused to the minimal promoter sequences of the human cytomegalovirus immediate to early gene 1 (hCMV IE1) promoter/enhancer. In the absence of Tc or Dox, tTA binds to the TRE and activates transcription of the target gene. This induction returns to basal levels upon administration of tetracycline (Fig. 9.19).

The *Tet-On* system works in exactly the opposite way. It is based on a reverse tetracycline controlled transactivator (rtTA) which is also a fusion protein made out of the TetR repressor and the VP16 transactivation domain. However, a four amino acid change in the TetR DNA-binding moiety alters rtTA's activity binding characteristics such that it can only recognize the tetO sequences in the TRE of the target transgene in the presence of the Dox effector. Thus in the *Tet-On* system, transcription of the TRE-regulated target is stimulated by rtTA only in the presence of *Dox*.

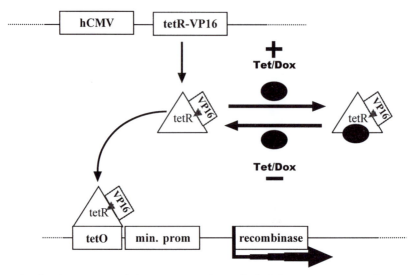

Fig. 9.19. The ubiquitous promoter hCMV triggers synthesis of a fusion protein tetR-VP16 where tetR is the tetracycline repressor and VP16 is the transcriptional activator domain (tTA). Binding of the fusion protein to the minimal promoter containing tetO elements activates transcription of the recombinase.

These *Tet-Off* and *Tet-On* systems can be used, for example, to design dominant gain-of-function experiments in which temporal control of transgene expression is required. In the absence of tetracycline, the transgene is expressed, while in its presence the transgene will not be expressed (Gossen and Bujard, 1992; Furth *et al.*, 1994).

Induction of mutations in ES cells by gene trapping. In an earlier section, we reported experiments where retroviruses were used as mutagenic agents in ES cells. The conclusions of these experiments were that retroviral vectors could be used as mutagenic agents, and when a mutation was induced, the retroviral insertion could be used as a tag to identify the mutated gene. Proviral tags are an advantage for positional cloning of mutated genes as soon as they are truly responsible for the observed mutation and are not closely linked to the mutated gene by chance. This unfortunately is not always easy to achieve especially when there are several such proviral insertions in the genome.

In order to improve the efficiency of recovering mutations that are likely to have phenotypic effects, the so-called gene trap approach was devised (Evans *et al.*, 1997). The principle of

this strategy is to transfect ES cells with transgenic vectors that lack some component needed for gene expression. Basically there are three types of vectors; the *enhancer*, the *promoter* and the *gene trap* vectors. The enhancer trap vectors contain a minimal promoter that requires the vector to insert near to a *cis*-acting enhancer element to produce expression of the reporter or marker gene (in general *lacZ*). Promoter-trap vectors are designed to be expressed only when they insert downstream of a promoter region and mutagenesis occurs when, for example, the transgene inserts into an exon. The gene-trap vectors contain a splice acceptor site immediately upstream of a promoterless reporter. In this case activation of transcription leads to the synthesis of a fusion transcript from the coding sequences of the trapped gene and the vector sequences. Other approaches have used a marker gene coupled to a suitable promoter but lacking a downstream polyadenylation signal. Here the marker gene is designed to be expressed after integrating into a host cell gene such that a fusion RNA product is made that utilizes 3′ host sequences in order to acquire a poly(A) tail. Of the different types of vectors designed for gene trapping, the gene-trap vectors appear to be the most efficient of all. Several groups have developed large collections of independently produced gene-trapped ES cell lines which have been made available to the community (Cecconi and Meyer, 2000; Hansen *et al.*, 2003; Stryke *et al.*, 2003). For an exhaustive review on this subject refer to Stanford *et al.* (2001).

Induction of mutations in ES cells with radiation or chemical mutagenesis. The induction of mutations in the mouse germline, with radiation or chemical mutagens, is an efficient way to annotate mammalian genes. However, a major drawback of this strategy is the high cost of breeding and/or the time necessary to collect the new mutations. In addition, these mutations are usually scattered over the entire genome, and represent a mixture of different types. Genotype-based screenings of production of targeted mutations have been developed to by-pass these drawbacks (Coghill *et al.*, 2002; Zan *et al.*, 2003; Smits *et al.*, 2004). Unfortunately this is insufficient when several alleles at a given locus are desired. An option will be a massive production of new mutations in ES cells using either radiation or ENU. You and colleagues (1997) were the first to report a method which generates radiation-induced deletions spanning up to several centiMorgans at defined regions. These deletions could be transmitted through the germline and were very useful for the systematic functional analyses. More recently a genotype-based screen for ENU-induced mutations has been adapted (Chen *et al.*, 2000; Munroe *et al.*, 2000). In a series of experiments focused on two loci of importance for mouse early development, *Smad2* and *Smad4*, Vivian and colleagues (2002) demonstrated that chemical mutagenesis in mouse ES cells, associated with high-throughput mutation detection, allows for the rapid identification of mutations in non-selectable genes of interest.

Getting information concerning transgenic mouse strains

Information concerning the many transgenic strains available worldwide can be obtained from several websites. The most important are:

- The T-base database: (http://tbase.jax.org/);
- The Frontier in Bioscience database: (http://www.bioscience.org/knockout/knochome.htm);

- The Mouse Genome informatics database: (http://www.informatics.jax.org/).

A full list of URLs for the databases referring to the lists of Gene Trapped and their locations is provided in the review by Stanford and colleagues (2001).

Other Strategies of Gene Inactivation

One of the new and promising strategies makes use of RNA interference (RNAi), where 21–23 nt small interfering RNAs (or siRNAs) induce transient silencing of more than 90% of a particular gene in mammalian cells (Hammond *et al.*, 2001). This strategy is briefly mentioned here because it is modern and certainly represents a promising tool for investigations in functional genomics. However, it will not be detailed in this chapter because it does not result from an alteration at the DNA level.

Conclusions

The first and maybe the most important comment is that the 'bottom-up' (or forward genetics) approach should be used to complement the 'top-down' (or reverse genetics) approach rather than as an alternative to it. The mutations that appear spontaneously and those that are induced by mutagenic treatments represent an enormous diversity that will probably never be replaced by mutations engineered in ES cells *in vitro*, even if it is true that in some cases the same phenotype is observed (Jaubert *et al.*, 1999). Furthermore, even if it is technically possible to design an experiment *in vitro* to produce virtually any single amino acid replacement in an essential protein, we still lack knowledge of protein biochemistry, in particular protein shaping and folding, to be able to design a missense mutation that will result in a deleterious phenotype. On the contrary, analysing in detail the genetic polymorphisms that can be observed in natural populations, including the deleterious effects of some alleles, is a rich source of information which is offered to geneticists almost for free. Many knockout mutations have either a very severe phenotype and are incompatible with normal embryonic development or have no obvious phenotype in traditional laboratory conditions, and in both cases they teach very little about the function(s) of the inactivated gene. On the other hand all mutations, spontaneous or induced, have an obvious phenotype since they are identified by this characteristic.

Further to these considerations, if it is true that the bottom-up approach has some invaluable advantages, one must recognize the tremendous capacities of the modern molecular techniques for the production of gene alterations. However these techniques also have some drawbacks that are difficult to by-pass. Firstly, they are applicable only to the mouse since ES cell lines, as we said, do not exist in other species. Secondly, they are expensive, frequently time consuming and even if they appear straightforward on paper, in reality their yield is often low and sometimes very low. Finally, they require almost as much breeding space as the strategies of positional cloning for the development and maintenance of the different strains generated, increasing the cost of the experiments further. The development of new techniques for chromosome engineering, termed recombineering (Liu *et al.*, 2003; Muyrers *et al.*, 2004), which make use of homologous recombination mediated by the phage Red proteins should facilitate, accelerate and reduce the cost of the generation of knockout mutations and transgene constructs.

Similarly, new techniques to detect ENU-induced mutations in F_1 males from DNA and cryopreserved spermatozoa should allow the identification of mutations in specifically selected (not targeted) genes and in any species of interest.

A second comment follows from the very basic question we addressed at the beginning of this chapter concerning the means that are available to scientists to assess gene function(s). In this chapter we chose to discuss *in vivo* or *in vitro* strategies based on genome alterations that are either spontaneous or induced. In fact much more information can be gathered using other valid approaches. Sequence homologies between functionally related genes in other species, analysis of spatial or temporal distribution of transcripts (micro-arrays) and analysis of the interactions between translated proteins with co-expressed proteins encoded in the same genome (double hybrids) can collectively aid in gene annotation. Based on the most recent data, we estimate that functional information was available for no more than 30% of mammalian genes. It is likely that the situation will change dramatically in the forthcoming years. Geneticists will then be ready to tackle another formidable challenge: understanding the nature and importance of the innumerable interactions that occur between genes, inside the mammalian genome. No doubt this will require a lot of work.

References

Ashby, J., Gorelick, N.J. and Shelby, M.D. (1997) Mutation assays in male germ cells from transgenic mice: overview of study and conclusions. *Mutation Research* 388, 111–122.

Babinet, C. and Cohen-Tannoudji, M. (2001) Genome engineering via homologous recombination in mouse embryonic stem (ES) cells: an amazingly versatile tool for the study of mammalian biology. *Anais da Academia Brasileira de Ciências* 73, 365–383.

Beier, D.R. (2000) Sequence-based analysis of mutagenized mice. *Mammalian Genome* 11, 594–597.

Beutler, B., Hoebe, K., Du, X., Janssen, E., Georgel, P. and Tabeta, K. (2003) Lps2 and signal transduction in sepsis: at the intersection of host responses to bacteria and viruses. *Scandinavian Journal of Infectious Diseases* 35, 563–567.

Biola, O., Angel, J.M., Avner, P., Bachmanov, A.A., Belknap, J.K., Bennett, B., Blankenhorn, E.P., Blizard, D.A., Bolivar, V., Brockmann, G.A., Buck, K.J., Bureau, J.F., Casley, W.L., Chesler, E.J., Cheverud, J.M., Churchill, G.A., Cook, M., Crabbe, J.C., Crusio, W.E., Darvasi, A. *et al.* (2003) The nature and identification of quantitative trait loci: a community's view. *Nature Reviews of Genetics* 4, 911–916.

Bode, V.C. (1984) Ethylnitrosourea mutagenesis and the isolation of mutant alleles for specific genes located in the T region of mouse chromosome 17. *Genetics* 108, 457–470.

Bode, V.C., McDonald, J.D., Guénet, J.L. and Simon, D. (1988) *hph-1*: a mouse mutant with hereditary hyperphenylalaninemia induced by ethylnitrosourea mutagenesis. *Genetics* 118, 299–305.

Bradley, A., Evans, M., Kaufman, M.H. and Robertson, E. (1984) Formation of germline chimaeras from embryo-derived teratocarcinoma cell lines. *Nature* 309, 255–256.

Breitman, M.L., Clapoff, S., Rossant, J., Tsui, L.C., Glode, L.M., Maxwell, I.H. and Bernstein, A. (1987) Genetic ablation: targeted expression of a toxin gene causes microphthalmia in transgenic mice. *Science* 238, 1563–1565.

Breitman, M.L., Bryce, D.M., Giddens, E., Clapoff, S., Goring, D., Tsui, L.C., Klintworth, G.K. and Bernstein, A. (1989) Analysis of lens cell fate and eye morphogenesis in transgenic mice ablated for cells of the lens lineage. *Development* 106, 457–463.

Brinster, R.L., Chen, H.Y., Trumbauer, M., Senear, A.W., Warren, R. and Palmiter, R.D. (1981) Somatic expression of herpes thymidine kinase in mice following injection of a fusion gene into eggs. *Cell* 27, 223–231.

Brinster, R.L., Allen, J.M., Behringer, R.R., Gelinas, R.E. and Palmiter, R.D. (1988) Introns increase transcriptional efficiency in transgenic mice. *Proceedings of the National Academy of Sciences USA* 85, 836–840.

Bulfield, G., Siller, W.G., Wight, P.A. and Moore, K.J. (1984) X chromosome-linked muscular dystrophy (*mdx*) in the mouse. *Proceedings of the National Academy of Sciences USA* 81, 1189–1192.

Capecchi, M.R. (1989) Altering the genome by homologous recombination. *Science* 244, 1288–1292.

Cecconi, F. and Meyer, B.I. (2000) Gene trap: a way to identify novel genes and unravel their biological function. *FEBS Letters* 480, 63–71.

Chapman, V.M., Miller, D.R., Armstrong, D. and Caskey, C.T. (1989) Recovery of induced mutations for X chromosome-linked muscular dystrophy in mice. *Proceedings of the National Academy of Sciences USA* 86, 1292–1296.

Charles, D.J. and Pretsch, W. (1987) Linear dose–response relationship of erythrocyte enzyme-activity mutations in offspring of ethylnitrosourea-treated mice. *Mutation Research* 176, 81–91.

Chen, Y., Yee, D., Dains, K., Chatterjee, A., Cavalcoli, J., Schneider, E., Om, J., Woychik, R.P. and Magnuson, T. (2000) Genotype-based screen for ENU-induced mutations in mouse embryonic stem cells. *Nature Genetics* 24, 314–317.

Church, D.M., Stotler, C.J., Rutter, J.L., Murrell, J.R., Trofatter, J.A. and Buckler, A.J. (1994) Isolation of genes from complex sources of mammalian genomic DNA using exon amplification. *Nature Genetics* 6, 98–105.

Coghill, E.L., Hugill, A., Parkinson, N., Davison, C., Glenister, P., Clements, S., Hunter, J., Cox, R.D. and Brown, S.D. (2002) A gene-driven approach to the identification of ENU mutants in the mouse. *Nature Genetics* 30, 255–256.

Coleman, D.L. (1973) Effects of parabiosis of obese with diabetes and normal mice. *Diabetologia* 9, 294–298.

Conlon, F.L., Lyons, K.M., Takaesu, N., Barth, K.S., Kispert, A., Herrmann, B. and Robertson, E.J. (1994) A primary requirement for nodal in the formation and maintenance of the primitive streak in the mouse. *Development* 120, 1919–1928.

Costantini, F. and Lacy, E. (1981) Introduction of a rabbit beta-globin gene into the mouse germline. *Nature* 294, 92–94.

Cox, R.D., Hugill, A., Shedlovsky, A., Noveroske, J.K., Best, S., Justice, M.J., Lehrach, H. and Dove, W.F. (1999) Contrasting effects of ENU induced embryonic lethal mutations of the quaking gene. *Genomics* 57, 333–341.

DeChiara, T.M. (2001) Gene targeting in ES cells. *Methods in Molecular Biology* 158, 19–45.

Duboule, D. (1998) Vertebrate hox gene regulation: clustering and/or colinearity? *Current Opinion in Genetics and Development* 8, 514–518.

Evans, M.J. and Kaufman, M.H. (1981) Establishment in culture of pluripotential cells from mouse embryos. *Nature* 292, 154–156.

Evans, M.J., Carlton, M.B. and Russ, A.P. (1997) Gene trapping and functional genomics. *Trends in Genetics* 13, 370–374.

Favor, J. (1986) The frequency of dominant cataract and recessive specific-locus mutations in mice derived from 80 or 160 mg ethylnitrosourea per kg body weight treated spermatogonia. *Mutation Research* 162, 69–80.

Favor, J. (1994) Specific-locus mutations tests in germ cells of the mouse: an assessment of the screening procedures and the mutational events detected. In: Mattison, D.R. and Olsham, A.F. (eds) *Male-mediated Developmental Toxicity*. Plenum Press, New York, pp. 23–36.

Favor, J., Sund, M., Neuhauser-Klaus, A. and Ehling, U.H. (1990) A dose–response analysis of ethylnitrosourea-induced recessive specific-locus mutations in treated spermatogonia of the mouse. *Mutation Research* 231, 47–54.

Fernandez-Gonzalez, A., La Spada, A.R., Treadaway, J., Higdon, J.C., Harris, B.S., Sidman, R.L., Morgan,

J.I. and Zuo, J. (2002) Purkinje cell degeneration (pcd) phenotypes caused by mutations in the axo-tomy-induced gene, Nna1. *Science* 295, 1904–1906.

Filippov, M.A., Hormuzdi, S.G., Fuchs, E.C., Monyer, H., Robertson, E., Bradley, A., Kuehn, M. and Evans, M. (2003) A reporter allele for investigating connexin 26 gene expression in the mouse brain. *European Journal of Neuroscience* 18, 3183–3192.

Friedrich, G. and Soriano, P. (1991) Promoter traps in embryonic stem cells: a genetic screen to identify and mutate developmental genes in mice. *Genes and Development* 5, 1513–1523.

Furth, P.A., St Onge, L., Boger, H., Gruss, P., Gossen, M., Kistner, A., Bujard, H. and Hennighausen, L. (1994) Temporal control of gene expression in transgenic mice by a tetracycline-responsive promoter. *Proceedings of the National Academy of Sciences USA* 91, 9302–9306.

Gordon, J.W. and Ruddle, F.H. (1981) Integration and stable germline transmission of genes injected into mouse pronuclei. *Science* 214, 1244–1246.

Goring, D.R., Rossant, J., Clapoff, S., Breitman, M.L. and Tsui, L.C. (1987) In situ detection of beta-galactosi-dase in lenses of transgenic mice with a gamma-crystallin/lacZ gene. *Science* 235, 456–458.

Gossen, M. and Bujard, H. (1992) Tight control of gene expression in mammalian cells by tetracycline-responsive promoters. *Proceedings of the National Academy of Sciences USA* 89, 5547–5551.

Gossler, A., Doetschman, T., Korn, R., Serfling, E. and Kemler, R. (1986) Transgenesis by means of blastocyst-derived embryonic stem cell lines. *Proceedings of the National Academy of Sciences USA* 83, 9065–9069.

Gossler, A., Joyner, A.L., Rossant, J. and Skarnes, W.C. (1989) Mouse embryonic stem cells and reporter con-structs to detect developmentally regulated genes. *Science* 244, 463–465.

Green, E.L. and Roderick, T.H. (1966) Radiation genetics. In: Green, E.L. (ed.) *Biology of the Laboratory Mouse*. Dover Publications, New York, pp. 165–185.

Gridley, T., Gray, D.A., Orr-Weaver, T., Soriano, P., Barton, D.E., Francke, U. and Jaenisch, R. (1990) Molecular analysis of the Mov 34 mutation: transcript disrupted by proviral integration in mice is con-served in *Drosophila. Development* 109, 235–242.

Gu, H., Marth, J.D., Orban, P.C., Mossmann, H. and Rajewsky, K. (1994) Deletion of a DNA polymerase beta gene segment in T cells using cell type-specific gene targeting. *Science* 265, 103–106.

Guénet, J.L. and Bonhomme, F. (2003) Wild mice: an ever-increasing contribution to a popular mammalian model. *Trends in Genetics* 19, 24–31.

Hammes, A. and Schedl, A. (2000) Generation of transgenic mice from plasmids, BACs and YACs. In: Jackson, I.J. and Abbott, C.M. (eds) *Mouse Genetics and Transgenesis: A Practical Approach*. Oxford University Press, New York, pp. 217–245.

Hammond, S.M., Caudy, A.A. and Hannon, G.J. (2001) Post-transcriptional gene silencing by double-stranded RNA. *Nature Reviews of Genetics* 2, 110–119.

Hansen, J., Floss, T., Van Sloun, P., Fuchtbauer, E.M., Vauti, F., Arnold, H.H., Schnutgen, F., Wurst, W., von Melchner, H. and Ruiz, P. (2003) A large-scale, gene-driven mutagenesis approach for the functional analysis of the mouse genome. *Proceedings of the National Academy of Sciences USA* 100, 9918–9922.

Harbers, K., Jahner, D. and Jaenisch, R. (1981) Microinjection of cloned retroviral genomes into mouse zygotes: integration and expression in the animal. *Nature* 293, 540–542.

Hasty, P., Ramirez-Solis, R., Krumlauf, R. and Bradley, A. (1991) Introduction of a subtle mutation into the Hox-2.6 locus in embryonic stem cells. *Nature* 350, 243–246.

Hasty, P., Abuin, A. and Bradley, A. (2000) Gene targeting, principles, and practice in mammalian cells. In: Joyner, A.L. (ed.) *Gene Targeting: A Practical Approach*. Oxford University Press, New York, pp. 1–36.

He, Y., Chen, H., Quon, M.J. and Reitman, M. (1995) The mouse obese gene. Genomic organization, pro-moter activity, and activation by CCAAT/enhancer-binding protein alpha. *Journal of Biological Chemistry* 270, 28887–28891.

Heintz, N. (2001) BAC to the future: the use of bac transgenic mice for neuroscience research. *Nature Review Neuroscience* 2, 861–870.

Hitotsumachi, S., Carpenter, D.A. and Russell, W.L. (1985) Dose-repetition increases the mutagenic effective-ness of N-ethyl-N-nitrosourea in mouse spermatogonia. *Proceedings of the National Academy of Sciences USA* 82, 6619–6621.

Hodgson, J.G., Smith, D.J., McCutcheon, K., Koide, H.B., Nishiyama, K., Dinulos, M.B., Stevens, M.E., Bissada, N., Nasir, J., Kanazawa, I., Disteche, C.M., Rubin, E.M. and Hayden, M.R. (1996) Human huntingtin derived from YAC transgenes compensates for loss of murine huntingtin by rescue of the embryonic lethal phenotype. *Human Molecular Genetics* 5, 1875–1885.

Hogan, N.C., Traverse, K.L., Sullivan, D.E. and Pardue, M.L. (1994) The nucleus-limited Hsr-omega-n transcript is a polyadenylated RNA with a regulated intranuclear turnover. *Journal of Cell Biology* 125, 21–30.

Hong, H.K., Noveroske, J.K., Headon, D.J., Liu, T., Sy, M.S., Justice, M.J. and Chakravarti, A. (2001) The winged helix/forkhead transcription factor Foxq1 regulates differentiation of hair in satin mice. *Genesis* 29, 163–171.

Hooper, M., Hardy, K., Handyside, A., Hunter, S. and Monk, M. (1987) HPRT-deficient (Lesch–Nyhan) mouse embryos derived from germline colonization by cultured cells. *Nature* 326, 292–295.

Hooper, M.L. (1992) *Embryonal Stem Cells*. Harwood Academic, Chur, 147 pp.

Houdebine, L.M. (2003) *Animal Transgenesis and Cloning*. Wiley, New York.

Hrabe de Angelis, M.H., Flaswinkel, H., Fuchs, H., Rathkolb, B., Soewarto, D., Marschall, S., Heffner, S., Pargent, W., Wuensch, K., Jung, M., Reis, A., Richter, T., Alessandrini, F., Jakob, T., Fuchs, E., Kolb, H., Kremmer, E., Schaeble, K., Rollinski, B., Roscher, A. *et al.* (2000) Genome-wide, large-scale production of mutant mice by ENU mutagenesis. *Nature Genetics* 25, 444–447.

Huxley, C., Passage, E., Manson, A., Putzu, G., Figarella-Branger, D., Pellissier, J.F. and Fontes, M. (1996) Construction of a mouse model of Charcot–Marie–Tooth disease type 1A by pronuclear injection of human YAC DNA. *Human Molecular Genetics* 5, 563–569.

Jaenisch, R. (1976) Germline integration and Mendelian transmission of the exogenous Moloney leukemia virus. *Proceedings of the National Academy of Sciences USA* 73, 1260–1264.

Jaenisch, R. (1988) Transgenic animals. *Science* 240, 1468–1474.

Jaenisch, R., Jahner, D., Nobis, P., Simon, I., Lohler, J., Harbers, K. and Grotkopp, D. (1981) Chromosomal position and activation of retroviral genomes inserted into the germline of mice. *Cell* 24, 519–529.

Jakobovits, A., Moore, A.L., Green, L.L., Vergara, G.J., Maynard-Currie, C.E., Austin, H.A. and Klapholz, S. (1993) Germline transmission and expression of a human-derived yeast artificial chromosome. *Nature* 362, 255–258.

Jaubert, J., Jaubert, F., Martin, N., Washburn, L.L., Lee, B.K., Eicher, E.M. and Guénet, J.-L. (1999) Three new allelic mouse mutations that cause skeletal overgrowth involve the natriuretic peptide receptor C gene (*Npr3*). *Proceedings of the National Academy of Sciences USA* 96, 10278–10283.

Johnson, F.M. and Lewis, S.E. (1981) Mutation-rate determinations based on electrophoretic analysis of laboratory mice. *Mutation Research* 82, 125–135.

Justice, M.J. and Bode, V.C. (1986) Induction of new mutations in a mouse t-haplotype using ethylnitrosourea mutagenesis. *Genetics Research* 47, 187–192.

Justice, M.J. and Bode, V.C. (1990) ENU-induced allele of brachyury (T^{kt1}) exhibits a developmental lethal phenotype similar to the original brachyury (T) mutation. *Journal of Experimental Zoology* 254, 286–295.

Justice, M.J., Zheng, B., Woychik, R.P. and Bradley, A. (1997) Using targeted large deletions and high-efficiency N-ethyl-N-nitrosourea mutagenesis for functional analyses of the mammalian genome. *Methods* 13, 423–436.

Justice, M.J., Noveroske, J.K., Weber, J.S., Zheng, B. and Bradley, A. (1999) Mouse ENU mutagenesis. *Human Molecular Genetics* 8, 1955–1963.

Katsuki, M., Sato, M., Kimura, M., Yokoyama, M., Kobayashi, K. and Nomura, T. (1988) Conversion of normal behavior to shiverer by myelin basic protein antisense cDNA in transgenic mice. *Science* 241, 593–595.

Kiernan, A.E., Erven, A., Voegeling, S., Peters, J., Nolan, P., Hunter, J., Bacon, Y., Steel, K.P., Brown, S.D.M. and Guénet, J.-L. (2002) ENU mutagenesis reveals a highly mutable locus on mouse chromosome 4 that affects ear morphogenesis. *Mammalian Genome* 13, 142–148.

Kile, B.T., Hentges, K.E., Clark, A.T., Nakamura, H., Salinger, A.P., Liu, B., Box, N., Stockton, D.W., Johnson, R.L., Behringer, R.R., Bradley, A. and Justice, M.J. (2003) Functional genetic analysis of mouse chromosome 11. *Nature* 425, 81–86.

Koike, S., Taya, C., Kurata, T., Abe, S., Ise, I., Yonekawa, H. and Nomoto, A. (1991) Transgenic mice susceptible to poliovirus. *Proceedings of the National Academy of Sciences USA* 88, 951–955.

Korstanje, R. and Paigen, B. (2002) From QTL to gene: the harvest begins. *Nature Genetics* 31, 235–236.

Kuehn, M.R., Bradley, A., Robertson, E.J. and Evans, M.J. (1987) A potential animal model for Lesch–Nyhan syndrome through introduction of HPRT mutations into mice. *Nature* 326, 295–298.

Kuhn, R., Schwenk, F., Aguet, M. and Rajewsky, K. (1995) Inducible gene targeting in mice. *Science* 269, 1427–1429.

Lakso, M., Sauer, B., Mosinger, B., Jr., Lee, E.J., Manning, R.W., Yu, S.H., Mulder, K.L. and Westphal, H. (1992) Targeted oncogene activation by site-specific recombination in transgenic mice. *Proceedings of the National Academy of Sciences USA* 89, 6232–6236.

Lalouette, A., Guénet, J.-L. and Vriz, S. (1998) Hotfoot mouse mutations affect the δ2 glutamate receptor gene and are allelic to Lurcher. *Genomics* 50, 9–13.

Lecuit, M., Vandormael-Pournin, S., Lefort, J., Huerre, M., Gounon, P., Dupuy, C., Babinet, C. and Cossart, P. (2001) A transgenic model for listeriosis: role of internalin in crossing the intestinal barrier. *Science* 292, 1722–1725.

Lee, J.T. and Jaenisch, R. (1996) A method for high efficiency YAC lipofection into murine embryonic stem cells. *Nucleic Acids Research* 24, 5054–5055.

Lewis, S.E. (1991) The biochemical specific-locus test and a new multiple-endpoint mutation detection system: considerations for genetic risk assessment. *Environmental Molecular Mutagens* 18, 303–306.

Lewis, S.E., Barnett, L.B., Sadler, B.M. and Shelby, M.D. (1991) ENU mutagenesis in the mouse electrophoretic specific-locus test, 1. Dose–response relationship of electrophoretically-detected mutations arising from mouse spermatogonia treated with ethylnitrosourea. *Mutation Research* 249, 311–315.

Lewis, S.E., Barnett, L.B. and Shelby, M.D. (1992) ENU mutagenesis in the mouse electrophoretic specific-locus test. 2. Mutational studies of mature oocytes. *Mutation Research* 296, 129–133.

Lira, S.A., Kinloch, R.A., Mortillo, S. and Wassarman, P.M. (1990) An upstream region of the mouse ZP3 gene directs expression of firefly luciferase specifically to growing oocytes in transgenic mice. *Proceedings of the National Academy of Sciences USA* 87, 7215–7219.

Liu, P., Jenkins, N.A. and Copeland, N.G. (2003) A highly efficient recombineering-based method for generating conditional knockout mutations. *Genome Research* 13, 476–484.

Mangiarini, L., Sathasivam, K., Seller, M., Cozens, B., Harper, A., Hetherington, C., Lawton, M., Trottier, Y., Lehrach, H., Davies, S.W. and Bates, G.P. (1996) Exon 1 of the HD gene with an expanded CAG repeat is sufficient to cause a progressive neurological phenotype in transgenic mice. *Cell* 87, 493–506.

Marshall, R.R., Raj, A.S., Grant, F.J. and Heddle, J.A. (1983) The use of two-dimensional electrophoresis to detect mutations induced in mouse spermatogonia by ethylnitrosourea. *Canadian Journal of Genetics and Cytology* 25, 457–466.

Martin, G.R., Stevens, M.E., Bissada, N., Nasir, J., Kanazawa, I., Disteche, C.M., Rubin, E.M. and Hayden, M.R. (1981) Isolation of a pluripotent cell line from early mouse embryos cultured in medium conditioned by teratocarcinoma stem cells. *Proceedings of the National Academy of Sciences USA* 78, 7634–7638.

Martin, N., Jaubert, J., Gounon, P., Salido, E., Haase, G., Szatanik, M. and Guénet, J.L. (2002) A missense mutation in *Tbce* causes progressive motor neuronopathy in mice. *Nature Genetics* 32, 443–447.

McDonald, J.D. (1995) Using high-efficiency mouse germline mutagenesis to investigate complex biological phenomena: genetic diseases, behavior, and development. *Proceedings of the Society for Experimental Biology and Medicine* 209, 303–308.

McDonald, J.D., Trischler, M., Stoorvogel, W. and Ullrich, O. (1994) The PKU mouse project: its history, potential and implications. *Acta Paediatrica Supplement* 407, 122–123.

Meisler, M.H. (1992) Insertional mutation of 'classical' and novel genes in transgenic mice. *Trends in Genetics* 8, 341–344.

Munroe, R.J., Bergstrom, R.A., Zheng, Q.Y., Libby, B., Smith, R., John, S.W., Schimenti, K.J., Browning, V.L. and Schimenti, J.C. (2000) Mouse mutants from chemically mutagenized embryonic stem cells. *Nature Genetics* 24, 318–321.

Muyrers, J.P., Zhang, Y. and Stewart, A.F. (2004) Techniques: recombinogenic engineering – new options for cloning and manipulating DNA. *Genomics* 83, 332–334.

Nadeau, J.H., Singer, J.B., Matin, A. and Lander, E.S. (2000) Analysing complex genetic traits with chromosome substitution strains. *Nature Genetics* 24, 221–225.

Nicolas, J.F. and Rubenstein, J.L. (1988) Retroviral vectors. *Biotechnology* 10, 493–513.

Nolan, P., Peters, J., Strivens, M., Rogers, D., Hagan, J., Spurr, N., Gray, I.C., Vizor, L., Brooker, D., Whitehill, E., Washbourne, R., Hough, T., Greenaway, S., Hewitt, M., Liu, X., McCormack, S., Pickford, K., Selley, R., Wells, C., Tymowska-Lalanne, Z. *et al.* (2000) A systematic genome-wide, phenotype-driven mutagenesis programme for gene function studies in the mouse. *Nature Genetics* 25, 440–443.

Overbeek, P.A., Chepelinsky, A.B., Khillan, J.S., Piatigorsky, J. and Westphal, H. (1985) Lens-specific expression and developmental regulation of the bacterial chloramphenicol acetyltransferase gene driven by the murine alpha A-crystallin promoter in transgenic mice. *Proceedings of the National Academy of Sciences USA* 82, 7815–7819.

Parvari, R., Hershkovitz, E., Grossman, N., Gorodischer, R., Loeys, B., Zecic, A., Mortier, G., Gregory, S., Sharony, R., Kambouris, M., Sakati, N., Meyer, B.F., Al Aqeel, A.I., Al Humaidan, A.K., Al Zanhrani, F., Al Swaid, A., Al Othman, J., Diaz, G.A., Weiner, R., Khan, K.T., Gordon, R. and Gelb, B.D. (2002) Mutation of TBCE causes hypoparathyroidism–retardation–dysmorphism and autosomal recessive Kenny-Caffey syndrome. *Nature Genetics* 32, 448–452.

Pereira, R., Khillan, J.S., Helminen, H.J., Hume, E.L. and Prockop, D.J. (1993) Transgenic mice expressing a partially deleted gene for type I procollagen (COL1A1). A breeding line with a phenotype of spontaneous fractures and decreased bone collagen and mineral. *Journal of Clinical Investigation* 91, 709–716.

Perou, C.M., Moore, K.J., Nagle, D.L., Misumi, D.J., Woolf, E.A., McGrail, S.H., Holmgren, L., Brody, T.H., Dussault, B.J., Jr, Monroe, C.A., Duyk, G.M., Pryor, R.J., Li, L., Justice, M.J. and Kaplan, J. (1996) Identification of the murine beige gene by YAC complementation and positional cloning. *Nature Genetics* 13, 303–308.

Peters, J., Ball, S.T. and Andrews, S.J. (1986) The detection of gene mutations by electrophoresis, and their analysis. *Progress in Clinical Biology Research* 209B, 367–374.

Pichel, J.G., Lakso, M. and Westphal, H. (1993) Timing of SV40 oncogene activation by site-specific recombination determines subsequent tumor progression during murine lens development. *Oncogene* 8, 3333–3342.

Pillers, D.A., Weleber, R.G., Green, D.G., Rash, S.M., Dally, G.Y., Howard, P.L., Powers, M.R., Hood, D.C., Chapman, V.M., Ray, P.N. and Woodward, W.R. (1999) Effects of dystrophin isoforms on signal transduction through neural retina: genotype–phenotype analysis of duchenne muscular dystrophy mouse mutants. *Molecular Genetics and Metabolism* 66, 100–110.

Plagge, A., Kelsey, G. and Allen, N.D. (2000) Directed mutagenesis in embryonic stem cells. In: Jackson, I.J. and Abbott, C.M. (eds) *Mouse Genetics and Transgenesis: A Practical Approach*. Oxford University Press, New York, pp. 247–284.

Pretsch, W., Favor, J., Lehmacher, W. and Neuhauser-Klaus, A. (1994) Estimates of the radiation-induced mutation frequencies to recessive visible, dominant cataract and enzyme-activity alleles in germ cells of AKR, BALB/c, DBA/2 and $(102 \times C3H)F_1$ mice. *Mutagenesis* 9, 289–294.

Rinchik, E.M. and Carpenter, D.A. (1999) N-ethyl-N-nitrosourea mutagenesis of a 6- to 11-cM subregion of the Fah–Hbb interval of mouse chromosome 7: completed testing of 4557 gametes and deletion mapping and complementation analysis of 31 mutations. *Genetics* 152, 373–383.

Robertson, E., Bradley, A., Kuehn, M. and Evans, M. (1986) Germline transmission of genes introduced into cultured pluripotential cells by retroviral vector. *Nature* 323, 445–448.

Russell, W.L., Kelly, E.M., Hunsicker, P.R., Bangham, J.W., Maddux, S.C. and Phipps, E.L. (1979) Specific-locus test shows ethylnitrosourea to be the most potent mutagen in the mouse. *Proceedings of the National Academy of Sciences USA* 76, 5818–5819.

Russell, W.L., Hunsicker, P.R., Carpenter, D.A., Cornett, C.V. and Guinn, G.M. (1982a) Effect of dose fractionation on the ethylnitrosourea induction of specific-locus mutations in mouse spermatogonia. *Proceedings of the National Academy of Sciences USA* 79, 3592–3593.

Russell, W.L., Hunsicker, P.R., Raymer, G.D., Steele, M.H., Stelzner, K.F. and Thompson, H.M. (1982b) Dose–response curve for ethylnitrosourea-induced specific-locus mutations in mouse spermatogonia. *Proceedings of the National Academy of Sciences USA* 79, 3589–3591.

Ryan, T.M., Townes, T.M., Reilly, M.P., Asakura, T., Palmiter, R.D., Brinster, R.L. and Behringer, R.R. (1990) Human sickle hemoglobin in transgenic mice. *Science* 247, 566–568.

Sauer, B. (1993) Manipulation of transgenes by site-specific recombination: use of Cre recombinase. *Methods in Enzymology* 225, 890–900.

Schedl, A., Beermann, F., Thies, E., Montoliu, L., Kelsey, G. and Schutz, G. (1992) Transgenic mice generated by pronuclear injection of a yeast artificial chromosome. *Nucleic Acids Research* 20, 3073–3077.

Schedl, A., Larin, Z., Montoliu, L., Thies, E., Kelsey, G., Lehrach, H. and Schutz, G. (1993) A method for the generation of YAC transgenic mice by pronuclear microinjection. *Nucleic Acids Research* 21, 4783–4787.

Schlager, G. and Dickie, M.M. (1967) Spontaneous mutations and mutation rates in the house mouse. *Genetics* 57, 319–330.

Schmezer, P. and Eckert, C. (1999) Induction of mutations in transgenic animal models: BigBlue and Muta Mouse. *International Agency for Research on Cancer-Research Publications* 367–394.

Shedlovsky, A., Guénet, J.L., Johnson, L.L. and Dove, W.F. (1986) Induction of recessive lethal mutations in the T/t-H-2 region of the mouse genome by a point mutagen. *Genetics Research* 47, 135–142.

Shedlovsky, A., King, T.R. and Dove, W.F. (1988) Saturation germline mutagenesis of the murine t region including a lethal allele at the quaking locus. *Proceedings of the National Academy of Sciences USA* 85, 180–184.

Silver, L.M. (1995) *Mouse Genetics: Concepts and Applications*. Oxford University Press, Oxford.

Skarnes, W.C., Auerbach, B.A. and Joyner, A.L. (1992) A gene trap approach in mouse embryonic stem cells: the lacZ reported is activated by splicing, reflects endogenous gene expression, and is mutagenic in mice. *Genes and Development* 6, 903–918.

Smith, D.J., Zhu, Y., Zhang, J., Cheng, J.F. and Rubin, E.M. (1995) Construction of a panel of transgenic mice containing a contiguous 2-Mb set of YAC/P1 clones from human chromosome 21q22.2. *Genomics* 27, 425–434.

Smithies, O., Gregg, R.G., Boggs, S.S., Koralewski, M.A. and Kucherlapati, R.S. (1985) Insertion of DNA sequences into the human chromosomal beta-globin locus by homologous recombination. *Nature* 317, 230–234.

Smits, B.M., Mudde, J., Plasterk, R.H. and Cuppen, E. (2004) Target-selected mutagenesis of the rat. *Genomics* 83, 332–334.

Stacey, A., Bateman, J., Choi, T., Mascara, T., Cole, W. and Jaenisch, R. (1988) Perinatal lethal osteogenesis imperfecta in transgenic mice bearing an engineered mutant pro-alpha 1(I) collagen gene. *Nature* 332, 131–136.

Stacey, A., Schnieke, A., McWhir, J., Cooper, J., Colman, A. and Melton, D.W. (1994) Use of double-replacement gene targeting to replace the murine alpha-lactalbumin gene with its human counterpart in embryonic stem cells and mice. *Molecular and Cellular Biology* 14, 1009–1016.

Stacey, D.W., Tsai, M.H., Yu, C.L. and Smith, J.K. (1988) Critical role of cellular ras proteins in proliferative signal transduction. *Cold Spring Harbor Symposia on Quantitative Biology* 53 Pt 2, 871–881.

Stanford, W.L., Cohn, J.B. and Cordes, S.P. (2001) Gene-trap mutagenesis: past, present and beyond. *Nature Reviews of Genetics* 2, 756–768.

Stryke, D., Kawamoto, M., Huang, C.C., Johns, S.J., King, L.A., Harper, C.A., Meng, E.C., Lee, R.E., Yee, A., L'Italien, L., Chuang, P.T., Young, S.G., Skarnes, W.C., Babbitt, P.C. and Ferrin, T.E. (2003) BayGenomics: a resource of insertional mutations in mouse embryonic stem cells. *Nucleic Acids Research* 31, 278–281.

Thomas, K.R. and Capecchi, M.R. (1987) Site-directed mutagenesis by gene targeting in mouse embryo-derived stem cells. *Cell* 51, 503–512.

Valancius, V. and Smithies, O. (1991) Testing an 'in–out' targeting procedure for making subtle genomic modifications in mouse embryonic stem cells. *Molecular and Cellular Biology* 11, 1402–1408.

Van Zeeland, A.A., Mohn, G.R., Mullenders, L.H., Natarajan, A.T., Nivard, M., Simons, J.W., Venema, J., Vogel, E.W., Vrieling, H., Zdzienicka, M.Z. *et al.* (1989) Relationship between DNA-adduct formation, DNA repair, mutation frequency and mutation spectra. *Annali dell'Instituto Superiore di Sanita* (ISDIS) 2003 25, 223–228.

Vivian, J.L., Chen, Y., Yee, D., Schneider, E. and Magnuson, T. (2002) An allelic series of mutations in Smad2 and Smad4 identified in a genotype-based screen of N-ethyl-N-nitrosourea-mutagenized mouse embryonic stem cells. *Proceedings of the National Academy of Sciences USA* 99, 15542–15547.

Vogel, E.W. and Natarajan, A.T. (1995) DNA damage and repair in somatic and germ cells *in vivo*. *Mutation Research* 330, 183–208.

Vogel, F. and Röhrborn, G. (1970) *Chemical Mutagenesis in Mammals and Man*. Springer, Berlin.

Wagner, E.F., Stewart, T.A. and Mintz, B. (1981a) The human beta-globin gene and a functional viral thymidine kinase gene in developing mice. *Proceedings of the National Academy of Sciences USA* 78, 5016–5020.

Wagner, T.E., Hoppe, P.C., Jollick, J.D., Scholl, D.R., Hodinka, R.L. and Gault, J.B. (1981b) Microinjection of a rabbit beta-globin gene into zygotes and its subsequent expression in adult mice and their offspring. *Proceedings of the National Academy of Sciences USA* 78, 6376–6380.

Waterston, R.H., Lindblad-Toh, K., Birney, E., Rogers, J., Abril, J.F., Agarwal, P., Agarwala, R., Ainscough, R., Alexandersson, M., An, P., Antonarakis, S.E., Attwood, J., Baertsch, R., Bailey, J., Barlow, K., Beck, S., Berry, E., Birren, B., Bloom, T., Bork, P. *et al.* (2002) Initial sequencing and comparative analysis of the mouse genome. *Nature* 420, 520–562.

Wilkie, A.O. (1994) The molecular basis of genetic dominance. *Journal of Medical Genetics* 31, 89–98.

Williams, R.W. (1999) A targeted screen to detect recessive mutations that have quantitative effects. *Mammalian Genome* 10, 734–738.

Wilsbacher, L.D., Sangoram, A.M., Antoch, M.P. and Takahashi, J.S. (2000) The mouse Clock locus: sequence and comparative analysis of 204 kb from mouse chromosome 5. *Genome Research* 10, 1928–1940.

Wong, E.A. and Capecchi, M.R. (1986) Analysis of homologous recombination in cultured mammalian cells in transient expression and stable transformation assays. *Somatic Cell Molecular Genetics* 12, 63–72.

You, Y., Bergstrom, R., Klemm, M., Lederman, B., Nelson, H., Ticknor, C., Jaenisch, R. and Schimenti, J. (1997) Chromosomal deletion complexes in mice by radiation of embryonic stem cells. *Nature Genetics* 15, 285–288.

Zan, Y., Haag, J.D., Chen, K.S., Shepel, L.A., Wigington, D., Wang, Y.R., Hu, R., Lopez-Guajardo, C.C., Brose, H.L., Porter, K.I., Leonard, R.A., Hitt, A.A., Schommer, S.L., Elegbede, A.F. and Gould, M.N. (2003) Production of knockout rats using ENU mutagenesis and a yeast-based screening assay. *Nature Biotechnology* 21, 645–651.

Zhang, Y., Proenca, R., Maffei, M., Barone, M., Leopold, L. and Friedman, J.M. (1994) Positional cloning of the mouse obese gene and its human homologue. *Nature* 372, 425–432.

Zheng, B., Sage, M., Cai, W.W., Thompson, D.M., Tavsanli, B.C., Cheah, Y.C. and Bradley, A. (1999) Engineering a mouse balancer chromosome. *Nature Genetics* 22, 375–378.

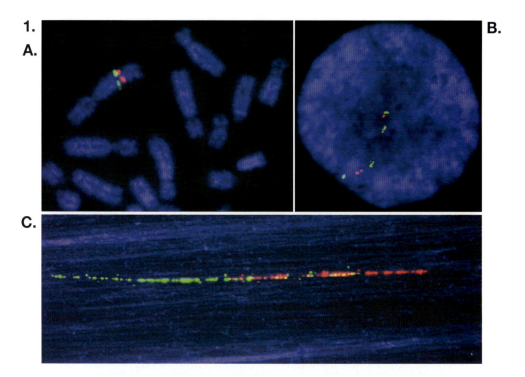

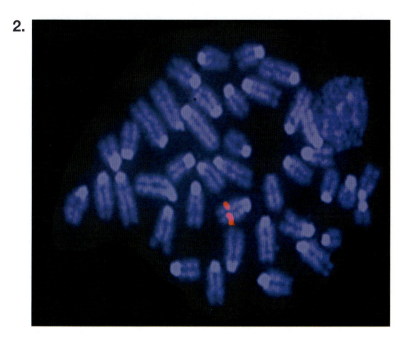

Plate 1 (from Chapter 2). Different resolutions of FISH analysis. **A** Two-colour FISH on metaphase chromosomes showing the location and physical order of two genes on the horse X chromosome – *TMSNB* (green) on Xq21 and *COL4A5* (red) on Xq22. **B** FISH on interphase chromatin showing relative order of three BAC clones from bovine major histocompatibility complex (BTA23). The red signal from one of the BACs lies in between green signals from two other BACs. **C** FISH on DNA fibres with two overlapping BAC clones from the horse Y chromosome contig map.
Plate 2 (from Chapter 9). Localization of a transgenic insertion by fluorescence *in situ* hybridization (FISH) (courtesy M.G. Mattei).

3.

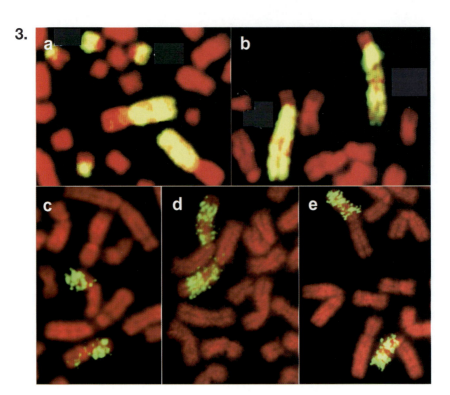

4. A.   **B.**

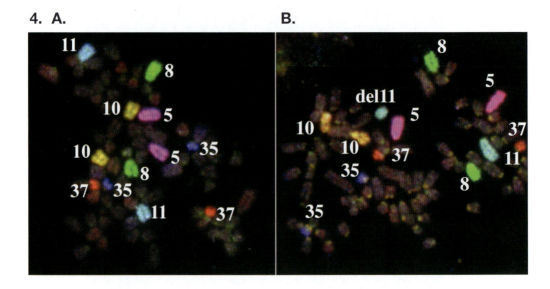

Plate 3 (from Chapter 14). Comparative chromosome painting between eutherian species that are closely (horse–donkey) and distantly (human–hippo) related. Top: (a, b): donkey chromosomes painted with horse chromosomes 3 and 4 paint. Bottom (c, d,e): hippo metaphase chromosomes painted with human chromosomes 2, 15 and 9 paints. Note the much stronger signal on chromosomes painted with probe from a closely related species (a, b).

Plate 4 (from Chapter 18). Seven colour chromosome painting in the dog. **A** and **B** show paint group 3 of a complete paint set for dog, balanced in 7 colours to allow examination of the entire karyotype in six hybridizations. (Three groups of six and three groups of seven colours are used. Group three is a six-colour set.) **B** shows the deletion of part of chromosome 11 in a fibrosarcoma (Milne and Sargan, unpublished data).

10   A Comparative Analysis of Mammalian Genomics: Prokaryote and Eukaryote Perspectives

M.I. Bellgard[1]* and T. Gojobori[2]*

[1]*Centre for Bioinformatics and Biological Computing, Murdoch University, Murdoch, Western Australia 6150, Australia;* [2]*Center for Information Biology, National Institute of Genetics, Mishima, Shizuoka-ken 411-8540, Japan*

Introduction	263
Bacterial Genome Structure Highlights	265
Horizontal gene transfer	267
Eukaryote Genomes in Perspective	268
G + C content, isochores, gene-rich regions	268
The transcriptome, non-coding RNA and systems biology	270
β-Tubulin genes	271
Major histocompatibility complex	272
Olfactory genes	272
Discussion	273
References	274

Introduction

Genomes of all extant living organisms, and not just mammalian species, share much genetic material as well as genomic structure and organization in common with each other. As a result, much of the current insights into evolutionary processes and mechanisms come from the analysis of the wealth of available molecular data. For instance, it is possible to identify and functionally characterize genes contained within any given species and find evolutionarily related genes (homologous genes) in most others. In broad terms this *in silico* analysis is referred to as comparative genomic analysis and is pivotal in deciphering and inferring the evolutionary relationships between species based on their molecular sequences. These studies continue to reinforce the profound statement by Theodosius Dobzhansky that 'nothing in biology makes sense except in the light of evolution' (Dobzhansky, 1973).

*Correspondence: m.bellgard@murdoch.edu.au; tgojobor@genes.nig.ac.jp

What we can interpret from the masses of molecular data that are continuing to become available through the numerous whole genomes sequence projects is that not only do we find 'gene repertoires' in common between species, but also a range of other genome characteristics that can form the basis for comparison. These characteristics can include gene order, conserved gene regions and rearrangements, nucleotide bias composition, repetitive elements, conserved intronic sequences and gene regulatory elements. It has been recently postulated that gene repertoires have more than likely been conserved even prior to the protestome/deuterstome divergence (for example, Rivera and Lake, 1992; Mineta et al., 2003).

By way of contrast, we find many differences between diverse and even closely related species given adaptive radiation and species divergence. While gene sequences between two diverse species might be inferred to be homologous (based on sequence similarity), it is highly likely that gene function will be different or perhaps differentially expressed. In fact, a given gene sequence may have multiple functions and splice variants even within a species (Imanishi et al., 2004). Thus, function determination and differentiation is not obvious via an exclusive comparative genomic analysis approach and the need for biological validation and systematic perturbation studies is essential (Davidson et al., 2002). Comparative genomic analysis can uncover orthologous regions (see Chapter 16) between species containing multicopy gene families. In some of these cases, we also find that each species has chosen a different progenitor gene as the basis for its lineage-specific 'version' of the multicopy family as in the case of olfactory genes (Dehal et al., 2001). Common to pairwise sequence comparison between species are many genome rearrangements that highlight the plasticity of genome sequences (Bellgard et al., 2004). For instance, bacterial genomes have been shown to have multiple rearrangements (Bellgard et al., 1999), based on an analysis of orthologous and paralogous genes. It is argued that genome plasticity has most probably contributed substantially to the dynamic evolution of genomes. The characteristic mosaic features of the Archaea genome that is comprised of both bacterial and eukaryal elements is described by investigating base compositional differences and similarities of this species' genes to either bacteria or eukarya. Certainly, genome and gene duplications have played an important role in genome evolution as was the case during the Cambrian explosion.

The field termed 'systems biology' involves the determination, analysis and integration of interrelationships of all elements in a biological system in response to genetic or environmental perturbations. The objective is to attempt to visually and mathematically model the system in order to understand the systems or emergent properties (Ideker et al., 2001). The motivation of this work is to understand the protein and gene regulatory networks of biological systems. Internationally, this field has achieved prominence as the likely future for life science and bioinformatics research. In this integration science, the need to develop new computational tools for gathering extensive genomic (and proteomic, and transcriptomic and functional genomic) data sets is essential and the dependence on computational tools for detailed comparative analysis is critical.

From a systems biology standpoint, three dimensions of time must be collapsed in order to begin to understand biological systems properly. As we explain, comparative genomic analysis has traditionally been viewed in the light of identifying genes shared between related species such as human and mouse, rice and wheat, or between bacterial organisms which may have diverged from a common ancestor over millions and hundreds of millions of years (dimension one). While the majority of genes between diverse species have been conserved over the course of evolution, gene regulatory elements start to emerge as important factors determining gene regulation, organism development (dimension two) as well as observed differences in physiology (dimension three) and phenotype. For example, while the genes involved in embryogenesis in a sea urchin and a starfish might be conserved over millions of years, it is the neighbouring regulatory genes and genomic regulatory elements that control the timing (as well as levels of expression) of genes activated within the cell (Davidson et al., 2002; Hinman et al., 2003). As a consequence, comparative analysis must be extended to a higher level of resolution in order to begin to fully understand the controlling gene sequences and related networks that define species and differentiate biological functions.

So, we might ask, what genes make significant contribution in terms of genome and species evolution and diversity? The immediate, and rather intuitive response might be that the genome could be characterized through gene and genome duplication. This mechanism facilitates the enrichment of gene number and possible differential function. Numerous examples of this are observed in the literature with representative coverage over all living organisms. Clearly, the rigorous work of Davidson and colleagues (Davidson et al., 2002) highlights the importance of regulatory gene networks and regulatory elements in species evolution, diversity and phenotype. The other important contributions come from nucleotide substitution, genomic rearrangement and transposition. This leads to the question, what of the role of gene loss and pseudogenes?

Pseudogenes are non-functional regions in the genome that have arisen as a consequence of accumulating mutations that result in either the premature termination of proteins during protein synthesis or the disruption of transcription (Babu, 2003). However, pseudogenes can also segregate into an intact form (Menashe et al., 2003). While it is perhaps a sizeable challenge in itself to attempt to characterize a genome in terms of duplication, rearrangements and regulatory elements, and to associate these with observed phenotypes, this becomes in most cases intractable when also considering the profound effects that gene loss and/or pseudogenes may have on the evolution and observed phenotype. If we accept that gene repertoires across species were established relatively early on in species evolution, it could be argued that gene loss is one of the more important evolutionary mechanisms to ultimately characterize extant species diversity and phenotype differences even between closely related species. This would certainly be relevant for species that could not undergo lateral gene transfer events and genome duplication in order to regain some or all their gene repertoire (Ochman and Moran, 2001). As we explore in subsequent sections,

comparative genomic analysis is able to provide us with some meaningful insights into the various mechanisms that have been described above.

In this chapter we attempt to place mammalian genomes in context with other species based on comparative genomic analysis. Bacterial and archaebacterial genome organization are first briefly explored followed by an investigation into eukaryote genome characteristics along with case studies on β-tubulin genes, the major histocompatibility complex region and olfactory receptor genes.

Bacterial Genome Structure Highlights

Bacterial genome sizes vary significantly compared with eukaryote genomes. They can range from around 580,000 bp as in *Mycoplasma genitalium* (Fraser et al., 1995) through to more than 10 Mbp in several cyanobacterial species. One of the smallest prokaryotes, the *Mycoplasmas*, can have anywhere from 470 protein-coding genes, including some 50 ribosomal proteins, one or two sets of rRNA genes (5S, 16S and 23S) and about 33 tRNA genes. Interestingly, this particular genome translates the codon TGA as a typtophan rather than a stop codon as in the 'universal' genetic code. This compact genome epitomizes the smallest gene complement required to sustain life; however, like other species such as *Buchnera* spp., it has a specialized symbiotic lifestyle. In the case of the *Buchnera* genomes, it has been argued that it is still experiencing a reductive process in its genome size (Sabater-Munoz et al., 2002).

It has been suggested that, as there is no correlation between genome size and phylogeny (Wallace and Morowitz, 1973), genome duplications have occurred frequently in the evolution of bacterial lineages and that larger bacterial genomes evolved from smaller genomes. Certainly, the identification of the putative ribonuclease HII (RNase HII) enzymes in the two *Mycoplasma* species (Bellgard and Gojobori, 1999a) and subsequent phylogenetic analysis that included species from archaebacteria and eukaryotes revealed an interesting observation with a range of intriguing and likely interpretations. Some bacterial species contained two copies of the enzyme while others species contained only one. Depending on which copy the species contained determined its position in the phylogenetic tree. For this particular gene, the expected taxonomic relationships between the particular species are not preserved (see Fig. 10.1). For instance, not all Gram-positive species were clustered together. One hypothesis is that early bacterial lineages contained multiple copies of RNase HII from which subsequent species either carried two copies or, as in the majority of observed cases, a particular species lost one of the copies during the course of evolution, thereby defining its position in the phylogenetic tree. Another possible theory is that lateral gene transfer (Ochman, 2001; Daubin and Perrière, 2003; Eisen and Fraser, 2003) events have played a role in restoring one or both copies.

Mycobacterium leprae is an example of a bacterial species that has accumulated a large number of non-functional (pseudogenes) genes within its genome. (Babu, 2003). It is postulated that repetitive DNA (about 2% of the genome) when compared with a closely related tubercle bacillus, *M. tuberculosis* H37Rv, has enabled dramatic recombination events that have resulted in loss of synteny, inversion and genome downsizing (Cole et al., 2001). In other studies of base composition it has been shown that these species have dramatic differences in guanine and cytosine (G + C) composition when compared with the estimated common ancestral base composition (Bellgard and Gojobori, 1999b). The role of loss of gene function has clearly had a dramatic effect on the phenotypes of *M. leprae*. In contrast, *Mycobacterium tuberculosis* is a high G + C

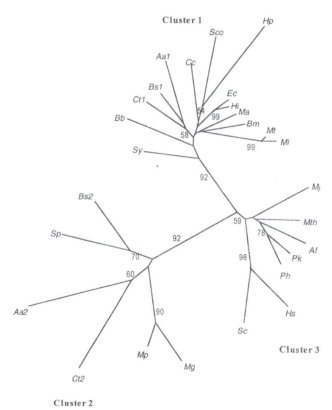

Fig. 10.1. Phylogenetic tree of ribonuclease HII genes in a wide range of species. The two eubacterial clusters are labelled cluster 1 and 2. *Mg* and *Mp* are contained within cluster 2. *Mj, Bs2* and *Aa2* are currently annotated as hypothetical ORFs in the public databases. Only bootstraps larger than 50 are shown in the figure. Reproduced with permission (Bellgard and Gojobori, 1999b).

member of the actinobacteria. *M. tuberculosis* has 250 genes involved in fatty acid metabolism, a much higher proportion than in any other organism. Comparative analysis was conducted and it was found that of the eight gene families examined, five of the phylogenies reconstructed suggest that the actinobacteria have a closer relationship with the proteobacteria than expected. Similar to the *Mycoplasma* spp. previously discussed, this is due to either an ancient transfer of genes or to deep paralogy and subsequent retention of the genes in unrelated lineages. These types of studies provide key insights into how species such as *M. tuberculosis* have developed their unique fatty acid synthetic abilities (Kinsella *et al.*, 2003).

Horizontal gene transfer

As previously mentioned, horizontal gene transfer (HGT) is the collective name for processes that permit the exchange of DNA among organisms of different species. There are three principal mechanisms that facilitate HGT in prokaryotes (reviewed in Syvanen and Kado, 1998). Genes can be horizontally transferred by means of transformation, conjugation, and/or transduction (Jain *et al.*, 2002). It has recently been recognized that HGT plays a significant role in genome evolution contributing to adaptation and ecological diversification (Karlin *et al.*, 1997; Lawrence and Ochman, 1998; Rivera *et al.*, 1998; Doolittle, 1999; Campbell, 2000; Ochman and Jones, 2000; Ochman, 2001; Gogarten *et al.*, 2002). Whereas prokaryotic and eukaryotic evolution was once reconstructed from a single 16S rRNA gene, the analysis of complete genomes is beginning to yield a different picture of microbial evolution, one that is wrought with the lateral movement of genes across vast phylogenetic distances (Lane *et al.*, 1988; Lake and Rivera, 1996; Lake *et al.*, 1999). The complete sequence of the methanogen *Methanococcus janaschii* revealed that its genome consisted of certain groups of genes being much more similar to eukaryotes than bacteria, while other groups of genes were much more closely related to their bacterial homologues (Bult *et al.*, 1996; Watanabe *et al.*, 1997; Bellgard *et al.*, 1999). HGT is also

widely prevalent in eubacteria and facilitates the ability of an organism to become pathogenic. The virulence genes can be identified in species-specific chromosomal regions referred to as 'pathogenicity islands' (Ochman, 2001; Jain *et al.*, 2002). Through gene loss and HGT, bacteria form intimate, and quite often mutually beneficial associations with a variety of multicellular organisms. Interestingly, comparative analysis of three strains of the endosymbiotic bacterium *Buchnera aphidicola* has revealed high genome stability with an almost complete absence of chromosomal rearrangements and HGT events during the last 150 million years. However, it is postulated that loss of genes involved in DNA uptake and recombination in the initial stages of endosymbiosis probably underlies this stability (Silva *et al.*, 2003); in particular, the loss of repeated elements that could act as putative sites for recombination such as phages, repeated sequences and transposable elements and the loss of important genes involved in recombination such as *recA* and *recF* (Tamas *et al.*, 2002).

However, the significant role of HGT in evolution is not without controversy. There are no data suggesting that eukaryotic genomes have originated as fusions between the genomes of Bacteria and Archaea. In addition, there is no doubt that HGT occurs and that it has important evolutionary consequences. The major issue requiring resolution appears to be whether or not there have been enormous amounts of HGT throughout the evolution of life. Unfortunately, to date, there is no rigorous proof of this which fuels the debate (Koonin, 2003; Kurland *et al.*, 2003; Lawrence and Hendrickson, 2003). It is also argued that an uncritical use of the bioinformatics techniques that are used to identify homologues in pairwise comparisons of archaea, bacteria and eukaryote is a systematic source of overestimates of HGT. A number of pieces of indirect evidence suggest a contrary major role for HGT. For instance, the estimate of the last universal common ancestor invariably leads to a *supergenome paradox* concept that can be rationalized with significant HGT. Other evidence is from higher numbers of shared genes in populations of very distinct prokaryotes inhabiting similar niches (Koonin, 2003).

Eukaryote Genomes in Perspective

The Human Genome Project (HGP) has provided the basis for determining the gene complement in addition to gene–disease associations and novel diagnostics derived from genes of interest (Collins et al., 2003). The HGP has also allowed the detailed analysis of the structure of genes in mammals with respect to the occurrence and distribution of repetitive and retro-transposable elements. The development of genomic-scale technologies such as high-throughput oligonucleotide synthesis and DNA sequencing, DNA microarrays, more efficient cDNA library production, whole-genome knockouts and scaled-up two-hybrid mapping provides the basis for more efficient analysis of other, model organisms (Collins et al., 2003). Completion, or near completion, of a number of eukaryote genome sequences has provided landmarks in a field characterized by rapid advances in defining the structure and function of eukaryote genetic material. In the HGP, the availability of other mammalian genome sequences provided the possibility of comparative analyses allowing the identification of conserved exon sequences to aid in identifying genes (Dehal et al., 2001a). The inclusion of comparative genomics in gene identification emphasizes the importance of intron–exon structure in defining DNA sequences that are conserved and more likely to code for orthologous genes in related species. The utilization of transcriptomics (expressed sequence tags (ESTs) and full-length cDNAs) and mapped chromosomal locations (Imanishi et al., 2004), as well as cDNA expression patterns (Galbraith, 2003), in particular the cis-acting sequence elements defining the detailed control of expression within the network of interacting genes in a given cell type (Brown et al., 2002; Davidson et al., 2002), will contribute to identifying orthologous/syntenic relationships. At the whole-genome level, detailed comparisons of human chromosome 19 (Dehal et al., 2001a,b) and chromosome 7 (Scherer et al., 2003) within the mouse genome demonstrated that against a stable order of conserved single-copy genes, certain lineages of genes have undergone dramatic changes in number.

Gene order in many eukaryote species is not random. In species such as human, flies and yeast there is clustering of co-expressed genes that cannot be explained as a trivial consequence of tandem duplication. It is postulated that chromatin-level gene regulation is the most likely explanation for regional co-expression in these species. However, in worms, such as C. elegans, approximately 15% of genes are incorporated into bacterial-like operons where genes within the same operon are transcribed together. This genome organization of C. elegans differs from the genomes of other eukaryotes and, interestingly, gene duplication that is not linked to existing operons is seen to be selectively favourable by allowing the evolution of new functions (Lercher et al., 2003).

G + C content, isochores, gene-rich regions

In eubacteria, the overall average guanine and cytosine content (G + C) differs substantially across species. It can range anywhere from 25 to 75% (Muto and Osawa, 1978). It is tempting to correlate bacterial G + C content with their phylogenetic relationship. However, we know that even relatively closely related species can have substantial differences in G + C content (cf. Mycoplasma species: Mycoplasma genitalium, Mycoplasma pneumoniae and Mycoplasma capricolium) (Bellgard and Gojobori, 1999b; Muto and Osawa, 1987). If the G + C content is examined in more detail, for example by examining the G + C content at the third codon position, it is possible to observe these differences even more clearly (Bellgard and Gojobori, 1999b, c; Bellgard et al., 2001). Between closely related species such as Mycobacterium species it is possible to identify the ancestral G + C content between the species and infer that M. leprae has lost G + C content compared with M. tuberculosis. It is also noted that M. leprae has a high proportion of pseudogenes, providing further evidence of its divergence from the common ancestor. This type of analysis can also be extended to within-species comparison (Bellgard and Gojobori, 1999c). In this study, it was discovered that the two completely sequenced strains of Helicobacter pylori are already showing G + C content differences. It is important to note that these differences cannot be noticed by examining average G + C content. Further analysis of codons reveals G + C content variation across the H. pylori genomes. There is one region in the genome that has a lower G + C content than the rest of the genome for both strains. Interestingly, this particular region contains the pathogenicity genes, known as the CAG pathogenicity island (Backert *et al.*, 2004). These are thought to be horizontally transferred into a *H. pylori* common ancestor. The detailed G + C analysis independently supports this hypothesis (see Fig. 10.2). The evolutionary mechanisms differentiating the substantial G + C content remain elusive. The G + C content could be a form of adaptation to the environmental conditions. For example, thermophilic bacteria inhabit very hot environments. Another possibility could be a correlation to UV radiation within the environment to G + C content within the species (Singer and Ames, 1970). An alternative to the selectionist views above are the mutationalist views where the G + C content of a given bacterial species is determined by the balance between the rate of substitution from G or C to T or A and the rate of substitution from A or T to G or C (Muto and Osawa, 1987; Sueoka, 1999a).

When the G + C content in eukaryotes is examined, some striking differences are found compared with bacteria. In general, the G + C content exhibits much smaller variation and even within eukaryotes such as plants, invertebrates and vertebrates, analysed separately, the G + C variation systematically reduces to roughly 35–45% (Sueoka, 1999b). Part of the

explanation for this could be from an historical perspective as vertebrates have not diverged long enough from one another to allow for considerable differences in G + C content to accumulate. However, as discussed previously in the context of the detailed analysis of G + C content in bacteria, vertebrates have even more significant within-species heterogeneity in G + C content. Such compositionally homogenous stretches of DNA may reflect that the genome is composed of a mosaic of these regions which have been termed isochores. These isochores are approximately 100–300 kb or even larger in size (Cuny *et al.*, 1981; Bernardi, 1985; Bernardi, 2001). The skewed distribution is composed of five normal distributions, corresponding to five distinct types of isochore (L1, L2, H2, H2 and H3, with GC contents of <38%, 38–42%, 42–47%, 47–52%, respectively. These isochores are correlated with gene-rich regions. The International Human Genome Sequencing Consortium (IHGSC) published a paper on the human genome and specifically examined the human genome to try to identify strict isochores (IHGSC, 2001). The consortium failed to find any strict isochores, which was to be expected as there is some confusion on the terminology (Bernardi, 2001). In addition, the consortium failed to agree that the human genome is a mosaic of isochores, but there is strong evidence for compositionally discontinuous sequence organization (Bernardi, 2001).

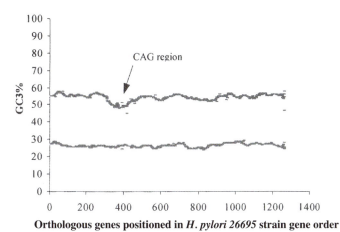

Fig. 10.2. G + C content bioinformatics analysis of the two strains of *H. pylori* species reveals the CAG pathogenicity island. GC3% refers to the content of G+C at the third base codon position for each gene. Reproduced with permission (Bellgard *et al.*, 2001).

The transcriptome, non-coding RNA and systems biology

The traditional view of biology is that there is a flow of genetic information from DNA to RNA to protein to protein interaction and through to metabolic pathway. As part of this view, genes generally code for proteins, fulfilling the majority of structural, catalytic and regulatory functions in all living cells. The recent high-throughput sequencing and annotation of numerous cDNA projects (such as human and mouse) indicate that additional variables need to be considered, namely, it is apparent that there is high putative splice variation and a relatively large number of non-coding RNA sequences (Imanishi *et al.*, 2004). These observations contribute significantly to the functional complexity of the mammalian transcriptome, representing a paradigm shift in the way we now view biological systems.

Analysing the full-length cDNAs of mouse (60,000) in addition to the publicly available mRNAs (44,000) revealed that a high proportion (41%) of the resulting transcript clusters showed evidence for alternative spliced forms; 79% of putative splice variants would produce altered protein sequences. In addition these estimates could be regarded as conservative as not all transcripts are represented in the current cDNA libraries. The importance of the mouse transcriptome is further highlighted as rare transcripts may occur in tissues that are not readily available in humans, such as the embryo (Rossant and Scherer, 2003).

Small interfering RNAs (siRNAs) and micro RNAs (miRNAs) are two types of ~22 nucleotide non-coding RNAs that play important roles as regulators of gene expression in eukaryotes (Allshire, 2002; Hutvagner and Zamore, 2002; Volpe *et al.*, 2002). siRNAs derive from the successive cleavage of long double-stranded RNA (dsRNA). They direct the destruction of corresponding mRNA targets during RNA inference in animals and plants (Reinhart and Bartel, 2002). It is thought to be responsible for post-transcriptional gene silencing, or co-suppression, a mechanism by which endogenous genes are silenced in the presence of a homologous transgene (Jorgensen, 1990).

MiRNAs are a rapidly growing family of 20–25 nucleotides (nt) RNA genes present in animals, fungi and plants that perform a variety of functions, including the regulation of gene expression. miRNAs act as post-transcriptional repressors of target transcripts or mediate RNA degradation via RNA interference (Aukerman and Sakai, 2003). They are processed from longer precursor transcripts that range in length from ~70 to 200 nt, and these precursor transcripts have the ability to form stable hairpin structures. miRNAs appear to regulate target genes by binding to complementary sequences located in the transcripts produced by these genes. Both animals and plants possess enzymes involved in processing miRNA precursors (Aukerman and Sakai, 2003). It is hypothesized that miRNAs, in many cases, may be involved in the regulation of important developmental processes. In animals, specific miRNAs are involved in development, such as temporal development (Lee *et al.*, 1993; Reinhart and Bartel, 2002), as well as others displaying developmentally regulated patterns of expression, both temporal and tissue specific (Lago-Quintana *et al.*, 2001, 2002; Lau *et al.*, 2001). There have been studies in both *Drosophila* and mammals that have associated specific developmental phenotypes with reduced expression of specific miRNAs (Brennecke *et al.*, 2003; Xu *et al.*, 2003). In plants, one striking observation about the putative target transcripts of plant miRNAs is that most of them encode members of transcription factor families that have been implicated in plant development patterning cell differentiation (Rhoades *et al.*, 2002; Aukerman and Sakai, 2003).

Given this paradigm shift, the role of transcribed retrotransposable elements (collectively referred to as retroelements) comes into question. Retroelements (which are covered in Chapter 11) have spread throughout eukaryote genomes by a process of transcription, retrotranscription, retrotransposition and insertion in various locations. Alu sequences are the largest family of short interspersed nucleotide elements (SINEs) in humans (Jurka *et al.*, 2002) with over 1 million copies of Alu present in the human genome. Alu sequences are derived from the 7SL RNA gene and are evident in all primate species. It is important to note that there is no known mechanism for the specific removal of Alu repeats. Other mammalian species have analogous incorporation of transcribed retroelements (for example, Kohnoe et al., 1987). In humans, a survey of recent literature highlights the potential role of Alu in transcription. For instance, Alu are involved in alternative splicing (Ganguly et al., 2003; Imanishi et al., 2004), involved in cryptic splicing resulting in a novel exon 2 product for the cathepsin B lysomal cystein protease (Berquin et al., 1997), are transcriptionally active which may eliminate transcriptional interference of neighbouring genes (Willoughby et al., 2000), form stable secondary structures (Sobczak and Krzyzosiak, 2002), and retrotransposition can be induced by genotoxic stress (Hagan et al., 2003). In addition, there are reports on specific tissues, such as thyroid, that possess Alu sequences as high affinity binding sites (Babich et al., 1999), mRNAs differentially expressed between benign and malignant breast tumours that are associated with Alu repeats (Liu et al., 2002), and higher processing rates of Alu-containing sequences in kidney tumours and cell lines with overexpressed Alu mRNAs (Vila et al., 2003).

β-Tubulin genes

β-Tubulin is a ubiquitous highly conserved eukaryotic protein that constitutes 10% of the total protein content of eukaryotes in the form of microtubules (Fong et al., 1984). A heterodimer with α- and β-tubulin subunits, microtubules perform a myriad of functions within eukaryotic cells. These include cell motility, cell secretion, nutrient absorption and formation of the cytoskeleton, acting as a substrate for intracytoplasmic transport of vesicles and organelles, and establishment of asymmetric neurone morphology (Lacey, 1988; Díaz-Nido et al., 1990; Robinson et al., 1991; Idriss, 2000; Vedrenne, 2002). A wide range of biochemical and molecular properties parallels this functional diversity. Microtubules within single cells can exhibit differences in stability, assembly and patterns of post-translational modification (Lopata et al., 1983; Murphy and Wallis, 1985).

Tubulin sequences have been used in molecular phylogeny and studies of early eukaryotic cell evolution. Tubulin is the target of many established and potential antiparasitic compounds (Fong et al., 1984; Lacey, 1988; Lubega et al., 1993; Katiyar and Edlind, 1994; Anthony and Hussey, 1999). The basis of tubulin's use as a chemotherapeutic target lies in the differences between mammalian and plant, fungal and algal tubulins. For example, the β-tubulins of Homo sapiens and Trypanosoma brucei rhodesiense differ in their amino acid composition by 15.4%. Phylogenetic analysis of parasitic β-tubulin amino acid sequences demonstrated that they are 'distinctly more similar to plant and algal tubulins than animal tubulins' (Caccio et al., 1997). The variable regions that characterize each isotypic class are conserved between different vertebrate species. This suggests that each isotype has been positively selected during evolution, which implies a functional role for the structural differentiation of these isotypes (Sullivan and Cleveland, 1986). In mammals, the class 4β-tubulin isotype is encoded by two variants with different expression properties.

Within the human genome, 15–20 β-tubulin genes have been identified. The majority of the identified human β-tubulin genes are pseudogenes, which are incapable of coding for a functional protein due to the presence of one or more mutations within the coding regions (Raff, 1984). Pseudogenes can be derived from the mRNA of its corresponding functional counterpart (Wilde et al., 1982). Human beta-one is very specifically expressed in erythroid cells, constituting structural components of erythrocytes. The beta-two subtype is abundant in testis, and present in lower levels in many tissues (Sullivan and Cleveland, 1984; Lewis et al., 1985; Lewis and Cowan, 1988). Human beta-three subtype is present in the central and peripheral nervous systems and is highly expressed during fetal and postnatal development, specifically in cerebellar and sympathoadrenal neurogenesis. In embryonic and 'adult-type' neuronal tumours, which express beta-three, neuronal differentiation occurs and cell proliferation is decreased. Increased expression of beta-three in these tumour lines is indicative of resistance (Katsetos et al., 2003).

The mouse beta-one subtype is expressed in erythroid cell lineages, forming the marginal band of platelets and fetal erythroblasts. Mouse beta-two is abundant in brain tissue. Mouse beta-three and beta-four are expressed in the same tissues as their human homologues. It has been demonstrated that single gene products are sufficient for essential microtubule construction in lower eukaryotes (Cleveland, 1983). More recently, conserved intronic sequences between diverse species have been identified (Stromback, 2003).

Major histocompatibility complex

The major histocompatibility complex (MHC) is a genomic region that contains a family of closely linked genes that are involved functionally with immune systems and immune responses. However, not all the vertebrate species, such as chicken and fish, have their MHC genes linked or contained in a complex on a single chromosome (Hansen *et al.*, 1999; Kaufman *et al.*, 1999). The draft genome sequences of humans and mice contribute greatly to our knowledge of MHC genetics, genomic organization, molecular evolution, gene function, protein structure and disease associations. Typically the human MHC is regarded as the reference standard for comparative genomic analysis, while the chicken MHC is considered to represent the minimal essential MHC within a non-mammalian vertebrate (Shand and Dixon, 2001; Kulski *et al.*, 2002).

Comparing MHC in humans with other primates reveals that the majority of genes appear to be highly conserved, although the genomic organization of the genes may be different. However, when the human MHC and a region from the chicken MHC (B locus) are compared, major structural and evolutionary divisions between the MHC of mammalian and non-mammalian vertebrates are found (Kulski *et al.*, 2002). The nucleotide diversity between three mammalian species, namely human, chimpanzee and macaque, was found to be 1.1% between human and chimp and 5.4% between human and macaque (Kulski *et al.*, 2002). As might be expected, there is a strong similarity (structurally) between mouse and rat MHC genomic organization, as well as to other mammals including human, non-human primates and swine. A characteristic difference between human and rodent is the olfactory receptor region (olfr block) of the mouse and rat, which we discuss further in the next section.

Interestingly, the analysis of avian lineages such as penguins, chickens, quails and blackbirds reveals differences in genomic organization while particular regions have remained conserved over a period of approximately 400 million years. This is in contrast to evolutionary origins of the eutherian mammals over a period of approximately 130 million years.

The complete MHC class II region of human, domestic cat and mouse was compared and revealed remarkable conservation of nucleotide sequence and gene organization given that they last shared a common ancestor around 80 million years ago. Interestingly, the cat MHC lacks the entire DQ subregion, being the first characterized mammalian species to lack this region. The loss of genes in this region could explain the immunological 'tolerance' such as the lack of cytotoxic antibody production (Winkler *et al.*, 1989; Yuhki *et al.*, 2003). Comparative analysis of interspersed repeat sequences in the MHC class II regions among these species indicated five types of interspersed repeats, namely SINEs, LINEs, long terminal repeat (LTR) transposons, short tandem repeats and DNA transposons, were common to all three species. There were, however, notable differences in the proportion of repeat sequences both within and across species. For instance, human SINE repeats were twice as dense as their counterpart SINEs in mouse and cat. Cats had 3–6 times fewer endogenous retroviral LTR sequences across their MHC compared with mouse or human, whereas mice showed a four- to eightfold reduction in DNA transposon elements relative to human and cat (Yuhki *et al.*, 2003).

Olfactory genes

The available sequence of both the human genome chromosome 19 and the related mouse DNA illuminates the function and evolutionary history of both genomes and provides us with a key to extrapolate on the overall evolution of mammalian genomes. The human chromosome 19 (HSA19) is approximately 65 Mb and contains approximately 1100 genes. HSA19 exists as a single, conserved linkage group in most primates (Ostrander *et al.*, 2000), and linkage within each respective HSA19 chromosome arm is well conserved in dogs, cats and cattle (Band *et al.*, 2000; Murphy *et al.*, 2000; Postlethwait *et al.*, 2000). Contained within HSA19 are 49 olfactory receptor loci distributed in four major clusters and 29 pheromone receptors (vomeronasal receptor gene, or VR) loci.

Olfactory receptors (ORs), are typically the first dedicated molecules with which odorants physically interact to arouse an olfactory sensation and constitute the largest gene family in vertebrates, including around 900 genes in humans and 1500 in the mouse. ORs are G protein-coupled, seven-transmembrane proteins that are responsible for binding odorants in the nasal epithelium. Interestingly, OR singletons are well conserved between human and mouse in a one-to-one mapping, however, lineage-specific changes continue at a rapid pace within gene families for which *in situ* duplication provides the major mode of expansion (Dehal *et al.*, 2001a). Of the known ORs in canines, approximately 18% are pseudogenes, compared with 63% in human and 20% in mouse (Carver *et al.*, 1998). Like in mouse, the canine OR repertoire appears to have expanded relative to that of humans, leading to the emergence of specific canine OR genes (Quignon *et al.*, 2003). In further detailed analysis by Quignon and his colleagues, canine OR gene clusters contain more genes than their orthologous human clusters and there are examples where particular clusters are centric on one chromosome in canines, whereas they scattered in multiple chromosomes in human. Interestingly, a number of these pseudogenes in human have been shown to segregate between an intact and pseudogene form. In this study, notably, non-African individuals had significantly fewer functional ORs than did African-American individuals. Thus, different evolutionary pressures may have shaped the chemosensory repertoire in different human populations (Menashe *et al.*, 2003).

OR genes are subject to unusually complex transcriptional regulatory mechanisms. Human MHC human leucocyte antigen (HLA)-linked OR genes not only function as odorant receptors, but are also suggested to play a role in the fertilization process. In testes of several mammalian species, including human, at least 50 OR genes are transcribed that could be involved in sperm development, sperm competition and chemotaxis (Volz *et al.*, 2003).

In comparing the VR genes between human and mouse, only one human VR gene appears to be functional. One of the more striking examples of primate-specific gene loss is that in human only one VR gene is consistent with the precipitous loss of pheromone receptor capacity in primates (Dehal *et al.*, 2001a). In contrast, in rodents, new VR genes have been generated through active rounds of gene duplication.

Discussion

In this chapter the characteristics of mammalian genomes in terms of genome and gene structure have been explored in context with other species. Base composition can differ dramatically across bacterial species in contrast to eukaryote species. In addition, eukaryote genomes exhibit significant heterogeneity within species where the high G + C content regions are known as isochores and usually correlate with gene-rich regions. Horizontal gene transfer has been acknowledged as a very important mechanism for shaping the adaptation and diversity amongst bacterial species. Understanding the transcriptome is paramount in order to obtain a holistic comprehension of the interaction of the various coding elements such as genes, regulatory elements and even retroelements that play critical roles in the observed phenotype differences between related species. Species display high levels of genome plasticity in terms of rearrangements, gene shuffling and even gene loss which is particularly striking in the bacterial species comparisons. Eukaryote genomes, through examining ubiquitous gene families and functional genome regions, display high levels of syntenic conservation. Interestingly, these studies make it possible to investigate gene loss and pseudogenes. The olfactory and pheromone receptors are striking examples across the animal kingdom of genes that are subject to gene loss, or become pseudogenes, or are lineage specific. The properties of these well conserved gene families highlight how gene loss can contribute to the rapid pace of species divergence.

Acknowledgements

The authors acknowledge the assistance of Rebecca Gardiner in formatting this manuscript and Gary Martin for proofreading.

References

Allshire, R. (2002) RNAi and heterochromatin – a hushed up affair. *Science* 297, 1818–1819.

Anthony, R.G. and Hussey, P.J. (1999) Dinitroaniline herbicide resistance and the microtubule cytoskeleton. *Trends in Plant Science* 4, 112–116.

Aukerman, M. and Sakai, H. (2003) Regulation of flowering time and floral organ identity by an MicroRNA and its *APETALA2*-like target genes. *The Plant Cell* 15, 2730–2741.

Babich, V., Aksenov, N., Alexeenko, V., Oei, S.L., Buchlow, G. and Tomilin, N. (1999) Association of some potential hormone response elements in human genes with the Alu family repeats. *Gene* 239, 341–349.

Babu, M. (2003) Did the loss of sigma factors initiate pseudogene accumulation in *M. leprae*? *Trends in Microbiology* 11, 59–61.

Backert, S., Schwarz, T., Miehlke, S., Kirsch, C., Sommer, C., Kwok, T., Gerhard, M., Goebel, U.B., Lehn, N., Koenig, W. and Meyer, T.F. (2004) Functional analysis of the CAG pathogenicity island in *Helicobacter pylori* isolates from patients with gastritis, peptic ulcer, and gastric cancer. *Infections and Immunity* 72, 1043–1056.

Band, M.R., Larson, J.H., Rebeiz, M., Green, C.A., Heyen, D.W., Donovan, J., Windish, R., Steining, C., Mahyuddin, P., Womack, J.E. and Lewin, H.A. (2000) An ordered comparative map of the cattle and human genomes. *Genome Research* 10, 1359–1368.

Bellgard, M. and Gojobori, T. (1999a) Identification of a ribonuclease H gene in both *Mycoplasma genitalium* and *Mycoplasma pneumoniae* by a new method for exhaustive identification of ORFs in the complete genome sequences. *FEBS Letters* 445, 6–8.

Bellgard, M. and Gojobori, T. (1999b) Inferring the direction of evolutionary changes of genomic base composition. *Trends in Genetics* 15, 254–256.

Bellgard, M. and Gojobori, T. (1999c) Significant differences between the G+C content of synonymous codons in orthologous genes and the genomic G+C content. *Gene* 238, 33–37.

Bellgard, M., Itoh, T., Watanabe, H., Imanishi, T. and Gojobori, T. (1999) Dynamic evolution of genomes and the concept of genome space. *Annals of the New York Academy of Sciences* 870, 293–300.

Bellgard, M., Schibeci, D., Trifonov, E. and Gojobori, T. (2001) Early detection of G + C differences in bacterial species inferred from the comparative analysis of the two completely sequenced *Helicobacter pylori* strains. *Journal of Molecular Evolution* 53, 465–468.

Bellgard, M., Ye, J., Gojobori, T. and Appels, R. (2004) The bioinformatics challenges in comparative analysis of cereal genomes – an overview. *Functional and Integrative Genomics* 4, 1–11.

Bernardi, G. (1985) Codon usage and genome composition. *Journal of Molecular Evolution* 22, 363–365.

Bernardi, G. (2001) Misunderstandings about isochors. Part 1. *Gene* 3, 3–13.

Berquin, I.M., Ahram, M. and Sloane, B.F. (1997) Exon 2 of human cathepsin B derives from an Alu element. *FEBS Letters* 419, 121–123.

Brennecke, J., Hipfner, D.R., Stark, A., Russel, R.B. and Cohen, S.M. (2003) *Bantam* encodes a developmentally regulated microRNA that controls cell proliferation and regulates the proapoptotic gene *hid* in *Drosophila*. *Cell* 113, 25–36.

Brown, C.T., Rust, A.G., Clarke, P.J.C., Pan, Z., Schilstra, M.J., De Buysscher, T., Griffin, G., Wold, B.J., Cameron, R.A., Davidson, E.H. and Bolourim, H. (2002) New computational approaches for analysis of cis-regulatory networks. *Developmental Biology* 246, 86–102.

Bult, C.J., White, O., Olsen, G.J., Zhou, L. and Fleischmann, R.D. (1996) Complete genome sequence of the methanogenic archaeon, *Methanococcus jannaschii*. *Science* 273, 1058–1073.

Caccio, S., La Rosa, G. and Pozio, E. (1997) The [beta]-tubulin gene of *Cryptosporidium parvum*. *Molecular and Biochemical Parasitology* 89, 307–311.

Campbell, A.M. (2000) Lateral gene transfer in prokaryotes. *Theoretical Population Biology* 57, 71–77.

Carver, E.A., Issel-Tarver, L., Rine, J., Olsen, A.S. and Stubbs, L. (1998) Location of mouse and human genes corresponding to conserved canine olfactory receptor gene subfamilies. *Mammalian Genome* 9, 349–354.

Cleveland, D.W. (1983) The tubulins: from DNA to RNA to protein and back again. *Cell* 2, 330–332.

Cole, S.T., Supply, P. and Honore, N. (2001) Repetitive sequences in *Mycobacterium leprae* and their impact on genome plasticity. *Leprosy Review* 72, 449–461.

Collins, F.S., Morgan, M. and Patrinos, A. (2003) The Human Genome Project: lessons from large-scale biology. *Science* 300, 286–290.

Cuny, G., Soriano, P., Macaya, G. and Bernardi, G. (1981) The major components of the mouse and human genomes. 1. Preparation, basic properties and compositional heterogeneity. *European Journal of Biochemistry* 115, 227–233.

Daubin, V. and Perrière, G. (2003) G+C3 Structuring along the genome: a common feature in prokaryotes. *Molecular Biology and Evolution* 20, 471–483.

Davidson, E.H. , Rast, J.P., Oliveri, P., Ransick, A., Calestani, C., Yuh, C., Minokawa, T., Amore, G., Hinman, V., Arenas-Mena, C., Otim, O., Brown, C.T., Livi, C.B., Lee, P.Y., Revilla, R., Rust, A.G., Pan, Z., Schilstra, M.J., Clarke, P.J., Arnone, M.I. *et al.* (2002) A genomic regulatory network for development. *Science* 295, 1669–1678.

Dehal, P., Predki, P., Olsen, A.S., Kobayashi, A., Folta, P., Lucas, S., Land, M., Terry, A., Ecale Zhou, C.L., Rash, S., Zhang, Q., Gordon, L., Kim, J., Elkin, C., Pollard, M.J., Richardson, P., Rokhsar, D., Uberbacher, E., Hawkins, T., Branscomb, E. and Stubbs, L. (2001a) Human chromosone 19 and related regions in mouse: conservative and lineage-specific evolution. *Science* 293, 104–111.

Dehal, P., Predki, P., Olsen, A.S., Kobayashi, A., Folta, P., Lucas, S., Land, M., Terry, A., Ecale Zhou, C.L., Rash, S., Zhang, Q., Gordon, L., Kim, J., Elkin, C., Pollard, M.J., Richardson, P., Rokhsar, D., Uberbacher, E., Hawkins, T., Branscomb, E. and Stubbs, L. (2001b) Homology-driven assembly of a sequence-ready mouse BAC contig map spanning regions related to the 46-Mb gene-rich euchromatic segments of human chromosome 19. *Genomics* 74, 129–141.

Díaz-Nido, J., Serrano, L., López-Otín, C., Vandekerckhove, J. and Avila, J. (1990) Phosphorylation of a neuronal-specific beta-tubulin isotype. *Journal of Biological Chemistry* 265, 13949–13954.

Dobzhansky, T. (1973) Nothing in biology makes sense except in the light of evolution. *American Biology Teacher* 35, 125–129.

Doolittle, W.F. (1999) Lateral genomics. *Trends in Genetics* 15, M5–M8.

Eisen, J.A. and Fraser, C.M. (2003) Phylogenomics: intersection of evolution and genomics. *Science* 300, 1706–1707.

Fong, D., Wallach, M., Keithly, J., Melera, P.W. and Chang, K.P. (1984) Differential expression of mRNA's for alpha- and beta-tubulin during differentiation of the parasitic protozoan *Leishmania mexicana. Proceedings of the National Academy of Sciences USA* 81, 5782–5786.

Fraser, C.M., Gocayne, J.D., White, O., Adams, M.D., Clayton, R.A., Fleischmann, R.D., Bult, C.J., Kerlavage, A.R., Sutton, G. and Kelley, J.M. (1995) The minimal gene complement of *Mycoplasma genitalium. Science* 270, 397–403.

Galbraith, D. (2003) Global analysis of cell type-specific gene expression. *Comparative and Functional Genomics* 4, 208–215.

Ganguly, A., Dunbar, T., Chen, P., Godmilow, L. and Ganguly, T. (2003) Exon skipping caused by an intronic insertion of a young Alu Yb9 element leads to severe hemophilia A. *Journal of Human Genetics* 113, 348–352.

Gogarten, J.P., Doolittle, W.F. and Lawrence, J.G. (2002) Prokaryotic evolution in light of gene transfer. *Molecular Biology and Evolution* 19, 2226–2238.

Hagan, C.R., Sheffield, R.F. and Rudin, C.M. (2003) Human Alu element retrotransposition induced by genotoxic stress. *Nature Genetics* 35, 219–220.

Hansen, J.D., Strassburger, P., Thorgaard, G.H., Young, W.P. and Du Pasquier, L. (1999) Expression linkage and polymorphism of MHC-related genes in rainbow trout, *Oncorynchus mykiss. Immunology* 163, 774–786.

Hinman, V.F., Nguyen, A.T. and Davidson, E.H. (2003) Expression and function of a starfish Otx ortholog, AmOtx: a conserved role for Otx proteins in endoderm development that predates divergence of the eleutherozoa. *Mechanisms of Development* 120, 1165–1176.

Hutvagner, G. and Zamore, D. (2002) RNAi: nature abhors a double-strand. *Current Opinion in Genetics and Development* 12, 225–232.

Ideker, T., Galistski, T. and Hood, L. (2001) A new approach to decoding life: systems biology. *Annual Reviews in Genomics and Human Genetics* 2, 343–372.

Idriss, H.T. (2000) Man to trypanosome: the tubulin tyrosination/detyrosination cycle revisited. *Cell Motility and the Cytoskeleton* 45, 173–184.

Imanishi, T. *et al.* (2004) Integrative annotation of 21,037 human genes validated by full-length cDNA clones. *Public Library of Science* 2 (6), 856–875.

International Human Genome Sequencing Consortium (IHGSC) (2001) Initial sequencing and analysis of the human genome. *Nature* 409, 860–921.

Jain, R., Rivera, M., Moore, J. and Lake, J. (2002) Horizontal gene transfer in microbial genome evolution. *Theoretical Population Biology* 61, 489–495.

Jorgensen, R. (1990) Altered gene expression in plants due to *trans* interactions between homologous genes. *Trends in Biotechnology* 8, 340–344.

Jurka, J., Krnjajic, M., Kapitonov, V., Stenger, J. and Kokhanyy, O. (2002) Active alu elements are passed primarily through paternal germlines. *Theoretical Population Biology* 61, 519–530.

Karlin, S., Mrazek, J. and Campbell, A.M. (1997) Compositional biases of bacterial genomes and evolutionary implications. *Journal of Bacteriology* 179, 3899–3913.

Katiyar, S.K. and Edlind, T.D. (1994) Beta-tubulin genes of *Trichomonas vaginalis*. *Molecular and Biochemical Parasitology* 64, 33–42.

Katsetos, C.D., Herman, M.M. and Mork, S.J. (2003) Class III beta-tubulin in human development and cancer. *Cell Motility and the Cytoskeleton* 55, 77–96.

Kaufman, J. *et al.* (1999) The chicken B locus is a minimal essential major histocompatibility complex. *Nature* 401, 923–925.

Kinsella, J., Fitzpatrick, D.A., Creevey, C.J. and McInerney, J.O. (2003) Fatty acid biosynthesis in *Mycobacterium tuberculosis*: lateral gene transfer, adaptive evolution, and gene duplication. *Proceedings of the National Academy of Sciences USA* 100, 10320–10325.

Kohnoe, S., Maehara, Y. and Endo, H. (1987) A systematic survey of repetitive sequences abundantly expressed in rat tumors. *Biochimica et Biophysica Acta*, 910, 93.

Koonin, E.V. (2003) Horizontal gene transfer: the path to maturity. *Molecular Microbiology* 50, 725–727.

Kulski, J.K., Shiina, T., Anzai, T., Kohara, S. and Inoko, H. (2002) Comparative genomic analysis of the MHC: the evolution of class I duplication blocks, diversity and complexity from shark to man. *Immunology Review* 190, 95–122.

Kurland, C.G., Canback, B. and Berg, O.G. (2003) Horizontal gene transfer: a critical view. *Proceedings of the National Academy of Sciences USA* 100, 9658–9662.

Lacey, E. (1988) The role of the cytoskeletal protein, tubulin, in the mode of action and mechanism of drug resistance to benzimidazoles. *International Journal for Parasitology* 18, 885–893.

Lagos-Quintana, M., Rauhut, R., Lendeckel, W. and Tuschi, T. (2001) Identification of novel genes coding for small expressed RNAs. *Science* 294, 853–858.

Lagos-Quintana, M., Rauhut, R., Yalcin, A., Meyer, J., Lendeckel, W. and Tuschi, T. (2002) Identification of tissue-specific microRNAs from mouse. *Current Biology* 12, 735–739.

Lake, J.A. and Rivera, M.C. (1996) The prokaryotic ancestry of eukaryotes. In: Roberts, D. McL., Sharp, P., Alderson, G. and Collins, M. (eds) *Evolution of Microbial Life* Symposium 54. Society for General Microbiology. Cambridge University Press, Cambridge, pp. 87–108.

Lake, J.A., Jain, R. and Rivera, M.C. (1999) Mix and match in the tree of life. *Science* 283, 2027–2028.

Lane, D.J., Field, K.G., Olsen, G.J. and Pace, N.R. (1988) Reverse transcriptase sequencing of ribosomal RNA for phylogenetic analysis. *Methods in Enzymology* 167, 138–144.

Lau, N.C., Lim, L.P., Weinstein, E.G. and Bartel, D.P. (2001) An abundant class of tiny RNAs with probably regulatory roles in *Caenorhabditis elegans*. *Science* 294, 862–864.

Lawrence, J.G. and Hendrickson, H. (2003) Lateral gene transfer: when will adolescence end? *Molecular Microbiology* 50, 739–749.

Lawrence, J.G. and Ochman, H. (1998) Molecular archaeology of the *Escherichia coli* genome. *Proceedings of the National Academy of Sciences USA* 95, 9413–9417.

Lee, R.C., Feinbaum, R.L. and Ambros, V. (1993) The *C. elegans* heterochronic gene *lin-4* encodes small RNAs with antisense complementarity to *lin-14*. *Cell* 75, 843–854.

Lercher, M.J., Blumenthal, T. and Hurst, L.D. (2003) Coexpression of neighboring genes in *Caenorhabditis elegans* is mostly due to operons and duplicate genes. *Genome Research* 13, 238–243.

Lewis, S.A. and Cowan, N. (1988) Complex regulation and functional versatility of mammalian alpha- and beta-tubulin isotypes during the differentiation of testis and muscle cells. *Journal of Cell Biology* 106, 2023–2033.

Lewis, S.A., Gilmartin, M.E., Hall, J.L. and Cowan, N.J. (1985) Three expressed sequences within the human beta-tubulin multigene family each define a distinct isotype. *Journal of Molecular Biology* 82, 11–20.

Liu, D., Rudland, P.S., Sibson, D.R. and Barraclough, R. (2002) Identification of mRNAs differentially-expressed between benign and malignant breast tumour cells. *British Journal of Cancer* 87, 423–431.

Lopata, M.A., Havercroft, J.C., Chow, L.T. and Cleveland, D.W. (1983) Four unique genes required for beta tubulin expression in vertebrates. *Cell* 32, 713–724.

Lubega, G.W., Geary, T.G., Klein, R.D. and Prichard, R.K. (1993) Expression of cloned beta-tubulin genes of *Haemonchus contortus* in *Escherichia coli*: interaction of recombinant beta-tubulin with native tubulin and mebendazole. *Molecular and Biochemical Parasitology* 62, 281–292.

Menashe, I., Man, O., Lancet, D. and Gilad, Y. (2003) Different noses for different people. *Nature Genetics* 34, 143–144.

Mineta, K., Nakazawa, M., Cebria, F., Ikeo, K., Agata, K. and Gojobori, T. (2003) Origin and evolutionary process of the CNS elucidated by comparative genomics analysis of planarian EST's. *Proceedings of the National Academy of Sciences USA* 100, 7666–7671.

Murphy, D.B. and Wallis, K.T. (1985) Erythrocyte microtubule assembly *in vitro*. Determination of the effects of erythrocyte tau, tubulin isoforms, and tubulin oligomers on erythrocyte tubulin assembly, and comparison with brain microtubule assembly. *Journal of Biological Chemistry* 260, 12293–12301.

Murphy, W.J., Sun, S., Chen, Z., Yuhki, N., Hirschmann, D., Menotti-Raymond, M. and O'Brien, S.J. (2000) A radiation hybrid map of the cat genome: implications for comparative mapping. *Genome Research* 10, 691–702.

Muto, A. and Osawa, S. (1987) The guanine and cytosine content of genomic DNA and bacterial evolution. *Proceedings of the National Academy of Sciences USA* 84, 166–169.

Ochman, H. (2001) Lateral and oblique gene transfer. *Current Opinion in Genetics and Development* 11, 616–619.

Ochman, H. and Jones, I.B. (2000) Evolutionary dynamics of full genome content in *Escherichia coli*. *EMBO Journal* 19, 6637–6643.

Ochman, H. and Moran, N. (2001) Genes lost and genes found: evolution of bacterial pathogenesis and symbiosis. *Science* 292, 1096–1098.

Ostrander, E., Galibert, F. and Patterson, D. (2000) Canine genetics comes of age. *Trends in Genetics* 16, 117.

Postlethwait, J.H., Woods, I.G., Ngo-Hazelett, P., Yan, Y.L., Kelly, P.D., Chu, F., Huang, H., Hill-Force, A. and Talbot, W.S. (2000) Zebrafish comparative genomics and the origins of vertebrate chromosomes. *Genome Research* 10, 1890–1902.

Quignon, P., Kirkness, E., Cadieu, E., Touleimat, N., Guyon, R., Renier, C., Hitte, C., Andre, C., Fraser, C. and Galibert, F. (2003) Comparison of the canine and human olfactory receptor gene repertoires. *Genome Biology* 4, R80.

Raff, E.C. (1984) Genetics of microtubule systems. *Journal of Cell Biology* 99, 1–10.

Reinhart, B. and Bartel, D. (2002) Small RNAs correspond to centromere heterochromatic repeats. *Science* 297, 1831.

Rhoades, M.W., Reinhart, B.J., Lim, L.P., Burge, C.B., Bartel, B. and Bartel, D.P. (2002) Prediction of plant microRNA targets. *Cell* 110, 513–520.

Rivera, M.C. and Lake, J.A. (1992) Evidence that eukaryotes and eocyte prokaryotes are immediate relatives. *Science* 257, 74–76.

Rivera, M.C., Jain, R., Moore, J.E. and Lake, J.A. (1998) Genomic evidence for two functionally distinct gene classes. *Proceedings of the National Academy of Sciences USA* 95, 6239–6244.

Robinson, D., Beattie, P., Sherwin, T. and Gull, K. (1991) Microtubules, tubulin, and microtubule-associated proteins of trypanosomes. *Methods in Enzymology* 196, 285–299.

Rossant, J. and Scherer, S.W. (2003) The mouse genome sequence – the end of the tail, or just the beginning? *Genome Biology* 4, 109.

Sabater-Munoz, B., Gomez-Valero, L., Van Ham, R.C., Silva, F.J. and Latorre, A. (2002) Molecular characterization of the leucine cluster in *Buchnera* sp. strain PSY, a primary endosymbiont of the aphid pemphigus spyrothecae. *Applied and Environmental Microbiology* 68, 2572–2575.

Scherer, S.W., Cheung, J., MacDonald, J.R., Osborne, L.R. *et al.* (2003) Human chromosome 7: DNA sequence and biology. *Science* 300, 767–772.

Shand, R. and Dixon, B. (2001) Teleost major histocompatibility genes: diverse but not complex. *Scandinavian Journal of Immunology* 2, 66–72.

Silva, F.J., Latorre, A. and Moya, A. (2003) Why are the genomes of endosymbiotic bacteria so stable? *Trends in Genetics* 19, 176–180.

Singer, C.E. and Ames, B.N. (1970) Sunlight ultraviolet and bacterial DNA base ratios. *Science* 170, 822–825.

Sobczak, K. and Krzyzosiak, W.J. (2002) Structural determinants of BRCA1 translational regulation. *Journal of Biological Chemistry* 277, 17349–17358.

Stromback, E. (2003) Comparative genome analysis of protozoan parasites via analysis of the structural and functional evolutionary relationships within the tubulin genes. Honours thesis, Murdoch University.

Sueoka, N. (1999a) Two aspects of DNA base composition: G+C content and translation-coupled deviation from intra-strand rule of A = T and G = C. *Journal of Molecular Evolution* 49, 49–62.

Sueoka, N. (1999b) Translation-coupled violation of Parity Rule 2 in human genes is not the cause of heterogeneity of the DNA G+C content of third codon position. *Gene* 238, 53–58.

Sullivan, K.F. and Cleveland, D.W. (1984) Sequence of a highly divergent beta tubulin gene reveals regional heterogeneity in the beta tubulin polypeptide. *Journal of Cell Biology* 99, 1754–1760.

Sullivan, K.F. and Cleveland, D.W. (1986) Identification of conserved isotype-defining variable region sequences for four vertebrate beta tubulin polypeptide classes. *Proceedings of the National Academy of Sciences USA* 83, 4327–4331.

Syvanen, M. and Kado, C. (eds) (1998) *Horizontal Gene Transfer*. Chapman and Hall, London, 474 pp.

Tamas, I. *et al.* (2002) 50 million years of genomic stasis in endosymbiotic bacteria. *Science* 296, 2376–2379.

Vedrenne, C., Giroud, C.E., Robinson, D.R., Besteiro, S., Bosc, C., Bringaud, F. and Baltz, T. (2002) Two related subpellicular cytoskeleton-associated proteins in *Trypanosoma brucei* stabilize microtubules. *Molecular Biology of the Cell* 13, 1058–1070.

Vila, M.R., Gelpi, C., Nicolas, A., Morote, J., Schwartz, S., Jr, Schwartz, S. and Meseguer, A. (2003) Higher processing rates of Alu-containing sequences in kidney tumors and cell lines with overexpressed Alu-mRNAs. *Oncology Reports* 10, 1903–1909.

Volpe, T., Kidner, C., Hall, I., Teng, G., Grewal, S. and Martienssen, R. (2002) Regulation of heterochromatic silencing and histone H3 lysine-9 methylation by RNAi. *Science* 297, 1833–1837.

Volz, A., Ehlers, A., Younger, R., Forbes, S., Trowsdale, J., Schnorr, D., Beck, S. and Ziegler, A. (2003) Complex transcription and splicing of odorant receptor genes. *Journal of Biological Chemistry* 278, 19691–19701.

Wallace, D.C. and Morowitz, H.J. (1973) Genome size and evolution. *Chromosoma* 40, 121–126.

Watanabe, H., Gojobori, T., Miura, K. and Watanabe, H. (1997) Bacterial features in the genome of *Methanococcus jannaschii* in terms of gene composition and biased base composition in ORFs and their surrounding regions. *Gene* 205, 7–18.

Wilde, C.D., Crowther, C.E., Cripe, T.P., Gwo-Shu Lee, M. and Cowan, N.J. (1982) Evidence that a human beta-tubulin pseudogene is derived from its corresponding mRNA. *Nature* 297, 83–84.

Willoughby, D.A., Vilalta, A. and Oshima, R.G. (2000) An Alu element from the K18 gene confers position-independent expression in transgenic mice. *Journal of Biological Chemistry* 275, 759–768.

Winkler, C., Schultz, A., Cevario, S. and O'Brien, S.J. (1989) Genetic characterisation of FLA, the cat major histocompatibility complex. *Proceedings of the National Academy of Sciences USA* 86, 943–947.

Xu, P., Vernooy, S.Y., Guo, M. and Hay, B.A. (2003) The *Drosophila* microRNA Mir-14 suppresses cell death and is required for normal fat metabolism. *Current Biology* 13, 790–795.

Yuhki, N., Beck, T., Stephens, R.M., Nishigaki, Y., Newmann, K. and O'Brien, S.J. (2003) Comparative genome organization of human, murine and feline MHC class II region. *Genome Research* 13, 1169–1179.

11 Elements and Mechanisms of Genome Change

R.J. O'Neill*, G.C. Ferreri and M.J. O'Neill

Department of Molecular and Cell Biology, U-2131, University of Connecticut, Storrs, CT 06269, USA

Introduction	279
Transposable Elements	280
Classification of transposable elements: class I and class II transposons	282
The genome shaping capabilities of TEs	282
Dynamic Rearrangements of Mammalian Genomes	284
Repeated Sequences and Genomic Changes	287
Chromosome evolution	287
Heterochromatin and centromeres	287
Duplications and Gene Families	288
Whole genome duplications	288
Segmental duplications	289
The role of duplications in mammalian genome evolution	291
Karyotypic Diversity, Reproductive Isolation and Speciation	292
References	293

Introduction

Dobzhansky and Sturtevant's (Dobzhansky and Sturtevant, 1938) seminal study 65 years ago classifying the chromosome rearrangements that distinguish two *Drosophila* species provided the first view of the molecular basis of species identity. Those pivotal observations launched decades of study of genome architecture from an evolutionary perspective. These studies have richly informed our understanding of developmental genetics, gene regulation, human genetic disorders and cancer and have greatly contributed to the Neo-Darwinian view of the divergence of species. Steadily throughout the last decade, with a sharp acceleration since the publication of the human genome sequence, molecular geneticists and bioinformaticians have described a eukaryotic genomic landscape that is fluid, dotted with evidence of both large-scale and fine-scale chromosome rearrangements. Recent findings in compara-

*Correspondence author: roneill@uconnvm.uconn.edu

© CAB International 2005. *Mammalian Genomics*
(eds A. Ruvinsky and J. Marshall Graves)

tive genomics have necessitated a re-evaluation of conventional models of chromosomal evolution, particularly in mammals.

The classical view is that the mammalian genome has evolved in static, conserved blocks (Nadeau and Taylor, 1984). The landscape of the eukaryotic genome had been considered relatively stable, characterized by regions of DNA conserved in gene content, order and sequence. Studies of the evolution of these 'conserved segments' (Nadeau and Taylor, 1984; O'Brien et al., 1999) have allowed evolutionary biologists to deduce phylogenetic relationships within Mammalia, as well as the evolutionary history of idiosyncratic portions of the genome, such as the Y chromosome (see Delbridge and Graves, 1999, for a review of Y chromosome evolution). Estimates of the number of conserved segments between the human and mouse genomes have been modified from approximately 180 (Nadeau and Taylor, 1984; Copeland et al., 1993; DeBry and Seldin, 1996; Gregory et al., 2002; Waterston et al., 2002) to 281 (Pevzner and Tesler, 2003a). The latter estimate takes into consideration homology not simply defined as segments of DNA conserved without disruption by rearrangement (Nadeau and Taylor, 1984), but as syntenic blocks that consist of 'short regions of similarity that may be interrupted by dissimilar regions and gaps' (Pevzner and Tesler, 2003a). This shift has largely been the result of refined sequencing efforts and the subsequent reanalysis of homologies between the human and mouse genomes (Pevzner and Tesler, 2003a,b). It is now realized that the mammalian genome has been shaped not only by large-scale rearrangements such as translocations, inversions, deletions, amplifications and duplications, but also by microrearrangements that occur across smaller distances within a single chromosome (Pevzner and Tesler, 2003b). When these rearrangements were viewed in a phylogenetic context (comparing syntenic blocks between two species across any given phylogenetic path) clustering of breakpoints was observed, thus giving rise to the concept of a mosaic genome in mammals, with regions prone to stability interspersed with fragile regions prone to rearrangement (Pevzner and Tesler, 2003b).

The molecular mechanisms responsible for these alterations, as well as the evolutionary forces behind their fixation within populations, have remained elusive. Remarkable insights, however, have been gained from examining the character of genetic elements involved in genome rearrangement. By carefully defining these elements and tackling the thorny issue of their evolutionary history a deeper understanding of the molecular mechanisms underlying genome plasticity may emerge. The second section of this chapter is a detailed catalogue of the types of mobile DNA present in the mammalian genome and the influence they have had on the evolution of gene function, chromosome structure and genome architecture. The third section of this chapter outlines several examples of mammalian clades that exemplify the dynamic nature of the genome through rapid karyotypic divergence. The fourth section focuses on repeated elements in the genome and their organization and involvement in chromosome structure and evolution, and the fifth section focuses on the role of duplication in genome evolution as well as gene function. Throughout this chapter, examples are given of rapid genome change in recently diverged species groups. This chapter concludes with a discussion of hypotheses regarding chromosome evolution and the process of speciation.

Transposable Elements

Transposable elements (TEs), or mobile DNA, are pervasive in mammalian genomes, accounting for approximately 50% of the human genome (Smit, 1999; International Human Genome Sequencing Consortium, 2001) (Table 11.1). They play an integral role in centromere structure (Kipling and Warburton, 1997; Schueler et al., 2001), allelic and non-allelic recombination, genome evolution (Wichman et al., 1992; DeBerardinis et al., 1998; O'Neill et al., 1998; Hughes and Coffin, 2001) and in disease-causing mutations in humans and mice (Kazazian, 1998). At the most basic level TEs can be considered very effective genomic parasites. They have colonized the genomes of all eukaryotes and many prokaryotes (Britten and Kohne, 1968; Singer, 1982; Coffin et al., 1997).

Table 11.1. Distribution and basic structure of different elements contained in the human genome.

Genomic element	Frequency	Structure
(a) Exons and introns[1]	~2%	
(b) SINE (Alu)[1–3]	13%	
(c) Non-LTR elements (L1)[1–3]	21%	
(d) LTR elements [1,3,4]	7.2%	
(e) DNA transposons[1,3–5]	3%	
(f) Simple sequence repeats[1]	3%	
(g) Processed pseudogenes[6]	0.5%*	
(h) Segmental duplications[1,7,8]	5.2%	

*Based on chromosome 22.

(a) Exons preceded by a promoter (P) and other regulatory regions are interrupted by introns and end in a poly(A) tail. (b) SINEs (Alu) are intronless and consist of a left (L-Alu) and right (R-Alu) monomer, separated by an A-rich linker, ending in a poly(A) tail and flanked by direct repeats (triangles) that are a duplication of the insertion site. (c) Non-LTR elements are flanked by target site duplications (triangles) and contain both 5′ and 3′ untranslated regions (UTRs), a poly(A) tail, and two open reading frames (ORFs) separated by an intergenic spacer. The endonuclease and reverse transcriptase enzymes are encoded in the ORF2. (d) LTR elements are flanked by target site duplications (triangles) and a 5′ and 3′ long terminal repeat (LTR) that carry promoter activity. These elements consist of overlapping coding regions for group-specific antigen (gag), protease (prt), polymerase (pol), which encodes reverse transcriptase and endonuclease, and envelope (env) proteins. (e) DNA transposons usually consist of a single open reading frame that encodes the transposase enzyme. (f) The simple repeats shown here are a pair of tandem repeats in a head-to-head orientation. (g) Processed pseudogenes are exonic sequences that carry intact UTRs (not shown) and a poly(A) tail. (Adapted from Prak and Kazazian, 2000.)

References: 1. International Human Genome Sequencing (2001); 2. Kazazian and Moran (1998). 3. Smit (1999); 4. Prak and Kazazian (2000); 5. Craig (2002); 6. Baily *et al.* (2002b); 7. Bailey *et al.* (2002a); 8. Bailey *et al.* (2001).

Classification of transposable elements: class I and class II transposons

TEs are generally partitioned into two classes based on whether they replicate through an RNA or a DNA intermediate (Sherratt, 1995; Robertson, 1996). The first class of TEs (class I), retroelements is by far the most well characterized in mammalian genomes (Auge-Gouillou et al., 1995; Hughes and Coffin, 2002; Neitzel et al., 2002). Retroelements in mammals are further subdivided into three categories: the long terminal repeat (LTR) retrotransposons, poly(A) or non-LTR transposons, and retrotranscripts. The first two contain an open reading frame coding for a reverse transcriptase while the last is mobilized through the action of this enzyme co-opted from another retroelement.

As the name suggests, LTR retrotransposons consist of elements that are flanked by long terminal repeat sequences. The most well characterized group of vertebrate LTR retrotransposons are endogenous retroviruses, accounting for ~7% of the human genome (Smit, 1999; Prak and Kazazian, 2000). Endogenous retroviruses are incapable of a complete replication cycle due to mutations in critical coding regions and, depending on the level of this degeneracy, are frequently transcriptionally silent. However, they can produce primary RNA transcripts, or even produce packaged virions that are capable of leaving the cell yet lack the ability to re-infect other cells (Doolittle et al., 1989; Levy, 1992).

Poly(A) or non-LTR retrotransposons comprise a second, large and diverse group of TEs. In mammalian genomes this group is typified by long interspersed nuclear elements (LINE-1 or L1). L1 elements are one of the most pervasive groups of elements, accounting for approximately 21% of the human genome (Smit, 1996; Kazazian and Moran, 1998; International Human Genome Sequencing Consortium, 2001), with similar abundance in other mammalian genomes (Smit, 1996, 1999; International human Genome Sequencing Consortium, 2001). The majority of human L1 elements are rendered incapable of further transposition because they are subject to the high mutation rate associated with reverse transcriptase-mediated transposition as well as epigenetic silencing (Dombroski et al., 1991).

Another group of retroelements that are closely tied to L1 elements are retrotranscripts. This group of non-autonomous elements includes short interspersed nuclear elements, or SINEs, and processed pseudogenes. The most prevalent SINE in humans, Alu elements, make up >10% (500,000 copies) of the genome (Deininger and Batzer, 1999; Smit, 1999; International Human Genome Sequencing Consortium, 2001). The primate Alu and the rodent B1 elements typify the structure and activity of mammalian SINEs (Table 11.1). The conserved secondary structure of these elements suggests that they possibly utilize L1-encoded endonucleases as vehicles for transposition. Although Alu and B1 elements do not integrate into the same target sites as their respective L1 elements, a similar process of in trans mobilization has been proposed (Jurka, 1997; Esnault et al., 2000; Wei et al., 2001). It has been estimated that transposition through L1-encoded sequences is responsible for one-third of the human genome (Coffin et al., 1997). Similar numbers have been suggested for mice, marsupials and other mammalian genomes (Burton et al., 1986; Dorner and Paabo, 1995). Recently it has been shown that L1 transposition enzymes also mediate transposition of processed pseudogenes, polyadenylated and spliced cDNA copies of the transcripts from endogenous protein coding genes (Esnault et al., 2000).

The second class of elements, known as class II DNA transposons, is not as well documented in mammalian genomes. This class of elements, which includes the Tc1/mariner family of elements, has been studied extensively outside of mammals (Berg and Howe, 1989; Craig, 2002). In these species such elements are active and responsible for many of the observed genome rearrangements and mutations (see Daviere et al., 2001 for an example). Current research indicates that class II DNA transposons in mammals are mostly 'fossilized' elements representing ancient, long inactive sequences (Robertson, 1996; Smit and Riggs, 1996).

The genome shaping capabilities of TEs

The realization that TEs are distributed ubiquitously in the genomes of all eukaryotes and most prokaryotes has led to a contentious debate regarding the evolutionary relationships that exist between TEs and their hosts. Phylogenetic analysis of transposable element proteins has established that eukaryotic TEs have co-evolved with their eukaryotic hosts, originating prior to the divergence of eukaryotes from prokaryotes (Doak *et al.*, 1994). This model of eukaryotic TE evolution suggests that DNA transposons (class II elements) and non-LTR retrotransposons (class I elements) co-evolved in eubacteria (Doak *et al.*, 1994; Eickbush, 1997). LTR retrotransposons would have then evolved as a result of a fusion between the two groups within a common ancestor (Eickbush, 1997). Tracing the phylogenetic path of TEs, however, has been complicated by the non-Mendelian inheritance patterns that have been observed in single lineages (Doolittle *et al.*, 1989; Doolittle and Feng, 1992). While these elements are inherited through the germline, they may also experience stochastic loss and horizontal transfer (Fig. 11.1), thereby creating extreme diversity in TE populations even among closely related species.

Although a large percentage of mammalian endogenous retroviral sequences are degenerate copies, a number of these sequences are intact and capable of transcription and transposition. This host-independent transposition reaction involves the excision, or replication, and re-insertion into a different location within a host genome through a 'cut and paste' or 'copy and paste' mechanism. The aforementioned mechanisms are responsible for the transposition of a distinct region of target DNA into a non-homologous site (Berg and Howe, 1989; Craig, 2002). The mechanism of transposition of TEs has been discussed in great detail in other sources (see Berg and Howe, 1989).

There are numerous examples of endogenous retroviral sequence expression in the human (Lower *et al.*, 1996; Andersson *et al.*, 2002), mouse (Thomas *et al.*, 1984) and marsupial (our unpublished data and O'Neill *et al.*, 1998) genomes. It is important to note that although some retroviruses may be expressed, they themselves may not be the target of the enzymes which they encode and could be working *in trans* to mobilize other retroelements (Coffin *et al.*, 1997). Sequence analysis of these active retroviruses has shown that they have diverged from their modern-day exogenous counterparts. Endogenous and exogenous retroviruses are obviously subject to different selective pressures. The selective forces influencing the activity of exogenous retroviruses are complex and do not necessarily depend on the fitness of the host. Endogenous retroviruses, on the other hand, have been subject to selective pressures that have largely rendered them compatible with the host (Coffin *et al.*, 1997).

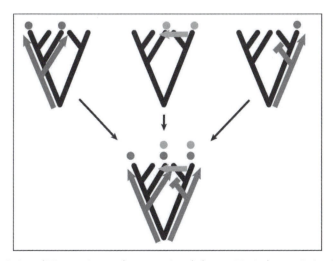

Fig. 11.1. Transmission of TEs superimposed on a species phylogeny. Vertical transmission, horizontal transfer and stochastic loss can create a complicated phylogenetic distribution (bottom) for TEs within a given species complex.

Despite the long association of retroelements with their host genomes, L1 elements are well-documented sources of deleterious mutations. It is estimated that there are approximately 40–80 full-length transposition-competent L1 elements in the human genome (International Human Genome Sequencing Consortium, 2001). Kazazian and co-workers (Kazazian *et al.*, 1988) were the first to show *de novo* insertion of an L1 element into a human gene resulting in disease (haemophilia A). L1 transposition in the human genome during gametogenesis and early development has also been associated with several genetic disorders such as Duchenne muscular dystrophy (Holmes *et al.*, 1994). In mice, L1 transposition has been shown to be responsible for an estimated 10% of spontaneous mutations (Kazazian and Moran, 1998). In addition to germline insertions in humans, L1 elements are also somatic mutagens, causing insertional disruption of tumour suppressor genes in several cases of cancer (Morse *et al.*, 1988; Miki *et al.*, 1992). SINE elements, with their close association with L1 elements, have also been associated with human disease, accounting for 17 known disease-causing mutations (Deininger and Batzer, 1999).

In contrast to the deleterious effects of transposition activity in the genome, TEs also play an adaptive role in the genome. L1 elements have been shown to integrate into double-stranded DNA breaks and mediate repair through reverse transcriptase activity in cell culture assays (Eickbush, 2002; Morrish *et al.*, 2002). Truncated TEs have also been co-opted by the mammalian genome for use as promoters and other transcriptional regulatory domains (Brosius, 1999). Variation in the activity of TEs, through such epigenetic modifications as cytosine methylation (Yoder *et al.*, 1997), has been implicated in generating novel phenotypic variation (such as the agouti locus in *Mus* (Whitelaw and Martin, 2001)) and genomic remodelling at the karyotypic level in marsupial hybrids (O'Neill *et al.*, 1998).

The genome shaping capabilities of TEs are evident not only in their patterns of activity, but also in their non-random distribution in mammalian genomes. A variety of biochemical and molecular genetic studies (as reviewed in Wichman *et al.*, 1992) have shown L1s to be preferentially associated with AT-rich, or G-banding, regions of the genome as well as the X chromosome, while SINEs are located in GC-rich, R-banding, regions. Other types of TEs have also been described that show a non-random distribution, such as the accumulation of the retroelement *Mys* on the sex chromosomes of *Peromyscus* spp. (Baker and Wichman, 1990). Wichman *et al.* (1992) have proposed several mechanisms, including sequence specificity, cell cycle-dependent replication timing and meiotic editing, to explain this non-random pattern of TE accumulation. While these mechanisms probably contribute to patterns of TE distribution, several other factors may be involved.

Lyon (1998), improving upon a previous model from Gartler and Riggs (Gartler and Riggs, 1983; Riggs, 1990), proposed that L1 elements on the X chromosome in mammals may act as 'booster elements' directing the spread of the *Xist* transcript in the conversion of one X chromosome in females to a facultative heterochromatic state, i.e. X inactivation. This theory relies on two assumptions: (i) there would be a disproportionately higher density of L1 sequences on the X chromosome than autosomes; and (ii) interaction between these sequences and the *Xist* RNA would direct the repeat-induced gene silencing system of the cell to initiate heterochromatization. In support of this hypothesis a higher density of L1 elements has been found on the X chromosome relative to autosomes in human (Bailey *et al.*, 2000) and in the bat species *Carollia brevicauda* (Parish *et al.*, 2002). The subset of L1s found on the human X are enriched for those active <100 million years ago and include an accumulation of these elements in Xq13–Xq21 (the X inactivation centre) (Bailey *et al.*, 2000). Recent experiments indicate that hypermethylation of L1 sequences by the DNA methyltransferase enzyme Dnmt3b on only one X chromosome may influence the spreading of *Xist* RNAs on the inactive X (Hansen, 2003).

Dynamic Rearrangements of Mammalian Genomes

From the ordering of genes in any given syntenic block to the location of functional components of the chromosome (e.g. centromeres), the entire genomic landscape offers a great deal of information regarding the evolutionary history of any given genomic region, as well as the potential for these regions to foster rapid and dramatic changes in the genome. For example, regional genomic architecture can affect genomic stability by influencing the frequency of non-allelic homologous recombination between low-copy repeats that are often specific to subregions of the genome (Lupski, 1998; Stankiewicz and Lupski, 2002a,b). Indeed, it is well known that the telomeric and centromeric portions of the genome are areas of rapid change and can contribute to significant changes in architecture. Telomeric and subtelomeric rearrangements in humans have led to the diversification of gene families (Trask et al., 1998) and often result in disease (see Mefford and Trask, 2002, for a review). Shifts in location of the centromere in the absence of any change in local DNA content, otherwise known as centric shifts or centromere emergence, have contributed to karyotypic changes in a short evolutionary period (Eldridge and Close, 1993; Ventura et al., 2001).

Several mammalian clades exhibit extraordinary amounts of chromosome diversity among closely related species. Interestingly, within some of these same clades certain species show relatively little chromosome change. Outlined below are several examples of species and races that have experienced rapid chromosome divergence.

The genus Muntiacus, the barking deer, is comprised of ten species, including one of the most recently discovered mammals on Earth, the giant muntjac, Muntiacus vuquangensis. Within this genus, Muntiacus muntjac, the Indian muntjac, and Muntiacus reevesi, the Chinese muntjac, exhibit extremely divergent karyotypes. The former species carries a diploid chromosome number of six in females and seven in males, while the latter carries a diploid chromosome number of 46 (Fregda, 1977). It has been estimated that no fewer than 20 rearrangements have occurred in the M. muntjac lineage (Baker and Bickham, 1980), including a series of tandem fusions (Hsu et al., 1975; Yang et al., 1997) (Fig. 11.2) as well as other complex rearrangements (Yang et al., 1997). Baker and Bickham described this dramatic change as 'karyotypic megaevo-

lution' (Baker and Bickham, 1980), in which linkage groups and G-band patterns are not congruent between two closely related species due to high frequency of different types of chromosome rearrangements. Karyotypic megaevolution is certainly not unique to deer. Baker and Bickham outlined the varied rates and types of chromosome changes that characterize 78 bat species across four families (Baker and Bickham, 1980). This study found that while some groups of species displayed no karyotypic divergence, others had experienced up to 36 different rearrangements through a pattern of karyotypic megaevolution that may have occurred over a very short time period.

Across Europe, the standard $2n=40$ telocentric karyotype of the house mouse, Mus musculus domesticus, can be found in small, dispersed populations. These populations are intermingled with over 40 different parapatric races of mouse that carry ~100 different types of Robertsonian (Rb) translocations (the fusion of two telo- or acrocentric chromosomes to form a metacentric chromosome with an internal centromere position) (Nachman et al., 1994). Remarkably, at least some of these populations have probably been established within only a few decades (Garagna et al., 1997). These races also exhibit a high frequency of whole arm reciprocal translocations (Garagna et al., 1995, 1997; Castiglia and Capanna, 1999; Britton-Davidian et al., 2000).

The genus Ctenomys contains more than 56 species (Reig et al., 1990) and exhibits one of the most diverse karyotypic ranges in mammals (Slamovits et al., 2001). These subterranean rodents (tuco-tucos) carry diploid numbers ranging from ten to 70. This diversity has been postulated to be the result of chromosomal speciation which has been facilitated by a population structure that, like the Mus musculus races, consists of small, isolated demes (Reig et al., 1990).

Marsupial chromosomes are among the most widely studied of any mammalian group, with over 50% of known marsupial species karyotyped (Hayman, 1977, 1990). Within marsupials, macropodines (the kangaroos and wallabies of the subfamily Macropodinae) have the most extensively studied and well-characterized karyotypes in terms of G-banding, chromosome rearrangements and homologies. They also carry a diverse array of karyotypes, with diploid

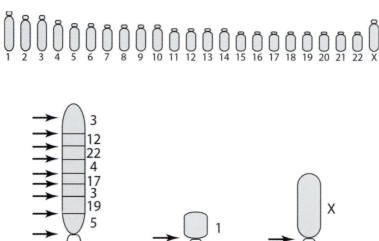

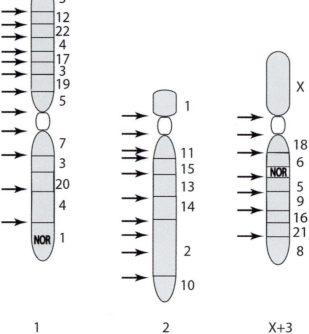

Fig. 11.2. The chromosome homologies of the Chinese muntjac (top), *Muntiacus reevesi*, on the chromosomes of the Indian muntjac, *Muntiacus muntjac* (bottom) as determined by chromosome painting (adapted from Yang *et al.*, 1997). The homologous chromosome segments are numbered. The arrows indicate the localization of a Chinese muntjac centromeric satellite sequence, corresponding to putative fusion points and probably representing remnant heterochromatin from the ancestral chromosome (Lin *et al.*, 1991; Yang *et al.*, 1997).

numbers ranging from 2n=10,11 in *Wallabia bicolor*, to 2n=22, found in several species and considered to be the ancestral karyotype for this subfamily (Rofe, 1979; Hayman, 1990).

Within macropodines, members of the genus *Petrogale* (rock wallabies) have undergone a recent and rapid explosion of chromosomal evolution (Eldridge and Close, 1993). With the exception of two subspecies, all 21 taxa within this genus exhibit distinct chromosomal complements (Sharman *et al.*, 1990). Centric fusions, centric shifts and inversions are characteristic of the majority of *Petrogale* taxa (see Eldridge and Close, 1993, for a review). It was found that

inversions were not responsible for the apparent mobility of the centromeres of this genus, but that the centromere location had shifted relative to the ancestral state (Eldridge and Close, 1993). Virtually every species within macropodines carries an X chromosome with the centromere or the X-linked NOR (nucleolar organizer region) in a different location. It has been hypothesized that centric shifts and inversions have been involved in the formation of this diversity of X chromosome morphology (Rofe, 1979; Hayman, 1990). Recently, a centric shift has also been identified in the X chromosome of the ringtail lemur (Ventura *et al.*, 2001).

Repeated Sequences and Genomic Changes

Many studies of chromosomal diversity in mammals have uncovered an association between the incidence of chromosomal rearrangement and the distribution of repeated sequences, both satellite sequences and TEs. A general pattern has emerged in which groups that show extensive karyotypic diversity exhibit a concomitant increase in the activity and/or abundance of repeated DNAs (Wichman *et al.*, 1991). Conversely, groups exhibiting karyotypic stability harbour fewer numbers of dynamic repeated DNAs (Bradley and Wichman, 1994).

Chromosome evolution

It has been suggested that the extremely high level of Robertsonian (Rb) translocations observed in races of *Mus musculus domesticus* (discussed above) has been facilitated by the interaction of minor satellite DNA sequences with a centromeric protein (Garagna *et al.*, 2001). Analyses of fusion points in these chromosomes indicate that the breakpoint lies within minor satellite DNA, which subsequently contributed to the symmetry of the newly formed Rb centromere. It has been proposed that the activity of the centromeric DNA-binding protein CENP-B facilitates these types of rearrangements (Kipling and Warburton, 1997; Garagna *et al.*, 2001). Interestingly, CENP-B shares sequence similarity to the *pogo* family of transposable elements (Tudor *et al.*, 1992) and may promote recombination, either inter- or intrachromosomally, through its putative ability to nick DNA adjacent to CENP-B boxes within the minor satellite repeats located at the centromeres of *Mus musculus domesticus* chromosomes.

In an analysis of the major satellite repeat sequence in *Ctenomys*, called RPCS, a sequence that shares similarity to the LTR sequences of retroviruses, Slamovits *et al.* found a positive correlation between the copy number of these elements and chromosome evolution in this group (Slamovits *et al.*, 2001).

Even between closely related species, amplifications and deletions were associated with extensive chromosome rearrangements while stability in RPCS copy number was found in clades that exhibit a stable karyotype.

Studies in marsupial hybrids between macropodine species have indicated a direct role for TEs in chromosome remodelling. In several *Petrogale* hybrids and *Macropus* hybrids, interchromosomal rearrangements were identified that involved the movement and amplification of pericentromeric DNA (O'Neill *et al.*, 2001). This activity was linked to a reduction in cytosine methylation levels, hypothesized as a form of genome defence and TE control (Yoder *et al.*, 1997), which resulted in the amplification of an endogenous retroviral sequence (O'Neill *et al.*, 1998).

Heterochromatin and centromeres

The term 'heterochromatin' refers to chromosomal DNA that remains condensed for most of the cell cycle (Balicek *et al.*, 1977). It is late-replicating DNA (Sumner, 1990) that is usually transcriptionally inactive, although examples of transcriptionally active domains localizing to heterochromatin have been identified in mammals (Sperling *et al.*, 1987; Neitzel *et al.*, 2002) (our unpublished data). Brown refined the definition of heterochromatin by separating it into two classes: facultative and constitutive (Brown, 1966). Facultative heterochromatin refers to DNA that is considered either condensed or epigenetically repressed without altering DNA content between either, such as in X inactivation in females. Constitutive heterochromatin refers to condensed DNA with an atypical composition.

C-band-positive constitutive heterochromatin had been considered rich in short, tandem arrays of repeated elements and poor in retroelements (Korenberg and Rykowski, 1988). Studies in rodents (Kuff *et al.*, 1986; Neitzel *et al.*, 2002) and marsupials (our unpublished data) have shown that these chromosome regions are far more complex in structure and do contain retroelements, some of which are still transcriptionally active. The link between heterochromatin and retroelements in general has remained controversial (Dimitri and Junakovic, 1999). It is unclear whether the satellite DNA sequences present in these regions act as sinks for the accumulation of retroelements as a protection against euchromatic 'damage' (i.e. a host defence mechanism) (Csink and Henikoff, 1998) or whether retroelements provide a functional component to heterochromatin (see Pimpinelli et al., 1995; Dimitri, 1997).

The full characterization of the heterchromatic DNA constituting the centromere of the human X chromosome (Schueler et al., 2001) was a major step forward in our understanding of mammalian centromeres. This study revealed that functional satellite sequences within the centromere contain evolutionarily young L1 elements while the more distal, non-functional satellites, contain signs of ancient retroelement activity. It was concluded that the age gradient revealed by L1 sequences showed that a monomeric satellite pre-dates the higher-order arrays of alpha satellites within this centromere. The implication that functioning centromeres are targets for recent retroelement integration is highly intriguing. However, it is unknown whether the basic structure of the centromere of the human X chromosome applies to other organisms.

The vast majority of centromere-associated sequences that have been identified to date have included arrays of repeated DNAs (Choo, 2000; Sullivan et al., 2001). Several authors have analysed centromeric DNA in an evolutionary context, although this has proved difficult due to the low conservation of these DNAs and their complex sequence evolution. An investigation of chromosome-specific alpha satellite DNA from the chimpanzee chromosome 4 led Haaf and Willard to postulate that chromosomally distinct sets of repeats have undergone species-specific amplification and/or concerted evolution (Haaf and Willard, 1997). Several chromosome-specific subsets of alpha satellite DNAs have also been identified in humans (Greig et al., 1989, 1993; Verma, 1999), suggesting that while this family of satellite repeats has been conserved in human evolution, there has been extensive chromosome-specific diversification.

Duplications and Gene Families

Evidence for dramatic change in the mammalian genome mediated by duplication and transposition of DNA is mounting from a variety of elegant comparative studies made possible by concerted genome sequencing efforts in several mammalian clades. Duplication and tandem insertion or transposition of single genes or larger DNA segments (segmental duplication), both herein referred to as duplicons, is now recognized as a major effector of karyotypic diversity in primates and rodents. Intriguing evidence from studies of the human genome suggests that gene duplication has also been a potent source of proteomic novelty. As of yet little is understood about the mechanisms underlying gene and segmental duplication and transposition. Like canonical transposable elements, duplicons show non-random distributions in mammalian genomes and are often linked to genomic instability. Recent examinations of the various scales of genetic duplication, from that of single genes to whole genomes, are presented below.

Whole genome duplications

The discussion of genetic duplication in mammalian genomes must begin with the arguments posed by Ohno in his influential book, *Evolution by Gene Duplication* (Ohno, 1970). Ohno sought to attribute the remarkably extensive and rapid ecological expansion of the vertebrates, 'the big leap', to the exploration of novel design space afforded by polyploidization of the vertebrate ancestral genome. Ohno's hypothesis has been co-opted into the more contentious 2R (two-rounds) hypothesis, positing two rounds of whole genome duplication in the establishment of the vertebrate lineage. The most compelling evidence in support of 2R was the discovery of the expansion of the Hox gene cluster from a single cluster in invertebrates to four in vertebrates (Schughart et al., 1989).

More recent discussion revolves around the question of whether the extensive paralogy observed in vertebrate genomes is the product of an ancient, whole-genome duplication (polyploidization, followed soon after by rediploidization and selective gene loss) or of a more piece-meal process of wide-scale, recurrent gene and segmental duplication. The task of resolving this issue is complicated not only by the remoteness in time of the origin of the vertebrate lineage but also by the high frequency of homogenization of duplicons by processes such as non-allelic recombination and gene conversion. Two recent studies applying different statistical approaches to identify and trace the evolutionary history of human gene families lend support to only one round of whole genome duplication (or nearly whole genome duplication) in the metazoan lineage giving rise to the vertebrates. Gu *et al.* analysed 749 gene families and showed evidence of three 'waves' of duplication in metazoan history (Gu *et al.*, 2002). An ancient wave of duplicative events pre-dates the split between *Drosophila* and human. The second wave, occurring prior to the divergence of ray-finned from lobe-finned fishes, shows a rapid increase in paralogous gene families consistent with a whole genome duplication event. The third wave was restricted to the mammalian radiation and is consistent with increased segmental duplication observed in primates (see next section). McLysaght *et al.* analysed 758 gene families and also found evidence for a large-scale duplication event soon after the origin of the chordate (McLysaght *et al.*, 2002). This analysis also provided a view of the history of the spatial relationship of duplicated regions by focusing on the lineage of groups of related genes travelling together, termed paralogons. Using this approach they identified 96 paralogons covering 44% of the human genome.

Segmental duplications

The majority of studies identifying, mapping and tracing evolutionary trajectories of gene and segmental duplications have focused on primates, for which extensive and broadly representative genome sequences are available. It is estimated that 5.2% of the human genome is duplicated (Bailey *et al.*, 2001, 2002a; International Human Genome Sequencing Consortium, 2001) with a tenfold enrichment for duplications in pericentromeric and subtelomeric regions (Bailey *et al.*, 2001). These enriched regions tend to harbour interchromosomal duplications, a process that involves the duplication of genetic material between two chromosomes, while euchromatic regions more often exhibit intrachromosomal duplications (Eichler, 2001; Bailey *et al.*, 2001). Many of the duplications found in the human genome have arisen in the last 35 million years of evolution (Bailey *et al.*, 2001, 2002b). Many region-specific duplicated segments in humans, termed low-copy repeats (LCRs), are prone to rearrangement through non-allelic homologous recombination. The effects of non-allelic recombination at these sites include micro-deletions, micro-duplications and inversions, and are a major contributory factor in several human diseases, termed genomic disorders (Stankiewicz and Lupski, 2002a,b). Recent analyses of the human genome sequence (Bailey *et al.*, 2002a) have shown a positive correlation between clustering of intrachromosomal segmental duplications and 24 human genomic disorders, and indicate a total of 169 regions that may be prone to instability (Table 11.2).

Comparative studies of segmental duplications among different mammalian groupings have given insight into the evolutionary history of subcompartments of the genome (see Horvath *et al.*, 2003, for an example) and may shed light on the formation of chromosomal rearrangements that accompany, facilitate or reinforce species divergence. Following the trajectory of duplicons is complicated by multiple rounds of duplication and rearrangement and becomes more difficult with increased evolutionary distance between species examined. Nevertheless, examples of interchromosomal duplication have been found in studies of paralogous regions between species of great apes. A pericentromeric DNA fragment found on the long-arm of human chromosome 21 (HSA21) was found to be the product of intrachromosomal duplication (Potier *et al.*, 1998) followed by interchromosomal transposition (to HSA2q, HSA13 and HSA18) in great apes after the divergence of orangutans (Golfier *et al.*, 2003).

Another example of the complex history of sequences that may have contributed to the formation of rearrangements during primate evolution can be found in the elegant studies of the evolution of the ancestral fusion site in HSA2q13–2q14.1 (Fan *et al.*, 2002a,b; Martin *et al.*, 2002; Mefford and Trask, 2002). Two

Table 11.2. Genomic disorders linked to low copy repeat (LCR)-associated genomic rearrangements.

Genomic disorder	OMIM[a]	Locus	Gene	LCR (kb)
Gaucher disease	230 800	1q21	GBA	14 (del)[b]
Familial juvenile nephronophthisis	256 100	2q13	NPHP1	45 (del)
Fascioscapulohumeral muscular dystrophy	158 900	4q35	FRG1	~100 (del)
Spinal muscular atrophy	253 300	5q11.2–q13.3	SMN	300 (inv[c]/dup[d])
Congenital adrenal hyperplasia III	201 910	6p21.3	CYP21	~25 (del)
Williams–Beuren syndrome	194 050	7q11.23	ELN/GTF21	>320 (del)
Glucocorticoid-remediable aldosteronism	103 900	8q21	CYP11B1/2	10 (dup)
Prader–Willi syndrome	176 270	15q11.2–q13	SNRPN, IPW?	400 (del)
Angelman syndrome	105 830	15q11.2–q13	UBE3A	400 (del)
Alpha-thalassaemia	141 800	16p13.3	Alpha-globin	4 (del)
Polycystic kidney disease	601 313	16p13.3	PKD1	50
Charcot–Marie–Tooth disease type 1A	118 220	17p12	PMP22	24 (del)
Hereditary neuropathy with liability to pressure palsies (HNPP)	162 500	17p12	PMP22	24 (dup)
Smith–Magenis syndrome	182 290	17p11.2	RAI1?	200 (del)
Neurofibromatosis type 1 (NF1)	162 200	17p12	NF1	85 (del)
Pituitary dwarfism	262 400	17q23.3	GH1	2.24 (del)
Cat eye syndrome	115 470	22q11		400 (inv/dup)
DiGeorge/velocardiofacial syndrome	188 400 192 430	22q11.2	TBX1	225–400 (del)
CYP2D6 pharmacogenetic trait	124 030	22q13.1	CYP2D6	2.8 (del/dup)
Ichthyosis	308 100	Xp22.32	STS	5–20 (del)
Hunter syndrome (mucopolysaccharidosis type II)	309 900	Xq28	IDS	3 (inv/del)
Red–green colour blindness	303 800	Xq28	RCP/GCP	39 (del)
Emery–Dreifuss muscular dystrophy	310 300	Xq28	EMD/FLN1	11.3 (del/dup/inv)
Incontinentia pigmenti	308 300	Xq28	NEMO	35.5 (del)
Haemophilia A	306 700	Xq28	FactorVIII	9.5 (inv)
Azoospermia(a)	415 000	Yq11.2	DBY/USP9Y	10 (del)
Azoospermia (c)	400 024	Yq11.2	RBMY/DAZ?	229 (del)

From Bailey et al., 2002a; Stankiewicz and Lupski, 2002a,b.
[a]OMIM (Online Mendelian Inheritance in Man database http://www3.ncbi.nlm.nih.gov/Omim).
[b]del, deletion.
[c]inv, inversion.
[d]dup, duplication.

chromosomes that have remained separate in other primates have fused to form human chromosome 2. Fan et al. (2002b) traced the history of the sequences present at this fusion site and found it has undergone a complex series of rearrangements and duplicative exchanges, including an intrachromosomal duplication pre-dating hominid divergence (16–20 million years ago (mya)). A duplicative exchange between two chromosomes (2 and 9) pre-dating the gorilla–chimpanzee–hominid split (14–8 mya) followed by an inversion at this site in the chimpanzee–hominid lineage

(8–6 mya). Two more duplications of material from HSA9q13 to HSA9p11.2 in the hominid lineage, and subsequently a fusion between duplicated segments to form HSA2 (Fan et al., 2002a) (Fig. 11.3).

Thus, genome architecture can act as a catalyst for chromosome rearrangements through the action of segmental duplication and nonallelic homologous recombination and points to a highly fluid genome in humans. The propensity towards intrachromosomal rearrangements in human chromosome evolution (Postlethwait et al., 2000; Pevzner and Tesler, 2003a,b) has

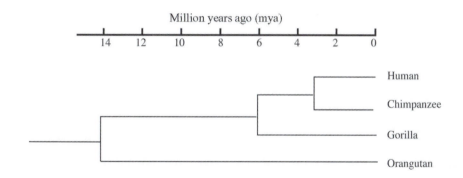

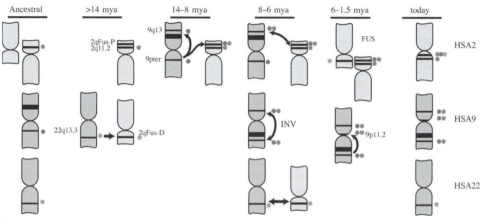

Fig. 11.3. The segmental duplication and chromosome rearrangements giving rise to human chromosome 2 (HSA2) mapped along the primate phylogeny (Fan *et al.*, 2002a). KEY: INV (inversion); FUS (fusion); chromosomes (ancestral and derived) HSA2, HSA9, HSA22; dots depict three sequences that have undergone rearrangement in the evolution of these chromosomes.

made delineating the complex history of human chromosomal segments across other mammalian lineages challenging. Studies of the conservation of breaks of synteny between human and mouse genomes found that 53% of evolutionary breakpoint rearrangements between these two species associate with segmental duplications (Armengol *et al.*, 2003). While it cannot be determined whether the duplications are associative or causative in these rearrangements, an intriguing correlation between karyotypic evolution in primates and segmental duplications is emerging (see Samonte and Eichler, 2002; Locke *et al.*, 2003). Additionally, Thomas *et al.* have identified recent (~5–7 mya) pericentromeric segmental duplications on mouse chromosomes 5 and 6 that have been implicated in a chromo-

some fission event as well as chimeric transcript formation, the derivation of novel satellite sequences and possibly centromere formation (Thomas *et al.*, 2003).

The role of duplications in mammalian genome evolution

Functional diversification frequently follows gene duplication when the duplicon produced from segmental duplication includes the intact exon–intron structure of a coding sequence(s). The resulting paralogue may diverge through processes such as positive selection (the selection for non-synonymous nucleotide substitutions), subfunctionalization (loss of some degree of pleitropy in one or both duplicates

(Prince and Pickett, 2002)) and domain accretion (the formation of chimeric transcripts which produce larger, multidomain proteins) (Eichler, 2001). Human gene families associated with the immune system, such as immunoglobulins (Ota *et al.*, 2000) and β-defensins (Hughes and Yeager, 1997; Hughes, 1999; Semple *et al.*, 2003) are examples of functional diversification following gene duplication. Recently, Johnson *et al.* identified a novel gene family that has undergone recent duplication, the *morpheus* genes, that have no known function but have undergone strong positive selection (Johnson *et al.*, 2001). An example of subfunctionalization of duplicate genes is growth hormone/placental lactogen in primates (Wallis, 2001). In this case the original copy retains ubiquitous expression while the duplicate is expressed only in placenta. Domain accretion would seem an abundant source for proteomic innovation by creating novel peptide domain combinations. Indeed, chimeric transcripts have been identified in the human genome that are the byproduct of fusion events between local genes and newly transposed, duplicated and unrelated exonic regions (see Eichler, 2001, for a review). Whether these chimeric transcripts encode functional peptides is as yet unknown.

Karyotypic Diversity, Reproductive Isolation and Speciation

In outbred populations, chromosomal rearrangements may act as post-mating reproductive isolating mechanisms by impairing the fertility of the offspring. This may occur as a rapid, 'single step' event requiring no other genetic or morphological differences between parent populations. The reduction in fertility of hybrid offspring, however, poses a serious problem to the theory of chromosomal speciation as there would be little chance for the fixation of novel rearrangements in a population (King, 1993). White argued that each rearrangement in species that was distinguished by multiple fixed rearrangements would only have a minor effect and would be fixed independently and sequentially (White, 1978). This fails to explain the occurrence of multiple rearrangements that are thought to

have reached fixation simultaneously (King, 1993). Baker and Bickham describe a model of peripatric speciation in which multiple events of one type of chromosomal rearrangement, monobrachial centric fusions, can become fixed between different founder populations and parental populations (Baker and Bickham, 1986). Under this model, there is little meiotic deficit between a single founder and parental population, however the meiotic impairment becomes pronounced between founder populations carrying different suites of fusions. They found their model to be applicable to several mammalian taxa, including *Mus* and *Rattus*.

Chromosome evolution in terms of the sites of chromosome breakage and the resulting type of chromosome rearrangement is undeniably a non-random process as numerous lineages are characterized by species or cytotypes that are distinguished by single or multiple rearrangements of the same type (King, 1993). White proposed that 'karyotypic orthoselection' could explain the tendency for the same rearrangement to reach fixation in different chromosomes of the same species, or complex of species (White, 1973, 1975, 1978). Through selection for an 'equilibrium' karyotype, karyotypic orthoselection could produce the reappearance of particular karyotypes within lineages (White, 1973). Extensive genome sequence comparisons are now pointing to structural features, such as centromeres, telomeres and fragile sites, which influence the number, position and type of rearrangement.

Karyotypic change can result from an increased mutation rate caused by genome destabilizing events, such as hybridization or exposure to environmental mutagens (Fontdevila, 1992). This may be attributed to the activity of TEs and other repeated DNAs. McClintock's work on *Zea mays* provided experimental evidence that the genome could produce multiple chromosomal rearrangements of the same type, in the same individual, simultaneously (collected papers, McClintock, 1987). Under varying levels of genomic stress, the activation of TEs resulted in varying degrees of reorganization, ranging from small segments of DNA to complex chromosomal translocations. There is now increasing evidence that chromosome instability can occur in species hybridization in mammals (*see* O'Neill *et al.*, 1998, 2001; Brown *et al.*, 2002, for examples). Not only may transposition bursts caused by hybridization greatly affect chromosome rearrangements, but the increased mutation rate provided by these bursts may also increase the rate of fixation and consequently the rate of speciation (Fontdevila, 1992).

TEs contribute significantly to the production of chromosomal rearrangements over extremely short time periods. However, the fixation of these novel rearrangements is a process that has been widely debated and is not well understood. *De novo* chromosome rearrangements are generally assumed to be at least slightly deleterious (i.e. underdominant or negatively heterotic) because of the disruption of pairing during meiosis. Population modelling and computer simulations indicate that fixation of rearrangements by stochastic processes may only occur under very restrictive conditions (Wright, 1941; Bengtsson and Bodmer, 1976; Lande, 1979; Hedrick, 1981; Chesser and Baker, 1986; Fontdevila, 1992). Other models that provide possible explanations for the fixation of chromosomal rearrangements include: positive selection for the homokaryotype (Hedrick, 1981), which proposes that rearrangements create new linkage groups that contribute to the accumulation of favourable combinations of genes (Sturtevant and Beadle, 1936); inbreeding which provides a positive effect when selection against the heterokaryotype is low, increasing the probability of underdominant fixation (Lande, 1979; Hedrick, 1981); and non-Mendelian segregation, which also can increase the probability of fixation of negatively heterotic chromosome rearrangements (White, 1978; Walsh, 1982; Lyttle, 1989; Pardo-Manuel de Villena and Sapienza, 2001; Malik and Henikoff, 2002).

Underdominance brought about by chromosome rearrangement is typified by decreased fertility or viability. The rapid fixation, therefore, of divergent karyotypes in populations is difficult to explain in the absence of cryptic adaptive significance for such changes. For protein-coding DNA sequences the legacy of natural selection can be found by comparing the rate of amino acid replacement mutation (K_a or d_N) with that of silent mutation (K_s or d_S). If K_a exceeds K_s it may be concluded that alteration of the peptide sequence is adaptive. Unfortunately, the signature of natural selection cannot be so easily discerned in instances of genome rearrangement. The dramatic restructuring, not just in mammals, of genomes in seemingly rapid bursts is remarkable given the apparent restrictions on neutral or underdominant fixation of these types of changes. As the whole genome sequences of more mammals become available, through comparative analysis we will gain a better understanding of the function of architectural features of the genome. With this information it may then be possible to gauge the adaptive significance of structural changes and how these changes have influenced the divergence of species.

References

Andersson, A.C., Venables, P.J., Tonjes, R.R., Scherer, J., Eriksson, L. and Larsson, E. (2002) Developmental expression of HERV-R (ERV3) and HERV-K in human tissue. *Virology* 297, 220–225.

Armengol, L., Pujana, M.A., Cheung, J., Scherer, S.W. and Estivill, X. (2003) Enrichment of segmental duplications in regions of breaks of synteny between the human and mouse genomes suggest their involvement in evolutionary rearrangements. *Human Molecular Genetics* 12, 2201–2208.

Auge-Gouillou, C., Bigot, Y., Pollet, N., Hamelin, M.H., Meunier-Rotival, M. and Periquet, G. (1995) Human and other mammalian genomes contain transposons of the mariner family. *FEBS Letters* 368, 541–546.

Bailey, J.A., Carrel, L., Chakravarti, A. and Eichler, E.E. (2000) Molecular evidence for a relationship between LINE-1 elements and X chromosome inactivation: the Lyon repeat hypothesis. *Proceedings of the National Academy of Sciences USA* 97, 6634–6639.

Bailey, J.A., Yavor, A.M., Massa, H.F., Trask, B.J. and Eichler, E.E. (2001) Segmental duplications: organization and impact within the current human genome project assembly. *Genome Research* 11, 1005–1017.

Bailey, J.A., Gu, Z., Clark, R.A., Reinert, K., Samonte, R.V., Schwartz, S., Adams, M.D., Myers, E.W., Li, P.W. and Eichler, E.E. (2002a) Recent segmental duplications in the human genome. *Science* 297, 1003–1007.

Bailey, J.A., Yavor, A.M., Viggiano, L., Misceo, D., Horvath, J.E., Archidiacono, N., Schwartz, S., Rocchi, M. and Eichler, E.E. (2002b) Human-specific duplication and mosaic transcripts: the recent paralogous structure of chromosome 22. *American Journal of Human Genetics* 70, 83–100.

Baker, R. and Bickham, J. (1980) Karyotypic evolution in bats: evidence of extensive and conservative chromosomal evolution in closely related taxa. *Systematic Zoology* 29, 239–253.

Baker, R.J. and Bickham, J.W. (1986) Speciation by monobrachial centric fusions. *Proceedings of the National Academy of Sciences USA* 83, 8245–8248.

Baker, R.J. and Wichman, H.A. (1990) Retrotransposon Mys is concentrated on the sex chromosomes: implications for copy number containment. *Evolution* 44, 2083–2088.

Balicek, P., Zizka, J. and Skalska, H. (1977) Length of human constitutive heterochromatin in relation to chromosomal contraction. *Human Genetics* 38, 189–193.

Bengtsson, B.O. and Bodmer, W.F. (1976) On the increase of chromosome mutations under random mating. *Theoretical Population Biology* 9, 260–281.

Berg, D.E. and Howe, M.M. (1989) *Mobile DNA*. American Society for Microbiology, Washington, DC, pp. xii, 972.

Bradley, R.D. and Wichman, H.A. (1994) Rapidly evolving repetitive DNAs in a conservative genome: a test of factors that affect chromosomal evolution. *Chromosome Research* 2, 354–360.

Britten, R.J. and Kohne, D.E. (1968) Repeated sequences in DNA. Hundreds of thousands of copies of DNA sequences have been incorporated in the genomes of higher organisms. *Science* 161, 529–540.

Britton-Davidian, J., Catalan, J., da Graca Ramalhinho, M., Ganem, G., Auffray, J.C., Capela, R., Biscoito, M., Searle, J.B. and da Luz Mathias, M. (2000) Rapid chromosomal evolution in island mice. *Nature* 403, 158.

Brosius, J. (1999) Genomes were forged by massive bombardments with retroelements and retrosequences. *Genetica* 107, 209–238.

Brown, J.D., Strbuncelj, M., Giardina, C. and O'Neill, R.J. (2002) Interspecific hybridization induced amplification of Mdm2 on double minutes in a *Mus* hybrid. *Cytogenetic and Genome Research,* 98, 184–188.

Brown, S.W. (1966) Heterochromatin. *Science* 151, 417–425.

Burton, F.H., Loeb, D.D., Voliva, C.F., Martin, S.L., Edgell, M.H. and Hutchison, C.A., 3rd (1986) Conservation throughout mammalia and extensive protein-encoding capacity of the highly repeated DNA long interspersed sequence one. *Journal of Molecular Biology* 187, 291–304.

Castiglia, R. and Capanna, E. (1999) Whole-arm reciprocal translocation (WART) in a feral population of mice. *Chromosome Research* 7, 493–495.

Chesser, R.K. and Baker, R.J. (1986) On factors effecting the fixation of chromosomal rearrangements and neutral genes: computer simulations. *Evolution* 40, 625–632.

Choo, K.H. (2000) Centromerization. *Trends in Cell Biology* 10, 182–188.

Coffin, J.M., Hughes, S.H. and Varmus, H. (1997) *Retroviruses*. Cold Spring Harbor Laboratory Press, Plainview, New York, pp. xv, 843.

Copeland, N.G., Jenkins, N.A., Gilbert, D.J., Eppig, J.T., Maltais, L.J., Miller, J.C., Dietrich, W.F., Weaver, A., Lincoln, S.E., Steen, R.G. *et al.* (1993) A genetic linkage map of the mouse: current applications and future prospects. *Science* 262, 57–66.

Craig, N.L. (2002) *Mobile DNA II*. ASM Press, Washington, DC, pp. xviii, 1204 (1232 for plates).

Csink, A.K. and Henikoff, S. (1998) Something from nothing: the evolution and utility of satellite repeats. *Trends in Genetics* 14, 200–204.

Daviere, J.M., Langin, T. and Daboussi, M.J. (2001) Potential role of transposable elements in the rapid reorganization of the *Fusarium oxysporum* genome. *Fungal Genetics and Biology,* 34, 177–192.

DeBerardinis, R.J., Goodier, J.L., Ostertag, E.M. and Kazazian, H.H., Jr (1998) Rapid amplification of a retrotransposon subfamily is evolving the mouse genome. *Nature Genetics* 20, 288–290.

DeBry, R.W. and Seldin, M.F. (1996) Human/mouse homology relationships. *Genomics* 33, 337–351.

Deininger, P.L. and Batzer, M.A. (1999) Alu repeats and human disease. *Molecular Genetics and Metabolism,* 67, 183–193.

Delbridge, M.L. and Graves, J.A. (1999) Mammalian Y chromosome evolution and the male-specific functions of Y chromosome-borne genes. *Reviews in Reproduction* 4, 101–109.

Dimitri, P. (1997) Constitutive heterochromatin and transposable elements in *Drosophila melanogaster*. *Genetica* 100, 85–93.

Dimitri, P. and Junakovic, N. (1999) Revising the selfish DNA hypothesis: new evidence on accumulation of transposable elements in heterochromatin. *Trends in Genetics* 15, 123–124.

Doak, T.G., Doerder, F.P., Jahn, C.L. and Herrick, G. (1994) A proposed superfamily of transposase genes: transposon-like elements in ciliated protozoa and a common 'D35E' motif. *Proceedings of the National Academy of Sciences USA* 91, 942–946.

Dobzhansky, T. and Sturtevant, A.H. (1938) Inversions in the chromosomes of *Drosophila pseudoobscura*. *Genetics* 23, 28–64.

Dombroski, B.A., Mathias, S.L., Nanthakumar, E., Scott, A.F. and Kazazian, H.H., Jr (1991) Isolation of an active human transposable element. *Science* 254, 1805–1808.

Doolittle, R.F. and Feng, D.F. (1992) Tracing the origin of retroviruses. *Current Topics in Microbiology and Immunology* 176, 195–211.

Doolittle, R.F., Feng, D.F., Johnson, M.S. and McClure, M.A. (1989) Origins and evolutionary relationships of retroviruses. *Quarterly Reviews in Biology* 64, 1–30.

Dorner, M. and Paabo, S. (1995) Nucleotide sequence of a marsupial LINE-1 element and the evolution of placental mammals. *Molecular Biology and Evolution* 12, 944–948.

Eichler, E.E. (2001) Recent duplication, domain accretion and the dynamic mutation of the human genome. *Trends in Genetics* 17, 661–669.

Eickbush, T.H. (1997) Telomerase and retrotransposons: which came first? *Science* 277, 911–912.

Eickbush, T.H. (2002) Repair by retrotransposition. *Nature Genetics* 31, 126–127.

Eldridge, M.D. and Close, R.L. (1993) Radiation of chromosome shuffles. *Current Opinion in Genetics and Development* 3, 915–922.

Esnault, C., Maestre, J. and Heidmann, T. (2000) Human LINE retrotransposons generate processed pseudogenes. *Nature Genetics* 24, 363–367.

Fan, Y., Linardopoulou, E., Friedman, C., Williams, E. and Trask, B.J. (2002a) Genomic structure and evolution of the ancestral chromosome fusion site in 2q13–2q14.1 and paralogous regions on other human chromosomes. *Genome Research* 12, 1651–1662.

Fan, Y., Newman, T., Linardopoulou, E. and Trask, B.J. (2002b) Gene content and function of the ancestral chromosome fusion site in human chromosome 2q13–2q14.1 and paralogous regions. *Genome Research* 12, 1663–1672.

Fontdevila, A. (1992) Genetic instability and rapid speciation: are they coupled? *Genetica* 86, 247–258.

Fregda, K. (1977) Chromosomal changes in vertebrate evolution. *Proceedings of the Royal Society of London* 199, 377–397.

Garagna, S., Broccoli, D., Redi, C.A., Searle, J.B., Cooke, H.J. and Capanna, E. (1995) Robertsonian metacentrics of the house mouse lose telomeric sequences but retain some minor satellite DNA in the pericentromeric area. *Chromosoma* 103, 685–692.

Garagna, S., Zuccotti, M., Redi, C.A. and Capanna, E. (1997) Trapping speciation. *Nature* 390, 241–242.

Garagna, S., Marziliano, N., Zuccotti, M., Searle, J.B., Capanna, E. and Redi, C.A. (2001) Pericentromeric organization at the fusion point of mouse Robertsonian translocation chromosomes. *Proceedings of the National Academy of Sciences USA* 98, 171–175.

Gartler, S.M. and Riggs, A.D. (1983) Mammalian X-chromosome inactivation. *Annual Reviews in Genetics* 17, 155–190.

Golfier, G., Chibon, F., Aurias, A., Chen, X.N., Korenberg, J., Rossier, J. and Potier, M.C. (2003) The 200-kb segmental duplication on human chromosome 21 originates from a pericentromeric dissemination involving human chromosomes 2, 18 and 13. *Gene* 312, 51–59.

Gregory, S.G., Sekhon, M., Schein, J., Zhao, S., Osoegawa, K., Scott, C.E., Evans, R.S., Burridge, P.W., Cox, T.V., Fox, C.A., Hutton, R.D., Mullenger, I.R., Phillips, K.J., Smith, J., Stalker, J., Threadgold, G.J., Birney, E., Wylie, K., Chinwalla, A., Wallis, J. *et al.* (2002) A physical map of the mouse genome. *Nature* 418, 743–750.

Greig, G.M., England, S.B., Bedford, H.M. and Willard, H.F. (1989) Chromosome-specific alpha satellite DNA from the centromere of human chromosome 16. *American Journal of Human Genetics* 45, 862–872.

Greig, G.M., Warburton, P.E. and Willard, H.F. (1993) Organization and evolution of an alpha satellite DNA subset shared by human chromosomes 13 and 21. *Journal of Molecular Evolution* 37, 464–475.

Gu, X., Wang, Y. and Gu, J. (2002) Age distribution of human gene families shows significant roles of both large- and small-scale duplications in vertebrate evolution. *Nature Genetics* 31, 205–209.

Haaf, T. and Willard, H.F. (1997) Chromosome-specific alpha-satellite DNA from the centromere of chimpanzee chromosome 4. *Chromosoma* 106, 226–232.

Hansen, R.S. (2003) X inactivation-specific methylation of LINE-1 elements by DNMT3B: implications for the Lyon repeat hypothesis. *Human Molecular Genetics* 12, 2559–2567.

Hayman, D.L. (1977) Chromosome number—constancy and variation. In: Stonehouse, B. and Gilmore, D. (eds) *The Biology of Marsupials*. Macmillan, London.

Hayman, D.L. (1990) Marsupial cytogenetics. In: Graves, J.A.M., Hope, R.M. and Cooper, D.W. (eds) *Mammals from Pouches and Eggs: Genetics, Breeding and Evolution of Marsupials and Monotremes*. Melbourne, CSIRO.

Hedrick, P. (1981) The establishment of chromosomal variance. *Evolution* 35, 322–332.

Holmes, S.E., Dombroski, B.A., Krebs, C.M., Boehm, C.D. and Kazazian, H.H., Jr (1994) A new retrotransposable human L1 element from the LRE2 locus on chromosome 1q produces a chimaeric insertion. *Nature Genetics* 7, 143–148.

Horvath, J.E., Gulden, C.L., Bailey, J.A., Yohn, C., McPherson, J.D., Prescott, A., Roe, B.A., de Jong, P.J., Ventura, M., Misceo, D., Archidiacono, N., Zhao, S., Schwartz, S., Rocchi, M. and Eichler, E.E. (2003) Using a pericentromeric interspersed repeat to recapitulate the phylogeny and expansion of human centromeric segmental duplications. *Molecular Biology and Evolution* 20, 1463–1479.

Hsu, T.C., Pathak, S. and Chen, T.R. (1975) The possibility of latent centromeres and a proposed nomenclature system for total chromosome and whole arm translocations. *Cytogenetics and Cell Genetics* 15, 41–49.

Hughes, A.L. (1999) Evolutionary diversification of the mammalian defensins. *Cellular and Molecular Life Sciences* 56, 94–103.

Hughes, A.L. and Yeager, M. (1997) Coordinated amino acid changes in the evolution of mammalian defensins. *Journal of Molecular Evolution* 44, 675–682.

Hughes, J.F. and Coffin, J.M. (2001) Evidence for genomic rearrangements mediated by human endogenous retroviruses during primate evolution. *Nature Genetics* 29, 487–489.

Hughes, J.F. and Coffin, J.M. (2002) A novel endogenous retrovirus-related element in the human genome resembles a DNA transposon: evidence for an evolutionary link? *Genomics* 80, 453–455.

International Human Genome Sequencing Consortium (2001) Initial sequencing and analysis of the human genome. *Nature* 409, 860.

Johnson, M.E., Viggiano, L., Bailey, J.A., Abdul-Rauf, M., Goodwin, G., Rocchi, M. and Eichler, E.E. (2001) Positive selection of a gene family during the emergence of humans and African apes. *Nature* 413, 514–519.

Jurka, J. (1997) Sequence patterns indicate an enzymatic involvement in integration of mammalian retroposons. *Proceedings of the National Academy of Sciences USA* 94, 1872–1877.

Kazazian, H.H., Jr (1998) Mobile elements and disease. *Current Opinion in Genetics and Development* 8, 343–350.

Kazazian, H.H., Jr and Moran, J.V. (1998) The impact of L1 retrotransposons on the human genome. *Nature Genetics* 19, 19–24.

Kazazian, H.H., Jr, Wong, C., Youssoufian, H., Scott, A.F., Phillips, D.G. and Antonarakis, S.E. (1988) Haemophilia A resulting from *de novo* insertion of L1 sequences represents a novel mechanism for mutation in man. *Nature* 332, 164–166.

King, M. (1993) *Species Evolution: The Role of Chromosome Change*. Cambridge University Press, Cambridge.

Kipling, D. and Warburton, P.E. (1997) Centromeres, CENP-B and Tigger too. *Trends in Genetics* 13, 141–145.

Korenberg, J.R. and Rykowski, M.C. (1988) Human genome organization: Alu, lines, and the molecular structure of metaphase chromosome bands. *Cell* 53, 391–400.

Kuff, E.L., Fewell, J.E., Lueders, K.K., DiPaolo, J.A., Amsbaugh, S.C. and Popescu, N.C. (1986) Chromosome distribution of intracisternal A-particle sequences in the Syrian hamster and mouse. *Chromosoma* 93, 213–219.

Lande, S. (1979) Effective deme sizes during long term evolution estimated from rate of chromosomal rearrangement. *Evolution* 33, 234–251.

Levy, J.A. (1992) *The Retroviridae*. Plenum Press, New York.

Lin, C.C., Sasi, R., Fan, Y.S. and Chen, Z.Q. (1991) New evidence for tandem chromosome fusions in the karyotypic evolution of Asian muntjacs. *Chromosoma* 101, 19–24.

Locke, D.P., Archidiacono, N., Misceo, D., Cardone, M.F., Deschamps, S., Roe, B., Rocchi, M. and Eichler, E.E. (2003) Refinement of a chimpanzee pericentric inversion breakpoint to a segmental duplication cluster. *Genome Biology* 4, R50.

Lower, R., Lower, J. and Kurth, R. (1996) The viruses in all of us: characteristics and biological significance of human endogenous retrovirus sequences. *Proceedings of the National Academy of Sciences USA* 93, 5177–5184.

Lupski, J.R. (1998) Genomic disorders: structural features of the genome can lead to DNA rearrangements and human disease traits. *Trends in Genetics* 14, 417–422.

Lyon, M.F. (1998) X-chromosome inactivation: a repeat hypothesis. *Cytogenetics and Cell Genetics* 80, 133–137.

Lyttle, T. (1989) Is there a role for meiotic drive in karyotype evolution? In: Giddings, L., Kaneshiro, K. and Anderson, W. (eds) *Genetics, Speciation and the Founder Principle*. Oxford University Press, New York.

Malik, H.S. and Henikoff, S. (2002) Conflict begets complexity: the evolution of centromeres. *Current Opinions in Genetics and Development* 12, 711–718.

Martin, C.L., Wong, A., Gross, A., Chung, J., Fantes, J.A. and Ledbetter, D.H. (2002) The evolutionary origin of human subtelomeric homologies – or where the ends begin. *American Journal of Human Genetics* 70, 972–984.

McClintock, B. (1987) *The Discovery and Characterization of Transposable Elements.* Garland Publishing, New York.

McLysaght, A., Hokamp, K. and Wolfe, K.H. (2002) Extensive genomic duplication during early chordate evolution. *Nature Genetics* 31, 200–204.

Mefford, H.C. and Trask, B.J. (2002) The complex structure and dynamic evolution of human subtelomeres. *Nature Reviews of Genetics* 3, 91–102.

Miki, Y., Nishisho, I., Horii, A., Miyoshi, Y., Utsunomiya, J., Kinzler, K.W., Vogelstein, B. and Nakamura, Y. (1992) Disruption of the APC gene by a retrotransposal insertion of L1 sequence in a colon cancer. *Cancer Research* 52, 643–645.

Morrish, T.A., Gilbert, N., Myers, J.S., Vincent, B.J., Stamato, T.D., Taccioli, G.E., Batzer, M.A. and Moran, J.V. (2002) DNA repair mediated by endonuclease-independent LINE-1 retrotransposition. *Nature Genetics* 31, 159–165.

Morse, B., Rotherg, P.G., South, V.J., Spandorfer, J.M. and Astrin, S.M. (1988) Insertional mutagenesis of the myc locus by a LINE-1 sequence in a human breast carcinoma. *Nature* 333, 87–90.

Nachman, M.W., Boyer, S.N., Searle, J.B. and Aquadro, C.F. (1994) Mitochondrial DNA variation and the evolution of Robertsonian chromosomal races of house mice, *Mus domesticus. Genetics* 136, 1105–1120.

Nadeau, J.H. and Taylor, B.A. (1984) Lengths of chromosomal segments conserved since divergence of man and mouse. *Proceedings of the National Academy of Sciences USA* 81, 814–818.

Neitzel, H., Kalscheuer, V., Singh, A.P., Henschel, S. and Sperling, K. (2002) Copy and paste: the impact of a new non-L1 retroposon on the gonosomal heterochromatin of *Microtus agrestis. Cytogenetics and Genome Research* 96, 179–185.

O'Brien, S.J., Menotti-Raymond, M., Murphy, W.J., Nash, W.G., Wienberg, J., Stanyon, R., Copeland, N.G., Jenkins, N.A., Womack, J.E. and Marshall Graves, J.A. (1999) The promise of comparative genomics in mammals. *Science* 286, 458–462, 479–481.

Ohno, S. (1970) *Evolution by Gene Duplication.* Springer, New York.

O'Neill, R.J., O'Neill, M.J. and Graves, J.A.M. (1998) Undermethylation associated with retroelement activation and chromosome remodelling in an interspecific mammalian hybrid. *Nature* 393, 68–72.

O'Neill, R.J.W., Eldridge, M.D.B. and Graves, J.A.M. (2001) Chromosome heterozygosity and *de novo* chromosome rearrangements in interspecific mammalian hybrids. *Mammalian Genome* 12, 256–259.

Ota, T., Sitnikova, T. and Nei, M. (2000) Evolution of vertebrate immunoglobulin variable gene segments. *Current Topics in Microbiology and Immunology* 248, 221–245.

Pardo-Manuel de Villena, F. and Sapienza, C. (2001) Female meiosis drives karyotypic evolution in mammals. *Genetics* 159, 1179–1189.

Parish, D.A., Vise, P., Wichman, H.A., Bull, J.J. and Baker, R.J. (2002) Distribution of LINEs and other repetitive elements in the karyotype of the bat *Carollia*: implications for X-chromosome inactivation. *Cytogenetics and Genome Research* 96, 191–197.

Pevzner, P. and Tesler, G. (2003a) Genome rearrangements in mammalian evolution: lessons from human and mouse genomes. *Genome Research* 13, 37–45.

Pevzner, P. and Tesler, G. (2003b) Human and mouse genomic sequences reveal extensive breakpoint reuse in mammalian evolution. *Proceedings of the National Academy of Sciences USA* 100, 7672–7677.

Pimpinelli, S., Berloco, M., Fanti, L., Dimitri, P., Bonaccorsi, S., Marchetti, E., Caizzi, R., Caggese, C. and Gatti, M. (1995) Transposable elements are stable structural components of *Drosophila melanogaster* heterochromatin. *Proceedings of the National Academy of Sciences USA* 92, 3804–3808.

Postlethwait, J.H., Woods, I.G., Ngo-Hazelett, P., Yan, Y.L., Kelly, P.D., Chu, F., Huang, H., Hill-Force, A. and Talbot, W.S. (2000) Zebrafish comparative genomics and the origins of vertebrate chromosomes. *Genome Research* 10, 1890–1902.

Potier, M., Dutriaux, A., Orti, R., Groet, J., Gibelin, N., Karadima, G., Lutfalla, G., Lynn, A., Van Broeckhoven, C., Chakravarti, A., Petersen, M., Nizetic, D., Delabar, J. and Rossier, J. (1998) Two sequence-ready contigs spanning the two copies of a 200-kb duplication on human 21q: partial sequence and polymorphisms. *Genomics* 51, 417–426.

Prak, E.T. and Kazazian, H.H., Jr (2000) Mobile elements and the human genome. *Nature Reviews of Genetics* 1, 134–144.

Prince, V.E. and Pickett, F.B. (2002) Splitting pairs: the diverging fates of duplicated genes. *Nature Reviews of Genetics* 3, 827–837.

Reig, O.A., Busch, C., Ortells, M.O. and Contreras, J.R. (1990) An overview of evolution, systematics, population biology, cytogenetics, molecular biology and speciation in *Ctenomys*. In: Nevo, E. and Reig, O.A. (eds) *Evolution of Subterranean Mammals at the Organismal and Molecular Levels*. Wiley-Liss, New York, pp. 97–128.

Riggs, A.D. (1990) DNA methylation and late replication probably aid cell memory, and type I DNA reeling could aid chromosome folding and enhancer function. *Philosophical Transactions of the Royal Society of London: B Biological Science* 326, 285–297.

Robertson, H.M. (1996) Members of the pogo superfamily of DNA-mediated transposons in the human genome. *Molecular and General Genetics* 252, 761–766.

Rofe, R.H. (1979) G-banding and chromosomal evolution in Australian marsupials. PhD thesis, University of Adelaide, Adelaide.

Samonte, R.V. and Eichler, E.E. (2002) Segmental duplications and the evolution of the primate genome. *Nature Reviews of Genetics* 3, 65–72.

Schueler, M.G., Higgins, A.W., Rudd, M.K., Gustashaw, K. and Willard, H.F. (2001) Genomic and genetic definition of a functional human centromere. *Science* 294, 109–115.

Schughart, K., Kappen, C. and Ruddle, F.H. (1989) Duplication of large genomic regions during the evolution of vertebrate homeobox genes. *Proceedings of the National Academy of Sciences USA* 86, 7067–7071.

Semple, C.A., Rolfe, M. and Dorin, J.R. (2003) Duplication and selection in the evolution of primate beta-defensin genes. *Genome Biology* 4, R31.

Sharman, G.B., Close, R.L. and Maynes, G.M. (1990) Chromosomal evolution, phylogeny and speciation of rock wallabies (*Petrogale*: Macropodidae). *Australian Journal of Zoology* 37, 351–363.

Sherratt, D.J. (1995) *Mobile Genetic Elements*. IRL Press, New York, pp. xiii, 179.

Singer, M.F. (1982) Highly repeated sequences in mammalian genomes. *International Reviews in Cytology* 76, 67–112.

Slamovits, C.H., Cook, J.A., Lessa, E.P. and Rossi, M.S. (2001) Recurrent amplifications and deletions of satellite DNA accompanied chromosomal diversification in South American tuco-tucos (genus *Ctenomys*, Rodentia: Octodontidae): a phylogenetic approach. *Molecular Biology and Evolution* 18, 1708–1719.

Smit, A.F. (1996) The origin of interspersed repeats in the human genome. *Current Opinion in Genetics and Development* 6, 743–748.

Smit, A.F. (1999) Interspersed repeats and other mementos of transposable elements in mammalian genomes. *Current Opinion in Genetics and Development* 9, 657–663.

Smit, A.F. and Riggs, A.D. (1996) Tiggers and DNA transposon fossils in the human genome. *Proceedings of the National Academy of Sciences USA* 93, 1443–1448.

Sperling, K., Kalscheuer, V. and Neitzel, H. (1987) Transcriptional activity of constitutive heterochromatin in the mammal *Microtus agrestis* (Rodentia, Cricetidae). *Experimental Cell Research* 173, 463–472.

Stankiewicz, P. and Lupski, J.R. (2002a) Genome architecture, rearrangements and genomic disorders. *Trends in Genetics* 18, 74–82.

Stankiewicz, P. and Lupski, J.R. (2002b) Molecular-evolutionary mechanisms for genomic disorders. *Current Opinion in Genetics and Development* 12, 312–319.

Sturtevant, A. and Beadle, G. (1936) The relations of inversion in the X chromosome of *Drosophila melanogaster* to crossing over and non-disjunction. *Genetics* 21, 554–604.

Sullivan, B.A., Blower, M.D. and Karpen, G.H. (2001) Determining centromere identity: cyclical stories and forking paths. *Nature Review Genetics* 2, 584–596.

Sumner, A. (1990) *Chromosome Banding*. Unwin Hyman, London.

Thomas, C.Y., Khiroya, R., Schwartz, R.S. and Coffin, J.M. (1984) Role of recombinant ecotropic and polytropic viruses in the development of spontaneous thymic lymphomas in HRS/J mice. *Journal of Virology* 50, 397–407.

Thomas, J.W., Schueler, M.G., Summers, T.J., Blakesley, R.W., McDowell, J.C., Thomas, P.J., Idol, J.R., Maduro, V.V., Lee-Lin, S.Q., Touchman, J.W., Bouffard, G.G., Beckstrom-Sternberg, S.M. and Green, E.D. (2003) Pericentromeric duplications in the laboratory mouse. *Genome Research* 13, 55–63.

Trask, B.J., Friedman, C., Martin-Gallardo, A., Rowen, L., Akinbami, C., Blankenship, J., Collins, C., Giorgi, D., Iadonato, S., Johnson, F., Kuo, W.L., Massa, H., Morrish, T., Naylor, S., Nguyen, O.T., Rouquier, S., Smith, T., Wong, D.J., Youngblom, J. and van den Engh, G. (1998) Members of the olfactory receptor gene family are contained in large blocks of DNA duplicated polymorphically near the ends of human chromosomes. *Human Molecular Genetics* 7, 13–26.

Tudor, M., Lobocka, M., Goodell, M., Pettitt, J. and O'Hare, K. (1992) The pogo transposable element family of *Drosophila melanogaster*. *Molecular and General Genetics* 232, 126–134.

Ventura, M., Archidiacono, N. and Rocchi, M. (2001) Centromere emergence in evolution. *Genome Research* 11, 595–599.

Verma, R.S. (1999) Evolution of the centromeric alpha-satellite DNA sequences of human chromosome 22. *Prenatal Diagnosis* 19, 590–591.

Wallis, M. (2001) Episodic evolution of protein hormones in mammals. *Journal of Molecular Evolution* 53, 10–18.

Walsh, J. (1982) Rate of accumulation of reproductive isolation by chromosome rearrangements. *American Naturalist* 120, 510–532.

Waterston, R.H., Lindblad-Toh, K., Birney, E., Rogers, J., Abril, J.F., Agarwal, P., Agarwala, R., Ainscough, R., Alexandersson, M., An, P., Antonarakis, S.E., Attwood, J., Baertsch, R., Bailey, J., Barlow, K., Beck, S., Berry, E., Birren, B., Bloom, T., Bork, P. *et al.* (2002) Initial sequencing and comparative analysis of the mouse genome. *Nature* 420, 520–562.

Wei, W., Gilbert, N., Ooi, S.L., Lawler, J.F., Ostertag, E.M., Kazazian, H.H., Boeke, J.D. and Moran, J.V. (2001) Human L1 retrotransposition: *cis* preference versus *trans* complementation. *Molecular and Cellular Biology* 21, 1429–1439.

White, M. (1973) *Animal Cytology and Evolution*, 3rd edn. Cambridge University Press, London.

White, M. (1978) *Modes of Speciation*. W.H. Freeman, San Francisco.

White, M.J. (1975) Chromosomal repatterning–regularities and restrictions. *Genetics* 79 (Supplement) 63–72.

Whitelaw, E. and Martin, D.I. (2001) Retrotransposons as epigenetic mediators of phenotypic variation in mammals. *Nature Genetics* 27, 361–365.

Wichman, H.A., Payne, C.T., Ryder, O.A., Hamilton, M.J., Maltbie, M. and Baker, R.J. (1991) Genomic distribution of heterochromatic sequences in equids: implications to rapid chromosomal evolution. *Journal of Heredity* 82, 369–377.

Wichman, H.A., Van den Bussche, R.A., Hamilton, M.J. and Baker, R.J. (1992) Transposable elements and the evolution of genome organization in mammals. *Genetica* 86, 287–293.

Wright, S. (1941) On the probability of fixation of reciprocal translocation. *American Naturalist* 75, 513–522.

Yang, F., O'Brien, P.C., Wienberg, J. and Ferguson-Smith, M.A. (1997) A reappraisal of the tandem fusion theory of karyotype evolution in Indian muntjac using chromosome painting. *Chromosome Research* 5, 109–117.

Yoder, J.A., Walsh, C.P. and Bestor, T.H. (1997) Cytosine methylation and the ecology of intragenomic parasites. *Trends in Genetics* 13, 335–340.

12 DNA Sequence Evolution and Phylogenetic Footprinting

E.T. Dermitzakis*[1,2] and A. Reymond[1]

[1]*Division of Medical Genetics, University of Geneva Medical School, Geneva, Switzerland;* [2]*Wellcome Trust, Sanger Institute, Hinxton, Cambridge, UK*

Introduction	301
Phylogenetic Relationships of Mammalian Species	302
Comparison of Sequences between Mammalian Species	304
Alignments of long genomic sequences	304
Global analysis of sequence evolution	304
Identification of functional genomic elements	306
Databases and programs for genome comparisons	307
Evolution of Protein-coding Genes	307
Evolution of nucleotides in a protein-coding sequence	307
Evolution of Regulatory Regions	309
Conserved Non-genic Sequences (CNGs) in Mammalian Genomes	311
Summary	312
References	312

Introduction

Extracting the complete functional information encoded in a genome is a major challenge in biological research. In an ideal situation, one would be able to mine this information directly from the ore of the DNA sequence itself without having to resort to experimentation. Comparative genome analysis of related species should provide a powerful and general approach to identify functional elements without previous knowledge of function. Non-functional sequences have diverged enough to allow functional sequence conservation signals

to stand out above the noise threshold. One advantage is that by increasing the number of species studied, one can raise the power of this approach. The term phylogenetic footprinting has been used extensively to indicate the presence of 'footprints' of strong negative selection on nucleotide sequences and the methodologies we use to detect them. Confident about the strength of phylogenetic footprinting, genome sequencing consortia are generating vast amounts of data to identify coding sequences, conserved non-coding regions and regulatory elements, and to get insight into the forces that shaped modern-day genomes.

*Correspondence: Emd4@Sanger.ac.uk

The elucidation of the evolutionary forces that govern sequence change over time has been the material of extensive research in the past few decades and many models accounting for different properties of nucleotide substitutions have been developed (e.g. Kimura, 1980; Tamura and Nei, 1993; see Yang, 1997 for applications). These models, however, were not tested on large amounts of data and as a result their validity and application remained unknown. This was a very important deficiency since we could not interpret the patterns of conservation or divergence in genomic regions of interest in the absence of the knowledge of what is the pattern of sequence evolution in non-functional regions.

We have recently seen the release of many vertebrate genomes such as the human (Lander et al., 2001; Venter et al., 2001), the fugu (Aparicio et al., 2002), the mouse (Mural et al., 2002; Waterston et al., 2002), the dog (Kirkness et al., 2003), and soon we will experience the availability of genomes from vertebrates such as the rat, the chimpanzee, the chicken, the cow, the rhesus macaque, the pig and others. The availability of these genomes provides for the first time ample information for comparative analysis of genomic sequences and deep understanding of sequence evolution. Whole genome sequence comparisons allow the identification of selectively constrained sequences by means of high conservation and detection of unusual patterns of evolution across multiple species.

The genes represent the best-known class of functional sequences. We have a relatively good understanding of the rules that govern sequence evolution in protein-coding sequences, and we are able to readily interpret the meaning and primary consequences of nucleotide changes in the open reading frame portion of their sequence (Kimura, 1977; Ohta, 1995). On the other hand, other functional classes of sequences are less well understood. Regulatory regions are redundant and fluid in their organization (Ludwig et al., 1998, 2000; Leung et al., 2000; Wasserman et al., 2000; Pennacchio and Rubin, 2001; Dermitzakis and Clark, 2002; Dermitzakis et al., 2003a; Elnitski et al., 2003). The multidimensional space of the rules for their evolution coupled with the small number of experimentally identified regulatory

regions contributes to a poor understanding of the consequences of the evolution of their sequence (Dermitzakis and Clark, 2002; Elnitski et al., 2003). Finally, there are probably other classes of functional sequences that we are currently unaware of, which makes our goal of understanding sequence evolution in mammalian genomes even more remote (Dubchak et al., 2000; Frazer et al., 2001; Dermitzakis et al., 2002; Mural et al., 2002; Waterston et al., 2002).

This chapter reviews studies and ideas of the last decades regarding DNA sequence evolution. More importantly, it is an attempt to put all these models and hypotheses into the framework of the vast amounts of data that have been recently available and see how well these models fit the data. We will start by giving an overview of the phylogenetic relationships of mammalian species and how these relate to the study of sequence evolution. Subsequently, a framework of platforms and databases that are useful for large-scale sequence analysis are presented. Then, we will present a global view and methodologies of sequence comparisons between mammalian species and describe how sequence conservation serves as a means for the identification of functional sequences in the human genome. Finally, we will focus on the evolution of the two best-characterized classes of functional sequences, the protein-coding genes and the regulatory regions, as well as the conserved non-genic sequences that have recently attracted a lot of attention.

Phylogenetic Relationships of Mammalian Species

Many studies have concentrated on the phylogenetic relationships of mammalian species (e.g. Kumar and Hedges, 1998; Murphy et al., 2001a,b). Although in most cases these studies have a natural history interest, the results of such studies will have a high impact on the understanding of sequence evolution as well. If we can accurately determine the relationships and the time of speciation of mammalian species we will be able to map the rate and pattern of evolution on their phylogenetic tree and understand the stepwise process by which sequence evolution occurs.

One of the caveats is that there is a certain amount of circularity in this approach. We are determining the phylogenetic relationships using DNA and protein sequences and subsequently we are mapping the characteristics of sequence evolution on the tree. How do we know whether the sequences we have used correctly illustrate the phylogenetic relationships? Which sequences do we use to build the relationships and which to study using the tree? To make the right choice one has to make certain assumptions about the pattern of evolution of DNA and protein sequences and then use them for the right purpose. Of course, the best way would be to use the whole genome to derive phylogenetic relationships and therefore average out any patterns of evolution that are contrary to the phylogenetic relationships and subsequently study each region of interest using the tree as given information. Unfortunately, this has not become possible yet since very few genomes are fully sequenced, but many studies have attempted to use large amounts of data for this reason (Kumar and Hedges, 1998). In the next few years we will experience the release of many mammalian genomes and phylogenetic approaches using whole-genome information.

The most widely accepted phylogenetic relationships of mammalian species that have been released or will be released in the next few years are shown in Fig. 12.1. From this figure it is evident that we will soon have a wide enough range of fully sequenced species to be able to reliably study patterns of evolution on the phylogenetic tree.

The phylogenetic information provides a large degree of power to analyse sequence evolution and detect patterns of nucleotide substitutions. One of the big advantages is that it enables one to put direction in sequence change events and scale their frequency and probability on the tree (Tajima, 1993). Moreover, one can detect species- or lineage-specific processes that are not obvious in multiple species comparisons by means of sequence alignment. Some sequence patterns (e.g. nucleotide substitutions) that are present in multiple species and in a multiple alignment would seem to be a result of a single evolutionary event in a common ancestor, but may actually be a result of more than one independent evolutionary event in different lineages. Such cases will carry very different interpretations for their impact in sequence function under the two alternative scenarios.

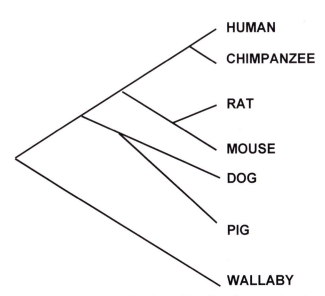

Fig. 12.1. A phylogenetic tree representing the relationships of vertebrate species whose sequences have been completed or will soon be completed. Phylogenetic relationships are based on the consensus of various studies and may not agree with all studies published to date.

Another aspect that is relevant to the phylogenetic relationships of mammalian species is which are the most informative species for the identification of functional genomic elements (O'Brien *et al.*, 1999). Depending on the level and pattern of selective constraint on the sequences, different combinations of species may be appropriate. Recent studies have utilized multiple species to select putatively functional sequences (Dubchak *et al.*, 2000; Dermitzakis *et al.*, 2002, 2003b; Boffelli *et al.*, 2003; Thomas *et al.*, 2003). However, some of these types of analysis have reduced power because the authors did not use the phylogenetic information in their models (Thomas *et al.*, 2003). The main source of the difference in the optimal choice of species is the different pattern of evolution for different classes of functional sequences. For example, it has been demonstrated that protein-coding sequences are conserved across a wider range of vertebrate species (as far as between human and fugu) than other functional sequences (Aparicio *et al.*, 2002; Thomas *et al.*, 2003), while non-coding functional sequences, such as regulatory regions, maintain a strong selective signal in shorter distances such as human, mouse, dog and marsupials (Loots *et al.*, 2000; Wasserman *et al.*, 2000; Dermitzakis and Clark, 2002) and thus analysis within these species will provide higher resolution. It is expected that implementing fine characteristics in a phylogenetic context will allow an efficient sub-classification of functional sequences even within the range of species where all of the functional elements are conserved (Boffelli *et al.*, 2003; Dermitzakis *et al.*, 2003b).

Comparison of Sequences between Mammalian Species

The availability of whole mammalian genomes allows the large-scale comparison of sequences and the elucidation of evolutionary forces that govern sequence change. Whole genome comparisons allow for a better understanding of the impact of variation of genome-wide parameters in genome evolution (Hardison *et al.*, 1997, 2003). Accounting for these parameters is a powerful way to distinguish conserved functional sequences from sequences that simply did not accumulate enough substitutions to diverge.

Alignments of long genomic sequences

One of the essential elements of large-scale genome comparisons is the use of reliable algorithms that can align the genomic regions of interest. Until recently only pairwise alignment methodologies were necessary for long regions since the availability of multiple species data was very limited. However, as mentioned above, many mammalian genomes are now being sequenced and several regions of the genome are being subjected to targeted sequencing in multiple species. For this reason the use of multiple species alignment programs is necessary.

Two of the algorithms that have been developed and are widely used in the alignments of large genomic regions are PIPMAKER and AVID (Schwartz *et al.*, 2000; Bray *et al.*, 2003). The first algorithm is based on BLASTZ (see also Chapter 15) and performs local alignments (Fig. 12.2), while AVID performs global alignments. Both programs were originally written to accommodate pairwise comparisons but have now been extended to handle multiple species (MULTIPIPMAKER and MAVID; Bray and Pachter, 2003; Schwartz *et al.*, 2003).

Global analysis of sequence evolution

Fine analysis of sequence evolution can be assisted substantially if one considers how characteristics of conservation and divergence are influenced by variation in genome characteristics such as nucleotide composition, repeat density, recombination, etc. (Hardison *et al.*, 2003). This will allow for the appropriate correction of nucleotide evolutionary rate using the variation of genomic parameters that can indirectly highlight functionally constrained sequences given the genomic context. In other words, one can map the expected neutral rate of nucleotide change in each genomic region and, based on indirect characteristics, select as functional sequences the ones that stand out as more conserved than the inferred neutral baseline.

To give an example, one of the characteristics that has been associated with genome conservation variation is repeat density. It has been shown that repeat density is higher in

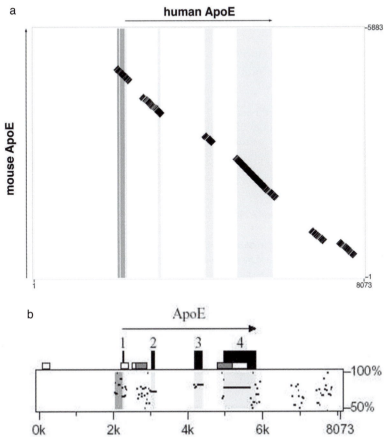

Fig. 12.2. (a) Dotplot of the human–mouse comparison of the ApoE region. (b) PIPMAKER representation in a scale of 50–100% conservation for the same region. Shading corresponds to the same genomic elements as in (a).

regions with low conservation, which provides a predictive model for genome conservation and possibly function (Chiaromonte *et al.*, 2001). It is reasonable to assume that regions that accumulated fixed repeat insertions will contain fewer functional genomic elements. Another possibility is that there are some regions of the genome that attract repeats more than others and consequently functional elements in those regions have been lost or 'translocated' in other, less repeat-dense regions.

Recombination is a genomic property that is also associated with sequence conservation (Hardison *et al.*, 2003; Hellmann *et al.*, 2003). G+C content is similarly associated with conservation and substitution rate (Hardison *et al.*, 2003). It is also known that G+C content is correlated with recombination (Fullerton *et al.*, 2001). This interdependence creates a problem, since we do not know what the cause is and what the effect is. The current thinking is that recombination and G+C content are causative of sequence conservation (Fullerton *et al.*, 2001). This problem illustrates one of the difficulties in trying to elucidate the cause and effect relationship of global genomic parameters, especially when so many data are involved that we have a lot of power to detect even small effects. This example shows the danger of attempting to assign a causal role simply by looking at genome correlations without independent biological support.

Nevertheless, these genome correlations are very useful even if they do not directly indicate cause and effect relationships. They can be used as predictive variables for sequence conservation. Also, when more in depth studies have characterized functional elements of the human genome in large genomic blocks, we may even be able to predict density and distribution of certain types of functional elements simply by looking at global characteristics such as G+C content, repeat density and recombination.

Identification of functional genomic elements

The ultimate goal of genomic comparisons is twofold. On the one hand, we want to be able to study the evolutionary processes that functional and non-functional sequences undergo so that we learn something about the biology of the species and the forces that contribute to its evolution. On the other hand, by studying examples of the phenotypic consequences of molecular evolution, we can speculate and design models for the potential effect of sequence changes in regions of unknown function.

So far we know very little about the actual fraction of the mammalian genome that is functional, but recent estimates based on sequence conservation patterns suggest that it is at least 5% (Waterston et al., 2002). Given that the protein-coding fraction of the genome is about 1.5%, there is a lot of room for the identification of additional functional elements, especially if we take into account that we have probably reached a plateau in the gene count and the number of currently unknown genes is probably small. We should also consider the fact that sequence conservation does not reveal the total fraction of the functional genome but simply the fraction of the genome that has *remained* functional within the group of species compared. Therefore, it is expected that an additional fraction will be species-specific or at least lineage-specific, and not conserved across large evolutionary distances such as human and mouse or across all vertebrate lineages. More importantly, this lineage-specific fraction can be either loss-of-function in one lineage or

gain-of-function in another lineage. These types of evolutionary changes will be very relevant when we start systematically thinking about the differences between species as a point of reference, instead of the similarities as we do now. Evolutionary comparisons of closely related species may reveal the genetic basis of differences between species such as human and chimp (Enard et al., 2002a,b; Clark et al., 2003).

The known functional attributes of the genome are very limited. The best known functional class is the class of protein-coding genes. The vast majority of nucleotide differences between species in coding sequences are silent. The silent substitution rate or the substitution rate in fourfold degenerate sites (sites in codons where all four of the nucleotides in one position code for the same amino acid) is commonly used as a means to estimate the neutral substitution rate of the genomic region (Hardison et al., 2003). However, there is a fraction of genes or gene regions that undergo diversifying selection in which the amino acid sequence changes as much or faster than the nucleotide sequence (Hughes and Nei, 1988; Wyckoff et al., 2000).

Another significant class of functional sequences is regulatory regions of gene expression. These regions are usually close to genes (within a few kb) but can be as far as hundreds of kb (Kleinjan and van Heyningen, 1998; Spitz et al., 2003). They consist of sets of transcription factor-binding sites and their DNA–protein interactions and the interactions between them and the promoter of the gene regulate the spatial and temporal expression of genes (Pennacchio and Rubin, 2001). The binding sites of transcription factors have usually redundant properties and are described with a probability weight matrix (PWM) that contains the information about the frequency of each of the four nucleotides in each of the positions of the site (Berg and von Hippel, 1987).

Non-coding RNA (ncRNA) genes are also considered a significant functional class of the mammalian genomes. While this class was mainly represented in the past by tRNAs, in the last few years we have witnessed the discovery of several hundred ncRNAs of various types. ncRNAs can be micro RNAs, small nuclear

RNAs and miRNAs (see reviews by Eddy, 1999, 2002). Each of these categories has its own characteristics and levels of conservation. Many of the ncRNAs have secondary structures which confer or facilitate their function. The conservation of this secondary structure is important and therefore most of the selective constraint is present in the sequences that pair to form these structures. Interestingly, there are often compensatory evolutionary changes in which two paired nucleotides can change almost simultaneously (in evolutionary terms) to create a new pair. This means that with two nucleotide substitutions the secondary structure can still be maintained, while that would not be possible with either of the two changes alone (Innan and Stephan, 2001). This pattern indicates that the evolution of ncRNAs with a functional secondary structure occurs in large steps of multiple nucleotide changes at a time rather than a slow and continuous process.

Another role of ncRNAs is to complement small regions of the transcript of a coding gene and via this interaction to regulate its rate of translation and/or its RNA stability (Carrington and Ambros, 2003). In this case the restriction on evolution is that the ncRNA maintains some level of complementation to the target mRNA. Most of the ncRNAs with this function (miRNAs) are small with their mature transcript being in the order of 20–25 ribonucleotides. This type of interaction predicts that there will be a co-evolution of the miRNA sequence with the target region of the mRNA. In addition, the small size of the miRNA suggests that there is a degree of stochasticity in the way that new miRNAs are created. A new interaction between an expressed random 25mer (a putative miRNA) and an mRNA, if advantageous, may be maintained and eventually fixed in the population. Such models generate very intriguing hypotheses for the evolution of genome regulation through birth and death processes of ncRNA.

Databases and programs for genome comparisons

The amount of data that has become available requires fast and efficient processing to make sense of it. Unfortunately, a single computer and a single person cannot process and obtain all the information and perform all types of analysis necessary to extract information from genome comparisons. For this reason, many databases have been developed that accommodate 'frequently asked' information that is necessary for the researcher to design experiments or study the biological significance of some sequences (see also Chapter 16). Most of these databases provide characteristics of the genome, such as genes, expressed sequence tags (ESTs), repeats, computational predictions and other information in the context of genome conservation. Such databases are the UCSC browser (genome.ucsc.edu), ENSEMBL (www.ensembl.org) and NCBI (www.ncbi.nlm.nih.gov). In these databases one can find information about the levels of conservation of genes and other non-genic regions between a number of species. Finally, we have recently seen the creation of a new database, GALA (http://globin.cse.psu.edu/gala/; Giardine *et al.*, 2003). This database provides information on the annotation of the human, mouse and rat genomes in the context of sequence conservation and allows the efficient extraction and primary analysis of this information, which can serve many subsequent analysis pipelines. The ability to efficiently extract these types of data will greatly facilitate the understanding of the mammalian genomes.

Evolution of Protein-coding Genes

Genes represent only a small proportion of the mammalian genomes but they are understandably the focus of great interest. Many innovations can occur in protein-coding genes such as evolution of the primary amino acid sequence, reorganization of their exons as well as the creation of a new gene from raw genetic material. All these issues will be discussed in this section.

Evolution of nucleotides in a protein-coding sequence

Proteins fall into various different categories of functional constraint and significance. There are also different selective constraints for different domains of proteins, which drive differential rates of evolution. The most abundant class of genes falls into the category of selectively constrained genes. However, there are genes which are generally less constrained and their sequence evolves either almost neutrally or faster than neutral due to positive selection. In highly constrained genes, nucleotide changes that alter the amino acids are disfavoured, but changes that are silent are mostly neutral. This type of selective constraint generates an excess of silent substitutions. On the other hand, about two-thirds of the potential changes in a protein-coding sequence result in the change of the amino acid (replacement). This biased pattern can be detected by computing the ratio of the rates of replacement over silent substitutions (K_A/K_S or Dn/Ds). For a selectively constrained sequence it is expected that the K_A/K_S ratio is significantly below 1. On the contrary, for a protein-coding sequence that is under positive selection the ratio will be significantly higher than 1. Initially, this statistic was used by Hughes and Nei (1988) to detect positive selection in the major histocompatibility complex (MHC) cluster.

Although the K_A/K_S ratio is a good indicator of the direction of selective constraint in a protein-coding sequence, some of the amino acid sequence properties are more subtle than can be expressed by calculating silent vs. replacement substitution rates. Some of the amino acid replacements have a stronger effect in the properties of the protein sequence than others. Amino acid changes that alter the amino acid to one with different biochemical properties are called radical whilst when the change is of small or no biochemical effect the change is called conservative. There are several properties by which we classify amino acid substitutions as radical vs. conservative (classified by charge, polarity or polarity and volume). Many methodologies have now implemented the conservative vs. radical criterion to detect an excess of conservative or radical changes and therefore detect negative or positive selection in order to tease apart the evolutionary forces that shape protein evolution (e.g. Zhang, 2000). In addition, there are other nucleotides that are not strictly part of the coding sequence but are also selectively constrained, such as the splicing signals. These are also usually conserved and good indicators of the presence of a gene structure.

Recent estimates suggest that the mammalian gene count may fall at the lower end of the 30,000–40,000 range (Waterston *et al.*, 2002). The sequencing of the human and mouse genomes allowed the creation of an improved catalogue of mammalian protein-coding genes (Lander *et al.*, 2001; Venter *et al.*, 2001; Waterston *et al.*, 2002). Comparative genomics coupled with experimental analysis can seriously improve gene discovery, especially for genes that are expressed at low levels or in very restricted patterns. In a recent comparative and experimental analysis of gene prediction programs the confirmed predictions that did not previously have experimental support (e.g. ESTs, cDNAs, etc.) showed a more restricted expression pattern than the average genes (i.e. mouse orthologues of human chromosome 21 genes) possibly explaining the absence of these predictions from transcript-based annotations (Reymond *et al.*, 2002; Guigo *et al.*, 2003). To further increase the proportion of novel predictions that are representing *bona fide* mammalian genes, the ratio of non-synonymous to synonymous substitution rates (K_A/K_S, see above) can be included in the new prediction pipelines, because, on average, experimentally verified Twinscan and SGP2 predictions were shown to have a lower K_A/K_S ratio (Guigó *et al.*, 2003).

Similarly, the exon structures and splice signals of more than 90% of exons of human–mouse RefSeq genes (i.e. genes with highly reliable structure annotated in the Reference Sequence database of NCBI) were shown to be highly conserved (Waterston *et al.*, 2002). These data suggest that a high quality annotation of the human coding sequences could be achieved through comparison of full genome sequences of a set of species covering the entire mammalian tree.

However, genome comparison might miss rapidly evolving genes, for example *Saccharomyces* species genome comparison overlooked a gene with less than 15% amino acid identity across the four analysed species (Kellis *et al.*, 2003). Duplication events are one of the forces which possess the potential to significantly alter the pace of gene evolution. Segmental duplication with a frequency, for example, of about one-tenth of the human genome (Bailey *et al.*, 2002), might complicate genome analysis by diverging from a strict 1:1

orthologous relationship. Numerous genes involved in sex and reproduction or in host defence and immunity were demonstrated to be rapidly evolving (Waterston *et al.*, 2002). This rapid change seems to be caused by selective forces rather than mutation rate, as the average rate of non-synonymous substitutions per site (K_A) is significantly higher while no differences appear in the synonymous substitution rate (K_S) of some mammalian genes.

It is easy to understand why a protein sequence will be highly constrained and not change, since natural processes are conservative in allowing already well-functioning components to change. For a protein to evolve fast there have to be forces that give a tremendous advantage to the new version vs. the old. As mentioned above, there are classes of genes that undergo positive selection due to the nature of their function. For example, immunity genes that participate in recognition of pathogens need to keep track of the evolving diversity of pathogens, thus participating in an 'arms race'. In addition, competition for sexual characteristics leads to extreme phenotypic characteristics, which are usually driven by fast-evolving genes. Finally, some new phenotypic attributes driven by a few changes can become extremely advantageous and are rapidly selected. Such an example is the *FOXP2* gene which, it has recently been speculated, could be associated with the gain of speech capability in humans (Enard *et al.*, 2002b). In this example, the authors showed that there is evidence for recent selective sweep in humans possibly driven by only a few amino acid changes that may facilitate speech capability in our species.

Evolution of Regulatory Regions

Another important class of functional sequences is the regulatory regions that control spatial and temporal levels and patterns of gene expression. Regulatory regions usually consist of sets of transcription factor-binding sites that are arranged along the sequence so that transcription factors interact with each other and with the promoter of the gene they regulate. These interactions modulate activation or suppression of expression of the gene of interest in a tissue- or stage-specific manner.

The evolutionary characteristics of regulatory regions are dependent on the organization of the functional DNA elements within them (Wray *et al.*, 2003). Transcription factor-binding sites are the functional units that are under selection and exert the function of the regulatory region. These binding sites are usually short in length, approximately 8–15 bp, and have redundant patterns of nucleotide organization. Binding sites are described by weight matrices in which each column describes the frequency of each of the four nucleotides in each position of the length of the binding site. Probability weight matrices provide a good metric of what types of nucleotide substitutions are conservative and thus do not seriously affect binding and what types are seriously affecting binding, in which case the nucleotide substitutions are radical (Berg and von Hippel, 1987).

Only a few studies have attempted to dissect the evolutionary characteristics of regulatory regions. In some early studies in *Drosophila*, Ludwig and colleagues (Ludwig and Kreitman, 1995; Ludwig *et al.*, 1998, 2000) have shown that there is substantial variation in fly regulatory regions and the pattern of their evolution appears to be compensatory. In other words, there is a lot of nucleotide divergence between species that is not reflected in phenotypic attributes controlled by the regulatory region (see model in Fig. 12.3).

Subsequently, a few studies have looked at the level of conservation of regulatory regions in mammals. Hardison (2000) showed that some regulatory regions are highly conserved between human and mouse providing lots of optimism in the field for an easy identification of them after the completion of the mouse genome. Wasserman *et al.* (2000) provided further evidence for this by looking at the pattern of conservation of experimentally verified and predicted transcription factor-binding sites in muscle-specific regulatory regions. On the contrary, the study by Dermitzakis and Clark (2002) showed that a large number of confirmed transcription factor-binding sites are not conserved above the levels of stochastic conservation and about 32–40% of them do not share function between human and mouse but are located in otherwise conserved regulatory regions. The discrepancy between the studies above shows the fluidity of the regulatory systems. In fact, all studies have described the

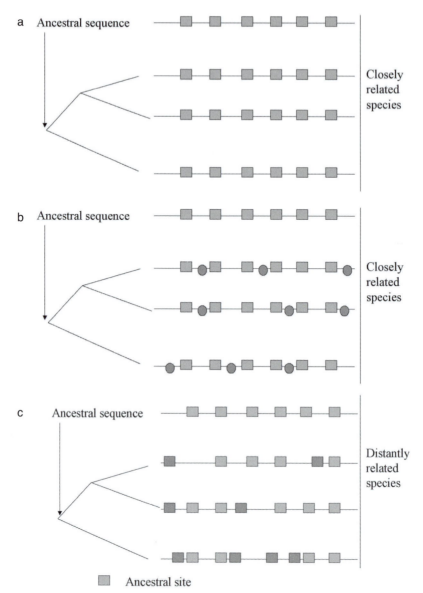

Fig. 12.3. Model for the evolution of regulatory regions. Closely related species have almost identical functional binding sites, which are highly conserved (a). Some other sequences are close to becoming functional binding sites (potential binding sites – circles) in the background and are a few nucleotide changes from that (b). In distantly related species some of the potential binding sites have replaced the previously functional ones (c).

correct pattern. It is true that many transcription factor-binding sites are present in highly conserved regulatory regions as Hardison and Wasserman and colleagues have observed but the actual binding site motifs are not conserved as much, but rather they undergo extensive turn-over as Dermitzakis and Clark (2002) have observed (see Fig. 12.4).

In a more recent study, Dermitzakis and colleagues (2003a) describe evolutionary characteristics of transcription factor-binding sites in *Drosophila* using five regulatory regions

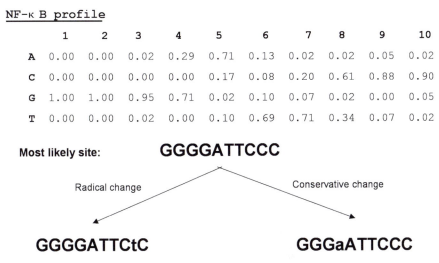

Fig. 12.4. Probability weight matrix for the NF-κB transcription factor. Below is the most likely binding site given the matrix and two scenarios of a conservative (no loss of binding ability) change and a radical change (partial loss of binding ability).

as models. By constructing models for prediction of binding sites, they study the potential evolutionary history of known and putative binding sites. These results indicate that although many of the binding sites are highly conserved between species, there is a lot of potential for new binding sites, through which many of the evolutionary changes are mediated.

One other important aspect is how much of the phenotypic variation within populations of the same species is associated with regulatory variation. This direction of research has been progressing slowly, mostly because our understanding of the consequences of nucleotide changes in regulatory regions is not adequate. In a recent study Rockman and Wray (2002) surveyed the literature for reports of variations in regulatory regions that have phenotypic effects as gene expression differences of disease phenotypes. To our surprise, there are a large number of reports that describe regulatory variation, which suggests that much of the phenotypic and/or disease variation is harboured in regulatory regions. Regulatory variation is more likely to be associated with multifactorial and quantitative phenotypes since regulatory changes exert quantitative differences and more subtle effects than changes in protein-coding sequences.

The pattern of evolution and polymorphism in regulatory regions will become a very important part in the studies of phenotypic variation and disease association studies in the next few decades (Wray *et al.*, 2003). Particularly the link to gene expression variation will provide a very useful endophenotype to bridge the big gap between genotype and disease (Cheung and Spielman, 2002). Whole-genome analysis of gene expression variation has been initiated in small population set-ups and it is anticipated that it will expand to more phenotype-driven studies and association with nucleotide polymorphism. This exercise will be very useful for the understanding of regulatory variation and mechanisms for the control of gene expression.

Conserved Non-genic Sequences (CNGs) in Mammalian Genomes

The last category of sequences that we should discuss in this chapter is the recently described class of conserved non-genic sequences (CNGs), which have been discovered after the comparison of the human and mouse genomes. These sequences appear to be abundant in numbers (~330,000 in the

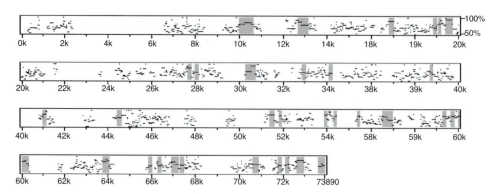

Fig. 12.5. PipMaker alignment of a gene-poor region of human chromosome 21. Blocks in grey indicate regions of the human genome that are at least 100 bp and at least 70% identity between human and mouse (conserved non-genic sequences: CNGs)

human genome – see Fig. 12.5 for example) and their level of conservation between human and mouse strongly suggests that they are functionally conserved (Dubchak *et al.*, 2000; Frazer *et al.*, 2001; Dermitzakis and Clark, 2002; Waterston *et al.*, 2002; Dermitzakis *et al.*, 2003b). Many hypotheses have been presented for their function but only a very small fraction of them have received a functional identity.

Most of the experimentally characterized CNGs have been designated as regulatory regions (Boffelli *et al.*, 2003). A few of them have been hypothesized to be structural elements that contribute to chromatin conformations. However, this picture is biased since the vast majority of CNGs have not been shown to belong to any particular functional category, most probably because we have not performed the right experiment. A recent study by Dermitzakis and colleagues (2003b) showed that the CNGs have evolutionary characteristics different from coding sequences and non-coding RNAs and resemble patterns of conservation of protein-binding regions, with alternate clusters of highly constrained and variable nucleotides. The elucidation of the function of CNGs will probably reveal new classes of functional elements as well as new ways that regulatory elements may function.

Summary

We are in the middle of a very exciting period. Soon, we will be able to analyse large amounts of sequence data and we are facing the unimaginable problem of too much data and poor understanding of how we can make sense of them. In the next decade we are bound to face surprises and big discoveries about the function of our genome and how it has evolved to this state.

References

Aparicio, S., Chapman, J., Stupka, E., Putnam, N., Chia, J.M., Dehal, P., Christoffels, A., Rash, S., Hoon, S., Smit, A., Gelpke, M.D., Roach, J., Oh, T., Ho, I.Y., Wong, M., Detter, C., Verhoef, F., Predki, P., Tay, A., Lucas, S. *et al.* (2002) Whole-genome shotgun assembly and analysis of the genome of *Fugu rubripes*. *Science* 297, 1301–1310.

Bailey, J.A., Gu, Z., Clark, R.A., Reinert, K., Samonte, R.V., Schwartz, S., Adams, M.D., Myers, E.W., Li, P.W. and Eichler, E.E. (2002) Recent segmental duplications in the human genome. *Science* 297, 1003–1007.

Berg, O.G. and von Hippel, P.H. (1987) Selection of DNA binding sites by regulatory proteins. Statistical-mechanical theory and application to operators and promoters. *Journal of Molecular Biology* 193, 723–750.

Boffelli, D., McAuliffe, J., Ovcharenko, D., Lewis, K.D., Ovcharenko, I., Pachter, L. and Rubin, E.M. (2003) Phylogenetic shadowing of primate sequences to find functional regions of the human genome. *Science* 299, 1391–1394.

Bray, N. and Pachter, L. (2003) MAVID multiple alignment server. *Nucleic Acids Research* 31, 3525–3526.

Bray, N., Dubchak, I. and Pachter, L. (2003) AVID: a global alignment program. *Genome Research* 13, 97–102.

Carrington, J.C. and Ambros, V. (2003) Role of microRNAs in plant and animal development. *Science* 301, 336–338.

Cheung, V.G. and Spielman, R.S. (2002) The genetics of variation in gene expression. *Nature Genetics* 32 (Supplement), 522–525.

Chiaromonte, F., Yang, S., Elnitski, L., Yap, V.B., Miller, W. and Hardison, R.C. (2001) Association between divergence and interspersed repeats in mammalian noncoding genomic DNA. *Proceedings of the National Academy of Sciences USA* 98, 14503–14508.

Clark A.G., Glanowski, S., Nielsen, R., Thomas, P.D., Kejariwal, A., Todd, M.A., Tanenbaum, D.M., Civello, D., Lu, F., Murphy, B., Ferriera, S., Wang, G., Zheng, X., White, T.J., Sninsky, J.J., Adams, M.D. and Cargill, M. (2003) Inferring nonneutral evolution from human–chimp–mouse orthologous gene trios. *Science* 302, 1960–1963.

Cliften, P., Sudarsanam, P., Desikan, A., Fulton, L., Fulton, B., Majors, J., Waterston, R., Cohen, B.A. and Johnston, M. (2003) Finding functional features in *Saccharomyces* genomes by phylogenetic footprinting. *Science* 301, 71–76.

Dermitzakis, E.T. and Clark, A.G. (2002) Evolution of transcription factor binding sites in mammalian gene regulatory regions: conservation and turnover. *Molecular Biology and Evolution* 19, 1114–1121.

Dermitzakis, E.T., Reymond, A., Lyle, R., Scamuffa, N., Ucla, C., Deutsch, S., Stevenson, B.J., Flegel, V., Bucher, P., Jongeneel, C.V. and Antonarakis, S.E. (2002) Numerous potentially functional but non-genic conserved sequences on human chromosome 21. *Nature* 420, 578–582.

Dermitzakis, E.T., Bergman, C. and Clark, A.G. (2003a) Tracing the evolutionary history of *Drosophila* regulatory regions with models that identify transcription factor binding sites. *Molecular Biology and Evolution* 20, 703–714.

Dermitzakis, E.T., Reymond, A., Scamuffa, N., Ucla, C., Kirkness, E., Rossier, C. and Antonarakis, S.E. (2003b) Evolutionary discrimination of mammalian conserved non-genic sequences (CNGs). *Science* 302, 1033–1035

Dubchak, I., Brudno, M., Loots, G.G., Pachter, L., Mayor, C., Rubin, E.M. and Frazer, K.A. (2000) Active conservation of noncoding sequences revealed by three-way species comparisons. *Genome Research* 10, 1304–1306.

Eddy, S.R. (1999) Noncoding RNA genes. *Current Opinion in Genetics and Development* 9, 695–699.

Eddy, S.R. (2002) Computational genomics of noncoding RNA genes. *Cell* 109, 137–140.

Elnitski, L., Hardison, R.C., Li, J., Yang, S., Kolbe, D., Eswara, P., O'Connor, M.J., Schwartz, S., Miller, W. and Chiaromonte, F. (2003) Distinguishing regulatory DNA from neutral sites. *Genome Research* 13, 64–72.

Enard, W., Khaitovich, P., Klose, J., Zollner, S., Heissig, F., Giavalisco, P., Nieselt-Struwe, K., Muchmore, E., Varki, A., Ravid, R., Doxiadis, G.M., Bontrop, R.E. and Paabo, S. (2002a) Intra- and interspecific variation in primate gene expression patterns. *Science* 296, 340–343.

Enard, W., Przeworski, M., Fisher, S.E., Lai, C.S., Wiebe, V., Kitano, T., Monaco, A.P. and Paabo, S. (2002b) Molecular evolution of *FOXP2*, a gene involved in speech and language. *Nature* 418, 869–872.

Flicek, P., Keibler, E., Hu, P., Korf, I. and Brent, M.R. (2003) Leveraging the mouse genome for gene prediction in human: from whole-genome shotgun reads to a global synteny map. *Genome Research* 13, 46–54.

Frazer, K.A., Sheehan, J.B., Stokowski, R.P., Chen, X., Hosseini, R., Cheng, J.F., Fodor, S.P., Cox, D.R. and Patil, N. (2001) Evolutionarily conserved sequences on human chromosome 21. *Genome Research* 11, 1651–1659.

Fullerton, S.M., Bernardo Carvalho, A. and Clark, A.G. (2001) Local rates of recombination are positively correlated with GC content in the human genome. *Molecular Biology and Evolution* 18, 1139–1142.

Giardine, B., Elnitski, L., Riemer, C., Makalowska, I., Schwartz, S., Miller, W. and Hardison, R.C. (2003) GALA, a database for genomic sequence alignments and annotations. *Genome Research* 13, 732–741.

Guigó, R., Dermitzakis, E.T., Agarwal, P., Ponting, C.P., Parra, G., Reymond, A., Abril, J.F., Keibler, E., Lyle,

R., Ucla, C., Antonarakis, S.E. and Brent, M.R. (2003) Comparison of mouse and human genomes followed by experimental verification yields an estimated 1,019 additional genes. *Proceedings of the National Academy of Sciences USA* 100, 1140–1145.

Hardison, R.C. (2000) Conserved noncoding sequences are reliable guides to regulatory elements. *Trends in Genetics* 16, 369–372.

Hardison, R.C., Oeltjen, J. and Miller, W. (1997) Long human–mouse sequence alignments reveal novel regulatory elements: a reason to sequence the mouse genome. *Genome Research* 7, 959–966.

Hardison, R.C., Roskin, K.M., Yang, S., Diekhans, M., Kent, W.J., Weber, R., Elnitski, L., Li, J., O'Connor, M., Kolbe, D., Schwartz, S., Furey, T.S., Whelan, S., Goldman, N., Smit, A., Miller, W., Chiaromonte, F. and Haussler, D. (2003) Covariation in frequencies of substitution, deletion, transposition, and recombination during eutherian evolution. *Genome Research* 13, 13–26.

Hellmann, I., Ebersberger, I., Ptak, S.E., Paabo, S. and Przeworski, M. (2003) A neutral explanation for the correlation of diversity with recombination rates in humans. *American Journal of Human Genetics* 72, 1527–1535.

Hughes, A.L. and Nei, M. (1988) Pattern of nucleotide substitution at major histocompatibility complex class I loci reveals overdominant selection. *Nature* 335, 167–170.

Innan, H. and Stephan, W. (2001) Selection intensity against deleterious mutations in RNA secondary structures and rate of compensatory nucleotide substitutions. *Genetics* 159, 389–399.

Kellis, M., Patterson, N., Endrizzi, M., Birren, B. and Lander, E.S. (2003) Sequencing and comparison of yeast species to identify genes and regulatory elements. *Nature* 423, 241–254.

Kimura, M. (1977) Preponderance of synonymous changes as evidence for the neutral theory of molecular evolution. *Nature* 267, 275–276.

Kimura, M. (1980) A simple method for estimating evolutionary rates of base substitutions through comparative studies of nucleotide sequences. *Journal of Molecular Evolution* 16, 111–120.

Kirkness, E.F., Bafna, V., Halpern, A.L., Levy, S., Remington, K., Rusch, D.B., Delcher, A.L., Pop, M., Wang, W., Fraser, C.M. and Venter, J.C. (2003) The dog genome: survey sequencing and comparative analysis. *Science* 301, 1898–1903.

Kleinjan, D.J. and van Heyningen, V. (1998) Position effect in human genetic disease. *Human Molecular Genetics* 7, 1611–1618.

Kryukov, G.V., Castellano, S., Novoselov, S.V., Lobanov, A.V., Zehtab, O., Guigo, R. and Gladyshev, V.N. (2003) Characterization of mammalian selenoproteomes. *Science* 300, 1439–1443.

Kumar, S. and Hedges, S.B. (1998) A molecular timescale for vertebrate evolution. *Nature* 392, 917–920.

Lander, E.S., Linton, L.M., Birren, B., Nusbaum, C., Zody, M.C., Baldwin, J., Devon, K., Dewar, K., Doyle, M., FitzHugh, W., Funke, R., Gage, D., Harris, K., Heaford, A., Howland, J., Kann, L., Lehoczky, J., LeVine, R., McEwan, P., McKernan, K. *et al.* (2001) Initial sequencing and analysis of the human genome. *Nature* 409, 860–921.

Leung, J.Y., McKenzie, F.E., Uglialoro, A.M., Flores-Villanueva, P.O., Sorkin, B.C., Yunis, E.J., Hartl, D.L. and Goldfeld, A.E. (2000) Identification of phylogenetic footprints in primate tumor necrosis factor-alpha promoters. *Proceedings of the National Academy of Sciences USA* 97, 6614–6618.

Loots, G.G., Locksley, R.M., Blankespoor, C.M., Wang, Z.E., Miller, W., Rubin, E.M. and Frazer, K.A. (2000) Identification of a coordinate regulator of interleukins 4, 13, and 5 by cross-species sequence comparisons. *Science* 288, 136–140.

Ludwig, M.Z. and Kreitman, M. (1995) Evolutionary dynamics of the enhancer region of even-skipped in *Drosophila*. *Molecular Biology and Evolution* 12, 1002–1011.

Ludwig, M.Z., Patel, N.H. and Kreitman, M. (1998) Functional analysis of eve stripe 2 enhancer evolution in *Drosophila*: rules governing conservation and change. *Development* 125, 949–958.

Ludwig, M.Z., Bergman, C., Patel, N.H. and Kreitman, M. (2000) Evidence for stabilizing selection in a eukaryotic enhancer element. *Nature* 403, 564–567.

Mural, R.J. Adams, M.D., Myers, E.W., Smith, H.O., Miklos, G.L., Wides, R., Halpern, A., Li, P.W., Sutton, G.G., Nadeau, J., Salzberg, S.L., Holt, R.A., Kodira, C.D., Lu, F., Chen, L., Deng, Z., Evangelista, C.C., Gan, W., Heiman, T.J., Li, J. *et al.* (2002) A comparison of whole-genome shotgun-derived mouse chromosome 16 and the human genome. *Science* 296, 1661–1671.

Murphy, W.J., Eizirik, E., Johnson, W.E., Zhang, Y.P., Ryder, O.A. and O'Brien, S.J. (2001a) Molecular phylogenetics and the origins of placental mammals. *Nature* 409, 614–618.

Murphy, W.J., Eizirik, E., O'Brien, S.J., Madsen, O., Scally, M., Douady, C.J., Teeling, E., Ryder, O.A., Stanhope, M.J., de Jong, W.W. and Springer, M.S. (2001b) Resolution of the early placental mammal radiation using Bayesian phylogenetics. *Science* 294, 2348–2351.

O'Brien, S.J., Menotti-Raymond, M., Murphy, W.J., Nash, W.G., Wienberg, J., Stanyon, R., Copeland, N.G., Jenkins, N.A., Womack, J.E. and Marshall Graves, J.A. (1999) The promise of comparative genomics in mammals. *Science* 286, 458–462, 479–481.

Ohta, T. (1995) Synonymous and nonsynonymous substitutions in mammalian genes and the nearly neutral theory. *Journal of Molecular Evolution* 40, 56–63.

Parra, G., Agarwal, P., Abril, J.F., Wiehe, T., Fickett, J.W. and Guigo, R. (2003) Comparative gene prediction in human and mouse. *Genome Research* 13, 108–117.

Pennacchio, L.A. and Rubin, E.M. (2001) Genomic strategies to identify mammalian regulatory sequences. *Nature Reviews of Genetics* 2, 100–109.

Reymond, A., Camargo, A.A., Deutsch, S., Stevenson, B.J., Parmigiani, R.B., Ucla, C., Bettoni, F., Rossier, C., Lyle, R., Guipponi, M., de Souza, S., Iseli, C., Jongeneel, C.V., Bucher, P., Simpson, A.J. and Antonarakis, S.E. (2002) Nineteen additional unpredicted transcripts from human chromosome 21. *Genomics* 79, 824–832.

Rockman, M.V. and Wray, G.A. (2002) Abundant raw material for *cis*-regulatory evolution in humans. *Molecular Biology and Evolution* 19, 1991–2004.

Schwartz, S., Zhang, Z., Frazer, K.A., Smit, A., Riemer, C., Bouck, J., Gibbs, R., Hardison, R. and Miller, W. (2000) PipMaker–a web server for aligning two genomic DNA sequences. *Genome Research* 10, 577–586.

Schwartz, S., Elnitski, L., Li, M., Weirauch, M., Riemer, C., Smit, A., Green, E.D., Hardison, R.C. and Miller, W. (2003) MultiPipMaker and supporting tools: alignments and analysis of multiple genomic DNA sequences. *Nucleic Acids Research* 31, 3518–3524.

Sorek, R. and Ast, G. (2003) Intronic sequences flanking alternatively spliced exons are conserved between human and mouse. *Genome Research* 13, 1631–1637.

Spitz, F., Gonzalez, F. and Duboule, D. (2003) A global control region defines a chromosomal regulatory landscape containing the HoxD cluster. *Cell* 113, 405–417.

Tajima, F. (1993) Simple methods for testing the molecular evolutionary clock hypothesis. *Genetics* 135, 599–607.

Tamura, K. and Nei, M. (1993) Estimation of the number of nucleotide substitutions in the control region of mitochondrial DNA in humans and chimpanzees. *Molecular Biology and Evolution* 10, 512–526.

Thomas, J.W., Touchman, J.W., Blakesley, R.W., Bouffard, G.G., Beckstrom-Sternberg, S.M., Margulies, E.H., Blanchette, M., Siepel, A.C., Thomas, P.J., McDowell, J.C., Maskeri, B., Hansen, N.F., Schwartz, M.S. *et al.* (2003) Comparative analyses of multi-species sequences from targeted genomic regions. *Nature* 424, 788–793.

Venter, J.C. Adams, M.D., Myers, E.W., Li, P.W., Mural, R.J., Sutton, G.G., Smith, H.O., Yandell, M., Evans, C.A., Holt, R.A., Gocayne, J.D., Amanatides, P., Ballew, R.M., Huson, D.H., Wortman, J.R., Zhang, Q., Kodira, C.D., Zheng, X.H., Chen, L., Skupski, M. *et al.* (2001) The sequence of the human genome. *Science* 291, 1304–1351.

Wasserman, W.W., Palumbo, M., Thompson, W., Fickett, J.W. and Lawrence, C.E. (2000) Human–mouse genome comparisons to locate regulatory sites. *Nature Genetics* 26, 225–228.

Waterston, R.H., Lindblad-Toh, K., Birney, E., Rogers, J., Abril, J.F., Agarwal, P., Agarwala, R., Ainscough, R., Alexandersson, M., An, P., Antonarakis, S.E., Attwood, J., Baertsch, R., Bailey, J., Barlow, K., Beck, S., Berry, E., Birren, B., Bloom, T., Bork, P. *et al.* (2002) Initial sequencing and comparative analysis of the mouse genome. *Nature* 420, 520–562.

Wray, G.A., Hahn, M.W., Abouheif, E., Balhoff, J.P., Pizer, M., Rockman, M.V. and Romano, L.A. (2003) The evolution of transcriptional regulation in eukaryotes. *Molecular Biology and Evolution* 20, 1377–1419.

Wyckoff, G.J., Wang, W. and Wu, C.I. (2000) Rapid evolution of male reproductive genes in the descent of man. *Nature* 403, 304–309.

Yang, Z. (1997) PAML: a program package for phylogenetic analysis by maximum likelihood. *Computer Applications in the BioSciences* 13, 555–556.

Zhang, J. (2000) Rates of conservative and radical nonsynonymous nucleotide substitutions in mammalian nuclear genes. *Journal of Molecular Evolution* 50, 56–68.

13 Evolution of the Mammalian Karyotype

Fernando Pardo-Manuel de Villena

Department of Genetics and Lineberger Comprehensive Cancer Center, Campus Box 7264, University of North Carolina at Chapel Hill, Chapel Hill, NC 27599-7264, USA

The Evolution of the Mammalian Karyotype: a Historical Overview	317
Evolutionary Trends in the Organization of the Mammalian Karyotype	318
Diploid number and fundamental number	319
Chromosome rearrangements	321
Chromosome rearrangements: effects on chromosome function	325
Comparative Studies	329
The ancestral karyotype	329
Rate of chromosome rearrangement	330
Karyotype orthoselection	333
Evolutionary Forces and Karyotype Evolution in Mammals	333
Phenotypic consequences of chromosome rearrangements	334
Mutation rate	336
Genetic drift	337
Natural selection	337
Karyotype and Speciation	340
References	342

The Evolution of the Mammalian Karyotype: a Historical Overview

The karyotype, the visual description of the complete set of chromosomes of a typical somatic cell of a eukaryotic organism, represents the highest level of organization of the genome. Autosomes are typically arranged in order of decreasing size with sex chromosomes placed at the end of the sequence. In most mammals, sex determination involves heterogametic XY males and homogametic XX females, although species with unusual sex determination systems have been reported (Fredga, 1972; Gardner, 1977; Benirschke *et al.*, 1980; Vassart *et al.*, 1995).

The genome not only contains a blueprint of an individual, but is also a complete and unique repository of the evolutionary history of its carrier. Characterization of the evolution of the genome is a powerful approach to solve basic evolutionary questions about natural history, speciation and adaptation (O'Brien et al., 1999). The ultimate goal is to describe and understand the evolutionary pathways taken by the ancestors of each mammalian lineage and the forces that shaped these processes.

Karotype evolution also has practical implications. For example, chromosome painting techniques allow the chromosomal position of genes in species with gene-poor maps to be inferred by projecting probes derived from one of the few mammalian species with high-density gene maps (O'Brien et al., 1999). These studies may also provide important information about the factors that influence genome instability, because chromosome rearrangements are the building blocks of karyotype evolution. Lastly, karyotype evolution provides avenues for the identification of functional constraints operating at mammalian centromeres and, therefore, an approach to address how mutation and selection had shaped these essential loci (Malik and Henikoff, 2002).

In 1938 Dobzhansky and Sturtevant proposed that 17 inversions were required to transform the Drosophila miranda karyotype into the Drosophila pseudoobscura karyotype (Dobzhansky and Sturtevant, 1938). This report represents the first comparative study addressing chromosome evolution. The existence of karyotype evolution in mammals has been known for decades, based on the high diversity of diploid and fundamental numbers among mammalian species and the presence of similar karyotypes, sparkled with few and distinctive chromosome rearrangements, among closely related species.

By the middle of the 20th century mammalian cytogenetics became an important tool in evolutionary and taxonomic studies. The landmark description of the human karyotype by Tjio and Levan in 1956 and a rapid succession of studies demonstrating the contribution of chromosomal abnormalities to human pathology highlighted the relevance of cytogenetic studies (Tjio and Levan, 1956). In another landmark study Mary Lyon proposed

X inactivation as the mechanism responsible for dosage compensation in mammals, providing the first and foremost example of chromosome biology unique to mammals. X inactivation explained the patterns of inheritance of X-linked traits, pioneered epigenetic studies in mammals and was key to the evolutionary views of S. Ohno.

The past three decades have witnessed the description of the karyotype of an increasing number of mammalian species and the development of new and powerful techniques to characterize the karyotype and the genome as a whole (Murphy et al., 2001a). Until the development of banding techniques only few features, mostly the position of the centromeres, allowed the comparison of karyotypes of different species. The development of more refined and effective DNA hybridization techniques, the generation of high-density and wide-coverage linkage maps and finally whole-genome sequencing projects, have dramatically increased both our understanding of the organization of the genome in individual species as well as the resolution at which the genome of different mammalian species may be compared.

Evolutionary Trends in the Organization of the Mammalian Karyotype

Mammals comprise 4600–4800 extant species including egg-laying mammals, marsupials and placental mammals classified into three subclasses, Monotremata, Marsupialia and Eutheria, respectively. Eutherians comprise 20 orders to which the vast majority of living species belong. The distribution of species richness among these orders is strongly skewed. Rodents account for approximately half of the species while other orders have only few extant species (Purvis and Hector, 2000).

The karyotype of more than a quarter of the extant mammalian species has been described (Pardo-Manuel de Villena and Sapienza, 2001a). However, there is considerable variability in the level of resolution at which the genome of a given species has been characterized (banding, chromosome painting, linkage maps, whole-genome sequence) and the number of individuals studied within a species. To avoid biases caused by this variability, comparative analyses based on the two most widely described parameters, the diploid number and fundamental number, are most useful to identify evolutionary trends. The fundamental number, FN, is the number of chromosome arms. However, there are limitations to what can be inferred from such studies, because only the subset of chromosome rearrangements that lead to changes in the number and/or gross morphology of chromosome (i.e. fusions, fissions, large and asymmetric pericentric inversions and heterochromatin expansions) can be detected.

Diploid number and fundamental number

Changes in diploid and fundamental numbers can be easily detected among all the mammalian species with known karyotypes. These changes may have a significant impact on important evolutionary processes such as meiotic drive and the level of genetic diversity that may be generated during meiosis. For example, the diploid number of possible segregations of chromosomes at meiosis is 2^N where N is the haploid number. On the other hand, the level of genome-wide recombination in mammals is strongly correlated with the fundamental number, and recombination determines the frequency at which different combinations of alleles at linked loci are recovered in the gametes (Dutrillaux, 1986; Pardo-Manuel de Villena and Sapienza, 2001b). These analyses also provide a good vantage point to confront one of the most challenging paradoxes of mammalian evolution, the abundance of opposing trends in the evolution of the mammalian karyotype. Furthermore, there is surprisingly little to no correlation between the directions of the trends and the phylogenetic relationships among mammalian species.

The distributions of the diploid and fundamental numbers in 1170 mammalian species are shown in Fig. 13.1. Only autosomes were used to plot the distribution of fundamental numbers to avoid bias due to morphological differences in the sex chromosomes and the existence of species with sex-dependent diploid numbers (Fredga, 1972; Gardner, 1977; Benirschke et al., 1980; Vassart et al., 1995). Both distributions are almost symmetric and

centred on the mean. The mean diploid number for all mammals is 43.3, ranging from $2n = 6$ in the Indian muntjac (*Muntiacus muntjack*) to $2n = 102$ in the South American rodent (*Tympanoctomys barrerae*). The wide range of diploid numbers observed in this distribution demonstrates that since the last common ancestor this parameter has undergone considerable changes in some mammalian species. Estimates of the diploid number of the hypothetical ancestral karyotype of placental mammals ranges from $2n = 44$ to $2n = 50$ (Chowdhary et al., 1998; Haig, 1999; Müller et al., 1999; Fronicke et al., 2003; Murphy et al., 2003; Richard et al., 2003). This value is remarkably similar to the mean diploid number among placental mammals shown in Fig. 13.1a ($2n = 44.7$). However, the fact that approximately 90% of mammalian species have a different diploid number indicates that the diploid number has changed in the majority of placental mammals. Based on this analysis it is possible to conclude that similar fractions of extant species have lower and higher diploid numbers than their common ancestor. Because chromosome fissions increase the diploid number while chromosome fusions decrease the diploid number, it follows that both types of chromosome rearrangement have been common themes during the evolution of the mammalian genome. Therefore, it would appear that there is no winner in the protracted controversy on whether fusions or fissions have predominated in the evolution of the mammalian karyotype (Todd, 1970; Redi et al., 1990; Qumsiyeh, 1994; Kolnicki, 2000). These observations provide the first example of the common theme of karyotype evolution, the presence of two opposing trends.

The distribution of the fundamental number of autosomes reinforces these conclusions (Fig. 13.1b). The mean fundamental number of autosomes is 61.8, ranging from $FN_a = 8$ to $FN_a = 198$. Therefore, there is a wide variety of fundamental numbers in mammals. The difference between the mean fundamental number of autosomes in placental mammals (64) and their ancestral karyotype (74) may suggest that most mammals have evolved towards a lower number of autosomes. However, this trend should be viewed with caution because the fundamental number depends on the posi-

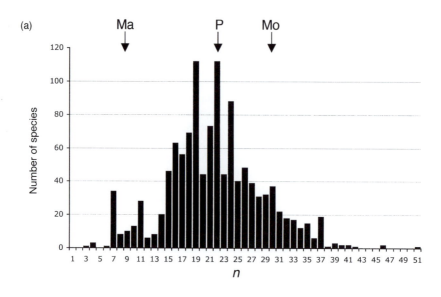

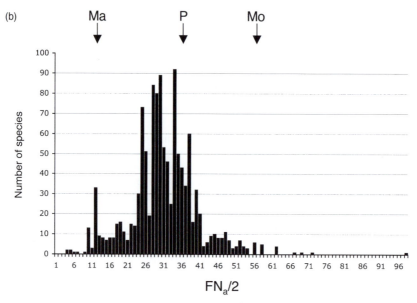

Fig. 13.1. Haploid number and fundamental number of mammalian species. (a) Haploid number of autosome (n) distribution among 1170 mammalian species. These species were selected to maintain a proportional representation of the species richness observed among eutherian orders and in the marsupial and monotreme subclasses. (b) Distribution of the fundamental number of autosomes (FN_a) among these same species. FN_a is equal to the number of autosome arms. Acrocentric and telocentric chromosomes are considered uni-armed, while metacentric and submetacentric chromosomes are bi-armed. Data are taken from Pardo-Manuel de Villena and Sapienza (2001). Arrows indicate the mean values for marsupials (Ma), placentals (P) and monotremes (Mo).

tion of the centromere, and recent studies suggest that it may be difficult to infer the position of these loci in the ancestral karyotype.

The shape of the two distributions shown in Fig. 13.1 might suggest that when these two parameters are considered together, most mammals will be clustered around the means defining an archetypical mammalian karyotype. This hypothesis can be tested plotting the diploid number and the ratio between the fundamental number and the diploid number ($FN_a/2n_a$, Fig. 13.2a). Inspection of Fig. 13.2a demonstrates that, in contrast to the expectations of the model, most mammals do not cluster in a single region. In fact, most mammals cluster in two widely separated regions while the rest are distributed across a broad region. These two regions are defined by opposite $FN_a/2n_a$ ratios, indicating that in most mammals there is a strong bias towards a particular chromosome morphology, either uni-armed or bi-armed, and that in only a minority of species do both types have similar contributions.

Figure 13.2 also suggests that the type of predominant chromosome morphology is correlated with the diploid number. This notion may be investigated by determining the average $FN_a/2n_a$ ratio in mammals with identical diploid numbers (Fig. 13.3a) and the average diploid number in species with similar $FN_a/2n_a$ ratios (Fig. 13.3b). Both analyses, taken together, indicate that lower diploid numbers are strongly correlated with karyotypes predominantly composed of bi-armed chromosomes and that higher diploid numbers are correlated with karyotypes composed predominantly of uni-armed chromosomes.

Comparisons of diploid number and fundamental number among mammals indicate that there is wide diversity and that most mammals had followed two opposing evolutionary trends, half of the species towards larger diploid numbers with mostly uni-armed chromosomes and half toward lower diploid numbers with mostly bi-armed chromosomes. Although these conclusions have been reached using placental mammals, consideration of the other two mammalian lineages, marsupials and monotremes, reinforces the presence of opposing evolutionary trends (Fig. 13.1).

Chromosome rearrangements

Karyotype evolution may be described as the accumulation of a particular set of chromosome rearrangements in a specific lineage.

Chromosome rearrangements are the source of the genetic diversity required for karyotype evolution. Rearrangements include translocations, inversions, deletions and duplications, as well as incompletely characterized processes such as centromere shifts and generation of extranumerary (B) chromosomes. Among the multiple criteria used to classify chromosome rearrangements (intra versus interchromosomal, effect on fitness, frequency, etc.), the impact of rearrangements on diploid number and fundamental number will be considered first due to the evolutionary trend correlating diploid and fundamental number described in the previous section.

Robertsonian translocations

Three types of chromosome rearrangements: fissions, fusions and generation of B chromosomes, are responsible for changes in diploid number. B chromosomes are comparatively rare among mammals as they have been described in fewer than 60 mammalian species (4.9% of species examined (Palestis *et al.*, 2003)). B chromosomes show variable frequencies within and between individuals, are composed of heterochromatin and segregate abnormally during mitosis and meiosis. B chromosomes are particularly frequent in phylogenetic groups such as the Muridae and Canidae families that have undergone high rates of chromosome rearrangements suggesting a causal link between both observations (Trifonov *et al.*, 2002). However, only recent work has begun to characterize the molecular nature and origin of B chromosomes in mammals (Yang *et al.*, 1999; Trifonov *et al.*, 2002; Nie *et al.*, 2003). These rearrangements will not be considered further in this section, with the exception of the role of meiotic drive in their maintenance (see below). For all practical purposes, fusions are viewed as translocations between two entire non-homologous chromosomes that result in a single rearranged chromosome, while fissions are the reverse process that generates two rearranged chromosomes from a single initial one. Robertsonian translocations are a subset of fusions and fissions in which the substrate of the rearrangements are whole chromosome arms. Therefore, in these chromosome rearrangements the initial and

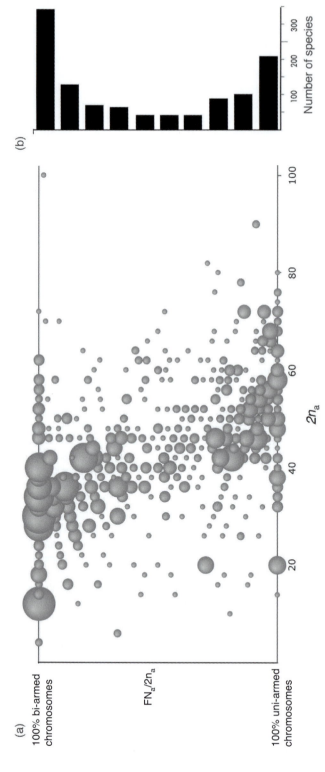

Fig. 13.2. Karyotype diversity in mammals. (a) Plot of mammalian species classified according to the diploid number of autosomes ($2n_a$) and average number of arms per chromosome. The ratio was calculated by dividing the fundamental number of autosomes (FN_a) by the diploid number of autosomes ($2n_a$). $FN_a/2n_a$ varies between 1 when all chromosomes are uni-armed (bottom line) and 2 if all autosomes are bi-armed (top line). Mammalian species are represented as bubbles and the size of the bubbles is proportional to the number of species observed. (b) Number of mammalian species classified according to the percentage of bi-armed chromosomes observed in their karyotypes.

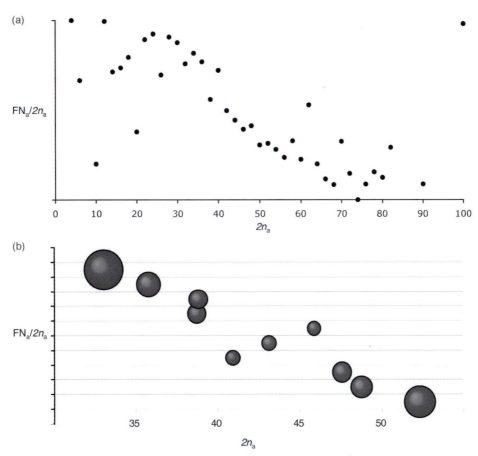

Fig. 13.3. Correlation between diploid number and fundamental number. (a) Average ratio of arms per autosome in mammalian species with identical diploid number of autosomes. (b) Average diploid number of autosomes in mammalian species with a similar ratio of arms per autosome. To simplify the analysis species were grouped into ten equal classes. The size of the bubbles is proportional to the number of species found within each category.

final state can be either a set of two non-homologous acrocentric chromosomes or a single metacentric. Robertsonian translocations are among the most easily recognized rearrangements and were essential in the identification of the genetic basis of inherited chromosomal disorders in humans. Heterozygous carriers for these rearrangements have been described in dozens of mammalian species. For these reasons they are probably the most intensively studied rearrangement in mammals and the consensus is that Robertsonian translocations have played a key role in the evolution of the mammalian karyotype (Qumsyieh, 1994).

When a Robertsonian translocation becomes fixed the diploid number decreases (fusion) or increases (fission) by two units, while the fundamental number remains unaffected. Therefore, if a karyotype is allowed to evolve only through Robertsonian translocations the diploid number and the $FN_a/2n_a$ ratio of all possible derived karyotypes may be easily determined (Fig. 13.4a). In this situation the fundamental number in the ancestor determines all possible outcomes and the number of possible different karyotypes is $((FN_a/2) + 1)/2$ (Fig. 13.4a). Other types of fusions and fissions also change the diploid number by two units, but they also lead to changes in fundamental number. Although these changes

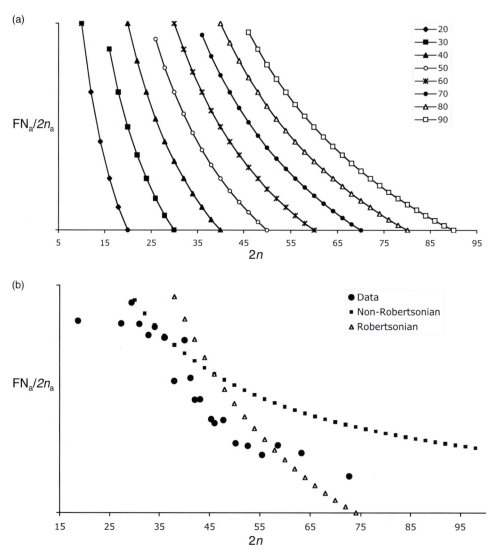

Fig. 13.4. The role of Robertsonian translocations in the generation of karyotype diversity in mammals. (a) Range of diploid numbers and ratios of arms per autosome that may be generated through Robertsonian translocations in mammalian species with different fundamental numbers. (b) Comparison between the range of diploid numbers and ratios of arms per autosome observed in mammals (data), with the possible ranges generated through non-Robertsonian fusions and fissions and Robertsonian translocations. The fundamental number of autosomes in the ancestral karyotype was 74 (Chowdhary et al., 1998; Haig, 1999; Müller et al., 1999; Fronicke et al., 2003; Murphy et al., 2003; Richard et al., 2003).

cannot be predicted with absolute confidence, they will be directly proportional to the changes in diploid number. Therefore, in these rearrangements there is less variation in the $FN_a/2n_a$ ratio than in Robertsonian translocations. In addition, these fusions and fissions may generate a larger number of different karyotypes (Fig. 13.4b).

Placental mammals are a monophyletic group and, therefore, the range of karyotypes that might be generated from a single ancestral karyotype through Robertsonian translocations or non-Robertsonian fissions and fusions may be determined and compared with the range of diploid number observed in mammals. As it is shown in Fig. 13.4b neither type of rearrangements can account, by themselves, for the full range of diploid numbers observed in placental mammals (Fig. 13.1a). However, both types of fusion and fissions combined with inversions and centromere shifts are enough to generate all known mammalian karyotypes. Although Robertsonian translocations cannot account for full diversity of diploid numbers, there is direct evidence, gathered from natural populations segregating for these rearrangements, that they play a key role in the generation of diploid number diversity. In fact, if the hypothetical karyotype of the ancestor of all placental mammals with 74 chromosome arms is allowed to evolve only by Robertsonian translocations, the distribution of karyotypes mirrors closely the diploid and the fundamental numbers observed in this group (Fig. 13.4b).

Inversions and reciprocal translocations

Since the discovery of the evolutionary role of inversions in *Drosophila* (Dobzhansky and Sturtevant, 1938), mounting evidence indicates that they represent the most frequent events in karyotype evolution. Because the ability to detect an inversion is related directly to its size, the total contribution of these rearrangements will be known only after the entire genome of multiple mammalian species is be sequenced. Large inversions detected through comparative cytogenetics have been known for decades (Yunis, 1976). Shorter inversions have sometimes been discovered serendipitously associated with dramatic reductions in recombination frequencies in heterozygous carriers of the 'normal' and inverted chromosomes (Silver, 1993). Comparative mapping based on locus-specific probes indicates that inversions have been common in many mammalian lineages (Goureau *et al.*, 2001; Goldamer *et al.*, 2002; Rogalska-Niznik *et al.*, 2002). However, only the direct comparison between aligned sequences of the completed genomes of humans and mice has begun to reveal the pervasive role of these rearrangements during evolution (Pevzner and Tesler, 2003). Using this approach it has become apparent that previous estimates (Lander *et al.*, 2001) of the number of rearrangements between these two species were too low (Pevzner and Tesler, 2003).

Reciprocal translocations have been observed at relatively high frequency in humans (Daniel *et al.*, 1989). However, these rearrangements are normally associated with strong reductions in fitness due to the generation of unbalanced gametes during meiosis. Therefore, it is expected that the role of reciprocal translocations in evolution is probably minor. Evidence for a special case of reciprocal translocations, known as whole-arm reciprocal translocations, WART (Hauffe and Pialek, 1997), has been reported recently in feral populations of mice (Castiglia and Capanna, 1999) and may play an important role in speciation in some instances.

Chromosome rearrangements: effects on chromosome function

Chromosomes have dual functions. On one hand, genetic information is encrypted within the DNA molecule. On the other hand, chromosomes are delivery systems that ensure the stable transmission of genetic information to the products of each cell division. Although rearrangements may affect the genetic information, for example by disruption of a specific gene, it is expected that these effects have little impact on karyotype evolution. In contrast, chromosome rearrangements may pose significant challenges to chromosome stability and faithful segregation during cell division and particularly during meiosis. For example, telomeres are specialized loci that cap the ends of the linear mammalian chromosomes and are essential to maintain their physical integrity. Rearrangements may lead to chromosomes that may initially lack such loci. However, several lines of evidence indicate that cells may easily add telomeric sequences and therefore restore physical integrity. On the other hand, chromosome rearrangements may severely compromise the essential roles that centromeres and recombination play in chromosome segregation.

The centromere

Centromeres mediate the faithful segregation of chromosomes during mitosis and meiosis. Centromeres show low levels of sequence conservation and extreme diversity in sizes among eukaryotes, ranging from 112 bp in *S. cerevisiae* to several megabases in humans composed mainly of thousands of copies of satellite repeats (Malik and Henikoff, 2002, Chapters 4, 11). In this situation it is difficult to obtain the complete and unambiguous sequence of a mammalian centromere. In the absence of sequence data, characterization has relied heavily on the characterization of trans-acting proteins and generation of artificial mammalian mini-chromosomes (Sullivan *et al.*, 2001; Malik and Henikoff, 2002). An additional complication is the use of the term centromere to refer to a sequence, a structure and a function. From an evolutionary point of view it is more informative to use a functional definition for centromeres. Therefore, the term centromere will be restricted here to functional loci that mediate equal and faithful chromosome segregation. Under this definition it is evident that chromosomes must have one, and only one, functional centromere. The terms centromeric sequence and centromeric regions will be used to refer to the satellite sequences and DNA–protein structures associated with mammalian centromeres.

Because chromosomes require one centromere for stable segregation, chromosome fusions and fissions pose specific challenges. For example, fusions may lead to the presence of two centromeric sequences in the rearranged chromosomes, while fissions may result in rearranged fragments lacking such sequences. If these rearranged chromosomes cannot establish a single centromere they will not be stably and faithfully transmitted and will play no role in karyotype evolution. However, if they are able to do so, they offer an approach to characterize the relationships between centromeric sequences and centromeres. Today the genetic, cytological and molecular evidence indicates that the centromeric sequences are neither necessary nor sufficient for centromeric function. For example, many Robertsonian translocations between two acrocentric human chromosomes have two centromeric sequences. However, there is only one functionally active centromere (detected using centromere-specific proteins that are required for the assembly of a functional kinetochore) (Earnshaw *et al.*, 1989; Page

et al., 1995; Sullivan and Schwartz, 1995). On the other hand, the discovery of human neo-centromeres, ectopic centromeres located at non-centromeric regions of a chromosome which in some cases completely lack satellite DNA (Amor and Choo, 2002), demonstrates that it is possible to establish a functional centromere in the absence of centromeric DNA. These observations, and additional results in yeast and *Drosophila*, support the hypothesis that centromeres in most eukaryotes are determined epigenetically rather than by primary DNA sequence (Sullivan *et al.*, 2001; Malik and Henikoff, 2002). In some instances, centromeric sequences might be inactivated (i.e. will lose function) while in other situations non-centromeric sequences are activated, becoming the locus for the assembly of the kinetochore.

These observations provide an explanation for how stable chromosomes may be generated by fusions and fissions. However, they also raise important questions on the methods used to infer the evolutionary history of mammalian centromeres. If centromeres are determined by the primary DNA sequence it should be possible to follow their evolutionary history using closely linked markers. However, if centromeres are epigenetically determined, then these inferences may be fraught with uncertainty, because it may be possible for centromeres to relocate to a different chromosome position without disturbing the surrounding sequences (chromosome shifts). A striking example of the phenomenon has been recently described in primates (Ventura *et al.*, 2001). This study compared the order of probes spanning the entire X chromosome between humans and two species of lemurs, in which the centromere occupies very different positions. They concluded that the centromere had relocated in at least two species without disturbing the order of any markers tested. This raises many question including how the position of the centromeres in the ancestral karyotype can be accurately inferred, what factors affect centromere shifts and the frequency and lineage-specific distribution of these events.

Independently of how centromere identity/activity is determined, it is evident that the function and location of these loci remain crucial. There is mounting evidence that selection has operated at the centromere with especial intensity and in unexpected ways (Henikoff et al., 2001). Both the centromeric sequence and the proteins involved in centromere function have evolved rapidly (Malik and Henikoff, 2001). Incompatibilities between these rapidly evolving centromeric components may be responsible for the organization of the centromeric region and most importantly for the reproductive isolation of emerging species. In addition, the position of the centromere determines the fundamental number, and this parameter determines the range of diploid number that may be generated from a given karyotype through Robertsonian translocations and affects recombination frequency.

Recombination and linkage maps

Although recombination plays many important roles in such basic processes as DNA repair and generation of genetic diversity, its role in ensuring the fidelity of chromosome segregation during the first meiotic division is, arguably, the most important (Chapter 1). Studies in most model organisms have shown that proper chromosome segregation requires recombination to take place in each chromosome. The physical manifestation of recombination during meiosis, the chiasma, is thought to be required for the proper orientation of homologous chromosomes in the spindle. In mammals, reduced recombination is associated with aneuploidy (Hassold et al., 2000; Hassold and Hunt, 2001). Several factors are known to affect recombination including the sex, sequence similarity and diploid number. However, these factors cannot explain the striking differences in genome-wide recombination observed among mammalian species. In 1986, Dutrillaux demonstrated that there is a strong positive correlation between the number of chiasma and the haploid number of chromosome arms (Dutrillaux, 1986). This observation was later expanded and confirmed in additional species and also using the length of the linkage map to estimate the recombination frequency (Pardo-Manuel de Villena and Sapienza, 2001b). These studies suggest that one recombination event per chromosome arm is required for proper segregation (Pardo-Manuel de Villena and Sapienza, 2001b).

Recently, the smallest linkage map in a mammal has been described (Zenger et al., 2002) and its size agrees with the prediction based on fundamental number (Pardo-Manuel de Villena and Sapienza, 2001b, and Table 13.1). Lengthy discussions about the possible causes of the diversity in map lengths among mammalian species are commonplace in the description of the linkage map. However, it is strange that among the multiple theoretical factors discussed, such a basic parameter as the fundamental number is rarely considered, despite the weight of supporting evidence. It would appear as if in the post-genomic era there is little room for a role for chromosome architecture in functional organization. Table 13.1 summarizes the characteristics of the linkage map in 13 mammalian species. Although the data support the presence of a direct correlation between FN and size of the linkage map, there are some species (horse and baboon) for which the length of the map appears to be smaller than expected. Although these might reflect real exceptions to the rule, the fact that most maps have consistently grown in size from earlier to later versions and the limited numbers of informative meiosis and markers (Table 13.1) suggest that more data need to be collected before arriving at a meaningful conclusion. It is also important to note that very different types of crosses have been used to estimate the map length (Table 13.1).

Because recombination depends on the fundamental number, some chromosome rearrangements will affect this basic process while others will not. Robertsonian translocations are predicted to have limited impact on recombination and to maintain at least one recombination per arm. Experimental evidence supporting both predictions has been reported recently in mouse (Dumas and Britton-Davidian, 2002). Lastly, chromosome rearrangements may have a profound impact on recombination in heterozygous carriers due to incomplete pairing between homologous chromosomes, changes in patterns of recombination along the entire chromosome and improper recombination between non-homologous sequences. In those instances, abnormal recombination is expected to lead to reduction in fitness.

Table 13.1. Length of linkage maps in mammalian species.

Subclass or order	Species	Type of cross	No. of markers	No. of meioses	*n*	FN	Length	Coverage	Reference
Perissodactyla	Horse	Intraspecific	140	89	32	46	6.79	–	Lindgren et al., 1998
		Intraspecific	359	61*	32	46	17.8	–	Swimburne et al., 2000
		Intraspecific	344	175	32	46	22.62	–	Guerin et al., 2003
Primates	Baboon	Intersubspecific	331	694*	41	42	19.75	–	Rogers et al., 2000
	Human	Intraspecific	5264	186**	23	41	36.99	+	Dib et al., 1996
		Intraspecific	8325	8 families	23	41	35	+	Broman et al., 1998
		Intraspecific	5136	1257	23	41	36.15	+	Kong et al., 2002
Carnivora	Dog	Intraspecific	150	200	39	40	8.842	–	Mellersh et al., 1997
		Intraspecific	276	218*	39	40	15.1	–	Neff et al., 1999
		Intraspecific	n.d	n.d.	39	40	20.4	–	www.fhcrc.org/science/dog_genome
	Cat	Interspecific	253	53*	19	38	20.4	–	Menotti-Raymond et al., 1999
		Interspecific	864	63*	19	38	26.46	–	Menotti-Raymond et al., 2003
Artiodactyla	Deer	Interspecific	621	351*	34	34	25.323	–	Slate et al., 2002
	Pig	Intraspecific?	128	200*	19	32	18.73	–	Ellegren et al., 1994
		Intraspecific	239	118*	19	32	18.37	–	Archibald et al., 1995
		Intraspecific	1042	94*	19	32	22.862	–	Rohrer et al., 1996
		Intraspecific	243	273	19	32	25.61	–	Mikawa et al., 1999
	Sheep	Intraspecific	232	140	27	31	20.75	–	Crawford et al., 1995
		Intraspecific	519	760	27	31	31.9	–	De Gortari et al., 1998
		Intraspecific	1093	128	27	31	36.32	–	Maddox et al., 2001
	Cow	Interspecific	313	180*	30	31	24.64	–	Bishop et al., 1994
		Interspecific	269	130 and 161	30	31	19.75	–	Ma et al., 1996
		Intraspecific	1250	223	30	31	29.9	–	Kappes et al., 1997
		Intraspecific	150	494*	30	31	27.64	–	Viitala et al., 2003
		Interspecific	417	602*	30	31	26.42	–	Kim et al., 2003
	Goat	Intraspecific	223	129	30	31	23	–	Vaiman et al., 1996
		Intraspecific	307	114	30	31	27.37	–	Schibler et al., 1998
Rodentia	Mouse	Interspecific	>6000	>500	20	20	14.5	+	www.jax.org
	Rat	Intraspecific	4736	91*	21	33	15.03	–	Steen et al., 1999
		Intraspecific	1620	136*	21	33	17.36	–	Bihoreau et al., 2001
Marsupialia	Tammar wallaby	Intraspecific	64	353	8	13	8.284	–	Zenger et al., 2002

The table summarizes the basic characteristics of the linkage maps in 13 mammalian species, classified according to the subclass or order to which they belong. The table provides the type of cross, the number of markers genotyped and the average number of informative meioses (a single asterisk denotes number of individuals typed rather than informative meiosis and two asterisks denote number of families). The haploid number of chromosomes (*n*) and the fundamental number (FN) are indicated. The raw length of the map is shown in Morgans. Based on the published data, maps are considered to cover more than 95% of the genome (+) or less (–). The table also provides the references or web site address for each map.

Comparative Studies

Comparative cytogenetics, comparative mapping and comparative genomics are powerful approaches to reconstruct the evolutionary history of all mammalian species by working backwards in time (Murphy *et al.*, 2001). All of them use the association between genomic segments as evolutionary characters and query whether such associations are observed between segments and among species. Comparative studies are based on the observation that genomic regions that are close to each other in the genome of one species tend to be close in other species, while genomic regions that are more distant in one species tend to be distant or are in different chromosomes in other species (Nadeau and Sankoff, 1998). The explanation is that processes that may separate genomic regions, i.e. chromosome rearrangements, are rare events. Under these assumptions it should be possible to reconstruct the ancestral karyotype by comparing the karyotypes present in living species and use rearrangements to determine phylogenetic relationships.

The main challenges faced by comparative studies include identifying conserved and rearranged regions of the genome among extant species, solving the combinatorial puzzle to rearrange genomes of related species, reconstructing the ancestral karyotype, estimating the rates and patterns of chromosome rearrangements in different lineages, explaining how rearrangements become fixed in some lineages and the functional consequences associated with these changes.

Until recently, most comparisons involved only few mammalian species belonging to closely related taxa and typically the human karyotype as the reference karyotype. Recently, there has been a considerable increase in both the number of species and the diversity of mammalian orders studied (Chowdhary *et al.*, 1998; Haig, 1999; Müller *et al.*, 1999; O'Brien *et al.*, 1999; Murphy *et al.*, 2001a; Frönicke *et al.*, 2003; Richard *et al.*, 2003; Yang *et al.*, 2003). However, the human karyotype retains central position in most comparisons (Murphy *et al.*, 2001a) based on the fact that it is the best characterized mammalian karyotype and that there is a legitimate interest in using com-

parative studies to increase our understanding of human physiology and disease. However, it is important to remember that the species included in comparative studies may bias the conclusions reached.

Comparative cytogenetics uses chromosome banding and DNA hybridization techniques to identify conserved regions. The development of cross-species fluorescent hybridization has allowed the easy visualization of chromosomal homologies among mammalian orders (Weinberg *et al.*, 1990; Weinberg and Stanyon, 1995). The type of probe used in these experiments varies from entire chromosomes to large-insert genomic clones and permits a precise identification of the extent of conservation and, in some cases, determination of the order within conserved fragments. However, the decrease in sequence conservation between distantly related species and the inability to resolve smaller rearrangements, mostly inversions, limits the usefulness of these techniques. This may lead to systematic undercounting of the number of chromosomal rearrangements between two species and, therefore, to deceptively low estimates of rates of chromosome rearrangements. Comparative mapping provides a more precise method to identify conserved segments and importantly also provides the order between consecutive genes (Nadeau and Sankoff, 1998). However, it is the alignment of whole genome sequences that will ultimately provide the greatest level of resolution. A striking example of the increased level of resolution provided by the later approach is the tenfold increase in the number of rearrangements detected between mouse and human (Pevzner and Tesler, 2003).

The ancestral karyotype

Reconstructing the ancestral karyotype is a complex process and there are still many methodological hurdles, including the lack of a consensus algorithm to track chromosome rearrangements. Although the approaches used are not new, the level of definition at which mammalian species may be compared has advanced dramatically in recent years. The reconstruction of the ancestral karyotype should make it possible to address important evolutionary questions, such as the presence of lineage-specific rates and modes of chromosome evolution. The hypothetical ancestral karyotypes of three mammalian groups, carnivores, primates and placental mammals, have been reported recently (Dutrillax and Couturier, 1983; Haig, 1999; Müller et al., 1999; O'Brien et al., 1999; Richard et al., 2003; Frönicke et al., 2003; Yang et al., 2003).

The reconstruction of the ancestral karyotype combines our knowledge of the associations between genomic segments observed among mammalian species with the basic evolutionary strategy of assuming that shared characteristics indicate common ancestry. The twist is that the goal here is to identify which of these associations were present in common ancestors of living species rather than using the shared versus non-shared associations to determine the phylogenetic relationships. When associations between pairs of specific genomic fragments are present in all species, then this situation is assumed to be ancestral. If associations are present only in a subset of species, determining whether this character is ancestral or derived relies on the analysis of the phylogenetic relationships between species sharing and not sharing the association. If sharing is observed in multiple independent branches then the association is assumed to represent the original situation. As a consequence, characters that are most informative to define the ancestral karyotype are those that would lead to the incorrect topology when used to generate a phylogenetic tree. Therefore, shared features among distantly related species are considered ancestral and are key to the reconstruction of the ancestral karyotype (Murphy et al., 2001a, also Chapter 14).

Rate of chromosome rearrangement

The rate of chromosome rearrangement is defined as the number of chromosome rearrangements that had occurred per unit of evolutionary time. The fossil record and molecular divergence are used to estimate evolutionary times. Calculating the number of rearrangements requires determining the shortest (most parsimonious (Palmer and Herbon,

1988)) path to convert the genome of one species into the genome of another. If only two species are considered, the rate of chromosome rearrangements represents the average number of rearrangements that have accumulated in both lineages in the time elapsed from the last common ancestor. On the other hand, the lineage-specific rate of chromosome rearrangements can be estimated when comparative data are available for more than two species. This involves reconstructing the ancestral karyotype, and then determining the ratio between the number of rearrangements that have occurred between the ancestor and each one of the extant species, and the time elapsed between them.

Multiple comparative studies have reported that the rate of chromosome rearrangements within mammals appears to be bimodal (Chowdhary et al., 1998; Burt et al., 1999; Haig, 1999; Müller et al., 1999; Cavagna et al., 2000). The rate of rearrangement observed in most comparative cytogenetics studies varies between one change per 10 million years to one change per million years. The lower rate, known as the basal rate of chromosome rearrangement, is very slow. Species with low rates of change include the cat (Carnivora), the dolphin (Cetartiodactyla), the squirrel (Rodentia) and man (Primates) (Fig. 13.5a). The presence of shared features among species belonging to phylogenetic groups that diverged more than 80 million years ago provides strong evidence that these features are ancestral. However, species with high rates of rearrangements have also been reported in these same groups, dog (Carnivora), muntjac (Cetartiodactyla), mouse (Rodentia) and gibbon (Primates) (Fig. 13.5a). Furthermore, the distribution of species with high and low rates does not follow any predictable pattern (Haig, 1999; Fig. 13.5a).

The presence of bimodal rates of chromosomal change and the distribution of these rates among phylogenetic groups are among the most striking characteristics of mammalian karyotype evolution. However, bimodality may be an artefact rather than a general characteristic of chromosome evolution. Bimodality may be caused by the small sample size, sampling bias and/or the methods used to reconstruct the ancestral karyotype. Compelling evidence

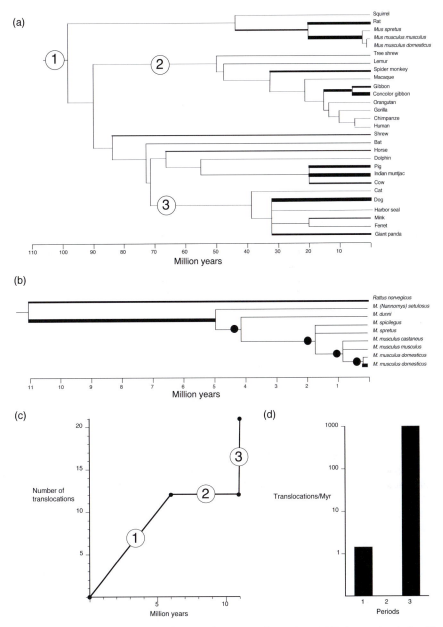

Fig. 13.5. Variable rates of rearrangement between lineages and over time. (a) Phylogenetic relationships and divergence times in selected mammalian species and position of the ancestors whose karyotypes have been recently described (Chowdhary *et al.*, 1998; Haig, 1999; Müller *et al.*, 1999; Frönicke *et al.*, 2003; Murphy *et al.*, 2001a,b, 2003; Richard *et al.*, 2003a,b; Stanyon *et al.*, 2003). (b) Phylogenetic relationships and divergence times in selected species of the *Mus musculus* lineage. In both figures the width of the horizontal lines is proportional to the rate of rearrangement observed in this lineages. Filled circles represent ancestors with a karyotype composed of 40 acrocentric chromosomes. (c) Number of translocations acquired in the lineage leading to populations of *Mus musculus domesticus* with 2*n* = 22 (Nachman and Searle, 1995; Britton-Davidian *et al.*, 2000) since its divergence from the rat (Stanyon *et al.*, 1999; Guénet and Bonhomme, 2003). The time scale has been divided into three periods according to the data shown in Fig. 13.3b. (d) Rate of rearrangement in the three periods shown in Fig. 13.3c.

that slow and high rates of chromosomal changes coexist in mammals can be obtained from lineages with many topologically well-defined branching nodes leading to species with well-characterized karyotypes. The house mouse, *Mus musculus*, is an excellent example. The *Mus* genus comprises more than a dozen species that diverged from the *musculus* lineage between 1.5 and 5 million years ago (Fig. 13.5b). Within the *Mus musculus* species several subspecies diverged from each other approximately 750,000 years ago. Lastly, several chromosomal races diverged from the *Mus musculus domesticus* branch 5000–10,000 years ago (Britton-Davidian *et al.*, 1989, 2000; Nachman *et al.*, 1994). All these mice have identical karyotypes composed of 40 acrocentric chromosomes, with the significant exception of these later chromosomal races in which the diploid number varies between 22 and 38 due to the accumulation of Robertsonian translocations (Nachman and Searle, 1995). One must conclude that the ancestral karyotype was composed of 40 acrocentric chromosomes and that the chromosomal races represent the derived situation. Therefore, during the last 5 million years no rearrangements occurred in the karyotype of six lineages, while in a very short period of time a karyotypic revolution takes place in some populations of the *Mus musculus domesticus* subspecies in Western Europe and northern Africa leading to a reduction in the diploid number by almost half.

The presence of lineage-specific rates should be used with caution when drawing general conclusions. For example, based on the presence of fast rates of rearrangements in the mouse and rat, high levels of rearrangements have been assumed to be typical in rodents. The slow rate of rearrangements reported recently in squirrels disprove this notion (Richard *et al.*, 2003b; Stanyon *et al.*, 2003). Although recent comparative studies frequently state that 'a slow rate is seen in most mammalian lineages', characterization of a large number of additional species would be required to determine the frequency of slow and fast rates in mammals. In addition, these species must reflect the species-richness of different phylogenetic groups. Note that rodents ($\approx$45% of extant mammalian species) are represented by only four species, mouse, rat and two types of squirrels, in

comparative cross-species fluorescent hybridization (Murphy *et al.*, 2001a; Richard *et al.*, 2003b; Stanyon *et al.*, 2003), while primates ($\approx$5% of extant mammalian species) are represented by more than 15 species.

Although these questions remain to be answered, the distribution of rates of rearrangement among mammalian species shown in Fig. 13.5a indicates that in addition to lineage-specific changes, variation in rates over time need to be considered. Lineage-specific rates offer a more precise picture than rates obtained comparing two extant species. However, they still represent an average change over time and assume constant rates over time. Again, chromosome evolution in the house mouse provides a useful system to test the presence and extent of variation in rates of rearrangement over time. The time elapsed between the divergence of mouse and rat and the present can be divided into three periods (Fig. 13.5b, c). During the 7 million years following the divergence of mouse and rat, the rate of change based on translocations has been estimated as one change per million (Stanyon *et al.*, 1999). In contrast, no rearrangement can be detected in the next 5 million years. In the last 10,000 years some populations of mouse acquired almost as many translocations as in the previous 12 million years. This analysis indicates that the rate of rearrangement is not constant, but rather varies dramatically within a single lineage (Fig. 13.5d). Interestingly, very low rates of rearrangement are observed for a long period of time despite the general consensus that mouse has high rates of change (Murphy *et al.*, 2001a). Lastly, the range of variation observed in rates of rearrangements is far wider than the averages reported on the basis of few distantly related species (Fig. 13.5d). These results suggest that karyotype evolution, in at least some lineages, involves long periods of stability interrupted by short intervals of dramatic change, which closely resembles the punctuated equilibrium hypothesis (Eldredge and Gould, 1972). If this pattern of variation occurs in a significant fraction of lineages it may have broad repercussions in the use of karyotypic characters in phylogenetic studies, in the evolutionary forces driving evolution and in the consequences of such changes from the viewpoint of speciation.

Karyotype orthoselection

The term karyotype orthoselection was coined by White (1978) to describe the acquisition of a series of rearrangements of particular type within a lineage. For example, among primates, pericentric inversions represent the preponderant type of rearrangement in the human lineage, whereas Robertsonian translocations have played the leading part among lemurs (Müller and Weinberg, 2001). Examples of intersubspecific karyotype orthoselection have been described in canids (Ward et al., 1987; Wurster-Hill et al., 1988; Nie et al., 2003), but the sequential accumulation of as many as nine Robertsonian fusions in the *Mus musculus domesticus* subspecies are among the most striking examples of karyotype orthoselection. In fact, over the past two decades an increasing number of examples of karyotype orthoselection have been reported in species belonging to the major mammalian orders. Random processes are unlikely to lead to the directionality that characterizes karyotype orthoselection.

The existence of karyotype orthoselection in mammals makes it useful to classify these forces into two classes according to whether they result in random changes or directional changes in the karyotype. Mutation and genetic drift are expected to result in the accumulation of random changes while the selective nature of adaptation and meiotic drive are likely to lead to directional changes.

The presence of karyotype orthoselection and opposing trends in karyotype evolution, in particular the enormous variations in the rate of rearrangements, raises questions regarding the use of karyotypic characters to determine phylogenetic relationships between mammalian species. These phylogenetic studies are based on the premise that chromosome rearrangements are rare enough that they must be regarded as independent events, unlike nucleotide substitutions that have a predictable rate of recurrence (White, 1978). Therefore, chromosomal rearrangements may be considered unique and may provide a powerful discriminating criterion in phylogenetic studies. Although there are many examples that support the hypothesis that closely related species show similar karyotypes, there

are also striking exceptions. In addition to the chromosomal races in *Mus musculus domesticus* discussed above, another example involves the Indian and Chinese muntjacs (*Muntiacus muntjack* and *Muntiacus reevesi*, respectively), two closely related species with dramatically different diploid numbers, $2n = 6$ and $2n = 46$, respectively. In fact many studies have reported that estimates of phylogenetic distance based on comparative cytogenetic parameters (i.e. index of homology) may lead to absurd conclusions, such as that humans are more closely related to the cat, a carnivore, than to the tree shrews, a primate (Cavagna et al., 2000). The presence of variable rates of chromosome change among lineages and over time and the likelihood that these variations may be driven by selective forces are the likely cause of these apparent contradictions. Although these observations do not preclude the use of karyotypic characters in phylogenetic studies, they are a cautionary note against their general use for comparisons between species with slow and fast rates and in the absence of additional supporting evidence.

Evolutionary Forces and Karyotype Evolution in Mammals

Using comparative studies it is possible to identify the mode and tempo of karyotype evolution in mammals. However, these approaches do not address directly the forces that drive this process. In particular, these studies ignore the phenotypic consequences of chromosome rearrangements in hybrid intermediates. However, reduced reproductive fitness is a general characteristic of these hybrids and a major potential obstacle to karyotype evolution (White, 1978). The following sections will briefly review the role of natural forces in karyotype evolution with particular emphasis on how such forces may contribute to overcome underdominance in hybrid intermediates. These forces could provide explanations for the opposing evolutionary trends affecting diploid and fundamental numbers, the extreme variation in rates of rearrangement over time and among species and the presence of karyotypic orthoselection.

Phenotypic consequences of chromosome rearrangements

Karyotype evolution requires the fixation of chromosome rearrangements. Intermediate hybrids, defined as heterozygous carriers of chromosome rearrangements, are a required step in the process of fixation. There is ample evidence that most chromosome rearrangements involved in karyotype evolution are generally deleterious when heterozygous but have normal fitness when homozygous (White, 1978). The lower fitness of heterozygotes relative to either homozygote is known as underdominance which is the consequence of germline defects and meiotic errors. Meiotic errors lead to the generation of unbalanced gametes that arise from incorrect segregation due to abnormal pairing and/or recombination and generation of deletion/duplication events within rearranged fragments. The degree of reduction in fitness defines the coefficient of selection (s) and depends on many factors, including the type and particular features of the rearrangement, the sex of the carrier and the species (Neri et al., 1983; Boue et al., 1985; Daniel et al., 1989). When a rearrangement is strongly underdominant (s > 0.2) it has an extremely low probability of fixation and, therefore, is irrelevant from an evolutionary standpoint. At the opposite extreme are rearrangements that have no effect on fitness and, therefore, behave as neutral mutations.

The reduction in fitness that characterizes underdominant mutations implies that selection plays a critical role in the fate of the variant allele. In the absence of additional forces, the fate of a rearrangement depends on the coefficient of selection against heterozygotes and the allele frequency (Hedrick, 1981). In

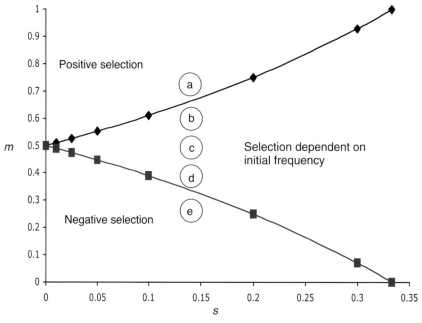

Fig. 13.6. (Above and opposite) Effect of meiotic drive on the probability of fixation of underdominant mutations. The graph on p.334 shows how the fate of underdominant mutations depends on the coefficient of selection (s) and the segregation ratio of the new variant during female meiosis (m). These two parameters define three sectors depending on the predicted outcome. The two lines define the levels of meiotic drive required to overcome a given level of underdominance and turn frequency-dependent selection into positive selection (dark line with diamonds) or negative selection (grey line with squares). Changes in allele frequency in underdominant mutations for five different levels of drive (a, m=0.7; b, m=0.6; c, m=0.5; d, m=0.4 and e, m=0.3) when s=0.15 are shown on the right side of the figure. q is the allele frequency of the new variant; Δq represents changes in allele frequency.

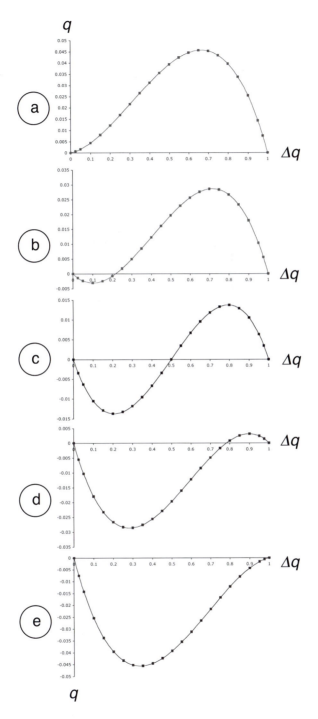

the simplest scenario the lower reproductive fitness of heterozygotes determines a fitness array defined by both types of homozygotes and the heterozygote that generates an unstable equilibrium (Hedrick, 1981). The point of unstable equilibrium is reached at the allele frequency at which the change in allele frequency for the new variant equals zero (Fig. 13.6c).

For allele frequencies below the point of unstable equilibrium, selection reduces the frequency of the variant allele and the variant is ultimately eliminated (Fig. 13.6c). For frequencies higher than the point of unstable equilibrium, selection increases the frequency of the variant allele and leads to its fixation. Therefore, variants are selected against when in the minority but selected for when in the majority. Because new variants are expected to be in the minority, this situation leads to a basic paradox. Chromosome rearrangements are known to occur but the presence of underdominance predicts that fixation of new variants should be extremely rare (White, 1978; Hedrick, 1981).

Five factors have been invoked to solve the basic paradox of chromosome evolution: mutation rate, genetic drift, inbreeding, adaptive selection and meiotic drive (White, 1978; Lande, 1979). Mutation, drift and inbreeding may affect the probability of fixation through changes in the allele frequency of the chromosome variant. On the other hand, adaptive selection and meiotic drive change the allele frequency at which the unstable equilibrium is reached. Although these factors will be considered separately, during evolution they will act together, further increasing the probability of fixation.

Mutation rate

For neutral mutations the fixation rate is equal to the mutation rate and is independent of the population size. In contrast, underdominant mutations have lower fixation rates and these rates depend on the coefficient of selection and the effective population size. The probability of fixation decreases as the population increases because the effective population size determines the allele frequency of the variant chromosome. The mutation rate for chromosome rearrangements in mammals has only been estimated accurately in humans. Although the frequency of chromosome rearrangements among newborns is very high (0.6–0.8%), many of the rearrangements are strongly deleterious ($s >$ 0.2) and are unlikely to contribute to karyotype evolution (Jacobs, 1981). On the other hand, the mutation rate for Robertsonian translocations is 4×10^{-4} per gamete per generation

(Jacobs, 1981). Interestingly, the mutation rate for each type of Robertsonian translocation is significantly different (13,14 and 14,21 represent 90% of all *de novo* rearrangements and most of them originate in the female germline (Jacobs, 1981; Page and Schafer, 1997)). This mutation rate may help to explain the fixation of weakly underdominant rearrangements in species with slow rates of rearrangements when the effective population sizes are small. However, this mutation rate cannot explain the observed fixation rates in mammals with fast rates of rearrangements, independently of the coefficient of selection and the effective population size. Obviously it may be possible that mutation rates are much higher in other species or for other types of rearrangements. Sequence comparisons provide evidence that some regions are more prone to breaks leading to chromosome rearrangements. Highly repetitive elements, transposons and Alu-mediated recombination, minisatellites and fragile sites have been found associated with chromosome breakpoints (Prak and Kazazian, 2000; Vergnaud and Denoeud, 2000). A recent study also shows that some genomic regions appear to be prone to undergo similar but distinct chromosomal rearrangements over and over again (Murphy *et al.*, 2003). However, the rapid changes in diploid number observed in *Mus musculus domesticus* have occurred in the absence of significant genetic divergence (Britton-Davidian *et al.*, 1989; Nachman *et al.*, 1994) and the mutation rates required for the fixation of Robertsonian translocations in these populations are extremely high. It has been reported that new Robertsonian fusions are more likely to occur in laboratory mouse stocks harbouring at least one of these rearrangements than in stocks with the normal karyotype (Nachman and Searle, 1995). Although promising, it is unclear to what extent this observation applies to the mutation rate in natural populations of mice in which Robertsonian translocations have become fixed and how steep increases in mutation rates will affect genomic instability in somatic tissues. The likelihood of the Robertsonian appears to be correlated with chromosome size rather than with specific genomic features (Gazave *et al.*, 2003). However, the effect of this apparent increase in mutation rate on the fixation rate would still be severely constrained by the coefficient of selection and the effective population size. Furthermore, variations in mutation rates do not predict, nor explain, the directional accumulation of a specific type of rearrangement in a lineage that is characterized by karyotype orthoselection.

Genetic drift

Because mammals live in subdivided populations where individuals have a limited number of potential mates, the size and breeding structure define the effective population size (Lande, 1979). In small populations genetic drift is a strong evolutionary force. This strength is compounded in the case of underdominant mutations because the fate of the mutation depends on its frequency. The stochastic processes that underlie genetic drift in small populations may occasionally lead to situations where the rearranged chromosome is in the majority and therefore likely to be fixed. Although drift may change the probability of fixation, this probability will be only a fraction of the probability of fixation for a neutral mutation (Hedrick, 1981; Nachman and Searle, 1995; Pardo-Manuel de Villena, 2003). Therefore, very small effective population sizes are required for drift to be solely responsible for the observed fixation rates and these sizes have little experimental support. Moreover, these small population sizes would lead to a higher risk of extinction and increased levels of inbreeding, a process that is associated with strong deleterious effects in mammals. Lastly, drift is a stochastic process that cannot explain karyotype orthoselection. Overall, genetic drift is expected to have played a significant, but secondary role in the fixation of chromosome rearrangements and it is not responsible for directional changes in the mammalian karyotype.

Natural selection

The operation of natural selection on genetic variants leads to directional changes in allelic frequencies over time. Natural selection may be positive or negative depending on the direc-tion of the change in allelic frequencies. In addition, natural selection may be adaptive if the directional change is driven by the differential fitness conferred by the genetic variants, or it may be non-adaptive if selection operates directly in the variants irrespective of the organismal fitness associated with them.

Adaptive selection

Adaptive selection may increase fixation rates if the new homokaryote (resulting from fixation of a rearrangement) has higher fitness than the old homokaryote. Fixation rates increase as a result of the lower unstable equilibriums that lead to a wider range of allelic frequencies for which fixation is the likely outcome. Adaptive selection cannot eliminate the unstable equilibrium (Hedrick, 1981). Importantly, large increases in fitness are required to significantly affect the probability of fixation. For example, increases in fitness for the new homokaryote in the same range as the coefficient of selection have very limited effects. Only when the fitness of the new homokaryote is several-fold larger than the coefficient of selection does this force have a significant impact. It is unlikely that any of the characters that have been proposed to be under selective pressure may increase fitness enough to have a significant impact on karyotype evolution. The most compelling ones, recombination and the generation of genetic diversity, propose that mammals select for different diploid and fundamental numbers to adjust the generation of genetic diversity at optimal levels (Burt and Bell, 1987; Qumsyeh, 1994). However, it is unclear why closely related species select for opposite levels of genetic diversity and no evidence for increased fitness has been reported. Therefore, adaptive selection is thought to have contributed little to the evolution of the mammalian karyotype.

Meiotic drive

Michael White (1978) and Max King (1991) have discussed extensively the putative role of meiotic drive in chromosomal evolution and several studies have modelled its effects on the probability of fixation of chromosome rearrangements (Hedrick, 1981). However, these landmark studies had only a limited impact in the field of karyotype evolution due to the ambiguous definition of drive, the lack of operational requirements for meiotic drive and the perception that there is little experimental evidence. As Max King has pointed out, meiotic drive was either viewed as 'an unnecessary complication for which there was no real need and less than significant evidence' or simply ignored. Independently of one's views on the impact of the karyotype in the process of speciation, the previous sections are a powerful reminder of the pressing need for an explanation of how underdominant mutations become fixed.

To discuss the role of meiotic drive it is first essential to define this term unambiguously. Meiotic drive was first defined by Sandler and Novistki (1957) as the failure of a heterozygote to produce both types of offspring at equal ratios. The authors explicitly excluded gametic selection and competition from their definition. However, for the next 40 years meiotic drive was consistently used to refer to gametic dysfunction systems as t-haplotype in mouse and Segregation Distorter in *Drosophila* (Ganetzky, 1999; Pardo-Manuel de Villena and Sapienza, 2001c). On the other hand, the use of meiotic drive to refer to the unequal recovery of alleles or chromosomes owing to alterations in the process of chromosome segregation during meiosis was limited. Lastly, meiotic drive may also be due to post-meiotic segregation or reciprocal crossover asymmetry (Jeffreys and Neumann, 2002). However, this latter type of drive affects only the vicinity of recombination hotspots and will not be considered further. Therefore, meiotic drive has been used to describe two very different phenomena, one meiotic and the other post-meiotic. It has been proposed to use meiotic drive to refer to any evolutionary force that results in distorted segregation ratios due to both meiotic and gametic processes (Ganetzki, 1999). In mammals, gametic drive is typically a male phenomenon while non-random segregation can only operate in females. Although many of the evolutionary consequences of the operation of drive are shared by both mechanisms, there are also significant differences (Ruvinsky, 1995; Pardo-Manuel de Villena and Sapienza, 2001c).

Systems of meiotic drive due to gametic dysfunction are fascinating. However, these systems are limited to specific genetic variants at a handful of specific loci and are independent from chromosome rearrangements. Therefore, they are unlikely to play a general role in the evolution of the karyotype. On the other hand, meiotic drive systems due to non-random segregation typically involve centromeres and are frequently associated with chromosome rearrangements. Therefore, this chapter will focus exclusively on non-random segregation. Non-random segregation is restricted to females because only they have an asymmetric meiosis that generates a single functional gamete. In this situation, any mechanism that preferentially segregates one chromosome or chromatid to the functional product of meiosis leads to drive. Once meiotic drive is restricted to non-random segregation systems it becomes possible to define three necessary and required conditions for drive to occur: (i) asymmetric cell division with respect to cell fate; (ii) functional asymmetry of the meiotic spindle; and (iii) functional heterozygosity at loci that mediate attachment of the driven chromosome to the spindle (Pardo-Manuel de Villena and Sapienza, 2001c).

Mathematical models demonstrate that meiotic drive acts by displacing the unstable equilibrium (Hedrick, 1981, and Fig. 13.6). In contrast with the evolutionary forces reviewed previously, drive may eliminate completely the unstable equilibrium (Hedrick, 1981). In this situation, if meiotic drive selects for the new variant, then this chromosome form behaves as if under positive selection and has an increased probability of fixation. On the other hand, if drive selects against the new variant, then this variant behaves as if under negative selection and it is eliminated. In both cases the fixation rate becomes less sensitive to the population size (Hedrick, 1981). Figure 13.6 shows the potential influence of meiotic drive in the probability of fixation of a chromosomal variant with different levels of underdominance. The segregation ratio (m) and the coefficient of selection against heterozygotes (s) define three distinct regions with different outcomes. When the segregation ratio is large enough to overcome the effect of underdominance, the variant behaves as if under positive selection (Fig. 13.6a,b). Negative selection for the variant is observed when the segregation ratio against the new variant reaches a certain value (Fig. 13.6d,e). When the unstable equilibrium is not eliminated then the direction of selection depends on the allelic frequency of the variant. In this latter case, three scenarios are possible, in the absence of drive the curve is symmetrical and the allele behaves as previously described (Fig. 13.6c). If $m > 0.5$ the unstable equilibrium is displaced towards lower allelic frequencies and, therefore, drift may easily tip the balance in favour of the variant chromosome leading to fixation. In contrast, if $m < 0.5$ the unstable equilibrium moves towards higher allelic frequencies, making it more difficult for stochastic events to overcome the effect of underdominance. Two other points are worth mentioning, drive may overcome underdominant mutations with $s < 0.33$, but the required levels of drive have only been reported in some insects and are unlikely to occur in mammals. However, modest levels of drive ($m < 0.7$) have a profound impact on the probability of fixation of chromosome rearrangements with coefficients of selection that are expected to play major roles in karyotype evolution ($s < 0.2$). In conclusion, mathematical models show that among evolutionary forces that may contribute to fixation of underdominant mutations, drive has the greatest potential.

Meiotic drive has been reported in several genetic systems and mammalian species, including Robertsonian translocations in both mouse and humans (Gropp and Winking, 1981; Ruvinsky *et al.*, 1987; Tease and Fisher, 1991; Aranha and Martin-DeLeon, 1994; Pacchierotti *et al.*, 1995; Pardo-Manuel de Villena and Sapienza, 2001a,d; Daniel, 2002), XO females (LeMarie-Atkins and Hunt, 2001), a chromosome 1 containing a homogeneously staining region (Agulnik *et al.*, 1991) and the *Om* locus in mouse (Pardo-Manuel de Villena *et al.*, 2000). The presence of drive in Robertsonian translocations is especially significant because these rearrangements play a major role in the evolution of the mammalian karyotype (Qumsyieh, 1994). In addition, these drive systems have a

unique set of characteristics in both mouse and humans including: (i) drive is independent of the chromosomes involved in the rearrangement; (ii) the direction of drive is constant within a species; (iii) the direction of drive observed for different types of chromosome abnormalities within a species is consistent with the unequal centromere rule; and (iv) the direction of drive is the same as the predominant chromosome form in the karyotype of each species (Pardo-Manuel de Villena and Sapienza, 2001a).

The presence and direction of drive stem from the functional asymmetry of the meiotic spindle acting on the unequal number of centromeres present in each side of the metaphase plate. In heterozygous carriers of Robertsonian translocations, at the first meiotic division one side of the metaphase contains the Robertsonian translocation with a single functional centromere (Earshaw *et al.*, 1985; Page *et al.*, 1995; Sullivan and Shwartz, 1995) while the other side of the spindle contains two acrocentric chromosomes with two centromeres. The presence of an unequal number of centromeres fulfils the requirement for functional heterozygosity at a locus that mediates attachment of the chromosomes subject to drive to the spindle. The fact that drive is independent of the chromosomes involved in the rearrangement confirms that the centromeres are the relevant chromosomal entity in non-random segregation.

When chromosome rearrangements result in unequal numbers of centromeres mouse consistently favours the segregation of the higher number of centromeres to the functional product of meiosis. That selection is applied to the centromeres rather than to any other feature is supported by the presence of drive in favour of the sole X chromosome in XO female mice (LeMarie-Atkins and Hunt, 2000). In contrast, humans favour the inclusion of the lower number of centromeres in the functional product of meiosis (Pardo-Manuel de Villena and Sapienza, 2001d; Daniel, 2003). These observations provide experimental evidence for meiotic drive for chromosome rearrangements in mammals. Importantly, the direction of drive in each species favours the predominant chromosome type found in their karyotype (i.e. acrocentric in mice and metacentric in humans), suggesting that drive has been instrumental in this process. An important consequence of the unequal centromere rule is that it predicts that most mammalian species should repeatedly select for the same chromosome morphology and, therefore, should have a strongly skewed distribution of uni-armed and bi-armed chromosomes in their karyotypes. These predictions are supported by the observed distribution of chromosome morphology in 1170 mammalian species (Pardo-Manuel de Villena and Sapienza, 2001a). Furthermore, the distribution of B chromosomes among mammalian species supports the hypothesis that meiotic drive operates to select higher number of centromeres in some species and lower number of centromeres in others because species with karyotypes composed mostly of acrocentric chromosomes have an excess of B chromosomes when compared with species with metacentric karyotypes (Palestis et al., 2003). Therefore, meiotic drive explains two of the characteristics of the mammalian karyotype (orthoselection and preference for uni-armed or bi-armed chromosomes) that cannot be explained by other forces. Although mouse and humans drive in opposite directions the level of distortion observed is similar, i.e. one chromosome form is preferentially transmitted to the functional gamete 60% of the time. This level of distortion will eliminate the unstable equilibrium for rearrangements with coefficient of selection lower than 0.092. For higher coefficients of selection this level of drive will displace the unstable equilibrium considerably (for example for $s = 0.2$ the unstable equilibrium will be reached at $q = 0.3$).

The fact that mouse selects for a higher number of centromeres while humans select for a lower number of centromeres indicates that the direction of drive reverses from time to time. Phylogenetic studies support the contention that reversal has occurred multiple times within each major phylogenetic lineage (Pardo-Manuel de Villena et al., 2001a). Opposing directions of selection and relatively frequent reversals may contribute to explain the bimodal rates of rearrangements reported in mammals and may play an important role in speciation.

Let us consider again the evolution of the karyotype in Mus musculus. Figure 13.5c shows that within this lineage periods of rapid change alternate with periods of stasis. Once the ancestral karyotype acquired 40 acrocentric chromosomes more than 5 million years ago, it has remained stable in most of its descendant species, including most populations of Mus musculus. Meiotic segregation analysis in carriers of Robertsonian translocations demonstrates that these rearrangements are preferentially segregated to the first polar body (Gropp and Winking, 1981; Ruvinsky et al., 1987; Tease and Fisher, 1991; Aranha and Martin-DeLeon, 1994; Pacchierotti et al., 1995). The coefficient of selection in these rearrangements and the level of drive will eliminate the unstable equilibrium and lead to negative selection. In other words, the all-acrocentric karyotype is maintained because Robertsonian fusions are selected against. On the other hand, it has been proposed that in some populations of the Mus musculus domesticus subspecies meiotic drive has been reversed, leading to positive selection for the rearranged chromosomes (Pardo-Manuel de Villena and Sapienza, 2001a). Although indirect evidence supports this hypothesis (Harris et al., 1986; Scriven, 1992), direct evidence is still lacking.

In conclusion, meiotic drive plays dual roles in shaping karyotype evolution. On one hand drive stabilizes the karyotype when the selection favours the old chromosome variant, while on the other hand drive increases the probability of fixation of chromosome rearrangements that are preferentially transmitted to the gametes during female meiosis (Pardo-Manuel de Villena, 2003). Although the latter situation has been the focus of most research, the former is important to address the diversity of rate of change in the mammalian lineage.

Karyotype and Speciation

There is a broad agreement that the establishment of a new species is often accompanied by changes in the karyotype (White, 1978;

King, 1993). Although most models of specia-tion require long periods of time (thousands or millions of generations) to develop reproduc-tive isolating mechanisms (RIMs), other modes, known as quantum speciation, may require only very short periods of time. Interestingly, most examples of quantum speci-ation involve changes in the karyotype. One form of quantum speciation is polyploidy, which involves the multiplication of entire chromosome complements, but this is never seen in mammals. Another form is the acquisi-tion of underdominant chromosome rearrangements leading to reduction in repro-ductive fitness of the hybrids. Some authors have favoured the idea that these chromo-some rearrangements may promote speciation in plants, insects and mammals (White, 1978; King, 1993; Piálek *et al.*, 2001). However, this idea remains controversial and is not favoured by many investigators (Piálek *et al.*, 2001). As these investigators have pointed out, several questions need to be addressed before chro-mosomal speciation can be a viable model. These include: (i) the mechanism by which rearrangements are fixed; (ii) the inverse cor-relation between potential of a rearrangement to act as a RIM and the likelihood that it will be fixed in a population; (iii) the requirements for geographical isolation; and (iv) the pace of the change.

Meiotic drive combined with genetic drift provides a satisfactory mechanism for fixation of underdominant rearrangements (Pardo-Manuel de Villena, 2003). However, the paradoxical relationship between fixation and RIM remains; if rearrangement acts as a RIM then it is unlikely that it would become fixed. Meiotic drive may overcome underdominance but then the isolat-ing potential of the rearrangement is reduced. Several models have been proposed to solve this paradox. Most of them involve geographical isolation and accumulation of several rearrange-ments or the action of positive selection (Piálek *et al.*, 2001; Navarro and Barton, 2003). Although these models may successfully explain the paradox, they imply that chromosomal spe-ciation is slow, very dependent on drift and do not explain the evolutionary trends observed in the mammalian karyotype.

Although meiotic drive by itself cannot be the cause of speciation, reversal of drive may play a significant role. It is predicted that reversal of drive should be a critical parame-ter determining the rate of rearrangement observed in different lineages and at different times within the same lineage. The longer the period during which the direction of drive has been stable in a lineage the lower will be the current rate of karyotype evolution because the mutation rate for rearrangement that dri-ves selected karyotypes is predicted to decrease. Rate of change in lineages that have undergone reversal of drive will be much faster, because the rearrangements that are selected for are expected to be much more frequent. A simplified model of reversal of drive is shown in Fig. 13.7. When reversal of drive occurs in some individuals of a popu-lation and these individuals are somehow geographically isolated, they will accumulate specific and directional changes at faster rates than the parental population. Although single rearrangements would not act as RIMs, multi-ple events would. Therefore, chromosome rearrangements that act as RIMs may become fixed very rapidly and hybrids would have severely reduced fitness. Importantly, in this situation multiple populations may fix differ-ent combinations of chromosomes and become isolated from each other even more rapidly than from the parental population, despite the fact that all of them descend from the single individual in which the direction of drive was reversed. The dichotomy between long periods of stasis and short but explosive periods of divergence, the directionality in the change, the different rates of rearrangement between and within lineages and the simulta-neous presence of multiple derived popula-tions with infertile hybrid intermediates agree with both theoretical models and experimen-tal observations (Eldredge and Gould, 1972; White, 1978; Britton-Davidian *et al.*, 2000; Piálek *et al.*, 2001). In conclusion, despite being out of favour for the last two decades, chromosomes play a crucial role in speciation and have much to contribute to our under-standing of this basic biological process.

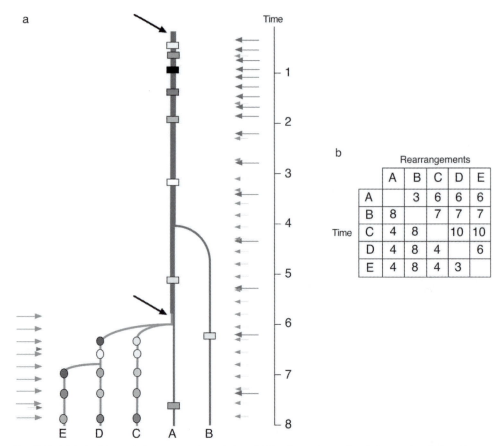

Fig. 13.7. Meiotic drive and chromosomal speciation. (a) The figure shows the phylogenetic relationships and chromosome rearrangements of five related species (A–E). Lineages in which meiotic drive favours a higher number of centromeres are shown in black and lineages in which drives favour a lower number of centromeres are shown in grey. An arbitrary time scale is shown on the right. The two large black arrows denote reversals of direction of drive. The distribution of black and grey arrows represents the mutation rates for these two types of rearrangements. Fixation of rearrangements leading to higher number of centromeres is depicted as rectangles and fixation of rearrangements leading to lower number of centromeres as circles. Different shades in rectangles and circles represent different rearrangements. (b) The table indicates the time of divergence and number of rearrangements between pairs of species.

References

Agulnik, S.I., Agulnik, A.I. and Ruvinsky, A.O. (1991) Meiotic drive in female mice heterozygous for the HSR inserts on chromosome 1. *Genetics Research* 55, 97–100.

Amor, D.J. and Choo, K.H.A. (2002) Neocentromeres: role in human disease, evolution and centromere study. *American Journal of Human Genetics* 71, 695–714.

Aranha, I.P. and Martin-DeLeon, P.A. (1994) Segregation analysis of the mouse Rb(6.16) translocation in zygotes produced by heterozygous female carriers. *Cytogenetics and Cell Genetics* 66, 51–53.

Archibald, A.L., Haley, C.S., Brown, J.F., Couperwhite, S., McQueen, H.A., Nicholson, D., Coppieters, W., Van de Weghe, A., Stratil, A., Wintero, A.K., Fredholm, M., Larsen, N.J., Nielsen, V.H., Milan, D., Woloszyn, N., Robic, A., Dalens, M., Riquet, J., Gellin, J., Caritez, J.C. *et al.* (1995) The PiGMaP consortium linkage map of the pig (*Sus scrofa*). *Mammalian Genome* 6, 157–175.

Benirschke, K., Ruedi, D., Müller, H., Kumamoto, A.T., Wagner, K.L. *et al.* (1980) The unusual karyotype of the lesser kudu, *Tragelaphus imberbis*. *Cytogenetics and Cell Genetics* 26, 85–92.

Bihoreau, M.T., Sebag-Montefiore, L., Godfrey, R.F., Wallis, R.H., Brown, J.H., Danoy, P.A., Collins, S.C., Rouard, M., Kaisaki, P.J., Lathrop, M. and Gauguier, D. (2001) A high-resolution consensus linkage map of the rat, integrating radiation hybrid and genetic maps. *Genomics* 75, 57–69.

Bishop, M.D., Kappes, S.M., Keele, J.W., Stone, R.T., Sunden, S.L., Hawkins, G.A., Toldo, S.S., Fries, R., Grosz, M.D., Yoo, J. and Beattie, C.W. (1994) A genetic linkage map of the cattle. *Genetics* 136, 619–639.

Boue, A., Boue, J. and Gropp, A. (1985) Cytogenetics of pregnancy wastage. *Advances in Human Genetics* 14, 1–57.

Britton-Davidian, J., Nadeau, J.H., Crosset, H. and Thaler, L. (1989) Genic differentiation and origin of Robertsonian populations of the house mouse (*Mus musculus domesticus* Rutty). *Genetics Research* 53, 29–44.

Britton-Davidian, J., Catalan, J., Ramalhinho, M.G., Ganem, G., Auffray, J.C., Capela, R., Biscoito, M., Searle, J.B. and da Luz Mathias, M. (2000) Rapid chromosomalvolution in island mice. *Nature* 403, 158.

Broman, K.W., Murra,y J.C., Sheffield, V.C., White, R.L. and Webe, J.L. (1998) Comprehensive human genetic maps: individual and sex-specific variation in recombination. *American Journal of Human Genetics* 63, 861–869.

Burt, A. and Bell, G. (1987) Mammalian chiasma frequency as a test of two theories of recombination. *Nature* 326, 803–805.

Burt, D.W., Bruley, C., Dunn, I.C., Jones, C.T., Ramage, A., Law, A.S., Morrice, D.R., Paton, I.R., Smith, J., Windsor, D., Sazanov, A., Fries, R. and Waddington, D. (1999) The dynamics of chromosome evolution in birds and mammals. *Nature* 402, 411–413.

Castiglia, R. and Capanna, E. (1999) Whole-arm reciprocal translocation (WART) in a feral population of mice. *Chromosome Research* 7, 493–495.

Cavagna, P., Menotti, A. and Stanyon, R. (2000) Genomic homology of the domestic ferret with cats and humans. *Mammalian Genome* 11, 866–870.

Chowdhary, B.P., Raudsepp, T., Froenicke, L. and Sherthan, H. (1998) Emerging patterns of comparative genome organization in some mammalian species as revealed by Zoo-FISH. *Genome Research* 8, 577–589.

Crawford, A.M., Dodds, K.G., Ede, A.J., Pierson, C.A., Montgomery, G.W., Garmonsway Hgm Beattie, A.E., Davies, K., Maddox, J.F., Kappes, S.W., Stone, R.T., Nguyen, T.C., Penty, J.M., Lord, E.A., Broom, J.E., Buitkamp, J., Schwaiger, W., Epplen, J.T., Matthew, P., Matthews, M.E., Ahulme, D.J., Beh, K.J., McGraw, R.A. and Beattie, C.W. (1995) An autosomal genetic linkage map of the sheep genome. *Genetics* 140, 703–724.

Daniel, A. (2002) Distortion of female meiotic segregation and reduced male fertility in human Robertsonian translocations: consistent with the centromere model of co-evolving centromere DNA/centromeric histone (CENP-A). *American Journal of Medical Genetics* 111, 450–452.

Daniel, A., Hook, E.B. and Wulf, G. (1989) Risk of unbalanced progeny at amniocentesis to carriers of chromosome rearrangements: data from the United States and Canadian laboratories. *American Journal of Human Genetics* 31, 14–53.

De Gortari, M.J., Freking, B.A., Cuthbertson, R.P., Kappes, S., Keele, J.W., Stone, R.T., Leymaster, K.A., Dodds, K.G., Crawford, A.M. and Beattie, C.W. (1998) A second-generation linkage map of the sheep genome. *Mammalian Genome* 9, 204–209.

Dib, C., Faure, S., Fizames, C., Samson, D., Drouot, N., Vignal, A., Millasseau, P., Marc, S., Hazan, J., Seboun, E., Lathrop, M., Gyapay, G., Morissette, J. and Weissenbach, J. (1996) A comprehensive genetic map of the human genome based on 5,264 microsatellites. *Nature* 380, 152–154.

Dobzhansky, T. and Sturtevant, A.K. (1938) Inversions in the chromosomes of *Drosophila pseudoobscura*. *Genetics* 23, 28–64.

Dumas, D. and Britton-Davidian, J. (2002) Chromosomal rearrangements and evolution of recombination: comparisons of chiasma distribution patterns in standard and Robertsonian populations of house mouse. *Genetics* 162, 1355–1366.

Dutrillaux, B. (1986) Le rôle des chromosomes dans l'evolution: une nouvelle interpretation. *Annals of Genetics* 29, 69–75.

Dutrillaux, B. and Couturier, J. (1983) The ancestral karyotype of Carnivora, comparisons with that of Plathyrrhine monkeys. *Cytogenetics and Cell Genetics* 35, 200–208.

Earnshaw, W.C., Ratrie, H. and Stetten, G. (1989) Visualization of centromere proteins CENP-B and CENP-C on a stable dicentric chromosome in cytological spreads. *Chromosoma* 98, 1–12.

Eldredge, N. and Gould, S.J. (1972) Punctuated equilibria: an alternative to phyletic gradualism. In: Schopf, T.J.M. (ed.) *Models in Paleobiology*. Freeman, Cooper & Co., San Francisco.

Ellegren, H., Chowdhary, B.P., Johansson, M., Marklund, L., Fredholm, M., Gustvasson, I. and Andersson, L. (1994) A primary linkage map of the porcine genome reveals a low rate of genetic recombination. *Genetics* 137, 1089–1100.

Fredga, K. (1972) Comparative chromosome studies in mongooses (Carnivora: Viverridae). *Hereditas* 71, 1–74.

Fronicke, L., Weinberg, J., Stone, G., Adams, L. and Stanyon, R. (2003) Towards the deliniation of the ancestral eutherian genome organization: comparative genome maps of human and the African elephant (*Loxodonta africana*) generated by chromosome painting. *Proceedings of the Royal Society of London: B Biological Sciences* 270, 1331–1340.

Ganetzki, B. (1999) Yuichiro Hizaizumi and forty years of segregation distortion. *Genetics* 152, 1–4.

Gardner, A.L. (1977) Chromosomal variation in *Vampyressa* and a review of chromosomal evolution in the Phyllostomidae (Chiroptera). *Syst Zool* 26, 300–318.

Gazave, E., Catalan, J., Ranalhinho, M.G., Mathias, M.L., Nunes, A.C., Dumas, D., Britton-Davidian, J. and Auffray, J.C. (2003) The non-random occurrence of Robertsonian fusion in the house mouse. *Genetics Research* 81, 33–42.

Goldammer, T., Kata, S.R., Brunner, R.M., Dorroch, U., Sanftleben, H., Schwerin, M. and Womack, J.E. (2002) A comparative radiation hybrid map of bovine chromosome 18 and homologous chromosomes in human and mice. *Proceedings of the National Academy of Sciences USA* 99, 2106–2111.

Goureau, A., Garrigues, A., Tosser-Klopp, G., Lahbib-Mansais, Y., Chardon, P. and Yerle, M. (2001) Conserved synteny and gene order difference between human chromosome 12 and pig chromosome 5. *Cytogenetics and Cell Genetics* 94, 49–54.

Gropp, A. and Winking, H. (1981) Robertsonian translocations: cytology, meiosis. Segregation patterns and biological consequences of heterozygosity. *Symposia of the Zoological Society of London* 47, 141–181.

Guénet, J.L. and Bonhomme, F. (2003) Wild mice: an ever-increasing contribution to a popular mammalian model. *Trends in Genetics* 19, 24–31.

Guerin, G., Bailey, E., Bernoco, D., Anderson, I., Antczak, D.F., Bell, K., Biros, I., Bjornstand, G., Bowling, A.T., Brandon, R., Caetano, A.R., Cholewinski, G., Colling, D., Eggleston, M., Ellis, N., Flyn, J., Gralak, B., Hasegawa, T., Ketchum, M., Lindgren, G. *et al.* (2003) The second generation of the International Equine Gene Mapping Workshop half-sibling linkage map. *Animal Genetics* 34, 161–168.

Haig, D. (1999) A brief history of human chromosomes. *Philosophical Transactions of the Royal Society of London: B Biological Sciences* 354, 1447–1470.

Harris, M.J., Wallace, M.E. and Evans, E.P. (1986) Aneuploidy in the embryonic progeny of females heterozygous for the Robertsonian chromosome (9.12) in genetically wild Peru-Coppock mice (*Mus musculus*). *Journal of Reproduction and Fertility* 76, 193–203.

Hassold, T. and Hunt, P. (2001) To err (meiotically) is human: the genesis of human aneuploidy. *Nature Review Genetics* 2, 280–291.

Hassold, T., Sherman, S. and Hunt, P. (2000) Counting cross-overs: characterizing meiotic recombination in mammals. *Human Molecular Genetics* 9, 2409–2419.

Hauffe, H.F. and Piàlek, J. (1997) Evolution of the chromosomal races of *Mus musculus domesticus* in Rhaetian Alps: the roles of whole-arm reciprocal translocation and zonal raciation. *Biological Journal of the Linnean Society* 62, 255–278.

Hedrick, P.W. (1981) The establishment of chromosomal variants. *Evolution* 35, 322–332.

Henikoff, S., Ahmad, K. and Malik, H.S. (2001) The centromere paradox: stable inheritance with rapidly evolving DNA. *Science* 293, 1098–1102.

International Human Genome Sequencing Consortium (2001) Initial sequencing and analysis of the human genome. *Nature* 409, 860–921.

Jacobs, P.A. (1981) Mutation rates of structural chromosome rearrangements in man. *American Journal of Human Genetics* 33, 44–54.

Jeffreys, A.S. and Neumann, R. (2002) Reciprocal crossover asymmetry and meiotic drive in a human recombination hot spot. *Nature Genetics* 31, 267–271.

Kappes, S.M., Keele, J.W., Stone, R.T., McGraw, R.A., Sonstegard, T.S., Smith, T.P.L., Lopez-Corrales, N.L. and Beattie, C.W. (1997) A second-generation linkage map of the bovine genome. *Genome Research* 7, 235–249.

Kim, J.J., Farnir, F., Savell, J. and Taylor, J.F. (2003) Detection of quantitative trait loci for growth and beef carcass fatness traits in a cross between *Bos taurus* (angus) and *Bos indicus* (Brahman) cattle. *Journal of Animal Science* 81, 1933–1942.

King, M. (1993) *Species Evolution. The Role of Chromosome Change.* Cambridge University Press, Cambridge, UK.

Kolnicki, R.L. (2000) Kinetochore reproduction in animal evolution: cell biological explanation of karyotypic fission theory. *Proceedings of the National Academy of Sciences USA* 97, 9493–9497.

Kong, A., Gudbjartsson, D.F., Sainz, J., Jonsdottir, G.M., Gudjonsson, S.A., Richardsson, B., Sigurdardottir, S., Barnard, J., Hallbeck, B., Masson, G., Shlien, A., Palsson, S.T., Frigge, M.L., Thorgeirsson, T.E., Gulcher, J.R. and Stefansson, K. (2002) A high-resolution recombination map of the human genome. *Nature Genetics* 31, 241–247.

Lande, R. (1979) Effective deme sizes during long-term evolution estimated from rates of chromosomal rearrangements. *Evolution* 33, 234–251.

LeMarie-Adkins, R. and Hunt, P.A. (2000) Nonrandom segregation of the mouse univalent X chromosome: evidence of spindle mediated meiotic drive. *Genetics* 156, 775–783.

Lindgren, G., Sandberg, K., Persson, H., Marklund, S., Breen, M., Sandgren, B., Carlstén, J. and Ellegren, H. (1998) A primary male autosomal linkage map of the horse genome. *Genome Research* 8, 951–966.

Ma, R.Z., Beever, J.E., Da, Y., Green, C.A., Russ, I., Park, C., Heyen, D.W., Everts, R.E., Fisher, S.R., Overton, K.M., Teale, A.J., Kemp, S.J., Hines, H.C., Guérin, G. and Lewin, H.A. (1996) A male linkage map of the cattle (*Bos taurus*) genome. *Journal of Heredity* 87, 261–271.

Maddox, J.F., Davies, K.P., Crawford, A.M., Hulme, D.J., Vaiman, D., Cribiu, E.P., Freking, B.A., Beh, K.J., Cockett, N.E., Kang, N., Riffkin, C.D., Drinkwater, R., Moore, S.S., Dodds, K.G., Lumsden, J.M., van Stijn, T.C., Phua, S.H., Adelson, D.L., Burkin, H.R., Broom, J.E. *et al.* (2001) An enhanced linkage map of the sheep genome comprising more than 1000 loci. *Genome Research* 11, 1275–1289.

Malik, H.S. and Henikoff, S. (2001) Adaptive evolution of Cid, a centromere-specific histone in *Drosophila*. *Genetics* 157, 1293–1298.

Malik, H.S. and Henikoff, S. (2002) Conflict begets complexity: the evolution of centromeres. *Current Opinion in Genetics and Development* 12, 711–718.

Mellersh, C.S., Langston, A.A., Acland, G.M., Fleming, M.A., Ray, K., Wiegand, N.A., Francisco, L.V., Gibbs, M., Aguirre, G.D. and Ostrander, E.A. (1997) A linkage map of the canine genome. *Genomics* 46, 326–336.

Menotti-Raymond, M., David, V.A., Lyons, L.A., Schäffer, A.A., Tomlin, J.F., Hutton, M.K. and O'Brien, S.J. (1999) A genetic linkage map of microsatellites in the domestic cat (*Felis catus*). *Genomics* 57, 9–23.

Menotti-Raymond, M., David, V.A, Chen, Z.Q., Menotti, K.A., Sun, S., Schäffer, A.A., Agarwala, R., Tomlin, J.F., O'Brien, S.J. and Murphy, W.J. (2003) Second-generation integrated genetic linkage/radiation hybrid maps of the domestic cat (*Felis catus*). *Journal of Heredity* 94, 95–106.

Mikawa, S., Akita, T., Hisamatsu, N., Inage, Y., Ito, Y., Kobayashi, E., Husumoto, H., Matsumoto, T., Mikami, H., Minezawa, M., Miyake, M., Shimanuki, S., Sugiyama, C., Uchida, Y., Wada, Y., Yanai, S. and Yasue, H. (1999) A linkage map of 243 DNA markers in an intercross of Gottingen miniature and Meishan pigs. *Animal Genetics* 30, 407–417.

Müller, S. and Weinberg, J. (2001) 'Bar-coding' primate chromosomes: molecular cytogenetic screening for the ancestral hominoid karyotype. *Human Genetics* 109, 85–94.

Müller, S., Stanyon, R., O'Brien, P.C.M., Ferguson-Smith, M.A., Plesker, R. and Weinberg, J. (1999) Defining the ancestral karyotype of all primates by multidirectional chromosome painting between tree shrews, lemurs and humans. *Chromosoma* 108, 393–400.

Murphy, W.J., Stanyon, R. and O'Brien, S.J. (2001a) Evolution of the mammalian genome organization inferred from comparative gene mapping. *Genome Biology* 2, REVIEWS0005. Epub 2001 Jun 05.

Murphy, W.J., Elzirk, E., Johnson, W.E., Zhang, Y.P., Ryder, O.A. and O'Brien, S.J. (2001b) Molecular phylogenetics and the origin of placental mammals. *Nature* 409, 614–618.

Murphy, W.J., Fronicke, L., O'Brien, S.J. and Stanyonn, R. (2003) The origins of human chromosome 1 and its homologs in placental mammals. *Genome Research* 13, 1880–1888.

Nachman, M.W. and Searle, J.B. (1995) Why is the house mouse karyotype so variable? *Trends in Ecology and Evolution* 10, 397–402.

Nachman, M.W., Boyer, S.N., Searle, J.B. and Aquadro, C.F. (1994) Mitochondrial DNA variation and the evolution of Robertsonian chromosomal races of house mouse, *Mus musculus*. *Genetics* 136, 1105–1120.

Nadeau, J.H. and Sankoff, D. (1998) Counting on comparative maps. *Trends in Genetics* 14, 495–501.

Navarro, A. and Barton, N.H. (2003) Chromosomal speciation and molecular diverge-accelerated evolution in rearranged chromosomes. *Science* 300, 321–324.

Neff, M.W., Broman, K.W., Mellersh, C.S., Ray, K., Acland, G.M., Aguirre, G.D., Ziegle, J.S., Ostrander, E.A. and Rine, J. (1999) A second-generation genetic linkage map of the domestic dog, *Canis familiaris*. *Genetics* 151, 803–820.

Neri, G., Serra, A., Campana, M. and Tedeschi, B. (1983) Reproductive risk for translocation carriers: cytogenetic studies and analysis of pregnancy outcome in 58 families. *American Journal of Medical Genetics* 16, 535–561.

Nie, W., Wang, J., Perelman, P., Graphodatsky, A.S. and Yang, F. (2003) Comparative chromosome painting defines the karyotypic relationships among the domestic dog, Chinese raccoon dog and Japanese raccoon dog. *Chromosome Research* 11, 735–740.

O'Brien, S.J., Menotti-Raymond, M., Murphy, W.J., Nash, W.G., Wienberg, J., Stanyon, R., Copeland, N.G., Jenkins, N.A., Womack, J.E. and Graves, J.A.M. (1999) The promise of comparative genomics in mammals. *Science* 286, 458–481.

O'Brien, S.J., Menotti-Raymond, M., Murphy, W. and Yuhki, N. (2002) The feline genome project. *Annual Review of Genetics* 36, 657–686.

Pacchierotti, F., Tiveron, C., Mailhes, J.B. and Davisson, M.T. (1995) Susceptibility of vinblastin induced aneuploidy and preferential segregation during meiosis I in Robertsonian heterozygous mice. *Teratogenesis Carcinogenesis Mutagenesis* 15, 217–230.

Page, S.L. and Schaffer, L.G. (1997) Nonhomologous Robertsonian translocations form predominantly during female meiosis. *Nature Genetics* 15, 231–232.

Page, S.L., Earnshaw, W.C., Choo, K.H. and Schaffer, L.G. (1995) Further evidence that CENP-C is a necessary component of active centromeres: studies of a dic(X;15) with simultaneous immunofluorescence and FISH. *Human Molecular Genetics* 4, 289–294.

Palestis, B.G., Burt, A., Jones, R.N. and Trivers, R. (2003) B chromosomes are more frequent in mammals with acrocentric karyotypes: support for the theory of centromeric drive. *Proceedings of the Royal Society of London B (Supplement)* DOI 10.1098/rsbl.2003.0084

Palmer, J.D. and Herbon, L.A. (1988) Plant mitochondrial DNA evolves rapidly in structure, but slowly in sequence. *Journal of Molecular Evolution* 27, 87–97.

Pardo-Manuel de Villena, F. (2003) *Human Karyotype: Evolution in Encyclopedia of the Human Genome.* Macmillan Publishers, Ltd.

Pardo-Manuel de Villena, F. and Sapienza, C. (2001a) Female meiosis drives karyotypic evolution in mammals. *Genetics* 159, 1179–1189.

Pardo-Manuel de Villena, F. and Sapienza, C. (2001b) Recombination is proportional to the number of chromosome arms in mammals. *Mammalian Genome* 12, 318–322.

Pardo-Manuel de Villena, F. and Sapienza, C. (2001c) Nonrandom segregation during meiosis: the unfairness of female. *Mammalian Genome* 12, 331–338.

Pardo-Manuel de Villena, F. and Sapienza, C. (2001d) Transmission ratio distortion in offspring of heterozygous female carriers of Robertsonian translocations. *Human Genetics* 108, 31–36.

Pardo-Manuel de Villena, F., De la Casa-Esperon, Bricoe, T.L. and Sapienza, C. (2000) A genetic test to determine the origin of maternal transmission ratio distortion: meiotic drive at the mouse *Om* locus. *Genetics* 154, 333–342.

Pevzner, P. and Tesle, G. (2003) Genome rearrangements in mammalian evolution: lessons from human and mouse genomes. *Genome Research* 13, 37–45.

Piàlek, J., Hauffe, H.C., Rodríguez-Clark, K.M. and Searle, J.B. (2001) Raciation and speciation in house mice from the Alps: the role of chromosomes. *Molecular Ecology* 10, 613–625.

Prak, E. and Kazazian, H.H. Jr, (2000) Mobile elements and the human genome. *Nature Reviews of Genetics* 1, 134–144.

Purvis, A. and Hector, A. (2000) Getting measure of biodiversity. *Nature* 405, 212–219.

Qumsyieh, M.B. (1994) Evolution of number and morphology of mammalian chromosomes. *Journal of Heredity* 85, 455–465.

Redi, C.A., Garagna, S. and Zuccotti, M. (1990) Robertsonian chromosome formation and fixation: the genomic scenario. *Biological Journal of the Linnean Society* 41, 235–255.

Richard, F., Lombard, M. and Dutrillaux, B. (2003a) Reconstruction of the ancestral karyotype of eutherian mammals. *Chromosome Research* 11, 605–618.

Richard, F., Messaoudi, C., Bonnet-Garnie, A., Lombard, M. and Dutrillaux, B. (2003b) Highly conserved chromosomes in an Asian squirrel (*Menetes bermorei*, Rodentia: Sciuridae) as demonstrated by ZOO-FISH with human probes. *Chromosome Research* 11, 597–603.

Rogalska-Niznik, N., Szczerbal, I., Dolf, G., Schlapfer, J., Schelling, C. and Switonski, M. (2002) Canine-derived cosmid probes containing microsatellites can be used in physical mapping of Arctic fox (*Alopex lagopus*) and Chinese raccoon dog (*Nyctereutes procyonoides procyonoides*) genomes. *Journal of Heredity* 94, 89–93.

Rogers, J., Mahaney, M.C., Witte, S.M., Nair, S., Newman, D., Wedel, S., Rodriguez, L.A., Rice, K.S., Slifer,

S.H., Perelygin, A., Slifer, M., Palladino-Negro, P., Newman, T., Chambers, K., Joslyn, G., Parry, P. and Morin, P.A. (2000) A genetic linkage map of the baboon (*Papio hymadryas*) genome based on human microsatellite polymorphisms. *Genomics* 67, 237–247.

Rohrer, G.A., Alexander, L.J., Hu, Z., Smith, T.P.L., Keele, J.W. and Beattie, C.W. (1996) A comprehensive map of the porcine genome. *Genome Research* 6, 371–391.

Ruvinsky, A.O. (1995) Meiotic drive in female mice: an assay. *Mammalian Genome* 6, 315–320.

Ruvinsky, A.O., Agulnik, S.I., Agulnik, A.I. and Belyaev, D.K. (1987) The influence of mutations on chromosome 17 upon the segregation of homologues in female mice heterozygous for Robertsonian translocations. *Genetics Research* 50, 235–237.

Sandler, L. and Novistki, E. (1957) Meiotic drive as an evolutionary force. *American Naturalist* 91, 105–110.

Schibler, L., Vaiman, D., Oustry, A., Giraud-Delville, C. and Cribiu, E.P. (1998) Comparative gene mapping: a fine-scale survey of chromosome rearrangements between ruminants and humans. *Genome Research* 8, 901–915.

Scriven, P.N. (1992) Robertsonian translocations introduced into an island population of house mice. *Journal of Zoology, London* 227, 493–502.

Silver, L.M. (1993) The peculiar journey of a selfish chromosome: mouse t-haplotypes and meiotic drive. *Trends in Genetics* 9, 240–245

Slate, J., van Stijn, T.C., Anderson, R.M., McEwan, K.M., Maqbool, N.J., Mathias, H.C., Bixley, M.J., Stevens, D.R., Molenaar, A.J., Beever, J.E., Galloway, S.M. and Tate, M.L. (2002) A deer (subfamily Cervinae) genetic linkage map and the evolution of ruminant genomes. *Genetics* 160, 1587–1597.

Stanyon, R., Yang, F., Cavagna, P., O'Brien, P.C.M., Bagga, M., Ferguson-Smith, M.A. and Weinberg, J. (1999) Reciprocal painting shows that genomic rearrangements between rat and mouse proceeds ten times faster than between humans and cats. *Cytogenetics and Cell Genetics* 84, 150–155.

Stanyon, R., Stone, G., Garcia, M. and Froenicke, L. (2003) Reciprocal painting shows that squirrel, unlike murid rodents, have highly conserved genome organization. *Genomics* 82, 245–249.

Steen, R.G., Kwitek-Black, A.E., Glenn, C., Gullings-Handley, J., Van Etten, W., Atkison, O.S., Appel, D., Twigger, S., Muir, M., Mull, T., Granados, M., Kissebah, M., Russo, K., Crane, R., Popp, M., Peden, M., Matise, T., Brown, D.M., Lu, J., Kingsmore, S. *et al.* (1999) A high-density integrated genetic linkage and radiation hybrid map of the laboratory rat. *Genome Research* 9, AP1–AP8.

Sullivan, B.A. and Schwartz, S. (1995) Identification of centromeric antigens in dicentric Robertsonian translocations: CENP-C and CENP-E are necessary components of functional centromeres. *Human Molecular Genetics* 4, 2189–2197.

Sullivan, B.A., Blower, M.D. and Karpen, G.H. (2001) Determining centromere identity: cyclical stories and forking paths. *Nature Reviews of Genetics* 2, 584–596.

Swinburne, J., Gerstenberg, C., Breen, M., Aldridge, V., Lockhart, L., Marti, B., Antczak, D., Eggleston-Stott, M., Bailey, E., Mickelson, J., Røed, K., Lindgren, G., von-Haeringen, W., Guérin, G., Bjarnason, J., Allen, T. and Binns, M. (2000) First comprehensive low-density horse linkage map based on two 3-generations, full-sibling, cross-bred horse reference families. *Genomics* 66, 123–134.

Tease, C. and Fisher, G. (1991) Two new X–autosome Robertsonian translocations in the mouse. I. Meiotic chromosome segregation in male hemizygotes and female heterozygotes. *Genetics Research* 58, 115–121.

Tjio, J.H. and Levan, A. (1956) The chromosome number of man. *Hereditas* 42, 1–6.

Todd, N.B. (1970) Karyotypic fissioning and canid phylogeny. *Journal of Theoretical Biology* 26, 445–480.

Trifonov, V.A., Perelman, P.L., Kawada, S.I., Iwasa, M.A., Oda, S.I. and Graphodatsky, A.S. (2002) Complex structure of B-chromosomes in two mammalian species: *Apodemus peninsulae* (Rodentia) and *Nyctereutes procyonoides* (Carnivora). *Chromosome Research* 10, 109–116.

Vaiman, D., Schibler, L., Bourgeois, F., Oustry, A., Amigues, Y. and Cribiu, E.P. (1996) A linkage map of the male goat genome. *Genetics* 144, 279–305.

Vassart, M., Seguela, A. and Hayes, H. (1995) Chromosomal evolution in gazelles. *Journal of Heredity* 86, 216–227.

Ventura, M., Archidiacono, N. and Rocchi, M. (2001) Centromere emergence in evolution. *Genome Research* 11, 595–599.

Vergnaud, G. and Denoeud, F. (2000) Minisatellites: mutability and genome architecture. *Genome Research* 10, 889–907.

Viitala, S.M., Schulman, N.F., de Koning, D.J., Elo, K., Kinos, R., Virta, A., Virta, J., Maki-Tanila, A. and Vilkki, J.H. (2003) Quantitative trait loci affecting milk production in Finnish Ayrshire dairy cattle. *Journal of Dairy Science* 86, 1828–1836.

Ward, O.G., Wurster-Hill, D.H., Ratty, F.J. and Song, Y. (1987) Comparative cytogenetics of Chinese and Japanese raccoon dogs, *Nyctereutes procyonoides*. *Cytogenetics and Cell Genetics* 45, 177–186.

Weinberg, J. and Stanyon, R. (1995) Chromosome painting in mammals as an approach to comparative genomics. *Current Opinion in Genetics and Development* 39, 792–797.

Weinberg, J., Jauch, A., Stanyon, R. and Cremer, T. (1990) Molecular cytotaxonomy of primates by chromosomal *in situ* hybrydization. *Genomics* 8, 347–350.

White, M.J.D. (1978) Chromosomal modes of speciation. In: *Modes of Speciation*. WH Freeman, San Francisco.

Wurster-Hill, D.H., Ward, O.G., Davis, B.H., Park, J.P., Moyzis, R.K. and Meyne, J. (1988) Fragile sites, telomeric DNA sequences, B chromosomes, and DNA content in raccoon dogs, *Nyctereutes procyonoides*, with comparative notes on foxes, coyote, wolf, and raccoon. *Cytogenetics and Cell Genetics* 49, 278–281.

Yang, F., O'Brien, P.C., Milne, B.S., Graphodatsky, A.S., Solanky, N., Trifonov, V., Rens, W., Sargan, D. and Ferguson-Smith, M.A. (1999) A complete comparative chromosome map for the dog, red fox, and human and its integration with canine genetic maps. *Genomics* 62, 189–202.

Yang, F., Alkalaeva, E.Z., Perelman, P.L., Pardini, A.T., Harrison, W.R, O'Brien, P.C.M., Fu, B., Graphodatsky, Ferguson-Smith, M.A. and Robinson, T.J. (2003) Reciprocal chromosome painting among human, aardvark, and elephant (superorder Afrotheria) reveals the likely eutherian ancestral karyotype. *Proceedings of the National Academy of Sciences USA* 100, 1062–1066.

Yunis, J.J. (1976) High resolution of human chromosomes. *Science* 191, 1268–1270.

Zenger, K.R., McKenzie, L.M. and Cooper, D.W. (2002) The first comprehensive genetic linkage map of a marsupial: the tammar wallaby (*Macropus eugenii*). *Genetics* 162, 321–330.

14 Comparative Gene Mapping, Chromosome Painting and the Reconstruction of the Ancestral Mammalian Karyotype

O.L. Serov,[1] B. Chowdhary,[2] J.E. Womack[2] and J.A. Marshall Graves[3]

[1]*Institute of Cytology and Genetics, Novosibirsk 90, Russia;*
[2]*Texas A&M University, College Station, Texas, USA;*
[3]*The Australian National University, Canberra, ACT 2601, Australia*

Contents

Introduction	350
Mammalian Relationships	350
Cytogenetic Comparisons	352
Comparative Mapping	353
Nomenclature and terms	354
Mapping human and other primates	355
Rodent maps	355
Carnivore maps	356
Eulipotyphla maps (Laurasiatheria)	357
Cetartiodactyla maps (Lauraiastheria)	358
Eutherian sex chromosomes	359
Marsupial and monotreme maps	359
Conclusions from comparative mapping	360
Comparative Chromosome Painting	361
Painting in eutherian mammals	362
Comparative chromosome painting in marsupials and monotremes	370
Reconstructing Ancient Mammalian Karyotypes	371
Reconstructing the ancient eutherian karyotype	372
Reconstructing ancient sex chromosomes	377
Conclusions	378
References	379

Correspondence: serov@bionet.nsc.ru; bchowdhary@cvm.tamu.edu; jwomack@cvm.tamu.edu; graves@rsbs.anu.edu.au

© CAB International 2005. *Mammalian Genomics*
(eds A. Ruvinsky and J. Marshall Graves)

Introduction

There are more than 4000 species of mammal in the world, of all shapes and sizes and habits, and occupying a great variety of ecological niches. However, they all evolved from a common ancestor and share a genome of about 3000 Mb that contains much the same set of genes. Comparing their genomes is an efficient way to discover genes and regulatory signals, to determine the function of genes and noncoding regions and correlate these with function and physiology, and to gain a better understanding of genome organization and evolution in mammals.

However, the enormous variation in mammalian karyotypes poses a major challenge. Eutherian karyotypes are very variable in their packaging, ranging from a diploid number of six huge chromosomes in the deer mouse to 102 small chromosomes in a South American rodent. Chromosome number is low and rather uniform in marsupials, with a basic karyotype of $2n = 14$, and monotremes are different again, having a few large chromosomes, and many small chromosomes, somewhat in the manner of birds and reptiles. Aligning these karyotypes represents a vast exercise in jigsaw puzzle solving.

The sex chromosomes are something of an exception at both extremes. The X is extremely conserved even between distantly related species, and the Y is extremely poorly conserved even between closely related species.

Methods for comparing mammalian genomes have developed enormously since the 1960s (Chapter 2), when it was possible only to compare chromosome sizes and centromere positions, and to construct maps of mammal genomes only by performing genetic crosses. The discovery of chromosome banding, the development of physical mapping techniques and ultimately comparative chromosome painting have led to an enormous increase in resolution and power, as described in detail in Chapter 13.

Thus early comparisons gradually evolved into a fully fledged branch of genetics – *comparative genomics* – that has become an integral part of genome analysis in various species. In this chapter we will focus on studies that used gene mapping and comparative painting to identify similarities across different mammalian genomes. We will then discuss how inferences and deductions from these studies help to deduce the likely karyotype of the common mammalian ancestor.

Mammalian Relationships

Mammals diverged from a branch of reptiles (synapsids) that left no other descendants. Thus mammals are equally distantly related to birds and reptiles, from which they diverged about 310 million years ago (mya).

There are three major groups of mammals that can be compared to deduce the form of ancient karyotypes. The mammalian subclass Theria (viviparous mammals) diverged from the egg-laying monotremes (subclass Prototheria) about 210 mya, then the two therian infraclasses Eutheria (placentals) and Metatheria (marsupials) diverged about 180 mya (Woodburne et al., 2003) (Fig. 14.1).

Of the ~4000 mammals, the vast majority are eutherians. There are about 3700 species in 19 extant orders, which are now divided into four major groups on molecular evidence (Murphy et al., 2001c; Springer et al., 2001; Novacek, et al., 2001). The oldest split at 103 mya separated off the Afrotheria, now thought to be a monophyletic group of very morphologically diverse animals (including elephants and aardvarks, tenrecs and golden moles). Xenarthra (sloths and armadillos) then diverged from the rest, which then split into two large groups. The Laurasiatheria, containing shrews, bats, carnivores, Perissodactyla (including horse) and Cetartiodactyla (whales and ungulates) diverged from Euarchontoglires containing rodents, tree shrews and primates. The close relationship between humans and rodents is a big surprise to geneticists and cytologists, who had long assumed from the multiple rearrangements between humans and mice that they belonged to widely divergent mammal groups.

This re-drawing of the boundaries has eliminated Edentata and Insectivora, which were shown to be polyphyletic by molecular data (Liu et al., 2001). For instance, the traditional view of Insectivora is that the group descended from a single common ancestor, and is composed of six families: Soricidae (shrews),

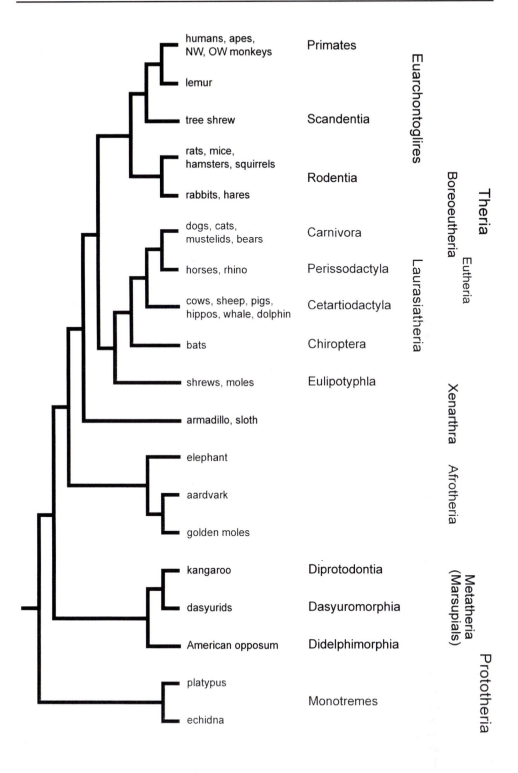

Fig. 14.1. Phylogeny of mammals (after Murphy *et al.*, 2001b,c).

Tenrecidae (tenrecs), Solenodontidae (solenodons), Talpidae (moles), Erinaceidae (hedgehogs and gymnures), and Chrysochloridae (golden moles). Recent molecular study of sequences of the mitochondrial 12S rRNA, tRNA-valine, and 16S rRNA genes demonstrated that golden moles and tenrecs belong to a new distinct order Afrosoricidae (Stanhope et al., 1998) of the Afrotheria. They are more closely related to elephants than to hedgehogs and shrews, which are now referred to as Eulipotyphla (Murphy et al., 2001b).

There are 270 known species of marsupials in seven orders, five of which live in Australasia and two in South America. The Australian species are monophyletic, and probably derived from a South American group before Australia split from Gondwana 80–55 mya (Amrine-Madsen et al., 2003). There are only three monotreme species in two Australasian families (the platypus and two spiny echidna species). Although they both fit the definition of mammals, bearing fur and suckling their young with milk, marsupials and monotremes are characterized by alternative reproduction strategies. The marsupial embryo develops in the uterus for only a short time and completes its development attached to a nipple, often protected by a pouch. Monotremes lay eggs.

In order to deduce the form of the genome of an ancestral eutherian, we need to compare the karyotypes of eutherians in different orders. Then to reconstruct an ancestral therian karyotype, we need to compare marsupials and eutherian ancestors. In order to deduce the karyotype of the ancestor of all mammals, we can compare this with the genome of monotremes, using chicken and reptiles as outgroups.

Cytogenetic Comparisons

Comparative cytogenetics represents the most traditional approach to comparing genomes across species. It came into existence following development of methods to obtain fixed chromosome preparations. Early comparisons were limited to comparison of chromosome numbers and shapes, defined by centromere and nucleolus organizer position. Even with these limitations, it was possible to judge that some species shared very similar karyotypes.

A major transformation in comparative cytogenetics occurred with the discovery of banding techniques, among which Q-, G- and R-banding approaches are the most significant. During the 1970s and 1980s these approaches played a critical role in the development of standard karyotypes and nomenclature in a variety of mammal species.

Banding studies also made it possible to compare chromosomes of closely and distantly related species. Meaningful comparisons were made mainly across closely related species, the karyotypes of which were either similar or had undergone relatively few detectable rearrangements. For instance, comparison of human and primate karyotypes showed that the chromosomes could be easily aligned, and the differences between the chromosomes of, for instance human and chimpanzee and gorilla were trivial (Dutrillaux et al., 1981). Sheep and cattle chromosomes also proved to be highly conserved within the limits of cytogenetic technologies (ISCNDB, 2000). Sheep ($2n = 54$) and cattle ($2n = 60$) karyotypes differ only in three metacentrics, whose arms correspond in banding pattern to individual cattle chromosomes. Similarly, karyotypes of felid species, for instance clouded leopard and African lion, are almost identical to that of the domestic cat (Tian et al., 2004).

However, karyotypes of some quite closely related species proved very difficult to align. For instance rat and mouse karyotypes took some effort to sort out (Stanyon et al., 1999; Helou et al., 2001; Nilsson et al., 2001), and dogs were difficult (Wayne et al., 1987). Even within karyotypically conservative groups such as primates, some species stood out as having atypical or highly rearranged karyotypes (e.g. gibbon).

Attempts to compare the chromosomes of more distantly related species, such as humans and mice, cats and cows, by their banding patterns were completely uninformative. It seemed that eutherian chromosomes had undergone so many changes during evolution that their genomes were scrambled nearly beyond recognition.

The eutherian X chromosome is exceptional in that, even at the cytogenetic level, it was seen to be particularly uniform in size. It maintained a relative length of 5% of the haploid chromosome length between even distantly related species, even in species with a few large chromosomes, or a lot of small ones (Ohno, 1969), an observation that became known as Ohno's Law. G-banding patterns were also very conserved between the X chromosomes of humans and apes (Dutrillaux, 1979), and even carnivores (Serov *et al.*, 1990).

In comparison with eutherian karyotypes, marsupial karyotypes are very straightforward, being obviously conserved between even the most distantly related groups. Chromosome number is low and rather uniform in all species, ranging from $2n = 10/11$ to $2n = 32$. Two striking modal numbers of 22 and 14 were identified (Hayman and Martin, 1969), and of these a 'basic' $2n = 14$ karyotype, with near identity of G-band patterns, can be recognized in each of the major marsupial groups (Rofe and Hayman, 1985; Hayman, 1990). This is certainly the ancestral karyotype at least of Australian marsupials and a South American order which probably gave rise to them (Amrine-Madsen *et al.*, 2003). Other marsupial karyotypes can be easily derived from this by simple fusions and fissions.

Monotreme chromosomes, too, are highly conserved across the two families. They are composed of a few large chromosomes, and many small chromosomes that are hard to distinguish and were at first thought to represent 'microchromosomes' like those of birds and reptiles (Bick, 1992). They are unique in containing several unpaired chromosomes that form a chain at male meiosis, presumably as the result of heterozygosity for translocations (Grützner *et al.*, 2003).

The basis of comparing genomes of different mammalian species gradually shifted during the 1970s from traditional cytogenetics to new molecular-based approaches. Construction of gene maps in several of these species is one of the factors that enabled this transition.

Comparative Mapping

Gene mapping by traditional means – measuring recombination between phenotypic markers – pre-dated cytogenetic comparisons by decades. However, it was of practical use only for the few mammal species with an enormous array of phenotypic markers and that could be captive bred over many generations. The only

complete linkage maps in the 1970s were those of mouse and rat. The development of anonymous DNA markers greatly widened the scope of linkage mapping. A complete linkage map of the human genome was constructed (Botstein *et al.*, 1980) and dense mouse and rat maps were developed (Jacob *et al.*, 1995; Dietrich *et al.*, 1996; Bihoreau *et al.*, 1997; Brown *et al.*, 1998). Linkage maps of several domestic species (e.g. cow, Barendse *et al.*, 1994, 1997; sheep, Maddox *et al.*, 2001; pig, Ellegren *et al.*, 1994; Rohrer *et al.*, 1996) and even exotic mammals that can be captive bred (e.g. tammar wallaby) were also constructed (Zenger *et al.*, 2002). One disadvantage of anonymous polymorphic molecular markers was that these could not usually be compared between species.

Before the advent of molecular markers, it was extremely difficult to map autosomes of all but a few model species. An exception was the X chromosome, whose genes could be recognized by the pattern of inheritance. Early scientific evidence for the conservation of gene arrangement came from comparative studies of the mammalian X chromosome (Ohno, 1967), which showed that the eutherian X was conserved in size and gene content, and some markers were conserved even between eutherians and marsupials (Richardson *et al.*, 1971).

A great advance to mammalian gene mapping was the development of somatic cell genetic mapping. Cell hybrids constructed between lines from different mammalian species were shown to preferentially lose chromosomes from one parent, and the patterns of segregation of markers and chromosomes could be correlated: markers that co-segregated were said to be syntenic (Ruddle, 1972). This made it possible to map genes in species that could not be easily captive bred, such as lions or whales, and provided physical maps of an array of mammalian species. One important feature was that this method enabled mapping of conserved coding genes that could easily be identified and compared across mammal species (Wakefield and Graves, 1996). For instance, early data demonstrated that the homologues of human chromosomes (HSA) 3 and 21 were syntenic in different placental (Threadgill *et al.*, 1991) and non-placental mammals (Maccarone *et al.*, 1992).

More recent still is the advent of efficient *in situ* hybridization of cloned sequences to chromosomal locations. Initially performed with homologous or heterologous radioactive probes to a particular gene, this method now usually employs homologous large insert clones that have been labelled with a fluorescent dye visible under a fluorescence microscope (fluorescence *in situ* hybridization, or FISH) (see Chapter 2 for details). Using combinations of these techniques, comparative mapping made huge advances in the last several years, such that it is now impossible to represent comparative maps on even an enormous spreadsheet, let alone a diagram such as those published previously (Wakefield and Graves, 1996; O'Brien *et al.*, 1999b).

Here we will review progress in gene mapping in particular mammal groups. In all these comparisons, the human map is taken as the standard. With limited financial and manpower resources, one of the most efficient ways to expand genetic information in the non-human species was to generate good comparative maps with regards to human, then to transfer the detailed information to the species of interest (e.g. see Rebeiz and Lewin, 2000; Murphy *et al.*, 2001a; Thomas *et al.*, 2002).

Nomenclature and terms

There are two types of maps that must be distinguished. Genetic ('linkage') maps are based on the frequency of recombination between markers amongst the offspring of parents with distinguishable alleles at two or more loci. In contrast, physical maps may be constructed either by direct observation of the physical location of a DNA probe, or by inference from segregation in somatic cell hybrids and radiation hybrids (RHS).

It is also useful to distinguish between two types of loci. Highly variable sequences (usually microsatellites) are particularly valuable for linkage mapping, but are of little use for comparative mapping. Coding genes and their products are the most valuable markers for comparative mapping, because they are conserved between species, orders or even classes, and homologies can be established (O'Brien and Graves, 1991).

It is critical for comparative gene mapping that the same markers are compared in different species. This makes it imperative to define homology – not an easy task. The usual definition of homology as identity by descent is not altogether helpful for comparative gene mapping, since duplication created many large gene families in vertebrate and even mammalian evolutionary history. Homologous genes in different species (such as alpha-globin in human and mouse) are referred to as orthologues. These must be distinguished from paralogues in the same species that resulted from duplication and divergence (e.g. human alpha- and beta-globin). Of many possible criteria for orthology discussed in detail in the Comparative Gene Organization Workshop (1996), by far the most useful is DNA sequence similarity, although there is no rigid rule for acceptable levels of percentage identity. Conserved map position is also an important criterion of orthology, and is particularly valuable in distinguishing between members of a gene family.

Comparison of gene maps between multiple species also requires some definition of terms, particularly since some have been widely and confusingly abused. Genes or other markers (phenotypic, protein or DNA) that co-segregate at meiosis in a cross are said to be linked. The term synteny (literally on the 'same strand') was coined by Ruddle as an alternative to linkage, to describe the co-segregation of markers on the same chromosome in a somatic cell hybrid. Because much of the early literature in comparative gene mapping describes 'conserved synteny' (Nadeau and Taylor, 1984), the word synteny is commonly adulterated in current literature to describe conservation between chromosomes of different species. Strictly speaking, *conserved synteny* refers to the association of two or more homologous genes on the same chromosome in two different species, regardless of gene order or interspersion with other segments. It is often useful to refer to a *conserved segment* where these markers are contiguous, and to *conserved order* where three or more homologous genes lie on one chromosome in the same order in different species (Comparative Gene Organization Workshop, 1996).

The human genome is a benchmark for all mammalian comparative gene mapping. This is largely for historical and practical reasons – the human gene map has, not surprisingly, received enormous funding and commanded most attention. We will briefly summarize progress in mapping in humans, rodents (Euarchontoglires), then carnivores, Eulipotyphla (shrew, no longer an insectivore) and cetartiodactyls (Laurasiatheria), as well as marsupials and monotremes.

Mapping human and other primates

The first human genes mapped were on the X chromosome, and consequently the first human linkage maps consisted of markers ordered on the X chromosome (review by McKusick and Ruddle, 1977). While a smattering of autosomal linkages emerged prior to the 1970s, nothing resembling a genome map was assembled until Ruddle (1972) and others (Goss and Harris, 1975) applied somatic cell genetics to gene mapping. Working primarily with human specific forms of biochemical markers identified by electrophoresis (Markert, 1968), these early pioneers of human gene mapping quickly constructed maps of specific human autosomes consisting of syntenic groups, i.e. groups of genes located on the same chromosome. These synteny maps generated from somatic cell genetics provided the foundation for comparative gene mapping with the human genome as the standard until the advent of molecular markers in the 1980s.

Synteny maps were complemented by cytogenetic maps, constructed by hybridizing gene-specific probes *in situ* to specific positions on individual chromosomes. This assigned all the syntenic groups to chromosomes and also provided the first data on linear order of genes within syntenic groups, albeit with only a small fraction of the genes.

Botstein *et al.* (1980) generated the first comprehensive linkage map of the human genome, employing restriction fragment length polymorphisms (RFLPs) generated by Southern (1975) blotting of restricted DNA and hybridization to anonymous molecular probes. The most polymorphic of these markers proved extremely valuable for mapping

Mendelian traits but were not generally useful for comparative mapping because they did not generally define conserved elements (genes) of the human genome. While the use of probes developed from coding regions provided this element of evolutionary conservation, the fragments identified were rarely sufficiently polymorphic for linkage mapping. The polymerase chain reaction (PCR) (Saiki *et al.*, 1985) and the discovery of microsatellites (Litt and Luty, 1989; Weber and May, 1989) largely replaced RFLPs as markers for the linkage map but once again identified highly polymorphic but rarely conserved elements in the human genome.

RH mapping resolved the dilemma of mapping gene order versus mapping conserved genes. Cox *et al.* (1990) revived a method conceived by Goss and Harris (1975) to produce whole chromosome maps of genes, a technique subsequently expanded to whole genomes (Walter *et al.*, 1994). RH maps, first in human and now in a variety of other mammals, have become the tools of choice for comparative mapping, short of whole genome sequencing.

Despite early interest in primate gene mapping by somatic cell genetics, the genomes of only a few other primates have been mapped, and the suite of loci mapped in different loci is idiosyncratic. A difficulty has been the lack of good family data for linkage analysis (Rogers and VandeBerg, 1998). There have been systematic studies of multigenerational pedigrees in baboons, which have led to the construction of a 10 cM linkage map with 330 markers. The situation has changed dramatically with the sequencing of the chimpanzee genome (Fujiyama *et al.*, 2003).

Rodent maps

Linkage maps in laboratory rodents first provided a basis for comparative maps almost a century ago. The first genetic linkage reported in a vertebrate species was between albino (c) and pink-eye dilute (p) in mice (Haldane *et al.*, 1915). Similar phenotypic variants, which proved to be mutations in homologous genes, were subsequently mapped in rats (Castle and Wachter, 1924: Feldman, 1924) and deer mice (Clark, 1936). The two loci were found to be linked in all three species with comparable recombination frequencies. This observation marked the birth of comparative gene mapping.

Although comparative mapping lay dormant for several decades, the mouse map roughly doubled in marker number each decade, thanks to the discovery and preservation of large numbers of mutants and intensive study by a generation of mammalian genetics pioneers. The rich history of the mouse gene map, as well as its current status, was summarized by Davisson et al. (1998). Linkage maps were not assigned to chromosomes until the 1970s. Eicher (1971) assigned the first linkage group (LGXII) to chromosome 19 and Roderick et al. (1976) finished the job with chromosome 16. This flurry of activity was precipitated by the advent of chromosome banding technologies and the innovative use of reciprocal translocations. Margaret C. Green annotated, edited, and maintained the mouse gene map throughout most of the 1950s, 1960s and 1970s.

Comparative mapping between mice and humans accelerated when the development of biochemical markers in the 1970s provided a set of common homologous markers for the two mammalian genomes. While Francke et al. (1977) and others developed somatic cell technologies in mice similar to those being used in humans, linkage mapping remained the mainstay of mouse genetics. The problem of gene monomorphism was largely overcome by the expeditious use of subspecies (Chapman, 1978) and interspecific hybrids in backcross linkage analysis (Avner et al., 1988), a technology that has extended into the molecular era (Copeland and Jenkins, 1991). The mouse map was enhanced by the development of recombinant inbred (RI) strains (Bailey, 1971; Taylor, 1989) which permit an accumulation of data in the form of strain distribution patterns of immunological, biochemical, molecular and other markers for which the progenitor strains carried different alleles. RHs were developed for mouse gene mapping by McCarthy et al. (2000) and have proved extremely valuable, along with the interspecific hybrid backcrosses, as a comparative mapping tool. The Mouse Genome Database (MGD), http://www.informatics.jax.org, presently lists more than 82,000 markers, 55,272 of which include genes or parts of genes, and 27,417 polymorphic DNA markers.

Although the Norway rat (Rattus norvegicus) was domesticated primarily for research use more than 150 years ago, its genetic map lagged far behind that of the laboratory mouse. As late as 1990, the rat map consisted of only about 70 loci on 13 linkage groups. As summarized by Levan et al. (1998), the rat map grew quickly with the advent of molecular markers in the 1990s, thanks to both linkage analysis of strain backcrosses (Jacob et al., 1995), and the development of somatic cell hybrid lines (Szpirer et al., 1984). An RH panel (McCarthy et al., 2000) has facilitated comparative mapping and the rat has grown into a fully fledged comparative genomic species, currently used by the NIHGR for whole genome sequencing. RatMap (http://www.Ratmap.gen.gu.se) lists more than 10,000 DNA markers, of which 1585 are genes.

Even the earliest sketchy maps of isozyme loci indicated that the mouse map shares only small regions of conserved synteny with the human map, and this is even more obvious now the maps of both species are so detailed. It was initially thought that this reflected a vast evolutionary distance between the species, reinforcing the opinion that primates and rodents are distantly related. This turns out to be far from the case, as they are both members of the Euarchontoglires (Murphy et al., 2001b). This closer relationship is demonstated by the rat genome, which shows considerably more conserved synteny with human. Further mapping confirms that the mouse has a particularly rearranged karyotype, which is not typical of other rodents. Indeed, most of the variation seems to have been introduced within a relatively short period of mouse evolution.

Carnivore maps

The carnivores are represented by two large phylogenetic branches: the Feliformia and the Caniformia; the first includes the families Felidae, Viverridae, Herpestidae and Hyaenidae, and the second includes the families Phocidae, Canidae, Ursidae, Otariidae,

Procyonidae and Mustelidae (Flynn and Nedbal, 1998; Bininda-Emonds *et al.*, 1999). At present, gene mapping data are available for four species: cat (2*n* = 38) (*Felis catus*; Felidae), dog (2*n* = 78) (*Canis familiaris*, Canidae) and fox (2*n* = 34) (*Vulpes vulpes*, Canidae) and American mink (2*n* = 30) (*Mustela vison*, Mustelidae).

The updated integrated map for cat is most advanced and comprises the linkage and RH, as well as the cytogenetic maps (Lyons *et al.*, 1997; Murphy *et al.*, 2000; Menotti-Raymond *et al.*, 2003). The cat map is also chromosomally well anchored, as a result of early efforts to map biochemical markers by somatic cell genetics, and covers about 60% of chromosome length. A total of 864 markers, including 585 gene loci and 279 microsatellites, are mapped in the cat genome. The cat linkage map contains 81 gene loci dispersed on all chromosomes except B1, and 248 microsatellites. The sum of the lengths of its linkage groups is 2646 with an average inter-marker spacing of 8 cM. The cat RH map includes 579 gene loci and 254 microsatellites and covers all feline chromosomes (Menotti-Raymond *et al.*, 2003). The cat map shows startling homology to the human map. Of the 23 human chromosomes, 16 are represented by a single cat chromosome, and the other seven by two cat chromosomes, representing a total of 30 segments homologous between the two species (O'Brien *et al.*, 1997).

Gene order is also highly conserved between cat and human. There are 110 conserved ordered regions common to cat and human genomes, one of the highest levels of conservatism with human shown by any non-primate mammalian genome (Menotti-Raymond *et al.*, 2003). However, gene mapping in cat reveals several small rearrangements: for instance, three small inserts carrying human markers of HSA1 lie in three different regions of feline chromosome A1 and a marker of HSA13q is located within the feline chromosome E3 between segments homologous to HSA7p and HSA16p (Menotti-Raymond *et al.*, 2003).

In the past few years, there have been significant advances in the development of a genome map of the dog (Neff *et al.*, 1999; Yang *et al.*, 2000a; Breen *et al.*, 2001a,b; Guyon *et al.*, 2003). The dog integrated map now includes detailed cytogenetic, RH and meiotic information. A 1500-marker RH map contains 1078 microsatellites, 320 dog gene loci, and 102 chromosome-specific markers (Breen *et al.*, 2001). Linkage analysis was performed on a panel of reference families, allowing one linkage group to be anchored to all 38 dog autosomes. Thus, each canine chromosome is identified by one linkage group and one or more RH groups. This updated integrated map, containing more than 1800 markers, covers approximately 90% of the dog genome (Breen *et al.*, 2001). More recently Spriggs *et al.* (2003) significantly infilled the RH and cytogenetic maps of the dog X chromosome. Moreover, regional localizations on 38 autosomes were determined for 266 chromosome-specific cosmid clones by FISH and, as a result, now chromosome localizations for 320 dog genes can be used for comparative mapping between and dog and human, as well as other mammalian species. The most recent achievement in dog mapping is the generation of a 1 Mb resolution RH map with 3270 markers, out of which 900 are genes and expressed sequence tags (ESTs) (Guyon *et al.*, 2003), adding power to comparative mapping between and dog and human, as well as other mammalian species.

Other carnivores for which there is significant map information are mink and fox, in which 127 and 35 gene loci, respectively, mark all autosomes and the X chromosome (Rubtsov, 1999; Serov, 1998; Kuznetsov *et al.*, 2003).

Eulipotyphla (Laurasiatheria)

Shrews have now been recognized to belong to the group Eulipotyphla, along with moles and hedgehogs, but distant from elephant shrews and golden mole which once shared the designation as Insectivora. Shrew species show many variations of karyotype. At present, gene mapping data are available for only one species, the common shrew (*Sorex araneus*) (2*n* = 20) (Serov *et al.*, 1998; Zhdanova *et al.*, 2003). The current shrew map contains 49 gene loci and four microsatellites (Zhdanova *et al.*, 2003). Eight of ten shrew chromosomes contain markers, but so far, none map to the largest *bc* and smallest *tu*.

Cetartiodactyl maps (Lauraiastheria)

Cetartiodactyls represent a recent taxonomic fusion of cetaceans, hippopotamuses, pigs and ruminants into a monophyletic group (Nikaido et al., 1999). Mapping in cetartiodactyls has included cattle, sheep, goats, pigs and deer, enterprises driven principally by the economic value of these species.

Although isolated linkages were reported early in cattle (Larsen, 1966; Hines et al., 1969; Gustavsson and Rendel, 1971), no genome level maps were produced prior to the development of hybrid somatic cells segregating cattle chromosomes (Heuertz and Hors-Cayla, 1981; Dain et al., 1984; Womack and Moll, 1986). Approximately 100 biochemical markers were placed on the somatic cell map by 1990, almost all with known human homologues (Womack, 1990). The molecular technologies of the 1980s produced an explosion in the number of markers on the map and also provided the basis for in situ hybridization which ordered some markers and anchored the rapidly expanding syntenic groups to chromosomes (Fries et al., 1986; Threadgill and Womack, 1988).

The development of microsatellite markers in cattle (Fries et al., 1990) led to the development of linkage maps (Barendse et al., 1994; Bishop et al., 1994) which were quickly expanded to several thousand markers (Barendse et al., 1997; Kappes et al., 1997). While these maps have been extremely valuable for mapping economically important traits, they consist primarily of non-conserved microsatellites and have done little for comparative mapping.

While somatic cell genetics provided 'cow on human' conserved syntenic segments, and Zoo-FISH with human paints identified conserved human segments on cow, neither answered questions regarding conservation of gene order between species. These answers came with the development of RH maps in cattle (Womack et al., 1997; Rexroad et al., 2000). Parallel RH maps of bovine and human genes proved highly effective in identifying intrachromosomal rearrangements within conserved syntenic groups, first with individual chromosomes (Yang and Womack, 1998) and then with whole genome approaches (Band et al., 2000). The latter map consists of more than 1000 markers, 638 of which have human orthologues. The 319 microsatellites on the map anchor the RH map to linkage maps and consequently to mapped traits. This map identified 105 conserved segments and accounts for 41 translocation events and 54 internal rearrangements relative to the human genome.

Higher resolution comparative mapping has been achieved by RH mapping bacterial artificial chromosome (BAC) end sequences selected for BLAST hits at 1 Mbp intervals on the human genome. The power of this approach was recently demonstrated by Larkin et al. (2003) with a cattle comparative map of human chromosome 11. The bovine genome is currently being sequenced by the NIHGR.

Sheep and cattle chromosomes are highly conserved within the limits of modern cytogenetic technologies (ISCNDB, 2000). Maps of the two genomes are correspondingly similar, within the limits of current mapping methods. As of 1994, only about 100 loci were mapped in sheep (Broad and Hill, 1994), primarily by somatic cell genetics (Burkin et al., 1993) and in situ hybridization (reviewed by Echard et al., 1994). Shortly thereafter, a linkage map of microsatellite markers was published (Crawford et al., 1995) providing a high level of coverage for trait mapping. It was possible to glean some comparative information relative to cattle, since many of the microsatellites were derived from cattle primers. This linkage map has subsequently been expanded to more than 1000 markers (Maddox et al., 2001). Noelle Cockett and co-workers have recently constructed a RH panel for sheep which will take comparative mapping to the next level of resolution. A genetic map of goat has also been established (Vaiman et al., 1996).

The river buffalo (Bubalus bubalis), $2n = 50$ is another bovid with chromosome arms conserved relative to cattle. Iannuzzi (2003) and others have mapped a variety of genes to buffalo chromosomes by in situ hybridization, generally confirming genomic conservation with cattle and sheep. A somatic cell panel has resulted in a synteny map (De Hondt et al., 1997; El Nahas et al., 2001) with more than 100 markers. Plans are in order for a RH buffalo map.

A linkage map of deer has contributed to an understanding of the evolution of the ruminant karyotype (Slate et al., 2002).

The development of the pig map has in many ways paralleled the development of the bovine map. Isolated linkages, primarily of blood group and protein polymorphisms, were reported in the 1960s (Andresen and Baker, 1964; Rasmusen, 1965). However, when reviewed by Fries *et al.* (1990), the map had only 18 loci assigned to chromosomes, including one linkage group of six markers. The pig was the first farm animal in which *in situ* hybridization was effectively used to map genes (Geffrotin *et al.*, 1984). When non-isotopic methods emerged, the number of chromosomal assignments in pigs escalated rapidly (reviewed by Chowdhary, 1998). Somatic cell genetics in pigs emerged in the 1990s (Rettenberger *et al.*, 1994; Zijlstra *et al.*, 1994b; Yerle *et al.*, 1996) quickly placing more than 200 markers on synteny maps. Multiple linkage maps have been developed in pigs consisting mostly of microsatellites (Archibald *et al.*, 1994; Marklund *et al.*, 1996; Rohrer *et al.*, 1996). The pig database in ArkDB (http://iowa.thearkdb.org) lists 6866 map assignments, 5039 by linkage and 1827 cytogenetic. Comparative mapping has benefited greatly from the development of RH panels replete with genes and ESTs (Yerle *et al.*, 1998; Hawken *et al.*, 1999; Rink *et al.*, 2002; Cirera *et al.*, 2003).

These detailed maps have demonstrated that, despite overall synteny shared between human and pig autosomes, there are numerous changes in gene order. For instance, gene order differences were detected between pig chromosome 13 and human 3, and between pig 5 and human 12, despite their shared synteny (Goreau *et al.*, 2001; Van Poucke *et al.*, 2003).

Eutherian sex chromosomes

Comparative mapping confirms that the X chromosome is exceptionally conserved between orders. The gene content of the X revealed by family studies and somatic cell genetic mapping in the 1970s and 1980s was completely invariant. Even the detailed comparisons now possible between sequenced genomes reveal that the gene content is largely identical. The sole exceptions are the two ends of the X chromosome, which have evidently undergone some terminal changes in or near the pseudoautosomal region (PAR) that is shared with the Y (Blaschke and Rappold, 1997; Graves *et al.*, 1998a; Charchar *et al.*, 2003).

Gene order on the X is also highly conserved even between humans and some voles, although differences in the relative distances suggested some rearrangements (Nesterova *et al.*, 1998). However, there have been many rearrangements of ten homology segments between the human and mouse X (DeBry and Seldin, 1996).

The Y chromosome, on the other hand, is poorly conserved between eutherian species. The Y is much smaller than the X, and has large stretches of heterochromatin. It contains few genes other than the sex-determining *SRY* – a mere 44 in humans – many of which have male-specific functions (Lahn and Page, 1997; Skaletsky *et al.*, 2003). Several genes on the Y chromosome in mice (e.g. *Ube1y* and *Ef1a*) do not lie on the human Y, and some genes (e.g. *RPS4Y*) on the human Y are missing from the rodent Y.

Marsupial and monotreme maps

Comparisons between the genomes of the three major groups of mammals relies completely on comparative gene mapping, because homology is too poor over this evolutionary distance for chromosome painting to work between marsupial or monotreme and eutherians.

Genetic mapping has been informative for two marsupial species that breed easily in captivity, and complete linkage maps have now been produced for the Australian tammar wallaby *Macropus eugenii*, and the Brazilian short-tailed grey opossum *Monodelphis domestica* (Zenger *et al.*, 2002; Samollow *et al.*, 2004). Some somatic cell genetic mapping was performed in the 1970s and 1980s using hybrids between rodent and marsupial cells. However, the hybrids were difficult to obtain and chromosomally unstable, and yielded few autosomal assignments. Increasingly, the tammar wallaby homologues of human genes have been cloned, and lambda or BAC clones cytologically mapped using FISH.

The small amount of gene mapping done in marsupials demonstrates at least some shared synteny with the human genome

(Samollow and Graves, 1998). For instance, seven genes spanning human chromosome 17 all map to linkage group 3 in *Mus musculus domestica*, and seven human chromosome 3p genes lie on chromosome 2q in the tammar wallaby.

Linkage studies in monotremes are all but impossible, since captive breeding is still an art practised by only two zoos in the world. Despite considerable effort, somatic cell genetic analysis has been limited by the instability of rodent–monotreme cell hybrids (Wrigley and Graves, 1988). The most successful means of mapping has been *in situ* hybridization. Using heterologous probes (usually human cDNA) to very conserved genes, physical locations on platypus and echidna chromosomes of cultured cells were obtained for many genes using radioactive *in situ* hybridization (Wrigley and Graves, 1984). More recently, several genes have been cloned from the platypus, and FISH using long lambda or BAC clones has been used to localize 22 platypus genes to autosomes. Such comparisons have shown that genes that are syntenic on human chromosome 21 lie in two clusters in monotremes as well as marsupials (Maccarone *et al.*, 1992).

Marsupials and monotremes have genome sizes in the range of eutherians, but their karyotypes are very distinctive, and it was initially expected that their genome arrangements would be scrambled beyond recognition. Even the tiny amount of data available at present shows that this is not the case.

Mapping marsupial and monotreme sex chromosomes has been particularly revealing. In marsupials, the basic X and Y are smaller than their eutherian counterparts, and there is no PAR (Toder *et al.*, 2000). In monotremes the X is large and obvious, but the Y remains a complete mystery. This is because the X is part of the translocation chain that contains several elements that appear to be male specific – which if any is a male-determining Y is not yet clear.

Comparative mapping by somatic cell genetics and FISH shows that the X is at least partly homologous in all three major mammal groups, demonstrating that mammalian sex chromosomes had a single origin in an ancestral mammal.

Genes on the long arm and pericentric region of the human X have homologues on the X in all marsupials, conforming to and extending Ohno's law of the conservation of the mammalian X. However, markers on the short arm of the human X (including pseudoautosomal genes) are autosomal in marsupials, most lying on the short arm of chromosome 5 in the tammar wallaby (Spencer *et al.*, 1990a,b; Graves, 1995; Wilcox *et al.*, 1996).

Comparative mapping of human Y genes shows that the marsupial Y chromosome, too, shares partial homology with the eutherian Y. Marsupial homologues of four human Y genes map to the marsupial Y. However, marsupial homologues of most human Y genes map to tammar chromosome 5p (Waters *et al.*, 2001).

The paucity of mapping data from these particularly interesting mammals is about to change dramatically with the commitment by NHGRI to sequence the genomes of the platypus *Ornithorhyncus anatinus*, and moves to sequence the tammar wallaby (*Macropus eugenii*) in Australia with support from NHGRI.

Conclusions from comparative mapping

Early comparisons at the cytogenetic level had painted a nihilistic picture of genome scrambling between mammal groups. Despite the complete lack of understanding of the basis for longitudinal differentiation of chromosomes, a common inference drawn from most of these studies was that banding similarities reflect similarities in gene content. Although attempts were made to show partial analogy between the banding and gene mapping data of evolutionarily diverged species such as human–mouse (Sawyer and Hozier, 1986), banding pattern has been found to be a poor predictor of gene content.

However, comparative mapping over the last three decades has revealed that the mammalian genome is much more conserved than was obvious from cytogenetic comparisons. Even the first comparative gene mapping, particularly between cat, bovine and human, showed a level of conservation that could not be appreciated using G-band comparisons.

The two decades of intensive mapping of hundreds of loci across more than 30 species since then has consolidated a picture of quite extraordinary genome conservation.

A giant jigsaw of the hundreds of genes mapped in more than 30 mammal species was constructed from the data available in 1998 and was presented as a poster (Wakefield and Graves, 1996). Very large regions of conserved synteny with the human genome were apparent when genes on the same chromosome were joined by vertical lines. For instance, the entirety of human chromosome 12 is represented by a single chromosome in the cat B4 and the bovine BTA5. It is evident from this comparative map that large regions have been conserved between human and primates, carnivores, artiodactyls, rodents and Eulipotyptila. The enormous explosion of mapping data in the last 5 years means that it is no longer practicable to present on a diagram, and the extent of conservation of synteny and gene order can now be best examined in the relational databases.

There are some exceptions to the picture of conservation of synteny that are not at all consistent with the evolutionary distance between species. For instance, within the primates, which are generally very conserved with respect to human, the gibbon stands out as having multiple breaks in synteny. It is evident that the dog map is far more fragmented with respect to human than is the highly conserved cat. Again, this suggests lineage-specific bursts of change.

Conserved synteny can give us an estimate of the number of breakpoints needed to transform one map into another. This is a minimum estimate, since internal rearrangements, such as those described in the pig, will only be detected when gene order is known. An even more dramatic representation of such conserved synteny can be obtained directly by chromosome painting, which can also detect interruptions of synteny within a chromosome.

Comparative Chromosome Painting

A variant of FISH is chromosome painting. This technique uses fluorescent probes derived from physically isolated whole chromosomes or chromosome regions to hybridize to the chromosomes of closely and distantly related mammal species. Whole chromosomes can be isolated by flow sorting or microdissection (Ferguson-Smith *et al.*, 1997), and chromosome regions by the latter technique. DNA is prepared from the isolated chromosome/region, and amplified by degenerate oligonucleotide-primed (DOP) PCR. The probe DNA is a cocktail of numerous sites from the originating chromosome or region, representing a large proportion of unique and repetitive sequences. Consequently, the signal observed is an aggregation of many hybridization sites, uniformly 'painting' the chromosome.

These chromosome paints may be applied to recognizing their chromosome/region of origin in the same species ('chromosome painting'), or homologous chromosomes or regions in closely or distantly related species ('comparative chromosome painting', 'cross-species painting' or 'Zoo-FISH'). Here we will use 'comparative painting'.

The probes used for chromosome painting ('paints') are usually prepared from flow-sorted chromosomes (e.g. Scherthan *et al.*, 1994; Ferguson-Smith, 1997), but may also be generated from microdissected chromosomes (Saitoh and Ikeda, 1997; Raudsepp and Chowdhary, 1999). For a long time, only human chromosome-specific libraries generated from flow-sorted chromosomes (Van Dilla and Deaven, 1990) were available as probes for comparative chromosomes. However, during the past 7–8 years, significant breakthroughs in the ability to resolve individual chromosomes of a large variety of species with complex karyotypes has enabled construction of chromosome paints for a wide range of mammals (Ferguson-Smith *et al.*, 1997; Gribble *et al.*, 2004).

In either case, the probe DNA is labelled (with biotin, digoxigenein or fluorescence moieties) and hybridized under suppression conditions (e.g. using a 1000-fold excess of repeat sequence-enriched DNA ($C_0t^{-}1$ DNA) to block hybridization to repetitive sequences), as described in detail by Chowdhary *et al.* (1996).

The first successful chromosome painting was on human chromosomes, using human whole chromosome paints, and this technique was subsequently extended to other primates. However, the breakthrough work showing chromosomal homology between species that diverged 60–80 mya was first reported by Scherthan *et al.* (1994). Now more than 40 evolutionarily distantly related mammalian species belonging to 11 orders have been analysed for whole karyotype comparisons using human whole chromosome paints. Additionally, karyotypes from a total of 65 closely related species have been compared using paints generated from ~15 species belonging to five eutherian orders.

In most cases, the paints have been used unidirectionally, i.e. paints from one species were hybridized to metaphase chromosomes of another species. However, in some cases bidirectional painting has been carried out. This involves use of paints from species 'A' on metaphase chromosomes of species 'B' and vice versa. Up to now, bidirectional painting has been done across 13 evolutionarily closely related species and between humans and 12 distantly related species. Bidirectional painting has emerged as a powerful tool to accurately compare genomes and even predict the likely location of orthologous genes between two species.

Here we will primarily discuss results from mammalian species that have been probed exclusively with human paints, as well as the reciprocal, where possible. In all these comparisons, the human genome has been taken as the standard for historical and practical reasons. Human paints were for a long time the only paints covering individual chromosomes. Practically, the rapid expansion of the human gene map provides a comparative map reference for other mammalian (or even non-mammalian) genomes. The gene content of a region in another species may be predicted by reference to the homologous human segment.

Here we will summarize progress in comparative painting within eutherian, marsupial and monotreme orders, then show how these data can be used with comparative gene mapping, to deduce the form of the genomes of ancestors of mammalian orders, infraclasses and ultimately the ancestor of all mammals.

Painting in eutherian mammals

Primates

The first comparative chromosome painting experiments were reported for primates (e.g. Wienberg *et al.*, 1990; Jauch *et al.*, 1992). Since then, data have expanded enormously. Comparative chromosome painting information is available for all human chromosomes in relation to the chromosomes of several primate species, depicting a high degree of molecular homology among the primate genomes (Wienberg *et al.*, 1994; Koehler *et al.*, 1995a,b; Arnold *et al.*, 1996; Consigliere *et al.*, 1996; Richard *et al.*, 1996; Sherlock *et al.*, 1996; Morescalchi *et al.*, 1997; Müller *et al.*, 1997; Wienberg and Stanyon, 1997). The use of whole chromosome paints cannot, of course, distinguish intrachromosomal arrangements (which are twice as frequent as interchromosomal rearrangements; Richard *et al.*, 2003), but paints derived from parts of chromosomes or specific loci may be used to detect changes in gene order. Several differences between human and gorilla chromosomes have been detailed by multicolour banding (Mrasek *et al.*, 2001).

Comparisons between human, chimpanzee and gorilla chromosomes revealed only a few rearrangements, whereas comparisons with chromosomes of Old World and New World monkeys showed more divergence, and with lemurs more divergence still. An account of the fusions and fissions that occurred during the evolution of the human karyotype could be put together from these comparisons. For instance, human chromosome 2 must be the result of a very recent fusion at 2q13 (Ijdo *et al.*, 1991) of two acrocentric chromosomes (chimpanzee chromosomes 12 and 13), because these are present as two separate blocks in all other primates and some non-primates (O'Brien *et al.*, 1999). Likewise, human chromosome 16, which is conserved between Old World monkeys, great apes and humans, was produced by Robertsonian fusion of two chromosomes in a Catarrhini ancestor (Misceo *et al.*, 2003). Fission of a single ancestral chromosome to produce human chromosomes 14 and 15 occurred further after the divergence of gibbons, and human chromosomes 3 and 21 were created by fission from a large ancestral chromosome that is present in lemurs, most other eutherian clades, and even in marsupials.

Chromosome painting has made it possible to chart the multiple chromosome rearrangement in species with highly derived karyotypes, such as gibbons (Muller *et al.*, 2003; Mrasek *et al.*, 2003), galagos (Stanyon *et al.*, 2002), proboscis monkey (Bigoni *et al.*, 2003) and a monkey with only 16 chromosomes (Stanyon *et al.*, 2003). Indeed the multiple rearrangements in the gibbon karyotype have made it possible to use sorted gibbon chromosomes to produce highly informative multicolour painting for human chromosomes. Non-Robertsonian fissions are responsible for the higher number of chromosomes in the African green monkey (Finelli *et al.*, 1999).

Comparisons between chromosomes have provided a wealth of information on primate phylogeny, and have enabled ancestral primate karyotypes to be deduced. For instance, comparative painting of howler monkeys (de Oliveira *et al.*, 2002) sorted out the genus *Alouatta*. Comparisons of five New World species identified rare genomic changes that could be used to clarify New World monkey phylogeny (Neusser *et al.*, 2001). 'Bar-coding' primate chromosomes using multicolour FISH across species enabled the ancestral hominid karyotype to be established (Muller and Wienberg, 2001).

Limited information is available for the Scandentia, which are now considered to be closely related to primates. Reciprocal painting between northern tree shrew and human chromosomes identified 40 conserved chromosome segments between the two species, similar to several other mammals. Very recently, Richard *et al.* (2003) analysed Chinese tree shrew chromosomes using paints from ten human autosomes. Half of the human chromosomes were conserved as a single block/chromosome in the Chinese tree shrew. Several common derived chromosomes link tree shrew and primate chromosomes.

Reciprocal painting between human, prosimian and tree shrew was used to define an ancestral 2n = 50 karyotype of all primates (Muller *et al.*, 1999; Haig, 1999).

Rodentia

Mouse chromosome-specific probes were generated with some difficulty by microdissecting the morphologically similar mouse chromosomes (Liechty *et al.*, 1995). These have been applied to many projects comparing the chromosomes of different mouse species. For instance, many tandem fusions have been identified in the Indian spiny mouse (Matsubara *et al.*, 2003), and complex rearrangements of a few chromosomes in the mouse lineage detected by comparisons with the African four-striped mouse (Rambau and Robinson, 2003).

Reciprocal painting between mouse and rat established homologies within 36 segments that are completely consistent with comparative gene mapping. Thirteen mouse paints hybridize to a single rat chromosome, but other mouse paints produce split signals in both rat and human (Grützner *et al.*, 1999; Guilly *et al.*, 1999). At least 14 translocations occurred in the 10–20 million years since the species diverged, implying that the rate of genomic rearrangement between rat and mouse was ten times faster than between human and cat (Stanyon *et al.*, 1999). Rat paints show three times less disruption compared with outgroups such as human. The conclusion is that rat is more similar to human than is mouse. Reciprocal painting between mouse and Chinese hamster (a member of the more distantly related Caviomorphs) produced 47 homologous segments from 19 mouse autosomal paints, confirming the recent rapid evolution of mouse chromosomes.

Recently, the chromosomes of three sciurid species (Indochinese ground squirrel, Asian squirrel and eastern gray squirrel) have been analysed using human paints. Twelve human paints hybridize completely to squirrel chromosomes (Stanyon *et al.*, 2003). The human autosomes corresponded to 33 conserved segments in Indochinese ground squirrel (Richard *et al.*, 2003a), 35 in Asian squirrel (Richard *et al.*, 2003b) and 39 in eastern gray squirrels (Stanyon *et al.*, 2003). Thus the karyotypes of squirrels are much closer to humans than are the chromosomes of rat and mice; and even of most of the non-rodent species analysed thus far. This confirms the conclusion that the ancestral rodent karyotype was not very different from the ancestral primate, but that mouse and rat karyotypes were recently greatly rearranged.

In early studies, painting of three human chromosomes (16, 17 and X) on to mouse chromosomes was reported (Scherthan *et al.*, 1994). Curiously, there is still no complete reciprocal painting between mouse and human, although results for some chromosome paints have been published (Ferguson-Smith, 1997, review). The painting results showed extensive rearrangement between the human and mouse karyotypes, in broad agreement with the available gene mapping data. The degree of precision visualized from the painting data is, however, not as accurate as is available today through human–mouse comparative gene mapping and sequence data (www.informatics.jax.org, www.thearkdb.org). Comparison of human and mouse genomes shows numerous small inserts marked by a few gene loci in large conserved syntenic associations (O'Brien *et al.*, 1999).

Lagomorpha

Lagomorphs (rabbits and hares) are the sister group of rodents. Rabbit whole chromosome paints were used to investigate the chromosomes of other rabbit and hare species, again revealing rapid chromosome evolution in the group (Robinson *et al.*, 2002). Painting rabbit ($2n = 44$) with human probes (Korstanje *et al.*, 1999) established 38 conserved syntenic segments, confirming extensive genome conservation with human. Whole chromosome paints obtained from male laboratory rabbit (*Oryctolagus cuniculus*; $2n = 44$) were used on human metaphase chromosomes, resulting in the detection of 40 homologous segments (Korstanje *et al.*, 1999). These findings, combined with comparisons of banding patterns, affirmed known homologies with humans for at least seven of the rabbit chromosomes. For other chromosomes the study provided corrections and even enabled detection of new segmental homology.

Chiroptera

Bats represent the second largest mammalian order, having almost 1000 extant species.

Application of human chromosome paints to the common long tongued bat (*Glossophaga soricina*; $2n = 32$) identified 41 conserved segments between the two species. Recently, chromosomes from several genera belonging to five bat families were analysed with human paints (Volleth *et al.*, 2002). This helped identify 25 evolutionarily conserved regions with human. It also clarified the phylogenetic relationship between Megachiroptera and Microchiroptera, which had been proposed to have different origins. In line with sequence comparisons, comparisons of painting patterns identified six synapomorphic features linking the two groups, strongly supporting monophyly of the bats.

Eulipotyphla

Of this new assemblage, incorporating shrews, hedgehogs and moles, only shrew species have been studied by comparative painting. Shrews are unusual in having multiple different Robertsonian rearrangements, so that their arms are given letter names.

Hybridization of human chromosome paints to the chromosomes of the common shrew (*Sorex araneus*) showed that 22 human autosomes are conserved in 32 segments on the shrew karyotype (Dixkens *et al.*, 1998). Fourteen large conserved regions were homologous to entire human chromosomes including the X. Eight human chromosomes were present as two segments in the shrew genome, and one was broken up into three segments (Dixkens *et al.*, 1998; Zhdanova *et al.*, 2003). Only ten breakpoints separate the shrew and the human karyotypes (Dixkens *et al.*, 1998), a result comparable with the conservation between human and cat (see O'Brien *et al.*, 1997).

These findings are reasonably consistent with the assignment of 40 of 49 gene loci (Matyakhina *et al.*, 1997). However, rearrangements of small segments and intrachromosomal inversions undetectable by Zoo-FISH analysis are revealed by comparative mapping. For instance, whereas human chromosome 17 painted only one large segment on the arm *h* of shrew chromosome *hn* (Dixkens *et al.*, 1998) and nine HSA 17p12–17qter loci map to this region, the MYH2 gene located on human pter17p12 is located on shrew chromosome *ik* (Zhdanova *et al.*, 2003). Interestingly, fission within this syntenic group was identified near, but not at, the same site in dog and fox karyotypes (Yang *et al.*, 2000a). In addition, the order of eight shrew genes within chromosome *hn* differs from that on human chromosome 17, probably reflecting small chromosomal rearrangements in Eulipotyphla.

Comparative gene mapping and chromosome-specific painting analysis of human and common shrew genomes revealed striking similarity in the organization of human chromosome 2 and the arm *d* of shrew chromosome *de*. However, it is known that HSA2 resulted from a very recent fusion of two chromosomes that are separated in other primates. The shrew arm *d* is homologous to HSA2, except for a small inversion at the putative fusion point in the human chromosome (Dixkens *et al.*, 1998; Zhdanova *et al.*, 2003), which is also present in other primate species and is a common polymorphism in human populations (Wienberg *et al.*, 1994). These two independent fusion events may therefore have occurred at sites prone to rearrangement.

Carnivora

Starting with the cat (Rettenberger *et al.*, 1995), nine carnivore species (cat, lion and palm civet in the Feliformia, and dog, red fox, domestic ferret, American mink, harbour seal and giant panda in the Caniformia) have been analysed using human paints. For others like Arctic fox and red fox, indirect deductions about homology with the human karyotype are based on the human–dog comparisons. Chromosome-specific paints for the cat, giant panda, red fox, dog and American mink have been developed and applied to the comparative analysis of genomic evolution of carnivores (Wienberg *et al.*, 1997; Nash *et al.*, 1998; Breen *et al.*, 1999; Yang *et al.*, 2000a; Graphodatsky *et al.*, 2002; Tian *et al.*, 2004).

Comparative gene mapping and reciprocal painting show that among carnivores there are two contrasting types of genomes. Felidae and Mustelidae show high levels of syntenic conservation, whereas Canidae and some of Ursidae have extensively rearranged

karyotypes (Wienberg *et al.*, 1997; Nash *et al.*, 1998; Breen *et al.*, 1999; O'Brien *et al.*, 1999a,b; Yang *et al.*, 2000a; Graphodatsky *et al.*, 2002; Menotti-Raymond *et al.*, 2003; Tian *et al.*, 2004).

Hybridizing cat chromosomes with human paints led to the detection of 30 conserved chromosomal segments between the two species. This astonishing conservation suggested that the human and cat karyotypes might represent the putative ancient mammalian founder karyotype. Reciprocal chromosome painting studies carried out later using paints prepared from flow-sorted cat chromosomes (Wienberg *et al.*, 1997) again identified 30 conserved segments. Fourteen of the 23 human chromosomes each hybridized with a single cat chromosome, and nine of the 19 cat chromosomes each corresponded to a single human chromosome, although gene mapping reveals several small rearrangements that are beyond the resolution of chromosome painting. Recently, some human paints were used to show that chromosomes in the lion and the palm civet are similar to those of the cat (Richard *et al.*, 2000). It was deduced that since divergence from a common ancestor ~80 mya, cats and humans have accumulated only 10–12 translocations.

The karyotypes of extant canids are the most extensively rearranged in mammals, except for the mouse (Yang *et al.*, 2000a; Tian *et al.*, 2004). Breen *et al.* (1999) used reciprocal painting between dog and human to identify 68 shared chromosomal segments whose homology was strongly supported by gene mapping data. Yang *et al.* (2000a) identified 90 homologous segments by painting dog probes on to human chromosomes, and 73 by painting with human probes on to dog chromosomes. The high number of conserved segments found in the dog is consistent with the large number of chromosomes ($2n = 78$), suggesting that the dog karyotype underwent rather complex chromosome rearrangements. Despite this, the dog karyotype maintains complete synteny for three human chromosomes (HSA14, 20 and 21). These findings received additional support from recent fine-scale painting of dog to human, as well as RH mapping (Sargan *et al.*, 2000). Painting between cat and dog revealed that seven fissions, 25 fusions and one inversion separate the cat and dog karyotypes (Yang et al., 2000b).

Although the red fox has a lower chromosome number than dogs, there are nearly as many (73) segments conserved with human (Yang et al., 2000a). Comparing the patterns of hybridization with human paints on dog and fox demonstrates considerable homology between dog and red fox karyotypes, and this is directly confirmed by recent gene mapping and comparative painting between red fox and human (Yang et al., 2000a). However, 26 fusions and four fissions are needed to reconstruct Vulpes vulpes fox karyotype from the dog karyotype (Yang et al., 2000a). These extensively rearranged karyotypes make it difficult to deduce an ancestral karyotype for Canidae (Fig. 14.2).

Hybridizing harbour seal chromosomes with human paints revealed 31 conserved syntenic segments (Frönicke et al., 1997) and a much higher degree of karyotype conservation with the cat. It was suggested that pinnipeds, felids and humans have maintained conserved karyotype complements despite several million years of independent evolution.

The largest carnivore family, the Mustelidae, includes the Mephitinae (skunks), Melinae (badgers), Lutrinae (otters), and Mustelinae (weasel, polecats, mink and martens) (Bininda-Emonds et al., 1999). Hybridizing American mink chromosomes with human paints showed that the karyotypes of the two species shared 34 chromosomal segments in an arrangement similar to that found in cat (Hameister et al., 1997). Hybridizing cat chromosomes with

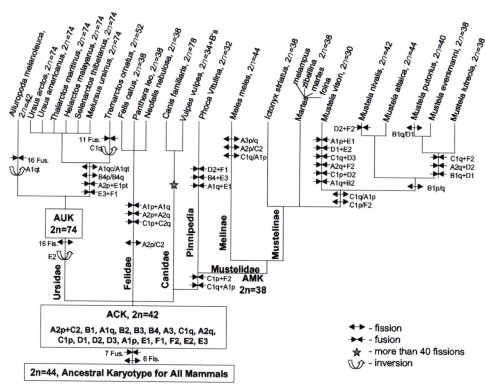

Fig. 14.2. The putative chromosomal evolution in Carnivorae order based on comparative Zoo-painting, G-banding and gene mapping data. The compilation scheme is established from those previously published for Felidae, Canidae, Mustelidae (Murphy et al., 2001a,b,c; Graphodatsky et al., 2002) and Ursidae (Nash et al., 1998; Tian et al., 2004) and adopted to domestic cat chromosome nomenclature. ACK, an ancestral carnivore karyotype, AMK, an ancestral mustelid karyotype, AUK (ancestral Ursidae karyotype). The first step is a transformation of an ancestral eutherian karyotype (Yang et al., 2003) to the ACK by seven fusions and six fissions.

mink paints delimits 21 homologous segments, implying only three rearrangements between the species. Chromosomal conservation between felid and mustelid genomes was confirmed by the demonstration that 49 out of 54 gene loci mapped to chromosome regions predicted from painting homology. (Kuznetsov *et al.*, 2003). The few exceptions, such as the presence of the GPT gene (on HSA8) in mink chromosome 14 (Khlebodarova *et al.*, 1995), reveal small rearrangements involving few genes.

The evolution of the small mink chromosome 14 (that initially was considered to be specific to mink or at least mustelids; Khlebodarova *et al.*, 1995) has been studied in detail. G-banding and chromosome painting identify this chromosome in more than 20 mustelid species in six genera, although some carried pericentric inversions (Hameister *et al.*, 1997; Graphodatsky *et al.*, 2002). The syntenic association on mink chromosome 14 forms part of the large metacentric chromosome in fox or the canine acrocentric (Rubtsov, 1998). Painting with human probes gave a complex pattern.

Human paints hybridized to the chromosomes of domestic ferret (*Mustela putorius furo*; $2n = 40$) identified 32 homologous autosomal regions (Cavagna *et al.*, 2000).

The chromosomes of three groups of the Ursidae (bears) are more highly rearranged than cats, but less so than dogs (Nash *et al.*, 1998; Tian *et al.*, 2004). Comparative painting of ursid genomes with human and feline paints revealed conserved whole chromosomes and segments, but some small terminal rearrangements. The $2n = 74$ karyotype common to six species of bear, including the Malayan sun bear (*Helartos malayanus*), is believed to closely resemble the ancestral ursid karyotype. This could have evolved from the $2n = 42$ putative ancestral carnivore karyotype by an inversion and 16 centric fissions (Fig. 14.2). Independent fusions of the acrocentric ancestral chromosomes have generated the unique karyotypes of the spectacled bear and the giant panda ($2n = 42$), which has 44 segments conserved with individual human chromosomes (Fig. 14.3). These relationships were indirectly verified by hybridizing cat chromosomes independently with human and panda whole chromosome paints (Rettenberger *et al.*, 1995; Weinberg *et al.*, 1997). Red panda chromosomes were painted with dog and cat paints to show that the karyotype is very similar to that of cat (Tian *et al.*, 2002).

Reconstruction of an ancestral carnivore karyotype has been considered in detail (Fig. 14.3). It is straightforward to deduce an ancestral mustelid karyotype of $2n = 38$, which is represented in three extant martens and distantly related mustelids such as striped and marbled polecats, otter, and sables (Graphodatsky *et al.*, 1989, 2002). The ancestral *Mustela* karyotype diverged from this ancestral mustelid karyotype by two centric fissions (Fig. 14.3). Eight whole chromosomes and 12 arms of the American mink correspond to the 20 autosomes of the carnivore ancestral karyotype (Graphodatsky *et al.*, 2002).

There is increasing evidence that the domestic cat karyotype closely resembles the ancestral carnivore karyotype (Murphy *et al.*, 2001; Graphodatsky *et al.*, 2002). An ancestral karyotype with $2n = 42$ can be derived from the domestic cat karyotype via three centric fissions and one fusion (Fig. 14.3). Combination of all these relationships has led to the reconstruction of the likely carnivore ancestral karyotype (Frönicke *et al.*, 1997; Breen *et al.*, 1999; Yang *et al.*, 2000a) and determination of its relationship with the ancestral karyotypes of other orders. The cat karyotype requires only one fission and three fusions from the ancestral carnivore karyotype, and the American mink karyotype can be derived by six fusions.

Perissodactyla

Among the Perissodactyls, horse, Hartmann's zebra, donkey and rhinoceros karyotypes have been analysed using human chromosome paints.

Painting horse chromosomes with human probes resulted in the detection of 43 homologous segments between the two species. No homology could be detected for some segments of the equine genome, but these have been filled in from the first generation radiation hybrid map (Rettenberger *et al.*, 1996; Lear and Bailey, 1997). Chaudhary *et al.* (1998) refined the observations for some of the human chromosomes.

Other equids examined using human paints are donkey (*Equus asinus*; 2n = 62) and Hartmann's zebra (*Equus zebra*; 2n = 32). Paints for six human autosomes were used to identify homologous regions on the chromosomes of the two species (Lear and Bailey, 1997; Raudsepp and Chowdhary, 1999). A detailed comparison of human and Hartmann's zebra chromosomes done by Richard *et al.* (2001) showed that individual human autosomes corresponded to a total of 46 conserved segments on the zebra chromosomes. Yang *et al.* (2003) demonstrated 48 conserved segments in Burchell's zebra. Robertsonian translocations have been identified between four equid species, using complete sets of horse, zebra and human paints (Yang *et al.*, 2003). Recently a refined genome-wide comparative map of horse and donkey with respect to human has been generated by reciprocal painting (Yang *et al.*, 2004). Chromosome paints of closely related species such as horse and donkey are very useful for identification of the homologous regions (Plate 3a,b).

Paints from Burchell's zebra and horse were recently used to probe the karyotypes of the white and black rhinoceros, which have 2n = 84 and 82, respectively (Trifonov *et al.*, 2003). Despite the high chromosome number of these species, the karyotypes were surprisingly conserved, with whole chromosomes shared, and many zebra and horse chromosomes recognizing two rhinoceros chromosomes.

Cetartiodactyla

At least five artiodactyl species have been probed using human paints. These include cattle, pig, sheep and river buffalo – and the fascinating muntjac group.

Pig was the first farm animal species whose genome was examined with human paints (Rettenberger *et al.*, 1995; Frönicke *et al.*, 1996). Both studies revealed 47 segments conserved with human, and conserved synteny was supported by available gene mapping data. The results were further refined by reciprocal chromosome painting (Goureau *et al.*, 1996), which detected homologies more efficiently, even though only 60% of the human genome could be painted using the pig paints. Despite the very conservative changes in karyotype revealed by whole chromosome paints, there is now much evidence at the cytological, as well as the gene map, level for small local rearrangements in the pig.

Painting between cattle, buffalo, sheep and goat confirms G-banding studies showing widespread homology of autosomes. Regional paints from microdissected bovine chromosomes were used to examine the X chromosome in detail, since G-band studies had suggested that deletions had occurred from the sheep and goat X. The missing segment was demonstrated to have been moved by inversion (Hassanane *et al.*, 1998).

The use of all human paints on cattle chromosomes (Hayes, 1995; Solinas-Toldo *et al.*, 1995; Chowdhary *et al.*, 1996) revealed 46–50

Table 14.1. Number and type of mapped markers in carnivores and insectivores .

Order/family/species	Number of gene loci (type I markers)	Number of microsatellites (type II markers)	Used methods for mapping*	Reference
Canidae (Canis familiaris)	320	1078	RH, LG, ISH	Breen et al., 2001
Vulpes vulpes	35		SHC	Rubtsov, 1998
Felidae (Felis catus)	585	279	RH, LG, ISH, SHC	Menotti-Raymond et al., 2003
Mustelidae (Mustela vison)	127		SHC, LG	Kuznetsov et al., 2003
Eulipotyphla (Sorex araneus)	49	4	SHC	Zhdanova et al., 2003

* RH, radiation hybrids; LG, linkage analysis; ISH, in situ hybridization; SHC, somatic cell hybrids.

syntenic segments conserved between the two species. A near complete coverage of the bovine karyotype was attained. The segmental homology was shown to be in close agreement with the available comparative gene mapping results (Solinas-Toldo *et al.*, 1995; Chowdhary *et al.*, 1996). Recently, detailed comparison of the whole human karyotype with that of buffalo (Iannuzzi *et al.*, 1998) was carried out using the whole set of human paints, revealing a total of 44 conserved segments, similar to those of cattle.

Initially, only limited Zoo-FISH data between human and sheep were available from hybridization of three human autosomes and the human X paints to sheep chromosomes (Chowdhary *et al.*, 1996). Recently, detailed comparison of the whole human karyotype with that of sheep (Iannuzzi *et al.*, 1999) using the whole set of human paints revealed a total of 48 conserved segments in a pattern closely resembling that of cattle. Hippo metaphase chromosomes painted by human chromosomes (Plate 3c,d,e) showed extended areas of homology, while intensity of painting was not as strong as between more closely related species.

The muntjacs are a group characterized by extraordinary differences in karyotype, from $2n = 47$ in the Chinese muntjac to $2n = 6$ enormous fused chromosomes in the Indian muntjac. The first indications of homology between Indian muntjac and human chromosomes was reported by Scherthan *et al.* (1994) using five human chromosome-specific paints, then a sixth was added (Sensi *et al.*, 1995). Later, human paints and telomere-specific probes were applied to Chinese and Indian muntjac chromosomes to investigate the role of tandem fusion events in the evolution of the muntjac karyotypes (Scherthan, 1995). Hybridization of the entire set of human paints to Indian muntjac chromosomes (Yang *et al.*, 1997) identified 48 autosomal segments conserved between the two species, as well as the relics of discarded centromere sequences. The findings supported the prevalent concept that the evolution of the Indian muntjac karyotype primarily involved fusion of huge blocks of entire chromosomes derived from a putative ancestral karyotype resembling that of the Chinese muntjac. Hybridization of a set of sheep paints to muntjac chromosomes identi-

fied 35 segments conserved between the two species, and supported the concept of multiple tandem fusions (Burkin *et al.*, 1997).

As yet, fin whale chromosomes have been hybridized only with paints from three human autosomes and X (Scherthan *et al.*, 1994). However, whole karyotype studies have been carried out between human and Atlantic bottlenose dolphin (*Tursiops truncatus*), resulting in the detection of 36 conserved chromosomal segments between the two species (Bielec *et al.*, 1998). This revealed a closer correspondence to human chromosomes than shown by any of the artiodactyls, and made it easier to relate the artiodactyl karyotypes to an ancestral mammalian karyotype.

Xenarthra

The nine-banded armadillo has recently been subjected to chromosome painting in an attempt to reconstruct the karyotype of an ancestral eutherian (Richard *et al.*, 2003). Nine chromosomes were fully conserved with human.

Afrotheria

Six orders are encompassed in this clade, which diverged from the rest of the eutherians about 105 mya. Elephant and aardvark have been subject to reciprocal chromosome painting using flow-sorted elephant and aardvark chromosomes, as well as human chromosomes (Frönicke *et al.*, 2003; Yang *et al.*, 2003). Once more, astonishing conservation between afrotherians and humans was observed, providing strong evidence for an ancestral eutherian karyotype of $2n = 44$ or 46. Nine elephant chromosomes are identical to their human counterparts and six to a chromosome arm, and seven are neighbour-joining segments conserved in other eutherians. Painting of elephant shrew chromosomes (Svartman *et al.*, 2004) defined 37 segments, and established 21 associations of human chromosome segments.

Summary of chromosome painting results

There is now a very good sampling of eutherian karyotypes that have been subjected to comparative chromosome painting. Painting confirms the surprising results of comparative mapping – that the eutherian karyotype is much more conserved than could be appreciated from comparisons of G-band patterns.

Comparative chromosome painting in marsupials and monotremes

The extraordinary conservation of marsupial chromosomes across all groups has greatly facilitated comparative studies between marsupial karyotypes. Marsupial chromosome evolution has been rather simple to sort out using chromosome banding and painting.

G-banding showed that marsupial karyotypes are highly conserved. Even the karyotypes of distantly related species are quite simply related. These relationships have largely been confirmed, with minor changes, by comparative chromosome painting. However, the decades-old debate about whether the conserved $2n = 14$ karyotype represents the original marsupial ancestor, or whether the original marsupial had a higher number remains unresolved, and will only be resolved by reference to an outgroup, not possible at this stage by cytogenetic methods.

The ubiquity of a virtually G-band identical $2n = 14$ karyotype suggested long ago that this represented the ancestral marsupial karyotype (Hayman and Martin, 1969; Rofe and Hayman, 1985), and the sequence homology of the six autosomes and X chromosome of this karyotype has been amply confirmed by comparative painting. This $2n = 14$ karyotype is present in all seven orders of marsupials, and is almost certainly the basic ancestral karyotype of the five Australian orders, which are monophyletic (Amrine-Masden et al., 2003), and also of a common ancestor with the $2n = 14$ South American order Paucituberculata.

The karyotypes of several Australian marsupials with higher or lower chromosome numbers have been compared with the $2n = 14$ species by comparative chromosome painting. Of particular interest is the order Diprotodontia, which includes kangaroos and possums. This order has the most diverse karyotypes and includes the species with the lowest, and the highest marsupial chromosome numbers. Cross-species painting shows obvi-

ous and simple homology of kangaroos with $2n = 10, 14, 16, 18$ and 22 to the $2n = 14$ basic painting (Glas et al., 1997; De Leo et al., 1999; O'Neill et al., 1999). This is compatible with the hypothesis (Hayman and Martin, 1974) that the chromosome number in this group increased to $2n = 22$ by multiple fission, then was reduced by a number of independent Robertsonian fusions. An example of comparative chromosome painting in marsupial species (Fig. 14.3) reveals the tandem fusion of two autosomes and centric fusion of the X chromosome and this fused chromosome.

Reciprocal painting was extended to distantly related marsupials including species from South America (Rens et al., 2001). Recent high-resolution work using paints from the species with the highest chromosome number ($2n = 32$) confirms this relationship, but shows also that several fusions and fissions occurred more than once (Rens et al., 2003).

Comparative painting between the Australian $2n = 14$ karyotype and karyotypes with higher chromosome numbers in two South American species of the basal marsupial order Didelphimorphia again shows simple and obvious homology. Arrangements of homologous blocks in the $2n = 20$ and $2n = 22$ species are clearly independent of arrangements within kangaroos with the same chromosome number. Since this marsupial order, too, contains species with the basic $2n = 14$ karyotype, there are two possibilities. Either the $2n = 14$ karyotype was the original ancestral marsupial karyotype, and the higher chromosome numbers were generated by fissions in this group, or the original ancestral number was higher ($2n = 22$), and the $2n = 14$ karyotype was generated by fusions. The latter possibility is favoured by the demonstration of interstitial telomere-specific sequences at projected fusion points in one of the $2n = 14$ didelphids (Svartman and Vianna-Morgante, 1999).

Monotreme chromosome paints have now been prepared (Rens et al., unpublished) for use in exploring homologies between the unpaired chromosomes in males (Grützner et al., 2004), and also the relationships between platypus and echidna chromosomes. Monotremes will be a vital outgroup to eutherians as well as marsupials, and will enable the

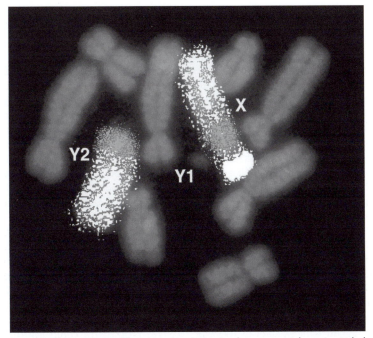

Fig. 14.3. An example of comparative chromosome painting in close marsupial species. Whole chromosome paints derived by flow-sorting chromosomes from the tammar wallaby (*Macropus eugenii*) were hybridized to chromosomes of the swamp wallaby *Wallabia bicolor*, which has the lowest recorded chromosome number in marsupials ($2n = 10$ in females and 11 in males). The tandem fusion of two autosomes, and centric fusion of the X chromosome and this fused autosome is apparent. A colour image of this metaphase can be seen on the front cover of this book (original by R. Toder, R. O'Neill and J. Graves).

deduction of an ancestral mammalian karyotype 210 mya.

Attempts to hybridize eutherian chromosome paints to marsupial and monotreme material have all been unsuccessful. The reverse – using marsupial and monotreme paints to hybridize to eutherian chromosomes – has also not been successful, with one important exception. When the sorted tammar X chromosome was used to hybridize to human cells, painting was observed on the X alone (Glas *et al.*, 1999). The regions hybridized were the long arm and the proximal region of the short arm, corresponding to the gene mapping data. Very recently, the sorted platypus X was used to hybridize to human cells, with similar hybridization patterns on to the human X (Kirby and Greaves, unpublished).

Thus painting between marsupial species, added to the cytogenetic data, to a large extent makes up for the paucity of gene mapping data for comparing the chromosomes of differ-

ent marsupial species. However, the inability to use painting to determine homologies between eutherian, marsupial and monotreme chromosomes will make it imperative to construct maps for marsupial and monotreme species in order to provide outgroups for eutherian comparisons, and to reconstruct ancient therian and mammalian karyotypes.

Reconstructing Ancient Mammalian Karyotypes

Combinations of comparative banding, painting and comparative gene mapping allow us to reconstruct ancestral mammalian karyotypes, to track chromosome evolution in mammalian lineages and derive the rules that govern chromosome change during mammal speciation.

Comparisons between closely related species, such as the primates, or the cats, can be used to deduce the karyotype of a common ancestor. The procedure is simply to spot chromosomes that have remained constant in all species (such as human chromosomes 11 and 17), so are likely to have been inherited from the common ancestor. It is then necessary to differentiate between rearrangements that are shared between distantly related species (likely to be ancestral) and shared derived characteristics that mark a recent expansion. We have described how this has been done for primates, rodents, carnivores and artiodactyls (Frönicke et al., 1997; Muller et al., 1999; O'Brien and Stanyon, 1999).

Combinations that are shared between distantly related mammals are more likely to be ancestral than to have occurred independently in several lineages, although it is necessary to distinguish between shared ancestral configurations and convergent rearrangements. Convergence is perhaps more common than is usually appreciated, with evidence of extensive breakpoint re-use from comparisons between mouse and human synteny blocks (Pevzner and Tesler, 2003) and painting comparisons across marsupials (Rens et al., 2003).

To reconstruct an ancient eutherian karyotype, the same logic has been applied to comparisons between eutherian orders. This has been surprisingly straightforward because of the unexpected degree of chromosome homology across even distantly related groups. The accumulation of cytogenetic, comparative mapping and comparative painting data of several eutherian species now makes it possible to reconstruct the karyotypes of an ancient eutherian. There is a satisfying degree of convergence in the estimates of an ancestral eutherian with about 46 chromosomes.

Reconstructing the ancient eutherian karyotype

Reconstructing an ancestral eutherian karyotype is like assembling a jigsaw puzzle with many missing pieces. Cytogenetic, gene mapping and comparative painting information – vital for tracing the karyotype of the common ancestor – is available for only a handful (fewer than 1%) of the ~4000 extant mammalian species. This arduous task is made more difficult by the wide variation in chromosome number, shape, size and gene content between mammalian orders, families, subfamilies and species.

Speculations about the ancient mammalian karyotype were first made long before detailed cytogenetic observations were available for more than a few eutherian species, and decades before the development of comparative gene mapping and chromosome painting. Based on cytogenetic measurements of chromosome numbers, sizes and arm ratios of fewer than 50 eutherian species, Todd (1970) suggested that the ancestral chromosomal complement for all mammals was essentially comprised of large mediocentric (i.e. metacentric, submetacentric and subtelocentric) chromosomes ($2n = 14$). These chromosomes underwent episodes of rearrangements – fission as well as fusions – that resulted in a huge range in diploid number spanning from $2n = 6$ in Indian muntjac to $2n = 102$ in a South American rodent. Soon thereafter, Matthey (1972) proposed a completely different number of $2n = \sim48 \pm 8$ chromosomes as the ancestral eutherian condition. The advent of G- and R-banding allowed better recognition of chromosomes and chromosome regions across species. For instance, R-banding comparisons spanning a range of primates led to the proposition of different versions of eutherian mammal karyotypes (Dutrillaux et al., 1980, 1981). The chromosome number ranged between 54 and 58, and the interpretations differed for the morphology of some of the chromosomes.

The development of comparative gene mapping from the early 1970s made it possible to discern conservation between chromosomes or regions across species or even orders even when cytogenetic similarity was unclear (O'Brien et al., 1985). Gene mapping information could be used to conduct chromosome segment comparisons across a range of evolutionarily distantly related species to discern the conservation of synteny. Unfortunately, this provided very little global information on the likely constitution of the ancestral karyotype because: (i) the mapping information was available on a very limited number of species and (ii) the information was not extensive/detailed enough to permit drawing conclusions regarding conservation of entire chromosome or chromosome arms.

The advent of Zoo-FISH in 1994 (Scherthan *et al.*, 1994) was therefore a major breakthrough in facilitating identification of evolutionarily conserved chromosomal segments across species because it provided direct molecular evidence of segmental similarities. Extended to a range of species in different mammalian orders, the technique has emerged as a major tool to draw inference of ancestral karyotypes for various mammalian families and orders, thereby permitting attempts to speculate on the chromosomal configuration in the ultimate mammalian ancestor. This is evident from the initial ground-breaking work of Rettenberger *et al.* (1995), Frönicke *et al.* (1996) and Chowdhary *et al.* (1996) and others who used human chromosome-specific paints to demonstrate a high degree of synteny conservation between the human (hominoid) and cat (felid), human and pig (suid), and human and cattle (bovid) karyotypes, respectively.

The cat and human karyotypes, with 36 and 44 autosomes respectively, were found to be very similar to a putative ancient mammalian founder karyotype. For the autosomes alone, 30 segments of conserved synteny were detected. The arrangement of these segments in the feline karyotype differs by only seven single chromosome breaks and one intrachromosomal inversion from their arrangement in humans, suggesting that during evolution to the human karyotype, the status quo of the ancestral configuration has been reasonably conserved for at least some 80 million years. Rettenberger *et al.* (1995) surmised that there has been no need to alter the well-balanced gene arrangement of the mammalian founder karyotype to get either the human or the cat karyotypes. Comparison of these data with conventional comparative gene mapping studies in mice, pig and cattle led to preliminary inferences that an ancestral karyotype probably had $2n = \sim 38$ chromosomes, a number intermediate between $2n = 14$ and 48, suggested by Todd (1970) and Matthey (1972), respectively.

One of the first organized attempts to reconstruct the putative ancestral mammalian karyotype using a combination of genome-wide comparative gene mapping and comparative painting data was made by Chowdhary *et al.* (1998). The authors compared several species (human, mouse, pig, cattle, Indian muntjac, cat, American mink, harbour seal and horse) belonging to four eutherian orders (Primates, Artiodactyla, Carnivora and Perissodactyla) to draw inferences on chromosomal regions that could fall into the category of 'ancestral'. On the basis of the set of chromosomal blocks conserved across the eight species, and assuming that maximum parsimony should disclose the probable ancestral chromosome complement, the authors proposed a chromosomal pool representing the likely karyotype of the common ancestor. This pool was generated by selecting those human chromosomes or chromosomal arms that were maintained as a single block or as a combination of two human chromosomes, in at least one of these non-primate species. Though Chowdhary *et al.* (1998) emphasized the deductions to be preliminary; the findings have since withstood subsequent tests by various research groups, and have served as a landmark for all future attempts at expansion of data aimed at refining the ancestral karyotype in mammals. The salient findings of their work are presented below. Any additions/refinements to this by other researchers are presented in later paragraphs and summarized in Table 14.2 and Fig 14.4a.

Three main categories of cross-species conservation in the ancestor were identified in relation to the human karyotype:

1. *Whole chromosome conservation:* these chromosomes generally show no *inter*chromosomal rearrangements across species studied through comparative painting, and are represented either as a single chromosome or as a whole chromosome arm in their karyotypes. HSA13, HSA17, HSA20, HSAX, and the corresponding homologous segments in the investigated species fell into this category.
2. *Conservation of large chromosomal segments including chromosomal arms:* these chromosomes represent homologues of human chromosomes that are, in general, primarily conserved as a single block or as two blocks representing the two arms of the chromosomes in most of the species. The chromosomes included homologues of HSA1, 2, 3, 4, 5, 6, 9, 11, 18 and 21 in different species.

Table 14.2. Chromosomal correspondence between the presumed ancestral karyotype for all mammals (AKM) (Yang et al., 2003) and human (HSA), between the presumed ancestral Carnivora karyotype (ACK)(Murphy et al., 2001; Graphodatsky et al., 2002) and human and cat (FCA), and between proposed ancestral Mustelidae karyotype (AMK) (Graphodatsky et al., 2002) and American mink (MVI).

AKM 2n = 44		ACK 2n = 42	HSA 2n = 46	FCA 2n = 38	MVI 2n = 30	AMK 2n = 38
1	HSA1p/1pq/19p	1	19p, 3, 21	A2p, C2	5	1
2	HSA8/4p	2	4, 8p	B1	6	2
3	HSA3/21	3	5	A1q	1q	15
4	+ 5 HSA5+ HSA6*	4	6	B2	1p	5
6	HSA15/14	5	14, 15	B3	10	7
7	HSA10/12pq/22q	6	10, 12p-q, 22q	B4	9	6
8	HSA7a	7	20, 2p-q	A3	11	8
9	HSA11	8	2q	C1q	4q	3q
10	HSA1q	9	7p-q	A2q	3q	10
11	HSA9	10	1p-q	C1p	2q	4q
12	HSA2p	11	10q	D2	2p	16
13	HSA8q	12	11	D1	7q	9
14	HSA10q	13	12q, 22, 18	D3	4p	11
15	HSA13	14	9	D4	12	12
16	HSA17	15	13	A1p	8q	3p
17	HSA18	16	17	E1	8p	18
18	HSA19q/16q	17	1q	F_1	13	13
19	HSA20	18	8q	F_2	3p	4p
20	HSA7b/16q	19	19q, 16q	E2	7p	17
21	HSA12qdis/22qdis	20	7q, 16p	E3	14	14
X		X	X	X	X	X

*Chromosomes 4 and 5 of the AKM and their HSA equivalents are artificially reunited in a single cell of the table for alignment of all cells.

3. *Conservation as neighbouring segments of two human chromosomes:* regions homologous to certain human chromosomes or chromosomal segments are contiguous/syntenic consistently in a number of evolutionarily diverged species suggesting that these combinations represent ancestral chromosomal conditions. It is hypothesized that recent fission events probably separated these ancestral combinations during human karyotype evolution (see below). The human chromosomal complement that comprised this group was: HSA3/HSA21, HSA12/HSA22, HSA14/HSA15 and HSA16q/HSA19q.

Beginning from this pool, that comprised human chromosome equivalents 1p, 1q, 2pter–q13, 2q13–qter, (3/21), 4, 5, 6, 7, 8, 9, 10, 11, (12/22a), 13, (14/15), (16q/19q), 16p, 17, 18, 19p, 20, 22b, X, and Y, likely rearrangements (mainly fusion and fission events) that led to the formation of the karyotype of each of the eight species were identi-

fied. The putative ancestral diploid karyotype comprised 48 chromosomes (including the sex chromosomes) – a number close to the earlier prediction of Matthey (1972).

The recent genome analysis projects initiated in several species (predominantly domesticated species: livestock and laboratory as well as pet/companion) have contributed substantially to comparative genomics. The key ways this has contributed to an improved understanding of the ancestral mammalian karyotypes are by: (i) precise demarcation of conserved synteny breakpoints across species and (ii) facilitating discovery of small conserved synteny segments that easily escape detection by comparative painting due to the limited resolution of the technique. Details regarding this are described elsewhere in this chapter. Of greater significance in augmenting our knowledge about the putative eutherian ancestral karyotype has been comparative painting experiments carried out in a range of

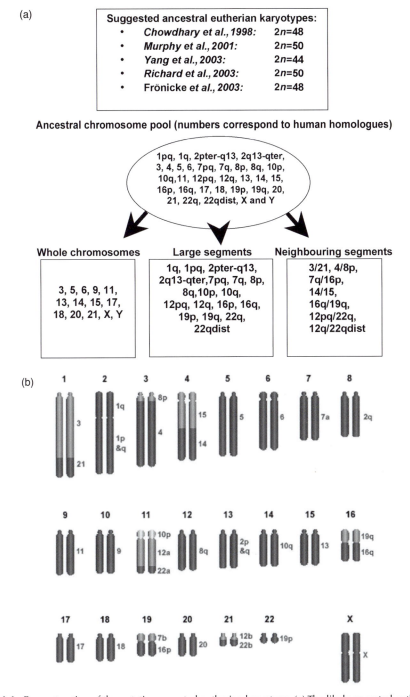

Fig. 14.4. Reconstruction of the putative ancestral eutherian karyotype. (a) The likely ancestral eutherian karyotype represented as homologues of human chromosomes and the three major conserved components it comprises. The conserved components, i.e. whole human chromosomes, large segments of human chromosomes and combination of neighbouring segments of human chromosomes, are commonly seen as conserved blocks in other evolutionarily diverged species. (b) The putative ancestral eutherian karyotype.

mammalian species belonging to different orders. These findings are significantly improving our insight into the probable ancestral eutherian chromosome configuration. Though several research groups have contributed to these developments (and these groups individually deserve credit for every step of advancement), we will focus on analysis from four of the recent publications that provide an overview of the information.

The first of these is a review by Murphy et al. (2001a) that describes the widespread synteny associations between species in nine eutherian orders (Primates, Scandentia, Carnivora, Perissodactyla, Cetardiodactyla, Chiroptera, Eulipotyphia Lagomorpha and Rodentia, with chicken and zebrafish as outgroups), pointing out that the number and the synteny associations of the reconstructed karyotype could be modified using fine mapping approaches. The authors agreed that the ancestral eutherian genome probably had a diploid number of ~50, but identified new conserved syntenies corresponding to human chromosomes 4+8p+4 and 7q+16p, and also refined some of the conserved chromosomal blocks.

The second set of articles (Frönicke et al., 2003; Yang et al., 2003) focus on comparative painting between the genomes of humans and their most distant eutherian relatives, the Afrotherians, which diverged ~95–105 mya (Eizirik et al., 2001). From comparative painting analysis of human, elephant and aardvark chromosomes, they deduced an ancestral eutherian karyotype of $2n = 44$ chromosomes. Though these authors consider this a major departure from the earlier deduction of 48–50 (Chowdhary et al., 1998; Murphy et al., 2001a), the differences seem primarily to be an outcome of refinement over previous works. Based on the studies carried out in elephants, Frönicke et al. (2003) also suggested that the comparative painting studies might have reached the limit of analysis.

Lastly, Richard et al. (2003), from the renowned group of B. Dutrillaux, recently conducted a detailed analysis using chromosome banding and/or comparative painting data from 29 species belonging to 11 mammalian orders to predict the ancestral eutherian karyotype. Applying the parsimony principle, the ancestral karyotype was considered to comprise 50 mostly acrocentric chromosomes. Ten human chromosomes and parts of nine others were considered ancestral, whereas seven ancestral chromosomes were considered to be disrupted in humans. These conclusions are consistent with the earlier deductions of Chowdhary et al. (1998), Murphy et al. (2001a), Frönicke et al. (2003) and Yang et al. (2003). It was proposed that, aside from intrachromosomal rearrangements, the human karyotype differed from the ancestral karyotype by only ten interchromosomal rearrangements and fissions. These conclusions are compared with the earlier deduction of an ancestral 54–58 chromosomes, based on comparisons of R-bands across species (Dutrillaux et al., 1981, 1982).

In conclusion, there is a building consensus that the number of chromosomes in the eutherian ancestor was around 46 ± 4. Researchers all agree that the number is not final. Obviously, the more evolutionarily diverse species are analysed, the greater will be the accuracy. Of increasing interest is the identification and analysis of break-points in human chromosomes that bear signatures of chromosome evolution. It will be a dream-come-true to be able to accurately trace back all the events that led to the formation of the human karyotype, beginning from the ancestral eutherian karyotype (Fig. 14.4b).

Comparison of this karyotype with ancestral marsupial karyotypes should enable reconstruction of an ancient therian karyotype that existed before the eutherian–marsupial divergence 180 mya, and comparison of this with monotreme karyotypes might enable us to reconstruct the karyotype of an ancestral mammal that lived more than 210 mya. However, attempts to use human paints on marsupials, and kangaroo and dunnart autosomal paints on human chromosomes have been completely unsuccessful, presumably because the homology between these distantly related sequences is too poor, or the homologous segments are too small. However, comparative painting between the tammar X chromosome and the platypus X on to humans produced very clear homology between the marsupial X and part of the human X (Glas et al., 1999; Kirby and Greaves, unpublished). This suggests either that the X chromosome sequences are more conserved than autosomal, or that homology blocks are larger, or both.

Reconstructing ancient sex chromosomes

Mammalian sex chromosomes appear to have evolved by a different set of rules than have autosomes, so are considered separately here. From the earliest gene mapping in mammals – even from the earliest cytogenetic comparisons between mammal karyotypes – it was obvious that the eutherian X chromosome was conserved in size and gene content, making up 5% of the haploid karyotype even in mammals with many small chromosomes, or a few large chromosomes (Ohno, 1969). The Y chromosome lies at the other end of the spectrum, being extraordinarily variable between even quite closely related species. Even in marsupials and monotremes, the X is homologous to a large region of the X of eutherians, and the Y with the eutherian Y.

There are two things that are special about mammalian sex chromosomes. One is that the Y chromosome carries the gene (*SRY*) that determines testis and therefore initiates male development. This accounts for the genetic poverty and variability of the mammal Y. The other is that the X chromosome is present in two copies in females but only one in males. Compensation for this dosage accounts for the unparalleled conservation of the X.

In mammals, sex is determined by an XX female:XY male chromosome constitution. The X and Y are almost completely differentiated, and pair only over a small pseudoautosomal region (PAR) at one or both tips. The eutherian X chromosome contains about 1500 genes, the Y only 49. Although there are about 30 genes shared between the X and Y, nearly all the genes on the X lack a partner in males. Gene dosage differences of X-borne genes between males and females are nullified by the inactivation of one or other X in somatic cells of females (Lyon, 1961). This whole-X inactivation is thought to have kept the X sacrosanct, since rearrangement would lead to disruption of a balanced gene dosage (Ohno, 1969). The non-conservation of genes in the PAR, or the region that was pseudoautosomal until recently

(reviewed in Graves *et al.*, 1998a) is understandable, since genes with a partner on the Y need not obey Ohno's Law (Graves, 1996).

The large X and small Y of mammals are thought to have differentiated in the last 300 million years from a pair of autosomes (Ohno, 1967). Other vertebrates do not share the mammalian X and Y chromosomes – in fact, the mammal sex chromosomes are represented by parts of chicken – and turtle – chromosomes 1, 4 and 12 (Nanda *et al.*, 1999; Graves and Shetty, 2001; Kohn *et al.*, 2004). This means that mammalian sex chromosomes got their start some time after the divergence of mammal-like reptiles 310 mya, and before monotremes diverged from therians 210 mya. Ohno's suggestion that sex chromosomes evolved from an ordinary autosomal pair is confirmed by the homology between the X and Y within the PARs, and also between about 30 Y-borne genes with homologues on the X.

The X and Y started on the path to differentiation as one member acquired a sex-determining locus. Other alleles with an advantage in males accumulated around this new testis-determining locus, and recombination was suppressed in the region (Charlesworth, 1991). Within the non-recombining region, numerous mutations and deletions accumulated, until the region lost most of its active genes and degraded progressively. The process of X–Y differentiation seems to have proceeded in fits and starts, as intrachromosomal rearrangements separated blocks of the Y from recombination with the X. This led to different dates of X–Y separation and therefore to different levels of homology in four different 'geological strata' (Lahn and Page, 1999b).

The creeping degradation of the Y can be observed by comparing the gene content, activity and function of Y-borne genes in different species. The extent of the PAR is different between species, being truncated on the human Y with respect to carnivores and ungulates. The spectrum of activity of Y-borne genes is also different in different species. For instance, many genes are like *ZFY* in humans, which is ubiquitously expressed and may complement its X-borne homologue *ZFX*, which is exempt from inactivation. However, in mice, *Zfy* is testis-specific and has a function in spermatogenesis, so its X-borne homologue is subject to X inactivation.

Many of the genes that have survived the process of degradation apparently did so by acquiring selectable male-specific functions. Ultimately, only genes that serve a selectable function in males will survive on the degrading Y. Even these have been lost in some lineages – indeed, the whole Y chromosome, including the *SRY* gene has been lost independently in the mole vole (Just *et al.*, 1995; Graves, 2002). If the Y continues to lose genes at its present rate, there will be nothing left of it in 9 million years (Graves, 2004).

Comparisons of X chromosome gene maps and painting patterns between the three major mammal groups show that only part of the eutherian X chromosomes is shared with the X of marsupials and monotremes. This could mean either that the original mammalian X was small (like the marsupial X) and received an addition in eutherians, or that it was larger like the eutherian X, and lost a large region in marsupials and monotremes. The first interpretation is favoured because markers on the short arm of the human X map to similar autosomal clusters in marsupials and the platypus. Since marsupials and monotremes diverged independently from eutherians, this must mean that an autosomal region was added to the X in the eutherian lineage after the divergence of marsupials 180 mya and before eutherian radiation 105 mya (Graves, 1996).

The eutherian X is therefore composed of an ancient X conserved region (XCR, at least 210 million years old) and a region XAR that was added 105–180 mya. Lahn and Page's 'geological layers' of the human X fit within the XCR (layers 1 and 2) and XAR (layers 3–5). Comparative mapping in chicken separates layer 1 and 2 (Kohn *et al.*, 2004).This recent addition explains why many genes on the short arm of the human X (including the PAR) act more like autosomal genes, being exempt from inactivation and having partners on the Y (Graves, 1995).

The Y, too, is composed of an ancient conserved region YCR defined by homology to genes on the marsupial Y, and an added region YAR defined by homology to genes on the marsupial autosomes. The ancient region of the human Y (corresponding to the ~10 Mb marsupial Y) is very tiny, containing only four known genes over a span of about 10 Mb. Most of the human Y chromosome is derived from the added region and, therefore, was not originally part of the Y chromosome (Waters *et al.*, 2001).

Thus a large autosomal region was added to the proto-X and -Y chromosome pair between 180 and 80 mya. The most likely scenario is that addition involved the PAR of an ancient sex chromosome pair – the region was added to one member, then recombined on to the other at meiosis, enlarging the PAR (Graves, 1996).

Thus comparative mapping and painting between distantly related mammals, and even non-mammal vertebrates, has clarified the origin and evolution of mammalian sex chromosomes, and provided some insight into the special rules that have formed the conservative X and the variable and degraded Y chromosomes in mammals.

Conclusions

Decades of cytogenetics, comparative gene mapping and most recently comparative chromosome painting have now provided many of the pieces of the jigsaw puzzle of mammal karyotype evolution. We can now begin to gain a picture of ancestral mammal chromosomes and the changes that have occurred in different lineages. Answers to the big questions of mammal karyotype evolution seemed very distant in the 1970s, when it was thought that mammalian karyotypes had rearranged beyond recognition. However, everything now tells us that the mammalian genome – indeed the vertebrate genome – has generally been very conserved in evolution.

Remarkably, conserved gene arrangements are apparent even in much more distantly related vertebrates. Birds, which came from a branch of reptiles that diverged from mammals 310 mya, show many shared syntenies of coding genes. For instance, eight HSA6 genes all map to chick chromosome 3 (Burt *et al.*, 1999). Even more remarkable is the conservation of synteny that is apparent in comparisons between humans and fish; for instance ten markers on HSA2q all lie in a single linkage group in the zebrafish LG9 (Postlethwait, 2000). As zebrafish sequence is assembled, it becomes apparent that mammalian chromosomes may have been parts of larger ancestral chromosomes that were broken up in the mammals – or at least in eutherians.

There are exceptions to this conservation in many lineages – gibbons among the conserved primates, dogs among the conserved carnivores, rock wallabies among the ultraconservative marsupials. Why do some karyotypes seem to be extremely stable and others very variable? Are there global changes that occur – for instance the invasion of new transposable elements (Ariga *et al.*, 1990), 'genomic stress' imposed by hybridization (O'Neill *et al.*, 1999) or changes in systems of repair – to destabilize a formerly well balanced karyotype (Chapter 13)?

Karyotype conservation means that information is readily transferred between different mammalian species. This has enormous implications for gene finding in map-poor species, as well as for the interpretation of the human gene sequence, since phenotypic traits may be mapped in one species and candidate genes spotted in another.

It also means that deducing the shape of an ancestral mammalian genome – maybe even an ancestral vertebrate genome – is not a vain hope. The construction of an ancestral eutherian karyotype is now a reality (a few disagreements about one or two fusions or fissions aside), thanks to the sharing of whole chromosomes, or large segments, across wide phylogenetic distances. These mild disagreements will soon be solvable by reference to an outgroup. Chicken gene maps are now detailed enough to provide many insights to mammalian chromosome evolution (Burt *et al.*, 1999), and the full sequence of the chicken genome is due to be released soon. Closer to home is the prospect of lining up this ancestral eutherian karyotype with the ancestral marsupial karyotype. At present there are insufficient gene mapping data for marsupial autosomes, but the autosome map will soon be fleshed out by linkage and physical mapping, as well as whole genome sequencing.

With this wealth of data and a new synthesis of karyotypic constancy and change, we can now approach other big questions of mammalian cytology, genetics and evolution. How relevant to gene function is chromosome location, interphase position and conformation? What forces select for karyotypes composed of big or small chromosomes? Telocentric or metacentric chromosomes? How is chromosome change – for instance centromeric position and heterochromatin distribution – coordinated in a karyotype? How does chromosome change relate to speciation? Mapping and sequencing mammal genomes over the next few years may provide the details to answer these questions.

References

Amrine-Madsen, H., Scally, M., Westerman, M., Stanhope, M.J., Krajewsky, C. and Springer, M. (2003) Nuclear gene sequences provide evidence for the monophyly of australidelphian marsupials. *Molecular Phylogenetics and Evolution* (1994) 28, 186–196.

Andresen, E. and Baker, L.N. (1964) The C blood group system in pigs and the detection and estimation of linkage between the C and J systems. *Genetics* 49, 379–386.

Archibald, A., Brown, J., Couperwhite, S., McQueen, H., Nicholson, D., Haley, C., Couppieters, W., van de Weghe, A., Stratil, A., Wintero, G., Gelderman, H., Davoli, R., Ruyter, D., Verstege, E., Groenen, M., Davies, W., Hoyheim, B., Keiserud, A., Andersson, L., Ellegren, H. *et al.* (1994) The PiGMap consortium linkage map of the pig (*Sus scrofa*). *Mammalian Genome* 6, 157–175.

Ariga, T., Carter, P.E. and Davis, A.E. (1990) Recombination between Alu repeat sequences that result in partial deletion within the C1-inhibitor family. *Genomics* 8, 607–613.

Arnold, N., Stanyon, R., Jauch A., O'Brien, P. and Wienberg, J. (1996) Identification of complex chromosome rearrangements in the gibbon by fluorescent *in situ* hybridization (FISH) of a human chromosome 2q specific microlibrary, yeast artificial chromosomes, and reciprocal chromosome painting. *Cytogenetics and Cell Genetics* 74, 80–85.

Avner, P., Amar, L., Dandolo, L. and Guénet, J.-L. (1988) Genetic analysis of the mouse using interspecific crosses. *Trends in Genetics* 4, 18–23.

Bailey, D.W. (1971) Recombinant-inbred strains. An aid to finding identity, linkage, and function of histocompatibility and other genes. *Transplantation* 11, L325–L327.

Band, M.R., Larson, J.H., Rebeiz, M., Green, C.A., Heyen, D.W., Donovan, J., Windish, R., Steining, C., Mahyuddin, P., Womack, J.E. and Lewin, H.A. (2000) An ordered comparative map of the cattle and human genomes. *Genome Research* 10, 1359–1368.

Barendse, W., Armitage, S.M., Kossarek, L.M., Shalom, A., Kirkpatrick, B.W., Ryan, A.M., Clayton, D., Li, L., Neibergs, H.L., Zhang, N., Grosse, W.M., Weiss, J., Creighton, P., McCarthy, F., Ron, M., Teal, A.J., Fries, R., McGraw, R.A., Moore, S.S., Georges, M., Soller, M., Womack, J.E. and Hetzel, D.J.S. (1994) A genetic linkage map of the bovine genome. *Nature Genetics* 6, 277–235.

Barendse, W., Vaiman, D., Kemp, S.J., Sugimoto, Y., Armitage, S.M., Williams, J.L., Sun, H.S., Eggen, A., Agaba, M., Aleyasin, S.A., Band, M., Bishop, M.D., Buitkamp, J., Byrne, K., Collins, F., Cooper, L., Coppettiers, W., Denys, B., Drinkwater, R.D., Easterday, K. *et al.* (1997) A medium-density genetic linkage map of the bovine genome. *Mammalian Genome* 8, 21–28.

Bennett, J.H., Hayman, D.L. and Hope, R.M. (1986) Novel sex differences in linkage values and meiotic chromosome behaviour in a marsupial. *Nature* 323, 59–60.

Bick, Y.A.E. (1992) The meiotic chain of chromosomes of monotremata. In: Augee, M. (ed.) *Platypus and Echidnas.* Royal Society of New South Wales, pp. 277–284.

Bielec, P.E., Gallagher, D.S., Womack, J.E. and Busbee, D.L. (1998) Homologies between human and dolphin chromosomes detected by heterologous chromosome painting. *Cytogenetics and Cell Genetics* 81, 18–25.

Bigoni, F., Stanyon, R., Koehler, U., Morescalchi, A.M. and Wienberg, J. (1997) Mapping homology between human and black and white colobine monkey chromosomes by fluorescent *in situ* hybridization. *American Journal of Primatology* 42, 289–298.

Bihoreau, M.T., Gauguier, D., Kato, N., Hyne, G., Lindpaintner, K., Rapp, J.P., James, M.R. and Lathrop, G.M. (1997) A linkage map of the rat genome derived from three F_2 crosses. *Genome Research* 7, 434–440.

Bininda-Emonds, O.R., Gittleman, J.L. and Purvis, A. (1999) Building large trees by combining phylogenetic information: a complete phylogeny of the extant Carnivora (Mammalia). *Biological Reviews of the Cambridge Philosophical Society (London)* 74, 143–175.

Bishop, M.D., Kappes, S.M., Keele, J.W., Stone, R.T., Sunden, S.L.F., Hawkins, G.A., Solinas Toldo, S., Fries, R., Grosz, M.D., Yoo, J. and Beattie, C.W. (1994) A genetic linkage map of cattle. *Genetics* 136, 619–639.

Blaschke, R.J. and Rappold, G.A. (1997) Man to mouse – lessons learned from the distal end of the human X chromosome. *Genome Research* 7, 1114–1117.

Bosak, N., Faraut, T., Mikawa, S., Uenishi, H., Kiuchi, S., Hiraiwa, H., Hayashi, T. and Yasue, H. (2003) Construction of a high-resolution comparative gene map between swine chromosome region 6q11→q21 and human chromosome 19 q-arm by RH mapping of 51 genes. *Cytogenetics and Genome Research* 102, 109–115.

Botstein, D., White, R.L., Skolnick, M. and Davis, R.W. (1980) Construction of a genetic linkage map using restriction fragment length polymorphism. *American Journal of Human Genetics* 32, 314–331.

Breen, M., Thomas, R., Binns, M.M., Carter, N.P. and Langford, C.F. (1999) Reciprocal chromosome painting reveals detailed regions of conserved synteny between the karyotypes of the domestic dog (*Canis familiaris*) and human. *Genomics* 61, 145–55.

Breen, M., Jouquand, S., Renier, C., Mellersh, C.S., Hitte, C., Holmes, N.G., Cherron, A., Suter, N., Vignaux, F., Bristow, A.E., Priat, C., McCann, E., Andrer, C., Boundy, S., Gitsham, P., Thomas, R., Bridge, W.L., Spriggs, H.F., Ryder, E.J., Curson, A., Sampson, J., Ostrander, E.A., Binns, M.M. and Galibert, F. (2001a) Chromosome-specific single-locus FISH probes allow anchorage of an 1800-marker integrated radiation-hybrid/linkage map of the domestic dog genome to all chromosomes. *Genome Research* 11, 1784–1795.

Breen, M., Switonski, M. and Binns, M.M. (2001b) Cytogenetics and physical chromosome maps. In: Ruvinsky, A. and Sampson, J. (eds) *The Genetics of the Dog.* CAB International, Wallingford, UK, pp. 299–328.

Broad, T.E. and Hill, D.F. (1994) Mapping the sheep genome: practice, progress and promise. *British Veterinary Journal* 150, 237–252.

Brown, D.M., Matise, T.C., Koike, G., Simon, J.S., Winer, E.S., Zangen, S., McLaughlin, M.G., Shiozawa, M., Atkinson, O.S., Hudson, J.R., Jr, Chakravarti, A., Lander, E.S. and Jacob, H.J. (1998) An integrated genetic linkage map of the laboratory rat. *Mammalian Genome* 9, 521–530.

Burkin, D.J., Morse, H.G., Broad, T.E., Pearce, P.D., Ansari, H.A., Lewis, P.E. and Jones, C. (1993) Mapping the sheep genome: production of characterized sheep × hamster cell hybrids. *Genomics* 16, 466–472.

Burt, D.W., Bruley, C., Dunn, I.C., Jones, C.T., Ramage, A., Law, A.S., Morrice, D.R., Paton, I.R., Smith, J., Windsor, D., Sazanov, A., Fries, R. and Waddington, D. (1999) The dynamics of chromosome evolution in birds and mammals. *Nature* 402, 411–413.

Castle, W.E. and Wachter, W.L. (1924) Variation of linkage in rats and mice. *Genetics* 9, 1–2.

Cavagna, P., Menotti, A. and Stanyon, R. (2000) Genomic homology of the domestic ferret with cats and humans. *Mammalian Genome* 11, 866–870.

Chapman, V.M. (1978) Biochemical polymorphisms in wild mice. In: Morse, H.C., III (ed.) *Origins of Inbred Mice*. Academic Press, New York.

Charchar, F.J., Svartman, M., El-Mogharbel, N., Ventura, M., Kirby, P., Matarazzo, M.R., Ciccodicola, A., Rocci, M., D'Esposito, M. and Graves, J.A.M. (2003) Complex events in the evolution of the human pseudoautosomal region 2 (PAR2). *Genome Research* 13, 281–286.

Charlesworth, B. (1991) The evolution of sex chromosomes. *Science* 251, 1030–1033.

Chaudhary, R., Raudsepp, T., Guan, X.Y., Zhang, H. and Chowdhary, B.P. (1998) Zoo-FISH with microdissected arm specific paints for HSA2, 5, 6, 16, and 19 refines known homology with pig and horse chromosomes. *Mammalian Genome* 9, 44–49.

Chowdhary, B.P. (1998) Cytogenetics and physical chromosome maps. In: Rothschild, M.F. and Ruvinsky, A. (eds) *The Genetics of the Pig*. CAB International, Wallingford, UK, pp. 199–264.

Chowdhary, B.P. and Raudsepp, T. (2001) Chromosome painting in farm, pet and wild animal species. *Methods in Cell Science* 23, 37–55.

Chowdhary, B.P., Frönicke, L., Gustavsson, I. and Scherthan, H. (1996) Comparative analysis of the cattle and human genomes: detection of ZOO-FISH and gene mapping-based chromosomal homologies. *Mammalian Genome* 7, 297–302.

Chowdhary, B.P., Raudsepp, T., Frönicke, L. and Scherthan, H. (1998) Emerging patterns of comparative genome organization in some mammalian species as revealed by Zoo-FISH. *Genome Research* 8, 577–589.

Cirera, S., Jorgensen, C., Sawera, M., Raudsepp, T., Chowdhary, B.P. and Fredholm, M. (2003) Comparative mapping in the pig: localization of 214 expressed sequence tags. *Mammalian Genome* 14, 405–426.

Clark, F.H. (1936) Linkage of pink-eye and albinism in the deer mouse. *Journal of Heredity* 27, 256–260.

Comparative Gene Organization Workshop (1996) First International Workshop (1996) Comparative genome organization of vertebrates. *Mammalian Genome* 7, 717–734.

Consigliere, S., Stanyon, R., Koehler, U., Arnold, N. and Wienberg, J. (1998) *In situ* hybridization (FISH) maps chromosomal homologies between *Alouatta belzebul* (Platyrrhini, Cebidae) and other primates and reveals extensive interchromosomal rearrangements between howler monkey genomes. *American Journal of Primatology* 46, 119–133.

Copeland, N.G. and Jenkins, N.A. (1991) Development and applications of molecular genetic linkage map of the mouse genome. *Trends in Genetics* 7, 113–118.

Cox, D.R., Burmeister, M., Price, E.R., Kim, S. and Myer, R.M. (1990) Radiation hybrid mapping: somatic cell genetic methods for constructing high-resolution maps of mammalian chromosomes. *Science* 250, 245–250.

Crawford, A.M., Dodds, K.G., Ede, A.J. Pierson, C.A., Montgomery, G.W., Garmonsway, H.G., Beattie, A.E., Davies, K., Maddox, J.F., Kappes, S.W., Stone, R.T., Nguyen, T.C., Penty, J.M., Lord, E.A., Broom, J.E., Buitkamp, J., Schwaiger, W., Epplen, J.T., Matthew, P., Matthews, M.E., Hulme, D.J., Beh, K.J., McGraw, R.A. and Beattie, C.W. (1995) An autosomal genetics linkage map of the sheep genome. *Genetics* 140, 703–724.

Dain, A.R., Tucker, E.M., Donker, R.A. and Clarke, S.W. (1984) Chromosome mapping in cattle using mouse myeloma/calf lymph node hybridomas. *Biochemical Genetics* 22, 429–439.

Davisson, M.T., Bradt, D.W., Merriam, J.J., Rockwood, S.F. and Eppig, J.T. (1998) The mouse gene map. *ILAR Journal* 39, 97–131.

Dawson, G.W. and Graves, J.A.M. (1984) Gene mapping in marsupials and monotremes I. The chromosomes of rodent–marsupial (*Macropus*) cell hybrids, and gene assignments to the grey kangaroo X chromosome. *Chromosoma* 91, 20–27.

DeBry, R.W. and Seldin, M.F. (1996) Human/mouse homology relationships. *Genomics* 33, 337–351.

de Gortari, M.J., Freking, B.A., Cuthbertson, R.P., Kappes, S.M., Keele, J.W., Stone, R.T., Leymaster, K.A., Dodds, K.G., Crawford, A.M. and Beattie, C.W. (1998) A second-generation linkage map of the sheep genome. *Mammalian Genome* 9, 204–209.

De Hondt, H.A., Gallagher, D., Oraby, H., Othman, O.E., Bosma, A.A., Womack, J.E. and El Nahas, S.M. (1997) Gene mapping in the river buffalo (*Bubalus bubalis* L.): five syntenic groups. *Journal of Animal Breeding and Genetics* 114, 79–85.

De Leo, A.A., Guedelha, N., Toder, R., Voullaire, L., Ferguson-Smith, M.A., O'Brien, P.C.M. and Graves, J.A.M. (1999) Comparative chromosome painting between three Australian marsupials with the 2n=14 putative ancestral marsupial karyotype. *Chromosome Research* 7, 509–517.

de Oliveira, E.H., Neusser, M., Figueiredo, W.B., Nagamachi, C., Pieczarka, J.C., Sbalqueiro, I.J., Wienberg, J. and Muller, S. (2002) The phylogeny of howler monkeys (Alouatta, Platyrrhini): reconstruction by multicolor cross-species chromosome painting. *Chromosome Research* 10, 669–683.

Dietrich, W.F., Miller, J., Steen, R., Merchant, M.A., Damron-Boles, D., Husain, Z., Dredge, R., Daly, M.J., Ingalls, K.A. and O'Connor, T.J. (1996) A comprehensive genetic map of the mouse genome. *Nature* 380, 149–152.

Dixkens, C., Klett, C., Bruch, J., Kollak, A., Serov, O.L., Zhdanova, N.S., Vogel, W. and Hameister, H. (1998) ZOO-FISH analysis in insectivores: 'evolution extols the virtue of the status quo'. *Cytogenetics and Cell Genetics* 80, 61–67.

Dutrillaux, B. (1979) Chromosomal evolution in primates: tenative phylogeny from *Microcebus murinus* (Prosimian) to man. *Human Genetics* 48, 251–314.

Dutrillaux, B., Viegas-Pequignot, E. and Couturier, J. (1980) [Great homology of chromosome banding of the rabbit (*Oryctolagus cuniculus*) and primates, including man (author's transl)]. *Annales de Genetique (Paris)* 23, 22–25.

Dutrillaux, B., Couturier, J. and Viegas-Pequignot, E. (1981) Chromosomal evolution in primates. *Chromosomes Today* 7, 176–191.

Dutrillaux, B., Couturier, J., Viegas-Pequignot, E. and Muleris, M. (1982) Cytogenetic aspects of primate evolution. *Progress in Clinical and Biological Research (New York, NY)* 103, 183–194.

Echard, G., Broad, T.E., Hill, D. and Pearce, P. (1994) Present status of the ovine gene map (*Ovis aries*); comparison with the bovine map (*Bos taurus*). *Mammalian Genome* 5, 324–332.

Eicher, E.M. (1971) The identification of the chromosome bearing linkage group XII in the mouse. *Genetics* 69, 267–271.

Eizirik, E., Murphy, W.J. and O'Brien, S.J. (2001) Molecular dating and biogeography of the early placental mammal radiation. *Journal of Heredity* 92, 212–219.

Ellegren, H., Chowdhary, B.P., Johansson, M., Marklund, L., Fredholm, M., Gustavsson, I. and Andersson, L. (1994) A primary linkage map of the porcine genome reveals a low rate of genetic recombination. *Genetics* 137, 1089–1100.

El Nahas, S.M., de Hondt, H.A. and Womack, J.E. (2001) Current status of the river buffalo (*Bubalus bubalis* L.) gene map. *Journal of Heredity* 92, 221–225.

Feldman, H.W. (1924) Linkage of albino allelomorphs in rats and mice. *Genetics* 9, 487–492.

Ferguson-Smith, M.A. (1997) Genetic analysis by chromosome sorting and painting: phylogenetic and diagnostic applications. *European Journal of Human Genetics* 5, 253–265.

Finelli, P., Stanyon, R., Plesker, R., Ferguson-Smith, M.A., O'Brien, P.C., Weinberg, J. (1999) Reciprocal chromosome painting shows the great difference in diploid number between human and African green monkey is mostly due to non-Robertsonian fissions. *Mammalian Genome* 10, 713–718.

Flynn, J.J. and Nedbal, M.A. (1998) Phylogeny of the Carnivora (Mammalia): congruence vs. incompatibility among multiple data sets. *Molecular Phylogenetics and Evolution* 9, 414–426.

Francke, U., Lalley, P.A., Moss, W., Ivy, J. and Minna, J.D. (1977) Gene mapping in *Mus musculus* by interspecific cell hybridization: assignment of the genes for tripeptidase-1 to chromosome 10, dipeptidase-2 to chromosome 18, acid phosphatase-1 to chromosome 12, and adenylate kinase-1 to chromosome 2. *Cytogenetics and Cell Genetics* 19, 57–84.

Fries, R., Hediger, R. and Stranzinger, G. (1986) Tentative chromosomal location of the bovine major histocompatibility complex by in situ hybridization. *Animal Genetics* 17, 287–294.

Fries, R., Vögeli, P. and Stranzinger, G. (1990) Gene mapping in the pig. *Advances in Veterinary Science and Comparative Medicine. Domestic Animals* 34, 273–303.

Frönicke, L., Chowdhary, B.P., Scherthan, H. and Gustavsson, I. (1996) A comparative map of the porcine and human genomes demonstrates ZOO-FISH and gene mapping-based chromosomal homologies. *Mammalian Genome* 7, 285–290.

Frönicke, L., Muller-Navia, J., Romanakis, K. and Scherthan, H. (1997) Chromosomal homologies between human, harbor seal (*Phoca vitulina*) and the putative ancestral carnivore karyotype revealed by Zoo-FISH. *Chromosoma* 106, 108–113.

Frönicke, L., Wienberg, J., Stone, G., Adams, L. and Stanyon, R. (2003) Towards the delineation of the ancestral eutherian genome organization: comparative genome maps of human and the African elephant (*Loxodonta africana*) generated by chromosome painting. *Proceedings of the Royal Society of London. Series B: Biological Sciences* 270, 1331–1340.

Fujiyama, A., Watanabe, H., Toyoda, A., Taylor, T.D., Itoh, T., Tsai, S.F., Park, H.S., Yaspo, M.L., Lehrach, H., Chen, Z., Fu, G., Saitou, N., Osoegawa, K., de Jong, P.J., Suto, Y., Hattori, M. and Sakaki, Y. (2003) Construction and analysis of a human–chimpanzee comparative clone map. *Science* 295, 131–134.

Geffrotin, C., Popescu, C.P., Cribiu, E.P., Boscher, J., Renard, C., Chardon, P. and Vaiman, M. (1984) Assignment of MHC in swine to chromosome 7 by *in situ* hybridization and serological typing. *Annales de Genetique* 27, 213–219.

Glas, R., De Leo, A., Reid, K., Ferguson-Smith, M.A., O'Brien, P.C.M., Westerman, M. and Graves, J.A.M. (1997) Chromosome painting in marsupials: genome conservation in the kangaroo family. *Chromosome Research* 7, 167–176.

Glas, R., Graves, J.A.M., Toder, R., Ferguson-Smith, M.A. and O'Brien, P.C.O. (1999) Cross-species chromosome painting between human and marsupial directly demonstrates the ancient region of the mammalian X. *Mammalian Genome* 10, 1115–1116.

Goss, S.J. and Harris, H. (1975) New method for mapping genes in human chromosomes. *Nature* 255, 680–684.

Goureau, A., Yerle, M., Schmitz, A., Riquet, J., Milan, D., Pinton, P., Frelat, G. and Gellin, J. (1996) Human and porcine correspondence of chromosome segments using bidirectional chromosome painting. *Genomics* 36, 252–262.

Goureau, A., Garrigues, A., Tosser-Klopp, G., Lahbib-Mansais, Y., Chardon, P. and Yerle, M. (2001) Conserved synteny and gene order difference between human chromosome 12 and pig chromosome 5. *Cytogenetics and Cell Genetics* 94, 49–54.

Graphodatsky, A.S., Sharshov, A. and Ternovsky, D.V. (1989) Comparative cytogenetics of Mustelidae. *Zoologicheski Journal* (in Russian) 68, 96–106.

Graphodatsky, A.S., Yang, F., Perelman, P.L., O'Brien, P.C.M., Serdukova, N.A., Milne, B.S., Biltueva, L.S., Fu, B., Vorobieva, N.V., Kawada, S.-I., Robinson, T.J. and Ferguson-Smith, M.A. (2002) Comparative molecular cytogenetic studies in the order Carnivora: mapping chromosomal rearrangements onto the phylogenetic tree. *Cytogenetics and Genome Research* 96, 137–145.

Graves, J.A.M. (1995) The origin and function of the mammalian Y chromosome and Y-borne genes – an evolving understanding. *BioEssays* 17, 311–320.

Graves, J.A.M. (1996) Breaking laws and explaining rules. *Nature Genetics* 12, 121–122.

Graves, J.A.M. (2002) The rise and fall of SRY. *Trends in Genetics* 18, 259–264.

Graves, J.A.M. (2004) The degenerate Y chromosome – can conversion save it? *Reproductive Fertility* and Development 16, 527–534.

Graves, J.A.M. and Shetty, S. (2001) Sex from W to Z – evolution of vertebrate sex chromosomes and sex determining genes. *Journal of Experimental Zoology* 281, 472–481.

Graves, J.A.M. and Westerman, M. (2002) Marsupial genetics and genomics. *Trends in Genetics* 18, 517–521.

Graves, J.A.M., Wakefield, M.J. and Toder, R. (1998a) Evolution of the pseudoautosomal region of mammalian sex chromosomes. *Human Molecular Genetics* 7, 1991–1996.

Graves, J.A.M., Disteche, C.M. and Toder, R. (1998b) Gene dosage in the evolution and function of mammalian sex chromosomes. *Cytogenetics and Cell Genetics* 80, 94–103.

Gribble, S., Ng, B.L., Prigmore, E., Burford, D.C. and Carter, N.P. (2004) Chromosome paints from single copies of chromosomes. *Chromosome Research* 12, 143–151.

Grützner, F., Himmelbauer, H., Paulsen, M., Ropers, H.H. and Haaf, T. (1999) Comparative mapping of mouse and rat chromosomes by fluorescence *in situ* hybridization. *Genomics* 55, 306–313.

Grützner, F., Deakin, J., Rens, W., El-Mogharbel, N and Graves, J.A.M. (2003) The monotreme genome: a patchwork of reptile, mammal and unique features. *Comparative Biochemistry and Physiology* 136, 867–881.

Grützner, F., Rens, W., Tsend-Ayush, E., El-Mogharbel, N., O'Brien, P.C.M., Jones, R.C., Ferguson-Smith, M.A. and Graves, J.A.M. In platypus a meiotic chain of ten sex chromosomes links bird and mammal sex determining systems. *Nature* (in press).

Guilly, M.N., Fouchet, P., de Chamisso, P., Schmitz, A. and Dutrillaux, B. (1999) Comparative karyotype of rat and mouse using bidirectional chromosome painting. *Chromosome Research* 7, 213–221.

Gustavsson, I. and Rendel, J. (1971) A translocation in cattle and its association to polymorphisms in red cell antigens, transferrins and carbonic anhydrases. *Hereditas* 67, 35–38.

Guyon, R., Lorentzen, T.D., Hitte, C., Kim, L., Cadieu, E., Parker, H.G., Quignon, P., Lowe, J.K., Renier, C., Gelfenbeyn, B., Vignaux, F., DeFrance, H.B., Gloux, S., Mahairas, G.G., Andre, C., Galibert, F. and Ostrander, E.A. (2003) A 1-Mb resolution radiation hybrid map of the canine genome. *Proceedings of the National Academy of Sciences USA* 100, 5296–5301.

Haig, D. (1999) A brief history of human autosomes. *Philosophical Transactions of Royal Society London B Biological Sciences* 354, 1447–1470.

Haldane, J.B.S., Sprunt, A.D. and Haldane, N.M. (1915) Reduplication in mice. *Journal of Genetics* 5, 133–135.

Hameister, H., Klett, C., Bruch, J., Dixkens, C., Vogel, W. and Christensen, K. (1997) Zoo-FISH analysis: the American mink (*Mustela vison*) closely resembles the cat karyotype. *Chromosome Research* 5, 5–11.

Hassanane, M.S., Chaudhary, R. and Chowdhary, B.P. (1998) Microdissected bovine X chromosome segment delineates homologous chromosome regions in sheep, goat and buffalo. *Chromosome Research* 6, 2213–2217.

Hawken, R.J., Murtaugh, J., Flickinger, G.H., Yerle, M., Robic, A., Milan, D., Gellin, J., Beattie, C.W., Schook, L.B. and Alexander, L.J. (1999) A first-generation porcine whole-genome radiation hybrid map. *Mammalian Genome* 10, 824–830.

Hayes, H. (1995) Chromosome painting with human chromosome-specific DNA libraries reveals the extent and distribution of conserved segments in bovine chromosomes. *Cytogenetics and Cell Genetics* 71, 168–174.

Hayman, D.L. (1990) Marsupial cytogenetics. *Australian Journal of Zoology* 37, 331–349.

Hayman, D.L. and Martin, P.G. (1969) Cytogenetics of marsupials. In: Benirschke, K. (ed.) *Comparative Mammalian Cytogenetics*. Springer-Verlag, New York, pp. 191–217.

Helou, K., Walentinsson, A., Levan, G. and Stahl, F. (2001) Between rat and mouse zoo-FISH reveals 49 chromosomal segments that have been conserved in evolution. *Mammalian Genome* 12, 765–771.

Heuertz, S. and Hors-Cayla, M.-C. (1981) Cattle gene mapping by somatic cell hybridization study of 17 enzyme markers. *Cytogenetics and Cell Genetics* 30, 137–145.

Hines, H.C., Kiddy, C.A., Brum, E.W. and Arave, C.W. (1969) Linkage among cattle blood and milk polymorphisms. *Genetics* 62, 401–412.

Iannuzzi, L., Di Meo, G.P., Perucatti, A. and Bardaro, T. (1998) ZOO-FISH and R-banding reveal extensive conservation of human chromosome regions in euchromatic regions of river buffalo chromosomes. *Cytogenetics and Cell Genetics* 82, 210–214.

Iannuzzi, L., Di Meo, G.P., Perucatti, A. and Incarnato, D. (1999) Comparison of the human with the sheep genomes by use of human chromosome-specific painting probes. *Mammalian Genome* 10, 719–722.

Iannuzzi, L. (2003) The river buffalo (*Baubalus bubalis*, 2n = 50) cytogenetic map: assignment of 64 loci by fluorescence *in situ* hybridization and R-banding. *Cytogenetics and Genome Research* 102, 65–75.

Ijdo, J.W., Baldini, A., Ward, D.C., Reeders, S.T. and Wells, R.A. (1991) Origin of human chromosome 2: an ancestral telomere–telomere fusion. *Proceedings of the National Academy of Sciences USA* 88, 9051–9055.

ISCNDB (2000) International system for chromosome nomenclature of domestic bovids. Di Berardino, Di Meo, G.P., Gallagher, D.S., Hayes, H. and Iannuzzi, L. (coordinator) (eds) *Cytogenetics and Cell Genetics* 92, 283–299.

Jacob, H.J., Brown, D.M., Bunker, R.K., Daly, M.J., Dzau, V.J., Goodman, A., Koike, G., Kren, V., Kurtz, T., Lernmark, A., Levan, G., Mao, Y., Pettersson, A., Pravenec, M., Simon, J.S., Szpirer, C., Szpirer, J., Trolliet, M.R., Winter, E.S. and Landers, E.S. (1995) A genetic linkage map of the laboratory rat, *Rattus norvegicus*. *Nature Genetics* 9, 63–69.

Jauch, A., Wienberg, J., Stanyon, R., Arnold, N., Tofanelli, S., Ishida, T. and Cremer, T. (1992) Reconstruction of genomic rearrangements in great apes and gibbons by chromosome painting. *Proceedings of National Academy of Sciences USA* 89, 8611–8615.

Jegalian, K. and Page, D.C. (1998) A proposed path by which genes common to mammalian X and Y chromosomes evolve to become X inactivated. *Nature* 394, 776–780.

Just, W., Rau, W., Akhverdian, M., Fredga, K., Graves, J.A.M., Lyapunova, E. and Vogel, W. (1995) Sex determination in *Ellobius lutescens* and *E. tancrei* in the absence of the Y chromosome and the *Sry* gene. *Nature Genetics* 11, 117–118.

Kappes, S.M., Keele, J.W., Stone, R.T., McGraw, R.A., Sonstegard, T.S., Smith, T.P.L., Lopez-Corrales, N.L. and Beattie, C.W. (1997) A second-generation linkage map of the bovine genome. *Genome Research* 7, 235–249.

Khlebodarova, T.M., Malchenko, S.N., Matveeva, N.M., Pack, S.D., Sokolova, O.V., Alabiev, B.Y., Belousov, E.S., Peremyslov, V.V., Nayakshin, A.M., Brusgaard, K. and Serov, O.L. (1995) Chromosomal and regional localization of the loci for *IGKC*, *IGGC*, *ALDB*, *HOXB*, *GPT*, and *PRNP* in the American mink (*Mustela vison*): comparison with human and mouse. *Mammalian Genome* 6, 705–709.

Kirsch, J.A.W., Lapointe, F.L. and Springer, M.S. (1997) DNA-hybridisation studies of marsupials and their implication for metatherian classification. *Australian Journal of Zoology* 45, 211–280.

Koehler, U., Bigoni, F., Wienberg, J. and Stanyon, R. (1995) Genomic reorganization in the concolor gibbon (*Hylobates concolor*) revealed by chromosome painting. *Genomics* 30, 287–292.

Koehler, U., Arnold, N., Wienberg, J., Tofanelli, S. and Stanyon, R. (1995) Genomic reorganization and disrupted chromosomal synteny in the siamang (*Hylobates syndactylus*) revealed by fluorescence *in situ* hybridization. *American Journal of Physical Anthropology* 97, 37–47.

Korstanje, R., O'Brien, P.C., Yang, F., Rens, W., Bosma, A.A., van Lith, H.A., van Zutphen, L.F. and Ferguson-Smith, M.A. (1999) Complete homology maps of the rabbit (*Oryctolagus cuniculus*) and human by reciprocal chromosome painting. *Cytogenetics and Cell Genetics* 86, 317–322.

Kuznetsov, S.B., Matveeva, N.M., Murphy, W.J., O'Brien, S.J. and Serov, O.L. (2003) Mapping of 53 loci in American mink (*Mustela vison*). *Journal of Heredity* 94, 386–391.

Lahn, B. and Page, D.C. (1997) Functional coherence of the human Y chromosome. *Science*, 278, 675–680.

Lahn, B.T. and Page, D.C. (1999a) Retroposition of autosomal mRNA yielded testis-specific gene family on human Y chromosome. *Nature Genetics* 21, 429–433.

Lahn, B.T. and Page, D.C. (1999b) Four evolutionary strata on the human X chromosome. *Science* 286, 964–967.

Larkin, D.M., der Wind, A.E., Rebeiz, M., Schweitzer, P.A., Bachman, S., Green, C., Wright, C.L., Campos, E.J., Benson, L.D., Edwards, J., Liu, L., Osoegawa, K., Womack, J.E., de Jong, J.P. and Harris, L.A. (2003) A cattle–human comparative map built with cattle BAC-ends and human genome sequence. *Genome Research* 13, 1966–1972.

Larsen, B. (1966) Test for linkage of the genes controlling haemoglobin, transferrin and blood types in cattle. *Royal Veterinary and Agricultural University, Copenhagen, Yearbook* 41–48.

Lear, T.L. and Bailey, E. (1997) Localization of the U2 linkage group of horses to ECA 3 using chromosome painting. *Journal of Heredity* 88, 162–164.

Levan, G., Stahl, F., Klinga-Levan, K., Szpirer, J. and Szpirer, C. (1998) The rat gene map. *ILAR Journal* 39, 132–137.

Liechty, M.C., Hall, B.K., Scalzi, J.M., Davis, L.M., Caspary, W.J. and Hozier, J.C. (1995) Mouse chromosome-specific painting probes generated from microdissected chromosomes. *Mammalian Genome* 6, 592–594.

Litt, M. and Luty, J.A. (1989) A hypervariable microsatellite revealed by *in vitro* amplification of a dinucleotide repeat within the cardiac muscle actin gene. *American Journal of Human Genetics* 44, 397–401.

Liu, F.-G.R., Miyamoto, M.M., Freire, N.P., Ong, P.Q., Tennat, M.R., Youmg, T.S. and Gugel, K.F. (2001) Molecular and morphological supertrees for eutherian (placental) mammals. *Science* 291, 1786–1789.

Lyon, M.F. (1961) Gene action in the X chromosome of the mouse. *Nature* 190, 372–373.

Lyons, L.A., Laughlin, T.F., Copeland, N.G., Jenkins, N.A., Womack, J.E. and O'Brien, S.J. (1997) Comparative anchor tagged sequences (CATS) for integrative mapping of mammalian genomes. *Nature Genetics* 15, 47–56.

Maccarone, P., Watson, J.M., Francis, D., Kola, I. and Graves, J.A.M. (1992) The evolution of human chromosome 21: evidence from *in situ* hybridization in marsupials and a monotreme. *Genomics* 13, 1119–1124.

Maddox, J.F., Davies, K.P., Crawford, A.M., Hulme, D.J., Vaiman, D., Cribiu, E.P., Freking, B.A., Beh, K.J., Cockett, N.E., Kang N., Riffkin, C.D., Drinkwater, R., Moore, S.S., Dodds, K.G., Lumsden, J.M., van Stijn, T.C., Phua, S.H., Adelson, D.L., Burkin, H.R., Broom, J.E., Buitkamp, J., Cambridge, L., Cushwa, W.T., Gerard, E., Galloway, S.M., Harrison, B., Hawken, R.J., Hiendleder, S., Henry, H.M., Medrano, J.F., Paterson, K.A., Schibler, L., Stone, R.T. and van Hest, B. (2001) An enhanced linkage map of the sheep genome comprising more than 1000 loci. *Genome Research* 11, 1275–1289.

Markert, C.L. (1968) The molecular basis for isozymes. *Annals of the New York Academy of Sciences* 151, 14–40.

Marklund, L., Johansson, M., Hoyheim, B., Davies, W., Fredholm, M., Juneja, R.K., Mariani, P., Coppieters, W., Ellegren, H. and Andersson, L. (1996) A comprehensive linkage map of the pig based on a wild pig–large white intercross. *Animal Genetics* 27, 255–269.

Matsubara, K., Nishida-Umehara, C., Kuroiwa, A., Tsuchiya, K. and Matsuda, Y. (2003) Identification of chromosome rearrangements between the laboratory mouse (*Mus musculus*) and the Indian spiny mouse (*Mus platythrix*) by comparative FISH analysis. *Chromosome Research* 11, 57–64.

Matthey, R. (1972) Chromosomes and evolution. *Triangle* 11, 107–112.

Matthey, R. (1973) The chromosome formulae of eutherian mammals. In: Chiarelli, A. and Capanna, E. (eds) *Cytotaxonomy and Vertebrate Evolution*. Academic Press, London, pp. 531–616.

Matyakhina, L.D., Koroleva, I.V., Malchenko, S.N., Bendixen, C., Cheryaukene, O.V., Pack, S.D., Borodin, P.M., Serov, O.L. and Searle, J.B. (1997) Chromosome location of sixteen genes in the common shrew, *Sorex araneus* L. (Mammalia, Insectivora). *Cytogenetics and Cell Genetics* 77, 201–204.

Mazeyrat, S., Saut, N., Mattei, M-G. and Mitchell, M.J. (1999) RBMY evolved on the Y chromosome from a ubiquitously transcribed X-Y identical gene. *Nature Genetics* 22, 224–226.

McCarthy, L.C., Bihoreau, M.T., Kiguwa, S.L., Browne, J., Watanabe, T.K., Hishigaki, H., Tsuji, A., Kiel, S., Webber, C., Davis, M.E., Knights, C., Smith, A., Critcher, R., Huxtall, P., Hudson, J.R., Ono, T., Hayashi, H., Takagi, T., Nakamura, Y., Tanigami, A., Goodfellow, P.N., Lathrop, G.M. and James, M.R. (2000) A whole-genome radiation hybrid panel and framework map of the rat genome. *Mammalian Genome* 11, 791–795.

McCarthy, L., Bihoreau, M.T., Kugawa, S.L., Browne, J., Watanabe, T.K., Hishigaki, H., Tsuji, A., Kiel, S., Webber, C., Davis, M.E., *et al.* (2000) A whole-genome radiation hybrid panel and framework map of the rat genome. *Mammalian Genome* 11, 791–795.

McKusick, V.A. and Ruddle, F.H. (1977) The status of the gene map of the human chromosome. *Science* 196, 390–405.

Menotti-Raymond, M., David, V.A., Chen, Z.Q., Menotti, K.A., Sun, S., Schaffer, A.A., Agarwala, R., Tomlin, J.F., O'Brien, S.J. and Murphy, W.J. (2003) Second-generation integrated genetic linkage/radiation hybrid maps of the domestic cat (*Felis catus*). *Journal of Heredity* 94, 95–106.

Misceo, D., Ventura, M., Eder, V., Rocchi, M. and Archidiacono, N. (2003) Human chromosome 16 conservation in primates. *Chromosome Research* 11, 323–326.

Morescalchi, M.A., Schempp, W., Consigliere, S., Bigoni, F., Wienberg, J. and Stanyon, R. (1997) Mapping chromosomal homology between humans and the black-handed spider monkey by fluorescence *in situ* hybridization. *Chromosome Research* 5, 527–536.

Mrasek, K., Heller, A., Rubtsov, N., Trifonov, V., Starke, H., Rocchi, M., Claussen, U. and Liehr,T. (2001) Reconstruction of the female *Gorilla gorilla* karyotype using 25-color FISH and multicolor banding (MCB). *Cytogenetics and Cell Genetics* 93, 242–244.

Mrasek, K., Heller, A., Rubtsov N., Trifonov, V., Starke, H., Claussen, U. and Liehr, T. (2003) Detailed *Hylobates lar* karyotype defined by 25-color FISH and multicolor banding. *International Journal of Molecular Medicine* 12, 139–146.

Muller, S. and Wienberg, J. (2001) "Bar-coding" primate chromosomes: molecular cytogenetic screening for the ancestral hominoid karyotype. *Human Genetics* 109, 85–94.

Muller, S., O'Brien, P.C., Ferguson-Smith, M.A. and Wienberg, J. (1997) Reciprocal chromosome painting between human and prosimians (*Eulemur macaco macaco* and *E. fulvus mayottensis*). *Cytogenetics and Cell Genetics* 78, 260–271.

Muller, S., Hollatz, M. and Wienberg, J. (2003) Chromosomal phylogeny and evolution of gibbons (Hylobatidae). *Human Genetics* 113, 493–501.

Muller, S., Stanyon, R., O'Brien, P.C., Ferguson-Smith, M.A., Plesker, R. and Wienberg, J. (1999) Defining the ancestral karyotype of all primates by multidirectional chromosome painting between tree shrews, lemurs and humans. *Chromosoma* 108, 393–400.

Murphy, W.J., Sun, S., Chen, Z., Yuhki, N., Hirschmann, D., Menotti-Raymond, M. and O'Brien, S.J. (2000) A radiation hybrid map of the cat genome: implications for comparative mapping. *Genome Research* 10, 691–702.

Murphy, W.J., Stanyon, R. and O'Brien, S.J. (2001a) Evolution of mammalian genome organization inferred from comparative gene mapping. *Genome Biology* 2, 1–8.

Murphy, W.J., Eizirik, E., O'Brien, S.J., Madsen, O., Scally, M., Douady, C., Teeling, E., Ryder, O., Stanhope, M.J., de Jong, W.W. and Springer, M. (2001b) Resolution of the early placental mammal radiation using Bayesian phylogenetics. *Science* 294, 2348–2351.

Murphy, W.J., Eizirik, E., Johnson, W.E., Zhang, Y.P., Ryder, O.A. and O'Brien, S.J. (2001c) Molecular phylogenetics and the origins of placental mammals. *Nature* 409, 614–618.

Nadeau, J.H. and Taylor, B.A. (1984) Lengths of chromosomal segments conserved since divergence of man and mouse. *Proceedings of the National Academy of Sciences USA* 81, 814–818.

Nanda, I., Shan, Z., Schartl, M., Burt, D.W., Koehlar, M. *et al.* (1999) 300 million years of conserved synteny between chicken Z and human chromosome 9. *Nature Genetics* 21, 258–259.

Nash, W.G., Wienberg, J., Ferguson-Smith, M.A., Menninger, J.C. and O'Brien, S.J. (1998) Comparative genomics: tracking chromosome evolution in the family Ursidae using reciprocal chromosome painting. *Cytogenetics and Cell Genetics* 83, 182–192.

Neff, M.F., Broman, K.W., Mellersh, C.S., Ray, K., Gregory, M. Acland, G.M., Aguirre, G.D., Ziegle, J.S., Ostrander, E.A. and Rine, J.A. (1999) Second-generation genetic linkage map of the domestic dog, *Canis familiaris*. *Genetics* 151, 803–820.

Nesterova, T.B. Duthie, S.M., Mazurok, N.A., Isaenko, A.A., Rubtsova, N.V., Zakian, S.M. and Brockdorff, N. (1998) Comparative mapping of X chromosomes in vole species of the genus *Microtus*. *Chromosome Research* 6, 41–48.

Neusser, M., Stanyon, R., Bigoni, F., Wienberg, J. and Muller, S. (2001) Molecular cytotaxonomy of New World monkeys (Platyrrhini) – comparative analysis of five species by multi-color chromosome painting gives evidence for a classification of *Callimico goeldii* within the family of Callitrichidae. *Cytogenetics and Cell Genetics* 94, 206–215.

Nikaido, M., Rooney, A.P. and Okada, N. (1999) Phylogenetic relationships among cetartiodactyls based on insertions of short and long interpersed elements: hippopotamuses are the closest extant relatives of whales. *Proceedings of the National Academy of Sciences USA* 96, 10261–10266.

Nilsson, S., Helou, K., Walentinsson, A., Szpirer, C., Nerman, O. *et al.* Rat–mouse and rat–human comparative maps based on gene homology and high-resolution zoo-FISH. *Genomics* 74: 287–298.

Novacek, M.J. (2001) Mammalian phylogeny: genes and supertrees. *Current Biology* 11, R573-R575.

O'Brien, S.J. and Graves, J.A.M. (1991) Report of the Comparative Gene Mapping Committee. *Human gene Mapping 11; Cytogenetics and Cell Genetics* 58, 1124–1151.

O'Brien, S.J. and Stanyon, R. (1999) Phylogenomics. Ancestral primate viewed. *Nature* 402, 365–366.

O'Brien, S.J., Nash, W.G., Wildt, D.E., Bush, M.E. and Benveniste, R.E. (1985) A molecular solution to the riddle of the giant panda's phylogeny. *Nature* 317, 140–144.

O'Brien, S.J., Seuanez, N.H. and Womack, J.E. (1988) Mammalian genome organization: an evolutionary view. *Annual Reviews in Genetics* 22, 323–351.

O'Brien, S.J., Cevario, S.J., Martenson, J.S., Thompson, M.A., Nash, W.G., Chang, E., Graves, J.A.M., Spencer, J.A., Cho, K.-W., Tsujimoto, H. and Lyons, L.A. (1997) Comparative gene mapping in the domestic cat (*Felis catus*). *Journal of Heredity* 88, 408–414.

O'Brien, S.J., Menotti-Raymond, M., Murphy, W.J., Wienberg, J., Stanyon, R., Nash, W.G., Copeland, N.G., Jenkins, N.A., Womack, J.E. and Graves, J.A.M. (1999a) The promise of comparative genomics in mammals. *Science* 286, 458–481.

O'Brien, S.J., Eisenberg, J.F., Miyamoto, M., Hedges, S.B., Kumar, S., Wilson, D.E., Menotti-Raymond, M., Murphy, W.J., Nash, W.G., Lyons, L.A., Menninger, J.C., Stanyon, R., Wienberg, J., Copeland, N.G., Jenkins, N.A., Gellin, J., Yerle, M., Andersson, L., Womack, J., Broad, T. *et al.* (1999b) Genome maps 10. Comparative genomics. Mammalian radiations. Wall chart. *Science* 286, 463–478.

Ohno, S. (1967) *Sex Chromosomes and Sex Linked Genes*. Springer Verlag, Berlin.

Ohno, S. (1969) Evolution of sex chromosomes in mammals. *Annual Reviews in Genetics* 3, 495–524.

Ohno, S., Becak, W. and Becak, M.L. (1964) X-autosome ratio and the behaviour pattern of individual X-chromosomes in placental mammals. *Chromosoma* 15, 14–30.

O'Neill, R.J.W., Eldridge, M.D.B., Ferguson-Smith, M.A., O'Brien, P.C. and Graves, J.A.M. (1999) Chromosome evolution in kangaroos (Marsupialia: Macropodidae). Cross species chromosome painting between the tammar wallaby and rock wallaby spp. with the 2n=22 ancestral macropodid karyotype. *Genome* 42, 525–530.

Pask, A., Renfree, M.B. and Graves, J.A.M. (2000) The human sex-reversing gene *ATRX* has a homologue on the marsupial Y chromosome. *Proceedings of the National Academy of Sciences USA* 97, 13198–13202.

Pevzner, P. and Tesler, G. (2003) Genome rearrangements in mammalian evolution: lessons from human and mouse genomes. *Genome Research* 13, 37–45.

Postlethwait, J.H., Woods, I.G., Ngo-Hazelett, P., Yan, Y.L., Kelly, P.D., Chu, F., Huang, H., Hill-Force, A. and Talbot, W.S. (2000) Zebrafish comparative genomics and the origins of vertebrate chromosomes. *Genome Research* 10, 1890–1902.

Rambau, R.V. and Robinson, T.J. (2003) Chromosome painting in the African four-striped mouse *Rhabdomys pumilio*: detection of possible murid specific contiguous segment combinations. *Chromosome Research* 11, 91–98.

Rasmusen, B.A. (1965) Linkage between the loci for C and J blood groups in pigs. *Vox Sanguinis* 10, 239–241.

Raudsepp, T. and Chowdhary, B.P. (1999) Construction of chromosome-specific paints for meta- and submetacentric autosomes and the sex chromosomes in the horse and their use to detect homologous chromosomal segments in the donkey. *Chromosome Research* 7, 103–114.

Rebeiz, M. and Lewin, H.A. (2000) Compass of 47,787 cattle ESTs. *Animal Biotechnology* 11, 75–241.

Rens, W., O'Brien, P.C.M., Yang, F., Solanky, N., Perelman, P., Graphodatsky, A.S., Ferguson, M.W.J. Svartman, M., De Leo, A.A., Graves, J.A.M. and Ferguson-Smith, M.A. (2001) Karyotype relationships between distantly related marsupials from South America and Australia. *Chromosome Research* 9, 301–308.

Rens, W., O'Brien, P.C.M., Fairclough, H., Harman, L., Graves, J.A.M. and Ferguson-Smith, M.A. (2003) Reversal and convergence in marsupial chromosome evolution. *Cytogenetics and Genome Research* 102, 282–290.

Rettenberger, G., Fries, R., Engel, W., Scheit, K.L.H., Dolf, G. and Hameister, H. (1994) Establishment of a partially informative porcine somatic cell hybrid panel and assignment of the loci for transition protein 2 (TNP2) and protamine 1 (PRM1) to chromosome 3 and polyubiquitin (UBC) to chromosome 14. *Genomics* 21, 558–566.

Rettenberger, G., Klett, C., Zechner, U., Bruch, J., Just, W., Vogel, W. and Hameister, H. (1995) ZOO-FISH analysis: cat and human karyotypes closely resemble the putative ancestral mammalian karyotype. *Chromosome Research* 3, 479–486.

Rettenberger, G., Bruch, J., Fries, R, Archibald, A.L. and Hameister, H. (1996) Assignment of 19 porcine type I loci by somatic cell hybrid analysis detects new regions of conserved synteny between human and pig. *Mammalian Genome* 7, 275–279.

Rexroad, C.E., III, Owens, E.K., Johnson, J.S. and Womack, J.E. (2000) A 12000 rad whole genome radiation hybrid panel for high resolution mapping in cattle: characterization of the centromeric end of chromosome 1. *Animal Genetics* 31, 262–265.

Richard, F., Lombard, M. and Dutrillaux, B. (1996) ZOO-FISH suggests a complete homology between human and capuchin monkey (Platyrrhini) euchromatin. *Genomics* 36, 417–423.

Richard, F., Lombard, M. and Dutrillaux, B. (2000) Phylogenetic origin of human chromosomes 7, 16, and 19 and their homologs in placental mammals. *Genome Research* 10, 644–651.

Richard, F., Messaoudi, C., Lombard, M. and Dutrillaux, B. (2001) Chromosome homologies between man and mountain zebra (*Equus zebra hartmannae*) and description of a new ancestral synteny involving sequences homologous to human chromosomes 4 and 8. *Cytogenetics and Cell Genetics* 93, 291–296.

Richard, F., Messaoudi, C., Bonnet-Garnier, A., Lombard, M. and Dutrillaux, B. (2003) Highly conserved chromosomes in an Asian squirrel (*Menetes berdmorei*, Rodentia: Sciuridae) as demonstrated by ZOO-FISH with human probes. *Chromosome Research* 11, 597–603.

Richard, F., Lombard, M. and Dutrillaux, B. (2003a) Reconstruction of the ancestral karyotype of eutherian mammals. *Chromosome Research* 11, 605–618.

Richard, F., Lombard, M. and Dutrillaux, B. (2003b) Reconstruction of the ancestral karyotype of eutherian mammals. *Chromosome Research* 11, 605–618.

Richardson, B.J., Czuppon, A.B. and Sharman, G.B. (1971) Inheritance of glucose-6-phosphate dehydrogenase variation in kangaroos. *Nature* 230, 154–155.

Rink, A., Santschi, E.M., Eyer, K.M., Roelofs, B., Hess, M., Godfrey, M., Karajusuf, E.K., Yerele, M., Milan, D. and Beattie, C.W. (2002) The first generation EST RH comparative map of the porcine and human genome. *Mammalian Genome* 13, 578–587.

Robinson, T.J., Yang, F. and Harrison, W.R. (2002) Chromosome painting refines the history of genome evolution in hares and rabbits (order Lagomorpha). *Cytogenetics and Genome Research* 96, 223–227.

Roderick, T.H., Davisson, M.T. and Lane, P.W. (1976) Personal communication. *Mouse Newsletter* 55, 18.

Rofe, R.H. and Hayman, D.L. (1985) G-banding evidence for a conserved complement in the Marsupialia. *Cytogenetics and Cell Genetics* 39, 40–50.

Rogers, J. and VandeBerg, J.L. (1998) Gene maps of nonhuman primates. *Ilar Journal* 39, 145–152.

Rohrer, G.A., Alexander, L., Hu, Z., Smith, T.P. and Wang, L. (1996) A comprehensive map of the porcine genome. *Genome Research* 6, 371–391.

Rothschild, M.F. and Plastow, G.S. (1999) Advances in pig genomics and industry applications. *AgBiotech Net* 1, 1–7.

Rubtsov, N.B. (1999) The fox gene map. *Institute for Laboratory Animal Research Journal* 39, 182–188.

Ruddle, F.H. (1972) Linkage analysis using somatic cell hybrids. *Advances in Human Genetics* 3, 173–235.

Saiki, R.K., Scharf, S., Falcona, F., Mullis, K.B. and Horn, G.T. (1985) Enzyme amplification of beta-globin genomic sequences and restriction site analysis for diagnosis of sickle-cell anemia. *Science* 230, 1350–1354.

Saitoh, Y. and Ikeda, J.E. (1997) Chromosome microdissection and microcloning. *Chromosome Research* 5, 77–80.

Sawyer, J.R. and Hozier, J.C. (1986) High resolution of mouse chromosomes: banding conservation between man and mouse. *Science* 232, 1632–1635.

Samollow, P. and Graves, J.A.M. (1998) Gene mapping in marsupials. *Institute for Laboratory Animal Research Journal* 39, 204–223.

Samollow, P.B., Kammerer, C.M., Mahaney, S.M., Schneider, J.L., Westenberger, S.J., VandeBerg, J.L. and

Robinson, E.S. (2004) First-generation linkage map of the gray, short-tailed opossum, *Monodelphis domestica*, reveals genome-wide reduction in female recombination rates. *Genetics* 166, 307–329.

Sargan, D.R., Yang, F., Squire, M., Milne, B.S., O'Bien, P.C. and Ferguson-Smith, M.A. (2000) Use of flow-sorted canine chromosomes in the assignment of canine linkage, radiation hybrid, and syntenic groups to chromosomes: refinement and verification of the comparative chromosome map for dog and human. *Genomics* 69, 182–195.

Scherthan, H., Cremer, T., Arnason, U., Weier, H.U., Lima-de-Faria, A. and Frönicke, L. (1994) Comparative chromosome painting discloses homologous segments in distantly related mammals. *Nature Genetics* 6, 342–347.

Sensi, A., Gruppioni, R., Bonfatti, A., Rubini, M., Giunta, C. and Fontana, F. (1995) Syntenic groups between human chromosome 9 and Indian muntjac chromosomes revealed by ZOO-FISH. *European Journal of Histochemistry* 39, 317–320.

Serov, O.L. (1998) The American mink gene map. *Institute for Laboratory Animal Research Journal* 39, 189–194.

Serov, O.L. and Rubtsov, N.B. (1998) Gene mapping in fur bearing animals: genetic maps and comparative gene mapping. *AgBiotech News and Information* 10, 179N–186N.

Serov, O.L., Zhdanova, N.S., Pack, S.D., Lavrentieva, M.V., Shilov, A.G., Rivkin, M.I., Matyakhina, C.D., Draber, P., Kerkis, A.Y., Rogozin, I.B. and Borodin, P.M. (1990) The mink X chromosome: organization and inactivation. *Progress in Clinical and Biological Research* 344, 589–618.

Serov, O.L., Matyakhina, L.D., Borodin, P.M. and Searle, J.B. (1998) The common shrew gene map. *Institute for Laboratory Animal Research Journal* 39, 195–202.

Sharp, P. (1982) Sex chromosome pairing during male meiosis in marsupials. *Chromosoma* 86, 27–47.

Sherlock, J.K., Griffin, D.K., Delhanty, J.D. and Parrington, J.M. (1995) Homologies between human and marmoset (*Callithrix jacchus*) chromosomes revealed by comparative chromosome painting. *Genomics* 33, 214–219.

Sinclair, A.H., Foster, J.W., Spencer, J.A., Page, D.C., Palmer, M., Goodfellow, P.N. and Graves, J.A.M. (1988) Sequences homologous to ZFY, a candidate human sex-determining gene, are autosomal in marsupials. *Nature* 336, 780–783.

Sinclair, A.H, Berta, P., Palmer, M.S., Hawkins, J.R., Griffiths, B.L., Smith, M.J., Foster, J.W., Frischauf, A.M., Lovell-Badge, R. and Goodfellow, P.N. (1990) A gene from the human sex-determining region encodes a protein with homology to a conserved DNA-binding motif. *Nature* 346, 240–244.

Skaletsky, H., Kuroda-Kawaguchi, T., Minx, P.J., Cordum, H.S., Hillier, L., Brown, L.G., Repping, S., Pyntikova, T., Ali, J., Bieri, T., Chinwalla, A., Delehaunty, A., Delehaunty, K., Du, H., Fewell, G., Fulton, L., Fulton, R., Graves, T., Hou, S.F., Latrielle, P. *et al.* (2003) The male-specific region of the human Y chromosome is a mosaic of discrete sequence classes. *Nature* 423, 825–837.

Solinas-Toldo, S., Lengauer, C. and Fries, R. (1995) Comparative genome map of human and cattle. *Genomics* 27, 489–496.

Southern, E. (1975) Detection of specific sequences among DNA fragments separated by gel electrophoresis. *Journal of Molecular Biology* 98, 503.

Spencer, J.A., Watson, J.M. and Graves, J.A.M. (1991a) The X chromosome of marsupials shares a highly conserved region with eutherians. *Genomics* 9, 598–604.

Spencer, J.A., Sinclair, A.H., Watson, J.M. and Graves, J.A.M. (1991b) Genes on the short arm of the human X chromosome are not shared with the marsupial X. *Genomics* 11, 339–345.

Spriggs, H.F., Holmes, N.G., Breen, M.G., Deloukas, P.G., Langford, C.F., Ross, M.T., Carter, N.P., Davis, M.E., Knights, C.E., Smith, A.E., Farr, C.J., McCarthy, L.C. and Binns, M.M. (2003) Construction and integration of radiation-hybrid and cytogenetic maps of dog chromosome X. *Mammalian Genome* 14, 214–221.

Springer, M.S. and De Long W.W. (2001) Which mammalian supertree to bark up? *Science* 291, 1709–1711.

Stanhope, M.J., Waddell, V.G., Madsen, O., de Jong, W., Hedges, S.B., Cleven, G.C., Kao, D. and Springer, M.S. (1998) Molecular evidence for multiple origins of Insectivora and for a new order of endemic African insectivore mammals. *Proceedings of the National Academy of Sciences USA* 95, 9967–9972.

Stanyon, R., Koehler, U. and Consigliere, S. (2002) Chromosome painting reveals that galagos have highly derived karyotypes. *American Journal of Physical Anthropology* 117, 319–326.

Stanyon, R., Bonvicino, C.R., Svartman, M. and Seuanez, H.N. (2003) Chromosome painting in *Callicebus lugens*, the species with the lowest diploid number (2n=16) known in primates. *Chromosoma* 112, 201–206.

Stanyon, R., Yang, F., Cavagna, P., O'Brien, P.C.M., Bagga, M., Ferguson-Smith, M.A. and Weinberg, J. (1999) Reciprocal painting shows that genomic rearrangements between rat and mouse proceeds ten time faster than between humans and cats. *Cytogenetics and Cell Genetics* 84, 150–155.

Svartman, M. and Vianna-Morgante, A.M. (1998) Karyotype evolution of marsupials: from higher to lower diploid numbers. *Cytogenetics and Cell Genetics* 82, 263–266.

Svartman, M. and Vianna-Morgante, A.M. (1999) Comparative genome analysis in American marsupials: chromosome banding and *in situ* hybridization. *Chromosome Research* 7, 267–275.

Svartman, M., Stone, G., Page, J.E. and Stanyon, R. (2004) A chromosome painting test of the basal eutherian karyotype. *Chromosome Research* 12, 45–53.

Szpirer, J., Levan, G., Thorn, M. and Szpirer, C. (1984) Gene mapping in the rat by mouse-rat somatic cell hybridization synteny of the albumin and α-fetoprotein genes and assignment to chromosome 14. *Cytogenetics and Cell Genetics* 38, 142–149.

Taylor, B.A. (1989) Recombinant inbred strains. In: Lyon, M.F. and Searle, A.G. (eds) *Genetic Variants and Strains of the Laboratory Mouse*, 2nd edn. Oxford University Press, New York, pp. 773–796.

Threadgill, D.S., Kraus, J.P., Krawetz, S.A. and Womack, J.E. (1991) Evidence for the evolutionary origin of human chromosome 21 from comparative gene mapping in the cow and mouse. *Proceedings of the National Academy of Sciences USA* 88, 154–158.

Threadgill, D.W. and Womack, J.E. (1988) Regional localization of mouse *Abl* and *Mos* protoonocogenes by *in situ* hybridization. *Genomics* 3, 82–86.

Thomas, J.W., Prasad, A.B., Summers, T.J., Lee-Lin, S.Q., Maduro, V.V., Idol, J.R., Ryan, J.F., Thomas, P.J., McDowell, J.C. and Green, E.D. (2002) Parallel construction of orthologous sequence-ready clone contig maps in multiple species. *Genome Research* 12, 1277–1285.

Tian, Y., Nie, W.H., Wang, J.H., Yang, Y.F. and Yang, F.T. (2002) [Comparative chromosome painting shows the red panda (*Ailurus fulgens*) has a highly conserved karyotype]. *Yi Chuan Xue Bao* 29, 124–127. Chinese.

Tian, Y., Nie, W., Wang, J., Ferguson-Smith, M.A. and Yang, F. (2004) Chromosome evolution in bears: reconstructing phylogenetic relationships by cross-species chromosome painting *Chromosome Research* 12, 55–63.

Todd, N.B. (1970) Karyotypic fissioning and canid phylogeny. *Journal of Theoretical Biology* 26, 445–480.

Toder, R. and Graves, J.A.M. (1998) *CSF2RA, ANT3* and *STS* are autosomal in marsupials: implications for the origin and evolution of the pseudoautosomal region of mammalian sex chromosomes. *Mammalian Genome* 9, 373–376.

Toder, R., Wakefield, M. and Graves, J.A.M. (2000) The minimal mammalian Y chromosome – the marsupial Y as a model system. *Cytogenetics and Cell Genetics* 91, 285–292.

Trifonov, V., Yang, F., Ferguson-Smith, M.A. and Robinson, T.J. (2003) Cross-species chromosome painting in the Perissodactyla: delimitation of homologous regions in Burchell's zebra (*Equus burchellii*) and the white (*Ceratotherium simum*) and black rhinoceros (*Diceros bicornis*). *Cytogenetics and Genome Research* 103, 104–110.

Vaiman, D., Schibler, L., Bourgeois, F., Oustry, A., Amigues, Y. and Cribiu, E.P. (1996) A genetic linkage map of the male goat genome. *Genetics* 144, 279–305.

Van Dilla, M.A. and Deaven, L.L. (1990) Construction of gene libraries for each human chromosome. *Cytometry* 11, 208–218.

Van Poucke, M., Yerle, M., Chardon, P., Jacobs, K., Genet, C., Mattheeuws, M., Van Zeveren, A. and Peelman, L.J. (2003) A refined comparative map between porcine chromosome 13 and human chromosome 3. *Cytogenetics and Genome Research* 102, 133–138.

Volleth, M., Heller, K.-G., Pfeiffer, R.A. and Hameister, H. (2002) A comparative ZOO-FISH analysis in bats elucidates the phylogenetic relationships between Megachiroptera and five microchiropteran families. *Chromosome Research* 10, 477–497.

Wakefield, M.J. and Graves, J.A.M. (1996) Comparative maps of vertebrates. *Mammalian Genome* 7, 715–716.

Wakefield, M.J. and Graves, J.A.M. (2003) The kangaroo genome: leaps and bounds in comparative genomics. *EMBO Reports* 4, 143–147.

Walter, M.A., Spillett, D.J., Thomas, P., Weissenbach, J. and Goodfellow, P.N. (1994) A method for constructing radiation hybrid maps of whole genomes. *Nature Genetics* 7, 22–28.

Waters, P., Duffy, B., Frost, C.J., Delbridge, M.L. and Graves, J.A.M. (2001) The human Y chromosome derives largely from a single autosomal region added 80–130 million years ago. *Cytogenetics and Cell Genetics* 92, 74–79.

Watson, J.M. and Graves, J.A.M. (1987) Gene mapping in marsupials and monotremes. V. Synteny between hypoxanthine phosphoribosyltransferase and phosphoglycerate kinase in the platypus. *Australian Journal of Biological Sciences* 41, 231–237.

Watson, J.M., Spencer, J.A., Riggs, A.D. and Graves, J.A.M. (1990) The X chromosome of monotremes shares a highly conserved region with the eutherian and marsupial X chromosomes, despite the absence of X chromosome inactivation. *Proceedings of the National Academy of Sciences USA* 87, 7125–7129.

Watson, J.M., Spencer, J.A., Riggs, A.D. and Graves, J.A.M. (1991) Sex chromosome evolution: platypus gene mapping suggests that part of the human X chromosome was originally autosomal. *Proceedings of the National Academy of Sciences USA* 88, 11256–11260.

Wayne, R.K. (1993) Molecular evolution of the dog family. *Trends in Genetics* 9, 218–214.

Wayne, R.K., Nash, W.G. and O'Brien, S.J. (1987) Chromosome evolution of the Canidae II. Divergence from the primitive carnivore karyotype. *Cytogenetics and Cell Genetics* 44, 134–141.

Weber, J.L. and May, P.E. (1989) Abundant class of human DNA polymorphisms which can be typed using the polymerase chain reaction. *American Journal of Human Genetics* 44, 388.

Wienberg, J. and Stanyon, R. (1997) Comparative painting of mammalian chromosomes. *Current Opinions in Genetics and Development* 7, 784–791.

Wienberg, J., Jauch, A., Stanyon, R. and Cremer, T. (1990) Molecular cytotaxonomy of primates by chromosomal *in situ* suppression hybridization. *Genomics* 8, 347–350.

Wienberg, J., Jauch, A., Lüdecke, H.J., Senger, G., Horsthemke, B., Claussen, U., Cremer, T., Arnold, N. and Lengauer, C. (1994) The origin of human chromosome 2 analysed by comparative chromosome mapping with a DNA microlibrary. *Chromosome Research* 2, 405–410.

Wienberg, J., Stanyon, R., Nash, W.G., O'Brien, P., Yang, F., O'Brien, S.J. and Ferguson-Smith, M.A. (1997) Conservation of human vs. feline genome organization revealed by reciprocal chromosome painting. *Cytogenetics and Cell Genetics* 77, 211–217.

Wilcox, S.A., Watson, J.M., Spencer, J.A. and Graves, J.A.M. (1996) Comparative mapping identifies the fusion point of an ancient mammalian X–autosomal rearrangement. *Genomics* 35, 66–70.

Womack, J.E. (1990) A comparative approach to the bovine gene map. In: Poli, G. and Beckmann, J. (eds) *Mappaggio del Genoma Bovino.* Societa Italiana Della Scienze Veterinarie, Milano, pp. 79–98.

Womack, J.E. and Moll, Y.D. (1986) A gene map of the cow: conservation of linkage with mouse and man. *Journal of Heredity* 77, 2–7.

Womack, J.E., Johnson, J.S., Owens, E.K., Rexroad, C.E., III, Schläpfer, J. and Yang, Y.-P. (1997) A whole-genome radiation hybrid panel for bovine gene mapping. *Mammalian Genome* 8, 854–856.

Woodburne, M.O., Rich, T.H. and Springer, M.S. (2003) The evolution of tribosphenv and the antiquity of mammalian clades. *Molecular Phylogenetics and Evolution* 28, 360–385.

Wrigley, J.M. and Graves, J.A.M. (1984) Two monotreme cell lines derived from female platypuses. (*Ornithorhynchus anatinus*; Monotremata, Mammalia). *In Vitro* 20, 321–328.

Wrigley, J.M. and Graves, J.A.M. (1988) Karyotypic conservation in the mammalian order Monotremata (subclass Prototheria). *Chromosoma* 96, 231–247

Yang, F., O'Brien, P.C., Weinberg, J. and Ferguson-Smith, M.A. (1997) A reappraisal of the tandem fusion theory of karyotype evolution of Indian muntjac using chromosome painting. *Chromosome Research* 103, 642–652.

Yang, F., O'Brien, P.C.M., Milne, B.S., Graphodatsky, A.S., Solansky, N., Trifonov, V., Rens, W., Sragan, D. and Ferguson-Smith, M.A. (2000a) A complete comparative chromosome map for the dog, red fox, and human and its integration with canine genetic maps. *Genomics* 62, 189–202.

Yang, F., O'Brien, P.C., Wienberg, J. and Ferguson-Smith, M.A. (1997) Evolution of the black muntjac (*Muntiacus crinifrons*) karyotype revealed by comparative chromosome painting. *Cytogenetics and Cell Genetics* 76, 159–163.

Yang, F., Graphodatsky, A.S., O'Brien, P.C., Colabella, A., Solanky, N., Squire, M., Sargan, D.R. and Ferguson-Smith, M.A. (2000b) Reciprocal chromosome painting illuminates the history of genome evolution of the domestic cat, dog and human. *Chromosome Research* 8, 393–404.

Yang, F., Alkalaeva, E.Z., Perelman, P.L., Pardini, A.T., Harrison, W.R., O'Brien, P.C.M., Fu, B., Graphodatsky, A.S., Ferguson-Smith, M.A. and Robinson, T.J. (2003) Reciprocal chromosome painting among human, aardvark, and elephant (superorder Afrotheria) reveals the likely eutherian ancestral karyotype. *Proceedings of the National Academy of Sciences USA* 100, 1062–1066.

Yang, F., Fu, B., O'Brien, P.C.M., Nie, W., Ryder, O.A. and Ferguson-Smith, M.A. (2004) Refined genome-wide comparative map of the domestic horse, donkey and human based on cross-species chromosome painting: insight into the occasional fertility of mules. *Chromosome Research* 12, 65–76.

Yang, Y.-P. and Womack, J.E. (1998) Parallel radiation hybrid mapping: a powerful tool for high-resolution genomic comparison. *Genome Research* 8, 731–736.

Yerle, M., Echard, G., Robic, A., Mairal, A., Dubut-Fontana, C., Riquet, J., Pinton, P., Milan, D., Lahbib-Mansais, Y. and Gellin, J. (1996) A somatic cell hybrid panel for pig regional gene mapping characterized by molecular cytogenetics. *Cytogenetics and Cell Genetics* 73, 194–202.

Yerle, M., Pinton, P., Robic, A., Alfonso, A., Palvadeau, Y., Delcros, C., Hawken, R., Alexander, L., Beattie, C., Schook, L., Milan, D. and Gellin, J. (1998) Construction of a whole-genome radiation hybrid panel for high-resolution gene mapping in pigs. *Cytogenetics and Cell Genetics* 82, 82–188.

Zenger, K.R., McKenzie, L.M. and Cooper, D.W. (2002) The first comprehensive genetic linkage map of a marsupial: the tammar wallaby (*Macropus eugenii*). *Genetics* 162, 321–330.

Zhdanova, N.S., Fokina, V.M., Balloux, F., Hausser, J., Volobouev, V., Serov, O.L., Borodin, P.M. and Larkin, D.M. (2003) Current cytogenetic map of the common shrew *Sorex araneus* L.: localization of 7 genes and 4 microsatellites. *Mammalia* 82, 285–293.

Zijlstra, C., Bosma, A.A., de Haan, N.A. and Mellink, C. (1996) Construction of a cytogenetically characterized porcine somatic cell hybrid panel and its use as a mapping tool. *Mammalian Genome* 7, 280–284.

15 Bioinformatics: From Computational Analysis through to Integrated Systems

M.I. Bellgard*

Centre for Bioinformatics and Biological Computing, Murdoch University, Murdoch, Western Australia 6150, Australia

Introduction	393
Comparative Genomic Analysis	395
Publicly Available Internet Resources	397
NCBI – the National Centre for Biotechnology Information	397
Other international bioinformatics resources	398
Errors within the data repositories	399
Bioinformatics Resources for Large-scale Genomic Analysis	400
Investigation of the diverse applications of bioinformatics tools and the need for developing bioinformatics pipelines	400
Sequence Alignment and Comparison	402
Large-scale sequence alignment	403
Multiple sequence alignment	405
Concluding Remarks	405
References	407

Introduction

The advent of large-scale genomic sequencing has resulted in masses of molecular sequences for a diverse range of organisms. There is a wealth of understanding to be gained from assimilating information by conducting extensive comparative genomic analysis. This information assists us in making reasoned interpretation on biological processes, evolutionary mechanisms as well as genomic and gene structure and characterization. More recently, the areas of functional genomics, transcriptomics, including the numerous international consortia for full-length cDNA projects, metabolomics, proteomics and structural biology are providing complementary data sources that provide further portals that assist in our understanding of the structure, function and evolution of genes and organisms from genome to phenome.

At a structural level, the definition of genes within a genome sequence is still a significant challenge that requires, ultimately,

*Correspondence: m.Bellgard@murdoch.edu.au

© CAB International 2005. *Mammalian Genomics*
(eds A. Ruvinsky and J. Marshall Graves)

the determination of sequences for full-length cDNAs for all the genes within a genome. Furthermore, as functions are assigned to genes in model organisms the bioinformatics challenge of transferring this information to other species comes into play. In the Human Genome Project (HGP) (Collins et al., 2003), the availability of other mammalian genome sequences provides the possibility of comparative analyses allowing the identification of conserved exon sequences to aid in identifying genes (Dehal et al., 2001). Human–mouse comparative genomics-based approaches to gene annotation (Flicek et al., 2003; Parra et al., 2003) include extending the classical sequence alignment approaches to maximize the score of amino acid alignments, combining the simultaneous sequence alignment with gene prediction and a separation of the sequence alignment and gene prediction processes. The inclusion of comparative genomics in gene identification emphasizes the importance of intron–exon structure in defining DNA sequences that are conserved and more likely to code for homologous (inferred evolutionarily related) genes in other species. In addition, the genome level analysis also provides a gene location within the chromosome and this may have implications for the rate of change in gene diversity since in both human–chimpanzee (Navarro and Barton, 2003) and even, for example, plant species such as rice–wheat (Akhunov et al., 2003) comparisons, genes in chromosome regions subject to structural change are more likely to undergo change themselves in evolutionary time.

The underlying capability enabling comparative genomic analysis is referred to as bioinformatics or computational molecular biology. Broadly speaking, bioinformatics is a multidisciplinary field that provides an enabling platform technology to investigate and interpret biological data. It relies on the integration of fields such as biotechnology, molecular biology, medical science, epidemiology, computer science, information systems and knowledge management, mathematics and statistics.

While the area of bioinformatics is relatively new, there have been a significant number of advances, which will be outlined below,

ranging from the computational aspects of bioinformatics such as sequence similarity, searching and alignment of molecular sequences, through to the development of robust and modular integrated systems that facilitate rigorous comparative genomic analysis for analysis and visualization. However, there are a number of bioinformatics challenges that must still be addressed. The size of the genomic repositories is nearly doubling every year. This has significant implications on the way bioinformatics analyses are carried out. There are a number of general characteristics common to most bioinformatics strategies for comparative analysis. For instance, it is widely acknowledged that there may be errors in the genomic sequences and their subsequent annotations, and that these can lead to error propagation (Bellgard and Gojobori, 1999; Pennisi, 1999; Kulski et al., 2003). Conflicting annotations (Carter et al., 2001; Carter and Bellgard, 2003) can be due to omissions, errors in the bioinformatics pipeline, oversights, or conducting analysis with out-of-date or alternative bioinformatics resources. In fact, given the pace of data submission, analysis can quickly become obsolete. Periodic releases of annotations of database 'freezes' for large genome sequencing projects can be significantly different from subsequent ones. However, the differences in information from one release to another are not usually documented in a convenient manner or not at all. These characteristics have implications on individual researchers who might be unaware that their current research might be invalidated (Bellgard and Gojobori, 1999; Carter et al., 2001; Bellgard, et al., 2003a; Kulski et al., 2003). Clearly, for information to be useful, it needs to be available at the right time or at the very least, a quality-managed approach to auditing the analysis process must be developed for scrutiny and inevitable reanalysis at a later date.

In this chapter, we examine bioinformatics issues that are relevant for an individual conducting a detailed comparative genomic analysis as well as exploring the broader issues that must be considered in order to address and anticipate current challenges that will impact on the future of this field. Thus, from one perspective, detailed issues covered will include repository sizes and the various analysis tools for comparative analysis. At the broader level, issues such as appropriately managing the *ad hoc* and ever changing nature of bioinformatics information and analysis will be addressed and a discussion on the role of developing bioinformatics integrated systems will be covered.

Comparative Genomic Analysis

Sequence similarity at both the nucleotide level and the amino acid level is the basis for defining the relationships between aligned sequence pairs or multiple sequences. Naturally, there are degrees of similarity such as 20%, 50% or 95% sequence identity, and this corresponds to a sequence similarity score. At the amino acid level, substitution matrices capture relationships between amino acids that could conservatively or non-conservatively be substituted for each other in sequence alignments. These matrices are in the form of scoring matrices and thus contribute to the alignment score between protein sequences. For a review of sequence alignment and sequence similarity see, for example, Altschul (1993), Apostolico and Giancarlo (1998) and Pearson (1996). The similarity score and the statistical significance are used to infer homology, i.e. whether the sequences share a molecular evolutionary relationship. Within a species, two homologous gene sequences are referred to as paralogous genes if they were the result of a gene, segmental or genome duplication. In the absence of gene duplication, orthologous genes can be strictly defined to be homologous genes from different species clearly related to each by a common (ancestral) evolutionary history, i.e. as a result of a speciation event. The precise identification of orthologous and paralogous genes between and within species is complicated due to the occurrence of duplications, polyploidy, integration of retroelements, transposition events in addition to multiple speciation events.

A corresponding definition of syntenic chromosomal regions between two species is the occurrence of chromosomal genomic regions containing clusters of orthologous regions. Comparative analyses have identified regions of synteny at both the macro and micro level between species. The significance of this is that highly syntenic regions are more likely to be conserved or at least have related phenotypes. Like paralogy and orthology, the identification of synteny is also difficult in many situations. Where whole genome sequences are available for comparative analysis, it is possible to interpret the homology/orthology/paralogy/synteny relationships. For example, as described in Chapter 10 of this book, a detailed relationship between human and mouse can be elucidated (Dehal *et al.*, 2001). However, in species where expressed sequence tags (ESTs) are predominantly available in one species, the task of identifying homology/orthology and synteny becomes more challenging. In some cases, the genomes may never be completely sequenced due to their immense size and the associated resource implications. For example, in the context of cereal bioinformatics where whole genome sequences are not available this becomes a significant challenge to bioinformatics. The utilization of transcriptomics (ESTs and full-length cDNA and mapped chomosomal locations) and functional genomics which provides gene expression patterns can assist in piecing together the syntenic relationship puzzle where one genome is used as the anchor (Bellgard *et al.*, 2004).

A recent study on the genomic and phylogenetic analysis of the *S100A7* (psoriasin) gene duplications on human chromosome 1q21 is a good example of the necessary level of detail to begin to fully characterize gene-encoding genomic sequences (Kulski *et al.*, 2003). The human *S100* gene cluster encodes the EF-hand superfamily of calcium-binding proteins and contains at least 14 family members. The genomic sequence containing the *S100A* gene cluster (approximately 260 kbp) revealed recent genome sequence duplication for only the *S100A7* gene region. These duplications are comprised of three distinct genomic regions of 33, 11 and 31 kbp, respectively. These regions in total contain at least five identifiable *S100A7*-like genes rather than the single *S100A7* gene presently identified in GenBank. In addition, these duplicated regions share a number of different retroelements (Malakowski *et al.*, 1994; Kulski *et al.*, 2003) including five Alu subfamily members that serve as molecular clocks which can be used to infer the timing of the genomic duplications, in this case approximately 30–40 mya. This study alone highlights the need to reanalyse carefully previously characterized genomic sequences. In this case four other *S100A7* genes not previously characterized were identified. An impact of this work would be on functional genomics studies through the use of genetic microarray chip technology. Here, the complex interplay of thousands of genes is revealed by simultaneously monitoring thousands of individual gene expression levels (Quackenbush, 2001; Wu, 2001). The functional role of the other four

S100A7-like genes must be carefully investigated as only a single *S100A7* has been spotted on the microarray chip. In this study, retroelements were used to assist in the inferred ageing of the genomic duplications. However, recent studies suggest that retroelements could also play a functional role. Hence, the characterization/annotation of genomic sequences is an ongoing, iterative, dynamic process rather than a perceived static view as presented in public repositories.

As an example of how quickly genome characterization can change, a region of human DNA genomic sequence from chromosome 22 is reanalysed using the Multiple (BLAST) Annotation System Viewer (MASV) system (Carter and Bellgard, 2003) (Fig. 15.1). The particular region shown highlights a number of new features

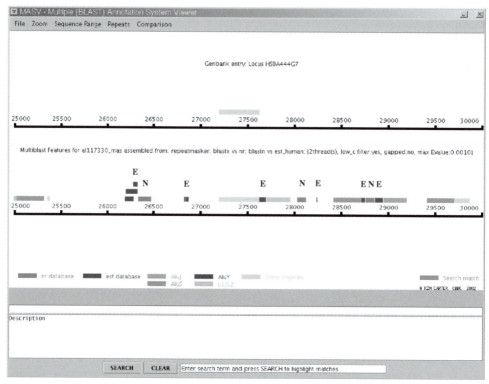

Fig. 15.1. Re-analysis of human DNA genomic sequence from clone RP11-444G7 (accession no. AL117330, 31,874 bp) on chromosome 22 which contains the CHK2 (protein kinase) gene. As can be seen in the figure that graphically shows the region from 25,000 bp to 30,000 bp, a number of protein matches (indicated by **N**), EST matches (indicated by **E**) and repetitive elements are identified (below) that were not previously identified in the GenBank entry (above). The screen shot was created using the MASV system (Carter and Bellgard, 2003). As this image is reproduced in black and white, EST (**E**) and NR (**N**) matches are explicitly shown; the rest are repetitive elements.

contained within the genomic sequence (below) compared with the original GenBank record (above). A number of protein matches, EST matches and repetitive elements are identified that were not previously identified. Naturally, the database version, the search program and parameters used will have impact on the features identified at any particular time point.

Publicly Available Internet Resources

Creating and maintaining repositories of bioinformatics data was originally managed by individual research groups conducting research in related areas. More recently, there are a selected few publicly funded bioinformatics resources that become portals for the majority of publicly available molecular data. (Please note, this section is intended to be complementary to Chapter 16 which has a genetic database focus. The context of this chapter is on the bioinformatics implications.) At these Internet web sites, various types of bioinformatics analysis can be conducted. These repositories contain both primary data and secondary (processed or interpreted information) data. These data and information can be interrogated via sophisticated search tools or a number of bioinformatics analyses can be conducted with them.

The International Nucleotide Sequence Database Collaboration is comprised of GenBank at the National Centre for Biotechnology Information (NCBI) (http://www.ncbi.nlm.nih.gov/collab/, the DNA DataBank of Japan (DDBJ) and the European Molecular Biology Laboratory (EMBL) (Stoesser *et al.*, 2001). These three organizations exchange data on a daily basis, and together have become the largest public sequence repositories in the world. While data are exchanged regularly, each site varies in its internal management of the data stored, but more importantly varies in how the data are made available to the public. These differences are in the form of visual presentation such as genome browsing and displaying matches between query sequences and repository hits through to different bioinformatics tools for conducting analysis. GenBank is the National Institutes of Health

(NIH) genetic sequence repository. It is an annotated collection of all publicly available DNA sequences (Benson, 2002). The impressive size of the repository is matched only by its growth rate, which has meant that since 1982 the database has doubled approximately every 14 months. Similarity searching against the nucleotide or protein sequences within GenBank is typically conducted against the non-redundant version of the repository. At this point it is important to comment on terminology used to describe these public repositories. The terms database, data set, databank and repository are often used interchangeably when referring to a data set of molecular sequences. For example, it is common to refer to the GenBank database or repository. This can typically cause confusion especially from a computer science perspective where the term database is reserved for the data management systems used to store the information, such as a relational database management system. However, it might be considered reasonable to refer to a sequence similarity search against one of these data sets by conducting what is commonly referred to as a nucleotide or a protein 'database' search. For the remainder of this chapter, only a databank or genomic repository or data set will be used interchangeably.

NCBI – the National Centre for Biotechnology Information

NCBI (US National Centre for Biotechnology Information Resources) is perhaps the best example of a public resource bioinformatics warehouse. It contains sophisticated linked resources for data set and literature searching and is a store of both primary and secondary data sets. For example, it links the NR (non-redundant) nucleotide and protein data sets, completed whole genome sequences, orthologous gene sets between species, single nucleotide polymorphism data sets and structural information, to name a few. These can be searched against using their system called Entrez. Entrez is the text-based search and retrieval system used for all of the major databases, including PubMed, Nucleotide and Protein Sequences, Protein Structures, Complete Genomes, Taxonomy, OMIM, and many others. Entrez is at once an indexing and retrieval system, a collection of data from many sources, and an organizing principle for biomedical information. It is important to note that most database searchers are typically against the NR nucleotide and protein sequence repositories. Historically, NR refers to the 'non-redundant' data sets. However, in practice the NR data sets are typically non-identical rather than non-redundant.

A typical bioinformatics analysis is a sequence similarity search against either a nucleotide or protein sequence repository using a tool known as BLAST (Altschul, 1990). The BLAST program provides the ability for a researcher with Internet access to search a molecular sequence against the up-to-date repository with all resources located at the NCBI website. There are a number of variations to this search. For instance, it is possible to search a nucleotide sequence against a nucleotide repository, or have it translated into all six reading frames and searched against a protein repository. There are other variations that tailor the search for query sequences that are short or just for comparing two sequences with each other rather than searching against a library of sequences. The results of this search will be graphically displayed and further searches and linking to

other related information such as protein structural information of literature via MED-LINE are available as significant integration of information has already been precomputed. There are a number of parameters that can be selected to ensure the search is sensitive to finding the correct matches while it is sufficiently selective to score highly true related matches from spurious ones (Pearson, 1996). There are sequence search techniques available other than BLAST, such as FASTA (Pearson and Lipman, 1988) and SSEARCH which implements the Smith-Waterman algorithm (Smith and Waterman, 1981) and others. Genbank records can be reanalysed and this may lead to identification of previously unknown features (Table 15.1, see section on large-scale genomic analysis).

Other international bioinformatics resources

The European Bioinformatics Institute (EBI) (http://www.ebi.ac.uk) and the DNA Database of Japan (DDBJ) (http://www.ddbj.nig.ac.jp) are the two other significant warehouses of international molecular sequence data. Together, NCBI, EBI and DDBJ represent the starting point for bioinformatics for molecular sequence submission, molecular database searching, and cross-linking with other information sources. The recently available NCBI Genome Workbench GBench (GBENCH,

Table 15.1. Selected GenBank records, their size and the year of submission: Z80998 (20,650 bp) 1999, AL117330 (31,874 bp) 2000, Z97184 (40,127 bp) 1999 and AL118506 (139,505 bp) 2001. These records were reanalysed to identify genomic features contained within them. As can be seen a number of features not previously characterized are identified (Carter and Bellgard, 2001).

Difference between assembled features using MASV vs. existing GenBank annotation	Z80998 20,650 bp	AL117330 31,874 bp	Z97184 40,127 bp	AL118506 139,505 bp
New repeats found	8	87	32	189
New NR matches (significant)	0	3	0	4
New ESTs found	6	8	32	64
Potential incorrect labels for repeat regions in existing (at approx. same position)	5	4	8	4
Potential incorrect labels for NR matches in Existing (at approx. same position)	0	0	14	5
Found same repeat, different start/end positions	8	3	22	11
Found same NR match, different start/end positions	0	2	24	26
Found same EST match, different start/end positions	0	0	1	0

2003) is a visualization application for bioinformatics analysis attempting to provide researchers with a more sophisticated environment to conduct sequence analysis. KEGG (Kyoto Encyclopaedia of Genes and Genome: http://www.genome.ad.jp/kegg/) is another useful resource containing both genomic and functional data as well as metabolic pathway links.

Genome data sets for specific organisms are usually maintained in special databases that combine not just the genome sequence (which is delivered to NCBI, etc.) and annotations but also other biological data related to the particular species such as rat, mouse and other species.

EMBOSS (http://www.hgmp.mrc.ac.uk/Software/EMBOSS/index.html) is an example of a free Open Source software analysis package specially developed for the needs of the molecular biologist. Within EMBOSS, extensive libraries are provided with the package to allow open source development. EMBOSS also integrates a range of currently available packages and tools for sequence analysis into a seamless whole. It is an excellent environment that integrates a number of different data and result formats transparently. The Sequence Manipulation Suite (http://www.sanbi.ac.za/~rmuller/SMS/) is another example of a useful bioinformatics analysis environment publicly available on the Internet.

Errors within the data repositories

While GenBank is undoubtedly a huge success, it should be noted that it is not without its problems. Surveys have revealed that as much as 2% of GenBank entries contain DNA produced by experimental procedures (Pennisi, 1999). Duplicate sequences also exist within GenBank, with no distinction made between correct and incorrect entries. Other problems are more systemic. For instance, the annotations on submitted entries are largely static, as they may only be updated by the scientist who originally submitted the sequence (Bellgard and Gojobori, 1999b; Carter and Bellgard, 2001; Benson et al., 2002; Kulski et al., 2003). With the large amount of new sequence data constantly

entering the database, it is safe to assume that older entries have missing links to new data. The problems with GenBank are widely known. One could argue, however, that the problems do not necessarily lie with GenBank but with how it is being used. GenBank is intended as a repository, which focuses on containing all public sequences (Benson et al., 2002), rather than a curated, non-redundant repository. More worrying than the current errors within GenBank are the systemic problems, particularly the fact that only submitting scientists can modify existing entries. This is said to be contributing to a large number of the errors within GenBank and these problems will only get worse as the amount of incoming data into public databases increases (Pennisi, 1999; Karp et al., 2001). Moreover, a number of data integrity-related problems have also been reported with these data repositories, such as confusion amongst gene synonyms and multiple representations of the same data (Apweiler et al., 2001), contamination with mitochondrial DNA sequence, vector contamination and sequencing errors (Brenner, 1999; Karp et al., 2001).

The NCBI resource provides alternative data set repositories, such as RefSeq, which is based on GenBank but aims to provide non-redundant, curated data: 'RefSeq sequences are derived from GenBank and provide non-redundant curated data representing our current knowledge of known genes.' (Refseq, 2003). While resources like RefSeq solve some of the problems researchers experience with GenBank, their solutions have so far only been applied to subsets of the GenBank data. By creating RefSeq, NCBI has proposed a solution to the problem of having multiple submitting authors with one submitting author, NCBI itself: 'RefSeq records are owned by NCBI and therefore can be updated as needed to maintain current annotation or to incorporate additional sequence information' (Refseq, 2003). There are clear bioinformatics challenges in this regard and the role of community-based moderation should be reconsidered given the advances in bioinformatics open source community (Stein, 2002) and the success of various online moderation systems (Hunter, 2003).

Bioinformatics Resources for Large-scale Genomic Analysis

Genome browsers are important for assimilating, visualizing and summarizing large amounts of information. Examples include DAS (Dowell et al., 2001), ENSEMBL (www.ensembl.org), the Human Genome Browser Gateway (http://genome.ucsc.edu/cgi-bin/hgGateway) MASV (Carter et al., 2001; Carter and Bellgard, 2003) and the SOE Genome Browser (Bellgard et al., 2004). MASV (Carter et al., 2001; Carter and Bellgard, 2003) system is a tool designed to aid in the annotation of genomic sequences. Unlike the current DAS viewers, MASV enables the researcher to conduct systematic searches and analysis as well as to compare and analyse differences in annotation, resulting from changes in databases, analysis program parameters and results. This provides a novel capability for individual users to conduct further bioinformatics analysis from the information obtained from third party sources.

These types of tools highlight the need to have interactive and flexible bioinformatics viewers and tools to track results in order to quality manage comparison of results at a later stage. The latter point is referred to as Bioinformatics Analysis Audit Trails (Bellgard et al., 2003a). Figure 15.1 is an example of an output of this viewer. The viewer shows the existing GenBank annotations along with the reannotations using up-to-date databases. As mentioned earlier, this analysis alone highlights the very fluid/dynamic nature of how changes in the genomic repositories can dramatically influence our interpretation of the genomic regions. Theses types of tools are essential for large-scale comparison of species such as human versus mouse.

Investigation of the diverse applications of bioinformatics tools and the need for developing bioinformatics pipelines

Related to the need to identify features contained within, and annotate, large genomic sequences, the BLAST (Altschul et al., 1990) sequence similarity search tool is extremely powerful. This is because, as more sequences

become available from diverse species, the more emphasis can be placed on homology-based annotation rather than prediction-based approaches such as GENSCAN (Burge and Karlin, 1997). When it is used to identify features within a particular genomic sequence, the search results often contain multiple 'hits' depending on the frequency of representation of the sequence in the database. For example, if one BLASTs a sequence that contains a commonly occurring mRNA for a gene as well as a small section from a less common neighbouring gene, against the NR data repository, chances are all the matching sequences in the repository (hits) will relate to the mRNA sequence. Matches with the neighbouring genes can invariably get swamped in the flood of hits for the larger more significant matches. If the aim is to characterize and annotate a query sequence as accurately as possible, it is important to retrieve as many different matches as possible rather than just find the best hits.

It is important to note here that this is a slight deviation from the basic use of BLAST and typifies the inherent ongoing customization of bioinformatics tools for various types of analysis. One solution might be to allow an 'infinite' number of hits in the BLAST report, which could be done within the existing Internet resources. However, this is not a viable solution. Another simple solution would be to find the first significant match in the BLAST result, mask out this region in the input sequence and BLAST the newly masked sequence against the database again. By repeating the BLAST process, it would be possible to record and mask out significant matches in an attempt to identify as many different hits as possible within the input sequence. With this concept in mind, a series of steps can be identified which form a 'pipeline' bioinformatics process. Thus, a simple program or script can be written that invokes the BLAST program repeatedly (Carter et al., 2001). This example is useful to highlight other important issues. One might wish to mask the query sequence for retroelements using programs such as REPEATMASKER (http://ftp.genome.washington.edu/cgi-bin/RepeatMasker) and CENSOR (Jurka et al., 1996) prior to conducting the BLAST search. However, there is a danger of masking genomic sequence especially given that retroelements have been shown to be transcribed, and thus should not be considered as 'junk'. Multiple pipelines are relevant in this situation (for instance, to mask the sequence or not after running it through a program to screen retroelements) to ensure no information is lost. The importance of visualization tools to compare the results of different pipeline analyses is integral (Carter and Bellgard, 2003). Figure 15.1 shows an example of this output of this system.

As discussed above, the public sequence repositories contain errors and the importance of re-analysis and verification of sequences and annotation, at a later date, becomes very important. This could be regarded as annotation iterative refinement. It is important to note that when a sequence is characterized, using a pipeline process, the accuracy is only as good as the quality of sequence, annotation information in the repository(s), bioinformatics tools (version) and the parameters employed. For the purposes of conducting research, the information at each stage of the 'pipeline' should be considered for storage in order to be assembled back together at the end of the process. The stored results can then be used for verification and comparison at a later date.

To demonstrate the value of this approach, the number of new retroelements found in a previously annotated genomic sequence is a simple example of how results can change in light of new information becoming available, such as a newer version of REPEATMASKER and/or updates to data sets it uses. As specific examples, Carter *et al.* (2001) conducted a study where they selected a number of human sequences from GenBank and reanalysed them (Table 15.1). There were examples of mislabelled retroelements, or differing start and end positions, as well as new retroelements found. While the latter might be expected, the fact that the GenBank entries do not contain sufficient detail as to what version of the programs were used, the parameters employed and the version of the data set searched against, makes it impossible to verify the results. These examples highlight the importance of keeping track of not only results, but database versions, software versions and parameters used in order to better compare results at a later stage. It is important to manage what could be referred to as annotation instability. Certainly, this has serious implications on up-to-date information in public repositories and relevance to other researchers conducting similar searches.

Researchers must be aware of the pitfalls of using database search tools like BLAST for characterizing genes. For example, in the same study, Carter *et al.* (2001) identified several issues relating to the sensitivity of BLAST in comparison with the existing GenBank entries. For instance, the version of BLAST used for searching is very important as it has been discovered that the older ungapped version of BLAST finds smaller exons whereas the gapped version does not. Another issue relating to BLAST sensitivity is that in some cases, the first BLAST match is not always the 'best' match in terms of description. When BLASTing commonly occurring sequences, containing mRNAs for example, BLAST will often return a number of different hits referring to the same region, but having different names, such as 'unknown' or 'hypothetical protein'. Sometimes these matches are listed in the BLAST results before the well-annotated descriptions. As an example, several of the coding sequences listed in the GenBank entry Z97184 relating to the Tapasin protein were not found. Instead what was found were NR matches of Tapasinas at (approximately) the same positions. Further examination of this revealed that Tapasinas is an alternatively spliced form of Tapasin, and that the Tapasin match was located deeper in the BLAST results. As this example shows, the first BLAST hit is not always the 'best' match, for this type of analysis.

The results discussed above illustrate not only the importance of storage, analysis and re-analysis of results, but the importance of tracking database versions, software versions and parameters. When researchers are basing or comparing their research projects with publicly available sequences, it would be desirable to have the most up-to-date and accurate representation of the sequences. In this dynamic process, the format and management of results and how they were obtained is critical in maintaining accuracy and data integrity within results.

Sequence Alignment and Comparison

Sequence alignment and comparison is unquestionably the most powerful computational tool available for comparative analysis of DNA and protein sequences. Sequence alignments are used to compare the sequences of genes and proteins with the aim of inferring structural, functional and evolutionary relationships among the sequences under study. The advent of automated DNA sequencing techniques and large genome sequencing initiatives has seen the number of gene and protein sequences in the public databases grow exponentially in recent years. The ability to compare a given sequence against all nucleotide and/or protein sequences *in silico* provides scientists with an ideal opportunity to rapidly identify their particular gene(s) and develop meaningful hypotheses to examine gene function in the laboratory. Improvements in the speed and sophistication of sequence alignment algorithms and computer performance have enabled scientists to keep pace with the growth of sequences in the public databases. However, it is now evident that with the availability of vast amounts of protein sequences, existing algorithms for sequence alignment may not be adequate.

The Smith & Waterman algorithm (Smith and Waterman, 1981) for sequence alignment is one of the most important techniques in computational molecular biology. The ingenious dynamic programming approach was designed to reveal the highly conserved fragments by discarding poorly conserved initial and terminal sequence segments. Gap penalties are an integral component of mainstream alignment programs as they restrict the initiation and extension of gaps, which attempt to model insertion and/or deletion (indels) events that have occurred in either or both sequences through the course of evolution. They can be modified in combination with using a substitution matrix. A substitution matrix defines the score for substituting one amino acid for another. This is used in alignment algorithms as well as to obtain the final alignment score. For aligning DNA sequences, only the identity matrix (exact match only) is used and substitutions between different nucleotides do not contribute to the score. As can be seen in Fig. 15.2 which shows the BLOSUM62 substitution matrix, some amino acids such as leucine (L) and valine (V) can be substituted with a positive score, where as cystine (C) cannot be conservatively substituted for any other amino acid (for a review, see Smith and Waterman, 1981; Pearson, 1996; Apostolico and Giancarlo, 1998). The scientific literature includes numerous empirical studies that each suggests generalized gap penalty parameters for 'typical' situations (Smith and Waterman, 1981; Altschul, 1991, 1993; Pearson, 1995; Barton, 1996). Even a zero gap penalty has been suggested (Roytberg et al., 1998; Morgenstern, 1999). However, it is now widely accepted that the use of a specific set of penalty parameters is not selective and/or sensitive enough for all types of alignments (Pearson, 1995; Apostolico and Giancarlo, 1998; Roytberg et al., 1998; Morgenstern, 1999; Arslan et al., 2001).

Bellgard et al. (2003) proposed an approach to pairwise sequence alignment that does not employ a gap penalty and refer to it as gap mapping. This algorithm is a modification of Smith and Waterman's dynamic programming approach in which a matrix of best partial match scores is populated via a recurrence relationship. Gap mapping uses an extra dimension on the matrix to allow the constraint of the number of gaps introduced. For a given k gaps, the alignment algorithm gives an alignment with the best score with at most k gaps. This approach was tested on a set of structurally aligned sequences and compared with other techniques and its performance was better (on average) than the other approaches. Surprisingly, this algorithm is a version of one proposed by Sankoff (Sankoff, 1972). The advantage of this type of approach is that it provides the researcher with the ability to evaluate more than one alignment for a given pair of sequences which typically are also optimal or near optimal alignments. This is in contrast to current approaches which give just one alignment and can mislead the researcher that this might be the only optimal or near optimal alignment. Figure 15.3 shows an example alignment of two amino acid sequences using the conventional Smith and Waterman algorithm and the corresponding series of alignments produced using the Gap Mapping approach (Bellgard et al., 2003b).

	A	R	N	D	C	Q	E	G	H	I	L	K	M	F	P	S	T	W	Y	V
A	4	-1	-2	-2	0	-1	-1	0	-2	-1	-1	-1	-1	-2	-1	1	0	-3	-2	0
R	-1	5	0	-2	-3	1	0	-2	0	-3	-2	2	-1	-3	-2	-1	-1	-3	-2	-3
N	-2	0	6	1	-3	0	0	0	1	-3	-3	0	-2	-3	-2	1	0	-4	-2	-3
D	-2	-2	1	6	-3	0	2	-1	-1	-3	-4	-1	-3	-3	-1	0	-1	-4	-3	-3
C	0	-3	-3	-3	9	-3	-4	-3	-3	-1	-1	-3	-1	-2	-3	-1	-1	-2	-2	-1
Q	-1	1	0	0	-3	5	2	-2	0	-3	-2	1	0	-3	-1	0	-1	-2	-1	-2
E	-1	0	0	2	-4	2	5	-2	0	-3	-3	1	-2	-3	-1	0	-1	-3	-2	-2
G	0	-2	0	-1	-3	-2	-2	6	-2	-4	-4	-2	-3	-3	-2	0	-2	-2	-3	-3
H	-2	0	1	-1	-3	0	0	-2	8	-3	-3	-1	-2	-1	-2	-1	-2	-2	2	-3
I	-1	-3	-3	-3	-1	-3	-3	-4	-3	4	2	-3	1	0	-3	-2	-1	-3	-1	3
L	-1	-2	-3	-4	-1	-2	-3	-4	-3	2	4	-2	2	0	-3	-2	-1	-2	-1	1
K	-1	2	0	-1	-3	1	1	-2	-1	-3	-2	5	-1	-3	-1	0	-1	-3	-2	-2
M	-1	-1	-2	-3	-1	0	-2	-3	-2	1	2	-1	5	0	-2	-1	-1	-1	-1	1
F	-2	-3	-3	-3	-2	-3	-3	-3	-1	0	0	-3	0	6	-4	-2	-2	1	3	-1
P	-1	-2	-2	-1	-3	-1	-1	-2	-2	-3	-3	-1	-2	-4	7	-1	-1	-4	-3	-2
S	1	-1	1	0	-1	0	0	0	-1	-2	-2	0	-1	-2	-1	4	1	-3	-2	-2
T	0	-1	0	-1	-1	-1	-1	-2	-2	-1	-1	-1	-1	-2	-1	1	5	-2	-2	0
W	-3	-3	-4	-4	-2	-2	-3	-2	-2	-3	-2	-3	-1	1	-4	-3	-2	11	2	-3
Y	-2	-2	-2	-3	-2	-1	-2	-3	2	-1	-1	-2	-1	3	-3	-2	-2	2	7	-1
V	0	-3	-3	-3	-1	-2	-2	-3	-3	3	1	-2	1	-1	-2	-2	0	-3	-1	4

Fig. 15.2. The BLOSUM62 amino acid substitution matrix showing the scoring relationships for substituting one amino acid for another. This is used to align sequences as well as for obtaining the overall score of the alignment.

Large-scale sequence alignment

Implicitly, sequence alignment is central to a vast range of comparative genomic sequence analysis and is even more critical now as sequencing efforts are focusing on strain-specific, individual differences. There are now major initiatives to align very large genomic sequences between human haplotypes and strains of cereals such as rice, as well as bacterial strains. These comparisons provide invaluable information, as there can be significant differences such as insertion/deletions of retroelements, nucleotide expansions/polymorphisms, translocations and gene copy number differences in one or the other sequence. These sequence differences make it difficult for computational alignment at the DNA level. The mere fact that it is possible to define these types of feature differences should suggest that alignment techniques could make use of this information.

Using any conventional approach to alignment sequences, such as the Smith & Waterman algorithm (Smith and Waterman, 1981) is not feasible. In fact, to align two very large genomic sequences is still a difficult problem. This is due to a number of reasons. First, repetitive elements and low complexity regions can lead to incorrect alignments (for instance, at least one-third of human DNA is made up of repetitive DNA). Secondly, it is difficult to optimize gap penalty parameters as sequences typically contain a substantial number of indels (Arslan *et al.*, 2001). Thirdly, the alignment process is computationally expensive. In addition, there may be other artefacts in the sequences, such as sequences containing large segment duplications, multicopy genes located in close proximity, etc. which can adversely affect a correct alignment. Thus, these techniques require some modification if they are to be useful for aligning very large sequences.

SANKOFF Alignment

```
<gap=1>
NLFVALYDFVASGDNTLSITKGEKLRVLGYNHNG---EWCEAQTKNGQGWVPSNYITPVN
KGVIYALWDYEPQNDDELPMKEGDCMTIIHREDEDEIEWWWARLNDKEGYVPRNLLGLYP
<gap=2>
-NLFVALYDFVASGDNTLSITKGEKLRVLGYNHNG--EWCEAQTKNGQGWVPSNYITPVN
KGVIYALWDYEPQNDDELPMKEGDCMTIIHREDEDEIEWWWARLNDKEGYVPRNLLGLYP
<gap=3>
-NLFVALYDFVASGDNTLSITKGEKLRVLGYNHNG-----EWCEAQTKNGQGWVPSNYITPVN
KGVIYALWDYEPQNDDELPMKEGDCMTII---HREDEDEIEWWWARLNDKEGYVPRNLLGLYP
<gap=4>
N--LFVALYDFVASGDNTLSITKGEKLRVLGYNHNG-----EWCEAQTKNGQGWVPSNYITPVN
KGVIY-ALWDYEPQNDDELPMKEGDCMTII---HREDEDEIEWWWARLNDKEGYVPRNLLGLYP
<gap=5>
-NLFVALYDFVASGDNTLSITKGEKLRVLGYNHNG-----EWCEAQTKNGQGWVPSNY--ITPVN
KGVIYALWDYEPQNDDELPMKEGDCMTII---HREDEDEIEWWWARLNDKEGYVPRNLLGLYP--
```

NB: With a gap=2, the alignment produced by Sank_al is the correct structural alignment.

Smith and Waterman Alignment

```
NLFVALYDFVASGDNTLSITKGEKLRVLGYNHNG--EWCEAQTKNGQGWVPSNYI
GVIYALWDYEPQNDDELPMKEGDCMTIIHREDEDEIEWWWARLNDKEGYVPRNLL
      **  *        *
```

NOTE: This alignment is close to correct structural alignment; however, it does not include entire sequences. Default parameters for the initiation gap penalty and gap extension penalty are −10 and −0.5, respectively.

Fig. 15.3. Alignment of two protein sequences P000520 and P04637 with known structural alignment. They share less than 25% identity in common. Alignments made use the BLOSUM62 substitution matrix, shown in Fig. 15.2.

A standard strategy to align very large sequences is to create a bioinformatics pipeline that subdivides the sequences into smaller sections and perform alignments on each section, and then manually (or semi-automatically) recombine the results. However, this leads to a number of further issues that need to be addressed. For instance, what size fragment is most efficient/effective when dividing the sequences, where should the sequences be divided to ensure that the edges are not missed out during alignment, how can we ensure that the reassembly process (of alignments) is accurate and reproducible? In addition, alignment outputs that are in different file formats from various programs can also add to the complexity of this approach. Other techniques filter out and remove repetitive sequences prior to conducting the alignment (Jiang and Zhao, 2000; Arslan et al., 2001).

There are currently a number of approaches to the alignment of large sequences: PIPMAKER/BLASTZ (Schwartz et al., 2000), MUMMER (Delcher et al., 2002) and SSAHA (Ning et al., 2001). Common to all three techniques is that features (such as repeats, genes, etc.) are only added to the visualization of the alignment and are not part of the alignment process. Thus, this information is treated as secondary information in the alignment process. However, sequences are aligned primarily to interpret and reason on the similarities/differences at the feature-level. For instance, determination of orthologous genes, repetitive elements and other useful markers; the degree of similarity/polymorphism; and insertion/deletion differences. Treating features as primary information in the alignment process. A useful property of this approach is that biologically relevant features are identified at the outset and can be used as part of the reasoning process as to how sequences should 'structurally' be aligned before aligning the details at the nucleotide level.

PIPMAKER is a webserver that employs BLASTZ (see below) as its underlying technique to produce alignments based on *pips*, which are non-gapped areas of identity. The display is either a dotplot analogue of these areas of similarity or a per cent identity. The output is a number of 'pdf' files emailed back. BLASTZ is the alignment technique currently employed by PIPMAKER, referred to as an experimental variant of the GAPPED BLAST program (Altschul *et al.*, 1997; Zhang *et al.*, 1998). It is has been designed for aligning two very long sequences. Software is available at (http://bio.cse.psu.edu). The output of BLASTZ is in a unique format, for which another software tool, called 'Laj' [http://bio.cse.psu.edu] can zoom in on a section of a sequence with views similar to the 'dotter' program (Sonnhammer and Durbin, 1995). MUMMER is a suffix tree-based algorithm available through TIGR (Delcher *et al.*, 2002). The suffix tree is a construct of similarities or Maximal Unique Matches (or MUMs) across two sequences. The sequences must be closely related in order to be efficient. Each MUM can only occur once and is, thus, unique. SSHA (Sequence Search and Alignment by Hashing Algorithm) is a software tool for very fast matching and alignment of DNA sequences. Sequences in a database are pre-processed by breaking them into consecutive k-tuples of k contiguous bases and then stored via a hash table at the position of each occurrence. Searching for a query sequence in the database is done by obtaining from the hash table the 'hits' for each k-tuple in the query sequence and then performing a sort on the results.

Bellgard and Kenworthy proposed the 'feature-based' pair-wise sequence alignment (FBSA) algorithm (Bellgard and Kenworthy, 2003). A database search will identify features within the sequences to be aligned which are then uniquely labelled. Once the original sequences are transformed into a string of features, the feature strings are then aligned. For example, two sequences, of 400,000 nucleotides in length can be reduced down to approximately 200 features in each (depending on the feature density). This drastically reduces computational time as instead of aligning two 400,000 base

pair sequences, one is now only aligning 200 symbols (features) in each string. Once the features are aligned, it is then essentially a matter of aligning the orthologous regions between the features. Each putative orthologous pair can be independently aligned, thus the algorithm naturally lends itself to computer parallelization. The original sequences can thus be described in terms of either contiguous feature strings that contain a feature (exactly) or those that are featureless according to some criteria. These regions will have defined boundaries. As part of this research, another tool FBPLOT that creates a feature-based dotplot is developed. Dotplots have been typically created for DNA and amino acid sequence comparisons. A traditional dotplot is shown in Fig. 15.4a, whereas a feature-based plot is shown in Fig. 15.4b. Apart from speed efficiency, the relationships between these two sequences are preserved.

Multiple sequence alignment

All of the discussions above on alignments have focused on pairwise alignment. It is often necessary to align multiple sequences such as: for evolutionary analysis (molecular phylogenetics (Li, 2000), to identify conserved regions between sequences from diverse species, as well as constructing substitution matrices. The most straightforward techniques are those with extensions to the dynamic programming methods, however the computational complexity inhibits the viability of these approaches. Alternative methods that employ heuristics have been developed, such as CLUSTALW (Higgins and Sharp, 1988). The basic approach is to make use of an approximate phylogenetic relationship between the sequences to be aligned and iteratively build up the alignment according to the phylogenetic relationship. T-Coffee (Notredame *et al.*, 2000) is a more recent approach to improve on the limitations of CLUSTALW and is shown to be more reliable than other approaches.

Concluding Remarks

The power of comparative genomic analysis cannot be overstated. However, in order to do

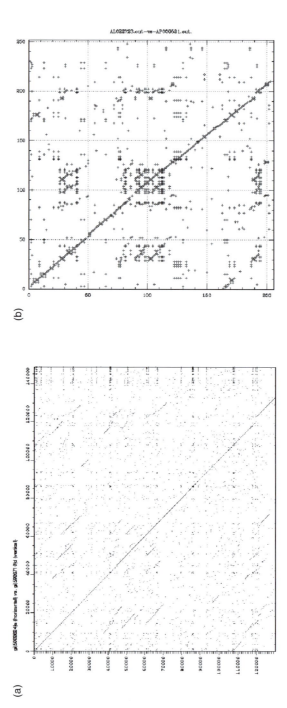

Fig. 15.4. Dotplot view of genomic sequences to be aligned. (a) Shows a dotplot (Dotter) of the original genomic sequences, and (b) shows a dotplot (FBPlot) of the features extracted out from these genomic sequences (Bellgard and Kenworthy, 2003). Features were extracted using REPEATMASKER (Smit and Green, 1997). These correspond to known retrotransposable elements in cereal species.

this effectively with attention to detail, there is a need to develop appropriate bioinformatics resources and strategies to conduct annotation and reannotation of genome sequences as well as to define syntenic relationships between species. These tools are often-times novel in nature but typically require some form of customization of existing resources combining bioinformatics analysis (alignment, database searching, molecular evolutionary analysis, etc.), integration and interoperability of local/third party data and information resources (Stein, 2002). In some cases, the form in which the data are presented (visualized) is essential to appropriate analysis and interpretation (Blomberg and Sunnerhagen, 2003). The Bio* projects (Stein, 2002), for example, provide a series of freely available open-source software for re-usable code libraries developed in the Perl, Java, Python and Ruby programming languages (Bioperl, BioJava, Biopython and Bioruby, respectively). The code libraries provide the basis for common bioinformatics tasks, such as manipulating DNA and protein sequences, and provide methods for importing and exporting data between data sources and among file formats. Seamless interoperability among online databases to scan and to aggregate data simply and reliably is becoming increasingly important and companies such as SUN, IBM and Microsoft are promoting a set of standards that include SOAP/XML, WSDL, UDDI and XSDL to be used by data providers (Stein, 2002).

If we consider a knowledge management approach to bioinformatics analysis it would inherently require an auditing and tracking mechanism as discussed earlier. For example, currently, it is a non-trivial task to compare the results of bioinformatics pipelines that are not precisely specified on the same genomic data. This might be best evident from the wide variation in estimates on the number of genes contained within the human genome (see, for example, Dowell *et al.*, 2001). Unless there is a clear and concise documented audit trail of the bioinformatics processes and a mechanism to validate them efficiently it is difficult to assess these differences which may result from differences in software parameters, choice of programs, versions of databases, or annotations in GenBank records and associated publications do not contain sufficient detail. Even if a publication provides sufficient detail and supplementary results are published on a WWW site, it is still a difficult task to reproduce bioinformatics analysis especially if there are many steps in the bioinformatics pipeline or if it requires significant interoperable resources. The bioinformatics 'process' is essential in order to obtain reliable/reproducible/up-to-date results.

Individual researchers significantly make use of third party, public bioinformatics Internet resources. However, typically there is no facility for them to quality manage and track their analysis in a sophisticated way except perhaps to keep a laboratory notebook with copious printouts. Laboratory information management systems (LIMS) are invaluable for biological sample tracking and analysis, but they do not provide functionality for bioinformatics analysis or data mining. The future role of bioinformatics project management that implements bioinformatics analysis audit trails (BAATs) will be critical to quality-managed bioinformatics analysis (Bellgard *et al.*, 2003a).

References

Akhunov, E.D., Akhunova, A.R., Linkiewicz, A.M., Dubcovsky, J., Hummel, D., Lazo, G., Chao, S., Anderson, O.D., David, J., Qi, L., Echalier, B., Gill, B.S., Miftahudin, Gustafson, J.P., La Rota, M., Sorrells, M.E., Zhang, D., Nguyen, H.T., Kalavacharla, V., Hossain, K. *et al.* (2003) Synteny perturbations between wheat homologous chromosomes caused by locus duplications and deletions correlate with recombination rates. *Proceedings of the National Academy of Sciences USA* 100, 10836–10841.

Altschul, S. (1991) Amino acid substitution matrices from an information theoretic perspective. *Journal of Molecular Biology* 219(3), 555–565.

Altschul, S. (1993) A protein alignment scoring system sensitive at all evolutionary distances. *Journal of Molecular Evolution* 36, 290–300.

Altschul, S., Gish, W., Miller, W., Meyers, E.W. and Lipman, D.J. (1990) Basic local alignment search tool. *Journal of Molecular Biology* 215, 403–410.

Altschul, S., Madden, T.L., Schaffer, A.A., Zhang, J., Zhang, Z., Miller, W. and Lipman, D.J. (1997) Gapped BLAST and PSI-BLAST: a new generation of protein database search programs. *Nucleic Acids Research* 25(17), 3389–3402.

Apostolico, A. and Giancarlo, R. (1998) Sequence alignment in molecular biology. *Journal of Computational Biology* 5, 173–196.

Apweiler, R., Kersey, P., Junker, V. and Bairoch, A. (2001) Technical comment to database verification studies of SWISS-PROT and GenBank by Karp *et al. Bioinformatics* 17 533–534.

Arslan, A., Egecioglu, O. and Pevzner, P. (2001) A new approach to sequence comparison: normalized sequence alignment. *Bioinformatics* 17, 327–337.

Barton, G. (1996) Protein sequence alignment and database scanning. In: Sternberg, M.J.E. (ed.) *Protein Structure Prediction – A Practical Approach.* IRL Press, Oxford.

Bellgard, M. and Gojobori, T. (1999a) Significant differences between the G+C content of synonymous codons in orthologous genes and the genomic G+C content. *Gene* 238, 33–37.

Bellgard, M. and Gojobori, T. (1999b) Identification of a ribonuclease H gene in both *Mycoplasma genitalium* and *Mycoplasma pneumoniae* by a new method for exhaustive identification of ORFs in the complete genome sequences. *FEBS letters* 445, 6–8.

Bellgard, M. and Kenworthy, B. (2003) FBSA: feature-based sequence alignment technique for very large sequences. *Applied Bioinformatics* (in press).

Bellgard, M., Hunter, A. and Kenworthy, B. (2003a) Microarray Analysis Using Bioinformatics Analysis Audit Trails (BAATs). Special Issue: *From Functional Genomics to Systems Biology, Proceedings of the French Academy of Sciences* (in press).

Bellgard, M., Gamble, T., Reynolds, M., Hunter, A., Trifonov, E. and Taplin, R. (2003b) Gap mapping: a paradigm for sequence alignment. *Applied Bioinformatics* 2, 531–535.

Bellgard, M., Ye, J., Gojobori, T. and Appels, R. (2004) The bioinformatics challenges in comparative analysis of cereal genomes – an overview. *Functional and Integrative Genomics* 4, 1–11.

Benson, D.A., Karsch-Mizrachi, I., Lipman, D.J., Ostell, J., Rapp, B. and Wheeler, A. (2002) 'GenBank'. *Nucleic Acids Research* 30, 17–20.

Blomberg, A. and Sunnerhagen, P. (2003) Visible trends in functional genomics. *Functional and Integrative Genomics* 3, 91–93.

Brenner, S.E. (1999) Errors in genome annotation. *Trends in Genetics* 15, 132–133.

Burge, C. and Karlin, S. (1997) Prediction of complete gene structures in human genomic DNA. *Journal of Molecular Biology* 268, 78–94.

Carter, K. and Bellgard, M. (2003) The multiple (BLAST) annotation system viewer (MASV). *Bioinformatics* 19, 2313–2315.

Carter, K., Oka, A., Tamiya, G. and Bellgard, M. (2001) Bioinformatics issues for automating the annotation of genomic sequences. *Genome Informatics Workshop* Tokyo, Japan, Dec 19–21, 2001, 204–211.

Collins, F.S., Morgan, M. and Patrinos, A. (2003) The Human Genome Project: lessons from large-scale biology. *Science* 300, 286–290.

Dehal, P., Predki, P., Olsen, A.S., Kobayashi, A., Folta, P., Lucas, S., Land, M., Terry, A., Ecale Zhou, C.L., Rash, S., Zhang, Q., Gordon, L., Kin, J., Elkin, C., Pollard, M.J., Richardson, P., Rokhsar, D., Uberbacher, E., Hawkins, T., Branscomb, E. and Stubbs, L. (2001) Human chromosome 19 and related regions in mouse: conservative and lineage-specific evolution. *Science* 293, 104–111.

Delcher, A.L., Phillippy, A., Carlton, J. and Salzberg, S.L. (2002) Fast algorithms for large-scale genome alignment and comparison. *Nucleic Acids Research* 30, 2478–2483.

Dowell, R., Jokerst, R., Day, A., Eddy, S. and Stein, L. (2001) The distributed annotation system. *BMC Bioinformatics* 2, 7.

Flicek, P., Keibler, E., Hu, P., Korf, I. and Brent, M.R. (2003) Leveraging the mouse genome for gene prediction in human: from whole-genome shotgun to a global synteny map. *Genome Research* 13, 46–54.

GBENCH (2003) http://www.ncbi.nlm.nih.gov/IEB/ToolBox/CPP_DOC/tools/gbench/gbench_manual.html

'Genbank release notes'. National Centre for Biotechnology Information, ftp://ftp.ncbi.nih.gov/genbank/gbrel.txt

Higgins, D.G. and Sharp, P.M. (1988) CLUSTAL: a package for performing multiple sequence alignment on a microcomputer. *Gene* 73(1), 237–44.

Hunter, A. (2003) Investigating the role of community-based review and moderation of bioinformatics analysis. Honours Thesis, School of Information Technology, Murdoch University.

Jiang, T. and Zhao, P. (2000) A heuristic algorithm for blocked multiple sequence alignment. In: *Proceedings of the 1st Annual IEEE International Symposium on Bioinformatics and Biomedical Engineering (BIBE 2000).* 8–10 November 2000; Arlington, VA, USA. 0–7695–0862–6/00. IEEE Computer Society Press, pp. 176–183.

Jurka, J., Klonowski, P., Dagman, V. and Pelton, P. (1996) CENSOR – a program for identification and elimination of repetitive elements from DNA sequences. *Computers and Chemistry* 20, 119–122. (ftp://ftp.ncbi.nlm.nih.gov/repository/repbase/SOFTWARE/)

Karp, P.D., Paley, S. and Zhu, J. (2001) Database verification studies of SWISS-PROT and GenBank. *Bioinformatics* 17, 526–532.

Kulski, J.K., Lim, C.P., Dunn, D.S. and Bellgard, M. (2003) Genomic and phylogenetic analysis of the S100A7 (psoriasin) gene duplications within the region of the S100 gene cluster on human chromosome 1q21. *Journal of Molecular Evolution* 56, 397–406.

Li, Z., Mouille, G., Kosar-hashemi, B., Raman, S., Clarke, B.C., Gale, K.R., Appels, R. and Morell, M.K. (2000) The structure and expression of the wheat starch synthase III gene. Motifs in the expressed gene define the lineage of the starch synthase III gene family. *Plant Physiology* 123, 613–624.

Malakowski, W., Mitchell, G.A. and Labuda, D. (1994) Alu sequences in the coding regions of nRNA: a source of protein variability. *Trends in Genetics* 10, 188–193.

Morgenstern, B. (1999) DALIGN 2: Improvement of the segment-to-segment approach to multiple sequence alignment. *Bioinformatics* 15, 211–218.

Navarro, A. and Barton, N.H. (2003) Chromosomal speciation and molecular divergence – accelerated evolution in rearranged chromosomes. *Science* 300, 321–324.

NCBI Handbook (2003) http://www.ncbi.nlm.nih.gov/books/bv.fcgi?call=bv.View..ShowTOC&rid=handbook.TOC&depth=2

Ning, Z., Cox, A.J. and Mullikin, J.C. (2001) SSAHA: a fast search method for large DNA databases. *Genome Research* 11, 1725–1729.

Notredame, C., Higgins, D. and Heringa, J. (2000) T-Coffee: a novel method for multiple sequence alignments. *Journal of Molecular Biology* 302, 205–217.

Parra, G., Agarwal, P., Abril, J.F., Wiehe, T., Fickett, J.W. and Guigo, R. (2003) Comparative gene prediction in human and mouse. *Genome Research* 13, 108–117.

Pearson, W. (1995) Comparison of methods for searching protein sequence databases. *Protein Science* 4, 1145–1160.

Pearson, W. (1996) Effective protein sequence comparison. *Methods in Enzymology* 266, 227–258.

Pearson, W. and Lipman, D.L. (1988) Improved tools for biological sequence comparison. *Proceedings of the National Academy of Sciences USA* 85, 2444–2448.

Pennisi, E. (1999) Keeping genome databases clean and up to date. *Science* 286, 447–450.

Quackenbush, J. (2001) Computational analysis of microarray data. *Nature Reviews of Genetics* 2, 418–427.

Refseq (2003) http://www.ncbi.nlm.nih.gov/LocusLink/refseq.html.

Roytberg, M.A. *et al.* (1998) Sequence alignment without gap penalties. Proceedings of the 1st International Bioinformatics Genome Regulation Structure Conference (BGRS'98). 24–31 August 1998; Novosibirsk, Altai Mountains, Russia. http://www.bionet.nsc.ru/bgrs/thesis/85/index.html

Sankoff, D. (1972) Matching sequences under deletion/insertion constraint. *Proceedings of the National Academy of Sciences USA* 69, 4–6.

Schwartz, S., Zhang, Z., Frazer, K.A., Smit, A., Riemer, C., Bouck, J., Gibbs, R., Hardison, R. and Miller, W. (2000) PipMaker – a web server for aligning two genomic DNA sequences. *Genome Research* 10, 577–586.

Smit, A.F.A. and Green, P. (1997) RepeatMasker. Accessed 15 September 2004. http://ftp.genome.washington.edu/RM/RepeatMasker.html

Smith, T. and Waterman, M. (1981) Identification of common molecular subsequences. *Journal of Molecular Biology* 147, 195–197.

Sonnhammer, E.L. and Durbin, R. (1995) A dot-matrix program with dynamic threshold control suited for genomic DNA and protein sequence analysis. *Gene* 167, GC1–GC10.

Stein, L. (2002) Creating a bioinformatics nation. *Nature* 417, 119–120.

Stoesser, G., Baker, W., van den Broek, A.E., Camon, E., Garcia-Pastor, M., Kanz, C., Kulikova, T., Lombard, V., Lopez, R., Parkinson, H., Redaschi, N., Sterk, P., Stoehr, P. and Tuli, M.A. (2001) The EMBL nucleotide sequence database. *Nucleic Acids Research* 29, 17–21.

Wu, T.D. (2001) Analysing gene expression data from DNA microarrays to identify candidate genes. *Journal of Pathology* 195, 53–65.

Zhang, Z., Berman, P. and Miller, W. (1998) Alignments without low-scoring regions. *Journal of Computational Biology* 5(2), 197–210.

16   Genetic Databases

V. Brusic and J.L.Y. Koh

Institute for Infocomm Research, 21 Heng Mui Keng Terrace, Singapore 119613

Summary	
Introduction	411
Genetic Databases	413
Catalogues of molecular databases	413
Overview of biological databases	415
Major sequence databases	415
Comparative genomics	417
Gene expression	417
Gene identification and structure	418
Genetic and physical maps	418
Genomic databases	419
Mutation databases	419
RNA sequences	420
Other databases	420
Retrieval systems	421
Searching Genetic Databases	422
Data Quality Issues	424
Concluding Remarks	425
References	426

Introduction

The second half of the 20th century brought revolutions in two fields – information technology and biotechnology. We have seen an explosive growth of these two technology fields, each of which has penetrated every pore of our daily lives. Information technology has provided hardware and software that enable data processing at unprecedented speed and efficiency. Biotechnology has provided methods and instrumentation for the analysis and manipulation of biological systems on a massive scale. The increased ability to measure

Correspondence: vladimir@i2r.a-star.edu.sg; judice@i2r.a-star.edu.sg

© CAB International 2005. *Mammalian Genomics*
(eds A. Ruvinsky and J. Marshall Graves)

molecular structures and to store and process these data resulted in ever growing biological databases and data repositories. Molecular biology databases show growth in number of entries, number of nucleotides or residues, and in the complexity of entries. In practice this translates to a rapidly growing number of database entries, longer sequences, and more fields added to individual records. In addition, new specialist databases are proliferating, providing researchers with increased number of views to the same data sets. The examples of growth of biological data are shown in Fig. 16.1. For the indicated period, the total number of both nucleotides and entries in GenBank (Benson et al., 2003) have grown on average at 75% per year. The number of residues in the Protein Information Resource (PIR, Wu et al., 2003) has grown on average at 35% and the number of their entries at 29%. The number of entries of protein structures in the Protein Data Bank (PDB, Westbrook et al., 2003) has grown at 25% annually since 1977. These numbers show that the growth of biological data is mainly fuelled by the growth of nucleotide data, particularly those pouring from the genome sequencing projects.

Since the structure of nucleic acids was discovered in 1953 (reprinted in Watson and Crick, 2003), the initial slow process of data accumulation accelerated to an explosive growth today. The 5386 base pairs of the bacteriophage ϕX174 constituted the first viral genome to be sequenced (Sanger et al., 1978). The 170 kb of the complete DNA sequence of Epstein–Barr virus was sequenced and published 6 years later (Baer et al., 1984). It took another decade before the first sequence of a free-living organism, Haemophilus influenzae, containing 1.8 million bases was published (Fleischmann et al., 1995). The sequence of some 3.5 billion bp of the human genome was published in 2001 (Lander et al., 2001), followed by publication of 3.2 billion bp of the mouse genome 1 year later (Waterston et al., 2002). A brief history of genomics is available in Roberts et al. (2001).

More than 1500 viral genomes, more than 110 bacterial and archaea genomes, and more than 20 eukaryotic genomes have been sequenced and are now accessible for analysis through the National Center for Biotechnology Information (NCBI) site (Wheeler et al., 2003). Sequencing of the human genome represents one of the great scientific milestones in the history of humankind. The large-scale sequencing employed in various genome projects was critically dependent on using computer facilities. The assembly, analysis, and annotation of genome sequences require significant computational infrastructure including molecular databases for storage and access to data and bioinformatics tools for the analysis of data. The databases and tools, collectively forming a significant proportion of bioinformatics resources, have enabled identification of homologues, comparative analysis of genomes, and genome annotation.

The database and database tools awareness among their primary users, biomedical researchers, is relatively low. Recent surveys (see Baxevanis, 2003) showed that only half of regular users of genome databases are familiar with the tools for accessing and the analysis of the data herein. Of human genome researchers only 11% regularly access the major human genome analysis tool ENSEMBL (Clamp et al., 2003) and 24% of users access it occasionally. The hands-on guide for human genome analysis tools has been made available to the public (Wolsberg et al., 2002) and similar guides have been developed at major genome repositories (NCBI, European Bioinformatics Institute – EBI, and University of California, Santa Cruz – UCSC).

A major purpose of this chapter is to address the main issues related to understanding genetic databases, list representative databases, and help raise awareness about the potential and possible pitfalls in using data from these databases. This chapter provides a description of the molecular database landscape focusing on nucleotide and genomics databases (genetic databases). We will first describe the types of genetic databases and provide starting points for readers interested in further exploration of the field. Secondly, we briefly discuss the multidatabase retrieval systems and tools for database searching. Thirdly, we discuss data quality issues which are important for adequate use of genetic databases. Finally we discuss data integration issues and provide our insight into the future.

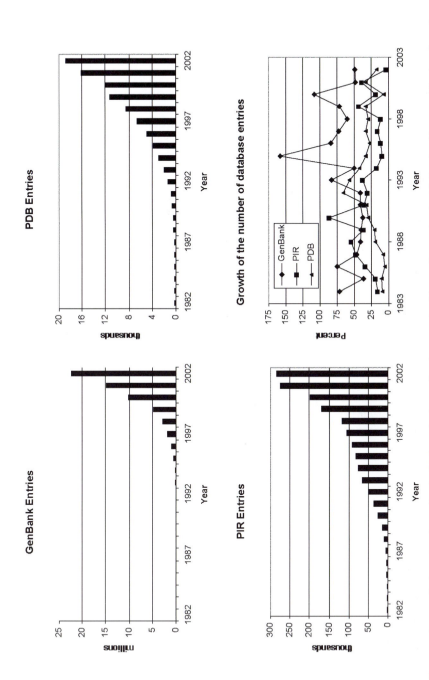

Fig. 16.1. The growth rates of selected major molecular biology databases showing growth of the total sequence length (nucleotides or residues) and number of entries. The examples include a major DNA database GenBank (Benson *et al.*, 2003), a major protein database PIR (Wu *et al.*, 2003), and a major protein structure database PDB (Westbrook *et al.*, 2003). Raw data were extracted from GenomeNet, University of Kyoto, Japan (www.genome.ad.jp/dbget/db_growth.html).

Genetic Databases

The main developments in the genomic field pertain to large-scale analysis, also known as genome-wide analysis. Major topics in this field are: genome and gene analysis, identification and annotation of genes, comparative genomics, single nucleotide polymorphism (SNP) analysis, study of full-length cDNAs, structural genomics, functional genomics and proteomics, applications for medical treatment, and bioinformatics for gene, genome, protein, and proteome analysis. Collectively, these international efforts produce large amounts of data. Access to databases is necessary for the extraction of existing information, experimental planning, and analysis and interpretation of experimental results. Bioinformatic tools that are available for database searching and the analysis of biological sequences are becoming more sophisticated. These tools enable users to quickly identify sequences of interest, analyse them, and provide substantial bibliographic, structural, functional, taxonomic or feature information. Tools for sequence comparison, motif searching, profiling and, more recently, functional analysis help researchers in identifi-

cation of biologically important sequence properties. Genetic databases have become crucial resources that facilitate these analyses.

Catalogues of molecular databases

There are two major catalogues of molecular databases: the annual database issue of *Nucleic Acids Research* (NAR, Baxevanis, 2003, nar.oupjournals.org), and the DBCAT – a catalogue of public databases at Infobiogen (www.infobiogen.fr/services/dbcat). These two catalogues draw attention to high-quality public databases and serve as starting points for searching relevant biological information. In August 2003, NAR listed a total of 399 databases and DBCAT listed 511 (Table 16.1). A large subset of NAR-listed databases gets published in the NAR database issue published on 1 January every year. The publication of databases adds value to the underlying data by setting minimum standards for content and updating through the peer review process. The actual number of databases is much higher, but the quality control for the broad population of databases is lack-

Table 16.1. Number of individual databases sorted by categories (as of September 2003) listed in the NAR (nar.oupjournals.org.) and DBCAT (www.infobiogen.fr/services/dbcat) catalogues.

NAR	Number	DBCAT	Number
Major sequence repositories	9	DNA	87
Comparative genomics	4		
Gene expression	19		
Gene identification and structure	27		
Genetic and physical maps	12	Mapping	29
Genomic databases	52	Genomics	58
Intermolecular interactions	7		
Metabolic pathways/cellular regulation	11		
Mutation databases	31		
Pathology	7		
Protein databases	56	Protein	94
Protein sequence motifs	20		
Proteome resources	6		
RNA sequences	26	RNA sequences	29
Retrieval systems and database structure	3		
Structure	34	Protein structure	18
Transgenics	2		
		Literature	43
Varied biomedical content	20	Miscellaneous	153
Total	399		511

ing and their use should be accompanied with due care.

Overview of biological databases

The major purpose of biological databases is to disseminate biological data and information, provide biological data in computer-readable form, and allow their analysis. A database needs to have at minimum one specific tool for database querying and data extraction. Web pages, books, journal articles, tables, text files and spreadsheet files cannot be considered as databases.

Databases may be general or specialized. The main purpose of general databases is to collect, organize, and disseminate annotated biological data to the public. The major challenge for general databases is coping with increased rates of data generation and providing reliable access to data. Consistent with their role, general databases have significant infrastructure including interfaces for data extraction and analysis, are centrally accessible, have standardized formats facilitating automation of searches, provide curation and quality assurance of entries, and are independently maintained and funded. On the other hand, general databases have limitations in providing quality control of content. They typically provide only basic annotation of features, and their entries do not always conform to specific standards and nomenclature in the fast changing fields. Additional problems are error propagation, obsolete, incomplete or redundant entries, and difficulty of synchronization between various data sources.

Specialist databases focus on a specific narrow domain and provide access to and the analysis of relevant data. The strengths of specialist databases are more detailed information; they are created and maintained by the domain experts and thus provide a high level of quality assurance of data, better compliance to standards and often have specialist tools. The advantages of the specialist databases come at the cost of generally irregular updates, low level of automation, lower reliability for access and currency, and funding uncertainty. In this chapter we will review the best known public repositories as well as other representa-tive databases. Additional databases can be found in the catalogues as well as in literature. While some databases can be classified into several categories, in this chapter we follow the classification defined in the NAR catalogue. The databases mentioned in this chapter are representative examples rather than a comprehensive listing. New databases are created, and some disappear over time. Major databases are well maintained and they are likely to provide ready access for users for years to come. Because of the distributed nature of the corpora of genetic databases, ready access cannot be guaranteed for all databases.

Major sequence databases

These databases represent general resources of nucleotide sequences. The main DNA sequence databases are DNA Data Bank of Japan, DDBJ (www.ddbj.nig.ac.jp), EMBL Nucleotide Sequence Database (www.ebi.ac.uk/embl.html), and GenBank (www.ncbi.nlm.nih.gov). The nucleotide databases, DDBJ, EMBL and GenBank focus on collecting, annotating, and providing access to the entries of DNA sequences and the related information. Data are collected, upgraded and exchanged between these three centres on a daily basis. DAD, TrEMBL and GenPept are the protein databases derived from the translations of all coding sequences of the three main nucleotides databases. A representative entry from GenBank is shown in Fig. 16.2. The main parts of the entry are the header, features and sequence. The header contains information on the entry itself, such as identifiers and versions, definitions, source organism, references and comments. The features part contains the description of the structural, functional and other physico-chemical properties of the sequence, protein translation, and their locations, if applicable. The sequence section contains the DNA sequence. The three major databases have the same overall header–features–sequence layout, but their field labels (schema) and corresponding fields (content) may differ. The individual databases may not have the same set of fields, which may lead to the loss of information when the entries are exchanged between databases.

```
NM_181867. Bos taurus interf...[gi:32480627] Links
LOCUS        NM_181867              579 bp    mRNA     linear    MAM 09-JUL-2003
DEFINITION   Bos taurus interferon induced transmembrane protein 3 (1-8U)
             (IFITM3), mRNA.
ACCESSION    NM_181867
VERSION      NM_181867.1  GI:32480627
KEYWORDS     .
SOURCE       Bos taurus (cow)
  ORGANISM   Bos taurus
             Eukaryota; Metazoa; Chordata; Craniata; Vertebrata; Euteleostomi;
             Mammalia; Eutheria; Cetartiodactyla; Ruminantia; Pecora; Bovoidea;
             Bovidae; Bovinae; Bos.
REFERENCE    1  (bases 1 to 579)
  AUTHORS    Pru,J.K., Austin,K.J., Haas,A.L. and Hansen,T.R.
  TITLE      Pregnancy and interferon-tau upregulate gene expression of members
             of the 1-8 family in the bovine uterus
  JOURNAL    Biol. Reprod. 65 (5), 1471-1480 (2001)
  MEDLINE    21526490
   PUBMED    11673264
COMMENT      PROVISIONAL REFSEQ: This record has not yet been subject to final
             NCBI review. The reference sequence was derived from AF272041.1.


FEATURES             Location/Qualifiers
     source          1..579
                     /mol_type="mRNA"
                     /db_xref="taxon:9913"
     gene            1..579
                     /gene="IFITM3"
                     /db_xref="LocusID:282255"
     CDS             49..489
                     /gene="IFITM3"
                     /note="16-kDa uterine protein; induced by interferon-tau;
                     interferon-induced protein 1-8U"
                     /codon_start=1
                     /product="interferon-induced transmembrane protein 3
                     (1-8U)"
                     /protein_id="NP_863657.1"
                     /db_xref="GI:32480628"
                     /db_xref="LocusID:282255"
                     /translation="MNRTSQLLLTGAHGAVPPAYEVLKEEHEVAVLGAPQSQAPLTTT
                     VININRSDTAVPDHIVWSLFNTIFMNWCCLGFVAFAYSVKSRDRKMVGDITGAQSYAST
                     AKCLNICSLVLGILLTVVLIVLVSNGSLMIVQAVSELMQNYGGH"
BASE COUNT       108 a      200 c      157 g      114 t
ORIGIN
        1 atctggaccg cagttgctca tctggactgc agttgctccg tccccaccat gaaccgcaca
       61 tcccagctct tactcactgg ggcccacggg gcggtgcccc cagcctatga ggtgctcaag
      121 gaggagcacg aggtggccgt gctgggggcg ccccagagcc aggcgcccct gacgaccacg
      181 gtgatcaaca tccgcagcga caccgccgtg cccgaccaca tcgtgtggtc cctgttcaac
      241 accatcttca tgaactggtg ctgcctgggc ttcgtggcat tcgcctactc tgtgaagtct
      301 agggaccgga agatggtcgg cgacatcact ggggcccaga gctacgcctc caccgccaaa
      361 tgcctgaaca tctgctccct ggtcctgggc atccttctga ctgtcgtcct catcgtcctc
      421 gtgtccaacg gctccctgat gatcgttcaa gcagtctccg agctcatgca aaactacgga
      481 ggccactagg cctgcccaaa agcccgaggc agtcgcccct ttccccgcag cctatccagg
      541 cacctgcccc cgtgaaataa aaggagggtt tgtgtgttg
//
```

Fig. 16.2. A representative GenBank entry.

ACTIVITY (util.bionet.nsc.ru/databases/ activity.html) is a database on DNA site sequences with known activity magnitudes, measurement systems and sequence–activity relationships, such as protein–binding sites.

ArrayExpress (www.ebi.ac.uk/arrayexpress) is a major microarray expression data repository that provides access to data in a standard format and facilitates sharing of the design of microarrays and of experimental protocols.

Comparative genomics

Comparative genomics is the study of genetics of one organism, such as human, by comparisons with model organisms such as bacterium (*Escherichia coli*), yeast (*Saccharomyces cerevisiae*), insects (fruit fly), vertebrates (fugu fish) or mammals (mouse). Interspecies comparison is a powerful methodology that can help infer function from genomic sequence. Comparative genomics helps decipher the differences between organisms. It serves as a powerful platform for understanding characteristics of individual species. Four comparative genomic databases have been listed in the NAR catalogue. These are COG – a database of clusters of orthologous groups of proteins from more than 40 genomes (www.ncbi.nlm.nih. gov/COG), CORG – a database of upstream regions of orthologous genes from human and mouse, focusing on gene regulatory regions (corg.molgen.mpg.de), Homophila – data relating human disease genes and their orthologues in the fruit fly (homophila.sdsc.edu), and XREFdb – cross-referencing of model organism genetics with mammalian phenotypes (www.ncbi. nlm.nih.gov/XREFdb).

Gene expression

Expressed sequence tags (ESTs) are sequenced cDNAs derived from a specified cell population. ESTs are transcribed under various developmental, environmental, or pathological states from various tissues and organs. An EST is a unique short segment of DNA within a gene which can be used in identification of full-length genes. While ESTs are typically considered as useful starting points for gene mapping, a careful analysis of these data can provide significant functional, structural, and evolutionary clues. A recent study (Okazaki *et al.*, 2002) indicated that at least one-third of the ESTs represent sequences that do not code for proteins.

Gene expression databases form a large category of databases which range from very general to very specialized. Major EST databases are STACK (www.sanbi.ac.za/ Dbases.html), TIGR gene indices (www.tigr.org/ tdb/tgi.shtml) and Unigene (www.ncbi.nlm. nih.gov/UniGene). Specialized databases focus on an organism, tissue, specific genome parts, gene regulatory networks, methodological data, or other relevant data (Table 16.2).

Table 16.2. Representative specialized gene expression databases.

Group	Description	dB name	URL
Organism	*Xenopus laevis*	Axeldb	www.dkfz-heidelberg.de/abt0135/axeldb.htm
	Drosophila	FlyView	pbio07.uni-muenster.de
	Mouse	GXD	www.informatics.jax.org/menus/expression_menu.shtml
	Medaka fish	MEPD	medaka.dsp.jst.go.jp/MEPD
Organ/ tissue	Kidney	Kidney Development dB	golgi.ana.ed.ac.uk/kidhome.html
	Dental	Tooth Development dB	bite-it.helsinki.fi
	Pancreas	EPconDB	www.cbil.upenn.edu/EPConDB
Location	3D-IMAGE	Mouse Atlas and Gene Expression dB	genex.hgu.mrc.ac.uk
Methods		Stanford Microarray dB	genome-www.stanford.edu/microarray
Regulatory networks	Signal transduction	TRANSPATH	www.biobase.de/pages/products/databases.html
	Gene regulation	TRANSFAC	www.gene-regulation.com
Other	Alternative splicing	ASDB	cbcg.lbl.gov/asdb

Gene identification and structure

Large quantities of genetic data have been accumulating, mainly from the genome projects. Identification of function and structure of genes is an important task for sequence-based function and structure prediction. Sequence analysis methods are better suited for prediction of protein structure and function than for DNA analysis (Bucher, 1999). Regulatory regions of genomes and biologically interesting sequence features are often poorly defined for nucleotide sequences. Some 30,000 human genes (Lander et al., 2001) and a similar number of mouse genes (Waterston et al., 2002) have been identified to date, but detailed annotation of human genes has not been completed. Alternative splicing and differential regulation result in even higher numbers of gene products. Gene identification and structure databases of a general nature include: protein-coding exon–intron sequence databases ExInt (intron.bic.nus.edu.sg/exint/exint. html) and EID (mcb.harvard.edu/gilbert/EID), the single nucleotide polymorphism – SNP Consortium database (snp.cshl.org), transcription regulatory regions in eukaryotes TRRD (www.bionet.nsc.ru/trrd), eukaryotic pol II promoters EPD (www.epd.isb-sib.ch), alternatively spliced mammalian genes AsMamDB (166.111.30.65/ASMAMDB.html), plant cis-acting regulatory elements, enhancers and repressors PlantCARE (oberon.rug.ac.be:8080/ PlantCARE/index.html), prokaryotic microsatellites MICdb (www.cdfd.org.in/micas), and open reading frames in viral genomes VIDA (www.biochem.ucl.ac.uk/bsm/virus_database/VIDA.html). The HvrBase (www.hvrbase.org) contains sequences from hypervariable regions and complete mitochondrial DNA, with related information, including anthropological data, on more than 10,000 individuals (humans, great apes and neanderthals).

Specialist gene identification databases may focus on an organism, or other gene features. Organism-centric databases include upstream regulatory information of Bacillus subtilis DBTBS (elmo.ims.u-tokyo.ac.jp/dbtbs), E. coli mRNA promoters PromEC (bioinfo.md. huji.ac.il/marg/promec), Caenorhabditis elegans WormBase (www.wormbase.org), S. cere-

visiae YIDB (www.EMBL-Heidelberg. DE/ExternalInfo/seraphin/yidb.html), and several human databases (for listing, see NAR and DBCAT). The feature-oriented databases include codon usage tables CUTG (www.kazusa.or.jp/codon), short tandem DNA repeats STRBase (www.cstl.nist.gov/div831/strbase) for identity testing, motifs in mRNAs that control translation and other biologically relevant mRNA regions Transterm (uther.otago. ac.nz/ Transterm.html), and composite elements of transcriptional regulation TRANSCompel (www.gene-regulation.com/ pub/databases.html#transcompel)

Significant redundancy is present in these databases, for example two databases contain information on plant cis-acting regulatory elements, and no less than four human gene databases. Because the content of databases varies in both number of entries and described features users are advised to search multiple databases to ensure access to maximum information available from public sources.

Genetic and physical maps

Viral genomes can be sequenced as a whole but larger genomes are usually studied in smaller portions. A cytogenetic map provides visualization of genetic markers spread over a chromosome. Radiation hybrid maps are produced by irradiation of cells, which breaks DNA, and the rescue of random fragments into a rodent cell. Overlapping clone maps are important for genome sequencing projects. The analysis of the rescued fragments for molecular markers can be used for identification of chromosomal positions. An overlapping clone map can be constructed from overlapping shotgun sequencing (Bankier, 2001) fragments. Cosmid clones, yeast (YAC) and bacterial (BAC) artificial chromosomes, respectively, can accommodate approximately 50, 1000 and 200 kb of DNA.

The Unified Database UDB (bioinfo.weizmann.ac.il/udb) contains an integrated map for each human chromosome, based on data from various radiation hybrid, linkage and physical mapping resources. UDB is a curated database in which the value-added information includes NCBI's genomic contigs, and repositioning of markers, genes and EST clusters in the sequenced regions. Physical map databases include HuGeMap (www.infobiogen.fr/services/Hugemap) and IXDB (ixdb.mpimg-berlin-dahlem.mpg.de). The radiation hybrid maps include GeneMap '99 (www.ncbi.nlm.nih.gov/genemap), G3-RH (www-shgc.stanford.edu/RH), RHdb (www.ebi.ac.ukRHdb) and GB4-RH (www.sanger.ac.uk/Software/RHserver/RHserver.shtml). The data on mapped human BAC clones at approximately 1000 kb intervals are available at GenMapDB (genomics.med.upenn.edu/genmapdb). Other specialized physical map databases include human cDNA clones homologous to *Drosophila* mutant genes DRESH (www.tigem.it/LOCAL/drosophila/dros.html), and comparative mapping in pig Genetpig (www.infobiogen.fr/services/Genetpig).

Genomic databases

Genomic databases form the largest category of nucleotide databases. Genomic data are the major component supporting the growth of biological databases. Various aspects of genomic data have been listed in the NAR and DBCAT catalogues. These range from databases of genome projects and approved gene symbols to mitochondrial genomes and genomes of specific organisms. A database of genome projects GOLD (igweb.integratedgenomics.com/GOLD) contains information on more than 150 (mainly prokaryotic) complete genomes and more than 600 ongoing genomic projects (360 prokaryotes and 240 eukaryotes).

Genomic databases may contain data on groups of organisms, such as Comprehensive Microbial Resource (www.tigr.org/tigrscripts/CMR2/CMRHomePage.spl), completely sequenced prokaryotic genomes EMGlib (pbil.univ-lyon1.fr/emglib/emglib.html), farm animals ArkDB (www.thearkdb.org), or crop plants CropNet (ukcrop.net). Examples of mitochondrial genomic databases include metazoan mitochondrial genes (bighost.area.ba.cnr.it/mitochondriome) or human mitochondrial genome MITOMAP (www.gen.

emory.edu/mitomap.html). Databases specializing in genomes of specific organisms include HIV (hiv-web.lanl.gov), *Bacillus subtilis* SubtiList (genolist.pasteur.fr/SubtiList), *Toxoplasma gondii* ToxoDB (toxodb.org), *Plasmodium falciparum* PlasmoDB (PlasmoDB.org), zebra fish ZFIN (zfin.org), *Arabidopsis thaliana* TAIR (www.arabidopsis.org), maize ZmDB (zmdb.iastate.edu), mouse MGD (www.informatics.jax.org), and human HOWDY (www-alis.tokyo.jst.go.jp/HOWDY), among others.

Other genome-related databases include an integrated database of human genes, maps, proteins, and diseases GeneCards (bioinfo.weizmann.ac.il/cards), organelle genome database GOBASE (megasun.bch.umontreal.ca/gobase/gobase.html), horizontally transferred genes in prokaryotes HGT-DB (www.fut.es/~debb/HGT), and several others.

Notably, genome databases may have a large level of redundancy. No less than four rice (approximately 430 Mb) genome databases are publicly available: INE (rgp.dna.affrc.go.jp/giot/INE.html), Oryzabase (www.shigen.nig.ac.jp/rice/oryzabase), International Rice Information System (www.iris.irri.org), and RiceGAAS (ricegaas.dna.affrc.go.jp). Although these databases have overlapping data, users are advised to search multiple databases to ensure access to maximum information.

Mutation databases

All living organisms show variation at phenotype as well as at genotype level. This has implications for polymorphism, allelic variation, and deleterious changes that may cause disease. Mutation data have been used for linkage to disease genes in diagnostic studies, in gene discovery, and in scanning of complete genomes (Cotton, 1999). The mutation databases may be of a general nature, such as single nucleotide polymorphisms dbSNP (www.ncbi.nlm.nih.gov/SNP), mutations in viral, bacterial, yeast and mammalian genes (info.med.yale.edu/mutbase), or contain information on allele frequencies and DNA polymorphism (alfred.med.yale.edu).

More specialized mutation databases focus on gene families, such as cytokines (www.bris.ac.uk/pathandmicro/services/GAI/cytokine4.htm) and G-protein coupled receptors (tinyGRAP.uit.no/GRAP), or focus on a single organism, such as a collection of T-DNA insertion transformants in *Arabidopsis thaliana* (genoplante-info.infobiogen.fr), and HIV reverse transcriptase and protease sequence variation (hivdb.stanford.edu/hiv).

A large proportion of mutation databases focus on diseases, such as Asthma Gene Database (cooke.gsf.de/asthmagen/main.cfm), Online Mendelian Inheritance in Man (www.ncbi.nlm.nih.gov/Omim), and chromosomal abnormalities in cancer (www.infobiogen.fr/services/chromcancer/). Alternatively, some databases focus on organs, such as mutations in human eye disease genes (131.113.190.126/mutview3/mutview/index_eye.html).

RNA sequences

RNA databases are another significant group containing more than 30 members. RNA molecules are important in protein synthesis and transport, catalysis, chromosome replication and regulation, and other biological functions of the cell. While most genes code for proteins, approximately 3% of genes code for RNA, such as rRNA and tRNA genes (Bucher, 1999). A variety of databases containing various aspects of RNA data are available from public resources. In addition to the ribosomal database project RDP-II (rdp.cme.msu.edu), at least four other rRNA databases have been constructed. Other examples include plant tRNA sequences PLMItRNA (bighost.area.ba.cnr.it/PLMItRNA), bacterial tmRNA TMRdB (psyche.uthct.edu/dbs/tmRDB/tmRDB.html), guide RNA (biosun.bio.tu-darmstadt.de/goringer/gRNA/gRNA.html), viroid and viroid-like RNA (penelope.med.usherb.ca/subviral), several non-coding RNA databases inclusive of Rfam (www.sanger.ac.uk/Software/Rfam), untranslated mRNAs UTRdb/UTRsite (bighost.area.ba.cnr.it/srs6/), and small RNA (mbcr.bcm.tmc.edu/smallRNA).

Other databases focus on structural or functional aspects of RNA, such as non-standard base–base interactions in known RNA structures NCIR (prion.bchs.uh.edu/bp_type), RNA functional sites SELEXdb (wwwmgs.bionet.nsc.ru/mgs/systems/selex), group II introns from bacteria and lower eukaryotic organelles (www.fp.ucalgary.ca/group2introns), RNA modification (medlib.med.utah.edu/RNAmods/), and RNA pseudoknots (wwwbio.leidenuniv.nl/~Batenburg/PKB.html).

Other databases

The databases discussed in this section are not necessarily genetic, but they contain information on gene products. We will only briefly mention some of these databases, however information extracted from them is often important for a better understanding of genetic and genomic data. Database users working on a particular problem should try to extract as much relevant information as possible.

More than 100 protein databases have been listed in the NAR and DBCAT catalogues. The most prominent protein database is SWISS-PROT/TrEMBL database (www.expasy.org). SWISS-PROT contains curated entries of protein sequences with annotations while TrEMBL contains entries of translated sequences from the EMBL database. The ExPASy (Expert Protein Analysis System) server also has numerous tools for peptide analysis, DNA to protein translation, sequence similarity searches, pattern and profile searches, prediction of post-translational modification and protein topology predictions, primary, secondary, and tertiary structure analysis, sequence alignment, and biological text analysis.

Other major protein databases include Protein Information Resource PIR (pir.georgetown.edu), Protein Data Bank PDB (www.pdb.org/pdb/) containing structural data of biological macromolecules, multiple sequence alignments and profiles of protein domains Pfam (www.sanger.ac.uk/Software/Pfam), biologically significant patterns and profiles (www.expasy.org/prosite).

Pathology databases contain human or mouse disease-related data. Examples include Tumor Gene Family Databases TGDBs (www.tumor-gene.org/tgdf.html), and Functional Immunology FIMM (sdmc.i2r.a-star.edu.sg:

8080/fimm). Intramolecular databases, such as the Database of Interacting Proteins DIP (dip.doe-mbi.ucla.edu) and MHC-binding peptides JenPep (www.jenner.ac.uk/Jenpep2) contain information on molecular interactions. Kyoto Encyclopedia of Genes and Genomes KEGG (www.genome.ad.jp/kegg) is a resource consisting of the PATHWAY database on molecular interaction networks, GENES database on genes and proteins generated by genome sequencing projects, and LIGAND database of chemical compounds and chemical reactions that are relevant to cellular processes. KEGG is a representative of metabolic pathways and cellular regulation databases. Several other databases with varied biomedical content have been listed in the NAR and DBCAT catalogues.

Retrieval systems

The major retrieval systems are associated with the providers of the three key DNA databases (DDBJ, EMBL and GenBank). These providers are the Japanese National Institute of Genetics (NIG, www.nig.ac.jp/index-e.html), European Bioinformatics Institute (EBI, www.ebi.ac.uk), and the US National Center for Biotechnology Information (NCBI, www.ncbi.nlm.nih.gov).

The three providers create public databases, develop database analysis tools, provide database analysis services, and disseminate biomedical information for better understanding of biological processes. They enable public access to a set of nucleotide, protein, genome, proteome, structure, microarray, literature and other specialized databases with a rich set of biological sequence analysis tools that can be used on-line. Users can download data using ftp access. These rich data sets enable comprehensive searches of multiple collections of databases. The Entrez retrieval system at NCBI facilitates access to, and analysis of 20 interlinked databases (Fig. 16.3). It provides tools for text searches, sequence similarity searches using the basic local alignment search tool

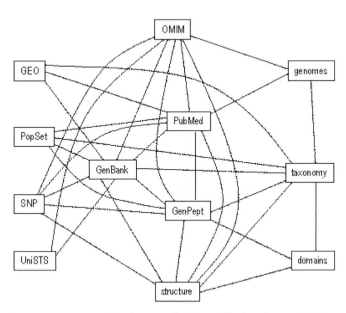

Fig. 16.3. The NCBI Entrez system and interface to a database collection. Some of the Entrez databases are not shown in this figure. The links show database connectivity. GenBank – nucleotide sequences; PubMed – publication abstracts; GenPept – protein sequences; OMIM – Mendelian inheritance and genetic disorders in humans; genomes – various genomes; taxonomy – names of organisms and taxonomic structure; domains – 3D structural domains; structure – 3D macromolecular structures, including proteins and polynucleotides; UniSTS – markers, or sequence-tagged sites (STS); SNP – single nucleotide polymorphism database; PopSet – DNA sequences selected for evolutionary studies; GEO – gene expression and hybridization array.

(BLAST, Altschul *et al.*, 1997), and 3D structure visualization. The NIG site enables database searches using BLAST, FASTA (Pearson and Lipman, 1988), and Smith–Waterman algorithm SSEARCH (Smith and Waterman, 1981). NIG also provides CLUSTALW for multiple sequence analysis bundled with tools for building of phylogenetic trees (Chenna *et al.*, 2003) and LIBRA I, a tool for compatibility analysis between protein 3D structure and its primary sequence (Ota and Nishikawa, 1999). The EBI site provides access to 60-odd analysis tools, grouped as similarity and homology searching, sequence analysis, protein function analysis, structure analysis and other tools.

Several additional retrieval systems or supporting resources are publicly available: hierarchical list of gene and protein names for data retrieval KEYnet (www.ba.cnr.it/keynet.html), transcription element search system TESS (www.cbil.upenn.edu/tess), and a database of rich links for data browsing, data analysis and database interconnection Virgil (www.infobiogen.fr/services/virgil). Virgil contains more than 40,000 rich links from five major databases: SWISS-PROT, GenBank, PDB, GDB and OMIM.

The University of California at Santa Cruz (UCSC) genome browser (genome.ucsc.edu) provides a convenient display of any requested portion of genomes at any scale, together with annotation tracks (known and predicted genes, ESTs, mRNAs, CpG islands, assembly gaps, chromosomal bands, genome comparisons, and some more). The genome browser analysis tools support text- and sequence-based searches that provide quick, precise access to any region of specific interest. The ENSEMBL genome browser at Sanger Institute, UK (www.ensembl.org) provides similar services for several genomes (human, rodent, fish, insect and worm).

Data integration systems mainly depend on two designs – federated databases or data warehouses (Durand *et al.*, 2003). Federated databases provide a software layer sitting on top of multiple data sources which are maintained independently. Data warehouses integrate data from multiple sources into a centralized repository. Several data integration systems have been developed (see Durand *et al.*, 2003). These include federated integration system

DiscoveryLink (Haas *et al.*, 2001), Kleisli (Chung and Wong, 1999), SRS (Zdobnov *et al.*, 2002) and TAMBIS (Stevens *et al.*, 2000). Data warehousing integration systems include Ensembl (Clamp *et al.*, 2003), Entrez (Wheeler *et al.*, 2003), and GenoMax (http://www.informaxinc.com/solutions/genomax/).

Searching Genetic Databases

A key issue for extracting value from databases is the ability to search and extract information at various levels. Common database search methods are by keyword matching, sequence similarity, motif searching, and class searching.

A standard basic search tool for most databases is a keyword or text matching facility. A keyword can be the name of the gene or protein (e.g. *lactalbumin*), species (e.g. *Homo sapiens, human*), a taxonomy term (e.g. *primates*), or a word from the reference title (e.g. *bee toxin*). Keywords can be combined into more complex logical expressions (e.g. *bee* OR *hornet* AND *toxin*). In an example from the authors' earlier research a query by the keyword 'snake' returned 1417 matches from GenBank, many of which were actually human, rat, mouse, or fruit fly sequences. The keyword '*serpentes*' – Latin name for snakes used in taxonomy – returned 4266 matches. Similarly '*snake* AND *toxin*' returned 250 matches, while '*serpentes* AND *toxin*' returned 443 matches. A good selection of keywords is therefore critical for getting correct and complete search results. Databases typically have hyperlinks that provide access to additional information related to the entry from other sources. Keyword searching will not provide any insight into these linked data sources and requires more sophisticated programming solutions.

Sequence similarity searches will return records containing sequences similar to the query sequence. Similar sequences often (but not always) have similar structure and, in turn, the same or similar biological function. The questions that can be answered by sequence analysis and sequence similarity searches are: What is the information contained in a biological sequence? How can we further analyse the sequence to gain additional knowledge? Does sequence similarity indicate any functional clues?

A large number of sequence similarity algorithms are available. The most popular algorithms are BLAST, FASTA or SSEARCH (see the 'Retrieval Systems' section for references). The results of database searching for sequence similarity will depend on the selected parameters. Usually, the results can be interpreted directly. Sometimes, interesting matches will be those where similarities are relatively low, but they will indicate similar structural properties or similar function. With distant homologues, the selection of search parameters, such as gap and gap length penalties may have significant effects on results. In addition, a filtering and masking facility, such as in a BLAST search, can be activated or disabled in a particular search. This may also result in different results when searching with a particular query sequence. If preliminary results indicate that there is a possibility of distant homology, users should attempt several search runs each with a different selection of parameters and filtering options. Sequence similarity algorithms have measures for the assessment of statistical significance of the matches. Most often, statistical significance is a good measure of the quality of similarity. Sometimes, the indicated statistical significance actually has no practical significance, particularly for short matches. Sometimes relatively low statistical scores are assigned to matches that have practical significance.

Motif searches and class searches are mainly used for protein databases, and are used less in genetic databases. Motif search refers to identification of a particular sequence pattern, such as TATA box (Juo *et al.*, 1996) in DNA sequence promoters. The class search refers to extraction of a complete sequence set belonging to a group or a family.

The extraction of comprehensive information available for an anonymous DNA sequence from multiple databases requires a multiple stops journey over the Internet. The starting point is typically one of the major DNA databases (GenBank, EMBL or DDBJ) or one of the expression databases. The journey will involve visits to tens of databases with an astounding amount of information collected. One strategy for gathering information about an anonymous sequence from various genetic databases is shown in Fig. 16.4. The first step includes identification of the sequence (or its nearest neighbours) in GenBank. One or more matches will enable access to relevant databases in Entrez (see Fig. 16.3). The SWISS-PROT database is a convenient entry to potential protein product analysis. The match will provide clues to structural and functional properties of the protein product of the query sequence. The links in SWISS-PROT include protein families, domains, motifs, and various genetic databases. The

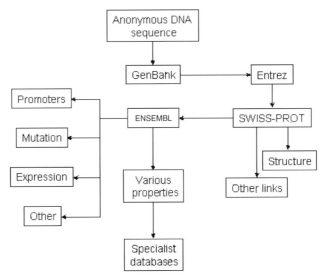

Fig. 16.4. A possible strategy for the analysis of an anonymous DNA sequence. The information retrieval strategy is explained in the main text.

ENSEMBL system will provide, if available, data on genetic sequence, chromosomal location, and various properties of the query sequence and its genomic counterpart. In addition, ENSEMBL enables comparative analysis across several genomes. The upstream region of the gene can be used for the analysis of promoters and regulatory elements via access to the relevant databases. Other databases, such as mutation or expression data sources, will provide additional information on the studied sequence. Various properties collected from the earlier steps in the analysis can be used for accessing specialist databases and extraction of additional information.

Data Quality Issues

Data quality generally refers to the 'fitness' of the data in the databases. Data quality can be assessed by measuring the agreement between data views presented by a system and the real world entity (Orr, 1998). According to this definition, a genomic database with data quality of 100% would have completely presented all real world genetic sequences. Although attaining perfect data quality is not possible, developers of genetic databases should strive for the highest level of quality achievable because quality data are crucial to the generation of accurate analytical results. Conversely, erroneous and noisy data may result in the extraction of inaccurate information and inaccurate analytical results.

Manual curation of the data is commonly used in many genetic databases to improve the quality of data originated from other public domain databases or directly submitted by individual researchers (Fredman, 2002). Using data analysis and visualization tools, curators inspect and correct the data for consistency, accuracy, completeness, correctness, timeliness, relevance and uniqueness. This process of improving data quality involving error correction, duplicate identification and restructuring data is also known as data cleaning. Although the curation at the database site helps eliminate a large proportion of errors in raw data, a significant number of errors and inconsistencies can be found in public databases. When using public databases for extraction of data, particularly for small data sets,

individual users are advised to perform curation of their data to ensure quality of data.

The use of a manual data cleaning process poses difficulties in sustaining data quality as genomic sequences continue to grow exponentially. Yet, an inspection of some of the discrepancies that can be found in genomic data explains why data cleaning issues are too complicated to be automated fully. For instance, a pair of DNA sequences differing by a few nucleotides can either be variants resulting from incorrect entry of the sequence or due to sequencing error. It is not easy to differentiate, especially when functional annotations from submissions are not expressed consistently. Curators need to cross-reference the annotations to the publications in an attempt to standardize the annotation and, if possible, impose use of controlled vocabulary.

In systems where manual curation is too expensive, automated pre-processing steps are adopted for improving the quality of data. When the prediction result of the analysis is highly sensitive to the noise in the data set, the pre-processing data cleaning steps are crucial for the accuracy of the analysis (Brusic et al., 1999; Teng, 1999). The focus on data quality improvement is seldom being emphasized in genetic databases, but there have been attempts to automate internal data cleaning. For example, the now discontinued Genomic Sequence Database (GSDB) identified the need for improved data quality and provided solutions for vector contamination in sequences and for identification of assemblies of dis-contiguous sequences (Harger et al., 1998). By removing erroneous sequences and organizing the set of sequence fragments into more meaningful groupings, they achieved better quality of the data to be used for gene level studies and genomic comparison studies.

Errors in genetic databases are common. We identified four general sources of errors: attribute, record, single-source database and multi-source database level. The attribute level errors are incorrect values of the fields (such as '0.9 gene' for gene name), misspellings, wrong abbreviations, or values placed in the wrong field (e.g. organism name inside the definition field). These errors occur most often because of the errors in the original data submission, or from automated systems for record processing.

The record level errors often result from conflicts or misplacement of fields within a record. Examples include wrong accession or version numbers, features entered as comments, or non-functional links inside the database. Single-source database level errors refer to conflicting or duplicate (redundant) entries within a single database. Examples include non-exact duplicates with identical content and identical partial or fragment sequences in separate records. The multi-source database errors occur because of data integration and source synchronization problems. A recent study of 221 scorpion toxins (Srinivasan *et al.*, 2002) extracted from public databases and literature identified significant sources of errors in the data. The data were extracted from GenBank (135), EMBL (65), DDBJ (65), SWISS-PROT (119), TrEMBL (20), PIR (62), and PDB (29), a total of 171 distinct entries, and literature (50 entries). None of the data sources had the complete data set. Additional functional and other annotations were extracted from the literature and manually assigned to 101 entries. The majority of data in public databases, therefore, lack important functional and structural data that are known and available in the literature. The types of errors identified were: wrong links between databases (one sequence), different names for the same sequence (11), different sequences for the same toxin name (15), no links between the same records in different databases (23), toxin names from journals not used in databases (30) and SWISS-PROT links to PDB structures of poor homology (51). These results indicate high error rates in molecular databases.

With increasing data mining and analytical projects that are dependent on the use of genomic databases, data quality is becoming an important factor to assess the useability of these databases. The need for quality data is tightly coupled with greater use of data cleaning techniques, which are likely to be more semi-automated in order to keep in pace with the growth in genomic information.

Concluding Remarks

Molecular databases are critical resources for life sciences. Searching molecular databases is important not only for rapid access to existing data and information, but also for new knowledge discovery. Knowledge discovery may be in the form of identifying crucial dependencies between molecular entities, understanding biological systems and processes, understanding effects of specific interactions, determining best experiments, minimization of cost and time, learning by summarization, and the use of predictive models.

Biological databases represent an invaluable resource in support of biological research. We can learn much about a particular molecule by searching databases and using available analysis tools. A large number of databases are available for that task. Some databases are very general, while some are specialized. For best results we often need to access multiple databases. Major types of databases are focusing on nucleotide, genome, protein, structure, pathways, molecular interactions, protein motifs, publication, and miscellaneous databases. Common database search methods include keyword matching, sequence similarity, motif searching, and class searching. The problems with using biological databases include incomplete information, data spread over multiple databases, redundant information, various errors, sometimes incorrect links and constant change. The trends are towards developing databases that are integrated with specialized tools and can be used for data mining and a higher level of knowledge discovery.

Database standards, nomenclature, and naming conventions are not clearly defined for many aspects of biological information. This makes information extraction more difficult. Retrieval systems help in extracting rich information from multiple databases. Examples include Entrez and other browsers. Formulating queries is a serious issue in biological database usage. The quality of results depends on the quality of the queries. Statistical measures in search algorithms indicate the quality of matches. Often the statistical and biological significance are related. Sometimes, however, matches of real biological significance have low statistical scores. Access to biological databases is so important that today virtually every molecular biological project starts and ends with querying biological databases.

The environment for genetic databases has dramatically changed since 1995. With the advent of the Internet, access to data and information has become practically unrestricted and it is no longer an exclusive privilege of the rich. The consequence is that molecular database user and developer communities have grown many-fold in the last 8 years and we are facing several major changes in the molecular database field. First, major established databases will grow into conglomerates comprising hundreds of integrated individual databases (see Fig. 16.3). A major challenge for conglomerated databases will be the management of the scale of these systems. Secondly, the retrieval systems, such as Entrez, will become increasingly sophisticated, providing seamless access to critical information from multiple sources.

They will keep integrating increasingly sophisticated analysis tools and will have to address access to diversified information sources. Thirdly, we will see the individualized systems which will enable researchers to build their own specialist databases that will contain detailed information currently not available in any of the public databases. Finally, both database development and user communities have become truly international which has resulted in an accelerated growth of specialist, well annotated databases. Currently, databases are heterogeneous, geographically dispersed, and have diverse data formats, access tools and analysis tools. In the near future, we will see the emergence of standards that will enable better automation and flow of information between various database centres.

References

Altschul, S.F., Madden, T.L., Schaffer, A.A., Zhang, J., Zhang, Z., Miller, W. and Lipman, D.J. (1997) Gapped BLAST and PSI-BLAST: a new generation of protein database search programs. *Nucleic Acids Research* 25, 3389–3402.

Baer, R., Bankier, A.T., Biggin, M.D., Deininger, P.L., Farrell, P.J., Gibson, T.J., Hatfull, G., Hudson, G.S., Satchwell, S.C., Seguin, C. *et al.* (1984) DNA sequence and expression of the B95-8 Epstein–Barr virus genome. *Nature* 310, 207–211.

Bankier, A.T. (2001) Shotgun DNA sequencing. *Methods in Molecular Biology* 167, 89–100.

Baxevanis, A.D. (2003) The Molecular Biology Database Collection: 2003 update. *Nucleic Acids Research* 31, 1–12.

Benson, D.A., Karsch-Mizrachi, I., Lipman, D.J., Ostell, J. and Wheeler, D.L. (2003) GenBank. *Nucleic Acids Research* 31, 23–27.

Brusic, V., Zeleznikow, J., Sturniolo, T., Bono, E. and Hammer, J. (1999) Data cleansing for computer models: a case study from immunology. *Proceedings of ICONIP99*, The Sixth International Conference on Neural Information Processing, IEEE, 603–609.

Bucher, P. (1999) Gene feature identification. In: Bishop, M.J. (ed.) *Genetic Databases.* Academic Press, San Diego, USA, pp. 135–164.

Chenna, R., Sugawara, H., Koike, T., Lopez, R., Gibson, T.J., Higgins, D.G. and Thompson, J.D. (2003) Multiple sequence alignment with the Clustal series of programs. *Nucleic Acids Research* 31, 3497–3500.

Chung, S.Y. and Wong, L. (1999) Kleisli: a new tool for data integration in biology. *Trends in Biotechnology* 17, 351–355.

Clamp, M., Andrews, D., Barker, D., Bevan, P., Cameron, G., Chen, Y., Clark, L., Cox, T., Cuff, J., Curwen, V. *et al.* (2003) Ensembl 2002: accommodating comparative genomics. *Nucleic Acids Research* 31, 38–42.

Cotton, R. (1999) Phenotype, mutation and genetic linkage databases and their links to sequence databases. In: Bishop, M.J. (ed.) *Genetic Databases.* Academic Press, San Diego, USA, pp. 39–52.

Durand, P., Medigue, C., Morgat, A., Vandenbrouck, Y., Viari, A. and Rechenmann, F. (2003) Integration of data and methods for genome analysis. *Current Opinion in Drug Discovery and Development* 6, 346–352.

Fleischmann, R.D., Adams, M.D., White, O., Clayton, R.A., Kirkness, E.F., Kerlavage, A.R., Bult, C.J., Tomb, J.F., Dougherty, B.A., Merrick, J.M. *et al.* (1995) Whole-genome random sequencing and assembly of *Haemophilus influenzae* Rd. *Science* 269, 496–512.

Fredman, D., Siegfried, M., Yuan, Y.P., Bork, P., Lehvaslaiho, H. and Brookes, A.J. (2002) HGVbase: a human sequence variation database emphasizing data quality and a broad spectrum of data sources. *Nucleic Acids Research* 30, 387–391.

Haas, L.M., Schwartz, P.M., Kodali, P., Kotlar, E., Rice, J.E. and Swope, W.C. (2001) DiscoveryLink: a system for integrated access to life sciences data sources. *IBM Systems Journal* 40, 489–511.

Harger, C., Skupski, M., Bingham, J., Farmer, A., Hoisie, S., Hraber, P., Kiphart, D., Krakowski, L., McLeod, M., Schwertfeger, J. *et al.* (1998) The Genome Sequence DataBase (GSDB): improving data quality and data access. *Nucleic Acids Research* 26, 21–26.

Juo, Z.S., Chiu, T.K., Leiberman, P.M., Baikalov, I., Berk, A.J. and Dickerson, R.E. (1996) How proteins recognize the TATA box. *Journal of Molecular Biology* 261, 239–254.

Lander, E.S., Linton, L.M., Birren, B., Nusbaum, C., Zody, M.C., Baldwin, J., Devon, K., Dewar, K., Doyle, M., FitzHugh, W. *et al.* (2001) Initial sequencing and analysis of the human genome. *Nature* 409, 860–921.

Okazaki, Y., Furuno, M., Kasukawa, T., Adachi, J., Bono, H., Kondo, S., Nikaido, I., Osato, N., Saito, R., Suzuki, H., *et al.* (2002) Analysis of the mouse transcriptome based on functional annotation of 60,770 full-length cDNAs. *Nature* 420, 563–573.

Orr, K. (1998) Data quality and systems theory. *Communication of the ACM* 41, 66–71.

Ota, M. and Nishikawa, K. (1999) Feasibility in the inverse protein folding protocol. *Protein Science* 8, 1001–1009.

Pearson, W.R. and Lipman, D.J. (1988) Imported tools for biological sequence comparison. *Proceedings of the National Academy of Sciences USA* 85, 2444–2448.

Roberts, L., Davenport, R.J., Pennisi, E. and Marshall, E. (2001) A history of the human genome project. *Science* 291, 1195. www.sciencemag.org/cgi/content/full/291/5507/1195

Sanger, F., Coulson, A.R., Friedmann, T., Air, G.M., Barrell, B.G., Brown, N.L., Fiddes, J.C., Hutchison, C.A., 3rd, Slocombe, P.M. and Smith, M. (1978) The nucleotide sequence of bacteriophage phiX174. *Journal of Molecular Biology* 125, 225–246.

Smith, T.F. and Waterman, M.S. (1981) Identification of common molecular subsequences. *Journal of Molecular Biology* 147, 195–197.

Srinivasan, K.N., Gopalakrishnakone, P., Tan, P.T., Chew, K.C., Cheng, B., Kini, R.M., Koh, J.L, Seah, S.H. and Brusic, V. (2002) SCORPION, a molecular database of scorpion toxins. *Toxicon* 40, 23–31.

Stevens, R., Baker, P., Bechhofer, S., Ng, G., Jacoby, A., Paton, N.W., Goble, C.A. and Brass A. (2000) TAMBIS: transparent access to multiple bioinformatics information sources. *Bioinformatics* 16, 184–185.

Teng, C.M. (1999) Correcting noisy data. *Proceedings of the Sixteenth International Conference on Machine Learning*, pp. 239–248.

Waterston, R.H., Lindblad-Toh, K., Birney, E., Rogers, J., Abril, J.F., Agarwal, P., Agarwala, R., Ainscough, R., Alexandersson, M., An, P. *et al.* (2002) Initial sequencing and comparative analysis of the mouse genome. *Nature* 420, 520–562.

Watson, J.D. and Crick, F.H. (2003) Molecular structure of nucleic acids. A structure for deoxyribose nucleic acid. 1953. *Rev Invest Clin.* 55, 108–109.

Westbrook, J., Feng, Z., Chen, L., Yang, H. and Berman, H.M. (2003) The protein data bank and structural genomics. *Nucleic Acids Research* 31, 489–491.

Wheeler, D.L., Church, D.M., Federhen, S., Lash, A.E., Madden, T.L., Pontius, J.U., Schuler, G.D., Schriml, L.M., Sequeira, E., Tatusova, T.A. and Wagner, L. (2003) Database resources of the National Center for Biotechnology. *Nucleic Acids Research* 31, 28–33.

Wolfsberg, T.G., Wetterstrand, K.A., Guyer, M.S., Collins, F.S. and Baxevanis, A.D. (2002) A user's guide to the human genome. *Nature Genetics* 32(suppl.), 1–79.

Wu, C.H., Yeh, L.S., Huang, H., Arminski, L., Castro-Alvear, J., Chen, Y., Hu, Z., Kourtesis, P., Ledley, R.S., Suzek, B.E., Vinayaka, C.R., Zhang, J. and Barker, W.C. (2003) The protein information resource. *Nucleic Acids Research* 31, 345–347.

Zdobnov, E.M., Lopez, R., Apweiler, R. and Etzold, T. (2002) The EBI SRS server – new features. *Bioinformatics* 18, 1149–1150.

17 Gene Predictions and Annotations

R. Guigó[1] and M.Q. Zhang[2]

[1]*Insitut Municipal d'Investigació Mèdica, Centre de Regulació Genòmica, Universitat Pompeu Fabra, Barcelona, Spain;* [2]*Cold Spring Harbor Laboratory, NY, USA*

Introduction	429
Ab initio Gene Prediction	430
Prediction of signals	430
Prediction of exons	433
Exon assembly into genes	434
Gene prediction programs	435
Comparative Gene Prediction	436
Genomic query against protein or cDNA target	436
Genomic query against genomic target	438
Prediction of selenoproteins	439
Accuracy of Gene Predictions	440
Measures of prediction accuracy	440
Accuracy in single gene sequences	441
Accuracy in large genomic sequences	442
Genome Annotation Systems	443
References	444

Introduction

After the genome of an organism is sequenced and assembled, comprehensive and accurate initial gene prediction and annotation by computational analysis have become the necessary first step towards understanding the functional content of the genome. This chapter describes typical computational methods for identification of protein-coding genes in a mammalian genome. The organization of a gene, as any other biological structure, is determined by functional and evolutionary constraints; all computational methods are therefore based on our experimental understanding of such constraints. Similar structure features would imply similar function and a tendency to be conserved through evolution. According to the central dogma, genetic information flows as DNA → RNA → protein (see Chapters 5 and 6). Namely, a gene is first transcribed into a pre-mRNA, this transcript is subsequently processed (i.e. capped, spliced and polyadenylated) and the mature mRNA transcript is then transported from the nucleus into the cytoplasm for translation into the gene product – a functional protein. An example of a eukaryotic gene structure and its corresponding mature mRNA transcript are depicted in Fig. 17.1. There are two types of genetic elements with respect to primary structure of proteins: short cis-regulatory elements (also called 'signals', mostly in non-coding regions) that control how the gene is expressed, and coding sequences (CDSs) that code for the gene product (protein). There are also RNA genes such as miRNAs and tRNAs, but they are not the focus of this chapter. Since the majority of the mammalian genes have introns, the key in gene prediction is to detect splice site signals and to locate CDSs. In the early 1980s, computer methods were developed to find genes either by detecting extended CDS regions based on codon usage (Staden and McLachlan, 1982) or coding periodicity (Fickett, 1982), or by detecting splicing signals (Mount, 1982; Staden, 1984). Later, more complex algorithms based on modern statistical (e.g. linear or quadratic discriminant analysis, or LDA/QDA), linguistic (e.g. hidden Markov model or HMM) or machine-learning (e.g. artificial neural networks or ANN) techniques have been developed to combine various sequence features for more accurate gene predictions. As more genomes become available, comparative genomics can offer even better prediction than ab initio methods for homologous genes. Although gene prediction tools have become more sophisticated, prediction accuracy is still far from satisfactory. Up to now, the total number of human genes, ranging from 24,500 to 45,000 (Pennisi, 2003), still cannot be estimated with certainty, and current mammalian gene counts are highly hypothetical in nature.

There is still a long way to go before gene prediction programs can produce accurate predictions of the gene content in mammalian genomes.

Ab Initio Gene Prediction

As most mammalian genes contain introns, here we mainly describe some basic methods for predicting coding exons of the intron-containing genes (see Zhang, 2002, for review and discussion of more general types of exons and their predictions).

Prediction of signals

There are four basic signals for any coding exon predictions: the translational start site (the START site), the 5′ splice site (5′ss, or donor site), the 3′ splice site (3′ss, or acceptor site), the branch site and the translational stop site (the STOP codon). The simplest measure for a signal site is so-called consensus (word pattern). For example, a STOP site may be any of TAA, TAG or TGA (the ratio among these three stop codons in mammalian genomes is about 1:1:2). The ATG site and the START site may be characterized by the Kozak consensus GCCGCCR-CCATGG (Kozak, 1987). Splice site consensus was first catalogued by Mount (1982) and later was refined with more data by Senapathy et al. (1990): donor site AG|GTRAGT, acceptor site (Y)nNCAG|G and branch site CTRAY (the average distance between the branch point 'A' and the 3′ss boundary, indicated by '|' in the

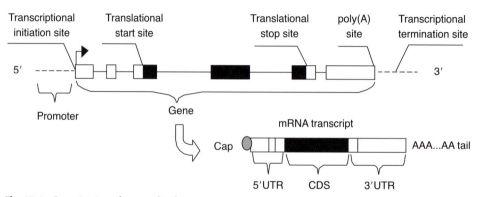

Fig. 17.1. Organization of a gene that has six exons.

acceptor site, is 26 bases). Senapathy *et al.* actually presented the splice signals by more quantitative frequency matrices (see Table 17.1), where each matrix element M_{ij} is equal to the frequency count of the base 'i' at position 'j' from the training set of aligned splice site sequences. If one defines $f_{ij} = M_{ij}/100$, and f_i as the background frequency of nucleotide i, then the popular log-odd scoring matrix (weight matrix model, or **WMM**) $S_{ij} = \log(f_{ij}/f_i)$ scores the signal as the sum of the scores over the bases within the signal.

Although useful in many cases, there are at least two main problems with any WMM approach. First, it assumes that bases at different positions are independent. There are many ways to incorporate base dependencies. One method is to assume that each base is correlated with its neighbouring bases (so-called Markov dependence or weight array model, **WAM**, (Zhang, 1993)). It is equivalent to extending the rows to doublets (Markov order-1: AA, AC AG, etc.) or to triplets (Markov order-2: AAA, AAC, AAG, etc.). However, more dependencies result in more parameters to be estimated and it requires much more training data. Another method is to apply a decision tree (so-called maximum dependence decomposition, MDM (Burge and Karlin, 1997)) to partition total training data into subsets so that splice site bases within each subset are approximately independent and hence can be modelled by a separate WMM. This also ignores the GC-content of the gene locus. It is well known that a mammalian genome has large variations in GC-content, often referred to as isochores (Bernardi, 1995),

therefore all signal compositions will be highly biased by the GC-content. More modern gene prediction tools use GC-content-specific signal models (Zhang, 1998). For example, the GC-specific splice site WMMs are shown in Table 17.2 with each element representing the frequency count obtained from both 'low' and 'high' GC genes (Zhang, 1998).

There are other ways one may further improve the accuracy of a signal prediction. For instance, one could model secondary structures of pre-mRNAs (e.g. Patterson *et al.*, 2002). However, the most effective way of improving signal site prediction is to combine flanking sequence features (for example, a donor score can be combined with upstream exon and downstream intron scores by LDA (Solovyev *et al.*, 1994)). Here we have only described statistical methods; there are also several machine-learning methods (e.g. Lapedes *et al.*, 1990; Degroeve *et al.*, 2002) in parallel to each of the statistical approaches.

Recently, Pertea *et al.* (2001) compared several leading prediction programs: NETGENE2 (Brunak *et al.*, 1991, http://genome.cbs.dtu.dk/services/NetGene2/), HSPL (Solovyev *et al.*, 1994, http://genomic.sanger.ac.uk/), NNSPLICE (Reese *et al.*, 1997, http://www.fruitfly.org/seq_tools/splice.html) GENIO (Mache and Levi, 1998, http://genio.informatik.uni-stuttgart.de/GENIO/splice/) and SPLICEVIEW (Rogozin and Milanesi, 1997, http://l25.itba.mi.cnr.it/~webgene/wwwspliceview.html) with their own GENESPLICER (Pertea *et al.*, 2001, http://www.tigr.org/tdb/GeneSplicer/gene_spl.html), and concluded that:

Table 17.1. Donor (from human), acceptor (from human) and branch point (from human, rat, chicken, plant and Drosophila) frequency matrices (Senapathy et al., 1990).

Donor frequency matrix

position	-3	-2	-1	|	+1	+2	+3	+4	+5	+6
A	28	59	8	|	0	0	54	74	5	16
C	40	14	5	|	0	0	2	8	6	18
G	17	13	81	|	100	0	42	11	85	21
T	14	14	6	|	0	100	2	8	4	45

Branch frequency matrix

position	-3	-2	-1	0	+1
A	1	0	39	99	11
C	76	8	15	1	45
G	2	0	42	0	6
T	21	91	4	0	38

Acceptor frequency matrix

position	-14	-13	-12	-11	-10	-9	-8	-7	-6	-5	-4	-3	-2	-1	|	+1
A	10	8	6	6	9	9	8	9	6	6	23	2	100	0	|	28
C	31	36	34	34	37	38	44	41	44	40	28	79	0	0	|	14
G	14	14	12	8	9	10	9	8	6	6	26	1	0	100	|	47
T	44	43	48	52	45	44	40	41	45	48	23	18	0	0	|	11

Table 17.2. GC-specific (low/high). donor, acceptor and branch point (all from human) frequency matrices (Zhang, 1998).

Donor

	−3	−2	−1	+1	+2	+3	+4	+5	+6
A	38/32	62/55	12/8	0	0	71/38	73/70	11/5	21/13
C	31/38	10/15	4/4	0	0	2/4	6/9	6/5	10/21
G	18/19	12/15	77/80	100	0	24/56	8/14	75/86	14/25
T	13/11	16/14	7/8	0	100	3/2	13/7	8/4	55/41

Acceptor

	−15	−14	−13	−12	−11	−10	−9	−8
A	15/10	14/8	13/7	11/8	10/6	10/6	11/4	12/8
C	24/41	21/42	20/41	22/40	21/38	22/43	25/42	28/46
G	10/15	12/14	10/14	9/13	10/13	9/12	10/13	10/14
T	51/34	53/36	57/38	58/30	59/43	59/39	54/41	50/32

	−7	−6	−5	−4	−3	−2	−1	+1	+2
A	13/8	11/7	10/6	26/19	7/2	100	0	26/21	24/19
C	28/49	25/54	22/45	25/38	55/82	0	0	11/13	15/21
G	8/10	5/8	5/8	15/26	1/0	0	100	50/58	20/29
T	51/33	59/31	63/41	33/17	37/16	0	0	13/8	41/31

Branch site

	−5	−4	−3	−2	−1	0	+1
A	25/15	25/15	0	0	39/16	100	18/7
C	19/36	22/38	66/88	2/5	24/31	0	33/56
G	15/24	17/22	0	0	32/51	0	3/12
T	41/25	36/25	40/12	98/95	5/2	0	46/25

Overall, NETGENE2 appears to be the best for donor site prediction, while for acceptor sites either GENESPLICER, NETGENE2 or HSPL perform comparably. One advantage of GeneSplicer for the latter task is that its thresholds can be adjusted by the user to vary the false negative and false positive rates.

The best average of error (false-positive + false-negative) rate for either donor or acceptor site prediction is about 5%. This may sound not bad if the search is restricted in a short (say, coding) region. If one uses any predictor to search a large region of a genome, the false-positive rate is unacceptable because for every true site there are hundreds of pseudo-sites. For example, if a large region has 40 true sites and 4000 pseudo-sites, one true site would be missed (2.5% false negatives) and 100 pseudo-sites would be predicted as true sites (2.5% false positives)! Since adjacent donor site and acceptor site are not independent (Zhang,

1998), this correlation can be explored for further eliminating false positives. For short introns, occurring mostly in lower eukaryotes, an intron is recognized by the interaction of splicing factors binding across the intron ends (hence 5'ss–3'ss correlation, Lim and Burge, 2001). In vertebrates, exons are much shorter; recognition of exons by the interaction of splicing factors binding across the exon ends (hence 3'ss–5'ss correlation, Zhang, 1998) is the key (Robbertson *et al.*, 1990). Therefore mammalian functional splice sites can only be effectively identified simultaneously through exon recognition.

An additional complication, dealt with poorly by current splice site prediction programs, is the presence of non-canonical sites. At least 1% of all introns do not conform to the canonical AG–GT boundaries (Burset *et al.*, 2000), and they are systematically ignored by splice site and gene prediction methods.

Prediction of exons

In addition to signal features, prediction of exons (here we only discuss internal coding exons which can be easily generalized to other types of exons. See Zhang (1998) for the complete classification of 12 types of exons in the mammalian genomes) should also require content features for both exon regions and flanking intron regions (typically <100 bp on each side). To discriminate CDS from introns, the best content feature is the frame-specific hexamer frequency score, so named because it captures codon-bias information and codon–codon correlation (Fickett and Tung, 1992; Guigó, 1998). There are many ways to construct such a score, both log-odd score $L_E(w,i) = log (f_E(w,i)/f_I(w))$ and preference score $P_E(w,i) = f_E(w,i)/[f_E(w,i) + f_I(w))$ are popular among exon finders, where $f_E(w,i)$ is the frequency of hexamer w in frame i calculated from known exon training data and $f_I(w)$ is the frequency of w calculated from known flanking introns. If the hexamer w is more likely to be found in an exon at the given frame than in an intron, L_E will be positive (P_E will be greater than 1/2). P_E is easier to use as it ranges from 0 to 1, and has been used in both LDA (HEXON: Solovyev *et al.*, 1994) and QDA (MZEF: Zhang, 1997) exon finders. In addition to coding measures, exon size is another important feature variable one must consider. For human internal coding exons, the size distribution is close to a log-normal distribution centred around 125 bp (Zhang, 1998). For intron regions, one may also construct similar hexamer scores: $L_I(w) = log (f_I(w)/f_E(w)) = -L_E(w)$ or $P_I(w) = f_I(w)/[f_E(w) + f_I(w)) = 1 - P_E(w)$, where $f_E(w)$ is the average of $f_E(w,i)$ over all three reading frames (i.e. $f_E(w) = (f_E(w,0) + f_E(w,1) + f_E(w,2))/3)$. These coding measures can only help with prediction of CDS regions; there has not been any effective method for predicting non-translated regions (UTRs) of an exon.

With a training set of exons and pseudo-exons (randomly selected AG–ORF–GT regions, also called a spliceable open reading frame, i.e. open reading frames bounded by the conserved 3'ss AG and 5'ss GT pair) and a set of feature variables $\boldsymbol{x} = (x_1, x_2, ..., x_k)$ (for example, $x_1 = $ 5'-flanking intron region score = average $L_I(w)$ over all hexamers in the 5'-flanking intron region, $x_2 = $ acceptor site score, $x_3 = $ maximum exon score = average $L_E(w,i)$ over all hexamers

in reading frame i and then take the maximum over i, $x_4 = $ exon size, $x_5 = $ donor site score, $x_6 = $ 3'-flanking intron region score, etc.), each training sample (a true exon or a pseudo-exon) can be represented by a point in the k-dimensional feature space. There are many statistical or machine-learning methods (most of them have been tried) that can be used to build an optimal (in the sense of minimization of false-positive and false-negative errors using cross-validation tests) discrimination function. This function is a decision surface (exon predictor) in the k-dimensional feature space that can best separate true exons from pseudo-exons. In the exon finder MZEF (Zhang, 1997), the quadratic discriminant analysis (QDA) is implemented as the core algorithm. Namely, the two training sample points (true exons and pseudo-exons) in the feature space are approximated by two separate k-dimensional Gaussian distributions characterized by the sample means and covariance matrices (estimated from the training data), the intersection of the two Gaussian distributions is a $(k–1)$-dimensional quadratic discrimination surface separating the two sample points in an optimal way (for a pedagogical introduction on DNA pattern discrimination, see Zhang, 2000). When the two Gaussians are assumed to have the same covariance matrix, the surface will become a hyper-plane and QDA will reduce to LDA, a method implemented in HEXON (Solovyev *et al.*, 1994) that is the exon-finder part of FGENEH (see below). Internal coding exon finding performance (Zhang, 1997) is shown in Table 17.3 using a data set of 570 mammalian single-gene sequences (Burset and Guigó, 1996; see section 'Accuracy of Gene Predictions', following).

Splice site identification helps exon recognition; internal coding exon measure can help further improve functional splice site selection. Indeed, Thanaraj and Robinson (2000) found that: (i) a high proportion of false-positive splice sites from computational predictions occur in the vicinity of real splice sites; and (ii) current algorithms are misled to predict wrong splice sites more often when the coding potential ends within ±25 nucleotides from real sites than when it ends at farther positions. Their integrated system (MZEF–SPC) with SPLICE PROXIMALCHECK (SPC) as a front-end filter for MZEF was able to eliminate two-thirds of the predicted false positives at the expense of losing one-tenth of predicted true positives. In fact,

Table 17.3. Accuracy of exon prediction methods.

Program	Nucleotide			Exon		
	S_n	S_p	CC	S_n	S_p	$\dfrac{S_n + S_p}{2}$
GRAIL2	0.79	0.92	0.83	0.53	0.60	0.57
FGENEH	0.83	0.93	0.85	0.73	0.78	0.76
MZEF	0.87	0.95	0.89	0.78	0.86	0.82

S_n is the proportion of coding nucleotides/exons predicted as coding;
S_p is the proportion of nucleotides/exons predicted as coding that are
actually coding; CC is a measure that summarizes both S_n and S_p (see
'Accuracy of Gene Predictions' section for detailed definitions).

these 'false-positive' splice sites could be alternative splice sites in a lot of cases. Exon extensions and truncations are commonly seen when an exon is skipped. Existing alternative exons obviously make it so much harder to predict the splice sites accurately, as there are quite a number of real splice sites. Over half of all genes are alternatively spliced, and over ten isoforms on average per gene, so the *in silico* 'false positives' may actually be 'false negatives' *in vitro*!

Exon assembly into genes

As exons are not independent, by splicing exons together to assemble a gene one can further eliminate false exon predictions by imposing translatability (i.e. adjacent exons must maintain the open reading frame). The main difficulty in exon assembly is the combinatorial explosion problem: the number of ways N candidate exons may be combined grows exponentially with N. The key idea of computational feasibility comes from dynamic programming (DP), which allows finding 'optimal assembly' quickly without having to enumerate all possibilities (Gelfand and Roytberg, 1993). DP is also used in GENEPARSE (Snyder and Stormo, 1993) to recursively search for exon–intron boundaries with signal and content measures obtained by a neural network. The FGENEH (Solovyev *et al.*, 1995a,b) algorithm incorporates 5'-, internal and 3'-exon identification linear discriminant functions and a dynamic programming approach for exon assembly. A more efficient dynamic programming algorithm was introduced by Guigó (1998), and it is incorporated in the program GENEID (Guigó *et al.*, 1992; Parra *et al.*, 2000).

A novel advance in gene prediction methodologies was the application of generalized hidden Markov models (HMMs), initially implemented in the GENIE algorithm (Kulp *et al.*, 1996; HMM was first used in a bacterial gene finder by Krogh *et al.* (1994) after its success in protein modelling). Soon after, it was implemented in the GENSCAN algorithm (Burge and Karlin, 1997) to predict multiple genes. Several other HMM-based gene prediction programs were developed later: VEIL (Henderson *et al.*, 1997), HMMGENE (Krogh, 1997) and FGENESH (Salamov and Solovyev, 2000). In a HMM approach, different types of structure components (such as exon or intron) are characterized by a state; a gene model is thought to be generated by a state machine: starting from 5' to 3', each base pair is generated by an 'emission probability' conditioned on the current state and surrounding sequences, and transition from one state to another is governed by a 'transition probability' which obeys all the constraints (such as an intron can only follow an exon, reading frames of two adjacent exons must be compatible, etc.). All the parameters of the 'emission probabilities' and the (Markov) 'transition probabilities' are learned (pre-computed) from some training data. Since the states are unknown ('hidden'), an efficient algorithm (called the VITERBI algorithm, similar to DP) may be used to select a best set of consecutive states (called a 'parse'), which has the highest overall probability compared with any other possible parse of the given genomic sequence (see Rabiner (1989) for a tutorial on HMMs). The reason these fully probabilistic state models have become preferable is that all scores are probabilities themselves and the weighting problem has become a matter of counting relative observed state frequency. It is *easy* to add more states (such as intergenic regions, promoters, UTRs, etc.) and transitions into HMM-based models to allow partial genes, intronless genes, *even* multiple genes or genes on different strands to be incorporated. These features are essential when annotating genomes or large contigs in an automated fashion. The advantage of modelling both strands simultaneously is that it avoids the prediction of overlapping genes on the two strands, which presumably are very rare in mammalian genomes, although in some cases, they may have a natural regulatory function.

Gene prediction programs

In Table 17.4, we have listed a number of the most popular gene finders available on the Internet, as well as other useful gene prediction resources. Benchmarks on the accuracy of these programs are described later. It has been reported that by integrating different programs

Table 17.4. Useful Internet gene prediction resources.

Weintian Li's bibliography on computational gene recognition	http://www.nslij-genetics.org/gene/
BCM Search Launcher: many gene finding programs	http://dot.imgen.bcm.tmc.edu:9331/seq-search/gene-search.html
PROMOTER RECOGNITION	
NNPP: ANN promoter prediction	http://www.fruitfly.org/seq_tools/promoter.html
PROMOTERINSPECTOR; MATINSPECTOR; FASTM; SMARTEST	http://genomatix.gsf.de/software_services/free_access/free_accounts.html
SIGNAL SCAN: TFD or TRANSFAC	http://bimas.dcrt.nih.gov/molbio/signal/
TESS: transcription factor site search	http://www.cbil.upenn.edu/tess/
TFBIND: transcription factor site search	http://tfbind.ims.u-tokyo.ac.jp/
CORE-PROMOTER: transcription start site prediction	http://argon.cshl.org/genefinder/CPROMOTER/
Dragon Promoter Finder	http://sdmc.lit.org.sg/promoter/promoter1_3/DPFV13.htm
Promoter: transcription start site prediction	http://www.cbs.dtu.dk/services/Promoter/
SPLICE SITE PREDICTION	
NETGENE2: splice sites in human, *C. elegans* and *Arabidopsis*	http://www.cbs.dtu.dk/services/NetGene2/
NNSPLICE: ANN, splice site prediction	http://www.fruitfly.org/seq_tools/splice.html
SPLICEPREDICTOR	@Normal:http://bioinformatics.iastate.edu/cgi-bin/sp.cgi
GENESPLICER	@Normal:http://www.tigr.org/tdb/GeneSplicer/gene_spl.html
SPLICEVIEW	@Normal:http://l25.itba.mi.cnr.it/~webgene/wwwspliceview_help.html
RNASPL	http://www.dl.ac.uk/CCP/CCP11/DISguISE/nucleotide_analysis/rnasplwww.html
GENE PREDICTION	
AAT: MZEF+homology	http://genome.cs.mtu.edu/aat.html
CDS: search coding region	http://bioweb.pasteur.fr/seqanal/interfaces/cds-simple.html
CRASA: EST-based gene finder	http://crasa.sinica.edu.tw/bioinformatics/introduce.html
DIOGENES: finding ORFs in short genomic sequences	http://www.cbc.umn.edu/diogenes/index.html
FGENESH: HMM (human, *Drosophila*, dicots, monocots, *C. elegans*, *S. pombe*); FGENESH ; HEXON; TSSW; TSSG; SPL; POLYAH	http://genomic.sanger.ac.uk/gf/gf.shtml http://searchlauncher.bcm.tmc.edu:9331/seq-search/gene-search.html HTTP://WWW.SOFTBERRY.COM/NUCLEO.HTML
FIRSTEF; Core_PROMOTER; CpG_PROMOTER; POLYADQ; JTEF; GENEID: hierarchical rules (human, *Drosophila*)	http://www.cshl.edu/mzhanglab/
GENEMACHINE: integrated gene finder	http://www1.imim.es/software/geneid/geneid.html#top http://genome.nhgri.nih.gov/genemachine/
GENEMARK: include HMM (many species)	http://dot.imgen.bcm.tmc.edu:9331/seq-search/gene-search.html
GENEPARSER: dynamic programming (DP)-ANN	http://beagle.colorado.edu/~eesnyder/GeneParser.html

Continued

Table 17.4. *Continued.*

	GENE PREDICTION *Continued*
GeneSeqer: EST-based gene prediction	http://bioinformatics.iastate.edu/cgi-bin/gs.cgi
GENEWISE2: DNA–protein alignment	http://www.cbil.upenn.edu/tess/
Genie: HMM (human, *Drosophila*)	http://www.fruitfly.org/seq_tools/genie.html
GENLANG: linguistic grammar	http://www.cbil.upenn.edu/genlang/genlang_home.html
GENOMESCAN: HMM+protein similarity (human, *Arabidopsis*, maize)	http://genes.mit.edu/genomescan/
GENSCAN: HMM (human, *Arabidopsis*, maize)	http://genes.mit.edu/GENSCAN.html
GRAIL: ANN (human, mouse, *Drosophila*, *Arabidopsis*)	http://compbio.ornl.gov/tools/index.shtml
HMM-GENE: HMM (human, *C. elegans*)	http://www.cbs.dtu.dk/services/HMMgene
MORGAN: tree; **VEIL:** HMM; **GLIMMER:** IMM (micro.)	http://www.tigr.org/~salzberg/
MZEF-SPC: MZEF+SpliceProximalCheck	http://industry.ebi.ac.uk/~thanaraj/MZEF-SPC.html
MZEF: QDA (human, mouse, *Arabidopsis*); **POMBE:** LDA	http://www.cshl.edu/mzhanglab/
ORFFINDER	http://www.ncbi.nlm.nih.gov/gorf/gorf.html
PREDICTGENES	http://cbrg.inf.ethz.ch/subsection3_1_8.html
PROCRUSTES: splice alignment	http://www-hto.usc.edu/software/procrustes/index.html
SLAM: dual HMM (human–mouse syntenic regions)	http://bio.math.berkeley.edu/slam/
TWINSCAN: Genscan+conservation sequence	http://genes.cs.wustl.edu/
WEBGENE: DP (human, mouse, *Fugu*, *Drosophila*, *C. elegans*, *Arabidopsis*, *Aspergillus*)	http://www.itba.mi.cnr.it/webgene/
XPOUND	ftp://igs-server.cnrs-mrs.fr/pub/Banbury/xpound/
YEASTGENE	http://tubic.tju.edu.cn/cgi-bin/Yeastgene.cgi
VEIL HMM	http://www.tigr.org/~salzberg/veil.html
EST_GENOME: aligns cDNA to DNA	http://www.well.ox.ac.uk/~rmott/est_genome.shtml
TAP: EST-based gene prediction	http://sapiens.wustl.edu/~zkan/TAP/
EbEST: EST-based gene prediction	http://rgd.mcw.edu/EBEST/
SGP1: comparative gene prediction	http://195.37.47.237/sgp-1/
SGP2: comparative gene prediction	http://genome.imim.es/software/sgp2/

together, one could achieve higher accuracy. For example, in DIGIT (Yada *et al.*, 2003), a Bayesian procedure and a HMM were used to combine the results of FGENESH, GENSCAN and HMMGENE to report better performance than the individual programs.

Comparative Gene Prediction

The rationale behind comparative or similarity-based gene prediction methods is that the regions in the genome sequence coding for proteins are generally more conserved during evolution that non-functional regions (Fig. 17.2). Essentially, there are two main classes of similarity-based approaches for gene identification: the comparison of the DNA query

sequence with a protein or cDNA sequence, or a database of such sequences, and the comparison of two or more genomic sequences. In both approaches query and target sequences may be from the same or different species.

Genomic query against protein or cDNA target

The backbone of similarity-aided or similarity-based gene structure determination is constituted by those methods that rely on a comparison of the query sequence with protein or cDNA sequences. Although mostly known as a database search program, BLASTX (Altschul *et al.*, 1990; Gish and States, 1993) illustrates

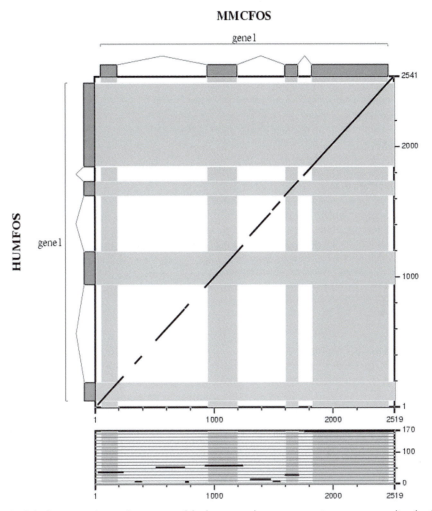

Fig. 17.2. Pairwise comparison using TBLASTX of the human and mouse genomic sequences coding for the FOS oncogene. The boxes at the top and on the left axis correspond to coding exons, while the diagonals indicate the conserved alignments. The score of the alignments (divided by 10) is given in the lower panel. Although conserved regions between the human and mouse genomic sequences coding for these genes fully include the coding regions, a substantial fraction of intronic regions are also conserved (although in general not as strongly). The TBLASTX output was post-processed to show a continuous non-overlapping alignment. The plot was obtained by the APLOT tool (Abril *et al.*, 2003).

the rationale behind this approach. With BLASTX a genomic query is translated into a set of amino acid sequences in the six possible frames and compared against a database of known protein sequences. The assumption is that those segments in the genomic query similar to database proteins are likely to correspond to coding exons. A similar assumption is behind the comparison of the genomic query against a database of cDNA sequences (such as ESTs), using BLASTN (Altschul *et al.*, 1990), FASTA (Pearson, 1999), or similar programs.

Such database search programs, however, are not dedicated gene prediction tools; they report exclusively matching sequences, but are not capable of automatically identifying start and stop codons or splice sites. Thus, after a database search and the identification of potential targets, additional tools are required to define exonic structures.

One approach is to use the top database match as target sequence, and obtain a so-called spliced alignment between this and the genomic query. In such an alignment, large gaps – likely to correspond to introns – are only allowed at legal splice junctions. A popular program (SIM4) to calculate such a spliced pairwise alignment has been developed by Florea et al. (1998). One of the first programs for the specific task to map ESTs on to a genomic sequence is EST_GENOME (Mott, 1997).

Splicing alignment algorithms with protein targets require a conceptual translation of the query sequence, computation of the alignment and post-processing, which includes the combinatorial problem of finding the best fit of a multi-exon structure to a related protein. Most such splicing alignment tools employ for this task dynamic programming techniques. Therefore, gene prediction in large-scale sequences may become extremely time- and space-demanding. This investment, however, is usually at the benefit of prediction accuracy. PROCRUSTES (Gelfand et al., 1996) and GENEWISE (Birney and Durbin, 1997) are powerful programs to predict genes based on a comparison of a genomic query with protein targets. GENEWISE is at the core of the ENSEMBL system. GENESEQER (Usuka and Brendel, 2000) is a similar spliced alignment program for plant genomes.

In an alternative approach, the results of a database search can be integrated in a more or less ad hoc way into the framework of a typical ab initio gene prediction program. In essence, these methods promote candidate exons in the query sequence for which similar known coding sequences exist. Indeed, the score of the candidate exon – initially a function of the score of the splice (start, stop) sites and of the coding potential of the exon sequence – is increased as a function of the similarity between the candidate exon and the known coding sequences. In this way, candidate exons showing similarity to known coding sequences are more likely to be included in the final gene prediction. In theory, this approach should produce predictions as accurate as pure ab initio programs when no similar target sequences exist, but more accurate ones (ideally, as accurate as those from splicing alignment tools) when such target sequences do exist. One popular example of this approach is the program GENOMESCAN (Yeh et al., 2001), an extension of GENSCAN over which it reports increased accuracy. CRASA

(Chuang et al., 2003) is a recently developed method that uses ESTs instead.

Genomic query against genomic target

With availability of genome sequences for an increasing number of eukaryote organisms, whole genome sequence comparisons are gaining popularity as a means of identifying protein-coding genes. Under the assumption that regions conserved in the sequence will tend to correspond to coding exons from homologous genes, a number of programs have been recently developed. The program EXOFISH (Crollius et al., 2000) was one of the first such programs developed and predicts human exons based on a comparison with a database of random sequences from *Tetraodon nigroviridis*, a puffer fish species. Later developments followed notably different approaches.

In one such approach (Pedersen and Scharl, 2002; Blayo et al., 2003), the problem is stated as a generalization of pairwise sequence alignment: given two genomic sequences coding for homologous genes, the goal is to obtain the predicted exonic structure in each sequence maximizing the score of the alignment of the resulting amino acid sequences. Both Blayo et al. (2003) and Pedersen and Scharl (2002) solve the problem through a complex extension of the classical dynamic programming algorithm for sequence alignment.

In a different approach, the programs SLAM (Pachter et al., 2002, http://baboon.math. berkeley.edu/~syntenic/slam.html) and DOUBLESCAN (Meyer and Durbin, 2002) combine sequence alignment pair hidden Markov models (HMMs) (Durbin et al., 1998) with gene prediction generalized HMMs (GHMMs; Burge and Karlin, 1997) into the so-called generalized pair HMMs. In these, gene prediction is not the result of the sequence alignment, as in the programs above, but both gene prediction and sequence alignment are obtained simultaneously.

A third class of programs adopts a more heuristic approach, and clearly separate gene prediction from sequence alignment. The programs ROSETTA (Batzoglou et al., 2000), SGP1 (from syntenic gene prediction, Wiehe et al.,

2001, http://195.37.47.237/sgp-1/), and CEM (from the conserved exon method, Bafna and Huson, 2000) are representative of this approach. All these programs start by aligning two syntenic sequences and then predict gene structures in which the exons are compatible with the alignment.

Although similarity-based gene prediction with homologous genomic sequences may produce high quality results (Miller, 2001), an obvious shortcoming is the need for two homologous sequences. Also, genes without a homologue in the partner sequence will escape detection. This is particularly problematic if species are compared where gene order or synteny is not preserved. Given only a single query sequence, it is therefore desirable to automatically search for homologous sequences or syntenic chromosome stretches in other species that are suited to be applied to similarity based programs. The programs TWIN-SCAN (Korf *et al.*, 2001, http://genes.cs.wustl.edu/query.html/) and SGP2 (Parra *et al.*, 2003, http://genome.imim.es/software/sgp2/) attempt to address this limitation. The approach in these programs is reminiscent of that used in GENOMESCAN (Yeh *et al.*, 2001) to incorporate similarity to known proteins to modify the GENSCAN scoring schema. Essentially, the query sequence from the target genome is compared against a collection of sequences from the informant genome (which can be a single homologous sequence to the query sequence, a whole assembled genome, or a collection of shotgun reads), and the results of the comparison are used to modify the scores of the exons produced by *ab initio* gene prediction programs. In TWINSCAN, the genome sequences are compared using BLASTN and the results serve to modify the underlying probability of the potential exons predicted by GENSCAN. In SGP2, the genome sequences are compared using TBLASTX (Gish, W., 1996–2002, http://blast.wustl.edu), and the results used to modify the scores of the potential scores predicted by GENEID.

TWINSCAN, SGP2 and SLAM have been successfully applied to the annotation of the mouse genome (Waterston *et al.*, 2002), and have helped to identify previously unconfirmed genes (Guigó *et al.*, 2003). Up to date predictions of these programs on the human and rodent genomes can be accessed through the UCSC genome browser and the ENSEMBL system.

Prediction of selenoproteins

The characterization of eukaryotic selenoproteins illustrates nicely the power of comparative gene prediction methods. In selenoproteins, the codon TGA is translated into a selenocysteine residue. Identification of the transcript encoding selenocysteine-containing proteins is particularly difficult even if the full-length cDNA of the gene is known, because computational gene prediction programs – including simple ORF finders – assume without exception that the TGA triplet codes for a stop codon. The alternative decoding of this codon is due to an mRNA structure, the seleno cysteine insertion sequence (SECIS), located at the 3' UTR of the selenoprotein-encoding genes. However, there is very little sequence conservation in the SECIS element, and searching for potential SECIS structures in eukaryotic genomes produces an overwhelming number of false positive hits. Castellano *et al.* (2001) developed a computational method to identify selenoprotein-encoding genes, which relies on the concerted prediction of SECIS structures and genes with in-frame TGA codons. In this regard, the fact that the region between the in-frame TGA codon and the real stop codon in selenoprotein genes exhibits the codon bias characteristic of protein-coding regions is of advantage to selenoprotein gene prediction. This approach led to the identification of the three selenoproteins so far identified in the *Drosophila* genome (Castellano *et al.*, 2001). When applied to mammalian genomes with lower coding density, however, the approach was less successful. In this regard, the availability of different mammalian and vertebrate genomes is essential for the characterization of mammalian selenoproteins. Comparative analysis helps in the identification of selenoproteins, primarily, at two different levels: (i) SECIS sequences are characteristically conserved between orthologous genes of species at the appropriate phylogenetic distance; and (ii) coding sequence conservation across a UGA codon between a query and a target DNA sequence may strongly suggest a Sec-coding function (Fig. 17.3). In contrast, if the conservation vanishes downstream of the UGA codon, this is often indicative of a stop codon function.

Conservation of SECIS sequences between human and mouse orthologous selenoproteins has been used in the search for mammalian selenoproteins (Kryukov *et al.* 2003), while conservation of UGA-flanking sequences as an indication of Sec-coding function has been used in the comparative analysis of the human and fugu genomes (Castellano *et al.*, 2004). Indeed, sequence comparisons of TGA interrupted predicted genes obtained independently in these two genomes has lead to the identification of SelU, a novel selenoprotein family (see Fig. 17.3).

Accuracy of Gene Predictions

One the first comprehensive comparative analyses of gene prediction programs was published by Burset and Guigó (1996), where a number of performance metrics were introduced to evaluate the accuracy of gene predictions. We describe first these measures, and discuss a few of the more important benchmarks concerning accuracy of gene predictions in mammalian

genomes. For a more detailed discussion and critical reviews of measures of gene prediction accuracy, see the papers by Bajic (2000) and Baldi *et al.* (2000). An extensive review on the accuracy of gene finding programs can be found in Guigó and Wiehe (2003).

Measures of prediction accuracy

To evaluate the accuracy of a gene prediction program on a test sequence, the gene structure predicted by the program is compared with the actual gene structure of the sequence. The accuracy can be evaluated at different levels of resolution. Typically, these are the nucleotide, exon and gene levels. These three levels offer complementary views of the accuracy of the program. At each level, there are two basic measures: sensitivity and specificity, which essentially measure prediction errors of the first and second kind. Briefly, sensitivity is the proportion of real elements (coding nucleotides, exons or genes) that

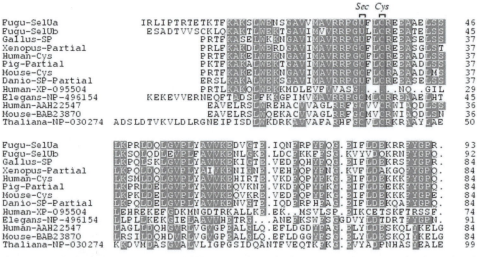

Fig. 17.3. Multiple sequence alignment of different members of SelU, a novel selenoprotein family predicted after computational analysis that involved genome sequence comparisons between human and fugu. SelU is a selenoprotein in fugu (the sequences of two members of the family, SelUa and SelUb are included in the alignment) but not in human, where the members of this family use cysteine (cys) instead of selenocysteine (sec, denoted by the letter U in the amino acid one letter code). After the prediction of SelU as a potential selenoprotein, exhaustive sequence similarity searches were performed against transcript sequences, and found members of this family across the whole eukaryotic spectrum; as the intial analysis suggested, SelU is a selenoprotein in some eukaryotic species, and it is not in others. Strong sequence conservation across the U residue (encoded by TGA, otherwise a stop codon) strongly suggests this to be a 'bona fide' selenoprotein, instead of a false positive. This selenoprotein was later verified by radioactive labelling of selenium. The pattern Sec-X-X-Cys is not uncommon in selenoproteins (adapted from Castellano *et al.*, 2004).

have been correctly predicted, while specificity is the proportion of predicted elements that are correct. More specifically, if TP are the total number of coding elements correctly predicted, TN the number of correctly predicted non-coding elements, FP the number of non-coding elements predicted coding, and FN the number of coding elements predicted non-coding, then, in the gene-finding literature, sensitivity is defined as $Sn=TP/(TP+FN)$ and specificity as $Sp=TP/(TP+FP)$. Both sensitivity and specificity take values from 0 to 1, with perfect prediction when both measures are equal to 1. Neither Sn nor Sp alone constitute good measures of global accuracy, since one can have high sensitivity with little specificity and vice versa. It is desirable to use a single scalar value to summarize both of them. In the gene-finding literature, the preferred such measure on the **nucleotide level** is the correlation coefficient defined as:

$$CC = \frac{(TP \times TN)-(FN \times FP)}{\sqrt{(TP+FN) \times (TN+FP) \times (TP+FP) \times (TN+FN)}}$$

CC ranges from -1 to 1, with 1 corresponding to a perfect prediction, and -1 to a prediction in which each coding nucleotide is predicted as non-coding and vice versa.

At the **exon level**, an exon is considered correctly predicted only if the predicted exon is identical to the true one, in particular both 5' and 3' exon boundaries have to be correct. A predicted exon is considered wrong (WE), if it has no overlap with any real exon, and a real exon is considered missed (ME) if it has no overlap with a predicted exon. A summary measure on the exon level is simply the average of sensitivity and specificity.

At the **gene level**, a gene is correctly predicted if all of the coding exons are identified, every intron–exon boundary is correct, and all of the exons are included in the proper gene.

Accuracy in single gene sequences

The results from the benchmark by Burset and Guigó (1996) are reproduced in Table 17.5. These authors evaluated seven programs in a set of 570 vertebrate single gene genomic sequences deposited in GenBank after January 1993. This was done to minimize the overlap between this test set and the sets of sequences which the programs had been trained on. Average CC for the programs analysed ranged

Table 17.5. Accuracy of gene predictions programs on single gene vertebrate sequences. Adapted from Burset and Guigó (1996).

| Program | Data set | Nucleotide | | | Exon | | | | |
		Sn	Sp	CC	Sn	Sp	$\frac{S_n + S_p}{2}$	ME	WE
FGENEH	ALLSEQ	0.77	0.88	0.80	0.61	0.64	0.64	0.15	0.12
	NEWSEQ	0.70	0.83	0.73	0.51	0.54	0.54	0.22	0.18
GENEID	ALLSEQ	0.63	0.81	0.65	0.44	0.46	0.45	0.28	0.24
	NEWSEQ	0.58	0.78	0.60	0.41	0.43	0.42	0.34	0.27
GENEPARSER2	ALLSEQ	0.66	0.79	0.65	0.35	0.40	0.37	0.29	0.17
	NEWSEQ	0.63	0.76	0.62	0.33	0.39	0.36	0.32	0.20
GENLANG	ALLSEQ	0.72	0.79	0.71	0.51	0.52	0.52	0.21	0.22
	NEWSEQ	0.63	0.73	0.63	0.39	0.44	0.43	0.29	0.25
GRAIL 2	ALLSEQ	0.72	0.87	0.76	0.36	0.43	0.40	0.25	0.11
	NEWSEQ	0.69	0.85	0.72	0.34	0.41	0.38	0.30	0.13
SORFIND	ALLSEQ	0.71	0.85	0.72	0.42	0.47	0.45	0.24	0.14
	NEWSEQ	0.65	0.79	0.65	0.36	0.39	0.38	0.29	0.19
XPOUND	ALLSEQ	0.61	0.87	0.69	0.15	0.18	0.17	0.33	0.13
	NEWSEQ	0.58	0.83	0.64	0.12	0.15	0.14	0.36	0.16
GENEID+	ALLSEQ	0.91	0.91	0.88	0.73	0.70	0.71	0.07	0.13
	NEWSEQ	0.88	0.87	0.84	0.68	0.64	0.66	0.10	0.15
GENEPARSER3	ALLSEQ	0.86	0.91	0.85	0.56	0.58	0.57	0.14	0.09
	NEWSEQ	0.83	0.89	0.82	0.50	0.53	0.51	0.17	0.09

from 0.65 to 0.78 at the nucleotide level, while the average exon prediction accuracy $((Sn+Sp)/2)$ ranged from 0.37 to 0.64. It cannot be ruled out, however, that some of the exons considered mispredicted may actually correspond to yet to be discovered splice isoforms of the gene.

Rogic et al. (2001) published a new independent comparative analysis of seven gene prediction programs. The programs were again tested in single gene sequences from human and rodent species. In order to avoid overlap with the training sets of the programs, only sequences were selected that had been entered in GenBank after the programs were developed and trained. Table 17.6 shows the accuracy measures averaged over the set of sequences effectively analysed for each of the tested programs. The programs tested by Rogic et al. (2001) showed substantially higher accuracy than the programs tested by Burset and Guigó (1996): average CC at the nucleotide level ranged from 0.66 to 0.91, while average exon prediction accuracy ranged from 0.43 to 0.76. Interestingly, DIGIT, a program that integrates the output of several gene finders, reports that the Rogic et al. data set gave a sensitivity of 0.80, and a specificity of 0.84, which is substantially larger than that of any individual program. (However, one needs to be cautious when optimizing the performance of a gene prediction program or system in one particular sequence data set, since this may lead to overtraining, i.e. to capturing the specificities of the sequences in the data set, rather than the generic features of coding sequences, that can be extrapolated to other data sets.)

Accuracy in large genomic sequences

While the paper by Rogic et al. (2001) represented a valuable update on the accuracy of gene finders, it suffered from the same limitation as did the previous work by Burset and Guigó (1996) and others: gene finders were tested in controlled data sets made of short genomic sequences encoding a single gene with a simple gene structure. These data sets are not representative of the genome sequences currently being produced: large sequences of low coding density, encoding several genes and/or incomplete genes, with complex gene structure. The exhaustive scrutiny to which the sequence of human chromosome 22 (Dunham et al., 1999) has been subjected through the vertebrate genome annotation (VEGA) database project at the Sanger Centre (http://vega.sanger.ac.uk/) offers, in this regard, an excellent platform to obtain a more representative estimation of the accuracy of current gene finders. However, VEGA uses GENSCAN and FGENES in the annotation pipeline, and may be biased towards these programs. Table 17.7 shows the accuracy of a number of ab initio and comparative gene finders in chromosome 22 when compared with the curated annotations from VEGA. Programs have been separated into categories: sequence similarity based, ab initio or comparative. As can be seen, accuracy suffers substantially when moving from single gene sequences to whole chromosome sequences. For instance, GENSCAN CC drops from 0.91 in the evaluation by Rogic et al. (2001) to 0.64 for chromosome 22. However, even more sophisticated gene find-

Table 17.6. Accuracy of gene prediction programs on single gene mammalian sequences. Adapted from Rogic *et al.* (2001)

| Program | Nucleotide | | | Exon | | | | |
	S_n	S_p	CC	S_n	S_p	$\frac{Sn+Sp}{2}$	ME	WE
FGENES	0.86	0.88	0.83	0.67	0.67	0.67	0.12	0.09
GENEMARK.hmm	0.87	0.89	0.83	0.53	0.54	0.54	0.13	0.11
GENIE	0.91	0.90	0.88	0.71	0.70	0.71	0.19	0.11
GENSCAN	0.95	0.90	0.91	0.70	0.70	0.70	0.08	0.09
HMMGENE	0.93	0.93	0.91	0.76	0.77	0.76	0.12	0.07
MORGAN	0.75	0.74	0.69	0.46	0.41	0.43	0.20	0.28
MZEF	0.70	0.73	0.66	0.58	0.59	0.59	0.32	0.23

Table 17.7. Accuracy of gene prediction programs on human chromosome 22.

| | Nucleotide | | | Exon | | | | |
| | | | | | | Sn + Sp | | |
Program	Sn	Sp	CC	Sn	Sp	2	ME	WE
Sequence similarity based								
ENSEMBL	0.74	0.83	0.78	0.75	0.80	0.77	0.18	0.13
FGENESH++	0.81	0.71	0.75	0.80	0.66	0.73	0.11	0.27
Ab initio								
GENSCAN	0.79	0.53	0.64	0.68	0.41	0.55	0.15	0.48
GENEID	0.73	0.67	0.70	0.65	0.55	0.60	0.21	0.33
Comparative								
SGP2	0.75	0.73	0.73	0.66	0.58	0.62	0.19	0.28
TWINSCAN	0.72	0.67	0.69	0.69	0.59	0.64	0.18	0.29

ers, such as ENSEMBL or FGENESH++, which use known cDNAs and RefSeq genes, respectively, are far from producing perfect predictions, with CCs around 0.75. These numbers strongly suggest that current mammalian gene counts are still of a highly hypothetical nature.

These examples demonstrate that the same program running on different data sets will often produce different results. Even running the same data set with the same software by different people (presumably with different ways of setting the parameters for different optimization purposes or perhaps with different ways of counting positives and negatives) can result in different statistics. There can be several additional caveats when reading such accuracy estimates. Upon closer scrutiny, some false positives have turned out to be true positives after more libraries and more sensitive technologies are used. One of the fundamental reasons that no gene prediction program can give accurate total gene number is that they are all based on training statistics, hence total gene number is an intrinsically free parameter (a normalization constant). This free parameter can be fixed (optimized) during the training process so that the total predicted gene number (or exon number) matches the total annotated (known) number in the training region. Therefore, if the training set is gene dense, the program will over-predict the total for the whole genome and vice versa. Another reason has to do with the large variation of the gene sizes as has been examined more care-

fully by a recent study (Wang *et al.*, 2003). It is shown that predictions of large multiple-exon genes and small single-exon genes are less reliable. The false positive is a particularly serious problem for genes longer than 100 kb as a long gene is more likely to be truncated into multiple shorter predicted genes (ESTs are a good way of identifying this). When the actual gene size is shorter than ~1 kb, both false-positive and negative predictions will rise sharply. In fact, when the predicted gene size is below 1 kb (GENSCAN), or below 10 kb (FGE-NESH), it can be expected that most of the exons will be false positives. Since less than 4.5% of genes are simple-exon genes (according to information derived from RefSeq), large multiple-exon genes are the main problem in *ab initio* gene predictions. Although internal exon size does not vary much, there is no typical exon number and there is no typical intron size either. It is interesting to note the mean gene size has been 'evolving' from 5–10 kb (before 1997), to 27 kb (draft human genome in 2001) and to 51–59 kb (finished chromosomes 14, 20 and 21 in 2003). The other three no less important reasons for not being able to obtain accurate gene number all relate to the definition of genes. The first is the pseudo-gene (non-functional gene) and paralogous gene (duplicated gene) problems that can complicate an accurate gene count. In fact, most of the single-exon gene prediction problems are caused by the pseudo-genes, and this is why unspliced ESTs are always looked upon as suspicious! The second is the overlapping, antisense or nested gene problems. The third is the alternative splicing problem. Many people define alternative transcripts as those that share exon regions; in principle this does not have to be the case. In fact some genes routinely produce transcripts containing no common exons (e.g. *Titin*). The bottom line is that we do not know all the rules about how genes are transcribed and processed, and thus current mammalian gene catalogues are of a highly hypothetical nature. Given there is always a trade-off between false positives and false negatives, it will depend on which errors will be most detrimental, Wang *et al.* (2003) suggest using ENSEMBL to control FPs (<7%, with default parameters) and using *ab initio* predictions to control FNs (<5%, with default parameters).

Genome Annotation Systems

Currently there are three popular public human gene annotation database systems, EBI and Sanger Institute ENSEMBL (Hubbard *et al.*, 2001; Clamp *et al.*, 2002; http://www.ensembl.org/), UCSC Genome Browser (Kent *et al.*, 2002; Karolchik *et al.*, 2003; http://genome.ucsc.edu/) and NCBI LocusLink (Maglott *et al.*, 2000; Pruitt *et al.*, 2003; http://www.ncbi.nlm.nih.gov/ LocusLink/). Each is associated with a native Internet web browser. Fortunately, they all use the NCBI assembly of the human genome (built 33 released on April 14, 2003, current at the time of writing).

Ensemble human genes are generated automatically by the ENSEMBL gene builder. For a number of chromosomes (currently chromosomes 6, 13, 14, 20 and 22), manual annotations are also available from Sanger Institute's VEGA curation system (http://vega.sanger.ac.uk/Homo_sapiens/). ENSEMBL genes are three basic types: those having full-length cDNA or proteins, those having high homology to proteins in other organisms and those GENSCAN-predicted genes matching to proteins/vertebrate mRNA and UniGene clusters. The basic gene-annotator engine (using protein homology to construct gene structure) is GENEWISE (Birney and Durbin, 2000). Currently ENSEMBL predict

24,500 human genes (Pennisi, 2003). These 'ENSEMBL genes' are regarded as being fairly conservative (with a low false positive rate), since they are all supported by experimental evidence of at least one form via sequence homology. Currently, the ENSEMBL project is attempting to add spliced EST information for identification of alternative transcripts and they are also incorporating comparative genomics to identify orthologues and syntenic regions.

UCSC Genome Browser provides a rapid and reliable display of any requested portion of genomes at any scale, together with dozens of aligned annotation tracks (known genes, predicted genes, ESTs, mRNAs, CpG islands, assembly gaps and coverage, chromosomal bands, mouse homologies, and more). Half of the annotation tracks are computed at UCSC from publicly available sequence data. The remaining tracks are provided by collaborators worldwide. Users can also add their own custom tracks to the browser for educational or research purposes. These customizable tracks have made the Genome Browser very popular. The basic annotator engine is BLAT (Kent, 2002) which allows rapid, reliable alignment of primate DNAs/RNAs or land vertebrate proteins on to the human genome, hence annotating the genome by sequence similarities. In its human gene prediction, in addition to ENSEMBL genes, it also displays 25,600 TWINSCAN genes, 32,400 GENEID genes, 39,800 FGENESH++ genes and 45,000 GENSCAN genes.

NCBI LocusLink has a rule-based genome annotation pipeline. Known genes are identified by aligning RefSq genes (http://www.ncbi. nlm.nih. gov/RefSeq/) and GenBank mRNAs to the genome using MEGABLAST (Zhang *et al.*, 2000). Transcript models are reconstructed by attempting to settle disagreements between individual sequence alignments without using an *a priori* model (such as codon usage, initiation, or poly(A) signals). Alternative mRNA models derived from the available mRNA and EST sequence data are grouped under the same gene when they share one or more exons on the same strand. Genes which produce non-coding transcripts are also annotated. These transcripts are annotated as 'misc_RNA' features. If the defining GenBank or RefSeq sequence aligns to more than one location on the genome, the best alignment is selected and annotation made on that contig. If they are of equal quality, both are annotated. Genes (and corresponding transcript and protein features) are annotated on the contig if the defining transcript alignment is $\geq 95\%$ identity and the aligned region covers $\geq 50\%$ of the length, or at least 1000 bases. Genes predicted by GENOMESCAN are annotated only if they do not overlap any model based on an mRNA alignment. GENOMESCAN predicted 38,600 human genes from the September 2000 GoldenPath human genome sequence.

In any case, the gene annotations provided by these systems are highly hypothetical, given the accuracy of currently available gene prediction programs. There is indeed a long way to go before automatic systems exist able to identify all genes in a given genomic sequence, provide the exhaustive catalogue of splice variants, and identify the sequence motifs involved in their regulation. A better understanding of what a gene is, and of the biological process involved in gene specification is certainly necessary to reach such a goal.

References

Abril, J.F., Guigó, R. and Wiehe, T. (2003) gff2aplot: plotting sequence comparisons. *Bioinformatics* 19, 2477–2479.

Altschul, S.F., Gish, W., Miller, W., Myers, E.W. and Lipman, D. (1990) Basic local alignment search tool. *Journal of Molecular Biology* 215, 403–410.

Bafna, V. and Huson, D.H. (2000) The conserved exon method. *Proceedings of the Eighth International Conference on Intelligent Systems in Molecular Biology (ISMB)* 8, 3–12.

Bajic, V. (2000) Comparing the success of different prediction software in sequence analysis: a review. *Briefings in Bioinformatics* 1, 214–228.

Baldi, P., Brunak, S., Chauvin, Y., Andersen, C. and Nielsen, H. (2000) Assessing the accuracy of predicition algorithms for classification: an overview. *Bioinformatics* 16, 412–424.

Batzoglou, S., Pachter, L., Mesirov, J.P., Berger, B. and Lander, E.S. (2000) Human and mouse gene structure: comparative analysis and application to exon prediction. *Genome Research* 10, 950–958.

Bernardi, G. (1995) The human genome organization and evolutionary history. *Annual Reviews in Genetics* 29, 445–476.

Birney, E. and Durbin, R. (1997) Dynamite: a flexible code generating language for dynamic programming methods used in sequence comparison. *Proceedings of the Fifth International Conference on Intelligent Systems in Molecular Biology (ISMB)* 5, 56–64.

Birney, E. and Durbin, R. (2000) Using Genewise in *Drosophila* annotation experiment. *Genome Research* 10, 547–548.

Blayo, P., Rouzé, P. and Sagot, M.-F. (2003) Orphan gene finding – an exon assembly approach. *Theoretical Computer Science* 290, 1407–1431.

Brunak, S., Engelbrecht, J. and Knudsen, S. (1991) Prediction of human mRNA donor and acceptor sites from the DNA sequence. *Journal of Molecular Biology* 220, 49–65.

Burge, C. and Karlin, S. (1997) Prediction of complete gene structures in human genomic DNA. *Journal of Molecular Biology* 268, 78–94.

Burset, M. and Guigó, R. (1996) Evaluation of gene structure prediction programs. *Genomics* 34, 353–367.

Burset, M., Seledtsov, I.A. and Solovyev, V.V. (2000) Analysis of canonical and non-canonical splice sites in mammalian genomes. *Nucleic Acids Research* 28, 4364–4375.

Castellano, S., Morozova, N., Morey, M., Berry, M.J., Serras, F., Corominas, M. and Guigó, R. (2001) In silico identification of novel selenoproteins in the *Drosophila melanogaster* genome. *EMBO Reports* 2, 697–702.

Castellano, S., Novoselov, S.V., Kryukov, G.V., Lescure, A., Blanco, E., Krol, A., Gladyshev, V.N. and Guigó, R. (2004) Reconsidering the evolutionary distribution of eukaryotic selenoproteins: a novel non-mammalian family with scattered phylogenetic distribution. *EMBO Reports* 5, 71–77.

Chuang, T.J., Lin, W.C., Lee, H.C., Wang, C.W., Hsiao, K.L., Wang, Z.H., Shieh, D., Lin, S.C. and Ch'ang, L.Y. (2003) A complexity reduction algorithm for analysis and annotation of large genomic sequences. *Genome Research* 13, 313–322.

Clamp, M. *et al.* (2002) Ensembl 2002: accommodating comparative genomics. *Nucleic Acids Research* 31, 38–42.

Crollius, H.R., Jaillon, O., Bernot, A., Dasilva, C., Bouneau, L., Fischer, C., Fizames, C., Wincker, P., Brottier, P., Quetier, F., Saurin, W. and Weissenbach, J. (2000) Estimate of human gene number provided by genome-wide analysis using *Tetraodon nigroviridis* DNA sequence. *Nature Genetics* 25, 235–238.

Degroeve, S., De Baets, B., Van De Peer, Y. and Rouze, P. (2002) Feature subset selection for splice site prediction. *Bioinformatics* 18 Suppl 2, S75–S83.

Dunham, I., Hunt, A.R., Collins, J.E., Bruskiewich, R., Beare, D.M., Clamp, M., Smink, L.J., Ainscough, R., Almeida, J.P., Babbage, A. *et al.* (1999) The DNA sequence of human chromosome 22. *Nature* 402, 489–495.

Durbin, R., Eddy, S., Crogh, A. and Mitchison, G. (1998) *Biological Sequence Analysis: Probabilistic Models of Protein and Nucleic Acids.* Cambridge University Press.

Fickett, J.W. (1982) Recognition of protein coding regions in DNA sequences. *Nucleic Acids Research* 10, 5303–5318.

Fickett, J.W. and Tung, C.S. (1992) Assessment of protein coding measures. *Nucleic Acids Research* 20, 6441–6450.

Florea, L., Hartzell, G., Zhang, Z., Rubin, G.M. and Miller, W. (1998) A computer program for aligning a cDNA sequence with a genomic DNA sequence. *Genome Research* 8, 967–974.

Gelfand, M. and Roytberg, M. (1993) Prediction of exon–intron structure by a dynamic programming approach. *BioSystems* 30, 173–182.

Gelfand, M.S., Mironov, A.A. and Pevzner, P.A. (1996) Gene recognition via spliced alignment. *Proceedings of the National Academy of Sciences USA* 93, 9061–9066.

Gish, W. and States, D. (1993) Identification of protein coding regions by database similarity search. *Nature Genetics* 3, 266–272.

Guigó, R. (1998) Assembling genes from predicted exons in linear time with dynamic programming. *Journal of Computational Biology* 5, 681–702.

Guigó, R. and Wiehe, T. (2003) Gene prediction accuracy in large DNA sequences. In: Galperin, M.Y. and Koonin, E.V. (eds) *Frontiers in Computational Genomics.* Caister Academic Press.

Guigó, R., Knudsen, S., Drake, N. and Smith, T. (1992) Prediction of gene structure. *Journal of Molecular Biology* 266, 141–157.

Guigó, R., Dermitzakis, E.T., Agarwal, P., Ponting, C.P., Parra, G., Reymond, A., Abril, J.F., Keibler, E., Lyle, R., Ucla, C., Antonarakis, S.E. and Brent, M.R. (2003) Comparison of mouse and human genomes followed by experimental verification yields an estimated 1,019 additional genes. *Proceedings of the National Academy of Sciences USA* 100, 1140–1145.

Henderson, J., Salzberg, S. and Fasman, K.H. (1997) Finding genes in DNA with a hidden Markov model. *Journal of Computational Biology* 4, 127–141.

Hubbard, D. *et al.* (2001) The Ensembl genome database project. *Nucleic Acids Research* 30, 38–41.

Karolchik, D., Baertsch, R., Diekhans, M., Furey, T.S., Hinrichs, A., Lu, Y.T., Roskin, K.M., Schwartz, M., Sugnet, C.W., Thomas, D.J., Webwe, R.J., Haussler, D. and Kent, W.J. (2003) The UCSC Genome Browser Database. *Nucleic Acids Research* 31, 51–54.

Kent, W.J. (2002) BLAT – the Blast-like alignment tool. *Genome Research* 12, 4656–4664.

Kent, W.J., Sugnet, C.W., Furey, T.S., Roskin, K.M., Pringle, T.H., Zahler, A.M. and Haussler, D. (2002) The human genome browser at UCSC. *Genome Research* 12, 996–1006.

Korf, I., Flicek, P., Duan, D. and Brent, M.R. (2001) Integrating genomic homology into gene structure prediction. *Bioinformatics* 17 Suppl 1, S140–S148.

Kozak, M. (1987) An analysis of 5'-noncoding sequences from 699 vertebrate messenger RNAs. *Nucleic Acids Research* 15, 8125–8132.

Kryukov, G.V., Castellano, S., Novoselov, S.V., Lobanov, A.V., Zehtab, O., Guigó, R. and Gladyshev, V.N. (2003) Characterization of mammalian selenoproteomes. *Science* 300, 1439–1443.

Krogh, A. (1997) Two methods for improving performance of an HMM and their application for gene finding. *Proceedings of the International Conference on Intelligent Systems in Molecular Biology* 5, 179–186.

Krogh, A., Mian, I.S. and Haussler, D. (1994) A hidden Markov model that finds genes in *E. coli* DNA. *Nucleic Acids Research* 22, 4768–4778.

Kulp, D., Haussler, D., Reese, M.G. and Eeckman, F. (1996) A generalized hidden Markov model for the recognition of human genes in DNA. *Proceedings of the International Conference on Intelligent Systems in Molecular Biology* 4, 134–142.

Lapedes, A., Barnes, C., Burks, C., Farber, R. and Sirotkin, K. (1990) Application of neural networks and other machine learning algorithms to DNA sequence analysis. In: Bell, G.I. and Marr, T.G. (eds) *Computers and DNA.* Addison-Wesley, New York, pp. 157–182.

Lim, L.P. and Burge, C.B. (2001) A computational analysis of sequence features involved in recognition of short introns. *Proceedings of the National Academy of Sciences USA* 98, 11193–11198.

Mache, N. and Levi, P. (1998) *GENIO– a Non-redundant Eukaryotic Gene Database of Annotated Sites and Sequences.* RECOMB-98 Poster, New York.

Maglott, D.R., Katz, K.S., Sicotte, H. and Pruitt, K.D. (2000) NCBI's LocusLink and RefSeq. *Nucleic Acids Research* 28, 126–128.

Meyer, I.M. and Durbin, R. (2002) Comparative *ab initio* prediction of gene structures using pair HMMs. *Bioinformatics* 18, 1309–1318.

Miller, W. (2001) Comparison of genomic DNA sequences: solved and unsolved problems. *Bioinformatics* 17, 391–397.

Mott, R. (1997) EST_GENOME: a program to align spliced DNA sequences to unspliced genomic DNA. *Computer Applications in the Biosciences* 13, 477–478.

Mount, S.M. (1982) A catalogue of splice junction sequences. *Nucleic Acids Research* 10, 459–472.

Pachter, L., Alexandersson, M. and Cawley, S. (2002) Applications of generalized pair hidden Markov models to alignment and gene finding problems. *Journal of Computational Biology* 9, 389–400.

Parra, G., Blanco, E. and Guigó, R. (2000) Geneid in *Drosophila. Genome Research* 10, 511–515.

Parra, G., Agarwal, P., Abril, J.F., Wiehe, T., Fickett, J.W. and Guigó, R. (2003) Comparative gene prediction in human and mouse. *Genome Research* 13, 108–117.

Patterson, D.J., Yasuhara, K. and Ruzzo, W.L. (2002) Pre-mRNA secondary structure prediction aids splice site prediction. *Pacific Symposium on Biocomputing* 2002, 223–234.

Pearson, W.R. (1999) Flexible similarity searching with the Fasta3 program package. In: Misener, S. and Krawetz, S.A. (eds) *Bioinformatics Methods and Protocols.* Humana Press, Totowa, New Jersey, pp. 185–219.

Pedersen, C. and Scharl, T. (2002) Comparative methods for gene structure prediction in homologous sequences. In: Guigó, R. and Gusfield, D. (eds) *Algorithms in Bioinformatics.* 2nd International Workshop WABI 2002. Lecture Notes in Computer Science. Springer-Verlag, Heidelberg, pp. 220–234.

Pennisi, E. (2003) Gene counters struggle to get the right answer. *Science* 301, 1040–1041.

Pertea, M., Lin, X. and Salzberg, S.L. (2001) GeneSplicer: a new computational method for splice site prediction. *Nucleic Acids Research* 29, 1185–1190.

Pruitt, K.D., Tatusov, T. and Maglott, D.R. (2003) RefSeq and LocusLink: NCBI gene-centered resources. *Nucleic Acids Research* 31, 34–37.

Rabiner, L.R. (1989) A tutorial on hidden Markov models and selected applications in speech recognition. *Proceedings of the IEEE* 77, 257–286.

Reese, M.G., Eeckman, F.H., Kulp, D. and Haussler, D. (1997) Improved splice site detection in Genie. *Journal of Computational Biology* 4, 311–323.

Robbertson, B.L., Cote, G.J. and Berget, S.M. (1990) Exon definition may facilitate splice site selection in RNAs with multiple exons. *Molecular and Cellular Biology* 10, 84–94.

Rogic, S., Mackworth, A.K. and Ouellette, F.B. (2001) Evaluation of gene-finding programs on mammalian sequences. *Genome Research* 11, 817–832.

Rogozin, I.B. and Milanesi, L. (1997) Analysis of donor splice signals in different organisms. *Journal of Molecular Evolution* 45, 50–59.

Salamov, A.A. and Solovyev, V.V. (2000) Ab initio gene finding in *Drosophila* genomic DNA. *Genome Research* 10, 516–522.

Senapathy, P., Shapiro, M.B. and Harris, N.L. (1990) Splice junctions, branch point sites, and exons: sequence statistics, identification and genome project. *Methods in Enzymology* 183, 252–278.

Snyder, E.E. and Stormo, G.D. (1993) Identification of coding regions in genomic DNA sequences: an application of dynamic programming and neural networks. *Nucleic Acids Research* 21, 607–613.

Solovyev, V.V. (2002) Finding genes by computer. In: Jing, T., Xu, Y. and Zhang, M.Q. (eds) *Current Topics in Computational Molecular Biology.* The MIT Press, Cambridge, Massachusetts, pp. 201–248.

Solovyev, V.V., Salamov, A.A. and Lawrence, C.B. (1994) Predicting internal exons by oligonucleotide composition and discriminant analysis of spliceable open reading frames. *Nucleic Acids Research* 22, 5156–5163.

Solovyev, V.V., Salamov, A.A. and Lawrence, C.B. (1995a) Prediction of human gene structure using linear discriminant functions and dynamic programming. In: Rawling, C. *et al.* (eds) *Proceedings of the Third International Conference on Intelligent Systems for Molecular Biology.* AAAI Press, Cambridge, UK, pp. 367–375.

Solovyev, V.V., Salamov, A.A. and Lawrence, C.B. (1995b) The gene-finding computer tools for analysis of human and model organisms genome sequences. In: Rawling, C. *et al.* (eds) *Proceedings of the 5th International Conference on Intelligent Systems for Molecular Biology.* AAAI Press, Cambridge, UK, pp. 294–302.

Staden, R. (1984) Computer methods to locate signals in nucleic acid sequences. *Nucleic Acids Research* 12, 505–519.

Staden, R. and McLachlan, A.D. (1982) Codon preference and its use in identifying protein coding regions in long DNA sequences. *Nucleic Acids Research* 10, 141–156.

Thanaraj, T.A. and Robinson, A.J. (2000) Prediction of exact boundaries of exons. *Nucleic Acids Research* 1, 343–356.

Usuka, J. and Brendel, V. (2000) Gene structure prediction by spliced alignment of genomic DNA with protein sequences: increased accuracy by differential splice site scoring. *Journal of Molecular Biology* 297, 1075–1085.

Wang, J., Li, S.T., Zhang, Y., Zheng, H.K., Xu, Z., Ye, J., Yu, J. and Wong, G.K. (2003) Opinion: Vertebrate gene predictions and the problem of large genes. *Nature Review Genetics* 4, 741–749.

Waterston, R.H., Lindblad-Toh, K., Birney, E., Rogers, J., Abril, J.F., Agarwal, P., Agarwala, R., Ainscough, R., Alexandersson, M. and An, P. et al., Mouse Genome Sequencing Consortium (2002) Initial sequencing and comparative analysis of the mouse genome. *Nature* 420, 520–562

Wiehe, T., Gebauer-Jung, S., Mitchell-Olds, T. and Guigó, R. (2001) SGP-1: prediction and validation of homologous genes based on sequence alignments. *Genome Research* 11, 1574–1583.

Yada, T., Takagi, T., Totoki, Y., Sakaki, Y. and Takaeda, Y. (2003) DIGIT: a novel gene finding program by combining gene-finders. *Pacific Symposium on Biocomputing* 2003, 375–387.

Yeh, R.F., Lim, L.P. and Burge, C.B. (2001) Computational inference of homologous gene structures in the human genome. *Genome Research* 11, 803–816.

Zhang, M.Q. (1993) A weight array method for splicing signal analysis. *Computer Applications in the Biosciences* 9, 499–509.

Zhang, M.Q. (1997) Identification of protein coding regions in the human genome based on quadratic discriminant analysis. *Proceedings of the National Academy of Sciences USA* 94, 565–568.

Zhang, M.Q. (1998) Statistical features of human exons and their flanking regions. *Human Molecular Genetics* 7, 919–932.

Zhang, M.Q. (2000) Discriminant analysis and its application in DNA sequence motif recognition. *Briefings in Bioinformatics* 1, 331–342.

Zhang, M.Q. (2002) Computational prediction of eukaryotic protein-coding genes. *Nature Reviews Genetics* 3, 698–709.

Zhang, Z., Schwartz, S., Wagner, L. and Miller, W. (2000) A greedy algorithm for aligning DNA sequences. *Journal of Computational Biology* 7, 203–214.

18 Genomic Research and Progress in Understanding Inherited Disorders in Humans and Other Mammals

D.R. Sargan[1] and A.I. Agoulnik[2]

[1]*Centre for Veterinary Science, University of Cambridge, Cambridge, UK;* [2]*Department of Obstetrics and Gynecology, Baylor College of Medicine, Houston, TX, USA*

General Introduction	449
Genetic disease burden in mammals	449
Current status of disease gene discovery	451
Other mammals as models for human inherited diseases	454
Methods to Define Disease-causing Genes	456
Gross chromosomal rearrangements	456
Single gene defects: the use of candidate genes	457
Mapping single gene defects: positional cloning	459
The investigation of polygenic diseases by association analysis in outbred populations	462
Mice and the Power of Modern Genetics	463
Human diseases and their modelling in mice. Large-scale mutagenesis	463
Gene targeting	465
Producing a complete set of gene knockout mutant animals for a whole mammalian genome	465
Diagnostic Applications of Mammalian Genomics – Ripening Fruit from the Genomic Tree	467
References	471

General Introduction

Genetic disease burden in mammals

Humans

Worldwide today, about 5% of children are born with a congenital or hereditary disorder and almost 40% of adults are treated for com- mon diseases with large inherited components during their lifetime. These include forms of mental illness and diseases such as cancer, car- diovascular disease, hypertension, asthma, diabetes and rheumatoid arthritis. In devel- oped countries, congenital and genetic disor- ders account for a nearly a quarter of deaths under the age of 4 years (WHO, 1999), although only a proportion of congenital diseases have a chromosomal or genetic origin. Amongst the most widespread single gene defects are haemoglobinopathies. As a result of heterozygote advantage conveyed by increased resistance to malaria, approximately 5% of the world population carry a haemoglobinopathy gene, and each year about 300,000 infants are born with major haemoglobin disorders. There are also huge health care costs. For example, between 30 and 40% of children's hospital beds in the UK are occupied by sufferers of inherited or congenital disorders (Polani, 1988). Individuals with diabetes mellitus in the USA have between 2 and 3 times the health care costs of others of similar age. This disease alone (with 177 million sufferers worldwide, and roughly 30% heritability) accounts for 8% of total health budgets in industrialized countries (WHO, 2002).

Domesticated species

Although controlled breeding of domesticated animals over time has reduced the occurrence of some forms of inherited disease, it also creates conditions in which recessive diseases can emerge. Single gene defects are spread rapidly by popular sire effects, in some cases combined with artificial insemination (AI). Thus, bovine leukocyte adherence deficiency (BLAD), citrullinaemia and uridine monophosphate synthase deficiency have all reached carrier frequencies of 5–15% in national dairy cattle herds. (The former, originating in a single popular bull in the early 1950s was probably carried by more than 1.25 million animals worldwide at the start of the 1990s (Shuster et al., 1992).) The mutation associated with porcine stress syndrome (malignant hyperthermia), which has advantageous effects on meat quality when heterozygous, was carried by an even higher proportion of many swine breeds (O'Brien et al., 1993). The prevalence of certain monogenic disorders within dog and cat breeds is also high, again driven by popular sire effects, inbreeding within small genetically isolated groups (the pure-bred or pedigree breeds) and probably also by co-selection with desired breed characteristics. For example, prevalence of copper toxicosis in Bedlington terriers in The Netherlands was 0.46 (Ubbink

et al., 2000) and in Britain 0.34 (Herrtage et al., 1987). (Breeding programmes making use of DNA diagnostics have since altered these prevalences.) Developmental and complex inherited diseases causing morbidity and mortality in later life are also important in pet as well as agricultural species. Cardiac, skeletal and other developmental defects and breed-associated cancers are prevalent in some dog and cat breeds, often causing such substantial breed-specific mortality that this is evidenced by reduction in average lifespan of the entire breed (Michell, 1999). For instance, the Newfoundland dog breed suffers subaortic stenosis with a breed incidence of greater than 10% (an at-risk odds ratio of 88:1 versus the general canine population). This disease has a mortality of 21% and morbidity of 33% within 3 years in this breed (Kienle et al., 1994).

Laboratory mice

For current inbred laboratory mouse strains the issue of inherited disease cannot be looked on as a sporadic or accidental happening in the same way that it is in other species. These strains are the survivors of many generations (in most cases >60) of brother × sister mating, are maintained in colonies using the same breeding system, and are considered isogenic. Hence all progeny are genetically identical and are of the normal phenotype for the strain except in the case of new mutations. For nearly all of these strains there has been temporary loss of fitness at some point in their creation and different strains differ in their fecundity and other aspects of fitness. Amongst commonly used strains, the fecundity of the relatively new FVB/N strain is nearly five times that of the least reproductively fit strain, BALB/cJ (Silver, 1995). It is for this reason, as well as others, that FVB/N has become the strain of choice for use in the production of transgenic animals. In many strains phenotypes which would be clearly deleterious outside the laboratory are maintained in the laboratory animals. Inheritance of these disease-like phenotypes can be examined by outbreeding. Examples include monogenic (e.g. retinal degeneration in C3H, CBA/J and NIH/Ola mice), digenic (e.g. glucose intolerance in KK mice), and polygenic inheritance (e.g. autoimmune haemolytic anaemia and nephropathy in NZY), whilst many strains show high incidence of particular types of tumour (http://www.informatics.jax.org).

Non-domesticated species

That inherited diseases also affect wild mammal populations is self-evident, although few data exist in most species. One might assume that occurrence rates are similar to those in human populations, or that in small, genetically isolated, or previously bottlenecked populations they might be considerably greater. This becomes important when considered in a conservation context. One good example of a known problem in a wild mammal population occurred in the Florida panther which suffered from inherited atrial septal defect ('hole in the heart') (Buergelt *et al.*, 2002) and cryptorchidism (Mansfield and Land, 2002). In this case introduction of Texan pumas to the Florida population is reducing occurrence of these conditions. In contrast, populations of at least two species (northern elephant seal and cheetah) appear to have recovered well from older genetic bottlenecks, but the cheetah is considered to remain vulnerable and is very difficult to breed successfully in captivity (O'Brien *et al.*, 1985). Inbreeding remains an issue in zoo populations of many species, where there are frequent reports of loss of genetic fitness. This governs the management of these species to a considerable degree (Frankham, 1999).

Current status of disease gene discovery

Database resources

Excellent general web sites compiling genetic diseases exist for man (On Line Mendelian Inheritance in Man, OMIM (http://www. ncbi.nlm.nih.gov/Omim)), all animals (On Line Mendelian Inheritance in Animals, OMIA (http://morgan.angis.su.oz.au/Databases/BIRX/ omia/)) and laboratory rodents (Mouse Genome Informatics, MGI (http://www.informatics.jax. org/)). In addition, specialist web sites exist for dogs (Inherited Diseases in Dogs, IDID (http://www.vet.cam.ac.uk/idid)) and for many human inherited diseases (see for

instance RetNet (http://www.sph.uth.tmc.edu/ Retnet/disease.htm), concerned with inherited diseases of the retina). The aforementioned OMIM database is a catalogue of all known human genes and genetic disorders; whilst the Human Gene Mutation Database (HGMD (http://www.hgmd.org, Stenson *et al.*, 2003)) constitutes a broad collection of data on germ-line mutations in nuclear genes specifically underlying or associated with human inherited disease. Both databases contain Internet links to a number of locus- or disease-specific mutation databases (for review see Claustres *et al.*, 2002). Another useful database is dedicated to the genetic abnormalities associated with polymorphisms and mutations of the human mitochondrial DNA (MITOMAP: a human mitochondrial genome database; http://www. mitomap.org). Table 18.1 has been compiled from information in OMIA, together with OMIM, MGI and IDID with removal of the more obvious production traits or non-disease traits, and of some repeated entries from the OMIA web site numbers.

Numbers of inherited diseases and of disease genes – monogenic disorders

The number of different disease conditions known to be inherited in different species is to some degree dependent on the level of medical/veterinary surveillance operating on the species (Table 18.1). The completion of human and murine genome sequences has led to an explosion in information about mutations in genetic diseases of these two species. As might be expected, the largest number of inherited defects and diseases has been described in humans. For monogenic disorders, many diseases have now been associated with a particular locus, and often with particular mutations. In mice, large-scale mutation and reverse genetic studies have been used to chart the phenotypes associated with many genes. Other species currently lag far behind. At the time of writing of this review the mutations in between 1400 and 1500 human genes which cause various diseases have been identified. In non-human mammals (excluding laboratory animals) the progress in identifying disorders/traits for which the causative mutations have been

Table 18.1. Number of described inherited disorders in various mammalian species. Except as stated (see footnote) this is adjusted from tabulated data in On Line Mendelian Inheritance in Animals (Sept, 2003), by removal of non-disease traits and probable repeat descriptions.

Species	No. of disorders	Single-locus disorders	Disorders/traits for which the causative mutation has been identified at the DNA level
Human	14,757[a]	12,367[a]	1,433[b]
Cat	250	36	7
Cattle	325	51	25
Dog	426	151 (97)[c]	33
Goat	41	7	4
Horse	168	17	4
Mink	18	3	0
Mouse[d]	NA	NA	3,620
Pig	179	28	6
Rabbit	44	11	3
Rat	30	3	0
Sheep	146	37	4

Human data are from: [a] On Line Mendelian Inheritance in Man (Sept 2003) or [b] the Human Gene Mutation Database. This is the number of unique single gene loci for which mutations have been connected to a phenotype. [c] The higher estimate is from IDID. The estimate in parentheses is from OMIA.
[d] The murine figure represents the number of genes for which the effects of polymorphisms on phenotype (if any) are available in MGI databases (Sept 2003). These experimental situations have not been recorded as inherited disorders.

identified at the DNA level is more modest (see Table 18.1). However, special features of population structure and of the economic usage of particular species may serve to inflate numbers of inherited diseases. Thus larger numbers of diseases have been described in dogs than cats despite the larger populations of cats in most western societies, perhaps because of the greater numbers of pure-bred dogs and the greater genetic stratification of the species into isolated breeding groups: the pure-bred breeds.

One hundred and eleven such mutated genes (including commercial and other traits as well as disease genes) are listed in the October 2003 update of the OMIA database for all mammalian species. The actual number of different mutated loci in humans for which a large animal disease model exists is even less, as many mutations in different species are of the same orthologous gene and are responsible for a similar phenotype. For example, mutations of the *agouti* locus encoding a peptide antagonist of the melanocyte-stimulating hormone receptor (MC1R) have been identified in human, mouse, rat, horse, pig and cow.

Table 18.2 shows an analysis of mutations by type of DNA lesion. The majority of known mutations (about 58%) are single base pair substitutions in the coding region of the gene,

Table 18.2. Type of mutations in the Human Gene Mutation Database. Mutations recorded in 1433 genes associated with human genetic disease, Oct 2003.

Mutation type	No. of entries
Micro-lesions	
Missense/nonsense	20,719
Splicing	3,423
Regulatory	322
Small deletions	5,856
Small insertions	2,292
Small indels	339
Gross lesions	
Repeat variations	63
Gross insertions and duplications	290
Complex rearrangements (including inversions)	390
Gross deletions	1,963
Total	35,657

whereas deletions account for about 22% of all changes and insertion/duplications constitute 7%. The remaining mutations are in the regulatory and splicing-relevant regions. It is interesting to note that only about 8% of all known mutations involve gross rearrangements of genomic DNA. This distribution apparently reflects technical difficulty both in analysis of regulatory mutations and in detection of big deletions, but, on the other hand, might indicate the true nature of the mutational process (Botstein and Risch, 2003).

Polygenic disorders

Single gene disorders probably account for less than one-third of the total load of inherited disease (0.5–1.4% of human live births) (Weatherall, 1991). The majority of the morbidity and mortality from inherited diseases in both human and agricultural species comes instead from multifactorial and polygenic diseases and chromosomal disorders. The investigation of multifactorial and polygenic traits is dealt with more extensively in Chapter 20. However, in considering population structure in domesticated, laboratory and wild animals it is worth noting here the effect of a reduced genetic diversity on the apparent complexity of inherited diseases. As detailed above, many laboratory mouse lines are isogenic, whilst breeds used in agriculture and more particularly companion animal pure-bred and pedigree breeds are rather inbred and likely to have fairly low numbers of founder animals. Hence they offer an opportunity to identify mutations which in humans are high-incidence, low-penetrance as single genes of large effect against a background which is well stratified and fairly genetically homogeneous. This is the situation for many domesticated animal breeds, as well as in gene knockout mice. In appropriate models the penetrance of such genes appears larger. Taking pedigree dogs as an example, such diseases include models of cardiac developmental defects (Patterson *et al.*, 1993), of diabetes (Kramer *et al.*, 1988), of inherited neoplasias (Moe and Lium, 1997), and many others. One interesting example of gene discovery showing the interplay between a domestic animal model and human disease is Birt–Hogg–Dube syndrome (multifocal renal cystadenocarcinoma

and nodular dermatofibrosis; Birt *et al.*, 1977). This had been noted as a rare familial sporadic neoplasia of humans with an estimated 15% penetrance, but was also observed as a breed-specific cancer of some lines of German shepherd dog segregating as a dominant gene with 100% penetrance (Moe and Lium, 1997). Full penetrance simplified mapping in the dog, localizing the disease gene to a portion of canine chromosome 5 with homology to human chromosome 17p11–12 (Jonasdottir *et al.*, 2000). Human geneticists armed with comparative synteny information were stimulated to further map the gene in human families and thus to identify a novel candidate gene (*BHD1*) showing mutations in affected individuals (Nickerson *et al.*, 2002). Subsequently the same gene has been shown to be mutated in dogs (Lingaas *et al.*, 2003). The ability to work on the same disease in both species facilitated the essential leaps that allowed gene identification. Thus genes contributing to human diseases with large heritable components, but which are difficult to map because of the large number of loci contributing to the disease in human populations, can be susceptible to mapping in other species. Mice will be considered in more detail below.

Chromosomal abnormalities and inherited diseases

Chromosomal abnormalities are an extremely important cause of genetic disease. Perhaps as many as 20% of all human conceptions have a chromosomal disorder, but most of these fail to implant or are spontaneously aborted so the birth frequency is 0.6% (Carr, 1969; Burgoyne *et al.*, 1991; McFadden and Friedman, 1997). Chromosomal numerical anomalies are the most common abnormality in humans, and it is likely that similar rates exist in cattle (Schmutz *et al.*, 1996), but estimates have tended to be lower in other agricultural species (sheep, pigs, horses (reviewed in King, 1990)). However such anomalies are strongly dependent on reproductive age in humans: one reason for the difference may be that many surveys of other species have concentrated on the young adults used in much agricultural production. In addition to well-documented environmental effects on aneuploidy, there are strong genetic components in its causation, easily seen in comparisons of rates of aneuploidy in gametogensis in different mouse strains. For instance, a mouse strain showing aberrant spindle formation (PL/J) is one model in which aneuploids are particularly frequent (Pyle and Handel, 2003).

Most chromosomal abnormalities are extremely serious for the individual concerned: leading to imbalances affecting large numbers of genes, and therefore reducing the fertility of the sufferer to zero. Hence most are not inherited. Inherited diseases caused by chromosomal abnormality are associated primarily with translocations and with chromosomal fragility syndromes. Translocation refers to the transfer of chromosomal material between chromosomes and usually occurs following failure to correctly repair breaks in one or two chromosomes. When there is no overall loss or gain of genetic material the individual, who is usually clinically normal, is described as a carrier of a balanced translocation. Problems occasionally arise when the original breaks happen to disrupt the coding or regulatory sequence of an active gene. A compilation of inherited balanced translocations with corresponding phenotypes is at the Mendelian Cytogenetics Network web site (http://www.mcndb.org/).

Although the carrier of a balanced reciprocal translocation is usually healthy, a proportion of gametes produced are unbalanced, that is, they have additional copies of translocation material from one chromosome, and reduced copies of the reciprocal material. When translocation chromosomes pair during meiosis I a cross-shaped quadrivalent is formed which allows homologous chromosomes to be in contact. When these chromosomes separate, at anaphase, the four chromosomes (two normal, two translocation) of the quadrivalent must segregate to the two daughter cells. Segregation of these chromosomes gives rise to six possible gametes only two of which are balanced, whilst four of the six gametes will result in chromosomally abnormal offspring. However, the actual risk for abnormal segregation, and thus eventually an abnormal offspring, is highly variable depending on the chromosomes involved, the size of the segments that are trisomic or monosomic, and the possibility of survival of unbalanced offspring *in utero*. One in 500 human

births are thought to carry a balanced reciprocal chromosome translocation and therefore have the possibility of propagating inherited disease through this route. About three-quarters of balanced translocations are inherited, in their turn, from parents, with the remainder arising *de novo*. Couples with spontaneous abortions are 20 times more likely to include a translocation carrier than the general population. Reciprocal translocations have been described in many commonly studied mammalian species.

Robertsonian translocations occur in the special case where acrocentric chromosomes have undergone centromere to centromere fusion. Thus the single most common translocation in humans is a centric fusion of chromosomes 13 and 14, followed by 14 and 21 centric fusions. Robertsonian translocations have also been reported in cattle (at least 10 different translocations involving 15 of the chromosomes, but notably t(1;29)); sheep (notably t(5;26) and t(8;11)); pig (notably t(13;17)); mouse, dog and goat but not so far in cat or horse. They can again produce normal, balanced, or unbalanced gametes. In some cattle breeds in particular, the abundant t rob(1;29) may be responsible for major losses of fertility. There is abundant evidence for speciation events in mammals (including rodentia, equidae and arteodactyls) in which Robertsonian fusions play an important part in maintaining genetic isolation.

Other mammals as models for human inherited diseases

The utility of inherited diseases of other mammals as models for human diseases depends both on the accuracy of the model in terms of the metabolic/physiological, anatomical, temporal and pathological aspects, and also on the ability to investigate, understand and manipulate the model. In the case of inherited disease, a genetic understanding is critical, and much of the remainder of this chapter will concentrate on this aspect. This is a two-way street: in the veterinary context it is often important to understand the limitations of a well-investigated human disease in modelling a veterinary one: human genomics is widely used as a source of candidate genes, biochemical insight, etc.

Amongst the mammals, the mouse offers models of genetic disease which are both very easily manipulated, and also (at least in mammalian terms) rapid and cheap to use. The genetic resources of murine models will be reviewed below. Here it is sufficient to say that transgenic and other genetic techniques allow investigation of the function of any human gene for which the murine equivalent can be unambiguously identified. In general this model is hugely successful, and it represents the zenith of achievement for disease geneticists. However, there are some disadvantages of the murine system in experimental, physiological and anatomical terms. It has become clear that our ability to assay subtle developmental and behavioural effects of a variety of genes knocked out in the laboratory mouse is limited. For instance, prion protein null (Prnp$^{0/0}$) mice have played a crucial role in investigation of the transmissible spongiform encephalopathies (TSEs); but although prion protein is under tight evolutionary conservation (Van Rheede *et al.*, 2003), and it is clear that abnormally folded forms of the protein have a central role in both transmission and pathogenesis of TSEs, its role in normal physiology is still under dispute. Several of the original *Prnp* gene knockout models show minor abnormalities in synaptic transmission and/or ataxia. However, these models also show altered expression of an adjacent gene encoding the prion-like protein Doppel (PrPLP/Dpl) and most phenotypic effects are now believed to be due to this. There is limited evidence for roles for prion protein itself in copper binding and regulation (Brown *et al.*, 1997) and in Purkinje cell electrophysiology (Herms *et al.*, 2001) but the former property is not conserved in evolution whilst the latter does not seem to be responsible for Purkinje cell loss in some Prnp$^{0/0}$ lines (Moore *et al.*, 2001; Rossi *et al.*, 2001). No changes in behaviour, learning or ageing processes have been convincingly demonstrated (Lipp *et al.*, 1998). More than 10 years of intensive study have failed to unequivocally clarify the normal function of this protein, which is clearly conserved in evolution despite the potential cost of spongiform degeneration. We may be pretty sure that we have overlooked something, and that we will be equally poor at recognizing function of other genes in future.

Physiological differences between humans and mice have been noted. An example drawn from retinal degenerations is seen in the effect of *MYO7A* gene mutation. In humans, mutations in the gene encoding myosin VIIa can cause Usher syndrome type 1b (USH1B), a disease characterized by deafness and retinitis pigmentosa. Myosin VIIa is also the gene responsible for the inner ear abnormalities at the shaker1 (*sh1*) locus in mice. To date, none of the *sh1* alleles examined have conveyed any signs of retinal degeneration, although a slight perturbation in electroretinogram at very high light intensities was noted in one study (Libby and Steel, 2001). However, differences are not clear-cut. Several human families have also been recorded with *MYO7A* mutation and non-syndromic deafness. Anatomical and temporal differences may be responsible for a non-exact correspondence between orthologous murine and human genetic diseases. One pertinent example is the differences in the X-linked, dystrophin-associated myopathies between the *Mdx* mouse and human Duchenne and Becker muscular dystrophy patients. Whilst mice, whose muscles have low loadings because of their small size, are able to survive into old age by rapidly turning over and regenerating dystrophin-lacking skeletal muscles, human male children lacking dystrophin go through a short period of muscular hypertrophy before relatively rapid and irreversible skeletal muscle loss, and death before age 20. The dog presents a better model than the mouse for some aspects of the human disease. However, even in this case the mouse offers valuable insights because of our ability to specifically target and knock out the other genes that interact with dystrophin (Durbeej and Campbell, 2002), allowing increased understanding of the whole system.

In therapeutic investigation it may be important to have a large and long-lived inherited disease model as an intermediate before human therapy where modelling delivery and long-term effects are important. In this respect dogs, with their wealth of naturally occurring inherited diseases, have been extensively used, but cats, sheep, pigs and other species have also had a role. In the next section we discuss the development of candidate genes and other strategies to articulate such models for both comparative and veterinary purposes.

Methods to Define Disease-causing Genes

Gross chromosomal rearrangements

The work of the disease cytogeneticist has been revolutionized over the last 5 years by a powerful suite of molecular tools. The analysis of human chromosomal rearrangements is still routinely begun using the G-, R- and Q-banding techniques on metaphases as described in Chapter 2. These are now supplemented by chromosome painting (fluorescence in situ hybridization, Chapter 14) which allows more rapid and complete analysis of complex karyotypes without the errors that can occur when using banding techniques. Methods for simultaneous labelling of all 24 human chromosomes such as the multiplex FISH (Speicher et al., 1996) and spectral karyotyping (Schrock et al., 1996) are available. Five fluorophores are used in combinations to assign different colours to each chromosome so that all chromosomes can be analysed in a single hybridization.

Chromosome painting resolves breakpoints only to the nearest band or around 5 million base pairs (Rens et al., 2001). However the same basic technology can be used with single bacterial artificial chromosome (BAC) or P1-derived artificial chromosome probes to improve resolution to roughly 100,000 base pairs in fully mapped species. This process is quite laborious as well as expensive as it requires the sequential hybridization of BACs at regular intervals down the chromosome or on either side of the translocation breakpoint to move in on breakpoint-spanning BACs. For use in rapid analysis of translocations in human, sets of subtelomeric probes are also available which allow rapid detection of the presence of reciprocal translocations involving any chromosome on a single slide.

FISH can now be supplemented by comparative genomic hybridization, using as targets metaphases or BACs or oligonucleotides which span the genome, spotted on to microarrays. This allows the analysis of amplifications and deletions with an accuracy of around or better than 100,000 base pairs. This methodology is beginning to be used to analyse whole genomes for aneuploidy (comparative genomic hybridization), largely in tumours. As yet it has received little attention as a means of analysing chromosome abnormalities in inherited or congenital disease, with a single published study including cases of DiGeorge syndrome and neurofibromatosis 2 syndrome (Fig. 18.1) (Buckley et al., 2002). However the future potential of the technique is enormous.

In the mouse, a similar range of techniques is being used to study chromosome aberrations, with the exception that genomic microarrays are not yet available. For veterinary and other mammalian species G- and other banding techniques have been the major source of our understanding of chromosome abnormalities. Yet in some domesticated and pet species these are particularly difficult to use: chromosomes are often less well defined or easily recognized than human ones. In these cases it would be useful to use FISH-based techniques in the analysis. Initial work in this area used centromeric (alphoid) satellite sequences in the examination of minute chromosomes and of Robertsonian translocations in cattle (Miyake et al., 1994; Tanaka et al., 2000). Subsequently a similar method was used as part of a study of reciprocal translocations in the pig (Pinton et al., 1998). Cloned individual genes have also been used to identify and characterize chromosome translocations in cattle (Iannuzzi et al., 2001a). Reciprocal chromosome painting methods to examine synteny relationships between a target species and a well mapped one such as human or mouse are described in Chapter 14. Chromosome painting methods can be extended to look at chromosomal abnormalities within a species, although relatively little work of this type has been published as yet. Human chromosome painting reagents (Zoo-FISH techniques) have been used in studies of chromosome translocation in cattle (Iannuzzi et al., 2001b) and horses (Lear and Layton, 2002). Ploidy levels and chromosome abnormalities have been examined using species-matched paints (two painted chromosomes only in each study) in cleavage embryos (Bureau et al., 2003) and blastocysts in cattle (Viuff et al., 1999) and oocytes in pigs (Vozdova et al., 2001). Chromosome translocations have been observed using species-matched chromosome paints in cattle (Basrur et al., 2001), pigs (WHO, 2002) and dogs (Schelling et al., 2001), and

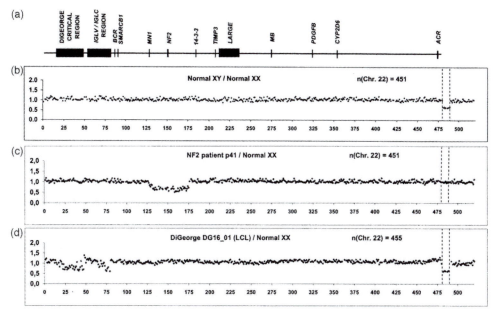

Fig. 18.1. Applications of a chromosome 22 microarray in the analysis of constitutional chromosome aberrations. 480 chromosome 22 loci are ordered from centromere (left hand side) to telomere (right hand side). Dots between the vertical broken lines indicate control loci from chromosome X (nine clones). Dots plotted after the chromosome X controls are derived from chromosomes other than X and 22 (31 clones). Each dot on a chart represents the average ratio between fluorescent signals for each locus normalized against normal female (XX) DNA ($n = 3$, y axis). (a) Positions of landmark features on chromosome 22 on the array, including the DiGeorge critical region (DGCR). (b) Normal male CGH profile. (c) CGH profile of a female NF2 patient, showing a large constitutional heterozygous deletion of ID132–177, encompassing the *NF2* gene. (d) Analysis of constitutional heterozygous deletions in a male patient affected with DiGeorge syndrome. A heterozygous deletion in this case extends from ID16 to ID47 (2.4 Mb), but is somewhat less clear than that in the NF2 case because the region contains many repeats. Adapted with permission from Buckley *et al.* (2002).

multicoloured paint sets are now being used for the latter in our own laboratory (Plate 4).

Single gene defects: the use of candidate genes

Historically, studies of the mutations causing genetic disease have concentrated on defining, cloning and sequencing individual genes in affected and normal individuals. These individual genes have been defined as 'candidates' for the mutation causing the disease either on functional grounds, through homology with a better understood disease in another species or through studies to define the position of the disease-causing mutation in the genome which show that it has close linkage to a particular gene (positional candidate).

Functional candidate genes

In the earliest studies of change of nucleotide sequence, candidate genes for study were derived from the known biochemistry and physiology of the defect. Thus the classical studies of the molecular basis of thalassaemia of Y.W. Kan and others began with knowledge that haemoglobin subunits were the proteins affected by mutation or reduced in expression in these diseases. From here it was possible to move to elegant studies showing the levels (Temple *et al.*, 1977), and eventually the sequences of purified normal and abnormal globin mRNAs (Chang and Kan, 1979).

This type of technique has guided most mutation-finding studies in both mouse and other domesticated and pet species up to the start of the 1990s. The retinitis pigmentosas (RPs), a group of blinding diseases seen in humans, lab rodents (where they are known as retinal degenerations (rds)) and domesticated and pet species (dogs, cats, horses and pigs, known as progressive retinal atrophies (PRAs)), provide a good example of how this applies in different groups. These diseases are characterized by bilateral, progressive retinal degeneration in which the photoreceptors either fail to mature, or show abnormalities in mature morphology or function and go on to die. For different human RP families or in different canine breeds with PRA, or different rd mice, it was observed that rates of disease progression are different. Breeding experiments showed that most forms of the disease are monogenic but that at least in dogs and mice, more than one locus is capable of causing disease. The dysplasia or degeneration of photoreceptors led to investigation of calcium and cGMP cascade in these diseases, and it was shown that there were associated early abnormalities in neural retinal cGMP concentration in several forms of the disease in dogs and mice (Aguirre *et al.*, 1978; Robb, 1979). Subsequently the phosphodiesterases present in photoreceptors were characterized and then their genes cloned. These were shown to be abnormal in mice (Bowes *et al.*, 1990; Pittler and Baehr, 1991), dogs (Suber *et al.*, 1993) and humans (McLaughlin *et al.*, 1993). At the same time rhodopsin had been mapped in the human genome and was swiftly confirmed as a positional candidate for some forms of dominant RP (Farrar *et al.*, 1990). This brought all members of the visual transduction cascade under suspicion and these have subsequently become a frequent target for functional candidate gene studies. Although most subsequent work in mouse and humans relied on mapping techniques, this has not been possible until recently in dogs (and is often not practical with clinical material). Further gene discoveries have been made without positional information (reviewed by Lin *et al.*, 2002), but there is often a great deal of wastage in chasing such candidates. In fact a recent survey of functional candidate gene cloning and analysis in PRA in different dog breeds unearthed 267 published studies, of which only seven were successful (Aguirre-Hernadez and Sargan, unpublished). Positional information hugely increases this success rate.

The type of functional information which can be collected has expanded in recent years to include information about the tissue distribution of gene expression and expression differences between normal and affected individuals. An early example of the latter type can again be drawn from retinal degenerations. The second rd-associated disease gene to be identified in the mouse (the gene causing rd2, previously retinal degeneration slow, rds) was recognized because of the loss of its transcript early in disease, as measured by subtractive hybridization selecting those retinal transcripts present in the normal mouse immediately after birth, but lost in rds retina (Travis *et al.*, 1989). One of these encodes a specialized structural protein of photoreceptors, rds/peripherin. Mutation in this gene has again been shown subsequently to be important in causing rd2-like diseases in other species, most particularly in humans. Currently, normal expression information on many genes has been collected by multiple techniques including tissue blotting (northern analysis) and serial analysis of gene expression (SAGE) mapping. A very large collection of human, murine, rat and macaque cDNA clones has been characterized and expression mapped by the IMAGE (Integrated Molecular Analysis of Genomes and their Expression) consortium (http://image.llnl.gov/). SAGE is based on generating and sequencing clones of concatenated (linked) short sequence tags derived from mRNA from the target cells or tissue (Velculescu *et al.*, 1995). By recognizing tags, individual to each gene, its transcript abundance can be estimated within each mRNA population. SAGE mapping data have been compiled for many human and murine genes to provide typical transcription profiles (http://www.ncbi.nlm.nih.gov/SAGE/). A further recent method to examine normal and disease populations of mRNAs is the use of cDNA microarrays. Currently these are commercially available for human, mouse and rat, but microarrays of cDNA from other species including cattle, pig and horse are already available in academic laboratories. Microarrays produce huge amounts of data, which must be interpreted with caution to distinguish variations caused by disease from noise or between-individual variation. Using either type of data set, or in conventional blotting studies, a candidate gene is expected to be expressed in tissues which show pathology in diseases in which it (the candidate) is mutated, and is likely to show variation in diseased tissues. But gene expression levels are aspects of phenotype that can be in cause or effect relationship with the presence of disease. The former relationship is much rarer than the latter, so that additional techniques are often needed to move from whole 'transcriptome' analyses to a small field of candidate genes.

Cloning candidate genes

Once candidate genes for a particular disease have been selected on functional or expression grounds, or by positional mapping, cloning and sequencing are expected to reveal the mutation causing the disease. Historically, candidate gene clones were often selected from full-length cDNA libraries cloned using complex assays of expression and function. More recently cloning of candidate genes in domesticated and wild mammalian species has relied on sequence information available from homologous genes in other species, largely human and mouse. Cloning strategies usually involve sequence alignment of the candidate in man, mouse and any other species for which relevant sequence information exists in order to select a pair of conserved oligonucleotides from which to amplify a portion of the target gene in the organism of interest. Where exon structure can be predicted from known human and murine gene structure, the oligonucleotides are selected from the same exon. A variety of primer selection software tools now exist allowing automatic selection of primers of particular length and base composition. Several of these can automatically check for repeats in primers and amplimer, etc. A typical collection of current tools is available at mol-biol.net (http://www.bioinformatics.vg/biolinks/bioinformatics/PCR%2520and%2520Primer%2520Design.shtml). These conserved oligonucleotides are used as PCR primers in the target species. The gene fragment can then be used as a probe to clone a full-length gene copy from cDNA or genomic libraries, or can be cloned as a gene fragment by standard methods.

Sources of sequence homology

In following such a cloning strategy some caution is needed. Homology (sequence similarity) in genes in two species has its sources either in orthologous, paralogous or convergent evolution. Orthologues and paralogues are both derived from the same ancestral gene. Orthologues are genes in two or more species derived from a common ancestor where the nearest common ancestor of the genes lies in the common ancestor of the two species under study. This gives rise to a set of sequences whose phylogeny is the same as the true phylogeny of the organisms from which the sequences were obtained. Often, but not always, orthologous genes retain the same function in two descendent species. Duplication of orthologues may have occurred in one or both species after speciation. Paralogues are always derived through one or more gene duplication events before speciation, and often have evolved new functions, either within a single descendant species (gene families) or between two new species. Thus many transcription factor and other developmental factor genes such as homeobox and paired homeobox genes, *WNT* genes, and others have formed multiple paralogues which differ between mammalian species; and care must be taken to recognize true orthologues. The level of homology (percentage identical residues) is generally higher between orthologues than between paralogues of the same evolutionary age, as the latter are either able to drift or undergo divergent evolution. A full study has been performed on nuclear receptor proteins (steroid/thyroid/retinoid hormone receptor proteins) showing similarities between mammalian orthologues of 80–100% at the amino acid sequence level, but much greater divergence between human paralogues (Garcia-Vallve and Palau, 1998). Paralogues evolve after gene or chromosome duplication events. The latter lead to low copy number repeats, in which gene order of paralogues adjacent to each other on the ancestral chromosome, but representing several different gene families, is often conserved (review Stankiewicz and Lupski, 2002). Such repeats, which are rapidly dispersed across chromosomes during evolution, are also common sites for translocation events, and can cause difficulties in the interpretation of comparative synteny maps, and potential confusion in candidate gene selection.

Convergent evolution occurs when selective forces cause similarity in function between two genes which come from different ancestral sequences. Such genes are analogues, but may not show high levels of overall sequence homology, because selection works on phenotype. Thus the *Ly49* genes of mouse and the *KIR* genes of humans have similar functions in major histocompatibility complex (MHC) class I recognition by natural killer (NK) cells, but are unrelated in sequence (Trowsdale *et al.*, 2001). Both *Ly49* and *KIR* genes form clusters of paralogues which are thought to interact with divergent MHC class I molecules. Members of both gene families are expressed singly on particular NK cell clones, and both families utilize a similar signal transduction pathway. Although only one of the two gene families exists in mouse and the other in human, both families co-exist in both cows and lower primates (baboons), showing that the study of diverse models can be crucial in working with candidate genes (Mager *et al.*, 2001; McQueen *et al.*, 2002). However, where function involves a specific molecular interaction, short stretches with sequence homology may have evolved (or been adopted by domain or exon swapping). In cloning a candidate gene for a disease one often has functional and sometimes positional information. If candidates are being cloned after are being discovered *in silico*, using tools such as BLAST, sequence homology can mislead. In general one is trying to clone an orthologue, but using the strategy outlined in the previous sub-section, cloning of paralogues is a frequent hazard, whilst in principle one might sometimes end up cloning an analogue.

Mapping single gene defects: positional cloning

Until the last two decades, despite some successes such as in thalassaemia, the complex nature of the symptoms observed in many diseases often rendered a prediction of the affected gene impossible. Forward genetics proved a slow process. But at the same time as the development of reverse genetic techniques, the development of comprehensive genetic maps has allowed the rapid advance of the forward genetic approach through the positional cloning of trait loci. Positional cloning has become the primary

means of gene identification in many studies of inherited disease, with reverse genetics the primary means of further investigation. This now classical method relies on an identification of the affected genes based solely on their chromosomal position. Five basic steps of positional cloning include: (i) localization of the gene on the genetic map using pedigree and genetic analysis; (ii) construction of the overlapping contig of cloned DNA fragments; (iii) identification of the genes residing within a critical genomic region; (iv) characterization of the candidate genes; and (v) confirmation of gene involvement through independent experiments. Progress in genomic sequencing and gene expression profiling outlined above now often abbreviates this process to the first and the last steps.

Localizing the gene: parametric and non-parametric mapping

Inherited disease genes can be traced to specific regions within the genome by their linkage to particular markers. Methods for linkage analysis can be divided into parametric and non-parametric (see for example Holmans, 2003). Parametric methods calculate a likelihood of linkage disequilibrium using routines and data in which the mode of transmission of all the loci involved is fully specified. Inheritance of alleles of a polymorphic locus is compared with inheritance of the disease trait by looking for the proportion of recombination between marker and trait, through completely specified meioses (i.e. gene flow and recombination is followed in fully sampled pedigrees). In non-parametric linkage analysis one looks for associations between particular alleles and the disease trait in specified sub-populations without the need to specify values for the parameters defining the transmission model (i.e. analyses do not attempt to model the mode of transmission – and are often described as model free). Techniques range from comparison of affected and non-affected sib pairs to general associations of particular alleles with disease within random samples of the whole population. A detailed description of methods and tools available for genetic mapping in relation to quantitative trait loci is available elsewhere in this book (Chapter 20). The tools described can be used with single or multiple loci. The same methods can be applied to inherited disease traits as to other traits. Here we review some more general considerations in mapping inherited disease loci in different mammalian species.

Ascertainment: trait onset and penetrance

A problem which must be overcome in mapping inherited disease traits is the correct ascertainment of individuals affected by the disease trait. Incorrect ascertainment compromises the power of linkage analysis. The loss of power caused by misclassification is greater in the case of parametric methods than association-based analyses.

Correct classification depends on both the accuracy of differential diagnosis and the penetrance of the condition at the age of sampling. Such problems are easily overcome in many monogenic, severe, congenital or juvenile onset conditions, but have increasing relevance in the context of the more complex and later onset diseases that are now being mapped in many laboratories. Studies of a late onset progressive retinal atrophy, for example, must be conducted on individuals old enough to have suffered disease, so a preliminary study must establish age of onset. One must also be aware of heterogeneity in the disease population: for instance late onset progressive retinal atrophies in both Labrador retrievers and miniature schnauzers have been suggested to occur in two forms, one with a more delayed onset than the other. For the miniature schnauzers at least, there is evidence of genetic heterogeneity (Lin et al., 1998; Zhang et al., 1998). Strategies which attempt to map a single gene will fail in such circumstances. Although several computational techniques now exist which allow multiple genes to be mapped, more cases will be required for success (Chapter 20), so the original experimental design must be altered.

Population sampling and population structures

Parametric linkage analysis follows allele and disease trait flow through multiple meioses. To maximize the information available it is important to have complete or near complete family groupings with parents and if possible grandparents of known phenotype and as many sibs as

possible. This situation is ideally managed in the laboratory environment, where colonies segregating the disease are established and where any animal in the pedigree can easily be sampled. With clinical or field samples this ideal is less easily attained, and the species under study has a large determining effect on what can be achieved. Human families are often very cooperative in studies, so that fairly full representation is obtained both of sibships and of surviving parents and grandparents. In general, pedigree information is pretty good (although studies over the last few decades suggest that rates of misreported paternity in present-day western populations are between 1.4% and 10% depending on the population under review (Macintyre and Sooman, 1991; Le Roux et al., 1992)). However, human families suffer from the disadvantage that family sizes in industrial countries are often small. When the number of children in the family is only one or two, this distorts segregation ratios (because families with carrier parents but no affected offspring are never ascertained). This can be allowed for in mathematical treatment of data, but it causes loss of power to find traits. It also causes additional complexity associated with chasing down a large number of small family units.

In cattle in the UK and many other intensive agricultural systems, a situation in some ways similar to the human one may apply. Pedigree records are good, whilst in this case it may also be relatively easy to trace parental samples (although in some agricultural situations male sibs may be difficult to sample). Because of the management systems used in the pig industry, in which multiple farrowings are performed with mixed semen lots and piglets are bulk reared post-weaning, it is less easy to perform sampling for parametric analysis in this species. A different set of problems becomes apparent in dealing with pet species: within the pet population, sibships are often rapidly dispersed, and weaker animals sometimes destroyed, whilst owners often see little incentive to cooperate in sampling of non-affected individuals. Furthermore, in pet populations there is often considerable inbreeding, creating genetic loops which reduce the power of parametric analysis. For those working with clinical case material in this field, it has proved very difficult to generate sample sets of sufficient power for mapping using parametric linkage analysis. Finally, it is clear that wild populations cannot easily be sampled for parametric analysis. In all these latter situations, forms of association analysis are more likely to succeed.

Association analysis works through various types of case–control study which compare polymorphic genetic markers within a disease group and a set of controls. The groups may be a set of affected and non-affected sib pairs or a group of known relatedness, but are often a more general group. The fundamental assumption of these studies is that the two series of subjects provide unbiased estimates of the corresponding marker distributions in affected and non-affected members of the entire population. Note that selection biases are easy to generate in populations: for instance, cases may be recruited through a nationwide contact scheme, but controls recruited locally. If there are local biases in frequency of particular alleles in the population (e.g. through founder effects) these will distort the analysis, leading to false associations between markers and disease and implicating the wrong chromosome regions.

Marker density and linkage disequilibrium in different species

It should be remembered that association analysis requires a dense genetic map and highly polymorphic markers to detect linkage. A particular problem in the case of many agricultural and pet species is that breeds act as genetically isolated and inbred groups. Within any one breed the level of genetic heterogeneity is reduced: we have already seen the advantage that this may give in the dissection of polygenic traits, but it has disadvantages in all forms of mapping – without polymorphism one cannot map. Furthermore, the popular sire effect and relatively low numbers of founders in many breeds lead to significant regions of linkage disequilibrium (in which particular alleles of different markers are found together in extended haplotypes in random individuals within the population). For instance in cattle in the UK dairy population, linkage disequilibrium was detected between pairs of loci in syntenic groups, extending to about 10 cM (Tenesa *et al.*, 2003), whilst within a large multigeneration outbred canine pedigree originating in crosses

of Labrador retriever with greyhound, linkage disequilibrium was detectable to 40 cM, and power was considered to be fully exploited using a screen of 1–2 markers per 10 cM (Lou *et al.*, 2003).

In real mapping studies, problems of homozygosity are reduced if an out-group can be found which carries the same disease. Thus, murine trait mapping studies by linkage analysis are often performed using a species intercross between *Mus musculus* and *Mus spretus* to generate heterozygosity at marker loci. In dogs, a classic study mapping the narcolepsy locus, begun with information derived from a breeding colony of Doberman pinschers, was supplemented by mapping of other narcoleptic dog groups including a second family of Dobermans, a family of dachshunds and a number of Labrador retrievers with the disease. This allowed reduction of the critical region to a size that could be physically mapped, and implicated a single gene (the hypocretin (orexin) receptor 2 gene *HCRTR2*) (Lin *et al.*, 1999).

The investigation of polygenic diseases by association analysis in outbred populations

Mutations in just under 1500 genes have been identified with human monogenic diseases and in just over 100 genes with naturally occurring genetic diseases in other mammals (Table 18.1). The majority of human disease genes have been identified by positional cloning based on a simple, single-gene hypothesis. This represents only a small portion (2.5–3%) of all human genes; those genes which when mutated allow fetal survival but present disease, and which are for a variety of reasons the easiest group of disease-causative genes to clone. Most of them are low-incidence, high-penetrance genes, as opposed to the genes for high-incidence, low-penetrance genetic variations, which form the genetic basis for complex disease phenotypes such as diabetes and hypertension. The latter genes are associated with 'polygenic' diseases (or complex traits), where disease occurrence is dependent on effects of alleles at multiple gene loci and of the environment. We shall not discuss the partitioning of variance in this chapter. Many excellent descriptions of this topic exist in the literature. Here we touch on the considerations from genomics for positional cloning of high incidence, low penetrance alleles.

Mapping of loci contributing to polygenic disease is just beginning. Several factors complicate the process such as locus and allelic heterogeneity, as well as difficulties in correctly assessing and stratifying complex phenotypes, which can consequently weaken the crucial phenotype–genotype link (Weiss and Terwilliger, 2000). The low contribution to variance from typical high incidence polygenic disease alleles (their low penetrance) adds to the noise within the system, making larger sample sizes necessary in mapping studies. Except where spread through popular sire, or similar effects, these high incidence alleles must often exist only in very short regions of linkage disequilibrium when considered across large outbred populations. They are likely to be of ancient origin and present on a variety of genetic backgrounds to have achieved high incidence in these populations, when their effect on phenotype is small. For instance, in contrast to the figures quoted for cattle and dog breeds above, linkage disequilibrium (LD) for SNP markers is detectable (on average) over only 0.1 cM in isolated Finnish human populations. If an allele is present on many chromosomes in this population, then there will be a similarly short region of LD around it. It might be anticipated that in less stable populations maintenance of LD will occur over even shorter distances (Varilo et al., 2003). As argued above, the very different population structures and levels of heterozygosity in most domesticated species means that very different considerations apply.

The noise within the system and the high marker densities required make mapping studies of polygenic disease in outbred populations very large. (Typically these are non-parametric studies involving hundreds of sib pairs or thousands of individuals grouped by phenotype.) Technologies such as pyro-sequencing, allele-specific priming or ligation-dependent PCR, mass spectrometric extension reactions (Sequenome), solid phase hybridization, etc., are now evolving for efficient and very cheap scoring of single nucleotide polymorphisms (SNPs). These suggest a route to examine these more complex diseases by performing association analysis utilizing a very high density SNP map to examine the haplotypes associated with disease. More than 6 million SNPs have already been described in humans, and only slightly lower densities in representative sequences of other species. SNP-based studies are now underway in many human diseases. Panels of SNPs can be selected to examine particular sets of loci in detail (for an example examining hypertension see Izawa et al., 2003). Where single studies have proved inconclusive, a rigorous statistical approach to meta-analysis of multiple studies has sometimes been more revealing. Examples from human heart disease include Chiodini and Lewis (2003) and Pajukanta et al. (2003).

Mice and the Power of Modern Genetics

Long before development of the modern tools of DNA manipulation, a number of human genetic diseases had been identified. Analysis of the mode of inheritance of many such diseases indicated an involvement of a single locus. As outlined above, before the 1980s, the only way to identify the causative gene was to study the biochemical processes affected by the disease. One looked for proteins controlling such processes and then moved towards the genes encoding these proteins. Thus the traditional order in which genetic disease is investigated is from phenotype to genotype: in conventional (or 'forward') genetics the geneticist tracks down the genetic difference that is responsible for a given phenotype. The development of methods for DNA cloning and manipulation allowed the introduction of an alternative approach: observation of the consequences of a deliberate and directed change in the genotype on the organism's phenotype. This approach is sometimes known as reverse genetics to distinguish it from conventional genetics. It has reached its pinnacle in recent applications to modelling human diseases in mice.

Human diseases and their modelling in mice. Large-scale mutagenesis

Mice are ideal animals for basic research applications because they are relatively easy to handle, they reproduce rapidly, and they can be genetically manipulated at the molecular level. For this reason a major aim of mouse genetics has become the modelling of all human diseases in the mouse. To increase the available pool of the mutant alleles in the mouse genome several methods have been employed. Historically, the first approach was linked with studying how low-dose radiation affected mammalian genomes. Irradiation caused large deletions and translocations, leading to advances in fine-structure mapping of the mouse genome. To increase the number of available mouse gene mutations researchers turned to chemical mutagenesis. Soon, ethyl-nitrosourea (ENU) emerged as a dominant frontrunner among different chemical compounds. ENU is an alkylating agent that mainly causes single base substitutions in DNA, and therefore allows for recovery of complete and partial loss-, and sometimes gain-, of-function alleles (see Chapter 9). Initial work over more than a decade was on a relatively small scale, often aimed at understanding mechanisms in mutagenesis but with useful mutant animals as a by-product. Recently, several big mutagenesis projects have been initiated in the USA, Europe, Japan and Australia which have catalogued and provided hundreds of new mutants that affect neurology and behaviour, haematopoiesis, clinical chemistry, immunology and allergy. Most recently such projects have set out to saturate the whole genome with chemical or radiation mutations (You *et al.*, 1997; Nolan *et al.*, 2000a). As with any mass screening approach, ENU mutagenesis requires lots of effort in breeding, identification and particularly in characterization of the mutants. Mutation events are random, so that all mice generated must be put through all phenotype screens in order to trap and describe mutations efficiently. For instance, mice from one project (Nolan *et al.*, 2000b) were put through a comprehensive phenotype assessment tool involving a battery of up to 42 simple tests to detect mutants in areas such as lower motor neurone/muscle function, spinocerebellar function, sensory function, neuropsychiatric function and autonomic function. A large fraction of progeny in the same series has also undergone two behavioural testing protocols: measurements of locomotor activity (LMA) to identify deficits in motor function and screens for abnormal

acoustic startle response and deficits in pre-pulse inhibition of the acoustic startle response. Mice were also subject to clinical chemistry screens (17 parameters). This battery of tests applied to over 26,000 mice produced about 500 mutants from a total of well over 1 million experimental observations made. Current lists of phenotypes from two of these projects are at their respective web sites (http://www.gsf.de/ieg/groups/enu-mouse.html, http://www.mut.har.mrc.ac.uk). Note that only dominant mutations are detected by the tests, which take place on the F_1 between mutagenized and normal mouse. Recessive mutations are also searched for using an F_2 to F_1 backcross on mutant mice generated by the Mouse Functional Genomics Research Group, RIKENGSC, but this is very laborious. Most significantly, neither ENU nor radiation mutation provide any tag or other advantages for mapping and isolation of the causative gene, whilst both may cause multiple mutations in a single genome.

Tagged mutagenesis

In 1981 an early step in the development of reverse genetics was the demonstration that genetic material could be successfully transferred into the genomes of newborn mice by injection of DNA into pronuclei of fertilized eggs (Wagner *et al.*, 1981). After integration into the mouse genome exogenous DNA becomes a part of the genetic material. Since then thousands of recombinant transgenic mouse models have been developed; the method has become a standard tool of DNA manipulation. Because neither ENU mutation nor radiation mutation leave any molecular tags associated with the incorporated mutation, identification of the affected gene can only be done by candidate gene screening or labour-intensive positional cloning. Hence mutagenesis of an adult is followed by an F_1 cross with a different mouse line to provide a polymorphic background. It is the offspring of this cross which are examined for phenotype. A less labour-intensive mutagenesis approach would require either direct targeting of the specific genes in the mouse genome, or random mutagenesis associated with creation of the molecular labels to allow rapid recovery of mutation sites. Both approaches were developed in the last 20 years.

Viral mutagenesis of embryonic stem cells

Most of the hundreds of different DNA constructs introduced into the mouse genome since 1981 include a cDNA or genomic DNA encoding a gene of interest, which gives the opportunity to study the effects of complementation of a disease phenotype by the gene product. Such constructs allow the experimental verification of gene therapy models whereby in cell culture the mutant gene has been complemented by the normal allele introduced through a transgene. Aberrant transgenic expression of the normal or mutated gene may cause specific phenotypic changes such as oncogenic transformation or developmental abnormalities. However, as a 'side product', many transgenic mouse experiments produce a new type of mutant. During construction of the transgenic line, the transgene integrates at random sites within the genome and in some cases causes disruption or deletion of a gene or its regulatory elements. Such mice display mutant phenotypes not associated with the transgene itself, but rather caused by rearrangements of the endogenous DNA. Significantly, the mutated site is now tagged by the transgene DNA, which provides an easy way to clone an adjacent DNA fragment and to identify the mutated gene. Based on this methodology transgene insertional mutagenesis programmes have been developed, generalizing aspects of 'reverse' genetic techniques. Transgenic DNA constructs have been engendered to contain enzymatic markers such as beta-galactosidase reporter constructs, and/or coat colour markers, such as a tyrosinase minigene for selection of the mutant carriers. Large-scale programmes have been performed using such reporter genes in retroviral vectors to infect embryonal stem (ES)cells, from which mouse lines are developed through injection of blastocysts with selected clones of cells (Capecchi, 1989). Such programmes effectively use random insertional mutagenesis to mutate all genes in the genome (Zambrowicz *et al.*, 1998). But they are more efficient than ENU mutagenesis, in that selection on the marker gene means that only lines containing it are analysed, and because the presence of the marker provides rapid methods for recovery of the mutation site. In fact many protocols recover and sequence the insertion site from cell culture before developing a mouse line from the transformed ES

cells (Zheng *et al.*, 1999). This is sensible as it has been estimated that investigation of phenotype costs over 20 times more than the creation and genotyping of random mutants (Capecchi, 2000). In one example, 18% of insertion sites were in recognized genes. A computerized database of transgenic and insertional mutants (TBASE) has been developed and linked to the on-line mouse databases maintained by the Jackson Laboratory (http://tbase.jax.org).

Gene targeting

A revolutionary step in analysis of hereditary diseases was achieved with the development of gene targeting technology. This approach combined the ability to culture and manipulate murine ES cells *in vitro*, the production of gene mutations through homologous recombination (knockout) occurring between an artificially designed DNA construct and an endogenous gene and, finally, creation of mutant animals carrying these constructs. To target an individual gene, a construct containing a selective marker flanked by sequences of the gene being targeted is transfected into ES cells (Fig. 18.2). After selection for stable marker uptake, clones of transfected cells are examined for homologous recombinants in which the target gene has been disrupted, using Southern blotting or PCR analysis. Only these homologous recombinants are used to populate blastocysts and create new mouse lines. Gene targeting allowed an analysis of the biological functions of one gene at a time. Hundreds of genes have been targeted during the last decade revealing their involvement in the aetiology of different diseases (see the Induced mutant resource web site (http://www.jax.org/ imr/notes.html)). An elaboration of this technique known as Cre–loxP technology, allows tissue- or effector-specific gene knockout or gene induction. CRE protein is a 38 kDa recombinase protein from bacteriophage P1 which mediates intramolecular (excisive or inversional) and intermolecular (integrative) site-specific recombination between loxP sites (a 13 bp inverted repeat with a minimum 8 bp spacer (see review article Sauer, 1993). The system was first developed in the 1980s as a means to efficiently manipulate yeast and plant genomes. It has been successfully

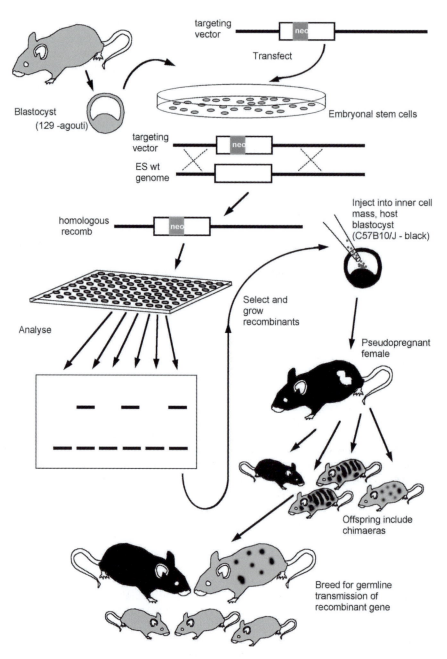

Fig. 18.2. Gene targeting using embryonal stem cell technology. Embryonal stem cells are derived from the inner cell mass of a blastocyst, and when reconstituted into the same environment go on to develop as normal constituents of any tissue. Initially embryonal stem cells are dispersed into tissue culture and transfected using a construct including a fragment of the target gene and a selective marker. Recombination takes place in a proportion of cells, knocking out the target gene. After selection and growth of clones containing the transfected marker, molecular characterization using Southern blotting or similar techniques is used to identify the clones in which homologous recombination has knocked out the target gene. Such clones are used to reconstitute a blastocyst, which is then reinserted into a pseudopregnant female. Chimaeric offspring are selected and interbred to produce monotypic transfectant offspring.

applied latterly in mouse gene targeting experiments to introduce recombination hotspots within chromosomes and specific genes. Modifications of this technique have been used to produce targeted mutations, transgenics, deletions, inversions and other chromosomal aberrations. Conditional mutants contain loxP sequences within the non-coding part of the otherwise intact gene. CRE protein expression is produced from a vector in which *Cre* cDNA under an experimentally regulated (for example by addition of tetracycline) or tissue-specific promoter induces deletion of the gene at a specific time of development or in specific tissues.

Producing a complete set of gene knockout mutant animals for a whole mammalian genome

Modifications of gene targeting and transgenesis approaches in ES cells, the so-called gene trapping techniques, allowed construction of large-scale libraries of clones with single mutated genes. Publicly funded projects have generated a reference library of thousands of gene trap sequence tags from insertional mutations in mouse ES cells (http://tikus.gsf.de; http://baygenomics.ucsf.edu;http://cmhd.mshri.on.ca/sub/genetrap.asp). Sequencing of gene fragments flanking the transgene provides instant access to the identity of the targeted gene. Creation of such library resources fuels the production of the mutant animals on a large scale. An idea of the increased efficiency of this process can be gained by a comparison. As of September 2003, the current version of the ENU mutation database at Harwell lists eight unmapped non-syndromic retinal degenerations or pigmentary disturbances (for all types – genes are not identified); whilst the database of transgenic and insertional mutants, T-base, lists ten photoreceptor-led retinal degenerations for all of which the responsible gene is identified and considerable pathogenesis information is available. However, a total of 15 loci associated with *naturally occurring* non-syndromic photoreceptor-led retinal degenerations have been described in mice (Chang *et al.*, 2002), of which only two overlap with those on T-base! This, plus the knowledge that over 30 non-syndromic photoreceptor-led retinal degeneration loci have

been mapped in humans suggests that there is still much work to do. In some other areas research is already more complete. The potential commercial application of animal disease models for development of therapeutics has received special interest from the pharmacological industry. The top 100 selling drugs have all already had their presumed target genes mutated in gene knockout mice. Although there are about 30,000–40,000 known genes in the mammalian genome, one can still predict that in the future every mouse gene will be mutated. However, even this will not complete murine geneticists' work, even on monogenic inherited disease. Many complete knockouts are fatal, so conditional mutants are needed which target each tissue with each gene. Table 18.3 gives a list of some promoters which use the *Cre* recombinase gene to direct tissue-specific gene knockouts in the central nervous system only: this is representative of the large number of mutants available in other tissues.

Already it has become clear that it is the interactions between genes (and thus the genetic background to mutations) which have large effects on phenotypes. For instance murine retinal degenerations rd1 and rd10 turn out to have mutations in the same gene, both of which abrogate functional gene product, but on different strain backgrounds. rd10 is a milder (slower) phenotype than rd1. The understanding of the networks that make up the transcriptome and the proteome, analysed using targeted gene disruption facilitated by the cDNA microarray and by network computation algorithms (Valenzuela *et al.*, 2003), are outside the scope of this chapter, but are referred to elsewhere in this book.

Diagnostic Applications of Mammalian Genomics – Ripening Fruit from the Genomic Tree

Genetic testing is now a well established branch of human medicine, employed routinely before and at birth in at-risk families for a large variety of inherited conditions and diseases. Depending on the specific type of genetic abnormality different techniques are used for testing. Genomic mutation associated with large-scale rearrangement of the chromosomes or their number has routinely been looked for

Table 18.3. Brain-specific *Cre* transgenic lines. Collected by the Dr Andras Nagy laboratory web site http://www.mshri.on.ca/nagy/ – September 2003. Used with permission.

Promoter	Specificity
Mox2	Epiblast
Pax3	Neural crest
Wnt1	Neural crest
Hoxa1	Floorplate, notochord, gut epithelium
Hoxb6	Extraembryonic mesoderm, lateral plate and limb mesoderm, midbrain/hindbrain junction
Engrailed-2	Mid–hindbrain constriction region, E9.5
Mouse promoter/enchancer D6	Neocortex and the hippocampus
Emx1 (knock-in)	Cerebral cortex and hippocampus
Foxg1 (BF-1) (knock-in)	Telencephalon, anterior optic vesicle, otic vesicle, facial and head ectoderm, olfactory epithelium, mid–hindbrain junction, and pharyngeal pouches
CamKII alpha	High levels in hippocampus, cortex and amygdala; lower levels in striatum, thalamus and hypothalamus
c-kit promotor	CA1, CA2 and CA3 regions of the hippocampus, the anterior region of the dentate gyrus, the ganglion cell layer of the retina
Neurone-specific rat enolase (NSE) promotor	High levels in developing and adult brain; occasionally low level in kidney
Nestin	Neuronal lineage
Mouse gonadotropin-releasing hormone (GnRH) promoter	Neurones
KA1	Most neuronal cells of CNS
Murine neurofilament-H (mNF-H) gene promotor	Neurones of the brain and spinal cord during the late stage of their development
CCDK-II and Thy-1-RU486 inducible	Cortex and hippocampus
hGFAP	Central nervous system; in neurones and ependyma
GluRepsilon3	Cerebellar granule cells
NMDA-type glutamate receptor GluRepsilon3 subunit gene	Cerebellar granule cells (Cre recombinase–progesterone receptor fusion (CrePR); inducible by antiprogestins)
L7/pcp-2	Cerebellar Purkinje cells
GFAP	Astrocytes and subpopulation of neural precursors
Myelin basic protein	Oligodendrocytes
CNP-Cre	Oligodendrocytes, Schwann cells
PO	Schwann cells
Krox20	Hindbrain, bones, PNS
Six3	Retina and ventral forebrain

using GTG banding. Widespread prenatal genetic testing and development of preimplantation genetic diagnosis (PGD, see Kuliev and Verlinsky, 2003; Verlinsky and Kuliev, 2003) have led to further improvement of FISH and molecular karyotyping. M-FISH and spectral karyotyping are now standard approaches in many clinical laboratories. These methods were successfully applied to resolve multiple chromosomal rearrangements in prenatal diagnostics (Buyse and Van den Veyver, 2001). Gene-specific FISH probes have been developed for detection of such disorders as cat-eye syndrome and Charcot–Marie–Tooth type 1A disease. Frequent causes of moderate and severe mental retardation are telomeric rearrangements (Knight and Flint, 2000). Use of chromosome-specific subtelomeric FISH probes enables identification of small chromosomal rearrangements not detected by classical karyotyping. Microdeletion syndromes such as Aniridia/Wilm's tumour (chromosome 11p13), Angelman's (15q11–13), Langer–Giedion (8q24), and others are detected using modifications of the FISH protocols with specific probes derived from the region of interest.

Most current genetic testing is performed on point mutations or small rearrangements within the causative gene. Although some diagnostic tests are still based on Southern blot analysis, the majority of assays are based on modifications of different PCR protocols and subsequent sequence or structural analysis of the amplified fragments. The easiest case for analysis is the presence of deletions. Unique in this sense is the Y chromosome, because all of its genes are in the hemizygous state in males. Deletion of the specific AZF (azoospermia factor) regions responsible for about 10% of all idiopathic infertility cases can be easily tested by a set of PCR assays (Repping *et al.*, 2003). The same is true for XY sex-reversal, caused by a deletion of the *SRY* testis-determining gene, found in several mammalian species (Morrish and Sinclair, 2002). The analysis of size difference of the different alleles due to the presence of an insertion/deletion or polymorphisms in repeat sequences is another simple case. For example, Huntington disease is caused by expansion of a polymorphic trinucleotide repeat $(CAG)_n$ located in the coding region of the gene for huntingtin. The range of repeat numbers is 9–37 in normal individuals and 37–86 in HD patients, which can be tested by sizing PCR fragments (Huntington's Disease Collaborative Research Group, 1993). Automation of the detection is achieved through the use of fluorescent PCR primers and capillary gel- and matrix-based sequencing machines. In many instances, the nucleotide substitution either creates or destroys a restriction endonuclease site. A classical example of such substitution is the defect in codon 6 of the beta-subunit of haemoglobin, resulting in the loss of an *Mst*II restriction site. This mutation causes sickle cell disease. A digest of the PCR fragment containing this site (restriction fragment length polymorphism analysis) easily identifies the mutant allele. Specific design of the primers across mutated sites enables allele-specific amplification for rapid testing for mutations associated with cystic fibrosis and various thalassaemias (e.g. Kanavakis *et al.*, 1995). The hybridization of the PCR fragments with specific oligonucleotides with sequences based on the mutant and wild type sequence provides a fast assay to detect the point mutation in Tay–Sachs disease, Gaucher's disease,

Canavan disease, and others. Current automated systems use single base extension reaction of PCR products. Based on primer extension technology, the reaction uses an unlabelled primer that anneals one base upstream of the target SNP. The reaction proceeds by adding a complementary labelled ddNTP at the SNP site, as a templated single base extension and examining the label of the nucleotide added (Shi, 2002). Application of microarrays for the analysis of point mutations and SNPs in genomic DNAs is currently under intensive development (Kolchinsky and Mirzabekov, 2002).

In many instances there are predominant mutant alleles over-represented in a specific population or race group. Approximately 70% of the mutations in cystic fibrosis (CF) patients have a specific deletion of 3 bp, which results in the loss of a phenylalanine residue at amino acid position 508 of the putative product of the CF gene. At the same time, a total of more than 1000 mutations has been described for the cystic fibrosis transmembrane conductance receptor (*CFTR*) gene. There are 27 exons of this gene which spread over 250 kb of genomic DNA. Breast cancer-associated1 gene (*BRCA1*) spans over 81 kb of DNA and contains 24 exons. There are more than 400 known allelic variants of the gene, mainly single nucleotide substitutions and small insertions/deletions. Such variability complicates genetic testing and requires high-throughput methods of DNA analysis capable of detecting any polymorphisms in a known sequence.

Single-stranded conformation polymorphism (SSCP) and heteroduplex analysis methods such as denaturing gradient gel electrophoresis (DGGE) are most commonly used in this type of diagnosis. They rely on changes in DNA secondary structure due to the sequence variations that cause differences in electrophoretic mobility. Heteroduplexes derived from the wild type and the mutant allele can be detected on denaturing gradient or non-denaturing gels, where they show differences in mobility from perfectly matching homoduplexes. This approach is commonly used for detection of many mutations in hard-to-analyse genes. Denaturing high-performance liquid chromatography (DHPLC) is a PCR scanning technique, which is also based on DNA duplex analysis, but does not require running the gels. An example of mutational screening in patients with cryptorchidism for sequence variations in the *GREAT* (G protein-coupled receptor affecting testicular descent) gene is shown in Fig. 18.3 (Gorlov *et al.*, 2002).

Currently, tests for about 1000 human genes and chromosomal abnormalities are established or in the research stage of development. A compendium of available human genetic tests can be found on the GeneTests web site (http://www.genetests.org). Over the next few years genetic testing will certainly become more routine in assessing the likely efficacy of particular therapies (see Chapter 19). It is also likely to become more widespread as a predictor for personal risks and avoidance strategies for the common degenerative, neoplastic and other complex diseases listed elsewhere in this chapter (although there has probably been some tendency to overplay what will be possible in this field).

Genetic diagnostic tests in agricultural and pet species have been made publicly available rapidly as mutations are discovered – more than three-quarters of the mutations in pet species enumerated in Table 18.1 are available as tests. Similarly, tests have been developed for virtually all the mutations reported in agri-cultural species. In human medicine the use of these tests is largely in counselling patients, parents or prospective parents. In veterinary medicine and agriculture, genetic testing can be used to motivate much more rapid changes in the genetic structure of populations. In companion animal medicine diagnostic tests are combined with advice to breeders which can be to some degree enforced in the issue of breed certificates by breed associations, kennel clubs, etc. With good breeder cooperation this allows rapid and complete elimination of diseases – for instance forms of progressive retinal atrophy are now close to elimination in the Cardigan Welsh corgi and Irish setter dogs. In agriculture in the developed world, tests for disease have become economic drivers in elimination of diseases mentioned previously. For both pet and agricultural species elimination may be complicated by the high incidence of the disadvantageous allele. The availability of the test can be used to ensure that carriers of a recessive disease are never crossed, so that the disease phenotype can be eliminated, whilst the gene frequency is reduced more slowly. In intensive and industrially organized agricultural systems, such as the pig industry in the developed world, DNA tests for inherited disease are

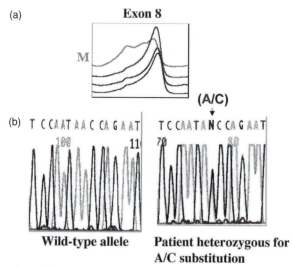

Fig. 18.3. Mutation analysis of the *GREAT* gene in cryptorchid patients. (a) DHPLC profiles showing sequence variations in exon 8. Three homozygous wild-type and one heterozygous mutant (M) profile is shown. The presence of the DNA heteroduplex can be seen as a double-pick profile. (b) Sequence variation detected by direct sequencing of the PCR fragment corresponding to exon 8 (right) and wild-type control (left). An A–C nucleotide substitution at position 664 is in heterozygous condition.

now combined with tests for production traits in characterizing the genetics of production stock. Particular genetic combinations can then be promulgated: in particular the use of AI allows sire traits to be promulgated to very large numbers of offspring.

DNA tests used in clinical practice or in commerce have very high sensitivity and specificity. Tests may look either for a haplotype associated with a disadvantageous allele or directly for the mutation. The former (linkage-based) tests have an intrinsic error rate due to recombination. (In some tests this has been unacceptably high, causing problems for all who engage in DNA testing: see for instance Haywood *et al.*, 2001.) In general, for those tests which detect the mutation directly, errors should arise only through problems in the audit chain rather than through lack of robust-

ness in the test's biochemistry. However, in cattle it has become clear that even mutation-detecting DNA tests are not infallible. The chimaerism of twinned cattle has been shown to cause wrong assignments of germline status in citrullinaemia testing, using leukocytes as DNA source (Healy *et al.*, 1994).

Such errors are very rare. In general the low error rate of DNA tests for inherited diseases, their lifelong predictive accuracy and the ability to observe carrier state even in recessive disease are all combined with rapidly reducing cost and increases in numbers of diseases that can be tested for from a single sample. These factors make it likely that agricultural and pet species will in future be subject to intensive genetic profiling in the same way as their owners may already choose to be, in an era of truly personalized medicine.

References

Aguirre, G., Farber, D., Lolley, R., Fletcher, R.T. and Chader, G.J. (1978) Rod–cone dysplasia in Irish setters: a defect in cyclic GMP metabolism in visual cells. *Science* 201, 1133–1134.

Basrur, P.K., Koykul, W., Baguma-Nibasheka, M., King, W.A., Ambady, S. and Ponce de Leon, F.A. (2001) Synaptic pattern of sex complements and sperm head malformation in X–autosome translocation carrier bulls. *Molecular Reproduction and Development* 59, 67–77.

Birt, A.R., Hogg, G.R. and Dube, W.J. (1977) Hereditary multiple fibrofolliculomas with trichodiscomas and acrochordons. *Archives of Dermatology* 113, 1674–1677.

Botstein, D. and Risch, N. (2003) Discovering genotypes underlying human phenotypes: past successes for mendelian disease, future approaches for complex disease. *Nature Genetics* 33 (Suppl.), 228–237.

Bowes, C., Li, T., Danciger, M., Baxter, L.C., Applebury, M.L. and Farber, D.B. (1990) Retinal degeneration in the rd mouse is caused by a defect in the beta subunit of rod cGMP-phosphodiesterase. *Nature* 347, 677–680.

Brown, D.R., Qin, K., Herms, J.W., Madlung, A., Manson, J., Strome, R., Fraser, P.E., Kruck, T., von Bohlen, A. and Schulz-Schaeffer, W. (1997) The cellular prion protein binds copper *in vivo*. *Nature* 390, 684–687.

Buckley, P.G., Mantripragada, K.K., Benetkiewicz, M., Tapia-Paez, I., Diaz De Stahl, T., Rosenquist, M., Ali, H., Jarbo, C., De Bustos, C., Hirvela, C., Sinder Wilen, B., Fransson, I., Thyr, C., Johnsson, B.I., Bruder, C.E., Menzel, U., Hergersberg, M., Mandahl, N., Blennow, E., Wedell, A. *et al.* (2002) A full-coverage, high-resolution human chromosome 22 genomic microarray for clinical and research applications. *Human Molecular Genetics* 11, 3221–3229.

Buergelt, C.D., Homer, B.L. and Spalding, M.G. (2002) Causes of mortality in the Florida panther (*Felis concolor coryi*). *Annals of the New York Academy of Sciences* 969, 350–353.

Bureau, W.S., Bordignon, V., Leveillee, C., Smith, L.C. and King, W.A. (2003) Assessment of chromosomal abnormalities in bovine nuclear transfer embryos and in their donor cells. *Cloning Stem Cells* 5, 123–132.

Burgoyne, P.S., Holland, K. and Stephens, R. (1991) Incidence of numerical chromosome anomalies in human pregnancy estimation from induced and spontaneous abortion data. *Human Reproduction* 6, 555–565.

Buyse, I.M. and Van den Veyver, I.B. (2001) Applied molecular techniques in reproductive genetics. *Reproductive Genetics* 12, 851–868.

Capecchi, M.R. (1989) The new mouse genetics: altering the genome by gene targeting. *Trends in Genetics* 5, 70–76.

Capecchi, M.R. (2000) Choose your target. *Nature Genetics* 26, 159–161.

Carr, D.H. (1969) Genetic factors in pregnancy wastage. *Medical Clinics of North America* 53, 1039–1050.

Chang, B., Hawes, N.L., Hurd, R.E., Davisson, M.T., Nusinowitz, S. and Heckenlively, J.R. (2002) Retinal degeneration mutants in the mouse. *Vision Research* 42, 517–525.

Chang, J.C. and Kan, Y.W. (1979) Beta-0-thalassemia, a nonsense mutation in man. *Proceedings of the National Academy of Sciences USA* 76, 2886–2889.

Chiodini, B.D. and Lewis, C.M. (2003) Meta-analysis of 4 coronary heart disease genome-wide linkage studies confirms a susceptibility locus on chromosome 3q. *Arteriosclerosis, Thrombosis and Vascular Biology* 23, 1863–1868.

Claustres, M., Horaitis, O., Vanevski, M. and Cotton, R.G. (2002) Time for a unified system of mutation description and reporting: a review of locus-specific mutation databases. *Genome Research* 12, 680–688.

Durbeej, M. and Campbell, K.P. (2002) Muscular dystrophies involving the dystrophin–glycoprotein complex: an overview of current mouse models. *Current Opinion in Genetics and Development* 12, 349–361.

Farrar, G.J., McWilliam, P., Bradley, D.G., Kenna, P., Lawler, M., Sharp, E.M., Humphries, M.M., Eiberg, H., Conneally, P.M., Trofatter, J.A. *et al.* (1990) Autosomal dominant retinitis pigmentosa: linkage to rhodopsin and evidence for genetic heterogeneity. *Genomics* 8, 35–40.

Frankham, R. (1999) Quantitative genetics in conservation biology. *Genetics Research* 74, 237–244.

Garcia-Vallve, S. and Palau, J. (1998) Nuclear receptors, nuclear-receptor factors, and nuclear-receptor-like orphans form a large paralog cluster in *Homo sapiens*. *Molecular Biology and Evolution* 15, 665–682.

Gorlov, I.P., Kamat, A., Bogatcheva, N.V., Jones, E., Lamb, D.J., Truong, A., Bishop, C.E., McElreavey, K. and Agoulnik, A.I. (2002) Mutations of the *GREAT* gene cause cryptorchidism. *Human Molecular Genetics* 11, 2309–2318.

Haywood, S., Fuentealba, I.C., Kemp, S.J. and Trafford, J. (2001) Copper toxicosis in the Bedlington terrier: a diagnostic dilemma. *Journal of Small Animal Practice* 42, 181–185.

Healy, P.J., Dennis, J.A., Nicholls, P.J. and Reichmann, K.G. (1994) Haemopoietic chimaerism: a complication in heterozygote detection tests for inherited defects in cattle. *Animal Genetics* 25(1), 1–6.

Herms, J.W., Tings, T., Dunker, S. and Kretzschmar, H.A. (2001) Prion protein affects Ca^{2+}-activated K^+ currents in cerebellar purkinje cells. *Neurobiology of Disease* 8, 324–330.

Herrtage, M.E., Seymour, C.A., White, R.A.S., Small, G.M. and Wight, D.C.D. (1987) Copper toxicosis in the Bedlington terrier, the prevalence in asymptomatic dogs. *Journal of Small Animal Practice* 28, 1141–1151.

Holmans, P. (2003) Nonparametric linkage. In: Balding, D.J., Bishop, M. and Cannings, C. (eds) *Handbook of Statistical Genetics*, 2nd edn. John Wiley & Sons, pp. 919–938.

Huntington's Disease Collaborative Research Group (1993) A novel gene containing a trinucleotide repeat that is expanded and unstable on Huntington's disease chromosomes. *Cell* 72, 971–983.

Iannuzzi, L., Molteni, L., Di Meo, G.P., De Giovanni, A., Perucatti, A., Succi, G., Incarnato, D., Eggen, A. and Cribiu, E.P. (2001a) A case of azoospermia in a bull carrying a Y–autosome reciprocal translocation. *Cytogenetics and Cell Genetics* 95, 225–227.

Iannuzzi, L., Molteni, L., Di Meo, G.P., Perucatti, A., Lorenzi, L., Incarnato, D., De Giovanni, A., Succi, G. and Gustavsson, I. (2001b) A new balanced autosomal reciprocal translocation in cattle revealed by banding techniques and human-painting probes. *Cytogenetics and Cell Genetics* 94, 225–228.

Izawa, H., Yamada, Y., Okada, T., Tanaka, M., Hirayama, H. and Yokota, M. (2003) Prediction of genetic risk for hypertension. *Hypertension* 41, 1035–1040.

Jonasdottir, T.J., Mellersh, C.S., Moe, L., Heggebo, R., Gamlem, H., Ostrander, E.A. and Lingaas, F. (2000) Genetic mapping of a naturally occurring hereditary renal cancer syndrome in dogs. *Proceedings of the National Academy of Sciences USA* 97, 4132–4137.

Kanavakis, E., Tzetis, M., Antoniadi, T., Traeger-Synodinos, J., Doudounakis, S., Adam, G., Matsaniotis, N. and Kattamis, C. (1995) Mutation analysis of 10 exons of the CFTR gene in Greek cystic fibrosis patients: characterization of 74.5% of CF alleles including 1 novel mutation. *Human Genetics* 96, 364–366.

Kienle, R.D., Thomas, W.P. and Pion, P.D. (1994) The natural clinical history of canine congenital subaortic stenosis. *Journal of Veterinary Internal Medicine* 8, 423–431.

King, W.A. (1990) Chromosome abnormalities and preganancy failure. *Advances in Veterinary Science and Comparative Medicine* 34, 229–250.

Knight, S.J. and Flint, J. (2000) Perfect endings: a review of subtelomeric probes and their use in clinical diagnosis. *Journal of Medical Genetics* 37, 401–409.

Kolchinsky, A. and Mirzabekov, A. (2002) Analysis of SNPs and other genomic variations using gel-based chips. *Human Mutation* 19, 343–360.

Kramer, J.W., Klaassen, J.K., Baskin, D.G., Prieur, D.J., Rantanen, N.W., Robinette, J.D., Graber, W.R. and Rashti, L. (1988) Inheritance of diabetes mellitus in Keeshond dogs. *American Journal of Veterinary Research* 49, 428–431.

Kuliev, A. and Verlinsky, Y. (2003) The role of preimplantation genetic diagnosis in women of advanced reproductive age. *Current Opinion in Obstetrics and Gynecology* 15, 233–238.

Le Roux, M.G., Pascal, O., Andre, M.T., Herbert, O., David, A. and Moisan, J.P. (1992) Non-paternity and genetic counselling. *Lancet* 340, 607.

Lear, T.L. and Layton, G. (2002) Use of zoo-FISH to characterise a reciprocal translocation in a thoroughbred mare: t(1;1 6)(q16;q21.3). *Equine Veterinary Journal* 34, 207–209.

Libby, R.T. and Steel, K.P. (2001) Electroretinographic anomalies in mice with mutations in *Myo7a*, the gene involved in human Usher syndrome type 1B. *Investigative Ophthalmology and Visual Science* 42, 770–778.

Lin, C.T., Petersen-Jones, S.M. and Sargan, D.R. (1998) Isolation and investigation of canine phosducin as a candidate for canine generalized progressive retinal atrophies. *Experimental Eye Research* 67, 473–480.

Lin, C.T., Gould, D.J., Petersen-Jones, S.M. and Sargan, D.R. (2002) Canine inherited retinal degenerations: update on molecular genetic research and its clinical application. *Journal of Small Animal Practice* 43, 426–432.

Lin, L., Faraco, J., Li, R., Kadotani, H., Rogers, W., Lin, X., Qiu, X., de Jong, P.J., Nishino, S. and Mignot, E. (1999) The sleep disorder canine narcolepsy is caused by a mutation in the hypocretin (orexin) receptor 2 gene. *Cell* 98, 365–376.

Lingaas, F., Comstock, K.E., Kirkness, E.F., Sorensen, A., Aarskaug, T., Hitte, C., Nickerson, M.L., Moe, L., Schmidt, L.S., Thomas, R., Breen, M., Galibert, F., Zbar, B. and Ostrander, E.A. (2003) A mutation in the canine BHD gene is associated with hereditary multifocal renal cystadenocarcinoma and nodular dermatofibrosis in the German Shepherd dog. *Human Molecular Genetics* 12, 3043–3053.

Lipp, H.P., Stagliar-Bozicevic, M., Fischer, M. and Wolfer, D.P. (1998) A 2-year longitudinal study of swimming navigation in mice devoid of the prion protein: no evidence for neurological anomalies or spatial learning impairments. *Behavioural Brain Research* 95, 47–54.

Lou, X.Y., Todhunter, R.J., Lin, M., Lu, Q., Liu, T., Wang, Z., Bliss, S.P., Casella, G., Acland, G.M., Lust, G. and Wu, R. (2003) The extent and distribution of linkage disequilibrium in a multi-hierarchic outbred canine pedigree. *Mammalian Genome* 14, 555–564.

Macintyre, S. and Sooman, A. (1991) Non-paternity and prenatal genetic screening. *Lancet* 338, 869.

Mager, D.L., McQueen, K.L., Wee, V. and Freeman, J.D. (2001) Evolution of natural killer cell receptors: coexistence of functional Ly49 and KIR genes in baboons. *Current Biology* 11, 626–630.

Mansfield, K.G. and Land, E.D. (2002) Cryptorchidism in Florida panthers: prevalence, features, and influence of genetic restoration. *Journal of Wildlife Diseases* 38, 693–698.

McFadden, D.E. and Friedman, J.M. (1997) Chromosome abnormalities in human beings. *Mutation Research* 396, 129–140.

McLaughlin, M.E., Sandberg, M.A., Berson, E.L. and Dryja, T.P. (1993) Recessive mutations in the gene encoding the beta-subunit of rod phosphodiesterase in patients with retinitis pigmentosa. *Nature Genetics* 4, 130–134.

McQueen, K.L., Wilhelm, B.T., Harden, K.D. and Mager, D.L. (2002) Evolution of NK receptors: a single Ly49 and multiple KIR genes in the cow. *European Journal of Immunology* 32, 810–817.

Michell, A.R. (1999) Longevity of British breeds of dog and its relationships with sex, size, cardiovascular variables and disease. *Veterinary Record* 145, 625–629.

Miyake, Y.I., Kawakura, K., Murakami, R.K. and Kaneda, Y. (1994) Minute fragment observed in a bovine pedigree with Robertsonian translocation. *Journal of Heredity* 85, 488–490.

Moe, L. and Lium, B. (1997) Hereditary multifocal renal cystadenocarcinomas and nodular dermatofibrosis in 51 German shepherd dogs. *Journal of Small Animal Practice* 38, 498–505.

Moore, R.C., Mastrangelo, P., Bouzamondo, E., Heinrich, C., Legname, G., Prusiner, S.B., Hood, L., Westaway, D., DeArmond, S.J. and Tremblay, P. (2001) Doppel-induced cerebellar degeneration in transgenic mice. *Proceedings of the National Academy of Sciences USA* 98, 15288–15293.

Morrish, B.C. and Sinclair, A.H. (2002) Vertebrate sex determination: many means to an end. *Reproduction* 124, 447–457.

Nickerson, M.L., Warren, M.B., Toro, J.R., Matrosova, V., Glenn, G., Turner, M.L., Duray, P., Merino, M., Choyke, P., Pavlovich, C.P., Sharma, N., Walther, M., Munroe, D., Hill, R., Maher, E., Greenberg, C.,

Lerman, M.I., Linehan, W.M., Zbar, B. and Schmidt, L.S. (2002) Mutations in a novel gene lead to kidney tumors, lung wall defects, and benign tumors of the hair follicle in patients with the Birt–Hogg–Dube syndrome. *Cancer Cell* 2, 157–164.

Nolan, P.M., Peters, J., Vizor, L., Strivens, M., Washbourne, R., Hough, T., Wells, C., Glenister, P., Thornton, C., Martin, J., Fisher, E., Rogers, D., Hagan, J., Reavill, C., Gray, I., Wood, J., Spurr, N., Browne, M., Rastan, S., Hunter, J. and Brown, S.D. (2000a) Implementation of a large-scale ENU mutagenesis program: towards increasing the mouse mutant resource. *Mammalian Genome* 11, 500–506.

Nolan, P.M., Peters, J., Strivens, M., Rogers, D., Hagan, J., Spurr, N., Gray, I.C., Vizor, L., Brooker, D., Whitehill, E., Washbourne, R., Hough, T., Greenaway, S., Hewitt, M., Liu, X., McCormack, S., Pickford, K., Selley, R., Wells, C., Tymowska-Lalanne, Z. et al. (2000b) Systematic, genome-wide, phenotype-driven mutagenesis programme for gene function studies in the mouse. *Nature Genetics* 25, 440–443.

O'Brien, P.J., Shen, H., Cory, C.R. and Zhang, X. (1993) Use of a DNA-based test for the mutation associated with porcine stress syndrome (malignant hyperthermia) in 10,000 breeding swine. *Journal of the American Veterinary Medical Association* 203, 842–851.

O'Brien, S.J., Roelke, M.E., Marker, L., Newman, A., Winkler, C.A., Meltzer, D., Colly, L., Evermann, J.F., Bush, M. and Wildt, D.E. (1985) Genetic basis for species vulnerability in the cheetah. *Science* 227, 1428–1434.

Pajukanta, P., Allayee, H., Krass, K.L., Kuraishy, A., Soro, A., Lilja, H.E., Mar, R., Taskinen, M.R., Nuotio, I., Laakso, M., Rotter, J.I., de Bruin, T.W., Cantor, R.M., Lusis, A.J. and Peltonen, L. (2003) Combined analysis of genome scans of Dutch and Finnish families reveals a susceptibility locus for high-density lipoprotein cholesterol on chromosome 16q. *American Journal of Human Genetics* 72, 903–917.

Patterson, D.F., Pexieder, T., Schnarr, W.R., Navratil, T. and Alaili, R.A. (1993) Single major-gene defect underlying cardiac conotruncal malformations interferes with myocardial growth during embryonic development: studies in the CTD line of keeshond dogs. *American Journal of Human Genetics* 52, 388–397.

Pinton, A., Ducos, A., Seguela, A., Berland, H.M., Darre, R., Darre, A., Pinton, P., Schmitz, A., Cribiu, E.P. and Yerle, M. (1998) Characterization of reciprocal translocations in pigs using dual-colour chromosome painting and primed *in situ* DNA labelling. *Chromosome Research* 6, 361–366.

Pittler, S.J. and Baehr, W. (1991) Identification of a nonsense mutation in the rod photoreceptor cGMP phosphodiesterase beta-subunit gene of the rd mouse. *Proceedings of the National Academy of Sciences USA* 88, 8322–8326.

Polani, P.E. (1988) The impact of genetics on medicine. Harveian Oration (Royal College of Physicians, 1990).

Pyle, A. and Handel, M.A. (2003) Meiosis in male PL/J mice: a genetic model for gametic aneuploidy. *Molecular Reproduction and Development* 64, 471–481.

Rens, W., Yang, F., O'Brien, P.C., Solanky, N. and Ferguson-Smith, M.A. (2001) A classification efficiency test of spectral karyotyping and multiplex fluorescence *in situ* hybridization: identification of chromosome homologies between *Homo sapiens* and *Hylobates leucogenys*. *Genes Chromosomes Cancer* 31, 65–74.

Repping, S., Skaletsky, H., Brown, L., Van Daalen, S.K., Korver, C.M., Pyntikova, T., Kuroda-Kawaguchi, T., De Vries, J.W., Oates, R.D., Silber, S., Van Der Veen, F., Page, D.C. and Rozen, S. (2003) Polymorphism for a 1.6-Mb deletion of the human Y chromosome persists through balance between recurrent mutation and haploid selection. *Nature Genetics* 35, 247–251.

Robb, R.M. (1979) Cyclic nucleotide phosphodiesterase activity in normal mice and mice with retinal degeneration. *Investigative Ophthalmology and Visual Science* 18, 1097–1100.

Rossi, D., Cozzio, A., Flechsig, E., Klein, M.A., Rulicke, T., Aguzzi, A. and Weissmann, C. (2001) Onset of ataxia and Purkinje cell loss in PrP null mice inversely correlated with Dpl level in brain. *EMBO Journal* 20, 694–702.

Sauer, B. (1993) Manipulation of transgenes by site-specific recombinations: use of Cre recombinase. *Methods in Enzymology* 225, 890–900.

Schelling, C., Pienkowska, A., Arnold, S., Hauser, B. and Switonski, M. (2001) A male to female sex-reversed dog with a reciprocal translocation. *Journal of Reproduction and Fertility, Suppl.* 57, 435–438.

Schmutz, S.M., Moker, J.S., Clark, E.G. and Orr, J.P. (1996) Chromosomal aneuploidy associated with spontaneous abortions and neonatal losses in cattle. *Journal of Veterinary Diagnostic Investigation* 8, 91–95.

Schrock, E., du Manoir, S., Veldman, T., Schoell, B., Wienberg, J., Ferguson-Smith, M.A., Ning, Y., Ledbetter, D.H., Bar-Am, I., Soenksen, D., Garini, Y. and Ried, T. (1996) Multicolor spectral karyotyping of human chromosomes. *Science* 273, 494–497.

Shi, M.M. (2002) Technologies for individual genotyping: detection of genetic polymorphisms in drug targets and disease genes. *American Journal of Pharmacogenomics* 2, 197–205.

Shuster, D.E., Kehrli, M.E., Jr, Ackermann, M.R. and Gilbert, R.O. (1992) Identification and prevalence of a genetic defect that causes leukocyte adhesion deficiency in Holstein cattle. *Proceedings of the National Academy of Sciences USA* 89, 9225–9229.

Silver, L.M. (1995) *Mouse Genetics: Concepts and Applications.* Oxford University Press.

Speicher, M.R., Gwyn Ballard, S. and Ward, D.C. (1996) Karyotyping human chromosomes by combinatorial multi-fluor FISH. *Nature Genetics* 12, 368–375.

Stankiewicz, P. and Lupski, J.R. (2002) Genome architecture, rearrangements and genomic disorders. *Trends in Genetics* 18, 74–82.

Stenson, P.D., Ball, E.V., Mort, M., Phillips, A.D., Shiel, J.A., Thomas, N.S., Abeysinghe, S., Krawczak, M. and Cooper, D.N. (2003) Human Gene Mutation Database (HGMD): 2003 update. *Human Mutation* 21, 577–581.

Suber, M.L., Pittler, S.J., Qin, N., Wright, G.C., Holcombe, V., Lee, R.H., Craft, C.M., Lolley, R.N., Baehr, W. and Hurwitz, R.L. (1993) Irish setter dogs affected with rod/cone dysplasia contain a nonsense mutation in the rod cGMP phosphodiesterase beta-subunit gene. *Proceedings of the National Academy of Sciences USA* 90, 3968–3972.

Tanaka, K., Yamamoto, Y., Amano, T., Yamagata, T., Dang, V.B., Matsuda, Y. and Namikawa, T.A. (2000) Robertsonian translocation, rob(2;28), found in Vietnamese cattle. *Hereditas* 133, 19–23.

Temple, G.F., Chang, J.C. and Kan, Y.W. (1977) Authentic beta-globin mRNA sequences in homozygous beta0-thalassemia. *Proceedings of the National Academy of Sciences USA* 74, 3047–3051.

Tenesa, A., Knott, S.A., Ward, D., Smith, D., Williams, J.L. and Visscher, P.M. (2003) Estimation of linkage disequilibrium in a sample of the United Kingdom dairy cattle population using unphased genotypes. *Journal of Animal Science* 81, 617–623.

Travis, G.H., Brennan, M.B., Danielson, P.E., Kozak, C.A. and Sutcliffe, J.G. (1989) Identification of a photoreceptor-specific mRNA encoded by the gene responsible for retinal degeneration slow (rds). *Nature* 338, 70–73.

Trowsdale, J., Barten, R., Haude, A., Stewart, C.A., Beck, S. and Wilson, M.J. (2001) The genomic context of natural killer receptor extended gene families. *Immunological Reviews* 181, 20–38.

Ubbink, G.J., Van den Ingh, T.S., Yuzbasiyan-Gurkan, V., Teske, E., Van de Broek, J. and Rothuizen, J. (2000) Population dynamics of inherited copper toxicosis in Dutch Bedlington terriers (1977–1997). *Journal of Veterinary Internal Medicine* 14, 172–176.

Valenzuela, D.M., Murphy, A.J., Frendewey, D., Gale, N.W., Economides, A.N., Auerbach, W., Poueymirou, W.T., Adams, N.C., Rojas, J., Yasenchak, J., Chernomorsky, R., Boucher, M., Elsasser, A.L., Esau, L., Zheng, J., Griffiths, J.A., Wang, X., Su, H., Xue, Y., Dominguez, M.G. *et al.* (2003) High-throughput engineering of the mouse genome coupled with high-resolution expression analysis. *Nature Biotechnology* 21, 652–659.

Van Rheede, T., Smolenaars, M.M., Madsen, O. and De Jong, W.W. (2003) Molecular evolution of the Mammalian prion protein. *Molecular Biology and Evolution* 20, 111–121.

Varilo, T., Paunio, T., Parker, A., Perola, M., Meyer, J., Terwilliger, J.D. and Peltonen, L. (2003) The interval of linkage disequilibrium (LD) detected with microsatellite and SNP markers in chromosomes of Finnish populations with different histories. *Human Molecular Genetics* 12, 51–59.

Velculescu, V.E., Zhang, L., Vogelstein, B. and Kinzler, K.W. (1995) Serial analysis of gene expression. *Science* 270, 484–487.

Verlinsky, Y. and Kuliev, A. (2003) Preimplantation diagnosis for aneuploidies using fluorescence *in situ* hybridization or comparative genomic hybridization. *Fertility and Sterility* 80, 869–870.

Viuff, D., Rickords, L., Offenberg, H., Hyttel, P., Avery, B., Greve, T., Olsaker, I., Williams, J.L., Callesen, H. and Thomsen, P.D. (1999) A high proportion of bovine blastocysts produced *in vitro* are mixoploid. *Biology of Reproduction* 60, 1273–1278.

Vozdova, M., Machatkova, M., Kubikova, S., Zudova, D., Jokesova, E. and Rubes, J. (2001) Frequency of aneuploidy in pig oocytes matured *in vitro* and of the corresponding first polar bodies detected by fluorescent *in situ* hybridization. *Theriogenology* 56, 771–776.

Wagner, T.E., Hoppe, P.C., Jollick, J.D., Scholl, D.R., Hodinka, R.L. and Gault, J.B. (1981) Microinjection of a rabbit beta-globin gene into zygotes and its subsequent expression in adult mice and their offspring. *Proceedings of the National Academy of Sciences USA* 78, 6376–6380.

Weatherall, D.J. (1991) *The New Genetics and Clinical Practice.* Oxford Medical Publications, OUP, Oxford.

Weiss, K.M. and Terwilliger, J.D. (2000) How many diseases does it take to map a gene with SNPs? *Nature Genetics* 26, 151–157.

You, Y., Bergstrom, R., Klemm, M., Lederman, B., Nelson, H., Ticknor, C., Jaenisch, R. and Schimenti, J. (1997) Chromosomal deletion complexes in mice by radiation of embryonic stem cells. *Nature Genetics* 15, 285–288.

Zambrowicz, B.P., Friedrich, G.A., Buxton, E.C., Lilleberg, S.L., Person, C. and Sands, A.T. (1998) Disruption and sequence identification of 2,000 genes in mouse embryonic stem cells. *Nature* 392, 608–611.

Zhang, Q., Acland, G.M., Parshall, C.J., Haskell, J., Ray, K. and Aguirre, G.D. (1998) Characterization of canine photoreceptor phosducin cDNA and identification of a sequence variant in dogs with photoreceptor dysplasia. *Gene* 215, 231–239.

Zheng, B., Mills, A.A. and Bradley, A. (1999) A system for rapid generation of coat color-tagged knockouts and defined chromosomal rearrangements in mice. *Nucleic Acids Research* 27, 2354–2360.

19 Pharmacogenomics

W.W. Weber and J.M. Rae

*University of Michigan School of Medicine, 1301b MSRB III, 1150 West Medical
Center Drive, Ann Arbor, MI 48109-0632, USA*

Introduction	478
Impact of Human Genome Project on Personalized Medicine	478
Switching from pharmacogenetics to pharmacogenomics	478
Edging towards personalized medicine	480
Profiting from human genetic diversity	481
Experimental Models	481
Gene targeting	481
QTL analysis	485
Simple eukaryotic organisms as experimental models	485
Genotyping and/or Phenotyping as Tools of Prediction of Individual Drug Response	487
Genotyping for human polymorphisms	487
Genotype predicts phenotype	487
Phenotype in the absence of genotype	489
Phenotyping	490
Genetic Criteria for Stratification of Patient Populations and Individual Assessment of	491
Treatment Risks	
DNA microarrays in the classification of haematopoietic tumours	492
DNA microarrays in the classification of solid tumours	493
Influence of Genomics on Drug Discovery and Development	494
Changing pharmaceutical perspectives	494
Applying genomics to drug development.	495
Engineered proteins and RNA aptamers as therapeutic targets	495
Inactivation mechanisms as therapeutic targets	496
G protein-coupled receptors as therapeutic targets	497
Simple eukaryotic organisms in drug discovery	497
References	498

Introduction

Pharmacogenomics has been much talked about recently at many national and international conferences, and widely publicized in the literature (Evans and Relling, 1999; Meyer, 2000; Evans and Johnson, 2001; Rana *et al.*, 2001; Roses, 2001; Weber and Smith, 2001; Xie *et al.*, 2001; Wolf and Smith, 2002; Evans and McLeod, 2003; Goldstein, 2003; Weinshilboum, 2003) and news media (Wade, 1997, 1999, 2000; Altman, 1998; Carr, 1998; Fisher, 2003; Pollack, 2003a). An abundance of information on the genetic diversity of human drug response has been collected over the past four decades that firmly fixes a person's unique genetic makeup as a major determinant of therapeutic failure and adverse drug reactions. However, much of the current wave of enthusiasm stems from the prospect that pharmacogenomics will feature prominently in drug discovery and drug therapy tailored to the individual.

Prior to the introduction of molecular genetics, investigators could only infer the existence of genetically polymorphic markers predictive of adverse drug reactions from familial Mendelian inheritance patterns of susceptible persons to explain such responses (Weber, 1997). When experimental pharmacogenetics began, such studies were performed to test the relevance of genetic variability to drug toxicity. These studies were retrospective in nature and focused on a single gene. During the 1980s, after biologists had learned how cells read information encoded in the genome, and had invented recombinant and allied techniques for manipulating DNA, adoption of these techniques in many laboratories dramatically quickened the pace of investigation, expanded their scope, and radically altered the way of doing pharmacogenetics. Investigators brought these advances into sharper focus by cloning and sequencing genes predictive of human drug response, expressing and characterizing the proteins they encode, and fixing their chromosomal location on the human genome.

Impact of the Human Genome Project on Personalized Medicine

Physicians know that genetic disorders in response to drugs and other exogenous substances are just as treatable as many other disorders they manage by restricting access to, or by replacing and/or removing the offending substance from the environment of susceptible persons. It follows that the occurrence of adverse responses to these substances among genetically susceptible persons might be averted or markedly curtailed if treatment is guided by risk profiles individualized for susceptibility to specific agents.

Switching from pharmacogenetics to pharmacogenomics

The date-line of discovery of selected human genetic polymorphisms is laid out in Table 19.1. This helps us see how the emergence of experimental pharmacogenetics in the 1950s and its growth throughout the 1980s relates to the switch to pharmacogenomics that began in the 1990s after the adoption of molecular genetics by many laboratories. It is evident that quite a few pharmacogenetic traits were identified and extensively characterized before molecular genetics tools were developed in the 1970s, and that the inheritance of numerous traits was described long before the responsible polymorphisms were described. It can also be seen that the drug-metabolizing enzymes (DMEs) were the first hereditary traits in drug therapy to be intensely scrutinized, and that comparatively few receptor polymorphisms were identified prior to the advent of molecular genetics.

The inception and maturation of the Human Genome Project was largely responsible for the rapid growth in pharmacogenetics, and the switch of pharmacogenetics to pharmacogenomics that occurred in the 1990s. This period also saw a flood of new gene discoveries, new approaches to the discovery and scoring of drug susceptibility loci, and technological advances to facilitate drug discovery and development. Aided by new technologies, investigators could examine simultaneously the structure and function of thousands of genes spanning a large fraction of the genome in many individuals and populations. The plethora of conferences, workshops, and educational programmes that ensued helped everyone keep up with advances, and encouraged scientists in academia and industry to

Table 19.1. Pharmacogenetics dateline.

Polymorphism	Inheritance discovery	Mutation discovery	Clinical effect
Serum cholinesterase	1957–60	1990–92	Succinylcholine sensitivity
G6PD deficiency	1958	1988	Drug-induced haemolysis
Potassium and sodium channelopathies	1957–60	1991–97	Long QT syndromes
N-Acetyltransferase (NAT2)	1959–60	1989–93	Isoniazid-induced neuropathy
Ryanodine receptor	1960–62	1991–97	Malignant hyperthermia, muscle rigidity, jaundice
Glucuronosyl transferase (UGT1A1)	1966–69	1992–?	Gilbert's syndrome; drug toxicity
Vasopressin receptor (AVPR2)	1969	1992	Vasopressin resistance
Aldehyde dehydrogenase (ALDH2)	1969	1988	Aversion to alcohol
Retinoic acid receptor PMLRARα	1970	1991–93	ATRA-responsive promyelocytic leukaemia
CYP2D6	1977	1988–93	Drug toxicity
Thiopurine methyltransferase (TPMT 3A)	1980	1995	Fatal drug-induced myelosuppression
CYP2C19	1984	1993–94	Drug toxicity
Glutathione transferase (GSTM1)	1986	1990	Chemical carcinogenesis
Aldolase B	1986	1988–95	Fructose-induced intolerance
Insulin receptor	1988	1988–93	Insulin-resistant diabetes
FMO III		1997	Fish malodour syndrome
5-Aminolavulinate synthase	1992	1994	Pyridoxine-responsive anaemia
Aldosterone synthase	1992	1992	Glucocorticoid remediable hypertension
CYP2C9 {8292} (Rettie, 1994)	1993–4	1993–4	Warfarin-induced bleeding
5-Lipoxygenase (ALOX5) {9275} (Drazen, 1999)	?	1999	Implicated in antiasthmatic therapy

form partnerships and networks for the betterment of testing pharmacogenomic initiatives.

The chronology (Table 19.1) also explains how genetic polymorphisms of the DMEs (pharmacokinetic variability) were mainly responsible for shaping the development of pharmacogenetics as a discipline. In contrast, the relative lag in progress of receptor pharmacogenetics (pharmacodynamic variability) is explained by the fact that, prior to the 1980s, receptors and transporters could only be defined by pharmacological differences. Pharmacokinetics and pharmacodynamics are the two basic subdivisions of pharmacology with the former pertaining to how the body absorbs, distributes, metabolizes and excretes drugs while the latter pertains to the kind of response they bring about, including both their beneficial and toxic effects. The pharmacological approach proved inadequate for pharma-

cogenetic analysis. The advent of molecular biology far surpassed previous efforts to establish receptor heterogeneity on a molecular plane and enabled well-defined receptor proteins to be produced in amounts sufficient for biochemical and pharmacological characterization. Subsequently, the pharmacogenetic analysis of receptors and transporters began to move ahead steadily.

The sample of traits listed in Table 19.1 is relatively small, but they are representative of dozens of additional polymorphic enzymes, receptors, transporters and other genetic markers of human drug response that have been characterized. In essence, each polymorphic marker is characterized by three types of information: the genetics (inheritance, allelic frequencies, and population variation); the molecular basis (genes responsible and their mutation spectrum); and the medical or biological significance of the trait. Broadly speaking, these markers fall into three groups: (i) those associated with altered transport, distribution and elimination of an agent; (ii) those resulting from adverse effects of an agent; and (iii) those associated with genetic variation of the drug target.

Among the genetic polymorphisms predictive of human drug response listed in Table 19.1, CYP2D6 (Kirchheiner *et al.*, 2001, 2002; Dalen *et al.*, 1997; Sallee *et al.*, 2000), CYP2C19 (Furuta *et al.*, 2002) and CYP2C9 (Taube *et al.*, 2000; Thijssen *et al.*, 2000; Brandolese *et al.*, 2001; Mamiya *et al.*, 2001) were among the first to attract the interest of clinical investigators as markers suitable for probing the concept of personalized medicine. CYP2D6 accounts for the clearance of >20% of all drugs, while CYP2C19 and CYP2C9 each account for clearance of an additional 15–20% of all drugs. CYP2C9 is also especially attractive as a model for testing this concept for the following reasons: (i) CYP2C9 metabolizes many drugs (warfarin, tolbutamide, phenytoin, glipizide, fluoxetine, losartan, celebrex and various non-steroidal anti-inflammatory drugs (NSAIDs) used in treatment of stroke, cardiovascular disease, diabetes, renal disease and hypertension; (ii) each of these drugs can exert severe, even lethal toxicity, and exhibit numerous clinically inhibitory drug interactions; (iii) between 5 and 40% of persons carry at least one copy of two CYP2C9 variants with reduced enzyme activity; (iv) the polymorphism is easy to detect and analyse; of three major allelic variants that occur in human populations, two confer the poor metabolizer phenotype on individuals; and (v) the methodology for detecting all three isoforms of CYP2C9 is established and validated.

The extensive legacy of pharmacogenetic polymorphisms makes a strong case for the structural and functional analyses of genomic diversity aimed at producing a complete catalogue of human genomic diversity (SNPs, deletions, insertions, repeats, and rearrangements). Molecular studies indicate that these polymorphisms are usually associated with only a limited number of important variants, raising the prospect that these genes and related genotypes (haplotypes) may be catalogued relatively quickly for many populations of interest (Stephens, 1999; Drysdale *et al.*, 2000). By

establishing associations between the unique genetic makeup of individuals and their responsiveness to specific drugs, foods and other exogenous substances, we expect to discover better therapies and improve prospects for personalized medicine.

Edging towards personalized medicine

Though implicit in the goals of pharmacogenetics set forth at the outset, the first explicit steps toward establishing the foundations of personalized medicine were initiated just a few years ago (Marshall, 1997a,b). The pharmaceutical industry has traditionally relied on population averages to develop new drugs and despite its capacity to produce an exquisite variety of therapeutics, made no attempt to differentiate patients according to differences in their response to these agents. Physicians were thus forced to decide about types of treatment and drug dosage empirically from information gathered on the basis of population averages instead of individual profiles.

Recently, the enormity of the pharmacoeconomic problems that may be associated with population averages as the approach to drug therapy has been clarified (Phillips *et al.*, 2001). Epidemiological studies reveal many approved drugs work as intended in only a fraction of patients, and may be accompanied by serious adverse reactions. Efficacious responses of patients for many drugs range from about 40 to 75%; the highest percentage, 80%, respond to Cox-2 inhibitors, and the lowest, 25%, respond to cancer chemotherapy (Spear *et al.*, 2001). Moreover, adverse drug reactions rank as a leading cause of hospitalizations and death (Lazarou *et al.*, 1998) and are estimated to cost more than US$177 billion annually in the USA (Ernst and Grizzle, 2001). Cost-effectiveness studies of the potential benefits of pharmacogenomics suggest that some types of diseases and drugs are more suited to individualized therapy than others (Lichter and Kurth, 1997; Phillips *et al.*, 2001). In their evaluation of these issues, Veenstra and Higashi have elucidated five criteria governing the cost-effectiveness of genotyping in conjunction with drug use: (i) avoidance of severe outcomes when genotype information is available; (ii) existing methods for measuring drug response are inadequate; (iii) established genotype–phenotype associations exist; (iv) availability of a rapid, accurate genetic test; and (v) relatively common occurrence of the variant genotype (Veenstra and Higashi 2000). The authors suggest that these criteria are likely to result in clinically useful and economically viable improvements in patient care.

Profiting from human genetic diversity

The hope of personalized medicine is that physicians will be able to use genetic information in a proactive rather than reactive manner in freeing patients of disease. By having a broad individual genotyping capability and an integrated rational therapeutic strategy of disease, and by profiling individuals for specific drugs, this hope should be an attainable objective, but only when all of these ends, including bioinformatic methods for analysing large data sets, are fully developed.

The body of evidence that has accumulated over the last four decades provides a starting point from which risk profiles suitable for medical practice and discovery of better therapies can be developed. Obviously, the construction of such profiles necessitates the collection of genomic data on a fairly large scale. As the Human Genome Project nears completion, such collections should not present any insurmountable technical problem. It is also obvious that the relationship of an individual's genotype (or haplotype) to his/her expressed phenotype for a specific drug must be established to complete the profile. The phenotypic effect of a defective pharmacokinetic mechanism is relatively easily identified, but our present understanding of pharmacodynamic (receptor-related) diversity is still relatively rudimentary. Establishing genotype–phenotype relationships, particularly for responses attributable to, or involving a major component of pharmacodynamic diversity, represents a major challenge to the construction of pharmacogenetic profiles.

Despite the availability of an abundance of pharmacogenetic evidence, its utilization in medical practice has been extremely limited, even when the relationship of genotype (or haplotype) to phenotype is reasonably firm (Tucker, 2000; Weber, 2001; Weber *et al.*,

2003). There appear to be several reasons for the lack of carry over: (i) many early pharmacogenetic studies did not assess clinically relevant end-points; (ii) objective criteria appropriate for assessing a clinically relevant outcome were lacking in those studies; (iii) the statistical power of early studies was low; (iv) most studies have failed to evaluate the effects of modifying genes; (v) suitable methodology for high-throughput phenotyping in a clinical setting is not available; and (vi) knowledge of physiological and pathophysiological factors that affect regulation of genetic markers of human drug response is deficient.

Experimental Models

Human studies of pharmacogenetic traits are often supplemented by studies in experimental models. Convincing models can aid in dissecting the genetic basis of the trait, identifying responsible mechanisms, and assessing its biological significance under experimental conditions that cannot be met in human experimentation for ethical or methodological reasons. Responders of known genotypes and phenotypes to a toxic chemical can be examined under carefully controlled conditions that reveal the pharmacological and toxicological consequences of a given trait. Even models that may be unsuitable for assessing new drug therapy may yet be excellent for elucidating the molecular basis and physiological mechanisms of human traits. Such studies can turn our thinking towards previously unsuspected pathways and mechanisms, or direct us to the acquisition of information that advances our understanding of the human condition. Of course, as the ultimate goal of any pharmacogenetic study is to further understanding of the human trait, the applicability of model findings to humans must be evaluated.

Gene targeting

Inbred mouse strains as well as recombinant, congenic and recombinant congenic inbred strains are favoured models for pharmacogenetic analysis (Weber, 1997). Shortly after recombinant techniques for manipulating DNA were introduced, investigators found that native chromosomal genes could be targeted and modified with exogenous DNA in a predictable manner (Capecchi, 2001; Smithies, 2001). Since then, 'gene-targeting' has been used to generate mice with 'knockout' alleles, over-expressed alleles, and 'humanized' transgenic mice, i.e. mice that have been modified by replacing a native mouse gene with its human counterpart. These techniques can be used to modify the pattern or property of a given gene including its transcription, mRNA, development, or the capacity of its products to interact with the products of other genes. Additional strategies have made possible the creation of mice with mutations that can be targeted to specific tissues (Kuhn *et al.*, 1995) and at specific developmental stages (Furth *et al.*, 1994).

Many experimental models have been created by gene targeting. Some concrete examples that are of pharmacogenomics interest are listed in Table 19.2. For instance, using a mapping study suggested the secretory phospholipase gene, *Pla2g2a*, located on chromosome 4 is a potential candidate for *MomI*, a strong modifier locus found in the same region of that chromosome (Cormier *et al.*, 1997). To test this hypothesis more directly, a transgenic mouse carrying a functional overexpressed

Pla2g2a allele was constructed on a B6 background. This transgene caused a reduction in tumour multiplicity and size comparable with that conferred by a single copy of *MomI*, indicating that this phospholipase can provide active resistance to intestinal tumorigenesis (Cormier *et al.*, 1997). The effect of the number of CF alleles on cholera toxin-induced intestinal secretion was examined in a knockout mouse model (Gabriel *et al.*, 1994). Homozygous knockout mice did not secrete any fluid in response to cholera toxin, while heterozygote mice expressed 50% of the CFTR protein and secreted 50% of the normal fluid and chloride ion in response to cholera toxin. This correlation suggests that CF might possess a selective advantage of resistance to cholera. PONI-knockout mice were produced by targeted disruption of exon 1 of the *PONI* gene. Results with these PONI null mice show their ability to inactivate organophosphate poisons was severely compromised, and when fed on a high-fat, high-cholesterol diet were more susceptible to atherosclerosis than their intact litter mates (Shih *et al.*, 1998). In another study, mice carrying a weakly expressed human leptin transgene indicated that dysfunctional regulation of the leptin gene can result in obesity with relatively normal levels of leptin, and that this

Table 19.2. Experimental models created by gene targeting.

Experimental model	System	Gene targeted	Reference
Resistance to intestinal tumorigenesis	Mouse	Phospholipase *Pla2g2a*	Cormier *et al.*, 1997
Resistance to cholera toxin	Mouse	*Cftr*	Gabriel *et al.*, 1994
Susceptibility to organophosphate toxicity; and atherosclerosis	Mouse	*Pon1*	Shih *et al.*, 1998
Obesity	Mouse	*Lep*	Ioffe and Moon, 1998
Long QT3	Mouse	*Scn 5a*	Nuyens *et al.*, 2001
Suppression of acute promyelocytic leukaemia	Mouse	PML$^{-/-}$ mutants crossed with human cathepsin G (hCG)-PMLRARa	Rego *et al.*, 2001
Deficiency of GABA degradation	Mouse	*Aldh5a1*	Hogema *et al.*, 2001
Organic cation transporter (Oct) defect	Mouse	*Oct1*	Jonker *et al.*, 2001; Wang *et al.*, 2002a
Acetylcholinesterase knockout	Mouse	*Ache*	Xie *et al.*, 2000b
Pregnenolone xenobiotic receptor	Mouse	*PXR*	Xie *et al.*, 2000a
α_{2C}-Adrenergic receptor defect	CHO cells	Human Del322–325 receptor variant	Small *et al.*, 2000
Tumorigenic cell line	Human DLD-1 cells	α_{2C}ARD$_e$\322–325	Torrance *et al.*, 2001

form of obesity is responsive to leptin treatment (Ioffe and Moon, 1998). Mutant mice heterozygous for a knock-in KPQ-deletion ($SCN5A^{\Delta/+}$) in the cardiac sodium channel show the features of LQT3 and spontaneously develop ventricular arrhythmias. Adrenergic agonists suppressed arrhythmias upon premature stimulation suggesting this mouse model may be useful for development of new treatments for the LQT3 syndrome (Nuyens et al., 2001).

In another study, a mouse model was constructed to determine whether promyelocytic leukaemia protein acts as a tumour suppressor. This model was constructed by crossing PML$^{-/-}$ mice with human cathepsin G (hcG)-PMLRARα transgenic mice. In this model, progressive reduction of the dose of PML resulted in a dramatic increase in the incidence of leukaemia, and an acceleration of leukaemia onset in PMLRARα transgenic mice. These results demonstrate PML acted as a tumour suppressor rendering cells resistant to proapoptotic and differentiating stimuli (Rego et al., 2001). Succinic semialdehyde dehydrogenase (ALDH5A1, encoding SSADH) deficiency is a defect in GABA degradation that manifests itself as 4-hydroxybutyric (γ-hydroxybutyric acid, GHB) aciduria. Aldh5a1-deficient mice constructed by gene targeting displayed ataxia and developed generalized seizures rapidly leading to death (Hogema et al., 2001). Therapeutic intervention with phenobarbital or phenytoin was ineffective whereas intervention with vigabatrin, or the GABA-receptor antagonist CGP 35348 prevented convulsions and significantly enhanced survival in mutant mice. This model may provide insight into pathological mechanisms of SSADH deficiency and may have therapeutic relevance for the human condition.

The effect of knocking out the organic cation transporter gene (Oct1) was explored in another mouse model (Jonker et al., 2001). Results show that Oct1 in itself is not essential for normal health and fertility of mice, but that it has an important role in the uptake and excretion of several organic cationic drugs and toxins by the liver and intestine (Wang et al., 2002a). This model may contribute to our understanding of mechanisms of drug transport and elimination, and provide insights for prevention of adverse effects of these substances. The acetylcholinesterase knockout mouse was constructed to explore the role of acetylcholinesterase in neural development (Xie et al., 2000b). Observations on these mice show that nullizygous mice were born alive and survived up to 21 days, but that physical development was delayed. The generally high levels of butyrylcholinesterase in tissues of nullizygous mice, including the motor-end plate, and additional observations, suggest that butyrylcholinesterase plays an essential role in these animals. To examine the significance of the pregnenolone xenobiotic nuclear receptor on the disposition of drugs subject to metabolism by CYP3A enzymes, transgenic mice containing a humanized form of the PXR mouse receptor were generated (Xie et al., 2000a). These mice were responsive to human-specific inducers of CYP3A such as the antibiotic rifampin. The exclusive profile of CYP3A inducibility exhibited by these mice suggests their potential usage in pharmacological studies and drug development.

The mouse has been used as the biological system of choice for the vast majority of experimental models in pharmacogenomics, but the last two entries in Table 19.2 use alternative systems for creating experimental models. The first of these utilizes Chinese hamster ovary (CHO) cells permanently transfected with constructs encoding the human wild-type α2C-adrenergic receptor or with a polymorphic deletion variant (Del322–325) of the wild-type α_{2C}AR (Small et al., 2000). The polymorphic α_{2C}AR variant lacks a sequence of four amino acids (Gly-Ala-Gly-Pro) in the third intracellular loop of the receptor that confers impaired coupling to multiple effectors. In this experimental model, the polymorphic receptor displayed markedly depressed coupling to Gi, significantly inhibiting adenylyl cyclase activity by epinephrine and norepinephrine as well as with synthetic agonists. The α_{2C}AR subtype has been identified in normal individuals, but the Del322–325 receptor variant is about ten times more frequent in African-Americans than in Caucasians (allele frequencies 0.381 vs. 0.040). Given these ethnogeographic observations, the experimental findings with this model suggest the α_{2C}AR locus may partially explain the individual variation in cardiovascular and central nervous system pathophysiology.

The last entry in Table 19.2 utilizes isogenic human cancer cell lines as an experimental model to identify compounds with gene-selective properties for high-throughput screening and drug discovery (Torrance et al., 2001). Rational screening for drugs with specificity towards cancers has been hindered because normal cells corresponding to cell types represented by common tumours are generally unavailable or do not exhibit growth properties comparable with those of tumours. To address this issue, Torrance et al. developed a drug screening strategy that exploits human cancer cells with endogenous alterations of specific genes (Torrance et al., 2001). The strategy is closely analogous to synthetic lethality screens in yeast wherein genes that elicit genotype-specific cell death are identified. In the present strategy, new compounds rather than genes are sought for genotype-specific effects. As shown in Fig. 19.1, a yellow fluorescent protein expression vector (shown as grey) was introduced into the colon cancer cell line DLD-1, and a blue fluorescent protein expression vector (shown as solid black) was introduced into an isogenic derivative in which the mutant K-ras allele had been deleted. Co-culture of the two cell lines allowed screening for compounds with selective toxicity towards the mutant Ras genotype. This strategy provides an approach to mine for therapeutic agents targeted to specific genetic alterations responsible for carcinogenesis.

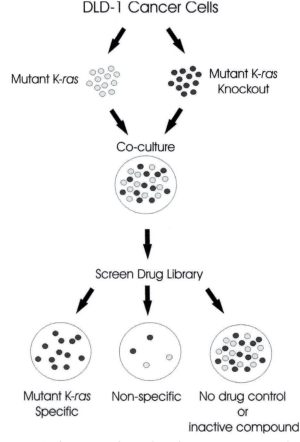

Fig. 19.1. Co-culture screening for compounds specific to the mutant K-ras gene (adapted from Torrance et al., 2001). The colon cancer cell line DLD-1 is transfected with a yellow fluorescent protein expression vector and the isogenic DLD-1 mutant K-ras knockout cells transfected with a blue fluorescent protein expression vector. The cells are co-cultured and screened with a drug library. Compounds specific to the mutant K-ras gene exhibit toxicity only towards cells with yellow fluorescent protein.

QTL analysis

Quantitative trait locus (QTL) mapping is another powerful analytical tool for genetic analysis, particularly for the dissection of complex genetic traits. This technique is aimed at mapping genes affecting quantitative or continuously distributed traits to broad chromosomal regions (Doerge, 2002). The mouse and rat are particularly well suited to systematic QTL mapping because of the availability of inbred lines and large families. Accurate construction of such maps requires multipoint linkage analysis of particular pedigrees, and computer packages specifically designed for this purpose have been described. However, the difficulty of QTL analysis is much greater than for a Mendelian disorder because the responsible genomic intervals are much greater, more difficult to define and the responsible variant more subtle to detect.

Examples of QTL mapping of pharmacogenomic interest include studies of loci controlling susceptibility to hypertension in the stroke-prone spontaneously hypertensive rat (Jacob *et al.*, 1991), susceptibility to intestinal neoplasia (Dietrich *et al.*, 1993), morphine preference (Berrettini *et al.*, 1994), airway hyper-responsiveness (De Sanctis *et al.*, 1999), morphine antinociceptive sensitivity (Hain and Belknap, 1999), and susceptibility to butylated hydroxytoluene-induced lung tumour production and pulmonary inflammation (Malkinson *et al.*, 2002).

Only a few of approximately 1000 QTLs have been identified at the molecular level because of the difficulties mentioned above. The combination of multiple perspectives on genome sequence, variation and function will probably be required to reveal molecular mechanisms of phenotypic variation (Waterston *et al.*, 2002). QTL identification will be advanced if genetic mapping can be combined with genomic sequence, gene expression array, and proteomic data. Conception and design of QTL experiments may be assisted if blocks of ancestral identity among mouse strains can be identified by high-density SNP maps and correlated with phenotypes. Also, testing of candidate genes will be facilitated through the construction of transgenic mice (Cormier *et al.*, 1997). The

availability of the complete genomic mouse sequence should enhance chances for success (Waterston *et al.*, 2002).

Simple eukaryotic organisms as experimental models

Human and mouse genomes have dominated the literature of genome science, but the annotation of the sequences of several non-mammalian eukaryotic organisms (yeast, fruit fly, nematode, zebrafish) provides additional opportunities to use comparative genomics for pharmacogenomics analysis and therapeutics (Hariharan and Haber, 2003).

Successful use of simpler organisms in medical research implies that important biological processes have remained essentially unchanged throughout evolution and that these processes are easier to dissect in these models. Additional advantages of these models for genetic studies are their short generation time and the fact that mutants responsible for specific phenotypes can be generated efficiently. Apart from the conservation of genomic sequences across species, the usefulness of a particular model is dictated primarily by its suitability for the study of specific cellular pathways. For example, yeast cells are particularly well-suited as a model for studies of cell cycle events and the effects of mutations on cell division; fly embryos for understanding genes that regulate organization of tissues and differentiation of cells; and nematodes for investigating the developmental fate of individual cells and for apoptosis (Hariharan and Haber, 2003). The zebrafish may greatly facilitate the study of genotype–phenotype relationships because the zebrafish genome is thought to contain a counterpart for almost every disease-causing gene in humans, and because the study of the consequences of aberrant gene expression can be readily studied at the level of the whole organism.

Already, clues to several human genetic defects of pharmacogenomic interest have been identified by genetic screening of these models. Thus, germline mutations in the human orthologue of the fly gene, *patched,* have been found as somatic mutations in most cases of sporadic basal cell cancer

(Hahn *et al.*, 1996; Johnson *et al.*, 1996). The human orthologue of the zebrafish gene, *ferroportin1,* is found to be mutated in certain cases of autosomal-dominant haemochromatosis (Montosi *et al.*, 2001; Njajou *et al.*, 2001). These models can also provide a means of defining cellular pathways by placing genes within a functional pathway — so-called modifier genes. Modifier genes often function as genes which function in the same pathway as a gene of interest. Modifier screening in flies and nematodes has shown that abnormal expansion of glutamines is causative of various inherited diseases of neurodegeneration such as Huntington's disease and spinocerebellar atrophy (Gusella and MacDonald, 2000). In flies, overexpression of the gene for α-synuclein causes degenerative changes in dopaminergic neurones and abnormalities in movement that may prove to be of interest in connection with Parkinson's disease (Feany and Bender, 2000).

Genetic screening in yeast, flies and nematodes might also provide new powerful approaches to discovery of therapeutic agents. Although genetic screening in these models has not led to any drugs in current use, two agents with therapeutic potential have been identified (Hariharan and Haber, 2003). One,

cyclopamine, is beneficial in treatment of basal cell cancer (Taipale *et al.*, 2000), and the other, sirolimus, may be beneficial in treatment of tuberous sclerosis. Cyclopamine causes prosencephaly in sheep, and a similar abnormality occurs in mice and humans lacking a *hedgehog* gene. In flies, cyclopamine interacts with *patched* and its partner (*smoothened*) to suppress the *hedgehog* signalling cascade. However, *patched* is known to be a target of mutations in the basal cell cancer syndrome raising the possibility that cyclopamine could be useful in treating this cancer (Fig. 19.2). The other example of genetic screening in flies concerns discovery of the potential therapeutic effect of sirolimus, also known as rapamycin, in treatment of the congenital anomaly, tuberous sclerosis. Rapamycin, an immunosuppressive antibiotic, was found to antagonize the function of TOR (target of rapamycin) kinase, a signalling molecule activated by several growth-promoting stimuli (Schmelzle and Hall, 2000). Tuberous sclerosis genes, *TSC1* and *TSC2*, restrict cell growth, while mutations in *TSC1* and *TSC2* cause excessive TOR kinase and augment cell growth. These observations provide a rationale for possible use of rapamycin in treatment of tuberous sclerosis (Kwiatkowski *et al.*, 2002).

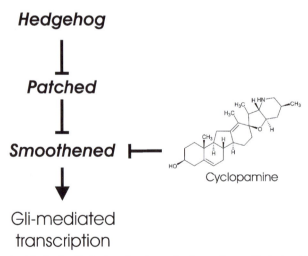

Fig. 19.2. Cyclopamine blocks the *Hedgehog* signal transduction pathway. *Hedgehog* blocks the inhibitory effects of *Patched* on *Smoothened* leading to Gli-mediated transcription, and the loss of *Patched* activity in tumours can lead to constitutive activation of *Smoothened*. Cyclopamine inhibits the *Hedgehog* pathway by blocking the activity of *Smoothened*.

Genotyping and/or Phenotyping as Tools to Predict Individual Drug Response

The completion of the first phase of the Human Genome Project (Lander *et al.*, 2001; Venter *et al.*, 2001) has ushered in a new era in biomedical research. One of the most promising and immediate impacts will be in the field of pharmacogenomics allowing a more rational approach to treatment by tailoring type of drug, dose and administration schedule to an individual's genotypic and phenotypic profile.

Genotyping for human polymorphisms

Many analytical techniques have been devised for mutation analysis enabling biologists to screen and score for molecular variations in genes, molecules and pathways relevant to human disease, including variations of pharmacogenomic interest (Weber and Cronin, 2000). All of these techniques are direct outgrowths of the molecular hybridization rules defined by Watson–Crick base pairing and have their origins in Southern's method (Southern, 1975), combining the specificity of restriction endonucleases with gel electrophoresis to probe genomic DNA for its sequence composition. The invention of the polymerase chain reaction (PCR) in the 1980s (Mullis and Faloona, 1987) permitted specific regions of DNA to be amplified enabling smaller and smaller quantities of DNA to be analysed more rapidly. Combining PCR with restriction endonuclease digestion provided a relatively simple and affordable way to determine clinically relevant polymorphisms or SNPs. This method, referred to as PCR–restriction fragment length polymorphism analysis (PCR-RFLP), is well within the capabilities of many laboratories and has been used in the majority of pharmacogenomic studies to date (shown in Fig. 19.3). A major disadvantage of this method is that it is not amenable to high-throughput screening due to its complexity and cost, and has limited potential in personalized medicine. The development of accurate high speed, nucleic acid sequencing technologies has made possible reliable and affordable high-throughput genotyping. The use of allele-spe-

cific probes or PCR with real-time fluorescence detection (Martin *et al.*, 2000) and primer extension assays using mass spectrometric detection (Ross *et al.*, 1998) are among recent advances. The specificity of the biochemistry combined with hybridization has gone far to improve the specificity of genotyping; however, each of these methods has the limitation of being able to test for only a single allele at a time. Multiplex genotyping methods allow the simultaneous analysis of multiple SNPs which greatly reduces the cost and time in pharmacogenomic studies and paves the way to personalized medicine. For example, Liljedahl *et al.* used a microarray-based genotyping system combined with primary extension to genotype hypertension patients for 74 SNPs in 25 genes to determine a genetic profile for response to antihypertensive drug treatment (Liljedahl *et al.*, 2003).

Genotype predicts phenotype

As mentioned above and in Table 19.1, there are many specific examples of genotypic profiles being used to predict a particular phenotype. Of these, *CYP2D6* polymorphisms rank among the highest in clinical relevance because they define three separate phenotypes (poor, extensive and ultrarapid metabolizers) that control the elimination of at least 25% of all prescribed drugs and a higher proportion of drugs that are most commonly prescribed. The expression and activity of *CYP2D6* are exclusively regulated at the genetic level (Zanger *et al.*, 2001; Bertilsson *et al.*, 2002) because this enzyme is not inducible and it has become the most extensively characterized polymorphic enzyme to date. More than 75 *CYP2D6* polymorphisms have been identified, although the most common six *CYP2D6* alleles appear to account for 95–99% of the detected phenotypes (Ingelman-Sundberg, 2001). SNPs that are known to change *CYP2D6* activity result in at least a tenfold decrease in the conversion of codeine to morphine (Caraco *et al.*, 1999), a sevenfold decrease in the clearance of desipramine (Brosen *et al.*, 1993), and up to as high as a 17-fold increase in the metabolic capacity of patients with duplicated alleles (Dalen *et al.*, 1998). Routine testing of

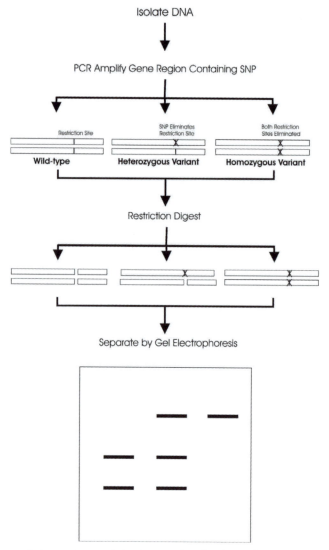

Fig. 19.3. Polymerase chain reaction–restriction fragment length polymorphism (PCR-RFLP) genotyping assay. Patient's genomic DNA is isolated and the genetic region containing the single nucleotide polymorphism (SNP) is amplified (PCR), the fragments are digested with a restriction endonuclease and separated by gel electrophoresis. The SNP eliminates the restriction site and the patient's genotype can be determined from the pattern of bands on the gel.

CYP2D6 genotypes has not yet been incorporated into clinical practice but it has made a significant contribution to oncology where the first examples of pharmacogenomic testing have been integrated into clinical practice.

The safe and efficacious administration of anticancer chemotherapy is one of the most challenging therapeutic specialties of pharma-

cology (Relling and Dervieux, 2001). Currently available chemotherapeutic drugs exhibit substantial variability in their pharmacokinetics and pharmacodynamics. These risks of variability are compounded by the narrow therapeutic range of these agents and the dangerous interactions that may occur on co-administration of other drugs. As a result, dose adjustment based on patient genotype is a logical extension of traditional oncological practice to monitor and adjust doses for individual patients based on other considerations.

A patient's genotype is important to safe and effective chemotherapy treatment as has been demonstrated in therapy with 6-mercaptopurine and 5-fluorouracil (Relling and Dervieux, 2001). The antimetabolite 6-mercaptopurine is used extensively in treatment of acute lymphoblastic leukaemia and its clearance is largely dependent on the polymorphic enzyme thiopurine methyltransferase (TPMT). Approximately one patient in 300 is functionally deficient for this enzyme and requires dose adjustments of about one-tenth of the standard to avoid serious acute and delayed intolerance to 6-mercaptopurine. Since administration of standard doses of 6-mercaptopurine to a TPMT-deficient person could result in severe and potentially fatal bone marrow suppression, genetic testing for this gene has now been incorporated into clinical practice (Evans et al., 2001). Similar considerations may apply to administration of 6-thioguanine and azathioprine, two other drugs whose responses are subject to TPMT deficiency. Patients harbouring certain polymorphisms in the gene for dihydropyrimidine dehydrogenase (DPD) are at risk of toxicity from standard doses of 5-fluorouracil (Wei et al., 1996). It is estimated that 1–3% of cancer patients are DPD deficient (Milano and Etienne, 1994) and when exposed to prolonged therapy with standard doses of 5-fluorouracil exhibit an altered pattern of drug metabolites and are susceptible to severe drug-related toxicity. As 5-fluorouracil occupies a pivotal role in the treatment of cancers of the gastrointestinal tract, breast, and head and neck, DPD deficiency is an important genetic marker.

Many of the most widely used chemotherapeutic agents including the anthracyclines, epipodophyllotoxins, vinca alkaloids, tamoxifen and taxanes as well as cyclophosphamide and ifosfamide (Kivisto et al., 1995; Iyer and Ratain, 1998) are metabolized by polymorphic cytochrome P450 DMEs. It is, therefore, quite possible that such polymorphisms might affect the response and toxicity profiles of patients treated with these agents. To date, in stark contrast to TPMT and DPD, studies are lacking to establish clear associations between CYP450 genotypes and response to these therapies.

Phenotype in the absence of genotype

While there are a few examples of genetic markers predictive of adverse drug response, for the vast majority of drugs no clear genotype–phenotype relationships are known. Originally, the lack of such relationships was believed to be due to insufficient information with clinical outcomes, but efforts to sequence entire candidate genes in hundreds of patients in the hopes of finding an allele that can be associated with a particular phenotype have not fulfilled that promise. There is no better example to illustrate this point than the CYP3A enzymes. Compared with other DMEs, the activity of these enzymes may have the largest impact on the inter-individual differences in the pharmacokinetics of drugs. The enzymes are predominantly expressed in the liver (Shimada et al., 1994) but also expressed in the gastrointestinal tract (Kolars et al., 1992). They possess a notoriously prolific and promiscuous metabolic activity that can oxidize a very wide range of endogenous and medicinal substrates. Furthermore, the expression and activity of CYP3A genes are highly variable with a 5- to 20-fold variability in drug clearance between individuals (Wilkinson, 1996), resulting in prescribing difficulties for up to 50% of all medicines (Benet et al., 1996). The fact that the genetic component of interindividual variability in CYP3A activity is between 60 and 90% (Ozdemir et al., 2000) may account in part for these findings.

Human CYP3A constitutes a family of enzymes with four genes: cytochromes P450 3A4, 3A5, 3A7 and 3A43 (Westlind et al., 2001). Although each may be relevant to drug disposition in adults, early studies focused on the CYP3A4 because it is the form predominantly expressed in human liver. Rebbeck et al. reported that a single SNP in the promoter of CYP3A4 resulted in higher clinical stage and grade prostate tumours (Rebbeck et al., 1998). One study showed this polymorphism was correlated with decreased risk for treatment-related leukaemia in patients treated with epipodophyllotoxins presumably as a result of altered metabolism (Felix et al., 1998) but another study indicated it resulted in only a small change in hepatic clearance when erythromycin or nifedipine were used as probes (Ball et al., 1999). These association studies set the bases for others to directly test the ability of this polymorphism to predict CYP3A4 activity. CYP3A5 has been considered less important than CYP3A4 because of its limited tissue distribution expression at lower levels, and a narrower range of substrate selectivity than CYP3A4 (Wrighton et al., 1990). However, the importance of this enzyme may have been underestimated as suggested by Kuehl et al. who, although they confirmed the limited expression of CYP3A5 in approximately a third of Caucasian patients, showed that up to 50% of African-Americans expressed the enzyme (Kuehl et al., 2001). Furthermore, individuals who express the CYP3A5 enzyme, do so at a level comparable with that of CYP3A4, suggesting that CYP3A5 may play a significant role in aggregate CYP3A activity. Kuehl et al. also identified a relatively frequent CYP3A5 genotype that correlated with activity: a SNP in the third intron that creates an aberrant splice site and results in a truncated, non-functional protein. Analysis of genomic DNA from a large number of patients with known drug metabolism phenotypes by these authors showed that CYP3A5 accounts for at least 50% of the total CYP3A protein in patients with at least one functional CYP3A5 allele. This study suggests that testing for this allele may help predict aggregate CYP3A activity, but its role will probably be minor.

Despite extensive study including numerous sequencing studies, a genetic test to predict aggregate CYP3A activity has eluded investigators (Watkins, 1994; Ball et al., 1999; von Moltke et al., 2000). The high homogeneity of the CYP3A genes and the existence of two pseudogenes (CYP3AP1 and CYP3AP2) has complicated efforts to demonstrate the presence of any CYP3A genetic variant that results in a clinically important change in activity (Eiselt et al., 2001). The absence of a clear genetic test has resulted in attempts to measure the phenotype directly.

Phenotyping

Measurement of CYP3A phenotypes could be done by studying the pharmacokinetics of a particular drug which exhibits high inter- or intra-patient CYP3A-dependent variability. However, collection and analysis of a series of blood or urine samples is unrealistic for most drugs in clinical practice. In 1989 Watkins et al. developed a practical way of estimating the activity of CYP3A . The test involves the intravenous injection of a radiolabelled [^{14}C]N-methyl erythromycin (Shou and Yang, 1990b). The rate of erythromycin, which is an indicator of CYP3A activity, can be determined by measuring the amount of $^{14}CO_2$ emitted in the breath. Data showing that patients pretreated with known CYP3A inducers dexamethasone or rifampin, or the CYP3A inhibitor triacetyloleandomycin exhibited increased and decreased expiratory $^{14}CO_2$ levels, respectively (Shou and Yang, 1990a) supported this concept. It was concluded that the 'erythromycin breath test' (EBT) can provide a reliable measure of CYP3A activity. Its rapidity, relative non-invasiveness, and the requirement for only a single measurement recommended its routine use for drug dosing adjustments. Hirth et al. used the EBT for dose adjustments to the anticancer drug docetaxel (Hirth et al., 2000) to estimate drug clearance and prevent over- and/or under-dosing. Unfortunately, the requirement for i.v. administration and radioactivity are drawbacks to the use of this test. So far, attempts to synthesize an orally available form of the drug have been unsuccessful (Paine et al., 2002). Additionally, the accuracy of the test in predicting clearance of orally administered CYP3A substrates (Kinirons et al., 1993) has been questioned.

The ability of the EBT to predict docetaxel clearance results from the fact that docetaxel is almost exclusively metabolized by CYP3A4 (Shou et al., 1998). In contrast, many clinically used drugs have more than one elimination pathway and, therefore, decreased metabolic activity resulting from genetic polymorphisms or enzymatic inhibition can compensate for other enzymes. In addition, the precise metabolic pathway for many drugs is not known. To get a more accurate determination of a patient's metabolic activity, investigators have administered cocktails of isoform-specific probe drugs and assayed for their pharmacokinetic parameters. For example, Dierks *et al.* (2001) showed that a cocktail of probe drugs, including midazolam, bufuralol, diclofenac, ethoxyresorufin, S-mephenytion, coumarin and paclitaxel can be used to simultaneously monitor the activity of seven clinically relevant cytochrome P450 enzymes (CYP3A4, CYP2D6, CYP2C9, CYP1A2, CYP2C19, CYP2A6 and CYP2C8) (Dierks *et al.*, 2001). Using fast gradient liquid chromatography tandem mass spectrometry to measure each probe drug's metabolism, this *in vitro* study could accurately predict the activity of the respective metabolic enzymes. While this study proved that a cocktail strategy could be used to predict drug metabolism, its clinical application has proven difficult. One reason is that some probe drugs cannot be administered to patients for ethical and safety reasons. For example, the anticancer drug paclitaxel is a suitable probe drug to assay for the activity of the polymorphic enzyme CYP2C8, but it has toxic properties that prevent its use *in vivo*. Other investigators have applied a similar concept to measure metabolic phenotype. Christensen *et al.* developed the 'Karolinska' cocktail for phenotyping five human cyctochrome P450 enzymes *in vivo* (Christensen *et al.*, 2003). The cocktail contains caffeine, losartan, omeprazole, debrisoquin and quinine suitable for accessing CYP1A2, CYP2C9, CYP2C19, CYP2D6 and CYP3A4 activities, respectively. The drugs are administered orally and their metabolism determined by assaying blood and urine samples using high performance liquid chromatography. This elegant study showed that by separating debrisoquin from the other cocktail drugs, they could successfully phenotype the patients for these clinically relevant enzymes using only two urinary collections and two blood samplings.

Phenotyping methods have thus far only been used in research settings and adequate methodologies for large-scale, high-throughput phenotyping are lacking. Additionally, there is the need for more adequate specific probe drugs that can better predict the metabolism of other drugs that fall into their category. These tools would greatly facilitate genotype–phenotype association studies as well as provide a means for predicting therapeutic response and toxicity in the absence of clear genetic tests.

Genetic Criteria for Stratification of Patient Populations and Individual Assessment of Treatment Risks

As the density of information derived from efforts to sequence, map and identify human genes increased, so did the demand for analytical tools capable of exploiting this information. In response, scientists developed one of the most significant technological advances of the late 20th century, DNA microarrays (Schena *et al.*, 1995). Their roots can be traced back to 1977 when the northern blot version of Southern's technique (Alwine *et al.*, 1977) established the link between genome sequence and mRNA expression. The power of DNA microarrays as experimental tools derives from the specificity and affinity of complementary base-pairing and allows for the gene-specific expression level determinations in complex mixtures (Brown and Botstein, 1999). Microarrays consist of discrete stretches of DNA sequences either in the form of complementary DNA (cDNA) or short (~25mers) oligonucleotide sequences immobilized or 'spotted' on solid supports, most often glass slides. To determine the relative abundance of genes in a biological system, total RNA is extracted, labelled with fluorescent dyes or radioactive nucleotides, and then hybridized with the DNA microarrays. After hybridization, signal measurements are made for each DNA spot, the intensity of which correlates directly with the expression of each specific gene. Originally, DNA microarrays were produced in small numbers of labs equipped with sophisticated equipment to produce and process the arrays. However, soon after microarray technology was realized, commercial enterprises responded swiftly to make it available to individual investigators (reviewed in 'The Chipping Forecast II', *Nature*, 2002).

Microarray technology has revolutionized the way fundamental biological questions are addressed in the post-genomic era. It is now possible to monitor expression of virtually the entire cellular transcriptome using DNA microarrays commonly referred to as 'chips'. In pharmacogenomics, they are being used to screen drug target candidates, and to score individuals and populations for polymorphisms that play a role in drug susceptibility. They are being applied to diagnose and classify diseases, elucidate developmental and physiological pathways, and track perturbations in gene activities and patterns of gene expression in cancer and other pathological conditions.

One of the first and most significant impacts of microarray technology has been in the field of oncology where it is changing the manner in which cancers are classified (Liu, 2003). This technology is being used to further define specific cancer subsets that are associated with the biological behaviour of tumours. The current morphologically based methods of cancer classification fail to completely account for clinical behaviour, and it is hoped that a 'molecular-based' taxonomy of cancers, based on global gene expression profiles, will provide a more accurate prognosis and better predict response to therapy (Chung *et al.*, 2002). Microarrays are being used to classify tumours according to their gene expression profiles in many types of cancer including breast, brain, colon, gastric, leukaemia, lymphoma, lung, kidney, ovarian and prostate as reviewed by Chung *et al.* (2002). In addition, since drugs are often targeted to particular gene products, monitoring the entire cellular transcriptome of tumours with DNA microarrays provides thorough drug target identification as well as potential candidate genes as new molecular targets for therapy.

DNA microarrays in the classification of haematopoietic tumours

The first applications of DNA microarrays to the diagnosis of cancer were made in leukaemia and lymphomas (Golub *et al.*, 1999; Alizadeh *et al.*, 2000). Samples of these cancers tend to be easily obtained and in relatively pure populations, making them well suited for DNA microarray analysis. Golub *et al.* used DNA microarrays to measure the expression profiles of 11 acute myeloid leukaemias (AMLs) and 27 acute lymphoblastic leukaemias (ALLs) previously classified by histological means (Golub *et al.*, 1999). Their results demonstrated the feasibility of mol-

ecularly classifying tumours based on their unique mRNA expression profiles and showed that tumours could be assigned to known classes (class prediction) based solely on these profiles. In addition, this approach could identify new cancer classes or subgroups within a class (new class discovery) that were otherwise undeterminable by the available histopathological methods. Yeoh *et al.* conducted a more extensive study on 360 ALL patients (Yeoh *et al.*, 2002). Since the intensity of treatment is tailored to each patient's risk of relapse, these investigators set out to determine whether gene expression profiles could better predict risk classifications. Their results not only showed that microarrays could accurately classify ALL tumours into the six previously known leukaemia subtypes, but also identified an additional distinct ALL subgroup based on its unique expression profile. Within each subgroup, the gene expression profile of the tumour could accurately identify those patients that would eventually fail therapy or develop therapy-induced secondary cancers. These results suggest that DNA microarray analysis could contribute to patient treatment decisions. Cheok *et al.* examined the treatment-induced changes in gene expression profiles in ALL blast in response to methotrexate and mercaptopurine given singly or in combination (Cheok *et al.*, 2003). These two antimetabolites, which are widely used in the treatment of ALL, kill tumours by distinctly different cellular mechanisms. Their results showed that the ALL subgroups share common pathways of genomic response to the same treatment and the changes in gene expression are treatment-specific.

Alizadeh *et al.* used the same approach as Golub *et al.* to classify diffuse large B-cell lymphoma (DLBCL), which is the most common lymphoid malignancy in adults (Alizadeh *et al.* 2000). Clinically and morphologically, DLCBL comprises a heterogeneous group of tumours. Approximately 40% of patients will respond to standard therapy but the remainder eventually succumb to their disease. Attempts to delineate these two groups of patients based on morphological analysis have largely failed and the current clinical prognostic models are far from perfect. Using DNA microarrays to characterize the gene expression profiles, Alizadeh *et al.*, identified two molecularly distinct forms of

DLBCL which had gene expression patterns indicative of different stages of B-cell differentiation (Alizadeh *et al.*, 2000). Patients whose tumour had expression profiles indicative of the germinal B-cell lineage had a significantly better overall survival rate. Similar results were shown by Shipp *et al.* who used a statistical analysis called the supervised learning prediction method to identify DLBCL patients responsive to standard chemotherapy (Shipp *et al.*, 2002). Their algorithm classified patients into two groups with strikingly different 5-year overall disease-free survival rates (Shipp *et al.*, 2002). Remarkably, their prognostic classification is based on the expression profile of only 13 genes.

It is clear that DNA microarrays can be used to classify non-solid tumours into distinct prognostic and predictive subgroups with diagnostic and therapeutic implications. Unfortunately, DLBCLs account for less that 10% of malignancies.

Application of microarray technology to other solid tumours has proven more difficult but this has not deterred investigators. A large number of studies using DNA microarrays to refine the classifications of solid tumours have been conducted in the hopes of better predicting clinical outcomes and responses to therapy. These studies are part of a rapidly expanding body of literature discussed in the next section.

DNA microarrays in the classification of solid tumours

The use of DNA microarrays to classify solid tumours by their molecular characteristics is inherently more difficult than that of non-solid tumours because they require surgical removal, they are a heterogeneous population containing normal stroma, endothelial and haematopoietic cells, and their cellular complement exhibits a great deal of heterogeneity consistent with the variability in their invasive and metastatic potential. Nevertheless, DNA microarrays have been used successfully to measure the genome-wide expression profiles in solid cancers including breast, brain, colon, gastric, lung, kidney, ovarian and prostate (Chung *et al.*, 2002). The majority of studies are retrospective in nature. They have

attempted to refine the classification of solid tumours and correlate gene expression profiles with traditional prognostic and predictive indicators as well as clinical outcomes.

Studies of melanoma and breast cancer exemplify some of the contributions of microarray technology to our understanding of these diseases. Primary melanomas have distinctly different clinical outcomes despite having very similar histopathological appearances. They are prime candidates for subset classifications based on DNA microarray measurements. Bittner *et al.* employed microarrays to measure the expression of 6971 unique genes in 31 primary melanomas and identified two biologically distinct groups (Bittner *et al.*, 2000). One subtype expressed genes that are associated with the *in vitro* invasive potential of melanoma cells suggesting that this molecular profile could predict the more aggressive clinical phenotype. Wang *et al.* attempted to confirm this finding in a prospective clinical study of 63 cutaneous melanomas (Wang *et al.*, 2002b). By following the history of the lesions, these investigators were able to correlate the tumour's gene expression profile with clinical outcomes. Their results confirmed the presence of the two distinct subgroups discovered by Bittner *et al.*, but the two subtypes did not appear to have divergent clinical outcomes (Bittner *et al.*, 2000). The study design by Wang *et al.* enabled the discovery of an interesting set of genes that were differentially expressed in pretreatment melanomas that responded to therapy. The results suggest that a melanoma's clinical response to immunotherapy is predetermined by its immunological configuration. A large prospective clinical trial is needed to properly test this hypothesis.

The high incidence of breast cancer in westernized countries and the significant morbidity and mortality associated with this disease has focused the attention of medical researchers and they soon realized that microarray technology could play a significant role in their attempts to improve diagnosis and treatment of this disease. Breast cancer classification has been refined, new subsets of cancer have been discovered, potential therapeutic targets have been identified, and treatment-induced changes in gene expression have been tracked and correlated with clinical outcomes (Chung et al., 2002; Mohr et al., 2002; Liu, 2003; Watters and McLeod, 2003).

Perou's original studies first showed the feasibility and usefulness of DNA microarrays to dissect and classify these cancers (Perou et al., 1999). Other studies using DNA microarrays to classify breast cancers and attempt to correlate the tumours' molecular profile with clinical outcomes soon followed (Perou et al., 2000; Gruvberger et al., 2001; Hedenfalk et al., 2001; Sorlie et al., 2001; West et al., 2001; van de Vijver et al., 2002; van't Veer et al., 2002; Hedenfalk et al., 2003; Huang et al., 2003). The study by Sorlie et al. showed that oestrogen receptor-positive and oestrogen receptor-negative breast cancers have distinctly different mRNA expression patterns (Sorlie et al., 2001), a result confirmed by others (Gruvberger et al., 2001). In addition, Gruvberger et al. showed that breast cancers can be divided into six specific subsets based on their unique molecular profiles and that each subset exhibits significant differences in their overall and relapse-free survival rates (Gruvberger et al., 2001; Sorlie et al., 2001).

The use of gene expression profiles to classify breast cancer has been shown in large clinical studies (van de Vijver et al., 2002; van't Veer et al., 2002). van't Veer et al. used DNA microarrays to define the gene expression profiles of low- and high-risk tumours by analysing a group (n = 78) of young, (<55 years) lymph-node-negative patients and identified 70 differentially expressed genes which accurately predict overall survival (van't Veer et al., 2002). Next they conducted a much more extensive study of 295 young (<53 years) stage I–II breast cancer patients who were both node-positive and node-negative (van de Vijver et al., 2002) and showed that patients whose tumours had a good prognosis signature were largely free of recurrence at the 10-year follow-up (85%) compared with patients whose tumours fell into the poor prognosis category (50%). It was not surprising that the poor prognosis signature strongly correlated with high histological grade and oestrogen receptor negativity both of which have been previously shown to be prognostic indicators. However, finding that the good prognosis and poor prognosis expression profiles did not correlate with axillary lymph node status was unexpected. This finding raises doubt about the long-held belief that the presence of tumour cells in lymph nodes near the breast indicates a more malignant form of breast cancer and therefore, poorer prognosis. Overall, classification of breast cancers according to their gene expression profile appears to outperform more traditional prognosticators of high-risk breast cancer.

Conclusion

The future promises that expression profiling of human tumours using DNA microarrays will become an integral part of cancer diagnosis and treatment. Retrospective studies conducted thus far have been performed for proof-of-concept and to establish optimal statistical approaches needed to analyse large amounts of information generated by DNA microarrays. In addition, these studies have shown that a tumour's molecular profile can accurately predict its tumour type and have discovered new classes or subgroups of cancer that were undeterminable by histological means. The predictive value of DNA microarray technology will only be determined by incorporating it into large prospective clinical trials.

Influence of Genomics on Drug Discovery and Development

Changing pharmaceutical perspectives

Until recently, the role of pharmacogenomics in drug discovery and development was viewed with scepticism by the pharmaceutical industry (Marshall, 1997a). However, opinions have gradually changed and now the elucidation of genetic diversity in human drug response is, in the long run, widely regarded as essential to rational, systematic design and evaluation of new, better therapeutic agents. By providing a better drug at the outset, pharmacogenomics would improve safety and efficacy, and diminish costly drug failures to protect a drug's market against erosion by competing products. It might even be possible to achieve a higher penetration of a selected population with a pharmacogenomically focused drug, and patients could be given a more effective drug dose.

The process of drug discovery and development begins with the selection of compounds exhibiting biological activity that correlate to a relevant drug target. Traditionally, the process starts with preclinical and clinical phases and proceeds sequentially to assess efficacy and toxicity of potential therapeutics before their approval and adoption for medical practice. History tells us that fewer than 500 validated targets have been exploited in the development of medicines currently in use, and that many of those are either agonists or antagonists of receptors, or are agents that inhibit or (less frequently) stimulate enzymes or protein–protein interactions (Drews, 2000).

However, with the maturation of the Human Genome Project coupled with advances in high throughput technologies for genomic and proteomic analyses, a much broader range of biological candidates is available for therapeutic intervention, greatly expanding the horizons for drug discovery (Broder and Venter, 2000). Interestingly, an early analysis indicates that chemically defined families nominally regarded as the major drug target classes (receptors) and select regulatory molecules (8.2%), transferase and oxidoreductase enzymes (4.1%), kinases (2.8%), ion channels (1.3%) and transporters (1.7%)) constitute only about 18% of the putative proteins predicted to be represented in the human genome (Venter *et al.*, 2001). A recent study of strategic trends in the drug industry (Drews, 2003) shows that productivity of the new chemical entities in the drug discovery pipeline of the pharmaceutical industry since 1995 has fallen short of expectations. The proportion of these entities, however, that originated in biotech firms, as a sign of the influence of genomics on drug discovery, has reached 20–25%, and is expected to increase to 50% within the next decade.

Until recently, the study of naturally occurring changes in single gene function in human subjects, or of functional changes of single genes in an appropriate cellular or animal model afforded the primary approaches to drug target validation. These are time-tested, powerful approaches to target validation, but they are labour intensive, slow and very expensive. Completion of the human genome sequence indicates there are some 30,000 genes, give or take a few thousand, in the sequence. By knowing the complement of

genes and taking advantage of advances in high-throughput technology, collection of drug target information on a genomic scale via high-throughput, global gene expression analysis and whole genome functional analysis is now possible (Nicolette and Miller, 2003). For example, the availability of microarray technology and the latest robotic equipment enables investigators to place the entire human gene complement of an individual on one microscope slide, and to study the expression patterns of thousands of genes simultaneously in cells or a tissue of their choice (Barrett and Kawasaki, 2003). Other non-array technologies for analysing gene expression are available (Zhang, 1997; Brenner *et al.*, 2000) but the simplicity and flexibility of microarray technology makes it an attractive option for simultaneous analysis of many genes. In principle, such technology enables researchers to study the expression and function of every gene in the human genome, and to greatly expand the number and diversity of therapeutic targets.

Applying genomics to drug development

By using modern genomic techniques, drug researchers seek to identify new drug targets for drug therapy, and attempt to validate these new targets. In the context of drug research, a validated target should satisfy the following criteria (Drews, 2003): (i) it should consistently lead to phenotypic changes that are in harmony with the desired therapeutic effect; (ii) its effect should be dose-dependent; (iii) the desired phenotypic change should be inducible in one or more relevant animal or experimental model; and (iv) the mechanism by which the target molecule brings about a particular phenotype should be known or determined.

The drug targets described below represent successful or promising applications of genomics to drug discovery. We have also described the extent to which each example satisfies (or fails to satisfy) the criteria listed above.

Engineered proteins and RNA aptamers as therapeutic targets

Engineered antibodies are one of the most visible applications of genomic information to drug discovery. These agents now represent 25–30% of biopharmaceuticals in clinical trials (Glennie and Van de Winkel, 2003; Hudson and Souriau, 2003). In 1997, rituximab (Rituxan®) for treatment of non-Hodgkin's lymphoma, was the first of a class of monoclonal antibodies to be approved for treating cancer (Wade, 1997; Marshall *et al.*, 2003). Rituxan® followed Reopro®, from Centocor Inc., an antibody for the prevention of blood clotting in heart patients approved in 1996 (Fisher, 2003). Rituxan® targets CD20, an antigen that resides on the surface of mature B cells and B-cell tumours. It then recruits the body's natural defences to attack and kill both malignant and normal mature B cells. Trastuzumab (Herceptin®) is another example of this class of biopharmaceuticals with indications for treating breast cancer. A promising report of the effectiveness of Avastatin®, another member of this class of agents, in treating colorectal cancer has recently appeared (Pollack, 2003a,b). Avastatin® works by blocking formation of blood vessels that supply oxygen and nutrients to the tumour. Some 400 antibody products are currently in clinical trial (Gura, 2002), and at least a dozen are currently on the market (Glennie and Van de Winkel, 2003; Marshall *et al.*, 2003).

Several additional engineered protein therapeutics are also currently on the market (Marshall *et al.*, 2003). Included among these are insulin lispro and insulin aspart for treating diabetes, etanercept for rheumatoid arthritis, darbepoetin for anaemia, and pegvasomant for acromegaly.

Finding that cells contain a bevy of RNA snippets was another genomics discovery with surprising therapeutic implications (Zaman *et al.*, 2003). These short segments of RNA — called aptamers — can bind tightly to proteins and block their function in a process known as RNA interference, or RNAi. Researchers are attempting to exploit RNAi to treat disease. This property plus their low molecular weight, stability, ease and low cost of preparation suggest therapeutic aptamers may ultimately be superior to monoclonal antibodies. One aptamer drug that has reached clinical trial has shown the ability to improve vision in some patients afflicted with aged macular degeneration, the leading cause of blindness in the elderly (Pollack, 2003c). Aged macular degeneration is characterized by blood vessel growth in the

back of the eye or leakage of blood vessels that results in total loss of vision. This agent exhibits dual capability in combating this disease by blocking vascular endothelial growth factor (VEGF) and vascular permeability.

Inactivation mechanisms as therapeutic targets

The discovery and development of imatinib (STI571, Gleevec or Glivec®) provides an example of how knowledge of a specific enzyme inactivation mechanism was used to design a therapeutic agent. Imatinib mesylate has specific indications for treating chronic myelogenous leukaemia. Chronic myelogenous leukaemia is a clonal haematological disorder that is characterized by a reciprocal translocation between chromosomes 9 and 22. The consequence of this exchange is the creation of the *BCR-ABL* gene which encodes a tyrosine kinase, a molecular pathogenic event that is linked to proliferation of leukaemic cells in affected patients. Imatinib binds to the inactive conformation of ABL to prevent proliferation of the cancer cells. The first patient was treated with imatinib in 1998, and within 3 years three large clinical trials showed the drug to be safe and effective in all stages of chronic myelogenous leukaemia. In 2001, the FDA approved imatinib and in 2002, it was approved with indications for treating gastrointestinal stromal tumours (Capdeville *et al.*, 2002).

Gleevec is extremely effective in treating chronic myelogenous leukaemia patients and causes remarkably few side effects. Unfortunately, some patients relapse and die because of Gleevec resistance (Gorre and Sawyers, 2002). However, additional inhibitors of BCR-ABL appear to be active against several mutations found in patients who develop Gleevec resistance (La Rosee *et al.*, 2002). Recently, investigators have tested such a compound as a prototype of a new generation of anti-BCR-ABL compounds that appear to be more effective than Gleevec (Huron *et al.*, 2003).

The success of imatinib suggests that strategies that target distinctive inactivation mechanisms provide a compelling approach to rational drug discovery.

G protein-coupled receptors as therapeutic targets

The GPCR family represents one of the most attractive therapeutic targets in the human genome. According to the analysis referred to earlier (Drews, 2000), cell membrane receptors, largely GPCRs, constitute the largest subgroup, with 45% of all therapeutic targets. Currently, 50–60% of all therapeutics in current use target GPCRs. *GPCRs*, represented by more than 600 genes, are one of the largest gene families in the genome (Venter *et al.*, 2001). These receptors conduct or participate in a wide variety of disorders associated with dysfunctional central and peripheral neurotransmission, as well as impaired autocrine, hormonal, and paracrine systems. Traditionally, therapeutic agents intended to modulate GPCR function are agonists or antagonists. GPCRs consist of membrane-bound molecules that are usually classified according to the drug ligand they bind; the major GPCR classes are those that bind bioaminergic, peptide, opioid, protease, chemokine, or fMLP (chemotactic) ligands or substrates. GPCRs that have been extensively studied or are currently of considerable interest include the $\alpha2$ adrenoceptor, dopamine receptor, V2 vasopressin receptor, the chemokine receptor, and the protease receptor.

A significant degree of individual variability is associated with therapeutic responses to GPCR agonists and antagonists. However, perusal of the prodigious GPCR literature indicates that pharmacogenomic studies on GPCRs are relatively scarce, and that the therapeutic relevance of GPCRs is often not well defined (Sadee *et al.*, 2001). Attempts to assess the relevance of specific and variant forms of *GPCR* alleles is often thwarted by a variety of factors: (i) cross-reaction of a single drug with multiple receptors; (ii) poorly defined ligand-binding receptor pockets that can accommodate drugs in different orientations and at alternative domains; (iii) the possibility of multiple receptor conformations with distinct functions; and (iv) multiple signalling pathways engaged by a single receptor.

On the other hand, insights into the nature and significance of GPCR variability that are of interest to drug discovery are advancing. Characterization of GPCR genetic variability

reveals that as much as 60% of this variability is attributed to genetic causes, and functional polymorphisms occur in multiple, potentially critical, genomic regions of GPCR genes. Studies of the $\beta2$ adrenoceptor, for example, suggest that such variability might have evolved through expression, ligand-binding, G-protein coupling or regulation, but the data suggest that variability developed most commonly in the transmembrane-spanning domains that are typical of ligand binding (Small and Liggett, 2003). The authors suggest that drug discoverers should be cognizant of ligand-binding SNPs and seek to delineate such variability early in the discovery process. They also note, however, that false non-synonymous polymorphisms of GPCR genes are frequently reported (68%), and caution against exclusive reliance on databases for selection of candidate GPCR polymorphisms for pharmacogenetic studies or disease associations (Small *et al.*, 2002).

Simple eukaryotic organisms in drug discovery

In addition to models for genetic studies of human disease (see 'Experimental models: simple eukaryotic organisms as experimental models'), the use of simple eukaryotic organisms as tools for discovery of better therapeutics has also been proposed (Marton *et al.*, 1998; Giaever *et al.*, 1999; Hughes, 2000). For instance, Giaever and colleagues have recently explored the possibility of genomic profiling yeast *Saccharomyces cerevisae* for drug sensitivities via induced haploinsufficiency (Giaever *et al.*, 1999). This approach is based on the idea that lowering the dosage of a single gene from two copies to one in diploid yeast yields a heterozygote that is sensitized to any drug that acts on the product of this gene. This 'haploinsufficient' phenotype serves to identify the gene product of the heterozygous locus as the drug target. This observation was exploited in a genomic approach to drug target identification as follows. First, a set of heterozygous yeast strains was constructed carrying deletions in genes encoding drug targets. Next each strain was grown in the presence of sublethal concentrations of the drug that directly targets the protein encoded by the heterozygous locus and analysed for drug sensitivity (e.g. reduced growth rate in the presence of the drug) on high-density oligonucleotide microarrays. A feasibility study on individual heterozygous strains verified six known targets. In each case, the result was highly specific as no sensitivity was exhibited when these strains were tested with other drugs. Additionally, parallel analysis of a mixed culture of 233 strains in the presence of the drug tunicamycin (a well-characterized glycosylation inhibitor) identified the known target and two unknown hypersensitive loci.

Today, drug discovery is largely driven by combinatorial chemistry followed by high-throughput screening of agents against a pre-selected target. The above method is advantageous in that no prior knowledge of the target is required, and only those targets that affect the fitness of the organism will be identified. The discovery that both drug target and hypersensitive loci exhibit drug-induced haploinsufficiency may help elucidate mechanisms underlying heterozygous disease phenotypes as well variable drug toxicities.

References

Alizadeh, A.A., Eisen, M.B., Davis, R.E., Ma, C., Lossos, I.S., Rosenwald, A., Boldrick, J.C., Sabet, H., Tran, T., Yu, X., Powell, J.I., Yang, L., Marti, G.E., Moore, T., Hudson, J., Jr, Lu, L., Lewis, D.B., Tibshirani, R., Sherlock, G., Chan, W.C. et al. (2000) Distinct types of diffuse large B-cell lymphoma identified by gene expression profiling. *Nature* 403, 503–511.

Altman, L.K. (1998) Drug shown to shrink tumours in type of breast cancer by targeting gene defect. *New York Times*, A12.

Alwine, J.C., Kemp, D.J. and Stark, G.R. (1977) Method for detection of specific RNAs in agarose gels by transfer to diazobenzyloxymethyl-paper and hybridization with DNA probes. *Proceedings of the National Academy of Sciences USA* 74, 5350–5354.

Ball, S.E., Scatina, J., Kao, J., Ferron, G.M., Fruncillo, R., Mayer, P., Weinryb, I., Guida, M., Hopkins, P.J., Warner, J. and Hall, J. (1999) Population distribution and effects on drug metabolism of a genetic variant in the 5' promoter region of CYP3A4. *Clinical Pharmacology and Therapeutics* 66, 288–294.

Barrett, J.C. and Kawasaki, E.S. (2003) Microarrays: the use of oligonucleotides and cDNA for the analysis of gene expression. *Drug Discovery Today* 8, 134–141.

Benet, L.Z., Kroetz, D.L. and Sheiner, L.B. (1996) Pharmacokinetics: the dynamics of drug absorption, distribution, and elimination. In: Hardman, J.G. (ed.) *Goodman and Gilman's The Pharmacological Basis of Therapeutics*, 9th edn. McGraw-Hill, New York, pp. 3–27.

Berrettini, W.H., Ferraor, T.N., Alexander, R.C., Buchberg, A.M. and Vogel, W.H. (1994) Quantitative trait loci mapping of three loci controlling morphine preference using inbred mouse strains. *Nature Genetics* 7, 54–58.

Bertilsson, L., Dahl, M.L., Dalen, P. and Al Shurbaji, A. (2002) Molecular genetics of CYP2D6: clinical relevance with focus on psychotropic drugs. *British Journal of Clinical Pharmacology* 53, 111–122.

Bittner, M., Meltzer, P., Chen, Y., Jiang, Y., Seftor, E., Hendrix, M., Radmacher, M., Simon, R., Yakhini, Z., Ben Dor, A., Sampas, N., Dougherty, E., Wang, E., Marincola, F., Gooden, C., Lueders, J., Glatfelter, A., Pollock, P., Carpten, J., Gillanders, E. et al. (2000) Molecular classification of cutaneous malignant melanoma by gene expression profiling. *Nature* 406, 536–540.

Brandolese, R., Scordo, M.G., Spina, E., Gusella, M. and Padrini, R. (2001) Severe phenytoin intoxication in a subject homozygous for CYP2C9*3. *Clinical Pharmacology and Therapeutics* 70, 391–394.

Brenner, S., Williams, S.R., Vermaas, E.H., Storck, T., Moon, K., McCollum, C., Mao, J.I., Luo, S., Kirchner, J.J., Eletr, S., DuBridge, R.B., Burcham, T. and Albrecht, G. (2000) In vitro cloning of complex mixtures of DNA on microbeads: physical separation of differentially expressed cDNAs. *Proceedings of the National Academy of Sciences USA* 97, 1665–1670.

Broder, S. and Venter, J.C. (2000) Sequencing the entire genomes of free-living organisms: the foundation of pharmacology in the new millennium. *Annual Reviews in Pharmacology and Toxicology* 40, 97–132.

Brosen, K., Hansen, J.G., Nielsen, K.K., Sindrup, S.H. and Gram, L.F. (1993) Inhibition by paroxetine of desipramine metabolism in extensive but not in poor metabolizers of sparteine. *European Journal of Clinical Pharmacology* 44, 349–355.

Brown, P.O. and Botstein, D. (1999) Exploring the new world of the genome with DNA microarrays. *Nature Genetics* 21, Supplement 33–37.

Capdeville, R., Buchdunger, E., Zimmermann, J. and Matter, A. (2002) Glivec (STI571, Imatinib), a rarionally developed, targeted anticancer drug', *Nature Reviews Drug Discovery* 1, 493–502.

Capecchi, M. R. (2001) Generating mice with targeted mutations. *Nature Medicine* 7, 1086–1090.

Caraco, Y., Sheller, J. and Wood, A.J. (1999) Impact of ethnic origin and quinidine coadministration on codeine's disposition and pharmacodynamic effects. *Journal of Pharmacology and Experimental Therapeutics* 290, 413–422.

Carr, G. (1998) The alchemists. The Pharmaceutical Industry. *The Economist* [February 21], 3–18.

Cheok, M.H., Yang, W., Pui, C.H., Downing, J.R., Cheng, C., Naeve, C.W., Relling, M.V. and Evans, W.E. (2003) Treatment-specific changes in gene expression discriminate *in vivo* drug response in human leukemia cells. *Nature Genetics* 34, 85–90.

Christensen, M., Andersson, K., Dalen, P., Mirghani, R.A., Muirhead, G.J., Nordmark, A., Tybring, G., Wahlberg, A., Yasar, U. and Bertilsson, L. (2003) The Karolinska cocktail for phenotyping of five human cytochrome P450 enzymes. *Clinical Pharmacology and Therapeutics* 73, 517–528.

Chung, C.H., Bernard, P.S. and Perou, C.M. (2002) Molecular portraits and the family tree of cancer. *Nature Genetics* 32, Suppl, 533–540.

Cormier, R.T., Hong, K.H., Halberg, R.B., Hawkins, T.L., Richardson, P., Mulherkar, R., Dove, W.E. and Lander, E.S. (1997) Secretory phospholipase Pla2g2a confers resistance to intestinal tumorigenesis. *Nature Genetics* 17, 88–91.

Dalen, P., Frengell, C., Dahl, M.L. and Sjoqvist, F. (1997) Quick onset of severe abdominal pain after codeine in an ultrarapid metabolizer of debrisoquine. *Therapeutic Drug Monitoring* 19, 543–544.

Dalen, P., Dahl, M.L., Ruiz, M.L., Nordin, J. and Bertilsson, L. (19980 10-Hydroxylation of nortriptyline in white persons with 0, 1, 2, 3, and 13 functional CYP2D6 genes. *Clinical Pharmacology and Therapeutics* 63, 444–452.

De Sanctis, G.T., Singer, J.B., Jiao, A., Yandava, C.N., Lee, Y.H., Haynes, T.C., Lander, E.S., Beier, D.R. and Drazen, J.M. (1999) Quantative trait locus mapping of airway responsiveness to chromosomes 6 and 7 in inbred mice. *American Journal of Physiology* 277, L1118–L1123.

Dierks, E.A., Stams, K.R., Lim, H.K., Cornelius, G., Zhang, H. and Ball, S.E. (2001) A method for the simultaneous evaluation of the activities of seven major human drug-metabolizing cytochrome P450s using an *in vitro* cocktail of probe substrates and fast gradient liquid chromatography tandem mass spectrometry. *Drug Metabolism and Disposition* 29, 23–29.

Dietrich, W.F., Lander, E.S., Smith, J.S., Moser, A.R., Gould, K. A., Luongo, C., Borenstein, N. and Dove, W. (1993) Genetic identification of MOM-1, a major modifier locus affecting Min-induced intestinal neoplasia in the mouse. *Cell* 75, 631–639.

Doerge, R.W. (2002) Mapping and analysis of quantitative trait loci in experimental populations. *Nature Review Genetics* 3, 43–52.

Drews, J. (2000) Drug discovery: a historical perspective. *Science* 287, 1960–1964.

Drews, J. (2003) Strategic trends in the drug industry. *Drug Discovery Today* 8, 411–419.

Drysdale, C.M., McGraw, D.W., Stack, C.B., Stephens, J.C., Judson, R.S., Nandabalan, K., Arnold, K., Ruano, G. and Liggett, S.B. (2000) Complex promoter and coding region Beta2-adrenergic receptor haplotypes alter receptor expression and predict *in vivo* responsiveness. *Proceedings of the National Academy of Sciences USA* 97, 10483–10488.

Eiselt, R., Domanski, T.L., Zibat, A., Mueller, R., Presecan-Siedel, E., Hustert, E., Zanger, U.M., Brockmoller, J., Klenk, H.P., Meyer, U.A., Khan, K.K., He, Y.A., Halpert, J.R. and Wojnowski, L. (2001) Identification and functional characterization of eight CYP3A4 protein variants. *Pharmacogenetics* 11, 447–458.

Ernst, F.R. and Grizzle, A.J. (2001) Drug-related morbidity and mortality: updating the cost-of-illness model. *Journal of the American Pharmaceutical Association* 41, 192–199.

Evans, W.E. and Johnson, J.A. (2001) Pharmacogenomics: the inherited basis for interindividual differences in drug response. *Annual Reviews in Genomics and Human Genetics* 2, 9–39.

Evans, W.E. and McLeod, H.L. (2003) Pharmacogenomics – drug disposition, drug targets, and side effects. *New England Journal of Medicine* 348, 538–549.

Evans, W.E. and Relling, M.V. (1999) Pharmacogenomics: translating functional genomics into rational therapeutics. *Science* 286, 487–491.

Evans, W.E., Hon, Y.Y., Bomgaars, L., Coutre, S., Holdsworth, M., Janco, R., Kalwinsky, D., Keller, F., Khatib, Z., Margolin, J., Murray, J., Quinn, J., Ravindranath, Y., Ritchey, K., Roberts, W., Rogers, Z.R., Schiff, D., Steuber, C., Tucci, F., Kornegay, N., Krynetski, E.Y. and Relling, M.V. (2001) Preponderance of thiopurine S-methyltransferase deficiency and heterozygosity among patients intolerant to mercaptopurine or azathioprine. *Journal of Clinical Oncology* 19, 2293–2301.

Feany, M.B. and Bender, W.W. (2000) A *Drosophila* model of Parkinson's disease. *Nature* 404, 394–398.

Felix, C.A., Walker, A.H., Lange, B.J., Williams, T.M., Winick, N.J., Cheung, N.K., Lovett, B.D., Nowell, P.C., Blair, I.A. and Rebbeck, T.R. (1998) Association of CYP3A4 genotype with treatment-related leukemia. *Proceedings of the National Academy of Sciences USA* 95, 13176–13181.

Fisher, L.M. (2003) Non-Hodgkin's lymphoma is treatable by a new drug. *New York Times Daily*, pp. A15.

Furth, P.A., St Onge, L., Boger, H., Gruss, P., Gossen, M., Kistner, A., Bujard, H. and Hennighausen, L. (1994) Temporal control of gene expression in transgenic mice by a tetracycline-responsive promoter. *Proceedings of the National Academy of Sciences USA* 91, 9302–9306.

Furuta, T., Shirai, N., Watanabe, F., Honda, S., Takeuchi, K., Iida, T., Sato, Y., Kajimura, M., Futami, H., Takayanagi, S., Yamada, M., Ohashi, K., Ishizaki, T. and Hanai, H. (2002) Effect of cytochrome P4502C19 genotypic differences on cure rates for gastroesophageal reflux disease by lansoprazole. *Clinical Pharmacology and Therapeutics* 72, 453–460.

Gabriel, S.E., Brigman, K.N., Koller, B.H., Boucher, R.C. and Stutts, M.J. (1994) Cystic fibrosis heterozygote resistance to cholera toxin in the cystic fibrosis mouse model. *Science* 266, 107–109.

Giaever, G., Shoemaker, D.D., Jones, T.W., Liang, H., Winzler, E.A., Astromoff, A. and Davis, R.W. (1999) Genomic profiling of drug sensitivities via induced haploinsufficiency. *Nature Genetics* 21, 278–283.

Glennie, M.J. and Van de Winkel, J.G.J. (2003) Renaissance of cancer therapeutic antibodies. *Drug Discovery Today* 8, 503–510.

Goldstein, D.B. (2003) Pharmacogenetics in the laboratory and the clinic. *New England Journal of Medicine* 348, 553–555.

Golub, T.R., Slonim, D.K., Tamayo, P., Huard, C., Gaasenbeek, M., Mesirov, J.P., Coller, H., Loh, M.L., Downing, J.R., Caligiuri, M.A., Bloomfield, C.D. and Lander, E.S. (1999) Molecular classification of cancer: class discovery and class prediction by gene expression monitoring. *Science* 286, 531–537.

Gorre, M.E. and Sawyers, C.L. (2002) Molecular mechanisms of resistance to STI571 in chronic myeloid leukemia. *Clinical Opinion in Hematology* 9, 303–307.

Gruvberger, S., Ringner, M., Chen, Y., Panavally, S., Saal, L.H., Borg, A., Ferno, M., Peterson, C. and Meltzer, P.S. (2001) Estrogen receptor status in breast cancer is associated with remarkably distinct gene expression patterns. *Cancer Research* 61, 5979–5984.

Gura, T. (2002) Magic bullets hit the target. *Nature* 417, 584–586.

Gusella, J.F. and MacDonald, M.E. (2000) Molecular genetics: unmasking polyglutamine triggers in neurodegenerative disease. *Nature Review Neuroscience* 1, 109–115.

Hahn, H., Wicking, C., Zaphiropolous, P.G., Gailani, M.R., Shanley, S., Chidabaram, A., Vorechovsky, I., Holmberg, E., Unden, A.B., Gillies, S., Negus, K., Smyth, I., Pressman, C., Leffell, D.J., Gerrard, B., Goldstein, A.M., Dean, M., Toftgard, R., Chenevix-Trench, G., Wainwright, B. and Bale, A.E. (1996) Mutations of the human homologue of *Drosophila* patched in the nevoid basal cell carcinoma syndrome. *Cell* 85, 841–851.

Hain, H.S. and Belknap, J.K. (1999) Pharmacogenetic evidence for the involvement of 5-hydroxytryptamine (serotonin)-1B receptors in the mediation of morphine antinociceptive sensitivity. *Journal of Pharmacology and Experimental Therapeutics* 291, 444–449.

Hariharan, I.S. and Haber, D.A. (2003) Yeast, flies, worms and fish in the study of human diseases. *New England Journal of Medicine* 348, 2457–2463.

Hedenfalk, I., Duggan, D., Chen, Y., Radmacher, M., Bittner, M., Simon, R., Meltzer, P., Gusterson, B., Esteller, M., Kallioniemi, O.P., Wilfond, B., Borg, A. and Trent, J. (2001) Gene-expression profiles in hereditary breast cancer. *New England Journal of Medicine* 344, 539–548.

Hedenfalk, I., Ringner, M., Ben Dor, A., Yakhini, Z., Chen, Y., Chebil, G., Ach, R., Loman, N., Olsson, H., Meltzer, P., Borg, A. and Trent, J. (2003) Molecular classification of familial non-BRCA1/BRCA2 breast cancer. *Proceedings of the National Academy of Sciences USA* 100, 2532–2537.

Hirth, J., Watkins, P.B., Strawderman, M., Schott, A., Bruno, R. and Baker, L.H. (2000) The effect of an individual's cytochrome CYP3A4 activity on docetaxel clearance. *Clinical Cancer Research* 6, 1255–1258.

Hogema, B.M., Gupta, M., Senephansiri, H., Burlingame, T.G., Taylor, M., Jakobs, C., Schutgens, R.B., Froestl, W., Snead, O.C., Diaz-Arrastia, R., Bottiglieri, T., Grompe, M., and Gibson, K.M. (2001) Pharmacologic rescue of lethal seizures in mice deficient in succinate semialdehyde dehydrogenase. *Nature Genetics* 29, 212–216.

Huang, E., West, M. and Nevins, J.R. (2003) Gene expression profiling for prediction of clinical characteristics of breast cancer. *Recent Progress in Hormone Research* 58, 55–73.

Hudson, P.J. and Souriau, C. (2003) Engineered antibodies. *Nature Medicine* 9, 129–134.

Hughes, T. R. e. al. F. S. H. (2000) Functional discovery via a compendium of expression profiles. *Cell* 102, 109–126.

Huron, D.R., Gorre, M.E., Kraker, A.J., Sawyers, C.L., Rosen, N. and Moasser, M.M. (2003) A novel pyri-dopyrimidine inhibitor of Abl kinase is a picomolar inhibitor of Bcr-abl-driven K562 cells and is effective against STI571-resistant Bcr-abl mutants. *Clinical Cancer Research* 9, 1267–1273.

Ingelman-Sundberg, M. (2001) Genetic susceptibility to adverse effects of drugs and environmental toxicants. The role of the CYP family of enzymes. *Mutation Research* 482, 11–19.

Ioffe, E., Moon, B., Connolly, E. and Friedman, J.M. (1998) Abnormal regulation of the leptin gene in the pathogenesis of obesity. *Proceedings of the National Academy of Sciences USA* 95, 11852–11857.

Iyer, L. and Ratain, M.J. (1998) Pharmacogenetics and cancer chemotherapy. *European Journal of Cancer* 34, 1493–1499.

Jacob, H.J., Lindpainter, K., Lincoln, S.E., Kusumi, K., Bunker, R.K., Mao, P., Ganten, D., Dzau, V.J. and Lander, E.S. (1991) Genetic mapping of a gene causing hypertension in the stroke-prone spontaneously hypertensive rat. *Cell* 67, 213–224.

Johnson, R.L., Rothman, A.L., Xie, J., Goodrich, L.V., Bare, J.W., Boniface, J.M., Quinn, A.G., Myers, R.M., Cox, D.R., Epstein, E.H., Jr and Scott, M.P. (1996) Human homolog of *patched*, a candidate gene for the basal cell nevus syndrome. *Science* 272, 1668–1671.

Jonker, J.W., Wagener, E., Mol, C.A.A., Buitelaar, M. *et al.* (2001) Reduced hepatic uptake and intestinal excretion of organic cations in mice with a targeted disruption of the organic cation transporter 1 (Oct1 [Slc22A1]) gene. *Molecular and Cellular Biology* 21, 5471–5477.

Kinirons, M.T., O'Shea, D., Downing, T.E., Fitzwilliam, A.T., Joellenbeck, L., Groopman, J.D., Wilkinson, G.R. and Wood, A.J. (1993) Absence of correlations among three putative *in vivo* probes of human cytochrome P4503A activity in young healthy men. *Clinical Pharmacology and Therapeutics* 54, 621–629.

Kirchheiner, J., Brosen, K., Dahl, M.L., Gram, L.F., Kasper, S., Roots, I., Sjoqvist, F., Spina, E. and Brockmoller, J. (2001) CYP2D6 and CYP2C19 genotype-based dose recommendations for antidepres-sants: a first step towards subpopulation-specific dosages. *Acta Psychiatrica Scandinavica* 104, 173–192.

Kirchheiner, J., Meineke, I., Muller, G., Roots, I. and Brockmoller, J. (2002) Contributions of CYP2D6, CYP2C9 and CYP2C19 to the biotransformation of E- and Z-doxepin in healthy volunteers. *Pharmacogenetics* 12, 571–580.

Kivisto, K.T., Kroemer, H.K. and Eichelbaum, M. (1995) The role of human cyotchrome P450 enzymes in the metabolism of anticancer agents; implications for drug interactions. *British Journal of Clinical Pharmacology* 40, 523–530.

Kolars, J.C., Schmiedlin-Ren, P., Schuetz, J.D., Fang, C. and Watkins, P.B. (1992) Identification of rifampin-inducible P450IIIA4 (CYP3A4) in human small bowel enterocytes. *Journal of Clinical Investigations* 90, 1871–1878.

Kuehl, P., Zhang, J., Lin, Y., Lamba, J., Assem, M., Schuetz, J., Watkins, P.B., Daly, A., Wrighton, S.A., Hall, S.D., Maurel, P., Relling, M., Brimer, C., Yasuda, K., Venkataramanan, R., Strom, S., Thummel, K., Boguski, M.S. and Schuetz, E. (2001) Sequence diversity in CYP3A promoters and characterization of the genetic basis of polymorphic CYP3A5 expression. *Nature Genetics* 27, 383–391.

Kuhn, R., Schwenk, F., Aguet, M. and Rajewsky, K. (1995) Inducible gene targeting in mice. *Science* 269, 1427–1429.

Kwiatkowski, D.J., Zhang, H., Bandura, J.L., Heiberger, K.M., Glogauer, M., el-Hashemite, N. and Onda, H. (2002) Mouse model of TSC1 reveals sex-dependent lethality from liver hemangiomas, and up-regula-tion of p70S6 kinase activity in Tsc1 null cells. *Human Molecular Genetics* 11, 525–534.

La Rosee, P., Corbin, A.S., Stoffregen, E.P., Deininger, M.W. and Druker, B.J. (2002) Activity of the Bcr-Abl kinase inhibitor PD180970 against clinically relevant Bcr-Abl isoforms that cause resistance to imatinib mesylate (Gleveec, STI571). *Clinical Cancer Research* 62, 7149–7153.

Lander, E.S., Linton, L.M., Birren, B., Nusbaum, C., Zody, M.C., Baldwin, J., Devon, K., Dewar, K., Doyle, M., FitzHugh, W., Funke, R., Gage, D., Harris, K., Heaford, A., Howland, J., Kann, L., Lehoczky, J., Levine, R., McEwan, P., McKernan, K. *et al.* (2001) Initial sequencing and analysis of the human genome. *Nature* 409, 860–921.

Lazarou, J., Pomeranz, B.H. and Corey, P.N. (1998) Incidence of adverse drug reactions in hospitalized patients: a meta-analysis of prospective studies. *Journal of the American Medical Association* 279, 1200–1205.

Lichter, J.B. and Kurth, J.H. (1997) The impact of pharmacogenetics on the future of healthcare. *Current Opinion in Biotechnology* 8, 692–695.

Liljedahl, U., Karlsson, J., Melhus, H., Kurland, L., Lindersson, M., Kahan, T., Nystrom, F., Lind, L. and Syvanen, A.C. (2003) A microarray minisequencing system for pharmacogenetic profiling of antihyper-tensive drug response. *Pharmacogenetics* 13, 7–17.

Liu, E.T. (2003) Classification of cancers by expression profiling. *Current Opinion in Genetics and Development* 13, 97–103.

Malkinson, A.M., Radcliffe, R.A. and Bauer, A.K. (2002) Quantitatve trait locus mapping of susceptibilities to butylated hydroxytoluene-induced lung tumour promotion and pulmonary inflammation in CXB mice. *Carcinogenesis* 23, 411–417.

Mamiya, K., Kojima, K., Yukawa, E., Higuchi, S., Ieiri, I., Ninomiya, H. and Tashiro, N. (2001) Phenytoin intoxication induced by fluvoxamine. *Therapeutic Drug Monitoring* 23, 75–77.

Marshall, A. (1997a) Getting the right drug into the right patient. *Nature Biotechnology* 15, 1249–1252.

Marshall, A. (1997b) Laying the foundations for personalized medicines. *Nature Biotechnology* 15, 954–957.

Marshall, S.A., Lazar, G.A., Chirino, A.J. and Desjarlais, J.R. (2003) Rational design and engineering of therapeutic proteins. *Drug Discovery Today* 8, 212–221.

Martin, E.R. *et al.* (2000) SNPing away at complex diseases: analysis of single-nucleotide polymorphisms around APOE in Alzheimer disease. *American Journal of Human Genetics* 67, 383–394.

Marton, M.L., DeRisi, J.L., Bennett, H.A., Iyer, V.R., Meyer, M.R., Roberts, C.J., Stoughton, R., Burchard, J., Slade, D., Dia, H., Bassett, J.D.E., Hartwell, L.H., Brown, P.O. and Friend, S.H. (1998) Drug target validation and identification of secondary drug target effects using DNA microarrays. *Nature Medicine* 4, 1293–1301.

Meyer, U.A. (2000) Pharmacogenetics and adverse drug reactions. *Lancet* 356, 1667–1671.

Milano, G. and Etienne, M.-C. (1994) Potential importance of dihydropyrimidine dehydrogenase (DPD) in cancer chemotherapy. *Pharmacogenetics* 4, 301–306.

Mohr, S., Leikauf, G.D., Keith, G. and Rihn, B.H. (2002) Microarrays as cancer keys: an array of possibilities. *Journal of Clinical Oncology* 20, 3165–3175.

Montosi, G., Donovan, A., Totaro, A., Garuti, C., Pignatti, E., Cassanelli, S., Trenor, C.C., Gasparini, P., Andrews, N.C. and Pietrangelo, A. (2001) Autosomal-dominant hemochromatosis is associated with a mutation in the ferroportin (SLC11A3) gene. *Journal of Clinical Investigations* 108, 619–623.

Mullis, K.B. and Faloona, F.A. (1987) Specific synthesis of DNA *in vitro* via a polymerase-catalyzed chain reaction. *Methods in Enzymology* 155, 335–350.

Nicolette, C.A. and Miller, G.A. (2003) The identification of clinically relevant markers and therapeutic targets. *Drug Discovery Today* 8, 37.

Njajou, O.T., Vaessen, N., Joose, M., Berghis, B., van Dongen, J.W.F., Beurning, M.H., Snijders, P.J.L.M., Rutten, W.P.F., Sandkuijl, L.A., Oostra, B.A., van Duijn, C.M. and Heutink, P. (2001) A mutation in SLC11A3 is associated with autosomal dominant hemochromatosis. *Nature Genetics* 28, 213–214.

Nuyens, D., Stengl, M., Dugarmaa, S., Rossenbacker, T., Compernolle, V., Rudy, Y., Smits, J.F., Flameng, W., Clancy, C.E., Moons, L., Vos, M.A., Dewerchin, M., Benndorf, K., Collen, D., Carmeliet, E. and Carmeliet, P. (2001) Abrupt rate accelerations or premature beats cause life-threatening arrhythmias in mice with long-QT3 syndrome. *Nature Medicine* 7, 1021–1027.

Ozdemir, V., Kalowa, W., Tang, B.K., Paterson, A.D., Walker, S.E., Endrenyi, L. and Kashuba, A.D. (2000) Evaluation of the genetic component of variability in CYP3A4 activity: a repeated drug administration method. *Pharmacogenetics* 10, 373–388.

Paine, M.F., Wagner, D.A., Hoffmaster, K.A. and Watkins, P.B. (2002) Cytochrome P450 3A4 and P-glycoprotein mediate the interaction between an oral erythromycin breath test and rifampin. *Clinical Pharmacology and Therapeutics* 72, 524–535.

Perou, C.M., Jeffrey, S.S., van de, R.M., Rees, C.A., Eisen, M.B., Ross, D.T., Pergamenschikov, A., Williams, C.F., Zhu, S.X., Lee, J.C., Lashkari, D., Shalon, D., Brown, P.O. and Botstein, D. (1999) Distinctive gene expression patterns in human mammary epithelial cells and breast cancers. *Proceedings of the National Academy of Sciences USA* 96, 9212–9217.

Perou, C.M., Sorlie, T., Eisen, M.B., van de, R.M., Jeffrey, S.S., Rees, C.A., Pollack, J.R., Ross, D.T., Johnsen, H., Akslen, L.A., Fluge, O., Pergamenschikov, A., Williams, C., Zhu, S.X., Lonning, P.E., Borresen-Dale, A.L., Brown, P.O. and Botstein, D. (2000) Molecular portraits of human breast tumours. *Nature* 406, 747–752.

Phillips, K.A., Veenstra, D.L., Oren, E., Lee, J.K. and Sadee, W. (2001) Potential role of pharmacogenomics in reducing adverse drug reactions: a systematic review. *Journal of the American Medical Association* 286, 2270–2279.

Pollack, A. (2003a) Genentech buoyed by results of new tack to colon cancer. *New York Times Daily*, C1–C2.

Pollack, A. (2003b) New weapons against cancer show promise. *New York Times Daily*[CLII], A1–A13.

Pollack, A. (2003c) RNA trades bit part for starring role in the cell. *New York Times Daily*, D1–D4.

Rana, B.K., Shiina, T. and Insel, P.A. (2001) Genetic variations and polymorphisms of G protein-coupled receptors: functional and therapeutic implications. *Annual Reviews in Pharmacology and Toxicology* 41, 593–624.

Rebbeck, T.R., Jaffe, J.M., Walker, A.H., Wein, A.J. and Malkowicz, S.B. (1998) Modification of clinical presentation of prostate tumours by a novel genetic variant in CYP3A4. *Journal of the National Cancer Institute* 90, 1225–1229.

Rego, E.M., Wang, Z.-G., Peruzzi, D., He, L.-Z., Cordon-Cardo, C. and Pandolfi, P.P. (2001) Role of promyelocytic leukemia (PML) protein in tumour suppression. *Journal of Experimental Medicine* 193, 521–529.

Relling, M.V. and Dervieux, T. (2001) Pharmacogenetics and cancer therapy. *Nature Reviews of Cancer* 1, 99–108.

Roses, A.D. (2001) Pharmacogenetics. *Human Molecular Genetics* 10, 2261–2267.

Ross, P., Hall, L., Smirnov, I. and Haff, L. (1998) High level multiplex genotyping by MALDI-TOF mass spectrometry. *Nature Biotechnology* 16, 1347–1351.

Sadee, W., Hoeg, E., Lucas, J. and Wang, D. (2001) Genetic variations in human G protein-coupled receptors: implications for drug therapy. *American Association of Pharmaceutical Sciences Online* 3, 1–27 (http://www.pharmsci.org)

Sallee, F.R., DeVane, C.L. and Ferrell, R.E. (2000) Fluoxetine-related death in a child with cytochrome P-450 2D6 genetic deficiency. *Journal of Child and Adolescent Psychopharmacology* 10, 27–34.

Schena, M., Shalon, D., Davis, R.W. and Brown, P.O. (1995) Quantitative monitoring of gene expression patterns with a complementary DNA microarray. *Science* 270, 467–470.

Schmelzle, T. and Hall, M.N. (2000) TOR, a central controller of cell growth. *Cell* 103, 253–262.

Shih, D.., Gu, L., Xia, Y.-R., Navab, M., Li, W.-F., Hama, S., Castellani, L.W., Furlong, C.E., Costa, L.G., Fogelman, A.M. and Lusis, A.J. (1998) Mice lacking serum paraoxonase are susceptible to organophosphate toxicity and atherosclerosis. *Nature* 394, 284–287.

Shimada, T., Yamazaki, H., Mimura, M., Inui, Y. and Guengerich, F.P. (1994) Interindividual variations in human liver cytochrome P-450 enzymes involved in the oxidation of drugs, carcinogens and toxic chemicals: studies with liver microsomes of 30 Japanese and 30 Caucasians. *Journal of Pharmacology and Experimental Therapeutics* 270, 414–423.

Shipp, M.A., Ross, K.N., Tamayo, P., Weng, A.P., Kutok, J.L., Aguiar, R.C., Gaasenbeek, M., Angelo, M., Reich, M., Pinkus, G.S., Ray, T.S., Koval, M.A., Last, K.W., Norton, A., Lister, T.A., Mesirov, J., Neuberg, D.S., Lander, E.S., Aster, J.C. and Golub, T.R. (2002) Diffuse large B-cell lymphoma outcome prediction by gene-expression profiling and supervised machine learning. *Nature Medicine* 8, 68–74.

Shou, M. and Yang, S.K. (1990a) 1-Hydroxy- and 2-hydroxy-3-methylcholanthrene: regioselective and stereoselective formations in the metabolism of 3-methylcholanthrene and enantioselective disposition in rat liver microsomes. *Carcinogenesis* 11, 933–940.

Shou, M. and Yang, S.K. (1990b) 1-Hydroxy- and 2-hydroxy-3-methylcholanthrene: regioselective and stereoselective formations in the metabolism of 3-methylcholanthrene and enantioselective disposition in rat liver microsomes. *Carcinogenesis* 11, 933–940.

Shou, M., Martinet, M., Korzekwa, K.R., Krausz, K.W., Gonzalez, F.J. and Gelboin, H.V. (1998) Role of human cytochrome P450 3A4 and 3A5 in the metabolism of taxotere and its derivatives: enzyme specificity, interindividual distribution and metabolic contribution in human liver. *Pharmacogenetics* 8, 391–401.

Small, K.M., Forbes, S.L., Rahman, F.F., Bridges, K.M. and Liggett, S.B. (2000) A four amino acid deletion polymorphism in the third intracellular loop of the human alpha 2C-adrenergic receptor confers impaired coupling to multiple effectors. *Journal of Biological Chemistry* 275, 23059–23064.

Small, K.M., Seman, C.A., Castator, A., Brown, K.M. and Liggett, S.B. (2002) False positive non-synonymous polymorphisms of G-protein coupled receptor genes. *FEBS Letters* 516, 253–256.

Small, K.M. e. al. S. J. C. and Liggett, S.B. (2003) Gene and protein domain-specific patterns of genetic variability within the G-protein coupled receptor superfamily. *American Journal of Pharmacogenomics* 3, 65–71.

Smithies, O. (2001) Forty years with homologous recombination. *Nature Medicine* 7, 1083–1086.

Sorlie, T., Perou, C.M., Tibshirani, R., Aas, T., Geisler, S., Johnsen, H., Hastie, T., Eisen, M.B., van de, R.M., Jeffrey, S.S., Thorsen, T., Quist, H., Matese, J.C., Brown, P.O., Botstein, D., Eystein, L.P. and Borresen-Dale, A.L. (2001) Gene expression patterns of breast carcinomas distinguish tumour subclasses with clinical implications. *Proceedings of the National Academy of Sciences USA* 98, 10869–10874.

Southern, E.M. (1975) Detection of specific sequences among DNA fragments separated by gel electrophoresis. *Journal of Molecular Biology* 98, 503–517.

Spear, B.B., Heath-Chiozzi, M. and Huff, J. (2001) Clinical application of pharmacogenetics. *Trends in Molecular Medicine* 7, 201–204.

Stephens, J.C. (1999) Single-nucleotide polymorphisms, haplotype, and their relevance to pharmacogenetics. *Molecular Diagnosis* 4, 309–317.

Taipale, J., Chen, J.K., Cooper, M.K., Wang, B., Mann, R.K., Milenkovic, L., Scott, M.P. and Beachy, P.A. (2000) Effects of oncogenic mutations in Smoothened and Patched can be reversed by cyclopamine. *Nature* 406, 944–945.

Taube, J., Halsall, D. and Baglin, T. (2000) Influence of cytochrome P-450 CYP2C9 polymorphisms on warfarin sensitivity and risk of over-anticoagulation in patients on long-term treatment. *Blood* 96, 1816–1819.

The chipping forecast II (2002) *Nature* 32, Supplement, 461–552.

Thijssen, H.H., Verkooijen, I.W. and Frank, H.L. (2000) The possession of the CYP2C9*3 allele is associated with low dose requirement of acenocoumarol. *Pharmacogenetics* 10, 757–760.

Torrance, C.J., Agrawal, V., Vogelstein, B. and Kinzler, K.W. (2001) Use of isogenic human cancer cells for high-throughput screening and drug discovery. *Nature Biotechnology* 19, 940–945.

Tucker, G.T. (2000) Advances in understanding drug metabolism and its contribution to variability in patient response. *Therapeutic Drug Monitoring* 22, 110–113.

van de Vijver, M.J., He, Y.D., van't Veer, L.J., Dai, H., Hart, A.A., Voskuil, D.W., Schreiber, G.J., Peterse, J.L., Roberts, C., Marton, M.J., Parrish, M., Atsma, D., Witteveen, A., Glas, A., Delahaye, L., van d., V, Bartelink, H., Rodenhuis, S., Rutgers, E.T., Friend, S.H. and Bernards, R. (2002) A gene-expression signature as a predictor of survival in breast cancer. *New England Journal of Medicine* 347, 1999–2009.

van't Veer, L.J., Dai, H., van de Vijver, M.J., He, Y.D., Hart, A.A., Mao, M., Peterse, H.L., van der, K.K., Marton, M.J., Witteveen, A.T., Schreiber, G.J., Kerkhoven, R.M., Roberts, C., Linsley, P.S., Bernards, R. and Friend, S.H. (2002) Gene expression profiling predicts clinical outcome of breast cancer. *Nature* 415, 530–536.

Veenstra, D.L. and Higashi, M.K.P.K.A. (2000) Assessing the cost-effectiveness of pharmacogenomics. *American Association of Pharmaceutical Sciences Online* 2, 1–11 (online article 29: http://wwwpharm-sci.org).

Venter, J.C., Adams, M.D., Myers, E.W., Li, P.W., Mural, R.J., Sutton, G.G., Smith, H.O., Yandell, M., Evans, C.A., Holt, R.A., Gocayne, J.D., Amanatides, P., Ballew, R.M., Huson, D.H., Wortman, J.R., Zhang, Q., Kodira, C.D., Zheng, X.H., Chen, L., Skupski, M. *et al.* (2001) The sequence of the human genome. *Science* 291, 1304–1351.

Von Moltke, L.L., Tran, T.H., Cotreau, M.M., Greenblatt, D.J. (2000) Unusually low clearance of two CYP3A substrates, alprazolan and trazolone, in a volunteer subject with wild-type CYP3A4 promoter region. *Journal of Clinical Pharmacology* 40, 200–204.

Wade, N. (1997) Researchers block cancer from developing defenses: tests targeting normal cells work in mice. *New York Times*, A15.

Wade, N. (1999) Tailoring drugs to fit the genes. *New York Times*.

Wade, N. (2000) Tiny beads allow close study of DNA. *New York Times Daily*, D3.

Wang, D.-S., Jonker, J.W., Kato, Y. *et al.* (2002a) Involvement of organic cation transporter 1 in hepatic and intestinal distribution of metformin. *Journal of Pharmacology and Experimental Therapeutics* 302, 510–515.

Wang, E., Miller, L.D., Ohnmacht, G.A., Mocellin, S., Perez-Diez, A., Petersen, D., Zhao, Y., Simon, R., Powell, J.I., Asaki, E., Alexander, H.R., Duray, P.H., Herlyn, M., Restifo, N.P., Liu, E.T., Rosenberg, S.A. and Marincola, F.M. (2002b) Prospective molecular profiling of melanoma metastases suggests classifiers of immune responsiveness. *Cancer Research* 62, 3581–3586.

Waterston, R.H., Lindblad-Toh, K., Birney, E., Rogers, J., Abril, J.F., Agarwal, P., Agarwala, R., Ainscough, R., Alexandersson, M., An, P., Antonarakis, S.E., Attwood, J., Baertsch, R., Bailey, J., Barlow, K., Beck, S., Berry, E., Birren, B., Bloom, T., Bork, P. *et al.* (2002) Initial sequencing and comparative analysis of the mouse genome. *Nature* 420, 520–562.

Watkins, P.B. (1994) Noninvasive tests of CYP3A enzymes. *Pharmacogenetics* 4, 171–184.

Watters, J.W. and McLeod, H.L. (2003) Cancer pharmacogenomics: current and future applications. *Biochimica Biophysica Acta* 1603, 99–111.

Weber, W.W. (1997) *Pharmacogenetics*. Oxford University Press, Oxford.

Weber, W.W. (2001) The legacy of pharmacogenetics and potential applications. *Mutation Research* 479, 1–18.

Weber, W.W. and Cronin, M.T. (2000) Pharmacogenetic testing. In: Meyers, R.A. (ed.) *Encyclopedia of Analytical Chemistry*. John Wiley & Sons, Chichester, UK, pp. 2–21.

Weber, W.W. and Smith, R.L. (2001) New directions in pharmacogenetics and ecogenetics. *Drug Metabolism and Disposition* 29, 467–614.

Weber W.W., Caldwell, M.D. and Kurthe, J.H. (2003) Edging toward personalized medicine. *Current Pharmacogenomics* 1, 193–202.

Wei, X., McLeod, H.L., McMurrough, J., Gonzalez, F.J. and Fernandez-Salguero, P. (1996) Molecular basis of the human dihydropyrimidine dehydrogenase deficiency and 5-fluorouracil toxicity. *Journal of Clinical Investigations* 98, 610–615.

Weinshilboum, R.W. (2003) Inheritance and drug response. *New England Journal of Medicine* 348, 529–552.

West, M., Blanchette, C., Dressman, H., Huang, E., Ishida, S., Spang, R., Zuzan, H., Olson, J. A., Jr, Marks, J.R. and Nevins, J.R. (2001) Predicting the clinical status of human breast cancer by using gene expression profiles. *Proceedings of the National Academy of Sciences USA* 98, 11462–11467.

Westlind, A., Malmebo, S., Johansson, I., Otter, C., Andersson, T.B., Ingelman-Sundberg, M. and Oscarson, M. (2001) Cloning and tissue distribution of a novel human cytochrome p450 of the CYP3A subfamily, CYP3A43. *Biochemical and Biophysical Research Communications* 281, 1349–1355.

Wilkinson, G.R. (1996) Cytochrome P4503A (CYP3A) metabolism: prediction of *in vivo* activity in humans. *Journal of Pharmacokinetics and Biopharmaceutics* 24, 475–490.

Wolf, R.C. and Smith, G. (2002) Pharmacogenetics. *British Medical Bulletin* 55, 366–386.

Wrighton, S.A., Brian, W.R., Sari, M.A., Iwasaki, M., Guengerich, F.P., Raucy, J.L., Molowa, D.T. and VandenBranden, M. (1990) Studies on the expression and metabolic capabilities of human liver cytochrome P450IIIA5 (HLp3). *Molecular Pharmacology* 38, 207–213.

Xie, H.G., Kim, R.B., Wood, A.J. and Stein, C.M. (2001) Molecular basis of ethnic differences in drug disposition and response. *Annual Reviews in Pharmacology and Toxicology* 41, 815–850.

Xie, W., Barwick, J.L., Downes, M., Blumberg, B., Simon, C.M., Nelson, M.C., Neuschwander-Tetri, B.A., Brunt, E.M., Guzelian, P.S. and Evans, R.M. (2000a) Humanized xenobiotic response in mice expressing nuclear receptor SXR. *Nature* 406, 435–439.

Xie, W., Stribley, J.A., Chatonnet, A., Wilder, P.J., Rizzino, A., Mccomb, R.D., Taylor, P., Hinrichs, S.H. and Lockridge, O. (2000b) Postnatal developmental delay and supersensitivity to organophosphate in gene-targeted mice lacking acetylcholinesterase. *Journal of Pharmacology and Experimental Therapeutics* 293, 896–902.

Yeoh, E.J., Ross, M.E., Shurtleff, S.A., Williams, W.K., Patel, D., Mahfouz, R., Behm, F.G., Raimondi, S.C., Relling, M.V., Patel, A., Cheng, C., Campana, D., Wilkins, D., Zhou, X., Li, J., Liu, H., Pui, C.H., Evans, W.E., Naeve, C., Wong, L. and Downing, J.R. (2002) Classification, subtype discovery, and prediction of outcome in pediatric acute lymphoblastic leukemia by gene expression profiling. *Cancer Cell* 1, 133–143.

Zaman, G.J.R., Michiels, P.J.A. and van Boeckel, C.A.A. (2003) Targeting RNA: new opportunities to address drugless targets. *Drug Discovery Today* 8, 297–306.

Zanger, U.M., Fischer, J., Raimundo, S., Stuven, T., Evert, B.O., Schwab, M. and Eichelbaum, M. (2001) Comprehensive analysis of the genetic factors determining expression and function of hepatic CYP2D6. *Pharmacogenetics* 11, 573–585.

Zhang, L. *et al.* (1997) Gene expression profiles in normal and cancer cells. *Science* 276, 1268–1272.

20 Genome Scanning for Quantitative Trait Loci

B.J. Hayes[1], B.P. Kinghorn[2]* and A. Ruvinsky[2]

[1]*Akvaforsk, Institute for Aquaculture Research, Agricultural University of Norway, Ås 1432, Norway;* [2]*Institute for Genetics and Bioinformatics, University of New England, Armidale, 2351 NSW, Australia*

Introduction	508
Principles of QTL Detection in Genome Scanning	509
What type of mutations cause variation in quantitative traits?	509
The number of genes underlying quantitative traits and the distribution of their effects	509
Detecting linkage between a single marker and a QTL	511
Two-marker QTL linkage mapping	512
Interval mapping for inferring location and effect of QTLs in the genome	513
The problem of significance testing with many marker tests	513
The problem of overestimation of QTL effects from genome scans	515
Precision of QTL mapping	515
Designs of QTL mapping experiments	515
Crosses of inbred or divergent lines	516
Case study: mapping of obesity QTLs in a cross between mouse lines divergently selected on fat content	516
Half-sib designs	517
Case study: mapping QTLs controlling milk production in cattle by exploiting progeny testing	518
Complex pedigree designs	519
Case study: QTLs influencing body-mass index reside on chromosomes 7 and 13: the National Heart, Lung and Blood Institute Family Heart Study	520
Comparison of case studies	521
Increasing the Power of QTL Detection Experiments	521
Selective genotyping	521
Selective DNA pooling	523
Strategies to Improve the Precision of QTL Mapping Experiments	523
Linkage disequilibrium mapping	523
Combined LD–LA mapping	529
Results from LD mapping in some mammalian species	530

*Sygen Chair of Genetic Information Systems.

© CAB International 2005. *Mammalian Genomics*
(eds A. Ruvinsky and J. Marshall Graves)

Meta-analysis   530
Using Information from Gene Expression Studies in QTL Mapping Experiments   530
Towards More Complex Genetic Models in QTL Detection   531
    Simultaneous detection of multiple interacting QTLs   531
Gene Identification   532
From Phenotype to Genotype or from Genotype to Phenotype?   532

Introduction

Understanding associations between phenotype and genotype is a key to understanding biology itself. We can observe high levels of structure and organization at all levels of phenotype, but the underlying genetic machinery has not been designed – it has fallen into place in a random fashion, and has then been directed by the forces of natural selection.

While growing knowledge narrows the gap between phenotype and genotype, a lack of linear relationships is quite obvious. There are numerous and clear indications that even the classical dogma of one gene–one protein does not work in many cases. However, understanding this general situation does not prevent numerous successful attempts to identify certain links between genes and traits, including quantitative traits.

This chapter relates to the *detection* and *location* of genes that cause some of the quantitative variation in phenotypes. This is based on associations between phenotypes and the marker loci that can be genotyped, and which themselves are unlikely to have any role in expression of the phenotypes. In ignorance of underlying mechanism, we refer to a detected genetic effect as a quantitative trait locus (QTL), and it remains a QTL until its true nature is discovered. In this chapter the major focus is on detecting the location of QTLs in the genome. Information on the rough location of QTLs can be exploited through marker-assisted selection, to increase the rate of genetic gain for important economic traits in livestock species. Identification of the causative mutation is more desirable, as the improvement in genetic gain with such information is greater than from linked markers. Additionally, without full characterization of the mutation, the possibility that the muta-

tion has undesirable pleiotropic effects on other traits cannot be ruled out. If the aim of the QTL detection experiment is disease diagnosis, genetic counselling, drug discovery or greater understanding of the biology underlying quantitative traits, identification of the causative mutation is the ultimate goal. Identification of mutations underlying variation in quantitative traits is difficult to achieve in a single experiment, rather QTL detection often occurs in a series of steps. The detection of a mutation responsible for a large proportion of the genetic variance in fat percentage in milk from dairy cattle demonstrates the possible stages required (Grisart et al., 2002). This article reported the identification of a single base pair mutation in the *DGAT1* gene, with major effects on milk yield and composition in cattle. The first step in identifying the mutation was a genome-wide linkage analysis (genome scan), which found that a region of chromosome 14 contained a QTL with a large effect on fat percentage (Georges et al., 1995). The confidence region surrounding this QTL was very large (about 20–40 cM), and contained so many (candidate) genes that could possibly be carrying the underlying mutation that it was impossible to select any of the genes with confidence. In the next step location of the QTL was narrowed to about 3 cM using linkage disequilibrium (LD) mapping (Riquet et al., 1999), which is discussed later in the chapter. LD mapping is often called fine mapping. The *DGAT1* gene was identified as a strong candidate in this interval, and subsequent sequencing detected a point mutation in this gene. The mutation caused a substitution from lysine to alanine in the *DGAT1* gene and further investigations showed this mutation to be associated with major effects on milk yield and composition (Thaller et al., 2003).

Identification of the *DGAT1* mutation is one of the first demonstrations in a mammalian species that the process of QTL mapping can be used to successfully identify a mutation underlying variation in a quantitative trait. The aim of this chapter is to describe in more detail each of the steps in this process, and to discuss strategies which have the greatest chance of success.

As more and more mutations that underlie variation in various traits are characterized, it is becoming apparent that the simple additive model of gene action often does not apply. For example the *Booroola*, *Inverdale* and *Javanese* genes for prolificacy in sheep (Piper and Bindon, 1982; Bradford and Inounu, 1986; Davis *et al.*, 1991); the double muscling gene in cattle (Charlier *et al.*, 1995); and the *Callipyge* muscling gene in sheep (Cockett *et al.*, 1994) all have non-additive gene action. Rather it seems that many genes operate in regulatory networks, often with complex interlocus interactions (see Chapter 8). To detect some of the mutations underlying variation in quantitative traits, it may be necessary to account for such complexity in models used for analysis of genotype and phenotype data. At the end of this chapter, we briefly discuss some recent attempts to include more complex genetic models in the analysis of QTL detection experiments.

Principles of QTL Detection in Genome Scanning

What type of mutations cause variation in quantitative traits?

The mouse and human genomes contain approximately 30,000 or more protein-coding genes, only 1% of which do not show homology (Waterston *et al.*, 2002). The coding part of the genome is relatively small ($\sim$1.5% of the genome) and highly conserved (see Chapter 4). Beside the protein-coding genes, the mammalian genome contains many functional sequences of other types, as well as numerous repeated sequences and pseudogenes, which evolve much faster and show less homology (see Chapters 4, 11 and 12). It is not known yet how much of the mammalian genome is likely to be essential for survival

and how much was accumulated as a result of random events having relatively small effect on phenotype. The latest estimates suggest that on average about 5% of the mouse genome is under selection (Waterston *et al.*, 2002). This is a surprisingly high estimate, given that only about 1.5% of the genome is protein coding, and probably implies that many non-protein-coding sequences are under selection control.

Given this, it seems that a large proportion of QTLs are likely to be generated by mutations outside protein-coding regions. An example is the callipyge trait in sheep, caused by what is probably a long-range control element (LRCE) within the *DLK1–GTL2* imprinted domain (Freking *et al.*, 2002; Georges *et al.*, 2003). If such QTLs are common, they could be mapped to a chromosomal location with some accuracy, but in many cases, may not be readily identified or functionally characterized for some time.

Obviously mutations in protein-coding regions can also give rise to QTL effects (Fujii *et al.*, 1991; Yu *et al.*, 2003) as well as mutations in non-coding regions, such as introns and promoters (Georges *et al.*, 2003). The QTL effect may also be due to mutations covering a conserved region of the genome. For example, Stam and Laurie (1996) evaluated the effect of different mutations in distinct parts of the *Adh* gene on variation in the level of alcohol dehydrogenase activity, and found at least three polymorphisms within the gene and its regulatory sites which contributed to enzyme activity (reviewed in Barton and Keightley, 2002). It is important to point out that the term QTL is also applied to mutations affecting traits, which are not strictly quantitative. As the Complex Trait Consortium put it recently, the distinction between Mendelian loci and QTLs is artificial (CTC, 2003).

The number of genes underlying quantitative traits and the distribution of their effects

Quantitative traits are affected by more than one gene. But how many genes, and what is the distribution of their effects? This question is relevant as it impacts on the best design of experiments for QTL mapping. Shrimpton and Robertson (1988) attempted to discern the number of genes causing variation in bristle number in *Drosophila*. They concluded at least 17 'genetic factors', or chromosome segments, were responsible for the difference in bristle number between a line with a high number of bristles and a line with a low number of bristles. However, they could not ascertain genes of effect less than 0.6 phenotypic standard deviations, and concluded their results were consistent with the hypothesis that quantitative characters are under the control of a few major genes supported by numerous genes with smaller effect.

Hayes and Goddard (2001) used the results of QTL mapping experiments in pigs and dairy cattle to derive the distribution of QTL effects. Figure 20.1 shows the Gamma distribution of QTL effects using their pooled results from the two species. Figure 20.1 also shows the importance of QTL of different effect, which peaks at the gamma distribution mean of 0.083 standard deviation units. This assumes that the frequencies of alleles at the QTL follow a distribution which would be expected if the quantitative trait was selectively neutral which of course may not be true (Hayes and Goddard, 2001). Meta-analysis of information from QTL mapping experiments in livestock led to a conclusion that for a typical trait approximately 20 of the largest QTLs would explain 90% of the genetic variance and approximately five QTLs would explain 50% of the variance in the traits used in their analysis (Hayes and Goddard, 2001). It was also found that in order to detect enough QTLs to explain 90% of the variance in a quantitative trait, a QTL detection experiment would have to detect QTLs with effects as small as 0.1 phenotypic standard deviations. The hypothesis that the majority of genetic variation in quantitative traits is controlled by a few genes of large effect is supported by a number of cases where the causative mutation underlying the QTL effect has been discovered, and the proportion of genetic variance the mutation causes estimated. The *DGAT1* mutation is responsible for approximately 43% of the genetic in fat percentage in Holstein-Friesian dairy cattle (Grisart *et al.*, 2002). In a genome scan for QTLs affecting obesity using a cross between mice lines divergently selected for fat percentage, four detected QTLs explained 4.9, 19.5, 14.4 and 7.3% of the F_2 phenotypic variance for fat percentage, respectively (Horvat *et al.*, 2000).

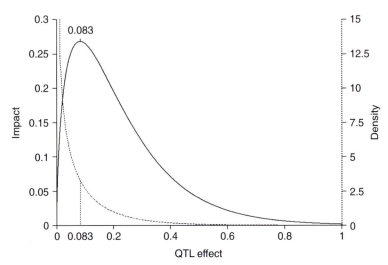

Fig. 20.1. Gamma distribution of the QTL effect, in phenotypic standard deviation units, from parameters pooled across dairy and pig results (gamma parameters $\alpha= 7.1$ and $\beta=0.59$, Hayes and Goddard, 2001). The dotted curve is distribution density, and the solid curve is the total impact or importance of QTLs of different effect, presented here as distribution density $\times$ QTL effect. Mean density and peak impact are at a QTL effect of 0.083 standard deviation units.

Hayes and Goddard (2001) assumed a simple additive model of gene action, and their conclusion that only a small number of the largest genes are required to explain the majority of genetic variance in a quantitative trait may not be valid if more complex genetic models were the reality. Barton and Keightley (2002) suggest that the effects of mutations may be quite transient due to epistasis. It may be that the QTL detected today have been found because of their importance in the prevailing genetic background. For example, redundant systems, which bypass the deleterious QTL mutations, evolve or are recovered, could render the QTL effects small. Of course, this phenomenon would be readily seen when QTL effects are estimated in breeds of domestic animal other than those in which they were discovered. Evidence in either direction is inconclusive: for the *DGAT1* mutation in dairy cattle effects on protein and milk yield are similar in both the Holstein-Friesian breed (in which the mutation was discovered) and Jersey cattle, one of the oldest dairy breeds. However, the effect of the mutation on fat yield is only half as large in Jerseys as it is in Holstein-Friesians (Spelman *et al.*, 2002). Another example is a mutation at the oestrogen receptor locus (*ESR*) which results in increased litter size in pigs (Rothschild *et al.*, 1991). This polymorphism was discovered in a crossbred population from the Chinese Meishan breed and commercial Large White animals. The size of the effect of the mutation was approximately one extra pig per litter in the Meishan crosses, and only 0.5

extra pigs per litter in Large White and Large White crosses (Rothschild *et al.*, 1994, 1996; Short *et al.*, 1997).

Detecting linkage between a single marker and a QTL

So how can the location in the genome of the QTL affecting the trait of interest be uncovered? To achieve this, in the absence of any knowledge of the location of such QTLs, one must use neutral molecular markers, which are randomly scattered throughout the genome. If variation at a marker genotype in the mapping population accounts for some of the phenotypic variation in the quantitative trait in the mapping population, then the marker may be linked to a gene containing a mutation that affects the quantitative trait.

Consider a very simple example where a male parent has a large number of progeny. For this example we will consider the alleles from the male parent only. Table 20.1 shows hypothetical marker genotypes and trait data for a sire and its progeny. For marker M1, one can deduce which sire marker allele has been inherited by each progeny except for progeny number 6. If progeny 6 were known to have a dam of genotype 2–4, then it could be concluded that progeny 6 had inherited allele 1 from the sire. The power to make inference about QTL position and effect from this data set is very weak indeed, but the small example makes for clearer illustration.

Table 20.1. Genotypes at marker loci *M1* and *M2* for a sire and nine of his progeny, sire allele inherited by each progeny and progeny merit for a hypothetical trait. Genotype 1–2 consists of allele 1 and allele 2 of the prevailing marker locus.

Individual	*M1* genotype	Sire allele inherited	*M2* genotype	Sire allele inherited	Progeny merit
Sire	1–2	–	1–2	–	–
1	1–3	1	2–5	2	9
2	2–3	2	1–3	1	11
3	1–4	1	1–2	?	11
4	1–1	1	2–6	2	9
5	2–5	2	1–1	1	11
6	1–2	?	2–4	2	9
7	2–4	2	2–2	2	10
8	1–4	1	1–5	1	10
9	2–2	2	1–3	1	11

Progeny inheriting M1 allele 1 from the sire perform on average 1 trait unit worse than progeny inheriting allele 2 (9.75 vs. 10.75). Assume for simplicity that this result holds true for very many progeny, and that this marker contrast of one trait unit ($c = 1$) is highly significant. The conclusion will follow that there is a QTL linked to marker M1. Where is this QTL? How big an effect does it have? If the QTL is at M1, with no recombination between M1 and the QTL, then the QTL effect is 1 trait unit.

However, if the QTL is located some distance from M1, then it must have a larger QTL effect in order to explain the observed marker contrast of 1 trait unit. If the recombination fraction between the QTL and M1 were 10% ($r = 0.1$), then 10% of progeny inheriting allele 2 would inherit the inferior QTL allele, and 10% of progeny inheriting allele 1 would inherit the superior QTL allele. The QTL effect would then be 1.25, giving an observed marker contrast of 90% times 1.25 plus 10% times -1.25, which equals 1.

In general, the QTL effect, the average effect of a QTL allele substitution, is estimated as $\alpha = c/(1 - 2r)$. Figure 20.2 shows the predicted QTL effect as a function of recombination fraction. With just one marker locus, we cannot unravel the QTL position and QTL effect.

Two-marker QTL linkage mapping

If the position of the markers relative to each other along the chromosome is known, the power of the QTL detection experiment to detect and locate QTLs can be increased by using information from flanking markers. Figure 20.3 illustrates an estimate of QTL position if the genotypes of flanking marker loci M1 and M2 are known. It can be noted from Table 20.1 that progeny inheriting allele 1 at locus M2 from the sire perform on average 1.5 trait units better than progeny inheriting allele 2 (10.75 versus 9.25). Moreover, in this sire, allele 1 of M1 seems to be linked to allele 2 of M2, and vice versa. Of the seven progeny for which the sire allele inherited can be deduced for both M1 and M2, only two progeny, numbers 7 and 8, break this trend. The recombination fraction between M1 and M2 is thus estimated at 2/7. This corresponds to a map distance of $d = 42.4$ cM, using the Haldane mapping function $d = -\frac{1}{2} \ln(1 - 2r)$; for detail see Chapter 1. Figure 20.3 illustrates the inference of map position of the putative QTL. Assuming that it is known that M1 is located on the map at 25 cM, and M2 is located 42.4 cM distally from M1, the QTL position is predicted to be at 56.3 cM, where the estimated QTL effects are equal for the two marker loci.

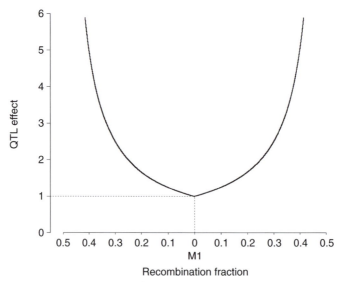

Fig. 20.2. Half-sibs inheriting alternative M1 marker alleles from their sire differ by 1 trait unit. This information is not sufficient to estimate QTL location. The further the QTL is from M1, the bigger its estimated effect must be to explain the marker contrast observed, as shown by the curves rising from coordinate (0, 1).

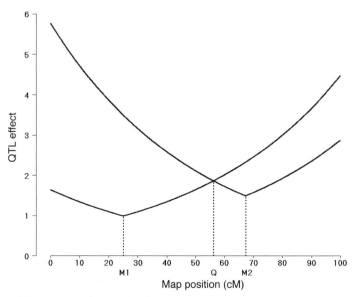

Fig. 20.3. The Haldane mapping function has been used to give a linear scale to compare different locations. With a second marker, M2, showing a marker contrast of 1.5 trait units, the QTL position (Q) is predicted at about 56.3 cM, with an allele substitution effect of 1.87 trait units.

Interval mapping for inferring location and effect of QTLs in the genome

In practice, a genome scan proceeds by evaluating support for existence of a QTL at regular intervals along the genome, for example every 1 cM, using information from the markers flanking each interval. If one of the markers flanking a particular interval is not informative (not segregating in the population for example), the closest informative marker is used. Maximum likelihood (ML) methods proceed by evaluating, for each putative QTL position, the log of the ratio of the maximum of the likelihood function under the null hypothesis of no segregating QTL to the maximum of the likelihood function under the alternative hypothesis of a QTL segregating at that position, given the marker and phenotype data. This provides a likelihood of odds (LOD) score for each putative QTL position (Fig. 20.4). The putative position with the highest LOD score is the most likely position of the QTL. The evaluation of the LOD score at many intervals along the genome is computationally intensive. An alternative is to use a regression approach, where the phenotype data are regressed on QTL allele probabilities at each putative location, as inferred from flanking markers (Haley and Knott, 1992). The regression approach appears often to give a high quality approximation to the ML method, and is much less computationally intensive (Haley and Knott, 1992). Lynch and Walsh (1998) gave an excellent description of statistical methodologies for QTL mapping.

The problem of significance testing with many marker tests

The significance threshold, which the LOD score must exceed, is usually set to control the number of false-positive results (a false positive occurs when a QTL is detected and there is no true underlying QTL effect at that position). A major issue in setting significance thresholds in interval mapping is the multiple testing problem. In interval mapping, many positions along the genome or a chromosome are analysed for the presence of a QTL. As a result, when these multiple tests are performed the 'nominal' significance levels for a single test do not correspond to the actual significance levels in the

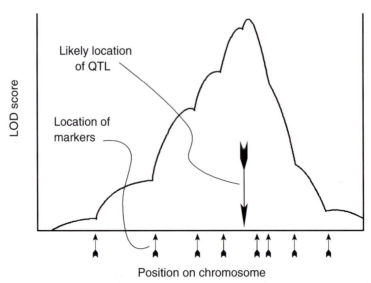

Fig. 20.4. Hypothetical LOD profile along a chromosome from a genome scan, showing the most likely QTL position.

whole experiment. For example, with a point wise (single test) significance threshold of 0.05, 5% of results at that location would be expected to be false positives. If 100 points along the chromosome are analysed (assuming independence of these points) five (100 × 0.05) false-positive results would be expected. To avoid numerous false positives, more stringent thresholds are required. The multiple testing problem in QTL mapping is even more complex because 'tests' on the same chromosome are not independent, as the markers are linked. Simple procedures, which assume independence between tests, such as the Bonferoni correction, will not give correct significance thresholds.

Churchill and Doerge (1994) proposed the technique of permutation testing to overcome the problem of multiple testing in QTL mapping experiments. Permutation testing is a method to set appropriate significance thresholds with multiple testing (e.g. testing many locations along the chromosome for the presence of the QTL). Permutation testing is performed by analysing simulated data sets that have been generated from the real one by randomly shuffling the phenotypes across individuals in the mapping population. This removes any relationship between genotype and phenotype, and generates a series of data sets corresponding to the null hypothesis. Chromosome or genome scans can then be performed on these simulated data sets. For each simulated data set the position in the genome yielding the highest value for the test statistic is identified and stored. The values obtained over a large number of such simulated data sets are ranked, yielding an empirical distribution of the test statistic under the null hypothesis of no QTL. The position of the test statistic obtained with the real data in this empirical distribution immediately measures the significance of the real data set. Significance of thresholds can then be set corresponding to 5% false positives for the entire experiment, 5% false positives for a single chromosome, and so on.

Another useful statistic for assessing the proportion of false-positive results from a QTL detection experiment is the false discovery rate (FDR). FDR is the expected proportion of detected QTLs that are in fact false positives (Weller et al., 1998; Storey and Tibshirani, 2003). The FDR can be used to set significance thresholds for each individual test, such that a pre-determined proportion (for example 5%) of significant results from the genome scan are in fact positions where no QTL are segregating.

The problem of over-estimation of QTL effects from genome scans

A consequence of setting significance thresholds for QTL detection is over-estimation of QTL effects. This occurs because the QTL effect is made up of both a true effect and an environmental error, and the QTLs which exceed the significance threshold are more likely to have received a favourable environmental error component. Georges *et al.* (1995) used simulation to demonstrate that the over-estimation of QTL effects is more pronounced for QTL of smaller true effect, and also that overestimation is more pronounced with smaller numbers of animals in the experiment. If the QTLs are to be used in marker-assisted selection in a livestock breeding programme, some strategy should be used to regress the estimate of the QTL effect to reflect the degree of overestimation that is likely to have occurred in the QTL detection experiment.

Precision of QTL mapping

The precision of positioning a putative QTL along a chromosome is usually expressed as an interval that contains the QTL with a level of statistical certainty, e.g. a 95% confidence interval (CI). One method to assign confidence intervals to QTL locations is based on the likelihood ratio test (Lander and Botstein, 1989; Zeng, 1994). The likelihood ratio test is performed at any position covered by markers across the whole genome. The location with the highest likelihood is the most likely putative QTL position. The CI is calculated by moving sideward (left and right) of the estimated position to the locations corresponding to a decrease in the LOD score of one or two units. The total width corresponding to a one- or two-LOD drop-off can then be considered as the 96.8 or 99.8% CI, respectively (derivation given in Mangin *et al.*, 1994). In the Lander and Botstein method, estimates of QTL position and its effects are approximately unbiased if there is only one QTL segregating on a chromosome (Zeng, 1994). Haley and Knott (1992) adopted a similar approach in a regression framework.

Darvasi and Soller (1997) proposed a formula for estimating the 95% CI for QTL location for simple QTL mapping designs under the assumption of a high density genetic map: CI = $3000/(kN\delta^2)$, where N is the number of individuals genotyped, δ^2 the substitution effect in units of the residual standard deviation, k the number of informative parents per individual, which is equal to 1 for half-sibs and backcross designs and 2 for F_2 progeny, and 3000 is about the size of the cattle genome in centiMorgans. For instance, given a QTL with a substitution effect of 0.5 residual standard deviations, and 1000 progeny genotyped, the 95% CI would be 12 cM.

A bootstrap method to determine approximate CIs for QTL position was proposed by Visscher *et al.* (1996). For data on N individuals, a bootstrap sample is created by sampling with replacement N individual observations from the data set. Each observation has marker genotype and phenotype. In the bootstrap sample some records can appear more than once. This process is repeated n times to generate n bootstrap samples. The Haley and Knott (1992) interval mapping method was used to detect QTLs from the bootstrap samples. The empirical central 90 and 95% CIs of the QTL position are determined by ordering the n estimates and taking the bottom and top fifth and 2.5th percentile, respectively. Another method to improve the mapping precision is to increase the marker density on the chromosome. In practice the effectiveness of this strategy is limited with linkage mapping, as a very large number of recombinants between closely spaced markers are required in order to refine the QTL position. LD mapping is a possible solution and will be discussed later.

Designs of QTL mapping experiments

For linkage experiments, power to detect and locate QTLs is derived from both the accuracy with which QTL effects are estimated, determining the ability of the experiment to separate the QTL effect from environmental and other effects, and the number of recombination events in the resource population. The more recombination events there are in the resource population, the more precise the estimation of the location of QTLs.

To a large extent the resource population which will be used in a QTL detection experiment depends on the reproductive biology and existing lines of the target species. In mice, breeding resources include numerous special inbred lines and strains (CTC, 2003) and dramatically exceed the QTL mapping capabilities of other mammalian species. Generally, breeding strategies which led to successful QTL mapping are based on backcrosses or quite often on intercrosses. An F_2 population created from the intercrosses of lines or breeds, in which the genes in question are segregating, is a powerful resource for QTL mapping. F_2 populations have also been used as a resource for QTL mapping in a number of mammalian species. In the livestock species, it is also possible to exploit the high reproductive capacity of males, and breed large half-sib progeny groups as a resource population. QTL detection experiments in humans are conducted in the existing populations, usually numbers of full-sib families, and also exploiting any genetic ties between the families. Large families are particularly attractive for such investigations (Dyer *et al.*, 2001). Each of these designs (F_2 half-sib, and complex pedigree) will be illustrated with an example.

Crosses of inbred or divergent lines

When two inbred lines are crossed, linkage disequilibrium is created in the F_1 population between the marker and QTLs that are fixed in the alternative lines. This LD can be exploited by randomly mating among the F_1s to produce F_2 animals, where all the markers and QTLs fixed for alternative alleles in the two parental lines will be segregating. If the lines are not inbred, but only divergent (for example different breeds), a proportion of the loci will not be homozygous in the parental lines, and the proportion of loci (both QTLs and markers) segregating in the F_2 population will be lower. Either ML or regression methods can be used to analyse phenotype and genotype data from F_2 crosses, with both methods giving similar results (Knott and Haley, 2000).

The power of a mapping experiment using F_2 populations from crosses of inbred lines is a function of the number of studied offspring, the size of the QTL effect, and the degree of the

QTL dominance. Lynch and Walsh (1998) give an expression for the number of F_2 progeny required to detect QTLs explaining a proportion r_{F2}^2 of the F_2 segregation variance with a power of $1 - \beta$:

$$n_{F2} = \left(\frac{1 - r_{F2}^2}{r_{F2}^2} \right) \left(\frac{z_{(1-|\alpha/2|)}}{\sqrt{1 - r_{F2}^2}} + z_{(1-\beta)} \right)^2 \left[1 + (k^2 / 2) \right]$$

where α is the significance level. For example, to detect a QTL explaining 5% of the total F_2 variance with a power of 90% and with $\alpha = 0.05$, with completely additive gene action ($k = 0$), 206 F_2 progeny would be required.

If the parental lines are not inbred, the power is also a function of the difference in allele frequencies at the marker and QTLs between the parental lines. The more similar the allele frequencies, the lower the power of the design to detect QTLs.

Case study: mapping of obesity QTL in a cross between mouse lines divergently selected on fat content

Horvat *et al.* (2000) performed a genome scan in an F_2 population created from parental outbred lines divergently selected for 53 generations for high and low fat percentage. There was a fivefold difference in fat percentage between the parental lines. Nineteen parents from each line were selected and mated to produce an F_1 population. One hundred and six F_1 individuals were selected, mated to produce 436 F_2 offspring, and phenotyped for total body fat percentage.

Seventy-one markers, on 20 chromosomes, were genotyped in the F_2 population. The majority of the markers were not fixed for alternative alleles in the parental line and, averaged across marker loci, 47% of F_2 progeny were fully informative (both alleles at a marker can be traced back to the parental line), 33% were partially informative (only one marker allele could be traced back to a parental line), and 21% were non-informative (neither allele could be traced back to a parental line).

The genotype and phenotype data from the F_2 progeny were analysed using an ML interval mapping procedure (Keightley *et al.*, 1998). At some locations the markers flanking a putative

QTL position were not informative. This was accounted for by considering the closest informative marker in the same linkage group in the analysis. Permutation testing (Churchill and Doerge, 1994) was used to set an experiment (or genome) wide significance 5% threshold for QTL detection. Four QTLs in the study whose effects exceeded the significance threshold were located on four different chromosomes. As an F_2 design was used, evidence for dominance interaction could be tested. Three QTLs had additive effects, and the fourth showed evidence for dominant effects. The QTLs accounted for 4.9, 19.5, 14.4 and 7.3% of the F_2 phenotypic variance for fat percentage.

The number of F_2 progeny (426) used in this experiment meant that the power of this design to detect QTLs was high. With 436 progeny, the smallest QTL that could be detected with 95% power would explain approximately 4% of the F_2 variance. Altogether, the QTLs accounted for 46.1% of the F_2 phenotypic variance for fat percentage, demonstrating the substantial power of the design. The effects of these QTLs are likely to be over-estimated, however, as the estimated effects of QTLs which receive a favourable environmental component are more likely to exceed the significance threshold (Georges *et al.*, 1995).

Half-sib designs

In species where the reproductive capacity of the male is much higher than that in the female, half-sib designs are a powerful tool for QTL detection. In the analysis of these designs, only the marker alleles of the sire are evaluated for linkage to putative QTLs.

Either regression or ML methods can be used to analyse the genotypic and phenotypic data from the half-sib families. Knott *et al.* (1994), Spelman *et al.* (1996) and Uimari *et al.* (1996) presented regression methodology for QTL mapping in half-sib populations. The assumptions underlying this method are that unrelated sires and unrelated dams are mated randomly, each dam has one progeny, the marker map is known, and phenotypic data are available on the offspring (Hoeschele *et al.*, 1997). Briefly, the procedure (Knott *et al.*,

1998) is to determine what the most likely linkage phase of genetic markers is for each sire; then the probabilities of inheriting each of the two sire gametes are calculated for fixed positions along the chromosome, conditional on the marker genotypes of the progeny. If marker genotypes are not informative (a sire is homozygous for the marker), this probability is 0.5. The phenotypes are then regressed on to the probabilities of inheriting the paternal gametes at each putative QTL position. Not all sires are expected to be heterozygous, and indeed, different heterozygous sires will generally have distinct linkage phases to adjacent markers. Therefore, the analysis should be carried out separately for each sire, with subsequent pooling of results, unless complex pedigree techniques are used. This allows the effect of the QTL (the gene substitution effect, b_i) to vary from one sire to another as if there were an infinite number of possible QTL alleles.

ML analysis has also been implemented for half-sib designs in outcross populations (e.g. Georges *et al.*, 1995; Mackinnon and Weller, 1995). The underlying genetic model in the ML analysis can be identical to that assumed in the regression analysis, that each sire has a unique QTL effect. Alternatively, a two-allele model can be assumed (Hoeschele *et al.*, 1997). In that case the likelihood for a QTL at a given position requires three parameters: the frequency of sires heterozygous at the QTL, the substitution effect of the QTL, and the residual variance (Knott *et al.*, 1996; De Koning *et al.*, 1999). The likelihood is maximized at fixed locations along the chromosome and compared to the likelihood of the null hypothesis of no QTL. Generally ML and regression methods give similar estimates of QTL location and effect from the analysis of half-sib data (Knott *et al.*, 1998).

The power of half-sib designs for the detection of QTLs is derived from both the number of sires and the number of half-sib offspring per sire. At least one of the sires used in the mapping experiment must be heterozygous at the QTL. The sire families must also be large enough to ensure that the difference between the effects of the two QTL alleles on the quantitative trait can be distinguished from environmental and other genetic effects. Weller *et al.* (1990) evaluated alternative designs for QTL detection (Table 20.2). Increasing the number of sires increases the chance of at least one sire being heterozygous at the QTL, so the power is increased. Very large family sizes are needed to detect small QTLs with any certainty.

If the total number of progeny is limited by cost, there is then a choice between a large number of sires, each with a small number of progeny, and a small number of sires, each with a large number of progeny. Which is a better design for QTL mapping? This question has been investigated using both deterministic predictions (e.g. Weller et al., 1990), and simulation (e.g. Hayes et al., 2001). Both approaches come to the same conclusion, which is to detect QTLs of medium–large size (e.g. effect of approximately 0.2 phenotypic standard deviations), very large half-sib families are needed (Fig. 20.5 and Table 20.2).

In Fig. 20.5, the criterion for measuring the success of the QTL mapping experiment is the total proportion of genetic variance explained by the QTL 'detected', that is, those QTLs which have effects exceeding a significance threshold. In this case the significance threshold was $P<0.05$ at the chromosome-wide level, set using permutation testing. Increasing the numbers of boars increased the chance of one or more boars being heterozygous for a QTL segregating in the population. However, for a given total number of progeny for the mapping experiment, increasing the number of boars decreased the progeny per boar and lessened the chance of the QTL effect to be statistically significant. Usage of five boars balanced these two phenomena and maximized the proportion of variance explained by detected QTLs, regardless of the total number of progeny in the experiment.

Case study: mapping QTLs controlling milk production in dairy cattle by exploiting progeny testing

Georges et al. (1995) performed a genome scan for QTLs affecting milk production in dairy cattle. The power of the design was increased by exploiting a feature of dairy cattle breeding, namely the use of progeny testing to evaluate the genetic merit for milk production of young bulls. In progeny testing, a large

,**Table 20.2.** Power of half-sib design to detect a segregating QTL.

	Number of		Size of QTL effects[1]		
Sires	Progeny per sire	Total progeny	0.1	0.2	0.3
5	200	1,000	0.03	0.18	0.50
	400	2,000	0.07	0.44	0.80
	600	3,000	0.12	0.64	0.90
	800	4,000	0.18	0.76	0.94
	1000	5,000	0.25	0.83	0.96
	2000	10,000	0.55	0.95	0.97
10	200	2,000	0.05	0.31	0.76
	400	4,000	0.11	0.70	0.96
	600	6,000	0.21	0.88	0.99
	800	8,000	0.32	0.95	0.99
	1000	10,000	0.43	0.97	0.99
	2000	20,000	0.81	0.99	0.99
20	200	4,000	0.07	0.56	0.95
	400	8,000	0.20	0.93	0.96
	600	12,000	0.38	0.99	0.99
	800	16,000	0.56	0.99	0.99
	1000	20,000	0.70	0.99	0.99
	2000	40,000	0.97	0.99	0.99

Weller et al. (1990); [1] The size of QTL effects = a/SD, where a is half the difference between the mean trait values for the two homozygotes, and SD is the residual standard deviation (the author's permission to use this table is kindly granted).

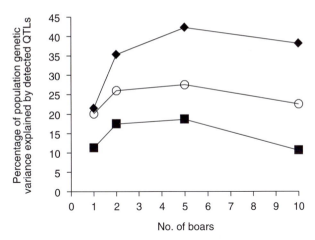

Fig. 20.5. Percentage of genetic variance explained by detected QTLs in genome scans with 1, 2, 5 or 10 boars and 500 (■), 1000 (○) or 2000 (♦) total progeny allocated to the mapping experiment.

number of daughters of a young bull are evaluated for their milk production, and the average of these records, the daughter yield deviation (DYD), is the phenotypic record of the young bull. These young bulls can be grouped according to their sire into half-sib groups. The young bulls are genotyped for markers, and a QTL mapping experiment performed. The extra power of this design, termed the granddaughter design, relative to a standard half-sib design, is derived from reduction of environmental noise in the phenotypes (DYDs) used in the analysis.

The resource population consisted of 1518 young bulls in 14 half-sib families. The 'phenotype', or DYD of each of the 1518 young bulls was derived from the milk production records (milk yield, protein yield, fat yield, protein %, fat %) of between 50 and 100 daughters per young bull. Each of the 1518 young bulls was genotyped for 159 microsatellite markers, spanning two-thirds of the bovine genome. An interval mapping approach using ML was used to analyse the genotype and phenotype data. No permutation testing was used to set significance thresholds, rather QTLs were taken as significant if the LOD score from the ML analysis was equal to or greater than three.

Five QTLs with LOD scores greater than or equal to three were detected. Three of these QTLs had significant effects on protein %, and two on fat yield and protein yield. Surprisingly, the detected QTLs explained more than 100% of the within sire variance for some traits, a clear result of overestimation of the QTL effect.

Complex pedigree designs

In humans, it is obviously not possible to breed large half-sib groups or to create crosses between inbred lines. Genome scans must be conducted within the population structure that pervades most societies, which is small family groups. If only the nuclear family is considered (parent–offspring), the power of these experiments is low. For example, to detect a QTL which explains 20% of the genetic variance for a trait, 8000 sib pairs (two sibs in each pair) would be required (Williams and Blangero, 1999)! Extra power can be gained in the genome scan if extended pedigrees can be genotyped, with more distant relationships than offspring–parent accounted for in the analysis, as this will include more recombination events to be traced in the experiment, and a large number of observations per QTL allele. In practice, the large number of missing marker genotypes reduces the power of complex pedigrees for QTL mapping.

Regression methods (George *et al.*, 1999), ML (Hoeschele *et al.*, 1997) and variance component methods have been proposed for the analysis of complex pedigrees. The variance component approach may be the most useful method of analysis due to its flexibility (George *et al.*, 2000). Such an analysis proceeds in two steps:

1. For each QTL position on the chromosome segment, calculate the (co) variance matrix associated with the QTL. This matrix is also called the **G** or **IBD** (identical by descent matrix), and has elements ij = Prob(QTL alleles i and j are identical by descent or IBD).
2. For each position considered in step 1, construct a linear model to estimate QTL variance and other parameters; test for the presence of a QTL.

Calculating the IBD matrix

The IBD matrix has the dimensions 2*the number of animals $\times$ 2*the number of animals, e.g. two QTL alleles for each animal in the pedigree. The IBD matrix traces the transmission of the alleles of the founder animals (those at the top of the pedigree) through the population. If the marker information was complete, and could be used to perfectly infer the transmission of QTL alleles, this matrix would contain 1s and 0s only. At the other extreme, if there is no marker information, the IBD matrix is characterized by many elements of 0.5, signifying equal probability of inheriting either allele from a parent.

The inference of QTL genotypes from marker genotypes is considerably more complicated in complex pedigrees than simple half-sib pedigrees, as marker alleles need to be tracked over multiple generations. This can lead to a large number of missing genotypes. These genotypes have to be inferred in some way, which has proved to be very difficult in populations with many inbreeding and or marriage loops (e.g. most livestock populations). Considerable effort has been invested in creating strategies to infer genotypes with missing marker information and complex pedigrees (e.g. Kerr and Kinghorn, 1996; Heath, 1997). Most strategies now use simulation-based methods, predominantly Markov Chain Monte Carlo (MCMC) approaches. A review of such methods is available in George *et al.* (2000).

Variance component approaches for estimation of QTL parameters

The linear model used to analyse data from complex pedigrees is (ignoring fixed effects): $Y = \mu + Zu + Zv + e$, where Y is a vector of observation, μ the overall mean, Z the design matrix relating animals to phenotypes, u the vector of additive polygenic effects, v the vector of additive QTL effects, and e the residual vector. The random effects u, v, and e are assumed to be distributed as follows: $u \sim (0, \sigma_u^2 A)$, $v \sim (0, \sigma_v^2 G)$, $e \sim (0, \sigma_e^2 I)$, where σ_u^2, σ_v^2, and σ_e^2 are the polygenic variance, the additive QTL variance, and the residual variance, respectively; A is the standard additive genetic relationship matrix, and G is a matrix whose ij element G_{ij} = Prob(QTL alleles i and j are identical by descent or IBD) is described above.

There is no simple approach to calculate the power of a complex pedigree design, though the power will obviously be some function of the total number of recombination events and observations per QTL allele. Simulation is one approach, where the actual pedigree to be analysed is simulated, with QTLs of known effect segregating. The simulated data can then be analysed to determine if the design is powerful enough to detect a QTL of that size.

Case study: QTLs influencing body-mass index reside on chromosomes 7 and 13: the National Heart, Lung and Blood Institute Family Heart Study

Body-mass index (BMI) is a complex trait, made up of component traits including amount of fat, lean mass and body build. Feitosa *et al.* (2002) conducted a genome scan in two populations, one of 1184 individuals from 317 sibships, and another of 3027 individuals in 401 three-generation families. In the first family, 243 markers were genotyped across the autosomal genome; in the second 404 markers were genotyped. A variance component approach was used to analyse the genotype and phenotype data.

This study is interesting because it allows a comparison of the power of the sibship design and the power of the extended pedigree approach. Both studies found strong evidence for a QTL affecting BMI on chromosome 7. With the multi-generational pedigree, a second QTL was below the significant level on chromosome 13 (LOD score of 2.6), while in the sibship pedigree the LOD score was only 1.0. Combining the two pedigrees in the same analysis was the most powerful design; in the combined analysis, the QTL on chromosome 13 had a LOD score of 3.2. This study also demonstrates that even with large numbers of individuals genotyped (1184 and 3027 individuals in each of the studies) the power to detect QTLs in human pedigrees was not high, as indicated by the comparatively low LOD scores.

Comparison of case studies

Although the traits in each of the case studies were different, as well as the analysis procedures, it is still informative to briefly compare the results. The ratio of number of QTLs found to the amount of genotyping is particularly interesting. In the F_2 mouse study, four significant QTLs that affected fat percentage were detected in a F_2 population of 436 individuals (Horvat et al., 2000). In the dairy cattle half-sib design, 1518 bulls were genotyped, and three QTLs were detected which affected protein percentage in milk (Georges et al., 1995). In the human study, 4201 individuals were genotyped, with two significant QTLs affecting BMI detected (Feitosa et al., 2002). Both the studies of Georges et al. (1995) and Feitosa et al. (2002) used criteria of LOD greater than or equal to three as the criteria for QTL detection, the lowest LOD score for a 'detected' QTL in the study of Horvat et al. (2000) was 3.2, so the test criteria are comparable across the studies. So the F_2 design returned the most QTLs detected per genotype. There are several reasons for this. The strategy of breeding the parental lines for 53 generations with divergent selection (selection for increased fat percentage in one line and selection for decreased fat percentage in the second line) maximized the probability that QTLs affecting fat percentage would be segregating in the F_2 population. Secondly, it is only the alleles from the two parental lines which are traced in the analysis, so there are a large number of observations

per allele, and allele effects are estimated accurately. This is necessary if the QTL effects are to be distinguished from environmental and other effects. Contrast this to the situation in the human population, where the founders of each pedigree are assumed to carry unique QTL alleles, so there are a very limited number of observations per allele. In the granddaughter design, there were on average 50 observations per grandsire allele. However the reduction of environmental noise in the experiment through the use of progeny testing increased the value of each of these observations, and the experiment was sufficiently powerful to detect QTLs explaining a considerable proportion of the genetic variance.

Increasing the Power of QTL Detection Experiments

QTL detection experiments are expensive. The cost of genotyping is often the largest cost in the experiment. Therefore strategies which either decrease the number of genotypings or return greater power to detect QTLs for the same number of genotypings are particularly valuable.

Selective genotyping

Selective genotyping is a method of QTL mapping in which the analysis of linkage between markers and QTLs is carried out by genotyping only individuals from the high and low phenotypic tails of the trait distribution in the population (Darvasi and Soller, 1992). In half-sib designs, the selective genotyping is usually done within each sire family. Individuals deviating most from the mean are considered to be most informative for linkage, as their genotypes at the QTLs can be inferred from their phenotypes more clearly than can those with average phenotypes (Fig. 20.6).

In fact, Darvasi and Soller (1992) demonstrated that it is not necessary to genotype more than 50% of a population to get maximum power from the design. For a constant number of genotyped progeny, selective genotyping can actually increase the power of the mapping experiment (Bovenhuis and

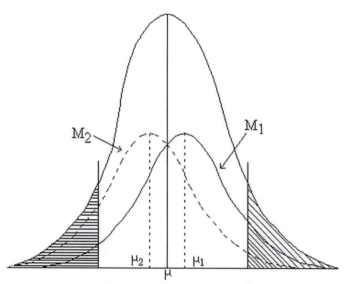

Fig 20.6. Distribution of the progeny of a M_1Q_1/M_2Q_2 sire, with high and low phenotypes for the production trait.

Meuwissen, 1996). Figure 20.7, from Bovenhuis and Meuwissen (1996), shows the power for different selected fractions as a function of the total number of animals with phenotypic records. The design consists of five sires with a large number of daughters. Other parameters are: heritability – 0.1, type I error – 0.05, QTL substitution effect – $0.2\sigma_p$, and

recombination fraction – 0. For a given number of animals genotyped, and no restrictions on the number of animals available for phenotypic trait evaluations, the power can be increased dramatically by using selective genotyping. This increase in the power results from the increased contrast between individuals carrying different marker genotypes. Nevertheless

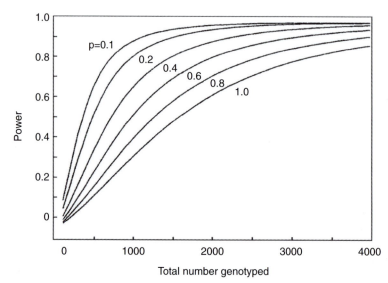

Fig. 20.7. The power of a daughter design (probability of detecting a QTL with effect of $0.2\ \sigma_p$), as a function of the number of individuals genotyped, for different selection fractions (p) (reproduced with permission from Bovenhuis and Meuwissen, 1996).

it is recommended that the selection not be lower than 10% in either tail (Bovenhuis and Meuwissen, 1996) because the data might contain outliers representing artefacts.

One drawback with selective genotyping is that the estimated QTL effect is severely biased upwards if only genotyped individuals are used to estimate the effect (Darvasi and Soller, 1992). This is a major problem if the QTLs are to be used in marker-assisted selection (MAS), as the overestimation of the QTL variance will erode the advantage of using the marker information. Methods have been derived to correct for this bias (Darvasi and Soller, 1992). Another drawback is that the power to detect QTLs for traits other than the one on which selective genotyping was performed is reduced. In livestock at least, one way to overcome this drawback is to use an index of the traits, which reflects their relative economic importance.

Various selective genotyping strategies have been proposed for human QTL detection experiments using sib-ships. Selective genotyping of discordant sib-ships is a strategy whereby only the sib-ships with the most extreme phenotypic differences between sibs are genotyped. The principle underlying this method is that the difference in phenotypes between pairs of sibs becomes larger as they share a decreasing number of alleles at a particular QTL identical by descent (IBD) from their parents (Chatziplis *et al.*, 2001), and in such families a QTL is more likely to be segregating. Other strategies, such as selection of sib-ships with the highest phenotypic variation within the sib-ship, have also been suggested (Chatziplis *et al.*, 2001).

Selective DNA pooling

A very efficient strategy to greatly reduce the number of genotypings was proposed by an Israeli mapping group (Darvasi and Soller, 1994; Lipkin *et al.*, 1998). In DNA pooling, the determination of linkage between a marker and a QTL is based on the distribution of parental alleles among pooled DNA samples of the extreme high and low phenotypic groups of offspring.

In theory, the amount of genotyping using selective DNA pooling is very low, with only two pools to be genotyped – or two pools per family if within-family contrasts are used to help reduce genetic differences between pools caused by specific parental contribution to different pools. In practice, the need to replicate the pools to increase the accuracy of results increases the number of genotypings required.

There are a number of difficulties to be overcome with selective DNA pooling, including accurate quantification of DNA from each individual going into the pool, and removal of 'shadow' ('stutter') bands in electropherograms when using microsatellites. A further complication is that the selective DNA pooling experiment has the power to detect QTLs affecting only the trait from which the pools of high and low phenotype progeny were made. The power to detect QTLs for other traits will be much lower, unless there is a high genetic correlation between the traits.

A large DNA pooling experiment has been carried out in Israeli-Holstein dairy cattle to detect QTLs affecting milk protein percentage (Lipkin *et al.*, 1998). Selective DNA pooling accessed 80.6 and 48.3%, respectively, of the information that would have been available through individual selective genotyping or total population genotyping. In effect, the statistical power of 45,600 individual genotypings was obtained from 328 pool genotypings. The experiment detected five QTLs with highly significant effects on protein percentage.

Strategies to Improve the Precision of QTL Mapping Experiments

Linkage disequilibrium mapping

The precision of a QTL detection experiment, the accuracy with which a QTL is positioned along the chromosome, is limited by the frequency of observable recombination in the genotyped progeny. Linkage experiments require a large number of progeny to position QTLs with any precision; such numbers are often beyond the reproductive capacity of the target species or resources of the experimenter. Yet to limit the number of genes which could harbour the causative mutation to a number, which is amenable to investigation by sequencing and other strategies, the location of the QTL must be narrowed to 1–3 cM. One strategy is to use historical recombinants, which have occurred in the resource population in the generations preceding the genotyped generation, using LD mapping.

Definitions of LD

The classical definition of LD refers to the non-random association of alleles between two loci (Hendrick, 2000). Consider two markers, A and B, which are on the same chromosome. A has alleles A1 and A2, and B has alleles B1 and B2. Four haplotypes of markers are possible A1_B1, A1_B2, A2_B1 and A2_B2. If the frequencies of alleles A1, A2, B1 and B2 in the population are all 0.5, then we would expect the frequencies of each of the four haplotypes in the population to be 0.25. Any deviation of the haplotype frequencies from 0.25 is LD, i.e. the genes are not in random association. As an aside, this definition serves to illustrate that the distinction between linkage and LD mapping is somewhat artificial – in fact LD between a marker and a QTL is required if the QTL is to be detected in either sort of analysis. The difference is:

- *linkage analysis* only considers the LD that exists within families or within a F_2 population, which can extend for tens of cM, and is broken down by recombination after only a few generations.
- *linkage disequilibrium* mapping requires a marker allele or marker alleles to be in LD with a QTL allele across the entire population. To be a property of the whole population, the association must have persisted for a considerable number of generations, so the marker(s) and QTLs must therefore be closely linked.

One measure of LD is D, calculated as (Hill, 1981):

$D = freq(A1\_B1)*freq(A2\_B2)-$
$freq(A1\_B2)*freq(A2\_B1)$

where freq (A1_B1) is the frequency of the A1_B1 haplotype in the population, and likewise for the other haplotypes. The D statistic is very dependent on the frequencies of the individual alleles, and so is not particularly useful for comparing the extent of LD among multiple

pairs of loci (e.g. at different points along the genome). Hill and Robertson (1968) proposed a statistic, r^2, which was less dependent on allele frequencies,

$$r^2 = \frac{D^2}{freq(A1)*freq(A2)*freq(B1)*freq(B2)}$$

where freq(A1) is the frequency of the A1 allele in the population, and likewise for the other alleles in the population.

These classical definitions of LD, while important and widely used, are not particularly illuminating with respect to the causes of LD, and may also not be especially useful for QTL mapping. For example, statistics such as r^2 consider only two loci at a time, whereas we may wish to calculate the extent of LD across a chromosome segment that contains multiple markers. An alternative multilocus definition of LD is the chromosome segment identity (equal to the chromosome segment homozygosity, CSH, introduced by Hayes et al., 2003). Consider an ancestral animal many generations ago, with descendants in the current population. Each generation, the ancestor's chromosome is broken down, until only small regions of chromosome which trace back to the common ancestor remain. These chromosome regions are identical by descent (IBD). Figure 20.8 demonstrates this concept.

The CSH, then, is the probability that two chromosome segments of the same size and location drawn at random from the population are from a common ancestor (i.e. IBD), without intervening recombination. CSH is defined for a specific chromosome segment, up to the full length of the chromosome. The CSH cannot be directly observed from marker data but has to be inferred from marker haplotypes for segments of the chromosome.

Why does LD occur?

LD can be a result of migration, mutation, selection, small finite population size or other genetic events that the population experiences (Lander and Schork, 1994; Hedrick, 2000). LD can also be created in livestock populations; in an F_2 QTL mapping experiment LD can be created between marker and QTL alleles by crossing two lines.

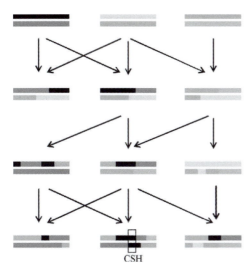

Fig. 20.8. Transmission of haplotypes through generations. The ancestor with a black-coloured chromosome has descendants. In each generation, this chromosome is broken down by recombination, until all that remains in the current generation are small conserved segments of the ancestor's chromosome. The chromosome segment homozygosity (CSH) illustrated requires some inbreeding. As a parameter, it is the probability that two chromosome segments of the same size and location drawn at random from the population are from a common ancestor.

In human and livestock populations, finite population size is generally implicated as a key cause of LD. LD due to cross-breeding or migration is large when crossing inbred lines but small when crossing breeds that do not differ as markedly in gene frequencies, and it disappears after only a limited number of generations (Goddard, 1991). Moreover, mutations are likely to have occurred many generations ago. While selection is possibly an important cause of LD, any impact is likely to be localized around specific genes, and have relatively little effect on the amount of LD 'averaged' over the genome.

Predicting the extent of LD with finite population size

There is a simple expectation for the amount of LD for a given size of chromosome segment and assuming finite population size is the only cause of LD. This expectation is (Sved, 1971):

$$E(r^2) = 1/(4N_e c + 1)$$

where N_e is the finite population size, and c is the length of the chromosome segment in Morgans. The CSH has the same expectation (Hayes et al., 2003). This equation predicts rapid decline in LD as genetic distance increases, and this decrease will be larger with large effective population sizes (Fig. 20.9).

If LD is predominantly a result of finite population size, then the extent of LD should be many times less in humans, where the effective population size is ~10,000 (Kruglyak, 1999), than in livestock, where effective population sizes can be lower than 100 (Riquet et al., 1999). In fact, this is what is observed. Significant LD in humans typically extends less than 5 kb (~0.005 cM), depending on the population studied (Dunning et al., 2000; Reich et al., 2001), while in cattle and sheep, considerable LD can extend up to 5–10 cM and longer (Farnir et al., 2000; McRae et al., 2001; Hayes et al., 2003).

In humans, there is some evidence that the extent of significant LD is greater in isolated populations, as a result of their smaller effective population size (though there is also some evidence that does not support this conclusion, e.g. Boehnke (2000). Shifman and Darvasi (2001) demonstrated that the level of LD, for markers spaced greater than 200 kb, was greater in isolated populations than outbred populations. As a result, less dense marker maps and less genotyping may be required to detect associations with QTLs in isolated populations than outbred populations. A further advantage of isolated populations is that genetic heterogeneity, which can be responsible for false positive results in QTL detection experiments, is reduced relative to outbred populations (Shifman and Darvasi, 2001).

Although LD that extends several hundred kb or even dozens of cM is advantageous in that the density of markers required for detecting QTLs can be reduced, compared with populations where LD extends only very short distances, the disadvantage of such high levels of LD is that the precision with which the QTL can be mapped may be limited. For example, if LD extends tens of cM, a marker located 0.5 cM from the gene may show a similar association to a QTL with a marker 0.25 cM from the gene.

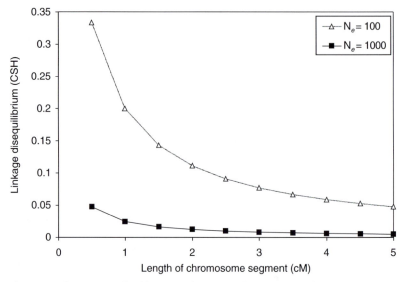

Fig. 20.9. The extent of LD as a result of finite population size, for increasing chromosome segment length, for N_e = 100 and N_e = 1000.

What type of markers are appropriate for detecting LD?

As population-wide LD extends over much shorter genetic distances than does LD within families, a denser marker map is required to detect and position QTLs using the LD approach.

For linkage analysis, microsatellite markers are favoured, due to their high polymorphism (many alleles in the population) and ease of amplification. A microsatellite marker is usually a dinucleotide repeat, e.g. ATATAT repeated many times. Microsatellites are nearly always neutral, having no effect on gene expression. The microsatellite marker maps are well developed in humans, mice and the major livestock species, and their position on the genetic map is known. In humans and mice the number of mapped microsatellites exceeds 10,000. In the major domestic mammals like sheep, cattle and pigs, the number of mapped microsatellites is well above 1000 and this continues to grow.

In livestock species, where LD extends for a considerable distance, microsatellite markers may be suitable for LD mapping. In humans, where significant LD extends for only very small distances, microsatellites are not as suitable for LD mapping, simply because they are

not sufficiently dense. Alternative markers are single nucleotide polymorphisms, or SNPs (see Chapter 4).

SNPs have the advantage that they occur very frequently throughout the genome, approximately 1 every 1000 bases (0.005 cM). It is possible that an SNP in the LD experiment may be the actual mutation causing the QTL effect. The disadvantage of SNPs is that they are not as polymorphic as microsatellites – they have a maximum of two alleles. About five SNPs are required to give the same amount of information as a single microsatellite (Dodds, 2003).

LD mapping analysis

There are relatively few methods that have been developed for detection of QTLs, rather than genes which have a qualitative effect, to exploit LD. Here we review a method outlined by Meuwissen and Goddard (2001).

The existence of LD implies there are small segments of chromosome in the current population which are descended from the same common ancestor. These IBD chromosome segments will not only carry identical marker haplotypes, if there is a QTL somewhere within the chromosome segment, the IBD chromosome segments will also carry identical QTL alleles. Therefore if two animals carry chromosomes which are likely to be IBD at a point on the chromosome carrying a QTL, then their phenotypes will be correlated. We can calculate the probability that the two chromosomes are IBD at a particular point based on the marker haplotypes and store these probabilities in an IBD matrix (**G**). If the similarities between animal phenotypes are proportional to **G**, there is evidence for a QTL at this position.

Before the IBD matrices can be calculated, the genotype data must be sorted into haplotypes (also called estimation of linkage phases). This can be done with Gibbs sampling, or following Mendelian rules.

We can infer the IBD status of two chromosome segments from their marker haplotypes (the CSH) as described above. For example, consider a chromosome segment which carries 10 marker loci and a single central QTL. Three chromosome segments were selected from the population at random, and were genotyped at the marker loci to give the marker haplotypes 11212Q11211, 22212Q11111 and 11212Q11211, where Q designates the position of the QTL. The probability of being IBD at the QTL position is higher for the first and third chromosome segments than for the first and second or second and third chromosome segments, as the first and third chromosome segments have identical marker alleles for every marker locus.

This is the basis for calculating an IBD matrix, **G**, for a putative QTL position from a sample of marker haplotypes (Meuwissen and Goddard, 2001). Element g_{ij} of this matrix is the probability that haplotype i and haplotype j carry the same QTL allele. The dimensions of this matrix are twice the number of animals $\times$ twice the number of animals, as each animal has two haplotypes. For computational purposes, this matrix is often condensed to an $n \times n$ matrix by adding the block of four elements pertaining to each individual and dividing by two. Meuwissen and Goddard (2001) described a method to calculate the IBD matrix based on deterministic predictions which took into account the number of markers flanking the putative QTL position that are identical by state, the extent of LD in the population based

on the expectation under finite population size, and the number of generations ago that the mutation is presumed to have occurred.

The important parameters are the number of markers and the length of the haplotype, as well as the effective population size discussed above. One way to gain some insight into the effect of the number of markers in the haplotype on the IBD coefficients in the **G** matrix is to investigate the proportion of identical marker haplotypes which carry the same QTL allele, by calculating the proportion of QTL variance accounted for by marker haplotypes (ρ). If each unique marker haplotype is associated with a single QTL allele, this proportion will be one. For example, in a simulated population of $N_e = 100$, and a chromosome segment of length 10 cM, the proportion of the QTL variance accounted for by marker haplotypes when there were 11 markers in the haplotype was close to one (Fig. 20.10). (Note that if the effective population size was larger, the proportion of genetic variance explained by a 10 cM haplotype would be reduced (Goddard, 1991).)

Now consider a population of effective population size 100, and a chromosome segment of 10 cM with eight markers. Two animals are drawn from this population. Their marker haplotypes are 12222111, 11122111 for the first animal, and 12222111 and 11122211 for the second animal. The putative QTL position is between markers 4 and 5 (i.e. in the middle of the haplotype). The **G** or IBD matrix could look like that shown in Table 20.3.

The probability is that the two identical haplotypes (animal 1 haplotype 1 and animal 2 haplotype 2) in the IBD matrix would be very similar to the ρ coefficient in Fig. 20.10 for eight markers from the above simulation. The following linear model can be used to estimate the QTL variance:

$$Y = Xb + Zu + Wv + e$$

where Y is a vector of phenotypic observations, X is a design matrix allocating phenotypes to fixed effects, b is a vector of fixed effect, Z is a design matrix relating animals to phenotypes, u is a vector of additive polygenic effects, W is a design matrix relating phenotypic records to QTL alleles, v is a vector of

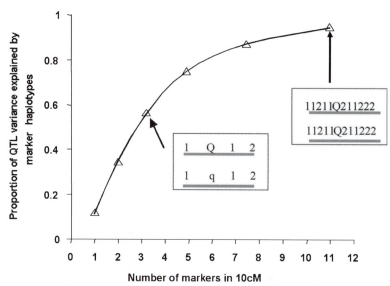

Fig. 20.10. Proportion of QTL variance explained by marker haplotypes with an increasing number of markers in a 10 cM interval. For three identical markers there is still a high probability that the QTL alleles are not identical by descent (IBD). indicated here by heterozygosity at the QTL. For 11 identical markers, which much better describe the haplotype, IBD is highly likely, indicated by the QTL homozygosity.

Table 20.3. Hypothetical QTL effects in animals carrying different haplotypes.

			Animal 1 Hap 1 12222111	Animal 1 Hap 2 11122111	Animal 2 Hap 1 12222111	Animal 2 Hap 2 11122211
Animal 1	Hap 1	12222111	1.00			
	Hap 2	11122111	0.30	1.00		
Animal 2	Hap 1	12222111	**0.90**	0.30	1.00	
	Hap 2	11122211	0.20	0.40	0.20	1.00

additive QTL effects and e is the residual vector. The random effects u, v, and e are assumed to be distributed as follows: $u \sim (0, A\sigma_u^2)$, $v \sim (0, G\sigma_v^2)$, $e \sim (0, \sigma_e^2 I)$, where σ_u^2, σ_v^2, and σ_e^2 are the polygenic variance, the additive QTL variance, and the residual variance, respectively; A is the standard additive genetic relationship matrix, and G is a matrix whose element G_{ij} is the probability that haplotypes i and j carry the same QTL allele (e.g. are IBD at the putative QTL position).

The precision with which the QTL variance, σ_v^2, is estimated will depend on both the number of unique haplotypes sampled from the population, and the number of observations per unique haplotype. The number of unique haplotypes sampled must be large enough to be representative of the population,

while the number of observations per unique haplotype determines the accuracy of estimating the haplotype effects. It is important to make a distinction here between the total number of haplotypes in the population, which will be twice the number of animals genotyped, and the number of unique haplotypes in the population, which is the number of different combinations of marker alleles that are present in the population. A unique haplotype can be represented many times in the population. If the marker haplotypes are to be used in marker-assisted selection, the accuracy of estimating the effect of a unique haplotype will determine the amount of improvement in the accuracy of selection as a result of using the marker information.

Combined LD–LA mapping

Authors investigating the extent of LD in both cattle and sheep were somewhat surprised/alarmed to find not only was LD highly variable across any particular chromosome, but there was even significant disequilibrium between alleles of loci located on different chromosomes (Farnir *et al.*, 2000; McRae *et al.*, 2001). It has been suggested that LD information be combined with linkage information to filter away any spurious LD likelihood peaks. This type of QTL mapping is referred to as LD–LA, for linkage disequilibrium–linkage analysis. The analysis proceeds by constructing an IBD matrix, which has two parts, a submatrix describing IBD coefficients between the haplotypes of founder animals, and a second matrix describing the transmission of QTL alleles from the founders to later generations of genotyped animals (Meuwissen *et al.*, 2002). A variance component approach can then be used to calculate the likelihood of QTLs at each putative position along the genome

The power of combining LD and LA information to filter out both spurious LD and spurious LA likelihood peaks was demonstrated in a study designed to map the QTLs for the twinning rate in Norwegian dairy cattle (Meuwissen *et al.*, 2002). Figure 20.11a is the likelihood profile from linkage only, Fig. 20.11b the likelihood profile from LD analysis only, and Fig. 20.11c the likelihood profile from combined LD–LA.

When LD-LA is performed, both linkage and LD information contribute to the likelihood profile. Any peaks due to LD or linkage alone are filtered from the profile. Using the LD–LA approach the QTL for the twinning rate was mapped to a 1 cM region (Meuwissen *et al.*, 2002).

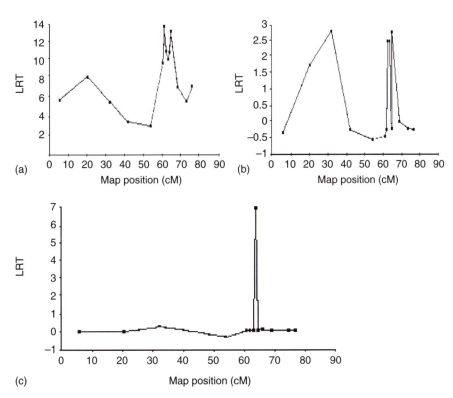

Fig. 20.11. Likelihood profile from linkage analysis (a), Linkage disequilibrium analysis (b) and combined linkage disequilibrium–linkage analysis (c) of marker data on chromosome 5 and twinning rate phenotypes in Norwegian dairy cattle. Meuwissen *et al.* (2002). LRT (Log-likelihood ratio test, 2* (Log Lik (QTL)-Log Lik (O))). Reproduced with permission from the authors.

Results from LD mapping in mammalian species

In dairy cattle, LD mapping has been an important step in identifying mutations underlying at least two QTLs affecting milk production traits. Riquet *et al.* (1999) refined the position of a QTL affecting fat percentage on chromosome 14 in dairy cattle to a 5 cM region using LD mapping. This allowed identification of a limited number of candidate genes, one of which (*DGAT1*) was subsequently found to harbour the mutation causing the QTL effect. Blott *et al.* (2003) used a combined LD–LA method to refine the position of a QTL affecting milk yield and composition on bovine chromosome 20 to a small chromosome segment. The genes in this segment included the bovine growth hormone receptor, and a mutation in this gene was later found to underlie the QTL effect.

In mice, the density of markers is sufficiently high to perform entire genome scans using LD methodology. Wang *et al.* (2002) conducted a whole-genome LD analysis to detect mouse lung tumour susceptibility QTLs in 25 strains of mice with known susceptibility to lung cancer, using 5638 genetic markers. As many as 63 markers were found to be significantly associated with lung tumour susceptibility, thus demonstrating the feasibility of using LD to map QTLs on a whole genome level.

The HapMap project (www.hapmap.org) in humans aims to go one step further, organizing marker locus information into a limited number of recognized or consensus haplotypes, which occur in the human population, to facilitate association studies in humans.

Meta-analysis

As genome scanning becomes more common, there is an increasing number of cases where the same trait and species were analysed in several different experiments, each of which may provide inconclusive or conflicting results. Key examples are milk traits in dairy cattle (Thomson *et al.*, 2003), and bi-polar disorder in humans (Segurado *et al.*, 2003). Ideally, the raw data sets involved would be merged and a single complete analysis carried out. However, differences in trait recording protocol, fixed effects to

be fitted, and markers used mean that this can only be done with considerable prior planning.

The alternative is to combine the results, rather than the data, in a meta-analysis. This allows integration of information from multiple studies and a balancing of factors that might be unproportionally presented in contributing data sets (Gu *et al.*, 2001). Meta-analysis can pool relatively weak signals from individual studies into stronger evidence for QTLs. Gu *et al.* (2001) defined three typical stages in meta-analysis: preparation, synthesis and interpretation. During preparation, the problem is defined, statistical inclusion/exclusion criteria are set, and potential contributing studies are found and assessed. Synthesis of the data can be problematic if different markers are used in different studies. In this situation, the results from each study were pooled according to chromosomal location (eg. Segurado *et al.*, 2003). As soon as the results of all studies are presented as additive LOD scores, an opportunity exists to combine the results. Alternatively, support for a QTL at a given chromosomal location can be ranked within each study and then accumulated across the studies (Segurado *et al.*, 2003). A test for heterogeneity of variances and its handling are essential during meta-analysis. The combined results of primary linkage studies can increase the statistical power to detect QTLs and to estimate their effects (Goffinet and Gerber, 2000; Etzel and Guerra, 2002; Levinson *et al.*, 2003).

Using Information from Gene Expression Studies in QTL Mapping Experiments

Development of QTL mapping approaches and microarray techniques eventually led to combining both methods, as Darvasi (2003) put it 'gene expression meets genetics'. This approach, the genetic linkage analysis of genome-wide expression patterns, could reveal gene–gene interactions in biochemical, metabolic, regulatory and developmental pathways (Jansen and Nap, 2001; Darvasi, 2003). Jansen and Nap (2001) proposed this merger, which involves expression profiling and marker-based fingerprinting of each individual of a segregating population, and exploits all the statistical tools used in the analysis of QTLs. After completion of phenotypic and genotypic description of individuals, expression levels are measured for appropriate individuals. The expression of every gene is treated as a quantitative trait. Statistical analysis allows identification of QTLs or genomic regions which affect the level of expression of particular genes.

Recently Schadt *et al.* (2003) used this approach to successfully identify a gene associated with obesity in mice. An F_2 population of mice produced by an intercross of inbred lines was profiled using a mouse gene oligonucleotide microarray. The expression values were treated as quantitative traits in a linkage analysis using autosomal markers. Among 23,574 genes represented in the microarray, 7861 showed differential expression in parental strains and not less than 10% of F_2 mice. Gene expression QTLs (so called eQTLs) with LOD scores greater that 4.3 seem to be responsible for about 25% of the transcription variation of the corresponding genes observed in this F_2 set. The authors identified two obesity-related genes within 2 cM of the peak of the eQTL located on chromosome 2. Similar methods have also been used to dissect transcriptional regulation in yeast (Brem *et al.*, 2002; Yvert *et al.*, 2003), identify candidate genes affecting ovariole number in *Drosophila melanogaster* (Wayne and McIntyre, 2002), and to dissect the genetic architecture of complex traits in maize (Consoli *et al.*, 2002). In general, such studies detect two categories of QTL; *cis*-acting modulators of single genes (i.e. the genes themselves or their promotors), and *trans*-acting modulators of many genes (for example a transcription factor). The studies suggest that combining gene expression data with linkage analysis can accelerate the identification of the causative mutations underlying QTL effects. This will be a valuable tool in unravelling at least part of the gene regulatory networks controlling gene expression.

Towards more Complex Genetic Models in QTL Detection

Simultaneous detection of multiple interacting QTLs

Many methods for the analysis of QTLs use a simple genetic model of additive or dominant gene action and a single QTL per chromosome. There are two consequences of adopting such a simple model. If there is more than one QTL per chromosome, the precision with which these QTLs can be located is relatively low. The LOD profile may appear as a single very broad peak, centred on a location between the two QTLs. In the case where the alleles of the QTL are in opposite phase in one or more of the parents, and the effects of the QTL are nearly equal, the mapping experiment may have no power to detect the QTL at all.

The second consequence of adopting an additive or dominance model in the analysis is that QTLs with more complicated gene action may not be detected. Epistatic gene action may account for at least some of the variation underlying quantitative traits, and for some traits, may be of considerable importance (Hood *et al.*, 2001). One approach to fitting multiple QTLs is composite interval mapping (Jansen, 1993; Zeng, 1994). In this case, the location and effect of putative QTLs are held constant in a second phase of analysis that aims to detect further QTLs. Another approach is to use Markov chain Monte Carlo methods to evaluate models with different numbers of QTLs on the chromosome (e.g. Heath, 1997). More complex genetic models with interaction between loci require a simultaneous search for two or more QTLs and accommodation of the interaction between the QTLs. The problem here is the large number of interactions which must be evaluated. A two-dimensional grid search has been used by a number of authors (Brockmann *et al.*, 2000; Fernandez *et al.*, 2000; Carlborg *et al.*, 2003). An alternative is to use an efficient search strategy such as a genetic algorithm in order to reduce the number of evaluations (Carlborg *et al.*, 2000).

A number of authors suggest that including epistatic interactions in the analysis of QTL mapping experiments allows QTLs to be detected which would not be significant under a purely additive model. Brockmann *et al.* (2000) detected a number of QTLs affecting growth and obesity traits in an F_2 mouse cross. They reported significant interaction between the QTL loci. One marker, *Igf1q1*, had no direct effect on fat accumulation, body weight, leptin and insulin levels, but did have significant effects on these traits when interacting with other loci. Overall, epistasis accounted for 64 and 63% of the phenotypic variance of body weight and fat accumulation, respectively, and for over 32% of muscle weight and serum concentrations of leptin, and insulin-like growth factor-I (IGF-I) in their F_2 population (Brockmann *et al.*, 2000). Carlborg *et al.* (2003) conducted a simultaneous search for multiple interacting QTLs for growth traits in a F_2 chicken population. They found 22 loci exceeding a 5% genome-wide significance threshold, nine of which were observed only as a result of their interaction effects. For hatch weight and growth from 1 to 8 days of age, 80 and 70%, respectively, of the total genetic variation was estimated to be the result of epistasis.

These two studies indicate the importance of including epistasis in the model underlying the QTL detection machinery. Although the QTLs with the largest effect on the quantitative trait were identified without an epistatic model in both cases, identification of a host of QTLs with small to moderate effects was only possible using a model which accounted for epistatic effects.

Gene Identification

Despite the large number of genome scans and considerable resources which have been devoted to QTL mapping studies, the number of mutations underlying the variation in quantitative traits which have been identified and characterized is still limited (Glazier *et al.*, 2002). In the livestock species, some of the genome scans which have been conducted have reasonable power, and some success has been achieved in narrowing the QTL to a small confidence interval using LD mapping (Meuwissen *et al.*, 2002). Rather, the limited number of causative mutations identified so far may perhaps be ascribed to the lack of knowledge at the genome level. To date, no projects sequencing the livestock genomes have been completed, and in some species, such projects have not even begun. So even if a QTL is narrowed down to a small chromosome segment, the genes in that segment, candidates for harbouring the underlying mutation, are not known, and have to be inferred from compara-

tive mapping. This is not always a simple procedure (see Chapters 2 and 14). For instance, if the QTL maps to a bovine chromosome segment, which corresponds to a break point in the comparative maps of human or mouse, a clear conclusion is not always possible.

In humans, QTL linkage mapping experiments must take place in existing populations, usually consisting of a number of small pedigrees. The vast majority of experiments conducted to date have not genotyped and phenotyped the thousands of such pedigrees required to detect QTLs of small to moderate effect. A further difficulty in humans is that even if QTLs are detected in a linkage experiment, the limited extent of LD in the human population makes it difficult to more precisely map the QTL. If significant LD only extends for tens of kb, extremely dense marker maps are required to detect significant associations between the markers and QTLs. Alternatively, the large number of gene candidates lying under the confidence interval from the linkage experiment can be evaluated for the causative mutation; high-throughput sequencing methodologies, mass array technologies and comprehensive databases of mutations make this increasingly possible.

Perhaps not surprisingly, the largest number of mutations underlying QTL effects have been identified and characterized in mice, where conditions for a powerful design for QTL detection are much more favourable (CTC, 2003). The complete genomic information in the mouse is also readily available (Waterston *et al.*, 2002).

From Phenotype to Genotype, or from Genotype to Phenotype?

The genome scanning approaches described so far proceed from phenotype to genotype, with QTL mapping followed by the candidate gene approach (Herrmann and Paul, 2002; or positional cloning, see Chapter 18). Over the last 10 years this type of methodology has led to the identification of about 1200 genes involved in human diseases (Botstein and Risch, 2003). These methodologies require significant resources and are sometimes called 'forward genetics' (McManus and Sharp, 2002).

With thousands of genes sequenced and identified, yet with little or nothing known about their function, 'reverse genetics' becomes profitable – proceeding from genotype to phenotype. The advent of small interfering RNA (siRNA) technology has made this approach much more powerful and affordable (McManus and Sharp, 2002; Chapter 5). This technology allows expression of any gene to be switched off, thus helping to identify its function.

Despite obvious advances in linking phenotypes and genotypes, complex phenotypes require a more integrated approach. This involves analysis of entire networks of biological processes. Different types of database and sophisticated bioinformatics tools are necessary for such analysis (see Chapters 15–17). Clearly these methods are particularly important for better understanding the nature of quantitative

traits. A recently published cross-genomic approach for systematic mapping of phenotypic traits to genes using a computational method develops a very promising direction of more close integration of bioinformatics and quantitative genetics (Jim *et al.*, 2004).

The validity of the process of identifying the mutations underlying variation in quantitative traits, beginning with a genome scan to identify QTLs, has been demonstrated (Grisart *et al.*, 2002; Blott *et al.*, 2003). Further advances in QTL mapping and identification will provide a great support for better understanding of genotype–phenotype relationships as well as having important practical implications in unravelling genetic disorders, developing pharmacogenomics (see Chapters 18 and 19) and animal breeding (Haley and Visscher, 2004). While success stories are limited at the time of writing, an optimistic future is anticipated.

References

Barton, N.H. and Keightley, P.D. (2002) Understanding quantitative genetic variation. *Nature Reviews of Genetics* 3, 1–21.

Blott, S., Kim, J.J., Moisio, S., Schmidt-Kuntzel, A., Cornet, A., Berzi, P., Cambisano, N., Ford, C., Grisart, B., Johnson, D., Karim, L., Simon, P., Snell, R., Spelman, R., Wong, J., Vilkki, J., Georges, M., Farnir, F. and Coppieters, W. (2003) Molecular dissection of a quantitative trait locus: a phenylalanine-to-tyrosine substitution in the transmembrane domain of the bovine growth hormone receptor is associated with a major effect on milk yield and composition. *Genetics* 163, 253–266.

Boehnke, M. (2000) A look at linkage disequilibrium. *Nature Genetics* 25, 246–247.

Botstein, D. and Risch, N. (2003) Discovering genotypes underlying human phenotypes: past successes for mendelian disease, future approaches for complex disease. *Nature Genetics* 33 (Suppl.), 228–237.

Bovenhuis, H. and Meuwissen, T. (1996) *Course Detection and Mapping of Quantitative Trait Loci*, 16–19 April, University of New England, Armidale, NSW, Australia.

Brem, R.B., Yuert, G., Clinton, R., and Kruglyak, L. (2002) Genetic dissection of transcriptional regulation in budding yeast. *Science* 296, 752–755.

Bradford, G.E. and Inounu, I. (1986) Prolific breeds of Indonesia. In: Fahmy, M.H. (ed.) *Prolific Sheep*. CAB International, Wallingford, UK, pp. 146–151.

Brockmann, G.A., Kratzsch, J., Haley, C.S., Renne, U., Schwerin, M. and Karle, S. (2000) Single QTL effects, epistasis, and pleiotropy account for two-thirds of the phenotypic F(2) variance of growth and obesity in DU6i $\times$ DBA/2 mice. *Genome Research* 10, 1941–1957.

Carlborg, O., Andersson, L. and Kinghorn, B.P. (2000) The use of a genetic algorithm for simultaneous mapping of multiple interacting quantitative trait loci. *Genetics* 155, 2003–2010.

Carlborg, O., Kerje, S., Schutz, K., Jacobsson, L., Jensen, P. and Andersson, L. (2003) A global search reveals epistatic interaction between QTL for early growth in the chicken. *Genome Research* 13, 413–421.

Charlier, C., Coppieters, W., Farnir, F., Grobet, L., Leroy, P.L., Michaux, C., Mni, M., Schwers, A., Vanmanshoven, P., Hanset, R. *et al.* (1995) The *mh* gene causing double-muscling in cattle maps to bovine chromosome 2. *Mammalian Genome* 6, 788–792.

Chatziplis, D.G., Hamann, H. and Haley, C.S. (2001) Selection and subsequent analysis of sib pair data for QTL detection. *Genetical Research* 78, 177–186.

Churchill, G.A. and Doerge, R.W. (1994) Empirical threshold values for quantitative trait mapping. *Genetics* 138, 963–971.

Cockett, N.E., Jackson, S.P. Shay, T.L., Nielsen, D., Steele, M.R., Green, R.D. and Georges, M. (1994) Chromosomal localization of the *callipyge* gene in sheep (*Ovis aries*) using bovine DNA markers. *Proceedings of the National Academy of Sciences USA* 91, 3019–3023.

Consoli, L., Lefeure, A., Zivy, M., de Vienne, D. and Damerval, C. (2002) QTL analysis of proteome and transcriptome variations for dissecting the genetic architecture of complex traits in maize. *Plant Molecular Biology* 48, 575–581.

CTC (2003) The nature and identification of quantitative trait loci: a community's view. *Nature Reviews of Genetics* 4, 911–916.

Darvasi, A. (2003) Gene expression meets genetics. *Nature* 422, 269–270.

Darvasi, A. and Soller, M. (1992) Selective genotyping for determination of linkage between a marker locus and a quantitative trait locus. *Theoretical and Applied Genetics* 85, 353–359.

Darvasi, A. and Soller, M. (1994) Selective DNA pooling for determination of linkage between a molecular marker and a quantitative trait locus. *Genetics* 138, 1365–1373.

Darvasi, A. and Soller, M. (1997) A simple method to calculate resolving power and confidence interval of QTL map location. *Behaviour Genetics* 27, 125–132.

Davis, G.H., McEwan, J.C., Fennessy, P.F., Dodds, K.G. and Farquhar, P.A. (1991) Evidence for the presence of a major gene influencing ovulation rate on the X chromosome of sheep. *Biology of Reproduction* 44, 620–624.

De Koning, D.J., Janss, L.L.G., Rattink, A.P., Van Oers, P.A.M., De Vries, B.J., Groenen, M.A.M., Van der Poel, J.J., De Groot, P.N., Brascamp, E.W.P. and Van Arendonk, J.A.M. (1999) Detection of quantitative trait loci for backfat thickness and intramuscular fat content in pigs (*Sus scrofa*). *Genetics* 152, 1679–1690.

Dodds, K.G. (2003) The number of markers required for parentage assignment. *Proceeding of the Association for the Advancement of Animal Breeding and Genetics* 15, 39–42.

Dunning, A.M., Durocher, F., Healey, C.S., Teare, M.D., McBride, S.E., Carlomagno, F., Xu, C.F., Dawson, E., Rhodes, S., Ueda, S., Lai, E., Luben, R.N., Van Rensburg, E.J., Mannermaa, A., Kataja, V., Rennart, G., Dunham, I., Purvis, I., Easton, D. and Ponder, B.A. (2000) The extent of linkage disequilibrium in four populations with distinct demographic histories. *American Journal of Human Genetics* 67, 1544–1554.

Dyer, T.D., Blangero, J., Williams, J.T., Goring, H.H. and Mahaney, M.C. (2001) The effect of pedigree complexity on quantitative trait linkage analysis. *Genetic Epidemiology* 21 (Suppl. 1), S236–S243.

Etzel, C.J. and Guerra, R. (2002) Meta-analysis of genetic-linkage analysis of quantitative-trait loci. *American Journal of Human Genetics* 71, 56–65.

Farnir, F., Coppieters, W., Arranz, J.J., Berzi, P., Cambisano, N., Grisart, B., Karim, L., Marcq, F., Moreau, L., Mni, M., Nezer, C., Simon, P., Vanmanshoven, P., Wagenaar, D. and Georges, M. (2000) Extensive genome-wide linkage disequilibrium in cattle. *Genome Research* 10, 220–227.

Feitosa, M.F., Borecki, I.B., Rich, S.S., Arnett, D.K., Sholinsky, P., Myers, R.H., Leppert, M. and Province, M.A. (2002) Quantitative trait loci influencing body-mass index reside on chromosomes 7 and 13: the National Heart, Lung and Blood Institute Family Heart Study. *American Journal of Human Genetics* 70, 72–82.

Fernandez, J.R., Tarantino, L.M., Hofer, S.M., Vogler, G.P. and McClearn, G.E. (2000) Epistatic quantitative trait loci for alcohol preference in mice. *Behavior Genetics* 30, 431–437.

Freking, B.A., Murphy, S.K., Wylie, A.A., Rhodes, S.J., Keele, J.W., Leymaster, K.A., Jirtle, R.L. and Smith, T.P. (2002) Identification of the single base change causing the callipyge muscle hypertrophy phenotype, the only known example of polar overdominance in mammals. *Genome Research* 12, 1496–1506.

Fujii, J., Otsu, K., Zorzato, F., de Leon, S., Khanna, V.K., Weiler, J.E., O'Brien, P.J. and MacLennan, D.H. (1991) Identification of a mutation in porcine ryanodine receptor associated with malignant hyperthermia. *Science* 253, 448–451.

George, A.W., Visscher, P.M. and Haley, C.S. (2000) Mapping quantitative trait loci in complex pedigrees: a two-step variance component approach. *Genetics* 156, 2081–2092.

George, V., Tiwari, H.K., Shu, Y., Zhu, X. and Elton, R.C. (1999) Linkage and association analyses of alcoholism using a regression-based transmission/disequilibrium test. *Genetic Epidemiology* 17 Suppl 1, S157–S161.

Georges, M. (2000) Extensive genome-wide linkage disequilibrium in cattle. *Genome Research* 10, 220–227.

Georges, M., Nielsen, D., Mackinnon, M., Mishra, A., Okimoto, R., Pasquino, A.T., Sargent, L.S., Sorensen, A., Steele, M.R., Zhao, X., Womack, J.E. and Hoeschele, I. (1995) Mapping quantitative trait loci controlling milk production in dairy cattle by exploiting progeny testing. *Genetics* 139, 907–920.

Georges, M., Charlier, C. and Cockett, N. (2003) The callipyge locus: evidence for the trans interaction of reciprocally imprinted genes. *Trends in Genetics* 19, 248–252.

Glazier, A.M., Nadeau, J.H. and Aitman, T. (2002) Finding genes that underline complex traits. *Science* 298, 2345–2349.

Goddard, M.E. (1991) Mapping genes for quantitative traits using linkage disequilibrium. *Genetics Selection Evolution* 23, 131s–134s.

Goffinet, B. and Gerber, S. (2000) Quantitative trait loci: a meta-analysis. *Genetics* 155, 463–473.

Grisart, B., Coppieters, W., Farnir, F., Karim, L., Ford, C., Berzi, P., Cambisano, N., Mni, M., Reid, S., Simon, P., Spelman, R., Georges, M. and Snell, R. (2002) Positional candidate cloning of a QTL in dairy cattle: Identification of a missense mutation in the bovine *DGAT1* gene with major effect on milk yield and composition. *Genome Research* 12, 222–231.

Gu, C., Province, M.A. and Rao, D.C. (2001) Meta-analysis for model-free methods. *Advances in Genetics* 42, 255–272.

Haley, C.S. and Knott, S.A. (1992) A simple regression method for mapping quantitative trait loci in line crosses using flanking markers. *Heredity* 69, 315–324.

Haley, C.S. and Visscher, P. (2004) DNA markers and genetic testing in farm animal improvement: current applications and future prospects. *Roslin Institute Annual Report* (in press).

Hayes, B.J. and Goddard, M.E. (2001) The distribution of the effects of genes affecting quantitative traits in livestock. *Genetics Selection Evolution* 33, 209–229.

Hayes, B.J., Bowman, P.J. and Goddard, N.E. (2001) Optimum design of genome scans to detect quantitative trait loci in commercial pig populations. *Proceedings of the 8th Conference of the Australian Pig Science Association, Adelaide*, pp. 144–145.

Hayes, B.J., Visscher, P.E., McPartlan, H. and Goddard, M.E. (2003) A novel multi-locus measure of linkage disequilibrium and it use to estimate past effective population size. *Genome Research* 13, 635–643.

Heath, S. (1997) Markov chain Monte Carlo segregation and linkage analysis for oligogenic models. *American Journal of Human Genetics* 61, 748–760.

Hedrick, Ph.W. (2000) *Genetics of Population*, 2nd edn. Jones and Bartlett. pp. 553.

Herrmann, S.M. and Paul, M. (2002) Studying genotype–phenotype relationships: cardiovascular disease as an example. *Journal of Molecular Medicine* 80, 282–289.

Hill, W.G. (1981) Estimation of effective population size from data on linkage disequilibrium. *Genetical Research* 38, 209–216.

Hill, W.G. and Robertson, A. (1968) Linkage disequilibrium in finite populations. *Theoretical and Applied Genetics* 38, 226–231.

Hoeschele, I., Uimari, P., Grignola, F.E., Zhang, Q. and Gage, K.M. (1997) Advances in statistical methods to map quantitative trait loci in outbred populations. *Genetics* 147, 1445–1457.

Hood, H.M., Belknap, J.K., Crabbe, J.C., Buck, K.J. (2001) Genomewide search for epistasis in a complex trait: pentobarbital withdrawal convulsions in mice. *Behaviour Genetics* 31, 93–100.

Horvat, S., Bunger, L., Falconer, V.M., Mackay, P., Law, A., Bulfield, G. and Keightley, P.D. (2000) Mapping of obesity QTLs in a cross between mouse lines divergently selected on fat content. *Mammalian Genome* 11, 2–7.

Jansen, R.C. (1993) Interval mapping of multiple quantitative trait loci. *Genetics* 135, 205–211.

Jansen, R.C. and Nap, J.P. (2001) Genetical genomics: the added value from segregation. *Trends in Genetics* 17, 388–391.

Jim, K., Parmar, K., Singh, M. and Tavazoie, S. (2004) A cross-genomic approach for systematic mapping of phenotypic traits to genes. *Genome Research* 14, 109–115.

Keightley, P.D., Morris, K.H., Ishikawa, A., Falconer, V.M. and Oliver, F. (1998) Test of candidate gene – quantitative trait locus association applied to fatness in mice. *Heredity* 81, 630–637.

Kerr, R.J. and Kinghorn, B.P. (1996) An efficient algorithm for segregation analysis in large populations. *Journal of Animal Breeding and Genetics* 113, 457–469.

Kinghorn, B.P., Kennedy, B.W. and Smith, C. (1993) A method of screening for genes of major effect. *Genetics* 134, 351–360.

Knott, S.A. and Haley, C.S. (2000) Multitrait least squares for quantitative trait loci detection. *Genetics* 156, 899–911.

Knott, S.A., Elsen, J.M. and Haley, C.S. (1994) Multiple marker mapping of quantitative trait loci in halfsib populations. *Proceedings of the 5th World Congress on Genetics Applied to Livestock Production*, Guelph, Canada, 21, 33–36.

Knott, S.A., Elsen, J.M. and Haley, C.S. (1996) Methods for multiple-marker mapping of quantitative trait loci in half-sib populations. *Theoretical and Applied Genetics* 93, 71–80.

Knott, S.A., Marklund, L., Haley, C.S., Andersson, K., Davies, W., Ellegren, H., Fredholm, M., Hansson, I., Hoyheim, B., Lundstrom, K., Moller M. and Andersson, L. (1998) Multiple marker mapping of quantitative trait loci in a cross between outbred wild boar and large white pigs. *Genetics* 149, 1069–1080.

Kruglyak, L. (1999) Prospects for whole-genome linkage disequilibrium mapping of common disease genes. *Nature Genetics* 22, 139–144.

Lander, E.S. and Botstein, D. (1989) Mapping mendelian factors underlying quantitative traits using RFLP linkage maps. *Genetics* 121, 185–199.

Lander, E.S. and Schork, N.J. (1994) Genetic dissection of complex traits. *Science* 265, 2037–2048.

Levinson, D.F. Levinson, M.D., Segurado, R. and Lewis, C.M. (2003) Genome scan meta-analysis of schizophrenia and bipolar disorder, part I: methods and power analysis. *American Journal of Human Genetics* 73, 17–33.

Lipkin, E., Mosig, M.O., Darvasi, A., Ezra, E., Shalom, A., Friedman, A. and Soller, M. (1998) Quantitative trait locus mapping in dairy cattle by means of selective milk DNA pooling using dinucleotide microsatellite markers: analysis of milk protein percentage. *Genetics* 149, 1557–1567.

Lynch, M. and Walsh, B. (1998) *Genetics and Analysis of Quantitative Traits*. Sinauer Associates, pp. 980.

Mackinnon, M.J and Weller, J.L. (1995) Methodology and accuracy of estimation of quantitative trait loci parameters in a half-sib design using maximum likelihood. *Genetics* 141,755–770.

Mangin, B., Goffinet, B. and Rebai, A. (1994) Constructing confidence intervals for QTL location. *Genetics* 138, 1301–1308.

McManus, M.T. and Sharp, P.A. (2002) Gene silencing in mammals by small interfering RNAs. *Nature Reviews of Genetics* 3, 737–747.

McRae, A.F., McEwan, J.C., Dodds, K.G., Wilson, T., Crawford, A.M. and Slate, J. (2001) Linkage disequilibrium in domestic sheep. *Genetics* 160, 1113–1122.

Meuwissen, T.H.E. and Goddard, M.E. (2001) Prediction of identity by descent probabilities from marker-haplotypes. *Genetics Selection Evolution* 33, 605–634.

Meuwissen, T.H., Karlsen, A., Lien, S., Olsaker, I. and Goddard, M.E. (2002) Fine mapping of a quantitative trait locus for twinning rate using combined linkage and linkage disequilibrium mapping. *Genetics* 161, 373–379.

Piper, L.R. and Bindon, B.M (1982) Genetic segregation for fecundity in Boorola Merino sheep. In: Barton, R.A. and Smith, W.C. (eds) *Proceedings of the World Congress on Sheep and Beef Cattle Breeding*. The Dunmore Press, Palmerston North, Vol. 1, pp. 395–400.

Reich, D.E., Cargill, M., Bolk, S., Ireland, J., Sabeti, P.C., Richter, D.J., Lavery, T., Kouyoumjlan, R., Farhadian, S.F., Ward, R. and Lander, E.S. (2001) Linkage disequilibrium in the human genome. *Nature* 411, 199–204.

Riquet, J., Coppieters, W., Cambisano, N., Arranz, J.J., Berzi, P., Davis, S.K., Grisart, B., Farnir, F., Karim, L., Mni, M., Simon, P., Taylor, J.F., Vanmanshoven, P., Wagenaar, D., Womack, J.E. and Georges, M. (1999) Fine-mapping of quantitative trait loci by identity by descent in outbred populations: application to milk production in dairy cattle. *Genetics* 96, 9252–9257.

Rothschild, M.F., Larson, R.G., Jacobson, C. and Pearson, P. (1991) Pvu II polymorphisms at the porcine estrogen receptor locus (ESR). *Animal Genetics* 22, 448.

Rothschild, M.F., Jacobson, C., Vaske, D.A., Tuggle, C.K., Short, T.H., Sasaki, S., Eckardt, G.R. and McLaren, D.G. (1994) A major gene for litter size in pigs. In: Smith, C., Gavora, J.S., Benkel, B., Chesnais, J., Fairfull, W., Gibson, J.P., Kennedy, B.W. and Burnside, E.B. (eds) *Proceedings of the 5th World Congress on Genetics Applied to Livestock Production* 21, 225–228.

Rothschild, M.F., Jacobson, C., Vaske, D.A., Tuggle, C., Wang, L., Short, T., Eckardt, G., Sasaki, S., Vincent, A., McLaren, D.G., Southwood, O., van der Steen, H., Mileham, A. and Plastow, G. (1996) The estrogen receptor locus is associated with a major gene influencing litter size in pigs. *Proceedings of the National Academy of Sciences USA* 93, 201–205.

Schadt, E.E., Monks, S.A., Drake, T.A., Lusis, A.J., Che, N., Colinayo, V., Ruff, T.G., Milligan, S.B., Lamb, J.R., Cavet, G., Linsley, P.S., Mao, M., Stoughton, R.B. and Friend, S.H. (2003) Genetics of gene expression surveyed in maize, mouse and man. *Nature* 20, 297–302.

Segurado, R., Detera-Wadleigh, S.D., Levinson, D.F., Lewis, C.M., Gill, M., Nurnberger, J.I., Jr, Craddock, N., DePaulo, J.R., Baron, M., Gershon, E.S., Ekholm, J., Cichon, S., Turecki, G., Claes, S., Kelsoe, J.R., Schofield, P.R., Badenhop, R.F., Morissette, J., Coon, H., Blackwood, D. *et al.* (2003) Genome scan meta-analysis of schizophrenia and bipolar disorder, part III: bipolar disorder. *American Journal of Human Genetics* 73, 49–62.

Shifman, S and Darvasi, A. (2001) The value of isolated populations. *Nature Genetics* 28, 309–310.

Short, T.H., Rothschild, M.F., Southwood, O.I., McLaren, D.G., DeVries, A., van der Steen, H., Eckardt, G.R., Tuggle, C.K., Helm, J., Vaske, D.A., Mileham, A.J. and Plastow, G.S. (1997) The effect of the estrogen receptor locus on reproduction and production traits in four commercial lines of pigs. *Journal of Animal Science* 75, 3138.

Shrimpton, A.E. and Robertson A. (1988) The isolation of polygenic factors controlling bristle score in *Drosophila melanogaster*. II. Distribution of third chromosome bristle effects within chromosome sections. *Genetics* 118, 445–459.

Spelman, R.J., Coppieters, W., Karim, L., Van Arendonk, J.A.M. and Bovenhuis, H. (1996) Quantitative trait loci analysis for five milk production traits on chromosome six in the Dutch Holstein-Friesian population. *Genetics* 144, 1799–1808.

Spelman, R.J., Ford, C.A., McElhinney, P., Gregory, G.C. and Snell, R.G. (2002) Characterization of the *DGAT1* gene in the New Zealand dairy population. *Journal of Dairy Science* 85, 3514–3517.

Stam, L.F. and Laurie, C.C. (1996) Molecular dissection of a major gene effect on a quantitative trait: the level of alcohol dehydrogenase expression in *Drosophila melanogaster*. *Genetics* 144, 1559–1564.

Storey, J.D. and Tibshirani, R. (2003) Statistical significance for genome wide studies. *Proceedings of the National Academy of Sciences USA* 5, 9440–9445.

Sved, J.A. (1971) Linkage disequilibrium and homozygosity of chromosome segments in finite populations. *Theoretical Population Biology* 2, 125–141.

Thaller, G., Kramer, W., Winter, A., Kaupe, B., Erhardt, G. and Fries, R.J. (2003) Effects of *DGAT1* variants on milk production traits in German cattle breeds. *Animal Science* 81, 1911–1918.

Thomson, P.C., Khatkar, M.S., Tammen, I. and Raadsma, H.W. (2003) Statistical considerations in meta-analysis of QTL reports: application to dairy cattle data. *Proceedings of the Association of Animal Breeding and Genetics* 15, 18–21.

Uimari, P., Zhang, Q., Grignola, F.E., Hoeschele, I. and Thaller, G. (1996) Analysis of QTL workshop I grand-daughter design data using least-squares, residual maximum likelihood and Bayesian methods. *Journal of Quantitative Trait Loci* 2, article no. 7.

Visscher, P.M., Thompson, R. and Haley, C.S. (1996) Confidence intervals in QTL mapping by bootstrapping. *Genetics* 143, 1013–1020.

Wang, D., Lemon, W.J. and You, M. (2002) Linkage disequilibrium mapping of novel lung tumor susceptibility quantitative trait loci in mice. *Oncogene* 3, 6858–6865.

Waterston, R.H. *et al.* (2002) Initial sequencing and comparative analysis of the mouse genome. *Nature* 240, 520–562.

Wayne, M.L. and McIntyre, L.M. (2002) Combining mapping and arraying: an approach to candidate gene identification. *Proceedings of the National Academy of Sciences USA* 99, 14903–14906.

Weller, J.I., Kashi, Y. and Soller, M. (1990) Power of 'daughter' and 'granddaughter' designs for genetic mapping of quantitative traits in diary cattle using genetic markers. *Journal of Dairy Science* 73, 2525–2537.

Weller, J.I., Song, J.Z., Heyen, D.W., Lewin, H.A. and Ron, M. (1998) A new approach to the problem of multiple comparisons in the genetic dissection of complex traits. *Genetics* 150, 1699–1706.

Williams, J.T. and Blangero, J. (1999) Power of variance component linkage analysis to detect quantitative trait loci. *American Journal of Human Genetics* 63, 545–563.

Yu, H., Pandit, B., Klett, E., Lee, M.H., Lu, K., Helou, K., Ikeda, I., Egashira, N., Sato, M., Klein, R., Batta, A., Salen, G. and Patel, S.B. (2003) The rat STSL locus: characterization, chromosomal assignment, and genetic variations in sitosterolemic hypertensive rats. *BMC Cardiovascular Disorders* 3, 3(1), 4 (19 pages).

Yuert, G., Brem, R.B., Whittle, J., Akey, J.M., Foss, E., Smith, E.N., Mackelprang, R. and Kruglyak, L. (2003) *Trans*-acting regulatory variation in *Saccharomyces cerevisiae* and the role of transcription factors. *Nature Genetics* 35(1), 57–64.

Zeng, Z.B. (1994) Precision mapping of quantitative trait loci. *Genetics* 136, 1457–1486.

21 Mammalian Population Genetics and Genomics

L. Chikhi[1] and M. Bruford[2]

[1]UMR 5174 Evolution et Diversité Biologique, Bat. IV R3 b2, Université Paul Sabatier, 118 Route de Narbonne, 31062 Toulouse cédex 4, France;
[2]Biodiversity and Ecological Processes Group, School of Biosciences, Cardiff University, Main Building, Cardiff CF10 3TL, UK

Introduction	539
Genetic diversity: Summarizing (or not) and Making Inferences	546
Genetic diversity within populations	546
Genetic diversity between populations	554
Individual-based approaches: inference of population structure with limited or without prior knowledge	559
Full-likelihood (Bayesian) and summary statistics (approximate Bayesian) methods	565
Extensions to Specific Applications	567
Population differentiation across the genome and selection	567
Contrasting patterns of molecular evolution in hominoids revealed by multiple marker studies	569
Livestock genetic diversity patterns	570
Conservation genetics: recent and ancient signals, and the importance of selection	571
Conclusions	573
References	574

Introduction

Population genetics aims to draw inferences on past demographic events or selection through the analysis of present-day populations (with the exception of ancient DNA studies). The first element of any such study requires populations to be sampled and typed or for data to be collected from the literature or from databases. In the second step the aim is to describe patterns of spatial and temporal variation and to answer the following questions: how is genetic diversity distributed within and among populations? Are the populations sampled genetically variable? Are their alleles or frequencies thereof significantly different from each other and do these differences follow particular geographical patterns? The third step is more complex: do the observed patterns (or lack thereof) favour a particular past demographic or selective scenario? Is it possible to quantify the relative importance of evolutionary factors (migration events, **genetic drift** (see Box 21.1) population size changes, selective pressures, mutation

*Correspondence: chikhi@cict.fr; BrufordMW@Cardiff.ac.uk

Box 21.1. Glossary (terminology).

Diversity measures

n_A: number of alleles.

S: number of segregating sites (= variable sites, see SNPs).

SNPs (single nucleotide polymorphisms): variable nucleotides sites (see segregating sites). SNPs are assumed to be bi-allelic. Most of population genetics was originally developed for bi-allelic loci and is thus directly applicable. SNPs are however often used jointly to define haplotypes, which re-creates multi-allelic loci.

H_o: observed heterozygosity, it is the proportion of loci that are heterozygous averaged across the sample. H_o measures genetic variability at the individual level, but not at the population level.

H_e: expected heterozygosity (=gene diversity) under random mating. It is a good measure of genetic diversity and is relatively little affected by sample size. For bi-allelic loci, such as SNPs, single-locus H_e can vary enormously even for large sample sizes, but this effect can be reduced when many loci are used, or when the two alleles have similar frequencies.
$H_e = 1 - \Sigma p_i^2 = \Sigma p_i p_k$ for $i \neq k$, where p_i is the frequency of allele i.
In a Wright–Fisher model, under mutation–drift equilibrium, the expected distribution of allele frequencies was derived by Ewens (1972), who therefore showed that H_e could be computed simply by knowing n_A.

Hardy–Weinberg (HW) equilibrium: principle stating that, in a large randomly mating population with non-overlapping generations, the *genotype* frequencies at one diploid locus (i) are the product of the *allele* frequencies at that locus, and (ii) remain constant between generations, hence the 'equilibrium', irrespective of the dominance of the alleles. The HW equilibrium holds provided that migration, mutation and natural selection do not affect the locus of interest. Another implicit assumption of the HW principle is that allele frequencies are the same in males and females. For a locus with n alleles, $A_1, A_2, ...A_n$, the frequencies of homozygotes A_iA_i and heterozygotes A_iA_j are obtained by expanding $(p_1 + p_2 + ... + p_n)^2$ and are thus p_i^2 and $2p_ip_j$, respectively. Note also that the HW equilibrium is reached after only one generation of random mating.

Panmixia: another term for random mating.

LD (linkage or gametic disequilibrium): measures the statistical association of alleles from different loci. If we consider two loci A and B, with alleles $A_1, ..., A_n$ and $B_1, ..., B_m$, respectively, a natural measure of statistical association between alleles A_i and B_j is $D_{ij} = p_{AiBj} - p_{Ai}p_{Bj}$, where p_G represents the frequency of allele/genotype G. If the two loci are independent, the frequency of genotype A_iB_j is simply the product of the respective allele frequencies and $D_{ij} = 0$ for all pairs of alleles. Thus, significant departures from zero indicate a possible correlation between alleles from the two loci. Different measures have been proposed to calculate D for multi-allelic loci. They have rather different properties (Lewontin, 1988) and some may therefore be better adapted to mapping while others to infer past demography.

$\theta = 4N_e\mu$, where N_e is the effective population size and μ is the mutation rate at the locus of interest.

π: mean number of nucleotide differences, is equivalent to H_e at the nucleotide level.

N_e: the effective population size is the size of an ideal population that would behave in a similar way to the real population. Depending on which aspect of genetic diversity one focuses on, different N_e have been defined (Crow and Kimura, 1970; Frankham, 1995). The concept of effective size is heavily used in population genetics, for instance to measure the relative importance of genetic drift and selection (Ohta, 1992). The long-term effective size represents the size of a demographically stable random mating population exhibiting the same amount of genetic diversity as the real population. The variance effective size is estimated by using the variation of allele frequencies over generations and therefore measures the recent behaviour of the population (Waples, 1989; Drummond, *et al.*, 2002; Berthier *et al.*, 2002; Laval *et al.*, 2003). Thus long-term and variance N_e can be extremely different. The concept of N_e while useful does not capture all aspects of the gene genealogy and should therefore be manipulated with care (see Box 21.2).

Continued

Box 21.1. *Continued.*

F_{ST}: used as a measure of genetic differentiation it actually measures fixation (the situation in which different populations have fixed different alleles). It varies between 0, when all populations have equal allele frequencies, and 1 when populations are fixed for different alleles. See text for discussion.

N_em: number of migrants exchanged between populations. It is expressed in terms of N_e, the effective population size of each population in a demographic model and m, the migration rate between populations. In the case of an infinite island model (see below), Wright expressed the relationship between N_em and F_{ST}, as $F_{ST} \sim 1/(1+4N_em)$.

Wright–Fisher model: a simple demographic model assuming neutrality, constant population size N_e and random mating. New allele frequencies are obtained by sampling alleles from an infinite pool of gametes using the parental allele frequencies. This is equivalent to a multinomial sampling scheme of size $2N_e$ (respectively N_e) from the parental allele frequencies for diploids (respectively, haploids). The Wright–Fisher model can be seen as a null model against which alternative models can be tested.

Mutation–drift equilibrium: describes the equilibrium state of a population when the increase of genetic diversity due to mutations is balanced by the loss due to genetic drift.

Infinite- and n-island models: demographic models in which an infinite number of (or n) island populations of size N_e are interconnected by gene flow at a rate m. The absolute number of migrants exchanged is thus N_em. This model assumes that migrants are equally likely to come from any of the populations considered, and there is consequently no idea of geographical structure. At equilibrium between migration and genetic drift, assuming the mutation rate is much smaller than the migration rate, Wright showed that $F_{ST} \sim 1/(1+4N_em)$.

Genetic drift: random variations of allele frequencies between generations; caused by the finite population size.

Admixture: demographic event during which a population is created by the contribution of two or more populations. In general admixture models are different from gene flow models in assuming one major event, rather than a continuous flow of migrants.

Ascertainment bias: bias introduced when data are collected in a non-random manner following a previous exploratory sampling scheme. For instance, the selection process of genetic markers in a particular species/population tends to favour polymorphic markers in this species/population. Future comparisons will tend to produce higher diversity estimates in the populations where the markers were isolated. This bias was invoked in humans, when comparing levels of diversity in different continents (Bowcock *et al.*, 1994). It was also observed in species comparisons. Ellegren *et al.* (1997) showed, using a reciprocal comparison between cattle and sheep, that the species in which microsatellite loci were isolated was more variable than the species to which they were later applied. Ascertainment bias is a serious problem in SNPs because SNPs are defined on the basis of very few individuals (e.g. Nielsen and Signorovitch, 2003).

Mutation models

ISM (infinite site model): assumes that every new mutation hits a non-mutated site. The different alleles can be seen as long DNA sequences for which mutations are recorded.

IAM (infinite allele model): assumes that every mutation creates an allele that was never observed before. Contrary to the previous case, the relationships between the different alleles are not specified.

SMM (stepwise mutation models): this is a family of mutation models that assume that mutations lead to incremental changes of a parameter of interest. Originally described by Ohta and Kimura (1973) to account for charge changes in allozymes, it has been mostly used to model changes in the number of repeats of microsatellites, notably after the work by Di Rienzo *et al.* (1994). Different SMMs exist (and different terminologies as well!). The single-step SMM (sometimes called SSM) assumes that all changes are of one repeat. There are different types of multiple-step model (e.g. the two phase model and the generalized stepwise model) which allow for changes by more than one repeat, the details of which can be found in Estoup *et al.*, 2002, and references therein). Some models allow infinite size alleles while others impose an upper boundary (e.g. Nauta and Weissing, 1996).

Continued

Box 21.1. *Continued.*

KAM (K-allele model): assumes that there are K possible allelic states and that mutations are equally likely to any other allele. The behaviour of the KAM is similar to the IAM for large K values.

Likelihood: assuming that some data, D, have been obtained and that a statistical model with parameters $(\theta_1, \theta_2, ...)$ can be used to model the process that generated the data, then it is possible to calculate the probability of observing the data for different parameter values. This conditional probability, $P(D \mid \theta_1, \theta_2, ...)$, is also called the likelihood $L(\theta_1, \theta_2, ... \mid D)$.

MCMC (Markov Chain Monte Carlo): set of methods used in statistical modelling. MCMC methods can be used to sample from statistical distributions from which it is otherwise difficult or impossible to sample. MCMC methods allow the user to explore the parameter space of a statistical model in such a way that by simply keeping track of parameter values visited, one obtains a sample from a statistical distribution of interest.

Bayesian inference, prior and posterior distributions: Bayesian inference rests on the application of Bayes formula to statistical problems. Bayes formula states that for two events A and B, we have $p(A|B) = p(AB)/p(B)$. Using the symmetric equation for $p(B|A)$, it is easy to show that:

$p(A|B) = p(A) \times p(B/A) / p(B) = p(A) \times p(B/A) / [p(A) \times p(B/A) + p(A^c) \times p(B|A^c)]$, where A^c is the complementary event to A.

The interpretation of this equation is important because it introduces the notions of prior and posterior probability. This equation can thus be expressed as follows.

$p(A|B)$ is the probability of A taking place *after* knowing that B has happened (*posterior* to B).

$p(A)$ is the probability that A happens *before* anything is known about B (i.e. *prior* to B).

$p(B|A)$ is the probability of B when A is known to have happened. This term represents how likely it is for A to have 'caused' B (i.e. it is the *likelihood* of B, assuming A took place).

The last term is a rescaling term corresponding to the probability of B. If B is rare, then $p(A|B)$ will be large.

This equation can be extended to probability densities (rather than events). In particular we could be interested in making inferences about a parameter (or a set of parameters) θ of a statistical model by using the information provided by the observation of the data, *D*. That would be described by the probability distribution function (pdf) of θ given the data: $p(\theta|D)$. The equation above is thus easily rewritten as:

$p(\theta|D) = p(\theta) \times p(D|\theta) / p(D)$.

The first right-hand term is the pdf of θ *before* the data have been obtained and is therefore called the *prior* as opposed to $p(\theta|D)$ which is the *posterior*. Practically, $p(\theta)$ summarizes the belief, knowledge or lack of knowledge about θ. The second term represents the probability of observing the data under the statistical model. Seen as a function of θ, $p(D|\theta)$ is $L(\theta)$ the likelihood function. The last term represents the probability of the data and is often impossible to evaluate but it is a constant *given* the data. As a consequence, this term can be ignored.

By taking a Bayesian (or full-likelihood) approach one considers that all relevant information about the parameter(s) is contained in the posterior pdf. Summary statistics are thus used for comparison with other methods (e.g. Chikhi *et al.*, 2001) but the use of summary statistics in comparison can be rather misleading. The Bayesian procedure requires that the user provide a prior on all parameters of the model. This involves some subjectivity but a lack of knowledge can be represented by a 'flat' prior on the parameter space. Consequently, the posterior will in fact be proportional to the likelihood function.

and recombination events, etc.) that led to the present-day patterns? Population genomics goes one step further by using large sets of genetic markers and, as opposed to summarizing data across loci, examines in detail the pat-terns of variation among loci in the genome. As Luikart *et al.* (2003) put it, there is a movement from 'genotyping' to 'genome typing'.

Therefore, population genomics is (or should be) more than population genetics with many markers. Inferential power will primarily come from the ability to understand the very manner in which genetic diversity is distributed in the genome. For instance, the recent discovery of 'blocks' of linkage disequilibrium in the human genome (e.g. Daly *et al.*, 2001; Goldstein, 2001; Jeffreys *et al.*, 2001; Gabriel *et al.*, 2002) may prove extremely fruitful to show how recombination events and demographic events can interact (Stumpf and Goldstein, 2003; Tisckoff and Verelli, 2003; Wall and Pritchard, 2003; Anderson and Slatkin, 2004). In fact, there is increasing awareness that concepts originally developed within population genetics do have implications and applications beyond the traditional scope of evolutionary studies. Such areas include pharmacogenomics (e.g. Wilson *et al.*, 2001, and Chapter 19) and association studies among others (see Goldstein and Chikhi, 2002, for a discussion).

While genomic maps may soon become available for a number of mammalian species, it is unlikely that they will be available for many species in the very short term. At the time of writing, only seven mammalian species have had their genome sequenced or in progress (human, chimpanzee, macaque, cattle, mouse, rat and dog) with sequence information only available for a few individuals. Most departments working on mammalian genomics are actually working on one or two of these species, which cannot objectively pretend to represent mammalian diversity (O'Brien *et al.*, 1999, 2001). There are between 4600 and 4800 different species of mammals world-wide, representing roughly 10% of all vertebrates. Mammal diversity is enormous whether we look at chromosome numbers (which vary from $2n = 6$ in the Indian muntjac to $2n = 102$ in the South American rodent (*Tympanoctomys barrerae*)), body size (from the massive 130 t blue whale to the tiny 2 g Kitti hog-nosed bat: more than seven orders of magnitude lighter!), reproductive systems (from the typical viviparous mammal to the egg-laying platypus), social behaviour (from colonial mole-rats through gregarious gnus and dolphins to solitary tigers and orang utans) and dispersal patterns (from female philopatry found in cercopithecoid primates and marmots to male philopatry in

chimpanzees) (McKenna and Bell, 1997; Di Fiore, 2003). Still, the number of anonymous markers covering the genome are becoming increasingly available for many mammals (microsatellite markers for 25 new mammalian species were described in *Molecular Ecology Notes* in 2003, including endangered species and marsupials) and methods to type increasing numbers of loci are becoming easier and cheaper by the day.

This chapter provides a review of both basic and advanced issues stemming from population genetics in order to show how methods and data have been (or might be) applied to mammalian species. Box 21.1 defines a number of terms which are highlighted in the text, the first time they appear. Many issues dealt with here are not mammal-specific but examples stemming from the class Mammalia were chosen whenever possible, in order to show how the extraordinary variation observed in this group can generate rather different patterns of genetic variation and how it can be used to increase our understanding of genomic constants. Some time is spent on specific studies in order to discuss interesting concepts. We have included a lengthy section on 'population genetics inference' This is important because too much weight is currently given to genetic data compared with other sources of biological information and some limitations should be kept in mind. Inference from genetic data is invaluable for instance to uncover social structure in mammals (e.g. Utami *et al.*, 2002, on orang-utans) but not error-free. Overall, in order to make *statistical statements* about demographic and selective parameters it is first necessary to find good ways to *represent* the genetic data, and to use whenever possible well-defined statistical models.

We therefore present how diversity observed both *within* and *between* populations can be summarized and show how it influences our understanding of evolutionary processes. In the last 20 years, significant progress has been made on the statistical front to develop efficient and powerful inferential methods (see reviews by Stephens, 2001; Beaumont, 2001, 2004). This is particularly true of the coalescent theory which has revolutionized population genetics and to which Box 21.2 is devoted. In parallel, large data sets have both accumulated and become available at an

Box 21.2. Coalescent theory: the shape of trees.

The coalescent theory focuses on the statistical properties of gene trees under different demographic scenarios (the standard coalescent assumes a Wright–Fisher model) and was originally developed by Kingman (1982a,b), Tajima (1983), Tavaré (1984), Hudson (1990) and many others (see reviews by Hudson, 1991; Donnelly and Tavaré, 1995; Fu and Li, 1999; Nordborg, 2001; Rosenberg and Nordborg, 2002). Our aim here is to present some properties of the standard coalescent with a discussion on the effect of some demographic events. For more thorough and mathematical issues, the reader will benefit much more from the previously cited reviews.

The principles of the theory can be loosely understood by noting that in any finite population, some individuals will by chance fail to produce offspring and that their genes may therefore not be passed on to the next generation. As time passes, more and more gene lineages become extinct. Looking backward in time, any present-day sample of a non-recombining gene (or DNA stretch) can be represented by a gene tree whose root is the most recent common ancestor (MRCA) of the *sample*. At each node (going backwards in time) genes are said to coalesce and coalescent events keep taking place until the MRCA is reached.

Given that evolution is a *forward* process affecting *populations* this tends to make forward simulations conceptually easier to grasp. However, by taking a retrospective or genealogical *sample*-based approach, the coalescent theory is more adapted to practical situations faced by population geneticists, where present-day *samples* are used to *infer the past* (Fu and Li, 1999) rather than predict the future of populations. Another reason for its success is that coalescent simulations are extremely fast (samples rather than whole populations are simulated).

One major result of the standard coalescent is that, in a sample of size n, the time T_n until the first coalescence (=the time during which there are n lineages) follows an exponential distribution of parameter $\lambda_n = n(n-1)/4N_e$, where N_e is the effective population size. The expectation of this time is therefore by definition of the exponential distribution $E(T_n) = 1/\lambda_n = 4N_e/n(n-1)$. Coalescent times are thus functions of both N_e and n. Once a coalescent event takes place, the number of lineages decreases to $n-1$ and the following coalescent time is sampled from the new exponential distribution with parameter λ_{n-1} and expectation $E(T_{n-1}) = 1/\lambda_{n-1} = 4N_e/(n-1)(n-2)$.

It is already possible to see a simple effect of this distribution on the structure of a coalescent tree. For a sample size of $n = 50$, the expected time until a coalescence will increase in expectation from $T_{50} \sim N_e/600$ (and similar values, around the tips of the tree) to $T_2 \sim 2N_e$, when only two lineages remain. Thus, coalescent times will vary by three orders of magnitude! Many coalescences will therefore take place around the tips for large sample sizes, whereas the time until the last two lineages coalesce will be rather long (as in Fig. 21.1, tree A). In fact, $E(T_2)$ will represent nearly half the 'height' of the gene tree. This 'height' is the age of the MRCA and can be computed by adding up the different coalescence times: $E(T_{MRCA}) = E(T_n) + E(T_{n-1}) + \ldots + E(T_2) = 4N_e(1 - 1/n)$. Thus, for reasonably large sample sizes, $E(T_{MRCA}) \sim 4N_e$ in a stable population, which is, as we just noted, twice $E(T_2)$.

All the results above assume that the population size is constant in time. If we assume that N_e has changed through time, then the theory can be extended: coalescent times are drawn from exponential distributions similar to those above except that N_e changes as time goes backward. For instance, a population increasing in size can be seen as having its N_e decreasing when we go back in time. One consequence is that the probability of coalescences is lower today than it was in the past. Thus, compared with a stable population size, coalescences will be moved backwards in time, and will concentrate around the MRCA. In other words, the tree will tend towards being 'star-like' (Fig. 21.1, tree B). On the contrary, in a declining population, coalescence times will be moved towards the present and the tree will have a 'comb-like' shape (Fig. 21.1, tree C). These are extreme cases where population size changes are monotonous, but they are helpful in interpreting gene trees, and may well approximate recent demographic histories of many species (recent colonizations, human-caused population crashes, etc.).

In populations, where N_e changes erratically through time (often on a longer evolutionary time scale), no clear predictions can be made on the tree structure. Similarly, the effect of population structure may also affect greatly both the tree structure and many aspects of genetic diversity patterns. This should explain why it is difficult to infer complex demographic histories, simply by using mtDNA trees or networks (see main text).

Continued

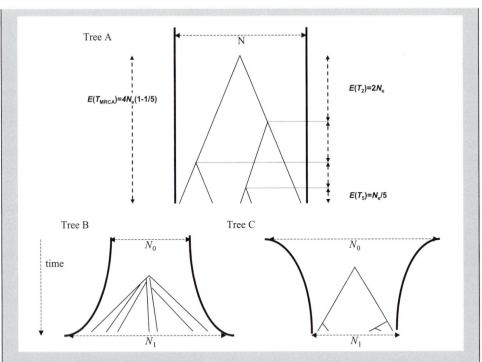

Fig. 21.1. Coalescent trees. Tree A is the tree expected under a simple Wright–Fisher model (population size constant, no selection). Tree B is expected under a population growth. Tree C is expected under a model of population contraction.

The coalescent theory provides results on the probability distribution of the times at which genes coalesce (e.g. Fig. 21.1, tree A). Such times are important because they influence the patterns of diversity in present-day samples. For instance if all nodes are close to the root (as in tree B), and assuming that the data simulated are DNA sequences, it is easy to see that mutations will hit the different branches in such a way that mutations observed in the final sequences will not be shared and thus appear independent (assuming an IAM model, see Box 21.1). Consequently, the distribution of pairwise differences between the sequences, will be Poisson like (Slatkin and Hudson, 1991, see main text). On the contrary, if most nodes are concentrated close to the tips (as in Fig. 21.1, tree C), then most mutations will take place in the few long branches and will thus be shared by many of the present-day sequences.

Overall, by taking a genealogical perspective, the coalescent theory has demonstrated that the structure of gene trees contains information on the demographic history of populations, information that summary statistics may not always contain (see main text). The coalescent is a highly stochastic process and averages given above have huge variances. Thus, independent neutral loci may exhibit extremely different behaviours despite having exactly the same demographic history. For instance, the T_{MRCA} of a stable population will vary enormously from locus to locus and large differences observed across loci should not be interpreted straightforwardly as implying different demographic histories or selective pressures. This is crucial when mtDNA and Y chromosome data are compared with each other or with nuclear loci. Different outcomes in present-day data are often interpreted as implying different demographic histories between males and females, without first accounting for that often large variance. This also means that gene trees based on linked loci (e.g. mtDNA and Y chromosome haplotypes) should be interpreted with great care as they essentially behave as (highly variable) single loci. See main text and the discussion on social structure and patterns expected (or unpredictable) for different markers.

The coalescent is a probability model that approximates real genealogies. It assumes for instance that $n(n-1)/4N \ll 1$ which is true if the sample size is small compared with N_e. So in theory the coalescent should not work for small populations (endangered species, bottlenecked populations). In practice, simulations show that the coalescent actually works very well even in such cases.

increasing pace (see Searls, 2000; Birney *et al.*, 2002, and Chapters 15–17 for database-related reviews). Computer software allows users not only to compute most useful statistics but also to perform simulations to statistically test complex demographic hypotheses (Box 21.3). There has been a gradual 'democratization' of analytical tools and this is good news. Still, most available inferential methods should not be used as 'black boxes' as they depend on models based on simplifying assumptions: a model accounting for population growth will usually ignore population structure or vice versa (e.g. Goldstein and Chikhi, 2002; Ptak and Przeworski, 2002). Due to these limitations, a number of '*ad hoc*' methods have also been developed and have had great success (e.g. Templeton, 2002). The apparent advantage of such methods is that they seem to uncover complex demographic histories. There is unfortunately little evidence that such methods, while useful for exploratory analyses, actually work as inferential tools (e.g. Knowles and Maddison, 2002; Beaumont, 2004; Felsenstein, 2004). We therefore argue at the outset that drawing conclusions from clearly specified models (even if they are simplistic) is preferable to using more loosely defined approaches. Overall, as evolutionary biologists trying to infer a past that cannot be replayed, we should keep in mind that: 'left to our own devices we are not very good at picking out patterns from a sea of noisy data. To put it another way, we are all too good at picking out non-existent patterns that happen to suit our purposes' (Efron and Tibshirani, 1993).

Genetic Diversity: Summarizing (or not) and Making Inferences

The aim of the following section is to take a bird's eye view on how genetic data can be described and how different measures can be used to improve both description and inference. The number of methods developed in recent years has increased at such a pace that it is impossible to review them all. Principles rather than specific mathematical details are thus presented here. The mathematically inclined reader will greatly benefit from reviews by Stephens (2001) or Beaumont (2001, 2003, 2004).

Genetic diversity within populations

Independent alleles and loci

Describing diversity within and between populations requires us to summarize genetic data (but see below: 'Full-likelihood methods'). Depending on the type and amount of genetic data available, different summary statistics have been proposed and are commonly used (see Box 21.1). Typically, these statistics have different properties and can therefore be influenced in different ways by demography and selection. For instance, population contraction tends to affect the number of alleles, n_A, much more than the expected heterozygosity, H_e (Nei *et al.*, 1975). This is because rare alleles, which will be eliminated after a population crash, do not influence H_e very much as allele frequencies are squared in the computation of H_e (see Box 21.1 and Fig. 21.2). Similarly, positive selection on some alleles will increase their frequency and reduce the proportion of rare alleles. Thus, the *joint* use of n_A and H_e can provide significant information on departures from demographic stability or from selective neutrality. However, this information became useful only after the work of Ewens (1972) and Watterson (1978). Ewens (1972) showed that for a demographically stable population and a neutral locus (or, using technical terms: under a **Wright–Fisher model** at **mutation–drift equilibrium**, see Box 21.1) the expected frequency distribution of alleles observed in a sample follows a statistical distribution now known as the Ewens sampling distribution. This means that n_A is a sufficient statistic to compute H_e (there is no need to know the allele frequencies). This expected equilibrium H_e value can then be compared with the *observed* H_e (computed using the *observed* allele frequencies, Box 21.1) to devise a statistical test which indicates that the data were not generated under a Wright–Fisher model. The results of this test are interpreted as signatures of either selection (Watterson, 1978) or ancient population bottlenecks (e.g. Watterson, 1986; Cornuet and Luikart, 1996; Luikart *et al.*, 1999).

The latter authors have used this result to devise statistical tests to detect population bottlenecks under different mutation models (Box 21.1) and have implemented them in the

Box 21.3. Population genetics software.

Below are given web pages of software mentioned in the text. When available, the papers describing the software itself or the methodology are given. Some pages containing links to other software are also provided.

Population structure
AIDA (Bertorelle and Barbujani, 1995): http://www.unife.it/genetica/Giorgio/giorgio.html
ARLEQUIN (Schneider *et al.*): http://lgb.unige.ch/arlequin/
DISTRUCT (Rosenberg, 2004): http://www.cmb.usc.edu/~noahr/distruct.html
GDA (Genetic Data Analysis) http://lewis.eeb.uconn.edu/lewishome/software.html
GENEPOP (Raymond and Rousset, 1995): http://wbiomed.curtin.edu.au/genepop/
GENETIX: http://www.univ-montp2.fr/%7Egenetix/genetix/genetix.htm
IBD (Isolation by distance) (Bohonak, 2002): http://www.bio.sdsu.edu/pub/andy/IBD.html
MULTILOCUS (Agapow and Burt, 2001) http://www.agapow.net/software/multilocus/
MSA, (Dieringer and Schlötterer, 2003) http://i122server.vu-wien.ac.at/MSA/MSA_download.html
POPGENE: http://www.ualberta.ca/~fyeh/
POPULATIONS (Olivier Langella): http://www.cnrs-gif.fr/pge/bioinfo/liste/index.php?lang=fr
PYPOP (Lancaster *et al.*, 2003): http://allele5.biol.berkeley.edu/pypop/
SAMOVA (Dupanloup *et al.*, 2002): http://web.unife.it/progetti/genetica/Isabelle/samova.html
SPAGEDI (Hardy and Vekemans, 2002): http://www.ulb.ac.be/sciences/lagev/spagedi.html

Admixture
LEA (Chikhi *et al.*, 2001; Langella *et al.*, 2001): http://www.cnrs-gif.fr/pge/bioinfo/lea/index.php?lang=en
ADMIX (Bertorelle and Excoffier, 1998): http://www.unife.it/genetica/Giorgio/giorgio.html
ADMIX2 (Dupanloup and Bertorelle, 2001): http://www.unife.it/genetica/Isabelle/admix2_0.html
LEADMIX (Wang, 2003): http://www.zoo.cam.ac.uk/ioz/software.htm#LEADMIX

Hidden structure and assignment methods
GENECLASS (Cornuet *et al.*, 1999): http://www.montpellier.inra.fr/CBGP/softwares/
GENECLASS2 (Piry *et al.*, 2004): same as Geneclass
NUCLEODIV (Holsinger and Mason-Gamer, 1996): http://darwin.eeb.uconn.edu/nucleodiv/nucleodiv.html
PARTITION (Dawson and Belkhir, 2001): http://www.univ-montp2.fr/~genetix/partition/partition.htm
SAMOVA (Dupanloup *et al.*, 2002): http://web.unife.it/progetti/genetica/Isabelle/samova.html
SPASSIGN (Pálsson, 2004): http://www.hi.is/~snaebj/programs.html
STRUCTURE (Pritchard *et al.*, 2000): http://pritch.bsd.uchicago.edu/
WHICHRUN (Banks and Eichert, 2000): http://www.bml.ucdavis.edu/whichrun.htm

Changes in population sizes
BOTTLENECK (Cornuet and Luikart, 1997): http://www.montpellier.inra.fr/CBGP/softwares/
LAMARC project (Joe Felsenstein's group): http://evolution.genetics.washington.edu/lamarc.html
MSVAR (Beaumont, 1999): http://www.rubic.rdg.ac.uk/%7emab/
MP_VAL (Garza and Williamson, 2001): http://www.pfeg.noaa.gov/tib/staff/carlos_garza/carlossoftware.html

Simulations
SIMCOAL: http://cmpg.unibe.ch/software/simcoal/
BOTTLESIM (Kuo and Janzen, 2003): http://www.arches.uga.edu/~chkuo/software/BottleSim.html
DNASP (Rozas and Rozas, 1995 ; Rozas *et al.*, 2003): http://www.ub.es/dnasp/
EASYPOP (Balloux, 2001): http://www.unil.ch/izea/softwares/easypop.html
SPLATCHE (Currat *et al.*, 2004): http://cmpg.unibe.ch/software/splatche

A few pages with lists of software
Evolution and Population Genetics web site: http://wbar.uta.edu
Software for population genetics analysis: http://www.biology.lsu.edu/general/software.html
Phylogenetics programs: http://evolution.genetics.washington.edu/phylip/software.html

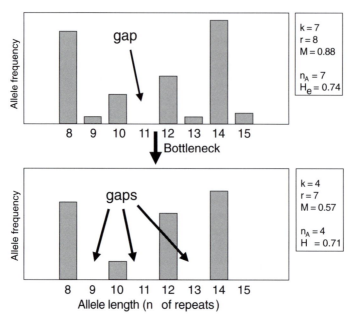

Fig. 21.2. Effect of a bottleneck on the gappiness of a microsatellite allele frequency distribution. After a poplation bottleneck, rare alleles are more likely to be eliminated.

BOTTLENECK software (Piry *et al.*, 1999; Box 21.3). It has been used on a number of mammalian species including the endangered northern hairy-nosed wombat known to have sharply decreased from thousands of individuals in the last centuries to less than 30 in the 1980s (Taylor *et al.*, 1994). Cornuet and Luikart (1996) demonstrated that a bottleneck was visible in Taylor *et al.*'s (1994) genetic data, apparently confirming the assumption that recent anthropogenic pressure was responsible for both the population crash and the consequent genetic signal.

Approaches using similar principles have also been developed for sequence data (e.g. Tajima, 1989a,b; Fu and Li, 1993). For example, Tajima's *D* is a rescaled (or 'normalized') difference between two estimates of $\theta = 4N_e\mu$ (Tajima, 1989a,b; see Boxes 21.1 and 21.4). One estimator uses **S, the number of segregating sites**, and the other uses the **average number of nucleotide differences, π** (Watterson, 1975; Tajima, 1983, Box 21.1). Values of *D* that are significantly different from zero indicate either selection or significant population size changes (but see Box 21.4 for a discussion on interpretations of Tajima's *D*).

Ultimately, these approaches focus on the

distribution of observed allele frequencies compared with those expected under different evolutionary scenarios. Such distributions can be summarized by n_A and H_e, but one could focus on other aspects of these distributions, as suggested, for instance by Garza and Williamson (2001). These authors studied the effects of demographic events on the gaps observed in allele frequency distributions for microsatellites (Fig. 21.2). They found that bottlenecked populations tend to exhibit more 'gappy' distributions than stable populations, because the range of allele size is not affected so much whereas rare alleles disappear (creating gaps). They developed a formal statistical test based on the ratio M = k/r, where k is the number of alleles and r the range (implemented in the MP_VAL software, Box 21.3). They found significant signal for population reductions in five out of 16 studies based on 13 mammalian species (including koala, northern wombat, and different canid, bear and seal species) known to have recently suffered from population size reduction. Note however that more complex demographic histories may not always leave simple genetic signatures. Indeed, Calafell *et al.* (1998) found that many microsatellite allele distributions in

Box 21.4. Tajima's *D* and inference of demography, population structure and selection.

Tajima's *D* (Tajima, 1989a) is a summary statistic estimated by rescaling the difference between θ_S and θ_{π}, two estimators of $\theta = 4N_e\mu$. The first is based on the number of segregating sites (*S*) and the second on the nucleotide diversity (π):

$$D = (\theta_{\pi} - \theta_S)/ [\text{Var}(\theta_{\pi} - \theta_S)]^{1/2}$$

Under a standard neutral model of population size constancy, the two estimators should provide similar values and *D* should be close to zero. Tajima (1989a) suggested that *D* could thus be used to test for selection. He studied the statistical properties of *D* by simulations and showed that *D*'s expected statistical distribution was not symmetrical (the expected value is slightly negative) and was closer to a beta distribution than to a normal distribution (Tajima, 1989a).

Tajima (1989b) showed that demographic events can affect *D*. In particular, population bottlenecks usually lead to a loss of rare alleles (Nei *et al.*, 1975) and to a reduction of *S*. Thus, smaller θ_S values will be expected. On the contrary, rare alleles have little effect on the computation of π and consequently of θ_{π}. Thus, population contractions tend to generate positive *D* values whereas population expansions create a bias towards negative values (Tajima, 1989b). Overall, values of *D* are influenced by the 'excess' or 'deficit' in rare alleles as compared with the neutral expectation (Ewens, 1972). Consequently, any factor that may change the importance of rare alleles will influence *D* values, and may be misinterpreted as signals for selection or ancient population size changes.

For instance Tajima (1995) showed that some forms of non-random sampling can push the expected distribution towards positive values, which might be interpreted as evidence for a population bottleneck. Bertorelle and Slatkin (1995) and Aris-Brosou and Excoffier (1996) showed that mutation rate heterogeneity could also lower S, which could either lead to apparent signals of population bottlenecks, or hide real population expansion signals by pushing *D* values in the opposite direction. Bertorelle and Slatkin (1995) found that 'even if only 2.5% of the sites are subject to frequent recurrent mutations, the confidence limits of Tajima's *D* may no longer be appropriate'. More recently, Ptak and Przeworski (2002) have shown that population structure can also lead to spurious effects on *D*. In particular, using real data from human population studies, they showed that the number of 'ethnicities' in the samples analysed was negatively correlated to *D*. This can be understood by noticing that more diverse samples tend to contain a higher proportion of rare alleles, when populations are differentiated. In other words, significant negative *D* values might be caused by population growth or population structure.

The situation is even more complex when time is considered. Indeed, depending on the severity of the population size change and the number of generations since that demographic event, *D* values may be transiently positive and negative. For instance, after a bottleneck, *D* values will first be positive, then become negative when the population grows again, before tending towards equilibrium (Tajima 1989b; Fay and Wu, 1999). This too can affect tests based on Tajima's *D*.

Thus, it appears that non-random sampling, heterogeneity in mutation rates, population ancient demography and population structure and sampling all influence *D* in rather complex ways.

Another effect was noted by Fay and Wu (1999) to explain apparent discrepancies between the negative *D* values observed at mitochondrial DNA data and the usually positive values estimated for nuclear loci. While these differences have been interpreted in terms of selective pressure. Hey (1997) and Harpending and Rogers (2000), showed by simulations that these differences were easily explained by the much smaller mtDNA effective size compared with autosomal loci. This difference led to disparate transient behaviours after a bottleneck, which had a greater effect on mtDNA diversity.

Overall, Tajima's *D* value should be interpreted with care as many factors appear to influence it. Fu and Li (1993) and others have suggested similar statistics, but there is no clear demonstration that they should work much better. For instance, Tajima (1995) showed that Fu and Li's (1993) *D** and *P** statistics were more influenced by non-random sampling than *D*. Despite its limitations, Tajima's *D* is currently the most computed statistic in studies interested in testing for selection or detecting demographic events (Przeworski *et al.*, 2000; Ptak and Przeworski, 2002) and it is likely to remain so for some time.

humans are bell-shaped, and have therefore few gaps, despite the well supported out-of-Africa bottleneck which characterized the human species (see below).

Based on single locus data (or linked loci from mitochondrial DNA or Y chromosome data sets), the previous tests have two limitations. First, they lack statistical power (this is particularly true for the Ewens–Watterson test, as it only uses allele frequency data). This means that non-significant results may miss a signal for population decline even when historical data show that this decline took place. For example, Côté et al. (2002) applied both Luikart et al.'s tests and Garza and Williamson's M test on isolated populations of Svalbard reindeer in Norway. Similarly, Waldick et al. (2002) applied the M test on the North Atlantic right whale. In both studies, genetic evidence for bottlenecks was not decisive, which is at odds with known demographical histories (in the Svalbard reindeer case, the situation is made more complicated by the manner in which loci were chosen to do the analysis, a point discussed by the authors). Secondly, these tests are ineffective in deciding whether significance is caused by selection or past demography. This can be problematic in interpreting genetic data. For example, patterns of diversity observed in human mtDNA and Y chromosome data are increasingly attributed to selection rather than to demographic expansions (e.g. Mishmar et al. (2003) for mtDNA and Repping et al. (2003) for Y chromosome data), but see Harpending and Rogers (2000) and Excoffier (2002).

These two limitations can be overcome by the use of large numbers of independent loci, which are becoming increasingly available with genomic data sets. Indeed, using independent loci provides an increase in both statistical power and the ability to detect one or more loci that have 'anomalous' behaviour (outlier loci, Beaumont and Nichols, 1996). The reason for this is that the demographic history is expected to have influenced all neutral loci in the same manner, whereas loci under selection will have a different behaviour. Thus, loci showing a genetic signal originally interpreted as due to a demographic event, might appear as outliers if the signal is actually caused by selection. Note that once outlier loci are identified one should always check whether selection is indeed the

most likely cause. The chances are that for most mammalian species, large numbers of loci will only be available through the use of 'anonymous methods' using RAPDs or AFLPs, for which the chromosome location may not always be known. Loci located on the sex chromosome might be outliers because of their smaller **effective population size, N_e**, rather than selection. Thus outlier search methodologies could be helpful in deciding whether mtDNA and Y chromosome data are indeed behaving differently because they are under selection or because of their different demographical histories.

This issue is particularly important for mammals where social structure and dominance systems typically influence male and female effective population sizes, migration rates, variance in reproductive success and a number of other demographic or genetic parameters (Chepko-Sade and Halpin, 1987; Chesser, 1991). For example, Hoelzer et al. (1998) showed the importance of female philopatry and variance in male reproductive success on increasing mtDNA N_e relative to autosomal N_e in macaques. More generally, the fact that in cercopithecine primates there usually is female philopatry and male dispersal (Di Fiore, 2003), whereas the opposite occurs in Hamadryas baboons and chimpanzees, allows us to make predictions about general patterns of genetic diversity in autosomal and mtDNA. Thus, mtDNA might appear as an outlier in cercopithecine primate data sets but not in Hamadryas baboons because of low female philopatry rather than selection (see also Box 21.2 on coalescence and single-locus variance). The role of selection can itself be complex. For instance, selection can play a significant role in maintaining genetic diversity in island populations at specific loci, as was shown in Soay sheep (Pemberton et al., 1996). In these islands, population crashes take place due to harsh environmental conditions, but despite the apparent low N_e, genetic variability has been maintained at loci apparently involved in resistance to intestinal parasites, among others.

Correlation of alleles within and between loci

The inferential methods discussed above do not account for the fact that alleles are not necessarily independent within or between loci.

Indeed, most methods available today do not use the information on *genotype* frequencies as they assume that populations are at **Hardy–Weinberg (HW)** equilibrium. For instance, Tajima's D can be computed on data from published work or databases for which only allele frequencies are available. Stratification of the data (for instance if different population samples are mixed) may then go unnoticed and lead to spurious significant (or non-significant) results (see Box 21.4). Ptak and Przeworski (2002) have recently shown that Tajima's D is negatively correlated to the number of 'ethnicities' in human population studies. In other words, they showed that apparent signals of population growth might be due to sampling non-random mating units. It should be kept in mind that this is always a problem with mtDNA data for which it is by definition impossible to test for HW equilibrium.

There are however a few studies (Pudovkin et al., 1996, and Luikart and Cornuet, 1999) which suggested using excess heterozygosity (when $H_o > H_e$) observed in progeny to estimate N_e. Similarly, population structure can also be uncovered if significant deficits in heterozygosity ($H_o < H_e$) are observed at many independent loci (the Wahlund effect). This can be important in detecting hidden social structure (see Goossens et al., 2001 on marmots for such an example) or the existence of genetically differentiated flocks in samples from breeds (see Byrne et al., submitted, for European breeds of sheep). There is, however, one reason why departures from HW equilibrium are unlikely to be very powerful: one generation of random mating is enough to restore HW proportions at all loci. Thus, **admixture** and other demographic events influencing HW proportions but taking place more than one generation before sampling will go unnoticed based on single-locus genotypes after one generation of random mating (but see Overall and Nichols, 2001 for a method to separate the effect of inbreeding and population structure).

The statistical association between alleles from different loci should on the contrary carry information on demographic events for much longer periods of time (e.g. Slatkin, 1994). It is typically measured by **linkage disequilibrium (LD**, also called gametic disequilibrium, Box 21.1). LD can be generated and main-

tained by a number of factors such as physical linkage, variation in recombination rates, selection, founder effects and bottlenecks (genetic drift), population growth, and admixture of populations (e.g. Slatkin, 1994; McKeigue, 1998; Wilson and Goldstein, 2000; Pritchard and Przeworski, 2001; Ardlie et al., 2002; Nordborg and Tavaré, 2002). However, in a large, randomly mating population, LD between alleles at neutral loci is expected to be mainly caused by physical linkage, and to decrease with recombination events.

If we call r the recombination rate per generation between the two loci and D_0 the initial LD, it is easy to show that the LD at generation t, D_t, will decrease geometrically $D_t = (1 - r)^t D_0$. For independent loci, $r = 0.5$. Thus, in cases where LD is created by drift or admixture, D_t is halved each generation, and decays much quicker than for tightly linked loci. In practice, the decrease is highly stochastic and high LD may be maintained for long periods and consequently extend over much larger distances than this equation would imply (Pritchard and Przeworski, 2001; Ardlie et al., 2002). The extent and distribution of LD are currently of great interest for fine mapping of complex genetic diseases and for its role in genome-wide association studies (see Chapter 21; Slatkin, 1994; Risch and Merikangas, 1996; Kruglyak, 1999; Abecasis et al., 2001; Reich et al., 2001; Pritchard and Przeworski, 2001). There is also increasing awareness that genome-wide patterns of LD can provide important insights into population-specific demographic histories (Stumpf and Goldstein, 2003; Wall and Pritchard, 2003). Indeed, in the last 20 years (following earlier work on and evidence from *Drosophila* studies) it has become clear that recombination rates vary enormously in the human and other genomes with regions having low recombination rates and others with so-called hotspots (reviewed in Nachman, 2003, see Chapter 1). With an average of slightly more than 1 cM/Mb in the human genome variations of more than one order of magnitude have been observed between regions (from <0.3 to >3.0 cM/Mb). The interpretation of such differences in terms of patterns of selection or of varying mutation rates is discussed in Nachman (2003) and Ellegren et al. (2003). Differences exist also between females and males both in average rates (1.7 vs. 0.9 cM/Mb, respectively) and in the genomic distribution of high and low rates. Centromeres also appear to recombine very little whereas recombination events are more common in telomeres (see also Chapter 1).

Recently, it has been suggested that hotspots and regions of low recombination are not simply the stochastic extremes in a model of a more or less uniform recombination rate, but rather correspond to what has been called the block-like structure of the human and other mammalian genomes (Daly *et al.*, 2001; Jeffreys *et al.*, 2001; Patil *et al.*, 2001; Gabriel *et al.*, 2002; Jeffreys and Neumann, 2002; Fig. 21.3). Indeed, in these studies the authors have found that SNPs separated by up to ~100 kb could exhibit high levels of LD. This was surprising because earlier work by Kruglyak (1999) had shown that in a model of population expansion similar to that thought to have taken place in humans, LD should not extend much over 3–5 kb around a particular marker. In fact, Kruglyak (1999) had also shown that by simply assuming a more recent

expansion and by reducing the initial population size, one could predict that LD could extend over much longer distances (~30 kb). Later results have repeatedly shown that LD patterns can extend over much more than 5 kb and depend on the population investigated. For instance, greater LD (and less genetic diversity) is found in European compared with African populations (Reich *et al.*, 2001). The strongest evidence for a block-like structure came after it was shown that regions where LD extended over long DNA stretches corresponded as well to regions with low recombination rates and were separated by recombination hotspots (Jeffreys *et al.*, 2001). These authors showed that the hotspot recombination rate was more than three orders of magnitude greater than the 'within-block' rate.

Recent discussion has focused on determining whether LD-blocks: (i) are found in different regions of the genome (they were originally found in regions of chromosomes 5, 6 and 21); (ii) are the same across populations; and (iii) there are objective ways to define such blocks (Stumpf, 2002). Overall, it seems that models

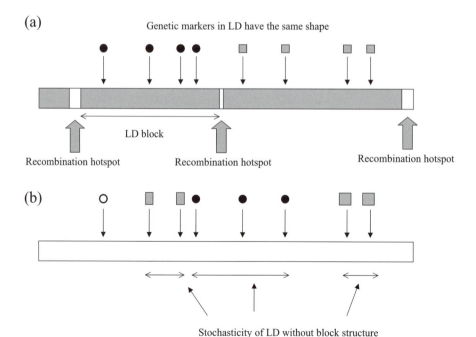

Fig. 21.3. Influence of LD blocks on patterns of LD along a stretch of DNA. (a) Genome with LD blocks. (b) Genome without LD blocks. Note that even without LD blocks the stochasticity can generate long stretches of DNA with markers in LD.

with a uniform recombination rate are unable to account for the patterns observed, as simulations have shown (Stumpf and Goldstein, 2003; Wall and Pritchard, 2003). However, the simulations show that the expected pattern of LD blocks is stochastic and that even when simulations explicitly assume an extreme variation in recombination rates, some regions may not exhibit a block-like structure depending on the demographic history of the population of interest (Stumpf and Goldstein, 2003). This led the authors to suggest that populations (and regions of the genome) could go through different stages from pre-block through block to post-block structure. These issues are made even more complicated by the problem of defining blocks, as noted by Stumpf (2002). Wall and Pritchard (2003) have proposed three criteria (the coverage, the hole and the overlapping blocks criteria) to define blocks and hotspots and applied them to both simulated and three real data sets including the data of Gabriel *et al.* (2002). Based on these criteria, they found that real data were only marginally block-like but could not be explained by a model with uniform recombination rate. They also noted that the estimated size of blocks is influenced by the distribution of SNPs, leading to perhaps imprecise assessment of these criteria (the number of SNPs is itself determined by the number of individuals sampled to define the SNPs). Interestingly, Anderson and Slatkin (2004) also used simulations that matched real data but did not reach the same conclusions. They used the data of Daly *et al.* (2001) and concluded that hotspots were only supported under a model of stable population, and not under a model of population growth. Moreover, a finer analysis of local recombination rates indicated that population growth was a better explanation for the pattern observed than recombination hotspots.

The role of selection in shaping LD along human chromosomes was investigated by Sabeti *et al.* (2002). These authors suggested that positive selection could be detected by analysing LD patterns around pre-defined loci or haplotype blocks for which specific haplotypes can be typed. The idea is that, in a population at mutation drift equilibrium, haplotypes are expected to have frequencies that correlate with both their age and the recombination rate of their genomic region. Population genetics theory has shown that frequent alleles are usually old whereas rare alleles can either be recent or old. Given that the number of recombination events increases with time, frequent haplotypes, being old, should not exhibit long-range LD. Thus, Sabeti *et al.* (2002) have proposed a test which they call the LRH test (for long-range haplotype) in which haplotypes showing both unusually high frequencies *and* long-range LD can be identified. By correcting for variation of local recombination rates and providing a visual representation of the pattern of LD, Sabeti *et al.*'s (2002) method appears to be very powerful at detecting selection and was applied to the regions containing the glucose-6-phosphate dehydrogenase gene and the CD40 ligand gene. The first is involved in resistance to malaria whereas the second is involved in the immune response to infectious agents. A comparison with other tests, which do not use the information on LD, such as the HKA (Hudson *et al.*, 1987) or the Tajima test (Tajima, 1989a), indicated that Sabeti *et al.*'s (2002) test was more powerful. It is important though to note that the method requires the information on haplotypes, which is going to be increasingly available thanks to the Haplotype Map project in humans, but may not be easily available for other mammalian species.

To conclude this section, recent years have seen a dramatic change in our understanding of patterns of LD in the human genome. After the work by Kruglyak (1999) it was believed that given the size of our genome, mapping would require more than 500,000 markers, a Herculean task, as it was seen at the time. The discovery of LD-blocks led to the conjecture that many fewer markers would be needed and that a limited number of SNPs would be enough to locate regions under selection or containing genes of interest. Now, it seems that studies done in some populations may not be valid in others and that the notion that LD extends over an average distance is a very misleading one (Ardlie *et al.*, 2002; Stumpf and Goldstein, 2003) as confirmed by both real and simulated data. One might consider this as problematic, but one could also consider that some populations may prove useful in isolating some genes while others, due to their unique demographic history, may also contribute in their own way. Overall, this is one more demonstration of human genetic diversity. On a wider perspective, comparison of mammalian genomes indicates the existence of important orthologous regions (see below) and it would be interesting to try and map LD-blocks across species. The use of patterns of LD to infer population history is still in its infancy, but we should note the recent work by Vitalis and Couvet (2001a,b,c) that uses a different measure of statistical association (the identity disequilibrium) to estimate N_e or migration rates between populations.

Genetic diversity between populations

Population differentiation: F_{ST} and related statistics

In the previous section we focused on measures of within-population diversity, but most data sets are usually taken from sets of populations, the structure of which is typically measured using **Wright's F_{ST}** (Wright, 1951; Box 21.1). F_{ST} was originally designed for bi-allelic markers and was later extended for multi-allelic markers (Nei's (1973) G_{ST}; Wright, 1978; Nei, 1987). Under an **infinite-island model**, Wright showed that F_{ST} can be used to estimate the number of migrants $N_e m$ exchanged by the populations (Slatkin, 1985, but see Whitlock and McCauley, 1999, for a critical review of this estimation procedure). This result together with the fact that F_{ST}: (i) is easy to estimate from allele frequency data, (ii) reaches migration–drift equilibrium quickly, (iii) is not too dependent on the theoretical demographic model (stepping-stone versus **n-island model**); and (iv) naturally arises in theoretical population genetics, is responsible for its popularity as a measure of population differentiation. Despite being an apparently good summary statistic, F_{ST} has a number of problems attached to it. First, there has been recurrent debate regarding the best way to estimate it (Wright, 1978; Nei, 1987). This is particularly problematic for multi-allelic loci, as different estimators weigh rare alleles in different manners leading sometimes to somewhat different estimates (Robertson and Hill, 1984; Weir and

Cockerham, 1984; Nei, 1987); a 'problem' which Raufaste and Bonhomme (2000, unpublished results) used to devise a statistical test of selective neutrality, implemented in the NEUTRALLELIX software (see Box 21.3). Overall, Weir and Cockerham's (1984) method remains the most popular because it is believed to be unbiased and based on a sound statistical model (but see Holsinger (1999) and Balding (2003) for other model-based approaches).

Secondly, it has become increasingly clear that absolute F_{ST} values are not comparable from study to study. While large F_{ST} values appear to always indicate high levels of genetic differentiation, the meaning of low F_{ST} values is unclear for hypervariable loci (Wright, 1978; Nagylaki, 1998; Hedrick, 1999; Balloux et al., 2000). The reason for this is that F_{ST}, despite being used as a measure of population differentiation, was originally introduced by Wright (1921) to measure the process through which genetic variation is lost, leading to fixation (hence the F). Unsurprisingly, fixation is reached much quicker for bi-allelic markers (such as allozymes or SNPs) than for multi-allelic loci (such as microsatellites). Thus, for highly variable loci, even if populations do not share alleles, they will by definition be far from fixation and will therefore not necessarily exhibit large F_{ST} values (not sharing alleles in bi-allelic loci would mean fixation of different alleles and thus correspond to an F_{ST} of 1). One consequence is that $N_e m$ estimates based on F_{ST} values from highly variable loci are likely to be artificially high. Strangely, though, many studies trying to compare F_{ST} estimates from different marker types (allozymes, microsatellites, and SNPs) have tended to generate not so different figures. In humans, F_{ST} values between 0.10 and 0.20 have repeatedly been observed (e.g. Lewontin, 1972; Barbujani et al., 1997; Jorde et al., 2000; Excoffier, 2002; Excoffier and Hamilton, 2003) indicating that most genetic variation (80–90%) is found within rather than between human groups. Tapio et al. (2003) found a more complex situation in sheep, where similar F_{ST} values were observed across loci in all breeds except those highly fragmented. Care should thus be taken when analysing mammal populations where both large and isolated populations are sampled and analysed together.

Thirdly, F_{ST} values provide little information on ancient demography. Indeed, there are many scenarios which may lead to similar F_{ST} values (e.g. extreme drift in populations recently isolated or low gene flow over longer periods), as we discuss further below on efficient summary statistics.

With the arrival of molecular data, a number of F_{ST}-like statistics have been proposed to use not only allele frequency information but also molecular information (i.e. molecular distances between alleles). For sequence or RFLP data, different methods using nucleotide information have been developed. Takahata and Palumbi (1985) and Holsinger and Mason-Gamer (1996) developed analogues of Nei's (1973) G_{ST} whereas others used information from haplotype phylogenies (Lynch and Crease, 1990; Hudson et al., 1992). The most popular approach, AMOVA (for analysis of molecular variance) was developed by Excoffier et al. (1992) and can be viewed as an extension of Weir and Cockerham's (1984) work, the latter being a special case of AMOVA when distances between alleles are equal. The AMOVA method is implemented in several software packages including the versatile ARLEQUIN software and allows the definition of hierarchical structures in a straightforward manner (Box 21.3). For microsatellite data, it is also possible to use molecular information such as the number of repeats to construct genetic distances between alleles. Different measures have been proposed and used to estimate population differentiation (R_{ST}: Slatkin, 1995; $(\delta\mu)^2$: Goldstein et al., 1995).

In principle, using molecular information should increase the statistical power to detect geographic patterns, however, the stochasticity of the underlying mutation model can actually lead to a loss of power. This is particularly true for R_{ST} and $(\delta\mu)^2$, which have been shown to have huge variances (e.g. Pérez-Lezaun et al., 1997; Calafell et al., 2000; Destro-Bisol et al., 2000). To overcome this problem, Hardy et al. (2003) have recently suggested a test to decide whether F_{ST} or R_{ST} should be used to measure differentiation. Note that in cases where most alleles become unique (for instance, when long polymorphic DNA sequences are used), the frequency information becomes meaningless, and all the information available is genealogical (Barton and Wilson, 1995).

Many mammals including domestic are characterized by having both large continental and small island populations (e.g. fox: Gilbert et al., 1990; reindeer: Côté et al., 2002; sheep: Byrne et al., 2004). Considering repeated samples from the large continental population independently might dilute the effect of the island samples (Tapio et al., 2003). The distribution of pairwise F_{ST} values is thus a useful way of identifying 'outlier' populations. Slatkin (1993) and others (e.g. Rousset, 1997) have also suggested the use of these pairwise F_{ST} values to determine whether genetic differentiation correlates with geographic distance. Such correlations cannot be tested using classical statistics as n samples will generate $n(n-1)/2$ pairwise values. The Mantel test, which takes a permutation approach on distance matrices, is therefore typically used for that (Mantel, 1967; Smouse et al., 1986) and is implemented in a number of softwares (Box 21.3). The latter approach can be used to study the most likely direction of gene flow in mammals. Indeed, different geographic distance matrices can be created allowing the user to determine the matrix that correlates best with that of genetic distances. For instance it was used by Goossens et al. (2001) on alpine marmots to determine whether gene flow was more likely along the valleys than across mountain tops.

Overall, F_{ST} and F_{ST}-like statistics appear to capture important aspects of genetic differentiation and are consequently very useful summary statistics, but the above-mentioned limitations should be kept in mind.

As noted above, the importance of social structure in determining patterns of diversity should not be underestimated. The typical female philopatry observed in cercopithecine primates is expected to generate higher F_{ST} values in mtDNA data than in autosomal loci for two reasons: (i) mtDNA has a lower N_e, which increases chances of random fixation; and (ii) female philopatry reduces mtDNA gene flow. However, the situation is never that simple with mammals in general and primates in particular. Indeed, these effects can be counterbalanced by male variance in reproductive success as shown by Hoelzer et al. (1998) in macaques. Making predictions is therefore often difficult and behaviour studies are often required to make sense of apparently chaotic genetic signals. For instance, F_{ST} is theoretically expected to increase at low population densities, because of a decrease in N_e. However, in some cooperative species such as callitrichine primates, it is the contrary that happens because higher densities correspond to limited territory availability. Juveniles tend to delay dispersal and stay in their natal group, where they usually do not breed (Di Fiore, 2003). Thus, higher densities tend to increase variance in reproductive success inside populations and limit gene flow between them.

The moment at which gene flow takes place is also crucial in species with male philopatry and female dispersal, because depending on whether females disperse before or after breeding, such dispersal may or may not favour dispersal of non-migrating male genes. While most models of population fission usually assume that individuals are separated at random, this does not appear to be the case in a number of primates (Di Fiore, 2003). In general, groups tend to be made of related individuals, which leads to higher F_{ST} values than expected if fission were a random process. Overall, it becomes increasingly clear that, for social species, behavioural studies are extremely important to interpret the genetic data (e.g. Chesser, 1991). They are also important in determining life-history patterns, which can then be used to make predictions, as in Tiedemann et al.'s (2000) simulation study. These authors demonstrated a higher effect of female than male migration using life-history parameters from the Asian elephant and the blue whale. The social structure also influences patterns of relatedness (e.g. de Ruiter and Geffen, 1998 in long-tailed macaque and Van Horn et al., 2004, in spotted hyenas), or the effective size in different groups of domestic cats (Kaeuffer et al., 2004).

Overall these studies demonstrate that the social structure can have a deterministic effect on patterns of genetic differentiation, but it should be kept in mind that part of the observed differences may also be caused by the stochasticity of the system. Differences between Y and mtDNA data should not always be interpreted in terms of male and female behaviour, as their smaller N_es could explain much of the sometimes large differences observed in population data (see Boxes 21.3

and 21.4). For example, larger F_{ST} values observed in Y chromosome data relative to mtDNA data in humans have been interpreted as an indication of greater female mobility in early human societies (Poloni et al., 1997; Seielstad et al., 1998). More samples would be needed to confirm whether the pattern is observed in all human groups and whether other factors (such as a large variance in male reproductive success) could not generate similar patterns.

Spatial patterns and genetic boundaries

Spatial patterns can also be analysed using specifically designed approaches such as spatial autocorrelation methods (Sokal and Oden, 1978; Bertorelle and Barbujani, 1995; Sokal et al., 1997). These methods require a reasonably large number of samples and have therefore been little used outside human populations and a few other species (e.g. Barbujani and Sokal, 1991, and Chikhi et al., 1998, on humans; Waser and Elliot, 1991, on kangaroo rats; Peakall et al., 2003, on the Australian bush rat). Their advantage over F_{ST}-based approaches is that the spatial component is integrated in the approach (in particular the fact that n spatial samples are not equivalent to n independent samples) and can be powerful in inferring large-scale demographic events (e.g. Sokal et al., 1997). Due to the natural stochasticity of demographic processes, Slatkin and Arter (1991) have argued that spatial autocorrelation may not be a very powerful inferential tool. This is true when very few loci are used (in which case few methods would be powerful anyway). Clearly, autocorrelation patterns should be interpreted with care, and one should concentrate on demographic signals over which all or most loci are in agreement. For example, Chikhi et al. (1998) found that all six loci analysed indicated a significant Europe-wide clinal pattern (a gradient of allele frequencies). This was unlikely to be caused by selection (as all loci agreed) and was in agreement with a hypothesized migration of people from the Near East during the Neolithic transition in Europe (Ammerman and Cavalli-Sforza, 1984).

Other spatially oriented methods include

wombling (Womble, 1951; Barbujani *et al.*, 1989), whereby regions of rapid genetic change ('genetic boundaries') can be detected or PC (principal component) analysis maps whereby the axes of PC analyses can be represented on a geographic map (e.g. Menozzi *et al.*, 1978; Cavalli-Sforza *et al.*, 1993). Both methods usually require a first step of 'interpolation' during which a regular grid is constructed based on the real data sets: in this step, artificial data such as allele frequencies, are created at the nodes by interpolation of the real data surrounding this area. This means that regions poorly sampled will be filled with non-existing (but hopefully representative) data which will be then used to define boundaries. This can be problematic (e.g. Sokal *et al.*, 1999).

An intermediate solution, avoiding the interpolation step, uses the Monmonier algorithm (e.g. Simoni *et al.* (2000) on humans, Townsend (2000) on sheep; Dupanloup *et al.* (2002) on roe deer). In this approach (Fig. 21.4), lines are drawn between the real samples in such a way that samples are connected to their nearest neighbour. The Monmonier algorithm results in a 'mapping' of the geographical area with triangles whose edges connect samples. Boundaries of high genetic change can then be determined by drawing a line across the edges following a simple

algorithm: (i) start from the border edge with the greatest genetic distance, (ii) extend this line by crossing adjacent edges exhibiting the highest genetic distances, (iii) stop when you reach a border.

These approaches are very interesting as exploratory tools, but unfortunately, little has been done to test the efficiency of the latter algorithm when genetic distances are similar at a bifurcation (i.e. what is the effect of choosing to draw the border across one of two edges having similar distance values given the uncertainty on genetic distance estimates?). It is therefore not clear how statistically valid such boundaries are (but see Dupanloup *et al.*, 2002, and below). Also, in cases where a population is the result of a re-introduction (or of an ancient long-distance migration event, or of admixture), the source and re-introduced populations may be separated by artificial boundaries because both populations are expected to be different from the populations currently in between. Overall, more should be done to improve and test these spatial methods, as the number of samples available for many species is going to increase.

Another recent trend, perhaps more developed in plant studies, has been to apply spatial autocorrelation methods to individuals rather than populations (e.g. Hardy and Vekemans, 2002, see the SPAGEDI software). These methods might be useful for solitary mammals for which defining populations may be rather difficult (if their geographic range is not too large).

Phylogeography: integrating geography and gene genealogies

Another set of approaches has also been developed to map gene phylogenies or gene networks to the geographical distribution of alleles (such as mtDNA haplotypes) and uncover ancient demographic events. For instance, the structure of gene trees (see below and Box 21.2) is known to contain information on population bottlenecks or selection. Mapping the gene tree to the geographical distribution of haplotypes might allow one to detect localized population expansions or long-term migration events. For example, observing very similar haplotypes at distant locations but not in intermediate populations

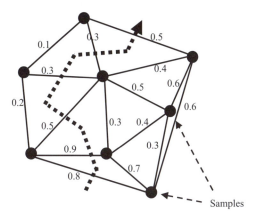

Fig. 21.4. Delauney triangles and genetic borders for a set of eight samples. The figures on the edges are genetic distances. The dashed line represents the genetic border drawn using the Monmonier algorithm (see text).

may be due to long-distance migration rather than to independent mutation events. These ideas were successfully applied in the 1980s (e.g. Bermingham and Avise, 1986; Avise, 1989; Ball et al., 1990). These authors used simulations to predict the kind of geographic patterns expected under different scenarios. Templeton (e.g. Templeton et al., 1995) attempted to formalize this type of approach, which he termed nested clade analysis (NCA). The principle is that a network of haplotypes can be constructed and clades of haplotypes identified. Templeton suggested a statistical test to determine whether the geographic distribution of these clades is consistent with chance and hence test for geographical structure. He then developed an interpretation key that could be applied to detect more specific demographic events.

Templeton's method has been applied to many species including a recent study on human genetic data (Templeton, 2002). While interesting in principle, NCA has a number of serious limitations and we would caution against its use in isolation. One basic problem is that the method does not account for the large and unavoidable stochasticity of the genealogical process (see Box 21.2). In other words, demographic events are inferred from the network in a deterministic manner assuming no statistical error. As a consequence, alternative and potentially valid evolutionary explanations different from those inferred using NCA are not taken into account. For instance, observing similar haplotypes in distant populations that are otherwise genetically distinct could also be due to ancient shared polymorphism. In the only test applied using simulated data sets, Knowles and Maddison (2002) found that NCA: (i) did not properly identify the clades simulated, (ii) identified clades that did not exist, and (iii) identified deterministic events that were actually created by the natural stochasticity of the demographic process. This is worrying because it is always possible to propose a scenario when applying NCA to real data sets. Interpretations based on NCA will therefore not be considered here.

A similar approach, termed median networks, was developed and applied to human populations by Bandelt et al. (1995). Such networks are often an extremely useful visual representation of complex and confusing data. Moreover, Bandelt et al. (2002) have recently developed a method that allows the detection of 'phantom' mutations. This method seems to be very powerful in identifying possible typing errors and could prove very useful for assessing large data sets. However, the same concerns arise regarding the use of median networks to infer demographic events.

The difficulty of linking geography and gene diversity in the frameworks presented above should not be interpreted as indicating that such a task is beyond hope (e.g. Emerson et al., 2001). The early work done by Bermingham et al. showed that it was possible to make simple phylogeographic inferences. In a number of these studies, work was carried out on divergent subspecies. This meant that enough time had elapsed for lineage sorting to have occurred. Simple demographic processes could thus be reasonably inferred. Similarly, the coalescent theory has clearly demonstrated that gene trees do contain significant information on demographic events (Box 21.2). Slatkin and Hudson (1991) were, to our knowledge, the first to study the effect of population growth on the properties of gene trees and DNA polymorphism and apply it to human populations. They simulated a population under exponential growth with mutations following a Poisson process. They showed that the expected gene trees are 'star-like' and that the distribution of pairwise differences between DNA sequences are Poisson-like. The reason for this can easily be understood, taking a coalescent approach (Box 21.2). An increasing population (with time going forward) can be seen as a shrinking population (going backwards in time). Consequently, for a sample of DNA sequences taken today, coalescence events occur at an increasing pace as we go backward in time. In other words, trees from expanding populations will tend to have most coalescent events concentrated close to the 'root' (the sample's MRCA), and will thus look star-like. Rogers and Harpending (1992) arrived at a similar conclusion but assuming a sudden population increase from an equilibrium population size of N_0 to N_1.

One recurrent question is whether signals observed at a particular locus are caused by demography or selection. Also, as Barton and

Wilson (1995) noted, gene genealogies are often influenced by ancient demographic events, making it difficult to estimate recent events or migration patterns. For instance, the signal for population growth found in human mtDNA by Slatkin and Hudson (1991) or by Rogers and Harpending (1992) was first interpreted as dating to 100,000–200,000 years ago, a period corresponding to the first appearance in the fossil record of anatomically modern humans. However, more recent expansion dates have also been suggested (between 40,000 and 60,000 years), a period which happens to correspond to the period during which a significant change in human behaviour is observed, with the first clear evidence of rituals and the use of more complex 'toolkits' (Klein, 2000). More recent dates again were found to be compatible with a past human expansion taking place between 12,000 and 250,000 years ago (e.g. Pritchard *et al.*, 1999). It thus appears that dates of population expansion would be compatible with the Neolithic transitions that took place around the world with the domestication of plants and animals. The reasons for such enormous variation in the dating of past human expansions are manifold. First, different studies have used different loci located in different chromosomes (from Y-linked to autosomal SNPs or microsatellites, through mtDNA data). Non-coding loci have tended to indicate a pattern of population expansion, whereas coding loci potentially under selection tended not to exhibit this expansion pattern. Secondly, the demographic and mutational models used are sometimes rather different. For microsatellites, accounting or not for recurrent mutations can lead to rather significant changes in population estimates as noted by Excoffier and Hamilton (2003) for the estimation of genetic diversity in human groups. Finally, recent years have seen an improvement in our ability to account for the stochasticity of the evolutionary process in the statistical inference. Thus, the increase in uncertainty seen in some estimates is also a reflection of the increasing awareness that very little can be said from genetic data alone and even less from single-locus studies (Goldstein and Chikhi, 2002, and see below).

To conclude on this section, we do believe that genetic data should be used to infer ancient demography, but that inference should be made with caution. Users should also be more critical of the sometimes very complex scenarios that can be constructed just by using mtDNA or Y chromosome data. Simulations of simple scenarios have clearly demonstrated that it is very difficult. The reader interested in the relationship between the gene and population/species tree issue may want to read the following papers and references therein: Pamilo and Nei (1988) at the species level; Barbujani *et al.* (1998) on the specific interpretation of gene trees in relation to the colonization of Europe; and Edwards and Beerli (2000) and Nichols (2001) at the species and population level.

Individual-based approaches: inference of population structure with limited or without prior knowledge

AMOVA and other F_{ST}-like-based approaches are powerful in detecting and analysing population structure, but some authors have expressed the concern that the population structure is defined *a priori* (by the sampling scheme for instance) rather than estimated from the data. This could lead to missing hidden patterns present in the data. Holsinger and Mason-Gamer (1996) have thus suggested combining an F_{ST} analogue using molecular information together with a clustering algorithm to define and test the hierarchical structure. This method is implemented in the NUCLEODIV software (Box 21.3). Recently, Dupanloup *et al.* (2002) have suggested an interesting approach that applies a maximization algorithm ('simulated annealing') to data that are explicitly spatial. Geographical location and molecular variance (as defined by Excoffier *et al.*'s 1992 AMOVA) are used to identify populations on the principle that molecular variance is maximized between and minimized within populations. This approach, implemented in the SAMOVA software (for spatial AMOVA, Box 21.3) was compared with the Monmonier algorithm presented above on both simulated and roe deer data by Dupanloup *et al.* (2002). They found that the Monmonier algorithm was, despite its simplicity, very efficient at detecting barriers in simulated data compared with the SAMOVA method. They also found that it tended to detect some poorly supported borders, whereas SAMOVA tended to be statistically more valid. In the roe deer data, one advantage of SAMOVA is that it allowed the grouping of populations which were geographically distant, a problem mentioned above for wombling and similar approaches. This allowed Dupanloup *et al.* (2002) to detect reintroduction events in the Ligurian sample, from Balkanic sources, whereas the Monmonier algorithm separated the Ligurian samples without linking it to Balkan samples.

Avoiding all *a priori* information is in practice difficult and different methods differ in how they try to uncover hidden structure in the data. The information on the samples is sometimes used (as in some assignment methods, below) or totally ignored (as in the model-based methods to infer hidden structure, below). However, most of these methods have in common that they are individual- rather than population-based. In the population genetics field, the first study to have explicitly taken this individual-based approach (for nuclear loci) is that of Bowcock *et al.* (1994). In this study, the authors developed a measure of genetic distance between individuals based on the proportion of shared alleles (see also Chakraborty and Jin, 1993) and then used it to build individual-based trees in order to determine whether geographical clustering would be visible. They found that with 30 microsatellites typed in 14 worldwide populations with approximately ten individuals each, they obtained good geographical clustering making it a very influential study. Interestingly, Bowcock *et al.* (1994) were also among the first to mention the problem of **ascertainment bias**, leading to biased estimates of genetic diversity. In their study, it was invoked to explain the observation that for allozymes and nuclear RFLP (but not for mtDNA or microsatellites), African samples were less diverse than most other populations, simply because these loci were first screened and selected for diversity in Europe. Note that ordination methods are to some extent also addressing the issue of finding 'hidden structure' in the data, but they are not specific to population genetics and shall not be addressed here.

Assignment methods

Assignment tests are individual-based methods that have recently become popular in the population genetics literature, particularly after the development of hypervariable loci for many species. They have been applied to both natural and domesticated mammalian species after the influential study by Paetkau *et al.* (1995) on bears. Assignment methods can be divided into **likelihood**- and distance-based methods. Likelihood-based methods rely on the calculation of the probability $P(G_i | \theta_j)$ that a multilocus genotype G_i (of a real or simulated individual) is observed in a population P_j for which allele frequencies are known (represented by θ_j). Loci and alleles are assumed to be independent (i.e. populations are at HW and linkage equilibrium) so that this probability (i.e. the likelihood $L(P_j)$) that an individual comes from P_j is simply computed as the product across loci of the single locus HW genotype frequencies. One typical problem with these calculations is that if an allele is not observed by chance in a particular population, the likelihood of the population will be zero, whatever the other loci might indicate (Fig. 21.5). Thus, a number of different *ad hoc* corrections have been proposed to account for unobserved alleles and/or missing genotypes and unequal sample sizes (Paetkau *et al.*, 1995; Cornuet *et al.*, 1999; Banks and Eichert, 2000). As there is no real statistical model behind these corrections, they should be used with caution. For instance, it has been suggested that the 'missing' alleles could be added to the excluded populations (to simulate the fact that the alleles could have been observed, had the sample size been slightly larger). However, assignment methods are typically used for hypervariable loci with many alleles, and this procedure might significantly change allele frequencies in the different populations. Another statistical problem related to the calculation of $L(P_j)$ is addressed by the 'leave-one-out' procedure, whereby allele frequencies are recalculated by excluding individuals from their population of origin (Efron, 1983, in Cornuet *et al.*, 1999). This method is applied to all individuals, including those having rare alleles (Paetkau *et al.*, 1995; Cornuet *et al.*, 1999). This is an *ad hoc* method and its consequences may vary widely from one data set to another.

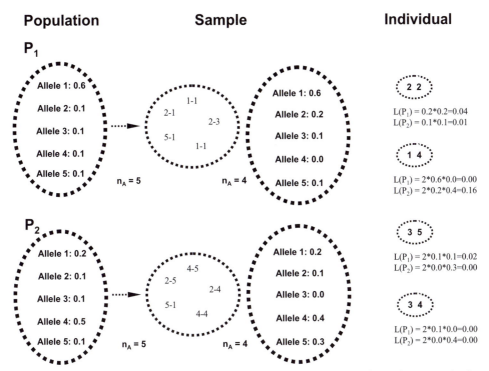

Fig. 21.5. Assignment methods. Two populations, P_1 and P_2, are represented together with one sample of size 5 from each. The likelihood is calculated for four single-locus genotypes. Even though both populations P_1 and P_2 have all alleles observed in the four individuals, the likelihoods are based on the frequencies calculated from the samples. Thus, some likelihood calculation can therefore be zero, leading to the rejection that the individual came from any of them (last individual).

The most statistically sound likelihood-based method was originally developed by Rannala and Mountain (1997) to detect migrants and was applied to human populations in Australia (the method also allowed researchers to determine if these individuals were 1st, 2nd or 3rd generation migrants). This method is set in a Bayesian framework and accounts for the sampling effect such that missing alleles are taken into account from the start. Simulations by Cornuet *et al.* (1999) indicated that it was the most efficient of all methods. It was recently extended by Wilson and Rannala (2003) to allow estimation of rates of recent immigration and has the property that it does not require populations to be at HW equilibrium. Wilson and Rannala (2003) applied their method to northwestern Canada grey wolf microsatellite data. They were able to demonstrate that migration patterns are not symmetrical among populations and that migration was unidirectional in some cases.

Distance-based assignment methods have been introduced recently by Cornuet *et al.* (1999). The rationale is similar to likelihood-based methods except that populations are 'chosen' with regard to the lowest genetic distance observed (rather than the highest likelihood). Three genetic distances were tested, namely Dc (the chord distance of Cavalli-Sforza and Edwards, 1967), D_{AS} (an allele shared distance by Chakraborty and Jin, 1993), and Goldstein *et al.*'s (1995) $(\delta\mu)^2$ distance. Details of how the two population distances (Dc and $(\delta\mu)^2$) are computed for individuals rather than populations can be found in Cornuet *et al.* (1999). Overall, the simulations performed indicated that distance methods were usually not as efficient as likelihood methods. Most previously cited assignment methods (plus a couple of other distance methods) have been implemented in the GENECLASS (Cornuet *et al.*, 1999), GENECLASS2

(Piry *et al.*, 2004) and WHICHRUN software (Banks and Eichert, 2000) (see Box 21.3).

Assignment methods can have many applications. We have seen above that they can help identify migrants and detect unidirectional movements (Wilson and Rannala, 2003). This is extremely important when monitoring populations of endangered species, and can be critical for island populations, to determine whether such populations are isolated from the rest of the species populations. Koskinen (2003) applied assignment methods and other clustering approaches to five dog breeds using ten microsatellites. He was able to show that the individuals were always assigned to their breed of origin and he suggested that, contrary to previous statements, breed identification could be achieved using genetic data. Similarly, Arranz *et al.* (2001) used 18 microsatellites on five Spanish sheep breeds and found very reliable results as well. Thus, in theory, genetic data might be very useful for individual identification in breeds or natural populations.

These methods also have applications in the detection of fraudulent meat of supposed guaranteed origin. In conservation issues, assignment methods have also be used to detect trafficked animals or identify the exact species of harvested animals (Manel *et al.*, 2002). This can be crucial for cetaceans when some species are protected while similar species are being harvested. Evans *et al.* (2001) applied assignment tests to identify potential hybrids in a hybrid zone between two macaque species, and Miller *et al.* (2003) used assignment methods to detect introgression of coyote in the endangered red wolf in North America and consequently determine strategies to minimize reproduction of hybrids and maximize reproduction of red wolves. Comstock *et al.* (2002) used assignment tests on African elephants and found that savannah elephants were not as highly identifiable as forest individuals (they had less specific alleles for instance), but they also identified potential hybrids. Interesting studies carried out on other non-mammalian species include differentiating stocks of exploited fish (Gum *et al.*, 2003, on the European grayling), identifying the origins of stock in captive breeding programmes (Burns *et al.*, 2003, on endangered Galapagos

tortoises) and identification of source populations in migratory species (e.g. Scribner *et al.*, 2003, on Canada geese).

Assignment methods are most powerful when the reference populations are clearly identified and genetically differentiated (i.e. when they exhibit high F_{ST} values). Thus, the results presented above should be interpreted with care and certainly not with excessive optimism. Indeed, the dog breeds used by Koskinen (2003) exhibited F_{ST} values of 0.18, and more loci would be required for lower F_{ST} values as simulations showed (Cornuet *et al.*, 1999). Similarly, the sheep study of Arranz *et al.* (2001) provided good results because the average F_{ST} was 0.07 and they used 18 microsatellite loci. For populations or breeds that exhibit lower differentiation, little work has been done, but preliminary results by Paulo *et al.* (unpublished) on Portuguese breeds indicate that 20 microsatellite loci are not enough to separate breeds with an average F_{ST} of 0.02. Overall, assignment methods have the advantage of being computationally very simple, and hence very well adapted to the huge data sets expected in the genomic era. We shall see that this is not the case with all methods.

Hidden structure: model-based approaches without prior sample information

Recently two major methods have been developed to uncover 'hidden structure' without using any *a priori* information regarding sampling origin (Pritchard *et al.*, 2000; Dawson and Belkhir, 2001; implemented in the STRUCTURE and PARTITION software, respectively). The underlying idea is to use a model-based approach to define the probability of generating the data assuming K hidden partitions (hopefully equivalent to the 'real' populations). In practice this probability can be difficult to estimate. However, if this probability can be approximated, it is possible to construct a **Markov Chain Monte Carlo algorithm (MCMC)** that will explore the parameter space defined by the model in a clever way. This exploration is performed in such a way that the parameter values sampled during the run are 'visited' in proportion to their probability of generating the data. In the general problem of uncovering hidden structure, parameters may include the number of partitions (Pritchard *et al.*, 2000; Dawson and Belkhir, 2001) or the proportion of genes coming from any of K partitions (Pritchard *et al.*, 2000; Falush *et al.*, 2003). In these methods, partitions are not predefined but constructed (estimated) during the analysis. In the Pritchard *et al.* (2000) method, the number of partitions has to be defined before the run and the analysis is performed for different K values using the STRUCTURE software. The average likelihood for the different K values can then be compared to estimate their relative probabilities.

In the Dawson and Belkhir (2001) approach the parameter space is defined by: (i) K, the number of possible partitions, and (ii) the distribution of individuals in the K partitions. During this process the likelihood of the data is estimated for the different K values and possible assignments of individuals to the K partitions. Thus, when the chain reaches equilibrium the different values of K have been sampled in proportion to their probability of generating the data. It therefore becomes possible to estimate the **posterior probability** distribution of K. The method also keeps track of the time during which individuals were in the same partition. A value of $K=1$ thus means that no hidden substructure is detected in the data set, and the maximum number of partitions allowed has to be specified beforehand. The current version of the PARTITION software allows the detection of up to 12 partitions (in practice this is very reasonable).

There are a number of differences between the two approaches. The STRUCTURE approach is more flexible and allows the user to either assign individuals to different populations or determine the proportion of their genes coming from the K potential populations. It is therefore able to deal with both hidden structure and admixture models. There is, however, a problem with it. If the STRUCTURE algorithm is run for very long, it is expected that individuals will have been randomly assigned to the different partitions in such a way that what was called 'partition 1' in the beginning of the run may become, say, 'partition 3' during the run. In other words, the fact that no *a priori* information exists on the partitions' genetic make-up makes it impossible to clearly define what partition 1 is. Consequently, long runs will tend

to attribute all individuals to the different partitions in proportion $1/K$. So, the STRUCTURE software is expected to work best when runs are not too long and the MCMC does not mix too well (see Box 21.1). This is rather strange as MCMC methods are usually expected to perform best when run longer and when mixing is good. Also, there is no definition of what a long or short run is, a general problem of any MCMC-based method. Thus, it is not necessarily easy to define the right time to terminate the analysis, so that individuals have been 'mixed' enough to be in the right partition, but not too long so that partitions have not started to be 'mixed too much'. Here the PARTITION method is clearly superior because it provides the proportion of time that different individuals have spent together after equilibrium is reached. This allows Dawson and Belkhir (2001) to propose a clustering method based on these posterior probabilities which can then be used to define partitions, if necessary. We applied the PARTITION method to genetic data from European sheep breeds (not published) and found that partition detected the existence of two flocks in the Soay sheep breed data. This was particularly impressive because the information on the two flocks had first not been given and the PARTITION analysis pointed to the need to have this information.

Individual-based analyses are appealing because they allow us to relax assumptions on the individuals' population or sample of origin (e.g. Estoup and Angers, 1998; Davies *et al.*, 1999). This led Luikart *et al.* (2003) to suggest that individual-based analyses should be used whenever possible. For instance, they suggested that it would be possible to use genetic distance measures to cluster individuals and consequently define populations without any *a priori* assumptions. However, while individual-based methods are becoming increasingly important (they may represent important alternatives for studying solitary species), it might be slightly misleading to consider them as being assumption-free. In a model of migration or admixture (which would potentially apply to many mammalian species: e.g. Evans *et al.*, 2001; Chikhi *et al.*, 2002) defining populations on the basis of genetic distances between individuals could actually lead to ignoring the very dynamic of interest. Note, however, that such clusters while potentially meaningless on their own could be used to identify migrants (they would cluster with their population of origin, but would be sampled in the population they migrated in). The key point here is that it is difficult to make any inference outside some underlying set of assumptions or model. The fewer assumptions one makes the more descriptive the analysis becomes. Population genetics teaches us that even when the model is correct, uncertain inference is the best we can get, whether we like it or not.

Another potential problem with individual-based approaches is that increased sampling instead of improving inference of structure may prevent it. This can be first seen with an imaginary example in which three well defined and distant populations are sampled (Fig. 21.6a). Clearly, we would expect individual-based methods to re-assign individuals to their populations of origin or hidden-structure seeking methods to infer that $K = 3$. Now assume that the sampling effort is increased and that ten samples covering the whole distribution are now genotyped; Fig. 21.6b gives a possible outcome of this increased effort. In such a case, it

is unclear what individual-based methods would be able to infer. This is problematic, as simple F_{ST}-based approaches would have no problem in demonstrating significant differentiation between samples. This situation is not as artificial as it may seem, as we found a similar situation with genetic data from European sheep breeds (Byrne et al., submitted). We also analysed the patterns of genetic differentiation of the Jersey cattle on the Island of Jersey (Chikhi et al., 2004) and found significant levels of population differentiation between parishes and farms. However, the PARTITION software was only able to detect one partition (with posterior probability of 0.98 for $K = 1$ and 0.02 for $K = 2$). These results are not necessarily contradictory. The island of Jersey being small, this indicated that gene flow had been much more intense than suggested by documented exchanges (we also found no spatial autocorrelation pattern). However, this problem may become crucial when isolation by distance is important, as it is for many species. The sampling scheme (within or between 'neighbourhoods') will then be important. This result is superficially counter-intuitive as it increases just

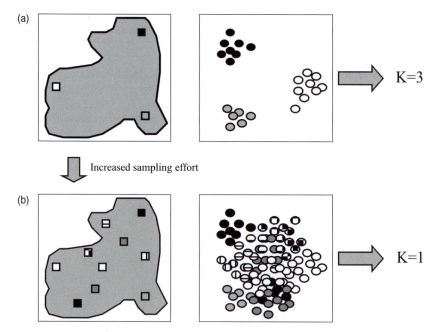

Fig. 21.6. Effect of the sampling efforts on methods using no *a priori* information on the origin of samples. In the first case (a), the methods are able to uncover three partitions. However, when the sampling effort increases (b), most samples overlap and cannot be separated.

because 'too many' samples were obtained (see Paulo *et al.*, unpublished on sheep breeds). As noted also by Excoffier (2002), the clusters found by the STRUCTURE method on 377 microsatellite loci by Rosenberg *et al.* (2002) using *c.* 1050 individuals are statistically reconstructed entities that do not necessarily correspond to real entities. They may be very useful in linking geography and clustering, as we noted above, but clearly, discarding non-genetic information *a priori* may be misleading. One possible solution to this problem may come from the development of new Bayesian methods that would be able to incorporate the sampling locations into the priors. Overall, it is very difficult to define steps to follow and general rules at this stage on whether to use individual or population-based approaches. We will need much more theoretical work and data before such rules can start to be set.

Full-likelihood (Bayesian) and summary statistics (approximate Bayesian) methods

We have seen above that there are many ways to summarize, sometimes visually, the genetic information collected from samples. Some of the methods allow the user to *detect* possible demographic or selective processes and others to *estimate* demographic parameters. Some are grounded in a sound statistical framework while others have more shaky foundations.

This naturally takes us back to the question of whether genetic data contain enough information to detect patterns and draw precise inferences. The conservative answer is unfortunately that there is often rather little information in genetic data and that it is currently impossible to infer complex demographic scenarios by genetic data alone. In other words, non-genetic and genetic data have to be used jointly to help discern usable or testable models (Goldstein and Chikhi, 2002). Fortunately, for many mammals, including humans, such external information is large enough to provide a limited number of alternative scenarios to test. But even in these cases, the interpretation of genetic data can be controversial (Barbujani and Bertorelle, 2001; Chikhi *et al.*, 2002; Richards, 2003, on recent origins of Europeans).

Felsenstein (1992) was among the first to provide a critical appraisal on how genetic data were used in population genetics. He noted that most of the useful information available in the gene phylogenies was discarded by most summary statistics. This has led a number of authors to develop full-likelihood methods that allow the user to utilize all the information present in the data. The principle is that instead of inferring, say, the migration rate between populations from F_{ST} values, it is the full allelic distributions in the different populations that can be used. Given that similar F_{ST} values can be obtained with huge numbers of allelic distributions, this represents a significant increase in information retrieval. The problem is that it is much easier to estimate the probability of obtaining a particular F_{ST} value than it is of obtaining particular allelic distributions.

Analytically, only a few demographic problems can currently be solved and even when this can be done, it is feasible to estimate these probabilities only for very small sample sizes (Griffiths and Tavaré, 1994; Beaumont and Bruford, 1999; Chikhi *et al.*, 2001; Beaumont, 2004). One could think of obtaining these probabilities by simulation. For instance, it is theoretically possible to simulate data for different demographic parameter values and count how often the sampled allelic distributions are observed. In practice this is impossible for most real data sets because these probabilities are so small that it would take huge amounts of time to obtain reasonable estimates of these probability values. Repeating the process for many parameter values is thus currently impossible. Fortunately a number of theoretical studies since the mid–late 1990s have shown that it is possible to estimate these probabilities by using recently developed statistical techniques such as importance sampling and Markov chain Monte Carlo (Griffiths and Tavaré, 1994; Kuhner *et al.*, 1995; Wilson and Balding, 1998; Pritchard *et al.*, 2000; Stephens and Donnelly, 2000; Beaumont, 2003, 2004). These methods remain highly computational but none the less much more feasible than trying to estimate the probabilities as suggested above and they are often set in a Bayesian framework (but see Kuhner *et al.*, 1995). For more technical details the interested reader is invited to turn to some very thorough reviews

(Stephens and Donnelly, 2000; Stephens, 2001; Beaumont, 2001, 2003, 2004).

Full-likelihood methods have been used for models of population size change (Beaumont, 1999, on humans and northern hairy-nosed wombat; Storz and Beaumont, 2002, on two bat species; and Storz et al., 2002, on savannah baboons), admixture (Chikhi et al., 2001, 2002; Pritchard et al., 2000, all on humans) and migration (Beerli and Felsenstein, 2001). The example of the northern hairy-nosed wombat, discussed above, is very illuminating in that it shows the potential increase in power that full-likelihood methods can bring. As we noted above, Luikart and Cornuet (1996) showed that the lack of genetic diversity observed by Taylor et al. (1994) in the northern hairy-nosed wombat was not due to a long-term small N_e, but rather to a recent demographic decline. This seemed to confirm the idea that the decline was due to recent and recorded human activities (Taylor et al., 1994). This was confirmed by the analysis of Garza and Williamson (2001) based on the gaps in microsatellite allele frequency distributions. However, Beaumont (1999) showed, using a more complex model, in which the start of the population contraction can be dated, that the decline had actually started much before the human-caused recorded population contraction. In other words, methods using reduced allelic information were able to detect a departure from mutation–drift equilibrium, which was itself an improvement on simply noticing a lack of genetic diversity, but which also led to erroneous interpretations. The use of all the information present in the allelic distribution allowed Beaumont (1999) to make much more precise (but still imprecise!) inferences. One interesting result of Beaumont's (1999) study is that the increase in power was also due to the use of monomorphic loci, which were discarded by previous analyses as being 'non-informative'. This study shows that care should be taken before interpreting signals of demographic events. It is all too easy to interpret such signals in relation to events that happen to be 'favoured'. Beaumont's method is implemented in the MSVAR software available from M. Beaumont's web page (Box 21.3).

This sets the problem of separating recent from ancient demographic events, which we discuss below. This problem is a recurring one in phylogeographic studies where signals of population growth or decline are automatically attributed to the last (and best studied) glaciation events. In practice, demonstrating that a population has been subjected to a bottleneck does not identify the bottleneck event in time. Given that the Pleistocene has seen up to ten cycles of glaciation alternating with interglacial periods, the former representing more than 90% of the duration, it is far from trivial to associate a demographic expansion to any of these individual events. Molecular clocks are not always precise enough to allow definitive associations with specific glaciation events. This is particularly true for recent events.

Despite their qualities, full-likelihood methods are highly computational and some of them at least meet a number of practical problems:

1. They cannot be used as a black box and often require adaptation to the data set at hand. For instance, MCMC methods require that only points sampled after convergence is reached be used for inference. In practice there is no way to know how much time this will require. This time is likely to change from one data set to another and some fiddling with the code might even be required.
2. Typically, outputs are complex and often require significant processing (but see STRUCTURE or PARTITION for more user-friendly methods).
3. They can be extremely slow even for typical population/conservation genetics data sets (i.e. 10–20 microsatellite loci typed for 200–300 individuals).

Thus, full-likelihood methods are very powerful and promising but do not seem to be adapted to data sets most likely to be produced in the genomic era, and sometimes, even with reasonable data sets.

This computational problem has favoured the development of so-called summary statistic methods (Pritchard et al., 2000). These methods try to find the right balance between slow full-likelihood methods and the less information-rich methods described at length in the preceding sections. The underlying idea is that there should be a way to efficiently summarize the allelic distribution by a limited number of statistics that are easy to evaluate. By doing that, one could simulate data and instead of checking whether the right allelic distribution is observed, compute the statistics of interest. Then instead of selecting parameter values only when there is an exact match between simulated and observed values, some threshold can be defined. These methods do not require the calculation of the likelihood, which is often highly computational (Beaumont et al., 2002; Marjoram et al., 2003). By testing varying combinations of summary statistics and of thresholds one can find a set that provides both precise estimates and a significant increase in speed, which allows its use on large data sets. There have been a number of improvements on summary statistics in the very last years. As they have been applied on different demographic models, these improvements can only be compared in terms of the criteria used to accept or reject the summary statistics values. In the original paper (Pritchard et al., 2000), the authors simply decided to keep simulations when the absolute difference between the observed and simulated statistics was below some arbitrary threshold. Recent work by Beaumont et al. (2002) has shown that the speed of the algorithm and the precision of parameter inferences can be significantly improved by using 'local' correlations between the summary statistics and parameter values. Excoffier and Hamilton (unpublished) have also made some similar improvements and applied them to models of population structure to estimate gene flow. Summary statistics approaches have been used to estimate and date population expansion (Pritchard et al., 2000), to estimate different parameters of genetic diversity (Beaumont et al., 2002). Overall, these methods are likely to become widely popular due to their ease of use.

Extensions to Specific Applications

In the first part of this chapter we dealt with population genetics issues showing how the access to increasing numbers of loci is naturally pushing population genetics methods towards population genomics. We also saw how patterns of LD make the first real step towards using genomic aspects of genetic diversity. In the following we address a few more specific issues, with even more explicit genomic views of genetic variability.

Population differentiation across the genome and selection

It seems reasonable to assume that unlinked neutral loci will represent independent replicates of the same demographic history. Hence, analysing variation among many independent loci should help identify outlier loci potentially under selection. Cavalli-Sforza (1966) is the first to have suggested the use of population differentiation to identify selection. His idea was that loci under balancing selection (maintaining similar frequencies between populations) should have lower F_{ST} values than neutral loci while loci under divergent selection (different alleles being more adapted in some populations/environments) should exhibit higher F_{ST} values. He applied his approach to a set of nine loci and 15 human populations, and suggested that selection might be required to explain that the R_0 allele of the Rh locus exhibited an F_{ST} of 0.38, a value more than an order of magnitude larger than that exhibited by the Kell blood locus. Lewontin and Krakauer (1973) were the first to try and quantify the statistical error associated with different F_{ST} values, but they made a number of simplifying assumptions which led to an underestimation of the real variance (Robertson, 1975a,b; Nei and Maruyama, 1975). Indeed, populations are neither spatially nor temporally independent, and this leads to erroneous rejection of neutrality (this is reminiscent of the stochasticity observed between loci due to coalescence, see Box 21.2).

The idea was therefore abandoned until Bowcock et al. (1991) suggested the use of coalescent simulations to obtain the distribution of F_{ST} values as a function of gene frequencies under reasonable demographic histories of human populations (Fig. 21.7). They were able to demonstrate the possible action of both stabilizing and disruptive selection on a number of loci. Beaumont and Nichols (1996) have consequently suggested the use of coalescent simulations to obtain the joint distribution of F_{ST} and H_e values under neutrality. Outlier loci could then be easily (and visually) identified as potential targets for selection. These methods have the limitation that the observed variance among loci is dependent on the demographic and mutational models

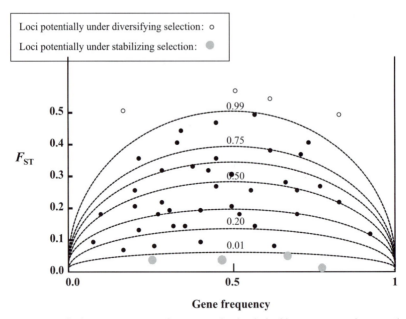

Fig. 21.7. Detection of selection using an outlier approach. The dashed lines represent the quantiles for the expected distribution for F_{ST} for different allele frequency under neutrality. Loci above the highest quantile are candidates for diversifying selection, whereas loci below the lowest quantile are possibly under stabilizing selection. Loci which are located between the highest and lowest quantiles (black dots) are less likely under diversifying or stabilizing selection. The figure was drawn after Bowcock et al. (1991). It was applied to binary markers and quantiles are therefore symmetrical around 0.5. In a binary marker the two alleles have the same F_{ST} value and it is irrelevant whether one or the other is represented. In this figure either the most frequent or less frequent allele is represented.

assumed (a simplified model of human history in Bowcock et al., 1991, and the n-island model and IAM mutation model in Beaumont and Nichols, 1996). The latter method was shown to be rather robust to different demographic models. Furthermore it was later improved to account for different mutational models (Flint et al., 1999). The availability of SNPs covering the whole genome of mammalian species would require that a two allele model be incorporated. Current implementations of the method (see M. Beaumont's web site, Box 21.3) will typically generate data sets with more than two alleles leading to a misleading joint distribution of H_e and F_{ST}. One way around this would be to filter the outputs to keep only bi-allelic loci, but this has not been implemented yet to our knowledge and would have to be done 'manually'.

Goldstein and Chikhi (2002) suggested that instead of using simulations, one could rather use the distribution based on SNPs that are known to be far from identified genes, and thus presumably neutral. SNPs of interest could then be compared with such empirical distributions. The advantage of this approach is that it should account for the (unknown) demographic history of the species of interest. This idea was first applied by Akey et al. (2002) and Soranzo et al. (unpublished data) for the human species, and could soon be applied to other mammals. Indeed, there are significant data suggesting that mammals share most of their orthologous genes and that many genetic markers such as microsatellites are conserved between close taxa such as primates (Blanquer-Maumont and Crouau-Roy, 1995; Clisson et al., 2000) but also among more distant taxa. Indeed, Jiang and Michal (2003) showed that among 1321 microsatellites identified in pig, 228 had homologous sequences in human.

If the genome of humans is block-like, as has been suggested recently (see above), then we would expect that a few SNPs could be enough to describe the genetic diversity along rather long DNA stretches (perhaps 5–100 kb depending on genome areas and population demographic history, e.g. Stumpf and Goldstein, 2003). One could apply this empirical approach to haplotypes rather than SNPs and uncover DNA stretches carrying genes under selection. This would open up fascinating discoveries, particularly when comparing different species (e.g. Murphy *et al.*, 1999, for the conservation of chromosome organization between human and cat but see also Murphy *et al.*, 2003, for the origin and multiple fission events that took place in the history of chromosome 1 across mammals). One could also use results obtained in humans to quickly identify orthologous regions from other mammals. For example, Jiang *et al.* (2002) showed that cattle and pig can share large amounts of contiguous autosomal markers with humans. Preliminary results indicate that 75% of the 24,567 annotated human genes align with dog sequences (Ostrander and Comstock, 2004). Regions possibly under selection in humans could therefore be identified in other mammals very rapidly. Conversely, disease-causing genes identified in domestic animals (particularly in dogs) might then help identify unknown genes in humans. Consequently, tissue-specific expression patterns could be compared as was done by Jiang *et al.* (2003) in human and pig tissues.

The identification of outlier loci should not hinder the complexity of both the demographic and selective history of genes. Genes recently influenced by selection (for instance, due to environmental changes posterior to the Neolithic transition in humans) might be undetectable under this type of approach. Another potential problem comes from the sampling scheme of populations, which influences average F_{ST} values. The effect of sampling very close populations repeatedly together with a few more divergent populations on the method proposed by Beaumont and Nichols (1996) has not been studied and it is not clear whether this could not lead to an increased difficulty in identifying outliers. Another way of using F_{ST} values to detect selection has recently been developed by Raufaste and Bonhomme (2000 and unpublished) and was mentioned above. This method could be applied to large sets of independent loci in parallel to the above

because it is likely to be influenced by different forms of selection.

Contrasting patterns of molecular evolution in hominoids revealed by multiple marker studies

Gagneux *et al.* (1999) examined phylogenetic diversity among African hominoids using 1158 mtDNA control region haplotypes, including 83 new sequences from western chimpanzees and bonobos. Analysis revealed different patterns of evolution among species and heterogeneity in species-level variation, with several chimpanzee and bonobo clade lineages retaining more mitochondrial variation than in extant *Homo sapiens*. Evidence was described that eastern chimpanzee and human clades experienced reduced effective population sizes, the latter apparently since the *Homo sapiens/H. neanderthalensis* split. The authors concluded that the patterns of mtDNA sequence variation seen in modern taxa reflect historical differences in ecological plasticity, female-biased dispersal, range fragmentation over differing periods of time, and competition among social groups. Further, a recent study of Y chromosome variation (Stone *et al.*, 2002) examined the population history of the genus *Pan*, which comprises chimpanzees and bonobos. The authors examined ~3 kb of Y DNA from 101 chimpanzees, seven bonobos and 42 humans, and identified ten informative SNPs associated with 23 polymorphisms in the genus. Nucleotide diversity was significantly higher on the Y chromosome of chimpanzees and bonobos than in humans. This great diversity in both chimpanzee species suggests that these species have substantially larger effective population sizes than humans, on an evolutionary scale, despite recent drastic population decrease of *Pan* sp. In addition, Stone *et al.* (2002) confirmed that Y chromosome lineages were distinct between bonobos and chimpanzees, and between chimpanzee subspecies. At the same time, Kitano *et al.* (2002) investigated nucleotide diversity in ten X-chromosomal genes where mutations are known to cause mental retardation in humans. For each gene, they sequenced the entire coding region from cDNA in humans, chimpanzees and orangutans and 3 kb of genomic DNA in 20 humans sampled worldwide and ten chimpanzees representing two 'subspecies.' As before, they found that nucleotide diversity was only half the level in humans compared with chimpanzees.

However, a recent study by Yu *et al.* (2003) has revealed a contrasting result for automosomal DNA. The authors sequenced 50 different non-coding, non-repetitive DNA segments from the nuclear genome of nine bonobos and 17 chimpanzees. Surprisingly, sequence diversity for bonobos was 0.078%, which is lower than humans (0.088%). The same segments had diversity of 0.092, 0.130 and 0.082% for East, Central, and West African chimpanzees, with an overall value of 0.132%. Such values are ~1.5 times higher than that for humans. The previously revealed larger difference in mtDNA diversity could, according to the authors, be due to a reduction in effective size of the human lineage after the human–chimpanzee divergence, since this would affect mtDNA diversity more than nuclear DNA. Finally, de Groot *et al.* (2002) examined major histocompatibility complex (MHC) class I gene intron variation in chimpanzees and showed that this species experienced a severe repertoire reduction at the orthologues of the HLA-A, -B and -C loci. The loss of variability apparently occurred before speciation of chimpanzees and has not obviously affected other gene systems. The authors inferred that the presumed selective sweep in the MHC class I gene may have resulted from a widespread virus infection. The authors further speculated that the fact that chimpanzees have a natural resistance to the development of AIDS might suggest that the selective sweep may have been caused by the chimpanzee-derived simian immunodeficiency virus (SIVcpz), the closest relative of HIV-1, or some similar virus. If this is the case, current chimpanzee populations could represent the offspring of AIDS-resistant animals, the survivors of a HIV-like pandemic that took place in the past.

Livestock genetic diversity patterns

A similar intensity of work has been carried out on the mammalian livestock domesticated by humans in recent millennia (Bradley and Cunningham, 1999; Ruvinsky and Rothschild, 1998; Bruford *et al.*, 2003). Mitochondrial DNA analysis has revealed that cattle (among other) domestication has been a more complex process than originally believed. For example, the two major distinct cattle types found today are zebu (humped cattle) and taurine. Some authors have classified the two types as separate species: *Bos indicus* (zebu) and *B. taurus* (taurine) cattle. However, the widely held view is that both types were merely differentiated forms that originated from a single domestication event. Taurine cattle from Europe and Africa, and zebu cattle from India and Africa were compared at the mitochondrial control region. The level of sequence divergence between the two mtDNA lineages was consistent with a most recent common ancestor dating to hundreds of thousands of years BP (before present) (Bradley *et al.*, 1996). However, unravelling evolutionary events since initial domestication has proved to be harder using mtDNA. For example, data showed that the African zebu only possesses taurine mtDNA (Loftus *et al.*, 1994) which seemed counterintuitive since it was thought that these zebu cattle originally came from Asia and the Middle East, transported by nomadic pastoralists. However, later work using microsatellites and Y chromosome DNA showed that the nuclear DNA of African zebu cattle was indeed most similar to that found in other zebu cattle populations. The reason for this difference was that mtDNA studies were not able to detect zebu genes (passed into the local taurine population via zebu bulls) because bulls are exclusively used in gene flow and introgression when managed by humans and male mitochondrial DNA alleles are not transmitted to descendent generations (MacHugh *et al.*, 1997) because mtDNA is inherited solely through the maternal line in mammals (but see Awadalla *et al.*, 1999, for a different claim).

It is problematic that comparisons of genomic diversity and its distribution are not straightforward in livestock due to the fact that the wild progenitor either is extinct, as in cattle, or highly endangered and inaccessible, as in sheep and goats for example. However, the origins of European taurine cattle were recently illuminated by a large-scale study of the variation in mtDNA control region variation in 400 cattle from Europe, Africa and the Near East. The study used ancient DNA analysis, which has been increasingly used to identify phylogenetic affiliations from the remains of organisms ranging from humans to bacteria. Skeletal remains found in the UK from four wild aurochs (putative cattle ancestors) were radiocarbon dated to 3720–7320 BP, and their phylogenetic relationships to modern cattle were determined. Surprisingly, they were found to be unrelated to UK domestic cattle, which were instead more similar to modern cattle in the Near East. These data indicate that at least the maternal origins of European cattle are consistent with the spread of cattle by pastoralists from the Fertile Crescent and not local, independent domestication (Troy *et al.*, 2001). This is an interesting and indirect indication that agriculture arrived in Europe with significant movement of human populations from the Near East as was suggested by Ammerman and Cavalli-Sforza, (1984, see also Chikhi *et al.*, 2002). Autosomal and sex chromosome analysis of these samples would be a crucial step forward in solving this longstanding problem.

Conservation genetics: recent and ancient signals, and the importance of selection

A common problem facing population geneticists centres around interpreting data from recent (often anthropogenically) isolated populations. Such populations can be unviable and may possess low amounts of genetic variation. In these cases, the management unit may often be a more applicable approach (e.g. O'Ryan *et al.*, 1998, for South African buffalo populations). Many such studies are carried out with the aim of identifying management units for translocating individuals to augment demographically unviable populations (Moritz, 1999). A major criterion identified by Moritz (1994) for defining separate 'management units' for special conservation measures is the possession of significant haplotype frequency differences at mitochondrial DNA regardless of the phylogenetic distinctiveness of the mitochondrial alleles.

Populations which have experienced radical changes in their habitat, are isolated from other populations and which are, or have recently been, small pose special problems when interpreting genetic data. Assignment of management unit status therefore must be carried out with caution. For example, the extreme demographic fluctuations which may be relatively common in small isolated populations are likely to produce substantial drift and/or inbreeding, accentuating allele frequency differences and may in turn result in a further loss of alleles, which will itself further increase the effects of inbreeding and increase the risks of long-term extinction due to genetic load (Lynch and Gabriel, 1990; Bickham *et al.*, 2000). On a shorter time scale this may also potentially result in the fixation of alleles which could *appear* to be locally unique.

Indeed, isolated populations often possess just a few mitochondrial alleles, and many isolated mammalian populations may have declined during the last 200 years. Inadequate genetic sampling may further produce patterns of apparent differentiation among populations (e.g. Sjögren and Wyöni, 1994, discuss the problems related to sampling too few individuals). The genetic structure often observed in endangered populations can result from recent demographic events as opposed to longer term divergence, potentially complicating translocation plans.

As an example, Barratt *et al.* (1999) found a large number of mitochondrial haplotypes in small, isolated populations of the red squirrel in the UK. The frequencies of these alleles were also extremely different, with many populations only containing alleles unique to the data set. Phylogenetic analysis revealed no phylogeographic pattern of diversity in different populations, either in the UK or in Western Europe. However, the red squirrel is known to have been extremely common, widespread and continuously distributed across Western Europe before deforestation for agriculture in the middle ages. A further complication is that it has been decimated following the introduction of the American grey squirrel in the 19th century. Many southern UK populations are today extremely small and isolated but have only been threatened for a few hundred years and may have indeed exchanged genes with neighbouring mainland European populations prior to isolation of the British Isles 9000 BP. The red squirrel apparently, therefore, has many diagnosable management units according to the mitochondrial data set. However, in the absence of phylogenetic structure in the populations analysed, and with the strong possibility of a purely demographic explanation of the data, one needs to consider whether it makes biological sense to manage the populations separately. Populations may possess large numbers of alleles due to long-term stability, rapid generation time and/or large effective population sizes. Population fragmentation and the sub-sampling of a diverse gene pool can instantaneously produce significant allele frequency differences, a pattern which carries no evolutionary signal or meaning.

One underlying issue is that of separating ancient from recent demographic events. As in the northern hairy-nosed wombat data, the possibility to separate these effects requires modelling that explicitly accounts for and estimates the time since the demographic event of interest. Beaumont's (1999) method was recently extended by Storz and Beaumont (2002) to provide absolute time estimates and applied to two bat species from the Indomalayan region, *Cynopterus sphinx* and *C. brachyotis*. The two species have different geographic distributions and have been potentially influenced to different extents by ancient climate changes. For both species, genetic data indicated strong support for a model of population decrease, with a more pronounced effect in *C. sphinx*. This came as a surprise because *C. sphinx* has a much wider range. The authors suggested that human-induced destruction of *C. sphinx*'s habitat might be the cause for the genetic signal. Another possibility is that both species suffered from reductions in tropical forests during the last glaciation and that *C. sphinx*'s more recent range expansion is not yet visible in the genetic data. The advantage of the hierarchical model developed by Storz and Beaumont (2002) is that it is possible to allow the mutation rates to vary among loci. Thus, it can provide indirect evidence of selection acting on some loci. It was also applied by Storz *et al.* (2002) to savannah baboon to test whether ancient demography in this species would be similar to that of other primates.

The major criticism which has been levelled at conservation genetics studies over the past 20 years (e.g. Laurenson *et al.*, 1995; O'Brien,

1998; Fraser and Bernatchez, 2001) is that neutral and quasi-neutral markers such as microsatellites, minisatellites, allozymes and anonymous SNPs tell us little about variation in adaptive traits. Indeed there is good reason to believe that adaptive variation is distributed very differently within the genomes of endangered species, which, having undergone demographic contraction, are more likely to exhibit non-additive variance in phenotypic traits, such as may be produced by dominance or epistasis (Armbruster *et al.*, 1998). Furthermore, single locus studies of candidate genes, although highly informative about certain selective pressures (e.g. de Groot *et al.*, 2002; O'Brien and Yukhi, 1999), are by their very nature isolated cases and do not benefit from the global perspective of whole genome analysis and multilocus comparisons. Recent studies on wild populations which look at whole genome variation (Slate *et al.*, 2002; Coltman *et al*, 2003) allow us the opportunity to examine the effects of selection on a genome-wide scale. The mapping of blocks of the genome under LD will allow us to discover which models of adaptive diversity and differentiation will best apply in endangered mammalian species, livestock breeds that are under threat (Macrae *et al.*, 2002; see Tishkoff and Verelli, 2003, for a review of human studies) and other demographically unstable populations.

Another problem arising in evolutionary studies in general and conservation genetics in particular is the problem of non-random sampling and in particular of ascertainment bias, already mentioned (this problem seems particularly important for SNPs because SNPs can be defined on the basis of as few as five individuals being typed – see Wakeley *et al.*, 2001; Nicholson *et al.*, 2002; Nielsen and Signorovitch, 2003). In the last 10 years, microsatellites have become the most used marker for studies of population genetics and conservation. However, microsatellite loci are often screened on the criterion that they should be polymorphic. Monomorphic loci are not always mentioned and it is difficult to determine to what extent published polymorphic loci are representative of the species' diversity. This type of non-random sampling has two different effects. First, it makes comparisons of diversity levels very difficult across species as the same H_e value may correspond to cases where many or very few loci were screened. Secondly, loci isolated in one species/population tend to be more variable than in subsequent species/populations (Ellegren *et al.*, 1997). These two effects can be problematic in devising recommendations in conservation biology or in the estimation of demographic parameters. For instance, we genetically typed the Island Jersey cattle breed to determine whether imports were needed to increase its genetic diversity (Chikhi *et al.*, 2004). Indeed, while this breed has been exported extensively becoming one of the most ubiquitous breeds, the island cattle were isolated with no recorded imports since 1789. It had been suggested that the breed was genetically depauperate but when the same loci were used for comparisons with different breeds, we found that the Island Jersey breed was no less variable than most other continental breeds. This led us to suggest that at this stage imports were probably not necessary. It should also be clear, that when recommendations are to be made, the importance of genetic data should not overshadow the work done in the field.

Conclusions

The importance of genetic data to infer population patterns, demographic histories, selection or social structure in mammals has been emphasized throughout this chapter. We have tried to show that new powerful methods are constantly developed and that the field is extremely lively. However, instead of pointing at all existing methods we have preferred to devote more space on fewer methods but more on their principles (see Beaumont, 2004).

Despite our belief that it is a very exciting period for population geneticists, we also tried to highlight problems we see related to the interpretation of genetic data. Indeed, one lesson that coalescent simulations taught us is that there is a tremendous amount of stochastic variation in the trees (and hence genetic data) that can be generated using the very same parameter values under a given demographic model (Box 21.2). As we have repeatedly written above, inference based on single locus data is bound to be of limited value *unless* it can

account and/or estimate this uncertainty, something that only very few methods can do (Pritchard *et al.*, 2000; Chikhi *et al.*, 2001, 2002 ; Wilson *et al.*, 2003).

To summarize, we probably can say that the sources of variance in any estimation procedure can be divided into the uncertainty due to: (i) the demographic (mutational, etc.) model used, (ii) the coalescent/genealogical process *conditional* on the model, and (iii) the sampling error *conditional* on the sampling process and the model. The third source is probably the easiest to identify and account for. It is also often the smallest. The second source of variation is usually well understood and can be rather large. It is rarely accounted for. This is changing and, in the genomic era, resampling among many independent loci would be a way to account for it (Goldstein and Chikhi, 2002). So, we can consider that in the very near future, this will be easily dealt with for most problems. The first source of variance is probably the most serious. It is also the one for which too little is currently being done. Unless one is fully aware of these limitations, the use of genetic data is bound to generate unending debate and disappointment among non-geneticists and geneticists alike.

Having said that, we should make it clear that we remain optimistic. The error caused by choosing inappropriate models can be reduced by using 'independent' information to choose among available models. There is also hope that new methods will, in the near future, help users choose among competing models while estimating parameter values or (better) posterior distributions of interest (Beaumont, 2004). Also, the increasing availability of multilocus data sets is already revolutionizing the handling and analysing of data. Better representations of the variability present in the genome are becoming available for many mammalian species. At the same time, the difficulty of handling huge data sets with methods that are very computer-intensive may resuscitate methods which were apparently not very powerful, but are easy to compute.

To conclude, one may ask what the real prospects are for bringing these approaches to bear for non-model organisms. Again, the situation may not be as bad as it first appears. As ungulate (*Bos, Ovis, Capra*), carnivore (*Canis,*

Felis), omnivore (*Sus*), rodent (*Mus*, *Rattus*) and primate (*Pan*, *Homo*, *Macaca*) genomes become densely and in some cases physically mapped, many markers may be transferable and syntenic regions of the genome may be rapidly mapped. Further, whole categories of genomic sequences (e.g. Caetano *et al.*, 1999; Banfi *et al.*, 2003) may be applicable across mammal taxa in ways previously unrealized. The potential application of generic and rapid ways to access microsatellites and SNPs (Nicod and Largiader, 2003) in combination with high throughput technologies (which are becoming cheaper by the day) will allow genomics to be applied to non-model organisms with increasing efficacy.

Acknowledgements

The authors would like to thank M. Beaumont for many useful discussions that helped us clarify a number of issues dealt with in this chapter. We also thank him for access to unpublished material. We are very grateful to C. Andalo, M. Beaumont, P. Géré, M. O'Hely, A. Ribéron and C. Thébaud for helpful comments on earlier drafts. Finally, we are extremely grateful to A. Ruvinsky for his patience in dealing with us, the job of editor is not a simple one, and we made it even harder. M. Trindade and D.Z. Chikhi's patience were also greatly appreciated.

References

Abecasis, G.R., Noguchi, E., Heinzmann, A., Traherne, J.A., Bhattacharya, S., Leaves, N.I., Anderson, G.G., Zhang, Y., Lench, N.J., Carey, A., Cardon, L.R., Moffatt, M.F. and Cookson, W.O. (2001) Extent and distribution of linkage disequilibrium in three genomic regions. *American Journal of Human Genetics* 68, 191–197.

Agapow, P.M. and Burt, A. (2001) Indices of multilocus linkage disequilibrium. *Molecular Ecology Notes* 1, 101–102.

Akey, J.M., Zhang, G., Zhang, K., Jin, L. and Shriver, M.D. (2002) Interrogating a high-density SNP map for signatures of natural selection. *Genome Research* 12, 1805–1814.

Ammerman, A.J. and Cavalli-Sforza, L.L. (1984) *The Neolithic Transition and the Genetics of Populations in Europe.* Princeton University Press, Princeton, New Jersey.

Ardlie, K.G., Kruglyak, L. and Seielstad, M. (2002) Patterns of linkage disequilibrium in the human genome. *Nature Reviews of Genetics* 3, 299–309.

Aris-Brosou, S. and Excoffier, L. (1996) The impact of population expansion and mutation rate heterogeneity on DNA sequence polymorphism. *Molecular Biology and Evolution* 13, 494–504.

Armbruster, P., Bradshaw, W.E. and Holzapfel, C.M. (1998) Effects of postglacial range expansion on allozyme and quantitative genetic variation of the pitcher-plant mosquito, *Wyeomyia smithii*. *Evolution* 52, 1697–1704.

Avise, J.C. (1989) Gene trees and organismal histories: a phylogenetic approach to population biology. *Evolution* 43, 1192–1208.

Awadalla, P., Eyre-Walker, A. and Smith, J.M. (1999) Linkage disequilibrium and recombination in hominid mitochondrial DNA. *Science* 286, 2524–2525.

Balding, D.J. (2003) Likelihood-based inference for genetic correlation coefficients. *Theoretical Population Biology*, 221–230.

Ball, R.M., Neigel, J.E. and Avise, J.C. (1990) Gene genealogies within the organismal pedigrees of random mating populations. *Evolution* 44, 360–370.

Balloux, F. (2001) EASYPOP (version 1.7) A computer program for the simulation of population genetics. *Journal of Heredity* 92, 301–302.

Balloux, F., Brünner, H., Lugon-Moulin, N., Hausser, J. and Goudet, J. (2000) Microsatellites can be misleading: an empirical and simulation study. *Evolution* 54, 1414–1422.

Bandelt, H.J., Forster, P., Sykes, B.C. and Richards, M.B. (1995) Mitochondrial portraits of human populations using median networks. *Genetics* 141, 743–753.

Bandelt, H.J., Quintana-Murci, L., Salas, A. and Macaulay, V. (2002) The fingerprint of phantom mutations in mitochondrial DNA. *American Journal of Human Genetics* 71, 1150–1160.

Banfi, S., di Bernardo, D., Boccia, A., Guffanti, A., Petrillo, M., Confalonieri, S., Mignone, F., Pesole, G., Missero, C., Paolella, G. and Ballabio, A. (2003) Analysis of human–mouse conserved genomic sequences (conserved sequence tags, CST) potentially involved in genetic diseases. *American Journal of Human Genetics* 73, 1022.

Banks, M.A. and Eichert, W. (2000) WHICHRUN (version 3.2): a computer program for population assignment of individuals based on multilocus genotype data. *Journal of Heredity* 91, 87–89

Barbujani, G. and Bertorelle, G. (2001) Genetics and the population history of Europe. *Proceedings of the National Academy of Sciences USA* 98, 22–25.

Barbujani, G. and Sokal, R.R. (1991) Genetic population structure of Italy. I. Geographical patterns of gene frequencies. *Human Biology* 63, 253–271.

Barbujani, G., Oden, N.L. and Sokal, R.R. (1989) Detecting regions of abrupt change in maps of biological variables. *Systematic Zoology* 38, 376–389.

Barbujani, G., Magagni, A., Minch, E. and Cavalli-Sforza, L.L. (1997) An apportionment of human DNA diversity. *Proceedings of the National Academy of Sciences USA* 94, 4516–4519.

Barbujani, G., Bertorelle, G. and Chikhi, L. (1998) Evidence for Paleolithic and Neolithic gene flow in Europe. *American Journal of Human Genetics* 62, 488–492.

Barratt, E.M., Gurnell, J., Malarky, G., Deaville, R. and Bruford, M.W. (1999) Genetic structure of fragmented populations of red squirrel (*Sciurus vulgaris*) in Britain. *Molecular Ecology* S12, 55–65.

Barton, N.H. and Wilson, I. (1995) Genealogies and geography. *Philosophical Transactions of the Royal Society of London B* 349, 49–59.

Beaumont, M.A. (1999) Detecting population expansion and decline using microsatellites. *Genetics* 153, 2013–2029.

Beaumont, M.A. (2001) Conservation genetics. In: Balding, D.J., Bishop, M. and Cannings, C. (eds) *The Handbook of Statistical Genetics*. John Wiley & Sons, New York, pp. 779–812.

Beaumont, M.A. (2003) Estimation of population growth or decline in genetically monitored populations. *Genetics* 164, 1139–1160.

Beaumont, M.A. (2004) Recent developments in genetic data analysis: what can they tell us about human demographic history? *Heredity* 92, 365–379.

Beaumont, M.A. and Bruford, M.W. (1999) Microsatellites in conservation genetics. In: Goldstein, D.B. and Schlötterer, C. (eds) *Microsatellites Evolution and Applications*. Oxford University Press, New York, pp. 165–182.

Beaumont, M.A. and Nichols, R.A. (1996) Evaluating loci for use in the genetic analysis of population structure. *Proceedings of the Royal Society of London Series B* 263, 1619–1626.

Beaumont, M.A., Zhang, W. and Balding, D.J. (2002) Approximate Bayesian computation in population genetics. *Genetics* 162, 2025–2035.

Beerli, P. and Felsenstein, J. (2001) Maximum-likelihood estimation of a migration matrix and effective population sizes in n subpopulations by using a coalescent approach. *Proceedings of the National Academy of Sciences USA* 98,4563–4568.

Bermingham, E. and Avise, J.C. (1986) Molecular zoogeography of freshwater fishes in the southeastern United States. *Genetics* 113, 939–965.

Berthier, P., Beaumont, M.A., Cornuet, J.M. and Luikart, G. (2002) Likelihood-based estimation of the effective population size using temporal changes in allele frequencies: a genealogical approach. *Genetics* 160, 741–751.

Bertorelle, G. and Barbujani, G. (1995) Analysis of DNA diversity by spatial autocorrelation. *Genetics* 140, 811–819.

Bertorelle, G. and Excoffier, L. (1998) Inferring admixture proportions from molecular data. *Molecular Biology and Evolution* 15, 1298–1311.

Bertorelle, G. and Slatkin, M. (1995) The number of segregating sites in expanding human populations, with implications for estimates of demographic parameters. *Molecular Biology and Evolution* 12, 887–892.

Bickham, J.W., Sandhu, S., Hebert, P.D.N., Chikhi, L. and Anthwal, R. (2000) Effects of chemical contaminants on genetic diversity in natural populations: implications for biomonitoring and ecotoxicology. *Reviews in Mutation Research*, 463, 33–51.

Birney, E., Clamp, M. and Hubbarb, T. (2002) Databases and tools for browsing genomes. *Annual Reviews in Genomics and Human Genetics* 3, 293–310.

Blanquer-Maumont, A. and Crouau-Roy, B. (1995) Polymorphism, monomorphism, and sequences in conserved microsatellites in primate species. *Journal of Molecular Evolution* 41, 492–497.

Bohonak, A.J. (2002) IBD (Isolation By Distance): a program for analyses of isolation by distance. *Journal of Heredity* 93, 154–155.

Bowcock, A.M., Kidd, J.R., Mountain, J.L., Hebert, J.M. and Cavalli-Sforza, L.L. (1991) Drift, admixture and selection in human evolution: a study with DNA polymorphisms. *Proceedings of the National Academy of Sciences, USA* 88, 839–843.

Bowcock, A.M., Ruiz-Linares, A., Tomfohrde, J., Minch, E., Kidd, J.R. and Cavalli-Sforza, L.L. (1994) High resolution of human evolutionary history trees with polymorphic microsatellites. *Nature* 368, 455–457.

Bradley, D.G. and Cunningham, P. (1999) Genetic aspects of domestication, common breeds and their origins. In: Fries, R. and Ruvinsky, A. (eds) *The Genetics of Cattle*. CAB International, Wallingford, UK, pp. 15–31.

Bradley, D.G., MacHugh, D.E., Cunningham, P. and Loftus, R.T. (1996) Mitochondrial diversity and the origins of African and European cattle. *Proceedings of the National Academy of Sciences USA* 93, 5131–5135.

Bruford, M.W., Bradley, D.G. and Luikart, G. (2003) Genetic analysis reveals complexity of livestock domestication. *Nature Reviews of Genetics* 4, 900–910.

Burns, C.E., Ciofi, C., Beheregaray, L.B., Fritts, T.H., Gibbs, J.P., Marquez, C., Milinkovitch, M.C., Powell, J.R. and Caccone, A. (2003) The origin of captive Galapagos tortoises based on DNA analysis: implications for the management of natural populations. *Animal Conservation* 6, 329–337.

Caetano, A.R., Lyons, L.A., Laughlin, T.F., O'Brien, S.J., Murray, J.D. and Bowling, A.T. (1999) Equine synteny mapping of comparative anchor tagged sequences (CATS) from human Chromosome 5. *Mammalian Genome* 10, 1082–1084.

Calafell, F., Shuster, A., Speed, W.C., Kidd, J.R. and Kidd, K.K. (1998) Short tandem repeat polymorphism evolution in humans. *European Journal of Human Genetics* 6, 38–49.

Calafell, F., Pérez-Lezaun, A. and Bertranpetit, J. (2000) Genetic distances and microsatellite diversification in humans. *Human Genetics* 106, 133–134.

Cavalli-Sforza, L.L. (1966) Population structure and human evolution. *Proceedings of the Royal Society of London Series B* 164, 362–379.

Cavalli-Sforza, L.L. and Edwards, A.W.F. (1967) Phylogenetic analysis: models and estimation procedures. *American Journal of Human Genetics* 19, 233–257.

Cavalli-Sforza, L.L., Menozzi, P. and Piazza, A. (1993) Demic expansions and human evolution. *Science* 259, 639–646.

Chakraborty, R. and Jin, L. (1993) A unified approach to study hypervariable polymorphisms: statistical considerations of determining relatedness and population distances. In: Pena, S.D.J., Chakraborty, R., Epplen, J. and Jeffreys, A.J. (eds) *DNA Fingerprinting: Current State of the Science*. Birkhauser, Basel, pp. 153–175.

Chepko-Sade, B.D. and Halpin, Z.T. (1987) Mammalian dispersal patterns. In: *The Effects of Social Structure on Population Genetics*. University of Chicago Press, Chicago, Illinois, USA, 342 pp.

Chesser, R.K. (1991) Influence of gene flow and breeding tactics on gene diversity within populations. *Genetics* 129, 573–583.

Chikhi, L., Destro-Bisol, G., Bertorelle, G., Pascali, V. and Barbujani, G. (1998) Clines of nuclear DNA markers suggest a largely neolithic ancestry of the European gene pool. *Proceedings of the National Academy of Sciences USA* 95, 9053–9058.

Chikhi, L., Bruford, M.W. and Beaumont, M.A. (2001) Estimation of admixture proportions: a likelihood-based approach using Markov chain Monte Carlo. *Genetics* 158, 1347–1362.

Chikhi, L., Nichols, R.A., Barbujani, G. and Beaumont, M.A. (2002) Y genetic data support the Neolithic demic diffusion model. *Proceedings of the National Academy of Sciences USA* 99, 10008–10013.

Chikhi, L., Goossens, B., Treanor, A. and Bruford, M.W. (2004) Population genetic structure of and inbreeding in an insular cattle breed, the Jersey, and its implications for genetic resource management. *Heredity* 92, 396–401.

Clisson, I., Lathuilliere, M. and Crouau-Roy, B. (2000) Conservation and evolution of microsatellite loci in primate taxa. *American Journal of Primatology* 50, 205–214.

Coltman, D.W., Pilkington, J.G. and Pemberton, J.M. (2003) Fine-scale genetic structure in a free-living ungulate population. *Molecular Ecology* 12, 733–742.

Comstock, K.E, Georgiadis, N., Pecon-Slattery, J., Roca, A.L., Ostrander, E.A., O'Brien, S.J. and Waser, S.K. (2002) Patterns of molecular genetic variation among African elephant populations. *Molecular Ecology* 11, 2489–2498.

Cornuet, J.-M. and Luikart, G. (1996) Description and power analysis of two tests for detecting recent population bottlenecks from allele frequency data. *Genetics* 144, 2001–2014.

Cornuet, J.-M., Piry, S., Luikart, G., Estoup, A. and Solignac, M. (1999) New methods employing multilocus genotypes for selecting or excluding populations as origins of individuals. *Genetics* 153, 1989–2000.

Côté, S.D., Dallas, J.F., Marshall, F., Irvine, R.J., Langvatn, R. and Albon, S.D. (2002) Microsatellite DNA evidence for genetic drift and philopatry in Svalbard reindeer. *Molecular Ecology* 11, 1923–1930.

Crow, J.F. and Kimura, M. (1970) *An Introduction to Population Genetics Theory*. Harper and Row, New York.

Daly, M.J., Rioux, J.D., Schaffner, S.F., Hudson, T.J. and Lander, E.S. (2001) High-resolution haplotype structure in the human genome. *Nature Genetics* 29, 229–232.

Davies, N., Villablanca, F.X. and Roderick, G.K. (1999) Determining the source of individuals: multilocus genotyping in nonequilibrium population genetics. *Trends in Ecology and Environment* 14, 17–21.

Dawson, K.J. and Belkhir, K. (2001) A Bayesian approach to the identification of panmictic populations and the assignment of individuals. *Genetical Research* 78, 59–77.

de Groot, N.G., Otting, N., Doxiadis, G.G.M., Balla-Jhagjhoorsingh, S.S., Heeney, J.L., van Rood, J.J., Gagneux, P. and Bontrop, R.E. (2002) Evidence for an ancient selective sweep in the MHC class I gene repertoire of chimpanzees. *Proceedings of the National Academy of Sciences USA* 99, 11748–11753.

de Ruiter, J.R. and Geffen, E. (1998) Relatedness of matrilines, dispersing males and social groups in long-tailed macaques (*Macaca fascicularis*). *Proceedings of the Royal Society of London Series B* 265, 79–87.

Destro-Bisol, G., Pascali, V. and Spedini, G. (2000) Application of different genetic distance methods to microsatellite data. *Human Genetics* 106, 130–132.

Dieringer, D. and Schlötterer, C. (2003) Microsatellite analyser (MSA): a platform independent analysis tool for large microsatellite data sets. *Molecular Ecology Notes* 3, 167–169.

Di Fiore, A. (2003) Molecular genetic approaches to the study of primate behaviour, social organization, and reproduction. *Yearbook of Physical Anthropology* 46, 62–99.

Di Rienzo, A., Peterson, A.C., Garza, J.C., Valdes, A.M., Slatkin, M. and Freimer, N.B. (1994) Mutational processes of simple-sequence repeat loci in human populations. *Proceedings of the National Academy of Sciences USA* 91, 3166–3170.

Donnelly, P. and Tavaré, S. (1995) Coalescents and genealogical structure under neutrality. *Annual Reviews in Genetics* 29, 401–421.

Drummond, A.J., Nicholls, G.K., Rodrigo, A.G. and Solomon, W. (2002) Estimating mutation parameters, population history and genealogy simultaneously from temporally spaced sequence data. *Genetics* 161, 1307–1320.

Dupanloup, I. and Bertorelle, G. (2001) Inferring admixture proportions from molecular data: extension to any number of parental populations. *Molecular Biology and Evolution* 18, 672–675.

Dupanloup, I., Schneider, S. and Excoffier, L. (2002) A simulated annealing approach to define the genetic structure of populations. *Molecular Ecology* 11, 2571–2581.

Edwards, S.V. and Beerli, P. (2000) Gene divergence, population divergence, and the variance in coalescence times in phylogeographic studies. *Evolution* 54, 1839–1854.

Efron, B. (1983) Estimating the error rate of a prediction rule: improvement on cross-validation. *Journal of the American Statistical Association* 78, 316–330.

Efron, B. and Tibshirani, R.J. (1993) *An Introduction to the Bootstrap*. Chapman and Hall.

Ellegren, H., Moore, S., Robinson, N., Byrne, K., Ward, W. and Sheldon, B.C. (1997) Microsatellite evolution – a reciprocal study of repeat lengths at homologous loci in cattle and sheep. *Molecular Biology and Evolution* 14, 854–860.

Ellegren, H., Smith, N.G.C. and Webster, M.T. (2003) Mutation rate variation in the mammalian genome. *Current Opinion in Genetics and Development* 13, 562–568.

Emerson, B., Paradis, E. and Thébaud, C. (2001) Revealing the demographic histories of species using DNA sequences. *Trends in Ecology and Evolution* 16, 707–716.

Estoup, A. and B. Angers, B. (1998) Microsatellites and minisatellites for molecular ecology: theoretical and empirical considerations. In: Carvalho, G. (ed.) *Advances in Molecular Ecology*. NATO Press, Amsterdam, pp.55–86.

Estoup, A., Jarne, P. and Cornuet, J.-M. (2002) Homoplasy and mutation model at microsatellite loci and their consequences for population genetics analysis. *Molecular Ecology* 11, 1591–1604.

Ewens, W.J. (1972) The sampling theory of selectively neutral alleles. *Theoretical Population Biology* 3, 87–112.

Excoffier, L. (2002) Human demographic history: refining the recent African origin model. *Current Opinion in Genetics and Development* 12, 675–682.

Excoffier, L. and Hamilton, G. (2003) Comment on 'Genetic structure of human populations'. *Science* 300, 1877.

Excoffier, L., Smouse, P.E. and Quattro, J.M. (1992) Analysis of molecular variance inferred from metric distances among DNA haplotypes: application to human mitochondrial DNA restriction data. *Genetics* 131, 479–491.

Falush, D., Stephens, M. and Pritchard, J.K. (2003) Inference of population structure from multilocus genotype data: linked loci and correlated allele frequencies. *Genetics* 164, 1567–1587.

Fay, J.C. and Wu, C.I. (1999) A human population bottleneck can account for the discordance between patterns of mitochondrial versus nuclear DNA variation. *Molecular Biology and Evolution* 16, 1003–1005.

Felsenstein, J. (1992) Estimating effective population size from samples of sequences: inefficiency of pairwise and segregating sites as compared to phylogenetic estimates. *Genetical Research* 59, 139–147.

Felsenstein, J. (2004) *Inferring Phylogenies.* Sinauer Associates, Sunderland, Massachussets.

Flint, J., Bond, J., Rees, D.C., Boyce, A.J., Roberts-Thompson, J.M., Excoffier, L., Clegg, J.B., Beaumont, M.A., Nichols, R.A. and Harding, R.M. (1999) Minisatellite mutational processes reduce Fst estimates. *Human Genetics* 105, 567–576.

Frankham, R. (1995) Effective population size/adult population size ratios in wildlife: a review. *Genetical Research* 66, 95–107.

Fraser, D.J. and Bernatchez, L. (2001) Adaptive evolutionary conservation: towards a unified concept for defining conservation units. *Molecular Ecology* 10, 2741–2752.

Fu, Y.X and Li, W.H. (1993) Statistical tests of neutrality of mutations. *Genetics* 133, 693–709.

Fu, Y.X. and Li, W.H. (1999) Coalescing into the 21st century: an overview and prospects of coalescent theory. *Theoretical Population Biology* 56, 1–10.

Gabriel, S.B., Schaffner, S.F. Nguyen, H., Moore, J.M., Roy, J., Blumenstiel, B., Higgins, J., DeFelice, M., Lochner, A., Faggart, M., Liu-Cordero, S.N., Rotimi, C., Adeyemo, A., Cooper, R., Ward, R., Lander, E.S., Daly, M.J. and Altshuler, D. (2002) The structure of haplotype blocks in the human genome. *Science* 296, 2225–2229.

Gagneux, P. (2002) The genus Pan: population genetics of an endangered outgroup. *Trends in Genetics* 18, 327–330.

Gagneux, P., Wills, C., Gerloff, U., Tautz, D., Morin, P.A., Boesch, C., Fruth, B., Hohmann, G., Ryder, O.A. and Woodruff, D.S. (1999) Mitochondrial sequences show diverse evolutionary histories of African hominoids. *Proceedings of the National Academy of Sciences USA* 96, 5077–5082.

Garza, J.C. and Williamson E. (2001) Detection of reduction in population size using data from microsatellite DNA. *Molecular Ecology* 10, 305–318.

Gilbert, D.A., Lehman, N., O'Brien, S.J., and Wayne, R.K. (1990) Genetic fingerprinting reflects population differentiation in the California Channel Island fox. *Nature* 344, 764–767.

Goldstein, D.B. (2001) Islands of linkage disequilibrium. *Nature Genetics* 29, 109–111.

Goldstein, D.B. and Chikhi, L. (2002) Human migration and population structure: what we know and why it matters. *Annual Reviews in Genomics and Human Genetics* 3, 129–152.

Goldstein, D.B., Linares, A.R., Cavalli-Sforza, L.L. and Feldman, M.W. (1995) An evaluation of genetic distances for use with microsatellite loci. *Genetics* 139, 463–471.

Goossens, B., Chikhi, L., Taberlet, P., Waits, L.P. and Allainé, D. (2001) Microsatellite analysis of genetic variation among Alpine marmot populations in the French Alps. *Molecular Ecology,* 10, 41–52.

Griffiths, R.C. and Tavaré, S. (1994) Simulating probability distributions in the coalescent. *Theoretical Population Biology* 46, 131–159.

Gum, B., Gross, R., Rottmann, O., Schroder, W. and Kuhn, R. (2003) Microsatellite variation in Bavarian populations of European grayling (*Thymallus thymallus*): implications for conservation. *Conservation Genetics* 4, 659–672.

Hardy, O.J. and Vekemans, X. (2002) SPAGeDi: a versatile computer program to analyse spatial genetic structure at the individual or population levels. *Molecular Ecology Notes* 2, 618–620.

Hardy, O., Charbonnel, N., Fréville, H. and Heuertz, M. (2003) Microsatellite allele sizes: a simple test to assess their significance on genetic differentiation. *Genetics* 163, 1467–1482.

Harpending, H. and Rogers, A. (2000) Genetic perspectives on human origins and differentiation. *Annual Reviews in Genomics and Human Genetics* 1, 361–385.

Hedrick, P.W. (1999) Highly variable loci and their interpretation in evolution and conservation. *Evolution* 53, 313–318.

Hoelzer, G.A., Wallman, J. and Melnick, D.J. (1998) The effects of social structure, geographical structure, and population size on the evolution of mitochondrial DNA: II. Molecular clocks and the lineage sorting period. *Journal of Molecular Evolution* 47, 21–31.

Holsinger, K.E. (1999) Analysis of genetic diversity in geographically structured populations: a Bayesian perspective. *Hereditas* 130, 245–255.

Holsinger, K.E. and Mason-Gamer, R.J. (1996) Hierarchical analysis of nucleotide diversity in geographically structured populations. *Genetics* 142, 629–639.

Hudson, R.R. (1991) Gene genealogies and the coalescent process. In: Futuyma, D.J. and Antonovics, J.D. (eds) *Oxford Surveys in Evolutionary Biology*, pp. 1–44.

Hudson, R.R., Kreitman, M. and Aguadé, M. (1987) A test of neutral molecular evolution based on nucleotide data. *Genetics* 116, 153–159.

Hudson, R.R., Slatkin, M. and Maddison, W.P. (1992) Estimation of levels of gene flow from DNA sequence data. *Genetics* 132, 583–589.

Jeffreys, A.J. and Neumann, R. (2002) Reciprocal crossover asymmetry and meiotic drive in a human recombination hot spot. *Nature Genetics*, 31, 267–271.

Jeffreys, A.J., Kauppi, L. and Neumann, R. (2001) Intensely punctuate meiotic recombination in the class II region of the major histocompatibility complex. *Nature Genetics* 29, 217–222.

Jiang, Z.H. and Michal, J.J. (2003) Linking porcine microsatellite markers to known genome regions by identifying their human orthologs. *Genome* 46, 798–808.

Jiang, Z.H., Melville, J.S., Cao, H., Kumar, S., Filipski, A. and Verrinder Gibbins, A.M. (2002) Measuring conservation of contiguous sets of autosomal markers on bovine and porcine genomes in relation to the map of the human genome. *Genome* 45, 769–776.

Jiang, Z.H., Zhang, M., Wasem, V.D., Michal, J.J., Zhang, H. and Wright, R.W., Jr (2003) Census of genes expressed in porcine embryos and reproductive tissues by mining an expressed sequence tag database based on human genes. *Biology of Reproduction* 69, 1177–1182.

Jorde, L.B., Watkins, W.S., Bamshad, M.J., Dixon, M.E., Ricker, C.E. *et al.* (2000) The distribution of human genetic diversity: a comparison of mitochondrial, autosomal, and Y-chromosome data. *American Journal of Human Genetics* 66, 979–988.

Kaeuffer, R., Pontier, D., Devillard, S. and Perrin, N. (2004) Effective size of two feral domestic cat populations (*Felis catus* L.): effect of the mating system. *Molecular Ecology* 13, 483–490.

Kingman, J.F.C. (1982a) On the genealogy of large populations. *Journal of Applied Probability* 19A, 27–43.

Kingman, J.F.C. (1982b) The coalescent. *Stochastic Processes and their Applications* 13, 235–248.

Kitano, T., Schwarz, C., Nickel, B. and Paabo, S. (2002) Gene diversity patterns at 10 X-chromosomal loci in humans and chimpanzees. *Molecular Biology and Evolution* 20, 1281–1289.

Klein, R.G. (2000) Archeology and the evolution of human behavior. *Evolutionary Anthropology* 9, 17–36.

Knowles, L.L. and Maddison, W.P. (2002) Statistical phylogeography. *Molecular Ecology* 11, 2623–2635.

Koskinen, M.T. (2003) Individual assignment using microsatellite DNA reveals unambiguous breed identification in the domestic dog. *Animal Genetics* 34, 1–5.

Kruglyak, L. (1999) Prospects for whole-genome linkage disequilibrium mapping of common disease genes. *Nature Genetics* 22, 139–144.

Kuhner, M., Yamoto, J. and Felsenstein, J. (1995) Estimating effective population size and mutation rate from sequence data using Metropolis–Hastings sampling. *Genetics* 140, 1421–1430.

Kuo, C.-H. and Janzen, F.J. (2003) BottleSim: a bottleneck simulation program for long-lived species with overlapping generations. *Molecular Ecology Notes* 3, 669–673.

Lancaster, A., Nelson, M.P., Single, R.M., Meyer, D. and Thomson G. (2003) PyPop: a software framework for population genomics: analyzing large-scale multi-locus genotype data. In: Altman, R.B. *et al.* (eds) *Pacific Symposium on Biocomputing 8* World Scientific, Singapore, pp. 514–525.

Langella, O., Chikhi, L. and Beaumont, M.A. (2001) LEA (Likelihood-based estimation of admixture) : a program to simultaneously estimate admixture and the time since admixture. *Molecular Ecology Notes* 1, 357–358.

Laurenson, M.K., Wielebnowski, N. and Caro, T.M. (1995) Extrinsic factors and juvenile mortality in cheetahs. *Conservation Biology* 9, 1329–1331.

Laval, G., SanCristobal, M. and Chevalet, C. (2003) Maximum-likelihood and Markov Chain Monte Carlo approaches to estimate inbreeding and effective size from allele frequency changes. *Genetics* 164, 1189–1204.

Lewontin, R.C. (1972) The apportionment of human diversity. *Evolutionary Biology* 6, 381–398.

Lewontin, R.C. (1988) On measures of gametic disequilibrium. *Genetics* 120, 849–852.

Lewontin, R.C. and Krakauer, J. (1973) Distribution of gene frequency as a test of the theory of the selective neutrality of polymorphisms. *Genetics* 74, 175–195.

Loftus, R.T., MacHugh, D.E., Bradley, D.G., Sharp, P.M. and Cunningham, P. (1994) Evidence for two independent domestications of cattle. *Proceedings of the National Academy of Sciences USA* 91, 2757–2761.

Luikart, G. and Cornuet, J.-M. (1999) Estimating the effective number of breeders from heterozygote-excess in progeny. *Genetics* 151, 1211–1216.

Luikart, G., Cornuet, J.-M. and Allendorf, F.W. (1999) Temporal changes in allele frequencies provide estimates of population bottleneck size. *Conservation Biology* 13, 523–530.

Luikart, G., England, P.R., Tallmon, D., Jordan, S. and Taberlet, P. (2003) The power and promise of population genomics: from genotyping to genome typing. *Nature Reviews* 4, 1–15.

Lynch M. and Crease T.J. (1990) The analysis of population survey data on DNA sequence variation. *Molecular Biology and Evolution* 7, 377–394.

MacHugh, D.E., Shriver, M.D., Loftus, R.T., Cunningham, P. and Bradley, D.G. (1997) Microsatellite DNA variation and the evolution, domestication and phylogeography of taurine and zebu cattle (*Bos taurus* and *Bos indicus*). *Genetics* 146, 1071–1086.

Manel, S., Berthier, P. and Luikart, G. (2002) Detecting wildlife poaching: identifying the origin of individuals using Bayesian assignment tests and multi-locus genotypes. *Conservation Biology* 16, 650–659.

Mantel, N. (1967) The detection of disease clustering and a generalized regression approach. *Cancer Research* 27, 209–220.

Marjoram, P., Molitor, J., Plagnol, V. and Tavaré, S. (2003) Markov chain Monte Carlo without likelihoods. *Proceedings of the National Academy of Sciences USA.* 100, 15324–15328.

McKenna, M.C. and Bell, S.K. (1997) *Classification of Mammals Above the Species Level.* Columbia University Press, New York.

McKeigue, P.M. (1998) Mapping genes that underlie ethnic differences in disease risk: methods for detecting linkage in admixed populations, by conditioning on parental admixture. *American Journal of Human Genetics* 63, 241–251.

Menozzi, P., Piazza, A. and Cavalli-Sforza, L.L. (1978) Synthetic maps of human gene frequencies in Europeans. *Science* 201, 786–792.

Miller, C.R., Adams, J.R. and Waits, L.P. (2003) Pedigree-based assignment tests for reversing coyote (*Canis latrans*) introgression into the wild red wolf (*Canis rufus*) population. *Molecular Ecology* 12, 3287–3301.

Mishmar, D., Ruiz-Pesini, E., Golik, P., Macauley, V., Clarck, A.G., Hosseini, S., Brandon, M., Easley, K., Chen, E., Brown, M.D., Sukernik, R.I., Olckers, A. and Wallace, D.C. (2003) Natural selection shaped regional mtDNA variation in humans. *Proceedings of the National Academy of Sciences USA* 100, 171–176.

Moritz, C. (1994) Defining evolutionary significant units for conservation. *Trends in Ecology and Evolution* 9, 373–375.

Moritz, C. (1999) Conservation units and translocations: strategies for conserving evolutionary processes. *Hereditas* 130, 217–228.

Murphy, W.J., Sun, S., Chen, Z.Q., Pecon-Slattery, J. and O'Brien, S.J. (1999) Extensive conservation of sex chromosome organization between cat and human. *Genome Research* 9, 1223–1230.

Murphy, W.J., Frönicke, L., O'Brien, S.J. and Stanyon, R. (2003) The origin of human chromosome 1 and its homologs in human placental mammals. *Genome Research* 13, 1880–1888.

Nachman, M.W. (2003) Variation in recombination rate across the genome: evidence and implications. *Current Opinion in Genetics and Development* 12, 657–663.

Nagylaki, T. (1998) Fixation indices in subdivided populations. *Genetics* 148, 1325–1332.

Nauta, M.J. and Weissing, F.J. (1996) Constraints on allele size at microsatellite loci: implication for genetic differentiation. *Genetics* 143, 1021–1032.

Nei, M. (1973) Analysis of gene diversity in subdivided populations. *Proceedings of the National Academy of Sciences USA* 70, 3321–3323.

Nei, M. (1987) *Molecular Evolutionary Genetics.* Columbia University Press, New York.

Nei, M. and Maruyama, T. (1975) Letters to the editors: Lewontin–Krakauer test for neutral genes. *Genetics* 80, 395.

Nei, M., Maruyama, T. and Chakraborty, R. (1975) The bottleneck effect and genetic variability in populations. *Evolution* 29, 1–10.

Nichols, R. (2001) Gene trees and species trees are not the same. *Trends in Ecology and Evolution* 16, 358–364.

Nicholson, G., Smith, A.V., Jónsson, F., Gústafsson, A., Stefánsson, K. and Donnelly, P. (2002) Assessing population differentiation and isolation from single-nucleotide polymorphism data. *Journal of the Royal Statistical Society Series B* 64, 695–715.

Nicod, J.C. and Largiader, C.R. (2003) SNPs by AFLP (SBA): a rapid SNP isolation strategy for non-model organisms. *Nucleic Acids Research* 31, e19.

Nielsen, R. and Signorovitch, J. (2003) Correcting for ascertainment biases when analyzing SNP data: application to the estimation of linkage disequilibrium. *Theoretical Population Biology* 63, 245–255.

Nordborg, M. (2001) Coalescent theory. In: Balding, D.J., Bishop, M. and Cannings, C. (eds) *Handbook of Statistical Genetics*. John Wiley & Sons, New York, pp. 179–212.

Nordborg, M. and Tavaré, S. (2002) Linkage disequilibrium: what history has to tell us. *Trends in Genetics* 18, 83–90.

O'Brien, S.J. (1998) Intersection of population genetics and species conservation – the cheetah's dilemma. *Evolutionary Biology* 30, 79–91.

O'Brien, S.J. and Yuhki, N. (1999) Comparative genome organization of the major histocompatibility complex: lessons from the Felidae. *Immunological Reviews* 167, 133–144.

O'Brien, S.J., Menotti-Raymond, M., Murphy, W.J., Nash, W.G., Wienberg, J., Stanyon, R., Copeland, N.G., Jenkins, N.A., Womack, J.E. and Marshall Graves, J.A. (1999) The promise of comparative genomics in mammals. *Science* 286, 458–481.

O'Brien, S.J., Eizirik, E. and Murphy, W.J. (2001) On choosing mammalian genomes for sequencing. *Science* 292, 2264–2265.

Ohta, T. (1992) The nearly neutral theory of molecular evolution. *Annual Reviews in Ecology and Systematics* 23, 263–286.

Ohta, T. and Kimura, M. (1973) A model of mutation appropriate to estimate the number of electrophoretically detectable alleles in a finite population. *Genetical Research* 22, 201–204.

O'Ryan, C., Harley, E.H., Bruford, M.W., Beaumont, M.A., Wayne, R.K. and Cherry, M.I. (1998) Microsatellite analysis of genetic diversity in fragmented South African buffalo populations. *Animal Conservation* 1, 85–94.

Ostrander, E.A. and Comstock, K.E. (2004) The domestic dog genome. *Current Biology* 14, R98–R99.

Overall, A.D. and Nichols, R.A. (2001) A method for distinguishing consanguinity and population substructure using multilocus genotype data. *Molecular Biology and Evolution* 18, 2048–2056.

Paetkau, D., Calvert, W., Stirling, I. and Strobeck, C. (1995) Microsatellite analysis of population structure in Canadian polar bears. *Molecular Ecology* 4, 347–354

Paetkau, D., Slade, R., Burden, M. and Estoup, A. (2004) Direct, real-time estimation of migration rate using assignment methods: a simulation-based exploration of accuracy and power. *Molecular Ecology* 13, 55–66.

Pamilo, P. and Nei, M. (1988) Relationships between gene trees and species trees. *Molecular Biology and Evolution* 5, 568–583.

Patil, N., Berno, A.J., Hinds, D.A., Barrett, W.A., Doshi, J.M., Hacker, C.R., Kautzer, C.R., Lee, C.H., Marjoribanks, C., McDonough, D.P., Nguyen, B.T.N., Norris, M.C., Sheehan, J.B., Shen, N., Stern, D., Stokowski, R.P., Thomas, D.J., Trulson, M.O., Vyas, K.R., Frazer, K.A., Fodor, S.P.A. and Cox, D.R. (2001) Blocks of limited haplotype diversity revealed by high-resolution scanning of human chromosome 21. *Science* 294, 1719–1723.

Peakall, R., Ruibal, M. and Lindenmayer, D.B. (2003) Spatial autocorrelation analysis offers new insights into gene flow in the Australian bush rat, *Rattus fuscipes*. *Evolution* 57, 1182–1195.

Pemberton, J.M., Smith, J.A., Coulson, T.N., Marshall, T.C., Slate, J., Paterson, S., Albon, S.D. and Clutton-Brock, T.H. (1996) The maintenance of genetic polymorphism in small island populations: large mammals in the Hebrides. *Philosophical Transactions of the Royal Society of London, Series B* 351, 745–752.

Pérez-Lezaun, A., Calafell, F., Mateu, E., Comas, D., Ruiz-Pacheco, R. and Bertranpetit, J. (1997) Microsatellite variation and the differentiation of modern humans. *Human Genetics* 99, 1–7.

Piry, S., Luikart, G. and Cornuet, J.-M. (1999) Bottleneck: a computer program for detecting recent reductions in effective population size from allele frequency data. *Journal of Heredity* 90, 502–503.

Piry, S., Alapetite, A., Cornuet, J.-M., Paetkau, D., Baudouin, L. and Estoup, A. (2004) GeneClass2: an assignment software to select or exclude populations as origins of individuals. *Molecular Ecology* (in press).

Poloni, E.S., Semino, O., Passarino, G., Santachiara-Benerecetti, A.S., Dupanloup, I., Langaney, A. and Excoffier, L. (1997) Human genetic affinities for Y-chromosome P49a,f/TaqI haplotypes show strong correspondence with linguistics. *American Journal of Human Genetics* 61, 1015–1035.

Pritchard, J.K. and Przeworski, M. (2001) Linkage disequilibrium in humans: models and data. *American Journal of Human Genetics* 69, 1–14.

Pritchard, J.K., Seielstad, M.T., Perez-Lezaun, A. and Feldman, M.W. (1999) Population growth of human Y chromosomes: a study of Y chromosome microsatellites. *Molecular Biology and Evolution* 16, 1791–1798.

Pritchard, J.K., Stephens, M. and Donnelly, P. (2000) Inference of population structure using multilocus genotype data. *Genetics* 155, 945–959.

Przeworski, M., Hudson, R.R. and Di Rienzo, A. (2000) Adjusting the focus on human variation. *Trends in Genetics* 16, 296–302.

Ptak, S.E. and Przeworski, M. (2002) Evidence for population growth in humans is confounded by fine-scale population structure. *Trends in Genetics* 18, 559–563.

Pudovkin, A.I., Zaykin, D.V. and Hedgecock, D. (1996) On the potential for estimating the effective number of breeders from heterozygote-excess in progeny. *Genetics* 144, 383–387.

Rannala, B. and Mountain, J.L. (1997) Detecting immigration by using multilocus genotypes. *Proceedings of the National Academy of Sciences USA* 94, 9197–9201.

Raufaste, N. and Bonhomme, F. (2000) Properties of bias and variance of two multiallelic estimators of Fst. *Theoretical Population Biology* 57, 285–296.

Raymond, M. and Rousset, F. (1995) Genepop (version 1.2), population genetics software for exact tests and ecumenicism. *Journal of Heredity* 86, 248–249.

Reich, D.E., Cargill, M., Bolk, S., Ireland, J., Sabeti, P.C., Richter, D.J., Lavery, T., Kouyoumjian, R., Farhadian, S.F., Ward, R. and Lander, E.S. (2001) Linkage disequilibrium in the human genome. *Nature* 411, 199–204.

Repping, S., Skaletsky, H. Brown L., van Daalen, S.K.M., Korver, C.M., Pyntikova, T., Kuroda-Kawaguchi, T., de Vries, J.W.A., Oates, R.D., Silber, S., van der Veen, F., Page, D.C. and Rozen, S. (2003) Polymorphism for a 1.6-Mb deletion of the human Y chromosome persists through balance between recurrent mutation and haploid selection. *Nature Genetics* 35, 247–251.

Richardson, G., Smith, A.V., Jónsson, F., Gústafsson, O., Stefánsson, K. and Donnelly, P. (2002) Assessing population differentiation and isolation from single-nucleotide polymorphism data. *Journal of the Royal Statistical Society B* 64, 695–715.

Risch, N. and Merikangas, K. (1996) The future of genetic studies of complex human diseases. *Science* 273, 1516–1517.

Robertson, A. (1975a) Letters to the editors: remarks on the Lewontin–Krakauer test. *Genetics* 80, 396.

Robertson, A. (1975b) Gene frequency distributions as a test of selective neutrality. *Genetics* 81, 775–778.

Robertson, A. and Hill, W. (1984) Deviations from Hardy Weinberg proportions: sampling variances and use in estimation of inbreeding coefficients. *Genetics* 107, 703–718.

Rogers, A.R. and Harpending, H. (1992) Population growth makes waves in the distribution of pairwise genetic differences. *Molecular Biology and Evolution* 9, 552–569.

Rosenberg, N. and Nordborg, M. (2002) Genealogical trees, coalescent theory and the analysis of genetic polymorphisms. *Nature Reviews of Genetics* 3, 380–390.

Rosenberg, N.A., Pritchard, J.K., Weber, J.L., Cann, H.M., Kidd, K.K., Zhivotovsky, L.A. and Feldman, M.W. (2002) Genetic structure of human populations. *Science* 298, 2381–2385.

Rousset, F. (1997) Genetic differentiation and estimation of gene flow from F statistics under isolation by distance. *Genetics* 145, 1219–1228.

Rozas, J. and Rozas, R. (1995) DnaSP, DNA sequence polymorphism: an interactive program for estimating population genetics parameters from DNA sequence data. *Computer Applications in the Biosciences* 11, 621–625.

Rozas, J., Sánchez-DelBarrio, J.C., Messeguer, X. and Rozas, R. (2003) DnaSP, DNA polymorphism analyses by the coalescent and other methods. *Bioinformatics* 19, 2496–2497.

Ruvinsky, A. and Rothschild, M.F. (1998) Systematics and evolution of the pig. In: Rothschild, M.F. and Ruvinsky, A. (eds) *The Genetics of the Pig*. CAB International, Wallingford, UK, pp. 1–16.

Sabeti, P.C., Reich, D.E., Higgins, J.M., Levine, H.Z.P., Richter, D.J., Schaffner, S.F., Gabriel, S.B., Platko, J.V., Patterson, N.J., McDonald, G.J., Ackerman, H.C., Campbell, S.J., Alshuler, D., Cooper, R., Kwiatkowski, D., Ward, R. and Lander, E.S. (2002) Detecting recent positive selection in the human genome from haplotype structure. *Nature* 419, 832–837.

Scribner, K.T., Malecki, R.A., Batt, B.D.J., Inman, R.L., Libants, S. and Prince, H.H. (2003) Identification of source population for Greenland Canada Geese: genetic assessment of a recent colonization. *Condor* 105, 771–782.

Searls, D.B. (2000) Bioinformatics tools for whole genomes. *Annual Reviews in Genomics and Human Genetics* 1, 251–279.

Seielstad, M., Minch E. and Cavalli-Sforza L.L. (1998) Genetic evidence for a higher female migration rate in humans. *Nature Genetics* 20, 278–280.

Simoni, L., Calafell, F., Pettener, D., Bertranpetit, J. and Barbujani, G. (2000) Geographic patterns of mtDNA diversity in Europe. *American Journal of Human Genetics* 66, 262–278.

Sjögren, P. and Wyöni, P.I. (1994) Conservation genetics and detection of rare alleles in finite populations. *Conservation Biology* 8, 267–270.

Slate, J., Visscher, P.M., MacGregor, S., Stevens, D., Tate, M.L. and Pemberton, J.M. (2002) A genome scan for quantitative trait loci in a wild population of red deer (*Cervus elaphus*). *Genetics* 162, 1863–1873.

Slatkin, M. (1985) Gene flow in natural populations. *Annual Reviews in Ecology and Systematics* 16, 393–430.

Slatkin, M. (1993) Isolation by distance in equilibrium and non equilibrium populations. *Evolution* 47, 264–279.

Slatkin, M. (1994) Linkage disequilibrium in growing and stable populations. *Genetics* 137, 331–336.

Slatkin, M. (1995) A measure of population subdivision based on microsatellite allele frequencies. *Genetics* 139, 157–162.

Slatkin, M. and Arter, H.E. (1991) Spatial autocorrelation methods in population genetics. *American Naturalist* 138, 499.

Slatkin, M. and Hudson, R.R. (1991) Pairwise comparisons of mitochondrial DNA sequences in stable and exponentially growing populations. *Genetics* 129, 555–562.

Smouse, P.E., Long, J.C. and Sokal, R.R. (1986) Multiple regression and autocorrelation extensions of the Mantel test of matrix correspondence. *Systematics and Zoology* 35, 627–632.

Sokal, R.R. and Oden, N.L. (1978) Spatial autocorrelation in biology. 1. Methodology. *Biological Journal of the Linnean Society* 10, 199–228.

Sokal, R.R., Oden, N.L. and Wilson, C. (1991) Genetic evidence for the spread of agriculture in Europe by demic diffusion. *Nature* 351, 143–145.

Sokal, R.R., Oden, N.L., Rosenberg, M.S. and DiGiovanni, D. (1997) Ethnohistory, genetics, and cancer mortality in Europeans. *Proceedings of the National Academy of Sciences USA* 94, 12728–12731.

Sokal, R.R., Oden, N.L. and Thomson, B.A. (1999) A problem with synthetic maps. *Human Biology* 71, 1–13.

Stephens, M. (2001) Inference under the coalescent. In: Balding, D.J., Bishop, M. and Cannings, C. (eds) *Handbook of Statistical Genetics*. John Wiley & Sons, New York, pp. 213–238.

Stephens, M. and Donnelly, P. (2000) Inference in molecular population genetics. *Journal of the Royal Statistical Society B* 62, 605–635.

Stone, A.C., Griffiths, R.C., Zegura, S.L. and Hammer, M.F. (2002) High levels of Y-chromosome nucleotide diversity in the genus *Pan*. *Proceedings of the National Academy of Sciences USA* 99, 43–48.

Storz, J.F. and Beaumont, M.A. (2002) Testing for genetic evidence of population expansion and contraction: an empirical analysis of microsatellite DNA variation using a hierarchical Bayesian model. *Evolution* 56, 154–166.

Storz, J.F., Beaumont, M.A. and Alberts, S.C. (2002) Genetic evidence for long-term population decline in a savannah-dwelling primate: inferences from a hierarchical Bayesian model. *Molecular Biology and Evolution* 19, 1981–1990.

Stumpf, M.P. (2002) Haplotype diversity and the block structure of linkage disequilibrium. *Trends in Genetics* 18, 226–228.

Stumpf, M.P. and Goldstein, D.G. (2003) Demography, recombination hotspot intensity, and the block-structure of linkage disequilibrium. *Current Biology* 13, 1–8.

Tajima, F. (1983) Evolutionary relationships of DNA sequences in finite populations. *Genetics* 105, 437–460.

Tajima, F. (1989a) Statistical method for testing the neutral mutation hypothesis by DNA polymorphism. *Genetics* 123, 585–595.

Tajima, F. (1989b) The effect of change in population size on DNA polymorphism. *Genetics* 123, 596–601.

Tajima, F. (1995) Effect of non random sampling on the estimation of parameters in population genetics. *Genetical Research* 66, 267–276.

Takahata, N. and Palumbi S.R. (1985) Extranuclear differentiation and gene flow in the finite island model. *Genetics* 109, 441–457.

Tapio, M., Miceikiené, I., Vilkki, J. and Kantannen, J. (2003) Comparison of microsatellite and blood protein diversity in sheep: inconsistencies in fragmented breeds. *Molecular Ecology* 12, 2045–2056.

Tavaré, S. (1984) Lines-of-descent and genealogical processes, and their application in population genetics models. *Theoretical Population Biology* 26, 119–164.

Tavaré, S., Balding, D.J., Griffiths, R.C. and Donnelly, P. (1997) Inferring coalescence times from DNA sequence data. *Genetics* 145, 505–518.

Taylor, A.C., Sherwin, W. B. and Wayne, R.K. (1994) Genetic variation of microsatellite loci in a bottlenecked species: the hairy nosed wombat (*Lasiorhinus krefftiz*) *Molecular Ecology* 3, 277–290.

Templeton, A.R. (2002) Out of Africa again and again. *Nature* 416, 45–51.

Templeton, A.R., Routman, E. and Phillips, C.A. (1995) Separating population structure from population history: a cladistic analysis of the geographical distribution of mitochondrial DNA haplotypes in the tiger salamander, *Ambystoma tigrinum*. *Genetics* 140, 767–782.

Tiedemann, R., Hardy, O., Vekemans, X. and Milinkovitch, M.C. (2000) Higher impact of female than male migration on population structure in large mammals. *Molecular Ecology* 9, 1159–1163

Tishkoff, S.A. and Verrelli, B.C. (2003) Role of evolutionary history on haplotype block structure in the human genome: implications for disease mapping. *Current Opinion in Genetics and Development* 13, 569–575.

Tishkoff, S.A., Dietzsch, E., Speed, W., Pakstis, A.J., Kidd, J.R. *et al.* (1996) Global patterns of linkage disequilibrium at the CD4 locus and modern human origins. *Science* 271, 1380–1387.

Townsend, S.J (2000) Genetic diversity and domestication in sheep (*Ovis*). Unpublished PhD Thesis, University of East Anglia, UK.

Troy, C.S., MacHugh, D.E., Bailey, J.F., Magee, D.A., Loftus, R.T., Cunningham, P., Chamberlain, A., Sykes, B.C. and Bradley, D.G. (2001) Genetic evidence for Near-Eastern origins of European cattle. *Nature* 410, 1088–1091.

Utami, S.S., Goossens, B., Bruford, M.W., De Ruiter, J. and Van Hooff, J.A.R.A.M. (2002) Male bimaturism and reproductive success in Sumatran orang-utans. *Behavioral Ecology* 13, 643–652.

Van Horn, R.C., Engh, A.L., Scribner, K.T., Funk, S.M. and Holekamp, K.E. (2004) Behavioural structuring of relatedness in the spotted hyena (*Crocuta crocuta*) suggests direct fitness benefits of clan-level cooperation. *Molecular Ecology* 13, 449–458.

Vitalis, R. and Couvet, D. (2001a) Two-locus identity probabilities and identity disequilibrium in a partially selfing subdivided population. *Genetical Research* 77, 67–81.

Vitalis, R. and Couvet, D. (2001b) Estimation of effective population size and migration rate from one- and two-locus identity measures. *Genetics* 157, 911–925.

Vitalis, R. and Couvet, D. (2001c) ESTIM 1.0: a computer program to infer population parameters from one- and two-locus gene identity probabilities. *Molecular Ecology Notes* 1, 354–356.

Wakeley, J., Nielsen, R., Liu-Cordero, S.N. and Ardlie, K. (2001) The discovery of single-nucleotide polymorphisms and inferences about human demographic history. *American Journal of Human Genetics* 69, 1332–1347.

Waldick, R.C., Kraus, S., Brown, M. and White, B.N. (2002) Evaluating the effects of historic bottleneck events: an assessment of microsatellite variability in the endangered, North Atlantic right whale. *Molecular Ecology* 11, 2241–2249.

Wall, J.D. and Pritchard, J.K. (2003) Assessing the performance of the haplotype block model of linkage disequilibrium. *American Journal of Human Genetics* 73, 502–515.

Wang, J. (2003) Maximum-likelihood estimation of admixture proportions from genetic data. *Genetics* 164, 747–765.

Waples, R. (1989) A generalized approach for estimating effective population size from temporal changes in allele frequencies. *Genetics* 121, 379–391.

Waser, P.M. and Elliott, L.F. (1991) Dispersal and genetic structure in kangaroo rats. *Evolution* 72, 771–777.

Watterson, G.A. (1975) On the number of segregating sites in genetical models without recombination. *Theoretical Population Biology* 7, 256–276.

Watterson, G.A. (1978) The homozygosity test of neutrality. *Genetics* 88, 405–417.

Watterson, G.A. (1986) The homozygosity test after a change in population size. *Genetics* 112, 899–907.

Weir, B.S. and Cockerham, C.C. (1984) Estimating F statistics for the analysis of population structure. *Evolution* 38, 1358–1370.

Whitlock, M.C. and McCauley, D.E. (1999) Indirect measures of gene flow and migration: $F_{ST} \neq 1/(4Nm + 1)$. *Heredity* 82, 117–125.

Wilson, G.A. and Rannala, B. (2003) Bayesian inference of recent migration rates using multilocus genotypes. *Genetics* 163, 1177–1191.

Wilson, I.J. and Balding, D.J. (1998) Genealogical inference from microsatellite data. *Genetics* 150, 499–510.

Wilson, I.J., Weale, M.E. and Balding, D.J. (2003) Inferences from DNA data: population histories, evolutionary processes and forensic match probabilities. *Journal of the Royal Statistical Society A* 166, 155–188.

Wilson, J.F. and Goldstein, D.B. (2000) Consistent long-range linkage disequilibrium generated by admixture in a Bantu–Semitic hybrid population. *American Journal of Human Genetics* 67, 926–935.

Wilson, J.F., Weale, M.E., Smith, A.C., Gratrix, F., Fletcher, B. *et al.* (2001) Population genetic structure of variable drug response. *Nature Genetics* 29, 265–269.

Womble, W.H. (1951) Differential systematics. *Science* 114, 315–322.

Wright, S. (1921) Systems of mating. *Genetics* 6, 111–178.

Wright, S. (1951) The genetical structure of populations. *Annals of Eugenics* 15, 323–354.

Wright, S. (1978) *Evolution and the Genetics of Populations.* University of Chicago Press, Chicago, Illinois.

Yu, N., Jensen-Seaman, M.I., Chemnick, L., Kidd, J.R., Deinard, A.S., Ryder, O., Kidd, K.K. and Li, W.H. (2003) Low nucleotide diversity in chimpanzees and bonobos. *Genetics* 164, 1511–1518.

Index

Note: page references given in **bold** refer to figures in the text; those in italics refer to tables or boxed material

aardvark 96, 369, 375
ablations, genetic 239
acetylcholinesterase knockout mouse *482*, 483
activation and repression domains 133–134
activator-recruited cofactor 124
ACTIVITY database 416
acute lymphoblastic leukaemia (ALL) 492
acute myeloid leukaemia (AML) 492
Adh gene 509
admixture *541*
adrenal hyperplasia, congenital *290*
advanced intercross lines 17
Aequora victoria 239
African-American race 273
Afrotheria 350, 369, 375
Agnatha 101
agouti locus 195, 452
Agouti Viable Yellow (A^vy) mouse 195
Agtpb1 gene 225
AIDS 570
Air RNA 138
alcohol dehydrogenase activity 509
aldehyde dehydrogenase 169
aldosteronism, glucocorticoid-remediable *290*
α_{2c}-adrenergic receptor defect *482*, 483
α-thalassaemia *290*
alteration vector 247
Alu sequences 182, 270–271, 282
amino acid sequence evolution 307–309
AMOVA (analysis of molecular variance) 555, 559
ancestral karyotypes 329–330, 364, 371–376
    carnivores 366, **367**, *378*
    marsupials 370, 379
    primates 364
aneuploidy 327
Angelman syndrome *290*
aniridia 134–135, 142
antibodies, engineered 495–496
ApoE region **305**
Arabidopsis thaliana 79, 139, 187
ArkDB 419
ARLEQUIN software 555

armadillo, nine-banded 369
ArrayExpress data repository 416
Artiodactyla 92, 98
ascertainment bias *541*
assignment tests 560–562
association analysis 462
AT-rich regions 284
auroch, wild 571
AVID 304
axial elements 2
AZF (azoospermia factor) regions 469
azoospermia *290*

B1 elements 282
BAAT (Bioinformatics Analysis Audit Trails) 400, 407
baboon 41, 327, *328*
    savannah 572
BAC contig maps 30
bacterial artificial chromosomes (BACs) 71–72, 92–93, 241, 456
bacterial genomes 264, 265–267, 273
    G+C content 268–269
    horizontal gene transfer 267
bacteriophage P1 cloning system 70–71
basal cell cancer syndrome 486
base pairs 94–96
basic Krüppel-like factor (BKLF) 133
basic-leucine zipper 133
bats 284, 285, 364, 572
Bayesian inference *542*, 565–567
B-cell lymphoma 492–493
B chromosomes 321, 340
bear, Malayan sun 367
Bedlington terriers 450
β-globin locus 122, 137, 209
β-tubulin 271
β-tubulin genes 271–272
bilaterians 202
Bio* projects 407

bioinformatics 531
    annotation and reannotation of sequences 405
    audit and tracking tools 400, 405
    public internet resources 397–399
    regulatory elements 89, 109–110
    resources for large-scale genomic analyses
        400–401
    sequence alignment and comparison 402–405
Bioinformatics Analysis Audit Trails (BAAT) 400,
        405
biological networks, analysis 212–213
birds 45, 109, 272, 378, 379
Birt-Hogg-Dube syndrome 451
BKLF (basic Krüppel-like factor) 133
BLAST 93, **101**, 398, 400, 401, 420, 421, 458
blastocyst 187
BLASTX 436–437
BLASTZ 304, 404, 405
BLAT 442
blepharophimosis ptosis epicanthus syndrome 96
blood cell development 135
BLOSUM62 substitution matrix 402, **403**
body-mass index (BMI) 520–521
bonobo 569
'booster elements' 284
bootstrap method 515
Borges, Jorge Luis 209
bottlenecks, population 546–548
BOTTLENECK software 546–548
boundaries, genetic 557
boundary (insulator) elements 204
bovine leukocyte adherence deficiency 450
brachyury (T) mutation 226, 230–231
breast cancer
    treatment 496
    tumour classification 493–494
breast cancer associated genes 235, 467
browsers, genome 400, 421–422, 444
Bubalus bubalis, see river buffalo
Buchnera spp. 265, 267
buffalo 368
    river 29, 37, 358

Caenorhabditis elegans 75, 79, 100, 105, 106, 268
    database 418
    RNA interference 138
callipyge locus **74**
callipyge phenotype 107, 509
camelids 99
cAMP-response element binding protein (CREB)
        136
Canavan disease 469
cancer
    classification using DNA microarrays 492–494
    drug screening 484
    serum profiling 165–166
    therapies 489, 495–496
canids 357, 365, 366
    see also dog
carboxyl-terminal repeat domain (CTD) 123
carnivores
    ancestral karyotype 366, **367**, 378
    chromosome painting 365–368
    comparative mapping 356–357
    see also named carnivores
Carollia brevicauda 284

cat 29, 37, 44, 272, 328, 330
    ancestral karyotype 378
    chromosome painting 365–366
    gene order 357
    inherited disorders 450, 452
    karyotype 365–366, 368, 373
        synteny with human 373
catalogues, molecular databases 414–415
cat-eye syndrome 290, 468
CAT gene 239
cattle
    chromosome painting 368
    comparative genome mapping 358
    conservation genetics 573
    FISH mapping 29
    genetic disorders 452, 461
    genetic diversity 570–571
    linkage map length 328
    QTL analyses 508, 510–511, 518–519, 523,
        529, 530
    RH mapping 42–43
    synteny mapping 35–36
Celera Genomics 90
cell signalling 205–207
cellular function genes 106
CEM program 438
CENP-B protein 287
CENSOR 398
centromeres 285, 288, 325–327
    unequal numbers 339–340
cetaceans 92, 99, 330, 358, 370
Cetartiodactyla 99, 350, 358, 368–369
    see also different cerartiodactylid species
CFTR gene 469
Charcot–Marie–Tooth type 1A disease 241, 290,
        468
cheetah 451
chemical mutagenesis 228–229, 253
chemotherapy agents 489, 490
chiasma distribution 4–6
    models 5
    sex differences in mammals 6–8
chiasmata 327
chicken 45, 109, 272, 379
chimpanzee 272, 288, 290, **291**, 362
    genome sequencing 92–93
    molecular evolution 569–570
    RH mapping 42
Chinese hamster ovary (CHO) cells 483
ChIP-chip technique 208–209
chiroptera 364
chloroquine 169
chordates, evolution 202, 289
chromatin 124–131, 183
chromatin immunoprecipitation (ChIP) 129–130
chromatin modifying enzymes 126–130
chromatin remodelling 124–126, 129–130,
        202–204
    and DNA methylation 187
    and gene expression 183–184
chromatography
    liquid chromatography 155–156, 164
    metal affinity 168
chromosome abnormalities 453–454
chromosome banding 25–26
chromosome diversity 285
chromosome duplication 91

chromosome mapping 24, 45–46
  approaches 27
  FISH 27–32
  future prospects 45–46
  RH mapping **33**, 39–45
  standard karyotypes and nomenclature 26–27
  synteny maps 32–39
chromosome painting 361–362, 456–457
  carnivores 365–368
  comparative 361–371
  eulipotyphla 364
  lagomorphs 364
  marsupials and monotremes 370–371
  primates 362–363
  rodents 363
chromosome rearrangements 284–286, 287,
      321–328
  analysis 456–457
  cat 357
  and disease 454
  effects on chromosome function 325–329
  genetic drift 337
  inversions and reciprocal translocations 325
  mutation rate 336–337
  and natural selection 337–340
  phenotypic consequences 334–335
  primates 362–363
  rates of 330–332
  and reproductive isolation/speciation 292–293,
      340–341, **342**
  Robertsonian translocations 285, 287, 321,
      323–325, 326, 333, 339, 369, 454
chromosome territories 109
chromosome walking 75
chronic myeloid leukaemia (CML) 496
ciliary neurotrophic factor 242
cis-regulatory elements 210
  identification 213–215
  prediction 430–462
citrullinaemia testing 471
civet, palm 366
Clock mutation 234
clone restriction fingerprinting 75
cloning systems 68–69
  bacterial artificial chromosome (BAC) 71–72
  bacteriophage P1 70–71
  comparison *70*
  fosmids 70
  P1-derived artificial chromosome (PAC) 72
CLUSTALW 422
CNGs (conserved non-genic sequences) 311–312
coalescent theory 543–546
coat colour 195
co-culture screening 484
coefficient of coincidence 9
COG database 417
ColE cosmid 69
collision-induced dissociation (CID) 157, 158, 160
colon cancer cell line DLD-1 484
colony hybridization 73–74
colorectal cancer 496
colour blindness, red–green *290*
comparative cytogenetics 25–26, 329–333,
      352–353
comparative genomics 88, 264, 304–307, 395–397,
      417
  databases and programs 307, 417

developmental genes 213
  functional element identification 306–307
  goal 306
  human–cat 37, 373
  human–mouse 92, **93**, 107–109, 268, 392
  human–rat 9
  long sequence alignments 304
  rat-human 93
  whole genome 304–306
complex pedigree design 519–521
Complex Trait Consortium 17–18, 509
complex trait mapping 16–18
Comprehensive Microbial Resource 419
conditional knockout (cko) mouse 249–251
confidence interval (CI) 515
connexin genes 248
conservation genetics 569–571
conservation 79, 91, 108, 280, 354, 360–361, 378
  categories 373
  chromosome segments 373
  chromosome structure 25–26
  gene order 44, 357
  genes 272, 307–309
  human–cat genomes 37, 373
  human–rat genomes 93
  regulatory regions 309–311
  and repeat density 304–305
  whole chromosome 373
conserved non-genic sequences (CNGs) 311–312
ConSite 108
consomic (congenic) strains 232–234
contig mapping 30
contigs 68, 90
  assembly 74–75
copper toxicosis 450
core promoter 119–122, 203–204
CORG database 417
cosmids 69–70
CpG dinucleotides, methylation 180–181, 183
CpG islands 89, 90–91, 92, 95, 122, 130
  human genome 180
  methylation 180
  X-inactivated genes 194
CREB-binding protein (CBP) 126–127
Cre–*lox*P strategy 234, 249–251, 465–467
CRE protein 465, 467
CRI-MAP program 10, 13
CropNet 417
cryptorchidism 470
Ctenomys spp. 285, 287
cyclopamine 484
Cynopterus spp. 572
CYP2C9 enzymes 480
CYP2D6 enzymes *290*, 480, 487–489
CYP3A enzymes 483, 489–490
cystic fibrosis 469
cytochrome P450 enzymes 489, 490–491
cytogenetics 24–25
  comparative 25–26, 329–333, 352–353
  mapping 27–32
cytokines 108–109
cytosine, methylation 189

DAD database 415
dairy cattle 508, 510–511, 518–519, 523, 529, 530
DataBank of Japan (DDBJ) 397, 398–399, 415

databases 397–399, 414–422
  catalogues 414–415
  comparative genomics 307, 417
  data quality issues 399, 400, 424–425
  diseases 420–421, 451
  gene expression/ESTs 207, 417
  gene identification and structure 418
  genetic and physical maps 418–419
  genomic 419
  growth in 412, **413**
  major sequence 415–416
  mutations 419–420
  pathology 420–421
  protein 157–158, 159, 415, 420
  purpose 415
  retrieval systems 421–422, 436–437
  RNA sequences 420
  searching strategies 422–424
  specialist 415, *417*, 418
  transgenic mouse strains 253
data integration systems 422
daughter yield deviation 519
DBCAT 414
deer 285, *328*
  roe 559–560
Delauney triangles 557
deletions 232
demographic events, inference 556–559
denaturing gradient gel electrophoresis (DGGE) 469
denaturing high-performance liquid chromatography (DHPLC) 469–470
Denys–Drash syndrome 142
development
  and DNA methylation 181
  epigenetic mark clearance and establishment 187–189
  gene expression 134–135
DGAT1 gene 508, 510–511
diabetes mellitus 450
diagnostic technology 467–471
differentially methylated regions (DMRs) 189
differentiation, genetic 554–556
diffuse large B-cell lymphoma 491–493
DiGeorge/velocardiofacial syndrome *290*, 456, **457**
digital differential display 207
digitation anormale (*dan*) mutation 243–244
DIGIT program 442
dihydropyrimidine dehydrogenase (DPD) 489
dinucleotide frequencies 94–95
diploid number 319–321, **322**, 333, 350, 375
diploid number:fundamental number ratio 321, **323**, **324**
Diprotodontia 370–371
diphtheria toxin 239
diseases
  animal models of human 240–241, 454–455
  databases 420–421, 451
  genetic, *see genetic disease*
  imprinted genes 192, 195–196
  and LCRs 289, *290*
  and LINEs 284
  serum profiling 165–166
  and trinucleotide repeats 95–96
dispersed elements 97–99
DMC1 gene product 3
DmD (mdx) locus 231
DNA amplification 76

DNA-binding domains, sequence-specific 131–134
DNA methylation 142, 179–180
  cellular machinery 180–182
  in development 188, 189
  functions 182
  and gene expression 130–131, 182–183, 187
DNA methyltransferases 181
DNA microarrays 208–209, 215, 491–492
  classification of haematopoietic tumours 491–493
  classification of solid tumours 493–494
  data use 215
  in QTL mapping 530–531
DNA pooling, selective 523
DNA testing, *see genetic testing*
DNA transposons **98**, 99, *281*
  Class II 282
  Class I (retroelements) 182, 195, 270, *281*, 282
Dnmt1 gene 181
Dnmt1 protein 181
Dnmt3a/3b proteins 181, 284
Doberman pinscher 462
Dobzhansky, Theodosius 263
docetaxel 490
dog 80–81, 105, 273
  breed identification 562
  breeds 80–81
  chromosome painting 365
  comparative genome mapping 357
  FISH mapping 29
  genetic disorders 450, *452*, 457–458, 461
  genetic testing 470–471
  linkage map length *328*
  as model of human disease 455
  RH mapping 44–45
  synteny mapping 37–38
dolphin 330
  Atlantic bottlenose 369
domestic animals
  FISH mapping 29
  genetic testing 470–471
  inherited diseases 450, *452*
  RH mapping 42–45
  synteny mapping 35–38
  see also livestock *and named domestic animals*
donkey 29
dosage compensation 137–138, 318
DOUBLESCAN 438
double-stranded breaks 3
double-stranded RNA (dsRNA) 138–140
down-stream promoter element (DPE) 121–122
Down syndrome 241
Drosophila 75, 79, 105, 106, 109
  development 134, 189
  karyotype evolution 318, 325
  quantitative traits 510, 531
  regulatory region evolution 309–311
drug development 494–498
  engineered proteins and RNA aptamers 495–496
  enzyme inactivation 496
  G protein-coupled receptor modulators 497
  simple eukaryotic models 497–498
  target identification 169
drug metabolism 478–479, 480, 487–491

drug response, prediction 487–491
drug responses, human diversity 480–481
drug targets, identification 169
Duchenne muscular dystrophy 284
duplications 91, 308–309
  chromosome 91
  genes 288–289
  role in genome evolution 291–292
  segmental *281*, 289–291, 308–309
  whole genome 288–289
dwarfism, pituitary *290*
dynamic programming 434
dystrophin associated myopathies 455

E-cadherin 240
echidna 360
Edentata 350–351
Edman degradation 168
Eed protein 193, 194–195
electroporation 71, 72
electro-retinogram (ERG) 231
electrospray ionization (ESI) 154–155
elephant 369, 375
  African 562
EMBL (European Molecular Biology Laboratory)
    397, 415
EMBOSS 399
embryo, transgenesis 236
embryonic stem (ES) cells 181, 242
  knockout mutations 245–247
  targeted homologous recombination 244–245,
    465–467
  viral mutagenesis 465
  X-inactivation **193**, 194
Emery–Dreifuss muscular dystrophy *290*
EMGlib database 419
encephalopathies, transmissable spongiform (TSEs)
    455
endangered species 562
endo16 gene 210
endomesoderm network 210
enhancers 119, **120**, 143, **203**, 204
enhancer trapping 241
ENSEMBL 443, 444
ENSEMBL Genome Browser 307, 400, 412, 422, 424
Entrez system 421–422
ENU (N-ethyl-N-nitroso-urea) 228–229, 234, 235,
    253
enzyme electrophoresis 33–34
enzyme inactivation therapies 496
epialleles, metastable 188, 195
epigenetic modifications 131, 179
  chromatin and histone modifications 183–186
  in development
    clearing and establishment in development
      187–189
    metastable epialleles 195
    parental imprinting 189–192
    X-chromosome inactivation 192–195
  DNA methylation 180–183
  inheritance 195
  interactions 186–187
Epstein–Barr virus 412

erythroid cells 135
erythroid Krüppel-like factor (EKLF) 132, 134
erythromycin breath test 490
ES cells, *see* embryonic stem (ES) cells
Escherichia coli 239
  database 418
ESI, *see* electrospray ionization
N-ethyl-N-nitroso-urea (ENU) 228–229, 234, 235,
    253
Euarchontoglires 350
euchromatin 96
eukaryotic genome, origins 267
Eulipotyphla 357, 364
European Molecular Biology Laboratory (EMBL)
    397, 415
evolution
  bacteria 267
  genes 307–309
  and genetic networks 216
  hominids 290, **291**
  hominoid 569–570
  parental imprinting 189
  role of horizontal gene transfer 267
  see also genome evolution; karyotype evolution
Evolution by Gene Duplication (Ohno) 288
Ewens sampling distribution 546
Ewens–Watterson test 550
EXOFISH program 438
exons *281*
  assembly into genes 434–435
  conservation 108, 308
  prediction 433–435
ExPASy (Expert Protein Analysis System) server
    420
experimental models
  human disease 240–241, 454–455
  pharmacogenomics 481–486
    drug discovery 497–498
    gene-targeting 481–484
    lower organisms 485–486, 497–498
    QTL analysis 485
expressed sequence tags (ESTs) 207, 417
  databases 417
eye formation 134–135

false discovery rate 514
Family Heart Study 520–521
fascioscapulohumeral muscular dystrophy *290*
FASTA 398, 422, 423
FASTS 159
FBPlot 405
'feature-based pair-wise sequence alignment (FBSA)
    algorithm 405
ferret 367
ferroportin1 gene 486
FGENEH program 434, *440*
FGENES 442
FGENESH++ 443
FGENESH 434
Fibre FISH 31
firefly luciferase gene 239
FISH, *see* fluorescence *in situ* hybridization
fish 45, 289, 378, 485, 486
  differentiation of stocks 562
fitness, and chromosome rearrangements 334–336
floxed 'stop' sequence 249–251
FLP-*Frt* system 250

fluorescence *in situ* hybridization (FISH) 27–32, 354, 456
　chromosomal assignment of linkage/syntenic groups 28–29
　complementing contig mapping/genome sequencing 30
　interphase/Fibre 31–32
　multicolour 30–31
　single locus mapping 29–30
5-fluorouracil 489
fly models 485–486
footprinting, phylogenetic 89, 301–302
'forward' genetics 222, 222–235, 532
　positional cloning 222–225, 460–462
fosmids 70
Fourier transform (FT) ion cyclotron analysers 157
foxes 357, 365, 366
FOXL2 gene 96
FOXP2 gene 88, 309
frame-specific hexamer frequency score 433
Frasier syndrome 142
F_{ST} statistic 541, 554–556
full-likelihood statistics *542*, 565–567
functional genomics, strategies 221–222, 532–533
fundamental number 319–321, **322**, 327, *328*
fundamental number:diploid number ratio 321, **323**, **324**

GABA degradation *482*, 483
GALA database 307
gametogenesis 2–4, 227, 327
gap mapping 402
GAPPED BLAST program 405
gap-repair cloning 235
GATA-1 transcription factor 209
Gaucher's disease *290*, 467
G-banding regions 284
GC content 305
　and crossing over 5–6
　eukaryotes 269
　mouse genome 92
　prokaryotes 268–269
GC-rich promoters 122
gel electrophoresis, two-dimensional 162–163, 167–168
GenBank 397, *398*, 415
　entries **416**
　errors 401
　growth 412, **413**
gene annotation systems 444–445
GeneCards 419
GENECLASS software 561–562
gene duplication 291–292
gene expression 118–119
　databases 207, 417
　developmental and tissue-specific 134–135
　'factories' 140–142
　regulation 89, 107–110, 134–137
　　artificial/inducible 143, 251–252
　　cascades 135–137
　　and cell signalling 205–207
　　chromatin remodelling 183–184
　　contemporary view **141**
　　and DNA methylation 130–131, 182–183, 187
　　investigation in transgenic mice 237–239
　　knowledge-based prediction 213–215

　long-range 109–110
　molecular networks 209–213, 216
　non-coding RNA (ncRNA) 137–140
　post-transcriptional 138, 140–142, 184–185
　promoter model 107–108
　selective 104
　whole genome studies 207–209
gene families 101–105, 289, 292
　inventories 105
　superfamilies 103
GENEID+ *441*
GENEID *441*, *443*
gene loss 265
　primate specific 273
GENEMARK.hmm *442*
gene order 268, 359
　conservation 44, 357
GENEPARSE 434
GENEPARSER2 *441*
GENEPARSER3 *441*
gene predictions 429–430
　ab initio 430–436
　accuracy 430, 440–443
　comparative 436–440
　programs 431–432, 435
general transcription factors (GTFs) 123, 202
genes 105–106
　for cellular function 106
　evolution 307–309
　number in mammalian genomes 79, 105–106
　structure 430
GENESEQER 438
gene silencing
　DNA methylation 183
　non-coding RNA *Air* 138
GENESPLICER program 431–432
gene-targeting 169, 244–245, 465–467, 481–483
GeneTest website 470
gene therapy 143
genetic disease
　animal models of human 454–455
　burden in mammals 449–451
　chromosomal abnormalities 453–454
　defining disease-causing genes 456–463
　polygenic 453, 462–463
　single gene 451–452, 457–460
　　mapping 460–462
　status of gene discovery 451–454
　tests 467–471
genetic drift 337, 341, *541*
genetic testing 467–471
　errors 471
　pet and agricultural animals 470
gene trapping 252–253, 467
gene trees *544–545*
GENEWISE 438
GENIE *442*
GENIO program 430
GENLANG *441*
genome, functional fraction 306
Genome Browser (UCSC) 422, 444
genome comparisons, *see* comparative genomics
genome components 94, **95**, *281*
genome duplication 202, 265
genome evolution 279–280
　mammalian 91–92, 94, 291–292
　protein-coding genes 207–209

regulatory regions 209–211
role of duplications 291–292
genome mapping 23, 67
    comparative 353–361, 360–361
    integrated maps 68, **69**
    linkage (genetic) maps 67
    nomenclature and terms 354–355, 459
    physical maps 24–25, 67–68
    radiation hybrid 24, 27, 39–45, 46, 67, 67–68,
       355
    see also chromosome mapping
genome plasticity 273
genome rearrangements 284–286
GENOMESCAN 438, 439
genome sequencing
    chimpanzee 92–93
    future challenges 88–89
    human 89–91
    mice 91–92
    past and present 88
    whole genome (WGS) 76–77, 90
       'chain termination method' 76
       costs and benefits 88
       novel methods/strategies 77–78
       ongoing projects 80
       shotgun approach 77, *78*, 90, 93
genome size 202
    bacteria 265
    mammals 79, 96
genome structure 96–105
Genome Workbench (GBENCH) 399
genomic (parental) imprinting 138, 182, 189–192
genomics, history 412
Genomic Sequence Database (GSDB) 424
genotype–phenotype relationship 80–81
genotyping, drug response prediction 487–491
GenPept 415
GENSCAN 442–443
genetic testing 467–471
geography, and genetic diversity 557–559
gibbon 330, 363
glucocorticoid receptor (GR) 207
goat *328*, 369, *452*
GOBASE 419
GOLD database 419
gorilla 42, 290, **291**, 362
G protein-coupled receptors 497
    affecting testicular descent (*GREAT*) 470
GRAIL 2 program *441*
great apes 42, 289, 290, **291**, 362, 570
green fluorescent protein 239
GRIP170 protein 207
growth factors 136
G_{ST} test 554, 555
GTFs (general transcription factors) 123, 202
Guthrie test 234

haemoglobin gene family 101
haemoglobinopathies 450
haemophilia A 284, *290*
Haemophilus influenzae 412
Haemophilus influenzae Rd genome 77
haemopoietic cancers 492–493, 496
Haldane function 13, 512, **513**
Haldane, J.B.S. 6, 9, 13
half-sib experiments 517–519, 521

haploid number 319–321, **322**, 327, *328*
HapMap project 17, 530, 553
HAPPY mapping 42
Hardy–Weinberg equilibrium *540*, 551
HAT medium 247–248
hCMV IE1 (human cytomegalovirus immediate early
    gene 1) promoter promoter 251, **252**
hedgehogs 364
hedgehog signalling cascade 486
Helicobacter pylori 268–269
hereditary neuropathy with liability to pressure
    palsies (HNPP) *290*
Herpes simplex virus 240
H_e *540*
heterochromatin 96, 139–140, 204, 287
    constitutive 287–288
    facultative 287
heterochromatin protein 1 (HP1) 128, 139–140
HEXON 433
HGB-DB 417
hidden Markov models 434, 438
histone acetyltransferases 124, *127*, **129**, 185
histone code **125**, 126–129
histone deacetylase enzymes 128, **129**, 183, 185
histone methyltransferases 124
histones 124, 183
    modification 183–185
       acetylation 124, 127–128, **129**, 185
       methylation 185–186
       phosphorylation 185
    modifications 186–187
history of genomics 412
HIV/AIDS 570
HMG-box gene family 101–103
HMMGENE 434, *442*
HNPP (hereditary neuropathy with liability to
    pressure palsies) *290*
Holliday junctions 3
homeodomains 132
homeogenes 238
hominid evolution 290, **291**
hominoids, molecular evolution 569–570
homologous recombination 243–245, 465–467
homology 354, 459
horizontal gene transfer (HGT), bacteria 267
hormone receptors 205–207
hormone responsive element (HRE) 205
horse 36–37, 327, *328*, 367–368
    FISH mapping 31
    inherited diseases *452*
    RH mapping 44
H_0 *540*
'house-keeping' genes 122
hox genes 202
*Hpa*II tiny fragment (HTF) islands 95
Hprt locus 242–245
HSPL program 429–430
human chromosome 2 (HSA2) 289–290, **291**
human chromosome 5q31 (HSA5q31) 108
human chromosome 9 (HSA9) 290
human chromosome 13 (HSA13) 94
human chromosome 16 (HSA16) 5, 362
human chromosome 18 (HSA18) 94, 109
human chromosome 19 (HSA19) 5, 94, 109, 273
human chromosome 21 (HSA21) 5, 7, 289
human chromosome 22 (HSA22) 5, 94, 442–443

human cytomegalovirus immediate early gene 1
    (hCMV IE1) promoter 251–252
human diseases
    burden of genetic 449–450
    genetic *452*
    genetic databases 420–421, 451
    imprinted genes 192, 195–196
    and LCRs 289, *290*
    and LINEs 284
    serum profiling 165–166
    transgenic models 241
    trinucleotide repeats 95–96
Human Gene Mutation Database (HGMD) 451,
    452
human genome
    β-tubulin genes 271
    CpG islands 180
    estimate of gene number 106
    MHC genes 272
Human Genome Project 79, 268, 394, 478–481,
    495
human genome sequencing 89–91
human–mouse genome comparison 92, **93**, 268,
    392
    regulatory sequences 107–109
Hunter syndrome *290*
Huntington disease 241, 469
HvrBase 418
hybrid cell panel, development 32–33
hygromycin B resistance 245
hypoxanthine, aminopterin and thymidine (HAT)
    medium 247–248
hypoxanthine phosphoribosyl transferase 242

IAM mutation model 568
ICAT (isotope coded affinity tagging) 164–165, **166**
ichthyosis *290*
identical by descent (IBD) matrix 520, 527
IGF2 gene 107
IHGSC (International Human Genome Sequencing
    Consortium) 269
IMAGE Consortium 458
imatinib 496
immunoglobulin gene superfamily 104
imprinted genes 138, 182, 189–192
    in mouse genome 189, *190–191*
imprinted X-inactivation 194–195
inbred strains, recombinant 18
incontinentia pigmenti *290*
infinite allele model *541*
infinite-models *541*
infinite site model *541*
Infobiogen 414
initiator element (Inr) 121
in ovo injection of DNA
    applications 237–241
    limitations of 241–242
    procedure 236–237
Insectivora 350–351
in situ hybridization 24, 27–28, 354
    fluorescence, *see* fluorescence *in situ*
        hybridization
    radioactive (RISH) 28, 29
    The Institute for Genomic Research (TIGR) 77
insulator (boundary) elements 204

integrated maps 68, **69**, 357
    databases 418–419
interchromatin compartment (IC) 109
interference 9–10
intergenic unique sequences 106–107
interleukins 108–109
internalin 240
International Human Genome Sequencing
    Consortium (IHGSC) 269
International Nucleotide Sequence Database
    Collaboration 397
International System for Cytogenetic Nomenclature
    (ISCN) 26–27
internet resources 397–399
    gene prediction 435–436
    see also databases
interphase mapping technique 31
intracisternal A particle (AP) family 182, 195
introns 212
    conservation 108
inversions 325
    use in mutagenesis 234
ion trap analysers 157
ISCN (International System for Cytogenetic
    Nomenclature) 26–27
Island Jersey cattle 564–565, 573
isochores 5–6, 91, 269, 273
isotope coded affinity tagging (ICAT) 164–165, **166**

Jackson laboratory 226
Jersey cattle 564–565, 573

K-allele model (KAM) *542*
kangaroos 285–286, 370–371
Karolinska cocktail 491
karyotype 26, 317
    ancestral 329–330, 371–376, *372–376*, *378*
        carnivores 366, **367**, *378*
        marsupials 370, *379*
        primates 364
    diversity of mammalian 350
    standard 26
    see also chromosome(s)
karyotype evolution 284–286, 318–319
    diploid number and fundamental number
        319–321, **322**
    evolutionary forces 333–340
    history of study 317–318
    orthoselection 333
    and reproductive isolation 292–293, 340–341,
        342
Kell blood locus 567
KEYnet 422
keyword matching 422
King, Max 338
KIR genes 460
knock-in strategy 249
knock-out mutations 483
    conditional 249–251
    ES cells 245–247
    whole genome 467
Kosambi function 14
Kyoto Encyclopaedia of Genes and Genome
    (KEGG) 399, 421

Labrador retriever 462
lacZ gene 239, 250
lagomorphs 29, 38, *452*
    chromosome painting 364
lampreys 101
Laurasiatheria 350, 357–359
'leave-one-out' procedure 560
Leishmania major 94
lemurs 326, 333
leptin gene 222–3, 239, 482–483
Lesch-Nyhan syndrome, mouse model 242–245
leucine zippers 133
leukaemia inhibitory factor 242
leukaemias 492, 496
libraries, chromosome-specific 75–76
library screening 73–74
likelihood *542*
likelihood of odds (LOD) 513, **514**
LINEs (long interspersed DNA (LI) elements) 97–98,
        182, 282
    and disease 284
    distribution 284
linkage disequilibrium 17, 54, 462, 463
    defined *540–541*
    patterns in human genome 551–554
linkage disequilibrium (LD) mapping 17, 515,
        523–528, 530
linkage mapping 68
    comparative in mammals 327, *328*, 353–361
    current status in mammals 15
    detection of linkage 10–13
    history 8–10
    male and female meiosis 4
    mapping functions 13–15
    purpose 15–16
    single gene diseases 460–462
    web sites *16*
lion 366
liquid chromatography 155–156
    coupled to MS analysis (LC-MS/MS) 164, 170
Listeria monocytogenes 240
Littlefield's HAT culture medium 247–248
livestock
    genetic diversity patterns 570–571
    see also named livestock species and groups
locus control regions (LCRs) 104, 122
LocusLink 444
LOD (likelihood of odds) 513, **514**
long interspersed nuclear elements (LINEs) 97–98,
        182, 282, 284
long range haplotype (LRH) test 553
long terminal repeat (LTR) elements *281*, 282, 283
low-copy repeats (LCRs) 289, *290*
'lower' animals
    pharmacogenomic analysis 485–486, 497–498
    transcriptional regulation 202
loxP sites 249–250
Lps2 mutation 234
LQT3 syndrome 483
LRH (long range haplotype) test 553
LTR (long terminal repeat) elements *281*, 282, 283
luciferase gene 239
lung adenocarcinoma 171
Ly49 genes 460
lymphomas 492–493, 496
lysine residues, methylation 186–187
lysine-threonine-serine (KTS) 142

macaque 272
macropodines 285–286
Macropus 287
Macropus eugenii, see tammar wallaby
major histocompatibility complex (MHC) 272
    genes 103
    peptide profiling 164
malaria parasite, proteomic analysis 164
MALDI (matrix-assisted laser desorption/ionization)
        154, **155**, 156, 168
MALDI-TOF mass spectrometer 157, 160
mammals
    comparative cytogenetics 25–26, 329–333,
        352–353
    genome size 79, 96
    genome structure 96–105
    phylogenetic relationships 266, 302–304,
        350–352
    see also wild mammals *and named*
        mammals/groups of mammals
Mantel test 555
map distance 13
map kinase inhibitor genes 171
map length, male/female comparison 8
mapping functions 13–15
marker-assisted selection 523
Markov Chain Monte Carlo 520, *542*, 562–563
marsupials
    ancestral karyotype 370, 379
    chiasma distribution 6–7
    chromosome painting **363**, 370–371
    gene maps 359–360
    karyotype 285–286, 350, 353
    phylogeny **351**, 352
    SCH hybrid analysis 38–39
    sex chromosomes 194, 360, 377
mass fingerprinting 156
mass spectrometry
    coupled with liquid chromatography (L-MS/MS)
        164
    instrumentation 154, **155**
    ion sources 154–156
    mass analysers 156–157
    spectrometer configuration and modes of
        operation 157, **158**
mass to charge (M/Z) ratio 154, 156
MASV 400
matrix-assisted laser desorption/ionization (MALDI)
        154, **155**, 168
matrix attachment regions (MARs) 204
maximum dependence decomposition (MDM) 431
maximum likelihood (ML) 513
MDscan *214*, 215
meat, fraudulent 562
6-mercaptopurine 489
MeCP2 protein 130
mediator complexes 124
MEDLINE 398
meiosis 2–4, 327
meiotic drive 337–341, **342**
    reversal 340, 341, **342**
melanoma 493
6-mercaptopurine 489
meta-analysis, QTL 530
meta-elements 96–107
metal affinity chromatography 168
metastable epialleles 188, 195
Methanococcus janaschii 267

methyl-binding protein (MBD) family 183
methyltransferase enzymes 181, 284
microarrays, *see* DNA microarrays
micro RNAs (miRNAs) 270, 307
microsatellites 95, 99–101, 358
 composition 100
 evolution 100–101
microtubules 271
migrants, detection 561–562
migration events 557–559
milk, yield and composition 508, 510–511,
 518–519, 523
minigenes 244–246
minisatellites 99
mink 38, 45, 357, *452*
 American 365, 366–367, *378*
mitochondrial DNA (mtDNA) 559
mitochondrial genomic databases 419
MITOMAP 419, 451
MLH1 (MutL homologue 1) foci 3–4, 7–8
mobile DNA, *see* transposable elements
modifier genes 486
molecular genetic networks 209–210
 analysis 212–213
 evolutionary impact 216
 mammalian gene 212
 reconstruction 210–211
moles 364
Moloney leukaemia virus (M-MuLV) 236, 243
MomI locus 482
monkey
 African green 363
 howler 363
Monmonier algorithm 557, 559–560
monoclonal antibodies 496
mononucleotide frequencies 94
monotremes
 chromosome painting 371
 genetic mapping 360
 karyotype 350, 353
 phylogenetic relationships **351**, 352
 X chromosome 377
MORGAN *442*
Morgan, T.H. 9
morpheus genes 292
motif-finding, computational 213–215
mouse 79, 91–92
 β-tubulin genes 272
 chromosome rearrangements 285
 epigenetic modifications 189, *190–191*, 195
 FISH mapping 29
 human genome comparisons 92, **93**, 107–109,
  268, 392
 imprinted genes 189, *190–191*
 inherited disease 450–451, *452*, 467
 karyotype evolution **331**, 332, 340
 microsatellites 99–100
 as model of human disease 455, 463–467
 mutagenesis 227–235
 obesity 222–223, 238, 516–517, 531
 olfactory receptor genes 103–104
 QTL mapping 17–18
 recombinant inbred strains 17–18, 356
 RH mapping 42
 sex determination 96
 transgenic, *see* transgenic mouse
Mouse Genome Database (MGD) 356

MPSS (massively parallel signature sequencing) 208
MP-VAL software 548
Mre11-Rad50-Nbs1 protein complex 3
mRNA expression profiling 169–171
mRNA population size 202
M test 548
mucopolysaccharidosis type II *290*
MudPIT (multidimensional protein identification
 technology) 156
MULTIMAP 41
Multiple (BLAST) annotation System viewer (MASV)
 400
'multiple loci' test 231
MUMmer algorithm 404, 405
Muntiacus spp., *see* muntjac
muntjac 330
 Chinese 285, **286**, 333, 369
 Indian 38, 285, **286**, 319, 333, 369
muscle development 135
muscular dystrophy
 Duchenne 284
 Emery–Dreifuss *290*
 experimental mutagenesis 231
 fascioscapulohumeral *290*
 models of 455
Mus musculus
 karyotype evolution **331**, 332, 340
 see also mouse
Mus musculus domesticus 285, 287, 340
 karyotype evolution **331**, 332, 333
Mustelidae 365–367
mutagenesis 225, 227–235
 at specific genome regions 232–234
 at specific loci 231–232
 chemical 228–229
 embryonic stem cells 242–253, 465
 genome-driven 235
 genome-wide 230–231
 homologous recombination 244–248
 human disease research 461–465
 phenotype-driven 234–235
 radiation 228
 tagged 464–465
mutation–drift equilibrium *541*
mutation models *541–542*
mutations
 databases 419–420
 insertional 241
 knock-out 245–247
  conditional 249–251
 and L1 elements 284
 point 247–248
 rate for chromosome rearrangements 336–337
 spontaneous 226–227
Mycobacterium
 leprae 266, 268
 tuberculosis 266–267, 268
Mycoplasmas 265
MyoD transcription factor 135
myopathy, dystrophin associated 455
myosin VIIa gene 455
MZEF *442*
M/Z (mass to charge) ratio 154, 156

National Centre for Biotechnology Information
 (NCBI) 397–398, 399, 412

Entrez system 421–422
LocusLink 444
National Heart, Lung and Blood Institute, Family Heart Study 520–521
National Institutes of Health (NIH) 395
natural selection 337
 adaptive 337
 karyotype evolution 337–341, **444**
 and population genetic differentiation 567–569
NCA (nested clade analysis) 558
N_e *540–541*
nematodes 485
 see also Caenorhabditis elegans
neomycin phosphotransferase (*neo^r*) minigene 345–346
neomycin resistance 345–346
nephronophthisis, familial juvenile *290*
nested clade analysis (NCA) 558
NETGENE2 program 431–432
network analysis 212–213
networks, *see* molecular genetic networks
neurofibromatosis type 1 *290*
neurofibromatosis type 2 456, **457**
neurological diseases 96
Neurospora crassa 187
Newfoundland dog 450
New World monkeys 101
n-island models *541*
NK cells 460
non-coding genes, number 106
non-coding RNA (ncRNA) 106, 137–140, 270–271, 306–307
non-Hodgkin's lymphoma 496
non-LTR elements *281*, *282*
nuclear compartmentalization 137
nuclear receptors 136–137, 205–207
Nucleic Acids Research 414
NUCLEODIV software 559
nucleosome core particle **125**, 183–184
NuRD complex 134

obesity, QTL 516–517, 531
 leptin gene 222–223, 238, 482–483
oestrogen receptor 206, 509
Ohno, S. 288, 376
Old World monkeys 103
olfactory receptor (OR) genes 103–105, 264, 272–273
oligonucleotide array technology 78
OMIA database 451, 452
OMIM database 108, 451, *452*
oncogenes 238
oogenesis 4
operon theory 107
orang-utan 289, **291**, 570
organic cation transporter gene (*Oct1*) 483
orthologues 354, 459
Orycteropus afer 96
Oryctolagus cuniculus, *see* rabbit
outlier approach 567–569
ovarian cancer 165–166

P1-derived artificial chromosome (PAC) 72
Pan 569–570
panda, giant 367–368

panmixia *540*
panther, Florida 451
paralogues 354, 459
parental imprinting 138, 182, 189–192
 X-inactivation 194–195
PARTITION software 562
patched gene 485–486
'pathogenicity islands' 267, 269
Pax6 gene 134
PCR, *see* polymerase chain reaction
peptide mass fingerprinting 156, 160–161
peptide sequencing, *de novo* 158–160
peripatric speciation 292
Perissodactyla 350, 367–368
 see also donkey; horse; zebra
permutation testing 514
Peromyscus spp. 284
Petrogale 286, 287
pharmacodynamics 479
 human diversity 480–481
pharmacogenomics 478
 drug discovery and development 494–498
 drug response prediction 487–491
 experimental models 481–485
 impact of Human Genome Project 478–481
pharmacokinetics 479
phenotyping, drug responses 490–491
phenylketonuria 234
pheromone receptor genes 273
phospholipase (*Pla2g2a*) gene 482
phosphoproteome 167–169
phosphorylation, histone 185
phosphotyrosines 168
Photinus puralis 239
phylogenetic footprinting 89, 301–302
phylogenies, gene, mapping 557–559
phylogeny
 mammals 266, 302–304, 350–352
 primates 363
phylogeography 557–559
physical maps 24–25, 27, 67–68
 databases 418–419
 see also chromosome mapping
pig 92, *328*, 450
 chromosome painting 368
 comparative genome mapping 358–359
 FISH mapping 31
 inherited disease 450, *452*
 QTL mapping 511
 RH mapping 43–44
 synteny mapping 36
pipelines, bioinformatics 400–401
PIPMAKER 304, **305**, **312**, 404–405
pituitary dwarfism *290*
pJC703 cosmid vector 69
Pla2g2a gene 482
plants 270, 418, 419
platypus 360, 371
pogo family 287
point mutations 247–248
Polb gene 249–250
polycomb group (PcG) proteins 193, 194–195
polycystic kidney disease *290*
polymerase chain reaction (PCR) 34, 74, 105
 in genetic testing 469–470
 library screening 74
 SCH analysis 34

polymerase chain reaction-restriction fragment length polymorphisms analysis (PCR-RFLP) 487, **488**
polyploidy 341
PONI gene 482
population genetics 539–540
  conservation/isolated populations 571–573
  differentiation across genome and selection 567–569
  diversity between populations 553–559
    population differentiation 553–559
    spatial patterns and genetic boundaries 556–559
  diversity measures *540–541*
  diversity within populations 546–554
    correlation of alleles within/between loci 550–554
    individual alleles/loci 546–550
  full-likelihood statistics *542*, 565–566
  hominoid molecular evolution 569–570
  individual-based approaches 557–565
  livestock species diversity 570–571
  mutation models *540–541*
  software *547*
  summary statistics *542*, 566–567
population genomics 542–543, 546
porcine stress syndrome 450
32P ortho-phosphate labelling 167–168
positional cloning 222–225, 231, 240, 458–460
  advantages 225
  limitations 225
  stages 223–225, 460–462
position effect variegation (PEV) 122
position weight matrix 215
possums 370
post-transcriptional gene expression regulation 138, 140–142, 184–185
Prader–Willi syndrome *290*
precursor ion scanning 168
pregnenolone xenobiotic nuclear receptor 483
pre-initiation complex (PIC) 119–120
primates
  ancestral karyotype 364
  chromosome painting 362–364
  comparative genome mapping 355
  gene loss 273
  karyotype evolution 326, 333
  phylogeny 363
  RH mapping 41–42
  SCH panel analysis 35
primates evolution 289–290, **291**
primitive ectoderm 188
principal component (PC) analysis maps 557
prion-like-protein Doppel 455
probability weight matrix (PWM) 306, **311**
PROCRUSTES 438
progressive motor neuropathy (*pmn*) mutation 225
progressive retinal atrophies (PRA) 457–458, 461, 470
promoters 89, 119–120
  core 120–122, 203–204
  defined 107
  human–mouse genome 107–108
  interspecific comparisons 107–108
  TATA-less GC-rich 122
promoter trapping 241
promyelocytic leukaemia protein 483

pronucleus, DNA injection 236–237
Protein Data Bank (PDB) 412, **413**, 420
protein expression profiling
  gel-independent 164–165
  two-dimensional gel electrophoresis 162–163, 167–168
protein identification 157–158
  databases 157–158, 159, 420
  de novo peptide sequencing 158–160
  peptide mass fingerprints 160–161
  sequence tags 160
  uninterpreted MS/MS database searching 161–162
Protein Information Resource (PIR) 412, **413**, 420
protein phosphorylation 167
proteins, engineered 496
proteome mining 169, **170**
proteomics 153–154
  combined protein and mRNA expression profiling 169–171
  defined 153
  future challenges 171–172
  protein expression profiling 162–165
  serum profiling 165–166
  sub proteome profiling 167–169
pseudoautosomal region (PAR) 376–377
pseudogenes 98, 106, 265, 273
  bacteria 266
  processed *281*, *282*
pulsed field gel electrophoresis (PFGE) 73
purine analogue drugs 169

quadrupole mass analysers 156
quaking (qk) locus 231
quantitative trait loci (QTL) 508
  gene expression 109–110
  number of genes and distribution of effects 509–511
  types of mutations causing 509
quantitative trait loci (QTL) mapping 16–18, 235
  detecting linkage to single marker 511–512
  detection of multiple interacting QTL 531–532
  experimental designs 515–516
    complex pedigree 519–521
    crosses of inbred/divergent lines 516–517
    half-sib 517–519
  gene identification 532
  improving precision 523–528
    combined LD–LA mapping 529
    linkage disequilibrium mapping 17, 523–528, 530
  increasing power of experiments 521–523
  meta-analysis 530
  pharmacogenetic traits 485
  principles 509–515
  statistical methods 513–515
  two-marker linkage 512, **513**
  use of gene expression data 530–531
quantum speciation 341
quelling 138
quinone reductase 2 169

rabbit 29, 38, *452*
Rad23b homologue **160–161**
RAD51 gene products 3

radiation hybrid (RH) mapping 27, 39–45, 46, 67–68, 355
  advantages 39
  characterization of hybrid cell lines 39–40
  data analysis 40–41
  databases 418–419
  humans 41
  non-human mammals 41–45
  non-mammalian vertebrates 45
  panel development **33**, 39
  panel genotyping 40
radiation mutagenesis 228, 253
rapamycin 486
rat 79
  FISH mapping 29
  gene maps 356
  human genome comparison 93
  inherited disease *452*
  linkage map length *328*
  mutagenesis 229
  Norway 356
  RH mapping 42
RatMap 356
Rattus norvegicus 356
reciprocal translocation 325
recombinant inbred strains 17–18, 356
recombination 2–4, 327
  frequency in males/females 6–8
  homologous 244–248, 465–467
  rate
    estimation 11–13
    variation 551–553
  see also meiosis
RefSeq 308, 399
'regular expressions' 214–215
regulatory sequences 264–265, 306
  bioinformatics 89, 109–110
  characterization using transgenic mice 238–239
  cis-acting promoter and enhancer sequences 119–122
  interspecific comparisons 107–109
  prediction 430–432
repeat density 304–305
REPEATMASKER 400
repetitive sequences 92, 97–105
  dispersed 97–99
  DNA methylation 182
  gene families 101–105
  and genomic changes 287–288
  satellite elements 99–101
  simple *281*
  unique 105–107
replacement vectors **245**, 246–248
reporter genes 238–239, 465
repression domains 133–134
reproductive isolation 292–293
  centromere changes 326–327
restriction fragment length polymorphisms (RFLP) 355
retinal atrophy, progressive 457–458, 461, 470
retinal degeneration, murine 457–458, 467
retinitis pigmentosas 457–458
retinopathy 169
retroelements 182, 195, 270, *281*, 282, 283, 287–288
retroviruses
  endogenous in genome 182, 283

mutagens in ES cells 243–244
  transgenesis 236
'reverse' genetics 222, 236–253, 463, 533
rhesus macaque 41
RHMAP 41
RHMAPPER 41
RH mapping, *see* radiation hybrid (RH) mapping
ribonuclease HII (RNase HII) enzymes 265–266
ribosomal genes 105
rice, databases 419
ricin 239
RIDGE (region of increased gene expression) 109
rituximab 496
river buffalo 29, 37, 358
RNA
  aptamers 495–496
  databases 420
  double-stranded (dsRNA) 138–140
  interference (RNAi) 138–139, **140**, 143, 253–254, 270, 496
  non-coding (ncRNA) 106, 137–140, 270–271, 306–307
RNA-induced silencing complex (RISC) 139, **140**
RNA polymerases 119
  polymerase II 121, 122–124
RNA splicing
  alternative 79, 142, 270
  databases 418
Robertsonian translocations 285, 287, 321, 323–325, 326, 333, 339, 368, 454
  mutation rate 336
rodents
  comparative gene mapping 355–356
  genome evolution 91–92
  karyotype evolution **331**, 332
  maps 355–356
  RH mapping 42
  synteny mapping 35
  see also named rodents
ROSETTA 436
RPCS satellite repeat 287
Russell, William 228, 229

Saccharomyces cerevisiae 3, 106
SAGE (serial analysis of gene expression) 207–208, 456
SAMOVA software 559–560
Sanger Institute 422, 442
Sank_al program 402, **404**
Sankoff alignment 402, **404**
satellite DNA 96, 99–101, 287, 288
Scandentia 363–364
SCH panels, *see* somatic cell hybrid (SCH) panels
SCP1 locus 9
sea urchin 210
segregation, non-random 338–341, **342**
SELDI (surface enhanced laser desorption ionization) 165–166
selection
  adaptive 337
  gene evolution 309
  karyotype evolution 337–341, **342**
  marker assisted 523
  and population genetic differentiation 567–569
Seleno Cysteine Insertion Element Sequence (SECIS) 439–440

selenoproteins, prediction 439–440
sequence alignment and comparison 304, 402–405
    dotplot 405, **406**
    large-scale 403–405
    multiple sequences 405
sequence evolution 302
    conserved non-genic sequences 312
    and mammalian phylogeny 302–304
    protein-coding genes 307–309
    regulatory regions 309–311
sequence homology, sources 459–460
sequence insertion vectors 246–247
Sequence Manipulation Suite 399
sequence similarity searches 422–423
sequential probability ratio test 10
serial analysis of gene expression (SAGE) 207–208, 458
serine residues, phosphorylation 185
serum profiling 165–166
sex chromosomes 350, 359
    marsupials and monotremes 194, 360, 377
    reconstruction of ancient 377–378
sex determination 96, 317–318
sex differences, chiasma distribution 6–8
SGP1 program 438
SGP2 program 439, *443*
sheep 29
    breed identification 562, 563
    callipyge phenotype 107, 509
    chromosome painting 368
    comparative genome mapping 358
    inherited disease *452*
    linkage map length *328*
    SCH panel analysis 36
short interfering RNAs (siRNAs) 138–139, **140**, 143, 270, 533
short interspersed nucleotide elements (SINEs) 98–99, 182, 270, 272, *281*, 282, 284
'shotgun' sequencing 77, *78*, 90, 93
shrews 357–358, 364–365
    Chinese tree 363–364
    common 38, 365
sickle cell disease 241, 469
signalling, and genome activity 205–207
signals, prediction 430–432
Signal Transducer and Activator of Transcription (STAT) family proteins 136
significance testing 513–514
SILAC technique 165
silencer elements **120**
simple animals, *see* lower animals
SINEs (short interspersed DNA elements) 92, 98–99, 182, 270, 272, *281*, 282, 284
single gene defects
    cloning candidate genes 459
    functional candidate genes 457–459
    mapping 460–462
Single Nucleotide Polymorphisms database 419
single nucleotide polymorphisms (SNPs) 80, *540*
single-stranded conformation polymorphism (SSCP) 469
sirolimus 486
SLAM 438, 439
Smith and Waterman algorithm 402, 403, **404**
Smith–Magenis syndrome *290*
SNP Consortium database 418
software

comparative genomics 307, 417
    gene predictions 431–432, 434
    population genetics *547*
somatic cell hybrid (SCH) panels 25, 32–33, 45–46
    analysis in mammalian species 34–39
    analysis methods 33–34
    construction 32–33
SORFIND *441*
Southern blotting 34
SOX genes 101–103, 107
SPAGEDI software 557
spatial autocorrelation 556–557
speciation 292–293
    and centromere changes 326–327
    karyotype changes 340–341, **342**
speech capability 309
spermatogenesis 4
spermatogonia 227
spinal muscular atrophy *290*
SPLICEVIEW 431–432
splicing, alternative 79, 142, 270
SPO1 (Sporulation 1) locus 3
squirrel 332
    red 571–572
SRY genes 96, 101
SSEARCH 396, 422, 423
STAT family proteins 136
statistical analysis packages, RH mapping 41
statistical methods
    population genomics *540–542*, 565–567
    QTL mapping 513–515
stem cells, *see* embryonic stem (ES) cells
stepwise mutation models (SMM) *541*
steroid hormone receptors 205–207
steroid hormones 136
'stop' sequence, floxed 250–251
STRUCTURE software 562–565
Sturtevant, A.H. 9
subfunctionalization 291–292
substitution matrix 402, **403**
succinic semialdehyde dehydrogenase deficiency 483
suffix trees 215
suffix trie 215
Suppressors of Cytokine Signalling (SOCS) family 136
surface enhanced laser desorption ionization (SELDI) 165–166
SWISS-PROT/TrEMBL database 420, 423
synaptonemal complex 2, 7–8
synteny maps 24, 27, 32
    domestic animals 35–38
    human 34–35, 355
    non-domestic animals 38–39
    non-human primates 35
    rodents 35
'systems biology' 264, 270

Tajima's D statistic 548, *549*, 551
tammar wallaby 6–7, *328*, **363**
tandem affinity purification (TAP) tag 170–171
tandem repeats, *see* microsatellites
TATA-binding protein (TBP) 120–121, 204, 205
TATA element 120–121
taurine cattle 570–571
Tay-Sachs disease 469
TBLASTX 439
T cell receptors 103

telomere 325
TESS 422
'Tet-Off' and 'Tet-On' expression systems 251–252
tetracycline-responsive promoter element (TRE) 251
TFIIB 122, 123
TFIID 122, 123, 204
TG microsatellites 95
TG repeats 99–101
thalassaemias *290*, 469
thermophilic bacteria 269
thymidine kinase 240
TIGR 77
time-of-flight mass analyser 156
tissue-specific gene expression 134–135
transcriptional activation 202–205
transcriptional regulation
  complexity 202, **203**
  knowledge-based predictions 213–215
  lower animals 202
transcriptional repressors 134
transcription factor-binding sites
  evolution 309–311
  identification 209, 306
transcription factors 119
  functions 131
  general/basal 122, 123
  hormone receptor interaction 207
  regulation 135–136
  sequence-specific DNA-binding 131–134
  tissue-specific 134–135
transcriptome, importance of 270
transgenes, expression 143
transgenesis 236
  induction of mutations 241
  in ovo injection
    applications 237–241
    limitations 241–242
    procedure 237
  viral infection of embryo 236
transgenic mouse
  databases and websites 253
  investigation of gene function and regulation 238–240
  as model of human disease 240–241
  pharmacogenetic models 481–484
  production of 236–237
transgenic rescue 223, 225
transmissable spongiform encephalopathies 453
transposable elements 280–284
  classification *281*, 282
  genome shaping capability 282–284
  transmission 283
transposons **98**, 99, *281*
trastuzumab 496
treble clef finger domains 132–133
TrEMBL database 415
trinucleotide repeats 95–96
trisomy 21, 241
Trypanosoma cruzi 235
tuberous sclerosis 486
tubulin 271
tufted (*tf*) locus 231
tunicamycin 498
twinning, dairy cattle 529
TWINSCAN 439, *443*
two-dimensional gel electrophoresis 162–163, 167–168

Unified Database (UDB) 418–419
unique sequences 105–107
upstream elements 108
Usher syndrome ype 1b 453

variance component approach 519–520
variegation 195
vectors
  alteration 247
  replacement **245**, 246–248
  retroviral 236, 252
  sequence insertion 246–247
VEIL 434
Venter, C. 77, *78*
Vertebrate Genome Annotation (VEGA) database project 442
vertebrates, evolution 202, 289
viral genomes, sequencing 412
viral mutagenesis 465
Virgil 422
vitamin D-receptor interacting protein (DRIP) 124
vomeronasal receptors 273

WAGR syndrome 142
Wahlund effect 551
Wallabia bicolor, see wallaby, swamp
wallaby 285–286
  rock 286
  swamp **363**
  tammar 6–7, *328*, **363**
WART (whole-arm reciprocal translocations) 325
websites
  cis-regulatory motif-finding 214
  transgenic mouse strains 253
weight array model (WAM) 431
weight matrix model (WMM) 431
WGS, *see* whole genome sequencing
whale, fin 369
WHICHRUN software 562
whole-arm reciprocal translocations (WART) 325
whole genome comparisons 304–306
whole genome duplications 288–289
whole genome sequencing (WGS) 76–77, 90
  'chain termination method' 76
  costs and benefits 88
  novel methods/strategies 77–78
  ongoing projects 80
  shotgun approach 77, *78*, 90, 93
wild mammals
  conservation genetics 571–573
  genetic disease 451
  synteny maps 38–39
  see also named wild mammals
Williams–Beuren syndrome *290*
Wilm's tumour 106
  suppressor gene (*WT1*) 106, 107, 142
WINNOWER algorithm 215
Wisconsin, University 235
WMM, *see* weight matrix model
wolf, red 562
wombling 556–557, 560
Wright–Fisher model *541*, 546
Wright's F$_{ST}$ 554–556
WT1 gene 106, 107, 142

X chromosome 350, 359
    ancient 376–377
    conservation 25, 352–353, 353, 359
    differentiation from Y chromosome 376–377
    inactivation 138, 182, 192–195, 284, 318, 376
      early events 192–193
      imprinted 194–195
      initiation 192
      late events 194
      marsupials and monotremes 360, 377
      morphological diversity 286
Xenarthra 350, 369
X-inactive specific transcript (*Xist*) RNA 138, 192,
    284
Xist 138, 192, 284
XPOUND program *441*
XREFdb 417
XY sex-reversal 469

Y chromosome 350, 359, 360, 469
    differentiation 376–377
yeast 106, 485, 497, 531
    mediator complexes 124
    mRNA and protein expression profiling 170–171
yeast artificial chromosomes (YACs) 72–73, 241
Yeast Genome Database 2

Zea mays 292
zebrafish 45, 378, 485, 486
zebras 367–368
zebu cattle 570
ZFY gene 377
zinc finger proteins 132, 143
zip1 gene 9
Zoo-FISH 26, 27, 372–373, 375
zoo populations 451

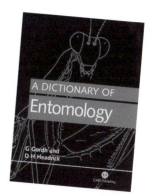